John Lindley, Thomas Moore

The treasury of botany

A popular dictionary of vegetable kingdom with which is incorporated a glossary of botanical terms

John Lindley, Thomas Moore

The treasury of botany
A popular dictionary of vegetable kingdom with which is incorporated a glossary of botanical terms

ISBN/EAN: 9783742838797

Manufactured in Europe, USA, Canada, Australia, Japa

Cover: Foto ©berggeist007 / pixelio.de

Manufactured and distributed by brebook publishing software (www.brebook.com)

John Lindley, Thomas Moore

The treasury of botany

Plate

RHODODENDRON DALHOUSIÆ
(AFTER HOOKER)

THE
TREASURY OF BOTANY:

A Popular Dictionary

OF

THE VEGETABLE KINGDOM;

WITH WHICH IS INCORPORATED

A GLOSSARY OF BOTANICAL TERMS.

EDITED BY

JOHN LINDLEY, Ph.D., F.R.S., F.L.S.

Emeritus Professor of Botany in University College, London;
Author of 'The Vegetable Kingdom'

AND

THOMAS MOORE, F.L.S.

Curator of the Chelsea Botanic Garden; Author of 'Index Filicum;' and
Co-Editor of 'The Gardeners' Chronicle.'

ASSISTED BY NUMEROUS CONTRIBUTORS.

ILLUSTRATED BY NUMEROUS WOODCUTS BY FITCH AND BRANSTON
AND STEEL ENGRAVINGS BY ADLARD.

IN TWO PARTS.—PART I.

NEW EDITION.

LONDON:
LONGMANS, GREEN, AND CO.
1870.

PREFACE.

THE OBJECT which it was proposed to accomplish by the publication of this TREASURY OF BOTANY, was to bring together, into the form of a Dictionary, a concise account of all the Plants concerning which a general reader was likely to seek for information; adding thereto, where practicable, longer notices of the more remarkable species, together with such popular matter as would give interest to the otherwise dry technical character of generic or specific descriptions. This information was to be diversified by woodcuts, and illustrated by views of scenery, representing the peculiar physiognomy of vegetation in different parts of the world.

The genera under which botanists have arranged the subjects of the Vegetable Kingdom are however, as is well known, so numerous, that they could not be all included in a moderate-sized textbook like the present, and hence selection became a necessity. In the choice of subjects, it has been made an especial object that none of the more important plants, whether in regard to their utility, their beauty, or their interest to botanical students, should be overlooked; although it is to be borne in mind that, in a progressive science like Botany, some omissions, such as of genera first made known or brought into notice during the progress of the work, must be inevitable.

What the TREASURY OF BOTANY really comprises, therefore, is a short history of those genera of plants which are known to possess especial interest on account of the medicinal qualities or the economical uses of their species, or by reason of their beauty or utility as garden plants; while to these two groups has been added a still larger one, comprising a selection of genera serving as representatives of the whole series of Natural Orders and their subdivisions. The space devoted to each separate genus is necessarily brief; and, except in the case of medicinal or economically valuable plants, of which a rather fuller account is given, the object has been to convey some notion of the characteristics of genera or families, rather than to attempt an enumeration, much less a description, of the species of which they consist. For that a massive Cyclopædia would have been necessary.

The remaining features comprised in the text of the TREASURY are—a Comprehensive Glossary of Botanical Terms, prepared by the late Dr. Lindley; an extensive selection of English Names of Plants, arranged, as far as possible, under their substantive terms; and a series of French Names both of genera and species, besides a large number of the Local Names of plants and vegetable productions in different countries throughout the world. In the Introduction, some excellent observations by Dr. Seemann, descriptive of the Plates, serve as illustrations of Phyto-Geography.

The style which has been aimed at, and as far as practicable realised, in the preparation of the several articles, is such that anyone, taking up the book in an idle hour, may be able to read a page without being reminded that he knows nothing about the plants which clothe our beautiful earth; and such also as, it is hoped, will induce in him a desire to increase his knowledge concerning them.

The Nomenclature-in-chief (that is to say, the names under which the articles pertaining to genera are written) is that of Lindley's *Vegetable Kingdom*, or Decandolle's *Prodromus*, so far as those books contain generic names forming the subject of an article.

The utility of this epitome of Botany depends very much on the able assistance which has been afforded, in carrying out the plan thus briefly sketched, by the various gentlemen, well-known in their several walks of the science, who consented to become contributors. The following list of their names, with the signatures adopted, will afford a sufficient guarantee of the value of their communications. The whole of the articles, of whatever kind, without signatures are editorial:—

Professor Balfour—[J. H. B.]
Rev. M. J. Berkeley—[M. J. B.]
Mr. A. A. Black—[A. A. B.]
Mr. W. B. Booth—[W. B. B.]
Professor Buckman—[J. B.]
Mr. W. Carruthers—[W. C.]
Mr. B. Clarke—[B. C.]
Professor Dickie—[G. D.]
Mr. W. B. Hemsley—[W. B. H.]
Mr. R. Heward—[R. H.]
Rev. C. A. Johns—[C. A. J.]
Dr. Masters—[M. T. M.]
Dr. Moore—[D. M.]
Dr. Seemann—[B. S.]
Mr. A. Smith—[A. S.]
Mr. J. T. Syme—[J. T. S.]
Mr. R. Thompson—[R. T.]
Mr. W. Thompson—[W. T.]

In the preparation of the Illustrations, the work has had the advantage of the admirable botanical and artistic talent of Mr. W. H. Fitch, by whom the very expressive though diminished woodcut figures have been drawn. These have been engraved with great fidelity by Mr. R. Branston; while the Plates, reduced by Mr. Adlard from well-known originals, are no less faithful as pictures of the aspects of vegetation in other lands.

The length of time which is taken up in the passage through the press of a book of so comprehensive a character, and into which, owing to the small type adopted, so great an amount of matter is compressed, has on this occasion been unhappily augmented by the complete failure of Dr. Lindley's health, which took place at an early stage of the progress of the work, and has not permitted him to witness its completion—for his labours, till lately so unceasingly devoted to the science of which he stood as one of the mightiest pillars, were stayed by the stroke of death shortly after this page had passed into the printer's hands. The plan of the book had, indeed, been perfected under his supervision, but he was unable to continue his editorial labours beyond letter C; and the superintendence of the subsequent portion has devolved entirely upon the writer of these sorrowing words, who is desirous of expressing not only his own keenly-felt sense of personal bereavement, but the still greater blow which has fallen on botanical science, by the loss of one of its ablest and most profound expositors.

In a book of so multifarious a character, it can scarcely be expected that mistakes do not occur—errors as to matters of fact as well as errors of the press, notwithstanding that both have been guarded against as far as possible. The Editor will be grateful to those readers who may be good enough to point out any such errata* that they may discover, with a view to their being corrected hereafter, should the patronage of the public lead to the issue of a subsequent edition.

T. M.

BOTANIC GARDEN, CHELSEA :
December 1865.

* One such, at p. 751 (line 3 of art. MELIACEÆ), may be here pointed out, where 'vinial' has been printed for 'renal.'

LIST OF PLATES.

I. Epiphytal Rhododendrons of the Himalaya . .	*Frontispiece*	
II. Vegetation of the Caroline Islands, on the Outskirts of Wood	*To face page*	190
III. Antiaris and Coffee Plantation in Java . .	,,	74
IV. Vegetation of Bamboos in Java	,,	120
V. Vegetation of New South Wales, near Port Jackson	,,	1239
VI. Forest on Guahan, one of the Marianne Islands .	,,	368
VII. Coral Reef in the Carolines	,,	1157
VIII. Vegetation of Teneriffe, with Succulent Euphorbias	,,	478
IX. Vegetation of Java—Tree-Ferns in the Foreground, a Forest of Amentaceæ in the Distance . .	,,	494
X. Swampy Forest, with Banyan Trees, in the Island of Ualan	,,	1139
XI. Vegetation of the Canary Islands—View in the Caldera	,,	69
XII. Vegetation of the Cinchona Forests of Peru, with Palms and Tree-Ferns	,,	294
XIII. Vegetation of Java	,,	840
XIV. Mountain Vegetation of Java	,,	856
XV. Vegetation of New Holland—an Acacia 'Scrub' .	,,	5
XVI. Vegetation of Kamtschatka, with tall Umbelliferæ—a Birch Forest in the Distance . .	,,	1169
XVII. Cactus Vegetation on the Banks of the Gila, New Mexico	,,	256
XVIII. Hyphæne or Doum Palm-Trees in Upper Egypt .	,,	512
XIX. Vegetation on the Ice-Cliffs in Kotzebue Sound, Arctic America	,,	1026
XX. Holy Cross Abbey, covered with Ivy . .	,,	572

ILLUSTRATIONS OF PHYTO-GEOGRAPHY.

The Plates of which the following pages furnish explanations, have been prepared with the view of showing some of the more remarkable aspects of the vegetation which clothes the surface of the earth in different parts of our planet. From these examples, which have been selected from a variety of sources, a tolerably adequate notion may be formed of the nature of the luxuriant and diversified leafage to be met with in tropical forests; while some knowledge may also be obtained of the quaint succulent vegetation which is scattered over the rocky arid wastes of the New World; of the scarcely more abundant, and much less developed, clothing to be found on arctic cliffs; of the peculiar tree-growth of the Australian continent; and of the characteristics of various other well-marked centres of plant life, the peculiar features of which are pointed out in the descriptive notice of each Plate.

EPIPHYTAL RHODODENDRONS OF THE HIMALAYA.

(Plate I.—*Frontispiece*.)

[Reference.—*Rhododendron Dalhousiæ*.]

The focus of the genus *Rhododendron* seems to be East Nepal and the Sikkim Himalaya mountains. It is there we find the species most numerous and their flowers of the greatest size and most brilliant tints. The genus chiefly prevails between 10,000 and 14,000 feet above the sea-level, its several species composing three-fourths of the vegetation above the forest region (12,000 feet). There Rhododendron wood supplies the native with fuel, and, from its tough nature and property of being easily worked, with many domestic utensils, poles for his tent, stools, saddle, bowl and spoon. The bark is used as that of the birch is in the Arctic regions, and the leaves serve as plates and wrappers for butter, curd, and cheese. It is the traveller's constant companion throughout every day's march; on the right hand and on the left hand of the devious path, the old trees and bushes are seen breast high or branching overhead, whilst the seedlings cover every mossy bank. At 13,000 feet the flanks of the snowy mountains glow with the blood-red blossoms of *Rhododendron fulgens*, whilst the beauty of *R. campanulatum* and the great elegance and delicacy of the white bells of *R. campylocarpum* excite the more admiration from their being found in such regions of fog and rain. Some kinds grow habitually as epiphytes, among them *R. Dalhousiæ* figured in our frontispiece, and one of the many noble introductions for which we are indebted to the labours of the indefatigable Dr. J. D. Hooker. *R. Dalhousiæ* is a slender straggling shrub, six to eight feet high, with oblong leaves, and white bell-shaped fragrant flowers with a delicate rosy tinge. It is generally growing, like many tropical orchids, amongst moss, with ferns and *Aroideæ*, upon the limbs of large trees, at from 6,000 to 9,000 feet above the sea, in a region of fogs, moisture, and rain, in sight of the snow-capped peaks of the Himalaya. [D. S.]

VEGETATION OF THE CAROLINE ISLANDS, ON THE OUTSKIRTS OF WOOD.

(Plate II.)

[REFERENCE.—a. *Artocarpus incisa*; b. *Caladium*; c. *Pandanus odoratissimus* in fruit; d. *Crinum*; e. Tree-Fern.]

This illustration introduces us into a valley of the island of Ualan, Caroline Archipelago, where, without much labour, the level land has been brought into a certain state of cultivation, being planted principally with those products of the island which furnish food. Bread-fruit trees, bananas, two gigantic species of *Caladium*, and the Tahitian sugar-cane grow here so intermingled that there is some difficulty in determining whether there has been any arbitrary plantation or not. The Bread-fruit tree (*Artocarpus incisa*) on the left-hand side is quite a young specimen, just beginning to bear fruit. The plantains and bananas of this place belong to four varieties, the specific types of which are *Musa paradisiaca* and *Musa sapientum*. Of the two larger Caladiums (fig. b) one is allied to, if not identical with, the well-known *C. macrorrhizum*, the root of which is used as an article of food. A third smaller species is the *Caladium esculentum*, the Kalo or Taro of the South Sea Islands. *Pandanus odoratissimus*, the Screw-pine, so called from its leaves being arranged like the windings of a screw, and its fruits having somewhat the outward appearance of pine-cones, is seen on the right-hand side of our illustration. Close to it will be seen the *Morinda citrifolia*, having a pale-green foliage and a whitish edible fruit of poor flavour. To the most prominent plants of this island belongs the widely-diffused *Dracæna terminalis*, commonly used for hedges. A fine *Crinum* with massive leaves grows isolated about the outskirts of the forests; and a *Maranta*, growing gregariously, abounds. Almost in the very centre of our picture are seen tree-ferns; and just above them the *Terminalia Catappa*, the horizontal branches of which form distinctly marked stories around the erect stems, imparting to the tree the aspect of a pine tree, and to the landscape a very peculiar and well-marked feature. [B. S.]

ANTIARIS AND COFFEE PLANTATION IN JAVA.

(Plate III.)

A glance at our illustration, taken from Blume's magnificent Rumphia, showing the *Antiaris toxicaria* or Upas tree of Java, surrounded by coffee plantations and other indications of human industry, at once disproves many of the exaggerated and fabulous accounts propagated about this famous poison plant by the early travellers. There is no sign of the extreme sterility of the ground in the vicinity of the poison trees, which was said for a considerable distance round to produce neither grass nor any other vegetable. Nor can it, with such surroundings, be true that, if the tree be pierced, those standing to windward would quickly be suffocated by its noxious effluvia, or that birds which fly over a recently wounded tree would meet the same fate. These and similar fables, Bennett and others have explained by transferring the odium to the marshy and unwholesome exhalations of parts of the Indian Archipelago to which state criminals, and especially those of the highest class, were sometimes banished, and where they speedily died of malaria, and not, as the vulgar believed, of the emanations of the Upas tree. The poisonous nature of the *Antiaris toxicaria*, stripped of all exaggeration, is, however, sufficiently powerful and deadly to make great precaution necessary. Dr. Horsfield had some difficulty in inducing the inhabitants of Java to assist him in collecting the juice which he required for his experiments, as they feared a cutaneous eruption and inflammation, resembling, according to the account they gave of it, that produced by the *Rhus vernix* of Japan, and the *Rhus radicans* of North America; but they were only affected by a slight heat and itching of the eyes. In clearing new grounds for cultivation, in which the Upas tree occurs, it is with difficulty the inhabitants can be made to approach the Upas, as they dread the cutaneous eruption which it is known to produce when newly cut down. But except when the tree is largely wounded, or when it is felled, by which a considerable portion of the juice is disengaged, the effluvia of which, mixing with the atmosphere, affect persons exposed to it with the symptoms just mentioned, the tree may be approached and ascended like the other common trees of the forests. [B. S.]

VEGETATION OF BAMBOOS IN JAVA.

(Plate IV.)

Those who compare the small meadow grasses of northern countries with the tall cocoa-nut trees of the tropics will probably think it fanciful that botanists should proclaim the plebeian grasses to approach in their external structure, as well as their internal organisation, nearest to the palms, termed by the illustrious Linnæus, the Princes of the Vegetable Kingdom. But this dictum of science will appear less fanciful when the huge bamboo, as the noblest representative of the grasses, with its tree-like trunk fit for fuel and building purposes, and its light feathery foliage, is placed by the side of some small rattan as the representative of the palms. Indeed grasses are regarded by many eminent botanists as a sort of palm of lower grade. In habit the two natural orders have much in common: their leaves are formed upon exactly the same plan, the only difference being that those of the palms are generally (not always) divided. Even the siliceous secretions so characteristic of grasses, are observable in rattans; whilst about their flowers, it may be said that those of the grasses are those of the palms, with the floral envelopes removed and only the bracts remaining. The group on the right-hand side of our plate affords a good illustration of the manner in which bamboos grow. They delight in humid localities, and are the ornament of most tropical rivers, often forming impenetrable thickets, the favourite retreat of wild animals. Their young shoots come up like asparagus, and in many Eastern countries are picked and preserved. The growth of the stem is rapid in the extreme. *Bambusa gigantea* was found to grow 25 feet 9 inches in length during the thirty-one days of July, 1833, when it was measured in the Calcutta Botanic Garden; and in the Botanic Garden at Glasgow the same plant was ascertained to rise one foot in twenty-four hours; so that an attentive observer could actually see a bamboo grow as plainly as he could see the movements of the hands of a watch. [B. S.]

VEGETATION OF NEW SOUTH WALES, NEAR PORT JACKSON.

(Plate V.)

[REFERENCE.—*a. Banksia; b. Xanthorrhœa.*]

The view here presented is that of Port Jackson as it was when the illustrious Banner visited it, rather than as it is at the present day, when Sydney has become a large magnificent city, and its wealthy inhabitants have scattered elegant villas and country-seats all over the neighbourhood, when thousands of *Araucarias* have been planted to give variety to the monotony of the Australian vegetation, and when foreign trees, shrubs, and weeds are as fast taking the place of native productions as the white race has usurped that of the black. Yet there is still a great deal of the original vegetation left. Even in Sydney itself, much that is seen in the parks and gardens consists of gum trees and other *Myrtaceæ*, which the hand of man has not planted. We need not go far from the town still to see Banksias with their thick coriaceous leaves and singular flower-heads so much like a grenadier's cap, or to come across the much more singular grass trees, with their charred trunks, grass-like leaves, and tall rod-like scapes of flowers. We can still revel amongst leafless *Acacias*, *Metrosideros*, strange forms of *Proteaceæ*, and Australian *Bignonias*; visit forests where the trees shed their bark instead of their leaves, and all the leaves are turned edgeways; or cast our eyes over large tracts of country still wearing the same evergreen, or rather brownish-green, mantle which it wore when Captain Cook and his naturalists first set foot on the shores of Port Jackson, then the unknown haunt of a few lawless savages, now the capital of Australia and the seat of the Governor-General. Unger, in his 'New Holland in Europe,' has shown us that at one period of the earth's history there flourished in Europe a vegetation very similar to that still beheld in Australia; but that the whole of it has been swept away, to make room for other vegetable forms, leaving no trace behind except what is recorded in the great stone-book of nature. Viewed in this light, the vegetation of New Holland is highly instructive. It is a faithful picture of what the aspect of the flora of our planet must have been ages ago; and on paying a visit to Australia, we are as it were transporting ourselves back to antehistorical periods. The effect which such an inspection produces on one's mind is very singular. It kindles within feelings of curiosity, but no sympathy. We delight in bright green foliage, sweet-smelling flowers, and fruits with some kind of taste in them. But we have here none of all that. The leaves are of a dull green colour, the flowers have no smell, and the fruits, without any exception, are tasteless and insipid. Not a single edible plant has the whole of Australia added to our tables, and Europeans who should have to rely upon what Australian vegetation can supply for their food, would have to share the melancholy fate of Burke and Wills when they tried to eke out their existence by eating the wretched Nardoo seeds of the Australian swamps. [B. S.]

FOREST ON GUAHAN, ONE OF THE MARIANNE ISLANDS.

(Plate VI.)

[REFERENCE.—*a. Ficus* with suspended roots; *b. Cycas circinalis; c. Cordia; d. Cycas circinalis*, old and branched; *e. Cerbera (tabulam?); f. Gigantic Ficus; g. Slender leaved Pandanus.*]

As far as the Mariannes are represented by Guahan, the most extensive and southernmost of these islands, they are at once distinguished from the more northern Caroline group by their dry climate, which imparts to the whole country the look of a steppe. The month of March, in which our illustration was taken, is evidently the dry season; everywhere is aridity, very few trees with fresh foliage are seen in the forest, and perhaps the third part of all is quite leafless. The sea-shores are either kept supplied with moisture by rivulets from the interior, and then overgrown with *Bruguiera* and other Mangroves; or they are sandy, and in that case distinguished by two forms very characteristic of this island—*Cycas circinalis*, very common hereabouts, and a shrubby pyramidal *Casuarina*, which is again met with in the upper steppes of the interior. Banks of coral surround the shores on all sides, making this larger island, as the high Carolines, appear like mountains risen in the centre of extensive coral plains. The plain, shown in our illustration, though destitute of springs, is nevertheless covered with fine tall trees, and, although thorny underwood abounds, is on the whole tolerably easy to penetrate. True, there are occasionally considerable thickets of luxuriant *Cycas*, as shown in the centre of our illustration, a few old trees of considerable height forming an agreeable contrast with this rather chaotic group of saplings. Amongst them are a few branching apparently very old specimens, as seen on the left of our plate. The forest trees include one distinguished by its slender growth and thick foliage (the leaves resembling those of the ash), which vernacularly is termed 'Pai-pai,' and esteemed on account of its extremely hard wood. The same remark applies to another tree of similar aspect, the leaves of which are, however, more like those of the myrtle, whilst the bark is pale yellow. A screwpine, *Pandanus*, though isolated, is rather common. It does not seem to differ essentially from *Pandanus odoratissimus*, and is conspicuous by its slender undulated branches, and long narrow leaves, of which there are comparatively few in each crown. Several species of *Cordia* exhibit their gigantic growth, and are in the dry season but sparingly clad with leaves; here and there their stem is surrounded by a network of creepers. But the most striking of all the trees is a huge species of Fig, the representative of the banyan in this place. It differs evidently in every respect from that of Ualan, the height of which it nowhere seems to attain. Its comparatively tall stem has the appearance of a gigantic bundle of sticks, the component parts of which must be considered as being curiously twisted around each other, and grown together into a compact mass. On the upper end of this rather conical bundle spreads out like an umbrella a crown formed of fantastically twisted branches, which has numerous fine leaves of a dark rather greyish-green. The tree seen on the right-hand side of the foreground is a smaller species of fig, the aerial roots of which have quite the look of creepers. Elegant ferns cover its branches. There is also a species of *Cerbera*, frequently met in the Caroline, Marianne, and Sassine Islands; it resembles in growth and in its leaves he *Terminalia Catappa*, but its principal ranches are more rectangular. [B. S.]

CORAL REEF IN THE CAROLINES.

(Plate VII.)

[REFERENCE.—a. Myrtaceous tree; b. *Scævola*; c. *Tournefortia*; d. Cocoa-nut Palms; e. *Artocarpus incisa*; f. *Tournefortia*; g. *Pandanus odoratissimus*.]

Imagine a chain of comparatively long narrow sandbanks, hardly elevated above the level of the ocean, having a general horseshoe-like outline, and sheltered against the waves by a coral reef surrounding the whole. Everywhere within the latter the water is shallow; the bottom, consisting of coral sand, is evidently rising and gradually becoming dry land, so that the open narrow channels crossing the long ridge of land and dividing it into several islands, will in time disappear. The present view represents one of these channels. Standing at the extremity of one island, we look across upon the other. On the right we have an expanded view of the reef, distant about 200 paces, and behind it the surf of the ocean; on the left we behold the basin of unequal depth, surrounded by the horseshoe-like chain, where the prospect is closed by a few islets of this selfsame chain. Such coral islands, but recently risen above the sea-level, exhibit no trace of that vegetation which establishes itself on the older ones. The first green appearing on the hitherto naked sand invariably consists of shrubby *Scævolas* with small white flowers, which afterwards form also the principal brushwood of the shores, and a specimen of which is represented in the centre of the foreground. The rich juicy foliage of this plant may be well suited to the formation of vegetable mould, in which afterwards a more diversified vegetation finds a home. Next follows a *Tournefortia*, common in all the islands of these seas, which assumes more the look of a small tree, and has a less bushy habit; the silvery grey colour of its leaves forms a strong contrast with the fresh light green of the *Scævola*. A young specimen of this exclusively littoral plant is seen on the right-hand side of the foreground, and an older one in the distance. Close by will be noticed the delicate foliage of another probably myrtaceous shrub peculiar to the outskirts of these forests; an old fully grown specimen of it is seen in the foreground to the left. In the outskirts of the forest at a distance are found, besides the exclusively littoral plants, other half-shrubby trees. Two specimens of *Pandanus odoratissimus*, so common in all these islands, will easily be recognised by their peculiar habit. Their trunks exhibit numerous crowns. On the right-hand side of the smaller specimen to the left are seen, besides the low *Scævola* and the just-mentioned *Myrtacea*, a species of *Hibiscus*, with cordate leaves and dark carmine-coloured flowers, which either occurs as a shrub or small tree; and above it a *Colophyllum*, which in other places becomes a stately forest tree, and has a dark green foliage. Immediately behind it rises an isolated cocoa palm, and more to the right a young specimen of *Barringtonia speciosa*, one of the most beautiful trees of this region, but which grows less freely in these coral islands. Groups of cocoa-nut palms, which suffer little underwood to spring up, show themselves here, and through these may be seen the other end of the forest, a proof of the limited extent of such an island as this. In its centre, where the accumulation of vegetable mould has been going on the longest, stately forest trees have already found a home, among them a large *Eugenia* with lanceolate leaves, and fruits about the size of a large plum of a pale green colour tinged with red, and several bread-fruit trees (*Artocarpus incisa*) of considerable height. [B. S.]

VEGETATION OF TENERIFFE, WITH SUCCULENT EUPHORBIAS.

(Plate VIII.)

Our illustration introduces us to a wild rocky glen, a barranco, on the east coast of Teneriffe, Canary Islands, where succulent cactus-like *Euphorbias* (*E. canariensis* and *piscatoria*), arborescent *Compositæ* and *Rubiaceæ* are the leading plants. The vegetation has a glaucous look. The *Euphorbia canariensis*, with its straight stiff branches, all springing from the root, is generally seen on the top of rocks and the very edges of precipices, imparting a peculiar feature to the landscape, and contrasting strangely with the tree-like *Kleinia nerifolia* with its long naked branches crowded at the top with tufts of leaves, or *Plocama pendula*, almost resembling a weeping willow, and seen in close proximity to a *Kleinia* in the lowermost right-hand corner of our illustration. *Pyrethrum crithmifolium*, *Conyza sericea*, and *Periploca lævigata*, are three plants also found in this spot. The whole may be taken as a fair illustration of the aspect of the vegetation of the coast region of the Canary Islands, where herbage is so scanty as to afford pasturage to only a few flocks of goats. [B. S.]

VEGETATION OF JAVA—TREE-FERNS IN THE FOREGROUND, A FOREST OF AMENTACEÆ IN THE DISTANCE.

(Plate IX.)

The traveller in Java, after emerging from the coast region and ascending to the height of 4,000 to 6,000 feet, experiences so great a change in everything surrounding him, that he can hardly believe himself to be in the same island. Instead of the sultry heat and clammy atmosphere, he now inhales a pure cool air which exercises a delightful reaction upon his spirits. Mountain streams of delicious coolness are met with at every step, and a bright verdure is spread over hill and dale. Our illustration introduces us to a view in the mountains of Java, where a large waterfall dashing over the edge of high perpendicular rocks, looking like a stream of silver from a distance, diffuses an unusual amount of moisture, and favours a great luxuriance of vegetation. Ferns, especially tree-ferns, those palm-like plants, unrivalled in their grace and beauty by any other members of the vegetable kingdom, are here plentiful, and attain often forty to seventy feet in height, their fronds measuring several yards in length. The background of this view is filled by a variety of amentaceous trees, chiefly species of evergreen oaks and *Castanopsis*, growing gregariously together as do their congeners in more temperate climates. [B. S.]

SWAMPY FOREST, WITH BANYAN TREES, IN THE ISLAND OF UALAN.

(Plate X.)

[REFERENCE.—*a.* Spiny *Cyperaceous* plant; *b.* *Ficus*, with root-covered trunk; *c.* Epiphytal *Freycinetia*; *d.* *Barringtonia amplegula*, young plant; *e.* *Cordia*; *f.* *Thamnopteris Nidus*, nestling on a big stem.]

A description of forest peculiar to the tropics. The adjacent ground, just above high-water mark, becomes inundated by the high tide forcing back the water discharged by rivers and rivulets. A soil thus periodically submerged is never quite dry, and only becomes somewhat firm by the gigantic roots of the trees occupying it. In Ualan, these swampy forests have a twofold character. Where the underwood consists of the *Hibiscus tiliaceus*, they are almost impenetrable; where this is wanting, there is, under the huge bower formed by the crown of large trees, a wider prospect. The underwood is composed of numerous small trees, the crowns of which have not been able to attain the height of the larger trees, and therefore remained undeveloped. The greater number of them belong to *Barringtonia aculangula*, the fine drooping flower-bunches of which are often seen on the ground. The stems are covered with epiphytal ferns; amongst them *Thamnopteris Nidus*, which imparts a striking character to the landscape. No less ornamental are the isolated *Freycinetias*, which in Ualan mostly grow epiphytically, and are shown quite in the foreground of the picture. The principal features of the plate are gigantic Fig-trees, such as are often met with in these forests. Those here represented may be assumed as having established, above the heads of other trees, a connection with each other by means of their branches, as is common in this kind of plant throughout India, where entire forests are formed, the stems of which are connected. These latter are the far famed banyan-trees, regarded in some places as sacred. Among the wonderful phenomena of the vegetable kingdom, as displayed in the tropics, they occupy the foremost place. The most striking peculiarity of these trees is their aerial roots, which, springing from the bark, grow downwards, often from a considerable height, but as soon as they touch the ground they enter it and form a new stem. They also have, in a prominent degree, a tendency to grow together as soon as their different parts come in contact with each other, by which is caused that extremely fantastic shape generally observed in these trees. The present species differs from other kinds of banyan, not only in its astonishing height (our illustration shows only the lower parts of the stems), but especially by its drooping aerial roots. These roots, appearing in bundles of tender, originally disconnected fibres, gradually grow together, and, after reaching the ground, increase in thickness. The new stem thus formed soon loses, more or less, all traces of its original formation. The height of the whole is so considerable that the crowns reach above that of other trees, and here and there form as it were a forest above a forest. The spectator, standing below, soon loses sight of the upper parts of the tree, and only notices accidentally the connection existing amongst trees which at first view would seem to be perfectly unconnected. It has been found impracticable to show in our plate the foliage of this tree; of the crown little was visible, and the leaves are small and of roundish shape. All the young saplings growing about here are those of the *Barringtonia acutangula*, which in these woods assumes an epiphytal character. [D. S.]

VEGETATION OF THE CANARY ISLANDS—VIEW IN THE CALDERA.

(Plate XI.)

[REFERENCE.—*a. Pistacia atlantica; b. Juniperus Cedrus; c. Phoenix dactylifera; d. Laurus indica; e. Pinus canariensis.*]

The Canary Islands are covered with a vegetation singularly characteristic of their geographical position. It is neither strictly tropical, nor typical of the temperate zone, but rather a blending of the forms most peculiar to either. A singular instance of this is presented to us in Webb's view of the Caldera. Surrounded by steep perpendicular walls of rock 4,000 feet high, that glen enjoys, like a garden conservatory, a temperature always uniform, allowing plants from all heights to flourish in company with each other—the Canarian cedar (*Juniperus Cedrus*) from the most elevated mountain ridge, and *Kleinia neriifolia* from the hot coast region. Here may still be witnessed the strange phenomenon of date-palms and pine-trees growing in the same spot harmoniously together. Leopold von Buch doubted the existence of this vegetative harmony, which had been mentioned by Viera, one of the earliest writers on the Canaries. But the fact is now placed beyond doubt by the united testimony of Berthelot, Webb, and Bolle. But the hand of man, even in this mysterious, almost inaccessible workshop of Nature, the way to which leads through so many dangers along yawning precipices, has not spared the 'Fawns of the wilderness' banished hither. It has allowed the fire to accomplish what the axe was not able to do. In September 1852, says Bolle, there stood only, on one inaccessible rock near the Barranco del Almendrero Amargos, surrounded by pine-trees, one solitary wild palm. Heine's conception of the longing of the two trees, so beautifully expressed in one of his elegies, had here found its realisation. [B. S.]

VEGETATION OF THE CINCHONA FORESTS OF PERU, WITH PALMS AND TREE-FERNS.

(Plate XII.)

The valley of San Juan del Oro represented in the accompanying engraving, is a continuation of the ravine of Sandia, in the Peruvian province of Caravaya. In this province great spurs run out from the main chain of the Cordillera, and gradually subside into the vast plains, covered with virgin forest and traversed by unvivable rivers, which extend to the Atlantic. These spurs form beautiful valleys, such as that of San Juan del Oro, which was once famous for its gold washings. Here torrents and cascades pour down on every side into the river flowing through the valley, and the mountain-sides are clothed with the richest subtropical vegetation. Here may be seen gigantic buttressed trees, festooned with creepers and fringed with graceful ferns and orchids; here are tall tree-ferns, bright-flowering melastomaceous shrubs, and numerous species of palms. Among them are the tall chonta, with its hard serviceable wood; the slender beautiful chinilla (*Euterpe*); the towering uuruma (*Iriartea*), with its roots shooting out from eight feet above the ground; and an *Astrocaryum* with thorny leaves, and a lofty stem thickly set with alternate rings of spines. But the prevailing vegetation of this valley, which is about 5,000 feet above the sea, consists of plants of the cinchonaceous order, with their graceful foliage and panicles of fragrant flowers. Among them are several species of the *Cinchona* which yields the inestimable bark of commerce. It was in these lovely Caravayan valleys that Mr. Clements Markham made a collection of cinchona plants for introduction into India, while he caused other collections of plants and seeds to be made in Northern Peru and Ecuador. Thus the cultivation of these precious quinine-yielding trees, which were until lately only met with growing wild in such valleys as that of San Juan del Oro, is now successfully established in our great Indian possessions. The cascarilleros or bark collectors are represented in the plate as engaged in packing the bark, previous to its being forwarded to the nearest depôt, on the backs of Indians. [B. S.]

VEGETATION OF JAVA.

(Plate XIII.)

[REFERENCE.—*a. Pandanus latifolius; b. Eriodendron indicum.*]

Perhaps one of the most singular genera of plants of the eastern hemisphere is that of the Screw-pines (*Pandanus*), so called from their long narrow sword-shaped leaves being arranged around the stem like the windings of a screw, and their fruits having the outward appearance of pine cones. In many instances their stem is branched and tree-like, and in several of our plates specimens of this mode of growth may be seen; but in some instances the stem is simple, and on the left-hand side of our present illustration will be noticed one of the finest and most robust species inhabiting the island of Java, the *Pandanus latifolius*. It grows here in company with feathery palms and the *Eriodendron indicum*; the latter easily recognised by its strictly horizontal branches, arranged in distinct whorls at certain intervals around the stem, and imparting to it the look of a coniferous tree. There are very few plants in the lower coast region of the tropics that have a similar habit. We can only recall *Terminalia Catappa*, and some *Myristicas*. [B. S.]

MOUNTAIN VEGETATION OF JAVA.

(Plate XIV.)

[REFERENCE.—*a. Rafflesia Rochusseni; b. Vanilla; c. Freycinetia; d. Selliguea.*]

Few spots on the globe support a more luxuriant and diversified vegetation than the island of Java. It is literally teeming with botanical treasures. Ferns and orchids, palms and oaks, bananas and nutmegs, vines and convolvuli, and an endless host of other plants of which not even the name has penetrated beyond the circle of scientific botanists, cover its surface. In the illustration before us, the artist has contrived to introduce us to a genus of plants which bears the most gigantic of flowers, the famous *Rafflesia*. Nature has equally divided her gifts by according to the New World the plant with the largest leaves (*Victoria regia*), to the Old World that with the largest flowers (*Rafflesia Arnoldi*); and it is not a little singular that both these plants, notwithstanding their prominence, have only been discovered in recent times. *Rafflesia Arnoldi* has flowers often three feet across, but, alas! it has no leaves. The gigantic flowers are seated on the stems of vines, different kinds of *Vitis (Cissus)*, from which they draw their nourishment parasitically. The species figured in our illustration is *Rafflesia Rochusseni*, not quite so large as the first-mentioned species, surrounded by excepting *Vanillas, Freycinetias*, ferns, and other mountain plants. A Dutch gardener, M. Teysmann, was the first who, by carefully observing the way in which *Rafflesias* grow, succeeded in cultivating and flowering them in the Botanic Garden at Buitenzorg in Java, and there is reason to hope that at no distant day we may grow in our hot-houses the largest-flowered plant of the eastern, as we do the largest-leaved plant of the western hemisphere. [B. S.]

VEGETATION OF NEW HOLLAND— AN ACACIA 'SCRUB.

(Plate XV.)

Among those plants which by beauty and elegance attract our attention, the *Acacias* occupy a prominent place. Few genera are richer in singular forms, or possess a greater number of truly ornamental species. Their graceful branches, airy foliage, and numerous often fragrant blossoms, have made them favourites with all those who are sensible to the charms of the vegetable kingdom. Especially the Acacias called *Phyllodineae*, are, by their habit, their curiously shaped and highly developed leaf-stalks, the absence of true leaves in old plants, and their diversified tints, even if destitute of flowers, objects of particular interest; and although the species with pinnated leaves do not rank so high in this respect, they are nevertheless not destitute of grace or beauty. The genus *Acacia*, though now considerably reduced, contains upwards of 300 species, a great number of which are peculiar to New Holland and the adjacent islands. Indeed, the *Phyllodineae* are almost exclusively Australian, only one species, *Acacia Koa*, being found north of the Equator in the Hawaiian or Sandwich Islands. The acacias are social plants; woods are often entirely formed of them, a fact which increases the commercial importance of several species. Our illustration introduces us to a grove of *Acacias* in Eastern Australia, the burying-ground of Milmeridien. 'On reaching the spot,' says Mitchell, in his *Three Expeditions into Eastern Australia*, 'the natives became silent and held down their heads. Nor did their curiosity restrain them from passing on, although I unfolded my sketch-book, which they had not seen before, and remained there half an hour for a purpose of which they could have no idea. The burying-ground was a fairy-like spot, in the midst of a "scrub" of drooping *Acacias*. It was extensive, and laid out in walks, which were narrow and smooth, as if intended only for "sprites," and they meandered in gracefully curved lines among the heaps of reddish earth which contrasted finely with the *Acacias* and dark *Casuarinas* around. Others girt with moss shot far into the recesses of the bush, where slight traces of still more ancient graves proved the antiquity of these simple hut touching records of humanity. With all our art, we could do no more for the dead than these poor savages had done.' (B.S.)

VEGETATION OF KAMTSCHATKA, WITH TALL UMBELLIFERS—A BIRCH FOREST IN THE DISTANCE.

(Plate XVI.)

[REFERENCE.—a, b, Tall species of Nettle; c, Angelica.]

Gigantic umbelliferous plants are more characteristic of the grassy plains of Kamtschatka than of any other part of the Bolschaja Reka district. The tallest among them are the *Heracleum dulce* (?), and a species of *Angelica* of surprising dimensions; it abounds in a few level valleys of the western slopes, principally in the district traversed by the Bannaja Reka, a tributary of the Bolschaja Reka, but is not met with again even in the neighbourhood of its real home. This stately herb is known throughout the country by the Russian name of 'Medweshie Koren' (Beor's root); its hollow stems are dark reddish in the autumn. Another plant is a tall, always gregariously growing nettle (*Urtica dioica*), which contributes an essentially characteristic feature to the country, but which does not occur anywhere in such masses as in these western districts. It is generally ten feet high. Its long stems yield a superior fibre for nettle yarn, which in former times was the only material the Kamtschatkans had for fishing-nets; lately it has in some measure been displaced, on the Kamtschatka river, by plantations of hemp, which also attains an astonishing height. In the western plains, however, it abounds in such quantities as to have preserved its place in the domestic economy of the inhabitants. The forest at the background consists of *Betula Ermanni*, edged by low willows. (B.S.)

CACTUS VEGETATION ON THE BANKS OF THE GILA, NEW MEXICO.

(Plate XVII.)

[REFERENCE.—a, *Cereus giganteus*.]

The *Cactaceæ* or family of Cactuses furnish perhaps the most singular plants of the present creation. Though a few species are found in the Old World, the bulk of this natural order is confined to the New; and in no part of America do we encounter a greater number both of species and individuals than in Mexico, California, and those countries until recently part and parcel of Mexican territory. In that region, but seldom visited by refreshing showers of rain, the Cactus is the leading plant, imparting a peculiar character to the landscape. The greater number of Cactuses are without leaves, and the generality present fleshy, globular, oblong compressed, or cylindrical stems, densely covered with bundles of spines. An estimate may be formed of the number of these spines, by stating that a single specimen of an *Echinocactus Visnaga*, the Toothpick Cactus, was found to have 51,000, and a *Pilocereus senilis*, the Old Man Cactus, 71,000. Yet the specimens on which they were counted were only such as had been brought over to England. The giants met with in Mexico and surrounding regions have hundreds of thousands of spines. The favourite haunts of Cacti are the mountain ridges which intersect or border the Mexican table'and, 5,000 to 7,000 feet above the sea. The view chosen as an illustration of this singular vegetation is a landscape on the banks of the river Gila, in the Colorado region of New Mexico, representing the largest form of all known *Cactaceæ*, the *Cereus giganteus*, which rises like a huge candelabrum amongst the rocks and ravines of that barren wilderness. In front is a specimen which, though already nearly sixty feet high, is still in vigorous health, and sending forth young side branches. On the right, a little towards the background, is a specimen in a decaying state, showing the form of a woody skeleton; and around them at short distances may be seen younger plants in various stages of development. A few *Opuntias* (Cactuses with compressed articulated stems), feathery *Mimoseæ*, the usual companions of

Cactus vegetation, affording shelter against the sun to the young plants, a couple of *Agaves* also typical of Mexico, and some herbs, are distributed over the soil, as yet the roving-ground of the wild Indian. Young plants of *Cereus giganteus* retain a globular shape for several years; and they begin to flower when about ten to twelve feet high. We have actual measurement of stems 45 feet high, so that there is nothing improbable in Colonel Emory's disputed statement that the *Cereus* attains 50 to 60 feet in height. The stem is thickest at or a little above the middle, and tapers upwards and downwards. It is mostly simple, but the older ones have often a few erect branches. Both stem and branches are ribbed, almost fluted like columns, and covered with bundles of spines. The flowers are produced in abundance near the summit, and the fruit has a crimson-coloured, sweet, but rather insipid pulp. [B. S.]

HYPHÆNE OR DOUM PALM-TREES IN UPPER EGYPT.

(Plate XVIII.)

As the traveller is leaving the lower and gradually ascending into the upper portion of Egypt, he meets with the most characteristic of African trees, in the shape of branched Palms, the famous *Hyphæne thebaica*. They are seen in their full beauty about the cataracts of the Nile, as represented in our plate. The contrast between these trees and the rest of the palm tribe is very great. Whilst most of the palms have a straight pole to which at the upper extremity a number of feathery leaves are attached, we have here a regularly branched tree, somewhat like a screw-pine or *Pandanus*, and large fan-shaped leaves, between which grow large bunches of light yellow fruits with a thick meaty rind, so much resembling in look and taste real gingerbread as to have conferred upon the palm the name of Gingerbread tree. In Cairo and other towns of Egypt these bunches are exposed for sale in the market-places, together with dates, figs, oranges, and other produce of the country. The wood of the tree is used for various domestic purposes; the seed of the fruit is eaten; and the kernels turned into beads for rosaries, and at Kano into toys. From the hieroglyphics we know that it was cultivated more than 4,000 years ago in and about Thebes; but though always a leading tree and a most striking object of the Egyptian landscape of the Upper Nile, the *Hyphæne* does not seem to have exercised, as far as we know, any decisive influence upon ancient Egyptian architecture, like the date palm (*Phœnix dactylifera*) and the Deleb palm (*Borassus æthiopum*), for instance, have done. The peculiar swelling of ancient columns is evidently copied from the trunk of the Deleb palm, of which a singular bulging out is one of the most striking characteristics; whilst the capitals of the column are, in many instances, slavish copies of the crowns of the date-palm, as may be seen in the ancient temple at Edfoo. The exact geographical range of *Hyphæne thebaica* has as yet to be ascertained. We know it extends considerably into Central Africa, but do not know exactly where it leaves off, as we have no botanical specimens to decide the question, and have to depend upon the information of travellers not able to discriminate between this species and those allied to it. It is certain, however, that there are more than one species of *Hyphæne*, and that some of them at least have a straight cylindrical and unbranched trunk, like that of the generality of palms. [B. S.]

VEGETATION ON THE ICE-CLIFFS IN KOTZEBUE SOUND, ARCTIC AMERICA.

(Plate XIX.)

The soil of the Arctic region is always frozen, and merely thaws during the summer months a few feet below the surface. But the thawing is by no means uniform. In peat it extends not deeper than two feet, while in other formations, especially in sand or gravel, the ground is free from frost to the depth of nearly a fathom. The roots of the plants, even those of shrubs and trees, do not penetrate into the frozen subsoil. On reaching it they recoil as if they touched upon a rock through which no passage could be forced. It may be surprising to behold a vegetation flourishing under such circumstances, existing independent as it were of terrestrial heat. But surprise is changed into amazement on visiting Kotzebue Sound, where, on the top of icebergs, herbs and shrubs are thriving with a luxuriance only equalled in more favoured climes. There, from Elephant to Eschscholtz Point, is a series of cliffs from 70 to 80 feet high, which present some striking illustrations of the manner in which Arctic plants grow. As may be seen in our plate, three distinct layers compose these cliffs. The lower, as far as it can be seen above the ground, is ice, pure ice, and from 20 to 50 feet high. The central is clay, varying in thickness from 2 to 20 feet, and being intermingled with remains of fossil elephants, horses, deer, and musk oxen. The clay is covered with peat, the third layer, bearing the vegetation to which it owes its existence. Every year, during July, August, and September, masses of the ice melt, by which the upper-most layers are deprived of support and tumble down. A complete chaos is thus created; ice, plants, bones, peat, clay, are mixed in the most disorderly manner. It is hardly possible to imagine a more stoneless [sic] aspect. Here are seen pieces of peat still covered with lichens, mosses, and saxifrages; there a shoal of earth with bushes of willow; at one place a lump of clay with *Smecies* and *Polygonums*; at another the remnants of the mammoth, tufts of its hair, and some brown dust which emits the smell peculiar to burial-places, and is evidently decomposed animal matter. The foot frequently stumbles over enormous osteological remains—some elephant's tusks measuring as much as 12 feet in length, and weighing more than 240 pounds. Nor is this formation confined to Eschscholtz Bay. It is observed in various parts of Kotzebue Sound, on the river Buckland, and in other localities, making it probable that a great portion of extreme North-Western America is, underneath, a solid mass of ice. With such facts before us, we must acknowledge that terrestrial heat exercises but a limited and indirect influence upon vegetable life, and that to the solar rays we are mainly indebted for the existence of those forms which clothe with verdure and gay flowers the surface of our planet. (B. S.)

HOLY CROSS ABBEY, COVERED WITH IVY.

(Plate XX.)

It is now, thanks to the indefatigable labours of Mr. H. C. Watson, an easy task to give a scientific man a clear idea of the nature and extent of the flora of our British Islands, by explaining to him that the whole territory is divisible into six zones of altitude, the super-arctic, the mid-arctic, the infer-arctic, the super-agrarian, the mid-agrarian, and the infer-agrarian, and yellow botanical provinces, the boundaries of which are founded upon physical and not upon political differences; and that the vegetation comprised in these divisions is composed of so-called Germanic, Scandinavian, Iberian, Boreal, and North-American types. This explanation, however, would convey but a vague notion of what the vegetation of the British Islands really looked like to one who had not had an opportunity of familiarising himself with the nature of the different zones, or the character of the types. To conjure up any idea of what the British flora really appears like, we should have to speak of waving corn fields, smiling meadows, shady lanes, mossy tombstones, jew-girt churches, gloomy pine woods, purple heather, and golden furze—objects which at once recall scenes and aspects of nature familiarised by the pen of the poet and the brush of the painter. For that reason we have chosen as one of the most characteristic features of the vegetation of the British Islands, Holy Cross Abbey covered with Ivy. This Ivy, it is true, is not peculiarly British, but diffused over the whole of Europe in several distinct varieties, some of which have white, some yellow, and some black berries. The yellow-berried Ivy is confined to the south of Europe, and is the plant with which in times gone by poets were crowned, and which played so prominent a part in the festivals held in honour of Bacchus. The black-fruited variety is much more common, and the one indigenous to our islands. Though it cannot claim the distinction of having encircled the heads of poets, it has furnished the theme of many a poet's song, and in no part of Europe does it thrive with such luxuriance as in the British Islands, especially in Ireland, where, favoured by a humid and mild climate, it ascends the tops of the highest trees, covers with its thick evergreen foliage rocks and walls, and gives a picturesqueness to many an old ruined castle or Gothic abbey. It has been mentioned that in remote times our European Ivy, *Hedera Helix*—at least the yellow-berried variety—was brought from the highlands of Asia; but the Ivy which flourishes in Nepal and throughout the Himalayas with such luxuriance is a species quite distinct, being covered with minute yellow scales instead of white stellary hairs, as our Ivy is. Our *Hedera Helix* is a strictly European plant, which may be said to attain in Britain its highest development, imparting to some of its landscapes a striking and characteristic peculiarity. (B. S.)

THE TREASURY OF BOTANY.

AARON'S BEARD. *Hypericum calycinum.*

ABACA. A name given in the Philippine Islands to *Musa textilis*, which yields Manilla hemp.

ABACOPTERIS. A name given by Fee to a group of the species of *Nephrodium*, in which the veins of the fronds are united in numerous superposed angles. [T. M.]

ABATIA. A small genus of *Lythraceæ*, consisting of Peruvian and Brazilian shrubs with greyish tomentum; opposite, shortly stalked, undivided crenate-serrate leaves, and terminal racemes of rather small, dull purplish, apetalous flowers, with numerous stamens. [J. T. S.]

ABBREVIATIONS. Signs to express particular attributes are largely employed by botanists. The following are those most in use:

♂ = male.
♀ = female.
☿ = hermaphrodite, or bisexual.
♂ - ☿ - ♀ = polygamous.
♂ ♀ = diœcious.
♂ - ♀ = monœcious.
♂ - ☿ - ♀ = triœcious.
① or ○ = annual.
② or ♂ = biennial.
♃ = perennial.
♄ = a tree or shrub.
∞ = an indefinite and considerable number of anything.
! placed after a person's name, indicates that an authentic specimen from that person has been seen.
* at the end of a citation, denotes that a plant is fully described in the place referred to.
v.v. = seen alive.
v.s. = seen in a dried state.
v.c. = seen cultivated.
v.sp. = seen wild.
′ ″ ‴ When these signs are placed after a number, they express a foot, an inch, or a line respectively; thus,

5′ = 5 feet.
5″ = 5 inches.
5‴ = 5 lines.

A very full account of all such signs is given in *Lindley's Introduction to Botany*, ed. 4, ii. 384.

ABELE TREE. *Populus alba.*

ABELIA. A small genus of ornamental shrubs, found in India, China, Japan, and Mexico, and belonging to the natural order *Caprifoliaceæ*. The species are of slender branching habit, bearing opposite leaves and terminal bunches of showy tubular flowers. The genus is distinguished by having an oblong calyx-tube, which is connate with the ovary, and terminated by a five-parted limb of foliaceous segments; a tubular funnel-shaped corolla, with a five-lobed spreading limb; four subdidynamous or nearly equal scarcely exserted stamens; a capitate stigma; and a three-celled ovary, of which two of the cells are many-ovuled, but abortive, and the other one-seeded and fertile, becoming a coriaceous berry. *A. floribunda*, which is a native of Mexico, is a very handsome freely-branching shrub, naturally rather straggling in habit, producing opposite, blunt, ovate, crenate leaves, which are smooth on the surface, and having large showy blossoms, which come from the axils of the leaves, at the ends of branches, so as to form a pendent leafy panicle. These flowers are a couple of inches in length, rich purple-red, tubular, the tube narrowing at the base and enlarging upwards, and finally spreading out into a limb of five nearly equal rounded lobes. *A. rupestris*, a native of China, on the Chamoo hills, has shorter tubular flowers, of a pale rose colour, and forms a lovely dwarf bush, loaded towards autumn with its ornamental blossoms. The few known species are rather objects of ornament than of utility. [T. M.]

ABELICEA. A genus of *Ulmaceæ*, containing a single species from Greece and Eastern Asia. It is so nearly related to *Planera* that it would perhaps be better to consider it as a section of that genus, separated from the true *Planeræ* by its smooth capsule and subsessile leaves. Both have alternate, ovate, crenate-serrate leaves, like the elm. The flowers are hermaphrodite, or polygamous from the non-development of parts. They occur in axillary fasciculate clusters, the inferior flowers of the fascicle being staminal, the superior hermaphrodite or rarely pistilline. There are four or five stamens. The ovoid ovary is one-celled and one-ovuled, and crowned with two spreading styles, which are stigmatose down the inner side. [W. C.]

ABELMOSCHUS. The name applied to a genus of plants of the mallow family

(*Malvaceæ*). The word is derived from the Arabic, signifying musk seed, and was given in allusion to the agreeable odour of the seeds of one species, *A. moschatus*, a native of Bengal. The seeds of this plant were formerly mixed with hair powder, and are still used to perfume pomatum. They possess cordial and stomachic properties, and are mixed with coffee by the Arabs. In the West Indies the bruised seeds, steeped in rum, are used both externally and internally as a remedy for snake-bites. *A. esculentus*, formerly called *Hibiscus esculentus*, a native of the West Indies, but naturalised in India, furnishes the Ochro or Gobbo pods, that are much used in thickening soups, for which their abundant mucilage well fits them. The young pods are gathered green and pickled like capers. The plant is cultivated in the south of France for the sake of its pods, which when ripe are of a conical shape, covered with hair, and about an inch in length. All the species of this genus furnish excellent fibre. The genus is botanically characterised by a deciduous ten-leaved involucel, a spathe-like, tubular, conical, five-toothed calyx, spreading petals, one-celled anthers, a style cleft into five divisions at the top, a capsule with five cells and five valves, whose edges are not bent inwards. [M. T. M.]

ABERRANT. Something which differs from customary structure. Also a group of plants which stands intermediate, as it were, between two other groups:—e.g. *Fumariaceæ*, which are by some regarded as an *aberrant* group of *Papaveraceæ*.

ABIES. In this genus of the cone-bearing family (*Coniferæ*) are included the plants commonly called Firs, in contradistinction to pines (*Pinus*). The firs are for the most part lofty trees, with small, narrow evergreen leaves, placed in two rows along the sides of the branches, or occasionally tufted. The flowers are unisexual; but the male flowers are borne upon the same trees as the female ones, and both kinds are produced in catkins. The mature female inflorescence constitutes the cone, which is usually of a cylindrical form, consisting of a number of woody scales overlapping each other, but not thickened at their points, as in pines. The species of fir are remarkable as timber trees, and for yielding turpentine, &c.

A. excelsa is the common or Norway Spruce Fir, which when well grown is a handsome tree, sometimes reaching the height of one hundred to one hundred and fifty feet. The leaves are of a dull green colour, of a four-cornered shape, and sharply pointed; the cones are cylindrical, pendulous, their scales bluntish, or slightly waved or toothed. The tree is a native of Norway, Russia, and the mountainous parts of Europe generally, thriving best on a damp soil. Its timber is much used under the name of white deal. From its trunk exudes a resin commonly called frankincense, which, when melted in water and strained, constitutes Burgundy pitch. The young leaf-buds or shoots of this and other species are boiled down in water to form essence of spruce, from which spruce beer is made.

A. balsamea is the Balm of Gilead Fir, an American tree of much smaller stature than the common spruce fir, with flat leaves, whitish beneath. Its cones are erect. It yields a pure form of turpentine, called Canada balsam, much used for optical purposes, and for preserving certain microscopic objects.

A. canadensis, or the Hemlock Spruce, is a native of North America and Canada, and, from its abundance and eminent beauty, is frequently referred to by the American poets under the name of the hemlock. The bark is much used for tanning purposes.

A. Picea, the Silver Fir, is so called from its leaves, which are whitish on their under surface, arranged in two rows, and have their points turned upwards. The cones are erect, of a greenish purple colour, their scales provided with long tapering bracts on their outer surface. The beauty of this tree is such that Virgil has applied to it the epithet *pulcherrima*, 'very beautiful.' It attains a height of 100 feet and upwards, and is a native of Central Europe and Northern Asia. Its timber is not so much prized as that of some other species, but is durable under water, and from its bark exudes a resin which, when purified, is known as Strasbury turpentine.

A. Larix is the common Larch fir. Its needle-shaped leaves are at first arranged in tufts, but subsequently become separated one from the other by the lengthening of the branch upon which they grow. They fall off at the approach of winter. The cones are small, erect, somewhat egg-shaped, but blunt-pointed, and the scales have irregular margins: for these reasons the larch is sometimes placed in a distinct genus, and called *Larix europæa*. The wood of the larch is much prized, and very durable; its bark is employed by tanners, and it, as well as the trunk, affords what is known as Venice turpentine, which differs from most other kinds of turpentine in not becoming hard by exposure to the air for a considerable period. A kind of sugary matter exudes from the larch in summer time, and is collected under the name of Manna of Briançon. The larch attains a great size, and forms a most beautiful object on the mountain sides in Switzerland and other Alpine districts of Europe, and is much cultivated in this country for the sake of its timber, while its pyramidal form, pendent branches, light green leaves, and purplish cones render it a very beautiful tree for ornamental purposes. Round some of the meres or lakes in Shropshire the larch is abundantly planted. Its leaves fall into the water, and become felted together into large ball-like masses by the agency of a peculiar species of *Conferva*. These larch balls may be met with of all sizes, from that of a marble to that of a child's head; they lie at the bottom of the lake, and are washed up round its margins.

A. Cedrus, or, as it is sometimes called, *Cedrus Libani*, is the well-known Cedar of Lebanon. It is principally distinguished from the larch by its evergreen leaves, and by its cones, which are from three to five inches in length, oblong, blunt, erect, and composed of numerous densely packed scales of a purplish-brown colour. They are not fully ripened till the third year, and remain on the tree for several years. The tree is a native of the mountains of Lebanon and of Taurus, where its majestic form and huge spreading branches render it a very prominent feature. A recent traveller in Syria, Mr. Urquhart, thus speaks of it. 'The trunk dividing at from ten to twenty feet from the ground, the branches contorted and snake-like, spreading out as from a centre, and giving to the tree the figure of a dome; the leaf-bearing boughs spread horizontally, the leaves or spiculæ point upwards, growing from the bough like grass from the earth. The leaves are thick and short, about an inch in length. The cones stand up in like manner, and are seen in rows above the straight boughs. The timber is in colour like the red pine.' The wood has been said to be very durable, but there is some reason to think that the wood of a species of Thuja has been mistaken for that of the cedar of Lebanon, which is not so indestructible as was once supposed. From the noble appearance that the tree presents, it is frequently met with in parks, &c., the habit or general appearance of the tree, and the arrangement of its branches, differing considerably in different individual trees. Many magnificent trees of this species are to be seen in Blenheim Park, Oxfordshire; but scarcely any two are alike in the disposition of their branches or the colour of the leaves.

A. Deodara, or *Cedrus Deodara*, the Deodar or Indian Cedar, differs from the cedar of Lebanon, in having the cones placed on short thick stalks; and the scales of the ripe cone fall off, instead of being persistent, as in the Lebanon cedar, while its leaves are longer and more distinctly three-sided than in that plant; but it is by no means certain that the two plants are really specifically distinct. The individual plants forming the species of this genus differ so remarkably in habit and general appearance one from the other, that great caution is necessary in dogmatising as to the distinctness of this or that form. The Indian or Deodar cedar is a native of Nepal and of the Himalayas, where it attains a height of from fifty to one hundred feet and upwards. Its timber is of great value from its durability, and it furnishes a turpentine which is much employed as a medicament by the natives in North-Western India. It was introduced into this country in the year 1822, and is now much cultivated as an ornamental tree, from its elegant form, gracefully pendent branches, and the glaucous hue of its foliage.

A. atlantica, the Algerian, or Mount Atlas Cedar, called also *Cedrus atlantica*, forms almost the entire vegetation of the upper mountainous regions of certain provinces of Algeria. According to M. Cosson, there is no doubt but that this is a mere variety of the Lebanon cedar, from which it differs in the length of its leaves. The form and size of the cones are too variable to constitute a point of distinction.

Several other species of this genus are grown in this country as ornamental trees, among which *A. bracteata* and *A. Douglasii*

Abies Douglasii (cone).

may be mentioned as particularly interesting species. [M. T. M.]

ABNORMAL. Opposed to usual structure. Thus, stamens standing opposite to petals, and nowhere else, as in Rhamnads, are abnormal, it being usual for stamens to be alternate with petals, if equal to them in number. Leaves growing in pairs from the same side of a stem, as in *Atropa Belladonna*, and flower-stalks adherent to the midrib of a bract, as in *Tilia*, are also abnormal.

ABO:.A radiata is a curious little orchid from New Grenada, differing from *Odontoglossum* and *Oncidium* in having a slender delicate caudicle, and solid pollen masses. The flowers are brown, with yellow streaks and a white lip.

ABOLBODA. A genus of *Xyridaceæ*, containing six or seven species of stemless plants, growing in tufts in the marshes of South America. This genus is nearly allied to *Xyris*, but differs from it in having the ovary and capsule always three-celled, while the predominant form in *Xyris* is one-celled, and, when otherwise, but imperfectly three-celled. The ovules also are attached to the central axis, while in *Xyris* they rise from parietal placenta. [W. C.]

ABORTIVE. Imperfectly developed; as abortive stamens, which consist of a filament only; abortive petals, which are mere bristles or scales.

ABRICOT SAUVAGE. A French name, used in the West Indies for the Mammee apple. Also applied in Cayenne to the fruit of *Couroupita guianensis*.

ABRODICTYUM. A name given by Presl to a very elegant species of *Trichomanes*, differing only in the form and arrangement of the cells of its tissue. *A. Cunningii* is now generally called *T. Smithii*. Van den Bosch has revived the name under the form of *Habrodictyon*. [T. M.]

ABROMA. The name given to a genus of the byttneriaceous family. They are small trees, with hairy lobed leaves, and terminal or axillary clusters of yellow or purple flowers. Their fruits are capsular, five-celled, with five membranous wings, and many seeds in each cell. They are natives of India, Java, and the tropical parts of New Holland. Three species are known; one of them, *A. augusta*, is the Wollut Comal, or Wullut Cumal, of the Benarlese. Its bark abounds with strong white fibres, which afford a good cordage. The plant grows quickly, and may be so managed as to afford three crops of cuttings in the year. The bark is separated from the shoots by maceration in stagnant water. From four to eight days is sufficient to effect this in hot weather, but in the cold season a much longer time is required. The fibre requires no artificial cleaning. Cord made from it, though not to be compared with that of hemp, is strong, and is not liable to be weakened, as many others are, by exposure to wet. [A. A. B.]

ABRONIA. A small genus of monochlamydeous plants, belonging to the order *Nyctaginaceæ*. In this genus the leading peculiarities of structure are a five-leaved involucre, surrounding a close head of many flowers; a coloured corolla-like silver-shaped perigone, having the tube inflated below, and the deciduous limb five-lobed, spreading, with obovate lobes; five hypogynous included stamens, connate at the base, and having oblong anthers; a simple style with a club-shaped stigma; and a one-celled ovary containing one erect ovule. There are but few species known, and these natives of N. W. America. *A. umbellata*, one of the best known, is a handsome dwarf trailing perennial herb, producing oppositely stalked, bluntly ovate, rather succulent leaves, and, from their axils, long-stalked, close umbels of pretty primula-like flowers, of a purplish rose colour. These flowers consist of a coloured calyx, the corolla being wanting; and they are very deliciously fragrant, especially towards evening. The other species are of similar character. They are not applied to any use. [T. M.]

ABRUPT. Suddenly terminating; as abruptly pinnated, when several pairs of leaflets are formed without any intermediate one at the end.

ABRUS. The name of a genus of plants of but one species, *A. precatorius*, and belonging to the pea tribe, of the order *Leguminosæ*. The plant was originally a native of India, but is now found in the W. Indies, the Mauritius, and other tropical regions. It is chiefly remarkable for its small egg-shaped seeds, which are of a brilliant scarlet colour, with a black scar indicating the place where they were attached to the pods. These seeds are much used for necklaces and other ornamental purposes, and are employed in India as a standard of weight under the name of Ratti. The weight of the famous Koh-i-noor diamond is known to have been ascertained in this way. The roots also are made use of in the same manner as the roots of the liquorice plant. The *Abrus* is of twining habit, with pinnate leaves, numerous stalked flowers in axillary clusters, a bell-shaped slightly four-lobed calyx, the upper lobe broadest, and a pale purple corolla, succeeded by an oblong compressed pod containing four to six seeds. [M. T. M.]

ABSINTHE. (Fr.) *Artemisia Absinthium*. — PETITE. *Artemisia pontica*.

ABUTILÆA. An E. Australian plant of the mallow family, not really distinct from *Abutilon*. [A. A. B.]

ABUTILON. A genus of *Malvaceæ* (mallow family), known by having a cup-shaped calyx without an involucre, an ovary of five carpels which open at the top, and are inseparably adherent one to the other by their inner angles. They are annual or shrubby plants, often very ornamental, inhabiting the W. Indies, Siberia, and even Piedmont. The flowers of one species, *A. esculentum*, are used as a vegetable in Brazil. *A. indicum* and *polyandrum*, Indian shrubs, furnish fibre fit for the manufacture of ropes. Their leaves contain a large quantity of mucilage. [M. T. M.]

A. striatum, *venosum*, *insigne*, and some others, are favourite garden plants, often seen in our gardens and greenhouses. They have palmately-divided or heart-shaped leaves, and axillary pendulous flowers, of which the petals converge so as to give them a semi-globular bell-shaped outline. They are of considerable size, yellow or white, beautifully veined with red. [T.M.]

ACACALLIS cyanea. Under this name Lindley describes a handsome Brazilian orchid with the habit of a *Huntleya* and large light blue flowers. It is distinguished from that genus by having a long narrow hypochil with a deep sac at its point, surrounded by a five-lobed border. It was found by Spruce, on trees near the Rio Negro.

ACACIA. A genus of shrubs or trees belonging to the *Mimosa* tribe of the leguminous family. Its principal points of distinction are the calyx, which is provided with four or five teeth, the corolla of four or five petals, the numerous stamens, and the pod, which is not divided into joints, and which does not contain a pulp. The great number of the stamens and the nature of the pod particularly distinguish this genus from the allied genus *Mimosa*. The flowers, which are small, are collected in large numbers in globular heads, or in long spikes. The true leaves are twice or thrice pinnated, and the small leaflets, being very numerous, confer a very elegant feathery appearance on the plants, but in many of the species, particularly those found wild in New Holland, the true leaves are seldom or never developed, but, to compensate for their absence, the leaf-stalk, which is usu-

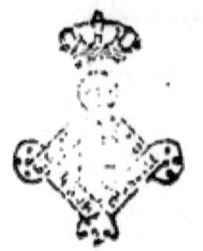

VEGETATION OF NEW HOLLAND: AN ACACIA SCRUB.

ally more or less cylindrical, and of small dimensions, becomes flattened out, and assumes a leaf-like appearance; these dilated leaf-stalks, or, as they are technically termed, *phyllodes*, fulfil the functions of the leaves, and are of very varied form in the different species. They are always so placed that their *edges* look upwards and downwards, so that by this means, as well as by the arrangement

Acacia argyrophylla (with phyllodes).

of the veins, they may be distinguished from true leaves, which have their *surfaces* looking upwards and downwards. It results from the singular position of these organs that the trees possessing them give but little shade, as the light is not intercepted in its passage to so great an extent as it is by the leaves of ordinary trees. This peculiar direction of the leaf is not confined to the acacias, but is also found in other Australian trees: e. g. *Eucalyptus*. At the base of the leaf-stalk of the acacias, where it joins the branch, are two small stipules, which are sometimes represented by spines.

The species of acacia are very numerous, and widely diffused in the warmer regions of the globe. *A. gummifera* is found in Mogador. *A. Nemu* is abundant in the environs of Nagasaki, Japan. *A. glandulosa* and *A. brachyloba* adorn the banks of the Mississippi. Some of the species are of great importance, as furnishing gum; others contain an abundance of tannin, which renders them useful for tanning purposes, and in medicine as astringent drugs. Gum arabic, or gum acacia, is an exudation from various species of acacia, such as *A. Verek, A. arabica, A. vera, A. Adansonii*, and others, for the most part natives of Arabia, Barbary, and the East Indies. Gum Senegal is a similar product from other species of the genus. Some Australian kinds called Wattle trees furnish gum. The drug known as catechu is prepared from various trees, but especially from *Acacia Catechu*, the wood of which is boiled down, and the decoction subsequently evaporated, so as to form an extract much used in medicine as an astringent. The bark of *A. arabica* is used in India for tanning leather, under the name of Babul bark; that of *A. Melanoxylon* is used for the same purposes in Australia. The pods of other species are likewise similarly employed in Egypt and Nubia. Many furnish excellent timber, and the flowers of one species, *A. Farnesiana*, yield a delicious perfume. The pods of *A. concinna* are used in India like those of the soap-nut for washing the head; the leaves also are acid, and used in cookery like those of tamarinds. The pounded seeds of *A. Niopo* are employed by certain of the Indian tribes on the river Amazon as the basis of a snuff, into the composition of which lime and the juice of a species of *Cocculus* also enter; its effects are to produce a kind of intoxication and invigoration of spirits. Many kinds are cultivated in greenhouses in this country, for the beauty of their flowers or for their foliage; some few even, such as *A. Julibrissin* and *A. lophantha*, will succeed out of doors in warm situations. The name *Acacia* is also commonly given to the locust-tree of North America, a very different plant. See ROBINIA, and MIMOSA. The aspect of an Acacia scrub, which is one of the characteristic features of Australian vegetation, is shown in Plate 15.

[M. T. M.]

ACACIA BLANC. (Fr.) *Robinia Pseud-Acacia*. — BOULE. *Robinia umbraculifera*. — DE CONSTANTINOPLE. *Acacia Julibrissia*. — DE SAINTE-HÉLÈNE. *Acacia testila*. — DE SIBÉRIE. *Caragana frutescens*. — PARASOL. *Robinia umbraculifera*.

ACACIA, BASTARD or FALSE. *Robinia Pseud-Acacia*, sometimes called the Locust-tree. — ROSE. *Robinia hispida*.

ACÆNA. A genus of the *Sanguisorbaceæ*. They are small herbs, mostly with woody stems. The leaves are unequally pinnate, the flowers small, white or purple, borne on scapes and arranged in terminal balls, or sometimes in spikes. Their calyces are often beset with slender spines which are furnished at their apex with reflexed bristles. *A. ovina* is a common weed in S. Australia and Tasmania, and is troublesome in grazing districts from the bristles of the fruit getting entangled in the wool of the sheep; it is also a pest to housewives from their adhering to linen exposed to dry on the grass; and, as well as many of the species, a common annoyance to travellers by catching their dress. A decoction of the leaves of *A. Sanguisorba* is used in New Zealand as tea and as a medicine. It is the Piri Piri of the natives. There are upwards of forty species in the genus, chiefly natives of the temperate regions of S. America; commencing in California, they extend through the Andes (where some of them reach the elevation of 15,000 feet) to Cape Horn; they attain their maximum in Chili. [A. A. B.]

ACALYPHA. A large genus of Spurgeworts (*Euphorbiaceæ*), comprising upwards of a hundred species, which are more or less distributed over all tropical and subtropical regions, attaining their maximum, however, in S. America. A goodly number are annual, but the great mass perennial shrubby plants, having much the appearance of nettles, and readily known in the family from their nettle-like leaves and the disposition of their flowers, which, usually of a green or reddish colour, and inconspicuous, are disposed generally in erect or drooping bracted spikes, which arise singly from the axils of the leaves or the end of the shoot, and vary in length from an inch to a foot, the upper portion of the spike bearing sterile, the lower fertile flowers, or the entire spike devoted to the one or the other. The sterile have a calyx of four triangular lobes, enclosing eight to sixteen stamens, whose curious flexuose anther cells are quite distinct from each other, and stand out nearly at right angles to their stalk. The fertile flowers have a calyx of three to five divisions, and a three-branched style, the branches deeply ramifying, crowning a three-lobed ovary, which, when ripe, is a three-celled and three-sided capsule of the size of a small pea. The Stringwood of St. Helena (A. rubra) is interesting as being one of a comparatively small number of plants now known to be extinct. It formed a beautiful small tree, and got its name of stringwood from the long spikes of reddish-coloured sterile flowers which hung in great profusion from the twigs. A. indica, an annual Indian weed, one to two feet high, with nettle-like leaves, and flower-spikes having toothed leafy bracts, has, according to Nimmo, roots which attract cats quite as much as do those of Valerian. This plant is the Upamarzi of Rheede, who says the root bruised in hot water is cathartic, and a decoction of the leaves laxative. [A. A. B.]

ACAMPE. Under this genus Lindley collects a few Indian and Chinese epiphytal orchids, formerly referred to Vanda, from which they differ in having small brittle flowers with a lip adnate to the edges of the column. They are of no interest except to botanists. The handsomest is Acampe (formerly Vanda) longifolia, a fine-looking species with small yellow flowers, occasionally met with in gardens.

ACANTHACEÆ (Acanthads). An order of monopetalous exogens, nearly related to scrophulariads, and for the most part tropical. In such regions they are extremely common, constituting a large part of the herbage. Nevertheless the genus Acanthus is found in Greece, and one species inhabits the United States. In a majority of cases Acanthaceæ are to be recognised by the presence of large leafy bracts, in the axils of which the flowers are partly concealed, and also by their calyx being composed of deeply imbricated sepals forming a broken whorl. But their most exact difference from other Orders of the Bignonal Alliance consists in the singular structure of their placenta, which expands into hard processes, which are most commonly hooked. In the form of their embryo they agree with bignoniads. They are of little importance to man. The greater part are mere weeds, but some are plants of great beauty, especially the species of Justicia, Aphelandra, and Ruellia. For the most part they are mucilaginous and slightly bitter; occasionally the bitterness increases, and they become pectoral medicines; some are dyers' plants. The genuine acanths, formerly called Branceursines, are emollients, as

Acanthus spinosus leaves growing round a pot; whence, as is said, the idea was derived of the Corinthian capital in Architecture.

also is Anisotes trisulcus, an Egyptian plant. About 1500 species are mentioned in books.

ACANTHE D'ALLEMAGNE. (Fr.) Heracleum Sphondylium.

ACANTHODIUM. A genus of acanthaceous plants, distinguished by Delile from the genus Acanthus by reason of its two-celled pod, each cell of which contains one compressed seed, the radicle or young root of which is placed near the scar of the seed, or that part where it is attached to the pod, whereas in Acanthus the rootlet is placed at a distance from the scar. The only species, A. spicatum, is a native of Egypt. It is provided with a very short stem, from which proceed three or four spikes of flowers, each provided with very spiny bracts. [M. T. M.]

ACANTHOGLOSSUM. An epiphytal orchid from Java, now merged in Pholidota.

ACANTHOLIMON. A genus of Plumbagineæ containing about forty species, most of which are natives of Persia, Asia Minor, and Greece. The technical characters are the union of the five styles at the base, and the capitate stigmas; but they are readily distinguished from their allies by their rigid, sharp-pointed leaves, which resemble those of juniper. The stems are very short, and much branched, so that the plants form dense prickly cushions on the rocks on which they grow; the flower-stalks are simple or forked; the spikelets in a spike which is generally lax; the calyx white, its limb surrounding the rose-coloured corolla like a frill. *A. glumaceum* is a very pretty garden rock plant. [J.T.S.]

ACANTHOPHIPPIUM. A genus of terrestrial orchids allied to *Bletia*, with large, fleshy tubular flowers, growing almost at the base of the leaves. These flowers are white or pink, and occasionally streaked with a deeper colour. The few species that are known come from the tropical regions of Asia.

ACANTHUS. The genus from which the order *Acanthaceæ* derives its name. The species of the genus are remarkable for the beauty of their foliage. The calyx consists of four unequal pieces, the two side ones being much smaller than the other two; the corolla is also irregular, and has but one lip; the stamens are four in number, one pair longer than the other; the anthers are one-celled, and covered with hairs; the capsule is two-celled, each cell containing two rounded seeds. *A. mollis* and *A. spinosus* both grow in Italy, Spain, and

Acanthus spinosus.

south of France, &c. The leaves of the latter plant are supposed to have furnished to Callimachus the model for the decoration of the capital of the columns in the Corinthian style of architecture. Both species are cultivated in this country, but are ill adapted to resist frost. [M.T.M.]

ACARPHÆA. A genus of the composite family (*Compositæ*) containing but one species, *A. artemisiæfolia*, a native of California. It is an herb with ragwort-like leaves three to four inches long, glandular above, and hoary beneath; the flower-heads few and stalked, the florets yellow. The name has reference to the absence of the chaffy pappus of *Chænactis*, and the chaffy receptacle of *Madia*, to both of which it is allied. The name has been by mistake printed *Acicarphæa*, instead of *Acarphæa*, in some books. [A.A.B.]

ACAULIS. Having a very short stem; literally stemless, but a plant without a stem cannot exist, unless it is a mere vesicle.

ACAULOSIA. A diseased condition of plants, in which the stem is imperfectly developed or wholly wanting. Its formation may moreover be retarded by the main powers of vegetation being directed to some other quarter, as in turnips to the formation of an enormous root. There may moreover be stemless varieties of some particular species; the primrose representing, for instance, a form of the cowslip in which the axis is reduced to little more than a point. The common hyacinth sometimes flowers imperfectly without any elongation of the stem, a state which arises from injury or decay of the roots; and from similar affections a like condition may be produced by heat. The stem of *Cnicus acaulis* is not developed in poor dry pastures, though it occasionally acquires a foot or more in length. The stunted growth of trees also may arise from a like cause, but is more frequently produced by actual injury, intentional or otherwise. [M.J.B.]

ACCRESCENT. Growing larger after flowering. The calyx of *Melanorrhœa*, which is small and green when in flower, becomes large and leafy when the fruit is ripe, and is therefore accrescent.

ACCUMBENT. Lying against anything; used in opposition to incumbent, or lying upon something; a term employed in describing the embryo of crucifers.

ACER. Under this name are included the Sycamore (not of Scripture) and the Maples, trees indigenous to the temperate regions of both the Old and New World, where they are either large-sized shrubs or moderate trees. They are mostly of rapid growth, and easily propagated. For these qualities, and for the beauty and variety of their foliage, the species are much planted in England for ornamental purposes, while in America one species has great economic value, being employed in the manufacture of sugar, a substance which is found more or less in the sap of all. The flowers, though they display no striking colours, attract the attention from their number, graceful arrangement, or the multitude of winged insects which, at a season when flowers are scarce, resort to them for food. The leaves are mostly lobed and toothed, in some species very large; and the seed-vessels (called *samara* by botanists), which are winged capsules, each containing a single seed and united by their bases into pairs, are strongly characteristic of the family.

The common sycamore, *A. Pseudo-platanus*, abundant as it is in England, and readily though it propagates itself by seed, is, on good grounds, supposed not to be indigenous, but to have been introduced from the European continent in the fourteenth century. Of a tree so well known it is unnecessary to give any description. Its uses are numerous. The wood is used for various articles of domestic furniture, musical instruments, and toys; as fuel it is said to be the most valuable of all woods, and it may be converted into excellent charcoal. From the sap collected in early

Acer Pseudo-platanus.

spring, sugar may be made, but not in remunerating quantities. The name Sycamore was given to it at an early period, from a supposition that it was the tree mentioned in the New Testament, which, however, as the etymology indicates, is a species of fig, (*sykon*, a fig, and *morea*, a mulberry-tree, resembling the former in its fruit, and the latter in its leaf), *Ficus Sycomorus*. In Scotland it is popularly known as the Plane.

The Maple (*A. campestre*) is a low hedge tree most conspicuous for the golden and purple tints of its foliage in autumn. The gnarled stems and knotted roots of this species have long been prized by turners and cabinet-makers for making choice articles of furniture. The wood also makes excellent fuel, and the best of charcoal. But the most important species of this family is the Sugar Maple (*A. saccharinum*), a native of North America. This tree forms extensive forests in Canada, New Brunswick, and Nova Scotia, and yields a saccharine juice in such abundance that maple-sugar is an important article of manufacture. It has been computed that in the northern parts of the two States of New York and Pennsylvania there are ten millions of acres which produce these trees, in the proportion of thirty to an acre. The season for tapping is in February and March, while the cold continues intense and the snow is still on the ground. A tree of ordinary size yields from fifteen to thirty gallons of sap, from which are made from two to four pounds of sugar. The tree is not at all injured by the operation, but continues to flourish after having been annually tapped for forty years without interruption. Greater facilities of intercommunication and the decreased cost of cane-sugar, which is far superior, have tended of late years greatly to check the manufacture of sugar from the maple. Old trees of this species are liable to a peculiarity of growth which gives to their timber the knotted structure known by the name of bird's-eye maple. The wood called 'curled maple' is obtained from old distorted trunks of *A. rubrum*, also a native of America. For an enumeration of other species, see *Loudon's Arboretum*. The common maple is the badge of the clan Oliphant.   [C. A. J.]

According to Mr. Hind, the ash-leaved maple, *Acer Negundo*, is tapped for sugar in the Red River settlement of Canada W.

ACERACEÆ (*Acera*; *Acerineæ*; the order of *Maples*). A natural order of trees and shrubs inhabiting Europe, the temperate parts of Asia, the north of India, and North America. The order is unknown in Africa and the southern hemisphere. The most important product is the sweet sap of some species, from which sugar is extracted. It is said, however, that their juices become acrid as the season advances. They yield a light useful timber. The bark of some is astringent, and yields reddish-brown and yellow colours. The order only contains three genera, and rather more than fifty species.

ACERANTHUS. A genus of *Berberidaceæ* containing a single species from Japan, a slender plant nearly allied to *Epimedium*, but having plain and not spurred petals.   [W. C.]

ACERAS. An *Orchis* without a spur, there being no other difference between the two genera, except that *Aceras* has only one pollen gland instead of two. The man-orchis, *Aceras anthropophora*, so called because of a fancied resemblance between its lip and the body of a man hung by the head, is common in meadows and grassy slopes all over Europe. It has greenish-yellow flowers bordered with red, a pair of oblong knobs or tubercles for its roots, and a heavy rather unpleasant odour. *Aceras hircina*, the lizard-orchis, is a much finer and rarer plant, with long spikes of dirty rose-coloured flowers, the middle lobe of whose lip has the form of a long, twisted strap; they emit an unpleasant odour like that of a goat. This species is occasionally found in chalky districts all over the temperate regions of Europe. Haller says that the bruised root increases the flow of milk in milch cattle. Other species occur in Asia, reaching as far as Gossain Than in the Himalayas; and one (*A. secundiflora*) found in Barbary and Madeira is occasionally seen living in the gardens of curious collectors.

ACEROSE. Needle-shaped; as in the leaves of heaths and pine trees.

ACETABULARIA. A beautiful genus of calcareous green-spored *Algæ*, the species of which resemble little umbrellas or such delicate gill-bearing fungi as *Coprinus plicatilis*. An erect articulated stem bears above a whorl of threads which are united laterally so as to form an umbilicate orbicular disk, from the centre of which arises a bunch of delicate branched threads. The most remarkable species is not uncommon in the Mediterranean, but none has yet been observed on our own coasts. {M. J. B.}

ACETABULUM. The receptacle of certain fungals.

ACHÆNE or ACHENE. Any small, brittle, seed-like fruit, such as Linnæus called a naked seed.

ACHANIA. The name given to a genus of plants of the mallow family (*Malvaceæ*), some of the species of which are cultivated in our stoves for the beauty of their flowers. They are shrubs inhabiting South America, Mexico, the West Indies, &c. The calyx is double, the outer of many pieces, the inner tubular and five-toothed; there are five petals with appendages at their base; the filaments are united into a spirally twisted tube, bearing the anthers on the summit; and the flowers are succeeded by a five-celled fruit. *A. Malvaviscus* is remarkable for the beauty of its scarlet axillary flowers, and its green, heart-shaped, sharply-pointed leaves. {M. T. M.}

ACHARIA. A genus of erect, slender, glaucous Cape herbs, belonging to the natural order *Papayaceæ*. They have alternate deeply trifid leaves, and axillary unisexual flowers, with a three-leaved involucel, and a campanulate trifid calyx. The male flower has three stamens, alternating with three scales. In the females the three scales surround the one-celled stipitate ovary. {W. C.}

ACHE. (Fr.) *Apium.* — DE MONTAGNE. *Levisticum officinale.* — DES CHIENS. *Æthusa Cynapium.*

ACHE'E. (Fr.) *Polygonum aviculare.*

ACHIAR. An Eastern condiment, formed of the young shoots of *Bambusa arundinaria*.

ACHILLEA. A name anciently given to a plant 'wherewith Achilles cured the wounds of his soldiers.' It is now applied to a family of plants belonging to the natural order of compound flowers. Most of the species have deeply-divided woolly leaves, and bear the flower-heads, which are white, yellow, or purple, in flat clusters (corymbs) at the extremity of the stem. Two species only are common in Great Britain: — *A. Ptarmica, Sneezewort,* an herbaceous plant, a foot high or more, bearing heads rather less in size than a daisy, which have the disk, as well as the ray, white. This is frequent in moist meadows, especially in the hill countries. It derives its name from its alleged property of exciting sneezing when pulverised, a virtue which it probably possesses, though not to an extent beyond that of many other plants undistinguished by special names. *A. Millefolium,* Milfoil or common Yarrow, is an herbaceous perennial, with tough upright stems, more or less woolly deeply-cut jagged leaves, and flat corymbs of flower-heads, containing very few florets, which are either white, pink, or, rarely, deep purple. Its properties are highly astringent, and it was anciently much prized as a vulnerary. The older English botanists called it Nose-bleed, "because the leaves being put into the nose caused it to bleed." Several foreign species are cultivated as border plants, and are conspicuous either by their flowers or hoary foliage. {C. A. J.}

ACHIMENES. An extensive genus of very handsome tropical and sub-tropical herbs, furnished with scaly underground tubers, by which they are perpetuated. They are much cultivated in hothouses on account of their ornamental character; and many new forms, developing greater variety and attractiveness than are to be found in the original kinds, have been obtained in the cultivated state. They belong to the order *Gesneraceæ*, and their most obvious peculiarities consist in a five-parted subequal calyx, the tube of which is joined with the ovary at its base; a funnel-shaped corolla, of which the tube is somewhat oblique, and gibbous behind at the base, and the limb spreading five-lobed and nearly equal; four didynamous included stamens inserted on the tube of the corolla, with the rudiment of a fifth; a simple style with a subcapitate obsoletely two-lobed stigma; and an ovary coherent with the base of the calyx, bordered by an annular or ring-formed glandular disk, one-celled, containing many ovules, which are attached to a of pair parietal placentæ. They have firmly erect stems; opposite, serrated, often hairy leaves; and axillary flowers, the pedicels of which are not unfrequently accompanied by little scaly, bulbiform tubers, like those produced at the base of the stem beneath the surface of the ground. The genus has been divided into several by modern botanists, but few of the proposed groups have been generally received. The principal of these new genera, in addition to *Achimenes* itself, — which is made to consist of erect herbs with axillary flowers, having a membranaceous entire glandular ring, and a two-cleft stigma, — are the following:

Kollikeria: dwarf herbs with a terminal racemose inflorescence, a membranaceous nearly entire glandular ring, and a stomatomorphous stigma.

Lockeria: erect herbs, with axillary or sub-panicled flowers, a thickened fleshy nearly entire five-angled glandular ring, and a two-cleft stigma.

Guthnickia: erect herbs, with axillary flowers, a thickened fleshy nearly entire five-angled glandular ring, and a stomatomorphous stigma.

Scheeria: erect herbs, with large axillary flowers, a thick fleshy subentire glandular ring, and stomatomorphous stigma.

Mandirola: erect herbs, with axillary or sometimes panicled flowers, having the glandular ring membranaceous, and composed of five crenatures or lobes, and a two-lobed stigma.

Tydæa: erect herbs, with axillary or somewhat panicled flowers, having the glandular ring composed of five distinct glands, and a two-cleft stigma.

Of these new genera *Tydæa* is the most distinct and the most generally accepted.

Achimenes as above restricted, consists of two distinct series, one of which is well represented by *A. coccinea*, a species found in Jamaica and Central America. This plant has slender, erect, branching stems a foot or rather more in height, furnished with small ovate, acute, serrated leaves, and axillary one- or few-flowered peduncles bearing small scarlet, somewhat salver-shaped flowers, having a broadish cylindrical or somewhat swollen tube, nearly equal at the base, and a spreading limb of five rounded segments. The other series is represented by the Mexican and Central American *A. longiflora*, in which the stems are also erect, about a foot and a half in height, with ovate, acute, serrated leaves, and axillary peduncles supporting one large flower, of which the tube is elongated, slender, curved, and deflexed, saccate at the base, and the limb very broad, plane, and lying in a direction oblique to the tube. The species are for the most part natives of Central America. They are not applied to any use, but are much prized for their ornamental properties. The mode of increase from the scaly tubers is very curious, every one of the scales, when separated, being capable of forming a new plant. The name *Achimenes* is also a synonym of *Artanema*, a genus of the order *Scrophulariaceæ*, its derivation is unknown. [T. M.]

ACHLAMYDEOUS. Having neither calyx nor corolla.

ACHOTE. The seeds of the Arnotto, *Bixa Orellana*.

ACH-ROOT. The root of *Morinda tinctoria*, used in India as a dye.

ACHYRACHÆNA. The generic name of a Californian annual of the composite family (*Compositæ*), nearly related to the better known and much prettier genera *Callichroa* and *Oxyura*, but differing from them in the nature of the pappus which crowns the cylindrical achenes, and consists of about ten very thin and membranous silvery scales, each about half an inch in length. The whole plant is clothed with soft white hairs, whence its specific name of *mollis*. The stems are seldom more than eight inches high, branched or simple, furnished with grassy leaves one to two inches long, and terminate in a single head of flowers half an inch across, with purple inconspicuous florets. The plant has also been called *Lepidostephanus madioides*. It has been cultivated at Kew. [A. A. B.]

ACHYRANTHES. A genus of *Amaranthaceæ*, found in the tropical and sub-tropical districts of the Old World and consisting of erect, procumbent, or sometimes climbing trees and shrubs, many of them being troublesome weeds in cultivated grounds. The flowers are in loose spikes, hermaphrodite, and have three spinous bracts. The calyx consists of five, rarely four, sepals. The stamens, the same in number as the sepals, are united at their bases into a cup. The one-celled ovary contains a single ovule, and has a simple style, and capitate stigma. The leaves are opposite. *A. aspera* and *fruticosa* are administered in India in cases of dropsy; *A. globulifera* is used in Madagascar as a remedy for syphilis. Upwards of thirty species have been described. Though natives of the Old World, three or four species have been accidentally carried to the United States, where they have rapidly spread, becoming perfectly naturalised. [W. C.]

ACHYROPHORUS. A genus of annual or perennial herbs belonging to the chicory group of the composite family, and only distinguished from *Hypochæris* by the feathery pappus-hairs being in a single instead of a double series. Of about twenty-five species four are S. European and Altaian; and one of these, *A. maculatus* is also common to Britain, but usually placed in *Hypochæris* in our floras. The remainder are entirely S. American, and chiefly extend from Chili southwards. A few, found in the Andes at elevations of 10,000 feet and upwards, are neat little stemless plants, with a spectie of linear or lance-shaped toothed or entire leaves, and nestling in their midst a large and handsome yellow flower-head often more than an inch across. One of this set, *A. sessiliflorus*, is called in N. Granada Chicoria de la tierra Caliente; and, according to Purdie, a decoction of its thick white tapering roots is employed in affections of the chest. In those species found at low elevations, the root-leaves are spreading, entire and grassy, or pinnatifid like those of our hawkbits (*Leontodon*), their surface smooth or hairy; the yellow flower-heads single on the ends of unbranched stalks, or the stalks branching and furnished with leaves at the points of forking. *A. aporgioides* and *A. Scorsonera* are known in Chili as Escorzoneras, and their tapering roots are eaten for their refreshing and purifying qualities, as those of the Spanish Scorzoneras (*Scorsonera Auspanica*) are in this country. [A. A. B.]

ACIANTHUS. A genus of Australian terrestrial orchids with solitary heart-shaped leaves and erect racemes of small green or dull purple flowers. They inhabit shady or damp places, and represent in the southern hemisphere the *Malaxis* and *Liparis* of the northern.

ACICARPHA. A genus of *Calycerace*, comprising seven species, all of them found in the provinces bordering on the river Plate. They are small herbs with toothed or entire leaves, and lateral or terminal heads of flowers which are enclosed in a spiny involucre. In general they are found in sandy or rocky soils, and may be considered as mere weeds. They do not appear to be applied to any useful purpose. *Acicarpha* embraces most of the members of the family found on the eastern side of the Cordillera. A name very similar to this, *Acicarphæa*, has by mis-spelling been given to *Acarphæa* in some books. [A. A. B.]

ACICULA. A bristle. The bristle-like abortive flower of a grass.

ACICULAR. Shaped like a needle.

ACICULATED. Marked by fine impressed lines, as if produced by the point of a needle.

ACIES. The edge of anything. The angles of certain stems.

ACINACIFORM. Scimetar-shaped; that is, curved, rounded towards the point; thick on the straighter side, thin on the convexity.

ACINETA. Noble epiphytal orchids from Central America, with angular pseudobulbs, membranous ribbed leaves, and large fragrant fleshy flowers in pendulous or occasionally erect racemes; some brownish-purple, others more or less yellow. The genus was founded upon the *Anguloa superba* of Humboldt and Bonpland, whose artist imagined the great drooping raceme to be erect, and otherwise misunderstood the true structure. Several species are known, but they are not very well distinguished; in all the lip is united to the column by a solid immoveable concave base, is three-lobed in various ways, and is furnished with a singular fleshy appendage rising from the middle in the form of a truncated body or of a mere horn. The species mentioned in books are *A. Humboldtii, Barkeri, chrysantha* (alias *densa*), *Warczewitzii, erythroxantha* (alias *chamæcyenoches*), *sella turcica,* and *cryptodonta,* all fine plants, and, with the exception of *sella turcica,* all in cultivation.

ACINODENDRON. A genus of Gronovius now reduced to *Sagræa,* and supposed by De Candolle to be the same as *S. guadalupensis.*

ACINUS. A bunch of fleshy fruits, as of currants or grapes. Now confined to the berries of such bunches.

ACIS. A small genus of bulbous plants, belonging to the order *Amaryllidaceæ,* and separated from *Leucojum,* from which they are distinguished by having a filiform style, and fleshy angular seeds. Both have a bell-shaped perianth, consisting of six nearly equal divisions, six stamens inserted in the epigynous disk, and an inferior three-celled ovary, containing numerous ovules. The species referred to *Acis* are plants of Southern Europe and Northern Africa, and are pretty subjects for bulb gardens. *A. rosea,* one of the nicest species, has a small round bulb, narrow blunt linear green leaves, and from one to three one-flowered scapes, blooming in succession, the flowers pendent, pale rose-coloured. The other species are *A. autumnalis* and *A. grandiflora.* [T. M.]

ACKAWAI NUTMEG. The fruit of *Acrodictidium Camara.*

ACKROOT or **AKROOT.** An Indian name for the Walnut.

ACLINIA. A supposed genus of Indian orchids founded by Griffith upon monsters of certain species of *Dendrobium,* in which regularity in the parts of the flower is substituted for their customary irregularity; the lip resembling the sepals and petals, and the column being triandrous, or nearly so. Five cases of the kind are recorded by Lindley, in the *Journal of the Linnean Society* (August 1858).

ACOCANTHERA. *Lycium cinereum.*

ACONIOPTERIS. A group of the *Acrostichea,* in which the parallel veins of the fronds are angularly united near the margin. It is now included in *Olfersia.* [T. M.]

ACONITE. *Aconitum.* — **WINTER.** *Eranthis hyemalis.*

ACONITUM. An important genus belonging to the order *Ranunculaceæ,* and botanically characterised by the calyx being not of a green colour, but blue or yellow, of five pieces, the upper of which is convex, and in form like a helmet. Within this are concealed two singularly shaped petals, formerly considered to be nectaries: the form of these bodies is somewhat like that of a hammer. There are also three other petals, very small and inconspicuous, though occasionally they also become hammer-shaped, like the two upper ones. The stamens are numerous; and the fruit consists of from three to five follicles. The plants constituting this genus are found in Europe and Northern Asia, and a few are natives of North America. One species, *A. napellus,* is said to have been found wild in Britain, but this is open to grave doubts. All the plants of this genus possess virulently poisonous properties; the roots of some of the Indian species produce the Bikh poison of Nepal, one of the most dangerous of poisons. The roots of *A. ferox* (supposed to be a variety of *A. napellus*) are used in the northern parts of Hindostan for poisoning arrows, with which tigers are destroyed. A tiger shot from a bow in Assam was found dead on the spot, so soon did the poison take effect. Several kinds are commonly cultivated in gardens, especially *A. napellus,* the fleshy roots of which have been occasionally used by mistake for horse-radish, and produced fatal results. This plant has a stem about three feet in height, with dark green glossy leaves, deeply divided in a palmate manner; the flowers are placed in erect clusters, and are of a dull blue colour. The roots, or more properly rootstocks, are of a tapering form, of a dark brown colour externally, and white internally; the younger roots, which are placed on either side of the older one, are of a lighter colour. The taste is bitter at first, but after a time numbness and tingling of the lips and tongue are perceived. The root has none of the acridity or pungency that fresh horse-radish possesses. The two plants are so dissimilar that it would seem impossible so terrible a mistake should be made, but it has generally arisen from taking the root of the aconite when the leaves and flowers, which are so unmistakable, have died away. The rootstock of the horse-radish is much larger than that of the aconite, not of a tapering form, dirty yellow externally, and the top or crown marked with transverse scars, indicating the position of the old leaves; its

Aconitum Napellus.

dour and taste are at first pungent and acrid. The venom of the aconite appears to depend upon the presence of an alkaloid called Aconitina, which is so extremely poisonous that so small a dose as one-fiftieth part of a grain has wellnigh produced fatal results. A tincture of aconite root, or a solution of the alkaloid, is occasionally used with much success as an application to relieve rheumatic pains, but it should be employed with the greatest caution.

Aconitum variegatum is also commonly cultivated; it has, as its name implies, flowers variegated with white and blue.

Aconitum Lycoctonum, or Wolfsbane, is a common plant in the Alps of Switzerland and Styria. Its leaves are palmate and

Aconitum Lycoctonum (flower).

hairy, of a dull yellowish green. Its flowers, which are borne in slightly branching clusters, are of a dull yellowish colour, and the shape of the upper sepal is that of an extinguisher, with a thick rounded knob at the extremity. This species does not possess such virulent properties as the others. [M. T. M.]

ACONTIAS. A genus of plants so named in allusion to the spots on the stem, which resemble those of a species of serpent so called. The genus belongs to the *Caladium* tribe of the arum family, and has tuberous rootstocks, lobed pedate leaves, green erect spathes, enclosing a spadix or fleshy spike, with female flowers at the lower portion, and male flowers at the upper. The species inhabit Brazil. [M. T. M.]

ACORE ODORANT. (Fr.) *Acorus Calamus*.

ACORIDIUM. A genus of cæspitose plants, natives of Manilla. They have slender stems, sheathed at the base, and bear diœcious flowers in a linear spike. They are too little known to refer them satisfactorily to their position, although they seem to be allied to *Burmanniaceæ* and *Xyridaceæ*. [W. C.]

ACORN. The fruit of the Oak or *Quercus* family. —SWEET. The fruit of *Quercus Ballota*.

ACORUS. The name of a genus of plants referred by some to the *Araceæ*, and by others to the *Orontiaceæ*. The most interesting plant of the genus is *Acorus Calamus*,

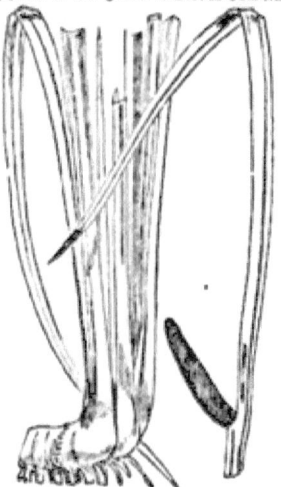

Acorus Calamus.

or sweet flag, a plant apparently known to the Greeks, though not to be confounded with the *Calamus aromaticus*, which, according to Royle, was a species of grass. The sweet flag grows in ponds, by the banks of rivers, and other wet places in England.

It is also found in the cooler parts of Europe, of India, and of North America. From the lower part of the thick jointed stem or rhizome, the plant sends down numerous roots, while from the upper surface it pushes upwards a number of lance-shaped leaves from two to three feet in length, sheathing at the base, also a long leaf-like stalk, from one edge of which, a foot or more above the rootstock, issues a spike of densely packed greenish flowers, each provided with a perianth of six pieces, enclosing six stamens and a three-celled ovary with a sessile stigma. All parts of the plant, but especially the rhizome, have a strong, aromatic, and slightly acrid taste; hence it has been used as a stimulant and mild tonic in medical practice, especially in some kinds of indigestion, and is said to be useful in ague. The rhizome is also used by confectioners as a candy, and by perfumers in the preparation of aromatic vinegar and other perfumed articles. *A. gramineus* is a much smaller plant, sometimes cultivated in gardens, especially the form with variegated leaves. [M. T. M.]

The *Acorus Calamus* imparts at once an aromatic taste, and an agreeable bouquet odour, to liquids in which it is infused. Professor Johnston states, in his *Chemistry of Common Life*, that it is used by the rectifiers to improve the flavour of gin, and is largely employed to give a peculiar taste and fragrance to certain varieties of beer. It abounds near the rivers of Norfolk, and, according to the same authority, the London market used to be principally supplied from this locality; as much as 40*l*. having been sometimes obtained for the year's crop of a single acre of riverside land, on which it naturally grows. It is still used, as "rushes," to strew the floor of Norwich Cathedral on certain festivals. [T. M.]

ACOTYLEDONS. A name often applied to CRYPTOGAMS, or flowerless plants, in consequence of their reproductive organs or spores, when germinating, having no seed-leaves or cotyledons. There is, however, no rule without an exception, and some lycopods present when young something very like cotyledons. Though Cryptogams have no true cotyledons, their spores produce, mostly by cell-division, a mass of threads, a leafy expansion, or a solid body, to which the name of false cotyledons (pseudo-cotyledons) has been given, and such productions, as the false cotyledons of mosses for example, have often been considered as distinct plants, belonging to a distinct natural order from the parent plant. Under this name are included all those plants called by Linnæus Cryptogamia, because he was unable to discover their organs of fertilisation, if they had any. They comprehend Sea-weeds, Fungi, Lichens, Mosses, Ferns, and their allies: which see. It is now known that all are multiplied by a sexual apparatus in structure wholly different from that of phænogamous plants, but in function the same. One very great peculiarity is that in the majority of the orders a true locomotive action is observable in the matter emitted by the male organs; and that in the higher orders, that is to say in Ferns, Lycopods, and Horsetails, the plant, properly so called, does not proceed directly from the spore or seed, but from a rudimentary intermediate organ called prothallium, on which the organs of fertilisation are formed, these organs not producing a spore or seed, but the very plant itself.

ACOTYLEDONOUS. Having no cotyledons, as in *Cuscuta*. But, in systematical botany, applied to what are now called spores, which were formerly thought to be embryos without cotyledons.

ACOUCHI RESIN. The inspissated juice of *Icica heterophylla*.

ACRADENIA. A plant found wild in the neighbourhood of Macquarrie Harbour, Tasmania, has been included in a genus of rutaceous plants of the above name. It belongs to the tribe *Boroniæ*, and its distinguishing characters are the following:— parts of the flower in fives; stamens free, not united together, perfectly smooth; anthers smooth, not provided with any appendage. The ovaries adhere closely together, and are everywhere clothed with a dense woolly covering, except that each bears at its summit a small gland, whence the name of the genus. When the fruit is ripe the inner shell remains firmly attached to the outer shell, instead of separating from it in two elastic valves, as in the neighbouring genera. *A. Franklinæ*, a species named after Lady Franklin, is a shrub now in cultivation at Kew, with opposite ternate leaves, which are thick, very rough and harsh on the upper surface, smooth on the under surface, and fragrant. [M. T. M.]

ACRÆA. A small genus of terrestrial orchids with fleshy fasciculate roots, and close racemes of yellowish, downy flowers. In general aspect the species resemble the European *Spiranthes*. The three or four that are described come from Central America.

ACRAMPHIBRYA. Plants that grow both at the point and along the sides, as endogens and exogens.

ACRIOPSIS. In Burmah, Borneo, and the Malay Archipelago, are found several epiphytal orchids with small reddish paniculate flowers, having their lip united firmly to the front of the column, from which it projects at right angles, in addition to which the column has two glandular arms, and is extended behind into a hood covering the anther. They belong to the vandeous sub-class, and are nearly related to the South American genus *Aspasia*. Three or four species are known, of which two, *A. densiflora* and *picta*, have been cultivated. Beyond the very curious structure of the parts of fructification, they have little to recommend them to notice.

ACROBRYA. A term used by Endlicher; synonymous with ACROGENS. [M. J. B.]

ACROCARPI. A division of mosses containing those species in which the female fruit terminates the branches. Unfortunately even in the same genus, as *Fissidens*, species with lateral and terminal fruit occur, so that the distinction is not without grave exceptions. [M. J. B.]

ACROCARPIDIUM. The plants constituting this genus of the natural order *Piperaceæ* are closely allied to those included in the genus *Peperomia*, from which they differ in habit and in the pseudo-pedicellate fruit. They are for the most part creeping plants, growing upon trunks of trees or mossy banks, with hairy or smooth, alternate roundish or kidney-shaped leaves, which have three or five prominent nerves; the flowers are placed in rings on long-stalked catkins with somewhat fleshy bracts; they have two distinct stamens, a simple stigma crowning the stalkless ovary, which latter ripens into a fruit so contracted at its base as to give an appearance as if it were placed on a stalk. They are natives of tropical America and the West Indies, and partake in some degree of the general cordial properties of the family to which they belong. *A. hispidulum* is made use of in the West Indies as a bitter and stomachic. Several kinds are cultivated in stoves as objects of curiosity or of botanical interest rather than for their beauty. They are best known under the old name of *Peperomia*. [M. T. M.]

ACROCHÆNE *punctata*. An epiphytal orchid from the Sikkim Himalaya, where it was found by Dr. Hooker at the height of 4,000 feet above the sea. It has an ovate pseudo-bulb, a long solitary coriaceous leaf, and an erect radical inflorescence. The flowers are straw-coloured, dotted with crimson. It is nearly related to *Sunipia*, with which it agrees in having a couple of long taper candicles for the pollen masses.

ACROCLINIUM. A beautiful genus of annual composites, at present represented in our gardens by the *A. roseum*, recently introduced from the Champion Bay district, Western Australia. Its flower-heads resemble those of the well-known *Rhodanthe Manglesii*, but are larger, and the habit of the plant is entirely distinct. It produces numerous erect unbranched stems a foot or more high, the primary one emitting two opposite shoots from its base, each of which in their turn throws out two additional ones, which again become the parents of others, until the plant assumes a bushy character. The stems are clothed with numerous linear, smooth, pointed leaves, and bear at the summit a single handsome flower-head an inch and a half in diameter, consisting of a bright yellow disk of tubular florets, surrounded by a many-leaved, imbricated involucrum, the innermost leaflets of which have spreading rose-coloured tips, presenting, as in *Rhodanthe*, *Helichrysum*, and other allied genera, the appearance of ray florets. The fruit, or, as it is popularly but incorrectly termed, the seed, is clothed with snow-white silky down, and is surmounted by a pappus of from fifteen to twenty feathery hairs or scales, flattened and connected at their base, and tipped with a yellow tassel-like brush, by which characters the genus is chiefly distinguished. The yellow colour of the disk is due less to the colour of the florets themselves than to the brush-like tips of the pappus hairs, which under a lens are very interesting objects. Four other species occur in the same locality, but do not appear to have been yet introduced. [W. T.]

ACROCOMIA. The name given to a genus of palms, in allusion to the elegant tufts of leaves at the summit of the stem. One species, *A. sclerocarpa*, grows almost all over South America, occurring in dry soil, rarely in woods. The tree belongs to the same tribe as the cocoa-nut palm; its trunk rises to twenty or thirty feet in height, and is sometimes swollen in the middle; the leaves are from ten to fifteen feet in length, pinnate, with from seventy to eighty leaflets on each side. The young leaves are eaten as a vegetable. It is cultivated in our hothouses. [M. T. M.]

ACROGENS. A large and most important division of CRYPTOGAMS, distinguished for the most part from THALLOGENS, as Fungusses, Seaweeds, and Lichens, by their herbaceous growth, the presence of leafy appendages which are frequently furnished with stomates, the different mode of impregnation, and the presence of vascular tissue. A few acrogenous Liverworts have the habit of Lichens, but differ totally in structure.

The most important distinction, however, undoubtedly is that the impregnation takes place somewhat after the manner of PHÆNOGAMS, by an impression made upon the contents of the embryonic sac, and not upon the spore itself, as is decidedly the case amongst Thallogens where the mode of impregnation has been ascertained, as in *Algæ*. In *Characeæ* alone the spore seems to be immediately impregnated, though even in this case it is uncertain whether impregnation does not take place before the spore is perfected.

In Mosses, Liverworts, and Ferns, the spore after germination produces at first either a web of threads, a solid mass, or a membranous expansion (*prothallium*). In the two former a distinct plant arises from the threads with frequently symmetrical leaves, and on these plants urn-shaped organs are produced (called *archegones*) analogous to pistils, which contain at their base a cell which, after impregnation, produces the proper fruit. In perennial species a fresh crop of archegones may be produced in two or three successive years, which require a distinct act of impregnation for the development of the capsules. In Ferns and their allies, on the contrary, the archegones give rise to a new plant, which for one or for many successive years produces a fresh crop of fruit without further impregnation. The result of impregnation in the two cases, then, is quite different. In Mosses the whole plant is, as to functions, a prothallium; in ferns, merely the membranous expansion immediately produced on the germination of the spores. Further details may be reserved for each successive group. In those species of Fungi, as *Puccinia*, *Podisoma*, &c., where a prothallium is produced, it has the nature of a spore, and germinates in the same manner.

As regards the tissues, it may be observed that the stem of many acrogens contains distinct vascular tissue. In *Jungermannia*, where such tissue is rare elsewhere, it almost universally accompanies the spores. In Mosses, as in *Sphagnum*, there are sometimes distinct spirals in the cells of the leaves. The vascular tissue in most of the higher cryptogams is scalariform; but in *Isoetes* and *Equisetum* it is annular, with transitions to short spirals, while in *Selaginella* and *Lycopodium* there is a transition from short spiral and reticulated cells to elongated cells, which may be called spiral vessels. In the stem of *Sphagnum* there is tissue closely resembling the glandular tissue of conifers. The spiral coats of the spores in *Equisetum* will be noticed hereafter. The impregnating bodies or spermatozoids have always flagelliform appendages, sometimes much more highly developed than in the spermatozoa of animals. The principal divisions of acrogens are:

1. CHARACEÆ. Spores solitary.
2. RICCIACEÆ. Capsules valveless, without spiral cells or elaters.
3. MARCHANTIACEÆ. Capsules dependent, containing elaters.
4. JUNGERMANNIACEÆ. Capsules erect, containing elaters.
5. MUSCI. Capsules mostly valveless, without elaters.

In these five orders the archegones give rise to the capsule.

6. FILICES. Capsules mostly with an elastic ring, but sometimes densely crowded and ringless.
7. OPHIOGLOSSACEÆ. Capsules ringless, bivalvate.
8. EQUISETACEÆ. Capsules dependent. Coat of spores spiral.
9. MARSILEACEÆ. Capsules multilocular.
10. LYCOPODIACEÆ. Capsules axillar, unilocular.

In these five orders the spores produce a prothallium bearing archegones which yield new plants and not capsules. For further details see *Berkeley's Cryptogamic Botany*, p. 421. [M. J. B.]

ACROGLOCHIN. A genus containing only a single species, *A. chenopodioides*, from Nepal. It has been referred by some to *Salsolaceæ*, because of the horizontal position of the seed, as in *Chenopodium*; but the dehiscent utricle seems to separate it from the true *Entsolaceæ* and join it to the *Amarantheæ*, with which, however, it does not perfectly agree, for in this order the seeds are vertical. It in fact occupies a position equally related to both these orders. The flowers are small, sessile, in axillary cymes. The calyx consists of five equal erect sepals. There are two stamens, and a unilocular ovary with a single ovule. The leaves are alternate, unequally dentate, and strongly reticulated below. [W. C.]

ACROLASIA. A genus of Chilian *Loasaceæ*, allied to *Mentzelia*, from which it differs in having a definite number of stamens (ten); white flowers, and sinuate-pinnatifid leaves, which are opposite below and alternate above. [W. C.]

ACRONIA. A spurious genus of orchids, now reduced to *Pleurothallis*. The only species was *A. phalangifera*, which proves to be identical with *Pleurothallis Mathewsi*.

ACRONYCHIA. A genus of rue-like plants (*Rutaceæ*), distinguished by a short four-parted calyx, four petals, eight stamens inserted on a disk; style short; stigma capitate, four-lobed; fruit berry-like, four-celled, each cell containing one seed. *A. Cunninghami*, an evergreen shrub, a native of Moreton Bay, is cultivated in this country. The flowers have a perfume like those of the orange, and the leaves abound in resinous or oily fluid of a powerful turpentine-like odour. [M. T. M.]

ACROPERA *Loddigesii*. A Mexican and Central American genus of orchids, consisting of about four species, growing on the bark of trees. They have the habit of *Maxillaria*, with fleshy pseudo-bulbs and a drooping radical inflorescence. Their name, which signifies a pouch at the point, was given them in consequence of there being a sac at the end of their labellum. The flowers are of some dull yellowish colour, with very small misshapen petals. The genus is very near *Gongora*, to which it is reduced by Reichenbach.

ACROPHORUS. A genus of polypodiaceous ferns of the group *Cystopterideæ*, distinguished by having its globose patches of fructification, situated mostly at the tips, rarely axillary in the forks of the veins, these sori being covered by suborbicular indusia affixed by their posterior side. They form a small genus, serving to unite the *Davallieæ* with the *Cystopterideæ*. The plants have creeping rhizomes, and very elegant membranaceous, pinnate or decompound, free-veined fronds, of which the divisions are either equal-sided or dimidiate. There are about a score of species, including those referred sometimes to *Leucostegia* and *Odontoloma*. The larger proportion of these are natives of India and the East. [T. M.]

ACROPHYLLUM. A genus of *Cunoniaceæ*, founded upon a Tasmanian plant allied to *Weinmannia*, but distinguished by the absence of a disk in the flowers. *A. venosum* is a small erect shrub with evergreen leaves placed in whorls of three; they are nearly sessile, oblong, cordate, acute, serrated, and smooth; the stipules are small and membranous; and the flowers are small, white tinged with red, in dense whorls round the upper part of the stem and branches. Above the whorls of flowers there is a terminal tuft of leaves, from which the genus takes its name. The sepals and petals are five each, and stamens ten. It was introduced into this country in 1836, and forms a very striking and handsome greenhouse shrub. [J. T. S.]

ACROPTERIS. A name sometimes given to *Asplenium septentrionale*, and a few other asplenioid ferns. [T. M.]

ACROSPIRE. The first leaf that appears when corn sprouts. It is a developed plumule.

ACROSTICHEÆ. A section of polypodineous ferns, in which the sori occupy almost or quite the whole fructiferous surface, and are not confined to distinct and determinate points of the veins. [T. M.]

ACROSTICHUM. A genus of polypodiaceous ferns, typical of the group *Acrosticheæ*, with which, in the wider sense, it is synonymous. As restricted by modern pteridologists, the name is chiefly confined to a somewhat variable subaquatic tropical fern found in different parts of the world, which is distinguished by having the veins of its fronds uniting everywhere in a close network of small meshes, and by the lowermost leaflets or pinnæ being sterile, and the upper ones fertile. The fertile parts, both in this genus and the rest of the *Acrosticheæ*, are entirely occupied by the densely packed spore-cases, which thus form universal or shapeless masses, without any special covering or indusium. The typical species is *A. aureum*, which, in one or other of its forms, is found in the West Indies, South America, Australia, the Pacific Islands and Eastern Archipelago, India, Mascaren Islands, Madagascar, South Africa, and Tropical Western Africa. It is a tall-growing plant, eight to ten feet high, with a thick rhizome or rootstock, and bold pinnated fronds, the upper pinnæ of which are smaller, and clothed with the dense mass of confluent spore-cases. The plant is generally found near the sea, in morasses or moist situations. There are very few other species retained in the genus, and these mostly of doubtful character. [T. M.]

ACROTOME. A genus of *Labiatæ*, containing three species, natives of Southern Africa. They are shrubs or herbaceous plants, with small opposite leaves. The flowers are in dense verticillasters in the axils on the upper portion of the stem. The calyx consists of a campanulate tube with ten nerves and five or ten teeth. The tube of the corolla is scarcely longer than the calyx; its upper lip is erect, entire, and slightly arching, the lower trifid, the middle lobe being largest. The stamens and style are included. This genus is nearly related on the one hand to *Leucas*, and on the other to *Marrubium* and *Sideritis*, but it is distinguished from all of them by its distinctly one-celled anthers. [W. C.]

ACROTRICHE. A genus of *Epacridaceæ*, found in the eastern and southern portions of Australia and Tasmania, and distinguished by having a bi-bracteate calyx; a funnel-shaped corolla, the segments of which are clothed at the apex with deflexed hairs, and five slightly exserted stamens, which are shorter than the lobes of the corolla. The fruit is a depressed globose berry. They are shrubs of dwarfish habit, the branches usually divaricate, and clothed with scattered ovate or lanceolate leaves. The flowers grow in short lateral or axillary spikes, and are white or pale red. The name of *Frebella fasciculiflora* has been proposed for *Acrotriche ramifiora*. [R. H.]

ACTÆA. A genus of plants so called from the resemblance borne by their leaves and fruit to those of the elder, in Greek *aktē*. The only British species, *A. spicata*, Baneberry, is of rare occurrence, and is found only in bushy, mountainous limestone districts in the north of England. It bears its flowers, which are white, slightly tinged with blush, in a spike. The berries are black and poisonous. The root has been used in nervous disorders, but is said to be a precarious remedy. It is sometimes called Herb Christopher, a name also formerly given to the flowering fern *Osmunda regalis*. It is indigenous to the greater part of Europe. Two American species are occasionally to be found in the gardens of the curious, introduced from their native country, where they are abundant in rocky mountainous districts, from Canada to Virginia, particularly about Lake Huron. These are considered valuable medicines by the natives, especially as a remedy against the bite of the rattlesnake; hence they are, with several other plants, sometimes known as the Rattlesnake herbs. [G. A. J.]

ACTINIOPTERIS. A genus of polypodiaceous ferns of the section *Asplenieæ*, and consisting of curious little plants like miniature fan-palms, by which appearance they may be known. The technical peculiarities of the genus among the *Asplenieæ*, consist in the simple, distinct indusia, free veins, and linear elongate sori, which are marginal on the contracted rachiform segments of the small flabelliform fronds. One of the species, *A. radiata*, is plentiful in Southern India; and both this and its ally, *A. australis*, occur in Africa. The former grows three to six inches high, and produces an erect tuft of fronds which have a roundish outline, and are divided inwards from the margin very much indeed like what occurs in the fan-palms. [T. M.]

ACTINODAPHNE. A name derived from Greek words signifying ray laurel, and applied to a genus of the laurel family (*Lauraceæ*). The plants are Indian trees with alternate leaves, sometimes clustered or whorled, feather-nerved or somewhat palmi-nerved. Flowers in clusters or tufts, the male and female sexes on different plants. The male flowers have nine fertile stamens, in three rows, those of the inner row having a gland on either side of its base. The style is thick, the stigma disk-shaped, the fruit berry-like, placed in the cup-shaped tube of the calyx. [M. T. M.]

ACTINOMERIS. A genus of perennial N. American and Mexican herbs of the composite family, closely allied to sunflowers (*Helianthus*), but differing in the compressed and winged,—instead of 4-sided and wingless,—achenes, which have a pappus of two smooth bristles. There are about eight known species, most of them tall branching herbs, with alternate or opposite ovate or lance-shaped serrate leaves, which are smooth or rough, often tapering to the base, and decurrent on the stem, thus giving it a winged appearance. The rayed flower-heads, disposed usually in a corymbose manner, are white or yellow, sometimes 1½ inches across, and not unlike those of some species of *Coreopsis*. The generic name alludes to the fewness or irregularity of the rays. A number of the species is cultivated in collections of herbaceous plants. [A. A. B.]

ACTINOPHLEBIA. A small group of cyatheaceous ferns, now included in *Hemitelia*. [T. M.]

ACTINOSTROBUS *pyramidalis*. A small shrub from Swan River, belonging to the coniferous order. The branches are three-cornered, and jointed like a *Callitris*, from which genus it differs in having six equal valves to its cones, and three winged seeds. It inhabits salt marshes.

ACTINOTUS. A genus of *Umbelliferæ*, containing three species, natives of the eastern districts of New Holland. It is nearly related to *Sanicula*, but differs from that and allied genera in having no petals. It is characterised also by a one-ovuled ovary, crowned by two styles; the fruit is ovate, villous, and marked with fine striæ. The leaves are alternate, petiolate, and deeply trisected. The umbels are simple and many-flowered, the flowers on short pedicels, and surrounded by a many-leaved large involucre, which gives the genus somewhat the appearance of belonging to the *Compositæ*. [W. C.]

ACULEUS. A prickle; a conical elevation of the skin of a plant, becoming hard and sharp-pointed: as in the rose.

ACUMINATE. A term applied to leaves or other flat bodies which narrow gradually till they form a long termination: if the narrowing takes place towards the base, it is so stated, *e.g.* acuminate at the base; if towards the point, the term is used without qualification.

ACUYARI WOOD. The aromatic wood of *Icica altissima*.

ADA *aurantiaca*. Under this name has been published a New Grenada epiphyte, found in the neighbourhood of Pamplona, at 8,500 feet above the sea. It has closely packed bright orange-coloured flowers, with much the same structure as *Brassia*, except that the lip is firmly consolidated with the base of the column.

ADAM and EVE. *Aplectrum hyemale*.

ADAM'S NEEDLE. The vulgar name for *Yucca*.

ADAMIA. A genus of the order *Saxifragaceæ*, related to *Hydrangea*, found in India, China, and other eastern countries. It has a short five-toothed calyx, a five to seven-petaled corolla, ten to twenty stamens, and a half-inferior ovary becoming a berry, which is many-seeded. *A. versicolor*, one of the most beautiful of the few-known species, is a native of China, and forms a dwarf smooth-branched shrub, furnished with largish opposite leaves, resembling those of *Hydrangea japonica*. The flowers are collected into a pyramidal panicle, nearly a foot in diameter; they are each six or seven-petaled, forming a pointed star, and while in bud are whitish, but they gradually change to purple and violet; they have twenty stamens. The berries are blue. Another species found in Nepal, *A. cyanea*, also bears blue berries. [T. M.]

ADANSONIA. This genus belongs to the natural family *Bombaceæ*. The *Adansonia* has, until lately, been considered the largest tree in the world, but it must now give place to the mammoth tree of California (*Wellingtonia gigantea*). Its height is from 40 to 70 feet, and not at all in proportion to the size of its trunk, which sometimes attains the great diameter of 30 feet. It soon divides into branches of great size, which bear a dense mass of deciduous leaves, somewhat like those of the horse-chestnut. The flowers are large, white, solitary, and pendent on long stalks, and when expanded are about 6 inches across. The fruit is an oblong woody capsule, covered with a short down, and from 8 inches to a foot and a half long, in appearance somewhat like a gourd; internally, it is divided into 8 or 10 cells, each cell filled with a pulpy substance in which the seeds are immersed.

A. digitata, the Baobab, Ethiopian Sour Gourd, or Monkey-bread, is a native of many parts of Africa. It has been found in Senegal and Abyssinia, as well as on the west coast, extending to Angola, and from thence across the country to Lake Ngami. It is cultivated in many of the warm parts of the world. It has been called 'the tree of a thousand years,' and Humboldt speaks of it as 'the oldest organic monument of our planet.' Adanson, whose name the genus bears, and who travelled in Senegal in 1794, has given an account of this tree. He made a calculation to show that one of them, 30 feet in diameter, must be 5,150 years old! He saw two trees, from 5 to 6 feet in diameter, on the bark of which were cut to a considerable depth a number of European names; two of these were dated, the one in the 14th, the other in the 15th century. In 1555, the same trees were seen by Thevet, another French traveller, who mentions them in the account of his voyage. Livingstone says of the tree, 'I would back a true *Mowana* (the name given to it in the neighbourhood of Lake Ngami) against a dozen floods, provided you do not boil it in salt water; but I cannot believe that any of those now alive had a chance of being subjected to the experiment of even the Noachian deluge.'

The bark of the Baobab furnishes a fibre which is made into ropes, and in Senegal woven into cloth. The fibre is so strong as to give rise to a common saying in Bengal; 'As secure as an elephant bound with a baobab rope.' The wood is soft, and subject to the attacks of a fungus which destroys its life, and renders the part affected easily hollowed out. This is done by the negroes, and within these hollows they suspend the dead bodies of those who are refused the honour of burial. There they become mummies, perfectly dry and well preserved, without any further preparation or embalmment.' Livingstone speaks of a hollow trunk, within which 20 to 30 men could lie down with ease. The leaves pounded constitute *Lalo*, which the Africans mix with their soups, sauces, &c., not as a relish, but to diminish the excessive perspiration, and keep the blood in a healthy state. 'The pulp of the fruit is slightly acid, agreeable, and often eaten; and the juice expressed from it constitutes a drink which is valued as a specific in putrid and pestilential fevers. Owing to this circumstance it forms an article of commerce.' The ashes of the fruit and bark boiled in rancid palm oil are used as a soap by the negroes.

The only other species of the genus is *A. Gregorii*. It is a native of the sandy plains of N. Australia, and is known as Sour gourd and Cream of tartar tree. It dif-

Adansonia Gregorii.

fers chiefly from *A. digitata* in its smaller fruit with a shorter foot stalk. The largest tree seen in Gregory's expedition was 29 feet in girth at 2 feet from the ground. The pulp of its fruit 'has an agreeable acid taste, like cream of tartar, and is peculiarly refreshing in the sultry climates where the tree is found. It consists of gum, starch, sugary matter, and malic acid.' [A. A. B.]

ADDER'S MOUTH. An American name for *Microstylis*.

ADDER'S TONGUE. The English name for *Ophioglossum*. — YELLOW. *Erythronium americanum*.

ADECTUM. A synonym of *Dennstædtia*, a handsome free-growing genus of ferns, related to *Dicksonia*. [T. M.]

ADELASTER (Gr. like something unknown). A name proposed for those garden plants which, having come into cultivation without their flowers being known, cannot be definitively referred to their proper genus. All Adelasters are therefore provisional names, to be abandoned as soon as the true names of the plants so called can be ascertained.

ADENANDRA. A genus of rutaceous or rue-like plants, so named on account of the presence of a small gland on the top of the stamens. They consist of small shrubs, natives of the Cape of Good Hope, and some of them are cultivated for the sake of their pink-coloured flowers. The genus is principally distinguished by its 5 sterile stamens, which are in form like the 5 fertile ones, but longer; both kinds tipped with a gland. The leaves are used for the same purposes as those of *Diosma* at the Cape. [M. T. M.]

ADENANTHERA. A genus of the pea family (*Leguminosæ*). The species are chiefly found in eastern India and the Malayan islands, and one is wild in Madagascar. They are trees or shrubs, with bipinnate or decompound leaves and spikes of small yellow flowers, the anthers of which are tipped with a stalked gland; and these gland-tipped anthers give rise to the generic name. *A. pavonina* grows to a great size in the East Indies, and yields a solid useful timber, called Red Sandal wood, a name which is also given to the wood of *Pterocarpus santalinus*. A dye is obtained by simply rubbing the wood against a wet stone; and this is used by the Brahmins for marking their foreheads after religious bathing. The seeds are of a bright scarlet colour, and are used by the jewellers in the East as weights, each seed weighing uniformly four grains. Pounded and mixed with borax, they form an adhesive substance. They are sometimes used as an article of food, and are frequently made into ornaments, such as bracelets, necklaces, &c. [A. A. B.]

ADEXOCALYMNA. The name given to a genus of *Bignoniaceæ*. The species are large climbers, and all of them natives of Brazil, where they scramble over trees, enlivening the forests with their clusters of bright-coloured yellow, orange, or pink flowers. Their stems are slender and often rough. Their leaves are ternate, or sometimes only binate; when this latter is the case, a tendril-like appendage takes the place of the third leaflet. Numbers of depressed circular glands are found on their surface, as well as on the calyx; and from this circumstance the genus receives its name *Adenocalymna*, which is composed of Greek words signifying gland and covering. The flowers are borne on long racemes; they are trumpet-shaped, and, intermixed with them, are large bracts, which fall off early. Some of the species are cultivated in our stoves for their beauty. [A. A. B.]

ADENOCARPUS. This is a genus of the pea family (*Leguminosæ*), composed of plants which are most of them extremely handsome, from their bearing profuse racemes of yellow flowers. The genus only differs from that of the common English broom (*Genista*) in having pods covered with glands: whence its name *Adenocarpus*, which is derived from two Greek words signifying gland and fruit. They are found in the Pyrenees, the Sierra Nevada, and in other parts of southern Europe, but chiefly at high elevations. One species is found in Madeira, and a few in the Canaries. Mr. Bunbury, in writing on the botany of the Peak of Teneriffe, says: 'To the region of the heath succeeds, as we ascend, that of the Codeso del Pico (*Adenocarpus frankenioides*). The limit of this plant is particularly well marked. For a little space it is intermixed with scattered and stunted bushes of the heath, but this soon thins out and disappears, and for miles the whole slope is covered with the *Adenocarpus* alone, as some of our commons and wastes in England are covered with Furze. It is in general a low compact rigid bush, peculiar in its multitude of short lateral branches, and the minute closely-crowded grey-green leaves; by no means a handsome plant when out of flower; but here and there, in sheltered spots, it assumes the character of a little tree. It is one of the most eminently social plants in the world.' Several species are cultivated in gardens. [A. A. B.]

ADENOPHORA. A genus of plants allied to *Campanula*, and like it bearing bell-shaped flowers, the chief mark of distinction being that the style of the present plant is surrounded by a cylindrical gland, whence its name (from the Greek *aden*, a gland, and *phero*, to bear). The plants of this family are perennial, rarely biennial herbs, with erect stems, alternate or somewhat whorled leaves, which below are broad and stalked, but gradually becoming narrower as they ascend the stem. The flowers are blue, stalked and drooping, and for the most part are situated towards the top of the stem, where, in some instances, they form a spike or cluster, while in others they are few in number. Most of the species are natives of Siberia, China, and Japan. One species, *A. liliifolia*, or *A. suaveolens*, is found in many countries of eastern Europe, and occurs also in France, Hungary, and Candia. In this the flowers are numerous, sweet-scented, and disposed in a loose pyramidal panicle. The root is thick and esculent, as are those of some of the other species. All are elegant border flowers, and are, therefore, worth cultivating in gardens. [C. A. J.]

ADENOPHORUS. A small group of ferns, in which the sori are terminal on the free veins, the receptacle at the apex of the simple costa-like or central veins being dilated or obovate. The fronds are small, very elegant in character, and bear glands over their surface. The species are now referred to *Polypodium*. [T. M.]

ADENOPUS. An imperfectly known genus belonging to the gourd family (*Cucurbitaceæ*). The male and female flowers are on different plants, and the female flowers are not at present known. The male flowers have a tubular five-toothed calyx; five petals inserted on the top of the calyx tubes, entire or slightly crisped at the margin; five stamens, in two parcels, attached to the middle of the tube of the calyx, with very short filaments and long wavy anthers. The leaves are palmately-lobed, stalked, with two glands at the extremity of the leaf-stalk. The plants are natives of Sierra Leone and Western Tropical Africa. [M. T. M.]

ADENOSMA. A genus of *Acanthaceæ*, containing eight or nine species, natives of Asia. They are annual herbaceous plants, having the odour of the Mints, with opposite leaves, and sessile flowers in the axils of the small leaves on the upper portion of the stem, so aggregated as to form a leafy spike. The genus is characterised by a five-partite calyx, a gaping corolla, four didynamous stamens, with anthers composed of two parallel cells. The long capsule is many-seeded. It differs from the allied genus *Ebermayera* in the gaping corolla, and in the structure of the anthers. [W. C.]

ADENOSTYLIS. A genus of the composite family, comprising but few species. They are perennial mountain herbs, with alternate stalked, cordate, or reniform leaves, which are smooth, or covered with a loose white cotton. Their flower-heads are numerous in terminal compact corymbs, with florets of a purple or white colour In appearance these plants are much like the common coltsfoot (*Tussilago Farfara*), but they differ from the coltsfoot in having all their florets fertile. They are all natives of mountain districts in southern Europe, the greater part of them being found in the Pyrenees, where they grow luxuriantly in stony places beside alpine rivulets. The leaves of *A. glabra* have been recommended in coughs. [A. A. B.]

ADESMIA. A large genus of the pea family (*Leguminosæ*), confined to the temperate parts of S. America. Commencing in the Bolivian Andes, they extend southward to Cape Horn; but are found in greatest numbers in Chili. They are annual or perennial; some of them shrubs four or five feet high, and most of them with alternate equally pinnate leaves terminated by a bristle. Their flowers are disposed in racemes at the apex of the branches, or solitary in the axils of the leaves, and are generally yellow with purple stripes. The pods are jointed, rough on the surface, and sometimes beset with feathery bristles. One of the species, *A. aphylla*, has its leaves reduced to mere scales; and in another, *A. trifoliata*, they are not unlike those of the common wood sorrel. *A. balsamifera*, a Chilian species called Jarilla, is a plant of great beauty when in flower; it yields a balsam which has a very pleasant odour, perceptible at a great distance. This balsam is said to be of great efficacy in healing wounds. A few of the species have their abortive flower-stalks converted into forked spines. There are upwards of fifty species. [A. A. B.]

ADHATODA. A genus of acanthaceous plants, consisting of herbs or shrubs with opposite leaves, and axillary spikes of flowers, each flower furnished with three bracts, the outer one of which is large and persistent, covering the calyx; the two inner ones smaller. The calyx is five-parted; the corolla two-lipped; the four stamens are inserted on the throat of the corolla; the anthers are two-celled, with a large connective, the lobes unequal, and the inferior ones often spurred; the filaments compressed, bent downwards; the style thread-shaped, bent downwards; and the capsule stalked, two-celled, four-seeded, bursting by two valves. *A. vasica*, the *Justicia Adhatoda* of Linnæus, is a common plant in India; its wood is soft, and its charcoal is excellent for the manufacture of gunpowder. The flowers, leaves, root, and especially the fruit, are considered as anti-spasmodic, and are given in cases of asthma and intermittent fever. The word *Adhatoda* is a latinised form of the native Malabar or Cingalese name. [M. T. M.]

ADIANTEÆ. A section of polypodiaceous ferns, in which the receptacles to which the spore-cases are attached, are placed on the under surface of the indusium itself, so that the fructification is, as it were, upside down, and is hence said to be resupinate. [T. M.]

ADIANTOPSIS. A small genus of elegant polypodiaceous ferns, of the section *Cheilanthea*, distinguished partly by their adiantoid aspect, but technically by having marginal punctiform sori terminal on the free veins, and covered by distinct orbicular indusia. The plants bear generally tufted stems, and small elegantly-divided fronds. The species are found in South America, the West Indies, and Africa. *A. radiata*, one of the best known of them, common in the West Indies and South America, grows about a foot high, from a tufted crown, the stipites shining black, and the fronds spreading out at top of the stipites into a radiate tuft of pinnate branches. The species are often seen in cultivation, on account of their small size and elegant character. [T. M.]

ADIANTUM. An extensive and much admired genus of polypodiaceous ferns, typical of the group *Adianteæ*. The species are scattered nearly over the whole world, but are most abundant in tropical countries. They have all black shining stipites, and mostly roundish or rhomboidal or lunately-curved pinnules, the fronds being very various in size and general character. The structure is very peculiar, unlike that of any other ferns. The sori are marginal, covered by indusia, which are either roundish and distinct, or become blended into a linear form, these two conditions respectively resembling the fructification seen in *Cheilanthes* and *Pteris*; but it is resemblance only, the fructification (spore-cases) being in the latter genera seated on the frond itself, and covered by the indusium, while in *Adiantum* they are not attached to the frond, but to the under side of the indusium, and are therefore turned upside down on to the surface of the frond. This structural peculiarity distinguishes *Adiantum* from all other ferns except *Hewardia*, which is known by having a reticulated venation, that of *Adiantum* being free. The genus is represented in the British Flora, by *A. Capillus-veneris*, the Maidenhair Fern, a very elegant plant, with a creeping scaly rhizome, and bipinnate fronds, the leaflets of which are between rhomboidal and wedge-shaped, margined with oblong sori, and more or less deeply lobed. This species is very extensively distributed in the temperate or tropical parts of Europe, Asia, Africa, and America, and not very materially varying in form, notwithstanding this wide range. Some species, as the *A. reniforme* of Madeira, have entire fronds; in others, as the *A. lunulatum* of India, they are pinnate; not a few species are, like our native one, bipinnate; and numerous others are tripinnate, or still more divided. *A. pedatum*, a very beautiful North American species, which has the fronds pedate, the divisions pinnate, and the pinnules halved oblong and lunate, incised along the upper edge; it is sometimes used in the preparation of capillaire. The species are great favourites in hothouses. [T. M.]

ADLUMIA. A climbing genus of fumeworts, consisting only of the *A. cirrhosa*, a pretty North American biennial, formerly known as *Corydalis fungosa*. It is distinguished from the other genera of the Order by the permanent cohesion of its four spongy petals into one piece, and by a many-seeded pod, splitting, when ripe, into two valves. Its chief attraction consists in its delicate pale green triply pinnate foliage, the twining footstalks of which act as tendrils; the small flesh-coloured blossoms are freely produced, but possess little beauty. The plant neither climbs nor flowers till the second year. [W. T.]

ADNATE. Grown to anything by the whole surface; when an ovary is united to the side of a calyx it is adnate.

ADONIS. A small genus of ranunculads, mostly European, comprising several popular border flowers, both annual and perennial. It is characterised by the absence of an involucre, a calyx of five sepals, a corolla of from five to fifteen petals, and numerous dry ovate carpels, pointed with the style, and grouped in a short spike or head; all the species have the foliage cleft into numerous linear segments, and produce but a single flower at the summit of each stem and branch. Of the annual section, eight or ten species are described, but only two are to be found in general cultivation: the *A. autumnalis*, and *A. æstivalis*, both indigenous plants, with small, deep crimson flowers, the latter having the petals much longer than the calyx; whilst those of the former scarcely exceed it. They are popularly known as Pheasant's Eye, and Flos Adonis. The perennial species are all showy, dwarf herbaceous plants, with long black fascicled roots, and large glossy yellow flowers. The best and most commonly cultivated species is the *A. vernalis*, a desirable and very effective early bloomer. [W. T.]

ADOXA. A small genus referred by Mr. Bentham to the *Caprifoliaceæ*, consisting of a single species, *A. Moschatellina*, the Tuberous Moschatel, found blooming in spring in woods and on shady banks in many parts of England, and extending through Northern and Central Europe and parts of Asia and North America, far into the Arctic regions. The genus is distinguished by bearing a calyx of two or three spreading lobes; a short-tubed corolla, with four or five spreading divisions; eight or ten stamens in pairs alternating with the divisions of the corolla, and inserted on a little ring at its base; three to five short styles united at the base; a three to five-celled ovary, with one ovule in each cell and maturing into a berry. The plant is a low herb, of four to six inches high, smooth, pale green, forming creeping half-buried runners, the leaves ternately divided, with broad deeply three-lobed segments, and the musky-scented flowers pale green in a little globular head at the top of the short leafy flower-stems. The upper flower in each head has generally a tetramerous arrangement of parts, two calyx lobes, four corolla lobes, and eight stamens; while the lateral ones have three calyx lobes, five corolla lobes, and ten stamens. The *Adoxa* has, until recently, been classed with the *Araliaceæ*. [T. M.]

ADPRESSED. Brought into contact with anything without adhering.

ADELPHIA. A fraternity — a Linnean term denoting a collection of stamens. *Monadelphia* = one such collection; *Diadelphia* = two such collections; and so on.

ÆCHMEA. A genus of *Bromeliaceæ*, having a six-parted perianth, of which the three outer sepaline divisions are equal, and much shorter than the inner petaloid ones. The flowers have six stamens, and an inferior three-celled ovary containing numerous ovules, and becoming a substitute berry. The species are found in tropical America, often epiphytal on the trunks of trees in the dense forests. They have strap-shaped or sword-shaped leaves, sometimes spiny at the margin; and from the centre of these is developed the flower scape, which is branched in a panicled manner and bears numerous flowers. *Æ. discolor*, one of the most striking of the species, has broad recurved leaves, which are dull green above and purplish beneath. The panicle is longer than the leaves, of a scarlet colour in the upper part, bearing the flowers distantly spiked along the branches. The flowers are without bracteoles, in which respect it is peculiar. The calycine segments are oblique and obtuse, coral-red below, blackish above, the petaline ones twisted, purplish. The unexpanded buds have a most striking resemblance to the seeds of *Abrus precatorius*, commonly called crab's eyes, and sometimes strung as beads. [T. M.]

ÆCHMOLEPIS. A genus of *Asclepiadaceæ*, containing a single species, a native of Angola. It is a shrub with ternate leaves, glabrous above, hoary and reticulated beneath. It is characterised by its filaments being connate at their base and distinct above, by having its anthers cohering at the apex, though free from the stigma, and by its twenty granular pollen masses. [W. C.]

ÆCIDIUM. A genus of *Fungi*, comprising a large number of parasites, which grow upon the living parts of plants. The reproductive organs or spores are nearly globose, arranged in little necklaces, which radiate from a thin cellular base, and, as they easily break off, form a little dust-like heap, which is white, yellow, orange, &c., according to the species. The whole mass is surrounded by a membranous coat or peridium, which sometimes bursts irregularly at the tip, but more frequently splits into a number of nearly equal lobes, which curl back, and have a very pretty appearance under the microscope. They grow on the leaves, petioles, fruit, or young shoots, sometimes producing but little constitutional derangement, but occasionally causing the adjacent parts to swell, or producing great distortion, as in a species which attacks the shoots of elder in North America. Sometimes the whole appearance of the plant is altered, as in one which commonly attacks species of *Epilobium*; while, again, at times, particular leaves only are affected, as in the garden and wood anemones where the outline is somewhat changed, and the substance is greatly thickened. Where the plant is only partially affected, the general health is not much impaired; but where the parasite is very vigorous, death may ultimately ensue. We are not aware that any species attacks our cereals. There has, however, been a very unjust charge brought against *Æcidium berberidis*, a beautiful species, which attacks the leaves, flowers, and young fruit of the berberry, as if it were the cause of mildew in wheat. Great, however, as are the changes which *Fungi* undergo occasionally in passing from one condition to another, there is not the slightest reason for imagining that the *Æcidium* is a transitional state of wheat mildew. It has its own mode of propagation, and passes through nearly the same phases of vegetation as the mildew, without affording a suspicion that it is not a perfect plant. The whole story has no doubt arisen from the *Æcidium* being common on the berberry in hedges surrounding wheat fields; and there is reason to believe the report is true, that wheat has been especially mildewed in the neighbourhood of the *Æcidium*. The peculiar situation, however, may be equally favourable to either parasite; and it is to be observed, that mildew is peculiarly prevalent in districts where the berberry is unknown, except as a garden plant. *Æcidia* attack phænogamous plants of various kinds, but they are far less frequent on endogens than exogens. Species occur in all parts of the world, but are more common in temperate regions. [M. T. B.]

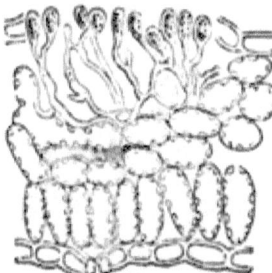

Æcidium rustlagiois

ÆGIALITIS. A genus of *Plumbaginaceæ*, containing a small number of Australian and Indian undershrubs, with thick articulated stems, and alternate-stalked ovate or roundish leaves of leathery texture; flowers in spikelets, arranged in branched spikes; calyx tubular-cylindrical; petals small and white; styles quite free and glabrous with awl-shaped stigmas. The seeds of *A. annulosa*, which grows in mangrove swamps, are said to germinate while on the plant. [J. T. S.]

ÆGICERACEÆ. This name has been given by Blume to *Ægiceras*, viewed as the representative of a natural order containing no other genus than itself. It is, however, generally included in MYRSINACEÆ, which see.

ÆGICERAS. A genus of *Myrsinaceæ*, differing from all the other genera in that family by its follicular fruit. The species, of which there are five, consist of small trees, inhabiting swampy shores in the tropical parts of India, the Indian Archipelago, and Australia, where they form impenetrable thickets like the mangroves (*Rhizophora*), in consequence of their seeds germinating while yet in the fruit, and sending down strong perpendicular roots into the mud, without separating from their parents. They have obovate entire dotted leaves, the upper surface of which is often covered with a saline incrustation, which, according to Blume, they scurvic. Their flowers are white, fragrant, in terminal or axillary umbels; the flower-stalks articulated at the base. *A. majus* is the only vegetation to be seen for miles along the coast of Sumatra. [A. A. B.]

ÆGILOPS. A genus of grasses allied to *Triticum*, or wheat grass. It occurs wild in the south of Europe and parts of Asia. Botanists have recognised as many as three species; but from recent experiments in the culture of *Ægilops*, there is reason to believe, not only that all the so-called species are referable to one, namely, *Æ. ovata*, but that the *Ægilops* is, in reality, the plant from which has originated our cereal wheats. Upon this subject will be found an interesting paper, translated from the French, in the *Journal of the Royal Agricultural Society* (vol. xv.), from which it would appear that M. Esprit Fabre, of Agde, has made the *Æ. ovata* the subject of experiment, and that from it he obtained the form known as *Æ. Triticoides*, the continued cultivation of which latter, for six years, resulted in the production of very respectable ears of wheat. The changes that occurred were a lessening in the numbers of the awns, and a gradual conforming of the chaff scales to those of wheat, a greater length and regularity of growth in the ear, an enlargement of the seed to that of the wheat, and a taller and more upright habit of growth of the whole plant. Both the experimental results, and the conclusions of M. Fabre have been doubted by some of the specific botanists, and we are, therefore, glad to have an opportunity of recording the result of our own experiments in this interesting matter. In 1854, we planted a plot with seed of *Æ. ovata*, from which was gathered seed for a second plot in 1855, leaving the rest of the first plot to seed itself, which it did, and came up spontaneously. This plot has since continued to bring forth its annual crop in a wild state, in which the spikes are short, and so brittle that they fall to pieces below each spikelet the moment the seed is at all ripe. The produce of the 1855 crop has, in the same manner, been cultivated year by year in different parts of the experimental garden of the Royal Agricultural College, and our crop for 1860 had many specimens upwards of two feet high, and with spikes of flowers containing as many as twelve spikelets. Our conclusions then are, that with us the *Ægilops* is steadily advancing; and we fully expect, in three or four years, to arrive at a true variety of cereal wheat. What too is confirmatory of this matter, is that the bruised foliage of the wild grass, and the cultivated wheat, emits the same peculiar odour, and, besides the *Ægilops*, is subject to attacks of the same species of parasites (blights), one examples of this year being much affected with the rust (*Uredo rubigo*), mildew (*Puccinia graminis*), and others. These, it

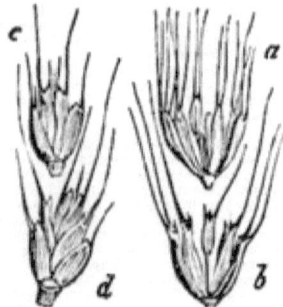

Ægilops:—a *Æ. ovata*; b *Æ. ovata triticoides*; c the same after four years' cultivation; d the same after five years' cultivation

would seem, are the effects of civilization; and it is not a little remarkable, that in this respect this grass should be so much like our field crops, which were particularly liable to blight in the straw and foliage during 1860. [J. B.]

ÆGINETIA. A genus of the broomrape family (*Orobanchaceæ*), found in India and the islands of the Indian Archipelago. They are annual, leafless, parasitical herbs, growing on the roots of various grasses, their stems from three inches to a foot high, bearing a solitary terminal flower, and having at their base a few scales. The whole plant is of a brown colour, except the flower, which is large and tubular; the tube white, and the limb rose, or altogether purple outside, and the throat yellow. The calyx is one-leaved and cleft in front. *A. indica*, 'prepared with sugar and nutmeg, is considered an antiscorbutic.' [A. A. B.]

ÆGLE. The name of a genus of plants belonging to the orange family (*Aurantiaceæ*). The fruit, known in India as the Bhel fruit, is the product of *Æ. Marmelos*. In appearance it is much like the orange. The thick rind of the unripe fruit possesses astringent properties, on which account it is used in India in cases of dysentery and diarrhœa. The ripe fruit has an exquisite flavour and perfume. Not only the fruit, but other portions of the plant are used for medicinal purposes; and a yellow dye is prepared from the rind of the fruit. The genus is distinguished by its numerous disunited stamens, from the orange (*Citrus*), to which, in other particulars, it is closely allied. [M. T. M.]

ÆGOCHLOA. A genus of dwarf, hardy annuals of the *Polemonium* family, chiefly Californian. Their tubular calyx, with unequal, rigid, multifid segments, and corolla with salver-shaped limb, distinguish them from the allied genera. The species have, for the most part, pinnate or pinnatifid clammy foliage, more or less spinous, and small-clustered pilia-like flowers. None of them are remarkable for their beauty, and, with the exception of *Æ. pungens*, are not known in British gardens. By some authors the species are classed under the genus *Navarretia*. [W. T.]

ÆGOPODIUM. An umbelliferous plant with smooth thrice-ternate leaves, unattractive white flowers, and an unpleasant odour when bruised: a common pest of orchards, shrubberies, and ill-kept gardens, where, by means of its creeping roots, or rather subterraneous stems, which are of rapid growth and singularly vivacious, it soon establishes itself when once admitted, and defies eradication, smothering all vegetation less rampant than its own, and disfiguring where it finds nothing to choke. Its old English names were Gout-wort or Gout-weed, Herb Gerard, Ash-weed, and English Master-wort. It was, at one time, accounted a specific for the gout, hence its specific name *Podagraria*; but though, like the rest of the umbelliferous tribe, partaking of aromatic properties, it is really of no more value in the pharmacopœia than in the garden. Linnæus says it is eaten in Sweden, boiled for greens when tender in the spring. It is a native of the whole of Europe to Caucasus and Siberia. Gerarde says of it: 'Herb Gerarde groweth of itselfe in gardens without setting or sowing, and is so fruitfull in its increase, that when it hath once taken roote, it will hardly be gotten out againe, spoiling and getting every yeere more ground, to the annoying of better herbes.' An Alpine species is found in Asia, which appears to possess all the bad properties of its congener. [C. A. J.]

ÆGOTOXICUM. A genus doubtfully placed in the spurgewort family (*Euphorbiaceæ*). There is but one species, *Æ. punctatum*, which is a native of Chili, and is described as a sombre-looking tree, forming immense woods. The leaves are opposite, shortly stalked, oblong, and entire, the upper surface smooth, the under covered with rusty scales. The flowers are male and female, the males alone on one tree, and the females on another; they are disposed in axillary racemes. The fruit is a one-seeded drupe, about the size of a pea. Planks and beams are made of the wood; and the fruits are said to be a powerful poison to goats. [A. A. B.]

ÆONIUM. The plants that are comprised in this genus of *Crassulaceæ* are much more generally known as species of *Sempervivum*, from which, however, they differ in their seed-vessels being partially sunk in the receptacle, and not regularly opening by their ventral suture, but only at the base and back by an irregular tearing. *Æ. arboreum* is well known to gardeners as the tree houseleek; its loose panicles, with a profusion of clammy yellow blossoms, are very elegant. *Æ. tabulæforme*, as well as others of the genus, is remarkable for the stem being so contracted that the leaves are closely packed in flat rosettes. Like other succulent plants, their tenacity of life is remarkable. They are natives of the Levant, Madeira, the Canaries, &c. [M. T. M.]

ÆQUALIS. This term signifies equality or similarity in size, and is also used in the sense of uniformity; thus, an equal umbel, is an umbel of which the florets are all alike.

AËRANTHUS *grandiflora*, is a Madagascar epiphytal orchid with distichous leaves, and large green solitary flowers at the end of weak, flexible scapes. It is sometimes seen in cultivation. From *Angræcum* it differs in having a lip articulated with the foot of the column.

AËRIDES. A large genus of tropical orchids, with distichous leaves, mostly channelled and unequally truncate, but sometimes terete. All the species inhabit the warmer parts of Asia. The flowers are usually among the largest of the order, of all tints except blue, and frequently extremely sweet. One of them, with small flowers, clings to the branches of trees in Sylhet with such long flat roots, resembling bands of a tape-worm, that it has gained the name of *Aerides tæniale*. This is not in cultivation.

AËROCYSTS. The air-cells of algals.

AËROPHYTES. Plants growing wholly in the air; such as epiphytal orchids, many lichens, bromeliads.

ÆRUA, or ŒRVA, a genus of *Amaranthaceæ*, consisting of shrubs and herbaceous plants from Africa, tropical Asia, and probably Central America. The plants have a more or less white tomentose appearance. The leaves are alternate. The minute flowers are in dense terminal or axillary spikes, and are hermaphrodite, with three concave persistent bracts. The calyx consists of five nearly equal, erect, and hairy sepals; the five stamens are united into a cup at their base; the ovary is one-celled, with a single ovule in each cell; the style is short, and the two stigmas are very minute. The fruit is a roundish utricle, included in the calyx, and the seed is vertical. Nineteen or twenty species have been described. [W. C.]

ÆRUGINOUS. Verdigris-coloured.

ÆSCHYNANTHUS. A beautiful genus of tropical plants of epiphytal habit, belonging to the cyrtandreous group of the *Gesneraceæ*. The peculiarities of structure are a tubulous five-cleft equal calyx, a monopetalous corolla with a curved tube dilated at the throat, and a two-lipped spreading five-lobed limb; there are four didynamous stamens, and the ovary is surrounded by a fleshy hypogynous glandular ring. The species inhabit the tropical parts of Asia, and are among the most gorgeous ornaments of hothouses in this country, many of them having been introduced to cultivation. They have mostly pendent stems, opposite fleshy leaves, and scarlet or orange-scarlet flowers. One of the finest species, *Æ. speciosus*, is of sub-erect habit, with ovate lanceolate acuminate fleshy leaves, and a terminal fascicle of from ten to twenty erect long-tubed flowers, of a rich orange-yellow below, passing into scarlet at the top, and marked on the face of the limb with yellow and black; the corolla tube in this species is narrow club-shaped, slightly curving towards the top, and the rich yellow of the throat, surrounded by a black band or zone, contrasting with the scarlet of the outer portion of the limb, produces a fine effect. This is a native of Java, as also is *Æ. longiflorus*, which has flowers of a similar shape, but of a deep crimson. The calyx, in these species, consists of narrow segments cut down nearly to the base; but in some other kinds, the calyx forms a deep vase-shaped tube, with a short slightly spreading limb. *Æ. Lobbianus* is one of these, with short elliptic leaves, and a large conspicuous purple calyx, covering half the length of the scarlet flower tubes. Another, *Æ. javanicus*, has the long tubular calyx green edged with red, and the flowers bright red, marked with yellow rays from the throat; these are both natives of Java. A still different form is met with in *Æ. tricolor*, a slender drooping Bornean plant, with ovate leaves, in which the calyx is cup-shaped with five short rounded lobes, and the flowers have a much shorter tube than in the foregoing. There are many other species, nearly all of ornamental character. [T. M.]

ÆSCHYNOMENE. A genus of the pea family (*Leguminosæ*), of which between thirty and forty species are known. They are herbs or small shrubs, with unequally pinnate leaves and half arrow-headed stipules. The flowers are disposed in axillary or terminal racemes, and are often of a bright yellow colour. The pods are jointed. The species are found in the greatest numbers in Brazil, but they are to be met with in most tropical countries. The pith-like stem of *Æ. aspera* is, on account of its extreme lightness, used in India (where it is called solah) for making hats, bottle-cases, swimming jackets, and for many other purposes where elasticity and lightness are required. To construct some of these articles, the wood is cut into thin slices and pasted together. It is sold in the bazaars of Calcutta, being brought from the neighbouring marshy places, where it grows to a great size. This substance is purchased by the natives, who use it for floats for fishing nets, and make fancy articles, as model temples, &c., from it. *Æ. montevidensis* is called the 'humming-bird bush' at Buenos Ayres, because that bird seems to take more delight in it than in any other flower. *Æ. viscidula*, a native of Florida, has sensitive leaves; and so also, as its name implies, has *Æ.*

sensitiva, which is common in the West Indies. [A. A. B.]

ÆSCULUS. The Horse-chestnut. The name Æsculus (from *esca*, food) was applied originally to a species of oak which, according to Pliny, was highly prized for its acorns; but how it came to be transferred to the horse-chestnut is very uncertain: perhaps, as Loudon suggests, it was given ironically, because its nuts bear a great resemblance, externally, to those of the sweet chestnut, but are unfit for food. *Hippocastanum* (the specific name of the common sort) is a translation of the modern name, which was given, Evelyn tells us, 'from its curing horses broken-winded and other cattle of coughs.' The Horse-chestnut is a tree of large size, frequently reaching a height of fifty or sixty feet, with an erect trunk and a broad pyramidal outline. It may be readily distinguished, even in the depth of winter, by its unusually large buds, set on the extremities of thick and heavy-looking branches, which are evidently destined to bear a weighty tuft of foliage and leaves. These buds are covered thickly with a gummy substance, which protects the tender interior from the cold and wet. As the sun gains power, the gummy covering melts and yields to the expanding pressure from within, and then the scales on which it is overlaid fall off, and the delicate green leaves are rapidly unfolded, encircling a conical mass of embryo flowers. In this stage the leaves present a singular appearance, drooping with their points towards the ground, as if not strong enough to assume a horizontal position. The buds expand very early in spring, but not prematurely, for within three or four weeks of their first unfolding they have attained their full length, amounting sometimes to eighteen inches. The leaves and flower-buds continue to increase in size until May, when the latter expand; and now the tree, having reached the meridian of its glory, stands forth prominently in all the gorgeousness of leaf and blossom. The downy covering, which was observable on the leaves in their early stage, has disappeared, and they have assumed instead a rich full green. Each leaf is composed of seven broad leaflets, unequal in size, which radiate from a common centre, a character of foliage different from that of any other British tree. The flowers, which grow in long cone-shaped clusters, are showy white, dashed with pink and yellow, destitute of perfume, but attractive to insects, and, as long as they continue in perfection, very beautiful. They soon, however, become tarnished, and the tree consequently loses much of its grace; yet it is still a fine tree, readily distinguished at a considerable distance by its tiers of large and massive foliage. Of the numerous flowers contained in every bunch, a few only mature their fruit; the rest drop off soon after they have begun to lose their beauty. The seed-vessels, which are set with short rigid prickles, attain their full size in October, when they fall off, and, splitting with even valves, disclose three cells, in each of which is contained a roundish polished nut, resembling the sweet chestnut in colour, but not, like it, terminating in a point. It rarely happens that all three nuts are perfected; frequently only two are developed, but the rudiments of all may be discovered. The nuts abound in farinaceous matter, but are too bitter to be fit for human food. They serve, however, as food for goats, sheep, and deer, and are sometimes boiled and given to poultry. Reduced to powder, and mixed with a third part of flour, they are said to make better paste than that composed of flour alone. The timber, owing to its rapid growth, is soft and of loose fibre, and is consequently of little value. The Horse-chestnut is supposed to be a native of Asia, probably of northern India, whence it was introduced into Europe about the middle of the sixteenth century. There is a very fine variety with deep rose-coloured blossoms; and in North America is found another species, the *Æ. ohioensis* or Buckeye, which is far inferior to the common sort in the beauty of its flowers. The tree sometimes called the Scarlet Horse-chestnut belongs to a closely allied genus, PAVIA, which see. [C. A. J.]

ÆSTIVAL. Of or belonging to the summer.

ÆSTIVATION. The manner in which the parts of a flower are folded up before the flower expands.

ÆTHALIUM. A genus of myxogastrous fungusses, inhabiting more especially stoves and garden-frames where a strong heat is kept up, and doing much damage, by first involving everything in a slimy mass, and then contaminating what it has not overrun, by its myriads of dust-like spores. The principal species, *Æthalium flavum*(If, indeed, the others are not mere varieties, differing only in colour), appears first under the form of a yellow cream-like mass, which is found to consist, when closely examined, of little wavy viscid strings; this at length swells, and produces abundant dark spores, collected in little heaps separated from each other by thin irregular yellow partitions; the outer surface is rough and scurfy. It sometimes occurs on leaves and rotten wood, in groves and forests, and is found in various parts of the world. The best way of getting rid of it, is dusting the plant, as soon as it appears, with quicklime or salt. This treatment must, however, be followed up perseveringly, as the growth is so rapid that the dusty stage, in which the lime or salt is of little use, may recur before a second application is made. [M. J. B.]

ÆTHERIA. A genus of terrestrial orchids found in the tropics of Central Asia, and nearly allied to *Goodyera*, from which it differs, indeed, in little except the presence of two callosities at the base of the lip. Five or six species are known.

ÆTHIONEMA. A genus of *Cruciferæ*, containing fifteen or sixteen species, chiefly natives of southern Europe and Central Asia, closely allied to the cress (*Lepidium*), but differing from it, as well as from *Thlaspi* and *Hutchinsia*, by having its four longer stamens winged and with a tooth; also from *Iberidella*, by having the placentas dilated at the base, and all the seeds attached to their lateral portions; and from *Iberis* by the petals being all equal. Some of the species, when in fruit, present a curious appearance, as the large dorsally-compressed and concave fringed pods are so closely imbricated, that the fruiting raceme resembles the fruit catkin of the hop, the individual pods representing the scale-like bracts. [J. T. S.]

ÆTHUSA. Under the name of Fool's Parsley, this plant is well known even to cottage gardeners. It is a common weed in cultivated ground, and is consequently likely to spring up uninvited in the parsley bed. When this happens, it runs a risk of being mistaken for true parsley, to the same natural order with which plant it belongs; this, however, can scarcely occur, except in an early stage of growth. It may then be distinguished by the bluish-green tint of its leaves, and by their fine subdivision. Being an annual, it comes into flower before parsley shows any indication of sending up a flowering stalk. By this unfailing criterion it may be discriminated when growing with the favourite pot-herb which it is supposed to simulate. By equally certain marks it may be distinguished from any other umbelliferous plant which approaches it in habit; each partial umbel, which helps to compose the general umbel of flowers termi-

Æthusa Cynapium.

nating the stalk, has at its base three approximate narrow pointed bracts or floral leaves, which hang down vertically. Its flavour and odour are unpleasant, and the seeds are very nauseous. The whole plant is said to be poisonous, and there are instances on record of persons having been made ill by eating it, even in the small quantities in which it is likely to have been present when mixed with parsley. Of its two names, *Æthusa Cynapium*, the former is derived from the Greek *aitho*, to burn, from its acrid properties; the latter, *kynos opion*, 'dog's parsley,' would seem to denote its worthlessness. [C. A. J.]

AFFINITY. A term in systematic botany, signifying that one thing resembles another in the principal part of its structure, as is the case with Crowfoots and Poppyworts.

AGALLOCHUM. The fragrant resinous heart-wood of *Aquilaria*; also called agila wood, aloes wood, and eagle wood.

AGALMYLA. A small genus of *Gesneraceæ* allied to *Æschynanthus*, having creeping stems, alternate leaves, and fascicles of axillary flowers. It differs in having the oblique five-lobed limb of the corolla scarcely two-lipped, and in having but two anther-bearing stamens. The species are tropical, inhabiting the islands of the Eastern Archipelago. *A. staminea* is a very handsome plant, epiphytal in habit, creeping and rooting on the trunks of trees, having robust stems, large fleshy gloxinia-like leaves, and axillary fascicles of from twelve to fifteen flowers, which are a couple of inches long, curved tubular, bright scarlet. The stamens are exserted an inch beyond the corolla. [T. M.]

AGAMÆ. A name sometimes given to cryptogams, resting on the supposition that they are asexual plants. [M. J. B.]

AGANISIA *pulchella* is an orchid with a creeping stem, throwing off at intervals rib-leaved pseudo-bulbs, from the base of which arise spikes of white or cream-coloured flowers. It is a native of Demerara, and has been figured in plate 33 of the *Botanical Register* for 1840.

AGANOSMA. A genus of *Apocynaceæ*, separated from *Echites*, with which it agrees, except that the coronet is cup-shaped or cylindrical, having its five parts so united that they appear only as lobes round the mouth of the cup, while in *Echites* the scales of the coronet are free or but slightly connate. The restricted genus contains eight or nine species, which are shrubs or creepers in the woods of India. Their large panicles of flowers have a showy appearance, and several have a fragrant smell. [W. C.]

AGAPANTHUS (literally Love-flower). A small genus of ornamental liliaceous plants, natives of South Africa, and long cultivated as ornaments of our greenhouses and terrace-gardens. The perianth in this family is tubular, with a short tube and six-parted spreading equal limb; there are six stamens inserted at the base of the limb, with somewhat declinate filaments:

the ovary is three-celled, with many ovules arranged in two series. The species form strong growing perennial herbs, with thick fleshy roots, and linear or somewhat lorate arching radical leaves, from among which springs the scape terminated by a large umbel of bright blue flowers. The species differ chiefly in size, in the breadth of their leaves, and in the intensity of colour in the flowers. The common one is called *A. umbellatus*. [T. M.]

AGAR-AGAR (or Agal-agal). The native name of the Ceylon Moss, *Gracilaria lichenoides*, a seaweed which is largely used in the East for soups and jellies. Another alga of equal excellence, *Gigartina speciosa*, is abundant on the coasts of the Swan River. The far-famed swallows' nests were formerly supposed to be formed of some seaweed abounding in gelatine; but it is now ascertained that they are formed from a peculiar secretion derived from the birds themselves. [M. J. B.]

AGARIC BLANC, (Fr.) *Polyporus officinalis*. — **CHAMPÊTRE**. *Agaricus campestris*.

AGARICINI. A group of *Fungi* agreeing with each other in having the hymenium or fructifying surface formed into distinct gill-like plates, the modifications of which, in combination with other circumstances, serve to distinguish the genera. The mushrooms and toadstools are familiar examples, in which the gills are highly developed. The chantarelle, on the contrary, presents a case in which they are reduced to mere veins. Sometimes the gills become hard and corky, as in *Lenzites*, of which a common species, *Lenzites betulina*, grows on old rails. [M. J. B.]

AGARICUS (*Agaric*). One of the largest and most important genera of *Fungi*, containing some of the highest forms which these plants are capable of attaining, of which the Common Mushroom is one of the most familiar examples. It is distinguished by the more or less fleshy substance of the hat-shaped receptacle, by being furnished on the under surface, whether supported by a stem or not, with gill-like plates, easily separable in the centre, as if composed of two membranes, the central substance consisting, not of subglobose cells, but of delicate filaments, and being immediately derived from the flesh of the cap or pileus.

The pileus may be either central or lateral, and, in a few instances, where the stem becomes at length obsolete, or is wholly wanting, it is attached to the substance on which it grows by the upper surface, in which case the gills become superior instead of inferior; directed, that is, towards the light, and not, as is usually the case, away from it. Where there is originally a very short stem, the pileus is at first in the usual position, but gradually turns over, so as to bring the gills towards the light. Sometimes the border of the pileus, which was at first resupinate, or having the gills on the upper side, turns over, so as to bring them into their normal position; in a very few instances alone, the whole plant is permanently resupinate.

The genus *Agaricus* is divided into five natural groups, according as the colour of the spores is white, pink, ferruginous, purple-brown, or black. These divisions, though presenting a few exceptional cases, are on the whole satisfactory, and, after a little experience, easy of determination. These groups are divided into sub-genera, according as they have a common wrapper or volva surrounding the whole plant, or a partial veil attached to the margin or forming a ring upon the stem; and then from various conditions of the stem and gills. Considering the fact that there are at least a thousand good species, it may readily be expected that some difficulties exist in the arrangement, and that the species are not always easily determined. Though, however, as in other parts of the vegetable kingdom, the limits of species are not easily defined, it may be asserted that no more certain species exist in the vegetable world, and that they are not to be considered as mere creatures of chance, without any stability. Many of them are of great beauty and elegance of form and colour, and are attractive from a thousand differences of sculpture, clothing, &c. They occur in all parts of the world, but abound most where the air is moist, with a tolerable degree of warmth. Some species afford the most delicious articles of food, while others are deleterious even when taken in small quantities. It is probable that the number of esculent species is far more numerous than is usually supposed; but as accidents are not unfrequent from confounding species altogether, or mixing poisonous kinds with those which are wholesome, they are far more neglected in this country than they deserve. It is impossible to give any positive rules for distinguishing those which are wholesome; but in general, where the taste of the raw agaric is not decidedly unpleasant, there is little danger, though even this is not without grave exceptions. With proper caution, the really useful kinds may readily be determined without the slightest risk. The common mushroom, however, is said to be poisonous in Italy, and as the bad properties depend upon the degree in which the poisonous alkali is developed — a circumstance which varies with climate and situation — even those species which are usually wholesome may at times prove deleterious.

Agarics grow in various situations. A vast variety affect dead wood, fallen leaves, and other matters while passing into a state of decay. Some affect the half-dead roots of grass, or large herbaceous plants, as the *Eryngo*. Many grow in pastures, or on the naked ground. Several occur only on dung or in highly-manured land; while a few inhabit principally stoves and other structures where the temperature is artificially kept up. Occasionally they appear under curious circumstances. In Naples, for instance, the grounds of coffee are placed in a heap in some subterranean place of moderate temperature, and an esculent species almost invariably makes its appearance. It is not, however, to be supposed that species which appear under such exceptional cases are creatures of spontaneous growth. They are generally mere altered forms of species which have usually a different habitat.

The word *acaric*, amongst the old herbalists, had a wider signification than it has now, and was applied to many of the corky fungoses. [M. J. B.]

AGARUM. A genus of olive-seeded *Algae*, distinguished from *Laminaria* principally by the frond being always perforated with roundish holes. These plants are peculiar to the northern parts of the Atlantic and Pacific Oceans, on the American and Asiatic shores. [M. J. B.]

AGASTACHYS. A Tasmanian genus of *Proteaceae*, containing only a single species, *A. odorata*, which has yellow apetalous flowers of four sepals and four stamens, one of which is attached by a short filament to the middle of each sepal; the style is filiform, rather shorter than the stamens, and bearing a two-lobed stigma. The flower-spikes are numerous, and, as the name implies, very handsome, from four to five inches in height, and crowded with flowers. The leaves are about two inches long, obtusely lanceolate, occasionally notched at the apex, with a smooth plane surface, subsessile, and rather thick in substance. [R. H.]

AGASYLLIS. A genus belonging to the umbelliferous order, and consisting of a single species, found in the Caucasus. It is a stout perennial herb, about three feet high, furnished with ternately decompound slightly downy leaves, having lanceolate, decurrent, serrate leaflets. The stems terminate in many-rayed umbels, without general, but with partial involucres of narrow leaflets. The flowers are small and white. The chief characters of the genus are an obsolete calyx margin, compressed oval fruit, with five primary obtuse ribs to each carpel, the two lateral ones shorter than the others, and the number of vittae eight to ten on the back, and five to six on the face of each carpel. [A. A. B.]

AGATHÆA. A genus of the composite family (*Compositae*), comprising twenty species, one of them, *A. abyssinica*, found, as its name implies, in Abyssinia, the others all natives of S. Africa. They are herbs or shrubs, with opposite, toothed or entire leaves, and solitary terminal flower-heads; the ray florets blue and pistilliferous, those of the disk yellow, and having both stamens and pistils. They are nearly allied to the well-known Michaelmas daisy (*Aster*), from which they differ chiefly in the pappus of their achenes consisting of one series of bristles. [A. A. B.]

AGATHELPIS. A genus of Cape under-shrubs, with alternate linear-filiform leaves and terminal flower-spikes, belonging to the natural order *Selaginaceae*. The genus is characterised by having a five-toothed tubular calyx, an elongated tubular corolla, two included stamens, and a bilocular ovary with a single ovule. By the abortion of one of the cells of the ovary, the fruit is a simple achene, covered by the persistent calyx. [W. C.]

AGATHOPHYLLUM. A name intended to express the good qualities of the leaves of the plants to which it is applied. The genus belongs to the laurel family, among which it may be known by its persistent calyx enclosing the fruit, and by its possessing nine stamens in three rows. The innermost stamens have, on either side of their base, a sessile awl-shaped gland or abortive stamen. The anthers are four-celled. One species, *A. aromaticum*, grows in Madagascar, where the natives use the leaves for a condiment. The fruit is aromatic, but encloses a kernel of an acrid caustic taste, known as Madagascar clove nutmegs. [M. T. M.]

AGATHOSMA. A genus of rutaceous plants, so named from their fragrance. They are natives of the Cape, and have regular flowers. The petals are divided, with long claws. They have ten stamens, five of which are fertile, with the anthers tipped by a small gland, and five sterile, dilated above into a petal-like mass, thread-shaped below. The fruit is two to three-celled, each cell containing two ovules placed side by side. *A. pulchella* is said to be made use of by the Hottentots to anoint their bodies, a process very distasteful to European noses. Some of the species are cultivated for their pretty white or purplish flowers. [M. T. M.]

AGATHOTES. A genus of plants of the gentian family, principally distinguished by its corolla, which is divided above into four pieces, while at the base are a number of small glandular pits, each protected by a fringed scale; and by the stamens, which are four in number, slightly connected together at the base. The dried stems of *A. Chirayta*, a native of the north of India, furnish a pure bitter, very similar in its properties to gentian, and used for like purposes under the name of Chiretta. By some this plant is referred to the genus *Ophelia*. [M. T. M.]

AGATI. A genus of the pea family (*Leguminosae*). *A. grandiflora* is the only species. It is a native of the East Indies and tropical Australia, but is commonly cultivated in tropical countries for the beauty of its flowers. It is a small slender tree twenty or thirty feet high, of rapid growth and short duration; its leaves alternate, abruptly-pinnate, with from eight to ten pairs of small leaflets. Flower stalks axillary, bearing from two to four large pea-like red or white flowers. The pods are about eighteen inches long, and as thick as a common quill. In India the flowers, pods, and young leaves are used by the natives in their curries; a juice is pressed from the flowers and used in curing dimness of vision; and the seeds are eagerly sought after by birds. The bark is powerfully tonic and bitter, and considered effective in small-pox. The wood is useless except for fuel. The tree, being a fast grower and sparingly clad with leaves, is used for the purpose of training the betel (*Piper Betel*). [A. A. B.]

AGAVE. A noble genus of *Amaryllidaceæ*, principally found in Mexico and other parts of South America. The species, of which several are known, are mostly of large size, with massive spiny-toothed fleshy leaves, forming a large spreading tuft, from the centre of which rises the tall flower scape, supporting a large compound inflorescence. The perianth is funnel-shaped, persistent, parted into a limb of six nearly equal divisions; the stamens are six in number, inserted in the tube of the perianth, and becoming exserted after the expansion of the flowers; the ovary is inferior, three-celled, with many ovules in two rows in the central angle of each cell. Some of the species become caulescent, and they are mostly long-lived plants, making comparatively slow progress in growth until the appearance of the flower stem, which, on the other hand, shoots up very rapidly. The best known species, *A. americana*, commonly called the American Aloe, affords a very good illustration of the family. This species is almost stemless; that is to say, its tuft of massive leaves is seated close to the ground, and they spread out on all sides so as to occupy considerable space. These leaves are very thick and fleshy, consisting of hard, firm pulpy matter intermixed with fibres; they are from three to six feet long, furnished with hard spines, both along the margins and at the point. These leaves are very durable, continuing to exist for many years. The plants are long in arriving at a mature or flowering age; indeed, so slow is their progress, under the artificial conditions in which they are placed in our gardens, as to have led to a popular though erroneous notion that they flower once only in a century. In reality they flower but once, the mature condition being attained in a longer or shorter period, ten to fifty or seventy years or more, according to the accelerating or retarding influences under which they are placed. Having, however, acquired full growth, the plant produces its giant flower-stem from the centre of the leaves, after which it perishes. New plants are formed around the base of the old one in the form of suckers. After the first appearance of the stem, it grows very rapidly, until a height of from fifteen to twenty or even forty feet is reached; and, towards the tip, a multitude of symmetrically-disposed horizontal branches are produced, at the ends of which branches are crowded bearing the numerous erect yellowish-green flowers, by which a sweetish liquid is secreted. The flowering plant remains for some weeks an object of interest, the flowers being durable and produced in succession.

The American Aloe appears to have been first introduced to Europe in 1561, at which date it is recorded as being in the possession of Cortusus. Parkinson, in 1640, relates that it was first brought into Spain, and from thence spread into all quarters, but is silent as to its being in England. A plant flowered in Paris in 1663. Mr. Versprit, of Lambeth, flowered one, twelve to fifteen feet high, about 1698, it being then a great rarity. Two were bloomed at Hampton Court about 1714. There is a wood engraving extant with the inscription '*Aloe americana quæ Sanderbuse floruit* 1662.' A plant flowered at Leipsic in 1700. Mr. Cowell, in 1729, flowered one at his garden in Hoxton; and this, he asserts, was the first seen in England, the others, mentioned above, not being the true American Aloe. There is a plate of this plant, by Kirkall, in mezzotinto, dated September 23, 1729. Another flowered at Eaton Hall, in 1737; a plate of it, engraved by Toms from a drawing by Badeslade, bearing date November of that year. This plant opened the crown for flowering on June 5th; the stem-bud appeared on the 15th, and grew five inches a day for some weeks; the flower branches were perfected in twelve weeks, and then it stood for a month while the buds were forming; the number of flowers was about 1,050. Two plants, about fifty years of age, flowered at Hampton Court in 1743, their respective heights being twenty-seven feet and twenty-four feet. The flower stems appeared on June 3rd, were in perfection in the middle of August, and continued blooming till the middle of October. A plant which flowered near Carlsbad in 1754 was twenty-six feet high, and produced twenty-eight branches, which bore above 3,000 flowers. Another flowered at Leyden in 1760, and a third at Friedricksberg, in Denmark, twenty-two feet high, with nineteen branches and more than 4,000 flowers. The tallest of which we have any account, was one that bloomed in the King of Prussia's garden, and this reached forty feet in height.

The species of *Agave* are not only ornamental in character, but are important on account of their uses and products. The plants themselves, with their hard, unyielding spiny leaves, form impenetrable fences, and they are used for this purpose in many parts. The roots as well as the leaves of *A. americana* and some allied species, especially the Pita plant, furnish a fibre (pita thread) which is extremely tough, and is useful for making twine and rope, and for various other purposes, such as paper-making. Humboldt describes a bridge of upwards of 130 feet span, over the Chimbo in Quito, of which the main ropes, four inches in diameter, were made of agave fibre. The fibre is separated by bruising the leaves, steeping them in water, and afterwards beating them.

The juice of the *Agave* leaves yields a

very useful succedaneum for soap. For this purpose the juice is expressed, and then the watery part is evaporated, either by artificial heat or exposure to the sun, until it is reduced to a thick consistence, when it may be made up into balls with the help of lye ashes. This soap lathers with salt water as well as fresh. A gallon of the juice yields about a pound of the soft extract. The roots of *A. saponaria*, a powerful detergent, are employed in Mexico for a similar purpose.

The most important product, however, of the *Agave*, and especially of *A. americana*, is the sap, which continues to flow for some time upon cutting out the inner leaves just before the flower scape is ready to burst forth. The plant is called Metl by the Mexicans, and Maguay de Cocuiza in Caraccas. Pittes, Arameti, Sequameti, and Maguey-meti, are varieties of this species, which is stated to be common everywhere in Equinoctial America, from the plains even to elevations of 9,000 to 10,000 feet. *A. mexicana*, a closely allied species, is sometimes called Maguei-meti, and also Manguai. According to Humboldt, the plant is extensively cultivated in the interior table-land of Mexico, and, indeed, extends as far as the Aztec language. *A. cempura* is Tiecometi or Manguei divinum; and in Cumana and Caraccas, *A. cubensis* is called Maguay de Cocuy.

The sap above referred to is of a sourish taste, and easily ferments, on account of the mucilage and sugar it contains, and in the fermented state is called pulque by the Spaniards. This vinous beverage, which resembles cider, has an odour of putrid meat, extremely disagreeable; but Europeans who have been able to overcome the aversion which the fetid odour inspires, prefer the pulque to every other liquor. A very intoxicating brandy, called Mexical or Aguardiente de Mauwey, is formed from the pulque. Royle states that the Government drew from the agave juice a net revenue of 166,497*l*. in three cities. The fresh leaves of *A. americana*, cut into slices, are occasionally used as fodder for cattle; and the centre of the flowering stem, split lengthways, is said to form no bad substitute for a European razor-strop, on account of the minute particles of silica in its composition. The leaves are also said to be used for scouring pewter. [T. M.]

AGDESTIS. A Mexican twining plant originally described by De Candolle, from a drawing of Mocino and Sessé's collection, and which has till lately been very little understood. Specimens recently examined have, however, shown that it forms a very distinct and somewhat anomalous genus of *Phytolaccaceæ*.

AGERATUM. A genus of composites, belonging to the *Eupatorium* tribe of the order, of which the *A. mexicanum*, a well-known occupant of the flower-border, with densely clustered lavender-blue capitules, may be taken as the type. Botanically,

it may be distinguished by its cup-shaped involucre of numerous imbricated linear leaflets, its naked receptacle, and its elongated angular fruit, crowned by a pappus of several awned scales, which are dilated at the base. The genus includes some other annual species in addition to the *A. mexicanum*, but none of them exceed, and few equal it in value for gardening purposes. The *A. conyzoides* very closely resembles it, and has recently appeared in gardens under the name of *Pholurera cœlestina*. The *A. angustifolium* and *A. latifolium* have white flowers, but are probably not in cultivation. There is a so-called white variety of *A. mexicanum*, but its flowers are really of a bluish cast. A few perennial species are comprised in the genus: they possess, however, but little general interest. The *Cœlestina ageratoides*, a half-hardy perennial, with blue ageratum-like flower heads, much employed in bedding, must not be confounded with the true *Ageratums*. [W. T.]

AGGLOMERATE. Heaped up; as the stamens in *Anona* and *Magnolia*, or the male flowers in a pine tree.

AGGREGATE. Several things collected together into one body; as the achenes in the fruit of a strawberry; the flowers of *Cuscuta*.

AGILA WOOD. The fragrant wood of *Aquilaria ovata*, and *A. Agallochum*.

AGLANDEAU. (Fr.) A kind of Olive.

AGLAOMORPHA. A genus of polypodiaceous ferns, of the group *Polypodiæ*, distinguished by having the veins of the fronds reticulated, with free included veinlets in the areoles, combined with the following peculiarities:—the free veinlets are divaricated: the fronds are naked, that is, not clothed with scales; they are articulated with the rhizome, and dimorphous, that is, certain sterile dwarfed oak-leaf-like fronds are produced as well as the larger fertile ones; and, finally, the fertile ones have the fertile segments, which are the upper ones, much narrower than the lower sterile ones. There is but one species, *A. Meyeniana*, a native of the Philippine Islands. [T. M.]

AGNOSTUS. A synonym of *Stenocarpus*.

AGNUS CASTUS. *Vitex Agnus-castus*.

AGRAPHIS 'The poets feign that the boy Hyacinthus, who was unfortunately killed by Apollo, was changed by that deity into a Hyacinth, which, therefore, was marked with the letters AI, alas! to express Apollo's grief. It is also feigned, that the same flower arose from the blood of Ajax when he slew himself; those letters being half the hero's name.'—*Note in Martin's Virgil*. The flower referred to is now supposed to be the Martagon lily, the spots on the petals of which sometimes run together so as to assume the required form; but the name *Hyacinthus* was given by the earlier botanists to a very different family, of which our common woodland plant, the wild Hyacinth or Blue-bell, was one. This, presenting no tracing of letters on its petals, even to the most imaginative eye, was named by Linnæus *H. non-scriptus*, or uninscribed Hyacinth. It has now been removed by Link into a distinct genus and named *Agraphis*, a Greek compound bearing the same meaning as *non-scriptus*. The wild Hyacinth, as it continues to be popularly called, is a liliaceous plant common in woods, and too well known to need any description. The blue-bell of Scotland, the harebell in poetry, is a totally different plant, *Campanula rotundifolia*. [C. A. J.]

AGRIMONIA. A family of herbaceous perennial plants with yellow flowers, belonging to the natural order *Rosaceæ*, among which they are distinguished by bearing their enclosed seeds in the hardened calyx, which is furnished on the outside with a circle of hooked bristles. The British representative of the genus, *A. Eupatoria*, is a common way-side plant, with interruptedly pinnate leaves, a scarcely branched stem about a foot and a half high, and an elongated spike of starry yellow flowers. When in fruit the calyx becomes inverted. The foliage is astringent and aromatic, and is an ingredient in several 'herb teas.' Its medicinal virtues, though far inferior to what they were anciently supposed to be, have retained for it a place in the repertory of herb collectors, who recommend it as tonic and astringent. It contains tannin, and will dye wool of a nankeen colour. A Canadian species is said to be used with success as a febrifuge. [C. A. J.]

AGRIMONY. *Agrimonia*. —, HEMP. *Eupatorium cannabinum*. —, WATER HEMP. An old English name for *Bidens cernua* and *B. tripartita*.

AGRIOPHYLLUM. A small genus of *Salsolaceæ*, containing two species, natives of Caucasian Siberia. They are annual plants, with alternate, sessile, entire leaves, and sessile axillary flowers in short squarrose spikes. The calyx, when present, consists of a single membranaceous sepal. There are three to five stamens, and two filiform styles. The fruit is a vesicular compressed capsule. [W. C.]

AGRIPAUME. (Fr.) *Leonurus Cardiaca*.

AGROSTEMMA. A genus of *Caryophyllaceæ*, of the tribe *Sileneæ*, founded by Linnæus, but now generally regarded as a section of the genus *Lychnis*, from which it only differs in the elongated segments of the calyx limb, in the petals being without a prominent scale at the base of the expanded portion, and in the capsule opening by valves alternate with and not opposite to the calyx segments. *Lychnis (Agrostemma) Githago*, the well-known weed Corn Cockle, with large, entire, purple petals, is the only species belonging to the section as it is now limited; the rest of the Linnean species being referred to the section *Coronaria*. [J. T. S.]

AGROSTIS. A genus of grasses, typical of the tribe *Agrostideæ*, and known by the English name of Bent grasses. The principal characters, which serve to distinguish this genus from its allies, are the flowers being single within the calyx glumes, and having short hairs at their base, and the upper glume being smaller than the lower. The species are numerous, no fewer than 171 being described in Steudel's *Synopsis Plantarum Graminearum*, and their range over the surface of the globe is also very extensive. The Falkland Islands, Nootka Sound, and Tasmania may be quoted as some of the outlying stations for the species of *Agrostis*. In the British Isles, the Bent grasses are of general currence on all damp pastures, as well as on dry waste ground. The Marsh Bent, *Agrostis alba*, is the once famous Florin grass of the late Dr. Richardson, who, by his writings on the subject, brought it prominently before the agricultural public, and caused it to be cultivated on a rather extensive scale, particularly in Ireland. It has not, however, been found to realise the expectations held concerning its worth, and, consequently, is not extensively grown at the present time. It is remarkable for having the long stems lying prostrate on the surface of the ground, and throwing out roots at their nodes or joints, by which means they frequently extend four feet or more from the main root of the plant without flowering. The Dog Bent, *Agrostis canina*, is the grass which sick dogs, and even cats, sometimes chew, for the purpose, it is supposed, of causing them to vomit. This species wants the inner glume or pale to the flower. *Agrostis pulchella*, a native of Quito, is cultivated in gardens, for the beauty of its elegant panicles of flowers, which, on being cut before they are fully ripe, remain a long time in a dry state, without much alteration in their appearance. Some of the foreign species of this genus are valuable as pasture grasses in the parts of the world where they grow spontaneously. [D. M.]

AGROSTOPHYLLUM. A genus of Java Orchids with fleshy stems, narrow leaves, and small flowers packed closely into terminal heads. Two or three unimportant species are known to botanists.

AGUILBOQUIL. A Chilian name for the berries of *Lardizabala biternata*.

AIAULT. (Fr.) *Narcissus Pseudo-Narcissus*.

AIGLANTINE. (Fr.) *Aquilegia vulgaris*.

AIGLE-IMPÉRIAL. (Fr.) *Pteris aquilina*.

AIGRELIER. (Fr.) *Pyrus torminalis*.

AIGREMOINE. (Fr.) *Agrimonia*; also *Aremonia agrimonioides*.

AIGUILLE DE BERGER. (Fr.) *Scandix Pecten-Veneris*.

AIL. (Fr.) *Allium sativum.* — À TOUPET. *Muscari comosum.* — DES BOIS. *Allium ursinum.* — D'ESPAGNE. *Allium Scorodoprasum.* — DORÉ. *Allium Moly.* — D'ORIENT. *Allium Ampeloprasum.*

AILANTUS. The *Vernis du Japon* of the French, *A. glandulosa* of botanists, is in its native countries, China and India, where it is called Ailanto, a tree of large size and handsome appearance, bearing numerous pinnate leaves from one to two feet long or more, and clusters of greenish flowers of a disagreeable odour. It is of rapid growth, making, when favourably situated, annual shoots from three to six feet in length. Its German name, Götterbaum, 'Tree of the gods,' is said to be a translation of *Ailanto*. French arboriculturists recommend that its lateral branches should be annually lopped off, when the main trunk will ascend perpendicularly and sustain a symmetrical spreading canopy. In France and Italy, it is

Ailantus glandulosus.

much valued as a tree for shading public walks, and is planted for that purpose along with the tulip-tree, horse-chestnut, plane, &c. Its leaves are not liable to be attacked by insects, which is a great recommendation; nevertheless they are the favourite food of the silk moth, *Bombyx Cynthia*; and they continue on the tree and retain their green colour till the first frosts of November, when the leaflets suddenly drop off, the leaf-stalks remaining on often a week or two longer. The wood is yellowish-white, satiny, and well suited for the purposes of the cabinet-maker. There are specimens, both in England and on the Continent, exceeding sixty feet in height. The name 'Japan varnish,' seems to have been applied to it through some mistake; probably from its having been mistaken for *Rhus succedaneum*. Other species are stove-plants. [C. A. J.]

AINSWORTHIA. A genus of *Umbelliferæ*, containing three species, natives of Palestine, having the habit of and nearly related to *Tordylium*, from which, however, it differs in the absence of the calyx teeth, and in having the margin of the fruit smooth. This genus was separated from *Hasselquistia* by Boissier, because of the breadth of the oleiferous vittæ in the fruit, and also from the characters of the calyx and fruit, which are the same in *Hasselquistia* as in *Tordylium*. [W. C.]

AIR PLANTS. A common name for *Arrides*. The name is also applied to Epiphytes, or plants which grow on trees and other elevated objects, not in the earth, and derive their nutriment from atmospheric moisture. They are to be distinguished from terrestrial plants, or those growing in earth, and from parasites, which derive nourishment directly from other plants on which they grow. [T. M.]

AIRA. A genus of grasses, belonging to the tribe *Aveneæ*, distinguished by having two perfect florets and frequently the rudiment of a third floret within the glumes. The pales are notched at the point, and bear short awns on the back, the awns being in most instances kneed or bent. The species are numerous, and have an extensive range of localities over the surface of the earth. Those that are natives of the British Isles are not held in great estimation for agricultural purposes, being of a coarse wiry nature.

The tufted Hair-grass, *Aira cœspitosa*, is one of the tallest-growing British grasses: indeed, under favourable circumstances, the culms, or stems frequently attain a height of six feet. In boggy land, the close growing tufts form what are called tussocks, which are found extremely useful for stepping on when walking over soft watery places. [D. M.]

AIRELLE. (Fr.) *Vaccinium.* — RAISIN D'OURS. *Vaccinium Arctostaphylos.* — ROUGE. *Vaccinium Vitis-idæa*.

AIROCHLOA. A name given to certain festuceous grasses, now generally referred to *Kœleria*. [T. M.]

AIROPSIS. A genus of grasses belonging to the tribe *Aveneæ*, distinguished from the genus *Aira* by the pales being partly attached, or adnate to the corn or seed. The majority of authors do not, however, consider this character, along with some others of minor importance, sufficient to separate it permanently from *Aira*, and, consequently, retain the species which Fries included under it, as a section of the genus *Aira*. The two British species, namely, *Aïropsis caryophyllea* and

A. præcox, are small elegant grasses, which flower in spring and the early part of summer, neither of them of much value as agricultural grasses, being only of annual duration, and loving to grow on dry barren sandy spots which produce little else besides them. [D. M.]

AITONIA. This name is applied to certain plants (usually referred, but with some doubt, to the family of *Meliaceæ*), in honour of Mr. Aiton, the former superintendent of Kew Gardens. The calyx is deeply divided into four divisions; the petals are four; the stamens eight, projecting from the corolla, their filaments united into a tube arising from beneath the ovary, which latter is surmounted by a thread-shaped style, terminated by an obtuse stigma. The fruit is membranous and triangular, of one cell, with several seeds attached to a central receptacle. *A. capensis* is a small shrub sometimes cultivated in this country. [M. T. M.]

AIZOON. A genus of plants referred by Endlicher to *Portulaceæ*, but separated from that order by Lindley on account of their want of petals and the small number of stamens, and formed, with some allied genera, into a distinct order, *Tetragoniaceæ*. The calyx is five-partite, and coloured on the inner surface. The stamens, about twenty in number, are inserted singly or in from three to five bundles in the base of the calyx. There are five subclavate stigmas; the ovary has five cells, each containing from two to ten ovules. The genus contains more than twenty species of prostrate herbaceous plants, very abundant in Southern Africa, and found sparingly also in Southern Europe, Northern Africa, and Arabia. The ashes of *A. canariense* and *A. hispanicum* abound in soda. [W. C.]

AJAX. A subdivision of the genus *Narcissus*, including the common Daffodil, and other species having a long trumpet-shaped coronet to the flowers. [T. M.]

AJONC, or AJONC MARIN. (Fr.) *Ulex europæus*.

AJOWAINS. The carminative fruits of some Indian species of *Ptychotis*. Also called *Ajwains*.

AJUGA. A genus of plants belonging to the labiate family, presenting nothing remarkable in appearance, nor possessing any properties which render it valuable. The species are all herbaceous, and the majority are annuals. The flowers either grow in whorls of six or more, or singly in the axils of the opposite leaves; sometimes contracted so as to resemble a spike, in other species more loosely, but in all cases accompanied by leaves or leaf-like bracts. Several species are furnished with stolons or runners. Of the four British species, the commonest is *A. reptans* (common Bugle), a woodland and hedge-side plant, rendered noticeable by the dull purple tinge of its upper leaves and bracts. A section of the family, named Ground Pines, is represented in Britain by *A. Chamæpitys*, a tufted spreading herb with three-cleft, very narrow hairy leaves, and yellow flowers dotted with red. Bugle was formerly held in high esteem for its vulnerary properties. 'Ruellius writeth that they commonly said in France, howe he needeth neither phisitian nor surgeon that hath Bugle and Sanickle, for it doth not onely cure woundes, being inwardly taken, but also applied to them outwardly.'—*Gerarde*. Other medical virtues assigned to the Bugle have as little foundation, in fact, as the above. [C. A. J.]

AKEBIA. A small genus of *Lardizabalaceæ*, distinguished by having separate male and female flowers; the former consisting of a three-leaved calyx of ovate-lanceolate, concave, nearly equal segments, six subequal free stamens in two rows, and the rudiments of six ovaries: the latter formed of three large roundish concave sepals, six to nine dwarfed abortive stamens, and from three to nine distinct oblong-cylindraceous ovaries, crowned by a short peltate stigma. The species are climbing plants of Japan and China, commonly cultivated in gardens, and also forming welcome half-hardy climbers in those of our own country. One of them, *A. quinata*, has its freely running stems furnished with very pretty leaves, consisting of three to five ovate or obovate entire obtuse emarginate leaflets; and from the axils of these leaves grow the racemes of dull-coloured fragrant flowers, of which the upper are smaller and sterile, the lower larger and fertile. Mr. Fortune found this plant in Chusan, growing on the lower sides of the hills in hedges; when climbing on other trees, its branches hung down in graceful festoons, attracting attention by the delightful fragrance of their flowers, the colour of which, a dark purplish brown, is not particularly showy. [T. M.]

AKA. The New Zealand *Metrosideros scandens*.

AKEE TREE. *Blighia* (or *Cupania*) *sapida*.

AKHROUT, INDIAN. *Aleurites triloba*.

AKRA. The name, in India, of the fodder Vetch, *Vicia sativa*.

AKUND. The *Calotropis gigantea* of India.

ALA. One of the lateral petals of a papilionaceous flower. Also a membranous expansion of any kind; as that round the seed of a bignoniad, from the summit or side of a seed-vessel, or on the angles of a stem. Formerly, the sail, but not now employed in that sense. The word is generally used in the plural form, alæ.

ALABASTRUS. A flower-bud.

ALAMANIA *punicea*. A little creeping Mexican orchid, scarcely distinct from *Epidendrum*. It has crimson flowers, with a small bar across the lip.

ALANGIACEÆ (*Alangiads*). A natural order of plants inhabiting tropical Asia. With the exception of the genus *Nyssa*, which is found in the United States, all are trees or shrubs with inconspicuous flowers, structurally similar to those of certain myrtles. Their fruit is succulent and eatable, but not agreeable to European tastes. The principal genera are *Alangium* and *Nyssa*. Eight or nine species are all that are known.

ALANGIUM. A genus of Indian trees, containing two, or perhaps three species, and belonging to the natural order *Alangiaceæ*. The leaves are alternate, exstipulate, entire, and reticulated on the under surface with transverse veins. The calyx is campanulate, five to ten-toothed; the petals, equal in number to the segments of the calyx, are linear and reflexed. The stamens are twice or four times as many as the petals, and have filaments which are very hairy towards the base, and bear adnate anthers. The ovary is coherent with the tube of the calyx, and somewhat crowned with its limb; it is one-celled, with one pendulous ovule. The single subulate style is expanded at the base into a coloured thick fleshy disk, covering the top of the ovary. The fruit, a fleshy one-seeded drupe, is edible but not palatable, being mucilaginous and insipid. The roots are aromatic, and the timber good and beautiful. Some of the branches occasionally become spinescent. The Malays believe the species of *Alangium* to have a purgative hydragogue property. De Candolle established the natural order *Alangieæ* on this genus, separating it from *Myrtaceæ* and other allied orders, because of its more numerous petals, adnate anthers, and one-celled fruit; and from *Combretaceæ*, on account of its adnate anthers, albuminous seeds, and flat cotyledons. [W. C.]

ALARIA. A genus of dark-spored *Algæ*, consisting of a very few species, confined to the colder regions of the North Atlantic and Pacific. The frond is from three to twenty feet long, of a membranous substance, but is furnished with a strong central nerve or rib, and is frequently much torn and split by the action of the waves; it is supported below by a short cylindrical stem, from the sides of which finger-shaped processes are given off, in whose outer coat the spore cases are immersed, supported on short pedunries, the contents of which are ultimately divided into four spores. We have a single species only upon our own coasts, *Alaria esculenta*, which is, however, well known by the Scotch under the name of Badderlocks, Henware, Honeyware, and Murlins, and is the best of all the esculent Algæ when eaten raw, the midrib and fruit-bearing appendages being the parts most in use. The name of Badderlocks, which has puzzled etymologists, is clearly a corruption of Balderlocks, or the locks of Balder, a Scandinavian deity to whom other plants have been dedicated. [M. J. B.]

ALASANDI or ARHAR. An Indian name for a common Eastern pulse, *Dolichos Catjang*.

ALATE. Furnished with a thin wing or expansion.

ALATERNUS. The common garden name of *Rhamnus Alaternus*, a well-known evergreen shrub.

ALBEFACTIO. A condition of plants induced by absence of light, commonly called Blanching, in which little or no chlorophyll is formed, the peculiar secretions are diminished, and the tissues are tender and unnaturally drawn out; and thus plants, which in a state of health are tough, unwholesome, and unfit for food, become palatable and wholesome. If light be restored, the plant may gradually recover its tone, but if it is absent for any great length of time death is sure to ensue. Some succulent plants, and those which have tubers, will sometimes survive the first season, but in general the confinement of a few months at the time of active growth is fatal. Flowers, when bleached, as of the phyllanthoid *Cacti*, sometimes recover their colour when exposed to light, but lilacs which are blanched for ornamental purposes remain white, though their leaves acquire a yellowish-green tinge. [M. J. B.]

ALBERTINIA. A genus of the composite family, containing about a dozen species. They are shrubs or small trees, with alternate, stalked, entire leaves attenuated at both ends, and either covered with short white hairs, or entirely smooth. Their flower-heads are arranged in compact globular bunches at the ends of the branches, each head containing from one to three florets. The hairs of the pappus are filiform, arranged in two or many series, and often rose-coloured. All of them are natives of Brazil. Their uses, if any, are not known. [A. A. B.]

ALBIZZIA. A genus of the leguminous family, related to *Acacia*. The name *Bezenna* was given by M. Richard to an Abyssinian tree, of which the flowers and fruits were unknown to him. Since then the plant has been found in flower, and proves to be a species of *Albizzia*. This plant, the *Albizzia anthelmintica*, is a small tree, with bipinnate leaves made up of one or two pinnæ, each of which bears three or four pairs of obovate, unequal-sided leaflets, about an inch long and half an inch broad. The flowers are in axillary stalked heads. The Abyssinian name of the plant is Besenna or Mesenna, and its bark is much used in that country in the treatment of tapeworm (*Tænia solium*), a pest to which the Abyssinians are much subject from their eating raw meat. [A. A. B.]

ALBUCA. A genus of African *Liliaceæ*, chiefly from the Cape of Good Hope, closely resembling *Ornithogalum*, but having the three inner segments of the perianth closed over the stamens, while the three outer ones are spreading; three of the six stamens are often sterile. They are bulbous plants, easily cultivated in the greenhouse when grown in pots with sandy peat earth; but they are not very ornamental, having green or yellowish flowers striped with white, and leaves more or less like those of the hyacinth. Seventeen or eighteen species have been in cultivation in this country. [J. T. S.]

ALBUMEN. The matter that is interposed between the skin of a seed and the embryo, or the vitellus, if there is one. It is, in reality, whatever substance is deposited in the cells of the nucleus during the growth of the seed.

ALBUMINOUS. Furnished with albumen when perfectly ripe. A term exclusively applied to seeds.

ALBURNITAS. A tendency to remain like alburnum. A disease of trees, when white rings of wood are interposed among heart-wood.

ALBURNUM. The sap wood of a tree: the younger wood, not choked up by sedimentary deposit, and therefore permeable to fluids.

ALCAMPHORA. A remedial preparation from *Croton perdicipes*.

ALCÉE DE LA FLORIDE. (Fr.) *Gordonia Lasianthus*.

ALCHEMILLA. A genus of herbaceous annual or perennial plants, belonging to the natural order *Rosaceæ*. All the species have lobed leaves, and inconspicuous yellow or greenish flowers. *A. vulgaris*, the common Lady's Mantle, is frequent in wet pastures and the borders of woods: the leaves are rather large, roundish, seven to nine lobed, plaited, and notched at the edges; the flowers, though small, are numerous, of a golden green colour, and collected into forked clusters. It often occurs in gardens, where it is valued more for the pleasant green of its foliage than for any showiness while in flower. Its properties are astringent, and slightly tonic; hence it comes within the province of the 'simpler.' *A. alpina* is a mountain species, found on the banks of rivulets in Scotland and the North of England. The leaves of this species are deeply divided into five oblong leaflets, and are thickly covered with lustrous silky hair. To this species probably belongs of right the not inappropriate name of 'Lady's Mantle,' which is shared in virtue of kin alone by its less daintily clothed relative. *A. arvensis* (Parsley-Piert) is a small annual plant, a few inches long, with jagged leaves, and tufts of minute green flowers growing in their axils. It grows abundantly in cultivated fields, and on hedge banks. *A. alpina*, and some of the foreign species, are well adapted for rock-work. [C. A. J.]

ALCORNOCO, or ALCORNOQUE BARK. The bark of several species of *Byrsonima*. The Alcornoque of Spain is the bark of the cork-tree.

ALDER. The common name for *Alnus*. —, BERRY-BEARING. *Rhamnus Frangula*. —, BLACK. An old English name for *Rhamnus Frangula*; also applied in America to *Prinos verticillatus*. —, RED. A name given at the Cape of Good Hope to *Cunonia capensis*. —, WHITE. A name given to *Platylophus trifoliatus* in South Africa; also to *Clethra alnifolia* in North America.

ALDROVANDA. A genus of *Droseraceæ*, containing a single species found in Southern Europe, growing in still water. This plant, *A. vesiculosa*, is remarkable for its curious leaves, which are in whorls of six to nine; they are pellucid, and inflated at the extremity, so as to form a vesicle, which acts as a float; the leafstalk is flat (not inflated), with four or five bristles at the extremity; the stems are only a few inches long, generally simple, with the whorls of leaves approximate; the flowers are white, and rather small and solitary, borne on longish slender peduncles, springing from the axils of the leaves. [J. T. S.]

ALE-COST. An old English name for *Pyrethrum Tanacetum*, commonly known as *Balsamita vulgaris*, the Costmary of gardens.

ALE-HOOF. An old English name for *Nepeta Glechoma*, the Ground Ivy.

ALEPYRUM. A genus of *Desvauxiaceæ*, containing three species of small tufted herbaceous plants, natives of the shores of New Holland. They have solitary or few terminal flowers, with two bracts; a single stamen; and six or eight ovaries, with simple styles to each. The genus differs from *Centrolepis* in wanting bracteoles, and in the spathe consisting of one or very few flowers. [W. C.]

ALETRIS. A genus of North American herbaceous *Hæmodoraceæ*, distinguished by the following features:—The perianth is half-inferior, tubular, with a six-cleft spreading or funnel-shaped limb; the six stamens are inserted into the base of the perianth segments, and have flat filaments and somewhat arrow-shaped anthers; the ovary is three-lobed, pyramidal, with a style composed of three connate bristles, distinct at the base, but joined at the top into a simple stigma; the capsule is pyramidal, three-celled, tricoccous, enclosed in the perianth, and opening at the point in three directions; and the seeds are numerous, minute, striated. *A. farinosa*, called Colic root and Star grass, is a dwarf perennial with somewhat distichous radical leaves, which are lance-shaped, ribbed, and sessile or somewhat sheathing at the base. The stem is simple, invested with remote scales, one to three feet high, terminating in a spiked raceme of short-stalked, white, oblong, bell-shaped flowers, the outer surface of which has a roughish frosted or mealy appearance. It is one of the most intense bitters known, and is used both as a tonic and a stomachic. [T. M.]

ALEURITES. A genus of the spurgewort family (*Euphorbiaceæ*). The only species, *A. triloba*, called the Candleberry tree, forms a tree of considerable magnitude, attaining the height of thirty to forty feet, and, though originally a native of the Moluccas and the S. Pacific Isles, is commonly cultivated in tropical countries for the sake of its nuts. The leaves are alternate, four to eight inches long, stalked, and without stipules, either oval acute and entire, or from three to five lobed, and, like all the young parts, covered with a whitish starry pubescence. The flowers are small and white, growing in clusters at the apex of the branches, the males and females together in the same cluster, the former being the most numerous. The fruit is two-celled, fleshy, roundish, and when ripe of an olive colour, its greatest diameter about two and a half inches; each cell contains one seed, in form something like a small walnut, the outer shell of which is very hard. The kernels, when dried and stuck on a reed, are used by the Polynesian Islanders as a substitute for candles; and as an article of food in New Georgia. They are said to taste like walnuts. When pressed they yield a large proportion of pure palatable oil, used as a drying oil for paint, and known as Country Walnut Oil and Artists' Oil. In Ceylon it is called Kekune Oil, and in the Sandwich Islands, where it 'is used as a mordant for their vegetable dyes,' Kukui Oil. In these islands alone about 10,000 gallons are annually produced. It has been imported to this country, but not to any considerable extent, and fetches about 20l. per imperial ton. The cake, after the oil has been expressed, is esteemed as a food for cattle, and also as manure. 'The root of the tree affords a brown dye, which is used by the Sandwich Islanders for their native cloths.' The plant is known in India under the name of Indian Akhrout. [A. A. B.]

ALEXANDERS. A common name for *Smyrnium Olusatrum*. Sometimes written Alisanders. —, GOLDEN. An American name for *Zizia*.

ALFA. The fibre of *Macrochloa tenacissima*, used in Algeria for paper-making.

ALFALFA. The Spanish name of Lucerne, *Medicago sativa*.

ALFREDIA. A genus of the composite family, founded on the *Cnicus cernuus* of old authors, which was cultivated in this country so long ago as 1760 by Miller in the Chelsea garden, and was figured by him in a publication illustrating his renowned *Gardener's Dictionary*. *A. cernua*, a native of Siberia, is a rank-growing, thistle-like plant, one to seven feet high, with stalked heart-shaped root-leaves nearly a foot long, having their scerate blades white underneath, and their footstalks crisped and prickly; the stem leaves are sessile and heart-shaped, except the uppermost, which are narrow lance-shaped. Each branch ends in a nodding yellow thistle-head, rather more than an inch across, containing numerous tubular florets, enclosed by an involucre of spinypointed and lacerated scales. From *Serratula* the genus differs in the pappus hairs (which crown the obovate streaked achenes) being bearded instead of rough, as well as in the long feathery tails of the anthers. Four species are known, all Siberian. [A. A. B.]

ALGÆ. A large and important tribe of cryptogams, far the greater part of which live either in salt or fresh water, a few only deriving their nourishment from the moisture contained in the surrounding air. Though many of them are confined to particular kinds of rocks, and have something resembling a root, it is not probable that they draw any important part of their nourishment from the substance on which they grow.

The higher *Algæ* have a distinct stem, from which arise variously shaped expansions, which often assume the semblance of leaves; but, though these are often strictly symmetrical, they never follow the spiral arrangement which is so marked in phænogams, and which exists even among mosses. In many the stem is quite obliterated, and the whole plant consists of an expanded membrane, consisting of one or more strata of cells, as the case may be. Frequently there is no expansion, and the whole plant, whether solid or fistulose, simple or branched, is everywhere more or less cylindrical. In other cases, again, it consists of a mere string of articulations; while in others, the whole is reduced to an adnate crust or a shapeless jelly, or to single cells. In one curious division, the frond, though often much divided, consists of a single cell only, however complicated, filled with endochrome. Whatever the colour of *Algæ* may be, it appears that they act upon the atmosphere in the same way as phænogams, that is to say, that they absorb carbonic acid and give out oxygen under the influence of light.

Algæ, whatever may be their outer form, or whatever their degree of complication, are cellular plants, in a very few instances only presenting anything like vessels, though the cell-walls themselves have frequently a spiral structure. The spores are often nothing more than the endochromes of cells, whether terminal, or chained together like the beads of a necklace, more consolidated than usual, and occasionally broken up into four or more distinct reproductive bodies. There are often two sorts of fruit upon the same or on different fronds, the one of which is regularly tetraspermous, the other variable in character, presenting often the appearance of a capsule perforated at the apex. Amongst the lower *Algæ* the spores are often furnished with one or more flagelliform processes, or with vibrating cilia, by means of which they move from place to place for a greater or less time, as if endowed with spontaneous motion, till they become attached and germinate. In most of the subdivisions sexual differences have been observed; the antheridia, or male organs, containing bodies often closely resembling the spermatozoa of animals. In some of the species fructification does not take place till the threads throw out little processes, by means of which a complete union with one another is established, the endochrome of the joint of one thread passing through their lateral tube and uniting with that of an opposite joint, and then forming a perfect spore.

In many of the lower *Algæ*, as indeed in some of the higher, reproduction takes place for an indefinite time by repeated subdivision of the original individual. At times, however, the proper fruit makes its appearance, and sometimes in such an anomalous form as to cause much perplexity.

Algæ are related on the one hand to fucuses, and on the other to lichens. Distinctive characters are more easily derived from their respective habits than from differences of structure.

The term *Algæ* had formerly a far wider range than at present, and it is now almost entirely confined to aquatic cryptogams. There is no English word which will comprise the whole. The most convenient, perhaps, is that of Hydrophytes, which, however, does not apply to the aerial species, and is objectionable because there are many plants with a submerged habit which are not *Algæ*.

Algæ are divided into three great classes, each of which comprises a number of very distinct groups, the more prominent of which will be noticed in their proper order. These three classes are characterised by the colour of their seeds, which correspond for the greater part with the general tint of the plants.

1. MELANOSPERMEÆ, or olive-spored.
2. RHODOSPERMEÆ, or rose-spored.
3. CHLOROSPERMEÆ, or green-spored.

The first of these comprises the olive-coloured species, which from their size and abundance are so conspicuous on our shores, or which float in dense masses, sometimes many leagues in extent, on the surface of the ocean. On our own coasts they attain the length occasionally of twenty feet or more, and in the genus *Laminaria* individuals are sometimes large enough to be a load for a man; but this is nothing to the size attained in the southern seas, or even in some parts of the northern hemisphere. Individuals of the genus *Macrocystis* attain a length of a hundred feet or more, and *Lessonia* forms submarine forests, the stems resembling the trunks of trees. Some of the lower species have nothing like leaves, and are reduced to mere articulated threads, or a shapeless mass.

The second class comprises those charming seaweeds, remarkable for their elegance of form, delicacy of texture, and brilliancy of colour, which attract the attention of all wanderers along the coast. These are often very abundant, but they seldom attain any considerable size, and some of them are as delicate as moulds.

The third class contains most of the smaller species, in which the frond seldom assumes the form of a membrane, but is more frequently reduced to a mere thread, or even to single articulations. A few only are conspicuous objects, amongst which the genus *Caulerpa* is most remarkable, affording on warm sandy coasts an abundant supply of nutritive food for turtles. Of the smaller and more obscure species, in which there is often no point of attachment, we have the most exquisite microscopical objects, exhibiting an almost inexhaustible variety of form and sculpture.

In the two latter classes, more especially, many species are so masked by calcareous matter as to present the appearance of corals, with which productions they have accordingly been arranged. A weak solution of hydrochloric acid, however, soon changes the fixed carbonate of lime into soluble chloride of calcium, and the structure and fruit are then unmasked and found to correspond with those of true *Algæ*. In *Diatomaceæ* silex instead of lime is imbedded in the substance of the cells.

Amongst the productions which appear upon rocks exposed to the action of the atmosphere, the lower *Algæ* are often the first to make their appearance. Even the cold surface of snow and ice produce the bright red *Algæ*, known under the name of Red Snow, while allied species appear on darker grounds. These gradually, by their decomposition, afford soil for higher growths.

The larger species of *Algæ* afford a useful though coarse article of food to men and domestic animals, not to mention the numberless tribes which they support in their own element. The Laver of our southwestern coasts is, however, considered by many an object of luxury, though, like olives, it is not in general relished at first. With use, however, it is esteemed by many a most acceptable condiment. Many of the rose-coloured *Algæ* abound in gelatine, and in consequence they are collected to make a fine kind of glue, or as a substitute for isinglass. Carrageen or Irish moss, which consists, in great measure, of common species of *Chondrus*, is a most useful article in cattle feeding, when boiled and mixed with other nutritious matters. Amongst the Chlorosperms, besides the Laver above mentioned, a species of *Nostoc* is much used as an ingredient in soup by the Chinese; but it seems not to have much to recommend it beyond the quantity of hassorin which it contains. *Durvillæa utilis* is employed for the same purpose in Chili. The siliceous coats of *Diatomaceæ*, of which the substance called Tripoli is entirely composed, form a capital substance for polishing, and the close parallel lines of extreme fineness, with which they are frequently grooved, make them very useful in microscopical researches as a test.

The larger *Algæ* were formerly much employed in the manufacture of kelp. More advanced chemical knowledge has, however, entirely suspended the practice, carbonate of soda being now obtained from other sources, to the great detriment of many of the proprietors on the sea-coasts of Scotland. They form also a very valuable manure, and it has lately been proposed by the writer of this notice to manufacture a portable manure from *Algæ* partially dried and then ground down with conical crushers, the pulpy mass being mixed with peat ashes and dried in strongly ventilated sheds.

Some of the lower *Algæ* approach, as before observed, very near to moulds, and in consequence many of these, when submerged and barren, have been assigned to *Algæ*. Such productions, however, as yeast, and other matters which occur in fermenting bodies, are now pretty well understood, and are referred to a more befitting place in the vegetable kingdom. It is very doubtful whether any true alga is parasitic on animals, those which have been supposed to be so, as *Sarcina*, &c., being in all probability *Fungi*. The curious productions which grow on fish and other aquatic animals, as *Leptomitus*, &c., are the only exception, if, indeed, these also should not be excluded. *Algæ* extend to the utmost limits of vegetation, and some of them are found at great depths in the sea. The limits of the distribution of species are not so extensive as in *Fungi*, though some have a very wide range. Many fossil species are described, but the nature of the greater part is obscure. [M. J. B.]

ALGAROBA BEAN. The fruit of *Ceratonia Siliqua*. Also applied to that of some South American species of *Prosopis*.

ALGAROVILLA. The seeds and husks of *Prosopis pallida*, a tannin material obtained from Chili.

ALHAGI. An Arabic name applied to a genus of *Leguminosæ*, characterised by having papilionaceous flowers in clusters, the pod stalked, woody, contracted between the seeds, but not dividing into separate joints. The plants are shrubby, with simple leaves and spiny flower stalks, and inhabit Southern Asia and Western Africa. A manna-like substance is produced from some of these plants in Persia and Bokhara, and is collected by merely shaking the branches. It is an exudation from the leaves and branches of the plant, only appearing in hot weather in the form of drops which soon harden by exposure to the air. Camels are very fond of it. *A. maurorum*, the plant mentioned as producing it, certainly does not do so in India. The secretion is supposed by some to be identical with the manna by which the Israelites were miraculously fed. [M. T. M.]

ALIAKOO An Indian tree, *Memecylon tinctorium*, whose leaves are used for dyeing yellow.

ALIBOUFIER. (Fr.) *Styrax officinale*.

ALISIER. (Fr.) *Pyrus Aria*. —, DE FONTAINEBLEAU. *Cratægus* or *Sorbus latifolia*. —, TRANCHANT or DES BOIS. *Pyrus torminalis*.

ALISMA. A family of aquatic plants, characterised by the parallel veins of their leaves, and their unimportant flowers of three lilac petals. *A. Plantago* grows commonly in still water, and bears large smooth, taper-pointed root-leaves on long stalks. These are thought to have some resemblance to the leaves of the plantain; and hence its name. The stem, which is leafless, is bluntly triangular, from two to three feet high, much branched in its upper part, and bearing numerous flowers in a loose pyramidal panicle or irregular cluster. The flowers, though not conspicuous, are singular from the unusual number of their petals; and the light spray-like subdivision of their stalks, joined to the vigorous habit of the leaves, claim for the plant a place among ornamental aquatics. The solid part of the root contains farinaceous matter, and, when deprived of its acrid properties by drying, is eaten by the Kalmucks. From some fanciful notion that the fearful disease hydrophobia could be counteracted by water-plants, *Alisma* was idly pitched on as a specific by empirics, but is now no longer in repute. Two other species occur in Britain: one of these, *A. natans*, is a floating plant, with larger flowers than the common water plantain ; the other, *A. ranunculoides*, is smaller in all its parts, and possesses no attractive qualities. [C. A. J.]

ALISMACEÆ (*Alismoideæ*), a small group of aquatic plants, with tripetaloid flowers and superior ovaries, each containing only one or two seeds. In some respects, although endogens, they much resemble ranunculaceous exogens, *Ranunculus parnassifolius*, having altogether the appearance of an *Alisma*. Although for the most part natives of the northern parts of the world, some species of *Sagittaria* and *Damasonium* inhabit the tropics. *Alisma* and *Sagittaria* have a fleshy rhizome, which is eatable; a species of the latter genus, *S. sinensis*, is cultivated for food in China, although its herbage is acrid. Various Brazilian *Sagittarias* are very astringent; and their expressed juice is even employed in the preparation of ink. The whole number of species does not exceed fifty, divided among the genera *Alisma*, *Sagittaria*, and *Damasonium*, which see.

ALK. A gum-resin obtained in North Africa from *Pistacia Terebinthus*.

ALKANET. The root of *Alkanna tinctoria*, which is used as a dye. Also applied in America to *Lithospermum canescens*.

ALKANNA. A genus of Mediterranean and Oriental *Boragineæ*, closely allied to *Lithospermum*, of which it perhaps ought to be considered a section, as it only differs by having the four small nuts which form the fruit contracted at the base. In habit it is, however, more like *Anchusa*, but the absence of scales closing the throat of the corolla, and the nuts not excavated at the base, are distinctive characters. The species are hispid or pubescent herbs, with oblong entire leaves and bracteated racemes, rolled up before the flowers expand. The corolla is rather small, between funnel and salver-shaped; usually purplish blue, but in some species yellow or whitish; the calyx enlarges in fruit. The root, which is often very large in proportion to the size of the plant, yields a red dye from the rind in many of the species. Alkanet, (*A. tinctoria, Anchusa tinctoria* of some authors, and *Lithospermum tinctorium* of others) is cultivated in Central and Southern Europe on account of this dye, which is readily extracted by oils and spirit of wine. It is employed in pharmacy to give a red colour to salves, &c., and in staining wood in imitation of rosewood, which is done by rubbing with oil in which the Alkanet root has been soaked. About eight or ten tons are annually imported from France and Germany. It is also said to be used in colouring some of the mixtures called by courtesy port wine; so it is to be feared that the whole quantity grown may not be applied to the legitimate purposes first mentioned. It is, however, perfectly harmless, which is so far satisfactory. [J. T. S.]

ALKE'KENGE. (Fr.) *Cardiospermum Halicacabum.* — JAUNE DOUCE. *Physalis pubescens.*

ALKEKENGI. The common Winter Cherry, *Physalis Alkekengi.*

ALLAMANDA. A genus of *Apocynaceæ* consisting of handsome climbing shrubs, found in Brazil and other parts of South America. They are well known in gardens, where they are prized for the gorgeous profusion of their rich golden flowers. The peculiarities of the genus reside in a small five-parted calyx; a large funnel-shaped corolla, having the tube narrow and cylindrical, the limb campanulate, and then spreading out into five obtuse lobes, the throat bearing five ciliated scales; five included stamens inserted in the throat, and a one-celled compressed ovary, containing numerous ovules. There are several species. *A. Aubletii*, one of the commonest found in Guiana, is a shrub with long trailing branches, bearing whorls of oblong-lanceolate leaves, and terminal or interpetiolar many-flowered panicles of large, showy, rich yellow flowers, of which the tube is an inch long or more; and the limb forms an irregular bell, about two inches long. Another still finer species, of similar habit, *A. Schottii*, a native of Brazil, has larger flowers, which are of a full yellow, funnel-formed, the lower half, or rather less, forming a narrow contracted tube, thence suddenly expanding into a campanulate faux (throat), of a deeper yellow inside; the limb of five rotundate spreading segments, often with a tooth or angle on one side. *A. scrifolia*, another Brazilian species, has a more compact shrubby habit of growth, broader, more oblong leaves, and a panicle of many flowers, which are really terminal, but by and by become lateral, from innovations, or young shoots, which grow past them and terminate also in clusters of flowers. The flowers of this species have a shorter tube and a longer faux or throat, and are deep yellow, streaked with orange. Though generally producing yellow flowers, the family yields, in the *A. violacea* described by Dr. Gardner, a species with flowers of a reddish-violet colour. The genus has, moreover, a medicinal reputation; the leaves of *A. cathartica* (perhaps not different from *A. Aubletii*, already mentioned) being considered a valuable cathartic in moderate doses, especially in the cure of painters' colic, though in over-doses it is said to be violently emetic and purgative. An infusion of the leaves is used in Surinam as a remedy for colic. [T. M.]

ALLANTODIA. A genus of polypodiaceous ferns, belonging to the *Asplenieæ*, among which they are distinguished by having the indusia simple and distinct; the veins of the frond reticulated, with free veinlets at the margin; and a vaulted or convex indusium. As thus defined, it includes one Indian species, *A. Brunoniana*, with pinnated fronds of large size. With this are sometimes associated various free-veined species, with short tumid sori, which are not distinct from *Asplenium*. [T. M.]

ALLELUIA. (Fr.) *Oxalis Acetosella.*

ALL-GOOD. An old English name for *Chenopodium Bonus Henricus.*

ALL-HEAL. *Valeriana officinalis.* —, CLOWN'S. *Stachys palustris.*

ALLIACEOUS. Having the smell of garlic.

ALLIAIRE. (Fr.) *Sisymbrium Alliaria*, often called *Erysimum Alliaria*, or *Alliaria officinalis.*

ALLIEZ. *Ervum Ervilia.*

ALLIGATOR WOOD. The timber of *Guarea grandifolia.*

ALLIONIA. A name given in honor of Charles Allioni, an Italian botanist, and applied to a genus of plants of the order *Nyctaginaceæ*. Some of them are cultivated as annuals in this country, though natives of central America. They are characterised by their flowers being placed within a three- or four-parted involucre; four free stamens arising from below the ovary, and included within the perianth, not projecting from it. The ovary is superior. [M. T. M.]

ALLIUM. A genus of bulbous plants of the lily family, remarkable for their pungent odour, having grassy or fistular leaves, and star-shaped six-parted hexandrous flowers, growing in an umbel at the top of the scape. The species are numerous, very few of them ornamental; but several are cultivated as esculents.

The Onion, *A. Cepa*, has been known and cultivated as an article of food from the very earliest period. Its native country is unknown, but it is believed to have originated in the East. In the sacred writings (Numbers xi. 5) we find it mentioned as one of the things for which the Israelites longed when in the wilderness, and complained to Moses. To show how much it was esteemed by the ancient Egyptians, we need only mention that Herodotus says in his time there was an inscription on the Great Pyramid, stating that a sum amounting to 1,600 talents had been paid for onions, radishes, and garlic, which had been consumed by the workmen during the progress of its erection. Even at the present day, the people of Western Asia, as well as the inhabitants of cold countries, are all large consumers of Onions, which, for culinary purposes, are more universally cultivated than almost any other vegetable. It is distinguished from other alliaceous plants by its larger fistular leaves, swelling stalk, and coated bulbous root. The uses to which it is applied are very numerous. From the time the plants are as large as an ordinary needle, until they attain the height of five or six inches, they are chopped and mixed in salads, which, according to the witty Sydney Smith, would not be perfect without them—

'Let onions, atoms, lurk within the bowl,
And, scarce suspected, animate the whole.'

When boiling and mature, they form an indispensable component in all soups and stews; at least, Dean Swift says —

'This is every cook's opinion—
No savoury dish without an onion;
But lest your kissing should be spoiled,
Your onions should be thoroughly boiled.'

The smaller-sized bulbs are highly prized for preserving in vinegar as a pickle. A number of varieties are cultivated, and esteemed in proportion to their being hardy, and good keepers.

The Under-ground, or Potato Onion, is supposed to be a variety of the common Onion, which it greatly resembles, but has the singular property of multiplying itself by the formation of young bulbs on the parent root, and thus produces an ample crop below the surface. Like the potato, its origin is not exactly known; but, from being sometimes called the Egyptian Onion, it is supposed to have been originally brought from Egypt about the beginning of the present century. In the West of England it is much cultivated, being quite hardy, productive, and as mild in quality as the Spanish onion.

The bulb-bearing Tree-Onion, *A. Cepa var. bulbiferum*, was introduced from Canada in 1820, and is considered to be a viviparous variety of the common Onion, which it resembles in appearance. It differs in its flower-stem being surmounted by a cluster of small green bulbs, instead of bearing flowers and seed. These bulbs are very similar to small Onions, and are said to be excellent in pickles, for which their diminutive size is a great recommendation.

The Welsh Onion is *A. fistulosum*. How this obtained the name of Welsh Onion it is impossible to say, as it is a native of Siberia and certain parts of Russia, where it is known as the Rock Onion, or Stone Leek, and regarded as an article of food. It has been cultivated in this country since 1629. It never forms a bulb like the common Onion, but has long tapering roots and strong fibres. From being very hardy, it is sometimes sown to furnish small green onions for spring salads.

The Leek, *A. Porrum*, is of great antiquity, and, although said to be a native of Switzerland, and to have been introduced in 1562, we think it is far more probable that, like the Onion, it originated in the East, mention being made in the sacred writings of both having been cultivated by the Egyptians in the days of Pharaoh. According to Pliny, Leeks were brought into great notice by the Emperor Nero, and the best were produced at Aricia, in Italy. Tusser and Gerarde, two of our earliest writers on gardening, speak of the Leek almost as if it were indigenous and in common use in their time. It is still very generally cultivated, not only in England, but more especially in Scotland and Wales, where it is esteemed as an excellent and wholesome vegetable. The whole plant, except the roots, is used in soups and stews. The stems are blanched by being planted deep for the purpose, and are much used in French cookery. The Leek, from time immemorial, has been regarded as the badge of Welchmen, who continue to wear it on St. David's day, in commemoration of a victory which the Welch obtained over the Saxons in the sixth century, and which they attributed to the Leeks they wore by the order of St. David to distinguish them in the battle.

The Shallot, *A. ascalonicum*, is a hardy bulbous perennial, native of Palestine, and more immediately of the neighbourhood of the once famous city of Ascalon, where Richard the First, King of England, defeated Saladin's army in 1192. It was first brought to this country in 1548. The bulbs are compound, separating into what are termed cloves, like those of garlic. They are used for culinary purposes, like onions, but are considered milder in flavour. In a raw state, they are occasionally cut very small and need to season chops or steaks; or mixed in winter salads. In French cookery, the Shallot is in great request, and several varieties are noticed by French writers, which have scarcely any other difference than that of the bulbs being larger or smaller than the ordinary size. They make an excellent pickle; and, by putting half a dozen cloves into a quart bottle of vinegar, an agreeable sauce may be formed.

The Garlic, *A. sativum*, is a hardy bulbous perennial, indigenous to the South of France, Sicily, and the South of Europe. It is stated to have been introduced in 1548, but appears to have been well known to the ancients. Homer makes it part of the entertainment which Nestor served up to his guest Machaon; and among the Greeks and Romans we are told it formed a favourite viand of the common people. Even at the present day, in many parts of the Continent the peasantry eat their brown bread with slices of Garlic, which give it a flavour they seem to relish. At Ovar, in Portugal, a great deal of this root is grown for exportation to Brazil. The bulb is compound, being composed of ten or twelve smaller bulbs, called cloves; and, although seldom employed with us, it is much used in Italian cookery for flavouring dishes, and is far more powerful for this purpose than any of the other species.

The common Chive or Cive, *A. Schœnoprasum*, is indigenous to Britain, having been found in Oxfordshire, as well as in Argyleshire, in the West of Scotland. It is perennial. The leaves, which rise from small slender bulbs, are about six or eight inches long, erect, awl-shaped and thread-like, and form dense tufts. They are generally cut off close to the ground, and used early in spring for salads, for which purpose they are much milder than onions or scallions—a name usually given to onions which have been sown thick for drawing, without forming bulbs. They are also used for seasoning soups, omelets, &c. In England they are little known; but in Scotland they are to be found in almost every cottage garden.

Rocambole, *A. Scorodoprasum*, is a native of Denmark and other parts of Europe, whence it was introduced in 1596. It is a hardy, bulbous-rooted perennial, with compound bulbs like garlic, but the cloves are smaller. It is used for nearly the same purposes as the shallot and garlic; and, although its flavour is considered more delicate than either, it is not much cultivated in this country. [W. B. B.]

ALLOBIUM. A genus of *Viscaceæ*, consisting of yellowish-green woody parasites on the branches of trees, with jointed, much-branched stems; thick firm persistent leaves, or only scales in their place; and small axillary spikes of flowers. The flowers are diœcious; the calyx is globular and three-lobed, each lobe in the male flowers bearing a transversely two-celled sessile anther; in the female flowers the calyx tube adheres to the ovary, which has a sessile obtuse stigma. The ovary contains a single pulpy seed, with a small embryo. The species of this genus are natives of America. [W. C.]

ALLOPLECTUS. A genus of *Gesneraceæ*, distinguished by having a free, coloured, five-leaved calyx; a funnel-shaped or club-tubulose corolla, with the tube gibbous at the base behind, and often ventricose in front above, the limb five-toothed or shortly five-cleft; four didynamous included stamens, with the rudiment of a fifth; and a free ovary surrounded by an annular disk. The genus consists of tropical American soft-wooded or sub-shrubby plants, of scandent habit, with opposite, fleshy, often unequal leaves, and axillary flowers which are solitary or aggregated, sessile or racemose. There are several species, most of which form desirable hot-house plants. *A. dichrous* is a Brazilian sub-shrub, of erect habit, with ovate-oblong entire leaves, having several flowers seated in their axils; these flowers consist of a large purple-red calyx of five triangular or cordate lobes, the three outer of which are larger and include the two inner, and of a large club-shaped tubular yellow hairy corolla, the colour of which contrasts strongly with that of the calyx. *A. concolor* is of similar habit, but has rather smaller flowers, of which both calyx and corolla are scarlet. The corolla in this latter plant is inserted at what appears to be the side of the tube near the base, and thus forms a blunt spur, whilst above it is remarkably ventricose on the upper side, with the mouth very oblique, as if the opening were at the side opposite to that by which it is affixed, thus producing a very singularly curved flower. *A. capitatus* is very distinct from the foregoing kinds, having tall stout red stems and large ovate leaves, from which the axils of the uppermost leaves are produced on short stalks, a few dense globular heads or umbels of flowers, having a very large blood-coloured calyx, and a comparatively small yellow tubular corolla. The most remarkable peculiarity of the genus among gesneraceous plants, is the large coloured calyx, which adds much to the beauty of the flowers. [T. M.]

ALLOSORUS. A genus of dwarf elegant polypodiaceous ferns, variously referred to the *Polypodieæ*, the *Cheilantheæ*, and the *Pterideæ*. They have punctiform sori at the apices of the free veins, and are without true indusia, the margin of the fronds being folded over the spore cases and somewhat altered in texture, so as to become indusia. Added to this, their fronds are dimorphous, the fertile and sterile being different in character, the former contracted by the involution of their margins, so that the divisions become pod-shaped or siliculiform. One of the species, *A. crispus*, is a native of England, and is found also throughout Europe and in North America. This is a pretty dwarf deciduous species, with bipinnate or tripinnate fronds. It is called the Rock Brake. There is another species, *A. Stelleri*, found in Siberia, India, and North America. The genus has a very close affinity with *Cryptogramma*. The name has been applied to various other ferns, especially to certain species which are more correctly referred to *Cheilanthes* and *Platyloma*. [T. M.]

ALLO] **The Treasury of Botany.** 42

ALLOUCHIER. (Fr.) *Pyrus Aria.*

ALLSEED. The common name for *Polycarpon*. Also sometimes applied to *Chenopodium polyspermum*, and *Radiola Millegrana*.

ALLSPICE. The fruit of *Eugenia Pimenta*. —, CAROLINA. *Calycanthus floridus*. —, JAPAN. The common name for *Chimonanthus*. —, WILD. *Dennzoin odoriferum.*

ALLSPICE TREE. The common name for *Calycanthus*.

ALLU'BODON, ALU'BO. The wood of *Calyptranthes Jambolana*, a common building material in Ceylon.

ALMEIDIA. The founder of this genus of rutaceous trees has devoted it to a Portuguese nobleman who assisted him in prosecuting his botanical researches in Brazil. The genus is allied to *Diosma*, but is known by its five equal, spoon-shaped petals, five fertile distinct stamens with flattened hairy filaments, an hypogynous cup-shaped disc, and a fruit opening by two valves. *A. rubra* is a handsome shrub with rose-coloured flowers, sometimes seen in hot-houses. [M. T. M.]

ALMOND. The fruit of *Amygdalus communis*; the Bitter and Sweet Almonds are the produce of different varieties of this species. —, AFRICAN. *Brabejum stellatifolium*. —, COUNTRY. The fruit of *Terminalia Catappa*. —, JAVA. *Canarium commune*.

ALMOND WORTS. An English name proposed for the group *Drupaceæ*.

ALNUS. A family of trees belonging to the natural order *Betulaceæ*, and all more or less approaching in character the common Alder, *A. glutinosa*. They inhabit most temperate countries of the northern hemisphere, and delight in a moist soil. The common Alder, in its young state, is a bushy shrub of a pyramidal form, heavily clothed with dark green leaves, which, as well as the young shoots, are covered with a glutinous substance. The leaves are stalked, roundish, blunt, jagged at the edge, shining above, and furnished at the angles of the veins beneath with minute tufts of whitish down. The flowers are of two kinds; the barren are long drooping catkins, which appear in the autumn and hang on the tree all the winter; and the fertile are oval, like little fir-cones, but are not produced till spring. When these ripen, the thick scales of which they are composed separate, and allow the seeds to fall, but remain attached to the tree themselves all the winter, and by them the tree may be distinguished when stripped of all its leaves. In young trees the bark is smooth, and of a dark purplish-brown hue, but in old trees it is rugged and nearly black. When allowed to attain its full growth, it reaches a height of forty or fifty feet, if the situation be favourable; but in the mountains and in high latitudes it does not rise above a shrub. The wood of the Alder is soft and light; and if exposed alternately to wet and dry, will scarcely last a year; but if kept entirely submersed, or buried in damp earth, no wood is more durable. By lying for a long time in peat bogs, it acquires a black hue, but from its softness will not take a good polish. The young branches are much used for the purpose of filling in drains, and are more durable than any other kind of brushwood. The charcoal is highly valued in the manufacture of gunpowder, for which purpose it is in some places largely planted. The colour of the wood when first cut is white, but by exposure it becomes of a bright orange-red, as is shown by the chips, which are left about where a tree has been felled. Several varieties are grown which differ from the typical species in having laciniated, lobed, or variegated leaves. Of the other species enumerated by Loudon, *A. cordifolia*, a native of Italy, is well adapted to this climate. It grows with rapidity, and is a most interesting and ornamental tree. The common Alder is the badge of the clan Chisholm. [U. A. J.]

ALOCASIA. A name applied to a section of the genus *Colocasia*, by some considered as a distinct genus. The species are natives of India, with peltate leaves springing from an erect root-stock; spathes glaucous, on short stalks. [M. T. M.]

A. metallica is a magnificent Bornean species, with very large cordate-ovate peltate leaves, having a rich bronze-coloured surface, and is a very conspicuous ornament of our hot-houses. The leaves look like great polished metal shields. [T. M.]

ALOE. A Latinised form of an Arabic name given to a genus of succulent plants of the lily family (*Liliaceæ*). The species of the genus vary very much in height, and in the appearance of their leaves and flowers, but are especially distinguished from allied genera by their having a stem, sometimes a very short one; permanent fleshy leaves; flowers arranged in erect spikes or clusters, each with a cylindrical perianth divided into six pieces, secreting nectar at the base; six stamens arising like the perianth from below the germen; a membranous fruit, consisting of three cells, each containing a great number of seeds. The species of Aloe are abundant in all warm countries, especially in the southern part of Africa and the Isle of Socotra, where 'the bristling aloes' give a character to their own to the landscape.

A. vulgaris, a native of the East and West Indies has been introduced into Italy, Sicily, Malta, and the Mediterranean region in general. The most important product of this genus is the drug known as aloes, which is the dried juice derived from the leaves of several species in the East and West Indies, Cape of Good Hope, and elsewhere. The finest kind of aloes is supposed to be derived from *Aloe socotrina*. The bitter resinous juice is stored up in greenish vessels, lying beneath the skin of the leaf, so that when the leaves are cut transversely, the juice exudes and is gradually evaporated to a firm consistence. The inferior kinds of aloes are prepared by pressing the leaves, when the resinous juice becomes mixed with the mucilaginous fluid from the central part of the leaves, and becomes proportionately deteriorated. In other cases the leaves are cut in pieces and boiled, and the decoction evaporated to a proper consistence.

The drug is imported in chests, in skins of animals, and sometimes in the cavity of large calabash gourds. It is largely used as a purgative, and in small doses as a tonic; the taste is peculiarly bitter and disagreeable, though the perfume of the finer sorts, when breathed on, is aromatic, and by no means so offensive as the taste. What is called aloes fibre seems rather to be the produce of an *Agave*, though it is stated that the negroes of Western Africa make nets and cords of the fibres of various species of Aloe.

Many of the species of Aloe are cultivated in this country, being extremely easy to grow, if planted in a dry soil and very little if any water supplied to them in the winter season. The thick leathery skin of the leaves prevents the internal moisture from escaping so readily, hence these plants retain their vitality for a long time under apparently adverse circumstances.

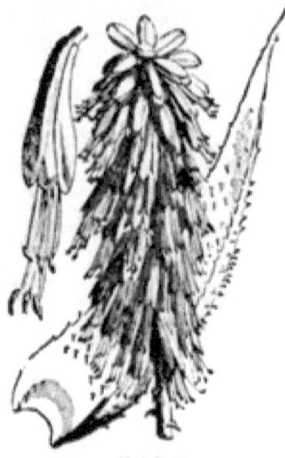

Aloe ferox.

Sailors sometimes bring home pieces of Aloe from the West Indies with a tarred cloth tied tightly round the cut end, so as to prevent the escape of the juices. Dr. Pereira mentions having had such a specimen suspended from the ceiling of his room for two years, and it was still living and growing when he wrote.

What is commonly known as the American Aloe is a species of agave much like an Aloe in general appearance, but particularly distinguished from it by the perianth being adherent to the ovary, or, as it is called, superior. [M. T. M.]

ALOE, AMERICAN. *Agave americana.* — FALSE. *Agave virginica.*

ALOE'S BEC DE CANNE. (Fr.) *Aloe*, or *Gasteria disticha.* — CORNE DE BE'LIER. *Aloe arborescens.* — LANGUE-DE-CHAT. *Aloe* or *Gasteria lingua* and *angulata.* — POUCE-E'CRASE', *Aloe retusa.*

ALOES-WOOD. The wood of *Aloexylon Agallochum.*

ALOEXYLON. The name given to a genus of the pea family (*Leguminosae*) which is said to grow on the high mountains of Cochin China. There is but one species, *A. Agallochum*, which is described as being a tree of about sixty feet in height, with simple, alternate, stalked, entire, lanceolate leaves, and terminal panicles of small flowers. The wood of this tree is one of the two woods known as Calambac, Lign-aloes or Eagle-wood. It yields the perfume the most esteemed by Orientals, who apply it to their clothes and apartments, and use it in medicine, in the treatment of paralytic affections. The perfume by some is said to originate by the concretion of oily particles into a resin; which action takes place in the centre of the trunk, and is occasioned by a disease which ultimately causes the death of the tree. This is, however, questioned by others. The wood is very valuable — selling at about 50*l.* per cwt. in Sumatra. It is sometimes used for inlaying in cabinet work. Some of the most precious jewels of East Indian manufacture are set in it; and, so highly is it prized, that it is considered equal to gold in point of value. The perfume derived from it is thought by some to be alluded to in the Bible, where it is said (Psalm xlv. 8), 'All thy garments smell of myrrh, *aloes*, and cassia.' The wood is said to retain its fragrance for years. Many conflicting statements have been published about this tree, and the *Aquilaria Agallocha*, which is also called Eagle-wood, and belongs to a very different family; and it is possible that some of the statements above given may apply to the *Aquilaria* rather than to the plant under consideration, which is, botanically, almost unknown. [A. A. B.]

ALONA. A small genus of South American *Nolanaceæ*, separated from *Nolana* by having several ovaries, with from one to six cells (not five, each of them four-celled, as in *Nolana*). Mr. Miers restricts the genus *Alona* to the species with woody stems and fasciculate terete or triquetrous leaves, as he finds that, in the allied genus *Sorema*, the way in which the carpels are combined varies in the same genus, and therefore cannot by itself be sufficient to make a generic distinction. The species have large handsome flowers, resembling those of the bindweeds. *A. cœlestis* is cultivated in this country for the beauty of its large pale blue flowers. It is a native of the coast of Chili. [J. T. S.]

ALONSOA. A small group of the *Scrophularineæ*, forming dwarf frutescent herbs, with opposite or ternately whorled serrated leaves, and axillary, submraceinose, pretty vermilion-coloured flowers. They are commonly cultivated in green-houses, and in the open air during summer. The characteristic features are a five-parted, sub-equal calyx; a resupinate corolla, having a very short tube, and a sub-rotate, five-cleft limb, of which the front, or upper lobe, is larger, and all rotundate; four didynamous declinate exserted stamens, affixed to the corolla tube; and a two-celled many-seeded ovary. The few species are mostly natives of Peru, where one of the species is famed for its anodyne and stomachic properties; and several of them have been introduced to our gardens. *A. linearis*, a pretty dwarf, bushy, suffruticose plant, of a span or foot in height, has the leaves, which are narrow linear, opposite or in threes, mostly fasciculate, from the non-evolution of the axillary shoots; and the branches terminate in racemes of curious, obliquely-rotate scarlet flowers, with a black spot at the base. In Peru it is called Ricaco and Ricarco, which mean Mask-flower. One of the best known species, *A. incisifolia* (sometimes called *Celsia urticæfolia*), is of a rather more vigorous habit, and has ovate, acute, deeply-toothed leaves, and elongated racemose panicles of scarlet and black flowers. *A. Warscewiczii*, a species more recently obtained from the mountains of Peru, is an erect, branched, sub-shrubby plant, with sub-cordate or ovate-lanceolate leaves, and racemes of scarlet flowers, without the black spot which is conspicuous in the other species already noticed. There are about a dozen described species. [T. M.]

ALOPECURUS. A genus of grasses called Fox-tail Grasses, of the tribe *Phleineæ*, distinguished from *Phleum*, to which some of the species are nearly allied, by having only one inner glume or pale to each flower, this bearing a long awn attached to the back portion of it. The species are mostly from temperate climates, and have an extensive range from their southern to their northern limits. Among those which are natives of Britain, the meadow Fox-tail Grass is one of the very best kinds, and forms a portion of all good pastures and meadows — particularly on limestone soils: the seeds are consequently sown in most instances as part of a mixture of grass seeds. It is one of the earliest kinds to flower in spring, and, when chemically analysed, is found to contain a large share of nutritive matter in its composition. The other species which are natives of Britain are of less agricultural value. The alpine Fox-tail Grass (*Alopecurus alpinus*) is one of the rarest native species, being much prized and eagerly sought after as a botanical rarity. [D. M.]

ALPHONSEA. Certain Indian plants are comprised in a genus bearing the above name, in honour of M. Alphonse de Candolle, the eminent botanist who has especially studied the natural order *Anonaceæ*, to which this genus belongs. Its principal characters are — petals valvate in the bud, nearly equal in size; stamens loosely imbricate. By these circumstances the genus may be distinguished from its nearest ally, *Saccopetalum*. The species comprise tall trees furnished with thick shining leaves, and small flowers, closely packed in tufts opposite to the leaves. [M. T. M.]

ALPINIA. A genus of plants deriving its name from Prosper Alpinus, an Italian botanist who lived in the 16th century. The genus belongs to the same natural family as the ginger (*Zingiberaceæ*), and is known by its thick, tuber-like, aromatic rhizomes; and by its flowers arranged in terminal spikes. Each flower has an outer row of three pieces, and an inner of four pieces, the lowermost of which is three-lobed. The filament is petal-like, and not prolonged beyond the two-lobed anther, as in some of the plants of this order. Stigma triangular, on a long style. The fruit is a somewhat fleshy capsule of three many-seeded cells. The species are natives of tropical America, the Indian Archipelago, etc. *A. Galanga* and other kinds furnish the aromatic stimulant root known as Galangale root, employed by the natives in cases of indigestion. The fruits of *A. alba* are known as ovoid China Cardamoms; others, as *A. nutans*, are remarkable for the exceeding beauty of their flowers, and are therefore cultivated in our stoves. [M. T. M.]

ALPISTE. (Fr.) *Phalaris canariensis*.

ALSINE. A genus of small caryophyllaceous herbs, generally distributed in temperate regions, and in alpine situations in warmer climates; closely resembling *Arenaria*, from which it differs by having the valves of the capsule equal in number to the styles, and not twice as many. The leaves are generally narrow, often subulate; the sepals strongly nerved; the petals white. Four species are natives of Britain: *A. verna*, a tufted perennial, with the petals longer than the calyx; *A. rubella*, a tufted alpine perennial, with short flower-stalks, and the petals not exceeding the calyx; *A. uliginosa*, also a tufted perennial, with the petals scarcely exceeding the calyx, but with long pedicels and smaller flowers than the last; and *A. tenuifolia*, a slender annual, not uncommon on wall tops and on dry commons in the South-east of England. [J. T. S.]

ALSODEIA. A genus of ornamental plants, belonging to the order *Violaceæ*, and inhabiting the islands of Madagascar and Timor. Some of the species are cultivated in this country. They are distinguished by their petals being all equal in size; by the absence of scales between the petals and stamens: the stamens spring from a disc surrounding the base of the ovary, free above; filament dilated, not narrowed into a claw. They are woody plants, with white flowers, and thickly beset with leaves; hence the name—from the Greek *alsodes*, leafy. [M. T. M.]

ALSOPHILA. A genus of cyatheaceous ferns, representing the *Alsophileæ*; often becoming magnificent umbrageous trees. Among the cyatheaceous ferns, which are known by the obliquity of the ring of their spore-cases, and by having an elevated receptacle, *Alsophila* is distinguished, primarily, by the absence of any indusium or cover to the sorus; and, secondarily, by producing only one sorus on each vein or venule. There are a considerable number of species, some of which have been imported for the decoration of our hot-houses. The species have bipinnate fronds, and a considerable number of them are found in the West Indies, South America, and Mexico, a few in Australia and the South Sea Islands, and several more in the East Indies and Malay Islands. *A. excelsa*—a native of Norfolk Island—is stated by Capt. King to grow to a height of eighty feet. 'The branches (fronds), which resemble those of the palm-tree in their growth, fall off every year, leaving an indentation on the trunk. The middle of the tree, from the root to the apex, consists of a white substance, resembling a yam, which, when boiled, tastes like a bad turnip: this the hogs feed on greedily. The outside of the trunk is hard wood, and full of regular indentations, from the top to the bottom.' Another tree of the same genus, cut down by Mr. Allan Cunningham, was fifty-seven feet long without the fronds; and Mr. Backhouse measured some forty feet high, crowned with magnificent circular crests of fronds. It is altogether a noble plant, having the stipes and main rachis of its fronds muricate, or rough, with small raised points. The fronds are bipinnate; the pinnules, or secondary divisions, oblong-lanceolate, acuminated, pinnatifid, with oblong acutish segments. *A. australis*, from the same region, is another fine species. In Tasmania, where Mr. Backhouse met with tree ferns in profusion, this species was seen with stems of all degrees of elevation up to twenty-five or thirty feet, some of them at the lower part as stout as a man's body, the whole length clothed with the bases of old leaves, which were rough, like the stems of raspberries, closely tiled over each other, and pointing upwards. Some of the larger fronds were thirteen feet long—making the diameter of the crest twenty-six feet. Some of the Indian species are also remarkable for their stature. There is preserved in the British Museum a stately trunk, forty-five feet long, of *A. Brunoniana*; and another of equal height, belonging to *A. gigantea*, is in the museum of the Linnæan Society of London. Some of the species are, however, without these elongated trunks, although all produce fronds of large size. [T. M.]

ALSOPHILE.E. A section of cyatheineous ferns, in which the sori have no cover. The plants referred to here are sometimes not easily distinguished from *Polypodium*, the compression of the spore cases being less marked, and the receptacle less obviously elevated than in the more typical species. [T. M.]

ALSTONIA. A genus of the periwinkle family (*Apocynaceæ*) differing from most others in the seeds having a tuft of silky hairs at each end, instead of at one end only; and from its nearest ally, *Blaberopus*, in the absence of the two nectary scales seen in the flowers of that genus. There are about a dozen species distributed over India, the Moluccas, tropical Australia, and West Africa. They are trees or shrubs with milky juice; opposite, often whorled, and entire leaves; small white flowers disposed in cymes at the ends of the branches, the corolla funnel-shaped with a flat border of five rounded lobes; and fruits consisting of two cylindrical pods (*follicles*) the thickness of a quill, and often a foot in length.

A. scholaris, called Devil-tree or Pali-mara about Bombay, is a widely-diffused plant in India and the Moluccas. It is a tree of fifty to eighty feet, with a furrowed trunk; oblong stalked leaves, three to six inches long, and two to four wide, disposed in whorls of four to six round the stem, their upper surface glossy, the under white, and marked with nerves running at right angles to the midrib. It has a powerfully bitter bark, which is used by the natives in India in bowel complaints, and its light wood is used in Ceylon for making coffins. The wood taken from near the root of what appears to be the same species in Borneo, is of a white colour, very light, and used for floats for nets, and household utensils, as trenchers, corks, &c. The genus bears the name of Alston, once Professor of Botany at Edinburgh. [A. A. B.]

ALSTRŒMERIA. A genus of very handsome amaryllidaceous plants, distinguished by having a six-parted regular subcampanulate perianth, of which the inferior segments are narrower, and two of them somewhat tubulose at the base; six stamens inserted with the perianth; a trifid stigma; and an inferior three-celled ovary with many horizontal ovules. They are tropical or extra-tropical herbs of South America, with fasciculate tuberous roots, and erect leafy stems, terminating in umbels of showy flowers. The numerous species, many of which have ornamented our gardens, are very similar in character. The leaves in this genus are, by the twisting of the petiole, resupinate; the upper surface, which is usually smooth, even, and destitute both of ribs and stomata, having the peculiar structure and performing the functions of the under surface. This curious economy in the leaves of *Alstrœmeria* was first pointed out by the late Robert Brown. Amongst the handsomest of the species may be mentioned *A. aurea*, an erect herb one to three feet high, with scattered, lanceolate, obtuse leaves, reversed, as is the case throughout the genus, by the twisting of the foot stalk. The flowers are produced several in a terminal umbel, the perianth consisting of three outer spathulate, deep orange-coloured segments, and three inner ones, which are narrower, lanceolate, acuminate, orange-coloured, the two upper of them marked with several dark red lines distributed over their surface. Quite distinct from this is *A. Flos Martini*, the St. Martin's Flower of Chili, which has an erect stem, linear acute leaves, and a perianth consisting of three outer cuneately obcordate yellowish-white segments, and an inner series of one short lower whitish lobe, and two upper oblong spathulate ones which are bright yellow in the upper half, and stained with irregular dark red spots, the spots becoming confluent towards the top. *A. Ligtu*, so named because, according to Feuillée, it is called Ligtu in Chili, is another very beautiful kind, in which the leaves are linear or linear-lanceolate, and the flowers, on corymbose two-flowered peduncles, are large, blush-coloured, with obovate emarginate sepaline divisions, the two upper petaline divisions narrow spathulate, yellow, striped with red below, and tipped with crimson. *A. psittacina* has the flowers a little hooded, rich crimson at the base, and at the tips green, spotted with purple. Another fine ornamental species is *A. Simsiana*, which has orange-scarlet flowers. The greater number of the species are natives either of Chili or of the Andes of Peru, a few being distributed in other parts of South America. The *A. pallida* furnishes in Chili a kind of arrowroot, which is prepared from its esculent roots. [T. M.]

ALTERNATE. Placed on opposite sides of an axis on a different level, as in alternate leaves. Placed between other bodies of the same or different whorls, as in an umbellifer, where the stamens are alternate with, that is between, the petals.

ALTERNATIVE. A term applied to æstivation, when of the pieces of a flower, being in two rows, the inner is so covered by the outer that each exterior piece overlaps half two of the interior row.

ALTHÆA. The Marsh-Mallow is, as the name implies, one of the *Malvaceæ*, and is distinguished by its flowers having an outer calyx of from six to nine pieces, and an inner one, partly divided above into five pieces. In other respects *Althæa* much resembles *Malva*. *A. officinalis*, the common Marsh Mallow, grows in marshes near the sea in this country, and also in Central and Southern Europe. The rootstock is perennial; the flowering stems are erect, branched, three or four feet high, covered with a soft velvety down, as also are the stalked, egg-shaped, cordate leaves, which are slightly notched at the margin, the lower ones five-lobed, the upper ones three-lobed. The flowers are of a pale rose colour, on short stalks, which spring from the axils of the upper leaves. The roots are much used, especially in France, under the name of Guimauve, to form demulcent drinks. *A. hirsuta* is a rare English plant, which has been probably introduced along with foreign agricultural seeds. It is an erect slender annual, much smaller than the preceding, with bluish flowers, and covered with long spreading stiff hairs. *A. rosea* is the origin of the hollyhock of gardens. It grows wild in China, also in the South of Europe. It possesses similar properties to the common marsh mallow, and is used for similar purposes in Greece. The leaves furnish a blue dye. Several species of *Althæa* are in cultivation, but the gay flowering shrub commonly called *Althæa frutex* is, properly speaking, a *Hibiscus*, *syriacus*. [M. T. M.]

ALTHÆA FRUTEX. The garden name for *Hibiscus syriacus*.

ALTHÆE. (Fr.) *Althæa officinalis*.

ALTHENIA. A genus of *Naiadaceæ*, containing a single species, a native of France. It is a slender tufted plant, growing in salt lakes, and resembling *Zannichellia*—except that that genus has male and hermaphrodite flowers; whereas, in *Althenia*, the flowers are diœcious; the male flowers being solitary, and below the female. [W. C.]

ALTINGIACEÆ. (*Liquidambars*, *Balsamaceæ*, *Balsamifluæ*.) A solitary genus, *Liquidambar*, represents this natural order, of which three species only are known—all trees of some magnitude—producing a fragrant resin called storax, or resembling that substance. They are nearly related to plane-trees and willows, from which they differ in having seed vessels with two distinct cells, instead of one; and seeds with broad membranous wings. See LIQUIDAMBAR and STORAX.

ALUM ROOT. The root of *Geranium maculatum*; also applied to some species of *Heuchera*.

ALUYNE. (Fr.) *Artemisia Absinthium*.

ALVEOLATE. Socketed, honey-combed; when a flat surface is excavated into conspicuous cavities, as in the receptacles of many Compositæ.

ALVIER, ALVIES. (Fr.) *Pinus Cembra*.

ALYSSUM. The generic name of several herbaceous annual plants with yellow flowers, belonging to the cruciferous tribe, and generally employed in decorating rock-work, or the open border. *A. saxatile*, a native of Transylvania, &c., popularly known as Gold-dust—in French, Corbeille d'Or—has somewhat woody, diffuse stems, lanceolate, hoary leaves, and numerous small flowers of a brilliant yellow colour, growing in dense clusters. These appear early in May, when flowers are scarce, and are consequently much prized. This species, like the rest of the family, thrives best in dry, somewhat stony ground; but may be made to grow anywhere. Several other species are cultivated, under the name of Madwort. [C. A. J.]

ALYSSUM, SWEET. *Glyce* (or *Koniga*) *maritima.*

ALYXIA. A genus of *Apocynaceæ*, containing sixteen species; natives of Australasia, Madagascar, and tropical Asia. They consist of evergreen trees or shrubs, with ternate, quaternate, or sometimes opposite, entire, and shortly-petiolate leaves. The flowers are fragrant (some species smelling like jasmine), axillary or terminal and solitary, or in cymes. The calyx is five-partite; the corolla is hypocrateriform, its long tube is swollen above the middle; the five included stamens, on short filaments, and with lanceolate anthers, are inserted on the dilated portion of the tube; there are two ovaries, with a single included style. While this genus has all the habit and the structure of the flowers of the true *Apocynaceæ*, it differs from the other genera of the order in having baccate, or sub-drupaceous fruits, in the shape of its seed, in its ruminated albumen, and in its erect embryo: in these two last particulars it agrees with *Anonaceæ*. The dark green foliage and fragrant flowers make the members of this genus an ornament in the conservatory, where they flower freely in the autumn. [W. C.]

AMADOU. A soft leathery substance, derived from *Polyporus fomentarius* and some other *Polypori*, and used for tinder, moxa, and other economical or medical purposes. It is prepared by cutting off carefully the cuticle and pores of the fungus, dividing it into convenient slices, beating them out, and steeping them in a solution of saltpetre. Occasionally, it is used to make coarse clothing, and then the latter process is omitted. The best Amadou is prepared in Germany, from *Polyporus fomentarius*, but *P. igniarius* and other species afford an inferior quality. The softer and more silky the substance of the fungus, the better the material. The fungus is generally collected from trunks of trees in the forests, where it is tolerably abundant; but attempts have also been made at cultivating it by collecting timber in proper situations, and watering it at proper intervals. The species occurs pretty generally in this country, but is not sufficiently frequent to make its collection a matter of interest. [M. J. B.]

Amadou is sometimes called German Tinder in the shops. The wood of *Hernandia guianensis* is used in a similar way in South America. [T. M.]

AMADOUVIER. (Fr.) *Polyporus igniarius.*

AMALAGA. *Chorica officinarum.*

AMANDE DE TERRE. (Fr.) *Cyperus esculentus.*

AMANDIER. *Amygdalus.* — **A LA MAIN,** or **DES DAMES.** *Amygdalus fragilis.* — **DE GEORGIE.** *Amygdalus nana.* — **SATINE.** *Amygdalus orientalis.* — **DU BOIS.** *Hippocratea comosa.*

AMANITA. A sub-genus of *Agaricus*, distinguished by its gills producing white spores, and the whole plant being covered at first by a distinct universal wrapper, or volva. It contains some of the most excellent and poisonous of Agarics—amongst the former being the Orange and *A. vaginatus*; and among the latter the Fly Agaric and *A. mrosus*. Some of the species have a distinct ring upon the stem; while others are wholly deficient in this ornament. The Fly Agaric (*A. muscarius*), with its vermilion pileum studded with white or yellow warts, and its stately stem, is the ornament of beech woods in most parts of the kingdom, and seldom fails to excite admiration, especially when illuminated by a strong stream of light. Several species—and especially those of Sikkim, where they abound—are amongst the largest of the Scaly Fungi. [M. J. B.]

AMANSIA. A lovely genus of rose-spored *Algæ*, mostly inhabiting the southern hemisphere, with a pinnate frond and generally involute tips. The frond is ribbed; the membrane formed of oblong six-sided cells, of equal length, arranged in transverse lines; the tetraspores are in marginal or superficial podshaped processes—generally in two rows; and the pyriform spores form a little fascicle at the base of the sub-globose capsules, which are perforated at the tip. Some species have almost exactly the habit of *Jungermannia*. [M. J. B.]

AMARACUS. (Fr.) *Origanum Dictamnus.*

AMARANTH, GLOBE. *Gomphrena globosa.*

AMARANTHACEÆ. (*Amaranthi*; *Polycnemeæ*.) Under this name are included about 500 species of weeds, or, occasionally, showy annual plants (very seldom undershrubs), with inconspicuous apetalous flowers, in almost all cases of a scarious texture, and most commonly with a white colour—although now and then pink, or orange, or intensely crimson. They are very nearly the same as dienopods, a still more weedy order. They occupy dry, stony, barren stations, or thickets upon the borders of woods, or even salt marshes; are much more frequent within the tropics than beyond them; and are unknown in the coldest regions of the world. Many of the species are used, with the addition of lemon-juice, as pot-herbs, on account of the wholesome mucilaginous qualities of the leaves. *Gomphrena officinalis* and *macrocephala* in Brazil, where they are called Para todo, Perpetua, and Raiz do Padre Salerma, are esteemed useful in all kinds of diseases, especially in cases of intermittent fever, colic, and diarrhœa, and against the bite of serpents.

AMARANTHUS. A genus of tropical annual plants, the type of a natural order, to which it gives its name—the Amaranthads. They are readily distinguished from the few other genera of the order by their three-bracted coloured calyx of three or five pieces, and their one-seeded fruit, splitting circularly round when ripe. The genus includes several handsome garden plants, the chief being the *A. caudatus*, popularly known as Love-lies-bleeding, and in France as *Queue de Renard* and *Discipline de Religieuse*, having long pendulous compound racemes of crimson flowers; the *A. hypochondriacus*, or Prince's Feather, with erect flower spikes and purplish foliage; the *A. speciosus*, or larger Prince's Feather, resembling the last, but differing by its more vigorous growth, and the *A. tricolor*, from China, are interesting species, more remarkable for the vivid colours of their foliage than for their flowers, which are insignificant. The last-named is much more tender than the other species; and, in the open air in this country, it is only in warm summers that its leaves assume the glowing tints to which the plant owes its specific name. In the gardens of the Southern United States, these hues are so richly developed as to have procured for it the popular appellation of Joseph's Coat. The plant known as Globe Amaranth belongs to another genus—*Gomphrena*. The name of this genus is often written *Amarantus*. [W. T.]

AMARANTINE. (Fr.) *Gomphrena globosa*.

AMARANTOIDE. (Fr.) *Gomphrena coccinea*.

AMARELLE. (Fr.) *Cerasus avium*.

AMARINIER. (Fr.) *Salix vitellina*.

AMARYLLIDACEÆ (*Narcissi*). A large natural order, consisting for the most part of bulbous plants, but occasionally forming a tall, cylindrical, woody stem, as in the genus *Agave*. They differ from Irids in having six introrse stamens, and from Liliaceous plants in their ovary being inferior. A few species of *Narcissus* and *Galanthus* are found in the North of Europe and the same parallels. As we proceed south they increase: *Pancratium* appears on the shores of the Mediterranean; *Crinum* and *Pancratium* in the West and East Indies; *Hæmanthus* is found for the first time with some of the latter on the Gold Coast; *Hippeastra* show themselves in countless numbers in Brazil, and across the whole continent of South America; and, finally, at the Cape of Good Hope, the maximum of the order is beheld in all the beauty of *Hæmanthus*, *Crinum*, *Clivia*, *Cyrtanthus*, and *Brunsvigia*. A few are found in New Holland, the most remarkable of which is *Doryanthes*. Poisonous properties occur in the viscid juice of the bulbs of *Buphane toxicaria* and *Hippeastra*; those of *Leucoium vernum*, the snowdrop and daffodil, and other kinds of *Narcissus* are emetic. Nevertheless the *Agave*, or American aloe, as it is called, has an insipid sweet juice. Others are deterrent, and a few yield a kind of arrow-root. Between 300 and 400 species are known.

AMARYLLIS. The type of the amaryllidaceous family, and formerly made to include a large number of species. It is now, however, generally limited in extent, and confined to those which have the tube of the perianth short, narrow funnel-shaped, and ribbed; the three petaloid filaments inserted at the base of the segments; the three sepaloid ones adhering to the mouth of the tube; the style declinate; the capsule obovate. They are handsome bulbous plants, with an autumnal flower-scape appearing before the leaves, which are biennial. The scape supports a many-flowered umbel of large stalked flowers, the anthers of which are incumbent, attached in the middle. The typical species is *A. Belladonna*, which is separated by some as a distinct genus. This plant is a native of the Cape of Good Hope, and is of vigorous habit, producing flower-scapes one and a half foot high, and large, showy, funnel-shaped flowers of a pale delicate rose beautifully pencilled with red, in the month of September, the flowers being succeeded by the leaves, which are ligulate or strap-shaped. *A. Josephinæ*, and *A. grandiflora*, sometimes placed in *Brunsvigia*, are referred hither by Herbert. Most of the plants called *Amaryllis* in gardens, e.g. *A. equestris*, are now referred to *Hippeastrum*; others, as *A. formosissima* to *Sprekelia*, *A. lutea* to *Oporanthus*, and *A. purpurea* to *Vallota*. The *A. Belladonna* has been said to be employed for poisoning in the West Indies, but this statement appears to be a mistake, and probably refers to some other plant of the same order, the Belladonna being a Cape plant. The name Belladonna Lily was given to the flower in Italy from the charmingly blended red and white of the perianth, resembling the complexion of a beautiful woman. [T. M.]

AMARYLLIS CANDE'LABRE, or GIRANDOLE. (Fr.) *Coburgia multiflora*.— BALTIMBANQUE. *Sprekelia Cybister*.— DE GUERNESEY. *Nerine sarniensis*.— DE VIRGINIE. *Zephyranthes Atamasco*— JAUNE. *Oporanthus luteus*.— REINE DE BEAUTÉ. *Sprekelia formosissima*.— VENE'NEUSE. *Buphane toxicaria*.

AMASONIA. A genus of the verbain family (*Verbenaceæ*) nearly allied to *Clerodendron*, and chiefly differing from that genus in its habit. The species enumerated are six, all of them natives of southern tropical America and the greater part of them found in Brazil. They are perennial dwarf shrubs, with alternate or opposite leaves, and terminal racemed panicles of flowers, each little group of yellow flowers being supported by a large scarlet-coloured or beautifully variegated bract, which bears on its outer surface a number of pellucid glands. They are well-deserving of cultivation, but seldom met with. [A. A. B.]

AMA-TSJA. Tea of Heaven; a kind of tea prepared in Japan from the leaves of *Hydrangea Thunbergii*.

AMAUROPELTA. A name given by Kunze to a West Indian Fern, supposed to have some affinity with the davallioid group, sometimes called *Saccoloma*. It is now referred to *Lastrea*. (T. M.)

AMBATCHA. *Arum abyssinicum*.

AMBER TREE. A common name for *Anthospermum*.

AMBERBOA. A genus of composites, several of the plants composing which have long been cultivated under the more familiar name of *Centaurea*, from which genus the present one differs only in a few obscure and minute characters of the fruit and pappus. The two best-known species are the *A. odorata*, or Yellow Sweet Sultan, and the *A. moschata* or Purple Sultan. Both are branching annuals, growing a foot or more high, with oblong pinnatifid foliage, and large terminal showy flower-heads. Those of the first are characterised by having the outer florets much longer than those of the centre, and the fruit is crowned with a short pappus of hairs. In the case of the latter species, *A. moschata*, the pappus is altogether wanting, and the florets of the circumference are scarcely longer than the central ones. The odour of this species is hardly suggestive of musk, as its name would imply, but is rather honey-like, differing but little from that of the Yellow Sultan, except in its greater intensity. (W. T.)

AMBLYOCARPUM. The generic name of a Persian weed of no beauty, belonging to the composite family, and closely related to *Carpesium*, but differing in the strap-shaped and female ray florets being in a single row, as well as in the achenes—which are five, angular, and without pappus—being beakless. The plant is called *A. inuloides* from its resemblance to our fleabane (*Inula Pulicaria*). Its lance-shaped leaves are, however, longer and smooth, not downy. The yellow flower heads are single at the ends of the twigs, and nearly half an inch across. (A. A. B.)

AMBLYOLEPIS. A Texan genus of composites, of which a single species, *A. setipera*, is in cultivation, and possesses some interest from the pleasing fragrance of its flowers, which they retain for many years when dried. This fragrance, which the seeds of the plant possess in a high degree, is doubtless due to the presence of coumarin, the chemical principle to which the well-known tonka bean, and the common vernal grass, *Anthoxanthum odoratum*, also owe their agreeable scent. The species in question is a dwarf, erect, branching annual, with entire, ovate, lance-shaped, stem-clasping leaves, two to three inches long, the branches being terminated by a single flower-head one and a half inch in diameter, with a ray of broadly wedge-shaped florets, and a disk of tubular ones, both being of a uniform orange-yellow colour. The involucre consists of about ten ovate, lance-shaped, spreading bracts, the receptacle is naked and conical, and the villous fruit is crowned by a pappus of five broad, blunt, transparent, colourless scales. (W. T.)

AMBORA. A genus of *Monimiaceæ*, consisting of trees from Madagascar and Mauritius, with entire evergreen leaves, and monœcious flowers, generally in racemes, though sometimes solitary, rising from the trunk or lower parts of the branches. The male flowers are scattered among the more numerous females. The stamens are numerous, with short filaments and bilocular anthers. There are many one-celled ovaries, each containing a single ovule. The fruit consists of many one-seeded drupes, enclosed in the enlarged calyx, which gives it a baccate appearance. The bark and leaves exhale an aromatic odour. (W. C.)

AMBORN. (Fr.) *Cytisus Laburnum*.

AMBOYNA WOOD. The beautifully mottled wood of *Pterospermum indicum*.

AMBRETTE JAUNE. (Fr.) *Amberboa odorata*. — MUSQUÉE *Hibiscus Abelmoschus*.

AMBRINA. A genus of plants belonging to the natural order *Chenopodiaceæ*. It comprises annual or perennial plants, with alternate, nearly sessile, cleft or sinuous leaves, covered, like the whole of the plants, with resinous spots. The flowers are clustered in heads, which are placed in the axils of the leaves, or in leafless or leafy terminal spikes. The genus is allied to *Chenopodium*, from which it differs in its obovate fruit, not depressed in the centre, and by the seeds being placed vertically in the seed vessel, not horizontally. From the genus *Blitum* it differs in the calyx, becoming of a pentagonal shape when it invests the fruit. All the species have an aromatic odour, and possess tonic and stimulant properties. *A. pinnatifida* is cultivated for the sake of its elegant and aromatic foliage. *A. ambrosioides*, or Mexican Tea, originally a native of North America, but long naturalised in the south of Europe, is used medicinally in the form of an infusion, having antispasmodic vermifuge and carminative properties. *A. anthelmintica* is common in the Southern States of America, where it is employed as a vermifuge. (M. T. M.)

AMBROISE. A name given in Jersey to *Teucrium Scorodonia*.

AMBROISINE. (Fr.) *Chenopodium ambrosioides*.

AMBROSIA. A genus of the composite family (*Compositæ*), chiefly annual coarse-habited weeds, with opposite, or alternate lobed, or dissected leaves and the flower-heads in racemes or in bundles in the axils of the leaves. The sterile and fertile flowers occupy different heads on the same plant. The sterile involucres, somewhat top-shaped, composed of seven to twelve scales, united into a cup, and containing five to twenty staminate flowers. The fertile ones top-shaped, closed, pointed, and usually with four to eight horns or tubercules near the top in one row, and containing a single flower composed of a pistil only. The species, of which there are about twelve, are pretty widely diffused, being found in India, tropical Africa, South Europe, and in North and South America, growing in fields and waste places. *A. artemisifolia* is very plentiful on the plains of the Saskatchawan and Red River; while *A. tenuifolia* is said to cover thousands of miles of the Pampas, south of Buenos Ayres, giving them a black appearance like that of the Scotch moors. *A. trifida* is called the Great Rag-weed in America, and *A. artemisifolia* the Roman Wormwood; indeed, all the species bear a great resemblance to the Wormwood (*Artemisia*). *A. maritima*, found in Italy and the Levant, is said to be tonic and resolutive; all its parts give out a sweet odour and have an aromatic taste, a little bitter, but agreeable. [A. A. B.]

AMBROSINIA. A genus of *Aroideæ*, containing a few species, natives of Sicily and Sardinia. They are small land plants, with tuberous, stoloniferous rhizomes, entire leaves, and a small spathe, inclosing a couple of scentless flowers, of which the uppermost has many monadelphous stamens perfectly destitute of a calyx, and a single unilocular ovary. They are referred by Endlicher to *Aroideæ*, but Lindley considers that the paucity of flowers in the spadix affords sufficient ground for establishing another order, which he calls *Pistiaceæ*, and which includes *Lemna*, *Pistia*, and some other allied genera. [W. C.]

AMELANCHIER, the Savoy name of the medlar, is given to a family of small trees, natives of Europe and North America, allied both to *Mespilus* and *Cotoneaster*. In British gardens they are cultivated for their flowers, which are white, abundant, showy, and produced early in the season; for their fruit, which ripens in June; and for the deep red or rich yellow hue which their foliage assumes in autumn. The common Amelanchier, *A. vulgaris*, has long been cultivated in England, where it sometimes attains the height of fifteen or twenty feet. It bears abundance of flowers, and its fruit, though not highly palatable, is eatable. This is a native of Southern Europe, where it grows in rocky mountainous woods. Of the American species, *A. Botryapium*, the Grape-Pear, bears sparingly small fruit of a purplish colour and of an agreeable sweet taste, which ripens in June, before that of any other tree. *A. ovalis*, considered by some to be merely a variety of the preceding, abounds, according to Dr. Richardson, in the sandy plains of the Saskatchawan. 'Its wood is prized by the Crees for making arrows and pipe-stems, and is thence termed by the Canadian voyageurs *Bois de flèche*. Its berries, about the size of a pea, are the finest fruit in the country, and are used by the Crees both in a fresh and dried state. They form a pleasant addition to pemmican, and make puddings very little inferior to plum-pudding.' [C. A. J.]

AMELLINGUE. (Fr.) A kind of Olive.

AMELLON. (Fr.) A kind of Olive.

AMELLUS. A genus of the composite family (*Compositæ*), containing twelve species, all of them natives of South Africa. They are herbs or shrubs, their lower leaves opposite, the upper alternate, oblong, entire or toothed, and hairy or canescent. Flower-stalks terminal, bearing a solitary head of flowers; the florets of the disc yellow, those of the ray blue. *A. Lychnitis* is cultivated in gardens. The flowers of it, and most of the species, are a good deal like those of the Michaelmas daisy (*Aster*), to which genus this is allied, differing chiefly in the opposite lower leaves, and in having the bristles of the pappus in a single series. [A. A. B.]

AMENTACEÆ. Under this name were once comprehended all apetalous unisexual plants, whose flowers grow in catkins, or amenta. Modern botanists find it more convenient to distribute them through several different orders, the chief of which are *Salicaceæ*, *Corylaceæ*, *Betulaceæ*, *Casuarinaceæ*, *Altingiaceæ*, *Myricaceæ*, which see. A forest of these amentaceous plants as they grow in the island of Java, is shown in Plate IX.

AMENTUM. A catkin. A deciduous spike of unisexual apetalous flowers, such as appears in the spring on the hazel and willow.

AMESIUM. A name once proposed to be given to *Asplenium septentrionale* and some allied species. [T. M.]

AMETHYSTEA. An insignificant Siberian genus of labiates, belonging to the *Ajuga* or bugle division of the order, and distinguished by the very short upper lip of its corolla, and the abortion of its upper pair of stamens. The only species, *A. cœrulea*, was formerly cultivated, but is now seldom met with, so many more deserving plants being available. It is a hardy annual, growing a foot or more high, with erect, square, branched stems; opposite, three-parted leaves; the segments oblong lance-shaped; and short terminal leafy racemes of very small pale-blue flowers, the corollas of which are scarcely longer than the calyx. As an ornamental plant, it is entirely worthless, but it possesses the merit of being slightly fragrant. [W. T.]

AMHERSTIA. A genus of the pea family (*Leguminosæ*), named in honour of the Countess Amherst. *A. nobilis* is the only species. It grows near Martaban, in the Malayan peninsula, and attains a height of about forty feet. When in flower, it is said to be 'one of the most superb objects imaginable, unrivalled in India or in any other part of the world.'

The leaves are equally pinnate, large, and, when young, of a pale purple colour. 'The flowers are large, scentless, and of a bright vermilion colour, diversified with three yellow spots, and disposed in gigantic ovate pendulous bunches.' The tree is cultivated in some of the larger English gardens; but, requiring so much space, is seldom met with in collections. The Burmese name of the plant is Thoca, and handfuls of the flowers are offered before the images of Buddha. [A. A. B.]

AMIANTHIUM. The name of a genus of North American plants, belonging to the same family — *Melanthaceæ* — as the *Colchicum* and *Veratrum*. The species have a widely-spreading petal-like perianth, without glands; six stamens attached beneath the ovary, with their anthers bursting outwardly; a capsule of three cells, which separate one from the other when ripe. One species, *A. muscætoxicum*, contains a narcotic poison which is injurious to cattle that browse on its foliage. Its bulbs, pounded and mixed with honey, are used as a fly poison. [M. T. M.]

AMICIA. A genus of the pea family (*Leguminosæ*), named in honour of Prof. J. B. Amici, of Modena, a distinguished microscopical observer. There are but two species known, one of them found in the vicinity of Loxa, and the other in the Cordilleras of Mexico, at an elevation of 5,000 to 8,000 feet. They are both straggling shrubs, having alternate pinnate leaves, with few leaflets. Their flower stalks are axillary or terminal, having at their base two large, kidney-shaped, coloured bracts. Two of the segments of the calyx are large compared with the others, and roundish in form. The pods are compressed, and jointed. All the parts of the plant are covered with pellucid, glandular dots, somewhat like those of St. John's wort. *A. zygomeris*, the Mexican species, is sometimes to be found in gardens, and is well worth cultivation, especially as it flowers late in autumn, or during the early part of winter. It is generally treated as a greenhouse plant; but, in the South of England, if planted out of doors in the spring, it generally flowers well in the autumn. The flowers are large, and of a pale yellow colour — about the size of an everlasting pea. [A. A. B.]

AMIDONNIER (Fr.) *Triticum dicoccum*, sometimes called *T. amyleum*.

AMMANNIA. A genus of inconspicuous herbs, of the order *Lythraceæ*, growing in wet places in the warmer regions of the globe; mostly glabrous annuals, with square stems, opposite entire leaves, and small axillary, nearly sessile flowers, often without petals. Several species have been introduced, but are more curious than beautiful. *A. vesicatoria* has acrid leaves, which, when bruised, are used by the native practitioners of India to raise blisters. [J. T. S.]

AMMI. A small genus of umbelliferous plants, with the habit of the carrot (*Daucus*), spindle-shaped roots, and many-parted leaves; it is remarkable for the large size of the outer petals of the umbel. As the name denotes (from the Greek *ammos*, sand), they affect sandy ground, but will thrive if sown in the common soil of the garden border. Common Bishop-weed, *A. majus*, is a native of the south and middle of Europe, Egypt, and the Levant, where it attains the height of three or four feet. Tooth-pick Bishop-weed, *A. Visnaga*, is so called on account of the use made in Spain of the rays or stalks of the main umbel. These, after flowering, shrink, and become so hard that they form convenient tooth-picks. When they have fulfilled this purpose, they are chewed, and are supposed to be of service in cleaning and fastening the gums; however this may be, the leaves have a pleasant aromatic flavour in the mouth. [C. A. J.]

AMMOBIUM. The *A. alatum*, the only known species, is a curious Australian annual of the composite order, remarkable for its winged stem. In a structural point of view, it is allied to the genera *Gnaphalium* and *Antennaria*, from which it differs but slightly, except in habit. The root-leaves are lance-shaped, with a long narrow foot-stalk, those of the stem and branches very small, and prolonged downwards in a narrow, wing-like form. The flower heads, which are of the dry everlasting character so common to plants of the Australian continent, are nearly an inch across, with a disk of tubular florets; a receptacle set with oblong, pointed, toothleted, chaffy scales; an involucre of imbricated leaflets, the inner series of which have membranous margins, and a four-angled, elongated fruit, furnished at the apex with four teeth, the two larger of which are terminated by a bristle. The plant remains some time in flower, and is not without a certain degree of interest; though, as an ornamental plant, it is almost superseded by the more recent species of *Helichrysum*, *Rhodanthe*, and *Acroclinium* from the same continent. [W. T.]

AMMOCHARIS. A genus of *Amaryllidaceæ*, in which the tube of the six-parted perianth is cylindrical, enlarged, the separate line divisions not imbricating thereon; the filaments of the stamens are adjusted almost equally at the base of the limb; the anthers are short, affixed in the middle; and the capsule is turbinate, three-celled. The leaves are vernal, and not sheathing. The genus is intermediate between *Crinum* and *Buphane*, differing from the first in its anthers, its filaments inserted just within instead of without the tube, its shorter limb and wider-mouthed tube, and its leaves not sheathing; and from the last, by the wider mouth of the tube of the perianth, the insertion of the filaments within the tube, and the more numerous ovules. The two species, sometimes referred to *Brunsvigia*, are South African. *A. falcata* has ligulate glaucous leaves, and a many-flowered umbel of greenish-white flowers, which afterwards become pinkish and finally rose-colour. (T. M.)

AMMONIACUM. A drug said to be obtained from *Dorema Ammoniacum*, and also from *Ferula tingitana*.

AMMOPHILA. A genus of grasses of the tribe *Arundineae*, inhabiting the sandy sea-shores of the coasts of Europe and North America, and extensively cultivated in many places, as in the eastern counties of England and in Holland, for preserving the sand-banks which prevent the inroads of the sea. In the northern parts of England, it is used for making table mats and basket work. It is the widely-creeping and matted rhizomes which serve to bind together the sand-banks on which it grows. The stems grow two or three feet high, and bear long, narrow, rigid involute leaves and a spiked cylindrical panicle, with laterally compressed spikelets. The glumes are nearly equal, and lance-shaped, stiff and chaffy. The flowering glumes, or outer pales, are the shorter, with a tuft of hairs outside, but the inner pales nearly equal them in length. The genus is nearly related to *Calamagrostis*, from which the inflorescence, the stiff glumes, and the absence of an awn to the flowering glume, serve to distinguish it. The only species, *A. arundinacea*, or *Psamma arenaria*, is variously called Maram, Marrum, Sea-reed, or Sea Matweed. (T. M.)

AMNIOS. The fluid that is produced within the sac which receives the embryo-rudiment and engenders it.

AMOMUM. A genus of aromatic herbs, belonging to the ginger family, *Zingiberaceae*. The root-stocks are jointed, creeping; the leaves placed in two rows, sheathing at the base, lance-shaped, and undivided at the margin. The flowers, in a spike or cluster, are provided with bracts, and but little raised above the ground; there is but one stamen, whose filament is prolonged beyond the two-celled anther, so as to form a more or less lobed crest; the capsule is three-celled, and opens, when ripe, by three pieces, so as to liberate the numerous small seeds. These plants are natives of India, the islands of the Indian Archipelago, etc. Their seeds are aromatic and stimulant, and form, with other seeds of similar plants, what are known as Cardamoms, of which there are many kinds.

Attare, Malaguetta Pepper, or Grains of Paradise, are the seeds of one, perhaps two, species of this genus, *A. Grana Paradisi* and *A. Meleguetta*. They are imported from Guinea, and have a very warm, slightly camphor-like taste. These seeds are made use of illegally to give a fictitious strength to spirits and beer, but they are not particularly injurious; although, from the very heavy penalty inflicted on brewers who have them in their possession, and on druggists who sell them to brewers (200l.

and 500l. respectively), it would seem as if such an opinion were entertained.

The large round China Cardamoms are supposed to be produced by *A. globosum*, the hairy round China sort by *A. villosum*, Java Cardamoms by *A. maximum*; but the botanical history of the plants producing

Amomum Grana Paradisi.

the various kinds of Cardamoms, Grains of Paradise, etc., is involved in much confusion and obscurity. Several species of the genus are in cultivation as ornamental stove plants. [M. T. M.]

AMOMUM. (Fr.) *Solanum pseudo-Capsicum*.

AMOREUXIA. A genus of *Cistaceae* containing two species from Mexico and New Granada. They are herbaceous plants, with the habit of *Maira*. The root is a large ligneous tuber. The leaves are alternate, on long petioles, and digitate-partite. The large flowers are in terminal racemes, and consist of five oblong persistent sepals, and five caducous obovate petals. The stamens are indefinite and arranged in two bundles, the one having very much longer and stouter filaments than the other. The ovary is ovate and trilocular, with many ovules attached to a central placenta. M. Planchon has joined this genus with *Cochlospermum* to form a small order *Cochlospermeae*, which he places near *Malvaceae* and *Zygophyllaceae*; but his reasons are not satisfactory for separating them from *Cistaceae*, with which they are more nearly allied. [W. C.]

AMORPHA. The flowers which belong to the natural order *Leguminosae*, though composed of petals unequal in size and irregular in form, have, for the most part, these organs symmetrically arranged, after the type of the pea and bean. In the present genera, however, the two pairs of petals, termed severally the wings and keel, are absent, the only representative of petals being the standard or vexillum, and hence its name *Amorpha*, 'deformed.' All the plants of the genus are deciduous shrubs, natives of North America. The leaves are pinnate with a terminal leaflet, covered with pellucid dots; and the flowers, of a blue-violet colour, are disposed in long spiked clusters, grouped at the tops of the branches. All the species are ornamental. The foliage is graceful; and the flowers, though individually small, are attractive from their numbers and colour, which is violet spangled with the golden anthers. As they only attain the height of a few feet, they are well adapted for small shrubberies, or the front of large ones, and thrive well in common garden soil. *A. fruticosa*, the commonest species in European gardens, was introduced to Britain in 1724 by Mark Catesby, who states that the inhabitants of Carolina at one time made a coarse sort of indigo from the young shoots. Hence it is sometimes called Bastard Indigo. [C. A. J.]

AMORPHOPHALLUS. A name given to a genus of plants of the araceous family, and used to indicate the exceedingly curious appearance of the plants, which are perennial, with tuberous rootstocks flattened on the upper surface. The leaves and spadices are solitary, invested below with imbricated scales. The spathe is spreading so as to fully expose the thick fleshy spadix, which is dilated and fungus-like at the upper extremity. The male flowers are placed above the females; their anthers are sessile and open by pores; the ovary has either two, three, or four cells, with erect ovules. These plants were formerly included in the genus *Arum*, from which they are distinguished by their spreading not convolute spathes; by their anthers opening by pores, not by longitudinal slits; by the numerous cells to the ovary; and by the solitary erect ovule, those of *Arum* being horizontal. They are natives of India and other parts of tropical Asia, where they are cultivated for the sake of the abundance of starch which is found in the rootstock. The presence of this starch, and especially the mode of preparation, deprives the roots of their otherwise acrid caustic properties. Dr. Wight says of *A. campanulatum*, that when in flower the fetor it exhales is most overpowering, and so perfectly resembles that of carrion, as to 'induce flies to cover the club of the spadix with their eggs.' [M. T. M.]

AMORPHOUS. Having no definite form.

AMOURETTE. (Fr.) *Briza media*; also *Saxifraga umbrosa*.

AMPELOPSIS. A North American genus of *Vitaceæ*, distinguished from *Vitis* and *Cissus*, to which it is closely allied, by the absence of the disk or expansion of the receptacle in a ring round the base of the ovary. *A. hederacea*, the Virginian Creeper, or American Ivy, is a shrubby climber, often planted in this country to cover walls, for which, from the rapidity of its growth, it is well adapted; the leaves, which have five large elliptical leaflets, turn red before they fall in autumn, when the plant presents a very beautiful appearance. The flowers are small and yellowish-green, in a many-flowered panicle. The tendrils are curious, adhering to supporting bodies by small sucker-like expansions which are formed at the apex of each of their divisions. [J. T. S.]

AMPELOPTERIS. A name proposed by Kunze for a few Indian ferns now referred to *Goniopteris*. [T. M.]

AMPHEREPHIS. A Brazilian genus of composites, of which *A. intermedia* is occasionally found in cultivation. It is a branched spreading annual, growing a foot or more high, with ovate, serrated foliage, and terminal flower heads, an inch across, composed wholly of tubular florets of a purple colour, and surrounded by a double series of leafy bracts or scales. It possesses few, if any, claims to general notice. [W. T.]

AMPHIBLESTRA. A genus of polypodiaceous ferns belonging to *Pterideæ*, and distinguished in this group by having the veins of the fronds compoundly reticulated, with free included veinlets, or little veins within the meshes or areoles. It is a coarse-looking fern of South America, with much the aspect of *Aspidium trifoliatum*, but having the pteroid linear marginal indusiate fructification. [T. M.]

AMPHIBOLIS. A genus of *Zosteraceæ*, formed to include a plant found in the Pacific Ocean and on the coast of New Holland; and considered by Endlicher and Kunth to belong to *Cymodocea*. The only species, *A. zosterifolia*, has branched annulated stems, and approximate alternate, linear, truncate, and bidentate leaves, with short truncated stipules. [J. T. S.]

AMPHICOME. A genus of the bignonia family (*Bignoniaceæ*). Two species are known, and both of them natives of the temperate regions of North Western India. They are perennial herbs, with alternate, unequally pinnate leaves, and toothed leaflets. The flowers are pink, tubular, and arranged in axillary or terminal racemes. The fruits are about the length and thickness of a crowquill, and their seeds are provided with a tuft of hairs at each end, this circumstance giving rise to the name of the genus—*amphi*, on both sides, and *kome*, a head of hairs. They are both in cultivation. *A. Emodi* is a remarkably handsome plant, and well deserves a place in choice collections; it is about one foot high, and the flowers, which are large for the plant, stand erect when expanded. *A. arguta* is about the same size, but it has smaller drooping flowers. [A. A. B.]

AMPHICOSMIA. A genus of cyatheaceous ferns, belonging to the section

Cyathea, in which group it is distinguished by having beneath the sorus, on the hinder side, a half-cup-shaped indusium, and by having the veins of the fronds free. The species are sometimes referred to *Alsophila*, to *Hemitelia*, or to *Cyathea*. Several species, chiefly South American, agreeing in having the half-cup indusium and free veins, are referred to the genus in *Index Filicum*; but there is also one species from the Cape of Good Hope, two from India, and one from New Holland. The typical species is *A. capensis*, found both at the Cape and in Java, a tree-fern growing twelve to fourteen feet high, and of which, according to Dr. Harvey, there is a noble forest in the woods on the east side of Table Mountain. The fronds of this are three times pinnate, and unarmed. [T. M.]

AMPHIDESMIUM. A genus of cyatheaceous ferns, closely related to *Alsophila*. They are distinguished — having obliquely-ringed spore-cases and naked sori with elevated receptacles—by producing two or three sori in different positions on the same vein, the veins in *Alsophila* bearing one only. They are also different in aspect, having bold pinnate fronds, which give them a noble appearance. The species *A. blechnoides* is found in various parts of South America and in the West Indies. [T. M.]

AMPHIGASTRIA. The so-called stipules of Scale-mosses, or *Jungermannias*.

AMPHIGENÆ. A name applied by Brongniart to Thallogens, implying that they are developed in every direction, without any distinct axis and appendages; and not especially at the apex, like ferns and mosses, to which he has applied the name of Acrogens, and which, is contradiction to Thallogens, are furnished with both axis and appendages. [M. J. B.]

AMPHILOCHIA. A genus of *Vochysiaceæ*, containing four species from Brazil. They are trees, with opposite petiolate and entire leaves, and glands at the base of the petioles. The flowers are in terminal racemes. The calyx consists of five coloured sepals, combined at the base, the upper being much the largest and spurred. The corolla has only a single petal, inserted in the base of the calyx between the two front sepals. There are two stamens, one on either side of the petal. The ovary is three-celled, with few ovules. [W. C.]

AMPHISARCA. A many-seeded many-celled superior indehiscent fruit, woody on the outside, pulpy within.

AMPHISTEMON, a genus of *Dioscoreaceæ*, formed by Grisebach by the subdivision of *Dioscorea* into many new genera. The section to which this name has been given, is separated from the others by having six short fertile stamens, which are inserted on the apex of the calyx tube. It contains eleven species of tropical, chiefly Brazilian, herbaceous plants. [W. C.]

AMPHITROPAL. When an ovule is attached by its middle, so that the two ends are equidistant from the point of insertion.

AMPLEXICAUL. Embracing; as when a leaf clasps a stem with its base.

AMPOULLEAU. (Fr.) A kind of olive.

AMPULLA. The metamorphosed flask-like leaves found on certain aquatics such as *Utricularia*: not different from *Ascidium*.

AMSINCKIA. A genus of the borage family, numbering seven species, found in Oregon, California, Mexico, and Chili. They are annual erect herbs, of little beauty; all their parts more or less clothed with rusty hairs. The stems, six inches to one and a half feet high, are furnished with alternate and entire linear, lance-shaped, or ovate leaves, one to five inches long, and terminate in one or more one-sided racemes of yellow funnel-shaped flowers, with a flat border of five rounded lobes. In the largest flowered species (*A. spectabilis*, from California), the corolla tube is three-quarters of an inch in length. The fruit consists of four triangular one-seeded nuts, their dorsal face smooth, or covered with warty excrescences. The seeds are remarkable, from having their cotyledons deeply biparted. [A. A. B.]

AMSONIA. A genus of *Apocynaceæ*, consisting of five species, natives of North America. They are perennial herbaceous plants, with alternate leaves, and pale blue flowers, in terminal panicled cymes. The calyx is small and five-parted; the corolla has the same number of long linear lobes; its narrow funnel-shaped tube is bearded inside, especially at the throat. There are five included stamens, with obtuse anthers, which are longer than the filaments; two ovaries, a simple style, and a rounded stigma, surrounded with a cup-shaped membrane. The two pods are long and slender, with many naked cylindrical seeds, in a single row. [W. C.]

AMYGDALOPSIS. A supposed genus of drupaceous plants formed on the Japanese *Prunus triloba*. Its distinctive character is having several carpels in each flower instead of one; probably a mere malformation, such as occurs in the peach and plum themselves.

AMYGDALUS. The name applied to the genus to which the Almond, the Peach, and the Nectarine belong. It is placed by botanists in the drupaceous subdivision of the rose family, and is especially known by the stone of the fruit, which encloses the kernel or seed, being coarsely furrowed or wrinkled, and by the leaves being folded in halves, not rolled round when young.

The Almond-tree (*A. communis*) appears to have been originally a native of Barbary and Morocco; but by long cultivation it has become distributed over almost the whole of the warmer temperate zones of the old world. It is a small tree, with oblong lance-shaped leaves, slightly saw-toothed at the margin. The flowers, which appear in spring, before the leaves, are solitary and of a beautiful pink colour. The fruit is a drupe, which is somewhat egg-shaped, downy externally; its middle portion tough and somewhat fibrous; its inner portion forming the hard wrinkled stone enclosing the seed within it. Many varieties of the Almond are cultivated, differing in the nature of their fruits; but the two principal are the Sweet and the Bitter Almond. The Bitter Almond has larger flowers than the sweet variety, and they are of a white colour. The styles are not longer than the stamens, and the seeds are bitter. The seeds of the Sweet Almond are much esteemed at the dessert table, in spite of their indigestibility. The bitter almonds, though occasionally used for flavouring purposes, should be employed in small quantities, as they contain a poisonous principle which is similar in its effects to prussic acid. The essential oil of almonds, which is much used as a flavouring ingredient by cooks and confectioners, is a most virulent poison: it contains prussic acid, and should therefore be employed with great care and in a diluted form, as in what is called in shops Essence of Almonds. It is curious that this oil does not exist naturally in the almond, but is formed by the chemical agency of water on some of its constituents.

A. persica is the botanical name given to the Peach, which is sometimes included in a separate genus (*Persica*), but it only differs from the almond in having a fleshy, not leathery, drupe. Instances have been cited of almonds having fleshy drupes, and thus assuming the character of the Peach. Three principal varieties of the Peach exist—clingstones, melters or freestones, and nectarines. The latter only differ from the peach in having smooth, not downy fruits; but both peaches and nectarines are occasionally met with on the same bough. The leaves of the Peach and Nectarine contain a small quantity of prussic acid, and have the taste and odour of bitter almonds. The fruits, taken in moderation, are as wholesome as they are delicious; but the kernels and blossoms contain prussic acid. The Peach is very extensively cultivated in America, but little attention is paid to the culture: the fruits are used in the manufacture of peach brandy, and for feeding hogs! [M. T. M.]

The common Almond-tree grows to the height of about twenty feet. The leaves closely resemble those of the Peach (*A. persica*), but the flowers are larger than those of that species. Its fruits, which are the Almonds of commerce, are well known. They seldom attain maturity in this country, in which, however, the tree is frequently to be seen, on account of its showy blossoms, which appear in great abundance very early in spring, when the season is not unusually cold; they often appear in February, and, in the mild winter of 1834, a standard almond-tree in the neighbourhood of London was in full flower in the end of January. De Candolle is of opinion that the Almond is a native of Persia, Asia Minor, Syria, and even Algeria. It is found growing spontaneously in many other countries, to which, however, it is not supposed to be indigenous, the plants met with having probably been derived from others introduced for the purpose of cultivation. In Palestine, it appears to have been cultivated from the earliest ages; for we find it enumerated among the best fruits of Canaan which were sent into Egypt as a present for Joseph, upwards of 3,500 years ago. The fruit of the Almond is of an ovate, somewhat curved, tapering form. It consists of a husk, which dries up and splits at maturity, exposing the stone, within which is the kernel, the only edible portion. There is a variety with bitter kernels, from which, like the sweet, oil can be extracted, but which are otherwise unfit for use, as they contain prussic acid in notable quantity. There are several varieties of the sweet-kernelled; some with hard, and others with comparatively tender, shells or stones. The most esteemed is the large thin-shelled, or Jordan Almond.

The Peach (*A. persica*) differs essentially from the Almond in the nature of the covering of the stone, which, instead of a dry husk, is fleshy, succulent, and delicious, when the fruit is ripened under favourable circumstances. The species comprises the Peach and Nectarine, the skin of the former being downy, and that of the latter quite smooth. They were supposed to be natives of Persia, and, on their introduction into the South of Europe, were called the *Malus persica*, or Persian apple. Professor De Candolle is, however, of opinion that China is the native country of the Peach. His reasons are, that if it had originally existed in Persia or Armenia, the knowledge and culture of so delicious a fruit would have spread sooner into Asia Minor and Greece. The expedition of Alexander is probably what made it known to Theophrastus, B.C. 322, who speaks of it as a Persian fruit. It has no name in Sanscrit; nevertheless, the people speaking that language came into India from the north-west, the country generally assigned to the species. Admitting this to be its country, how can it be explained that neither the early Greeks, nor the Hebrews, nor the people who speak Sanscrit,—and who have all sprung from the upper region of the Euphrates, or from parts communicating with it,— had grown the Peach-tree? On the contrary, it is very possible that the stones of a fruit tree cultivated from all antiquity in China, may have been carried across the mountains from the centre of Asia into Cashmere, or Bokhara and Persia; for the Chinese had discovered this road at a very remote period. This importation must have been made between the time of the Sanscrit emigration and the intercourse of the Persians with the Greeks. The cultivation of the Peach-tree, once established at this point, would easily extend on one side towards the west, and on the other, by Cabul, towards the North of India. In support of the supposition of a Chinese origin, it may be added that the Peach-tree was introduced from China into Cochin-China, and that the Japanese call it by the Chinese name, *Too*. In the Japanese encyclopedia it is stated to be a tree from western countries, which applies to China with regard to Japan, or rather to the interior of China relatively to its eastern coast; the statement having been taken from a Chinese author. The Peach is mentioned in the books of Confucius, 5th century before the Christian era; and the antiquity of the knowledge of the fruit in China is further proved by the representations of it in sculpture and on porcelain. The above are some of the arguments adduced by De Candolle against the commonly-received opinion that the Peach originated in Persia: for the full investigation of the subject, we must refer the reader to his *Géographie Botanique*, according to which excellent authority the conclusion is that China is the native country of this esteemed fruit. That it is there cultivated extensively, and to great perfection, is certain. The Flat Peach of China was introduced into this country more than thirty years ago. It is figured in the *Transactions of the Horticultural Society* (iv. 512, t. 19); and, more recently, a very large variety was brought from Shanghae by Mr. Fortune, which has the usual form exhibited by those cultivated in Europe.

In the South of France, and in other Continental countries possessing a similar climate, Peach-trees ripen their fruit very well as standards in the open air; but at Paris they require a wall; and, with this assistance, they also succeed very well in the southern parts of England, but in the northern the aid of fire-heat, and the protection of glass, are necessary. In America, the Peach grows almost without any care—extensive orchards, containing from 10,000 to 20,000 trees, being reared from the stones. At first the trees there make rapid and healthy growth, and in a few years bear in great abundance; but they soon decay, their leaves becoming tinged with yellow, even in summer, when they should be green. This is owing to their being grown on their own roots; for when that is the case in this country, the trees present a similar appearance. They require, therefore, to be budded on the plum or on the almond. Some doubts have been entertained as to whether the Peach is not the same species as the Almond. They appear, however, to maintain their respective characters sufficiently distinct, unless artificially or by accident they are crossed with each other. The possibility of this being effected was successfully tried by Mr. Knight; and the circumstance of their crossing readily proves their close affinity. He fertilised an almond blossom with pollen from a Peach blossom. An almond was the result; but from its kernel he raised a tree which bore peaches of fair size and round form, with succulent melting flesh, of tolerably good quality, better, indeed, than some seedlings of the Peach itself.

The varieties of Peaches and Nectarines are very numerous, and would be difficult to distinguish, were it not for a classification formed from certain characters afforded by the fruit, leaves, and flowers. In some varieties the fruit has firm flesh, adhering to the stone; such are termed clingstones. Others have melting flesh, parting readily from the stone; these are called melters or freestones. The leaves are either glandless, or are furnished with globose, or with reniform glands at their bases. And in some the flowers are large, in others small. Formerly the Peaches and Nectarines, known in Europe, had all bitter kernels; but sweet-kernelled varieties have of late years been introduced from Syria. The following are some of the best varieties of Peaches: Noblesse, Royal George, Acton Scot, Grosse Mignonne, Bellegarde, Late Admirable, and Walburton Admirable. Of Nectarines, the Violette Hative, Pitmaston Orange, Downton, Elruge, Impératrice, and Balgowan are amongst the most esteemed sorts. [R.T.]

AMYLACEOUS GRANULES. Grains of starch.

AMYLIDEÆ. Cells in algals, secreting starch.

AMYLUM. Starch; that organised granular matter of plants which iodine stains violet or blue.

AMYLOID. A substance analogous to starch, but becoming yellow in water after having been coloured blue by iodine.

AMYRIDACEÆ. (*Terebintaceæ*, *Burseraceæ*, *Amyrids*.) With the appearance of oranges, and sometimes with the dotted leaves of that order, these plants differ in their fruit, forming a shell whose husk eventually splits into valve-like segments. In general, moreover, the petals have a valvate æstivation. The genera collected under this name are by no means perfectly known, and demand a scrupulous revision. The tropics of India, Africa, and America exclusively produce the species. Their resinous juice is of great importance, forming an ingredient in frankincense and other preparations demanding a fragrant combustible matter. See AMYRIS, BURSERA, BOSWELLIA, BALSAMODENDRON, ICICA, and CANARIUM.

AMYRIS. A genus of trees belonging to the order *Amyridaceæ*, known by their unequally pinnate leaves, and by their solitary ovary, which contains two pendulous ovules. The plants are natives of tropical America and India, and are remarkable for yielding resinous products. It is supposed that the resin called Elemi is produced from some species of *Amyris*, such as *A. Acrandra* and *A. Plumieri*, though there is much doubt, not only as to the plant or plants producing the drug, but even as to whence it is imported. Indian Bdellium, or False Myrrh, is obtained from *A. commiphora*; it is a gum resin, possessing properties somewhat similar to those of the Myrrh, but is not so highly valued. *A. balsamifera* yields some descriptions of the wood which is called Lignum Rhodium. *A. toxifera* is poisonous. See BALSAMODENDRON. [M. T. M.]

ANABAINA. A genus of green-spored *Algæ*, the species of which consist of necklace-shaped threads, of which some of the articulations are much larger than the rest. They either form a shapeless scum on the surface of pools, or roundish patches on the bare soil. They never develop a distinct solid frond like that of *Nostoc*. One of the species, *A. bicheniformis*, is extremely common in gardens where the ground has been much trodden, as amongst raspberry bushes. The threads are a pretty object under the microscope, the large articulations being reproductive. One or two closely allied *Algæ*, as, for example, *Aphanizomenon*, are remarkable for being suspended in the water in which they grow, and giving to it a green tint. [M. J. B.]

ANABASIS. A genus of *Salsolaceæ*, consisting of trees and shrubs, natives of Central and Eastern Asia, and of the eastern shores of the Mediterranean. They are jointed plants, generally aphyllous, or with leaves small and opposite. The flowers are sessile and single, or in a glomerulus, hermaphrodite, and furnished with two bracts. There are five sepals, and the same number of stamens inserted in the receptacle; between these and united to their bases are five minute scales or staminodies. The ovary is unilocular and uniovulate, and the style double and divaricate. There are seventeen species. [W. C.]

ANACAMPSEROS. A genus of undershrubs from the Cape of Good Hope, referred to the order *Portulacaceæ*. They are succulent plants with crowded, imbricated, sessile, ovate-trigonous terete or subglobose leaves with stipules cut into five segments, often hair-like. Flowers large, white, rose, purple, or yellow, with twelve to twenty stamens; peduncles in some species very short, in others elongated, simple or branched. Several species are cultivated in greenhouses. [J. T. S.]

ANACAMPTIS. A genus of orchids, established by A. Richard for the *Orchis pyramidalis*, which differs from the rest of the genus by two small plates or appendages at the base of the labellum. A South European species in which these two plates are united into one horse-shoe-shaped appendage has since been added, and some botanists unite both species with *Gymnadenia* or with *Aceras*, but in a more natural arrangement they would be retained in *Orchis*. The *Anacamptis* (or *Orchis*) *pyramidalis* is not uncommon in central and southern Europe, extending eastward to the Caucasus.

ANACARDIACEÆ (*Terebintaceæ*, Cassuvieæ, Spondiaceæ, Anacardia, Terebinthi). When trees or bushes have a resinous, milky, often caustic juice, dotless leaves, and small inconspicuous flowers, with an ovary containing a single ovule suspended at the end of an erect cord, it is pretty certain that they belong to this order, of which more than 100 species are described, inhabiting the tropics both north and south of the equator, but not known to occur in Australia. *Pistacia*, and some kinds of *Rhus*, inhabit temperate latitudes. Among the products of the order are the Mango fruit, and that called in the West Indies the Hog Plum; the nuts named Pistachios and Cashews, the Black Varnish of Burmah and elsewhere, Mastich, Fustic, &c. These varnishes are extremely acrid, and produce dangerous consequences to persons who use them incautiously. See MELANORRHŒA, MANGIFERA, SPONDIAS, RHUS, ANACARDIUM, SCHINUS, &c.

ANACARDIUM. A genus of woody plants, from which the family to which they belong derives its systematic name, *Anacardieæ*. The plants of this group are chiefly remarkable for their kidney-shaped fruit, which is placed on the end of the thickened fleshy pear-like receptacle. *A. occidentale*, a plant cultivated in the West Indies and other tropical countries, produces the fruits known as Cashew Nuts. It is a large tree, somewhat like a walnut-tree in appearance, but with oval, blunt,

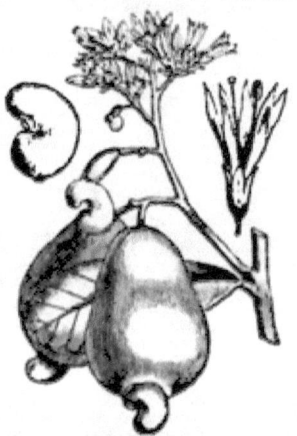

Anacardium occidentale.

alternate leaves; the fragrant rose-coloured flowers are borne in panicles. The stem furnishes a milky juice, which, as it dries, becomes black and hard, and is used as a varnish. A gum is also secreted by this plant, having qualities like those of gum arabic. It is imported into this country from S. America, under the name of Cadjil gum, and is used in S. America by bookbinders, who wash their books with it, to keep away moths and ants. The thickened receptacle has an agreeable acid flavour, with some degree of astringency; the fruit at its extremity is kidney-shaped, of an ash colour; the shell of the fruit consists of three layers, the outer and inner of which are hard and dry, but the intermediate layer contains a quantity of black, extremely acrid, caustic oil, which gives rise to severe excoriation of the lips and tongue in those who attempt to crack the nut with their teeth. This oil is sometimes applied to the floors of houses in India, to protect them from the attacks of white ants. The acrid matter is destroyed by heat, hence the kernels are roasted before being eaten, and then become wholesome and agreeable. The process of roasting has to be carefully conducted, the acridity of the fumes being so great as to produce severe inflammation in the face of persons approaching too near. [M. T. M.]

ANACHARIS. A submerged aquatic, belonging to the natural order *Hydrocharidaceæ*, having long, much-branched stems, small pellucid leaves, which are usually inserted in whorls of three, or less frequently four, and inconspicuous flowers; an American plant, which made its appearance in several remote parts of Britain almost simultaneously about the middle of the nineteenth century. How it was introduced is unknown, and it is equally a mystery by what means it traversed the wide tracts of country which separate the various stations in which it first appeared; for, as it is diœcious, and pistilliferous plants alone have found their way to this country, it perfects no seeds. It easily propagates itself from a small portion either of stalk or root, and is of wonderfully rapid growth; hence it has in many instances destroyed the beauty of ornamental pieces of water, impaired navigation in not a few inland canals, and interferes with the working of water-mills, by choking the outlets of reservoirs, especially towards the close of summer, when its debris are often brought down by the current in large quantities. Great efforts have been made to eradicate it in various places, but with imperfect success. It is greedily eaten by swans and some other water-birds; but even this service is not without its disadvantages, since portions of the stem, torn off but not consumed by these seeming allies, are carried away by the current, and transplanted elsewhere. In some places it is said to have almost disappeared, from having exhausted of its specific nutriment the soil in which it was rooted; but whether the ground will not, after a certain lapse of time, recover its productiveness, and bear a fresh crop of *Anacharis Alsinastrum*, remains to be seen. [C. A. J.]

Anacharis Alsinastrum.

ANACHASTE *sanguinea* is a terrestrial orchid, found at the sources of the Amazon river by Warczewitz, a Polish traveller. It has the habit of *Epidendrum maculatum*, with rich blood-red or rose-coloured flowers, having somewhat the form of a *Comparettia*, without the spur. It is a native of the forest of Talaneay, on the borders of the river Chanchan. Living plants were introduced by Mr. Skinner in 1853. The genus is nearly related to *Cochlioda*.

ANACYCLUS. A genus of the composite family (*Compositæ*), comprising eight species, which are found chiefly on the coasts of countries bordering on the Mediterranean sea. They are herbs with alternate, pinnati-lobed, much-cut leaves, and terminal solitary flower-heads; the ray florets white. In appearance they much resemble Chamomiles (*Anthemis*), and are chiefly distinguished from them by their achenes being bordered with a membranous wing. *A. Pyrethrum* grows in Barbary, Arabia, and Syria, and is cultivated in many places for the sake of its roots, which are used in medicine, and are called Pellitory of Spain. They are imported by the French from Africa, in pieces about the size of the finger. These, when fresh, if applied to the skin, cause a singularly cold sensation, immediately followed by heat. It is very pungent, and causes the saliva to flow freely. Sometimes it is given in toothache, but is seldom taken inwardly. 'When chewed, it causes a pricking sensation in the lips and tongue, and a glowing heat.' [A. A. B.]

ANADENIA. The name given to a number of plants belonging to the large family of *Proteaceæ*, but which are now generally included in the genus *Grevillea*. They are shrubs chiefly of West Australia, with variously cut and lobed leaves.
[A. A. B.]

ANADYOMENE. A genus of calcareous green-spored *Algæ*, consisting of branched or dichotomous articulated threads, which are laterally confluent with each other, so as to form a more or less fan-shaped membrane. They are most beautiful objects under a low power of the microscope, and might suggest elegant designs for the silversmith. The species, which are few in number, occur in the warmer seas, and consequently we have none upon our own coast. Some species of *Cladophora*, of which we possess so many, give a good idea of the nature of the filaments. [M. J. B.]

AN.ECTOCHILUS. A genus of terrestrial orchids, nearly allied to *Goodyera* and *Ætheris*, with creeping slender-jointed rhizomes, one or two radical leaves, and spikes of white or yellow flowers. Some of the species have the leaves traversed by beautiful silver or golden veins, on a rich green or purplish ground; hence they have become favourites in the gardens of the curious. There are other tropical terrestrial orchids, with similarly veined leaves, belonging to other genera, one of the commonest of which is *Physurus pictus*.

ANAGALLIS. An interesting genus of primworts, consisting of dwarf trailing herbaceous plants, annual or perennial roots, angular stems, opposite leaves in pairs or threes, and pretty axillary blue or red flowers. The Pimpernels, by which name the species are popularly known, are easily distinguished from the rest of the order by their conspicuous wheel-shaped, five-parted corolla, five stamens with bearded filaments, and especially by their many-seeded globular pods, opening when ripe by a transverse fissure all round, the top falling off like a lid. With the exception of *Centunculus*, an obscure weed, no other genus of the order presents this feature. Every one is familiar with the common red Pimpernel (*A. arvensis*). The *A. indica*, with blue flowers, scarcely differs from it, except in colour and the larger size of its blossoms. Of much greater interest are the Italian Pimpernel (*A. Monelli*), with still larger flowers and of stronger habit, and the *A. fruticosa*, a barbary species, with handsome light red tinted corollas, which, unlike those of the other species, remain expanded even in the absence of sunshine. It is probable that most of the varieties now cultivated in gardens, among which may be mentioned *A. Phillipsii*, with deep blue, and *A. Parksii*, with red flowers, are either hybrids or mere semiual variations of these two species. [W. T.]

ANAGYRIS. A genus of the pea-flower tribe, of the leguminous family. One species only is known, a large bush, with trifoliolate leaves, entire elliptical bracts, and axillary racemes of yellow flowers, much like those of the Laburnum. The pod is narrow, compressed, and curved backwards; and from this circumstance the genus is named— *ana*, signifying backwards, and *gyros*, a circle. It is found in the South of France, Spain, and other countries bordering on the Mediterranean Sea. The seeds are kidney-shaped, violet in colour, and are said to be poisonous, like those of the laburnum. [A. A. B.]

ANALOGY. Resemblance to a thing in form, but not in function; or in function, but not in form. Corresponding with a thing in many points, but differing in more, or in points of more importance. Thus, the flowers of *Potentilla* and *Ranunculus* are analogous.

ANAMIRTA. A genus of plants inhabiting Malabar, Ceylon, and the Eastern Isles of India, and belonging to the *Menispermaceæ*. The flowers are unisexual and dioecious—that is, the male blossoms are borne on different plants from the female blossoms. The sepals are six in number; there are no petals; the stamens are numerous, but united into one parcel, forming a globular head. The ovaries of the female flower are three in number, attached to a short thick hemispherical receptacle; they become succulent and drupe-like in the fruit. The most important, if not the only plant of this genus, is the *A. Cocculus*, the plant which produces the fruits known as *Cocculus indicus*. It is a climbing plant, with ash-coloured, corky bark. The leaves are stalked, more or less heart-shaped, smooth above, pale beneath, and provided with tufts of hairs at the junctions of the nerves, the larger of which radiate from the base of the leaf. The flowers are borne in pendulous panicles. The fruits are roundish or kidney-shaped; the outer coat is thin and dry, of a dark brown or black colour and wrinkled appearance; within this is a white hard shell, divided into two pieces; this encloses the whitish seed, which is very oily, of a crescent-like shape, and much smaller than the fruit, so that it never entirely fills up the cavity.

Cocculus indicus is imported from the East Indies, and is used for adulterating porter, though, very properly, a heavy penalty is inflicted upon brewers detected in so doing, and upon druggists who supply brewers, as it contains an acrid irritant poison, called picrotoxin. It is used to poison fish, and to increase the intoxicating properties of porter, being employed in the shape of a black extract. Its effects are to produce giddiness, convulsions, and insensibility. It has been occasionally used externally to destroy vermin, and in some skin diseases. [M. T. M.]

ANANASSA. A genus of tropical *Bromeliaceæ*, having rigid foliage, with sharp spines along the edge, distinguished among the inferior-fruited genera of the order, chiefly by its berries being consolidated with the bracts into a compound or syncarpous fruit, which is edible.

The Pine Apple, *A. sativa*, is generally believed to have derived its name from the great resemblance which the fruit bears in its form to the cone of some species of the pine or fir tribe. It is universally acknowledged to be one of the most delicious fruits in existence. Three hundred years ago it was described by Jean de Lery, a Huguenot priest, as being of such excellence that the gods might luxuriate upon it, and that it should only be gathered by the hand of a Venus. It is stated to be a native of Brazil, and having been carried from thence to the West, and afterwards to the East Indies, cannot be regarded as indigenous to the tropical parts of Asia, Africa, and South America. It first became known to Europeans in Peru, where it is called *Nanas*, and under this name it was described in 1555 by Andre Thevet, a monk, who says it was often preserved in sugar. The plant is biennial, not unlike an *Aloe*, but the leaves are much thinner, and of a hard fibrous texture, with numerous short sharp spines on the edges. The fruit is produced on a short stem which rises from the centre of the plant, and bears a scaly central spike, surmounted by a number of small spiny leaves called the crown. This conical body, after flowering, gradually enlarges and eventually becomes the rich and succulent Pine Apple we so highly prize. Besides being the first of dessert fruits, it is made into marmalades and various confectioneries, and is used to flavour rum. The earliest account of Pine Apples being seen in England, is that of a present of some having been received by the Protector Cromwell. We next find them noticed by the celebrated Evelyn, from whose Diary we subjoin the following extract:—'August 9, 1661. I first saw the famous Queen Pine brought from Barbados and presented to His Majesty' (Charles II.); again under date of July 19, 1668, he observes, 'I was at a banquet which the King gave to the French Ambassador. Standing by His Majesty at dinner in the presence, there was of that rare fruit called the King Pine, growing in Barbados, in the West Indies. The fruit of them I had never seen. His Majesty cutting it up was pleased to give me a piece from his own plate to taste of; but in my opinion it falls far short of those ravishing varieties of deliciousness ascribed to it.' It has been conjectured that from the crowns of these Pines, Mr. Rose, the royal gardener, succeeded in raising plants, and that one of the latter might have produced the fruit he is represented, in a well-known picture, as presenting on his knee to King Charles II. as the first Pine Apple grown in England. It is just possible that such might have been the case, but, unless in the picture above alluded to 10f which a copy is in the possession of the Horticultural Society), we have no evidence to show that the Pine Apple was then cultivated in the royal gardens, or at any other place in this country, until many years afterwards. For its introduction into Europe we are indebted to M. Le Cour, a Dutch merchant, who about the middle of the seventeenth century made an attempt to cultivate it in his garden at Driehock, near Leyden. After a great many trials he at last hit upon a plan by which he obtained a sufficient degree of heat to produce fruit equally good, though not so large, as that produced in the West Indies. According to the best authorities, the first plants introduced into England were brought from Holland by the Earl of Portland in 1692. Twenty years afterwards we find Pines successfully cultivated by Sir Matthew Decker, in his garden at Richmond; and to this gentleman the honour has usually been ascribed of having first fruited the Pine Apple in Britain, about the year 1712. From that time to the present every possible means that art and ingenuity could devise for the culture of this fine fruit have been adopted, and in no other instance, perhaps, has the care and skill of the gardener been attended with more signal success. Pine Apples having been produced in this country far surpassing in size and flavour the very best of those matured in a tropical climate. The difficulties which formerly attended the cultivation of the Pine Apple have disappeared since the mode of heating hot-houses with hot water was introduced, and handsome fruit weighing from six to twelve pounds, are by no means uncommon; but the heaviest on record, we believe, was grown in 1826 by Dixon, gardener to John Edwards, Esq., Rheola, Neath, Glamorganshire, and weighed fourteen pounds twelve ounces! The most remarkable experiment, however, that has been made in pine growing was one by Barnes, gardener to Lady Rolle, at Bicton, in Devonshire, who, in September 1845, cut some excellent fruit of four and five pounds weight from plants that had been exposed in the open air during the whole of the summer. Pine Apples are no longer a novelty, large quantities being annually imported and sold at a cheap rate in the principal towns throughout the kingdom. Like most of our cultivated fruits, they vary in quality and appearance; no less than fifty-two sorts being described in the *Transactions of the Horticultural Society* (2 ser. i. i). The greater number have been introduced from abroad; but several have originated from seed in England. That which is now so commonly imported from the Bahamas is a sort called the Providence, one of the least valuable of the race.

[W. B. B.]

ANANDRÆ. A name sometimes given to Cryptogams on the supposition that they have no male organs. See ASEXUAL PLANTS. [M. J. B.]

ANANDRIA. A genus of the composite family, and of the tribe *Mutisiaceæ*, in which the florets are two-lipped. *A. Bellidiastrum* is a stemless herb of Siberia and Japan, having rosettes of stalked lyrate toothed leaves, covered with white down underneath, and arising from their midst a flower scape bearing a single head about half an inch across, containing numerous white or purple florets. When the plant is in flower the leaves are seldom more than three inches long, but, when mature, are from five to six inches. The flower scape also lengthens after the flowers wither, and is often upwards of a foot in length. The naked receptacle and broad-beaked achenes terminating in a small cavity, are the characters which distinguish the genus from *Peridicium*, to which this plant was formerly referred. It has been cultivated in this country, and is the only species of its genus. [A. A. B.]

ANAPAUSIA. A genus of polypodiaceous ferns belonging to the *Acrosticheae*, among which they are distinguished by having a portion of their fronds wholly fertile, and the veins of their fronds compoundly reticulated, with free divaricate veinlets in the areoles. The species are mostly found in the West Indies and South America, and form coarse-growing herbaceous plants with compound fronds, a portion of which are entirely sterile, and the remainder somewhat contracted and covered with the fructification. *A. vespertilio*, a Javanese species, has coriaceous lunately bilobed sterile fronds, and linear-lanceolate fertile ones. [T. M.]

ANARRHINUM. A genus of biennial, or perennial herbaceous plants, natives of Southern Europe, Northern Africa, and Syria, belonging to *Scrophulariaceae*, and containing seven species. The radical leaves are generally rosulate, the cauline opposite or alternate, sometimes both on the same plant. The flowers are small in spike-shaped racemes. The calyx is deeply five-fid, the corolla tubular and bilabiate, the upper lip erect, then reflexed, the lower patent. The stamens are included, four being fertile, and the fifth sterile and undeveloped. The ovary is bilocular, with many ovules. [W. C.]

ANARTHRIA. A genus of *Restiaceae*, containing five species of perennial plants, with flattened simple or branched stems, indigenous to the eastern shores of New Holland. The flowers are dioecious, and have six glumes. The male flower has three free stamens, with bilocular anthers; the female has three styles. The capsule is three-lobed and tri-locular, with a single seed in each loculament. [W. C.]

ANASARCA. A condition of plants analogous to dropsy, though not always attended by extravasation. In extremely wet weather the tissues get gorged with fluid, and as the vegetative powers are generally lowered by the decrease of temperature, the contents of the cells are badly supplied, and, in consequence, their walls, unconsolidated, become subject to decay, which is soon exhibited in a variety of untoward symptoms. Fruit, in consequence, which has been produced in a wet season, is notoriously subject to decay, except compensated, as in the late unusual summer, by a high state of the hygrometer, a circumstance which may perhaps account for the extremely small quantity of decay which has been experienced in our autumnal fruits. In some cases, as in elms, there is sometimes direct extravasation, and then the fluid accumulates, and at length forces its way through the bark, producing permanent ulcers. [M. J. B.]

ANASTATICA. A genus of *Cruciferae*, consisting of a single species, the Rose of Jericho (*A. Hierochuntina*), a small annual growing in the arid wastes of the extra-European Mediterranean region, from Syria to Algeria. The stem is short, branched in a corymbose manner at the top; the leaves obovate, with stellate hairs, the lower ones entire, the upper remotely toothed; the flowers are small and white, forming spikes along the branches; the fruit is a short pouch, with two cardike projections at the top, and divided by a transverse partition within into two cells, in each of which there is a seed. This plant is interesting on account of its hygroscopic properties: when the plant is in flower, the branches spread rigidly, but when the seed ripens, the leaves wither and drop, and the whole plant becomes dry, each branch curls inwards, until the plant presents the appearance of a little ball of wickerwork at the top of the unbranched part of the stem. In this state it is soon loosened from the soil, and carried about by the wind, and often blown into the sea: when this happens, or the plant is otherwise wetted, the branches unbend, and the pods begin to open by splitting longitudinally, so that, when thrown on shore by the waves, the circumstances are favourable for the production of fresh individuals in a locality remote from the original place of growth. The plant retains its property of expanding when moistened, and again curling up when dry for a long time. Specimens, collected ten years ago, exhibit the phenomenon as perfectly as ever. In Palestine it is called 'Kaf Maryan,' or Mary's Flower; and there is a tradition that the plant expanded at the birth of the Saviour. [J. T. S.]

ANASTOMOSIS. The angle formed by the union of veins, or of their branches.

ANATHERUM. A group of grasses, the species of which are now included in the genus *Andropogon*. [D. M.]

ANATROPA. A generic name given to a small, succulent, herbaceous plant from Syria, belonging to the natural order *Elatineaceae*. Except that it has stipules, it differs in no respect from *Tetradiclis*, to which it is consequently generally referred. [W. C.]

ANATROPAL. When an ovule is turned down upon itself, so that the foramen, or true apex, points to the base, and the chalaza is at the apex.

ANBURY. A gouty nodular condition of certain roots, as turnips, arising from the presence of grubs. It must not be confounded with dactylorhiza, which is a very different affection, and entirely independent of the attacks of insects. [M. J. B.]

ANCHIETEA. P. Anchietea, a Brazilian writer on plants, is commemorated in the name of this genus of violetworts, *Violaceæ*. The species are shrubs, with undivided leaves, and small white flowers in axillary tufts. The calyx consists of five unequal divisions, not prolonged at the base; petals, four, unequal, the hindermost one large, and prolonged into a spur; filaments very short; anthers fixed together, membranous at the top, two of them prolonged by means of their connectives into the spur of the petals. Fruit large, capsular, membranous, and bladder-like. Of *A. salutaris*, a creeping bush, with a nauseous taste, and a smell of cabbage, the root is used as a purgative by the Brazilians, and as a remedy in skin diseases. *A. pyrifolia* is an ornamental stove-creeper, with white flowers. [M. T. M.]

ANCHUSA. A genus of *Boragineæ*, consisting of rough or hispid plants, most of which are natives of Southern Europe and the East, resembling *Lithospermum* and *Alkanna*, but with the nuts which form the fruit hollowed out at the base, and the corolla, which is from funnel-shaped to salver-shaped, closed by five scales at the throat; flowers purple, blue or yellowish, in scorpioid racemes, which are generally bracteated, and in pairs. Three species occur in Britain, but two of them appear to be doubtful natives. The evergreen Alkanet, *A. sempervirens*, has broad ovate leaves, those of the root large, and on long stalks; with sky-blue flowers in short twin racemes on short stalks, from the axils of the leaves. The common Alkanet, *A. officinalis*, has narrow oblong leaves, and deep purple flowers, in several racemes, at the top of the stems. The small Bugloss, *A. arvensis*, is often separated from the other species, under the name of *Lycopsis arvensis*, on account of the tube of the corolla being curved; it is a very bristly annual, with small pale blue flowers, and narrow oblong leaves, and is a common weed on cultivated ground. The first two species are often cultivated in gardens, but the biennial European *A. paniculata*, or Italica, is the most common in flower borders, as its flowers are as large as a fourpenny piece, and bright azure blue. *A. capensis* is from the Cape of Good Hope. *A. tinctoria*, the Alkanet, is now called *Alkanna tinctoria*. [J. T. S.]

ANCIPITOUS. Two-edged, as the stem of an Iris.

ANCISTROCLADEÆ. Under this name Planchon proposes to form a new natural order, out of the solitary genus *Ancistrocladus*—which see

ANCISTROCLADUS. A genus of climbing plants, inhabiting the East Indies. Its prominent characters are the branches, some of which are curved and hook-like; the alternate, stalked, leathery leaves unprovided with stipules; inflorescence a panicle; ten stamens in one row, five shorter than the others, all slightly adherent one to the other at their base; the anthers have a slightly prolonged crest at the top; the one-celled ovary has a single ovule at its base, and ripens into a sort of nut, crowned by the persistent calyx. The genus is by Planchon referred to a new order, *Ancistrocladeæ*, more nearly allied to *Dipterocarpaceæ* than to any other group. "It combines," he says, "with the venation of *Nepenthes*, the leaves of *Lophira*, the stamens and staminodia of certain malpighiaceous plants, the adherent calyx of *Dipterocarpus*, the adherent ovary of *Symplocos*, the hook-like branches of *Hugonia*, and a peculiar fleshy fungus-like embryo, with a cylindrical rather long radicle, and a disc-shaped cotyledonary mass." [M. T. M.]

ANCOLIE. (Fr.) *Aquilegia*.

ANDER. In names formed from the Greek—the male sex or stamens. *Monander*—having one stamen.

ANDERSONIA. A genus of squarrose-leaved shrubs, belonging to the natural order *Epacridaceæ*, containing several species, remarkable for the great beauty of their flowers and the singular structure of their leaves. They are natives of New Holland. The flowers are terminal, and solitary or in spikes. The calyx is coloured, five-partite, with two or more bracts. The corolla is subcampanulate, hypocrateriform, and five-lobed, the lobes bearded at their base. The five hypogynous stamens just appear beyond the throat of the corolla. The existence of a nectary, consisting of five scales, separates this genus from *Sprengelia*, with which it is otherwise identical. The ovary is five-celled, with many ovules in each cell. The fruit, a berry, is esculent. [W. C.]

ANDIRA. A genus of the pea-flower tribe, of the natural family *Leguminosæ*. About twelve species are known, all of them trees of moderate height, with alternate unequally pinnate leaves, about one foot long, of five to ten pairs of leaflets. The stipules are sometimes large and persistent, or they are small and fall early. The flowers are often showy, and are disposed in axillary or terminal panicles—the reddish lilac of their petals contrasting well with the often dark purple branches of the flower-stalks. The pod is one-seeded, drupaceous, and somewhat like a plum in appearance. All the species are natives of tropical America, but a variety of *A. inermis* is found in Sierraambola. This species is called in the West Indies the Cabbage-tree. Its bark is anthelmintic, but requires great care in its administration, being powerfully narcotic. It has a sweetish taste, but a disagreeable smell, and is given in the form of a powder, decoction, or extract. The decoction is generally preferred, and is made by boiling an ounce of the dried bark in a quart of water, until it assumes the colour of Madeira wine. The effects of an over-dose are vomiting, delirium, and fever; the antidote for this is lime-juice or castor oil. The powder, administered in doses of three or four grains, purges like jalap. The bark is known as Bastard Cabbage Bark, or Worm Bark; formerly it was used as a medicine in English practice, but its use is now obsolete. *A. refusa*, a Brazilian species, has purple flowers, having an odour of oranges, with a slight aroma: the fruit is said to smell like the tonka bean. Most of the species are beautiful objects when in flower. [A. A. B.]

ANDRACHNE. A genus of spurgeworts, (*Euphorbiaceæ*) comprising ten species; distributed over the Mediterranean region, N. India and China, one species occurring in Arkansas. They are perennial herbs of no beauty, with erect or prostrate stems, furnished with alternate, shortly-stalked leaves, having rounded or ovate and entire blades of a pea-green colour, varying from a quarter of an inch to two inches in length; while the small greenish flowers —which are sterile and fertile on different plants—are borne singly, or two to four together, in the axils of the leaves. The sterile flowers have a calyx of five to six sepals, a like number of petals, five entire or bifid glands opposite the petals, and five stamens, slightly united below by their filaments, and surrounding an abortive ovary. The fertile flowers have a like calyx and corolla, rather larger glands, and an ovary surmounted by a three-branched style, each of the branches forked. The fruits are trilobed capsules, of the size of peas, with three cells, and two seeds in each cell. [A. A. B.]

ANDREÆA. A genus of mosses named after Andreæ, a Hanoverian Doctor. It is remarkable for having a capsule which splits into four or sometimes six valves, which, however, do not expand as in *Jungermanniæ*, but adhere at the apex to which the columella is attached. The capsule is always sessile, even to maturity, but is at length supported by the elongated base or pseudopodium. In consequence of this arrangement, the calyptra does not burst as soon as in most other mosses, as it is ruptured by the swelling of the capsule and not by the elongation of the pedicel. The species are Alpine or sub-Alpine, occurring on quartzose or granitic, never on calcareous rocks, and are found in cold or temperate regions of either hemisphere, or at considerable heights in warmer countries, as in the Himalayas. Four species in which the leaves have a central nerve, occur in this country, while in the others the leaves are nerveless. [M. J. B.]

ANDREÆACEÆ. A natural order, or, according to some, a distinct tribe of mosses. They do not, however, differ essentially, being distinguished mainly by the longitudinal splitting of the valves at maturity. It consists of but two genera, *Andreæa* just described, and *Acroschisma*, an Antarctic genus, distinguished by the cylindrical capsule splitting into four or eight valves at the apex only. All the species are of a peculiar dark hue, and the leaves, which are of a close texture, are of a beautiful yellow or golden brown under the microscope. [M. J. B.]

ANDROCYMBIUM. A genus of *Melanthaceæ*, containing three or four species from the Cape of Good Hope. They are plants with tunicated bulbs, and simple, short, subterraneous stems, crowned with from two to four ovate, lanceolate, or linear leaves, and having one to nine flowers in short spikes, hid in coloured foliaceous bracts. [W. C.]

ANDRŒCEUM. The male system of a flower. The stamens taken collectively.

ANDROGLOSSUM. A genus founded on a single species, *A. reticulatum*, a native of the Island of Hongkong. It seems to belong to the natural order *Ebenaceæ*; the arrangement of the stamens being precisely the same as in that order; but it differs from it in the structure of the ovary, the carpels being almost if not quite distinct, and the ovules, two in each carpel, horizontally attached to the axis, instead of being solitary and erect from the base. The calyx is five-partite; the corolla five petaled; the five stamens are opposite to and inserted in the petals; there are two styles. The fruit from the abortion of one of the carpels, is simple, spherical, and unidrupaceous, with a crustaceous covering. [W. C.]

ANDROGYNOUS. A term applied to such kinds of inflorescence as consist of both male and female flowers.

ANDROMEDA. A genus of *Ericaceæ*, consisting of shrubs and trees having various habits, and a wide geographical range; but found chiefly in boreal districts, or at considerable heights on mountains in North America, Europe, and Asia. The genus has a calyx of five nearly or partly distinct sepals, valvate in the early bud, but very soon separate or open. The corolla is ovate or campanulate, five-toothed, and deciduous. The stamens are ten in number, with the anthers fixed near the middle; the cells generally opening by a terminal pore. The ovary is five-celled, with many ovules in each cell. The style is simple. The fruit is a dry capsule, superior, globular, five-celled and five-valved, and loculicidal, the dissepiments being from the middle of the valves. The genus is by some modern botanists limited to the single British species, *A. polifolia*; but it is usually extended to include a very large number of species. So extended, it is divided into the following sub-orders:—1. *Andromeda proper*, boreal herb-like plants, with calyx five-cleft, corolla sub-globose, filaments bearded, anthers having a slender ascending awn, and seed smooth. 2. *Cassiope*, Arctic and Alpine under-shrubs, with calyx five-parted, without bracts, corolla campanulate, anthers fixed by the apex, and having a long recurved awn, stigma truncate, and seeds smooth and wingless. 3. *Cinnamomea*, boreal shrubs, with calyx five-parted, and having two bracts, corolla tubular, anthers without awns, and seeds flattened and wingless. 4. *Zenobia*, North American shrubs, with calyx five-toothed, corolla campanulate, deeply five-parted, anthers two-awned, and seeds angular. 5. *Leucothoe*, North American shrubs, with calyx five-parted, corolla tubular, five-toothed, anthers naked, or with one or two awns, and stigmas broadly capitate. 6. *Pieris*, Nepal trees or shrubs, with calyx five-parted, corolla tubular or ovate, anthers two-awned, and stigmas truncate. 7. *Agarista*, tropical American evergreen shrubs, with calyx five-parted, corolla ovate, anthers with two bristles, stigmas capitate, and seeds angular.

The members of this genus are more or less narcotic. *A. polifolia*, the only British species, but found also in peat bogs throughout the north of Europe, Asia, and America, is an acrid narcotic, and proves fatal to sheep. Similar properties have been observed in the United States in *A. mariana* and other species. The shoots of *A. ovalifolia* poison goats in Nepal. Dr. Horsfield states that a very volatile heating oil, with a peculiar odour, used by the Javanese in rheumatic affections, is obtained from one of the species. [W. C.]

ANDROPHORE. The tube formed by monadelphous filaments, as in mallow.

ANDROPOGON. An extensive genus of grasses, typical of the tribe *Andropogoneæ*, which, according to Steudel's *Synopsis Plantarum Graminearum*, contains 438 species. The flowers are polygamous. The species are mostly natives of the warmer parts of the globe, especially South America; none of them British. Among the more interesting kinds, the Sweet-scented Lemon Grass, *A. Schœnanthus*, may be mentioned, which is a native of Malabar, and well-known in British gardens as a stove conservatory plant. The fresh leaves, when bruised, emit a delightful odour, and when roasted are used in India for medicinal purposes. One of the grass oils is yielded by a species of this genus. (See *Brewster's Journal*, ix. 333.) *A. muricatus*, called in India khus, is employed there for making covers for palanquins, and screens, &c. (see *Lindley's Vegetable Kingdom*, p. 115)—The roots are woven neatly into screens or mats, and suspended before the doors or windows, so that the breeze, in passing through them, is cooled and receives a portion of its healthy elasticity, while a slight but very agreeable fragrance is diffused around. Mrs. Calcott in the *Scripture Herbal*, considers *A. Calamus aromaticus* to be the Sweet Cane of Isaiah, the Sweet Calamus of Exodus, the Calamus of the Canticles and of Ezekiel. Dr. Royle also considered the plant of that name described by Dioscorides to be the Sweet Cane and 'rich aromatic reed from a far country' of Scripture. Steudel does not enumerate this species under the genus of *Andropogon*, nor give any synonym of it. [D. M.]

ANDROSACE. Mountain plants, with flowers much resembling the primrose, from which they differ principally in having the mouth of the corolla contracted. They are found on the mountains of Europe, from Siberia to the Pyrenees, and are well adapted for growing on rockwork, though not easy of culture. The leaves, which vary in shape in the different species, are tufted, and grow close to the ground; the flowers are either white or pink, and grow on a scape of leafless stalk, in umbels; in one species, *A. Vitaliana*, they are solitary. Some of the species are annuals, and some perennials; one only is a biennial. They vary in height from two to six inches. [C. A. J.]

ANDROSÆMUM. Under the name of *A. officinale*, the Tutsan, *Hypericum Androsæmum*, is sometimes separated from the rest of the *Hypericeæ*, from which it differs chiefly in having a berry-like capsule; but the genus is not generally adopted. [T. M.]

ANDROSTEMMA. A genus of *Hæmodoraceæ*, nearly allied to *Conostylis*, containing a single species, from Swan River Colony, New Holland. It is a rushy plant, of no beauty, although its flowers are an inch and a half long, for they are green, and buried among the leaves. [W. C.]

ANDROUS, in the composition of names derived from the Greek, refers to the stamens; thus, *monandrous* signifies having one stamen, &c.

ANDRYALA. A family of evergreen herbaceous plants, belonging to the natural order *Compositæ*, growing to the height of about half a foot, and having yellow flowers. Two species are found in dry stony fields about Nice, near the Var; the others are natives of Madeira and the North of Africa, and, when cultivated in this country, require the protection of a greenhouse. [C. A. J.]

ANEE. (Fr) *Alnus glutinosa.*

ANEILEMA. A genus of *Commelynaceæ*, the flowers of which have a six-parted unequal perianth, the three outer divisions of sepals persistent, and the three inner deciduous; six stamens, of which three are dissimilar, scarcely polliniferous; no involucre. It is principally distinguished from *Commelyna* by the latter circumstance. There are several species, natives of New Holland and India. [T. M.]

ANEMIA, often erroneously written *Aneimia*. A genus of schizæaceous ferns, belonging to the section *Schizæeæ*, in which it is distinguished by having the fructifications paniculate on distinct fronds, or on lateral branches of the fronds, and the veins free. The separate branches of fructification produce the appearance of a flowering plant, with a spicate inflorescence. There are numerous species of South America and the West Indies, some of which are of a very ornamental character, and much prized in gardens. One of the most beautiful is *A. adiantifolia*, a species having the barren branch triangular and tripinnatifid, and the two fertile branches erect, rising from its base, and bearing a cylindrical spike of small fertile segments. Several species, with a similar arrangement of the parts, have the sterile branch pinnate. Other species, as, for example, *A. millefolia* and *A. bonifolia*, have the fertile parts distinct, rising from the base. One species, *A. Wightiana*, is found on the Neilgherry Hills of India, and another, *A. Dregeana*, is met with in South Africa and Natal. [T. M.]

ANEMIDICTYON. A genus of schizaeaceous ferns, distinguished from *Anemia* only by the reticulated venation of its fronds. It is consequently included in that genus by those who do not admit the generic importance of the venation in ferns. The principal species, *A. Phyllitidis*, occurs in various forms in the West Indies and South America, and is a fine herbaceous species, with pinnate sterile branches, and tall, compactly-panicled fertile ones. [T. M.]

ANEMIOPSIS. A genus of the small family *Saururaceae*, peculiar to California and New Mexico, and represented by a single species, *A. californica*, a semi-aquatic perennial herb, with stalked and nearly smooth root-leaves, like those of the sorrel, *Rumex Acetosa*, but of a much thicker texture. The flower scape, nearly a foot in length, and exceeding the root leaves, bears near its middle a leafy bract (which often produces in its axil a young plant), and terminates in a compact cone of small green flowers, surrounded by an involucre of six oblong petal-like leaves of a white colour, spotted with red, so that the whole head has some resemblance to an anemone flower. The flowers are destitute of calyx and corolla, have six to eight stamens seated on the top of an ovary, which is one-celled, with three bundles of ovules hanging from the top of the cell, and crowned with three short styles. The plant is in cultivation in this country. [A. A. B.]

ANEMONE. A large genus of *Ranunculaceae*, generally distributed in temperate regions — most numerous in alpine situations in the warmer districts. They have tuberous or thickened root-stalks and root-leaves, often ternately divided or cleft. The stem, or rather scape, is leafless, and often unbranched, having an involucre below the flower, formed by a whorl of three (rarely two) bracts; when the scape is unbranched, there is only one involucre, when branched, each flower has one, and the branches spring from the interior of the involucres, together with the peduncle, which bears the central flower. The flowers are handsome, for, though the petals are absent in single flowers, the sepals are brightly coloured, especially on their inner faces. The flowers very readily become double by the conversion of the numerous stamens into narrow petals; this is often seen in gardens, but occurs even in wild plants of some species found in the South of Europe — a very unusual circumstance.

The genus forms three groups, or sections. (1) *Anemone proper*, or *Anemanthus*, of Endlicher, in which the carpels terminate in a short point (not a feathery tail), and the involucre is remote from the flower. (2) *Hepatica*, with the carpels as in *Anemanthus*, but the involucre close to the flower, resembling a calyx. (3) *Pulsatilla*, in which the carpels end in a long feathery tail, formed by the persistent styles, which elongate after the flower fades. Of the first section, three species occur in Britain. The Wood Anemone, *A. nemorosa*, is the only one truly native; it has white flowers, sometimes tinted with purple on the outside. *A. ranunculoides*, a common European plant, naturalised in a few stations in Britain, has bright yellow flowers, otherwise like the wood Anemone. *A. apennina*, a native of Southern Europe, also naturalised in a few British localities, has the flowers bright blue on the inside of the sepals, which are narrow, and more numerous than in the other two; the root-stalk is also shorter and thicker. The last two are often cultivated in gardens, as well as the more showy Japan Anemone, *A. japonica*, which has ternate leaves, branched flowering stems, and large purplish-red flowers. The Star Anemone, *A. hortensis*, or *stellata*, has ternate leaves, with the segments not finely divided, unbranched flower-stalks, and star-like flowers, smaller than those of the Japan Anemone and very variable in colour; and the Poppy Anemone, *A. coronaria*, which, like the last, is a native of the Mediterranean region, has ternate leaves, with the divisions cut into fine segments, unbranched flower-stalks, and large flowers, with broad sepals, very variable in colour — scarlet, purple, blue, whitish, striped, or with an eye of a different hue from the rest of the flower. The last two, and especially the Poppy Anemone, are florists' flowers. [J. T. S.]

ANEMONE, RUE. *Thalictrum anemonoides*.

ANEMONOPSIS. A genus of *Ranunculaceae*, containing a single species, *A. macrophylla*, a native of Japan. It is allied to *Helleborus*, having three to five follicles to form the fruit. The flower, however, resembles in aspect that of an *Anemone*, whence the name. The calyx has nine sepals; the corolla ten petals; and the leaves are three or four times ternately divided, resembling an *Actaea*. [J. T. S.]

ANEMOSIS. The condition known in timber by the name of wind shaken. A trunk which is apparently sound externally, proves, when felled, to have given way in the direction of the concentric layers of which it is composed, so that the connection between them is more or less completely broken. This occurs in many kinds of exogenous timbers, and is no less common in foreign woods than in those of native growth, being, as it is supposed, due to the pressure of extremely violent gales. This, however, is very doubtful, the effect being more probably due to frost or lightning. Wind, however, may be injurious to trees without producing absolute fractures or separation of parts, by causing too rapid evaporation, and in consequence chilling the tissues to such a degree as to retard development, or induce an unhealthy condition, or temporary sterility. [M. J. B.]

ANESORHIZA. A genus of *Umbelliferæ*, containing seven or eight species of biennial or perennial herbaceous plants, with one or more fusiform roots: natives of the Cape of Good Hope. The root of *A. capensis*, known vulgarly as Ang-wortel, is used as an esculent. The radical leaves are petiolate, and two or three times pinnatisect; those of the stem are scale-like. The umbels are few or many radiate; in some species being as few as three, in others as many as twenty-eight radiate. The limb of the calyx is five-toothed and persistent. The petals are elliptical and acuminate. The involucres and involucels are many-leaved, the margins of the leaves being often scarious. [W. C.]

ANETHUM. The name applied to a genus of umbelliferous plants, which is distinguished by the absence of involucre to the umbel, by the absence of the limb or upper part of the calyx, by the fruit being flattened from back to front, provided with a membranous border or wing, and with six ridges, three on each half of the fruits. In each of the furrows, between these ridges, is placed a broad channel, or vitta, filled with volatile oil. The Common Dill, *A. graveolens*, which in appearance resembles the fennel, is cultivated in herb gardens in this country for the sake of its fruits, and is a native of the south of Europe, Egypt, the Cape of Good Hope, &c. The fruits, or as they are commonly but erroneously called, the seeds, when distilled with water, furnish an oil on which the carminative effects of the plant depend. It is generally used in the form of dill water, to relieve flatulence in children, and to prevent the griping properties of some purgative medicines. The plant and the fruit are used as condiments in the East. It is supposed to be the plant which is called Anise in the New Testament narrative. [M. T. M.]

ANETIA. A genus of *Homalineæ*, containing a single species, from tropical Africa. It is a shrub with alternate shortly petiolate leaves, and small cinereous flowers in branched spikes. It is nearly allied to *Homalium*, but differs from it in having a double series of stamens and glands, fifteen of each, that is three times the number of the calyx segments, and in having five diverging styles. [W. C.]

ANETIUM. A genus of polypodiaceous ferns, sometimes referred to the *Acrostichæ*, but more closely allied to the *Hemionitideæ*, among which it is distinguished by having the veins of the fronds uniform, reticulate, and the sori sporadic, or dispersed, sometimes reticulated following the veins, and in some places distributed on the surface, but everywhere partial, as if scattered. The species *A. citrifolium*, is a West Indian and South American plant, producing narrow ish simple fronds. [T. M.]

ANETTE. (Fr.) *Lathyrus tuberosus*.

ANEURA, ANEUREÆ. The name of a division and genus of frondose *Jungermanniaceæ*. *A. multifida* and *A. pinguis*, found on the margins of ponds, on the walls of wells, and in similar damp situations, associated with *Pellia*, differ from that genus in having more divided and irregular ribless fronds, with the fruit marginal and ventral, the capsule oval or oblong, and the sisters attached to the tips of the valves. *Aneura* is the only genus of the division *Aneureæ*, a name indicative of the ribless fronds. [M. J. B.]

ANFRACTUOSE. Twisted or sinuous, like the anther of a cucumber.

ANGELICA. A family of umbelliferous plants, the several species of which inhabit Europe and America. The name was given as a record of the angelic virtues possessed by some of the species; for not only was it a singular remedy against poison, the plague, all kinds of infection, and malaria, but it was invaluable against witchcraft and enchantments. The British species, *A. sylvestris* is a tall and stately plant, five or six feet high, with a polished stem, most frequently tinged with purple and covered with a glaucous bloom like that of a plum, much branched, bearing large compound leaves covered with a bloom like that of the stem, and at the extremity of each branch a large convex umbel of white flowers tinged with pink. It is common on the banks of rivers in withy holts and other watery places, and may often be distinguished at a considerable distance by the large egg-shaped expansions of the leaf stalks, which serve as an instrument for the as yet unsuspended flowers. The plant is now little regarded for any intrinsic virtues it may possess; but it forms a picturesque addition to the landscape, and may be made very useful in the garden by cutting the hollow stalks into convenient lengths and placing them about in the shrubs as traps for earwigs. Candied Angelica is made from the stalks of an allied plant, *Archangelica*. The appearance of a tall species of this genus growing in Kamtchatka is shown in plate 16. [C. A. J.]

ANGELICA. The garden name for *Archangelica officinalis*.

ANGELICA TREE. *Aralia spinosa*.

ANGELICO. *Ligusticum actæifolium*.

ANGELIQUE. (Fr.) *Archangelica officinalis*. — DE BOHEME. *Trochiscanthes nodiflorus*. — EPINEUSE. *Arabia spinosa*.

ANGELONIA. A genus of Scrophulariaceæ distinguished by its five-cleft or five-parted calyx; its short-tubed corolla, with fornicate throat, and somewhat two-lipped limb; the upper lip very blunt, two-lobed, the lower one longer, three lobed with the middle lobe saccate at the base; four didynamous included stamens, and a two-celled many-ovuled ovary. The species, which are rather numerous, inhabit South America, and form erect or procumbent herbs, with opposite leaves alternate on the upper part of the stems, and solitary one-flowered axillary or racemose peduncles. *A. salicariaefolia* is a pretty perennial species, with tallish stems, lance-shaped leaves, and a long racemose inflorescence consisting of light purple flowers. *A. angustifolia* is another species of similar character, but dwarfer; it has smooth, narrow, lance-shaped leaves, and dense, terminal, erect racemes of deep violet-coloured flowers. The species have no particular use, but are ornamental. [T. M.]

ANGIOPTERIDEÆ. A section of the marattiaceous division of marattiaceous ferns, in which the spore cases are free, and set close together, face to face, in two opposite contiguous lines. [T. M.]

ANGIOPTERIS. A noble genus of marattiaceous ferns, representing the group *Angiopterideæ*. The genus is eastern, being common in India, Ceylon, and the islands of the Eastern Archipelago. There are probably but few species, though they have been very much extended by De Vriese, the author of a monograph of the family. The differences observed perhaps rather indicate varieties than species. The plants form a large round massive rhizome or rootstock, covered with the great scale-like bases of the fronds, and from this solid mass rise up the stout stipites, supporting the very large bipinnated fronds, the pinnules of which are articulated on the rachides. The genus is known by its spore-cases being destitute of any elastic or jointed ring, and by having oblong distinct dorsal sori longitudinally bivalved, the spore-cases being separate though crowded into two opposite linear series. The original species, *A. evecta*, was found in the Society Isles. *A. angustifolia*, a Philippine Island plant, is described as having a cylindrical caudex, three feet high; the other species, so far as they are known, have the caudex of a depressed globular form. These plants form noble objects when cultivated in our hot-houses, but require much space. [T. M.]

ANGIOSPERMS. In modern classifications all exogens are divided into those whose seeds are enclosed in a seed-vessel, and those with seeds produced and ripened without the production of a seed-vessel. The former are ANGIOSPERMS, and constitute the principal part of the species; the latter are GYMNOSPERMS, and chiefly consist of conifers and cycads.

ANGLE-POD. A common name for *Gonolobus*.

ANGOLA WEED. *Roccella furfuracea*.

ANGOPHORA. A genus of New Holland plants, belonging to the myrtle family, *Myrtaceæ*. They are large trees, with large, opposite, not dotted leaves. The flowers, which are of a white or yellow colour, are arranged in a corymbose manner, and have their calyx divided into five or six segments, a circumstance that serves to distinguish them from the members of the genus *Eucalyptus*; their petals are free, the stamens are distinct; the fruit is dry, dehiscent, many-celled, containing several seeds, which are not winged. Some of the species furnish a dark-coloured astringent gum. [M. T. M.]

ANGRÆCUM. The tropics of Africa, and its islands, some parts of the West Indies, and the Cape Colony, yield this remarkable genus of orchids, one of whose species produces the largest flowers known in the order — the sepals and petals and prodigious spur extending to the length of more than a foot in *A. sesquipedale*, a native of Madagascar. But although others approach this, yet the species are for the greater part small-flowered, and little better than weeds. A whole section of them consists of leafless plants, clinging to the branches of trees by their flat bands, representing roots, as in *Aerides funalis*. The finest species, after Madagascar, are from the West of Africa, where they assume some extraordinary forms. None are found except in the hottest latitudes.

ANGUILLARIA. A genus of *Melanthaceæ*, containing a few species from New Holland. They have the leaves and roots, and the general habit of *Melanthium*, with the structure of *Ornithoglossum*. [W. C.]

ANGULOA. A genus of very remarkable terrestrial orchids, inhabiting the forests of tropical America. They have broad, ribbed leaves, and short leafy scapes, bearing a single large fleshy flower, either white, yellow, or spotted with crimson, on a pale yellow ground. One of them grows in the Equator, at the height of 7,000 feet above the sea. Six or seven species are cultivated in this country.

ANGURIA. Under this name are included certain plants belonging to the gourd family, *Cucurbitaceæ*. They are natives of South America, and have lateral tendrils, male and female flowers distinct, but on the same plant; the male flowers provided with two distinct, not united stamens; and the fruit a gourd. Some of the species are cultivated. [M. T. M.]

ANHALONIUM. A genus of South American Cactaceæ, containing two species of napiform plants. The genus approaches *Mamillaria* in the arrangement and structure of its flowers, and has by some been made a section of that genus; but its fruit and seed unite it on the other hand with *Echinocactus*. [W. C.]

ANIA. A genus of terrestrial orchids of little interest, related to *Bletia*. They have plaited radical leaves, and flowers in spikes. Only two are known, inhabiting the hottest parts of Asia.

ANIBA. A Guiana plant, described by Aublet under this name, but which has not since been recognised. It is probably some genus of the Laurinaceous order.

ANIGOSANTHUS. A curious and handsome genus of *Hæmodoraceæ*, distinguished by having its woolly, tubular, elongated, often-curved perianth connate with the ovary, but at length deciduous; the limb six-cleft, and turned to one side; six stamens inserted in the throat; a filiform style and simple stigma; and a three-celled ovary, containing numerous ovules. The species, which are not very numerous, are herbs of the Australian continent, producing linear-ensiform leaves, slightly sheathing at the base, and a tall flower-scape, supporting a branching subcorymbose head, or short raceme of large and often showy flowers. The outer surface of the perianth, and the upper part of the flowering stem, are clothed with a peculiar short dense pile of branching coloured hairs, which are very curious objects when slightly magnified. Several of the species have found their way to our greenhouses, where they form desirable plants, on account of their distinct and peculiar, and not unornamental aspect when in flower. The flowers last a considerable time. One of the best-known species is *A. Manglesii*, a perennial tufted-growing plant, with glaucous green leaves, a foot to eighteen inches long, and an erect branched stem, clothed with a short thick crimson felt of the branched hairs above alluded to. The flowers are arranged on the branches in short terminal spiked racemes, and are two to three inches long, curved, clothed with velvety hairs, which, for the greater part of the length of the tube, are of a bright green colour, and on the peduncles, as well as the swollen base of the perianth enveloping the ovary, are rich crimson. Another species well known in gardens, *A. coccinea*, has a dichotomously-forked inflorescence, and flowers of a dull crimson below, and green towards the tips. In another beautiful species, *A. pulcherrimus*, in which the inflorescence is branched and loaded with flowers, the colour of the short velvety hairs on the flowers is bright yellow, while those on the stems are scarlet, curiously branched, on a yellow ground. *A. tyrianthinus*, again, has the paniculated branches and copious flowers clothed with dense tomentum of the richest Tyrian purple; while in *A. fuliginosus*, which has been called a flower of mourning, the upper parts of the stem, and the lower parts of the flower, are downy, as if covered with black velvet. These species are all from the Swan River district. [T. M.]

ANIME. A resin procured from *Hymenæa Courbaril*.

ANIS. (Fr.) *Pimpinella Anisum.* — **ÉTOILÉ**, or **DE LA CHINE**, *Illicium anisatum*. — **DES VOSGES**. *Carum Carvi*.

ANISADENIA. A genus of *Frankeniaceæ*, containing a single species, from Nepal, a plant having the appearance of *Triumfetta*, with a simple erect stem, bearing a number of alternate, entire, membranaceous leaves at its summit. The calyx and corolla consist of five parts. The petals are unguiculate. There are five filiform fertile, and five short barren stamens. The ovary is sessile, and trilocular, with two ovules in each cell. There are three filiform styles. [W. C.]

ANISE. *Pimpinella Anisum.* —, **STAR**. *Illicium anisatum*.

ANISEED TREE. A common name of *Illicium*.

ANISOCALYX. A genus of *Scrophulariaceæ*, containing a single species, found on the margins of streams in the island of Hong Kong. The calyx is unequally five-partite; the corolla is deciduous, and nearly equally five-partite. There are four didynamous stamens, scarcely longer than the corolla, with oblong purple and bilocular anthers; the style has a simple capitate stigma. [W. C.]

ANISOCHÆTA. A genus of the composite family (*Compositæ*). There is but one species, which is a native of Caffreland. It is a sub-climbing shrub, with alternate, ovate, coarsely-toothed leaves, and terminal panicles of flower-heads. The genus is nearly related to the *Ageratum* of our gardens, but differs in habit, as well as in the scales of the pappus and in the form of the achenes. It is a plant of no beauty. [A. A. B.]

ANISOCHILUS. A genus of *Labiatæ*, consisting of annual or perennial herbaceous plants, natives of Asia, chiefly of India, and containing nine species. The verticillasters are in ovate-oblong, or cylindrical terminal spikes, compact and imbricate; the floral leaves are bract-like. The calyx is ovate, swollen below, contracted above. The tube of the corolla is bent down after leaving the calyx; the throat is inflated; the upper lip is three or four-fid, the lower lip is longer, entire, and concave. There are four stamens, and a bifid style. [W. C.]

ANISODUS. A genus of plants belonging to the *Solanum* family, or by Miers referred to *Atropaceæ*. Its name is derived from its calyx, which is irregularly five-toothed, a circumstance which distinguishes it from *Hyoscyamus*, or the henbane genus. *A. luridus*, a Nepalese plant, is common in cultivation; it has a tap-shaped root, alternate leaves, which are stalked, oval, somewhat woolly on their under surface; the greenish-yellow, bell-shaped flowers are borne on axillary flower-stalks, and the fruit bursts by a transverse crack, like that of the henbane. [M. T. M.]

ANISOMELES. A genus of *Labiatæ*, containing eight species, natives of South-eastern Asia, the Mauritius, and tropical Australia. They are herbaceous plants, having the habit of *Stachys*. The terminal verticillasters are dense and many-flowered, or the axillary are few-flowered; the calyx is ovate, tubular, and five-toothed. The corolla is the same length as the tube of the calyx; its upper lip is erect, oblong, and entire; the lower lip has the two lateral lobes ovate and obtuse, the middle one emarginate and bifid. The stamens, four in number, are longer than the upper lip. The style is bifid at the apex. *A. malabarica* has the reputation of being a tonic and febrifuge, and is so used by the natives of India. [W. C.]

ANISOMERIA. A genus of *Phytolaccaceæ*, containing a single species from Chili. It is so nearly related to *Phytolacca* that it is generally considered as a sub-genus, differing from the true *Phytolacca* in the inequality of the lobes of the calyx, the ascending stamens, and the absence of an elevated central axis, leaving the ovaries free at their inner edge. [W. C.]

ANISOMEROUS. When the parts of a flower are unequal in number. The same as *Unsymmetrical*.

ANISOPTERA. A genus of *Dipterocarpeæ*, containing six species of trees, natives of the Islands of the Eastern Archipelago. They are nearly related to *Dipterocarpus*, but differ from it in having alternate leaves, and in the stamens, twenty-five in number, having their short filaments united together at the base. [W. C.]

ANISOTES. A genus of *Lythraceæ*, founded on the Brazilian *Lythrum anomalum*, which differs from *Lythrum* in having irregular flowers, the upper pair of petals being much larger than the rest, and the stamens only six. [J. T. S.]

ANISOTOMA. A genus of *Asclepiadaceæ* from the Cape of Good Hope, consisting of climbing herbs, with heart-shaped or kidney-shaped leaves, and small lateral umbels of subsessile flowers having a downy corolla. [T. M.]

ANISOTOME. A genus of *Umbelliferæ*, containing three or four species, natives of Auckland and Campbell Islands. They are amongst the largest and noblest plants of the natural order to which they belong, attaining a height of six feet, and bearing large umbels of rose-coloured or purple flowers. The stem is strong, erect, and furrowed. The leaves are large, petiolate, and two or three-pinnate. The flowers are diœcious, with the calyx margin five-lobed, and one or more of the lobes longer and more lanceolate than the remainders. In the male flower there are two rudimentary abortive ovaries, with the styles as mere points on their inner margins. In the female flowers the ovaries are conical, and terminate in long stout recurved styles, capitate at the extremity. The furrowed seeds are covered with a blackish testa. The whole plant of *A. lavida*, when bruised, emits an aromatic smell. [W. C.]

ANNOTINOUS. A year old. *Rami annotini* are branches one year old.

ANNUAL. Flowering and fruiting in the same year when raised from seed.

ANNULAR. Having the form of a ring, as in certain embryos.

ANNULATE. Surrounded by rotated rings or bands, or by scars to that form.

ANNULUS. A ring, as that which surrounds the apercure of a fern, or the peristome of a moss; or the membrane remaining round the stipe of an agaric when the cap has expanded. In the latter case, it is a membranous or filamentous veil, inserted on the one hand round the stem, and on the other into the edge of the pileus, so as to cover the organs of reproduction.

ANODA. A genus of *Malvaceæ* differing little, if at all, from *Sida*, except in the fact that its peduncles are not jointed, from which circumstance also its name has been derived. The species are natives of tropical America, north of the equator, and are herbaceous plants with solitary violet or yellow flowers. Some of them are in cultivation. [M. T. M.]

ANOMALOUS. Irregular, unusual, contrary to rule.

ANOMATHECA. A small genus of pretty iridaceous bulbs, inhabiting South Africa. The genus is distinguished by having a hypocrateriform perianth with a different triquetrous tube constricted at the throat, and a six-parted limb of oblong spreading segments, of which the three hinder ones are appendiculate; three unisecund stamens inserted in the throat of the perianth, and having short filaments, a filiform style bearing three narrow linear stigmata; and an inferior roundish ovate ovary, three-celled, containing many ovules. *A. cruenta* is a very pleasing dwarf plant, often seen in gardens. This produces a stem six to twelve inches high, furnished at the base with two-ranked, narrow, sword-shaped leaves, branched above, and terminating in a subsecund spike of flowers, of which the long slender tube is whitish, and the limb rich carmine crimson; the three lower segments have also a deep blood-coloured basal spot. There are one or two other species. [T. M.]

ANONA. A South American and West Indian genus of shrubs and trees, from which the name of the order to which they belong, *Anonaceæ*, is derived. The separate characters are a calyx of three minute sepals, united at the base; a corolla of six petals in two rows; the stamens numerous with linear, two-celled anthers, surmounted by an oval crest; numerous ovaries placed on a rounded receptacle and partly united together, becoming completely fused when mature into a many-celled, fleshy, oval or rounded fruits.

Several species of this genus are cultivated in tropical countries for the sake of their fruits. The Sour-sop of the West Indies is the fruit of *A. muricata*. It is of considerable size, often weighing upwards of two pounds; it is greenish and covered with prickles, the pulp is white, and has an agreeable slightly acid flavour. The

Anona squamosa.

Sweet-sop is the fruit of a tree, *A. squamosa*, native of the Malay Islands, but extensively cultivated in the East and West Indies. The fruit is ovate, covered with projecting scales, the rind is thick, but encloses a luscious pulp, concerning which, however, tastes differ; it appears to be highly esteemed by the Creoles, while the Europeans think lightly of it. The fruit grown in the Indian Archipelago is said to possess a finer flavour than that grown in the West Indies. The leaves of this plant have a heavy disagreeable odour, and the seeds, according to Royle, contain an acrid principle, fatal to insects, on which account the natives of India use them powdered and mixed with the flour of gram (*Cicer arietinum*) for washing the hair. The Cherimoyer of Peru is the fruit of *A. Cherimolia*, which is nearly allied to the preceding. The fruit is somewhat heart-shaped and scaly on the exterior, and is reputed by the Creoles as being the most delicious fruit in the world. A verdict which Europeans do not confirm. The common Custard Apple, or Bullock's Heart, is an eatable fruit produced by *A. reticulata*, a native of the West Indies, but cultivated in the East Indies also. Its yellowish pulp is not so much relished as that of the other kinds. In addition to their fruits, the plants of this genus are remarkable for their fragrant leaves and aromatic properties. The wood of *A. palustris* is so soft and compressible that it is made use of in Jamaica in place of cork; the fruit is called the Alligator Apple, but is not eaten, as it contains a narcotic principle. [M. T. M.]

ANONACEÆ (*Anona*, *Anonads*, *Glyptospermæ*) form an important natural order of tropical trees, remarkable for the powerful aromatic qualities of some of the species. They are nearly allied to magnoliads, differing mainly in the want of stipules, and in having an albumen ruminated like a nutmeg. In most species, moreover, the æstivation of the petals is valvate, so that the flowers, being formed on a ternary plan, the buds are three-sided pyramids. Some bear eatable pulpy fruit, like the Cherimoyers and Custard Apples; in others it is dry, aromatic, and pungent, like pepper; in all there seems to be present a stimulating quality, which renders them unsafe as articles of food, or as condiments, except in small quantities. The timber of some is extremely elastic, as lancewood, and occasionally is intensely bitter. See XYLOPIA, UVARIA, GUATTERIA, ANONA, MONODORA, etc.

ANONYMOS. A name occasionally given by the older botanists to various plants which they could not readily compare with any one that had a name already. It has been entirely rejected from the modern nomenclature.

ANOPLANTHUS. A genus of the broomrape family (*Orobanchaceæ*). They are annual, leafless, parasitical herbs, growing on the roots of various plants; seldom more than one foot in height, and the whole plant of a brown or purple colour. The flower-stalks are naked above and scaly below, bearing a single terminal flower; the corolla with a curved tube about an inch long, or short, and somewhat bell-shaped. There are five species known, three of them found in North America; the other two, which have large scarlet flowers, are natives of Asia Minor. *A. uniflorus*, a N. American species, is called one-flowered Cancer root. [A. A. B.]

ANOMIA. A genus of *Umbelliferæ*, containing a single species from Candia—a biennial, erect, herbaceous plant, with a fusiform root, obtusely-trifid leaves, and white hermaphrodite flowers. It is nearly related to *Smyrnium*, from which, however, it is separated by the want of involucres and involucels. [W. C.]

ANOTTA, or ARNOTTO. *Bixa orellana*.

ANREDERA. A genus of *Basellaceæ*, containing a single species, a native of the West Indies and Peru. It is a climbing herbaceous plant, with alternate petiolate leaves, and pedicellate flowers, arranged in many-flowered simple axillary spikes. The calyx consists of five membranaceous sepals. The five stamens rise from the base of the sepals. The ovary is uniovular, and uniovulate, with a short style, and three long slender stigmas. [W. C.]

ANSELLIA africana. A very fine orchid, found growing on oil-palm trees in the island of Fernando Po. It has a tall stem, not unlike a sugarcane; broad strap-shaped leaves, and great drooping panicles of greenish flowers, blotched with purple. There is also a plant of this genus found at Natal, and called *Ansellia gigantea*; but it does not seem to be distinct from the plant of the West Coast.

ANSÉRINE. (Fr.) *Chenopodium*; also, *Potentilla anserina*.

ANT TREE. *Triplaris Bonplandiana*, the habitation of a species of ant (*Myrmica*).

ANTENNARIA. A family of herbaceous evergreen perennials, belonging to the natural order *Compositæ*, and distinguished by the dry, coloured, chaffy scales encircling each head of flowers, of which the stamens and pistils are on different plants. *A. margaritacea*, the Pearly Everlasting (*Gnaphalium* of Linnæus), is a native of North America, where it grows in some districts in great profusion. It has long been a favourite garden plant in this country, and, having escaped from cultivation, has in some places thoroughly established itself as a denizen. Gathered just before their prime, the flowers retain their form and lustrous pearly hue for an indefinite period; before they are often laid by to be added to winter bouquets, or, having been previously dyed of various colours, to be employed in decorating rooms. On the Continent, under the name of *Immortelles*, they are much used in the construction of wreaths, to be placed as votive offerings on the graves of the departed, and renewed on the anniversaries of their saints' days. *A. dioica* is a British species, not unfrequent in hilly and mountainous districts. It is a much smaller plant than the preceding, from five to six inches high, with decumbent stems, cottony leaves, and white or rose-coloured flowers. [C. A. J.]

ANTENNARIA (bis). The black web-like masses which hang down from the ceilings of wine vaults, and from thence cling to the casks and bottles, forming the pride of the wine-merchant, are derived from a fungus of this genus, belonging to the race of sac-bearing moulds, *Physomycetes*. Other species creep over living leaves, entering them with a black felt, and hindering both the proper access of light and their especial functions of breathing and perspiration. The threads of which the mass is composed are either even, or swollen into joints, like necklaces, and the fruit arises from swollen portions of the threads, a miniature plant being sometimes produced within the swellings or sporangia, without distinct spores. Since many of the species are succeeded by a *Coryneum*, it is possible that the greater part are only imperfect or transitional forms of that genus. *Antennaræ* are far more common in warm than in cool climates, and are the pest of orange groves and coffee plantations. They seem frequently to accompany different species of *Coccus*, from whose exudations they probably derive their nutriment. (M. J. B.)

ANTHEMIS. The genus of plants to which the Chamomile belongs. It forms part of the composite family, among which it may be known by its involucre, consisting of a number of overlapping scales, with membranous margins; by the absence of pappus, or feathery hairs; by its outer florets — flowers of the ray, as they are called — being in one row, ligulate, or strap-shaped, containing pistils only; while those of the centre, or disc, are numerous, tubular in form, and contain both stamens and pistils. The receptacle on which the flowers are placed is convex, and covered with little chaffy scales or bracts, which stand up between the florets.

The Chamomile, *A. nobilis*, is a native of Britain. Its stems are procumbent or erect, much-branched, leafy, furrowed, and hollow in the interior. The leaves are downy, pinnately divided into narrow segments. The bitterness of the Chamomile is due to a principle which possesses tonic properties. The aromatic fragrance is due to the presence of an essential oil, which is of a light blue colour when freshly extracted. Both these ingredients exist in larger quantities in the central yellow florets than in the outer white ones; hence the wild Chamomile is preferred for medical purposes, as in the cultivated variety the flowers are apt to become double by the conversion of the yellow tubular central florets into white strap-shaped ones like those of the ray. Owing to its stimulant tonic properties, it is much used in certain cases of weak digestion, and occasionally as an emetic, in the form of an infusion. *A. tinctoria* furnishes a yellow dye. *A. Cotula* is a common weed in the South of England, where it is called Stinking May-weed. The leaves differ from those of the true Chamomile in being quite smooth, not downy. The plant is covered with glands, which emit a powerful and disagreeable perfume, and cause swelling of the hands of persons employed to pull the plant up as a weed. [M. T. M.]

ANTHER. The case which contains the pollen of a plant; the terminal hollow of a stamen.

ANTHERICUM. An extensive genus of *Liliaceæ*, distinguished by its six-leaved, equal spreading, or campanulately connivent perianth; six hypogynous stamens with short filaments; a filiform declinate style, with an obtuse subcapitate stigma; and a three-celled ovary, containing numerous ovules. The species consist of herbs having fleshy fasciculate roots, radical filiform or linear lanceolate leaves, sometimes fleshy, often hairy, and flower scapes bearing racemes or panicles of white flowers. They are found indigenous in the middle and south of Europe, in New Holland, and in South Africa. The species are ranged in three groups, viz.:— 1. *Anthericum proper*, in which the perianth is spreading, and the stamens smooth. 2. *Bulbine*, in which the perianth is spreading and the stamens are all, or each alternate one, bearded. 3. *Czackia*, in which the perianth is campanulately connivent, and the stamens smooth. *A. Liliastrum*, which belongs to the last group, is a pretty plant with broadish grassy leaves, and a scape one and a-half foot high, bearing several largish white sweet-scented flowers, marked on each segment with a green dot. This is called St. Bruno's Lily. [T. M.]

ANTHERIDIA (Antheridia). A term applied to the male organs in cryptogams, which, though analogous to anthers in phænogams, cannot be considered as precisely the same organs, or, in strict botanical language, their homologues. They are of very various character and variously seated in the different sections. While amongst the higher cryptogams, in ferns and some other allied sections, they consist of little sacs, seated on the threads or membrane immediately derived from the spores, producing bodies altogether analogous with the spermatozoa of animals: in *Selaginella*, *Pilularia*, &c., they appear under the guise of a second smaller spore, which after a time produces the spermatozoids; and in mosses they appear on the plant itself, once only, or more frequently, according as it is annual or perennial. In *Characeæ* the antheridia are of a very complicated structure, as described under that head.

In the lower cryptogams there is no less difference in the character of the antheridia. Amongst the rose-spored and dark-spored *Algæ* there are several distinct types; amongst the green-spored we have sometimes free bodies, analogous to the small spores of *Selaginella*. The spermatozoids are not, however, spiral, as in the higher cryptogams, though still endowed with active motion by means of delicate appendages. Amongst fungi and lichens we have again a totally different type, the spermatozoids, or spermatia as they are called, not having active motion or external motile appendages. Amongst the lower kinds it is possible that they may be merely representative, without any active functions, and in some cases, probably, they exist merely as a sort of conidia. The whole subject is, however, at the present obscure, as far as *Fungi* and lichens are concerned. Organs which are sometimes called antheridia, are occasionally so denominated without any sufficient grounds, as for example the prominent cells on the gills of the bottled toadstools, *Coprini*. [M. J. B.]

ANTHEROMANIA. An unnatural development of stamens. This may take place without any detriment to the plant. As the petals are multiplied, the stamens are usually multiplied; but where the stamens are epigynous, the addition will sometimes materially disturb the normal construction of the flower. In some double flowers, the apparent multiplication of the petals is really a multiplication of the stamens, and as such petaloid stamens are generally devoid of anthers, the fertility of the plant may in consequence be affected. [M. J. B.]

ANTHISTIRIA. A genus of grasses of the tribe *Andropogoneæ*. The flowers are polygamous, the hermaphrodite being sessile, or very shortly stipitate, whilst the male flowers are pedicellate. The species are all foreign, and many of them only grow spontaneously in tropical countries. One of the most interesting is the Kangaroo grass of Australia, *A. australis*, which supplies a considerable portion of food to the cattle in some parts of that country, and is consequently, valuable to the colonists. *A. arundinacea* grows from six to twelve feet high, in the East Indies, according to Roxburgh. [D. M.]

ANTHOCARPOUS. Composed of flowers and fruit blended into a solid mass, as in the pine apple.

ANTHOCERCIS. A genus of plants allied to *Salpiglossis* and belonging to the same natural family, *Scrophulariaceæ*. The plants are shrubs, natives of the extratropical parts of New Holland. Some of them are cultivated in greenhouses. The calyx is five-parted, equal; the corolla bell-shaped, its tube contracted at the base, its limb five-parted; there are four perfect stamens, and a rudiment of the fifth. [M. T. M.]

ANTHOCERIDEÆ, ANTHOCEROS. A small natural order and genus of liverworts, distinguished by the capsule, which is threaded by a linear columella, bursting longitudinally on one or both sides, and by the fronds being without the pores which are so conspicuous in *Marchantiaceæ*. It consists of about three genera only, which occur in different parts of the world, one of which, *Carpobolus*, is remarkable for the absence of the spiral threads or elaters in the capsule, that are so characteristic of the tribe. Two species *Anthoceros lævis* and *A. punctatus* occur in this country on the ground, principally in the eastern counties. None of the species bear anything like the true leaves of the higher *Jungermanniæ*. [M. J. B.]

ANTHOCHORTUS. A genus containing a single species of *Restiaceæ*, a native of the Cape of Good Hope. It is a little known herbaceous plant with filiform leafless stems. The flowers are diœcious. The male has a six-parted infundibuliform and purple-coloured calyx, and three stamens; the female is unknown. [W. C.]

ANTHOCYANE. The blue colouring matter of plants.

ANTHODISCUS. A genus of *Rhizobolaceæ*, containing a single species, a tree from Guiana. It has alternate or opposite trifoliate leaves, and flowers in racemes, each flower having a short pedicel with two bracts. The small persistent calyx is cup-shaped, and obscurely five-lobed. The five coriaceous petals are concave and coherent. The stamens are very numerous, and inserted with the petals on a hypogynous disc; they have filiform and glandular filaments, and bilocular anthers. The ovary is globose with many loculaments, each containing a single ovule. There are as many oblong incurved styles as there are cells in the ovary. [W. C.]

ANTHODIUM. The head of flowers, or capitulum of compositæ.

ANTHOGONIUM gracile. A terrestrial orchid from the north of India, with long, narrow, grassy, leaves. The flowers form pretty complete tubes of a crimson colour, at the end of slender scapes.

ANTHOLYSIS. The retrograde metamorphosis of a flower; as when carpels change to stamens, stamens to petals, petals to sepals, and sepals to leaves, more or less completely.

ANTHOLYZA. A small genus of showy Irideæ, having a tubulose perianth, of which the limb is six-parted, unequal, the upper segment being much the longest, straight, spoon-shaped, the two lateral ones spreading and ascending, and the three lower once very small; there are three stamens, and three stigmata, and a three-celled ovary, containing numerous ovules. They are herbs with bulb-tubers or corms, and are allied to Gladiolus. A. splendens, one of the pretty species sometimes seen in gardens, has the corms about as large as a hazel nut, a stem two to three feet high, bearing at the base, long linear or linear-ensiform strongly-nerved leaves, and terminated by a many-flowered spike of distichous flowers, the tubular portion of which is slender at the base, and triangularly gibbous about the middle the limb being bright scarlet. The flowers appear to be long tubular, with a pair of expanded wings. A. Cunonia, another well-known species, has the flowers scarlet, yellow towards the base of the tube, and arranged in a scorund manner, instead of being distichous on the spike. There are a few other species, all South African. [T. M.]

ANTHOPTERUS. Under this name are comprised certain plants belonging to the order Vacciniaceæ. Their prominent characters are a calyx tube provided with five wings, a corolla with a tube similarly winged, and ten stamens, united together into a membranous tube. [M. T. M.]

ANTHOPTOSIS. Most flowers are mere temporary organs, which, when they have performed their functions, are destined to fall. In many cases, however, the flowers fall before impregnation has taken place, or shortly after, involving with them the pistil, and so inducing sterility. This may arise from various causes, as excess or want of proper moisture, but more frequently from late frosts or cold winds. The disease amongst grapes known by the name of 'coulure,' is of this description. This, however, arises frequently from poverty of sap, and may be prevented by ringing, provided the weather be not very unfavourable. In many instances the fall of the flower naturally follows impregnation, and cannot be regarded as a disease; indeed, the time of its fall seems to depend upon the process of fertilisation, for even in cases where the flowers naturally fade very rapidly, their duration may be prolonged by preventing the access of pollen to the style. [M. J. B.]

ANTHOXANTHINE. The yellow colouring matter of plants.

ANTHOXANTHUM. A genus of grasses of the tribe Phalaridæ. The few species which belong to the genus are all from temperate parts of the globe, and there is only one British, namely, the sweet-scented Vernal grass, A. odoratum. It is distinguished from its allies by having membranous, awnless glumes, compressed and connate below; pales, one to each flower, bearing an awn on its back part. This grass is rather remarkable botanically, by having flowers diandrous, i. e. with two stamens to each, there being the normal number in grasses; hence, in accordance with the Linnæan system, it is included in a different class from most of the other grasses. It forms a large proportion of many meadows and pastures, but is not considered a first-class species, having a less quantity of saccharine matter and more mucilage than some other kinds in its composition. The peculiar odour which well-saved new hay gives out, is supposed to be principally emitted from this grass, hence the English name. [D. M.]

The fragrant resinous principle which occurs in this grass, and is called coumarin, is a widely-diffused natural perfume, being found, according to Professor Johnstone, in the Tonka Bean (Dipterix odorata), the Faham Tea-plant of the Mauritius (Angræcum fragrans), the common Sweet Woodruff (Asperula odorata), the Sweet-scented Vernal grass (Anthoxanthum odoratum), the common Melilot (Melilotus officinalis), and the blue, or Swiss Melilot (Melilotus cæruleus). 'It is the same odour,' he continues, 'therefore, which gives fragrance to the Tonka Bean, to the Faham Tea of the Mauritius, to our Melilot Trefoil, and to sweet-smelling hay-fields. In Switzerland the blue Melilot is mixed with particular kinds of scented cheese, and the coumarin it contains gives to that of Schabzieger its peculiar well-known odour.' The vapour of coumarin is stated to act powerfully on the brain; and it is not improbable that hay fever, to which many susceptible people are liable, may be owing to the presence of this substance in unusual quantities during the period of haymaking. [T. M.]

ANTHRISCUS. A genus of umbelliferous plants, with thin, finely-divided leaves, and small, inconspicuous white flowers, arranged in umbels. Two species only are cultivated—the Chervil, A. Cerefolium, for flavouring salads, &c., and the Parsnip Chervil, A. bulbosus, for its roots as a vegetable. Chervil is an old-fashioned pot-herb, having been cultivated by Gerarde in his garden in Holborn in 1596. It is a native of various parts of Europe, and occasionally met with in England in waste places. The young leaves, when about two inches high, are considered fit for gathering, and are then used in soups and stews, to which they impart a warm aromatic flavour. They are also used with mustard and cress in small salads, but are not much in demand in this country. The French and Dutch, however, have scarcely a soup or salad in which Chervil does not form a part; and, as a seasoner, it is by many preferred to parsley. There is a curled leaved variety of this plant, which makes a very handsome garnish for dishes, and is on this account more esteemed than the common sort. The *Cerfeuil frisé* of the French is very similar to the last, the only difference being in the leaves, which are even more frizzled than those of plants raised from seed saved in England. Care must be taken not to confound this plant with *Anthriscus vulgaris*, the common rough Chervil, which bears so great a resemblance to it as to have deceived some Dutch soldiers, who gathered it, when in England, in 1745, and put it into their soups, by which several of them were poisoned. The Parsnip Chervil, sometimes called *Chærophyllum bulbosum*, is a native of France, and, although known to gardeners since its introduction to this country in 1726, it is only within the last few years that attention has been directed to its culture as an esculent vegetable. In size and shape the root attains the dimensions of a small Dutch carrot. It is outwardly of a grey colour, but when cut the flesh is white, mealy, somewhat nutty in a raw state, and by no means unpleasant to the taste. When boiled, the flavour is intermediate between that of the chestnut and potato, in consequence of which it has been recommended for cultivation as a substitute for the latter root. [W. B. B.]

ANTHURIUM. A genus of plants of the *Arum* family, or by some referred to *Orontiaceæ*. The name is derived from two Greek words, signifying flower-tail, and is given in allusion to the inflorescence, which is a spike somewhat like a tail. The plants are better known under the old name of *Pothos*. They comprise several tropical plants, natives of Central or Tropical America, for the most part growing upon trees, or in their forks, and hence called *epiphytes*, in contradistinction to parasites, which not only grow upon other plants, but also derive their nourishment from them. From the root-like stems the leaves arise; these are of varied shape, in some entire, in others palmate or digitate, sometimes with swollen leaf-stalks, but in all invested below by a small sheath. The stem also gives off numerous aerial roots, like those of the common ivy, but on a larger scale. The flowers contain both stamens and pistils, enclosed within a perianth. The ovary is two to four-celled. The flowers thus constituted are densely packed upon a cylindrical often almost sessile spadix or spike, at the base of which is a large bract or spathe, which becomes bent backwards as the flowers come to maturity. [M. T. M.]

ANTHYLLIS. A genus of plants belonging to the natural order *Leguminosæ*, herbaceous or shrubby, having a permanent calyx, which, after flowering, becomes inflated; petals almost of equal length; and a pod always hidden by the calyx, and containing one or two seeds. The only British species is *A. Vulneraria*, so called from its supposed property of staunching the blood of wounds, which virtue it probably possesses to the same extent with many other plants having equally downy leaves. Its popular name is Kidney Vetch, or Lady's Fingers, and it is frequently met with in dry pastures, especially such as are chalky, or near the sea. The leaves are rather large, of a bluish tinge, hairy, pinnate, with the terminal leaflet largest. The flowers are most commonly yellow, and grow in crowded heads, which are disposed in pairs, with large deeply-lobed bracts beneath each; the calyx is of a delicate straw colour. In some of the mountain stations — especially at the Lizard, in Cornwall — the colour of the flowers varies to a remarkable extent, yellow, cream-coloured, white, purple, and crimson being found all growing together. Of the shrubby species, *A. Barba-Jovis* is an evergreen shrub, a native of the South of Europe. This also has pinnate leaves and yellow flowers, and the whole plant has a silvery appearance, from which it derived its name of Jupiter's Beard and the Silver-bush. 'The elegance of this shrub did not escape the ancients; and Pliny mentions its beauty, adding that it dislikes water, and that it makes a very elegant ornament for gardens, when clipped into a round shape. It is one of the finest shrubs that can be planted against a conservatory wall. It will grow in any light soil.' [C. A. J.]

ANTIARIS, the arto-carpaceous genus of plants to which the Upas-tree of Java belongs. The stamens and pistils are in separate flowers, on the same tree. The male flowers are numerous, and enclosed within a hairy involucre, formed of several fleshy divisions, rolled inwards. The calyx is in three or four pieces, and encloses an equal number of stalkless anthers. The female flower has an adherent calyx of several leaves, and is terminated by a long two-parted style. It contains a single suspended ovule, and becomes converted when ripe into a succulent drupe-like fruit. The female flowers are solitary, placed in the axils of the leaves, side by side with the heads of male flowers.

The Upas-tree, when pierced, exudes a milky juice, which contains an acrid virulent poison, called *antiaris*. Most exaggerated statements respecting this plant were circulated by a Dutch surgeon about the close of the last century. The tree was described as growing in a desert

ANTIARIS & COFFEE PLANTATION IN JAVA.
(AFTER BLUME)

tract, with no other plant near it for the distance of ten or twelve miles. Criminals condemned to die were offered the chance of life if they would go to the Upas-tree and collect some of the poison. They were furnished with proper directions, and armed with due precaution, but not more than two out of every twenty ever returned. The Indian surgeon, Focrsch, states that he had derived his information from some of those who had been lucky enough to escape, albeit the ground around was strewn with the bones of their predecessors; and such was the virulence of the poison, that 'there are no fish in the waters, nor has any rat, mouse, or any other vermin been there; and when any birds fly so near this tree that the effluvia reaches them, they fall a sacrifice to the effects of the poison.' Out of a population of 1,600 persons, who were compelled, on account of civil dissensions, to reside within twelve or fourteen miles of the tree, not more than three hundred remained in less than two months. Focrsch states that he conversed with some of the survivors, and proceeds to give an account of some experiments that he witnessed with the gum of this tree, these experiments consisting principally in the execution of several women, by direction of the Emperor! Now, as specimens of this tree are cultivated in botanic gardens, the tree cannot have such virulent properties as it was stated to have; moreover, it is now known to grow in woods with other trees, and birds and lizards have been observed on its branches. It occasionally grows in certain low valleys in Java, rendered unwholesome by an escape of carbonic acid gas from crevices in the ground, and which is given off in such abundance as to be fatal to animals that approach too closely. These pestiferous valleys are connected with the numerous volcanoes in the island. The craters of some of these emit, according to Reinwardt, sulphureous vapours in such abundance as to cause the death of great numbers of tigers, birds, and insects; while the rivers and lakes are in some cases so charged with sulphuric acid, that no fish can live in them. So that doubtless the Upas-tree has had to bear the opprobrium really due to the volcanoes and their products: not that the Upas is by any means innocent, for severe effects have been felt by those who have climbed the tree for the purpose of bringing down the branches and flowers. The inner bark of the young trees, which is fabricated into a coarse garment, excites the skin the most horrible itching. It clings to the skin, if exposed to the wet before being properly prepared. The dried juice, mixed with other ingredients, forms a most venomous poison, in which the natives dip their arrows. A view of one of these trees in the midst of coffee plantations, will be found in Plate 3.

A species of *Antiaris*, called also *Lepurandra saccidora*, furnishes the natives of Bombay with sacks, which are made by beating the cloth-like bark, and peeling it off from the felled branches, leaving a small portion of wood to form the bottom of the sacks. They are used to hold rice. Specimens of these may be seen in the Kew Museum. [M. T. M.]

Antiaris innoxia.

ANTICAL. Placed in front of a flower, the front being reckoned as the part most remote from the axis. Thus, the lip of an *Orchis* is antical.

ANTIDAPHNE. A genus of *Loranthaceæ*, containing a single species parasitic on the trees of the primeval forests of Peru, found chiefly on laurels. It has alternate, obovate, and entire leaves. The flowers are monœcious, arranged in small axillary spikes. The male flowers have a simple, three-lobed calyx, three stamens, with petaloid filaments, and bilocular anthers. The calyx of the female flowers is simple, with an entire margin; the ovary is unilocular, and uniovulate, with a subsessile capitate stigma. [W. C.]

ANTIDESMA. A genus of the natural family *Stilaginaceæ*. Upwards of thirty species are known, all of them natives of tropical India, Africa, and Australia, and their islands. They are trees or shrubs, with alternate, simple, entire leaves, and spicate inflorescence. The flowers are inconspicuous, the males and females on the same plant. The fruit is a one-seeded drupe about the size of a pea. The bark of *A. Bunius*, which is a native of Java and the adjacent isles, affords a fibre from which ropes are made. The fruits are of a bright red colour, ripening into an intense black, with a sub-acid taste. They are used in Java for preserving, principally by Europeans, bringing about twopence per quart.

The Treasury of Botany.

The leaves are used as a remedy against snake bites, and in syphilitic affections. The wood, when immersed in water, becomes black and as heavy as iron. All the parts of the plant have a bitter taste. The berries of *A. dioandrum* are eaten in India by the natives, as well as those of *A. pubescens*, also a native of India; and its bruised leaves are used in native practice, and applied in the form of a poultice to ulcers and tumours. [A. A. B.]

ANTIGRAMMA. A genus of polypodiaceous ferns of the *Asplenium* section, belonging to that series in which the sori are contrived in pairs, with the indusia opening face to face (scolopendrioid). In that series it is known by having the veins of its fronds reticulated, and its sori parallel and oblique. The genus comprises a couple of Brazilian species, having simple fronds. [T. M.]

ANTHIRILEA. A genus of cinchonaceous plants inhabiting Mauritius and Bourbon, consisting of shrubs with leaves arranged in whorls of three. The flowers are borne on forked peduncles, and have a calyx which is short, bell-shaped, and four-toothed; and a tubular corolla, with four sessile anthers attached to its interior. The fruit is succulent externally, and contains a kernel with two one-seeded cells. The name of these plants is expressive of their valuable properties in arresting hemorrhages, and as astringents. [M. T. M.]

ANTIRRHINUM. A genus of *Scrophularineæ*, containing fourteen species, natives chiefly of the Mediterranean region, though some are found in California. They are annual or perennial, rarely shrubby herbaceous plants, with the lower leaves often opposite, and the upper ones alternate. The flowers are commonly showy, and solitary in the axils of the upper leaves, or form lax terminal racemes. The calyx is five-partite. The corolla has a broad tube, saccate and slightly protruding below the calyx on the lower side, but not spurred as in *Linaria*. The throat is closed by a large, projecting, bearded palate, which gives to the flower a resemblance to the face of an animal or a mask, whence the name, meaning "snout-like." The capsule is two-celled, oblique, and opening by two or three pores at the top. The seeds are truncate.

The genus has been divided into three sections. 1. *Orontium*, annual plants with penninerved entire leaves and compressed seeds. 2. *Antirrhinastrum*, perennial plants with penninerved entire leaves and obovate seeds. And 3. *Asarina*, with palminerved lobed leaves and ovoid oblong seeds. Two species are found in Britain; the larger, *A. majus*, has probably escaped from gardens, and is found on old walls and in clefts of rocks; *A. Orontium* occurs in the corn-fields of the south of England and Ireland. [W. C.]

ANTITROPAL. The same as Orthotropal.

ANTJAR. The poisonous *Antiaris toxicaria*; also called *Antschar*.

ANTROPHYUM. A genus of polypodiaceous ferns, belonging to the *Hemionitideæ*, distinguished by having the veins of the fronds uniform and reticulated, and the fructification, which is linear, and usually immersed in a shallow groove, also more or less, but only partially, reticulated. It is a group of very distinct aspect, though in technical characters coming close to *Hemionitis*, in which latter, however, the lines of spore-cases are more completely joined together into a network, and superficial on the frond. There are several species, all simple fronded, found in various parts of the tropics of both hemispheres. [T. M.]

ANYCHIA. A genus of knotworts, *Illecebraceæ*, near to *Paronychia*, but differing in the absence of petals as well as of awns to the calyx leaves. There are two known species, both North American weeds of no beauty. The Forked Chickweed, *A. dichotoma*, is a slender herb four to ten inches high, with capillary, many-times-forked branchlets, bearing minute flowers in the forks, and opposite oblong leaves rather more than half an inch in length, accompanied with stipules like those of buckwheats (*Polygonaceæ*). [A. A. B.]

AOTUS. A genus of Australian and Tasmanian shrubs, belonging to the pea-flowered section of the leguminous family, containing ten species. They are slender plants, with heath-like leaves, arranged in whorls round the stem, therein in a whorl. The flowers are small, bright yellow, with short stalks, and the calyx is destitute of the two small bracts which are found on those of the allied genera; this gives rise to the name of the genus—*Aotus*, signifying without ears. *A. gracillima*, a native of West Australia, is a favourite greenhouse plant. It is a slender shrub, with copious yellow flowers, which are so thickly set on the stems as to hide the leaves from view. One or two other species are in cultivation, but many very pretty species have yet to be introduced. The pods of most of the species are not larger than a grain of barley, and contain each two seeds. [A. A. B.]

APACITIS. A Japanese tree, very imperfectly described by Thunberg, and not since recognised.

APALANCHE VERT. (Fr.) *Prinus verticillatus*.

APALANTHE. A generic name given by Planchon to a few species of *Anacharis*, separated from that genus because of their having hermaphrodite flowers. It has, however, been found that one of three species, *A. Schweinitzii*, is the same as *Anacharis (Elodea) Planchonii*, which, although its flowers are generally dioecious, yet frequently bears fertile flowers with three to six stamens; sometimes merely short sterile filaments without anthers or with imperfect ones, and sometimes with oblong almost sessile anthers. It is probable that, when the other species are more carefully examined, no foundation will be found for the separation of *Apalanthe*. [W. C.]

APARGIA. This name is used in some English books for one or two species of Hawkbit, *Leontodon*, called *A. hispida* and *A. autumnalis*, the latter of which is sometimes referred to *Oporina*. [T. M.]

APATURIA. Terrestrial leafless orchids, from the continent of India, Ceylon, and China. They are of no interest, their pallid flowers hanging down from the side of a rather long spike. One of the marks by which they are most easily known is having, along with a structure similar to that of *Bletia*, stems covered by thin membranous scales.

APEIBA. A genus of the lime-tree family, *Tiliaceæ*, containing twelve species. They are trees or shrubs with alternate, stalked, entire or serrate leaves, which are covered on both surfaces with starry pubescence, and having at the base of their footstalk two stipules, which fall early. The peduncles are terminal and opposite the leaves, much-branched and many-flowered; the flowers yellow or greenish, interspersed with bractese. Their fruits are woody, roundish, and often covered with tubercules, or stiff prickles. The species are found in Mexico, the West Indies, and Southern tropical America. The fibrous bark of *A. Petoumo* is known in Panama as Cortesa, and is used for making cordage, being strong, tough, and distinguished from other Indigenous fibres by its whiteness. The wood of *A. Tibourbou* being light and soft, is used in Brazil for making the raft-boats called jangadas. Its fruit, in size and appearance, is much like that of the Spanish chestnut. *A. aspera* has a flattened circular fruit, with rough points, resembling the cup of an acorn, cosy closed at the top. [A. A. B.]

APERA. A genus of grasses of the tribe *Agrostideæ*. As defined by Adanson and Beauvois, a few species only are referable to this genus, which in more modern works will be found described under the genera *Agrostis*, *Muhlenbergia*, and *Vulpia*. The principal characters depended on to separate it from *Agrostis* are the lower glume being smaller than the upper, and the presence of a rudimentary second floret, beside the perfect floret. The British species, *Apera Spica-venti*, Wind-bent Grass, is one of the prettiest of English grasses, the light feathery panicles of inflorescence, with the long awns attached to the glumes, seldom failing to attract attention from even those who are little in the habit of observing plants. Although of small importance as an agricultural species, it is valued for the beauty of its flowers, which remain long on the rachis, and form a handsome drawing-room ornament, even in their natural state, but particularly so when dyed crimson, green, or any other bright colour. [D. M.]

APETALON. A minute leafless orchid, found beneath the shade of Bamboos in Coorg.

APETALOUS. Having no petals. Also extended to plants that have neither calyx nor corolla.

APHANES. A synonym of *Alchemilla*.

APHANOSTEPHUS. A genus of the composite family, numbering three species, which are found in Texas and New Mexico. They are related on the one hand to the daisy (*Bellis*), from which they differ in the presence of a pappus to the achenes, and on the other to the Australian genus *Brachycome*, from which the rounded and striate, instead of flattened, achenes at once distinguish them. They are much-branched annual plants, six inches to a foot high, having linear or spathulate, toothed or entire, more or less hoary leaves, and slender twigs, terminating in a single stalked flower-head about half an inch across, the rays pink or white, the disc yellow. *A. ramosissimus*—is quite a pretty plant in cultivation, producing a great abundance of flower-heads, with white rays, tinged with pink underneath; and it lasts through the summer. [A. A. B.]

APHELANDRA. This name is applied in consequence of the flowers of the plants of this genus having one-celled anthers. They are small shrubs, natives of tropical America. The inflorescence consists of four-sided spikes, with slightly membranous bracts, handsome reddish or scarlet flowers, with a gaping two-lipped corolla, the lower lip divided into three lobes, the central one of which is much larger than the lateral ones; the stamens are four in number, one pair longer than the other (didynamous), the anthers one-celled; the capsule is sessile roundish, two-celled, each cell containing two compressed seeds. *A. cristata* is a remarkably handsome show plant, with fine ovate pointed leaves, and showy spikes of blossom. It was formerly referred to the genus *Justicia*, and belongs to the order *Acanthaceæ*. *A. aurantiaca* is scarcely less handsome than the other. [M. T. M.]

APHELEXIS. A genus of Madagascar plants, belonging to the composite family (*Compositæ*), having much resemblance to the everlasting flowers (*Helichrysum*), and differing from these chiefly in the hairs of the pappus. Five species are known, all of them having very small leaves, which are closely pressed to the stem, like those of the club-moss. The flower-heads are either large, solitary, and of a pink colour, or small, yellow, and two or three together at the ends of the branches. The plants known in gardens as *Aphelexis*, and so commonly cultivated in greenhouses, are natives of the Cape, not of Madagascar, and are generally placed in the genus *Helipterum*. [A. A. B.]

APHELIA. A genus of *Desvauxiaceæ*,

consisting of a small sedge-like plant, *A. cyperoides*, from the southern shores of New Holland, which grows in small tufts, with short thread-like leaves and naked stems, a few inches high, at the top of which are short two-ranked spikes of glumaceous flowers, the lower glumes frequently empty; the outer glume of each flower is much acuminate, the upper glume shorter, keeled at the base. [J. T. S.]

APHLŒA. A genus of the *Flacourtia* family, containing but few species. They are small trees, with much-cut, serrate, or entire alternate leaves, and axillary, solitary, or fascicled flowers, without petals. From all the allied genera, they are distinguished by their single one-sided placental line, to which the ovules are attached. They are natives of Madagascar or the Mauritius, some of them varying much in their foliage, entire or pinnatifid leaves being found on the same plant. *A. theiformis* has an emetic bark. [A. A. B.]

APHYLLÆ. A name applied to that portion of cryptogamic plants comprehended under the term *Thallogens*, in consequence of the greater part of them being destitute of such modifications of leaves as occur in mosses, ferns, &c. Some seaweeds, or Algæ, indeed, have leaf-like organs, but these differ in many respects from leaves, and are mere expansions of the common stem. [M. J. B.]

APHYLLANTHES. A genus of *Liliaceæ*, consisting of a single species, found in the south of Europe. It is a perennial, slender, rush-like herb, leafless, the scapes having membranous sheaths at the base, like those of the rush, and hence terminated by a small head of numerous blue flowers. The perianth is six-parted, spreading at the apex, connivent into a short tube at the base; six stamens, with thread-like filaments, are inserted above the base of the perianth; the filiform style is terminated by a three-lobed stigma; and the ovary is three-celled, with a solitary basal ovule in each cell. The scapes appear like grassy leaves, but are seen to be tipped by the glumaceous scales which protect the blossom-buds. [T. M.]

APIACEÆ (*Umbellifera*, *Umbelliferæ*). Under this name is collected a very large number of plants inhabiting for the most part, in the northern regions of the northern hemisphere, woods, bogs, marshes, and dry places. As we approach the equator they become less and less known, and in the southern hemisphere, are comparatively rare. They all have a double diagynous inferior ovary, separating when ripe into two similar parts, vulgarly called seeds, surrounded by a superior calyx, which is generally scarcely, and often not at all observable; five separate petals; five intervening epigynous stamens; and two styles proceeding from what is not very correctly termed a double epigynous disk. Hemlock, carrot, parsley, and parsnip are familiar examples. Although the order numbers at least 1,500 species, divided nearly 300 genera, not a tree is known among them; a very few only attaining the condition of woody bushes. Many are important as producing articles of food; many are poisons; most are merely unimportant weeds; a few, like *Astrantia*, are furnished with gay colours, and thus become objects of decoration. One of them, *Bolax Glebaria*, forms huge tussocks in the Falkland Islands, resembling haycocks. Of the harmless species, in which, with a little aroma, there is no considerable quantity of acrid watery matter or gum-resinous secretion, must be most particularly named celery, fennel, samphire, parsley, and the roots of the carrot, parsnip, and skirret (*Sium Sisarum*). The root of *Eryngium campestre* and maritimum, vulgarly called Eryngo, is sweet, aromatic, and tonic. The aromatic roots of *Meum athamanticum* and *Mutellina* form an ingredient in Venice treacle. Angelica root, belonging to *Archangelica officinalis*, is fragrant, sweet when first tasted, but leaving a glowing heat in the mouth; others are gum-resinous, as the species of *Ferula*, yielding Asafœtida, the fœtid odour of which is supposed to be owing to sulphur in combination with a peculiar essential oil. For aromatic and carminative fruits, the most celebrated are anise (*Pimpinella Anisum*), dill (*Anethum graveolens*), caraway (*Carum Carvi*), and coriander (*Coriandrum sativum*). Besides these, great numbers of less note are also employed for the same reason, the chief of which are the ajwains or ajowans of India (species of *Ptychotis*), honewort (*Sison Ammomum*), whose fruits smell of bugs, and cummin (*Cuminum Cyminum*), now only used in veterinary practice. Among poisons, hemlock (*Conium maculatum*) holds the first place. *Anthriscus vulgaris* and *sylvestris* are not so dangerous. *Æthusa Cynapium*, *Œnanthe crocata*, *Œ. Phellandrium*, *Cicuta maculata*, and *C. virosa* are other fatal species. See Plate 16.

APICRA. A division of the genus *Aloe*, sometimes regarded as distinct, and comprising, along with *Haworthia*, a group of species of very different aspect from the great cylindrical or tubular-flowered aloes more commonly associated with the name. The genera are dwarf or acaulescent plants, with very crowded leaves and slender flower-scapes, bearing erect greenish-white flowers, which consist in the *Apicra* series of a regular cylindrical perianth, having short, spreading, conformable limb segments. A considerable group of species is referred hither. [T. M.]

APICULATE. Terminating abruptly in a little point.

APIOS. An elegant climbing plant belonging to the natural order *Leguminosæ*, having pinnate leaves, with a terminal leaflet, and lateral clusters of brownish-purple sweet-scented flowers. It is a native of North America, from Pennsylvania to Carolina, on the mountains, in hedges, and among bushes. In this country it grows freely in common garden soil, and is easily increased by tubers. It requires to be supported like peas. The tubers, though small, are numerous, farinaceous, and eatable. [G. A. J.]

APIOSPERMUM. A genus of *Pontaceæ*, consisting a single species, a native of the marshes of Cuba. The genus has been separated from *Pistia*, with which it agrees, except that its spadix is continued beyond the whorl of stamens, and its seeds are smooth. [W. C.]

APIUM. A genus of umbelliferous plants consisting of but few species, one of which is the well-known Celery, *A. graveolens*; and the other the common Parsley, *A. Petroselinum*, which occupies a spot in almost every garden.

The Celery, in its wild state, is found in marshy places and ditches near the sea coast in various parts of England. It is a biennial; and as grown in its native ditches the whole plant has a strong taste and smell, and is acrid and dangerous to eat. Such, however, are the wonderful changes effected by cultivation, that this rank, coarse, and more than suspicious plant has by degrees been transformed into the sweet, crisp, wholesome, and most agreeable of our cultivated vegetables. In Italy and the Levant, where it is much grown, but not blanched, the green leaves and stalks are used as an ingredient in soups. In this country they are always blanched and used raw as a salad, or dressed as a dinner vegetable. They are also sometimes made into an agreeable conserve. There are two kinds of Celery; the red and white-stalked, of both of which there are many sub-varieties. The seeds, when bruised and tied into a bag, form an excellent substitute for flavouring soups when Celery cannot be procured.

Celeriac or turnip-rooted Celery, is a variety of the preceding, obtained by cultivation. It is very seldom grown in this country; but in France, and more especially in Germany, it is commonly employed as a vegetable, and is considered hardier than Celery, and capable of being preserved for use much later in spring. It is excellent for soups, in which slices of it are used as ingredients, and readily impart their flavour. With the Germans it is also a favourite salad; the roots being prepared by boiling until a fork will pass readily through them, and when cold eaten with oil and vinegar.

Parsley, which is sometimes called *Petroselinum sativum*, is a hardy biennial plant; and although so common as to be naturalised in some parts of England and Scotland, was originally introduced from Sardinia, of which it is a native, in 1548. It is a well-known seasoning herb, and is in constant demand throughout the year for a variety of culinary purposes, such as sauces, soups, &c., and for garnishing various dishes. Among the ancient Greeks and Romans, Parsley always formed a part of their festive garlands, on account of retaining its colour so long; and Pliny states that in his time there was not a salad or sauce presented at table without it. The ancients supposed that its grateful smell absorbed the inebriating fumes of wine, and by that means prevented intoxication; but however this may be, we believe nothing is more effectual than the eating a leaf or two of Parsley to take off the smell and prevent the after-taste of any dish that has been strongly flavoured with onions. In Cornwall it is much esteemed and largely used in parsley pie, which are peculiar to that part of England. If dried and preserved in bottles from which the air is excluded, it will retain its flavour for a long time, and be found extremely useful for seasoning omelets and similar dishes. The curled-leaved Parsley is always preferred for use as being more ornamental than the common sort, of which it is nothing more than a variety obtained and continued by careful cultivation.

Hamburgh Parsley, *A. Petroselinum* var. *fusiforme*, is a variety of the preceding, and may be used for the same purposes; but it is chiefly grown for the sake of its long spindle-shaped roots, which are dressed and served at table as a separate dish like those of the parsnip. [W. B. B.]

APLECTRUM. A genus of melastomaceous shrubs, from the Moluccas, with opposite, stalked, elliptical-oblong, entire, five-ribbed leaves, and flowers in axillary and terminal panicles, with four petals and eight stamens; fruit, a sub-globose berry. [J. T. S.]

APLECTRUM hyemale belongs to a distinct race from the foregoing. This plant, which bears in the United States the names of Putty-root and Adam-and-Eve, is a terrestrial orchid, allied to the genus *Corallorhiza*, and inhabiting woods in rich mould, but rare. It forms tubers an inch in diameter, and sends a foot high, bearing a few dingy green flowers. Owing to its tubers containing a large quantity of very adhesive mucilage, which is employed in mending broken porcelain, it has gained in the United States the name of Putty-root. The solitary leaf is broad and ribbed, like that of a *Veratrum*.

APLOCARYA. A genus of South American scrubby shrubs, of the order *Nolanaceæ*, with fleshy leaves, separated from *Nolana* on account of the five ovaries being free, and the fruit of five separate nuts. [J. T. S.]

APLOPHYLLUM. The plants constituting this genus of the rue family (*Rutaceæ*) are perennials or small shrubs, with simple, alternate, dotted leaves, no stipules, and bearing yellow or white flowers in panicled cymes. They are distinguished from rue (*Ruta*) by their simple, undivided leaves, whence also they derive their name, as well as by the parts of the flower being arranged in fives, not in fours. They are natives of S. Europe, etc. [M. T. M.]

APLOTAXIS. A genus of the composite family (Compositae), chiefly found in the alpine and temperate regions of the Himalayas, one only being known in Siberia. Upwards of twenty species are recorded. They are herbs from one inch to three feet in height, varying much in appearance, those growing in the high alpine regions being very dwarf, the taller species being found at much lower elevations, and some of them not unlike burdocks, but the scales of their involucres are not hooked, as in that genus. The hairs of the pappus being in a single series, give rise to the name of the genus. A. *gossypina* is found in Kumaon, at an elevation of from 16,000 to 18,000 feet. The plant, altogether not higher than two or four inches, has its leaves densely clothed with long cottony hairs, which form an admirable covering to protect it from the cold to which it is exposed. A. *Lappa*, the root of which is the Costus of the ancients, is found on the mountain slopes of the Cashmere Valley, at an altitude of 8,000 to 9,000 feet. It is a gregarious herb, six to seven feet high, with an annual stem and perennial root, which is thick and aromatic. The leaves lyrate-pinnatifid, and about two feet long. The flower-heads two to three, sessile, and the florets of a purple colour. Dr. Falconer (from whose account the following is abridged) described the plant under the name of *Aucklandia Costus*. In Cashmere the plant is called Koot, in Bengal Putchuk, and the Arabic name is Koost. It is gathered largely, the greater portion being taken on bullocks, sent through the Punjab to Bombay, and there shipped for the Red Sea, the Persian Gulf, and China. A portion of it finds its way to Calcutta, through Hindostan. The roots are dug up in September and October, cut into pieces, two to six inches long, and exported without further preparation. The quantity collected amounts to about 200,000 lbs. per annum; the cost of collecting and transport to a mercantile depot in Cashmere is said to be 2s. 6d. per cwt., but when it reaches Canton it is sold for 47s. 6d. per cwt. The root is used by the Chinese as an aphrodisiac, and for burning as incense in their temples. In Cashmere the root is only employed for protecting bales of shawls from the attacks of moths; and the stems of the plants are suspended from the necks of children, to avert the evil eye. [A. A. B.]

APOCARPOUS. Having the carpels, or at least their styles, disunited.

APOCYN GOBE-MOUCHE. Fr. *Apocynum androsæmifolium*.

APOCYNACEÆ. (*Contortæ*, *Vincæ*, *Apocyneæ*, *Dogbanes*.) A natural order of corolliferal exogens, with a superior ovary, free epipetalous stamens, a pulley-shaped (trochlear) stigma, and unequal-sided lobes of the corolla, on which last account Linnæus called them *contorted*, or *twisted-flowered plants*, the corolla having some resemblance to a Catharine-wheel firework in motion. Most of the species inhabit tropical countries; the northern forms are the *Vinca*, or Periwinkle, *Nerium*, or Oleander, and a few more. In several the species form a poisonous, acrid, milky secretion, which renders them dangerous; but others are mild enough in their action to be useful in medicine, and in a few cases the milk is bland enough to form a palatable beverage. Some yield the gum-elastic Caoutchouc (see **VAHEA**); while some *Hancornias* and *Carissas* produce an eatable, and, as travellers say, a pleasant fruit. See **TANGHINIA**, **TABERNÆMONTANA**, **HANCORNIA**, &c. The commoner forms in cultivation are those of *Allamanda*, *Parsonsia*, *Vinca*, and *Tabernæmontana*. About 600 species are known, distributed through about 100 genera.

APOCYNUM. A genus of *Apocynaceæ*, containing four species of perennial herbs, with upright branching stems, opposite, mucronate-pointed leaves, a tough fibrous bark, and small, pale, and terminal or axillary flowers on short pedicels. The calyx is five-parted, and the corolla campanulate, five-cleft, bearing five triangular scales in the throat opposite the lobes. The five stamens, inserted on the very base of the corolla, have the filaments flat and shorter than the arrow-shaped anthers, which converge around the ovoid obscurely two-lobed stigma, and slightly adhere to it by their inner face. The fruit consists of two long, slender and coriaceous follicles, containing numerous ovoid seeds, compose, with a long tuft of silky down at the apex. From the fibrous bark of *A. cannabinum* (commonly called Indian Hemp), and *hypericifolium*, the Indians prepare a substitute for hemp, of which they make twine, bags, fishing-nets, and lines, as well as linen for their own use. The members of the genus afford by incision a milky juice, which, when sufficiently dried, exhibits the properties of India-rubber.

A. androsæmifolium, the Fly Trap of North America, is cultivated as an object of curiosity in this country. The five scales in the throat of the corolla of this plant secrete a sweet liquid, which attracts flies and other insects to settle on them; the scales are endowed with a peculiar irritability, the cause of which has not been accurately determined, but which causes them to bend inwards towards the centre of the flower, when touched, and to retain the unlucky flies as prisoners. Numbers of dead flies may be seen in the several flowers of this plant; the movement of the scales probably serves to scatter the pollen on the stigma. These plants are more or less poisonous and acrid, and produce emetic and diaphoretic effects. They are widely distributed over the temperate parts of both hemispheres, and a few are in cultivation, but possess no great beauty. [M. T. M.]

APODANTHES. One of the genera of *Rafflesiaceæ*, characterised by unisexual flowers, a four-cleft calyx, which is provided with two bracts, petals inserted on the ovary. The male flowers are not known; the female flowers have a half-superior ovary which, when mature, becomes a fleshy fruit with a four-cornered cavity, containing several seeds with a hard-pitted covering. The plants are natives of Guiana. (M. T. M.)

APODYTES. A genus of *Olacineae*, containing a single species from Port Natal, South Africa. It is a tree or shrub, with alternate, exstipulate, petiolate, and entire leaves. The flowers are in loosely branched terminal racemes. The calyx is small, five-toothed, and persistent. The corolla consists of five oblong linear petals, rising from the receptacle. The stamens are five in number, alternating with the petals, and, by thin dilated filaments, uniting the petals together through two-thirds of their length. The ovary is free and unilocular, with two ovules. The style is excentric, and kneed at the base; the stigma is minute. The fruit is a drupe, one-celled, and one-seeded by abortion, of a peculiar kidney shape, with a fleshy protuberance from the hollow side. (W. C.)

APONOGETON. A genus of aquatic plants belonging to the *Juncagineae*, and remarkable for producing its flowers in conjugate or bifidate spikes at the ends of the flower scapes. The flowers consist of several (six to eighteen) stamens with subulate filaments, and are destitute of both calyx and corolla, the conspicuous part of the inflorescence being a double row of large white bracts, at the base of which the minute apetalous flowers are seated. *A. distachyon*, a very handsome, deliciously fragrant water-plant, a great favourite in gardens, has been well figured and described from vigorous, well-developed specimens in *Paxton's Flower Garden* (il. t. 40) by Dr. Lindley, who writes:—"In appearance it resembles a pondweed (*Potamogeton natans*), except that it is of a clear green colour without any tinge of brown. Its bulb or corm is described as being as large as a hen's egg. The leaves float on the surface of the water, are oblong, about eighteen inches long when full grown, flat, and have three distinct veins running parallel with the main rib. When young their sides are rolled inwards. The flowers are placed on a forked inflorescence, originally included within a taper-pointed calyx-parted spathe (cup), which is forced off as they advance in size. When fully formed each fork of the inflorescence is very pale green, and is bordered by two rows of large, ovate-oblong, oblique, ivory-white bracts, in the axils of which stand the minute flowers. The latter are bisexual, and destitute of both calyx and corolla. Twelve hypogynous five stamens, with dark purple anthers, surround from four to six distinct carpels, each of which has a short curved style, a simple minute stigma, and six erect anatropal ovules. After flowering the bracts and inflorescence grow rapidly, acquire a deep green colour, and soon resemble tufts of leaves, among which lie in abundance large membranous indehiscent beaked carpels, containing about four seeds each, and readily tearing at the sides." This species is common at the Cape of Good Hope, where it bears the name of Water Uintjies. The flowering tops are, according to Mr. Bunbury, sometimes used in the colony both as a pickle and as a substitute for asparagus. Marlyn states that the "bulbs" are eaten roasted. There are one or two other species from South Africa. The Indian *A. monostachys* is now referred to *Spathium*. (T. M.)

APOO. (Fr.) *Urtica*.

APOPHYSIS. A name given to a swelling, often hollow, or of extremely loose texture, at the base of the capsules in several mosses. It is developed extremely in the natural order *Splachnei*, where it often exceeds in size the true capsule. It attains its maximum in *Splachnum luteum* and *rubrum*, where it is a most conspicuous object, hanging down like an umbrella or the vesicle of *Aesophera*. In *Oedipodium* almost the whole of the stem consists of apophysis, which is confluent at once with it and the capsule. (M. J. B.)

APOROCACTUS. A genus of *Cactaceae*, distinguished from *Cereus* by the elongated narrow tube of the perianth, and its obliquely ampling, somewhat two-lipped limb, and also by the graduated insertion of the lower stamens, all of which are exserted, and the upper ones longer than the rest. It includes a few species known in gardens under the name of *Cereus*. *A. Baumanni*, sometimes called *Cereus Treedici*, is a handsome, cylindraceous, erect-stemmed succulent plant from Buenos Ayres, having a many-angled stem, and numerous rich orange-crimson, slender-tubed flowers, curved at the base so as to be inserted obliquely, and also curving in an opposite direction at the mouth, which is slightly spreading. *A. flagelliformis*, the *Cereus flagelliformis* or Creeping Cereus of gardens, is a well-known plant, with long, slender, pendent stems, producing a profusion of narrow rose-coloured flowers, 'so beautiful, and produced in such great plenty, that this may be placed in the first class of exotic plants" in point of ornament. It is a native of Peru. (T. M.)

APORUM. A division of the great genus *Dendrobium*, distinguished by having fleshy equitant leaves. The flowers are small, and have no beauty.

APOSTASIA. Among the forests of Malacca, Burmah, and Assam are found two species of Endogens, with low stems, covered with grassy leaves, and terminal panicles of small yellow flowers, which throw an unexpected light upon the structure of the curious order of orchids. The calyx and corolla consist each of three narrow equal pieces. The anthers—two or three—are distinct; the style is perfectly free from the stamens, and the ovary is three-celled; so that the synandrous structure of orchids wholly disappears. Another genus nearly related is *Neowiedia*, a Borneo plant resembling a dwarf palm, with dense spikes of resinous flowers, and a three-winged ovary, terminating in a narrow neck.

APOSTASIACE.E. This is a very small group, bordering on the limits of the vast orchidaceous order, from which it differs mainly in its stamens not being synandrous, but distinct from each other and from the style. It stands near the genus *Corysanthes*, some of the reputed species of which, now called *Acianthus*, have a three-celled ovary. The flowers of all the known species are small and inconspicuous, while the leaves are strongly marked by stout parallel nerve veins, as in *Curculigo*, or any similar plant.

APOSTAXIS. Unusual discharge of the juices of plants. This may arise merely from an extreme abundance of fluid, which is in consequence discharged, as in Indian summer of the vine, from the point, or serrated top of the leaves. If, however, it is exhibited say which flows out, either from injury or weakness of the tissues, the effect may be injurious. And this is exactly the case in what is called gum-mosis; a condition which may be induced artificially, by allowing water to drop constantly over a branch. This always precedes from injured or diseased tissues, and is with difficulty arrested when once set up, and, if so, is the certain forerunner of fatal canker. In some cases, as in the transvaalit plants, the gum is organised, and is derived apparently from the medullary rays. In conifers, a flow of resin is often attended with the same fatal results as gumming in plums and other allied plants. In this case it seems to arise generally from root-enfeeblement and a consequent check of circulation. (M. J. B.)

APOTHECIA. The shields of lichens; from being disks arising from a thallus, &c., containing spires.

APPENDAGES. Leaves and all their modifications are appendages of the axis. Hairs, prickles, &c., are appendages of the part which bears them. A name applied to processes of any kind.

APPENDICULA. A genus of inconspicuous orchids, inhabiting tropical Asia. They have tufted stems, clothed with oblong distichous leaves, bearing at the end a few green flowers. About twenty species are known, one only of which has been in cultivation in Europe.

APPETIT. (Fr.) *Allium Schœnoprasum*.

APPLE. *Pyrus Malus*. —, ADAM'S. A variety of the Lime, *Citrus Limetta*. —, ALLIGATOR. The fruit of *Anona palustris*. —, BALSAM. The fruit of *Momordica Balsamina*. —, CHERRY. The Siberian Crab, *Pyrus baccata*. —, CUSTARD. The fruit of *Anona reticulata*; also a common name for the family of *Anonæ*. —, DEVIL'S. The fruit of *Mandragora officinalis*. —, EGG. The fruit of *Solanum esculentum*. —, ELEPHANT

The fruit of *Feronia elephantum*. —, KANGAROO. The fruit of *Solanum laciniatum*. —, KAU. The name, in south Africa, of a fruit supposed to belong to *Dospyros*. —, LOVE. The fruit of *Lycopersicum esculentum*. —, MAD, or JEW'S. The fruit of *Solanum æthiopicum*. —, MAMMEE. The fruit of *Mammea americana*. —, MANDRAKE. The fruit of *Mandragora officinalis*. —, MAY. *Podophyllum peltatum*. —, MONKEY. A West Indian name for *Clusia flava*. —, OAK. A spongy excrescence, formed on the branches of the oak-tree. —, of PERU. The fruit of *Nicandra physaloides*. —, of SODOM. The fruit of *Solanum sodomeum*. —, OTAHEITE. The fruit of *Spondias dulcis*. —, PERSIAN. A name given to the peach, when first introduced into Europe. —, PINK. *Anamassa sativa*. —, PRAIRIE. *Psoralea esculenta*. —, ROSE. The fruit of *Eugenia malaccensis*, *E. aquea*, *E. Jambos*, and others. —, STAR. The fruit of *Chrysophyllum Cainito*. —, THORN. *Datura Stramonium*. —, WILD BALSAM. *Echinocystis lobata*.

APPLE BERRY. A colonial name for *Billardiera*.

APPLE-TREE, MALAY. *Eugenia malaccensis*.

APPLEWORTS. An English name proposed for the order *Pomaceæ*.

APRICOT. *Prunus Armeniaca*; formerly sometimes written *Apricock*. —, WILD. *Mammea americana*.

APTANDRACEÆ. Out of the genus *Aptandra*, Mr. Miers has proposed to form a natural order, bearing this name. Only one species is known, a tree with alternate leaves and minute flowers, a native of the banks of the river Amazon. It is usually referred to Hamelieæ. Its great feature is having anthers opening by reflexed valves, as in *Lauraceæ*.

APTERIA *setacea*. An obscure North American plant related to *Burmannia*, but destitute of wings to the fruit.

AQUIFOLIACEÆ. (*Ilicineæ*, *Hollyworts*). The common holly-tree is the type of a small natural order of shrubs and trees, with rotate monopetalous flowers, a definite number of epipetalous stamens, and a fleshy fruit. The species may be said to possess in general emetic qualities, variously modified in various instances. Birdlime is obtained from the bark of the common holly, and the beautiful white wood is much esteemed by cabinet-makers for inlaying. A decoction of *Ilex cassinioria*, called Black drink, is used by the Creek Indians at the opening of their councils, and it acts as a mild emetic. But the most celebrated product of the order is Maté, of Paraguay tea, the dried leaves of *Ilex paraguayensis*; which see.

AQUILARIA. The Eagle-wood, or Agallochum of the ancients, is produced from certain species of this genus; hence the name. The genus gives its name to the order *Aquilariaceæ*, and is characterised by a top-shaped leathery calyx, downy externally, whose limb is divided into five small oblong, reflexed segments; from the throat of the calyx project ten woolly scales, which adhere to the whole length of the interior of the calyx tube, and alternate with the ten stamens, the filaments of which also adhere for nearly their whole length to the calyx tube, and are attached to the back of the anthers below their middle. The ovary is two-celled, each cell containing a single ovule, suspended from the placentas; these ovules are flat on one side, convex on the other, and winged, the wing being prolonged downwards into a horn-like process; the ovary is surmounted by a short style, terminated by a large round stigma, which is depressed in the centre. *A. Agallocha*, a large tree, inhabiting Silhet, and provided with alternate lance-shaped stalked leaves, furnishes an odoriferous wood, called Aloes-wood, or Eagle-wood, supposed to be the aloes-wood of Scripture. The wood contains an abundance of resin, and an essential oil, which is separated, and highly esteemed as a perfume. The Orientals burn it in their temples for the sake of its slight fragrance, on which account also it was used in the palace of Napoleon the First. It has been prescribed in rheumatic affections in Europe, either but inferior kinds of this wood are said to be furnished by species of *Excæcaria* and *Brucea*. [M. T. M.]

AQUILARIACEÆ (*Aquilariads*) consist of fragrant tropical Asiatic trees, with small spetalous flowers, resembling those of a *Rhamnus*. Only ten species are known, of which the most important is the genus *Aquilaria*; which see.

AQUILEGIA. A genus of *Ranunculaceæ*, widely distributed over the temperate regions of the northern hemisphere. It is generally considered to consist of many species, but the authors of the *Flora Indica* believe that the greater number of these are merely varieties. It is distinguished by the curious structure of the flowers, which have five flat, elliptical coloured sepals, alternating with as many spurred petals; the spurs are very large, and produced backwards into hollow tubes, like a cornucopia with the mouth downward, and are frequently curved round towards the central axis of the flower at the extremity. The fruit consists of five follicles, with numerous seeds. In cultivation, double varieties occur, which have a series of spurred petals, with the spurs included in those of the exterior ones, like a nest of crucibles. Stellate varieties also occur, which have the petals flat, and destitute of a spur. The flowers are drooping, unless *A. parviflora*, which Ledebour describes with the flowers perfectly erect, be an exception. The five-spurred petals with incurved heads have been compared to five doves, the sepals representing the wings, and to this the English name Columbine refers. The leaves are ternate, the root-leaves twice or thrice-ternate. *A. vulgaris*, the Common Columbine, is apparently native in Britain. It has the flowers usually purplish blue, but in cultivation they vary much, being dark purple, dull reddish, or white. *A. alpina* has much larger flowers and shorter spurs, and stamens hardly exceeding the petals. *A. canadensis* has scarlet and yellow flowers, with very long slender straight spurs, and very long stamens. The species are quite hardy in the open border. [J. T. S.]

ARABETTE (Fr.) *Arabis*.

ARABIS. An extensive genus of annual or perennial herbaceous plants, belonging to the natural order *Cruciferæ*, and bearing white, or (rarely) purple flowers. For the most part they are under a foot in height, the root-leaves are stalked, but the upper ones clasp the stem, and all are more or less thickly set with forked hairs. They inhabit various countries, but the British species possess little interest. The name *Arabis* was probably given to the genus because most of the species delight in stony or sandy soil, such as that of Arabia is presumed to be; Wallcress, the English name, has similar reference to the natural place of growth. Many species are well adapted for rock-work, and others are equally fitted to be grown as border flowers, as they bloom earlier than most garden plants. The genus being closely allied with others, some confusion exists as to the names severally assigned to the plants which it contains. Some species are described under the names of *Turritis* and *Cardamine*. [C. A. J.]

ARAÇA. A name given to the fruit of some Brazilian *Eugenias*.

ARACEÆ (*Aroideæ*, *Arads*) are incomplete plants of the Endogenous class, with numerous naked unisexual flowers, closely packed upon a spadix, shielded when young by the hooded leaf called a spathe, as is seen in the common wake-robin (*Arum maculatum*). They are common in tropical countries, but rare in those with a cold or temperate climate. Botanists have mixed them with Orontiads, from which their hermaphrodite flowers distinguish them. Most have tuberous roots (*cormus*), but some acquire the stature of little trees, the most interesting of which is the Dumb-cane, a species of the genus *Dieffenbachia*. The acrid poisonous qualities which have given rise to the latter name are characteristic of the order. Nevertheless the whole contain starch in such abundance that it may be separated in the form of arrowroot or used as food in its combined state; only, however, after careful washing to remove the acrid juices. Thus, the common spotted *Arum* was eaten with usin time of scarcity, and yields a kind of arrowroot, and the *Colocasias* are grown everywhere in hot countries as common field crops. See all these names. Among the peculiarities of the order is to extend the end of the spadix into a soft, cellular, enlarged process, which is the growing point of the flower-branch, and analogous to the succulent receptacle of the strawberry, the dry core of the raspberry, the spongy excrescence called the oak apple, and even the stiff hard spine of the *Gleditschia*. Scarcely more than 200 species are known. The appearance presented by this very distinct race of plants, is shown in plate 7, in which a group of *Caladiums* is seen at fig. 5.

ARACHIS. A genus of leguminous plants, remarkable for the peculiar structure of its calyx, and the habit of thrusting its fruit into the ground. M. Poiteau, in the *Annales des Sc. Nat.*, 1853 (xix. 265), gives a good description and figure of *A. hypogœa*. The principal characters of the genus are the immensely long tube of the calyx, whose limb is two-lipped; the corolla papilionaceous and yellow; and eight stamens united into one parcel. The ovary is very small, and placed at the bottom of the very long calyx tube; it contains two ovules, and is terminated by a very long style, thickened at its extremity, and covered with hairs at the place where it comes in contact with the stamens. After the fall of the flower, the ovary, which is very small, is gradually raised upon a stalk which in time attains a length of two to three inches, and in its growth curves downwards, so that at length the small ovary at its extremity is thrust into the ground. When this happens, the ovary begins to enlarge, and ripens into a pale yellowish wrinkled slightly curved pod, often contracted in the middle, and containing two seeds. Should the ovary by some accident not be enabled to thrust its pods into the ground, it withers and does not attain perfection. The plant was originally a native of the West Indies and West Africa, but is now cultivated in warm climates, preferring a light sandy soil. The seeds which are of the size of a pea are eaten as food, but are chiefly valuable for the quantity of oil they produce when pressed. The oil is used as a substitute for that of olives, to which it is equal in quality. The plant might with much advantage be extensively cultivated in Australia and others of our colonies for the sake of its excellent oil, while the herbage would form valuable forage for cattle, who eat it greedily. The pods are known in this country as Ground Nuts. The peculiarity of thrusting the fruit into the soil to effect its maturation there, is not confined to this genus, but exists also in the allied genus *Voandzeia*, a native of Surinam, where its seeds are eaten, like those of the *Arachis*, as peas by the negroes. [M. T. M.]

ARACHNIODES. A doubtful genus of ferns, supposed to belong to the *Polypodium* or *Aspidium*. The veins of the fronds are free, and the sori are said to have an arachnoid or cobweb-like involucre covering them. The only recorded species is a native of Java. [T. M.]

ARACHNIS (from the Greek; a spider. A Javanese orchid, of epiphytal habit, whose name has been derived from the extraordinary resemblance of its flowers to a huge spider. The plant has flowers five inches in diameter, of a lemon colour, with great purple spots; they grow as many as twelve together, on a long loose spike arising from one side of a strong scrambling stem. They are said to have the most delicate smell of musk, but so penetrating withal that a single spike will scent an entire meeting-hall. Kaempfer, however, asserts that this odour resides exclusively in the ends of the sepals and petals, which are broader at the end than elsewhere; and he says that if they are cut off all fragrance ceases. The plant has had several names, as *Epidendrum flos aeris*, *Aerides arachnites*, *Renanthera arachnites*, and *Arachnis moschifera*. It is called in Java, Katong ging. Undoubtedly it is one of the most remarkable plants of its remarkable order, and it is not a little surprising that it should never have been introduced into Europe.

ARACHNITIS is a name given to the spider *Ophrys*.

ARACHNOID. Resembling cobweb in appearance; seeming to be covered with cobweb, in consequence of the entanglement of long white hairs.

ARACHNOTHRIX. A genus of plants closely related to *Rondeletia*, from which it differs in having the corolla four-parted, with its tube and throat smooth. The anthers are placed towards the top of the tube of the corolla on very short filaments. The plants are covered with a more or less cobweb-like clothing of hairs, hence the name. It belongs to the natural order *Cinchonaceæ*. [M. T. M.]

ARADS. An English name for the *Araceæ* or Arum family.

ARALIA. This genus is the type of the order *Araliaceæ*, and consists of trees, shrubs, and herbs of rather striking character, found in North America, and in New Zealand, Japan, and the East. The flowers are inconspicuous, collected in umbels, the umbels not unfrequently raised in large compound panicles. The calyx has a very short superior limb, which is entire or five-toothed; the corolla consists of five petals inserted on the margin of the epigynous disk; the stamens are five in number, alternating with the petals; and the ovary is inferior, five to ten-lobed, with a solitary pendulous ovule in each cell, and becomes a berry-like drupe. The foliage is very various in character, but generally of an ornamental aspect; sometimes simple, entire, or lobed, sometimes digitate, pinnate, twice ternate, bipinnate, or superdecompound. Some of the species have smooth, and others prickly stems. One of the former, *A. racemosa*, grows three to four feet in height, with a divaricately-branched herbaceous stem, bearing coun-pound leaves, the petioles of which are tripartite, each division bearing from three to five ovate or heart-shaped serrated leaflets. This plant is called Spikenard in North America, and is highly esteemed as a medicine. The roots of *A. nudicaulis*, another North American herbaceous species, were formerly imported and sold for sarsaparilla; and they are stated to be used by the Crees, under the name of Rabbit-root, as a remedy against syphilis, and also as an application to recent wounds. *A. spinosa*, one of the prickly species, is a small, simple-stemmed tree, eight to twelve feet high, the stems and leafstalks both prickly, the leaves doubly and triply pinnated with ovate serrate leaflets, and the panicle much branched, downy, bearing numerous umbels of flowers. This is known in America under the name of Angelica tree, and the berries are used in an infusion of wine or spirits for relieving rheumatic pains and violent colic. The tincture has also been found to relieve toothache. The Rice Paper plant of China has been referred to this genus by Sir W. J. Hooker, under the name of *A. papyrifera*. This plant grows in the deep swampy forests of the island of Formosa, and apparently there only, forming a small tree, branching in the upper part, the younger portions of the stem, together with the leaves and inflorescence, covered with copious stellate down. The full-grown leaves are sometimes a foot long, cordate, five to seven-lobed, of a soft and flaccid texture. The panicles of flowers come from the extremities of the stem and branches, rising above them, and then becoming pendulous, one to three feet long, bearing the numerous capitate umbels of small greenish flowers. The stems are filled with pith of very fine texture, and white as snow, which when cut forms the article known as rice paper. Large quantities of the stems are 'taken in native crafts from Formosa to Chinchew, where they are cut into thin sheets for the manufacture of artificial flowers.' A lengthened account of this interesting plant will be found in *Hooker's Journal of Botany*. [T. M.]

ARALIACEÆ (*Araliads, Ivyworts*) form a small natural order closely approaching umbellifers, from which they in reality differ in little, except their fruit always consisting of more carpels than two, and having no double epigynous disk. They are also more generally arborescent, many of them being trees or large shrubs, and very few herbs. Several are conspicuous for their broad noble foliage. The species are found in the tropical and sub-tropical regions of the world; and in some of the coldest, as in Canada, the north-west coast of America, and Japan. *Aralia polaris* even occurs in Lord Auckland's Islands, in 50° south latitude. They have aromatic qualities, usually slight, but occasionally intense. One of them forms a soft white spongy pith, which when cut into thin plates and flattened becomes the so-called Rice Paper plant of the Chinese. See HEDERA, ARALIA, PANAX, GUNNERA, ADOXA, &c.

ARA-ROOT. The same as Arrow-root, which see.

ARAR TREE. *Callitris quadrivalvis*, formerly called *Thuja articulata*, a great coniferous tree, which yields gum sandarac.

ARATICU DO MATO. A Brazilian name for *Anona spinifera*.

ARAUCARIA. A genus of *Coniferæ*, consisting of lofty evergreen trees, with verticillate spreading branches, covered with stiff, flattened, pointed leaves, usually imbricate, but more or less spreading. The spikes of male flowers are cylindrical and terminal; each anther divided into from six to twenty cells. The ripe cones in the females are large, globular, terminal, densely imbricated with numerous woody scales, sometimes winged, each bearing a single adnate seed, and many of them usually barren. There are five or six species known, all from the southern hemisphere. *A. imbricata* is the species commonly planted in this country, and the only one which will bear our climate without protection. It is a native of the mountains of Southern Chili, where it forms vast forests, attaining a great height, and supplying a hard and durable timber. The seeds are also edible when fresh. The leaves are very spreading, vertically flattened, broadly lanceolate, very stiff, with long pungent points, and attain a couple of inches in length. The cones, sessile at the extremities of the branches, are of the size of a child's head. *A. brasiliensis* forms large forests in south tropical Brazil. It much resembles the Chilian species, but is rather more elegant in growth and of a better colour. It is occasionally planted in Southern Europe, where it succeeds better than the *A. imbricata*, but is too tender for this country. *A. excelsa*, the Norfolk Island pine, attains the height of 200 feet. The leaves are much shorter than in the two preceding species, and but slightly flattened, and the scales of the cone are broadly winged with a hooked point. It will not bear the open air in our climate, but forms a conspicuous object in lofty conservatories. It has been considered by some botanists as forming, with two Australian species, a distinct genus under the name of *Eutassa* or *Eutacta*.

ARBOL DE CORAL. A Mexican name for *Piscidia Corallodendron*. — DEL CERA. A South American name for *Elæagia utilis*. — DE LECHE. The Cow Tree, *Brosimum Galactodendron*. — DE ULE. A Mexican name for *Castilloa elastica*.

ARBOR JUDÆ. A common name for *Cercis* or Judas tree. — VITÆ. The common name for *Thuja*.

ARBOUSIER. (Fr.) *Arbutus*.

ARBRE A' CRAPELET. (Fr.) *Melia Azederach*. — A' FRANGES. *Chionanthus virginica*. — A' LA CIRE. *Myrica cer-*

tera. — A' PERRUQUE. *Rhus Cotinus.*
— A' SUIF. *Arctosgos adipova.* — A'
POIVRE. *Vitex Agnus-castus.* — AUX
ANEMONES. *Calycanthus floridus.* — AUX
FRAISES. *Arbutus Unedo.* — AUX QUA-
RANTE ECUS. *Salisburia adiantifolia.*
— A'RGENT. *Leucodendron argenteum.*
— DE CASTOR. *Magnolia glauca.* — DE
JUPITER. *Abrus hipartitum.* — DE
NEIGE. *Chionanthus virginica.* — DE
SOIE. *Acacia Julibrissin.* — DE STE.
LUCIE. *Cerasus Mahaleb.* — DE VIE.
Cupressus thuroides, Thuja occidentalis,
and *Thuja orientalis.* — DU VOYAGEUR.
Ravenala madagascariensis, sometimes
called *Urania speciosa.* — SAINT. *Melia
Azederach.*

ARBUTUS. A genus of *Ericaceæ,* consisting of trees and shrubs, natives of
Southern Europe, the Canary Islands,
North America and Chili. Twenty-five
species have been described. They have
alternate, entire, or toothed evergreen
leaves. The pedicellate and bracteate flowers
are in terminal panicles or racemes; the
corolla is white or reddish; the calyx
inferior, and consisting of five small sepals.
The deciduous corolla is globosely or
ovately campanulate, with a small contracted five-cleft and reflexed border.
It encloses the ten stamens, which have
flattened filaments, and anthers compressed at the sides, opening by two
terminal pores, and attached below the
apex, where two reflexed awns are produced. The ovary has five cells, with many
ovules in each. There is a single style
with an obtuse stigma. The fruit is a
globular fish-coloured berry, rough with
granular tubercles, and containing five
many-seeded cells. The berries are edible
though not agreeable. *A. Unedo* is called
the strawberry tree from its fruit resembling a strawberry at a distance. When
eaten in quantities this fruit is said to be
narcotic. A wine is made from it in
Corsica, but it has the same property as
the fruit. In Spain both a sugar and a
spirit are obtained from it. The bark and
leaves of the same plant are used as
astringents; in some parts of Greece they
are employed for tanning leather. This
species grows abundantly on the rocks at
Killarney. It is cultivated as an ornamental shrub, and as it ripens its fruit the
second year, it is peculiarly beautiful in
October and November, being covered at
the same time with blossoms and ripe
fruit. [W. C.]

ARBUTUS, TRAILING. An American
name of *Epigæa repens.*

ARCEUTHOS. (Gr.) *Juniperus oxycedrus* and *Juniperus phœnicea.*

ARCHANGEL. A common name for
Lamium and *Galeobdolon;* also applied to
Archangelica officinalis.

ARCHANGELICA. A genus of umbelliferous plants, whose stems and leaves have
a very powerful and agreeable aromatic
smell. The *Angelica, A. officinalis,* is the

only species grown for culinary or medicinal
purposes. *Angelica* is a hardy biennial, from
three to six feet high, found in **Eastland** in
moist situations, but believed to be originally a native of the northern parts of
Europe. It has been in cultivation since
1568. The leaf stalks were formerly blanched and eaten like Celery. They have,
however, been ceased to be so, and are
now in request for the use of confectioners,
who make an excellent sweetmeat with
the tender stems, stalks, and ribs of the
leaves, candied with sugar. The seeds and
leaves are powerfully aromatic, and are
used in country places for their supposed
medicinal properties. [W. B. B.]

ARCHEGONE. A term applied to the
from-celled cellular sacs which occur in
the higher of acrogenous cryptogams, and
which are analogous to the pistils of
phenogams. They contain at the base of
their cavity a sac which is analogous to
the embryo-sac of phenogams, and which is
impregnated by the agency of spermatozoids. Within this latter sac, either the
young plant as in ferns, or the capsule as
in mosses, is formed by means of cellular
division. [M. J. B.]

ARCHILL, or ORCHIL. A colouring
matter obtained from various species of
lichens, especially *Roccella tinctoria.*

ARCTIUM. One of the familiar plants,
which, without culture or management,
flourishes in nearly all climates and every
kind of soil. To the agriculturist it is
best known as a troublesome weed, always
ready to make its appearance in neglected
ground, growing rapidly, and with its
large spreading leaves choking all other
vegetation; to the artist it affords a bold
and striking foreground for his landscapes;
and to the school boy its heads of flowers,
under the name of burs, offer an ever welcome supply of means for playing practical
jokes. The Burdock is of no utility to man,
as no domesticated animals, except, it is
said, the ass, will eat its leaves; though it
is a question whether it might not be sown
with advantage as a cover for pheasants in
places where there is a difficulty in raising
underwood. It was formerly commended
for its medicinal virtues, and was prescribed
for rheumatic affections. Some writers
too speak of its excellence as a culinary
vegetable. The stems, they say, should
be gathered young, stripped of their rind,
and treated as asparagus. When burnt
the ashes afford a large quantity of alkaline
salt. There is but one British species of
Burdock, of which modern botanists reckon
two varieties, *A. Bardana,* with a cottony
substance investing the heads; and *A.
Lappa,* which is destitute of this appendage. By some continental authorities,
Lappa is made the name of the genus, and
the two plants are described as distinct
species, *L. tomentosa* and *L. major.* A third
species, *L. minor,* grows on the continent
of Europe, but does not occur in Britain.
The name *Lappa* is derived from the
Celtic *llap,* a hand, from its prehensile pro-

perties. *Arctium* is from *arctos*, the Great Bear, from the rough character of the plant. [C. A. J.]

A. Lappa, under the name of Gobo, is cultivated in Japan as a vegetable.

ARCTOCALYX. A genus of *Gesneraceæ*, consisting of half-shrubby plants inhabiting the mountains of Mexico. They have ovate-lanceolate or elliptical leaves, axillary flowers of an unusually large size for the order, and of a bright orange spotted with brown or purple. The calyx is large, tubulous-bellshaped, fifteen-nerved, and five-toothed. The corolla is funnel-bell-shaped, and the ovary sunk into the calyx, and surrounded by a glandular disk. *A. Endlicherianus* is not uncommon in our gardens. [B. S.]

ARCTOSTAPHYLOS. A genus of *Ericaceæ*, consisting of procumbent shrubs, with small deciduous or persistent leaves, and rather small lunulate flowers, two or three together, in very short terminal racemes. It is very nearly related to *Arbutus*, differing from it in having a glabrous berry with five stones, and each stone being one-seeded. The genus has been recently very much limited. Ten species have been separated and placed under the generic title of *Comarostaphylis*, having as their distinctive characteristic a drupaceous fruit, with a stone hard for its albuminous stones, and a single seed in each cell. Five more species have been removed to a new genus *Daphnidostaphylis*, which is characterised by having the ovary placed on a ten-angled, hypogynous disk, and containing six to ten cells. The restricted genus *Arctostaphylos*, containing only the two species found in Britain, has the ovary without true dissepiments. The three genera have all alike a five-partite, persistent, and hypogynous calyx, a five-lobed reflexed corolla, inserted on the calyx, and ten stamens. The two species are natives of the northern regions of both the old and the new world. The whole plant of *A. uva-ursi* is astringent; it has been used for tanning leather. The berries form a favourite food of grouse and other game. The plant is a valuable medical astringent, used to check an excessive secretion of mucus, as in urinary and bronchial affections, and even in calculus. The *Arctostaphylos alpina*, or Black Bearberry, is the badge of the clan Ross. [W. C.]

ARCTOTHECA. A genus of the composite family, consisting of two species, both perennial herbs, peculiar to Southern Africa, and found in sandy spots near the sea. *A. repens*, which has been cultivated in this country, is a branching plant, six inches to a foot high, with pinnatifid leaves, about six inches long, covered underneath, as are also the stems, with a white tomentum, and bearing solitary yellow flower heads, nearly three-quarters of an inch across, borne on long naked stalks. The ray florets are strap-shaped and neuter, those of the disc perfect; involucral scales in many series; pappus none; achenes wingless and four-sided. [A. A. B.]

ARCTOTIS. A genus of composites, all natives of the Cape of Good Hope, with savoy orange-coloured flower heads, of which several species occur in gardens. It has a honey-combed receptacle set with bristles, oval grooved achenes crowned with several broad membranous scales, and an involucre of numerous imbricated leaflets with chaffy margins. The *A. speciosa* is not unfrequently cultivated under the name of *A. breviscapa*, as a half-hardy border annual, though our English summers are too short and too cool to bring it to the perfection it attains in its southern home. It would probably be seen to more advantage if treated as a tender biennial. As grown in our gardens, it is a dwarf, tufted, tomentose plant, with numerous, short, prostrate stems, proceeding from the crown of the root; three-nerved leaves, varying from oval and entire to lyrately pinnatifid, in the latter case with a large terminal lobe; and large terminal flower-heads, with a brownish disk and orange-coloured ray, expanded only in fine weather during the middle of the day. Many of the species are greenhouse perennials, which would succeed in the open ground in summer in warm situations, and some of them would be desirable additions to our gardens. Of this section, probably the only attainable species is the *A. grandiflora*, with handsome deep-orange-coloured capitules, four inches in diameter, and silvery-grey pinnatifid foliage, blooming freely during the whole summer, in a sunny situation and dry soil. [W. T.]

ARCTURIA. A section of the genus *Drosera*, of which the Tasmanian *D. Arcturi* is the type. This has three undivided styles with thickened stigmas; the stem is short, and not bulbous, with narrow leaves passing insensibly into the leaf-stalk; scape with a single white flower. [J. T. S.]

ARCTURUS. A genus of *Scrophulariaceæ*, established by Bentham, but subsequently abolished by him; the name being retained to characterise a group of the genus *Craisa*, in which the anthers are attached by their middle. [W. C.]

ARDISIA. This is a large genus of the family *Myrsineæ*, containing upwards of 100 species. They are evergreen shrubs or small trees, with alternate, rarely opposite, leaves covered with transparent dots. Their flowers are white or rose coloured, and arranged in panicles, the branches of which are often of an intense rose colour, thereby adding greatly to the beauty of the plant. The flower-stalks are often disposed in little umbellets on the branches of the panicle. They are found in India, the islands of the Indian Ocean, and America. The bark of *A. colorata* is known in Ceylon as Dun, and is used in native practice in bowel complaints, fevers, and externally for healing ulcers. It is tonic and astringent. *A. solanacea*, a native of India, is to be met with in some gardens; the juice of its berries is of a beautiful red colour, which, when put on paper, changes to a durable brown. A. *crenata*, a native of China and Penang, is a beautiful dwarf bush often cultivated in greenhouses. The leaves are always green, and in the winter season, if well managed, the plant is covered with a mass of scarlet berries, much like those of the holly. A. *primulifolia*, a native of Hong Kong, is only about six inches high, and has thin leaves like those of the common primrose. A number of species are in cultivation in English gardens. [A. A. B.]

ARDISIADS. An English name for the order *Myrsinaceae*.

AREC. (Fr.) *Areca oleracea*.

ARECA. The generic name applied to certain species of palms, characterised by having a lofty stem, pinnated leaves whose stalks are rolled up into a cylinder at their base, a double spathe enclosing the flowers, which are loose upon a branched spadix, and are unisexual. The male flowers have a six-parted perianth; the female flower contains six rudimentary stamens, and a superior one-seeded ovary which ripens into a drupe-like fruit with a fibrous rind.

A. *Catechu* is a handsome tree, cultivated in all the warmer parts of Asia for the sake of its fruits, which are of the size of a hen's-egg, of a reddish yellow colour, and with a thick fibrous rind, within which is the seed. This is known under the name of areca nut, pinang, and betel nut, and is about the size of a nutmeg, but conical in shape, flattened at the base, brownish externally, and mottled internally like a nutmeg. These nuts are cut into narrow pieces, which are rolled up with a little lime in leaves of the betle pepper. The pellet is chewed, and is hot and acrid, but possesses aromatic and astringent properties. It tinges the saliva red, and stains the teeth, and is said to produce intoxication, when the practice of chewing it is begun. The effects seem to be as much due to the other ingredients as to the areca nut. So addicted are the natives to the practice, that Blume tells us, 'they would rather forego meat and drink than their favourite areca nuts, whole ship loads of which are annually exported from Sumatra, Malacca, Siam, and Cochin China. The practice is considered beneficial, rather than otherwise. In this country the charcoal of the nuts is used as tooth powder, for which it is well adapted by its hardness. A sort of Catechu is furnished by boiling down the nuts of this palm to the consistence of an extract, but the greatest quantity of the drug called Catechu used in this country is the produce of *Acacia Catechu*. The flowers of the tree are very fragrant, and used on festive occasions in Borneo, where they are considered a necessary ingredient in medicines, and charms employed for healing the sick. In Malabar another species, A.

Dicksoni, is found wild, and furnishes a substitute for the true betel nut to the poorer classes.

A. *oleracea* is the Cabbage Palm which is found in abundance in the West Indies. It derives its name from the bud which terminates its lofty stem. This bud consists of a great number of leaves densely packed, so that the inner ones are of a white colour, and delicate flavour, and serve as a vegetable. The noble trees are destroyed for the sake of this luxury; and it is related that in the cavity formed by the removal of the 'cabbage' a kind of beetle deposits its eggs, from which maggots are produced which are articles of diet, much relished by the negroes of Guiana. [M. T. M.]

AREGMA. A remarkable genus of parasitic *Fungi*, which abound on several species of *Rosaceae*. Their first appearance is that of some yellow *Uredo*, in which condition the fruit is not distinguishable from that of the genus just mentioned, but, after a time, cylindrical dark multiseptate bodies are produced on long bulbous stalks, forming a sort of spore-shaped prothallus, the articulations of which terminate, and produce at length the true spores. Nothing is more common than the *Aregma* of the Rose and Bramble (A. *Rosae* & A. *rubi*), which afford interesting objects for the microscope, and food for much reflection, from their peculiar mode of reproduction. [M. J. B.]

AREMONIA. A name altered from *Agrimonia*, and now applied to an evergreen herb belonging to the natural order *Rosaceae*. It grows about a foot high and bears irregularly pinnate downy jagged leaves, of which the upper leaflets are largest, those of the stem in threes; the flowers are small, yellow, and grow in tufts. The plant is a native of Italy and Carniola. [C. A. J.]

ARENARIA or Sandwort. A genus of *Caryophyllaceae*, belonging to the tribe *Alsineae*, consisting of small herbs, distinguished from the others of the tribe by having the styles generally three; the capsule opening by twice as many valves as there are styles; at last splitting down to the base; the seeds without an appendage; and the petals not cleft into two segments. The species are extensively distributed; three occur wild in Britain: A. *serpyllifolia*, which is a common annual plant, with the petals not exceeding the calyx. Some authors consider we have two species included under this name, and separate from the common form, A. *leptoclados*, which is a much more slender plant, with softer capsules. A. *ciliata*, a perennial found on Ben Bulben, in the West of Ireland, has the petals much longer than the calyx, and the leaves fringed with hair. A. *norvegica*, also a perennial, from the Orkney and Shetland Islands, is closely allied to the last, but the leaves are not fringed. A. *trinervis* is sometimes placed in the genus *Moehringia*, as the seeds have an

appendage; it, however, accords ill with the other species of that genus. [J. T. S.]

AREOLATE. Divided off into distinct spaces usually more or less angular. The skin of a pine is areolate.

ARETHUSA bulbosa is a small swamp plant, belonging to the order of orchids, with a one-leaved scape, terminated by a single very handsome rosy-purple sweet-scented flower. It is found exclusively in North America, and is the only species of its genus. Other supposed species belong to *Pogonia*.

ARGALOU. (Fr.) *Paliurus aculeatus*.

ARGANIA. A genus of plants belonging to the family of *Sapindaceæ*. The calyx has ten sepals, in two rows; the throat of the corolla has five scales or abortive stamens, alternating with the five fertile stamens; anthers opening outwardly; style awl-shaped. *A. Sideroxylon* is the Argan tree of Morocco, in certain provinces of which it grows in woods. It is a spiny evergreen tree, with a trunk of considerable size, ten of low stature. It gives off branches at a few feet from the ground, which incline downwards till they rest on the earth; at length, at a considerable distance from the stem, they ascend. A tree mentioned in the *Journal of Botany* for April 1854, measured 18 feet only in height, while the circumference was as much as 220 feet. The fruit is an egg-shaped or roundish drupe, dotted with white. These fruits are much relished by all ruminating animals, who, in chewing the cud, eject the hard seeds, from which a valuable oil is extracted. The culture of the plant for the sake of its oil has been recommended in Australia and certain parts of Cape Colony subject to droughts. The wood is very hard, and so heavy as to sink in water. [M. T. M.]

ARGEL, or ARGHEL. A Syrian name for *Solenostemma Arghel*, the leaves of which are common in Egyptian senna.

ARGEMONE. The name of a genus of the Poppy family, *Papaveraceæ*, thus characterised: sepals 2-3; petals 4-6; stamens numerous; stigmas 4-7, radiating, sessile, or elevated on a very short style; capsules obovate, opening at the top by a number of little valves. *A. mexicana*, a native of Mexico, has become widely distributed over the globe, abounding in roadsides, and waste places in proximity to human habitations. The seeds possess acrid, narcotic, and purgative properties, and are employed as a substitute for ipecacuanha. They also contain an oil which has been recommended—as what has not?—as a remedy for cholera. The yellow juice of the plant is used in ophthalmia. [M. T. M.]

ARGENTINE. (Fr.) *Cerastium tomentosum*; also *Potentilla anserina*.

ARGEMONE. The same as *Argemone*.

ARGOPHYLLUM. A small genus of Tasmanian *Escallonieæ*, with alternate stalked ovate undivided leaves, silky and silvery on the under side. The flowers are very small, in terminal many-flowered corymbose or paniculate cymes. A curious cup from which the stamens rise, is cut into comb-like teeth. [J. T. S.]

ARGOUSIER. (Fr.) *Hippophae*. — DU CANADA. *Shepherdia canadensis*.

ARGYLIA. A genus of *Bignoniaceæ*, containing eight species, from Chili, of perennial herbs, with fleshy roots, an angled stem, petiolate palmate and alternate leaves, and white or purplish axillary flowers. The genus has a five-partite calyx; a tubular corolla, with a five-lobed limb; four didynamous stamens; and a bilocular ovary. [W. C.]

ARGYREIA. A genus of the natural order *Convolvulaceæ*, having large handsome flowers, with a bell-shaped corolla, into the base of the tube of which the stamens are inserted. The ovary is two-celled, each cell containing two seeds, unless, as often happens, some of them become abortive. Fruit berry-like, indehiscent. They are natives of tropical Asia. The leaves of *A. bracteata* and *A. speciosa* are used in India as a poultice in cases of scrofulous disease of the joints, and as a rooting application in headaches. The root of *A. malabarica* possesses purgative properties. Two or three species are in cultivation. They are climbing plants with white or purple flowers, and much resemble *Ipomœa*. [M. T. M.]

ARGYROLOBIUM. A genus of the pea-flower tribe of the natural family *Leguminosæ*, containing upwards of forty species, all of them herbs or dwarf shrubs with trifoliolate stalked or nearly sessile leaves, having two stipules at their base, and generally covered with silvery hairs. The flowers are yellow, solitary or racemed, with bracts. About thirty species are found in South Africa, a number in the countries bordering on the Mediterranean, and some extending through Affghanistan into N.W. India. The name of the genus has reference to the pod which is often clad with silvery hairs. [A. A. B.]

ARGYRORCHIS. An obscure terrestrial orchid from Java, with pinkish flowers. It appears to be a mere *Peloria* of *Macodes Petola*.

ARHYNCHIUM lobbsum, an epiphyte from tropical Asia, with dull green and brown flowers, referred by the younger Reichenbach to *Braseltera bilinguis*. Its lip is so constructed as to look as if composed of two tongues laid one upon the other.

ARIL, ARILLUS. A body which rises up from the placenta, and encompasses the seed like the mace in nutmeg, and the red sac in *Euonymus*.

ARILLODE. A false aril; a coating of the seed proceeding from its own surface, and not from the placenta.

ARIOPSIS. A curious genus of plants belonging to the *Aroideæ*, and similar to the genus *Arum* in appearance, hence the name. The species were formerly included in the genus *Arisarum*, and consist of small Indian herbs with inconspicuous stems, globular rhizomes, and sinuate heart-shaped petiolate leaves, on long stalks. The spathe is nodding, boat-shaped, adherent to the lower part of the spadix, on the upper part of which the male flowers are placed in little depressions; each little cavity contains six abortive anthers, bursting by one pore. The female flowers at the lower portion of the spadix, consist of obliquely ovate ovaries with three to five stigmas. The fruit is like a berry, but somewhat dry, angular, one-celled, with four to five placentas, and several seeds placed in two rows along each placenta. *A. peltata* is sometimes met with in cultivation as an object of curiosity. [M. T. M.]

ARISÆMA. The plants of this genus of the *Aroid* family have tuber-like rootstocks, from which proceed pedate, pedate, palmate, or more rarely undivided leaves. The spathe is rolled round the spadix at the base; the spadix has unisexual flowers below, its upper part covered with rudimentary flowers; the anthers are provided with distinct filaments; the ovaries are numerous, and contain 2-6 ovules, and are terminated by very short styles. The tuberous rootstocks of two species are used by the natives of Sikkim Himalaya, as food; they are beaten into a pulp with water, and allowed to ferment, a process which destroys their acridity. The *Dragon-root*, or Indian turnip of America, is the tuber of *A. atrorubens*, which furnishes a kind of starch. [M. T. M.]

ARISARUM. A genus of plants of the *Arum* family, closely allied to *Arisæma*. The lower part of the spadix has unisexual flowers, but no rudimentary ones, and is naked at the top; the ovaries are few in number, and have a distinct style. The plants are herbaceous, with a tuberous or branching and creeping rootstock, heart-shaped or spear-shaped leaves, on long stalks, and livid purple spathes. They are natives of Southern Europe and the Mediterranean region. [M. T. M.]

ARISTA. The awn or beard of corn, or any such like process.

ARISTATE. Furnished with an arista.

ARISTOLOCHIACEÆ (*Aristolochieæ, Asarineæ, Patchouriaceæ, Birthworts*). In the tropical parts of both hemispheres, and occasionally beyond those limits, occurs a race of plants with singularly inflated irregular flowers, consisting of a calyx only, of a dull dingy colour, varying from yellow to shades of chocolate, purple, or brown, and often emitting an offensive odour. A hot summer appears to be one condition of their existence, with a few exceptions, the most striking of which are *Asareæ*, little stemless plants, with in Europe and North America, and the *Aristolochia Clematitis*, which has become as it were naturalized in England. The wood of these plants, when they have any, consists of parallel plates, held loosely together by soft medullary processes. The ovary is inferior, with many ovules, and for the most part consists of six cells, the number three being, as in Endogens, characteristic of the floral apparatus of the order. In medicine these plants are slightly aromatic stimulating tonics, useful in the latter stages of low fever; the taste is bitter and acrid; the odour strong and disagreeable. They are also said to be sudorific, emmenagogue, parturitive, and diuretic. The principal genera are ARISTOLOCHIA and ASARUM, which see.

ARISTOLOCHIA. A remarkable genus of plants belonging to the family *Aristolochiaceæ*, and characterized by the possession of a calyx of some other colour than green, of an irregularly tubular form, inflated at its lower portion, and adherent at its base to the ovary. The stamens are six in number, and adhere to the solitary style; the fruit is a six-celled capsule with numerous seeds. The wood of these plants differs much in appearance from that of Exogenous trees or shrubs in general, as it consists of radiating plates of wood, surrounding a pith and encircled by the bark, not disposed in rings.

The plants of this genus are for the most part shrubs, generally climbing round the branches of trees. They are abundant in tropical South America, while a few species are distributed throughout North America, Europe, and India. One species *A. indica*, is common to India and to New Holland. The flowers of some of the kinds are remarkable for the oddity of their form, and for their large size. Humboldt mentions one, *A. cordata*, as growing on the shady banks of the Magdalena, and having blossoms measuring four feet in circumference, and which the Indian children sportively draw on their heads as caps. *A. Clematitis*, the common Birthwort, is found in this country, but generally in the neighbourhood of old ruins, as if it had at some time been cultivated in the gardens attached to such buildings, probably for medical purposes, as an aid in parturition. It is a low growing shrub, with slender erect greenish furrowed stems, stalked heart-shaped leaves, in the axils of which the yellowish trumpet-shaped flowers are produced in clusters. Others of the species had formerly a similar reputation, such as *A. rotunda* and *A. longa*.

A. Serpentaria is the Virginian Snakeroot, furnishing the drug known as serpentary, which is esteemed in the southern states of America as a cure for the bite of the rattlesnake or of a mad dog. Its effects, when given in large doses, are a feeling of sickness, purging, and subsequently increased fulness of the pulse; hence it is still occasionally used as a stimulant in fevers.

The roots of other species are used in the United States for the same purposes, as *A. Serpentaria* and *A. tomentosa*; and several kinds are employed in Brazil for their stimulant properties. Many of these plants, besides those above mentioned, are said to be useful in effecting the cure of snake bites, not only in tropical America, but also in the West Indies, Hindostan, and Egypt. It is stated that the Egyptian jugglers use some of these plants to stupify the snakes before they handle them; and Jacquin relates that the juice of the root of *A. anguicida*, if introduced into the mouth of a serpent, so stupifies it, that it may be handled with impunity. If the reptile be compelled to swallow a few drops, it perishes in convulsions; hence it is perhaps, on homœopathic principles, that the root is affirmed to be an antidote to snake bites. *A. bracteata* and *A. indica* are both used for similar purposes in India, so that there is the concurrent testimony of the natives of different quarters of the globe as to the peculiar property of these plants. The two kinds just named are bitter plants, used as purgatives and vermifuges, and for other purposes in India.

In Central America one or more plants, called Guaco by the natives, are held in high esteem for the cure of snake bites. It is conjectured, with much probability, that the Guaco is some species of *Aristolochia*. So satisfied are the natives of Peru, Central America, and Mexico of its extraordinary medicinal powers and specific virtues in cases of snake bite, that every Indian or Negro who has to traverse the country, invariably has a supply of this friendly plant in a dry or prepared state, to meet any accident that may befal him, by inadvertently placing his foot upon one of these dreaded and deadly foes of mankind. Mr. Temple, to whose account of this plant, published in the *Journal of the Society of Arts* for the year 1835, we are indebted in drawing up this notice, states that he employed the Guaco in four cases of snake bite with complete success. He also gives a strange account of the way in which the Guaco is reported to have been first discovered, the substance of which is as follows:—A traveller passing through a forest observed two formidable snakes engaged in deadly encounter; after a short time one was severely bitten and fled from the scene of conflict, until it reached a cowpine plant, of the leaves of which it partook with avidity—that plant was the Guaco. He secured the reptile, and brought away the plant the leaves of which it had eaten. The snake, although bitten by one of a most deadly species, quite recovered. Another report, as probable as the other, is that snakes have been observed carefully to avoid localities where the plant grows. Many persons are so firmly persuaded that the snake will not approach the Guaco, that when travelling in the bush, they carry a small piece of the root of the plant in their pocket. So then, this wonderful plant prevents the access of snakes, stupefies them, and kills them if they do come, and cures them if bitten by a fellow snake, and likewise cures human beings bitten by these venomous reptiles. There can be no doubt of the partial truth of some of these statements, and hence, not only the botanical history, but the medical properties of Guaco, demand accurate investigation.

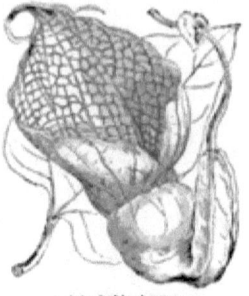

Aristolochia gigantea.

Several different kinds of *Aristolochia* are cultivated in hot-houses for the singularity, and in some cases for the handsome appearance, of their flowers, albeit their colour is usually of a dingy hue. The flowers of *A. ploscera* act as a sort of fly-trap; the flowers are bent in the middle, and lined with hairs pointed downwards, so that insects in easy but escape impossible to the unlucky insect, which thus, it may be unconsciously, aids in the ripening of the seed. *A. Sipho*, a native of the Alleghany mountains, is cultivated as a climbing plant, out of doors, for the sake of its large, heart-shaped leaves; the flower is curved like a siphon, hence its name; it has also received the name of Pipe-vine, from a resemblance in the form of the flowers to that of a tobacco-pipe.   [M. T. M.]

ARISTOTELIA. A genus of the natural family *Tiliaceæ*. There are four species known: shrubs or small trees with opposite or ternate stalked leaves, which vary much in size and form. The flowers are small and white, arranged in axillary fascicles or in racemes; the berries roundish, varying its colour from pink to black, and is also from that of a small pea to a cherry. The wood of *A. Macqui*, a native of Chili, is used for making musical instruments, its tough bark forming the strings. The berries are acid but eatable, and a wine is made from them by the Chilians which is given in malignant fever. "It was employed by Bomley in Chili during the plague of 1791, with boasted success." Two species are found in N. Zealand; the berries of *A.*

racemose, the Mako-Mako of the natives, are eaten. The genus is named in honour of the Greek philosopher.   [A. A. B.]

ARJOONA. A genus of *Olacaceæ*, containing three species of under-shrubs or herbaceous plants, with alternate rigid and acute leaves, and flowers in spikes at the end of the branches. The calyx consists of a cylindrical five-lobed tube. The stamens, five in number, have slender filaments and oblong anthers. The inferior ovary is fleshy and contains three ovules. There is a thread-like style, and three short linear stigmas. The fruit, included in the persistent bracts and bracteoles, is one-seeded. The species are natives of South America.   [W. C.]

ARMARINTHE. (Fr.) *Cochrys*.

ARMATURE, ARMS. Any kind of defence consisting of spines, prickles, &c.

ARMENIACA. The Apricot, *Prunus Armeniaca*.

ARMENIACUS. A native of Armenia; but more generally used to signify apricot-coloured.

ARMERIA. Thrift. A genus of *Plumbaginaceæ*, with narrow often grass-like leaves and naked scapes, terminating in compact heads, almost like those of the scabious, surrounded by an involucre of bracts, the two outermost of which have the bases produced downwards, forming a cylindrical sheath or tube, enclosing the upper part of the scape. The flowers are rose-coloured, purple or white. Two species are included in the British Flora, but one of them, *A. plantaginea*, which has the leaves three or five-nerved, broader towards the end, is only found in Jersey. The other, *A. vulgaris*, is the common Sea-Pink or Thrift, and occurs on all the coasts and many of the mountains of the British Islands: the leaves are narrow and parallel-sided. This plant is often cultivated in gardens, where it is sometimes used to supply the place of box-edging, for which its compact (tufted growth makes it very suitable.   [J. T. S.]

ARMILLARIA. A sub-genus of *Agaricus*, distinguished from other white-spored groups by its partial ring-like veil, without any universal volva, which remains attached to the stem. *Agaricus melleus*, a species common on almost every rotten stump in autumn, is the most prominent example met with in this country. This is frequently eaten abroad under the name of Halimasch; but it is very acrid, and causes a strong constriction of the throat when eaten raw. It would not be an acceptable article of food in this country, even were it free from danger.   [M. J. B.]

ARMOISE. (Fr.) *Artemisia*.

ARMORACIA. The Horse-Radish, *Cochlearia Armoracia*.

ARNEBIA. A small genus of oriental and North African *Boraginaceæ*, allied to *Lithospermum*, but having the style bifid at the apex, and the stigmas often forming four lobes. The species are small and very hispid or bristly, with pale yellow or purplish-blue flowers.   [J. T. S.]

ARNICA. The name applied to a genus of the composite family, distinguished by the following characteristics:—Involucre bell-shaped, of two rows of bracts; outermost florets strap-shaped, containing pistils only with rudimentary stamens; central ones tubular, five-toothed, containing both stamens and pistils; the tube of the corolla hairy; style with long arms covered with downy hairs; fruit cylindrical, tapering at each end, ribbed, hairy, and surmounted by a pappus, consisting of close rigid rough hairs arranged in one row.

A. montana, the Mountain Tobacco, is a native of Central Europe. Its roots and leaves possess powerfully acrid properties, but in small doses it is employed as a stimulant in low fevers and other conditions of debility, also in paralytic affections; externally it is much used as a tincture applied to bruises, wounds, and sprains. It promotes the speedy absorption and removal of the effused blood. The peculiar properties seem especially due to a resinous substance called *arnicin*, and to a volatile oil.   [M. T. M.]

ARNOLDIA. The name of a section of the genus *Dimorphotheca*, which see. The same name was applied by Blume to a Java plant, which is now placed in the genus *Weinmannia*.

ARNOSERIS. Nipplewort. An insignificant native annual weed belonging to the tribe *Cichoraceæ*, of compound flowers. It grows from six to eight inches high, with a branched leafless stem, the upper branches being hollow and singularly swollen upwards so as to assume a club-shaped form. The flowers are small and yellow. By Smith, Hooker, and others, it is placed in the genus *Lapsana*, from which it was separated by Gertner on account of the fruit being crowned with a pappus of many short entire broad scales; in *Lapsana* the fruit is naked.   [C. A. J.]

ARNOTTO or **ANATTO.** *Bixa Orellana*.

ARONICUM. A genus of the composite family, closely allied to, and only differing from, *Doronicum*, in all the achenes being furnished with a pilose pappus, instead of those of the disc only. There are four known species, all of them pretty perennial herbs restricted to mountain districts in Central Europe and Asia. They have stems varying from three inches to two feet high, terminating in one or more yellow-rayed flower-heads, sometimes two inches across; the root-leaves are stalked, heart-shaped or oblong, and toothed, with a smooth or downy surface, and those of the stem sessile and arranged alternately—not opposite as in the nearly allied genus *Arnica*. *A. Clusii* is a neat little Alpine species, three to five inches high, frequently met with in collections of Alpine plants. [A. A. B.]

ARBOPHYLLUM. Under this name are collected about four species of epiphytal Orchids, inhabiting Mexico and New Grenada. They have slender bulb-like stems, invested with stout sheaths and one or two narrow leathery leaves at their tip. The flowers are collected in close cylindrical spikes, are small, somewhat violaceous, and have a rich deep crimson colour. One of the species, *A. cardinalis*, is as much as three feet high. One or two species exist in gardens, where they are valued for their elegant manner of growth. *A. alpinum* is the hardiest, inhabiting Mount Totonicapan, at the elevation of 10,000 feet above the sea-level, where it rides on the branches of the Mexican alder, in a region where oaks refuse to grow.

ARRABIDEA. A genus of *Bignoniaceæ*, composed of about twenty South American, chiefly Brazilian species, all of which are climbing shrubs, having, when young, pinnate or trifoliate, when old bifoliate leaves, generally furnished with tendrils. The genus may be readily distinguished from all other *Bignoniaceæ* by having by far the smallest flowers in the order, the corolla being, in some instances, only three to four lines long; also by its stamens, four of which are fertile, whilst the fifth is sterile and of equal breadth with the rest. The calyx is cup-shaped; the corolla hypocrateriform; the fruit a dehiscent, smooth, flattened capsule, linear in shape, and having a septum placed parallel with the valves of the latter. The flowers, though small, are arranged in large terminal panicles, and render the *Arrabideæ* ornamental objects. The leaves of several have a deep fuse or purplish tint, and are used for dyes. One of these species is *A. rosea*, from which a purplish colour is extracted in the forests of Rio Negro, and imported to Europe. The doubtful *Bignonia Chica*, probably also a congener, furnishes, by boiling its leaves in water, a red fecalent substance, which is quickly precipitated by adding some juices of the bark of an unknown tree, called Arayana; the Indians use it for painting their body red. It is also an article of importance to dyers. In nature it approaches the resins, but contains some peculiar properties; it gives an orange colour to cotton. *Bignonia*(?) *Chica*, termed "Chica" in the Chinese districts, is probably identical with the "Carajuru" in the isthmus of Panama: it is known as "Hojita de teñir," and used for dyeing Spanish hammocks. Silk-worms fed with the leaves are stated to produce red silk. [B. S.]

ARRACACHA. A name applied by the natives of the northern parts of South America to several kinds of plants, possessing tubers or tuberous roots, but, botanically speaking, confined to a genus of umbelliferous plants allied to the hemlock. Its principal distinguishing characteristics are—limb of the calyx entire;

petals ovate or lance-shaped, purplish, with the point bent inwards; fruit turgid, compressed from side to side, wingless, surmounted by the thickened bases of the style; albumen curved. *A. esculenta* is cultivated in the cooler mountainous districts of Northern South America, where the roots form the staple diet of the inhabitants. The plant is somewhat like the hemlock (*Conium maculatum*), but its leaves are broader, its stem not spotted, and its flowers are of a dingy purple colour; the roots are large and divided into several fleshy lobes of the size of a carrot, which when boiled are firm and have a flavour intermediate between a chestnut and a parsnip. Trials have been made to cultivate the plant in this country, but the climate has not been found suited for it. It might be tried in some of our colonies with advantage. [M. T. M.]

The name Arracacha is also given to one of the tuber-bearing species of Oxalis, *O. crenata*. [T. M.]

ARBÊTE-BŒUF. (Fr.) *Ononis procurrens*.

ARRHENATHERUM. A genus of grasses of the tribe *Avenaceæ*, distinguished chiefly by having two florets within the glumes, the lower of which is abortive. The only British species is the tall Oat-grass, *A. avenaceum*, which in many instances forms a very considerable portion of good meadows and pastures. Although a large growing species, and one which cattle appear to like, it is found, on being chemically analysed, to be low in nutritive properties compared with some other kinds, consequently, it is mostly cultivated as a portion in mixtures of grasses, and never alone as a crop. For this purpose it is useful in assisting the weaker stemmed kinds to stand upright while ripening. The few species which were formally included under this genus, as defined by Beauvois, will be found described under the genus *Avena* in Steudel's *Synopsis*. [D. M.]

ARROCHE ÉPINARD. (Fr.) *Atriplex hortensis*. — **FRAISE.** *Blitum capitatum*. — **POURPIER.** *Atriplex portulacoides*.

ARROW-GRASS. A common name for *Triglochin*. The name Arrow-grass is also applied to the *Juncaginaceæ* generally.

ARROW-HEAD. *Sagittaria sagittifolia*.

ARROW-ROOT. A pure kind of starch obtained from various plants, and employed for dietary and other purposes. That called Bermuda or West Indian Arrow-root is obtained from *Maranta arundinacea*. Brazilian Arrow-root or Tapioca meal is obtained from *Manihot utilissima*. Chinese Arrow-root is said to come from the tubers of *Nelumbium speciosum*. East Indian Arrow-root is obtained from different species of *Curcuma*. English Arrow-root is the starch obtained from the tubers of the potato, *Solanum tuberosum*. The seeds of *Dioscorea* furnish a kind of Arrow-root in Mexico. Oswego Arrow-root is obtained in

America from Indian corn, *Zea Mays*. A kind of Arrow-root, called *Tous les mois*, which comes from the West Indies, is supposed to be the produce of *Canna edulis*, *C. Achiras*, and probably of other species. That of the Sandwich Islands comes from *Tacca oceanica*. Though the name Arrowroot is that applied to the produce of various plants, it is more particularly associated with that of the *Maranta*. The word is a corruption of the name Araroot. [T. M.]

ARROW-WOOD. An American name for certain species of *Viburnum*, as *V. dentatum*, *pubescens*, etc.

ARRUDEA. A genus of the mangosteen family (*Clusiaceæ*), differing from *Clusia*, to which it is most nearly related, in having a many-leaved calyx, a larger number of petals and stamens, as well as a stalked stigma. Three species are known; two of them found in Brazil, the other in Surinam. Their leaves are opposite, smooth and leathery, and their flowers solitary, stalked at the ends of the branches, and sometimes as large as those of a camellia. *A. chusioides*, a Brazilian species, is said to be a small tree, from the branches of which a viscid gum exudes; while *A. rosea*, the Surinam species, is said to grow on the trunks of other trees which it clasps with its long stringy roots so tightly as eventually to kill them. As in the other species, a gum exudes from the stems, which sometimes are upwards of forty feet long, while their greatest thickness is two and a half feet. The genus is named in honour of M. Arruda de Camara, who wrote on fibrous plants of Brazil. [A. A. B.]

ARTABOTRYS. A genus of plants belonging to the family of *Anonaceæ*. Its name is derived from the hook-like form of the flower stalks, by the aid of which the fruit is hung or suspended. The prominent characters of the genus are: hooked woody flower stalks; three sepals, coherent at the base; six petals in two rows, all of the same shape, and so placed in the flower bud that they touch by the margins only, hollowed at the base, and constricted around the ovaries; numerous densely packed stamens; ovaries indefinite in number, each containing at the base two erect ovules. The plants constituting this genus are shrubs or climbing plants, natives of India and the Indian Archipelago chiefly, but one is found in the western part of tropical Africa. *A. odoratissima* is cultivated as an ornamental shrub, and for the sake of its fragrant flowers, throughout the East, and also in hot-houses in this country. The leaves of certain kinds are highly esteemed in Java, against cholera, their value being probably dependent on the warm aromatic principle pervading them. [M. T. M.]

ARTANEMA. A genus of *Scrophularineæ*, synonymous with *Achimenes* of Vahl. It is characterized by a five-parted subequal calyx; a funnel or bell-shaped corolla, bearing four scales inside the tube, and having a four-cleft somewhat two-lipped limb, the upper segment of which is broader; four didynamous stamens inserted in the tube of the corolla, the hinder ones shorter; a simple style with a bilamellate stigma; and a two-celled ovary containing many ovules. The species are glabrous herbs of India and the East, and have opposite leaves, with terminal racemes of flowers. *A. fimbriatum* is an ornamental species, sometimes seen in gardens. [T. M.]

ARTANTHE. The name of a genus of plants belonging to the pepper family (*Piperaceæ*). They are woody plants with jointed stems, rough leaves, and spikes of flowers opposite the leaves. The flowers are perfect with peltate or hooded bracts. *A. elongata*, formerly called *Piper angustifolium*, furnishes one of the articles known by the Peruvians as Matico, and which is used by them for the same purposes as cubebs, the produce of a nearly-allied plant; but its chief value is as a styptic, the rough leaves of the plant having the power of staunching blood. The under surface of the leaf is rough, traversed by a network of projecting veins, and covered with hairs; hence its effect in stopping hæmorrhage is probably mechanical like that of lint, cobweb, and other commonly-used styptics. It has also been employed internally to check hæmorrhages, but with doubtful effect. Its aromatic bitter stimulant properties are like those of cubebs, and depend probably on a volatile oil, a dark green resin, and a peculiar bitter principle called *maticin*. *A. adunca* is made use of in Brazil for its pungent aromatic stimulant qualities, as well as for its aperitive effects. Other plants appear to furnish leaves having similar properties, and called by the same name by the Columbians. See EUPATORIUM. [M. T. M.]

ARTEMISIA. A genus of plants commonly called Wormwood, belonging to the tribe *Senecioneæ* of the Compositæ. The Wormwoods are shrubby or herbaceous plants with their leaves usually much divided and frequently of a grey colour. The flower heads are small, borne in panicles, and provided with an involucre of overlapping bracts; the florets are as long as the involucre, yellow or greenish, either all tubular and five-toothed, or the central ones tubular, five-toothed and barren, and the outer ones filiform or three-toothed, female and fertile; the florets are placed on a receptacle without scales, and the fruits are obovate and not provided with a pappus. The genus is widely distributed over the temperate and warmer temperate regions of the globe, and most of them are remarkable for their strong odour and bitter taste. Three or four species grow wild in this country. In certain of the Western states of North America, as Utah, Texas, New Mexico, &c., are large tracts almost entirely destitute of other vegetation than that afforded by certain kinds of *Artemisia*, which cover vast plains, and give them an universal greyish green hue. The plants are known under various names by the trappers, who find the snarled and interlacing branches an almost insurmountable obstacle to man or horse. The plants, moreover, are of no value as forage. The few wild animals that feed on them are said to have their flesh rendered of a bitter taste in consequence. The *Artemisias* also abound in the arid soil of the Tartarian steppes, and in other similar situations.

The Common Wormwood, *A. Absinthium*, is found wild in some parts of Britain and cultivated in cottage-gardens. It possesses aromatic bitter and tonic properties, and was formerly much employed as a vermifuge. The active properties of the plant, and probably those of the other kinds used for like purposes, depend on a volatile oil, a peculiar bitter principle called absinthine, and an acid called absinthic acid. What is called salt of wormwood is an impure carbonate of potash, obtained from the ashes of wormwood.

A large number of the species possess similar properties to those found in the common wormwood, and are hence used for the same purposes in various parts of the world. The flower stalks and heads of several species of *Artemisia* are sold by herbalists under the name of Wormseed; they are chiefly imported from the Levant, and are the produce of plants growing in Syria, Persia, and Barbary. Others imported from India are employed as vermifuges. *A. Moxa* is said by Dr. Lindley to be the plant used by the Chinese and Japanese in the formation of their Moxa, a small pellet of combustible material, placed on the skin and burns there so as to produce a sore. It is used for the same purposes, and on the same principle as a blister, but it is exceedingly painful and now very rarely employed. Some of the species of *Artemisia* growing in Switzerland are used in the manufacture of the bitter aromatic Extrait d'Absinthe.

The Southernwood of gardens, *A. Abrotanum*, sometimes called by country people Old Man, is a shrub with finely divided greyish green leaves, which have a fragrant aromatic odour, said to be disagreeable to bees and other insects. The plant is a native of the South of Europe.

The Tarragon, *A. Dracunculus*, differs from the majority of its fellows in that its leaves are undivided; they are narrow and lance-shaped, of a bright green colour, and possess a peculiar aromatic taste, without the characteristic bitterness of the genus. The plant is a native of Siberia. [M. T. M.]

ARTHANITA. (Fr.) *Cyclamen europæum*.

ARTHROBOTRYS. A name proposed for a small group of Indian ferns, now referred to *Lastrea*. [T. M.]

ARTHROCNEMUM. A genus of Chenopodiaceæ, separated from *Salicornia* to receive *S. fruticosum* and a few other species, which differ from the restricted *Salicornia* in having the flowers hidden in the articulations of the branches, and not concealed in excavations in the axis. The calyx also is trigonous or tetragonous, with three to five teeth, and without wing or appendage. The seed has a distinctly double integument, while in *Salicornia* it is single. Otherwise the two genera agree. They have perfect flowers, without scales; one or two stamens; two styles; and an ovate one-celled and one-seeded ovary. The species are found in the salt marshes of all parts of the world. *A. fruticosum* is abundant on the British coasts. [W. C.]

ARTHROLEPIS. The name given to a genus of the composite family (Compositæ). There is but one species known, a perennial herb, native of Syria, a foot high, with alternate linear pinnatisect leaves, the segments very small and closely overlapping each other. The flower-heads are single at the ends of the branches; the ray florets yellow. All the parts of the plant are covered with a white mealy pubescence. It is nearly related to the Chamæmelies (*Anthemis*) and the Millfoils (*Achillea*); differing from the first in its winged achenes, from the second in its single flower-heads, and from both in the pointed scales of the involucre. The name of the genus is derived from this latter circumstance. [A. A. B.]

ARTHROLOBIUM. An unimportant genus of leguminous plants distinguished from the equally unpretending *Ornithopus*, by the heads of flowers being destitute of a floral leaf, or bract, at the base. There are two European species, one of which, *A. ebracteatum*, grows in the Channel isles and in Scilly. It is a small plant with prostrate stems, pinnate leaves, and minute cream-coloured flowers veined with crimson, growing in heads of four or five, and succeeded by as many-jointed and curved pods, which together bear a singular resemblance to a bird's foot. [C. A. J.]

ARTHROPHYLLUM. A genus of the *Bignonia* family, containing five species, all of them shrubs or small trees, found in Madagascar and the islands of Eastern tropical Africa. Their leaves are compound, opposite, or alternate, and very peculiar in structure; indeed, in four of the species no true leaves may be said to be developed, but their petioles, or leafstalks are winged and leaf-like, with two to four joints, the segments between the joints being wedge-shaped, and the terminal one acute. In *A. Thouarsianum* leaflets are produced from the joints of the petiole. Their flowers are generally large and tubular, disposed in racemes or corymbs from the ultimate forkings of the branches. *A. madagascariense* is cultivated in England. Its flowers are pink in colour, large and tubular, the limb of the corolla five-lobed, with crisped margins. The name *Arthrophyllum* signifies jointed leaf. It has been changed by some authors to *Phyllarthron* because the name *Arthrophyllum* is given also to a genus of the *Aralia* family. [A. A. B.]

ARTHROPODIUM. A genus of Australian and New Zealand *Liliaceæ*, allied to *Anthericum*, with grass-like radical leaves, fasciculate roots, and small purplish or white flowers in lax racemes or panicles; the filaments of the stamens are clothed with fine short hairs for half their length. A few species are cultivated in our greenhouses. [J. T. S.]

ARTHROPTERIS. A name proposed for a few tropical ferns distinguished by having a jointed stalk. They are referred severally to the genera *Lastrea*, *Nephrolepis*, and *Polypodium*. [T. M.]

ARTHROSTEMMA. A genus of tropical American *Melastomaceæ*, which have little resemblance to each other, but agree in having the parts of the flower in fours, the anthers curved at the base, the ovary bristly at the apex. Some of the species are handsome, resembling the *Rhexias*. A few of the species are cultivated in our stoves and greenhouses. [J. T. S.]

ARTHROTAXIS. A genus of *Coniferæ*, consisting of much branched evergreen trees of no great height, with short, thick, densely imbricated leaves, closely covering the branches. The male flowers form very short terminal spikes with two-celled anthers. The ripe cones are also terminal, sessile, small, and globular, with almost woody peltate scales, each bearing three to six inverted seeds. There are only three species known, all natives of Tasmania.

ARTICHAUT. (Fr.) *Cynara Scolymus.*

ARTICHOKE. *Cynara Scolymus.* —, JERUSALEM. *Helianthus tuberosus.*

ARTICULUS. A joint; a place where spontaneous or easy separation takes place.

ARTILLERY PLANT. *Pilea serpyllifolia*, and *herniarifolia*.

ARTOCARPACEÆ (*Artocarpus, Artocarpads*). A group of apetalous trees, not unlike the plane trees of Europe, but for the most part inhabiting the tropics and always the warmer parts of the world. They abound in a milky juice, and have for the most part their female flowers collected into fleshy masses or heads. Moreover, they have stem sheathing convolute stipules like those of a fig tree. The more important genera are *Artocarpus* and *Antiaris*.

ARTOCARPUS. This name, signifying Bread-fruit, is applied to the genus of trees furnishing the well-known fruit of that name. It gives the name to the order *Artocarpaceæ*, and is distinguished by having its male or stamen-bearing flowers borne on long club-shaped spikes, and the pistil-bearing ones in round heads. The male flowers have a tubular calyx of two parts containing a single stamen; the female flowers have a simple ovary, containing a single ovule, and surmounted by a style with two stigmas curved downwards. These female flowers soon grow together, and form one large fleshy mass, which becomes the fruit, which is thus formed exactly in the same way as the mulberry is, but in the bread-fruit farinaceous matter takes the place of the sugar and vegetable jelly of the former.

The Bread-fruit tree of the South Sea Islands (*A. incisa*) is a moderate-sized tree, whose young branches are marked with ring-like scars indicating the spot where the large convolute stipules have been placed. The leaves are large rough dark green, divided into lobes, something like those of a fig tree. The fruit is roundish, of the size of a melon, rough on the exterior, marked with hexagonal knobs, or in some of the varieties smooth and of a green colour. The pulp in the interior is whitish, and of the consistence of new bread. It is roasted before it is eaten, but has little flavour. The best varieties contain no seeds, the tree being propagated by shoots that spring from the roots. The tree contains a viscid milky juice, containing caoutchouc, which is used instead of glue, and for caulking the canoes of the South Sea Islanders, who

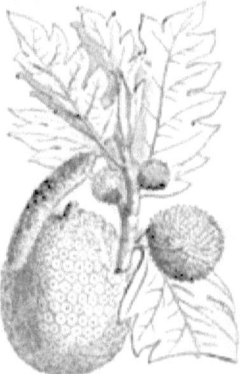

Artocarpus incisa.

make use also of the timber of the tree, which is soft, and is said to attain a mahogany colour by exposure. The bark of the young tree is also fabricated into a coarse cloth. In the South Sea Islands, the bread-fruit constitutes the principal article of diet, which is prepared by baking it in an oven heated by hot stones. The plant is now cultivated in the West Indies, but does not equal the plantain as an article of food. The history of its introduction into those islands is associated with the well-known incidents of the mutiny of the crew of the 'Bounty,' which had been sent out under the command of Lieut. Bligh to procure bread-fruit trees, at the earnest request of Captain Cook, and the naturalists who accompanied him in his voyages. The first attempt was frustrated by the mutiny of the crew after the plants had been procured, and all promised well. A second attempt, four years subsequently, proved successful.

The Jack, *A. integrifolia*, is a native of the Indian Archipelago; it produces a fruit like that of the above-named kind, but the leaves are not at all lobed or divided. It is cultivated in Southern India and all the warm parts of Asia. The fruit is a favourite article of food among the natives, as also are the roasted seeds. The timber is much used for making furniture; it is at first of a pale colour, but subsequently becomes dark, approaching to a mahogany tint. Bird-lime is manufactured from the juice. *A. hirsuta*, a native of Malabar, possesses similar qualities. *A. indica* is shown in Plates 2a and 7c.
[M. T. M.]

ARUM. A genus of plants of the family *Araceae*, known by having a large spathe, whose edges are rolled over one another at the base. The flowers are unisexual, and placed on a fleshy spike or spadix, which is club-shaped and destitute of flowers at the summit, but at the lower portion bears male flowers or stamens, consisting merely of anthers, opening by lateral slits, unprovided with filaments, and not invested by a perianth. Between the male flowers and the female are a number of abortive flowers having the appearance of threads or hairs. Quite at the base of the spadix the female flowers are placed; these consist merely of ovaries, containing two to six ovules, and ripening into a berry-like fruit.

The common Arum of the hedges (*A. maculatum*), or, as it is commonly called, Lords and Ladies, or Wake Robin, is an extremely acrid poisonous plant, but by drying, or the agency of heat, the acrid matter is destroyed. It has a whitish rootstock, from which proceed ovate arrow-shaped green leaves, frequently marked with dark blotches and placed on long stalks; the spathe is green. The spadix is of various shades of purple, or more rarely of a dull yellowish colour. When the fruits are ripe, they are of a brilliant red colour, and very conspicuous, as not only the tops of the spadix, but also the investing spathe wither, and fall off as the fruit attains to maturity. From the tubers of this plant, in the Isle of Portland, a starch called Portland Arrowroot was formerly extensively prepared by pounding the tubers with water, and then straining. On allowing the strained liquid to stand the starch was deposited. Care was necessary from the extremely acrid nature of the plant. Indeed, Mr. A. Irvine, in his *Illustrated Handbook of British Plants*, records that many hours' boiling did not suffice to dissipate the acrid matter.

A. italicum has been found to grow in the Isle of Wight. It is a much larger plant than the common *A. maculatum*, the leaves, with white veins, have the lobes at the base spreading more widely than in that species, and the spathe is rolled backwards at the point as the flowers ripen.

The root of *A. montanum* is used in India to poison tigers. The roots of *A. lyratum* furnish an article of diet to the natives of the Circar mountains. They require, however, to be carefully boiled several times, and dressed in a particular manner, to divest them of a somewhat disagreeable taste.

All the species of *Arum*, and those of allied genera, such as *Arisaema*, *Colocasia*, *Caladium*, *Amorphophallus* and others, possess a similar combination of extremely acrid properties, with the presence of a large quantity of farina, which can be separated from the poisonous ingredient by heat or water, or by merely drying in some instances.

A. Dracunculus is commonly cultivated in gardens for the sake of its large palate leaves, its spotted stem, and purple spadix. The smell is fetid, and apt to produce headache. The *Arums* have been made use of in experimenting upon vegetable heat, as by reason of the investing spathe, the heat generated by the flowers does not so easily escape as in other plants, and its degree can the more readily be ascertained. Moreover, it appears that these plants really do generate more heat than other flowers; for instance, a difference of more than 50° is recorded between the temperature of the air and that of the flowers of *A. cordifolium*.
[M. T. M.]

ARUM, ARROW. *Peltandra virginica*—, WATER. *Calla palustris*.

ARUM D'ETHIOPIE. (Fr.) *Richardia æthiopica*.

ARUNDINA. Reed-like terrestrial orchids, with slender stems and narrow-ribbed leaves. The flowers are large, thin, richly coloured with rose or purple, but very fugitive. Three or four species are known, all inhabitants of tropical Asia. The genus is nearly allied to *Bletia*.

ARUNDINARIA. A genus of grasses belonging to the tribe *Bambuseae*. The species are either of a shrubby or arborescent nature, with strong jointed stems, resembling those of the bamboo cane. They are mostly from the warmer parts of the globe, and in some instances attain a great size, where they grow spontaneously. *A. falcata* is one of the hardiest kinds, being able to bear the cold of ordinary winters in Britain, especially in the southern counties of England and Ireland. In the county of Cork, several brakes were planted of this plant about the year 1848, and in ten years after the canes had reached a height of from sixteen to twenty feet, being about a foot in circumference at the base. The joints on the stems are nearly of equal growths, and owing to their regularity the plantations present a curious appearance, to those, particularly, who have not seen tropical vegetation. In Nepal and on the slopes of the great Himalayan range of mountains, they are used in many instances by the inhabitants in the way of thatch to cover their dwellings, for which purpose it is valuable, resisting, as it does, the effects of weather a long time, owing to the large amount of silica contained in the joints and on the cuticle of the stems.

Another important species is *A. Schomburgkii*, a native of Guiana, where the straight canes attain a height of sixteen feet and upwards, with a diameter of one to one and a half foot at their base. It is this plant, which chiefly furnishes the tubes to the native Indians, from which they blow their poisoned arrows, which after being dipped in the deadly woorali poison, act with such fatal effect on the victims they are aimed at. [D. M.]

ARUNDO. A genus of grasses typical of the tribe *Arundineæ*. This genus, as now defined by Steudel and other authors, excludes the British species, which were formerly included in it; they will be found in the genera *Psamma* and *Phragmites*. *A. Donax* is one of the most important kinds, and may be seen occasionally cultivated in British gardens, for the ornamental effect it produces when growing in groups. The stems attain a height of eight to ten feet in this country; but in Spain and other parts of the south of Europe they grow much taller. The leaves are broad, of a fine siliceous green colour, and in one variety they are beautifully striped in different colours, similar to those of the common ribbon-grass of gardens (*Phalaris arundinacea variegata*). The reeds are sometimes used in making bagpipes and some other musical instruments. Mrs. Callcott, in the *Scripture Herbal*, considers it probable that *A. Donax* is one of the plants alluded to in Scripture as the Reed, especially when the original word is 'keneh.' The canes being long, straight, and light, make admirable fishing-rods, and excellent arrows; the latter quality being of great importance to the warlike Jews, after they began to practise archery with effect. The heroes of Homer made their arrows of this reed (*Iliad* 11.), and the tent of Achilles was thatched with its leaves. [D. M.]

ARVORE DE PAINA. A Brazilian name of *Chorisia speciosa*.

ASA DULCIS. A drug held in high repute among the ancients, supposed to be the produce of *Thapsia garganica*.

ASAFŒTIDA. A drug formed of the concreted milky juice of *Narthex*, and of various species of *Ferula*.

ASAGRÆA. A Mexican genus of plants belonging to the colchicum family, *Melanthaceæ*. The single species of this genus, *A. officinalis*, furnishes the Cebadilla seeds from which the alkaline poison veratrine is prepared. The plant is bulbous, with long linear grass-like leaves, and a long bracteate cluster of flowers; which have a six-parted perianth; six stamens, three shorter than the remainder; anthers heart-shaped, becoming shield-shaped, and bursting vertically; and fruit consisting of three lance-shaped pointed follicles, of thin papery consistence, and containing a number of winged seeds. The seeds, called Cebadilla seeds, were formerly used to destroy vermin, but are now employed in the preparation of veratrine, an alkaline substance, of a powerfully irritant poisonous nature, occasionally made use of in neuralgia and rheumatic affections. It has been given internally, but from its doubtful action and dangerous nature, it is now rarely if ever employed. [M. T. M.]

ASARABACCA. The common name for *Asarum*.

ASARUM. A genus of the order *Aristolochiaceæ*, known by its bell-shaped three-cleft perianth, twelve stamens inserted at the base of the style, and with the connective of the anthers prolonged into an awl-shaped process. The fruit is a six-celled capsule, surmounted by the persistent limb of the calyx. The species of this genus are dispersed over Europe, and the temperate parts of Asia, and North America.

A. europæum is the Asarabacca of herbalists; it is said to be found wild in Westmoreland and other places in the north of England. It is a low growing plant, with a creeping rootstock, from which proceed a number of roots, and also two rounded kidney-shaped stalked leaves; between them is placed the dull brownish flower. The roots and leaves are acrid and somewhat aromatic, they contain a volatile oil, a bitter matter, and a substance like camphor. Asarabacca was formerly used as a purgative and emetic, and also to promote sneezing, but it is now rarely used, having been supplanted by safer and more certain remedies. *A. canadense* is sometimes met with in gardens; it greatly resembles the European plant, but has larger leaves provided with a short spine. [M. T. M.]

ASARINEÆ. A synonym of *Aristolochieæ*.

ASCARINA. A genus of *Chloranthaceæ*, founded by Forster on a single species from the Society Islands. It is a tree with opposite, petiolate, and serrate leaves. The flowers are diœcious and inferior, on tax spikes. The male flower consists of a single stamen with a short filament, and a large oblong quadrivalvate anther. The ovary is globular-truncate, one-celled, and one-seeded. The stigma is sessile, depressed, and obsoletely three-lobed. [W. C.]

ASCENDING. Directed upwards; as the stem, which is the ascending axis. Rising upwards with a curve, from the horizontal to a vertical position, as many stems, simply rising upwards.

ASCI. The name of the fruit-bearing cells in the important division of *Fungi*, called *Ascomycetes*. These may be thread-shaped, cylindrical like little sausages, ellipsoid, or subglobose. In the latter case they are mostly few in number, and are occasionally reduced to one in each cyst or perithecium, as in *Sphærotheca*, to which genus belongs the felted mildew of Rose-leaves and the Hop mildew. [M. J. B.]

The term *Asci* is also applied to spore-cases, consisting of a long or roundish cell containing spores. These are characteristic of lichens.

ASCIDIUM. A pitcher; various modifications of leaves containing, or capable of holding fluid, such as are found in *Sarracenia*, *Nepenthes*, *Cephalotus*, or even *Utricularia*.

ASCLEPIADACEÆ. (*Asclepiadeæ, Asclepiads*, *Apocyneæ* in part.) Among monocotalous exogens with a superior ovary, the very large natural order which bears this name is known by its pollen being collected in the form of waxy masses or bags, derived from the separate inner lobes of the anther cells, and by the fruit consisting most commonly of a pair of divaricating follicles. The species differ from *Apocynaceæ* or *Dogbanes* in the peculiar structure of the staminal apparatus: the stamens in the latter order being distinct, the pollen powdery, the stigma not particularly dilated, and all these parts distinct the one from the other. But in *Asclepiads* the whole of the sexual apparatus is consolidated into a single body, the centre of which is occupied by a broad disk-like stigma, and the grains of pollen cohere in the shape of waxy bodies attached finally to the five corners of this stigma, to which they adhere by the intervention of peculiar glands.

Fully 1000 species are known, for the most part inhabiting the tropics of the Old and New Worlds. Two genera only are found in northern latitudes, one of which, *Asclepias*, has many species, and is confined apparently to North America; the other, *Cynanchum*, is remarkable for extending from 50° north latitude to 32° south latitude. *A. Stapelia* is found in Sicily. They vary extremely in appearance; many being leafless succulents, like *Stapelia*; others, and they are the more numerous, consisting of twiners, like *Hoya*; while another portion consists of upright herbaceous plants, such as *Asclepias* and *Vincetoxicum*; a few are tropical trees. As a general rule the species are poisonous; an acrid milk which pervades all their parts being eminently emetic and purgative.

The genera *Stapelia*, *Hoya*, *Asclepias*, *Vincetoxicum*, *Ceropegia*, *Periploca*, are good examples of the orders. The manner in which the ovules of these plants are fertilised by the pollen is among the most curious phenomena known in plants. Instead of the grains of pollen falling on a viscid stigmatic surface, and then producing tubes of impregnation, the tubes are formed inside the pollen bags, whence they ultimately find their way by a spontaneous emission, and reach the surface of the stigma without being projected upon it, conducted by some inherent vital power. For a full account of this extraordinary fact, see Lindley's *Introduction to Botany*, 4th edition.

ASCLEPIAS. From this genus the order *Asclepiadaceæ* takes its name. Its characters are as follow:—The corolla consists of five petals, bent downwards towards the stalk; within the petals are five curious boat-shaped processes or cups, forming what is called the coronet, and from each of these cups a curved horn-like body projects; within these are five stamens, whose filaments are united into a pentangular tube bearing five anthers, which adhere to the five-angled stigma; the pollen is also remarkable in being aggregated into two separate parcels, suspended on two threads from a sort of gland, but this is a peculiarity not confined to the plants of this genus; the fruit consists of a pair of follicles, which opening, disclose a number of seeds provided with a tuft of glossy silk-like hairs.

The genus consists of herbaceous plants with a milky juice, and which are for the most part natives of America. Several species are cultivated for the sake of their showy flowers. All of them are more or less poisonous. *A. curassavica* is employed in the West Indies as an emetic, and goes by the name of Ipecacuanha; the drug truly so named, however, is derived from a very different plant; see CEPHAELIS. *A. tuberosa*, the butterfly-weed, has mild, purgative properties, and promotes perspiration and expectoration. *A. syriaca*, a plant misnamed, as it is a native of America and Canada, is frequently to be met with in gardens; its dull red flowers are very fragrant, and the young shoots are eaten as asparagus in Canada, where a sort of sugar is also prepared from the flowers, while the silk-like down of the seeds is employed to stuff pillows. Some of the species furnish excellent fibre, which is woven into muslins, and in certain parts of India is made into paper. Some one of the species of *Asclepias* is thought to be the Soma plant so often alluded to as an object of prayer and praise by the ancient natives of India, in the Sanskrit Vedas, which are of a not less remote antiquity than the thirteenth century B.C., while some place them so far back as twenty centuries B.C. 'The bruised stem and leaves of the Soma plant yield a juice which, by standing, ferments into an intoxicating liquor, which is supposed to gratify the gods, and animate them to extraordinary exploits. The elevation of the plant to the rank of a deity can only have originated in a stage of semi-barbarism, in the same way as we can imagine that ardent spirits might have won the adoration of the North American Indians when first introduced among them. See Max Müller's *History of Sanskrit Literature*. [M. T. M.]

ASCOBOLUS. A genus of ascomycetous Fungi, distinguished from *Peziza* by its shooting out its asci when mature. The species grow almost exclusively on the dung of various animals. One, however, is found on decayed leather; and another, which is perhaps a doubtful species, occurs on cover leaves. The sporidia are often of a beautiful colour and form exquisite objects under the microscope. Few Fungi are more common than *Ascobolus furfuraceus*, on old cow-dung. [M. J. B.]

ASCOMYCES. A small genus of *Fungi* of the most simple construction, remarkable principally for the effect they have upon the plants upon which they are developed. The whole plant consists of a stratum of club-shaped cells filled with sporidia, with scarcely any filamentous or cellular base developed in the shape of a white powder on the surface of the leaves, which are generally swollen and distorted, as is especially in the case with blistered peach-leaves, when attacked by *Ascomyces deformans*. The asci are either accompanied by naked spores which sprout like the cells of yeast, or else the sporidia, when set free, are propagated after the same fashion. The genus occurs on the leaves of trees, or sometimes of herbaceous plants, as *Trimudia carpens*. The most obvious examples besides those mentioned above are the *Ascomyces* of the walnut and pear, which trees are, however, far less deformed by it than the peach. [M. J. B.]

ASCOMYCETES. A large division of *Fungi* distinguished by their fruit being contained in hyaline sacs (asci), and not situated at the top of certain privileged cells as in the mushrooms and allied *Fungi*. The asci are placed parallel to each other, barren threads or nore intervening, and are packed into a thin stratum, which equally with the fructifying stratum of mushrooms is called the hymenium. This may be entirely exposed, or may be included in an especial organ called a perithecium. The asci are for the most part colourless, and vary from mere threads to globose sacs. The sporidia or fructifying bodies which they contain are generally definite in number and multiples of two. Their most usual number is eight, but when they are very large these are reduced to four or two, or even one; and in other cases their number is greatly increased, so as in particular instances to be indefinite as far as our powers of observation go. In a particular condition a large proportion of these fungi produce also naked spores on distinct plants, and occasionally naked spores and asci are produced upon the same hymenium. The distinction from sporigerous *fungi* is not therefore as definite as might be wished, though the group is strictly natural. Some of the species approach the lichens so nearly as to be scarcely distinguishable. It is said that asci have been lately found on the gills of one at least of the higher fungi, *Agaricus melleus*, which is largely consumed abroad under the title of hallimasch, though justly neglected here. This, however, wants confirmation, and an assurance that some parasite is not in question. Some of the moulds again produce fruit containing a single spore, or a number of asci; but whether these moulds are true allies of the *Ascomycetes* or not is at present doubtful. The Morel is one of the most familiar examples of the division, and one of the most highly organised. The Truffle belongs to the same division, though so different at first sight from its near allies. [M. J. B.]

ASCOPHORA. A genus of vesicular moulds (*Physomycetes*), differing from *Mucor* principally in the head being at length flaccid and hanging over the top of the stem like a cap or bonnet. The Bread-mould belongs to this genus, and there are one or two more species of some consequence. A singular fact about some of the species is that the fruit upon the sides of the stem is different from that at the apex, retaining its globular form, and containing sporidia of a different size. *Ascophora elegans* is a most beautiful object, from the repeated and regular forked branching of the lateral threads, each division of which is terminated by a fertile vesicle. The Bread-mould is easily cultivated, and the whole developement of the plant in consequence easily traced. Other species, like *A. elegans*, may be cultivated on rice paste under a bell glass, and are interesting objects of study. [M. J. B.]

ASCYRUM. A genus of the St. John's wort family (*Hypericaceae*), numbering five species, all of them American, with a distribution from the N. United States southwards to S. Grenada. All the species have been cultivated in Britain, and one of them (*A. Crux Andreae*) is called St. Andrew's Cross, from the circumstance of the four pale yellow petals approaching each other in pairs, so that they appear like a cross with equal arms. Collectively, they are called St. Peter's worts, to distinguish them from St. John's worts. [A. A. B.]

The species are all under-shrubs, resembling the St. John's worts in general appearance, having opposite sessile leaves, sprinkled below with black dots, and large terminal yellow flowers, singly or three together. The genus is characterised by the tetramerous (four-part) arrangement of the calyx and corolla; the two exterior sepals of the persistent calyx much larger than the inner pair; the deciduous petals cruciate and widely-spreading; the stamens indefinite, from nine to a hundred, with slender filaments, and ovoid two-celled anthers; the ovary ovoid, one-celled and two to four-lobed, with as many styles, and numerous ovules; and the capsule enveloped in the enlarged calyx. [W. C.]

ASEROË. A genus of phalloid *Fungi*, distinguished by the bold rays of the receptacle. The species, which may probably be reduced to three, are of a delicate pink or green. They vary greatly in the degree to which the rays are divided. Like others of the group, they are very fetid when fresh. They are confined to the islands of the southern hemisphere. The genus de-

Aserōe pentactina.

mands more especial notice here, as one of the species appeared in great perfection some years since in one of the stoves at Kew. [M. J. B.]

ASEXUAL PLANTS. This term was once applied to cryptogams, but since the discoveries which have been made during the last thirty years, it is no more applicable to them than to phænogams. Sexual organs have now been discovered in every branch of cryptogams. Amongst *Fungi* alone they are still obscure in several divisions, but if such genera as *Leptostroma* really belong to *Fungi*, of which there is little doubt, there is even amongst them the same type as amongst the higher cryptogams.

It is, moreover, singular that the impregnation of cryptogams comes nearer the type of that in animals than in phænogams. Their spermatozoids resemble closely those of animals, and indeed are often more complicated. Amongst *Fungi* alone, and lichens, which nearly approach *Fungi*, they are mere cells, without motion, analogous to pollen-grains, though they do not germinate like them, at least, so far as has been observed at present. [M. J. B.]

ASH. The common name for *Fraxinus*. —, BITTER. A West Indian name for *Simaruba excelsa*. —, CAPE, *Ekebergia capensis*. —, HOOP. *Celtis crassifolia*. —, MANNA or FLOWERING. *Ornus europæa*. —, MOUNTAIN. The Rowan tree, *Pyrus Aucuparia*. —, POISON. *Rhus venenatum*. —, PRICKLY. *Xanthoxylon fraxineum*. —, RED. *Alphitonia excelsa*.

ASH-WEED. An old English name for *Ægopodium Podagraria*.

ASPALATHE. (Fr.) *Caragana frutescens*.

ASPALATHUS. A large genus of S. African shrubs or under-shrubs belonging to the pea-flowered tribe of the leguminous family. Their leaves are commonly heath-like, often three together (ternate), or sometimes tufted, that is, a number of additional small leaves grow from nearly the same point. The flowers are terminal, racemed, or spiked, and generally yellow, but sometimes bluish purple, red, or white. About 150 species are known, but there is not much of interest about them; many of them have spiny-pointed leaves, and are not unlike dwarf furze bushes. [A. A. B.]

ASPARAGUS belongs to the natural order *Liliaceæ*, and represents the sub-order *Asparageæ*, which are lilies with succulent fruit. The genus consists of many species, but only one is cultivated, the common Asparagus, *A. officinalis*, so well known for its powerful properties as a diuretic, which are ascribed to the presence of a peculiar principle called *asparagin*. See Lindley's *Vegetable Kingdom*, p. 205.

The Common Asparagus is a native of several places in Britain near the sea; such as the Isle of Portland, and Kynance Cove, near the Lizard, Cornwall. In the southern parts of Russia and Poland the waste steppes are covered with this plant, which is there eaten by horses and cattle as grass. It is also common in Greece, and was formerly much esteemed as a vegetable by the Greeks and Romans. It appears to have been cultivated in the time of Cato the Elder, 200 years B.C.; and Pliny mentions a sort that grew in his time near Ravenna, of which three heads would weigh a pound.

In this country Asparagus is reckoned among the oldest and most delicate of our culinary vegetables; and in its cultivated state the whole plant has a very graceful appearance. It is noticed by Gerarde in 1597; and in 1670 forced Asparagus was supplied to the London market. At Mortlake, Battersea, and other places near London, where the soil is suitable, Asparagus is extensively cultivated, and by skilful management is brought to a higher degree of perfection, perhaps, than in any other part of the world. The part of the plant which is used is about six or eight inches of the young shoot, which is considered to be fit for cutting when it has emerged two or three inches out of the ground, and has a firm, compact, roundish point, of a fine green colour, slightly tinged with purple. In preparing Asparagus for table, its delicate flavour is rather deteriorated than improved by the additions which skilful cooks deem necessary for it and other vegetables. It is usually boiled and served alone with melted butter and salt, or on toasted bread with white sauce; and the smallest heads are sometimes cut into small pieces and served as a substitute for green peas. Its virtues are well known; as a diuretic it is unequalled; and for those of sedentary habits who suffer from symptoms of gravel, it has been found very beneficial, as well as in cases of dropsy.

Prussian Asparagus, which is brought to the markets in Bath, is not a species of *Asparagus* at all, but consists of the spikes, when about 8 inches long, of *Ornithogalum pyrenaicum*, which grows abundantly in hedges and pastures in that locality.

[W. B. B.]

ASPASIA. Under this name are collected a few species of epiphytal orchids of the Vandeous suborder, with a lip half united to the column. They have broad oval thin pseudobulbs, and flowers mottled with purple on a violet ground. The most important species is *A. epidendroides*, a plant from Central America. The other species are from the tropics of the same continent.

ASPEN, or ASP. *Populus tremula*.

ASPERGE. (Fr.) *Asparagus officinalis*.

ASPERGILLUS. A genus of filamentous moulds, characterised by the hyaline or brightly-coloured jointed thread being swollen at the apex, and there studded with radiating cells, each of which produces a necklace of spores. The most common species, *A. glaucus*, distinguished by its globose echinulate spores, is one of the *Fungi* which produce the well-known blue mould, and whose spores form occasionally a part, with other common species, of the substance called yeast. It is distinguished from the genus *Rhinotrichum*, formerly associated with it, by the fertile radiating cells, which in *Rhinotrichum* are replaced by a few sporules. There is some reason to believe that there is a second form of fruits in *Aspergillus*, which constitutes the genus *Eurotium*, but this requires confirmation. [M. J. B.]

ASPERIFOLIÆ. An old name for what are now called Borageworts, or *Boraginaceæ*, derived from the remarkable roughness of the leaves of the greater part of the species.

ASPERUGO. A genus of *Boraginaceæ*, consisting of a single species, *A. procumbens*, which occurs in Britain as well as the whole of Europe and a great part of Central Asia. It is an annual plant with trailing dichotomous stems, which, as well as the aborate leaves, are rough with curved bristles; the flowers are very small, bluish-purple. The calyx is curious; it enlarges as the fruit ripens, and takes the form of two large puckered valves, which are triangular and marked with prominent veins; these valves are applied flat to each other, and enclose the fruit which has the structure common to the order. [J. T. S.]

ASPERULA. A family of herbaceous plants with square stems, whorled leaves and four-cleft flowers, which are either pure white, white tinged with purple externally, or more rarely blue or yellow. Many of the species are ornamental, and well fitted by their habit for the decoration of rock-work. *A. odorata*, Woodruff or Woodrowel (so called from the resemblance between its whorled leaves and the rowel of a spur), is a common woodland plant, conspicuous in May and June by its brilliantly white flowers, and at other seasons by its bright green leaves, arranged in a star-like form round the stem. The flowers are sweet-scented, but the plant derives its name from the fragrance of its leaves. This is not perceptible while the herbage is fresh, but after being gathered a short time it gives out the perfume of new hay, and it retains this property for years. Woodruff is a useful plant in shrubberies, increasing rapidly and thriving under the shade of most trees, even the beech. It is used in Germany to impart a flavour to some of the Rhine wines. *A. Cynanchica*, a small trailing plant with slender stems, very narrow leaves, four in a whorl, and small white flowers delicately tinged with pink, occurs on chalky downs in many parts of Britain. It owes its specific name, and its popular name Quinsy-wort, to its supposed virtues in curing quinsy. In the time of Gerarde, *A. arvensis*, a species with blue flowers, grew in 'many places of Essex, and divers other parts, in sandie ground.' It was also found during the present century in a slate-quarry in Devonshire, but has disappeared. Most of the foreign species are hardy and may be raised from seed. [C. A. J.]

ASPHODEL. *Asphodelus*. — BOG, or LANCASHIRE. *Narthecium ossifragum*. — FALSE. An American name for *Tofieldia*. — SCOTCH. *Tofieldia palustris*.

ASPHODELUS. The Asphodel. A genus of *Liliaceæ*, distinguished by having a six-leaved equal spreading perianth; six hypogynous stamens, of which the alternate ones are shorter, the filaments dechiate; a filiform or subclavate style, with a capitate three-lobed stigma; a three-celled ovary, with three collateral ovules in each cell. The species are perennial herbs of Southern Europe, with fleshy, fasciculated roots; radical, subulate, triquetrous, or linear-lanceolate leaves; and a simple or branched scape bearing the white flowers in close racemes. There are several species, *A. albus* is a common garden plant, formerly called King's Spear; and this, and *A. ramosus*, which is probably only a branched variety of it, are very ornamental plants. It is stated on the authority of Symonds to cover large tracts of land in Apulia, an ancient province of Italy, and to afford good nourishment to sheep. [T. M.]

ASPHYXIA. Plants, like animals, require free access to atmospheric air, and if this is impeded, or the air is loaded with noxious gases, a greater or less degree of mischief is sure to follow. Death may not be the immediate consequence, but the tissues may be so impaired that there is only a short respite. The communication between these tissues is carried on in phænogams, and many cryptogams by means of the stomata. If, therefore, these are clogged on, the proper degree of aeration cannot take place, and since the same apertures are the safety-valves for the discharge of superabundant moisture or gases which have performed their office, the whole system becomes gorged, and the proper functions impeded. In such cases a true Asphyxia or suffocation takes place; and the same effect may be produced by the air-passages being filled with gummy matter, or their apertures covered by parasitic fungi, as *Antennaria*, *Capnodium*, *Cladosporium*, &c. Plants may also be drowned by a few days' immersion in water, though in some instances there is a provision by which such an effect is altogether prevented. [M. J. B.]

ASPIC. (Fr.) *Lavandula Spica*.

ASPIDIEÆ. A section of polypodineous ferns, in which the sori are punctiform, or dot-like, and covered either by reniform or peltate indusia. [T. M.]

ASPIDISTRA. A genus of *Liliaceæ* found in China and Japan. They are stemless glabrous herbs, with oblong-lanceolate striate leaves, and radical one-flowered peduncles, bearing a single dull purple flower. The perianth is bell-shaped, six to eight-cleft, with spreading segments; the stamens six to eight, inserted in the tube of the perianth; the ovary small, cylindrical, three to four celled, with two ovules in each cell; the style short, thick, continuous with the ovary, terminated by a large discoid, radiate, lobed stigma. Three or four species are known. [T. M.]

ASPIDIUM. The name formerly given to a group of polypodiaceous ferns, including all those in which the dot-like or punctiform sori were covered by a roundish cover or indusium. In this sense it is synonymous with the modern section *Aspidica*. It is now, however, generally divided into a greater or lesser number of genera, according to the views of individual pteridologists. The smallest amount of division is adopted by those who separate the group into two, having the indusia respectively peltate or reniform, the first being then called *Aspidium*, and the second *Nephrodium*. Those who subdivide more extensively, and separate the free-veined from the net-veined species, restrict the name to a few typical kinds having the indusium orbicular and peltate, and the veins of the fronds compoundly reticulated, with free included veinlets, which are divaricate or variously directed. Thus limited it comprises about a dozen species, with as many more doubtful ones, chiefly found in India and the east, but also occurring in South America and the West Indian Islands. The majority of the species are strong-growing pinnate ferns, with the pinnæ sometimes lobed. The typical *A. trifoliatum* is sometimes seen cordate and undivided, a stunted condition, caused probably by the depressing influences under which the plants are grown; sometimes three-leaved, which seems to have been the form originally described;

and sometimes decidedly pinnate, with the pinnæ more or less deeply lobed, all these forms being sufficiently developed to become fertile. These facts clearly show the variableness to which the species of ferns are liable. *A. singaporianum*, a simple-fronded species, has the fronds very remarkably narrowed at the base, and is furnished with very numerous evenly arranged sori.

ASPIDOSPERMA. A genus of *Apocynaceæ*, consisting of about twenty-five species of trees, from tropical America. The leaves are alternate, and mostly entire. The flowers are small and arranged in solitary or numerous dichotomously-branched cymes, at the ends of the branches. The calyx is five-partite, the corolla is hypocrateous, sub-infundibuliform and five-lobed. The five inclined stamens are inserted in the middle of the corolla tube; they bear ovate sub-sessile anthers. There are two ovaries, with many ovules attached to the ventral suture. The fruit is a double, rarely a single follicle, compound, obovate, and woody, with numerous membranaceous seeds. The wood of this genus is valuable. *A. excelsum*, called by the colonists Paddle-wood, is remarkable for its singularly fluted trunk, composed of solid projecting radii, which the Indians use as ready-made planks. [W. C.]

ASPLENIDICTYON. A synonyme of *Hemidictyum*, a genus of large growing asplenium-like ferns. [T. M.]

ASPLENIEÆ. A section of polypodineous ferns, in which the simple linear or oblong sori are parallel with the veins, and oblique to the midrib, produced on one side of the vein, and covered by indusia of the same form. The modern group, *Asplenieæ*, is nearly synonymous with the genus *Asplenium* of the elder and some modern writers. [T. M.]

ASPLENIUM. A genus of polypodiaceous ferns established by Linnæus, and, as originally defined, synonymous with the modern group *Asplenieæ*, including the *Scolopendrieæ*, and *Diplazieæ*. In this sense it included all the ferns with lines of fructification lying parallel, or nearly so, on the disk of the frond (not marginal). The group is now considerably subdivided, and the name *Asplenium* restricted to those species in which the veins of the frond are free, the sori are linear or oblong, and lying obliquely on the parts of the frond, and the indusia are simple and distinct. Even thus reduced it is a very extensive family, found in all parts of the world, mostly evergreen, numbering about 200 species, of which nine are natives of Great Britain. As might be anticipated in so large a family, the species are exceedingly varied, especially as to division, some being simple, others lobed, or pinnate, or bipinnate, or tripinnate, or even decompound; and while some are delicately membranaceous in texture, others are of a stouter herbaceous character, and some are thick and leathery. *A. Adiantum nigrum*, the

Black Maidenhair Spleenwort, one of the commoner British species, has bipinnate fronds; another common one, A. Trichomanes, the common Maidenhair Spleenwort, has pinnate fronds; while A. septentrionale, one of the rarer native species of northern habitats, has the fronds reduced to the appearance of two or three forked rigid ribs. Some of the exotic species are very beautiful in form; and many of them are cultivated on account of their beauty in our gardens and hothouses. Several species have the very singular property, strongly developed, of bearing little buds on their surface, from which young plants are formed even while they are retained upon the parent frond. The genus has been named Asplenium, or Spleenwort, on account of some supposed potency in the plants over diseases or affections of the spleen; but, as in many other instances, these virtues are both fanciful and fabulous. The principal genera, separated from Asplenium by modern pteridologists, are Diplazium, Athyrium, Thamnopteris, Hemidictyum, Allantodia, Ceterach, and Callipteris. [T. M.]

ASSAGAY-TREE. *Curtisia faginea.*

ASSARACUS. A subdivision of the genus *Narcissus*, including N. *capax* and N. *reflexus*, in which the segments of the perianth are semi-reflexed, and the coronet poculiform, about equalling the perianth segments. [T. M.]

ASTELIA. A genus of sedge-like Juncaceæ, from the islands of the Southern Ocean, with polygamous-dioecious flowers, having a perianth of glumaceous texture, as in the rush; ovary three-celled or one-celled by the incompleteness of the partitions; fruit berry-like; stem very short; leaves broadly-linear, hairy, very silky at the base. A. *alpina* has leaves three-quarters of an inch broad, and an extremely short flattened scape, crowned by a dense panicle of rather large chestnut-coloured flowers. The leaves of this species, which grows on the sand-hills of the coast of Tasmania, are edible, and are said to have a nutty flavour. [J. T. S.]

ASTEMON. A genus of labiates, found in Bolivia; it is related to *Colebrookia*, from which it differs in having a non-plumose calyx, and a five-lobed, not four-lobed, corolla. The calyx is tubuloso-campanulate, with five subequal lanceolate acuminate teeth. The corolla tube is as long as the calyx, the limb short and nearly equally five-lobed, and the throat bearded. The stamens, described as wanting, are in reality reduced to four small distant stalkless anthers, adnate to the corolla-tide. A. *grareolens* is a shrub six to eight feet high, having an unpleasant odour; the leaves oblong-lanceolate, attenuately acute, green and smoothish above, whitish and tomentose beneath; and the flowers small, white, in a terminal panicle, which is still more densely clothed than the stems and leaves with white tomentum. [T. M.]

ASTEPHANUS. A genus of Asclepiadaceæ, containing thirteen species, natives chiefly of the Cape of Good Hope and Madascasar, but found also sparingly in America. They are climbing or decumbent under-shrubs, with small opposite leaves, and interpetiolar umbels, consisting of a few small and generally white flowers. The calyx consists of five acute sepals; the corolla is campanulate and has no squames within the tube—the character by which this genus is distinguished from *Metastelma*. There are ten small pendulous masses of pollen. [W. C.]

ASTER. A well-known genus of the composite family, numbering nearly 200 species, which are distributed sparingly over Europe, Asia, and S. America, but occur in great abundance in N. America, where three-fourths of them are indigenous. They are perennial (rarely annual) herbs with alternate and simple entire or toothed leaves, and panicled, racemed, or corymbose star-like flower-heads, having an involucre of numerous imbricated scales, enclosing many florets, the outer row strap-shaped and pistil-bearing, those in the centre tubular, and all having more or less flattened achenes crowned with a pappus of numerous capillary bristles. From their time of flowering, Asters are often called Michaelmas Daisies and Christmas Daisies, some of them continuing in flower in the open air in mild seasons up to the latter period; and for this reason they are valuable garden plants, because there are few things but themselves which flower so late in the year. The Seaside Aster, A. *Tripolium*, is the only British species. It is a pretty plant, six inches to two feet high, with linear or lance-shaped smoothish and fleshy leaves, and stems terminating in corymbs of purple-rayed flower-heads, rather more than half an inch across. It occurs pretty generally over all the British as well as European coasts. The Alpine Aster, A. *alpinus*, is the type of a small group which inhabit Alpine regions alone. It is found on the mountains of Central Europe, Asia, and N. America, growing from three inches to a foot high, the stem furnished with lance-shaped or linear leaves, one to two inches long, and terminating in a blue-rayed flower-head, one to two inches across. The remainder are mostly branching plants, from one to ten feet high, with heart-shaped, willow or heath-like leaves, and starry flower-heads, always with the central tubular florets yellow, and the ray varying from white to lilac blue or purple. There is a great sameness about many of the species, and they are most difficult to determine. We can only name as some of the more showy kinds A. *spectabilis*, A. *Novæ-Angliæ*, A. *versicolor*, and A. *turbinellus*, all North American; A. *alpinus*, from the Sikkim Himalaya; and the Italian Star wort, A. *Amellus*, from S. Europe. [A. A. B.]

ASTER, CAPE. *Agathea amelloides.* —

CHINA, *Callistephus chinensis*. —GOLDEN. A common name for *Chrysopsis*. — WHITE-TOPPED. An American name for *Sericocarpus*;

ASTERACEÆ (*Compositæ, Synantheræ*.) This is the largest natural order of plants, the species occurring in all parts of the world, and in all places, and forming a total equal to about a tenth of the whole vegetable kingdom. They are recognised by their monopetalous flowers, growing in close heads (*capitula*), and having at once an inferior one-celled ovary, and stamens whose anthers cohere in a tube (i.e. are syngenesious). De Candolle states as the result of his examination of their natural habit, that out of 8323, 1,229 were annuals, 745 biennials, 2,491 perennials, 1,264 undershrubs from 1 to 3 feet high, 566 shrubs from 4 to 15 feet high, 72 small trees, 4 large trees above 25 feet high, 81 woolly plants of which nothing further was known, 126 twiners or climbers, and 1,501, about which nothing certain could be ascertained. Of these 247 grow in the South Sea Islands, 2,234 in Africa, 1,427 in Asia, 1,042 in Europe, and 3,500 in America; the Cape of Good Hope possessed 1,540, Mexico 725, Brazil 722, United States and Canada 678, the Levant 610, the Continent of India 481, north and middle Europe 447, Europe in the Mediterranean 395, Australia 294. But these numbers greatly require rectification. The uses of the order, real or imaginary, are very numerous and conflicting. Some are tonic and aromatic like wormwood (*Artemisia Absinthium*, and others); or vermifuges like those other *Artemisias* known in foreign pharmacy as *Semen-contra*, or *Semencine*. A few are powerful rubefacients, as pellitory of Spain (*Anacyclus Pyrethrum*), and various kinds of *Spilanthes* which excite salivation. *Arnica montana* is powerfully narcotic and acrid. Similar evil qualities belong to *Crepis lacera*, a most venomous species, said to be no infrequent cause of fatal consequences to those who, in the South of Europe, incautiously use it as salad. Nor are *Hieracium virosum* and *H. suaveolens* altogether free from suspicion. Some species of *Pyrethrum* have the power of driving away fleas. Many yield in abundance a bland oil when their achenes or "seeds" are crushed: such are the sunflower (*Helianthus annuus*), the til or ramtil (*Verbesina sativa*), largely cultivated in India, and *Madia sativa*. A purgative resin is obtained from some allies of the thistles; others, as *Aucklandia Costus*, now referred to *Aplotaxis Lappa*, have aromatic roots, and are looked upon by Orientals as aphrodisiacs. Finally, under the name of artichoke, succory, scorzonera, endive, salsify, and lettuce, we have some of our most harmless and useful esculents. Botanists adopt various modes of classifying this immense mass of species; but all are subordinate to the four following capital groups, viz :—CICHORACEÆ: florets all ligulate; CORYMBIFERÆ: florets tubular in the disk; CYNARACEÆ: florets all tubular, with an articulation below the stigma; LABIATIFLORÆ: florets bilabiate.

ASTERANTHOS. A genus represented by a single African shrub, whose true affinities have not yet been established, and may be one of the survivors of an extinct world. A natural order (provisional to some extent), *Napoleoneæ*, has been created to receive it, and the allied genus *Napoleona*, both possessing sufficient characters to separate them from every known family. This genus has alternate, ovate-lanceolate, entire, and shortly petiolate leaves, and solitary axillary flowers. It has a short campanulate, many-toothed calyx, adherent to the ovary, and a simple many-lobed corolla. The indefinite stamens are inserted at the base of the corolla, and have filiform filaments, and oblong anthers. The inferior ovary has a simple style, and an obtusely six-lobed stigma. [W. C.]

ASTÈRE D'AFRIQUE. (Fr.) *Agathæa amelloides*.

ASTEROLINUM. A genus of the primrose family, with a single species, *A. stellatum*, found in S. Europe and Asia Minor, chiefly on the sea coast. It is a little erect or decumbent annual, one to three inches high, with opposite linear leaves, one-sixth of an inch in length, bearing in their axils single stalked minute flowers, whose greenish-white corollas are nearly hidden by the calyx. The minute corollas and few-seeded capsules are the chief distinguishing marks between this genus and *Lysimachia*, in which the plant was placed by Linnæus with the name *Lysimachia linum-stellatum*, alluding to its flax-like leaves and starry flowers. [A. A. B.]

ASTEROSTEMMA. A genus of *Asclepiadeæ*, having the following distinguishing characters: the coronet of the stamens five-lobed, fleshy, short, coloured, and cup-shaped, its lobes crescent-like, or three-toothed, opposite to the anthers, which latter are terminated by a membranous crest; the pollen masses are erect. *A. repandum* is a climbing shrub inhabiting Java. [M. T. M.]

ASTEROSTIGMA. A genus of the family *Araceæ*, comprising one or two Brazilian species, which have a tuberous rootstock, from which arise the leaves and also the stalked spadix, encircled by a spreading purplish spathe. The male flowers are at the upper part of the spadix; the anthers open by a terminal pore. The ovaries which are placed at the lower part of the spadix surrounded by abortive stamens, are three or four-celled, each cell containing a single erect ovule. The style is short and terminated by a flattened stigma, which is divided into three or four segments, each of which is again divided into two, giving to the stigma that star-like or radiating appearance denoted by the name of the genus. [M. T. M.]

ASTRAGALUS. A genus of perennial

plants belonging to the papilionaceous subdivision of the leguminous family. They have woody roots; unequally pinnate leaves; flowers in axillary clusters; a tubular or bell-shaped calyx, with five teeth; a corolla with the standard larger than the wings, and a blunt keel; stamens in two parcels; a curved fruit or legume, divided into two cavities by the projection inwards of the hinder wall of the fruit. They have compound leaves, and frequently spiny branches. There is a large number of species distributed all over Europe, Central and Northern Asia, North America, the Andes, penetrating into the Arctic regions, ascending high Alpine summits, and abundant in the hot rocky districts of the Mediterranean region. A great number are cultivated in this country, and three species are found wild. One, *A. glycyphyllos*, has long stems, trailing on the ground like those of a pea, large leaves, and yellow flowers. The other two are humble plants with small leaves: *A. hypoglottis* has purplish flowers and erect pods, rather longer than the calyx; and *A. alpinus*, which is only found in the Clova mountains, has pendulous pods, which are three times the length of the calyx.

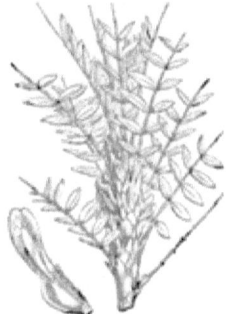

Astragalus gummifer.

The gum-like substance called Tragacanth is the produce of several species of *Astragalus* growing in Persia, Asia Minor, Kurdistan, etc. The gum exudes naturally from the bark in the same way that gum exudes from the bark of cherry or plum trees. The seeds of *Astragalus boeticus* are used in place of, and sometimes mixed with, coffee; the plant is cultivated for those seeds in certain parts of Germany and Hungary. [M. T. M.]

ASTRANCE. (Fr.) *Astrantia.*

ASTRANTIA. A genus of umbellifers, natives of Europe and Western Asia, containing some ten or twelve species. They are perennial herbs, with black, aromatic roots, palmatilobed, petiolate, radical leaves, and few generally sessile stem-leaves. The umbels have a leaf-like involucre, and few umbellules containing of many white or rose-coloured, often polygamous flowers, surrounded by a conspicuous involucel, consisting of many membranaceous, whitish or slightly-coloured leaves. The tube of the calyx is tuberculated, and the limb has five foliaceous lobes. The petals are conniveut, oblong-obovate, and divided from the middle. The fruit is compressed; the mericarp, without vittae, has five prominent toothed juga, with smaller juga in the intervening spaces. [W. C.]

ASTRAPÆA. A genus of small trees, belonging to the natural family *Byttneriaceae*. Their leaves are alternate, stalked, cordate, and from three to five-lobed, generally hairy or viscid, and having at their base large stipules. The peduncles are axillary, long, pendulous or erect, bearing on their apex an umbel of large sessile flowers, enclosed in a leafy involucre. *A. Wallichii* has large leaves and bright red flowers, nearly as large as those of some rhododendrons; the leaves are sometimes one and a half feet long, and of the same breadth. It has long been in cultivation in England, is an elegant plant, and the drooping heads of flowers give it a peculiar appearance. *A. viscosa* has erect peduncles and much smaller flowers, which are white with a pink centre. The bark of *A. cannabina* is used in Madagascar (where all the species are found) for making cords. [A. A. B.]

ASTROLOMA. A genus of *Epacrideae*, containing eight species, natives of New Holland and Tasmania. They are low under-shrubs, often prostrate, with crowded alternate linear or obovate-lanceolate and mucronate leaves; and with solitary axillary flowers of a reddish colour. The calyx is five-partite, with four or more bracteoles. The corolla is hypogynous and tubular, distended above the middle, and with a short five-cleft limb; it has five bundles of hairs in the inside near its base. There are five included stamens, with linear filaments, and oblong simple anthers. The disk is hypogynous and cyathiform. The ovary is five-celled, with one ovule in each cell; the style is simple, and the stigma capitate. The fruit is a drupe, composed almost entirely of solid putamen with five cells. [W. C.]

ASTRONIA. A genus of melastomaceous trees, from the Moluccas, with the habit of *Melastoma*. The leaves are opposite, long-stalked, three-ribbed, oblong acuminate; the flowers small, purplish, in terminal panicles; petals five or six; stamens ten or twelve; fruit a three or four-celled many-seeded berry, branches four-angled, and, as well as the peduncles and calyx, with dark scales. *A. papetara* has subacid leaves, which are cooked as a sauce to fish. [J. T. S.]

ASTROPHEA. A section of the genus *Passiflora*, characterized by the absence of cirrhi and involucre, by its ten partite calyx and five stamens. The species are South American trees. [W. C.]

ASUL. The Arabic name of *Tamarix orientalis*. Also, an Indian name for *Tamarix furas*, a nut-gall tree.

ASYSTASIA. A genus of acanthaceous plants, natives of the East Indies, the warmer and temperate parts of Asia and Africa. They are herbaceous, or shrub-like, with slender branches; axillary or terminal clusters of flowers, which are blue or yellowish, and handsome, with a regularly five-parted calyx, a somewhat funnel-shaped corolla, with a limb divided into five nearly equal segments; stamens four, united in pairs at their base; anthers two-celled; cells parallel, thickened or provided with an appendage at the base; stigma capitate two-lobed. Capsule compressed, slender, and seedless below; above somewhat four-cornered, two-celled, four-seeded. Seeds with a prominent angle at the base. [M. T. M.]

ATACCIA. A genus of the small order *Taccaceae*. The tube of the perianth is in these plants connate with the ovary, and the six-parted limb has the inner segment larger reduced and persistent. The flowers contain six stamens, inserted at the base of the segments of the limb, and having broad filaments concave above; the style is short, thick, three-furrowed, with a capitate three-lobed stigma; the ovary is sub-three-celled, containing numerous ovules, and becomes a semi-three-celled many-seeded berry. There are few more remarkable-looking plants than *A. cristata*, sometimes met with in gardens under the incorrect name of *Tacca integrifolia*. It has a short conical underground caudex, or rhizome, and produces from this caudex three or four large oblong acuminate purplish-green stalked leaves. The scape is about as long as the leaves, erect, stout, angled, dark purple, terminated by a large four-leaved involucre, of which the two outer leaflets are dark purple, opposite, sessile, and spreading; and the two inner much larger, placed side by side, green with a deep purple base and stalk. The numerous flowers form a drooping one-sided umbel; the perianth dark purple, with a turbinate six-angled tube, and a six-parted limb suddenly reflexed, the segments arranged in an outer smaller, and an inner larger series, the rim of the mouth forming a crenated ring. This plant is a native of the islands of the Malayan Archipelago, and one or two other species are Indian; besides which, Sir W. J. Hooker mentions one a native of Demerara. Though remarkably curious in structure, these plants are of no known utility. [T. M.]

ATALANTIA. A genus of aurantiaceous plants, known by their undivided leaves, few stamens, united below into a tube, and one ovule in each cell of the ovary. The trees and shrubs of this genus are natives of the East Indies. The wood of *A. monophylla*, a native of Coromandel, furnishes a heavy closely grained yellow wood, suitable for cabinet work. [M. T. M.]

ATAXIA. A genus of grasses of the tribe *Phalarideae*, with the inflorescence in thyrsoid panicles; spiculae three flowered; inferior flower male, with two pales, intermediate neuter with one pale, and the terminal hermaphrodite, triandrous; glumes unequal. Steudel describes four species, none of which are British. *A. Horsfieldi*, a native of Java, has the peculiar property of emitting, when bruised, a similar odour to that given out by the English sweet-scented vernal grass, which is supposed to result from the presence of a portion of benzoic acid (?) in their tissues. [D. M.]

ATCHAR or ACHIAR. A condiment prepared from *Bambusa arundinacea*.

ATHANASIA. A genus of yellow-flowered composites, consisting chiefly of greenhouse evergreens of shrubby habit, from Southern Africa, and a single annual species from Barbary, the *A. annua*, formerly much more cultivated than at present on account of the long duration of its flowers. It attains a height of about two feet or more, with diffusely branched furrowed stems; pinnatifid fleshy foliage, with linear segments; and corymbs of clustered flower-heads on long foot-stalks, the florets being all tubular. Though of somewhat rambling habit, the small amount of care it requires, and the lasting character of its blossoms, render it deserving of some attention. The cut flowers preserve their freshness for a long period. The genus derives its name from the Greek 'Αθανασια, signifying immortality, in allusion to the unfading nature of its flowers. [W. T.]

ATHERANDRA. A genus of *Asclepiadaceae*, containing two species, from the Moluccas; climbing shrubs, with slender branches, ovate and opposite leaves, and few flowered axillary cymes. The calyx consists of five lanceolate sepals, and the corolla of as many linear-lanceolate lobes. The filaments are free above, and the anthers are appressed to the stigma, and more or less connate among themselves. There are twenty granular masses of pollen. [W. C.]

ATHEROSPERMACEÆ. (*Plumæ Nudæ.*) A small natural order of trees from Australia and Chili, deriving their English name from their aromatic nuts, being furnished with a permanent style, clothed with long hairs. Only three genera are known; *Atherosperma*, *Laurelia*, and *Doryphora*, which see. Their flowers are insignificant. They are placed by Lindley in the monospermal alliance of dicheous exogens.

ATHEROSPERMA. A genus of *Monimieae*, containing a single species from New Holland. It is an aromatic tree with four-cornered branches, opposite leaves, and pedicellate axillary solitary flowers, with two deciduous bracteoles. The flowers are monœcious. The male flower is campanulate, with a very short tube, and eight lobes; and the ten to twelve fertile stamens are mixed with scale-shaped barren ones. The calyx of the female flower is the same as in the male; the ovaries numerous, sessile, and distinct, with one cell and one ovule. [W. C.]

ATHYRIUM. A genus of polypodiaceous ferns of the section *Asplenieæ*, closely allied to *Asplenium*, with which some have united it. It agrees with that genus in the peculiarities of having free veins, and simple distinct indusia; but the sori are innate or more or less horse-shoe-formed (hippocrepiform), this distinguishing character being most strongly developed in the sori which are placed more immediately in the neighbourhood of the principal veins. The species, several in number, found in various parts of the world, are mostly deciduous in habit; and one of them, *A. Filix fœmina*, the Lady Fern, is plentiful in Great Britain, where it assumes a great variety of beautiful forms, which will be found described in Moore's *Handbook of British Ferns*, or more fully, accompanied in many instances, by figures, in the Octavo *Nature-Printed British Ferns*. The Lady Fern has bipinnate or tripinnate fronds of delicate texture, and of a remarkably elegant plumy character. [T. M.]

ATLEE GALL. A gall nut produced abundantly by *Tamarix orientalis*, which is called *Bil* by the Egyptians. It is filled with a deep scarlet liquid.

ATRACTYLIS. A genus of prickly-leaved thistle-like plants belonging to the composite family (Compositæ). They are perennial or annual herbs, from three inches to one and a half foot high, with toothed or pinnatifid leaves, their margins often spiny. The flower-heads are terminal, solitary, or three to four together, having a double involucre, the scales of the external one pinnatifid and leafy, with spinous teeth, standing apart from the inner involucre, and forming, as it were, a fence round the flower head. The scales of the inner involucre are ovate or lanceolate, terminating in a spinous point. The florets are generally of a pink colour. One species is found in North China, and all the others are natives of the Mediterranean region, abounding in Algeria, and growing chiefly in arid desert places. [A. A. B.]

ATRAGENE. A genus of somewhat woody-stemmed *Ranunculaceæ*, of climbing habit, differing from *Clematis* only by having petals, which, however, are small and pass gradually into stamens. The sepals are large and coloured, usually purplish as in the solitary flowered species of *Clematis*; the leaves opposite, compound, ternate, the leaf-stalks twining round supporting bodies. They occur in the temperate regions of both the New and Old Continents, in the northern hemisphere. One species, *A. alpina*, is not uncommon in gardens. [J. T. S.]

ATRAPHAXIS. A genus of Asiatic and Cape of Good Hope *Polygonaceæ*, forming low shrubs with rigid much-branched often spiny stems, and small entire oblong leathery leaves; stipules shrubling, with a small free appendage on each side; flowers fasciculate near the end of the branches, on short peduncles; calyx coloured, four-leaved, the two inner divisions largest, conspicuously so in fruit (they are probably really petals); stamens six; styles two; fruit a small lens-shaped nut, included between the two large ovate deltoid calyx segments. A few species are cultivated as greenhouse plants, but their scrubby stems and small flowers present few attractions. *A. spinosa* is a dwarf hardy shrub. [J. T. S.]

ATRIPLEX. Orache. A genus of *Chenopodiaceæ*, with the foliage covered with a granular mealiness. The Oraches are chiefly distinguished by the two hearts or small leaves, enclosing the fruit, and enlarging after flowering; they are frequently dotted with large-coloured warts, which give them a peculiar appearance. The genus possesses several species, which are very variable in form, according to soil and situation. They inhabit waste places or road banks by the sea shore, rarely occurring inland, with the exception of the *Atriplex patula*, which accompanies arable cultivation, especially in wet sandy clays. The following are British species:—*A. portulacoides*, Purslane Orache, a straggling branched sub-shrub; *A. patula*, Common Orache, of which there are several varieties, distinguished by a more or less upright habit, and leaves of various gradations, from lancet to halberd-shaped; *A. pedunculata*, Stalked Orache, which has the fruits always on a pedicel or footstalk. [J. B.]

The Garden Orache, or Mountain Spinach (*A. hortensis*), is a tall erect growing hardy annual plant, a native of Tartary, introduced in 1548. It is not much cultivated in this country, but in France, under the name of *Arroche*, it is grown to some extent for the sake of its large and somewhat succulent leaves, which are either used alone as spinach, or mixed with sorrel, for the purpose of correcting its acidity. The quality of the spinach yielded by Orache is far inferior to that of the common spinach (*Spinacia oleracea*), or even of the New Zealand spinach (*Tetragonia expansa*); but its leaves being produced abundantly during summer, it is occasionally found useful for culinary purposes. There are several varieties of this plant cultivated, but they do not differ in any other respect, excepting in the colour of their stems and leaves, which vary from pale green to a red or lurid purple, and are very ornamental. The seeds are said to be so unwholesome as to excite vomiting. See Lindley's *Vegetable Kingdom*, p. 573. [W. B. B.]

ATRIPLICEÆ. A synonyme of *Chenopodiaceæ*.

ATROPA. A genus of plants of the natural order *Solanaceæ*; or by Miers made the type of a new family called *Atropaceæ*. The genus is known by its five-parted calyx; its bell-shaped corolla formed of five united petals; five stamens adhering to the lower part of the tube of the corolla, with their anthers opening by long slits; a two-celled ovary and succulent fruit, each containing several seeds.

The Deadly Nightshade (*A. Belladonna*) is found wild in Southern Europe and Western Asia, also in this country, frequently on chalky soils, and not uncommonly in the vicinity of ruins. Though the stems die down annually, they spring from a perennial rootstock, and form in summer time a bushy plant, with stalked egg-shaped entire leaves of a dull green colour, and a peculiar heavy smell. The flowers are borne on short drooping flower-stalks in the axils of the leaves, or in the forks of the stem; they have a widely-spreading bell-shaped calyx, deeply divided into five-pointed segments, and a bell-shaped corolla, somewhat less than an inch in length, and of a dull purplish-brown colour, but whitish or yellowish at the lower portion (uppermost as the flower hangs on the bush). The berry is of a dark shining black colour, about the size and form of a cherry, of a sweet or mawkish taste, and placed at the bottom of the permanent spreading calyx. All parts of the plant are poisonous. It is supposed to have been the plant which produced such remarkable and fatal effects on the Roman soldiers during their retreat from the Parthians under Mark Antony, as recorded in Plutarch's life of Antony. Buchanan relates the destruction of the army of Sweno the Dane, when it invaded Scotland, by the berries of this plant. They were mixed with the drink which the Scots, according to the terms of the truce, were to supply to the Danes.

When taken in large or poisonous doses, Belladonna produces a peculiar form of delirium, widely-dilated pupils, great thirst and dryness of the mouth, and ultimately coma and death. The poisonous principle is an alkaloid called *atropin*, which exists in all parts of the plant, and is of a frightfully poisonous nature. Belladonna is much used in medicine in small doses in the shape of an extract; this and the alkaloid atropin are also used as an external application. Belladonna is employed as a sedative to allay pain and spasm, and to relieve incontinence of urine, for which purpose it has a remarkable effect. It is frequently smeared round the eye in cases where it is necessary to dilate the pupil, this being one of the peculiar effects of Belladonna. It is said by homœopathists to act as a preventative of scarlet fever, as the use of Belladonna causes dryness and redness of the throat, such as also occurs in scarlet fever; hence, on the principle of like curing like, the use of Belladonna is recommended for this disease. It has been recently discovered that quantities of Belladonna, which would seriously affect adults, can be taken with impunity by children, and also that the action of Belladonna and of Opium are so mutually antagonistic that the one may be employed as an antidote to the other. Valuable as Belladonna is as a remedy, it is obvious that it should never be employed except by a duly qualified person. In cases of poisoning by Belladonna, the stomach-pump and emetics should be had recourse to as speedily as possible.

Atropa Belladonna.

The mandrake was formerly referred to this genus, but is now included in the genus *Mandragora*. [M. T. M.]

ATROPAL. An ovule which never alters its original position; same as *Orthotropal*.

ATTALEA. The name of a genus of lofty palms, natives of tropical South America. The leaves are large and pinnate. The fruits hang in large clusters, each nut consisting of three cells, and containing as many seeds, a circumstance which serves to distinguish the genus from all its allies.

A. funifera, called by the Brazilians Piassaba, yields a fibre of much value, derived from the decaying of the cellular matter at the base of the leaf-stalks, and the consequent liberation of the fibrous portions. This fibre is much used in Brazil for the purpose of rope-making, and in this country is employed for making brooms to sweep the streets. A fibre, having the same name, is also produced from another palm called *Leopoldinia Piassaba*.

The seeds of *A. funifera* are known as Coquilla nuts; they are three or four inches long, oval, of a rich brown colour, and very hard in texture; hence they are much used in turnery for making the handles of doors, umbrellas, &c. The seeds of *A. compta*, the Pindova Palm of Brazil, are eaten as a delicacy; the leaves of the same plant are used for thatching, for making hats, &c. A. *speciosa* and A. *cerifera* furnish nuts, which are burnt to dry the juice of *Siphonia elastica*, which furnishes India rubber. A. *Cohune*, a native of Honduras, produces nuts called Cohoun nuts, which yield a valuable oil. [M. T. M.]

ATTRAPE-MOUCHE. (Fr.) *Apocynum androsæmifolium*; also applied to *Arum crinitum*, *Drosera muscipula*, *Lychnis Viscaria*, and *Silene muscipula*.

ATWISHA. An Indian poison, supposed to be *Aconitum ferox*.

AUBAINE ROUGE. (Fr.) A kind of wheat.

AUBEPINE. (Fr.) *Cratægus Oxyacantha*.

AUBERGINE. (Fr.) *Solanum esculentum*, sometimes called *S. Melongena*.

AUBOUR. (Fr.) *Cytisus Laburnum*.

AUBRIETIA. A section of the cruciferous genus *Farsetia*, from which it is separated by having the valves of the oval pod convex and not flattened. The outer sepals bulging at the base, and the shorter stamens with a tooth on the filaments, distinguish it from the allied genera. The species are low diffuse plants, with leaves somewhat like those of a stock in miniature, or rather those of *Arabis albida*; the short flowering stems bear few flowers. A. *deltoidea*, a native of the eastern Mediterranean region, is a pretty early-flowering plant, often introduced on rockwork. [J. T. S.]

AUCKLANDIA. A name given to an Indian composite plant, which proves to be identical with *Aplotaxis*, which see.

AUCUBA. A genus of evergreen shrubs, referred to the order *Cornaceæ*, and distinguished by their diœcious flowers, of which the males have a small four-toothed calyx, a four-petaled corolla, and four short stamens alternating with the petals; and the females have, instead of the stamens, an inferior one-celled ovary, surrounded by a fleshy epigynous disk, the style short thick turned at the base, the stigma orbicular, and the ovary containing a single ovule. The fruit is a one-seeded berry. The *Aucuba japonica* is a well-known shrub of vigorous habit, highly prized for its capability of enduring and even thriving in the atmosphere of towns and cities. It forms a dense roundish bush, furnished with large glossy leathery leaves of an elliptic form, remotely serrated, and in our common garden form conspicuously blotched with pale yellow, the green-leaved type having been only lately introduced. The flowers are inconspicuous. Several variegated varieties are known. In another species, of more recent discovery, A. *himalaica*, the leaves are wholly green. [T. M.]

AUDOUINIA. A genus of *Bruniaceæ*, containing a single species, from the Cape of Good Hope. It is a small shrub with erect branches, spirally arranged imbricate leaves, and purple flowers in a terminal oblong capitulum. The calyx tube is short and adherent to the ovary, the limb being deeply five-partite. The corolla consists of five spreading unguiculate petals. The five stamens are shorter than the petals and alternate with them. The ovary is slightly three-lobed, and three-celled, with two ovules in each cell. There is a single trigonous style, with three small papillæform stigmas. [W. C.]

AUGEA. An annual glabrous fleshy herb, with the aspect of a *Mesembryanthemum*, but with small inconspicuous green flowers, without petals, ten short stamens, and a ten-celled superior ovary. It forms a genus of *Zygophylleæ*, and is a native of sandy saline wastes in the Cape Colony.

AUGUSTIA. A genus of begoniads, separated by some modern botanists from *Begonia*, and consisting of succulent tuberous plants found at the Cape of Good Hope. The staminate flowers have two, the pistillate five sepals; anthers small, elliptical, lengthened into an obtuse cone; filaments long, not united; style persistent, its branches furnished with a continuous papillose band, making two spiral turns; placentas split lengthwise, their transverse section ovate-oblong; seed vessel with three nearly equal wings. There are four species known, viz.—A. *Bergii*, A. *Coffra*, A. *saffrutescens*, found at the Cape, and A. *natalensis* at Port Natal. The genus is named after Dr. August of Berlin. [J. H. B.]

AUGUSTINIA major (or *Bactris major* of Jacquin) is the only known representative of a genus of palms inhabiting Venezuela and New Grenada, and bearing an edible fruit of a pleasant acid flavour. It grows from twelve to twenty feet high, and its cane-like trunks, several of which spring from the same root, form thick bushes, quite impenetrable on account of the spines with which the plant is clad. The leaves are pinnate. The inflorescence, enclosed in a double spathe, is axillary; the flowers are monœcious; and the fruit is a dark violet-coloured smooth drupe, about the size of a pigeon's egg. [B. S.]

AULACOSPERMUM. A genus of *Umbelliferæ*, containing two species of perennial, glabrous, herbaceous plants, with bipinnate leaves, natives of Altai. The limb of the calyx is five-toothed or obsolete. The petals are ovate and entire. The fruit is ovate and slightly compressed; each mericarp has five longitudinal winged ridges, with intervening vittate furrows; the commissure is plain. [W. C.]

AULAYA. A genus of *Scrophularineæ*, containing eight species, natives of the Cape of Good Hope. They are parasitic herbaceous plants, having the habit of *Orobanche*, with imbricate scale-like leaves, and gaudy flowers. The calyx is campanulate and five-cleft, with two bracteoles. The tube of the corolla is clavate at the base, and inflated upwards; the limb is spreading and five-cleft. There are four included didynamous stamens, inserted in the base of the tube. The anthers have two cells, the one being perfect, ovate-acuminate, the other abortive, longer and subulate. The ovary is two-celled, and contains many ovules; and the style is terminated by a clavate involute stigma. [W. C.]

AULNE. (Fr.) *Alnus glutinosa.* — NOIR. *Rhamnus Frangula.*

AUNÉE. (Fr.) *Inula Helenium.*

AURANTIACEÆ. (*Aurantia, Citronworts.*) The orange, lemon, and similar fruits, are produced by trees belonging to a natural order bearing this name. They are all bushy or woody plants, having the leaves filled with transparent oil-cysts, giving them a dotted appearance, a definite number of hypogynous stamens, and a fruit more or less pulpy. Less than 100 species are known. The various genera are almost exclusively found in the East Indies, whence they have in some cases spread over the rest of the tropics. Mention is made of a wild orange of Brazil, which has a mawkish sweet taste, but must have been introduced. The *Skimmias* are remarkable among so tender a race for the hardiness of their constitution.

AURICULA. *Primula Auricula,* a favourite garden flower.

AURICULARINI. An order of hymenomycetous *Fungi,* distinguished by the hymenium being destitute of gills, pores, prickles, or other decided prominences. In a few species of one genus only there are a few obscure folds or papillæ. The nobler species have the hymenium inferior, as in the mushrooms, &c., but as the order contains a great mass of a low condition of organisation, very many of the species are permanently glued to the substance from which they spring. But even amongst these there is a tendency to become free at the margin, and to reflect it so as to take the hymenium away from the light. Several of the species are amongst the most common of fungi. The yellow *Stereum hirsutum* grows on almost every oak log, and the purple *Stereum purpureum* on every fallen poplar. The dark Indigo-blue *Corticium ceruleum*, so common on damp rotten rails, is said to be occasionally phosphorescent. The order is distinguished from the *Tremellini* by the expanded horizontal hymenium, which is, besides, more definite, and formed after the same type as that of the higher orders of the family, whereas in the *Tremellini* the fructifying cells or sporophores are of unequal length. There is, moreover, in the higher *Auricularini* a distinct pileus, while in the *Tremellini*, with the exception of *Hirneola* and *Exidia,* where there is often a distinct barren outer coat, as in *Peziza,* the whole surface, even in the highest species, bears fruit. No plant of the order is known to have any economical use. [M. J. B.]

AURICULATE. Having a pair of small round lobes or ears, as is the case with many leaves.

AURONE FEMELLE. (Fr.) *Santolina Chamæcyparissus.* — MÂLE. *Artemisia Abrotanum.*

AVA. A kind of pepper, called *Macropiper Methysticum.* The name is also given to a spirit distilled in the Sandwich Islands from the root of a species of *Cordyline.*

AVANT-PÂQUES. (Fr.) *Tulipa sylvestris.*

AVENA. Oat grass. A genus distinguished by large membranaceous outer pales, enclosing from two to three florets, each armed with a bent more or less twisted awn.

Meadow species: *A. pubescens,* Downy Oat Grass; leaves downy, with soft hairs; a common meadow-grass in limestone pastures, which should be included in the seeds for such situations. *A. pratensis,* Narrow-leaved Oat Grass; leaves hard and rigid; a denizen of moors and poor clays. Its specific name is inappropriate as its favourite habitat is seldom worthy of the name of meadow. *A. alpina,* Great Alpine Oat Grass; a larger and coarser form than the preceding, of which it is probably a mountain variety. *A. flavescens,* Yellow Oat Grass; flowers small yellow; an upland pasture grass of considerable merit.

Agrarian species: *A. strigosa,* Bristle-pointed Oat; seeds much like those of Corn Oats, the awned inner pales with two long bristly points; occasionally met with in corn-fields, where it has probably been introduced with foreign seed. *A. fatua,* Wild Oat; awn much bent, the lower half twisted, the inner pales covered with stiff hairs. These peculiarities give the seed so much the appearance of a fly, that the rustics often make use of it in trout fishing, and as the twisted awn uncoils when it comes in contact with the water, the fish is deceived by its apparent struggling; this property of the awn has likewise caused it to be used as a hygrometer; it is a common weed in clay soils.

The two latter species have lately attracted considerable attention from their connection with agriculture. Dr. Lindley, in an article in Morton's *Cyclopædia of Agriculture*, suggested that the cultivated Oat 'is a domesticated variety of some wild species, and may be not improbably referred to *Avena strigosa*;' but perhaps, after all, the *A. strigosa* may be but a variety of *A. fatua*, from the cultivation of which it has been shown that Cereal or crop Oats may be grown, in illustration of which we here give a short account of our own experiments.

In 1852 we sowed a plot of the seeds of *A. fatua,* collected in 1851; they grew well, but were scarcely different from the wild plant, except in a tendency to an increased plumpness of grain. The produce of this crop was preserved throughout the winter, and sown in a different part of the garden in the spring of 1853; we repeated the process with successive crops in 1854 and 1855, in each of which we noted an increase of tendencies in the following direction: 1. a gradual decrease in the quantity of hairs on the palea; 2. a more tumid grain, in which the palea were less coarse and the awns not so strong and rigid; 3. a gradual increased development of kernel or flour. The produce again sown in 1856 had so far advanced, that we collected poor, but still decided samples of what are known as the Potato and Tartarian forms of Oat. These we have gone on improving until, in 1860, we had a quarter of an acre each of good white Tartarian and Potato Oats, as a farm crop, which had been derived from the wild example. This is the more interesting, because farmers have always stated, especially on the poor lias clays of Gloucester and Worcester, that they could not grow oats without leaving behind a quantity of wild or weed oat; and our subsequent inquiries have convinced us that shed oats in some situations do really degenerate into wild oats, and the first stage in the process of degeneration will be observed in an accession of hairs at the base of the grain, which good cereal oats never possess.

We may then view the different forms of crop Oats, as induced varieties from the *A. fatua*. In cultivation, it would appear that the best and plumpest oats are grown in North Britain; here they make a good meal, which is much used as human food, Oatmeal, 'parritch' being indeed an article of diet far more nourishing than the potato, which is the more usual food of the southern. [J. B.]

AVENS. The common name for *Geum*.

AVERRHOA. A genus of *Oxalidaceæ*, consisting of a few small trees, originally from the Moluccas and Ceylon, but cultivated throughout India. They have evergreen alternate pinnated leaves, somewhat like those of the ash, or rather the sumach, and small purplish flowers in racemes. The fruit is like a gurken in shape, very acid, but pleasant when made into syrup, candied, or pickled. The leaves are slightly sensitive. *A. Bilimbi*, the Bilimbing, has many pairs of leaflets, and the flowers produced from the trunk. *A. Carambola*, the Caramba, has only from two to five pairs of leaflets, and the flowers produced from the branches. [J. T. S.]

AVET. (Fr.) *Abies pectinata*.

AVICENNIA. A genus of the verrain family, *Verbenaceæ*. The plants composed in this genus are called White Mangroves, and, like the true Mangroves, are found in the tidal estuaries of most tropical countries. They are small trees, with opposite evergreen leaves, which are oblong, entire, and covered beneath with a white pubescence. Their flowers are inconspicuous, and arranged in closely-packed terminal bunches. Their roots stand out of the mud in which they grow, overarching each other in erect-arched masses, and sending up Asparagus-like shoots from their underground parts. *A. tomentosa* is in great reputation in Rio for tanning. The native washermen of India (dhobies) make a preparation from the ashes of the wood, which they use in washing of cleaning cotton cloths. The green fruits boiled with butter form poultices, used in native practice. In N. S. Wales the wood is valued for stone-masons' mallets, on account of its toughness. *A. nitida* is called Courida in British Guiana. The wood is used for the foundations of buildings and underground work, on account of its power of resisting damp; exposed to the atmosphere it soon perishes. The bark is used for tanning in the W. Indies. [A. A. B.]

AVIGNON BERRIES. The yellow dye-berries of the Buckthorn, *Rhamnus infectorius*.

AVOCATIER. (Fr.) *Persea gratissima*.

AVOINE. (Fr.) *Avena sativa*. — A' CHAPELET. *Avena bulbosa*. — DE MONGRIE. *Avena orientalis*.

AWL TREE. The Indian Mulberry, *Morinda citrifolia*.

AWL-WORT. The common name for *Subularia*.

AWN. The beard of corn, or any such slender process.

AXIL, AXILLA. The angle formed between the axis and any organ that grows from it; the base of a lateral ascending organ, on the upper side.

AXILE, AXIAL. Of or belonging to the axis.

AXILLARY. Growing in the axil of anything.

AXIS. The stem, including the root; or any centre round which leaves and other organs are arranged. The stem is called the ascending axis, the root the descending axis. —, ACCESSORY. An axis of a second rank; secondary to some principal axis. —, APPENDAGES OF THE. All the leafy or thin expansions that grow upon a stem, such as leaves, and the parts of a flower.

AYAPANA. The sudorific *Eupatorium Ayapana*, which is said to be a valuable remedy for the bites of poisonous snakes.

AYART. (Fr.) *Acer opulifolium*.

AYDENDRON. A genus of tropical American trees of the laurel family, *Lauraceæ*. They have a funnel-shaped, six-parted perianth, containing twelve stamens in four rows; the nine outer stamens have anthers, the three innermost are sterile; of the fertile stamens the three innermost have glands on each side at the base, and their anthers open outwardly; the remainder have no glands, and their anthers open inwardly. The fruit is succulent, at first concealed within the base of the perianth, which afterwards falls off, leaving only a portion surrounding the base of the fruit.

Cajuputy beans are the seeds of *A. Cajumary*, and are esteemed in Brazil as tonics and stimulants in cases of weak digestion. [M. T. M.]

AYER AYER. The esculent fruit of some species of *Lansium*.

AYLMERIA. A genus of *Paronychiaceæ*, consisting of two species of Australian annuals, with much-branched stems, opposite or verticillate leaves, small scarious stipules, and terminal corymbose cymes of rose-coloured or purple flowers on long stalks. [J. T. S.]

AZADIRACHTA. A genus of the order *Meliaceæ*, represented by an Indian tree with unequally pinnated leaves, whose leaflets are oblique. The young shoots are smooth, not covered with down as in the allied genus *Melia*. The flowers are small, white, borne in axillary panicles; they differ from those of *Melia* in having a three-celled ovary, and a three-lobed stigma, and also in the fruit, which is purple when ripe, of the size of a small olive, one-celled, one-seeded.

The bark of *A. indica* is used in India as a tonic, the root as a vermifuge, and the leaves as an application to glandular swellings, bruises and rheumatism. They have also been employed successfully in some forms of skin disease. From the fruit an acrid oil is obtained for burning in lamps, and for dyeing cotton cloths. A stimulant gum exudes from the bark. The seeds are used as a poison for insects, and mixed with water as a hairwash. A kind of toddy is said to be prepared from the young trees. [M. T. M.]

AZALEA. A genus of *Ericaceæ*, established by Linnæus, and including many plants which have since been separated and arranged under different genera. So conflicting are the opinions of botanists as to the set that should retain the original Linnæan name, that it seems in danger of being lost altogether. Some seek to retain it for *A. procumbens*, as the only plant to which it is truly applicable, and propose the name *Anthodendron* for the showy shrubs so well known in our gardens as *Azaleas*; whereas others, because of the almost universal application of the name to these plants, and to prevent unnecessary confusion in the synonymy, have given the name *Loiseleuria* to the small genus containing the single species, *A. procumbens*, and retained the original name for the showy American shrubs. This course being adopted generally by continental botanists, as well as by many in Britain and America, it seems better to consider the genus as so limited.

Azaleas are upright shrubs with alternate and obovate or oblong deciduous leaves, which are entire, ciliate, and mucronate, with a glandular point. The flowers are large and showy, often glandular, and glutinous outside; they rise in umbelled clusters from large scaly-imbricated terminal buds. The calyx is five-parted, often minute. The corolla is funnel-shaped, with five spreading lobes. The stamens are five in number, with long exserted filaments, and short ovate anthers, opening by terminal pores. The ovary is five-celled, with many ovules: the style is simple. The pod is five-celled and five-valved, and contains many scale-like seeds. There are about twenty species, natives of North America and Asia. They are largely cultivated as ornamental shrubs, on account of the abundance of their flowers, and the fragrant smell of most of the species. Some possess dangerous narcotic qualities. Pallas was of opinion that *A. pontica* was the plant from whose flowers the bees of Pontus collected the honey that produced the extraordinary symptoms of poisoning, described as having attacked the Greek soldiers, in the famous retreat of the Ten Thousand. Xenophon says that after eating it, the men fell stupefied in all directions, so that the camp looked like a battle-field covered with corpses. The natives are aware of the deleterious qualities of the plant. Cattle and sheep which browse on its leaves are poisoned. [W. C.]

AZAREBO. (Fr.) *Cerasus lusitanica.*

AZARA. A genus of Chilian shrubs, belonging to the *Flacourtia* family, having twin or solitary alternate leaves, generally toothed and varying in form from egg-shaped to almost linear. Their flowers are small and yellow, destitute of petals, and arranged in axillary bundles. A few of the species are in cultivation in English gardens, and can be grown outside with the protection of a wall in the southern counties. *A. Gilliesii* is the most handsome species of the genus. Its leaves are evergreen and somewhat like those of the holly, bearing in their axils roundish fascicles of yellow flowers. About a dozen species are known. The leaves of many have a bitter taste. [A. A. B.]

AZAROLE. The fruit of *Cratægus Azarolus.*

AZEDARACH. *Melia Azedarach.*

AZEROLIER. (Fr.) *Cratægus Azarolus.*

AZOLLA. A very curious genus belonging to the marsileaceous division of the pseudo-ferns. Its habit is that of a floating pinnately-branched *Jungermannia*, with two or four-ranked imbricating leaves; but its fructification is totally different, and is nearer to that of *Salvinia* than of any other genus, and with which it forms a distinct section or order, according to the views of authors. Indeed, its peculiarities are such that it has been sometimes supposed to constitute a distinct order by itself. The species float upon the water, forming green or reddish patches, which are frequently several yards across, throwing down rootlets on the under side, stronnest which are situated, principally in the axils of the leaves, the organs of fructification. These are twofold:—1. Thin membranous ones bearing on a short cylindrical axis, springing from the base, stipitate globose cysts, filled with angular bodies, which are furnished either with curious arrow-headed or root-like appendages. These armed granules are doubtless the antheridia, though their spermatozoids have not yet been discovered. 2. Ovate sporangia, divided within by a transverse partition, which incloses below, a granulous or at length pulverulent mass, and gives off from its centre above a column fringed at the apex with a tuft of hair, and having attached to it from three to nine dependent spores, which are at length exposed by the separation of the upper half of the sporangium at the above-mentioned partition. The species occur from Australia and New Zealand as far as New York. One has been found in Western Africa by Vogel. It has been supposed that the differences in the antheridia and the number of spores, accord with the geographical distribution of the species, which may accordingly be separated into two genera; but this is at present more than doubtful. [M. J. B.]

BABEER. A Syrian name for *Papyrus*.

BABIANA. A genus of bulbous-tuberous *Iridaceæ*, found in South Africa, and having two-ranked sword-shaped plicately-nerved leaves, and flower-stems terminated by a loose subsecund or two ranked spike of flowers, which consist of a funnel-shaped tube, with a dilated throat, and a six-parted regular or somewhat two-lipped limb of nearly equal segments; they are furnished with three stamens, and the three-celled many-ovuled ovary is terminated by a filiform style, dividing at top into three conduplicate wedge-tongue-shaped undivided stigmas. The flowers are large and showy, and in some of the species sweet-scented. There are upwards of thirty species, many of which have been in cultivation in this country, and some are still occasionally met with, but, like many others, they have been undeservedly neglected in the rage for novelties which distinguishes the present age, so that they are less frequently seen than they deserve to be among the ornaments of our greenhouses. *B. plicata*, which may be taken as an illustration of the genus, is a slender plant, of six inches to a foot high, everywhere pubescent, with oblong lanceolate leaves, and pale violet-coloured flowers, the lower segments of which are streaked with yellow in the middle, and spotted with brown at the base; these flowers have the odour of cloves. [T. M.]

BABINGTONIA. This genus of *Myrtaceæ* was named after Professor Babington, a well-known English botanist. It is allied to the genus *Backea*, but differs from it in the stamens being collected in groups of three, opposite the petals. The anthers also are placed directly on the top of the filaments, and open by pores. The style seems to be a direct prolongation of the placenta; it protrudes through a hole in the top of the ovary, and does not even touch the carpels. *B. camphorosma* is a graceful greenhouse shrub, with white or pinkish flowers, and has been introduced from New Holland. [M. T. M.]

BABOOL. The Indian name for the gum-bearing *Acacia arabica*.

BABOUNY. A name used in Egypt for the flower-heads of *Santolina fragrantissima*, a substitute for chamomiles.

BACCA. A berry; that is to say, a succulent seed-vessel, filled with pulp, in which the seeds nestle, as in *Solanum*. — CORTICATA. A berry having a rind; as an Orange. — SICCA. A fruit which is a berry when unripe, but becomes a dry body when ripened. — SPURIA. Any fleshy fruit, which is not a true bacca or berry; as the juniper, strawberry, raspberry, &c.

BACCATE. Having a pulpy texture; a term only applied to the parts of a flower or fruit.

BACCATE SEEDS. Seeds with a pulpy skin.

BACCAULARIUS. Such a fruit as that of the mallow; viz. several one or two seeded dry carpels cohering round an axis.

BACCHARIS. A large and natural genus of the composite family, distinguished from its allies by having male flowers only on one plant, and the females on another. Upwards of 200 species are known. They are herbs, shrubs, or sometimes small trees, many of them smooth and covered with a resinous substance, which gives to the leaves a glossy appearance. The latter are generally alternate, rarely opposite, and vary much in form. In one section of the genus they are three-nerved, and ovate or lanceolate in form; in another, one- or three-nerved, and wedge-shaped; in a third they are very small, or absent altogether; while in a fourth the stems are winged and leaf-like, performing the functions of the leaves, which are small or almost absent. The flower-heads are arranged in various ways, and the florets are generally white in colour. The species are confined to the New World, and are found, in greater or less number, from the United States to the extreme south of the continent. Many of them are found at an elevation of 13,000 feet above the sea level in the Andes, and a few of them reach the snow limit. Immense tracts are covered on the plateaus of the Cordillera with plants of this genus, and shrubby groundsels, taking the same place there that the heaths do on our moors in Peru and Bolivia, the shrubby species are known by the names of Tola, or Chilca, and by the latter name in S. Grenada and Chili. The resinous species are almost universally used as firewood for ovens. An infusion of the winged stems of *B. trimera* is used by the Brazilians as a sudorific and tonic; while another, also with winged stems, *B. microcephala*, is used in Parana for curing rheumatism by putting bushes of it in warm baths. A bitter is extracted from *B. genistelloides*, which is held in great reputation in Brazil when used with a specific aroma in cases of intermittent fever. Horses devour this herb with avidity, and it is further reckoned of great service in curing chronic diseases in that animal. *B. Douglasii* is remarkable as being found in California, and appearing again in Chili, without being found in any intervening place.   (A. A. B.)

BACHE. A South American name for *Mauritia flexuosa*, an economical species of palm.

BACHELOR'S BUTTONS. A garden name for the double-flowered variety of the buttercup, *Ranunculus acris*.

BACILLARIA. A genus of diatomaceous *Algæ* consisting of a single species, which occurs on our coasts, known by its linear rectangular articulations, which are at first joined by the longer sides into a straight tabular series, and then slip over each other so as to make oblique series. The articulations or frustules, individually, are not so beautiful in respect of structure as many others of the group; the chief point of interest consisting in the curious manner in which the articulations or frustules incessantly slip backward and forward over each other, with a more or less isochronal motion, yet so as always to adhere to each other. The whole mass is thus in motion, though the several groups of frustules, of which it is composed, may be moving in opposite directions. An obstacle, says Mr. Smith, is not evaded but pushed aside; or, if sufficient to avert the onward course, the latter is detained for a time equal to that which it would have occupied in its forward progress, and then retires from the impediment as if it had accomplished its full course. The motion is about one two-hundredth of an inch per second.   (M. J. B.)

BACILLE. (Fr.) *Crithmum maritimum.*

BACILLI. The separable moving narrow plates, of which the genus *Diatoma* is composed.

BACILLUS. The little bulbs found on the inflorescence of some plants; a term rarely employed.

BACKHOUSIA. One or two showy-flowered myrtaceous plants have been considered to form a new genus, named in honour of Mr. James Backhouse, who has travelled much in Australia and South Africa, and otherwise contributed to advance botanical science. The principal characters of the genus are: the tube of the calyx covered with dense hairs, the five segments of the limb large, whitish, and petal-like; the petals themselves small and comparatively inconspicuous; the stamens very numerous, and longer than the calyx or the corolla; the ovary adherent below to the tube of the calyx, but free at its upper portion, very hairy on the exterior, the interior containing several seeds in each of its two compartments. *B. myrtifolia* is a small tree, with opposite ovate pointed leaves, and starred corymbs of whitish flowers, and is cultivated as a greenhouse plant.   [M. T. M.]

BACTRIDIUM. A very curious genus of *Fungi*, of rather doubtful affinity, but supposed to belong to the division *contomycetes*, and to be allied to *Corynæum*. The plant consists almost entirely of oblong septate hyaline spores, which radiate from a little dot-like receptacle. The spores in our most conspicuous native species, *B. flavum*, which occurs in this country, although but rarely, on dead elm stumps, are of a pale yellow. We have a species from Venezuela, with enormous spores, one-sixtieth of an inch long, which afford an interesting microscopic object under a low magnifying power; in this the spores, which seem en masse, are of a pale fawn colour.   [M. J. B.]

BACTRIS. A genus of slender palms, natives of the West Indies, Brazil, and other tropical countries on the eastern side of South America; generally growing in low marshy places, or inundated tracts of land, upon the banks of rivers, and on the sea coast. There are about forty species, but very few of them attain anything like the majestic proportions of the generality of palms, the majority having thin reed-like stems, not much exceeding the height of a man. A few, however, grow to a height of forty or even fifty feet, with trunks averaging about four inches in diameter. Almost all of them are armed with sharp black or brown spines, several having their stems encircled with bands of them, placed at short intervals all the way up, whilst others have them only at their summits; and, as they usually grow together in large masses, and throw up numerous suckers from their creeping roots, they offer a really formidable and often impassable barrier both to man and beast. Their flower-spikes are produced either from the apex of the trunk or from the bases of the leaves, and while young are enclosed in a double sheathing spathe, which, in nearly all the species, is densely covered with short black spines. The male and female flowers are borne upon the same spike, and are yellow, green, or rose-coloured; the males have a three-parted thin calyx, and three fleshy petals, and contain from six to twelve stamens; the females have a cup-shaped or cylindrical calyx and corolla, three-toothed at the apex, and they contain a triangular ovary, with three sessile stigmas. Their fruits are generally small, seldom exceeding a pigeon's egg in size, and frequently not larger than a pea, mostly of a bluish black colour, having a thin coating of white fibrous pulp surrounding a hard black stone, which has three small holes at the top, and contains a single seed. Their leaves do not fall away from the trunk like those of many other palms, but remain attached long after they have withered, hanging down and concealing the trunk; they are nearly always pinnate, and from two to eight feet long; in a few species, however, the leaves are nearly entire, or merely divided into two broad sharp-pointed lobes.

B. Maraja, the Marajah Palm of Brazil, grows upon the banks of the Amazon and other rivers. It is the largest species of the genus, its trunk attaining a height of fifty feet. It is thickly armed with spines, and has a succulent rather acid but agreeably tasted fruit, from which a vinous beverage is prepared. *B. minor* has a stem about twelve or fifteen feet high, and seldom more than an inch in diameter. It is common in Jamaica and some parts of tropical South America, growing in open places in the vicinity of woods. Its stems are used for walking-sticks, and are said to be sometimes imported into this country under the name of Tobago canes. [A. S.]

BADAMIE. An Indian name for oil of almonds.

BADDERLOCKS. *Alaria esculenta*.

BA'DEK. A fermented liquor prepared in Java from rice.

BADGER'S-BANE. *Aconitum meloctonum*.

BADHAMIA. A genus of gelatinous puffballs (*Myxogastres*), named after the late Dr. Badham, remarkable for its spores being contained in little groups in distinct hyaline sacs or asci; whereas in most of the immediately allied fungi they are naked. The species were formerly referred to *Physarum*. Other instances of asci occur in the same division, as in the genus *Enerthenema*, separated from *Stemonitis*. The most common species, perhaps, is *B. hyalina*, which is known by its delicate peridia as well as by its long confluent yellowish stems. [M. J. B.]

BADIANE. (Fr.) *Illicium*.

BADIERA. A genus of the milkwort family (*Polygalaceæ*), which includes three species, all of them natives of the West Indian Islands. They are woody plants with evergreen leaves, and axillary corymbs of white or greenish-yellow flowers, differing chiefly from the common milkworts (*Polygala*), in having a large oily aril to the seed which fills the upper part of the cell, and in the anthers opening inwards by an oval partitioned slit. The bark of *B. diversifolia* is acrid and bitter, like that of the Lignum Vitæ, and is called Bastard Lignum Vitæ, in Jamaica, on this account. [A. A. B.]

BADIOUS. Chestnut-brown.

BADULA. A genus of the Myrsine family, of which seventeen species are enumerated. They are evergreen shrubs or small trees, with smooth entire dotted leaves, having short and broad foot-stalks. Their flowers are numerous, disposed in axillary or terminal panicles, and either white, dotted or streaked with pink, or entirely of a pink colour. The fruits are small scarlet or black berries, containing few seeds. They are nearly related to *Ardisia*, and differ chiefly from that genus in their short round-headed stigma, and few seeds. Their distribution is unusual, one being found in the Philippine Isles, a considerable number in Mauritius, Bourbon, and Madagascar, but the greatest number in the West Indies, Peru, and Brazil. [A. A. B.]

BÆA. A small genus of *Gesneraceæ* consisting of herbaceous plants, with short stems or entirely stemless, and crowded leaves. The calyx is five-partite and persistent; the corolla is campanulate, the tube scarcely as long as the calyx, while the subdilabiate limb is five-partite with roundish lobes. There are two fertile stamens with very short filaments, and large cordate-ovate anthers. The lanceolate ovary is one-celled, with two parietal placentæ. The capsule is elongated and pod-shaped, and the two valves, after dehiscence, are spirally twisted to the right. The oblong seeds are numerous and very small. This genus differs from *Streptocarpus* chiefly in the length of the corolla tube. [W. C.]

BÆCKEA. The name of a genus of plants belonging to the *Myrtaceæ*. The flowers are sessile or stalked; the limb of the calyx five-cleft, persistent, its tube top-shaped; petals five, longer than the stamens, which are from five to ten in number, and distinct; stigma capitate, capsule many-seeded. The plants are small shrubs, with opposite leaves and white flowers. They are natives of New Holland and China. Some of them are in cultivation as pretty greenhouse plants. [M. T. M.]

BÆOMETRA. Certain bulbous plants, belonging to the order *Melanthaceæ*, are so called. From the bulbs or corms arise narrow sheathing leaves and spikes of flowers, each of which latter has a six-parted petal-like spreading deciduous perianth, into the base of the segments of which the six stamens are inserted. The ovary is somewhat triangular, and terminated by three recurved spreading stigmas; it ripens into a cylindrical capsule, its three compartments separating one from the other at the top, so as to liberate the numerous seeds, which are of a compressed four-cornered shape, arranged in two lines along the inner edge of each compartment. They are all natives of South Africa. [M. T. M.]

BÆOMYCES. A small genus of Lichens, distinguished amongst *Lecideineæ* by their subglobose terminal fruit, which is supported by a short unbranched stem. The disc is generally bright-coloured, as rose, chestnut, &c. *B. roseus* and *B. ericetorum*, which abound in heaths, are often taken at first-sight for fungi. [M. J. B.]

BÆRIA. A genus of composites, allied to *Callichroa*, of which but a single species is known, the *B. chrysostoma*, from California. It is a pretty dwarf annual, of slender erect habit, with downy stems about a foot high; linear opposite entire leaves and solitary terminal bright yellow flowers, an inch across. Botanically, the genus is distinguished by an involucre of about ten leaflets, arranged in two series, a conical naked receptacle, and an elongated fruit without pappus. It differs from *Callichroa*, not only in its general habit, but also by its smaller flower-heads, and the oblong-pointed form of the ray florets, the florets of the latter being wedge-shaped. [W. T.]

BAGASSA. An imperfectly-known genus of *Artocarpaceae*, comprising one or more species of trees, with opposite leaves, deciduous stipules, and orange-shaped fruit, consisting of egg-shaped, pointed achenes, clustered around a thick central receptacle. This fruit is eaten in Guiana, where the tree is a native. [M. T. M.]

BAGUENAUDIER. (Fr.) *Colutea arborescens.* — D'ETHIOPIE. *Sutherlandia frutescens.*

BAJREE. *Penicillaria spicata*, a bread-corn cultivated in India.

BALANITES. The name given to a thorny shrub or small tree, with a very forbidding aspect, growing almost always in dry barren places. Its leaves grow in pairs (binate), the leaflets oval, or oblong, stalked, and pubescent when young. The flowers are small, greenish, white and fragrant, arranged in short axillary racemes. The fruit is oval, about one and a half inch long, and when ripe of a greyish colour. The plant is a native of many parts of India, Egypt, Senegambia, and the W. coast of Africa. The leaves in the Egyptian variety are slightly acrid and anthelmintic, and the bark is used by the Hyots in India as a medicine for their cattle. The young fruits are purgative, but when ripe are edible, and formed into an intoxicating drink by the negroes on the W. coast of Africa. In India the nut, which is very hard, is employed in fireworks. A small hole being drilled in it, and the kernel taken out, it is filled with powder and fired, bursting with a loud report. An oil, called by the negroes Zachun, is obtained from the seeds, and the wood, which is yellow, hard, and durable, is used in Africa for household work. The place of the plant in a natural arrangement is somewhat doubtful, some authors placing it with *Olax*, others with *Amyris*, while a few think it should constitute a separate order. [A. A. B.]

BALANOPHORACEÆ. (*Cynomorineae*.) A small natural order, consisting of about thirty species, of singular-looking succulent leafless plants, usually highly coloured, of various shades of yellow or red; all parasites on roots, and rising from an inch or two to about a foot above ground. Their colour and consistence, the absence of all leaves, excepting in some species, imbricated scales of the colour of the rest of the plant, and the greatly reduced structure of the flowers, had induced some botanists to consider them as cryptogams allied to fungi; but their structure is now much better understood, and has been fully described, especially by Dr. J. D. Hooker. He has shown them to be most nearly connected with *Balanophora*, and to have no real affinity with *Rafflesiaceae*, *Orobanchaceae*, or any other root parasites, which assume sometimes a similar colour and consistence. The flowers are, in nearly all the species, unisexual, of very simple structure, and produced, in considerable numbers, in compact terminal heads or cones; the small perianth, usually simple and inferior in the females, more or less three-cleft or six-cleft in the males, is in some species wholly wanting; the stamens, usually few, are very variable in number and form; the ovary has one or two styles, and always a single cavity with one pendulous ovule.

The *Balanophoreae* are natives of hot climates, in various parts of both the New and the Old World, one species only, the *Cynomorium coccineum* or *Fungus melitensis* of old authors, being found as far north as the southern shores of the Mediterranean. They have been distributed into fourteen genera. The most remarkable for the size or beauty of the species, or for the use made of them, are *Sarcophyte*, *Lophophytum*, *Ombrophytum*, *Langsdorfia*, and *Cynomorium*.

BALANOPHORA. Singular leafless parasitical plants, giving their name to the order *Balanophoraceae*. These plants are found on the roots of oaks, maples, vines, and other trees in tropical countries, especially in mountainous districts. One species is found in Australia. Some of the Himalayan species cause the formation of large knots on the roots of oaks and maples, which are much sought after by the natives for the manufacture of wooden cups, in general use throughout the Himalaya and Thibet. Some of the species, as *B. elongata*, furnish wax in great abundance, which is used for making candles in Java. [M. T. M.]

BALANSÆA. A genus of *Umbelliferae* or *Apiaceae*, consisting of one species, inhabiting North Africa. It has a tuberous root, large broadly cut leaves, and hermaphrodite flowers. Each half of the fruit is compressed laterally, elongated into a conical 'stylopod,' terminated by an erect style, and marked by five prominent thread-like ridges, in the intervals between which, in the rind, run solitary channels, or 'vittae,' filled with volatile oil, while in the commissure are two such channels; albumen furrowed. [M. T. M.]

BALANTIUM. A name proposed for a genus of Ferns, now considered synonymous with *Dicksonia*. It is represented by the *Dicksonia Culcita* of Madeira. [T. M.]

BALAUSTA. The fruit of the pomegranate.

BALAUSTION. A Greek word for the pomegranate, but applied by Sir W. Hooker to another genus of *Myrtaceae*. B.

The Treasury of Botany.

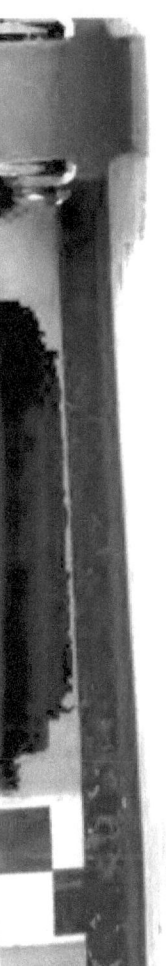

puickervissum, the only known species, is described as being one of the most lovely of plants. It is a shrub, with a thick twisted stem, numerous slender branches crowded together in places, and short linear leaves, dotted with resinous fragrant dots. The flowers are abundant, placed towards the end of the branches, in shape and colour like those of the dwarf pomegranate, but drooping on short stalks, with two small coloured bracts beneath the calyx, which has a cup-shaped tube, adherent by the base only to the ovary; the remainder is detached, and the limb divided into five ovate membranous spreading segments; the petals are five, oval, slightly larger than the calyx, and of a deep scarlet or crimson colour. The stamens are very numerous inserted in one row at the throat of the calyx; anthers inserted on the awl-shaped filaments by their backs, the lobes parallel, splitting by a long cleft. Ovary small, top-shaped, with three compartments, each containing about six ovules, placed one over the other in a double line; style thread-shaped; stigma rather dilated. A native of South-western Australia. [M. T. M.]

BALD-MONEY, or BAWD-MONEY. *Meum athamanticum*.

BALFOURIA. A genus of the natural family *Apocynaceæ*, consisting of shrubs with opposite narrow sickle-shaped leaves. The flowers are arranged in cymes at the end of the branches, or from their side, and have a five-parted calyx, a funnel-shaped corolla, with its limb divided into five straight equilateral lobes, having also at its throat a small coronet or tube with a wavy margin. The five stamens are inserted on the throat of the corolla, from which they protrude. The anthers are arrow-shaped, provided with a sharp point or mucro, and adherent to the sessile stigma; style thread-shaped; ovaries with two compartments. They are natives of tropical Australia, and have not yet been introduced into cultivation. [M. T. M.]

BALISIER. (Fr.) *Canna indica*.

BALLOTA. A family of herbaceous plants belonging to the labiate order, among which they are distinguished by the strongly ten-ribbed salver-shaped calyx. They are natives of the temperate regions of the Eastern hemisphere, and are remarkable for nothing but their strong offensive odour, on account of which they are for the most part rejected by cattle; hence the name from the Greek *ballo*, to reject. *B. nigra*, Black Stinking Horehound, a common wayside perennial, has stout-branched stems, ear-shaped wrinkled leaves, and whorls of numerous dull purple flowers. The whole plant is as offensive in odour as it is unattractive in appearance, and suffers little from being generally covered with dust. It is mostly found growing near towns and villages, and has accompanied our colonists to many remote countries. In Scotland, according to Don, it is an universal remedy in disorders incident to cattle. French *Ballote*; German *Zahntee*. [C. A. J.]

BALLOTE. (Fr.) *Ballota nigra*.

BALM. *Melissa officinalis.* —. BASTARD. The common name for *Melittis*. —. FIELD. *Calamintha Nepeta*. —. HORSE. An American name for *Collinsonia*. —. MOLDAVIAN. *Dracocephalum moldavicum*. —. MOLUCCA. The common name for *Molucella*. — of GILEAD. A resinous product of *Balsamodendron gileadense*, and *B. Opobalsamum*, called sometimes Balm of Mecca; also, a garden name for *Dracocephalum canariense*; also, an American name for *Populus candicans*. — of GILEAD (AMERICAN). A resin obtained from *Abies rorosa*. — of MECCA. The same as Balm of Gilead, a resinous product of *Balsamodendron*.

BALOGHIA. The name given to a tree of the spurgewort family (*Euphorbiaceæ*), which attains the height of twenty to thirty feet, and has opposite entire oblong leaves, which are stalked, and have at their base two membranaceous stipules which fall early. The flowers are numerous, and are disposed in terminal cymes, the male having a calyx of five divisions, five petals longer than the calyx, and a large number of stamens, their stalks united at the base, while the calyx and corolla of the female flowers are as in the male, and their ovary is three-celled, each cell containing one ovule. *B. lucida* is the only species known, and it is found in Norfolk Island, where it is called Blood Wood, as also in the colony of Queensland, in N. Holland. The wood is close-grained, impregnated with a resinous substance, and burns readily in a green state. A blood-red sap oozes from the trunk when cut, and is obtained in the following manner in Norfolk Island: "A knife, similar to a farrier's, is used, but stronger, fixed upon a handle four to five feet long, which enables the workman to reach high up the trunk of the tree. A perpendicular incision is made through the bark, an inch wide at the surface, but tapering to a point near the wood, and from eight to ten feet long, forming the main channel through which the sap flows to the base of the tree, where a vessel is placed for its reception; branch channels are cut on each side of the main one, leading obliquely into it, six or eight inches apart, and extending nearly two-thirds round the trunk. The sap generally flows from these channels for about twelve hours, when it is collected. The quantity produced by each tree varies; sometimes about a pint, but on an average about a gill." The sap forms an indelible paint, and was formerly used in the island for marking bags, blankets, and other articles. [A. A. B.]

BALSAM. A name given to various gum-resinous or oleo-resinous vegetable substances. —. BAYRE. A product of *Balsamodendron pubescens*. —. CANADIAN. A product of *Abies balsamea*. —. CARPATHIAN. A product of *Pinus Cembra*. —. COPALM. A product of *Liquidambar styraciflua*. —. GARDEN. *Impatiens Bal-*

sensis, sometimes called *Balsamina hortensis*. —, HUNGARIAN. An oleo-resinous product of *Pinus Pumilio*. — of ACOUCHI. A product of *Icica Aroucichini*. — of CAPAIVA. An acrid product of various species of *Copaifera*. — of MARLA. A product of *Verticillaria acuminata*. — of PERU. A product of *Myrospermum peruiferum*. — QUINQUINO. A product of *Myrospermum pubescens*, sold as White Balsam. — of TOLU. A product of *Myrospermum toluiferum*. — of UMIRI. A product of *Humirium floribundum*. —, TAMACOARE. A product of a Brazilian species of *Caraipa*. —, WHITE. The Balsam of Quinquino.

BALSAM HERB. A garden name for *Justicia comata*.

BALSAM SEED. A garden name for *Myrospermum*.

BALSAM TREE. A common name for *Clusia*, and *Balsamodendron*.

BALSAM WEED. An American name for *Gnaphalium polycephalum*, a plant used in the manufacture of paper.

BALSAM WOOD. A garden name for *Myroxylon*.

BALSAMINACEÆ. (*Hydroceræ*: the family of Balsams.) The large genus *Impatiens* (Balsam), and a single species separated from it under the name of *Hydrocera*, included by Jussieu in the *Geraniums* family, have been raised to the rank of a distinct order, on account of the remarkable irregularities in the flowers, which have been variously explained by different botanists. The sepals and petals, all coloured, consist usually of six pieces, two outer ones, small, flat and oblique, the next large, hood-shaped, ending below in a conical spur; the fourth opposite to it, small, but yet very broad and concave, the two innermost very oblique, and more or less divided into two unequal lobes. It has been a matter of much dispute which of them should be considered as sepals and which as petals. It has now, however, been proved by the examination of some Asiatic species, where there are two additional small sepals, and especially of the *Hydrocera*, where the flowers are less irregular, that the two outer pieces, and the large spurred one, with the two occasional additional ones, are the sepals, that the two innermost lobed pieces consist each of two united petals, and that the broad concave one is the fifth petal, thus bringing the structure more in conformity with that of true *Geraniaceæ*, with which Balsams agree also in their ovary, and in the fruit which, in bursting open, leaves the attachment of the seeds adhering to the persistent axis. The *Balsaminaceæ* may therefore be again considered as a tribe only of *Geraniaceæ*.

BALSAMINA. A name sometimes given to the garden Balsam, and some few species resembling it in habit, but which are, however, more usually and correctly referred to *Impatiens*. [T. M.]

BALSAMINE. (Fr.) *Impatiens Balsamina*.

BALSAMITA. A genus belonging to the natural order *Compositæ*, belonging to that group in which the florets are all hermaphrodite, and distinguished by having a naked receptacle, no pappus, and an imbricated involucre, &c. One species of this genus, grown for culinary purposes, the common Costmary or Alecost (*Balsamita vulgaris*), is a native of Italy, from whence it was introduced in 1568. It is a creeping-rooted hardy perennial, from two to three feet high, remarkable for the strong balsamic odours of its leaves, which are roundish, oblong, and toothed, and were formerly put into ale and negus, hence its old English name of Alecost; whilst that of Costmary indicates that it is the Costus or aromatic plant of the Virgin. Although common in every cottage garden, it is almost entirely discarded from the plants that are grown for culinary purposes; and even in France it is only used occasionally to mix in salads. The plant is the *Pyrethrum Tanacetum* of Linnæus. [W. B. B.]

BALSAMOCARPON belongs to the pea family (*Leguminosæ*), and is a native of the province of Coquimbo, in Chili, where it is common in dry hilly places. There is but one species, *B. brevifolium*, the Atareribo of the Chilians. It is a shrub with undivided elongated branches, having many tubercules; from these the leaves proceed, and are accompanied with two or three short spines. The leaves are simply pinnate, and are not more than half an inch long, the leaflets six in number and very small. The flowers are large, yellow, and arranged in few-flowered clusters at the ends of the branches, their calyces covered with long glandular hairs. The pods are thick, short and sessile, and are remarkable for being almost entirely transformed into a cracked resinous substance, which is astringent, and used commercially for dyeing black and making ink. [A. A. B.]

BALSAMODENDRON. A word, as the name implies, applied to certain balsam-bearing trees, of the natural order *Amyridaceæ*. Their foliage is generally scanty, pinnated, and the branches frequently spiny. The flowers are small, green, axillary, often uniserial, with a four-toothed persistent calyx, four narrow petals bent inwards, and eight stamens, inserted with the petals beneath a circular cup-shaped disc, from which arise eight small lobes, which alternate with the stamens. The fruits are small, oval and drupe-like, with four sutures. The nut is thick and hard, two-celled or sometimes one-celled by abortion; each cell contains one seed.

B. Myrrha, a plant growing wild in Arabia Felix, is supposed to yield some of the gum resin known as Myrrh. *B. gileadense* and *B. Opobalsamum* are stated to produce Balm of Gilead, or Balm of Mecca, sometimes called Opobalsamum. A gum resin obtained by incision into the bark, and considered by the ancients as a panacea for almost all the ills that flesh is heir to. *B. Kafal*, one of the plants supposed to yield Myrrh, has a red resinous wood, which is a common article of sale in Egypt. *B. africanum*, a species found in Abyssinia and Western Africa, yields a resin known as African Bdellium, and the Indian drug of the same name is the produce of another species of this genus, *B. Roxburghii*, or of the closely allied one *Amyris*. Bdellium is like myrrh in its properties, but is not considered so good; it is moister than myrrh, not brittle, and has not so agreeable an odour. It is rarely used in this country.

B. Mukul yields a resin known in Scinde under the name of Googul, and in Persia as Mukul. The late Dr. Stocks has shown that this is identical with the Bdellium of Dioscorides and of the Scriptures. The tree producing it is abundant in Scinde, in rocky ground, and the resin is collected by making incisions into the tree and letting the resin fall on the ground, hence it is mixed with much dirt and many impurities. The resin has cordial and stimulant properties. It is given as a medicine to horses in Cabul; it is also used as a plaster for boils. It is burnt as incense, and is mixed by builders with the mortar used in the construction of houses, when durability is an object. A similar resin with the same native name is obtained in other parts of India, from other species of the genus.

Balsamodendron Mukul.

B. pubescens, according to Dr. Stocks, furnishes Bayce Balsam, which is brittle, but tasteless and inodorous. The bark of this tree peels off in thin layers like that of the Birch. As is so frequently the case, there is considerable doubt as to the plants producing these several gum-resins, though it is agreed on all hands that the plants, whatever their species may be, belong to this genus; nay, it is not unlikely that more than one species may furnish the same kind of resin. *B. zeylanicum* is cultivated in this country as an ornamental stove plant. [M. T. M.]

BALSAMORRHIZA. A genus of the composite family (Compositæ). Seven species are enumerated, all of them dwarf perennial herbs, with chiefly radical leaves, which are heart-shaped with long stalks, or pinnatifid. Their stems are simple, usually bearing a solitary flower-head, which is about two inches in diameter, having the appearance of a small sunflower. All the florets are yellow in colour, and the greater part of the species are covered with a whitish pubescence. They are found on the west side of the Rocky Mountains, in Oregon and California. The thick roots of *B. Hookeri*, which is found on gravelly banks of the Columbia river, yield a copious pellucid resin, which has a powerful turpentine-like odour, while those of *B. incana* and *heliantheumoides* are eaten by the Indians in Oregon. They are cooked on hot stones, and have a sweet and rather agreeable taste. The name is given from the occurrence of a balsamic resin in the roots of some of the species. [A. A. B.]

BAMBOO. The common name of *Bambusa*. Bamboo-canes are the stems of different kinds of *Bambusa*. —, SACRED, of the Chinese: *Nandina domestica*.

BAMBUSA. A genus of grasses, typical of the tribe *Bambuseæ*. This tribe is remarkable among those belonging to the great family of grasses, in consequence of the gigantic size some of its species attain. The flowers are hexandrous, more rarely triandrous, and are produced in panicled spikelets. Occasionally some are neuter, and others male only. Steudel describes thirty-three species, which are all natives of warm countries, and have an extensive range over the surface of the globe. It is *B. arundinacea*, which is generally considered to be the species the largest and best canes are produced from, but frequent errors regarding it no doubt occur, and the canes of other species are mistaken for it. *B. vulgaris*, with culmi inermes, appears to be the species which is generally cultivated in British gardens, whereas the *B. arundinacea* is described with culmis spinosus. In the East and West Indies the canes frequently grow from fifty to sixty feet high; and even in this country they have been known to grow forty feet in one season, in some of the large Palm-houses. The finest known species is, perhaps, *B. latifolia*, a native of the Orinoko, which produces much thicker and larger canes in every way than those of *B. vulgaris* or *B. arundinacea*. A fine plant of the large sort is growing in the Botanic Garden at Berlin.

The variety of purposes to which the Bamboo is applied is almost endless. The Chinese use it, in one way or other, for nearly every thing they require. The sails of their ships, as well as their masts and rigging, consist chiefly of Bamboo, manufactured in different ways. Almost every article of furniture in their houses, including mats, screens, chairs, tables, bedsteads, and bedding, are made of the same material. (*See Library of Entertaining Knowledge.*) A similar extensive use of the hollow reed is made in Japan, and also in

VEGETATION OF BAMBOOS IN JAVA.

Java, Sumatra, and other eastern countries. Although the Bamboo grows spontaneously, and more profusely in nearly all the immense districts included in the southern portion of the Chinese empire, the people do not rely on the beneficence of nature, but cultivate the gigantic reed with much care. They have treatises and whole volumes solely on this subject, laying down rules derived from experience, and showing the proper soils, the best kinds of water, and the seasons for planting and transplanting the useful production. (*Ibid.*) A view of the Bamboo vegetation of Java, is given in Plate 4. [D. M.]

BANANA, or WISE-MEN'S BANANA. *Musa sapientum.*

BANANIER. (Fr.) *Musa.*

BANNETTE. (Fr.) *Dolichos melanophthalmus.*

BANARA (including *Ascra, Bosca, Kuhlia,* and *Pineda*). A genus of *Samydaceæ,* confined to the tropical parts of America, and consisting of about fifteen species, all of which are either small trees or shrubs, with ovate leaves, and paniculate, racemose or fasciculate flowers. The calyx is four to five cleft; the petals from four to five in number; the stamens disposed in several rows, inserted in a perigynous disk, and indefinite. Uses unknown. [B. S.]

BANDAKAI. The fruits of *Abelmoschus esculentus.*

BANDED. Marked with cross-bars of colour.

BAND-SHAPED. Narrow and very long.

BANDALA. The strong outer fibre of *Musa textilis*, from which Manilla white rope is made.

BANDOLIER fruit. The berries of *Zamomia indica.*

BANEBERRY. The common name of *Actæa spicata.*

BANG. A narcotic preparation from the leaves of the Hemp, *Cannabis sativa.*

BANGIA. A genus of *Algæ*, which deserves notice as connecting the filamentous with the membranous series, the perfect plant of *B. atropurpurea*, closely resembling very young examples of the common *Porphyra*, which produces its layer of our oil shops. Like *Porphyra* its place is doubtful, as it has almost equal claims to be ranked amongst the green and rose-spored genera. Both, however, are usually placed amongst *Chlorosperms*. *B. atropurpurea* is a common species on old jetty piles, &c., and is a pretty microscopic object. We do not consider such species as *B. velutina* belonging to the same section. See ULVA and PRASIOLA. [M. J. B.]

BANISTERIA. A name applied to a genus of the natural family *Malpighiaceæ,* consisting of trees or shrubs, frequently climbing, with simple stalked leaves, often provided with glands on the stalks. The flowers have a five-parted calyx, also provided with glands at its base externally; the petals are furnished with long stalks or claws; there are ten stamens, frequently somewhat coherent at the base; three styles, often leaf-like at their extremities; and three carpels, each containing one seed, and terminating in a simple membranous wing. The seed-leaves or cotyledons are thick and unequal. The plants are natives of Brazil and the West Indies; several are in cultivation for the sake of their pretty yellow flowers, and, in some instances, fine foliage. [M. T. M.]

BANKSIA. A genus of *Proteaceæ*, established by the younger Linnæus, and named in honour of Sir Joseph Banks. It is distinguished by having four-parted apetalous flowers, the anthers of which, four in number, are sub-sessile and attached one to the concave apex of each sepal; the style is filiform or subulate, with a clavate or cylindrical stigma. The seed-vessel, which is termed a follicle, is large and woody, and contains large winged seeds which are generally black. The genus is peculiar to Australia and Tasmania. In the former colony it is very generally distributed throughout the extratropical portion, while only two intertropical species have been discovered, viz.—*B. compar* at Keppel Bay, on the east coast, and *B. dentata* at Arnheim's Land, on the north coast, and at Endeavour River, on the north-east coast. There are upwards of fifty species known, of which only a few become trees. Mr. C. Fraser mentions having seen a specimen of *B. grandis* which he considered to be fifty feet in height, and with a stem two and a half feet in diameter. The other arborescent species are *B. littoralis, B. cylindrostachya, B. australis, B. prionotes, B. Menziesii* and *B. ilicifolia*. The remainder are more generally shrubs of from fifteen to twenty feet in height, though in some instances, as *B. nutans, B. pulchella* and *B. sphærocarpa*, of much humbler growth. The foliage is remarkable for its harsh rigid coriaceous character, and the leaves are generally dark green on the upper surface, and clothed with a white or rufous down beneath, their margins being either deeply serrated or only spinous, rarely entire. Their form is singularly variable, thus in *B. Menziesii* they are small, refracted and sharp pointed; in *B. spinulosa* and *B. ericifolia* they are linear, three to four inches in length, and about an eighth of an inch in breadth. *B. latifolia* is distinguished by having lanceolate leaves, nearly a foot long and three inches broad, covered with a rich rufous down on the underside. *B. Solandri* has broad ovate leaves, deeply sinuated. *B. speciosa* and *B. Victoriæ* have scurious leaves; fourteen inches covered with whitish down beneath. *B. dryandroides* and *B. Brownii* have very elegant foliage, the latter bearing very much the appearance of a species of *Mimosa*. *B. ærvina* is remarkable for its large head of deep red flowers. One species, *B. integrifolia,* is named the Honeysuckle by the

Australian colonists, in consequence of the great quantity of honey which the flowers contain. These plants, from their handsome and peculiar foliage, have always been great favourites in gardens. The appearance of the *Banksias* in their native habitats is shown in a view of the vegetation of New South Wales, taken near Port Jackson, which forms the subject of Plate 5. [B. H.]

BANQUOIS. A name given in the Mauritius to a species of Screw-pine, *Pandanus vacoa*, the leaves of which are used for making sacks.

BANYAN-TREE. *Ficus indica.*

BAOBAB-TREE. *Adansonia digitata.*

BAPHIA. A genus of leguminous plants (*Fabaceæ*: *Cæsalpineæ*), four species of which are described in botanical works, all of them natives of the coast of western tropical Africa. They are either trees or shrubs, with unequally pinnate leaves. Their flowers are produced in clusters, upon short stalks at the bases of the leaves, each flower having two small bracts underneath its calyx; they have a sheathing calyx which splits along the underside, and is either entire or five-toothed; their corolla is papilionaceous; and they have ten free stamens, all fertile. The fruit is a narrow flattened straight or stalk-shaped pod, of a leathery texture, and having its edges slightly thickened; it contains numerous seeds, and splits open when ripe.

B. nitida, which produces the Camwood or Barwood of commerce, is a tree growing to the height of forty or fifty feet. It has shining green leaves, composed of two pairs of leaflets, with an odd one, and its yellow flowers bear some resemblance to those of the common laburnum of our gardens. About 300 or 400 tons of the wood of this tree are annually imported from Sierra Leone, being collected from various parts of the coast between that place and Angola. In 1858 the imports were 464 tons, valued at 13,832*l*. It usually comes in trimmed logs, about four feet in length and a foot in diameter, but sometimes, though rarely, in the form of balls or cakes, made of the roughly powdered wood. It is of a deep red colour, and yields a brilliant but not permanent dye; with a mordant of sulphate of iron it produces the red colour of the English Bandana handkerchiefs, and dyers generally employ it for much the same purposes as the better known Brazil-wood. The native women on the West coast of Africa use the pounded wood for painting their bodies; amulets are also made of it, and it is used in their Fetish ceremonies. (A. S.)

BAPTISIA. American herbaceous plants belonging to the order *Leguminosæ*, among which they are distinguished by their two-lipped calyx, by their petals, which are equal in length, their deciduous stamens, and swollen pod, which is supported by a stalk, and many-seeded. All the species are herbaceous, and, with one exception, *B. perfoliata* (in which the leaves are simple and entire), have trifoliate leaves. They grow from one to two feet high, and bear blue or yellow flowers, either solitary or in clusters. They are ornamental border flowers, and being perennial may be increased by division of the roots. One species, *B. tinctoria*, a native of dry hilly woods from Canada to Carolina, has been used as Indigo by dyers, and from this the name (from the Greek *bapto*, to dye) was given to the genus. The root and leaves are said to possess astringent and antiseptic properties. The species most frequently cultivated are *B. australis* (French, *Baptisie de la Caroline* or *Podalyre*), a pretty border plant, with large blue flowers, tinged on the keel with greenish white, and arranged in a long cluster; and *B. minor*, a smaller plant with blue or white flowers. [C. A. J.]

BARANETZ, or BAROMETZ. *Cibotium Barometz*, called the Scythian Lamb. *Baran* is Russian for Lamb.

BARBA JOVIS. *Anthyllis Barba Jovis.*

BARBACENIA. A genus of monocotyledonous plants, related to *Velloziea*, and referred with some doubt to the order *Hæmodoraceæ*. It consists of perennial herbs, with simple or dichotomously branched stems, which sometimes attain two or three feet, or sometimes more, in height, and are furnished at the ends with spirally disposed firm spreading narrow acute-keeled leaves, from amongst which issue one-flowered scapes, which are usually clothed with glandular and resiniferous hairs, especially towards the top. The flowers are large and generally showy, and consist of a funnel-shaped perianth, resinously-hairy on the outside, the base of the tube confluent with the ovary, and the limb spreading, of six equal segments; there are six included stamens, having piano-compressed filaments, which are three-toothed at the apex, the middle tooth being the smaller and bearing the anther. The ovary is three-celled, containing numerous ovules affixed to the central angles of the cells, and becomes a cylindraceous three-cornered capsule. The style is triquetrous, three-parted, and the stigma is capitate, three-cornered. There are upwards of a dozen species, all South American, and nearly all found in Brazil, where they occur in hot dry mountain regions, lying between 14° and 20° S. lat. *B. discandrine*, found in the southern parts of British Guiana by Sir R. Schomburgk, is stated to grow from ten to twelve feet high. *B. purpurea*, one of the most familiar species, is frequently met with in hothouses, and affords a very good illustration of the family. This has a short dichotomous striated stem, bearing numerous linear acuminate rigid leaves, sheathing at the base, and minutely spiny-toothed at the margin. The flower-stalks are longer than the leaves, one-flowered; the flowers erect, rich violaceous purple, with lanceolate segments, the three inner of which are broader and more erect than the outer three, which are narrower and spreading. It is a plant of ornamental character. B. spumosa is similar in habit, but is dwarfer, with a more scaly stem, and smaller reddish flowers. Between these species some very showy hybrids have been raised in gardens. [T. M.]

BARBADOS PRIDE. *Poinciana pulcherrima.*

BARBAREA. Winter-cress. A herb held in some repute in the days when the field or brook furnished the only salads, but banished from the table by vegetables of better flavour. The common species, *B. vulgaris*, sometimes called Land-cress, by way of distinction from water-cresses, to which its leaves bear a distant resemblance, is a weed frequently seen in gardens and waste grounds, where the soil is damp. In winter and early spring it is a tuft of pinnate glossy leaves, of a dark green hue, sending up in May an erect leafy stalk, having numerous yellow flowers, which are succeeded by largish four-angled pods. *B. præcox*, Early Winter-cress, is a smaller plant of similar habit; it is well-distinguished by the slender divisions of its upper leaves and its very narrow pods. This, though common enough in the West of England, is considered a relic of cultivation. A variety of the common species is sometimes cultivated for the sake of its double flowers, under the name of Yellow Rocket Herb (French, *Julienne jaune*). The French name of the wild plant is *Barbarée* or *Herbe de St. Barbe*; German, *Winter-kresse*. (C. A. J.)

BARBATE. Having long weak hairs in one or more tufts.

BARBE-DE-BOUC. (Fr.) *Spiræa Aruncus.* — DE CAPUCIN. *Nigella damascena.* — DE CHÈVRE. *Eryngium campestre*, also *Spiræa Aruncus.* — DE JUPITER. *Anthyllis Barba Jovis*, also *Centranthus ruber.*

BARBEAU. (Fr.) *Centaurea Cyanus.* — JAUNE. *Centaurea Amberboi.* — MUSQUE, *Centaurea Moschata.* — VIVACE. *Centaurea montana.*

BARBELLÆ. The hairs of the pappus of composites, when they are short, stiff, and straight.

BARBELLULÆ. Small conical spine-like processes of the pappus of composites, as in *Aster*.

BARBERRY. The Berberry, *Berberis vulgaris.*

BARBON. (Fr.) *Andropogon.*

BARBS. Hooked hairs.

BARBULA. The inner row of fringes or teeth in the peristome of each moss as *Tortula*; also the name of a genus of mosses.

BARBYLUS. An imperfectly known genus, belonging to the natural family *Amyrideæ*. Its describer speaks of the single species, *B. jamaicensis*, as a tree inhabiting Jamaica, with a rough bark, alternate pinnate leaves, and the flowers in racemes. The calyx is bell-shaped, four to five-cleft; the corolla, with four or five petals, arising from the margin of the calyx; stamens eight to ten, arising from the bottom of the calyx; ovary five; style and stigma simple; capsule with three two-seeded compartments. [M. T. M.]

BARCLAYA. A singular genus of *Nymphæaceæ*, not much resembling ordinary water-lilies in appearance, though botanically allied to them. It consists of aquatic plants with tuber-like root-stocks, whence the leaves and flowers spring. The calyx is composed of five distinct sepals; the corolla is tubular at the base, and united below to a disc surrounding the ovary, the limb being divided into five red-coloured petals; stamens numerous, in several rows, inserted on the tube of the corolla, the upper ones sterile; the anthers are without appendages. Fruit adhering to the fleshy disc, composed of several carpels, with radiating stigmata. Each compartment of the fruit contains several seeds, which are albuminous internally, and externally covered with thick bristles. These curious plants are natives of the East Indies, and are especially remarkable for the calyx consisting of distinct sepals detached from the ovary, while the petals are united together below, and are attached to a disc in which the ovary is immersed, so as to give an appearance as though it were inferior, which, however, is not the case. [M. T. M.]

BARDANA. The Burdock, *Arctium Bardana* or *Lappa tomentosa.*

BARDANE. (Fr.) *Arctium Lappa* or *Lappa major.*

BARDANETTE or B. FAUX. (Fr.) *Echinospermum Lappula.*

BARK. All the outer integuments of a plant beyond the wood, and formed of tissue parallel with it. The only true bark is that of Exogens. In Endogens, False bark, also called Cortical Integument, stands in place of bark, from which it is known by the fibrous tissue of the wood passing into it obliquely.

BARK. The officinal name given to the cortical layers of various plants, used chiefly for medicinal and tanning purposes. The name is, par excellence, applied to the Peruvian or Cinchona barks, the source of quinine. Of these there are many varieties, namely:— Calisaya, Royal Yellow, Cinchona Calisaya; Light Calisaya, *C. boliviana*, *scrobiculata*; Peruvian Calisaya, *C. scrobiculata* β. *Delondriana*; Carabaya, Ash, Jaen, *C. ovata*; Dark Jaen, *C. villosa*; Hard Carthagena, *C. cordifolia*; Woody Carthagena, *C. Cordiauxiana*; Spongy Carthagena, Coquetta, Bogota, *C. lancifolia* α. *condaminea* δ.; Crown, *C. Calisaya*; Select Crown, *C. chahuarguera*; Ashy Crown, *C. macrocalyx*, *rotundifolia*; Fine Crown, *C. crispa*; Loxa Crown, *C. Condaminea*; Wiry Crown, *C. hirsuta*; Cinnamon, *C. coccinea*;

Cusco, Arica, *C. pubescens*; Red Cusco, St. Ann's, *C. scrobiculata*; Huanuco, Grey, *C. micrantha, glandulifera, nitida*; Original Loja, *C. uritusinga*; Negrilla, *C. Asterophylla*; Red, *C. conglomerata*; Genuine Red, *C. succirubra*; Spurious Red, *C. magnifolia*. The principal sorts are sometimes classed thus:—GREY BARKS: Crown or Loxa, *C. Condaminea, scrobiculata, macrocalyx*; Lima, Huanuco, Silver, *C. micrantha, lanceolata, glandulifera*, and probably *purpurea*. RED BARKS: *C. nitida*. YELLOW BARKS: *C. Calisaya, macrantha, Condaminea, lancifolia*. RUSTY BARKS: *C. hirsuta, macrantha, ovalifolia*, and probably *purpurea*. WHITE BARKS: *C. ovata, pubescens, cordifolia*. For a complete account of the medicinal cinchona barks, see Mr. Howard's splendid volume, entitled *The Nueva Quinologia of Pavon*.

The following Barks are also employed officinally or economically:—, ALCORNOCO or ALCORNOQUE. The astringent bark of several species of *Byrsonima*; or, according to some authorities, of *Bowdichia virgilioides*. —, ANGOSTURA. The febrifugal bark of *Galipea Cusparia* or *G. officinalis*. —, BABUL. The astringent bark of *Acacia arabica*. —, BASTARD CABBAGE. The bark of *Andira inermis*: same as Worm Bark. —, BASTARD JESUITS. The bark of *Ilex frutescens*. —, BONACE. The bark of *Daphne tinifolia*. —, CANELLA. The stimulant aromatic bark of *Canella alba*. —, CARIBEAN. The astringent bark of *Exostema caribæum*. —, CASCARILLA or SWEET WOOD. The aromatic bark of *Croton Cascarilla* and *C. pseudochina*. —, CHINA. The febrifugal bark of *Buena hexandra*. —, CONESSI. The astringent bark of *Wrightia antidysenterica*. —, CULILAWAN. The aromatic stimulant bark of *Cinnamomum Culilawan*. —, ELEUTHERA. The aromatic bark of *Croton Cascarilla*. —, FALSE ANGOSTURA. The bark of *Strychnos nuxvomica*. —, FRENCH GUIANA. The febrifugal bark of *Portlandia hexandra*. —, JESUITS. The same as Peruvian Bark. —, JURIBALI. An astringent bark of Demerara, supposed to be the produce of some cedrelaceous plant. —, MELAMBO. The aromatic febrifugal bark of some species of *Galipea*, or one of its allies. —, MEZEREUM. The acrid irritant bark of *Daphne Mezereum*. —, MONESIA. The bark of some S. American *Sapotaceæ*. —, MURUXI. The astringent bark of *Byrsonima spicata*, used by the Brazilian tanners. —, NIEPA. The febrifugal bark of *Samadera indica*. —, PANOOOCCO. The sudorific bark of *Swartzia tomentosa*. —, QUERCITRON. The yellow dye bark of *Quercus tinctoria*. —, QUILLAI. The bark of *Quillaia saponaria*, used as a substitute for soap. —, STRINGY, of Tasmania, *Eucalyptus robusta*. —, SWEET WOOD. The same as Cascarilla Bark. —, NINE. An American name for *Spiræa opulifolia*. —, WHITE WOOD. The same as Canella Bark. —, WINTER'S. The tonic aromatic bark of *Drymis Winteri*. —, WORM. The bark of *Andira inermis*, formerly used as an anthelmintic.

BARKERIA. A small genus of beautiful orchids, from Mexico and Central America, differing little from *Epidendrum* except in the column being bordered by a broad membranous wing. About half-a-dozen species are known, of which *B. spectabilis*, called in Guatemala Flor de Isabel, is the finest. It is one of the votive offerings of the Catholics in that country.

BARKLYA *syringifolia*, the only species of a genus belonging to the section of the pea family bearing regular flowers, is a large tree, with alternate simple coriaceous leaves, which have long stalks, and are in form like those of the lilac (*Syringa*), but have seven radiating nerves. The flowers are golden yellow, very numerous, and disposed in axillary or terminal racemes. The pods are stalked, about half an inch long, thin, and containing few seeds. The tree has been lately introduced into English gardens. It is a native of Eastern Australia, near the Brisbane river. The genus bears the name of Sir Henry Barkly, governor of the colony of Victoria.
[A. A. B.]

BARLERIA. A large genus of herbs or shrubs, natives of the tropical regions of both the Old and New Worlds, and belonging to that division of the *Acanthaceæ* in which the corolla lobes are imbricate or two-lipped in the bud, and not contorted, and the seeds are inserted on hooked retinacula. The flowers of this genus are axillary, or in terminal spikes or heads, and have herbaceous or pungent bracts. The calyx has four sepals, the two outer being larger than the others; the corolla has a long tube, and five nearly equal spreading lobes. Of the four stamens the upper pair are sometimes abortive; the anthers are linear and parallel. The two-celled ovary has two ovules in each cell; the style is entire with a truncate stigma. The capsular fruit is acuminate. The allied genera all have a distinctly two-lipped corolla, and are thus easily distinguished.
[W. C.]

BARLEY. The common name for *Hordeum*, a genus of corn-producing grasses. Pearl Barley is the grain of the common Barley deprived of its hard integuments.

BARNADESIA. A genus of the composite family, belonging to that section of the order which has two-lipped corollas. All the species are spiny bushes, furnished with entire generally elliptical or lanceolate pointed leaves, each having at its base two spiny stipules. The flower-heads are terminal and elongated. The florets and often the involucres are purple or pale pink in colour. The pappus is feathery, and the scales are clothed with silky hairs. *B. rosea* has delicately rose-coloured florets, which are covered with silky hairs, and is a favourite plant in the tropical houses of English gardens, being a very free bloomer. The species, nine in number, are natives of tropical S. America. The genus is named in honour of Michael Barnades, a Spanish botanist.
[A. A. B.]

BARNARDIA. A genus of *Liliaceæ*, containing rather small bulbous plants, resembling *Scilla*, natives of China and Japan. They have linear cuspidate radical leaves, and scapes bearing small pink flowers in racemes; filaments winged and dilated at the base; ovary three-celled, each cell with one ovule erect from the base — which distinguishes it from its allies. *B. scilloides* is a pretty frame plant. [J. T. S.]

BAROMETZ. *Cibotium Barometz.*

BAROSMA. This name has been applied to a genus of *Rutaceæ*, on account of the heavy powerful odour that the species possess. The genus is botanically characterised by an equally five-parted calyx; five oblong petals; ten stamens, of which five are sterile and petal-like, alternating with the five shorter fertile stamens; the style of the same length as the petals; and the ovary five-celled. The species are small evergreen shrubs, with opposite or alternate simple dotted leathery leaves, in the axils of which the flowers are placed on stalks. They are all natives of the Cape of Good Hope, where the leaves, which have a rue-like smell, are used by the Hottentots to perfume themselves with!

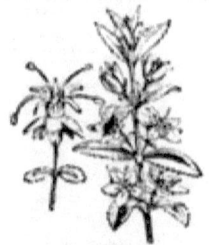

Barosma crenulata.

They also use a tincture of the leaves as an application to wounds, and in urinary diseases. Several species are used by the Hottentots under one common name of Bucku. The Bucku leaves of commerce are produced chiefly from *B. crenulata*, *B. crenata*, and *B. serratifolia*. Bucku leaves are much used in medicine as a stimulant and tonic, and appear to have a special effect in chronic diseases of the bladder, their action probably being dependent on the powerfully-smelling volatile oil which they contain. [M. T. M.]

BARRALINGUE. (Fr.) A kind of olive.

BARRAS. The French name of the resinous exudation of *Pinus maritima*, the basis of Burgundy Pitch.

BARRENWORT. The common name for *Epimedium*.

BARRINGTONIACEÆ. (*Barringtoniads*.) A small family consisting of about five-and-twenty species, usually considered as forming a tribe of *Myrtaceæ*, with which they agree in the structure of their ovary and perianth, and in the very numerous perigynous stamens, turned towards the bud. They differ chiefly in the presence of albumen in the seed. Their leaves are also alternate, not dotted, and often serrate; but these characters occur also occasionally in true *Myrtaceæ*. They are all trees or shrubs, inhabiting the tropics in the New and the Old World, some of them bearing large flowers of considerable beauty. The principal genera are *Barringtonia* and *Careya* in the Old World, and *Gustavia* in the New.

BARRINGTONIA. This genus consists of trees, some of them of large dimensions, with alternate opposite or whorled leaves, often of large size and generally obovate in form, their margins toothed or entire. The flowers are in spikes or racemes, generally large and handsome, and in colour pink, scarlet, or white. The stamens are very numerous and form a conspicuous feature in the flower, from the great abundance of yellow anthers; and their filaments, being slightly united at the base, fall off in the form of a ring when the flower fades, and have the appearance of a painter's brush. Their fruits are one-seeded, fleshy, more or less four angled, and in the larger-flowered species about two inches in length, tapering towards the base. They are found in many parts of India, but in the greatest numbers in the Malayan peninsula and the islands of the Indian Ocean; two species are present in N. Australia, and one grows on the banks of the Zambesi river in East Africa. Without exception they are beautiful objects when in flower.

The bark of a number of the species has narcotic qualities. *B. acutangula*, an Indian species, grows to a large size, and bears some resemblance to an oak in its branching. It furnishes a solid durable wood, useful for ordinary purposes; and from the leaves an extract or juice is obtained which, when mixed with oil, is used in native practice for eruptions of the skin. The kernels, powdered and prepared with sugar and butter, are used in diarrhœa; mixed with milk they promote vomiting. Young plants of this species are shown in Plate 10, figure d.

B. speciosa, a native of the Moluccas, and one of the handsomest of the genus, attains the height of forty or fifty feet, with a circumference of ten to fourteen feet; it is generally found near the sea. From its seeds a lamp-oil is expressed; mixed with bait they are used to inebriate fish, in order to facilitate their capture.

The root of *B. racemosa* has a bitter taste, and is used by Hindoo practitioners on account of its aperient and cooling qualities. The seeds and bark are also used in native medicine, the latter is of a reddish colour, and is said to possess properties akin to those of quinine (*Cinchona*). The pulverised fruit is used as snuff, and, combined with other remedies, is applied externally in diseases of the skin. The genus

Barringtonia speciosa.

was dedicated to the Hon. Daines Barrington, the English antiquary. [A. A. B.]

BARROWIA. A genus of *Asclepiadaceæ*, containing a single species, from the Cape of Good Hope. It is a slender branched and climbing plant, with oblong-lanceolate leaves, and three or more white flowers on interpetiolar peduncles. The calyx is five-parted, with lanceolate erect sepals. The funnel-shaped corolla is slightly swollen at the base, and the limb is cleft into five lanceolate spreading divisions. The gynostemium is included, and has the sinuous staminal corona attached to its base. The ovoid pollen masses are attached to a small corpuscle by slender processes, and have a projecting pellucid apex. The stigma is five-sided, with a slightly projecting central cone. [W. C.]

BARTERIA. A tropical African shrub, with alternate glabrous entire or crenate leaves, and rather large sessile axillary flowers, forming a genus of *Passifloreæ*, allied to *Smeathmannia*, but differing chiefly in the stigmas being consolidated into one large terminal capitate mass, exceeding the ovary in diameter, and in the fruit, which is said to be an indehiscent berry the size of a pigeon's egg.

BARTHOLINA. This is one of a singular race of terrestrial orchids, peculiar to the Cape Colony, with solitary shaggy leaves, small white flowers, and a great lip cut into narrow strips, resembling the teeth of a comb. They have been grown in this country, but perish after having been imported for a year.

BARTLINGIA. A genus of *Chamælauciaceæ*, founded on an undershrub found in Eastern Australia. It has slender fastigiate branches; alternate shortly-stalked obovate retuse entire glabrous leaves, with immersed glands secreting oil; short stipules; and the flowers clustered at the apices of the branches; calyx with five bony segments; petals five, small and scale-like; stamens ten; ovary free, with one style; ovules two. [J. T. S.]

BARTONIA. A showy genus of annual North American Loaseæ, of which the *B. aurea*, a Californian species, is one of the best known. The most important features of the genus are, a cylindrical or club-shaped calyx tube, with a five-parted persistent limb; five or ten flat spreading equal petals; numerous stamens; and a capsule having the seeds arranged in two rows on each of the parietal placentæ, opening at the summit when ripe. The *B. aurea* is a succulent branched spreading plant, of a greyish-green aspect, growing two feet high, with lanceolate pinnatifid roughish foliage, and large lustrous golden yellow blossoms in terminal clusters, expanding only in the middle of the day. When in perfection, it is really a splendid plant, and may be made to contribute greatly to the gaiety of the borders; for, although its habit and foliage are less attractive than those of some other annuals, in size and brilliancy of blossoms it is inferior to none. There are several other species peculiar to the Western and North Western States, of which the most remarkable is the *B. ornata*, with very large white flowers, figured many years since in the *Botanical Magazine*, under the name of *B. decapetala*, from dried specimens, but apparently unknown in England in the living state. By some botanists the genus *Bartonia* is not considered distinct from *Mentzelia*. [W. T.]

BARTRAMIA. A genus of mosses, included in the order *Bryaceæ*.

BARTSIA. Unpretending annuals, belonging to *Scrophularineæ*, and distinguished from *Rhinanthus* (Yellow Rattle) by having the upper lip of the corolla arched, and not laterally compressed. *B. Odontites* is a common weed by waysides and in corn-fields, growing from six to eight inches high, with an erect branched stem, bearing many one-sided clusters of inconspicuous dull purple flowers; the foliage is scanty, and the whole plant roughish, and tinged more or less with purple. A less common species is *B. viscosa*, which grows in marshes and damp pastures to the height of six to twelve inches, and bears numerous brightish green leaves, which are narrow, cut at the edges, and taper to a point; it is very common in many parts of Devon and Cornwall, where it sometimes grows two feet high. The flowers are solitary, imbedded among the leaves, and much larger than in the last. The whole plant is singularly clammy to the touch. *B. alpina* is a rare species, found only in rocky mountainous pastures in the north. All the species turn black in drying. [C. A. J.]

BARU. A woolly material, found at the base of the leaves of *Saguerus saccharifer*, sometimes called *Aronga saccharifera*.

BARWOOD. An African dye wood, produced by *Baphia nitida*.

BARYA. A genus of begoniads, established by Klotzsch, and consisting of herbaceous plants, found on the mountains of Peru. The staminate flowers have four, and the pistillate five sepals; anthers elliptical and short; filaments united; style persistent, with elongated branches, surrounded by an interrupted papillose band, making five spiral turns; placentas stalked, with two lamellae. There is one known species, namely, *B. monadelpha*, gathered by Ruiz, under the name of *Begonia monadelpha*, near Muña, in Peru. The genus is named after Dr. Ant. de Bary, a patron of botany. [J. H. B.]

BASAL. Growing at the base of anything, as ovules at the base of an axile placenta.

BASELLACEÆ. (*Basellads*.) A small family, chiefly distinguished from *Chenopodiaceæ* by what has been called a double calyx, and perigynous stamens; but the so-called order calyx consists merely of the two bracts, which are here admitted to the perianth, instead of being free, or at some distance from it; and more or less perigynous stamens occur also in other chenopodiaceous genera. *Basellaceæ* have therefore been now re-united with that family as a tribe. They are mostly herbaceous climbers, with more or less succulent leaves, and small inconspicuous flowers. The perianth is usually thick and fleshy, and the style is three-cleft, whilst in true *Chenopodieæ* it is more frequently (but not always) only two-cleft. There are sixteen or seventeen species, all tropical, and they have been distributed into six genera, of which the most important are *Basella*, *Boussingaultia*, and *Anredera*.

BASELLA. A genus of climbing plants, belonging to the order or tribe *Basellaceæ*. The simple ovary becomes converted into a membranous fruit, which is adherent to the inner part of the persistent calyx, and contains a single seed, with little or no albumen, and an embryo, coiled up spirally, like a watch-spring. *B. alba* and *B. cordifolia* are cultivated in the East Indies as pot herbs, and are used as a substitute for spinach. *B. rubra*, a variety of *B. cordifolia*, yields a rich purple dye, but it is difficult to fix. These plants are grown in India over trellis-work, where the cucumber shoots and leaves form an agreeable protection from the sun. Some of the species have tuberous roots. *B. alba* is in cultivation, and might with advantage be more frequently grown from a suspended basket, as its appearance when in bloom is elegant. [M. T. M.]

BASIBRACTEOLATE. A term applied chiefly to the involucre of a composite, when it is surrounded at the base by a distinct order of bracts, as in dandelion.

BASIDIA. Little elevations found among fungals, consisting of a single cell, having one or more points at its apex, each bearing a spore.

BASIDIOSPORES. The spores which stand upon the basidia.

BASIFIXUS. Attached by the base.

BASIL, BUSH. *Ocymum minimum.* —, SWEET, or BASILICUM. *Ocymum Basilicum*, an aromatic pot-herb. —, WILD. *Calamintha Clinopodium*.

BASILARIS. Seated at the base of anything.

BASILIC COMMUN. (Fr.) *Ocymum Basilicum*. — DE LA CHINE. *Plectranthus rugiflorus*. —, PETITE. *Ocymum minimum*. —, ROMAIN. *Ocymum Basilicum*.

BASIL-THYME. *Calamintha Acinos*.

BASINERVED. When the ribs of a leaf all spring from its base, as in most Melastomads.

BASISOLUTE. A term applied to leaves which, like those of *Sedum* and *Echeveria*, are extended downwards below their true origin.

BASSIA. A genus of the natural order *Sapoteæ*, consisting of tropical trees, with alternate entire leaves, and whitish axillary stalked flowers, having a calyx of four or five sepals, a fleshy corolla, tubular below, but divided at its limb with eight segments. Stamens numerous; ovary terminated by a tapering style, and containing six to eight compartments, of which three or four undergo an arrest of growth, so that the pulpy fruit does not contain more than three or four one-seeded cells.

B. butyracea, the Indian Butter tree, or Phulwara, is a native of Nepaul, and the Almorah hills. From its seeds when bruised and pressed is squeezed out a fatty substance of the consistence of hog's lard and of a white colour. It is used to adulterate ghee, and is considered serviceable in rheumatism, and as an application to the hair. It makes good soap, and is adapted for burning. It is soluble in warm alcohol, and does not become rancid when kept, but is completely melted at a temperature of 120°. From the juice of the flowers a kind of sugar is prepared.

B. latifolia, the Mahwah tree of Bengal, furnishes a hard and strong timber used for the wheels of carriages, &c. The flowers are sweet-tasted and are eaten raw, and they are also largely made use of in the distillation of an ardent spirit like whisky, which is consumed in great quantities by the natives of Guzerat, &c. When fresh it is very deleterious to Europeans. The seeds yield an oil used for lamps, in the manufacture of soap, and for culinary purposes; but it is thick, coarse, and only used by the poorer classes. The flowers are stated to collect the sweetly tasting flowers of this plant, and dry them to store as a staple article of food; and hence, 'in expeditions undertaken for the punishment or subjection of these tribes when unruly, their

Bassia trees are threatened to be cut down by the invading force, and the threat most commonly ensures the submission of the tribes.'—(*Gibson*.)

The flowers of *R. longifolia* are roasted and eaten in Malabar and Coromandel; they are also bruised and boiled to a jelly. The leaves as well as the milky juice of the unripe fruit are used medicinally. The bark contains a gummy juice which exudes and is used in rheumatism; the bark itself is likewise employed as an astringent, and as a remedy for the cure of the itch. The seeds furnish an oil like that of the other kinds, but of an inferior quality.

The Shea tree or Butter tree of Africa, whose seeds produce the tallow butter, mentioned by Mungo Park in his travels, is a species of this genus, *B. Parkii*, or of the closely allied one, *Lucuma*. The seeds are boiled in water to extract the latter from them. This fatty substance is of a white colour, and agreeable taste, and keeps well, hence it is an important article of commerce in Sierra Leone. Some of the species of this interesting genus are in cultivation. [M. T. M.]

BASSIN D'OR. (Fr.) *Ranunculus repens.*

BASSINET. (Fr.) *Ranunculus repens.*

BASSORA GUM. A partially soluble gum of uncertain origin, supposed to be the produce of a *Cactus* or *Mesembryanthemum*.

BASS-WOOD. The American Lime or Linden, *Tilia americana*.

BAST. A strong woody fibre, much used for brooms, brushes, &c., obtained from the leaf-stalks of *Attalea funifera*, and of *Leopoldinia Piassaba*. Also, the inner bark of the lime tree, of which the Russian mats used in gardens, are made. —, CUBA. The fibrous inner bark of *Paritium elatum*, much used for tying up cigars, and in gardens for tying plants, as also is the bast of the lime tree.

BATARREA. A genus of *Fungi* allied to the puffballs. Its most striking characteristics are a thick gelatinous volva, a tall rigid stem, and a hemispherical cap-shaped peridium. Some of the filaments, moreover, have a spiral structure, a very rare circumstance amongst *Fungi*. The British species is extremely rare, and occurs on sandhills, for the most part near the sea, or amongst the vegetable soil in hollow trees. The habit is that of *Phallus*, and the volva with its intermediate gelatinous coat is precisely the same. The early stages of this plant has not been observed since the true structure of the hymenium in the higher fungi has been ascertained; but there can be little doubt that it resembles that of the true puffballs. [M. J. B.]

BATATE. (Fr.) *Batatas edulis*, formerly *Convolvulus Batatas*.

BATATAS. A genus of Bindweeds (*Convolvulaceæ*), of which about twenty species are described, mostly natives of tropical America. They are creeping or twining, herbaceous or shrubby plants. Their flowers have a bell-shaped corolla, enclosing the stamens, and a four-celled ovary, with a single style and a two-lobed capitate stigma.

The most interesting species is *B. edulis*, the tuberous roots of which, under the name of Sweet Potato, are extensively used in many warm countries in the same way that we use common potatoes. The plant has a creeping or sometimes twining stem five or six feet long, and either running along the ground, or rambling over other shrubs. Its leaves are about six inches long and heart-shaped at the base; and its flowers resemble those of the common *Convolvulus*, of a pale purple colour, and arranged in threes or fours on a stalk. This plant has been so long cultivated and naturalised in various tropical countries, that its precise origin is somewhat obscure, but probably it is indigenous to both hemispheres. The first mention of it is said to be by an author named Pigafetta, who went to Brazil in 1519, and found it in use among the Indians as an article of food. It was soon afterwards introduced into Spain, where it is still cultivated. The roots were known in England before the introduction of the common potato, with which they were frequently confounded by early writers. They were imported in considerable quantities from Spain and the Canary Islands, and, when steeped in wine, or made into sweetmeats, were supposed to have the effect of restoring decaying vigour. At the present day sweet potatoes are largely cultivated in many tropical and sub-tropical countries; such for instance as India, China, Japan, the Malayan Archipelago, &c., in the east; and in the west, very generally throughout tropical America, also in Texas, Alabama, Carolina, and other Southern States of America, extending even as far north as New York, where, however, they are not found to be a profitable crop: they are also grown to a small extent in the south of

Batatas edulis.

Europe, and more extensively in the Canary Islands, Madeira, and North Africa. There are several varieties, some having white roots and others red. The roots grow to a very great size; according to Crawfurd they sometimes attain the enormous weight of fifty pounds in Java; but in the United States the general weight is from three to twelve pounds each tuber, and the yield per acre is estimated at from 200 to 300 bushels. They have an agreeable sweetish taste, and contain rather more flesh-forming matters than the common potato, considerably more sugar, and a slight excess of starch.

B. Jalapa has large tuberous roots and creeping stems like the last. The leaves of this species are heart-shaped, of a deep green upon the upper surface, and covered with a white woolly down beneath; and the flowers are either white or rose-coloured, and very showy. It is a native of Mexico, growing commonly in the vicinity of the town of Xalapa, whence the specific name *Jalapa* is derived. It was formerly supposed to produce the jalap of the Pharmacopœia, but that drug is now known to be derived from another plant of the same natural order; the roots of *B. Jalapa*, however, possess purgative properties, and are probably sometimes substituted for true jalap.

B. paniculata has thick smooth twining stems and large hand-shaped leaves; and its flowers are very handsome and of a fine purple colour. It is a native of India, Java, New Holland, Mauritius, West Africa, Guiana, Brazil, &c.; and is the species commonly cultivated for food in Western tropical Africa. From the seeds of a species of this genus is obtained what is called Natal Cotton, a textile material resembling true Cotton. [A. S.]

BATEMANNIA. *Colleyi* is an inconspicuous orchid with dull brownish-purple flowers, from Demerara. It differs in little from *Maxillaria*, excepting having an anther-bed with a membranous border. Some other plants bearing this name belong to the genus *Galeottia*.

BATHMIUM. A name given by Link to a group of large-growing ferns now included in *Aspidium* and *Sagenia*. [T.M.]

BATIDEÆ. The *Batis maritima*, a low shrubby succulent plant, with opposite leaves, abundant in the salt-marshes on the sea-coasts of the West Indies, has much puzzled botanists as to its real affinities. They have therefore, in compliance with a custom now very prevalent in similar cases, endeavoured to solve the problem by supposing it to constitute a family by itself, giving it the plural name of *Batideæ*. The habit of the plant, and the small green flowers half buried in a succulent spike, give it a great general resemblance to *Salicornia*; but the ovary having (according to Torrey) four cells with one erect ovule in each, differs materially from that of *Chenopodiaceæ*. The flowers are in unisexual cylindrical spikes. The males, solitary, under imbricated bracts, have a two-lobed calyx and four stamens, alternating with as many minute scale-like petals, or rather staminodia.

The females have a two-lobed sessile stigma, without either perianth or stamens. The seed has no albumen, and the embryo is but little curved; yet Grisebach is probably right in proposing once more to include it among the *Chenopodiaceæ*, with which it accords so well in outward appearance.

BATIS. This genus of Exogens has a structure so anomalous that it has been separated as a distinct order, *Batideæ*. The leading features have been just explained; in addition to which it may be remembered that the naked ovaries adhere to each other in the form of a short green four-rowed cone. Each ovary consists of two carpels, the stigmas being only two; but it is four-celled, with one ovule in each cell, in consequence of the dorsal rib of each carpel being inflexed so as to form a partition, the partition passing between the two ovules, making the two-celled ovary four-celled, with one ovule in each cell. The ovule is erect from the base of the cell. The seed contains no albumen, and the embryo has an inferior radicle. The position of the *Batideæ* in the natural system is a question of much interest with botanists, who have assigned it widely different stations. There is no doubt, however, that it has some relation with the *Callitrichaceæ*, and in common with that order shows some resemblance to the *Caryophyllaceæ*. The writer has also succeeded a very near affinity with the *Verbenaceæ*, which have the same kind of four-celled ovary. (*Trans. Linn. Soc.* xxii. 411). The plant is sometimes used in making West Indian pickles, and its ashes yield large quantities of barilla. [H. C.]

BATODENDRON. A name applied to a genus of *Vacciniaceæ*, more commonly considered as a section of the genus *Vaccinium*. It is known by its flowers being borne on long thread-like stalks, in leafy or leafless clusters; the corolla five-lobed, spreading, bell-shaped; filaments hairy; anthers provided with two long awns bent backwards; fruit berry-like, almost tasteless, its cells few-seeded by abortion. The species are all natives of America and Mexico. [M.T.M.]

BÂTON BLANC or ROYAL, (Fr.) *Asphodelus ramosus*. —, DE JACOB. *Asphodelus luteus*. —, DE SAINT JEAN. *Polygonum orientale*.

BATRACHOSPERMEÆ, BATRACHOSPERMUM. A division and genus of green-spored *Algæ*. The division contains two distinct groups, in one of which the frond is cartilaginous, solid or hollow, with the outer coat cellular; in the other it is made up of a central articulated axis coated with close-packed descending threads, covered with whorls of necklace-like branchlets at regular intervals. In both, the fruit consists of chains of spores, which are external in the latter, internal in one genus, at least, of the former. *Batrachospermum* belongs to the second group. Its species which are highly gelatinous, vary from black to nearly violet, but are never of a deep rose as in the analogous rose-spored *Algæ*. *B. mosiliformis* is a common inhabitant of our rivulets, where it is found attached to stones or roots, and never fails to obtain admiration when closely examined. Most of the species grow in fresh water. The genera of the first group will be noticed under *Lemanea*. [M.J.B.]

BATSCHIA. A genus of *Montispermaceæ*, allied to *Abuta* and *Anelasma*, but differing from both in its three to five-nerved leaves, which are almost glabrous. The genus is imperfectly known, Mr. Bentham is of opinion that it should be merged with *Abuta*. The plants included in it are natives of Darien. The name has been applied to certain boraginaceous plants now included under *Lithospermum*; and also to a genus of *Leguminosæ*, now referred to *Humboldtia*. [M.T.M.]

BAUDRIER DE NEPTUNE. (Fr.) *Laminaria saccharina*.

BAUERACEÆ. The genus *Bauera*, belonging to the *Hydrangeæ* family (or tribe of *Saxifrageæ*), has by some botanists been thought to possess distinctive characters sufficient to establish it as a separate family under the name of *Baueraceæ*, which has not, however, been generally adopted.

BAUERA. A genus of *Hydrangeaceæ*, consisting of small shrubs found in Australia. They have opposite sessile trifoliate leaves, with oblong undivided leaflets, and no stipules; and the flowers are handsome nodding, rose-coloured or purple, axillary, solitary on rather long stalks, rarely terminal and clustered; calyx with six to ten segments, adhering to base of ovary; corolla of six to ten petals; stamens numerous; styles two. [J.T.S.]

BAUHINIA. This genus of leguminous plants (*Fabaceæ*; *Cæsalpineæ*) was selected by Plumier to bear the name of two brothers, John and Caspar Bauhin, celebrated botanists of the sixteenth century, in consequence of most of the species having their leaves composed of two lobes, which are either quite separate, or, more frequently, joined together by a portion of their inner margins, and which Plumier thought symbolical of the united labours of the two Bauhins in the cause of science. The numerous species are extensively diffused throughout the tropics, particularly in Brazil and India. They are generally climbers, frequently attaining a gigantic size; some few, however, form trees or large shrubs. Their flowers are produced either simply or in racemes opposite the leaves, and have a calyx with a cylindrical tube split on one side, or rarely 5-parted; five unequal spreading petals, inserted along with the stamens into the top of the calyx-tube; ten stamens, which are either joined together at the base or distinct, a portion of them being sometimes barren; and a long-

stalked ovary, which ultimately becomes a many-seeded two-valved pod.

B. tomentosa is a native of Ceylon, where it forms a small tree, growing about fifteen feet high, and having pale yellow flowers spotted with crimson, which has given rise to the superstitious idea that they are sprinkled with the blood of St. Thomas, hence the tree is called St. Thomas' tree. Its leaves are composed of two oval, blunt-topped leaflets joined together for more than half their length, and hairy on the under side. The dried buds or flowers, and also the leaves, are employed by the native Indian doctors as a remedy against dysentery.

B. Vahlii is the Maloo climber of India, a plant whose gigantic shrubby stems often attain a length of 300 feet, and climb over the tops of the highest trees of the forest, twisting so tightly round their stems that they not unfrequently strangle and cause death, the stems ultimately decaying and leaving a sheath of climbers standing in their place. The young shoots and leaves are covered with a rust-coloured scurf, and are furnished with tendrils. The leaves are very large, often more than a foot in diameter, composed of two oval-shaped lobes joined together for about half their length, and heart-shaped at the base. The flowers are snowy-white, and arranged in racemes. The exceedingly tough fibrous bark of this species is employed in India for making ropes, which, from their great strength, are used in the construction of the suspension bridges across the river Jumna. The bark of another Indian species is used for making the shoe-matches used with native guns.

B. variegata is a small tree of about twenty feet in height, a native of India, China, and the Moluccas Islands, and now naturalised in some of the West India Islands. It has two broadly egg-shaped leaflets joined for about one-third their length; and its rosy-white flowers are produced in twos upon a forked stalk. The wood of this species is of a dark colour, and forms one of the many woods called Ebony; the bark is used medicinally in India, and also for dyeing and tanning leather. [A.S.]

BAUME A' COCHON, or A' SUCHIER. (Fr.) A balsamitous resin produced by *Hedyosmia balsamifera*. —, A' SALADE. *Mentha viridis*. —, COQ, or GRANDE. *Balsamita vulgaris*, or *Pyrethrum Tanacetum*. — DE CANADA. *Abies balsamea*. — DE PEROU. *Myroxoacarum peruiferum*. — DE PEROU FAUX. *Melilotus cærulea*. — DE VANILLE. A liquid which exudes from the Vanilla. — DES JARDINS. *Mentha rosea*. — FORMICAT. A balsamineous product contained in galls borne by the Elm, *Ulmus campestris*, in Italy, France, Persia, &c. — SAUVAGE. *Mentha rotundifolia*.

BAUMIER. (Fr.) *Populus balsamifera*. — DE GILE'AD. *Abies balsamea*.

BAWCHEE SEED. An oil seed, the produce of *Psoralea corylifolia*.

BAWDMONEY, or BALDMONEY. *Meum Athamanticum.*

BAXTERA. A little known genus of *Asclepiadaceae* from Brazil, containing a single species, an erect shrub with opposite cordate elliptical leaves, and reddish flowers in terminal umbels. The calyx is five-parted. The tube of the corolla is dilated below, and the limb is five-cleft; the staminal corona consists of five fleshy leaves. The anthers are terminated by a membranaceous appendage. [W. C.]

BAY, INDIAN or ROYAL. *Laurus indica.* —, LOBLOLLY. *Gordonia lasianthus.* —, RED. *Laurus carolinensis.* —, ROSE. *Epilobium angustifolium.* —, SWEET, or BAY-TREE. *Laurus nobilis;* also an American name for *Magnolia glauca.*

BAYBERRY. *Myrica cerifera.*

BDELLIUM, AFRICAN. A gum resin obtained from *Balsamodendron africanum;* also applied to that obtained from *Ceradia furcata.* —, INDIAN. A resinous product of *Balsamodendron Roxburghii,* or *Amyris Bdellium.*

BEAD TREE. The common name for *Melia.*

BEAKED. Ending in a long sharp terete, or angular point.

BEAK-SEDGE. A common name for *Rhynchospora.*

BEAM TREE. *Pyrus Aria.*

BEAN. The common name for *Faba.* —, BOG. The Buckbean, *Menyanthes trifoliata.* —, CUJUMARY. The tonic seed of *Aydendron Cujumary.* —, EGYPTIAN, or PYTHAGOREAN. The fruit of *Nelumbium speciosum.* —, FRENCH. *Phaseolus vulgaris.* —, HARICOT. The seed of *Phaseolus vulgaris.* —, INDIAN. An American name for *Catalpa.* —, KIDNEY. The common name for *Phaseolus,* especially for those kinds cultivated as esculents. —, LIMA. An American name for *Phaseolus lunatus.* —, LOCUST. The pod of *Ceratonia Siliqua.* —, MOLUCCA. The seed of *Guilandina Bonducella.* —, ORDEAL, of Old Calabar. The seeds of *Physostigma venenatum.* —, OX-EYE. The seed of *Mucuna urens.* —, PICHURIM. A commercial name for the cotyledons of *Nectandra Puchury.* —, SACRED. The common name for *Nelumbium.* —, SAJUFA. *Soja hispida.* —, ST. IGNATIUS'. The seed of *Ignatia amara,* alias *Strychnos Ignatia.* —, SCARLET RUNNER. *Phaseolus multiflorus.* —, TONKA, or TONQUIN. The seed of *Dipterix odorata.* —, UNDERGROUND KIDNEY. *Arachis hypogaea.* —, WATER. An English name for the family of *Nelumbiaceae.* —, WILD. An American name for *Apios.*

BEAN CAPER. The common name for *Zygophyllum.*

BEAN TREE, Swedish. *Pyrus intermedia.* —, of Australia. *Castanospermum australe.*

BEAN TREFOIL. The common name for *Anagyris;* sometimes also applied to *Menyanthes trifoliata,* and anciently to *Cytisus Laburnum.*

BEAR-BANE. *Aconitum arctophonum.*

BEARBERRY. *Arctostaphylos.*

BEARBIND. The common name for *Calystegia.*

BEAR'S BREECH. *Acanthus.*

BEAR'S EAR. *Primula Auricula.*

BEAR'S FOOT. *Helleborus foetidus.*

BEASTS BANE. *Aconitum theriophonum.*

BEATONIA. A genus of bulbous *Iridaceae,* containing a single species, *B. purpurea,* found in Mexico. This has a simple flower stem about six inches high, a few plaited leaves, and one or two terminal flowers, which are crateriform, rosy purple, whitish and dotted with purple near the base. The perianth is very irregular, almost as in *Tigridia;* the filaments cylindraceously connate, and the style slender, with a three-lobed stigma, the lobes of which are split and recurved. [T. M.]

BEATSONIA. A genus of *Frankeniaceae,* scarcely differing from *Frankenia,* but having two styles, with globular stigmas, and a two-valved capsule, while in *Frankenia* there are three styles and a three-valved capsule. *B. portulacifolia* is one of the few plants indigenous to the Island of St. Helena. [J. T. S.]

BEAUCARNEA. A name lately given to a genus of Agave-like filaceous plants, till recently known in gardens under that of *Pincenectitia,* which, it is said, originated in a misspelling on a garden label of *Prognectia,* the name of a genus of screwpines, with which these have no affinity whatever. The few known species are natives of Mexico. They are arborescent stems, remarkable for the large bulbiform swelling which, from the earliest stages, forms at its base; these support a spreading terminal crown of long narrow leaves. The inflorescence in *B. recurvata* is a large terminal panicle, a yard or more in height, bearing a multitude (4,000 to 5,000) of small white fragrant flowers. The genus is very nearly allied to *Dasylirion,* being, like it, dioecious, but it differs in having the segments of the perianth more decidedly biseriate and unequal, in its more truly panicled inflorescence, and in the remarkable bulbiform base of its Dracaena-like stems. The species are: *B. recurvifolia* (*Pincenectitia tuberculata*), which has the leaves channelled and recurved; *B. stricta* (*P. glauca*), which has plain glaucescent leaves, straighter and more erect; and *B. gracilis* (*P. gracilis*), which has very straight sharp erect leaves, slightly channelled in front. [T. M.]

BEAUFORTIA. A genus of *Myrtaceae,* consisting of shrubs with opposite sessile leaves. The showy flowers have a calyx with a top-shaped tube, and a limb divided into five acute segments; a corolla of five petals; stamens united into four or five parcels placed opposite the petals, the anthers attached by their base; style thread-shaped. The fruit is a capsule. These handsome flowering shrubs are natives of New Holland, and some of them are in cultivation in greenhouses. [M.T.M.]

BEAUMONTIA. Under this name are included some very handsome flowering shrubs of climbing habit belonging to the order *Apocynaceae*. The leaves are opposite. The flowers are white, large, borne in terminal or axillary corymbs, and have a calyx of five broad spreading coloured sepals, and a distended bell-shaped corolla, with a short limb divided into five erect nearly equilateral divisions. The stamens are placed on the top of the tube of the corolla, and alternate with the five lobes of the hypogynous disc; anthers arrow-shaped, adhering to the thick oblong two-cleft stigma. Ovary two-celled. Fruit a follicle, with many hairy seeds. These plants, especially *B. grandiflora*, are remarkable for their handsome flowers; they are natives of the East Indies, and are cultivated as stove plants in this country. [M.T.M.]

BEAVER TREE. *Magnolia glauca.*

BECK-BEAN. *Menyanthes trifoliata.*

BECKEA. A small group of South African dwarf shrubs, of the order *Bruniaceae*, closely related to *Brunia*, differing in having a smooth instead of hairy calyx, included instead of exserted stamens, and in the fruit being crowded by the persistent calyx, the petals and stamens falling away. The species have by some been referred to *Phylica.* [T. M.]

BECKERA. A genus of grasses belonging to the tribe *Paniceae*. The few species which belong to it are all natives of Abyssinia, save one, *B. nubica*, an annual, which grows wild in Nubia. [D. M.]

BECKMANNIA. A genus of grasses belonging to the tribe *Phalarideae*. The inflorescence is in close compressed spikelets; spicules two-flowered; glumes unequal, navicular, slightly stalked at the base, obtuse, or rather obovate; palea two, nearly equal. One species only is described, *Beckmannia erucaeformis*, a native of Japan, and a very elegant grass, which proves hardy in Britain. [D.M.]

BECLARDIA. A synonyme of *Cryptopus*.

BEDAGORA. A Brazilian name for the seeds of *Cassia occidentalis*, which are used as a substitute for coffee.

BEDEGUAR. (Fr.) Sweet-briar Sponge. A sponge-like gall found on the Eglantine and other roses.

BEDFORDIA. A genus of the composite family, which scarcely differs from *Senecio*. Two species are known, both Tasmanian shrubs or small trees, with alternate lanceolate or linear leaves, which are always above and covered underneath with a white tomentum, as are the branches and flowerheads. The latter are axillary and solitary, or few together, and have no strap-shaped florets. *B. salicina*, the Dogwood of Tasmania, has beautifully marked wood, suitable for cabinet-work, and is sometimes to be met with in English gardens. The genus is named in honour of the late Duke of Bedford, a great patron of horticulture and botany. [A. A. B.]

BEDSTRAW. The common name for *Galium*. It is also applied to *Desmodium Apurines*.

BEEBERRY. The Guianian name of *Nectandra Rodiei*.

BEE-DE-CIGOGNE. (Fr.) *Erodium ciconium*, —, DE GRUE. *Geranium Robertianum*.

BEECH. The English name of *Fagus sylvatica*, —, of Australia. *Trifoenia australis*, —, of New South Wales. *Monotoca elliptica*, —, BLUE or WATER. *Carpinus americana*, —, SEA-SIDE. A name used in Jamaica for *Exostema caribbeum*.

BEECH-DROPS. An American name for *Epiphegus*, —, FALSE. An American name for *Hypopitys lanuginosa*.

BEECH-MAST. The fruits of the Beech tree, *Fagus sylvatica*.

BEEFINGS. Apples prepared by being oven-dried and pressed flat.

BEE-FLOWER, or **BEE ORCHIS.** *Ophrys apifera.*

BEEFSUET TREE. *Shepherdia argentea.*

BEEFWOOD. The common name for *Casuarinas*. Also applied, in N. S. Wales, to *Stenocarpus salignus*, and in Queensland to *Banksia compar*.

BEET. The common name of the esculent *Beta vulgaris*, —, CHARD or SICILIAN. *Beta Cicla*.

BEFARIA. A genus of *Ericaceae*, containing about twenty species of small and often glutinous shrubs, natives of the Alpine districts of Peru and Mexico. They have alternate often crowded entire coriaceous leaves, and bear flowers, generally of a purple colour, in terminal racemes of corymbs. The calyx is 6-7 cleft; the corolla consists of as many petals; the double series of stamens have filiform filaments, and smooth two-celled anthers dehiscing by oblique pores at the apex. The ovary has six or seven cells, each containing many ovules; the style filiform, often long and exserted; and the stigma capitate. The capsular fruit dehisces septicidally. The plants of this genus are extremely beautiful; they grow at a great height on the mountains of South America, often at the very extreme of vegetation. The genus is nearly related to *Rhododendron* from which it differs in its petals being all distinct, overlapping each other, and not being united into a tube. [W. C.]

BEGGAR-TICKS. An American name for *Bidens frondosa* and *B. connata*.

BEGONIACEÆ (*Begoniads*.) A natural order of dicotyledonous plants, belonging to the monochlamydeous sub-class of De Candolle. Lindley places the order in his Cucurbital (Cucumber) Alliance. The order contains herbaceous plants or succulent under-shrubs. The leaves have an oblique form, and are placed alternately on the stem, having stipules at their base. The flowers have no petals, but consist of a single perianth, usually pink-coloured, which is placed above the ovary or well-vessel. Some flowers have stamens only, others pistils only; in the former, the perianth has from two to four divisions; in the latter from two to eight. The stamens are numerous and are collected in a head. The stigmas are three, and the fruit is winged with three divisions. Some of the plants produce buds which are easily detached so as to constitute living plants. The plants are common in the East and West Indies, and South America; a few occur in Madagascar and South Africa. They are said to possess bitter and astringent qualities, and some have been used in the cure of fluxes of various kinds. The succulent acid stalks of several species are employed as potherbs like rhubarb. There are two divisions of this order: 1. *Stephanocarpeæ*, having a persistent style; 2. *Gynanocarpeæ*, having a deciduous style. In the order there are, according to Klotzsch, forty-two genera and about one hundred and ninety species; Illustrative genera:—*Begya, Begonia, Diploclinium, Evoltia, Mezeria, Girondia*, and *Protecha*. [J. H. B.]

BEGONIA. The genus whence the natural order *Begoniaceæ* derives its name. It consists of herbaceous plants found in the East and West Indies, Brazil and other parts of South America, and in Mexico. The staminate flowers have four and the pistillate five sepals. Anthers oblong, with an obtuse connective elongated at the apex; filaments short and not united. Style persistent, its branches surrounded by a continuous papillose band, which makes two spiral turns. Placentas stalked and bilamellar, split lengthwise. There are forty-seven known species. The genus is named after Michel Begon, a Frenchman, who promoted the study of botany. The plants receive the name of Elephant's-ear from the form of their leaves. The stalks of some of the species are used in the same way as rhubarb. [J. H. B.]

BEHEN. *Silene Behen, Cucubalus Behen*, now *Silene inflata*, and *Serratula Behen*.

BEHEN BLANC. (Fr.) *Silene inflata*. —ROUGE. *Centranthus ruber*.

BEJARIA. A synonyme of *Befaria*, a genus of ericaceous plants related to *Rhododendron*.

BELANGERA. A genus of Brazilian trees belonging to *Cunoniaceæ*, with opposite stalked leaves having from three or five serrated leaflets; stipules caducous; acumens simple, axillary; calyx six-parted; petals none; stamens numerous, on a perigynous disk; ovary five. [J. T. S.]

BELIS. A synonyme of the coniferous genus *Cunninghamia*.

BELLADONNA. A name sometimes given to a group consisting of certain species of *Amaryllis*, of which *A. Belladonna* is the type. The genus is not generally adopted, and indeed, according to Herbert, its type is also the type of the Linnæan *Amaryllis*. [T. M.]

BELLADONNA. *Atropa Belladonna*, the Deadly Nightshade.

BELLADONE. (Fr.) *Atropa Belladonna*.—D'AUTOMNE. *Amaryllis Belladonna*.—DE ROUEX, or D'ETÉ. *Hippeastrum vittatum*.

BELLE DAME. (Fr.) *Atriplex hortensis*.

BELLE-DE-JOUR. (Fr.) *Convolvulus tricolor.*—DE NUIT. *Mirabilis Jalapa.*—D'ONZE HEURES. *Ornithogalum umbellatum*.

BELLENDENA. A genus of proteaceous plants. The only species, *B. montana*, is a native of Tasmania, where it has been found on Mount Wellington. It bears a short spike of apetalous flowers of four sepals, with four free stamens rather shorter than the sepals, and a filiform style; its fruit is a nut. The plant is of humble growth, eight to ten inches in height, having crowded short-stalked wedge-shaped leaves, obsoletely three nerved with a three-toothed apex. [B. H.]

BELLERIC. The astringent fruit of *Terminalia Bellerica*.

BELL-FLOWER. The common name for *Campanula*; also applied to *Canarina Campanula*.

BELLEVALIA. A genus of *Liliaceæ*, consisting of a few bulbous plants found in the Mediterranean region and in temperate Asia. They have the habit of some of the larger species of grape hyacinth (*Muscari*), but are distinct by having their perianth divided half way down into six folded lobes, expanding to form a prismatic bell. From the true hyacinths they differ by the perianth having an angular and not a circular section. The few leaves are radical, broadly linear; the flowers small, whitish or violet tinged with green. [J. T. S.]

BELLIDIASTRUM. A genus of the composite family, containing but one species, *B. Michelii*, which is found in the Alps of Central and Southern Europe. The plant without close examination might be readily taken for a common daisy (*Bellis*), but can be easily distinguished from that genus by the presence of a copious pappus of rough hairs, the daisy having no pappus at all. [A. A. B.]

BELLIS. The Daisy; the favourite flower of the poets of nature, from Chaucer to Burns; the first which children learn to call by name, and thenceforth love for evermore. The 'bonnie gem' of the latter poet was regarded by 'the firste fynder of our fayre language' with such deep feeling, that the reader cannot but be pleased to have the opportunity of culling its characters from the following extracts:—

'And leaning on my elbowe and my side
The longe dale I shope me for to' abide,
For nothing ellis, and I shall not lie,
But for to lookin upon the Daisie,
That well by reson men it calle maie
The Daisie, or els the eye of the daie,
The emprise, and the floure of flouris alle.

. . . . .

'When that the sunne out of the south gan west,
And that this floure gan close and gon to rest
For darknes of the night, the which she drede,
Home to my house full swiftly I me spede
To gone to rest, and erly for to rise
To seene this floure to sprede as I devise.

He falls asleep, and, in his dream, the Queen of Love appears, 'clad in roiail habite grene,' with a fret of gold on her head:—

'And upon that a white coroune she bere
For all the worlde sight as a Daisie
Icrouned is, with white levis lite,
So were the flowrins of her crowne white,
For of a perle fine orientall
Her white coroune was imakid all,
For which the white coroune above the grene
Ymade her like a Daisie for to sene
Considrid eke her fret of gold above.'

. . . . .

'The flour
Icrownid is with white, as men made ar,
And Mars gaave her a coroun red purir,
Instede of rubies set among the white.'

And as the Queen of Love was crowned with the flowers which the poet did 'love and drede,' so the fairest land he had ever seen was

'a launde of white and grene;
The groundë was grenë, ypoudrid with Daisye.'

The daisy again fills a prominent place in the 'Floure and the Leafe,' where the band of knights and ladies

'founde a tuft that was
All oversprad with flouris in compas,
Whereto they enclined everichoone
With grete reverence, and that full humbly;
And at the last mere this began anon
A lady for to sing right womanly
A bargaret in praising the Daisie,
For (as methought) among her notis swete,
She said, "Si douce est la Margarete!"'

Marguerite, the reader need scarcely be informed, is the French for 'a pearl,' and 'a daisy.'   [C. A. J.]

BELLIUM. The species of this genus are five in number — one of them, *B. cordifolium*, found in Spain, and the others in Southern Italy, and the neighbouring islands. They are nearly related to the common daisy (*Bellis perennis*), and all of them are plants of a very similar appearance, but they differ in having a pappus of six to eight broad scales, torn at the apex, alternating with an equal number of long scabrous bristles.   [A. A. B.]

BELLWORTS. An English name for the group *Campanulaceæ*. The term Bellwort is also used in America for *Uvularia*.

BELLYING. When a round body is more prominent on one side, or at one point, than at another.

BELMONTIA. A genus of *Gentianaceæ*, which may be recognised by the calyx, which is more or less deeply divided, five-parted, with prominent angles or wings; a regular five-cleft corolla, with an almost cylindrical tube; five stamens included within the corolla; a stigma, with two roundish club-shaped lobes; and a two-celled capsule, whose placenta ultimately divides into four pieces. The plants are natives of Southern Africa.   [M. T. M.]

BELOTES. The Spanish and Italian name for the edible nuts or acorns of *Quercus Gramuntia*.

BELOANTHERA. A genus of *Mesembryaceæ*, founded on a herb from Java, which has a procumbent rooting stem, nearly sessile alternate leaves, a glandular hairy calyx, five very small white petals, five stamens, and two deflexed styles, with violet stigmas.   [J. T. S.]

BELOPERONE. A considerable genus of *Acanthaceæ*, from tropical America, containing many species of beautiful shrubs, with large purple or blue flowers, borne on short secund axillary or terminal spikes, and having the bracts frequently brightly-coloured. The calyx is deeply five-cleft, the corolla gaping, the upper lip concave, the lower trifid. The two stamens are inserted in the corolla tube; the anthers are spurred at the base. The two-celled ovary has two ovules in each cell; the stigma is subulate. The lower half of the capsule is unguiculate, without seeds, the upper portion containing four coloured discoid seeds.   [W. C.]

BELOTIA. A genus of the lime-tree family (*Tiliaceæ*), and nearly allied to *Grewia*, but differing from that genus in having a two-celled capsule, with many seeds in each cell; while the fruit of *Grewia* is from four to eight-celled, each cell containing but one seed. Only one species, *B. grewiæfolia*, is known. This is found in Mexico and Cuba. It is a small tree, having the younger branches thickly clothed with dense starry hairs. Its leaves are alternate, stalked, elliptical in form, and acute, the upper surface nearly smooth, and the lower covered with white starry pubescence. The flowers are white, almost the size of those of the lime tree, and arranged in racemes in the upper axils of the leaves. The genus bears the name of Dr. Charles Belot, a distinguished surgeon of Havannah. [A. A. B.]

BELVEDERE. (Fr.) *Kochia scoparia*.

BELVISIACEÆ. A group of three or four tropical species, whose affinities had been long misunderstood, owing to the imperfect specimens which had been obtained. They are now, however, better known, and have been shown to constitute a small family closely allied to *Myrtaceæ*, or even a tribe only of that family, of which they have the inferior several-celled ovary, the numerous stamens turned inwards in the bud, the fruit and the seeds. They differ in their plaited petals, united into a rotate lobed corolla, and in their stamens, united in connective rings, of which the outer ones are converted into barren staminodia. They are all shrubby or arborescent, with alternate leaves and axillary almost sessile flowers. They form two genera: *Napoleona* (unwarrantably altered by freemen, for political reasons, to *Belvisia*, from tropical Africa; and *Asteranthos*, from North Brazil.

BELVISIA. A group of ferns, separated by Mirbel from *Acrostichum*, as formerly understood. The species referred to it have little or no affinity, according to modern systems of classification, but are included severally in *Acrostichum*, *Leptochilus*, *Gymnopteris*, *Hymenolepis*, and *Schizæa*. The characters relied on were the fructification, occupying the whole space between the edge of the frond and the costa, so as to form a line on each side, and covered by a recurved membrane, attached to the edge of the frond; but the proposed species have only an external resemblance, even in these particulars. The name *Belvisia* is also a synonym of *Napoleona*. [T. M.]

BENÇAO DE DEOS. The Brazilian name for the esculent flowers of *Abutilon esculentum*.

BENCOMIA. The species of this genus, which belongs to that section of the rose family called *Sanguisorbeæ*, are both found in Teneriffe. They are low perennial shrubs, with unequally pinnated leaves and pertinately toothed stipules. Their flowers are male and female, on different plants, and arranged in long catkin-like bracted spikes. The tube of the calyx, when mature, has the appearance of a berry, and encloses two to four achenes. The genus is nearly allied to that of the garden burnet (*Poterium*), but differs from it in having dioecious flowers, as well as in the long spikes. *B. caudata* was introduced to English gardens in 1779, but is seldom to be met with. The flowers are greenish, tinged with purple, and very small. The genus bears the name of Bencomi, the last king of Teneriffe. [A. A. B.]

BENGAL ROOT. An old name for the roots of the Yellow Zedoary, *Zingiber Cassumunar*.

BENINCASA. This name is applied to a genus of cucurbitaceous plants, in honor of Count Benincasa, an Italian nobleman. The plants are herbs with hairy musk-scented leaves, climbing stems, and simple tendrils. The flowers are solitary, yellow, polygamous, and monoecious. They have a calyx with short wide tubes, wavy and dentate at the margin; a corolla, with five roundish spreading lobes, wavy at the margin; stamens in three bundles, diverging, rudimentary in the females, the filaments short and wide, and the anthers very irregular in shape; stigma thick and irregular; fruit ovate, cylindrical, downy, the seeds thickened at the margin. The unripe fruits of *B. cerifera*, the White Gourd of India, are universally employed by the natives in their curries. [M. T. M.]

BENJAMIN TREE. *Styrax Benzoin*. The name is also sometimes applied to *Ficus Benjamina*, and to *Benzoin odoriferum*, the *Laurus Benzoin* of Linnæus.

BENJAMIN-BUSH. An American name for *Benzoin odoriferum*.

BENNET, HERB. *Geum urbanum*.

BENOÎTE COMMUNE. (Fr.) *Geum urbanum*.

BENTHAMIA. A genus of epigynous exogenous plants, belonging to the natural order *Cornaceæ*, distinguished by having the calyx segments small, four-toothed; petals four, fleshy and cup-shaped; stamens four; style one; fruits small drupes growing together and forming a large berry resembling the fruit of *Arbutus*, and red when ripe. The leaves are opposite, and bear no inconsiderable resemblance to those of the Cornelian cherry. *B. fragifera* was introduced to English gardens about the year 1825, and is now to be found in some good collections; but being a native of northern India, it is rather tender, and frequently hurt during severe winters, unless protected, especially in the midland and northern counties. [D. M.]

BENTS. A common country name for the dried stalks or culms of various grasses occurring in pastures, especially those of *Agrostis* and *Cynosurus*.

BENZOIN. A genus of *Lauraceæ*, inhabiting the damp shady woods of North America, and found also in Nepal. It has dioecious involucrated flowers; the males with a calyx of six equal segments; six stamens, and nine stamens in three rows, and females smaller than the males, with fifteen to eighteen sterile stamens, amongst which smaller spathulate bodies are dispersed. The ovary is one-celled with a single ovule, and the style short with a two-lobed stigma. There are in the male flowers two or three rows of glands, six to

BANZ] The Treasury of Botany. 135

nine in number, with a uniform compressed head, alternating with the rows of stamens. The fruit is succulent, seated on the permanent six-cleft calyx. *B. odoriferum*, which furnishes an aromatic stimulant tonic bark, is a bush of eight to ten feet high, with oblong or elliptic wedge-shaped leaves, and small yellow flowers on naked kernels appearing before the leaves. The berries are said to have been used in the United States during the American war, as a substitute for allspice; and they yield an aromatic stimulant oil. [T. M.]

BENZOIN. The gum-resinous or balsamic exudation of the Benjamin tree, *Styrax Benzoin*. A milky juice resembling benzoin is also obtained from *Terminalia Benzoin*. —, FALSE. A name given in Bourbon to *Terminalia mauritiana*, which furnishes a resinous gum resembling benzoin.

BEQUETTE. (Fr.) *Delphinium Ajacis*.

BERAH. *Caladium costatum*, which is said to supply an edible root to the natives.

BERBERIDACEÆ (*Berberids*.) A natural order of Exogenous plants, belonging to the Thalamifloral subclass of De Candolle. Lindley includes the order in his Berberal Alliance along with vineworts and fumeworts. The plants of this family are shrubs or herbaceous perennials, with alternate compound leaves, which are often spiny. Sepals three, four, or six in a double row; petals equal to sepals in number, or twice as many; stamens equal in number to petals and opposite to them; anthers having two lobes, each opening by a valve, which rolls up from the bottom to the top. Ovary solitary and one-celled; stigma orbicular. Fruit, either a berry or a capsule, with one, two, or three seeds. These plants are found in the mountainous parts of the Northern Hemisphere, and of South America as far as the Straits of Magelhaens. They are common in the northern provinces of India. None occur in Africa, Australasia, or the South Sea islands. They possess acid, bitter, and astringent qualities. The fruit of some is used as a preserve, and sometimes eaten in a fresh or dried state. Oxalic acid occurs in some of the species. The stem and bark of several barberries are used in dyeing yellow. The astringent substance called Lycium of Dioscorides is supposed to be furnished by the root of various species of barberry; and a preparation of a similar kind is much used as a febrifuge in India. The pinnate-leaved barberries in cultivation form the subgenus *Mahonia*. In the order there are twelve genera and 110 species. See *Berberis*, *Epimedium*, *Leontice*, *Nandina*, *Diphyleia*, and *Jeffersonia*. [J. H. B.]

BERBERIS. The typical genus of the family Berberidaceæ. It consists of shrubs found chiefly in the temperate parts of Europe, Asia, and America, having the leaves simple or pinnate, the primary ones sometimes abortive or changed into simple or variously divided spines, with a tuft of smaller secondary leaves in their axils. The flowers consist of a six to nine-leaved deciduous calyx of coloured sepals, disposed in two or three series; six hypogynous clawed petals opposite the interior row of sepals, and having two glands inside at the base; six stamens opposite the petals, and opening by valves; and a one-celled ovary containing two to eight erect ovules, and surmounted by a peltate stigma on a very short style. This grows into a one-celled ovary containing from one to eight seeds. The pinnate-leaved species are sometimes separated under the name of *Mahonia*; these have the glands at the base of the petals frequently obsolete. [T. M.]

The common Berberry, or Barberry, *B. vulgaris*, forms a deciduous shrub, attaining the height of eight or ten feet. It is found wild in Britain as well as most other parts of Europe, and is also commonly met with in a wild state in North America, and particularly in New England; but it is very doubtful whether the plant is really indigenous to that continent or was carried there by the early settlers, and disseminated through the agency of birds. The distinguished botanists Torrey and Gray affirm that it was introduced; and this opinion is strengthened by the fact of the species not being found in Iceland and Labrador, nor in the eastern parts of Siberia. The Berberry forms a compact bush, composed of numerous very spiny shoots springing from the base, which are covered with a whitish bark, the wood itself being of a fine yellow. The leaves are small, obovate, toothed, and ciliated on the margin, and of a pleasant green; the flowers are yellow, appearing in May. The berries are in pendulous racemes; their colour is generally bright red, but in some varieties they are purple or yellowish-white. Occasionally plants are met with, the berries of which have no seeds; but such do not constitute a permanent variety, for stoneless berberries are only found on old plants; and it has been proved that young suckers taken from them and planted in fresh soil, fruit with perfect seeds. The fruit is less sold for use in its natural state; but it makes excellent refreshing preserves, for which Rouen is particularly celebrated. It is likewise candied; and when sewn is sometimes pickled in vinegar. The inner bark affords a bright yellow dye. The roots, which are of deep yellow colour throughout, boiled in an alkaline ley, yield a yellow dye, used in Poland for colouring leather. [R. T.]

The bark of the Berberry, of which a decoction was made, was formerly much celebrated as a remedy in jaundice, but it has long since been discarded from modern practice, as its claims as a medicinal plant only rested upon the doctrine of similitudes, which assumed that nature when she made a plant, impressed upon it some sign to point out its curative properties to those who properly sought such knowledge. In this way it was supposed that as the patient's skin in jaundice is yellow, so the yellow bark of the berberry indicated it as a remedy for this diseased condition.

Another popular notion with respect to this shrub is, that it is the cause of blight or rust in corn. This has arisen from the circumstance that the berberry is itself frequently attacked by a species of epiphyte—the *Æcidium Berberidis*, in which the leaves appear to be covered with spots of a brightish red colour, whilst wheat is subject to another epiphyte, the *Uredo rubigo* or rust. There has, however, been no connection traced between these two, and there can be no doubt that the peculiarity of colour is at the bottom of both the popular errors now described; at all events, with regard to the last, we can point to fields and districts where rust is common on wheat, and yet there is no berberry near, while in other spots close under a berberry hedge, this disease of wheat has scarcely been heard of. [J. B.]

BERBERRY. *Berberis vulgaris*, sometimes called Barberry. —, OPTHALMIC. *Berberis Lycium*.

BERCE (Fr.) *Heracleum Sphondylium*.

BERCHEMIA. Twining, or erect, deciduous shrubs, belonging to the order *Rhamnaceæ*. *B. volubilis* is a native of Carolina and Virginia, in deep swamps near the coast. According to Pursh it ascends the highest trees of *Taxodium distichum*, in the Dismal Swamp, near Suffolk, in Virginia; and it is known there by the name of Suppo Jack. The stems twine round one another, or any object which they may be near. The flowers are small and of a greenish-yellow colour; and in America they are succeeded by oblong violet-coloured berries. It will grow in any common soil, and is well adapted for bowers or trellis-work. It rarely, however, exceeds the height of eight or ten feet in this country, owing probably to the neglect of planting it in wet peaty soil. It was introduced in 1714. The other species are not much cultivated. [C. A. J.]

BERGAMOT. *Mentha citrata* or *odorata*. — WILD. An American name for *Monarda fistulosa*.

BERGAMOTTE. (Fr.) The Lime, *Citrus Limetta*.

BERGERA, one of the genera of *Aurantiaceæ*, is so named in honour of a Danish botanist. The genus consists of a few species of small trees with pinnate leaves, small white flowers in terminal panicles, with a five-cleft calyx, five spreading petals, ten stamens with ovate anthers, and filaments flattened at the base. The fruit is one-celled and one-seeded. *B. Kœnigi* is known in India as the Curry-leaf tree, as the natives flavour their curries with its aromatic fragrant leaves. The leaves, root, and bark are likewise used medicinally. The wood is hard and durable, and from the seeds a clear transparent oil, called Simbolee oil, is extracted. [M. T. M.]

BERGIA. An unimportant family of herbaceous plants belonging to the order *Elatinaceæ*. All the species are natives of the East Indies, Java, or the Cape of Good Hope, where they grow in moist places, such as rice fields, which are irrigated the greater part of the year. Dr. Wight says, that in India the little *B. ammannioides* bears a Tamul name equivalent to Water-fire, which, as Lindley observes, seems a curious coincidence with the word Water-pepper, given in English to *Elatine*, and seems to indicate a popular belief in these plants possessing some acridity. [C. A. J.]

BERGSMIA. A genus of the *Flacourtia* family, containing but one species, *B. javanica*, which is a native of Java, and is described as being a large tree with alternate or opposite stalked leaves, which are entire, from three to five inches long, and lanceolate in form, their upper surface smooth, and of a whitish colour beneath. The flowers are arranged in axillary racemes and have a three-parted calyx, five petals, and four or five stamens, the stalks of which are united. The fruit is not known. The genus bears the name of C. A. Bergsma, a professor of botany in Holland. [A. A. B.]

BERLANDIERA. A genus of the composite family, nearly related to *Silphium*, but differing in the wingless achenes of the ray florets, which are arranged in a single series, and are adherent to the large interior involucral scales. The five known species, distributed over the S. W. states, Texas, and Mexico, are pretty perennial herbs one to two feet high, the stems slightly branching above and terminating in solitary stalked yellow-rayed flower-heads an inch or more across, while the alternate heart-shaped ovate-oblong, or in one case pinnatifid, leaves have notched margins, and are clothed underneath with a white velvety down. The involucral scales are foliaceous (like those of the *Dahlia*) and in three series; the strap-shaped ray-florets have a pistil only; the tubular disc ones are sterile and enveloped by the dilated chaffy scales of the receptacle; and the flattened obovate achenes have a pappus of two short awns. The genus bears the name of M. Berlandier, an American botanist, who collected largely in Texas. [A. A. B.]

BERLE. (Fr.) *Sium*.

BERMUDIENNE A' PETITES FLEURS. (Fr.) *Sisyrinchium Bermudianum*.

BERNHARDIA. A synonyme of *Psilotum* and *Tmesipteris*.

BERRIED. The same as Baccate.

BERRY. See Bacca.

BERRYA. A genus of the lime tree family (*Tiliaceæ*). But one species, *B. Ammonilla*, is known; it is a tree with alternate stalked heart-shaped leaves, which are smooth and have from five to seven nerves radiating from the base. The flowers are white and very numerous,

...inal or axillary panicles; ...ved, splitting irregularly ...divisions, and about half ...which are five in num-...ng in shape. The fruit is ...le ornamented with six ...are covered with silky ...one to four in each cell, ...bout rigid hairs which ...in and produce as much ...those of the Cowitch ...The tree is a native of the Philippine ...ds and Ceylon. In the ...nes one of the largest ...ber trees for building ...considered the best wood ...king oil casks. Being ...is employed in the con-...Masoola boats of Madras. ...larger quantities under ...make wood. The native ...looncilla. The genus is ...of the late Dr. Andrew ...danist. [A. A. B.]

BERTIA. A genus of European and ... cruciferous plants, allied ... which it ought to be con-...n, as it merely differs ...the sepals bulging at ...bifid, and the valves of ...without a nerve; in ... a nerve. From Alyssum, ...frequently associated, it ...bifid petals and swollen ...ually more elongated. ... common European plant, has ...ves, and its stems and ...with close white striate ...

[J. T. S.]

BERTHOLLETIA. A genus of the coun-... in honour of Baron ... for with Mr. P. B. Webb ...k on the flora, and ...ies. But one species ... a small shrub with ... rigid entire leaves, ...tle point. The flower ...in dense corymbs at ...nches, and the florets ... those of the ray-...the disk having both ... The pappus is white, ...ets of the plant are ...pubescence, which gives ... flowers. The plant is ... ds districts of India, ...le, as well as in Sene-... [A. A. B.]

BERTHOLLETIA. A genus of Lecythi-... only one species, B. ...lds the Brazil nuts ...is known. This tree ...ans, Youchebs, and ... forests on the banks ... the Negro, and like-...adas on the Orinoco, ...ll it "Juvia." The tree ...most majestic in the South ...attaining a height of ... a smooth cylindrical

trunk about three or four feet in diameter, and seldom having any branches till near the top. It has bright green leaves about two feet long and six inches wide, entire or undivided, and placed alternately upon the branches. The flowers have a two-parted deciduous calyx, six unequal cream-coloured petals, and numerous stamens united into a broad hood-shaped mass, three at the base being fertile and the upper ones sterile. The fruit is nearly round, and about six inches in diameter, having an extremely hard shell about half an inch thick, and containing from eighteen to twenty-four triangular wrinkled seeds, which are so beautifully packed within the shell that when once disturbed it is impossible to replace them. When these fruits are ripe they fall from the tree and are collected into heaps by troops of Indians called cushnalcros, who visit the forests at the proper season expressly for this purpose; they are then split open with an axe and the seeds (which are what we call Brazil nuts) taken out and packed in baskets for transportation to Para in the native canoes. Brazil nuts form a considerable article of export from the port of Para (whence they are sometimes called

Bertholletia excelsa.

Para nuts), about 50,000 bushels being annually sent to this country alone. Besides their use as an article of dessert, a bland oil, used by watchmakers and artists, is obtained from them by pressure. And at Para the fibrous bark of the tree is used for caulking ships, as a substitute for oakum. [A. S.]

BERTOLONIA. A genus of Melastomaceæ, containing dwarf or procumbent herbs, natives of the dense forests of Brazil. Leaves opposite, stalked, heart-shaped, ciliated, with five or more ribs; flowers nearly sessile in cymes; petals five, white rose-coloured or purplish; stamens ten; ovary free, three celled; capsule three-winged. B. maculata is a pretty little creeping hot-house plant with spotted leaves, purple beneath. [J. T. S.]

BERTYA. A genus of the Spurgewort family (*Euphorbiaceae*) composed of a number of small resinous shrubs much like the rosemary in appearance and habit. Their leaves are alternate, crowded on the branches, and linear in form, their margins entire and curved backwards. Their flowers are solitary in the axils of the leaves, those on the lower part of the stem male, the upper ones female; they are small, have no corollas, and without being looked for would be easily passed by. The leaves of most of them are covered with minute starry hairs of a white colour. Five species are known, all natives of Eastern Australia and Tasmania. The genus is named in honour of Count L. de Lambertye, a patron of horticulture in France, and the name is shortened to Bertya because of an already existing genus called *Lambertia*. [A. A. B.]

BERULA. Under this name is included a small section of the Linnæan genus *Sium*, *B. angustifolia*, particularly known as the Narrow-leaved Water-parsnip, is a native of the whole of Europe and a great part of Asia, growing in ditches and rivulets. The roots are fibrous, and send out stolons from the crown; the leaves are smooth, pinnate, and unequally lobed and serrated; the flowers, which are small and white, are either terminal or grow in umbels opposite the leaves, and are all stalked. [C. A. J.]

BESCHORNERIA. A genus of agave-like amaryllidaceous plants, related to *Littæa* and *Pourcroya*, from which they differ in having tubulose flowers, the former also differing in its exserted stamens, and the latter in its habit. The flowers consist of a deeply six-parted perianth, the segments of which are linear-spathulate, tubulously connivent, sometimes slightly spreading at the point. There are six stamens which are about as long as the perianth; and an inferior and somewhat club-shaped ovary, terminated by a long slender style and small stigma. *B. tubiflora* is a stemless plant with a tuft of linear sword-shaped acuminate leaves, and an erect scape supporting a many-flowered raceme of purplish-green flowers. *B. yuccoides*, another species of considerable beauty, has also a radical tuft of thickish lanceolate pointed leaves, a foot and a half long, and a tall slender curved scape three to four feet high, the upper half of which forms a drooping panicle of slender branches of the same rich coral-red, springing from deep rose-coloured bracts, and supporting a short pendent raceme of bright green flowers. It is indeed a most striking plant, the coralred scape and panicle, the graceful slender drooping branches, and the racemes of large pendent greenflowers, which in shape are not much unlike those of some long-flowered *Fuchsia*, but of a dark yellow green tinged with red, rendering it very ornamental, the more so as it continues a long time in blossom. The species are natives of Mexico. [T. M.]

BESENNA. The Abyssinian name of *Albizia anthelmintica*, and at one time adopted as the scientific name of the plant, which was then imperfectly known.

BESHAN. The Balm of Mecca, *Balsamodendron Opobalsamum*.

BESLERIA. A genus of erect branching undershrubs, abundant in the forests of tropical America, and belonging to the section of *Gesneraceæ*, which have albuminous seeds. They have opposite petiolate and fleshy leaves, with prominent nerves and veins, and axillary peduncles, with one or many flowers. The calyx is free, five-cleft, and reddish coloured; the corolla is campanulate, and sub-equally five-cleft. The four didynamous stamens, along with the rudiment of the fifth, are inserted in the tube of the corolla; the anthers are two-celled. The one-celled ovary is free, surrounded at its base by a fleshy ring without glands, and has two two-lobed parietal placentæ, to which are attached numerous anatropal ovules; the style is simple and bifid; the fruit is a berry filled with the fleshy placentæ, and numerous small obovate seeds, with very short cotyledons. [W. C.]

BESSERA. A genus of liliaceous bulbs, found in Mexico, the species of which have narrow linear leaves, and umbel-bearing flower scapes. The perianth is bell-shaped, six-parted, furnished with six stamens, which are connate below into a cylindrical coronet, and having a sessile ovary, containing numerous ovules, lying in two rows in the cells, the style being terminated by a capitate-depressed obscurely three-lobed fringed stigma. *B. elegans*, one of the best known of the few species, produces a pair of radical leaves, which are erect for two-thirds of their length, and then become pendulous; they are one to two feet long, semicylindrical, smooth and hollow. The scape, also smooth and fistular, is solitary, erect, taller than the leaves, and bears a terminal umbel of from three to sixteen drooping flowers of an orange-red colour, and having a turbinately bell-shaped tube, a moderately spreading limb of six nearly equal oblong-obtuse segments, and a cluster of six green stamens on long red filaments projected considerably beyond the limb, and united at the base for half their length into a six-ribbed tube. [T. M.]

BETA. A genus belonging to the natural order *Chenopodiaceæ*, a group comprising various genera of coarse weedy-looking plants, among which *Beta* is one of the most remarkable, on account of its roots and leaves being valuable both for culinary and agricultural purposes. It is a genus with hermaphrodite flowers, in which the five-parted urceolate perigone becomes hardened at the base, the segments merely shrivelling up. There are five stamens inserted in a fleshy ring opposite the limb segments, and the depressed one-celled ovary becomes a one-seeded utricle.

The Common Beet, *Beta vulgaris*, is a native of the South of Europe, and although cultivated by the ancient Romans, and much esteemed by epicures, it was not introduced into this country until 1656. It is a hardy biennial, with large erect succulent leaves, generally of a deep reddish-purple colour; but that for which it is most valued is its fleshy roots, which vary in form from that of a carrot to a flat round turnip. The long-rooted sorts are preferred; they are usually about a foot or more in length, and from two to four inches in diameter at the top, from which they taper to a blunt point, and are prized in proportion to their being wholly of a deep blood-red colour when cut. In France and Germany, beet root is far more extensively used than in England; and, when properly dressed, it is generally considered to be a wholesome and nutritious vegetable. Boiled and sliced, it is eaten cold, either by itself or mixed in salads. It is also excellent with vinegar as a pickle, and is capable of being made into a conserve. There are many varieties in cultivation, which do not differ materially from one another, except in the colour of their roots and leaves.

A variety of Beet is grown on the continent, under the name *Betterave à Sucre*, from which sugar nearly equal to that from the cane is extracted, as well as a powerful spirit.

The White, or Sicilian Beet, *Beta Cicla*, as its name imports, is a native of Sicily, near the sea coast, as well as the shores of Spain and Portugal, from whence it was introduced in 1570. It is a biennial, and is grown solely for its leaves, which are either put into soups, or dressed like spinach. In France they are often mixed with sorrel, to lessen its acidity. The ancient Greeks used to eat the leaves of beet in preference to lettuce, and blanched them by laying a tile over the plant, as some gardeners do at the present day to blanch endive.

The large white, or Swiss Chard Beet, *Beta Cicla* var., is a very distinct variety, remarkable for the thick midribs and stalks of its large upright leaves. It is the *Poireé à Carde* of the French, with whom it is a favourite vegetable, when stewed and served up in the same manner as sea kale, or asparagus; but unless it is properly dressed, it has a peculiar earthy taste, and on this account it is not generally relished in this country.

The Sea Beet, *Beta maritima*, a perennial, which grows wild on the sea coast in various parts of Britain, is occasionally used as spinach or greens in situations where it is plentiful. [W. B. B.]

The Sea Beet is chiefly remarkable for the chances which it undergoes in cultivation, as from it have been produced the different varieties of Garden Beet, and Mangold Wurzel. If we examine the wild plant, we find some specimens in which the roots and foliage are highly tinctured with a purple colour, whilst others incline to a yellowish-green hue. These two varieties are the initiatives of the red, and the white beet, and also of the red, orange, and white mangold wurzel. With respect to those forms which are cultivated for their roots, the size to which they have been brought is remarkable; but it should be observed, with regard to the white beet, which is cultivated for the midrib of the leaves, that the roots are usually much forked, and, indeed, are not greatly better in form than those of the woody wild examples, which, however, by being cultivated with a view to the root alone, attain a compact shape and large size. [J. B.]

BETEL. The fruit of *Areca Catechu*.

BETOINE, (Fr.) *Stachys Betonica*. —, AQUATIQUE. *Scrophularia aquatica*.

BETONICA. An old Linnæan name for various plants, now referred to *Stachys*, *B. officinalis*, or *Stachys Betonica*, is the Wood Betony of the herbals. [T. M.]

BETONY. The English name for the species sometimes separated from *Stachys* under the name *Betonica*; also *Trocrium betonicum*. —, WATER. *Scrophularia aquatica*. —, WOOD. *Stachys Betonica*.

BETTE. (Fr.) *Beta maritima*.

BETTE-RAVE. (Fr.) *Beta vulgaris*.

BETULACEÆ. (*Birchworts*.) A natural order of Exogenous plants, belonging to the monochlamydeous sub-class of De Candolle, and to the amental or catkin-bearing alliance of Lindley. They are trees or shrubs, having alternate, simple, stipuled leaves, often with the primary veins running straight from the midrib to the margin. The flowers are in catkins, some having stamens only, others pistil only; and they have scales in place of a perianth or floral envelope. In the alder, however, there is a four-leaved perianth. The stamens are opposite the scales. The ovary is two-celled, with a single pendulous ovule in each cell; stigmas two. The fruit is dry, does not open, is one-celled and one-seeded. The plants are found in the woods of Europe, Northern Asia, the Himalayas, and North America; they also inhabit the mountains of Peru and Columbia, and the Antarctic regions. They are usually timber trees, with deciduous leaves. Their bark is used as an astringent for gargles, and for dyeing and tanning; it also possesses tonic qualities, and is occasionally employed as a substitute for paper, and for making boats. *Betula alba* is the common birch. Its sap contains sugar, and, by fermentation, yields a kind of wine. The empyreumatic oil of the birch has been recommended in various affections; it is used in the preparation of Russian leather, and gives to it a peculiar odour. The alder, *Alnus glutinosa*, grows in moist places; the wood resists well the action of water, and has been used for the piles of bridges. The Rialto of Venice is built on alder piles, and so are many houses in Amsterdam. Sabots are made of the wood. There are two genera, *Betula* and *Alnus*, and upwards of sixty species. [J. H. B.]

BETULA. The Birch. Trees or shrubs inhabiting high latitudes in the northern hemisphere, or, when found in temperate regions, growing principally in rocky mountainous situations. They are characterised by slender, often drooping branches, which are covered by a smooth durable bark; by small leaves possessing little succulency, and in their mature astringent and aromatic; and by having their fructification at the same time, with the leaves in catkins of two kinds, barren and fertile, both on the same tree.

The Common Birch, *B. alba*, pronounced by the poet Coleridge—

'most beautiful
Of forest trees, the Lady of the Woods'—

is remarkable for its lightness, grace, and elegance, nor less so for its hardiness; standing in no need of protection from other trees in any stage of its growth, and living on the bleak mountain side and other exposed situations from which the sturdy oak shrinks with dismay. It is a native of the colder regions of Europe and Asia. Throughout the whole of the Russian empire it is more common than any other tree, being found in every wood and grove from the Baltic sea to the Eastern ocean, and frequently occupying the forest to the exclusion of all other arboreous plants. It grows from Mount Etna to Iceland: in the warmer countries being found at a high elevation among the mountains, and varying in character according to the temperature. In Italy it forms little woods at an elevation of 6,000 feet. On some of the highlands of Scotland it is found at the height of 3,500 feet. In Greenland it is the only tree, but diminishes in size according to the decreased temperature to which it is exposed. It is a tree of rapid growth, especially when young; and as it is little affected by exposure, it forms an excellent nurse for other trees. The soil which it prefers is turf over sand, and in such situations it attains maturity in about fifty years; but it seldom exceeds fifty feet in height, with a trunk from twelve to eighteen inches in diameter. The bark possesses the singular property of being more durable than the wood which it encircles. Of this the peasants of Sweden and Lapland take advantage, and, shingle-like it tiles, cover their houses with it. The wood is white shaded with red, and, if grown in a very cold climate, it lasts a long while. The highlanders of Scotland employ it for all purposes for which wood is available; the branches are used as fuel in the distillation of whisky; the spray for thatching and for smoking hams and herrings; the bark for tanning leather; and the leaves for bedding. In Russia, an oil is extracted from the bark, which is used in the preparation of Russian leather, to which it not only imparts a fragrant odour, but renders it durable, preventing it from becoming mouldy, and repelling insects. The variety known as *B. pubescens* differs from the common species only in having the branches pendulous, smoother, and more slender.

B. nana is found in Scotland, and in all the northern countries of continental Europe and America. It is a low wiry shrub, rarely exceeding three feet in height, with numerous round notched leaves, which are beautifully veined.

The Paper Birch, *B. papyracea*, so called from the brilliant white colour of the bark of young trees, is an American species no less valuable than the common birch, and attains a far larger size. By the Indians and French Canadians the durability of the bark is turned to good account. The Canadians select a tree with a large and smooth trunk; in the spring two circular incisions are made quite through the bark several feet from each other. Two vertical incisions are then made on opposite sides of the tree; after which a wooden wedge is introduced, by which the bark is easily detached. These plates are usually ten or twelve feet long, and two feet nine inches broad. To form a canoe, they are stitched together with the fibrous roots of the Canadian spruce. The seams are coated with resin. Great use is made of these in long journeys into the interior of the country; they are very light, and are easily carried on the shoulders from one lake or river to another. A canoe calculated for four persons weighs from forty to fifty pounds. Some are made to carry fifteen passengers. Numerous other species of birch are known to botanists, all of which approach more or less in character those described above.—French *Bouleau*, German *Birke*.

Plate 14, which is a view in Kamtschatka, represents a birch forest as seen in the distance. [C. A. J.]

BEURRE. A general name applied to a class of dessert pears, which have their flesh of what is called a buttery texture, as indeed the name itself indicates.

BEUREE. (Fr.) *Hesperis matronalis*.

BEUREE DE SPERGULE. (Fr.) *Spergula arvensis*.

BEYCHE SEED or NUT. A Siamese name for *Strychnos nux-vomica*.

BEYRICHIA. A genus of *Scrophulariaceæ*, containing a few species of herbaceous plants from Brazil and Guiana. They have opposite ovate leaves, and axillary flowers, on very short pedicels, either lax or in dense leafy spikes. The calyx is five-parted, the upper segment being ovate, and the four lower ones narrow. The upper lip of the corolla is emarginate, the lower is slightly trilobed, the palate is prominent. There are four stamens, two of which are frequently sterile. The capsule dehisces septicidally or loculicidally, and contains numerous small seeds. The genus is divided into two sections: *Achetaria*, having two sterile stamens and septicidal dehiscence of the capsule; and *Deppostreum* with the four stamens fertile and dehiscent loculicidally. In habit the species of this genus have very much the appearance of *Acanthaceae*. [W. C.]

BHABHUR or **BHABHUR.** An Indian name for the silky leaves of *Eriophorum cannabinum*, used for making cordage.

BHADLEE. *Panicum pilosum*, a breadcorn cultivated in India.

BHANG. An intoxicating drug obtained in the East from the Hemp, *Cannabis sativa*.

BHEL. The Indian name of the fruit of *Ægle Marmelos*.

BI. In compound words=twice.

BIACUMINATUS. Having two diverging points.

BIARTICULATUS. Two-jointed.

BIARUM. One of the numerous new genera of *Araceæ* proposed by Schott on comparatively slight grounds. This genus differs from *Arum* in its spathe being tubular at the base, with the limb spreading. The female flowers, moreover, have a distinct style, and the fruit contains only one ovule. The plants, which are much like the species of *Arum*, are natives of the south of Europe. [M. T. M.]

BIAURITE. Having two little ears. See also *Auriculate*.

BIBACIER. (Fr.) *Eriobotrya japonica*.

BIBIRI. The Greenheart Tree, or Beebeeree of Guiana, *Nectandra Rodiei*.

BICALLOSE. Having two callosities, as the lip of many orchids.

BICARINATE. Having two elevated ribs or keels on the under side, as in the pales of many grasses.

BICEPS. A term sometimes applied to the keel of a papilionaceous corolla when the ungues of the two petals of which it is composed, are distinct.

BICONJUGATE. When each of two secondary petioles bears a pair of leaflets.

BICONJUGATO-PINNATE. When each of two secondary petioles is pinnated.

BICORNELLA. A genus of little known Madagascar orchids nearly related to *Habenaria*. They have long almost leafless stems, terminated by a few orchid-like flowers.

BICORNES. A name originally given by Linnæus to a group of genera, corresponding nearly to the heath family (*Ericaceæ*) taken in its most extended sense. It has been lately revived by Klotzsch and others for the designation of a class to consist of *Vacciniaceæ*, *Ericaceæ*, *Epacridaceæ*, and the smaller families or tribes included in or closely allied to them.

BICORNIS, BICORNUTE. Having two horn-like processes.

BICONDONA. A genus of *Apocynaceæ* remarkable, as the name implies, for having in the throat of the salver-shaped corolla, a double row of scales, each row consisting of ten, and the upper series being placed in pairs, before each lobe of the limb of the corolla, the lower row alternate with them. The five filaments are very short; the stigma is two-parted; and the fruit is like a berry. The only species is a shrub with erect branches, thick leaves, and flowers in axillary and terminal cymes, and is a native of New Caledonia. [M. T. M.]

BICRURAL. Having two legs or narrow elongations, as the lip of the man-orchid.

BICUIBA. *Myristica Bicuiba*.

BIDENS. A somewhat extensive genus of herbaceous compound flowers, growing both in the old and new world, well marked by the pericarp having, instead of a pappus, from two to five rigid awns which are rough with minute deflexed points. The British species of Bur-marigold, *B. cernua* and *B. tripartita*, are not unfrequent on the borders of ponds and streams. They grow from one to two feet high, and may be distinguished while in flower by their button-like dingy-yellow flowers, which are surrounded at the base by an involucre of long bracts. The former has its flowers drooping; the latter has tripartite leaves. Neither of them is remarkable except for the tenacity with which the fruits cohere by their serrated awns to any penetrable substance to which they may happen to attach themselves. The foreign species possess little interest. French *Bident*; German *Zweyzahn*. [C. A. J.]

BIDENTATE. Having two teeth.

BIDIGITATO-PINNATE. Same as Biconjugato-pinnate.

BIDUOUS. Lasting two days only.

BIDWILLIA. A genus of Australian and Peruvian liliaceous bulbs, with paniculate or racemose white flowers, only differing from the asphodels (*Asphodelus*) by having the filaments of the stamens thickened upwards. The leaves are linear, more or less glaucous; the roots fasciculate with knobbed ends. *B. glaucescens* is a native of the table land called New Zealand, in Australia. [J. T. S.]

BIEBERSTEINIA. This name commemorates the botanical services of Marschall von Bieberstein, a Russian naturalist. It is applied to a genus of *Rutaceæ*, or, according to some authors, of *Zygophyllaceæ*. The species are herbaceous plants, with pinnately divided leaves; flowers in terminal racemes, with five sepals and five petals; the stamens with filaments dilated at the base, between which and opposite to the petals are placed five round glands; ovaries five, distinct at the base and at the summit, but cohering in the middle, the five thread-shaped styles proceeding from the inner side of the ovaries near their base, and uniting at the top into a single cup-shaped five-lobed stigma; fruits membranous, one-seeded. The species are natives of Persia, and the Altai and Himalayan mountains. [M. T. M.]

BIENERIA. A genus of terrestrial orchids, proposed by the younger Reichenbach, but hardly distinct from *Chloræa*.

BIENNIAL. Requiring two years to form its flowers and fruit, and then dying: growing one year, and flowering, fruiting and dying the next.

BIFARIOUS, BIFARIAM. Arranged in two rows. This term is frequently applied to flowers and to ovules.

BIFARIOUSLY IMBRICATED. Overlapping in two rows.

BIFERUS. Double bearing; producing flowers or fruit twice in the same season.

BIFIDUS. Split half way down into two parts.

BIFOLIOLATE. Having two leaflets only to a leaf.

BIFOLLICULUS. A double follicle.

BIFLORUS. Bearing two flowers on the same footstalk. Also flowering twice in the same year.

BIFORATE. Having two pores or apertures.

BIFORINES. Oblong cells, with an aperture at each end, through which raphides are expelled.

BIFRENARIA. A name given to those Maxillaria-like plants which have two frena or caudicles to their pollen masses instead of four. The species are all from the tropics of America.

BIFRONS. Growing on both surfaces of a leaf. Also appearing equally like two different things. A term seldom used.

BIFURCATE. Twice-forked; having two pairs of diverging horn-like arms.

BIG. The common Bere or four-rowed Barley, *Hordeum vulgare*.

BIGAMEA. A Ceylon plant generally considered as belonging to *Combretaceæ*, but referred by Planchon to a separate order, *Ancistrocladeæ*; from *Ancistrocladus*, which is equivalent to *Bigamea*. It is a shrubby-stemmed climber, whose main stem is short and terminated by a tuft of wedge-shaped leaves; from this fascicle springs a branch, which towards its apex bears short alternate patent branches with terminal tufts of lanceolate leaves. The flowers have five petals, five stamens, an inferior ovary, a pyramidal style with three stigmas; fruit a pear-shaped drupe covered by the five-parted calyx limb. [J. T. S.]

BIGANDELLE. (Fr.) *Cerasus vulgaris*.

BIGARADE. The bitter or Seville Orange, *Citrus vulgaris*.

BIGARADIER. (Fr.) *Citrus vulgaris*.

BIGARREAUTIER. (Fr.) *Cerasus avium*.

BIGEMINATE. Same as biconjugate.

BIGEMINOUS. In two pairs; as the placentæ of many plants.

BIGENERS. Mule plants obtained by crossing species of different genera. This kind of hybridism has been said to be impossible; Kœlreuter in particular adduced examples of failure in the attempt; but modern experiments seem to show the possibility of such a union.

BIGLUMIS. Consisting of two of the scales called, among grasses, glumes.

BIGNONIACEÆ (*Bignoniads*; the Trumpet-flower family.) A natural order of dicotyledonous or exogenous plants belonging to the subclass Corolliflora of De Candolle, and to the Bignonial Alliance of Lindley, which includes the Sesamworts, acanthads, and gesnerworts. The order contains trees or twining or climbing shrubby plants, with usually opposite compound leaves, and showy often trumpet-shaped flowers. Calyx divided or entire, sometimes in the form of a spathe; corolla usually irregular, four to five lobed, and with a swollen portion below its mouth; stamens five, unequal, one generally, two occasionally, abortive. Ovary having two cavities, surrounded by an expansion at its base; ovules attached to the central part of the ovary. Fruit a two-valved often pod-like capsule, divided by a spurious expansion of the placenta; seeds generally numerous and winged; embryo without albumen, and having broad leafy cotyledons. The plants are found in the tropical regions of both hemispheres, but predominate in the eastern. They extend in America from Pennsylvania in the north, to Chili in the south. They do not occur wild in Europe. The plants produce abundance of showy finely-coloured flowers. Some yield dyes; others supply timber. Among them are medicinal agents used in chest affections, and for worms. There are 46 genera and 462 species described. Illustrative genera: *Bignonia*, *Colosanthes*, *Catalpa*, *Eccremocarpus*, *Jacaranda*, *Spathodea*, *Tecoma*. [J. H. B.]

BIGNONIA. The order *Bignoniaceæ* takes its name from this genus, which was itself so called in honour of the Abbé Bignon, librarian to Louis XIV. The species of Bignonia are remarkable for the beauty of their flowers, and hence many are cultivated in this country. *B. capreolata*, one of the handsomest species, is a native of North America, but capable of being grown in warm places in this country as an ornamental climbing plant. The other species are for the most part natives of the warmer regions of the western hemisphere. The botanical characteristics of the genus are a bell-shaped calyx, slightly wavy at the margin; an irregular bell-shaped corolla; five stamens, two long and two short of which are fertile, and one sterile; stigma divided into two lamellæ;

capsule like a long pod, with the partition between its two compartments parallel with the valves or walls of the pod; the seeds arranged in two rows and provided with a membranous wing. The wood of some of the climbing species is arranged in four divisions, so as to present a cross-like appearance when cut. The leaves are pinnate or sometimes consist of only two opposite leaflets. The flowers are borne in panicles, and are of various colours, but always handsome-looking. The fruit of most species is either unknown, or but superficially described. [M. T. M.]

The *Bignonias* are ascendant tendrilled plants, frequently climbing to the tops of the highest trees, their flexible stems, twisted like ropes, sometimes passing from tree to tree, descending to the ground at intervals, taking fresh root, and again ascending other trees; in some of the Brazilian forests they are so numerous as to render them almost impassable.

B. alliacea, the Garlic shrub, or *Liane à l'ail* of the French, is a native of Guiana and the West Indies, and is so called in consequence of the powerful odour of garlic emitted by its bruised leaves and branches. It is a square-stemmed climber, with leaves composed of elliptical leathery leaflets, joined together in pairs; its flowers are large and white.

B. Kerere is a climbing shrub with smooth angular stems. Its leaflets are in pairs or threes upon a single stalk, of an elliptical form, and rather hairy upon the under surface; the flowers are about two inches long, downy, and of a yellow colour. The natives of French Guiana, where this plant is indigenous, use the tough flexible stems as a substitute for ropes; and from strips of them they weave various kinds of baskets, and broad-brimmed hats which protect them from both sun and rain.

B. (?) Chica is the most useful species of the genus. It is a native of Venezuela, New Grenada, and Guiana, and has long climbing stems, which reach to the tops of the trees, where they divide into numerous small branches which support themselves by means of their tendrils. Its leaves consist of eight leaflets arranged in pairs conjugate, each pair having a tendril betwixt them, and possessing a separate stalk branching from the central leaf-stalk; the leaflets are oval. The funnel-shaped flowers are arranged in loose drooping panicles, and are of a violet colour; they produce a long flattened pod-like fruit, containing numerous winged seeds. A red pigment called Chica on the Orinoco, and Carajuru on the Rio Negro, is obtained by macerating the leaves of this plant in water, and is greatly used by the natives for painting their bodies, so much so that M. Humbolt, in speaking of the natives of the Orinoco, says:—'To form a just idea of the extravagance of the decoration of these naked Indians, I must observe, that a man of large stature gains with difficulty enough by the labour of a fortnight, to procure in exchange the chica necessary to paint himself red. Thus we say, in temperate climates, of a poor man, "he has not enough to clothe himself;" you hear the Indians of the Orinoco say, "that a man is so poor, that he has not enough to paint half his body." See ARRABIDÆA. [A. S.]

BIHAI. *Heliconia Bihai*.

BIJUGUS. A pinnate leaf with two pairs of leaflets.

BIKH or BIKHMA. The poisonous root of *Aconitum ferox*.

BILABIATE. A corolla divided into two separate parts or lips, placed one over the other, as in sage, bugle, and similar plants.

BILAMELLATE. Consisting of two plates, as many placentæ, stigmas, &c., or bearing two vertical plates, as the lip of some orchids.

BILBERRY. The fruit of *Vaccinium Myrtillus*, sometimes called Whortleberry, Whorts, or Hurts in country places.

BILIMBI TREE. *Averrhoa Bilimbi*.

BILLARDIERA. A genus of shrubs belonging to the *Pittosporaceæ*, natives of Australia and Tasmania, with twining stems and alternate leaves; peduncles solitary from the apex of the branches, one-flowered, pendulous; calyx of five subulate sepals; petals five, combined into a tube below, generally yellow, occasionally blue or purple; stamens five; style thread-like, stigmas lobed; berry elliptical or cylindrical-ovoid, two-celled, many-seeded; pulp generally resinous. *B. mutabilis*, however, is said by Backhouse to have pleasant subacid fruit which at first is green, and at last amber-coloured. *B. longiflora* has pretty blue berries. [J. T. S.]

BILLBERGIA. A genus of *Bromeliaceæ*, so called in honour of a Swedish botanist. It is characterised by a superior three-parted calyx; corolla of three convolute petals, scaly at the base; stamens inserted into the base of the perianth; style thread-shaped; stigmas linear convolute; fruit berry-like. The flowers are generally very elegant, bluish-red or yellow, borne on light panicles; the leaves are harsh and rigid. These plants are found growing on trees in tropical America, and being capable of living without contact with the earth, they are hung on balconies, &c., in South American gardens, where they are much prized for the beauty and fragrance of their flowers. Many species are cultivated for ornament in our stoves. A yellow dye is extracted from the root of *B. tinctoria* in Brazil. [M. T. M.]

BILOBE'S. Divided into two lobes.

BILSTED. An American name for *Liquidambar Styraciflua*.

BIMESTRAL. Existing for two months only.

BINUS. Lasting two years.

BINATE, BINUS. In pairs. Also the same as Bifoliolate.

BINATO-PINNATE. The same as Bipinnate.

BINDWEED. The common name for *Convolvulus*, especially *C. arvensis*; also applied to *Smilax aspera*. —, BLACK. *Polygonum Convolvulus*.

BINDWITH. A name applied to *Clematis*.

BINI. Two together; twin.

BINIFLORUS. Bearing flowers in pairs; a term seldom used.

BINODAL. Consisting of two nodes or articulations, and no more.

BIOTA. A generic name proposed for the *Thuja orientalis* and *T. pendula*, which differ from the *T. occidentalis* and other American species in not having wings to the seeds. The genus is not, however, generally adopted.

BIOTIA. Formerly considered as a distinct genus from that of the Michaelmas daisy (Aster), but now united with it. The species are perennial herbs, one to three feet high, their root leaves large, on long stalks, and heart-shaped in form; those of the stem, ovate or oblong and narrowed towards the base into a winged footstalk; their flower-heads arranged in terminal corymbs, and very like those of the asters. The species are found in Canada, and the United States, and one occurs in Manchuria. [A. A. B.]

BIPALEOLATE. Consisting of two small scales or paleæ, as in grasses.

BIPARTITE. Divided nearly to the base into two parts.

BIPENTAPHYLLOUS. Having from two to five leaflets.

BIFID. Same as Bicrural.

BIPINNATE, BIPINNATISECTED. When the primary and secondary divisions of a leaf are pinnated.

BIPINNATIFID, BIPINNATIPARTED. When both the primary and secondary segments of a leaf are pinnatifid.

BIPINNATIPARTITO-LACINIATE. Being bipinnatifid with the divisions laciniated.

BIPINNULA. A small genus of terrestrial orchids related to *Arethusa*, with fleshy fascicled roots, consisting of little except starch and gum. The flowers are large, racemose, greenish-yellow, and most remarkable for having the lateral sepals broken up into tufts of exquisitely beautiful fringes. Two species occur in Chili, and one in the Argentine States, near Buenos Ayres.

BIPLICATE. Having two folds or plaits.

BIPOROSE. Opening by two round holes.

BIRADIATE. Consisting of two or more rays as in certain umbels.

BIRCH. The common name for *Betula*. —, WEST INDIAN. *Bursera gummifera*.

BIRCH CAMPHOR. A resinous substance obtained from the bark of the black Birch, *Betula nigra*.

BIRCHWORTS. A name given by Lindley to the betulaceous order.

BIRDLIME. A preparation of the bark of the Holly, *Ilex Aquifolium*; also obtained from the viscid berries of the Mistletoe, *Viscum album*.

BIRD-PLANT, MEXICAN. *Heterotoma lobelioides*.

BIRD'S-BILL. *Trigonella ornithorhynchos*.

BIRDS-EYE. *Adonis autumnalis*. —, AMERICAN. *Primula pusilla*.

BIRDS-FOOT. The common name for *Ornithopus*, sometimes called Bird's-foot Vetch; also applied to *Euphorbia Ornithopus*.

BIRDS-HEAD. The common name for *Ornithocephalus*.

BIRDS-NEST. *Neottia Nidus-avis*; also applied to *Thesmopteris* or *Asplenium Nidus*. —, YELLOW. *Monotropa Hypopitys*.

BIRDS-NEST PEZIZA. The common name for the species of *Cyathus* and *Nidularia*.

BIRDS-TONGUE. The common name for *Ornithoglossum*; also applied to *Senecio paludosus*.

BIRIMOSE. Opening by two slits, as most anthers.

BIRTHROOT. An American name for *Trillium erectum*.

BIRTHWORT. The common name for *Aristolochia*.

BISAILLE. (Fr.) *Pisum arvense*.

BISCUIT ROOTS. A name given in Oregon to the tuberous roots of some umbelliferous plants allied to *Ferula*.

BISCUTELLA. A genus of herbs belonging to the *Cruciferæ*, natives of central Europe, the Mediterranean region, and central Asia. Often hispid, with erect rigid stems, frequently corymbosely branched at the summit; leaves oblong, entire, or pinnatifid, very variable in this respect even within the limits of a single species; racemes short, elongated in fruit; flowers rather small, yellow; pouch flattened, with the partition narrow and the valves orbicular, flattened and winged, breaking away from the axis when the seeds are ripe; seeds one in each valve, and contained in it when they fall off. *B. lævigata* is a common subalpine plant of central Europe, &c., very variable in appearance, and remarkable for its curiously-shaped seed-vessels, which are notched both at the base and apex. Some of the species have them notched only at the base. [J. T. S.]

BIENNIAL. Arranged in two rows not on opposite sides of an axis; as on a flat surface.

BISERRATE. When serratures are themselves serrate.

BISEPTATE. Having two partitions.

BISH or BISHMA. The poisonous root of *Aconitum ferox*.

BISHOP'S CAP. An American name for *Mitella*.

BISHOPWEED. *Ægopodium Podagraria*; also applied to the *Bunium Ammi* of Linnæus; and as a common name to the genus *Ammi*. —, MOCK. An American name for *Discopleura*.

BISTORT. *Polygonum Bistorta*.

BITCHWOOD. The timber of *Pisidia erithagyroxus*; much esteemed in Jamaica for making the naves of wheels.

BITERNATE. When the principal divisions of a leaf are three, each of which bears three leaflets.

BITTEN. Terminated irregularly and abruptly; applied to leaves and roots.

BITTER-BLAIN. A name given by the Dutch Creoles in Guiana to *Tovmidelia diffusa*.

BITTER KING. *Soulamea amara*.

BITTER-SWEET. *Solanum Dulcamara*; also an American name for *Celastrus scandens*.

BITTER WOOD. *Xylopia glabra*; also used in gardens for the genus *Xylopia*.

BITTERWORT. An old name for *Gentiana lutea*.

BIVITTATE. Having two vittæ.

BIXACEÆ, BIXINEÆ. A name sometimes given to the order of plants, more generally called FLACOURTIACEÆ (which see).

BIXA. A name applied by the Indians of Darien to the plant producing the Arnotta of commerce, and adopted by botanists for the genus of *Flacourtiaceæ*, to which it belongs. There are four species known, all of them natives of tropical America, and forming small trees, with entire leaves marked with numerous pellucid dots. Their flowers are produced in large bunches at the ends of the young branches; and have a calyx consisting of five sepals, which alternate with five wart-like swellings on the stalk, and likewise with the five petals; numerous hour free stamens, and a long style terminating in a two-lobed stigma. Their fruit has a dry prickly husk, which splits into two pieces, each bearing numerous seeds attached in a perpendicular row on their inside.

B. Orellana is a small tree growing about twenty or thirty feet high, having broad heart-shaped pointed leaves, and bunches of rose-coloured flowers. Its fruit is heart-shaped, rather more than an inch long, of a reddish-brown colour, and covered with stiff prickles. The seeds have a thin coating of red waxy pulp, which forms the substance called arnotta; it is separated by throwing the freshly-gathered seeds into a tub of water, and stirring them until the red matter is detached, when it is strained off and evaporated to the consistency of putty

Bixa Orellana.

In this state it is made up into rolls and wrapped in leaves, and is then known as flag or roll arnotta; but when more thoroughly dried, it is made into cakes and called cake arnotta. In South America arnotta is greatly used by the Caribs and other tribes of Indians for painting their bodies; paint being almost their only article of clothing. In this country it is used for colouring cheese, inferior chocolates, &c.; and by the Dutch for colouring butter. It is also used by silk-dyers; and by varnish-makers for imparting a rich orange tint to some kinds of varnish. [A. S.]

BLACKBERRY. The Bramble, *Rubus fruticosus*, and its numerous varieties.

BLACKBURNIA. A genus of *Xanthoxylaceæ*, consisting of trees with alternate pinnate leaves, and flowers in panicles. The parts of the flower arranged in fours; ovary solitary, on a short stalk, one-celled, one-seeded, with a short style and simple stigma; capsule tough, partly two-valved. These trees, inhabiting Norfolk Island and the East Indies, resemble the species of *Pteles*, but are known by their simple stigmas and wingless fruit. *B. pinnata* is occasionally cultivated. [M. T. M.]

BLACK DRINK. A decoction of *Ilex vomitoria* used by the Creek Indians.

BLACK JACK. An American name for *Quercus nigra*.

BLACK NONESUCH. *Medicago lupulina*.

BLACKTHORN. *Prunus spinosa*.

BLACK-WOOD. An Indian furniture wood obtained from *Dalbergia latifolia*; also a name for that of *Melanoxylon*, —, of New South Wales. *Acacia Melanoxylon*.

BLACKWELLIA. A genus of *Homalineae*, named in honour of Elizabeth Blackwell, the author of a forgotten herbal; technically it is characterised by having an adherent top-shaped calyx, whose limb is divided into from five to fifteen divisions, glandular at the base or in the centre; stamens opposite the petals; ovary conical above, with three to five styles; fruit a one-celled many-seeded capsule. The species are small trees, natives of India, Mauritius, and China. *B. padiflora*, a greenhouse shrub, much resembles the common *Prunus Padus* in appearance; and there are also other species of the genus in cultivation. [M. T. M.]

BLADDER-GREEN. A colour obtained from the berries of *Rhamnus catharticus*.

BLADDER KETMIA. *Hibiscus Trionum*.

BLADDER-POD. The common name for *Physolobium*.

BLADDER-SEED. The common name for *Physospermum*.

BLADDERWORT. The common name for *Utricularia*.

BLADDERY. Inflated like an animal bladder; as the fruit of the Bladder Senna, *Colutea arborescens*.

BLADE. The lamina or expanded part of a leaf.

BLAERIA. A genus of *Ericaceae*, containing many heath-like shrubs, from the Cape of Good Hope, with opposite and ternate leaves, and terminal clusters of flowers. The calyx is four-parted, the persistent corolla is campanulate, sometimes a little expanded below. The four stamens are inserted below the hypogynous disc. The ovary is four-celled, with many ovules in each cell. The capsule is globular with four rounded angles. The habit and structure of the members of this genus are the same as in *Erica*, from which they differ only in having four instead of eight stamens. [W. C.]

BLAKEA. A genus of trees or shrubs belonging to *Melastomaceae*, natives of tropical America, with opposite petiolate three or five-nerved leathery leaves, glabrous and shining above, often covered with short rust-coloured wool beneath as well as the peduncles, which are axillary and one-flowered; flowers large, handsome, rose-coloured, the bell-shaped calyx with four or six broad scales at the base; petals six; stamens twelve to sixteen, anthers cohering, opening by a pair of pores at the apex, shortly spurred at the base; ovary half-inferior, six-celled; style thread-like; fruit a six-celled berry with numerous seeds. *B. quinquenervis* of Guiana has an edible yellow fruit. [J. T. S.]

BLANC D'EAU. (Fr.) *Nymphaea alba*. —, DE HOLLANDE. *Populus alba*.

BLANCHETTE. (Fr.) *Valeriana Locusta*.

BLANCHING. A whitening of the usually green parts of plants, to which the term Albefaction is applied.

BLANCOA. A genus of hæmodoraceous plants, consisting of dwarf stemless herbs, with the aspect of a *Barbacenia*, having equitant hoary falcate acuminate leaves as long as the furfuraceous scape, which latter supports two or three large nodding flowers, both flowers and peduncles being clothed on the outside with plumose hairs. The perianth is elongately bell-shaped or subulate, with an erect six-toothed equal limb, and is furnished with six sub-sessile anthers. The species *B. canescens* is found in the Swan River Colony. [T. M.]

BLANDFORDIA. A genus of *Liliaceae*, consisting of very handsome perennial herbs, having linear elongate striate radical leaves, dilated and somewhat sheathing at the base; others shorter and more distant, appearing on the flower stem, which is simple with a many-flowered raceme at the top. The flowers are solitary on recurved pedicels, and have a tube funnel-shaped six-cleft regular perianth, with ovate acutish segments, six equal stamens, scarcely exserted, and a free long-stalked narrow three-celled ovary, terminated by a filiform style and obtuse stigma. Several species, natives of New Holland and Tasmania, are known. *B. marginata* has rigid sub-erect leaves, scabrous along the margin, and lengthened racemes of pendulous, conically funnel-shaped flowers, which are of a deep rich coppery red outside, yellow within and at the edges of the rounded petaline divisions, which at the back terminate in a sharp orange-coloured point. In *B. nobilis* the leaves are very narrow and entire, and the flowers are ventricosely funnel-shaped, subtubulate, red with the upper half yellow. *B. grandiflora* has rigid erect leaves serrated at the point and short racemes of pendulous ventricosely funnel-shaped flowers, which are red with the upper half yellow, and have retuse petals. In *B. Cunninghamii* the leaves are weakish, spreading, quite entire and smooth, the flowers pendulous, conical, inflated at the apex, subumbellate, reddish throughout, the segments all acute, and the stamens somewhat exserted. They are all handsome plants, and some one or other of them may not unfrequently be met with in our greenhouses. [T. M.]

BLASTEMA. The axis of an embryo, comprehending the radicle and plumule, with the intervening portion. Also the thallus of a lichen.

BLASTIDIA. Secondary cells generated in the interior of another cell.

BLASTUS. The plumule.

BLASTHEMANTHUS. A tree found near the Amazon river, has been considered by

Planchon to belong to a new genus of *Ochnaceæ*, to which he gives the above name. Its botanical characters are interesting; the chief are: a double calyx, each of five overlapping pieces; five petals; twenty glands in one row, exterior to the ten stamens; anthers prolonged into a leaf-like process, opening by two pores, the stamens after flowering turned to one side of the flower; ovary placed on a very short stalk, three to five-celled, and many-seeded. The alternate oblong leaves have cartilaginous stipules inserted on to the branch above the insertion of the leaf. [M. T. M.]

BLAZE′, or BLANZE. (Fr.) A species of *Triticum*.

BLAZING STAR. A North American name for *Liatris squarrosa*, and *Chamælirium luteum*.

BLE′. (Fr.) *Triticum vulgare*. — BARBU. *Triticum turgidum*. — D'ABONDANCE. *Triticum compositum*. — DE BARBARIE. *Polygonum Fagopyrum*. — DE MIRACLE. *Triticum compositum*. — DE TURQUIE. *Zea Mays*. — DE VACHE. *Melampyrum arvense*. — NOIR. *Polygonum Fagopyrum*. —, TURQ. *Triticum compositum*.

BLEABERRY. The Bilberry, *Vaccinium Myrtillus*; sometimes also applied to the Bog Whortleberry, *Vaccinium uliginosum*.

BLECHNIDIUM. A genus of polypodiaceous ferns, closely related to *Blechnum*, from which it differs only in the veins being reticulated instead of free. The only species, *B. melanopus*, is a native of India, and is a moderate-sized pinnatifid fern, with falcate segments, having a general resemblance to the common garden *Blechnum occidentale*. As its trivial name indicates, the stipes or stalk of the frond is black. [T. M.]

BLECHNOPSIS. A name proposed by Presl for certain species separated from *Blechnum*, namely, *B. orientale*, *cartilagineum*, *brasiliense*, &c. It is not adopted by other pteridologists. [T. M.]

BLECHNUM. A considerable genus of polypodiaceous ferns belonging to the group *Lomarieæ*. They are plants with simple pinnatifid or pinnate fronds, of which the fertile ones are sometimes more or less contracted. They are distinguished by having the sori linear, lying parallel with and more or less approximate to the midrib, and therefore theoretically distant from the margin, but sometimes becoming at the same time sub-marginal by the contraction of the fronds. These sori are covered by linear indusia, which are attached along that side of the receptacle which is nearest the margin of the frond, and open along the inward side, or that which is nearest to the midrib. The veins, as seen in the sterile fronds, where they are uninterrupted by the development of the fructification, are free, that is, they branch out from the costa, and become forked as they extend towards the margin, without coming in contact with each other; but in the fertile fronds they are combined within the margin, and generally near the base by the receptacle which runs transversely to them. Leaving out of view *Blechnidium*, which is distinguished from *Blechnum* only by the reticulation of its veins, its nearest ally is *Lomaria*, which indeed presents sometimes so little difference that the same plants are in some cases differently referred to either genus by different authors, or even by the same author in different publications. The proper distinction between the two consists in the fructification of *Lomaria* being marginal, and that of *Blechnum* within the margin, and this irrespective of the contraction of the fronds, which latter feature has sometimes been taken as the mark of *Lomaria*.

The species of *Blechnum* range under two divisions, in one of which, represented by the Indian *B. orientale*, the sori is placed very near the costa, and in the other, represented by our native *B. Spicant*, it becomes sub-marginal from the contraction of the fronds. The former group is the more typical. *B. orientale* is a tall growing and very handsome fern found throughout India and the East. It has a short caudex, which is clothed with long narrow glossy scales. The fronds, which are often three feet long or more, are pinnated, the pinnæ sometimes a foot long, elongately linear, tapering to a narrow point. *B. Spicant* is a humbler plant, producing horizontal pectinately pinnatifid sterile fronds, and erect fertile ones, with narrower or contracted segments. The genus contains a considerable number of species, which are abundant in tropical countries, a large proportion of them being found in the northern parts of South America, in the West Indian Islands, in India, and in the various islands of the Eastern sea. A few species occur in Australasia, at the Cape of Good Hope, and in Chili; and our native, *B. Spicant*, is found throughout Europe, in Madeira and the adjacent islands, in the Caucasian regions, and in Kamtschatka. [T. M.]

BLECHUM. A genus of herbaceous plants of the order *Acanthaceæ*, abundant in tropical America, and occurring also in India and Madagascar. The flowers are in large axillary or terminal spikes; they spring from the axils of broad herbaceous imbricated bracts. The calyx is deeply five-cleft; the corolla is funnel-shaped, with a long tube and a small regular five-fid limb; the four included didynamous stamens are inserted in the middle of the tube; the anthers consist of two oval parallel cells; the ovary is two-celled, with four or more ovules in each cell; the style is simple, and the stigma bifid. The ovate capsule is two-celled, with eight or more roundish seeds. [W. C.]

BLEEKERIA. This name has been applied to a tree, native of New Holland, and the island of Ceram, in honour of Dr. Bleeker, a distinguished student of the natural history of India, especially of the fishes of that country. The genus is one of the apocynaceous family, characterised by a calyx without glands, a salver-shaped corolla with a slightly distended tube, and no scales at its throat. Filaments adherent to the tube of the corolla for some distance, hairy; anthers linear, slender, with the connective prolonged for a short distance beyond the lobes. Ovaries two, small, roundish, each containing two ovules, placed one over the other; style short; stigma almost globular below, tapering above and hairy, slightly two-lobed at the point. Fruit of two fleshy purple drupes, or one by abortion, with a hard woody inner shell. (M. T. M.)

BLENNOSPORA. *B. Drummondii* is the name given to a little West Australian plant which belongs to the cudweed section of the composite family. It is seldom taller than three inches, and is altogether covered with loose woolly hairs. Its leaves are alternate, and linear in form. The flower heads, of a brown colour, are arranged in dense terminal clusters, each of the heads containing but two florets. The generic name refers to the cellular coating of the achene becoming gelatinous when moistened. (A. A. B.)

BLEPHARÆ. The teeth or fringes belonging to the peristome of an urn-moss.

BLEPHARIS. A genus of *Acanthaceæ*, natives of Asia and Africa. They are creeping herbaceous plants, with verticellate unequal leaves, and axillary spikes in which the lower bracts are sterile and closely imbricated, while the two terminal bracteoles contain a single flower. The calyx is four-parted, of which the upper and lower divisions are broadest, and the lower bidentate. The corolla is one-lipped, its anterior portion being trifid, and the posterior tridenticulate. The four stamens are sub-didynamous, the anthers on the longer pair of filaments one-celled, while the shorter filaments bear two-celled anthers. The two-celled ovary has two ovules in each cell; but the capsullary fruit contains sometimes only two seeds, from the abortion of two of the ovules. (W. C.)

BLEPHAROCHLAMYS. A name synonymous with MYSTROPETALON (which see). (M. T. M.)

BLEPHILIA. A genus of the mint family, *Labiatæ*, peculiar to the United States, and nearly related to horse-mints (*Monarda*), but the calyx tube has thirteen instead of fifteen nerves, and is naked in the throat, while the throat of the corolla, which are much smaller than in *Monarda*, are more markedly dilated. There are two species, *B. hirsuta* and *B. ciliata*, the former with long stalks to the leaves, the latter with nearly sessile leaves; and both with the habit, appearance, and odour of our own mints (*Mentha*). The purplish flowers are disposed in axillary or terminal globular whorls, surrounded with coloured bracteas, which, like the calyx-teeth, are fringed with hairs. To this fringe the generic name, derived from the Greek, signifying eyelash, refers. (A. A. B.)

BLETIA. A large genus of terrestrial orchids chiefly from tropical America, where they inhabit swampy places. They have narrow grass-like leaves, and purple or whitish flowers in long terminal racemes, in almost all cases handsome enough to claim the notice of gardeners. Very few species occur in the Old World, among which is *B. hyacinthina*, cultivated in China for the sake of its fragrance. In their manner of growth they are much like *Cymbidiums*.

BLETTING. That kind of change in tissue which results in the formation of a brown colour, without putrefaction, as in the fruit of the medlar. The term *Hyposaphria* is applied to this change.

BLEWITS. The popular name in some parts of England for *Agaricus personatus*, a species which is frequent in rich meadows in autumn, and is known by its pale bistre-coloured or purplish convex fleshy pileus, pallid gills, and thick stem, tinged more or less with violet. It is sometimes exposed for sale, but is a fungus of inferior quality for the table. It is in general believed to be wholesome; but in a case of poisoning from the use of fungi at Cambridge, some years since, the principal part of the stew consisted of this species. Dr. Badham, however, speaks highly of it, when not sodden with water, and suggests that the name is a corruption of Blue Hats. (M. J. B.)

BLIGHIA. A genus of *Sapindaceæ*, named in honour of Captain William Bligh, of H.M.S. Bounty, who, in the year 1787, was appointed to convey the bread-fruit and other trees from Tahiti to the West Indies. It consists of only one species, *B. sapida*, which produces the Akee fruit. This plant is a native of Guinea; but it has been introduced into and is now common in the West Indies and South America. It forms a small tree about thirty feet in height, having compound leaves consisting of three or four pairs of broadly lance-shaped downy leaflets. Its flowers are produced in racemes from the axils of the leaves. They have a calyx consisting of five pieces; five white petals bearing a large two-lobed scale near the base on their inside; eight stamens; and a short style bearing three stigmas. The fruit is fleshy, and of a red colour tinged with yellow, about three inches long by two in width, and of a three-sided form; when ripe it splits open down the middle of each side, disclosing three shining jet-black seeds, seated upon and partly immersed in a white spongy substance called the aril. This aril is the eatable part of the fruit, and in tropical countries, where it comes to perfection, it is said to possess an agreeable sub-acid taste, very grateful to the palate; but fruits ripened in the hothouses of this country have not been found to possess such good qualities, their taste resembling that of a chestnut. A small quantity of semi-solid fatty oil is obtainable from the seeds by pressure. [A. S.]

BLIGHT. This word is used by cultivators with great latitude, and is extended to those diseases of corn, and other objects of field and garden cultivation which depend upon the presence of parasitic fungi. It is best, however, as far as cereals are concerned, to confine it to that diseased condition of corn in which the plant dies prematurely without bringing any fruit to perfection. This often depends upon some kind of fungous spawn attacking the roots, and we believe arises in many cases from undecomposed remains of the last year's crop, which encourage the growth of fungi, the threads of which spread to the living roots, and gradually impair their vigour, and ultimately cause death. The notion that the gloomy turbid state of the atmosphere or the haze, so common in sultry weather, which depends upon differences of temperature between the earth and the air, is caused by the presence of some blighting substance which attacks plants, giving rise to noxious insects and fungi, is founded on popular error. [M. J. B.]

BLIMBING. The Bilimbi tree, *Averrhoa Bilimbi*.

BLINKS. *Montia fontana*.

BLITE. (Fr.) *Amaranthus Blitum*. — SAUVAGE. *Chenopodium polyspermum*.

BLITE. *Amaranthus Blitum*. —, SEA. A common name for *Sueda*. —, STRAWBERRY. The common name for *Blitum*.

BLITUM. A singular genus of chenopods, remarkable for the succulent fruit-like character assumed by the calyx of several species after flowering. The flowers themselves are inconspicuous, and quite elementary in their structure, consisting of a three-cleft calyx, one stamen, and an ovary containing a single vertical seed, and crowned by two styles; all these organs being, however, so small as to be scarcely discernible without the aid of a lens. After the fertilization of the ovary is effected, the calyx gradually increases in size, and at length becomes fleshy and filled with a red-coloured juice, swelling around the membranous capsule, but not entirely concealing it. The flowers being produced in clusters, the resulting compound fruit is sufficiently conspicuous, and from its supposed resemblance to a small strawberry, has arisen the popular name of Strawberry Blite, applied to two plants of this genus. The fruit of the strawberry differs, however, essentially in its character from that of the Blites, consisting as it does of a fleshy succulent receptacle, the calyx itself undergoing no transformation. The structure of the fruit in *Blitum* more closely resembles that of the common mulberry, *Morus nigra*, in which the matured ovary is completely enclosed in a succulent berry-like calyx. Two species of *Blitum*, both European, are cultivated in gardens; *B. capitatum* which has an ascending branched stem, triangular sinuate foliage, and terminal clusters of flowers and fruit; and *B. virgatum*, which has long rod-like shoots, and rather smaller foliage than that of the preceding species, with axillary flowers and berries. The fruit of both species, though insipid, is said to have been formerly employed in cookery. The leaves have a spinach-like flavour, and may be used as a substitute for it. [W. T.]

BLOODFLOWER. The common name for *Hæmanthus*.

BLOODRAIN. Many of the tales of the descent of showers of blood from the clouds which are so common in old chronicles, depend upon the multitudinous production of infusorial insects or some of the lower *Algæ*. To this category belongs the phenomenon known under the name of Red Snow. One peculiar form, which is apparently virulent only in very hot seasons, is caused by the rapid production of little blood-red spots on cooked vegetables or decaying fungi, so that provisions which were dressed only the previous day are covered with a bright scarlet coat, which sometimes penetrates deeply into their substance. This depends upon the growth of a little plant which has been referred to the *Algæ*, under the name of *Palmella prodigiosa*, but which seems rather to be one of those conditions of moulds which under various colours are so common on paste and other culinary articles, to which they seem to bear the same relation as yeast globules do to *Penicillium* and other Fungi. The spots consist of myriads of extremely minute granules, and though they are propagated with great ease, at present no one has been able to follow up their evolutions. In damp weather fresh meat is covered with little colourless gelatinous or creamy spots, which are clearly of the same nature. One curious point about the fungous bloodrain is, that when cultivated on rice paste, little spots spring up on the surface of the paste, apart from the main patch, which look just like blood spirted from an artery, and therefore increase the illusion. The colour of the Bloodrain is so beautiful that attempts have been made to use it as a dye, and with some success; and could the plant be reproduced with any constancy, there seems little doubt that the colour would stand. On the same paste with the Bloodrain we have seen white, blue, and yellow spots, which were not distinguishable in structure and character. We refer for further information to Dr. H. O. Stephen's article in *Taylor's Annals of Natural History*, which is suggestive if not conclusive. [M. J. B.]

BLOODROOT. *Sanguinaria canadensis*, and *Geum canadense*.

BLOODROOTS. A name applied by Lindley, to the hæmodoraceous order.

BLOODWOOD, of Jamaica. *Gordonia Hæmatoxylon*. —, of Norfolk Island. *Baloghia lucida*. —, of Queensland. Eucalyptus paniculata. —, of Victoria. Eucalyptus corymbosa.

BLOOD-TREE. *Croton gossypifolium*

BLOODWORT. *Sanguinaria canadensis*; also an old name for *Rumex sanguineus*.

BLUEBELL. *Hyacinthus nonscriptus* —, SCOTCH. *Campanula rotundifolia*, the 'Bluebell of Scotland.'

BLUEBERRY. An American name for *Vaccinium*.

BLUEBOTTLE. *Centaurea Cyanus*.

BLUE HEARTS. An American name for *Buchnera*.

BLUE MOULD. A name commonly given to *Aspergillus glaucus* when growing upon cheese. In some cases its presence is thought so desirable from its inducing a particular condition of the curd that pains are taken to inoculate cheeses with the mould. It is not clear that the *Aspergillus* is the only mould to which the name has been applied. Cheese is generally eaten in such small quantities that if the mould has any deleterious properties they are not experienced; but it is believed that this or some allied mould when produced abundantly in dried sausages or rolled bacon, known under the name of Italian cheese, has often produced disastrous and even fatal results. It has, however, been supposed that some decomposition in the meat may have taken place before the mould had made its appearance, and that the bad effects were due to this rather than to the fungus. Similar effects have been produced by eating bread made of damaged flour, which was overrun with fungi a few hours after it was taken from the oven. It is asserted that in either case a considerable quantity of the fungus has been collected and swallowed without producing any evil consequences, and that the poisonous quality must therefore be ascribed to the meat or bread itself, and not to the fungus. As the question is of some consequence, further experiments are desirable. [M. J. B.]

BLUE TANGLES. An American name for *Vaccinium frondosum*.

BLUE-WEED. An American name for *Echium vulgare*.

BLUET. (Fr.) *Centaurea Cyanus*. —, DU LEVANT. *Centaurea moschata*.

BLUETS. An American name for *Vaccinium angustifolium*; also for *Hedyotis cærulea*.

BLUMEA. A large genus of the composite family, annual or perennial weeds, found in the tropical or sub-tropical countries of the Old World, the greater part in India, and the islands of the Indian ocean, a few in Africa, and still fewer in Australia. Above one hundred species are enumerated. Their leaves are alternate, and vary much in size and form; in the greater portion they are oblong, sessile, cordate at the base, and toothed, and, as well as the stems, clothed with villous hairs; in others they are large pinnatifid and smooth, somewhat like those of a sow-thistle; while in a third group they are dilated at the base and prolonged down the stem, so as to give to it a winged appearance. The flower-heads are generally arranged in loose panicles or corymbs, their florets all tubular, those of the ray female, those of the disc male, and either purple or yellow in colour.

B. scandens, a native of Borneo, and a few others, are long scrambling shrubs, often found growing among brushwood; they have pretty purple flowers. *B. aurita* and *B. lacera*, both Indian species with small yellow flowers, have a strong turpentine smell, and are used by the natives in cases of dyspepsia. *B. balsamifera*, when bruised, smells strongly of camphor, and *B. aromatica* has, as its name implies, a sweet aromatic smell. [A. A. B.]

BLUMENBACHIA. A curious genus of loaseads, comprising several species of climbing annuals, of which two only are known to be in cultivation, *B. insignis* and *B. multifida*. In habit and inflorescence the genus closely resembles true *Loasa*, having like it flowers with five spreading boat-shaped petals; the stamens placed opposite to them in five distinct parcels, arranged horizontally within them when the flower first expands, but ultimately becoming erect; and alternating with these five fleshy concave scales to which sterile filaments are attached. In the structure of their fruit, however, the two genera differ very materially; that of *Blumenbachia* being of a roundish spongy character, spirally striated, splitting to the base when ripe into ten pieces, five of which are real valves, having the black wrinkled seeds imbedded in their substance, three on each side, the alternate five thinner pieces being the dissepiments or partitions of the fruit. *B. insignis* has an angular-branched stem, clothed with hairs, some of which are simply glandular, and others of a stinging character, with opposite palmately-lobed or deeply-pinnatifid foliage, and flowers produced singly from the axils of the upper leaves, on long footstalks, which are at first erect, but ultimately drooping. The blossoms of this species are pure white, an inch across, with compressed keeled petals, furnished with a large serrated tooth on each side. *B. multifida* is a plant of much stronger growth, more hispid with stings, and with much larger five-parted leaves, longer two-bracted flowerstalks, and broader obtuse petals. Both species are natives of the southern parts of South America, and are not known to possess any sensible properties. [W. T.]

BLUSHWORT. The common name for *Æschynanthus*.

BLYSMUS. A genus of cyperaceous plants, belonging to the tribe *Scirpeæ*. The inflorescence is in more or less compound compressed spikes; spikelets with two to eight flowers, which are all hermaphrodite; stamens three; styles cleft. Four species are described, two of which are natives of Britain, namely, *B. rufus* and *B. compressus.* The former occurs frequently in salt-marshes, near the sea-coasts of England and Ireland; but the latter is rather rare, particularly in Ireland. *B. brevifolius* is a native of India. [D. M.]

BLYTTIA. A genus of grasses belonging to the tribe *Agrostideæ*. Only one species is described, *B. suaveolens*, which is the *Cinna pendula* of *Steudel's Synopsis*, and a native of Norway. [D. M.]

BLYXA. A small genus of stemless aquatic plants found in India and Madagascar, belonging to *Hydrocharidaceæ* and allied to *Vallisneria*. They have linear leaves, which as well as the flowers are submerged. The flowers are diœcious, produced from a tubular spathe split at the end, which has several stalked flowers in the male plant, but only a single sessile one in the female; the perianth has three calyx-like outer segments, and three linear oblong petaloid inner ones, but in the female flower these are at the top of a long tube, adhering to the inferior ovary at the base; stamens three to eight; berry one-celled, as in *Vallisneria*. [J. T. S.]

BOATLIP. The common name for *Scaphyglottis*.

BOAT-SHAPED. Having the figure of a boat in miniature, with its keel.

BOBUA. A genus which was for a long time known only from a short and imperfect description, and was generally placed in the *Combretum* family, but is now generally allowed to be a species of *Symplocos* (*S. spicata*). It is a small tree, all its parts of a yellowish-green colour, and retaining that colour in a dried state. The leaves are alternate, stalked and oblong; the flowers small, white or yellowish-coloured, and borne on short axillary spikes. The fruits are hard, small, and in form like a miniature pitcher, and are sometimes seen strung like beads, and used as necklaces by native children. The plant is common in India and Ceylon. [A. A. B.]

BOCAGEA. One of the genera of *Anonaceæ*, characterised by the calyx, which is either divided, of three segments, or entire and cup-shaped; the petals are six in number, distinct; the stamens definite in number, opposite to the petals. Ovaries three, one-celled, and containing five to eight ovules; styles free or none. Fruit berry-like, of from one to three carpels, which are on short stalks and contain only three seeds, the remainder being arrested in their growth; the seeds are horizontal and provided with an arillus. The species are trees inhabiting Brazil. [M. T. M.]

BOCCONIA. An interesting genus of *Papaveraceæ*, so named in honour of a Sicilian botanist. The calyx consists of two cream-coloured or pinkish sepals; corolla none; stamens eight to twenty-four; style bifid; capsule not jointed, but two-valved, and containing from one to four seeds. The species have their flowers in graceful clusters, and the foliage is also elegant. *B. frutescens* and *B. integrifolia*, natives of the West Indies and Mexico, are in cultivation. *B. cordata*, a hardy species, is a native of China. [M. T. M.]

BŒCKHIA. A genus containing a few sedge-like plants from the Cape of Good Hope, belonging to the natural order *Restiaceæ*. The rhizome is creeping, throwing up slender simple rigid stems with small membranous sheaths. The flowers are unisexual with six small glumes, and are arranged in pairs or in terminal spikes. The male flowers have three stamens; the female a two-celled ovary with one ovule in each cell, and two plumose stigmas; fruit a hard nut containing one seed. [J. T. S.]

BŒNNINGHAUSENIA. A genus of *Rutaceæ*, nearly allied to *Ruta* itself, but distinguished by its flat entire oblong petals; the ovaries also are placed on a thread-like column or stalk, which projects from a short cup-shaped disc. The species are natives of the East Indies. [M. T. M.]

BŒRHAAVIA. This genus of *Nyctaginaceæ* commemorates a famous Dutch physician and naturalist, a cotemporary and patron of Linnæus. The plants are herbs, widely distributed over the tropical and warmer regions of the globe. The flowers have no involucre; their perianth is in two divisions, the lower portion cylindrical, black, persistent, the upper funnel or bell-shaped, coloured, deciduous, five-lobed at the top; stamens one to three, more rarely four, arising from a ring placed beneath the ovary. Ovary very small at the base of the perianth. Fruit within the enlarged hardened base of the perianth, frequently five-ribbed. The root of *B. procumbens*, a troublesome weed in India, is given as a laxative and vermifuge. Others are used as emetics, and for other medicinal purposes. Several species are in cultivation, but have no particular beauty to recommend them. [M. T. M.]

BŒHMERIA. This genus of the order of nettleworts (*Urticaceæ*) contains numerous species distributed throughout the tropics and subtropics of both hemispheres. They are herbaceous plants or shrubs, closely allied to true nettles (*Urtica*), but differing from them in not having stinging hairs. The male and female flowers are produced in separate spikes on the same plant; the males having a four-parted calyx and four stamens, the females a tubular calyx divided into four teeth at the top, and a slender style with hairs along one side. Several of the species yield valuable fibres. The most interesting of them is *B. nivea*, the Tchou-ma of the Chinese, the Rheea of Assam, and the Chinese Grass-cloth plant of English writers. It is a small shrubby plant about three or four feet high, throwing up numerous straight shoots, which are about as thick as the little finger and covered with short soft hairs. Its leaves grow upon long hairy footstalks, and are broadly heart-shaped, about six inches long by four broad, terminating in a long slender point, and

Böhmeria nivea.

having their edges cut like a saw. They are of a deep green colour on the upper side, but covered on the under side with a dense coating of white down, which gives them an appearance, like that of frosted silver. The beautiful fabric known in England as Grasscloth, and rivalling the best French cambric in softness and fineness of texture, is manufactured from the fibre obtained from the inner bark of this shrub, which is a native of China and Sumatra, and has long been cultivated in those countries and also in India, where it has recently been recognised as identical with the Rheea of Assam. The Chinese bestow an immense amount of care and labour upon its cultivation and the preparation of its fibre; they obtain three crops of the stems annually, the second being considered the best. To obtain the fibre the bark is stripped off in two long pieces and carefully scraped with a knife, so as to get rid of all useless matter, after which it is softened and separated into fine filaments, either by steeping it in hot water or holding it over steam. The fibre is of different degrees of fineness according to the age of the plant, and the part of the bark from which it is taken; the inner bark of young quickly grown stems yielding the beautifully fine delicate fibre from which the best fabrics are manufactured, while the outer portion affords a coarse fibre only fit for making ropes, canvass, &c. Experiments made with the view of testing the strength of this fibre have proved it to possess nearly double the tenacity of Russian hemp.

B. Puya, which is a native of Nepal, very closely resembles the preceding both in its botanical characters and general appearance. It is, however, rather taller, growing as high as six or eight feet, and its leaves are of a different form, being broadly lance-shaped, and terminating in a sharp point; but they have serrated edges, and are silvery on the under side as in the last. This plant is called Pooah or Puya in Sikkim and Nepal, and its fibre has long been in use among the natives; but they have hitherto employed clay or mud in its preparation, which greatly deteriorates its value. When properly prepared it is very strong, and makes good cordage and sailcloth. Of the other species of this genus we may mention that the inner bark of *B. albida* is used in the Sandwich Islands for making cloth; and *B. caudata* is employed medicinally in Brazil. [A. S.]

BOIS A' BALAIS. (Fr.) *Betula alba*. —A' LARDOIRE. *Evonymus europœus*. — DOUTON. *Cephalanthus occidentalis*. — CUIR. *Dirca palustris*. — D'ARC. *Maclura aurantiaca*. — DE CHINE. *Murraya exotica*. — DE CHYPRE. *Cordia Gerascanthus*. — DE COCHON. *Hedwigia balsamifera*. — DE COLOPHANE. *Bursera paniculata*. — D'HUILE. *Erythroxylon hypericifolium*. — DE LOSTEAU. *Antirhœa verticillata*. — DE LETTRES. *Brosimum Aubletii*. — DE MAI. *Cratœgus Oxyacantha*. — DE PALIXANDRE. The Rosewood of the cabinet-makers, obtained from some Brazilian species of *Triptolomea*. — DE PERDRIX. *Heisteria coccinea*. — DE SAINTE LUCIE. *Prunus Mahaleb*. — DE ROSE. *Licaria guianensis*. — GENTIL or JOLI. *Daphne Mezereum*. — ROUGE. *Guarea grandifolia*. — TAN. *Byrsonima spicata*.

BOISDUVALIA. A small genus of North American onagrads, separated by Spach from *Œnothera*, from which it differs chiefly in the four stamens, which are opposite the petals, being shorter than the alternate ones, and in the rosy or pinkish colour of the corolla; the flowers of the true *Œnotheras* being either white or yellow. Only two species are known, *B. densiflora* and *B. concinna*, both of annual duration. The former is an erect woolly slightly-branched plant, with linear-lanceolate pointed toothed leaves, and is remarkable for having the axillary buds of the main stem, which usually produce but a single flower, developed into a short branch bearing a small corymb of flowers; it has little beauty to recommend it. *B. concinna* is of trailing habit, with small ovate lanceolate leaves and pretty pink flowers in terminal leafy spikes. [W. T.]

BOISSIELLE. (Fr.) *Bossiœa Scolopendra*.

BOJERIA. A genus of one species (*B. speciosa*) belonging to the composite family, and found in Madagascar. It is a shrub about ten feet high, the stems towards the apex covered with dense rusty hairs. The leaves are alternate, entire, ovate or lanceolate in form, and clasping the stem by their base, nearly smooth above, and densely tomentose beneath. The flower-heads are single from the apex of the branches, and about one inch in diameter, having numerous purple tubular florets, all of them containing both stamens and pistil. The genus bears the name of M. Bojer, Professor of Botany in the Mauritius. [A. A. B.]

BOLBITIS. A name proposed for certain acrostichaceous ferns, now referred to *Pœcilopteris*. [T. M.]

BOLBOPHYLLUM. A very extensive genus of orchids of small stature growing on trees or overrunning the ground among mosses. Their leaves are usually solitary on fleshy pseudo-bulbs; and their flowers are small and inconspicuous in racemes or small capitules. Some, however, have fleshy deeply-coloured flowers in dense spikes. In structure they differ little from dendrobes except that the column is terminated by two conspicuous lateral bristles or teeth. Nearly one hundred species are known from the tropics of both the Old and New World. The focus of the genus is Africa and Asia.

BOLDOA. The name given to a small Chilian tree belonging to the *Monimia* family. It has opposite short-stalked ovate leaves, which are entire and rough on the surface. The flowers in little axillary racemes, the males and females on different plants. The centre of the male flower is occupied by a great many stamens, and that of the female by from two to nine ovaries, which when ripe are succulent drupes, about the size of haws, and very aromatic, as are all the parts of the plant. The bark is serviceable to tanners, and the wood is preferred before any other in the country for making charcoal; while the fruits are eaten. The tree is known in Chili as Boldu, whence the generic name. The origin of the specific name *fragrans* is evident. [A. A. B.]

BOLDU. A genus of *Lauraceæ*, consisting of Chilian shrubs, with hermaphrodite flowers in axillary panicles. The calyx is six-cleft, rotate, with persistent thick segments; the three inner stamens have on either side at their base a sessile gland; the anthers are two-celled. Boldu is besides the Chilian name for *Boldoa fragrans*. [M. T. M.]

BOLETS. (Fr.) *Boletus*.

BOLET DU MÉ'LÈ'ZE. (Fr.) *Polyporus officinalis*.

BOLETUS. A genus of hymenomycetous Fungi, distinguished by the hymenium consisting of tubes separable from each other, as well as from the pileus or cap. In a few instances the tubes are separable from the pileus in the more fleshy *Polypori*, but never so completely from each other as in this genus. All the species have a strong stem, and in a few this is furnished with a ring. They are numerous and often difficult of determination. Some of the poisonous, while *B. edulis* is most people an excellent ar It is not much used in this co Hungary it is preferred to th which is regarded generally w The most poisonous species cognised by the red orifice of t with the exception just men are not more than one or tw species. One of the most c about these fungi is, that species the flesh from whi turns instantaneously to b vided. It is believed that thi the action of ozone on the ju has sometimes been cultivat in its native woods.

BOLIVARIA. A genus of family confined to South Br They are small woody plants two feet high, with opposi three-lobed leaves and axillar yellow flowers, either singl three together, and not unl the jasmine, but smaller. T two-lobed cartilaginous capsu part of which falls off in t cap when the seeds are ripe. bears the name of Bolivar, th liberator of S. America; it i with MENODORA (which see).

BOLTONIA. A genus of belonging to the composite peculiar to North America, extend from Canada southv Southern states. They are st branched perennial herbs, wi pale green sessile leaves, and of flower-heads with white or very much like Michaelmas d to which genus they might at be referred; but they differ i of the ray and disc florets help and consisting of numerous in often with two to four longe *B. glastifolia* has been cultiv land. The genus is dedicated an English botanist.

BOMAREA. A genus of am plants closely related to *Alst* which it is principally disti its twining habit, and some c the capsule or fruit, which is valvate, splitting from t three parts, and in *Bomarea* and coriaceous, with a dehisce lid. The species are rather nu are all South American, foun on the Peruvian Andes, a fe met with in Mexico, Quito the greater part of them elevated situations. *B. Salsi* pretty twining plant, with sm and umbels of purple flowers long, having a dark eye-like base of the two upper segm nearly equal perianth, and a the lowest. This spotting h to the name *oculata*, under sometimes been known.

aspect of the species is similar. *B. edulis*, a West Indian species, produces tubers which are eaten in St. Domingo like those of the Jerusalem artichoke are in this country. [T. M.]

BOMBACEÆ. The Silk-cotton family, a group of Thalamifloral dicotyledons or Exogens belonging to Lindley's malval alliance, and usually considered as a sub-order of STERCULIACEÆ. [J. H. B.]

BOMBAX. Derived from the Greek word *bombyx*, signifying raw silk, and applied to a genus of large soft-wooded trees belonging to the order of sterculiads (*Sterculiaceæ*), the fruits of which contain a beautiful silky substance attached to their seeds, and to which the name of Silk-cotton has been appropriately given. There are about a dozen species, almost entirely confined to the tropical regions of America, one species only being a native of Western Africa. Several Indian species, however, were formerly included in the genus, but they are now separated under the name of *Salmalia*; and the West Indian tree, commonly called *B. Ceiba* or Goi-tree, is the same as *Eriodendron anfractuosum*. Their flowers are produced either singly or in clusters upon the trunk or old branches, and are generally large and of a white or greenish colour: they have a short calyx shaped like the cup of an acorn, and a corolla of five pieces joined together at the bottom; their stamens are arranged in five or more bundles, which are connected together at the base into a short cylindrical tube, the filaments being divided into two branches near the top, each bearing an anther; and they have a shield-like stigma with five angles and furrowed sides. Their fruit is a large woody capsule, containing numerous seeds arranged in five cells, each seed being surrounded by a quantity of beautiful silky hairs, and when ripe it bursts into five pieces, allowing the escape of the seeds, which are then wafted about by the wind.

B. Munguba is a smooth-stemmed tree about eighty or one hundred feet high, commonly found on the banks of the Amazon river and the Rio Negro, where the natives call it Munguba. It has large smooth leaves deeply cut into eight divisions radiating from a centre, and large white or greenish flowers arranged in twos or threes on the branches. Its fruit is about eight inches long by four wide, and of a clear brick-red colour. The silk-cotton surrounding its seeds is of a light brown colour, and, although exceedingly beautiful, it has not hitherto been employed for any purpose more important than stuffing cushions; but it is to be hoped that a better use will some day be found for it.

B. pubescens is called Embirussu in the province of Minas Geraes, in Brazil. This species does not attain the great height of the preceding, being generally only about twenty-five or thirty feet high. It has a smooth trunk covered with a very tough fibrous bark, which the Brazilians use for making ropes. The leaves are variable in shape; those on the lower part of the branches being hand-shaped, that is, cut into five radiating divisions, whilst those higher up on the branches have only three divisions: they are of a leathery texture and covered on the under side with star-like hairs. The large flowers are clothed with white silky down. [A. S.]

BOMBYCINE. Silky, feeling like silk; this term is not applied to hairiness of any sort.

BONAVERIA. A genus of the pea-flower family (*Leguminosæ*), consisting of a single species, *B. securigera*, formerly placed in the genus *Coronilla*, with which it accords entirely in habit, but differs in the form of the pod, which is about four inches long by a quarter of an inch wide, flattened, thickened at both margins, and not jointed distinctly between the seeds. In *Coronilla*, on the contrary, the pod is nearly cylindrical, and distinctly jointed. The plant grows in South Europe, and is a smooth pea-green herb a foot or more high, with unequally pinnate leaves five or six inches long, made up of many pairs of wedge-shaped leaflets; the yellow flowers are borne in an umbellate manner at the end of a long naked stalk, the umbels being about half an inch across. It is often seen in collections of herbaceous plants, and is frequently called *Securigera Coronilla*. [A. A. B.]

BONA-NOX. *Ipomœa Bona-nox*; *Argyreia* or *Rivea Bona-nox*; *Smilax Bona-nox*.

BONAPARTEA. A genus of *Bromeliaceæ*, named in honour of Napoleon I., and consisting of plants with tufted narrow rigid leaves, which are convolute at the base; hermaphrodite flowers protected by bracts, and arranged on a simple or cone-like or branched scape; sepals spirally twisted, either all equal in size, or the two hinder ones larger, all more or less adherent at the base; petals convolute at the base, forming a tube, linear-lance-shaped and spreading at the top; stamens hypogynous, distinct, the filaments thread-shaped, the anthers sagittate, protruded beyond the corolla. The ovary is superior with a thread-like style and three linear fringed stigmas coiled up spirally. The fruit is an ovate capsule, dehiscing by three valves, which expose a central column bearing the numerous seeds each provided with a hair-like appendage. Two species are in cultivation, one especially, *B. juncea*, a graceful plant, from its elegant drooping grass-like leaves. The same name has also been applied to a genus of *Amaryllidaceæ*, now included under *Littæa*. [M. T. M.]

BONATEA. Under this name are collected many species of terrestrial orchids, with the oblong fleshy roots of our wild *Orchis*. The genus is perhaps not distinct from *Habenaria*, from which it is only separated by an excessive enlargement of the upper lip of the stigma. The true lip is always divided to the very base into thread-like lobes. The flowers appear to be in all cases greenish, verging on yellow or white.

BONE-SEED. The common name for *Osteospermum*.

BONESET. *Eupatorium perfoliatum*.

BONGARDIA. A genus of the berberry family, but not at all like a berberry in appearance. One species only (*B. Rauwolfii*) is known, and it is a small stemless plant, with a tuberous underground rootstock, somewhat like a small potato, from the upper part of which spring four or five long-stalked pinnatisect leaves. The flower stalk is slightly branched and panicled, and the flowers small, golden yellow, with three to six calyx leaves, and six petals, each of which has a little pit at its base, like that of the buttercups. The genus comes near to that of the lion-leaf (*Leontice*), but differs in the pit at the base of the petals, and in having a dilated stigma. The plant is a native of Greece, Syria, and Persia, extending to Affghanistan and Scind. It was noticed as early as 1573 by Rauwolf, who spoke of it as the true *Chrysogonum* of Dioscorides. The Persians roast or boil the tubers, and use them as an article of food, while the leaves are eaten like sorrel. [A. A. B.]

BONHOMME. (Fr.) *Narcissus pseudo-Narcissus*; also *Verbascum thapsiforme*.

BONNE DAME. (Fr.) *Atriplex hortensis*.

BONNET D'ÉLECTEUR or DE PRÊTRE. *Cucurbita Melopepo*; also *Euonymus europæus*.

BONNAYA. A small genus of *Scrophulariaceæ*, found in tropical and subtropical Asia. They are annuals, usually glabrous, with opposite leaves, and flowers in the axils or in terminal racemes. The calyx has five distinct narrow sepals; the upper lip of the corolla is erect and two-lobed, the lower is larger, spreading and three-lobed. The two upper stamens alone are fertile, the lower pair, inserted at the base of the lower lip of the corolla, are represented by the linear obtuse filaments. The style is filiform with a dilated generally two-lobed stigma. The linear capsule is longer than the calyx. [W.C.]

BONNETIA. A genus of the tea family (*Ternstrœmiaceæ*), composed of a few Brazilian and Peruvian shrubs or small trees, with sessile spathulate entire leaves, having prominent parallel veins: they are generally crowded at the ends of the branches, which are marked with prominent scars where they have fallen. The flowers are numerous and panicled, or single, and as large as those of *Camellia*; generally white in colour, and composed of a five-leaved calyx, five petals, a large number of stamens, a three-parted style, and a one-celled ovary, which becomes when ripe a three-celled capsule containing many seeds. The leaves of *B. paniculata*, a Peruvian species which attains the height of twenty or thirty feet, have an aromatic smell when bruised. [A. A. B.]

BONPLANDIA. A genus of *Rutaceæ*, now generally merged in GALIPEA (which see). [M. T. M.]

BONTIA. A genus of *Myoporaceæ*, containing a single species from the West Indies. It is a small evergreen tree, in habit so like the olive as to have been named *Olea sylvestris*. The leaves are alternate lanceolate and sub-entire, and the flowers are solitary or in pairs on axillary pedicels. The calyx is divided into five ciliated imbricated lobes, two being exterior. The corolla is tubular and bilabiate. The four didynamous stamens are shorter than the corolla. The ovoid ovary is two-celled, each cell being almost divided by an incomplete secondary septum; there are two ovules in each cell. The baccate drupe has eight hard seeds. [W. C.]

BONUS HENRICUS. Good King Henry, *Chenopodium Bonus Henricus*.

BOOPIS. A genus of the *Calycera* family comprising a few annual or perennial herbs, some of them stemless and with entire leaves, others branching with pinnatisect leaves, and a habit not unlike that of the chamomile. Their flower-heads are stalked and terminal, containing many white or yellow florets enclosed by a membranous toothed involucre. The genus is readily distinguished from its allies by the absence of spiny points to the calyx leaves, and the nature of the involucre. The species, eight in number, are found in the Cordilleras of Chili, the neighbourhood of Buenos Ayres, and also in the extreme south of the continent. The generic name is derived from the Greek *bous*, an ox, and *ops*, an appearance; the flowers having somewhat the appearance of an ox-eye. [A. A. B.]

BOOREE. An Indian name for the inflammable pollen of a species of *Typha*.

BOOR-TREE or BOUNTRY. A Scotch name for the Elder, *Sambucus nigra*.

BOOTIA. A genus of the natural order *Hydrocharidaceæ*, found in the margin of the river Irrawadi in Ava. The leaves are all radical, some of them submersed, elongate linear lanceolate, others cordate, floating, with long petioles and a scape rising out of the water; flowers diœcious from a tubular inflated spathe, which is toothed at the apex, and includes many stalked male flowers or a single sessile female one. Perianth with three outer oblong calyx-like divisions, and three inner obovate petaloid ones. As usual in the order, these segments are in the female flower at the top of a tube adhering to the ovary at the base. Stamens twelve; ovary with nine parietal placentas. [J. T. S.]

BOQUILA. *B. trifoliata*, the only known species, is a small diœcious trailing shrub found in Chili in the neighbourhood of Valdivia, and there called Baquti-blanca, whence the generic name. Its leaves are alternate, with three entire or slightly-toothed leaflets which are glossy above and pea-green underneath. The flowers are white and solitary, or sometimes two or four, in the axils of the leaves: the calyx and corolla each of three membranous leaves; the male flowers containing six stamens and the females three or four ovaries, which when ripe are berries about the size of a pea, and with few seeds. The few seeds and membranous floral leaves distinguish the genus from *Lardizabala*, to which it is allied. [A. A. B.]

BORA. A common Indian pulse, *Dolichos Cajan*, or *Cajanus bicolor*.

BORAGE. *Borago officinalis*.

BORAGEWORTS. A name applied by Lindley to the boraginaceous family.

BORAGINACEÆ. (*Borageworts; Asperifoliæ*.) A natural order of Corolliflora dicotyledons or Exogens belonging to Lindley's echial alliance. Herbs or shrubs with round stems, alternate rough leaves, and spirally-coiled inflorescence; calyx four to five divided, persistent; corolla generally regular and five-cleft; stamens five, inserted in the corolla, and alternate with its divisions; ovary four-lobed with a style arising from the base of the lobes. The fruit consists of distinct achenes without albumen. The order was called formerly *Asperifoliæ* from the rough leaves of the plants. Natives of the northern temperate regions principally. They abound in the southern part of Europe, the Levant, and middle of Asia. They are less frequent in high northern latitudes, and they nearly disappear within the tropics. Demulcent mucilaginous qualities pervade the order. Some yield dyes, as alkanet (*Anchusa tinctoria*); others are used for potherbs, as comfrey (*Symphytum officinale*), which is employed as a substitute for spinach. The common borage (*Borago officinalis*), when steeped in water, imparts coolness to it, and is used in the beverage called cold-tankard. The leaves of *Mertensia maritima* have the taste of oysters, so that it is called the oyster-plant. The species of *Myosotis* receive the name of Forget-me-not. There are fifty-eight known genera, and 688 species. Illustrative genera:—*Cerinthe, Echium, Borago, Lithospermum, Cynoglossum, Myosotis, Symphytum, Anchusa, Omphalodes*. [J. H. B.]

BORASSUS. There are only two species of this magnificent genus of palms, both having separate male and female flower-spikes on distinct trees: the males in cylindrical branching catkins, composed of a number of scales closely packed and overlapping each other, from amongst which the flowers only partially emerge; the female spikes seldom branched, and their scales not so closely packed as those of the male.

B. flabelliformis is the Palmyra Palm. The parts of this tree are applied to such a multitude of purposes that a poem in the Tamil language, although enumerating 801 uses, does not exhaust the catalogue. It is widely distributed throughout the tropical parts of Asia, generally growing in low sandy tracts of land near the sea-coast, and forming lofty trees with straight and almost cylindrical trunks from sixty to eighty or even one hundred feet high, and about two feet in diameter. Like all endogenous trees, it has the hardest part of its wood towards the outside of the trunk, and the older the tree the harder this wood becomes; so that, while the wood of young trees is almost worthless, that of centenarians is very valuable on account of its hardness, weight, and durability. The leaves of the Palmyra are from eight to ten feet long, including the stalk, and of a nearly circular form, consisting of seventy or eighty ribs, radiating from a centre and plaited like a half-open paper fan; in old trees they form a large round head at the summit of the trunk. These leaves are employed by the natives for a variety of useful purposes; houses are thatched with them; matting for floors and ceilings is platted from strips of them, also bags and baskets of all kinds, hats and caps, umbrellas and fans, and a host of minor articles; they likewise, in common

Borassus flabelliformis.

with those of the Tallpot palm, supply the Hindoo with paper, which he writes upon with a stylus. A most important product, called toddy or palm-wine, is obtained from the flower-spikes in the following manner: as soon as a spike makes its appearance among the leaves, a toddyman ascends the tree, and securely binds it with thongs so that it cannot expand; he then for three successive mornings beats the lower part of the spike with a short baton, and on the four following mornings, in addition to the beating, he cuts a thin slice off the end: on the eighth day the sap or toddy begins to flow, and is collected in an earthenware jar tied on the end of the spike. A tree continues to yield toddy for four or five months, the toddyman ascending the tree every morning to empty the jar, and at the same time to cut a fresh slice off the end of the spike. Palm toddy is intoxicating, and when distilled yields strong arrack. Very good vinegar is also obtained from it; but its most important product is jaggery, or palm sugar, large quantities of which come to this country. The fruits of this palm are about the size of a child's head, and are produced in bunches of fifteen or twenty together. They have a thick coating of fibrous pulp, which the natives roast and eat, or make into a jelly. But the most singular use of this palm is the consumption of the young seedlings as an article of food; these are cultivated for the market, and either eaten in a fresh state or after being dried in the sun, or else they are made into a very nutritious kind of meal.

B. æthiopum is a native of the central part of tropical Africa, occurring from the Niger on the west to Nubia on the east. It forms a large tree resembling the Palmyra in general appearance, but having a curious bulging out or swelling in its stem at about the middle of its height. Its leaves and fruits are used by the Africans for the same purpose as those of the Palmyra by the Asiatics, and its young seedlings are likewise used for food; but the custom of extracting toddy does not appear to be known in Africa. [A. S.]

BORBONIA. A genus of the pea-flowered section of the leguminous family, numbering thirteen species, all of them natives of South Africa. They are small shrubs with simple alternate many-nerved leaves. The flowers, arranged in axillary or terminal few-flowered racemes, are yellow, as in those of the common broom, and much like them but smaller. The pods are linear compressed and often covered with long soft hairs, and contain few seeds. The segments of the calyx are equal, and the upper petal or vexillum is hairy; these two characters distinguishing the genus from its allies. *B. crenata* has roundish leaves which embrace the stem by their base, and terminal racemes of pretty yellow blossoms. *B. parviflora* has many-nerved sharp-pointed leaves like those of the butcher's-broom (*Ruscus*). The genus was named in honour of Gaston de Bourbon, Duke of Orleans, son of Henry IV. of France, a patron of botany. [A. A. B.]

BORECOLE. A loose or open-headed variety of the cabbage, *Brassica oleracea*, cultivated in gardens under the name of Kale.

BORKHAUSIA. A family of compound flowers allied to the hawkweeds and dandelion. Several species are described as inhabiting Southern Europe, all of which are annuals. Two are natives of Great Britain, but are of rare occurrence. The group to which they belong are very difficult of discrimination, and will scarcely admit of a satisfactory popular description. They have yellow flowers, and leaves somewhat like those of the dandelion. *B. fœtida* has an unpleasant odour, in which a flavour of bitter almonds can be distinguished. *B. rubra*, an Italian species, is cultivated as a border plant; it has compound leaves and large flowers of a delicate rose colour or sometimes white. (French, *Barkhausie*.) [C. A. J.]

BORONIA. This name was applied in honour of Francis Borone, an Italian attendant of Dr. Sibthorp, of *Flora Græca* celebrity, to a genus of *Rutaceæ*. The genus is known by a four-cleft persistent calyx; four ovate persistent petals; eight stamens, of which the four opposite the sepals are fertile, the remaining four abortive, with filaments studded with hairs and bent inwards; four styles, erect, approximate or fused together; carpels four to two-valved, combined within into a four-celled capsule; seeds few in each cell, flattened. The species are shrubs, natives of New Holland, with opposite pinnate leaves and pretty pinkish or whitish flowers. Many of them are in cultivation as elegant greenhouse shrubs. [M. T. M.]

BORYA. The same as *Forestiera*.

BOSCIA. Louis Bosc was a French professor of agriculture; and in his honour this genus of *Capparidaceæ* was named. The plants have four sepals disunited or joined together at the base only; petals none; stamens twelve to twenty; berry globose, stalked, one-seeded; leaves simple. The species are natives of Africa. *B. senegalensis* is in cultivation as an ornamental stove-plant. [M. T. M.]

BOSEA. A genus consisting of a shrub from the Canary Islands, of which the natural order is doubtful, but most generally supposed to be *Chenopodiaceæ*. The leaves are alternate, exstipulate, shortly-stalked, elliptical-acuminate, and shining; racemes axillary and terminal, the flowers small polygamous-dioecious; perianth five-cleft, membranous, with two bracts; stamens five; ovary one-celled; drupe sub-globose, fleshy; embryo with foliaceous cotyledons. [J. T. S.]

BOSSED. Circular and flat, with a prominent centre, like the Highland target; as in the fruit of *Patiurus australis*.

BOSSIÆA. A genus of Australian shrubs or small herbs belonging to the pea-flowered section of the leguminous family. Their stems are round or compressed, often when compressed without leaves; the leaves are simple, of various forms, and the flowers are axillary and solitary, always yellow, the base of the vexillum or the keel generally blotched or veined with purple. The genus differs from its allies in its alternate leaves and compressed pods, the margins of which are thickened but not winged. It is named in honour of M. Bossieu Larmartinière, a French botanist, who accompanied La Peyrouse in his voyage round the world. Many of the species are highly ornamental, and no greenhouse collection of any pretensions is to be found without some of them. Among the leafless species in cultivation are *B. scolopendra* and the sword-branched *B. ensata*: both these, however, when in a seedling condition, have true leaves. Amongst the leafy species the choicest are the slender-stemmed *B. tenuicaulis*, with ovate acute leaves and very numerous yellow flowers streaked with purple; *B. lanceolata*; and *B. disticha*, a Swan river species, with ovate acute leaves arranged in a two-ranked manner. [A. A. B.]

BOSTRYCHIA. A genus of rose-spored *Algæ* belonging to the natural order *Rhodomelaceæ*, and remarkable at the same time for the curled tips of the fronds, and their amphibious habit like that of *Lichina*. *B. amphibia* occurs on our coasts as high as the Wash, extending from thence to Spain; it grows attached to the base of marine phænogamous plants, which are covered only at high water. Several species grow in the United States in similar situations or on the margins of tidal rivers, and others are found nearer the equator and in the Southern hemisphere. They do not serve in the structure of the frond, but their habit and general character are so alike that it is better not to separate them. [M. J. B.]

BOSWELLIA. A genus of the family *Amyridaceæ*, consisting of trees with compound leaves; and white flowers in clusters, each with a small five-toothed persistent calyx, and five petals spreading widely, inserted, as are also the ten stamens, beneath a cup-shaped fleshy disc, which is larger than the calyx; the filaments of the stamens are persistent, but the anthers fall off. Ovary sessile, with a long style, terminated by a three-lobed stigma. The fruit is triangular, three-celled, and bursts by the separation of the three component leaves one from the other; the seeds are winged. These trees are remarkable as furnishing a gum resin. That of *B. glabra* is used in India in place of pitch, and as a medicine, both externally and internally. The Hindoos employ it as incense in their religious ceremonies.

B. thurifera, a tree common in Coromandel, also known as *B. serrata*, furnishes the resin known as Olibanum, which is supposed to have been the Frankincense of the ancients. It is rarely used in medicine, but is an astringent and stimulant, and is employed for its grateful perfume as incense in Roman Catholic churches. African olibanum, a drug rarely met with in this country, has been conjectured with much probability to be the product of a species of *Boswellia*, probably *B. papyrifera*, a tree so named on account of its bark, which peels off in thin white layers, capable of being used for packing purposes. The two first-mentioned species are in cultivation in our stoves. [M. T. M.]

BOTHRENCHYMA. The pitted, or dotted, or so-called porous tissue of plants.

BOTROPHIS. A genus of *Ranunculaceæ*, synonymous with *Macrotis*, containing a North American herb allied to *Cimicifuga*, from which it differs by having only one carpel (very rarely two), which becomes a solitary follicle in fruit. This distinguishes it from the berry-bearing *Actæa*. The leaves are twice or thrice ternate, with large oval leaflets irregularly cut; the stem is about from three to eight feet high, with long racemes of white flowers, of which the central one is by far the longest: sepals petaloid, white, soon dropping off; petals, or rather abortive stamens, very small with long claws; stamens numerous, white, and very conspicuous; seeds seven or eight in the follicle. The flowers are very fetid, and the large knotted root-stocks, which have a nauseous astringent and bitter taste, are considered in the United States to be a remedy for the bite of the rattlesnake. The only species rejoices in several names both generic and specific. [J. T. S.]

BOTRYCHIUM. A genus of ophioglossaceous ferns, distinguished by having the fructifications in a compound or rachiform panicle, forming a separate branch of the frond. The spore-cases in this group have no jointed band or ring surrounding them, as in the generality of ferns, but are fleshy, coriaceous, and burst vertically in two equal hemispherical valves. The fronds spring from a short erect fleshy rhizome, and are variously pinnatifid, pinnate, or ternately decompound, the sterile and fertile branches being always separate, and the spore-cases ranged in two rows on the ultimate divisions of the latter. The genus, which consists of about a dozen species, is found in all parts of the world excepting Africa, and extends from the tropical to the arctic regions, and over both the eastern and western hemispheres. The common British species, *B. Lunaria*, called Moonwort, is a dwarf fleshy-looking plant, having the sterile branch pinnate with lunate leaflets, and the fertile branch panicled with sessile distinct globular spore-cases. *B. simplex* is a smaller and less divided plant found in North America and the north of Europe. Another species, *B. virginicum*, of which somewhat varied forms are found in North and South America and in India, is much larger in size and more compound in structure than either of the foregoing; the sterile branch being ternate, then bipinnatifid, with the segments again inciso-pinnatifid, and the fertile branch larger and bipinnate or tripinnate. [T. M.]

BOTRYDIUM. A genus of green-spored *Algæ* belonging to the division *Siphoneæ*, in which it is remarkable for the predominance of the large capsule over the vegetative part, which consists only of a few threads, that like roots penetrate the soil, the capsules being the only part externally visible. *B. granulatum* occurs in little vesicular strata on the sides of ponds, but not very commonly. [M. J. B.]

BOTRYOGRAMMA. A synonyme of *Llava*.

BOTRYOPSIS. A genus of *Menispermaceæ*, briefly described by Mr. Miers, in the *Annals of Natural History*. The male flowers have six petals; the female flowers six ovaries, with an embryo without albumen, and curved so as to resemble a horse-shoe; cotyledons large, thick, curved; radicle superior. The plants are natives of the Organ mountains of Brazil. [M. T. M.]

BOTRYOPTERIS. A synonyme of *Helminthostachys*.

BOTRYOSICYOS. A name apparently implying a resemblance in the plant to which it is applied, to a grape vine and a gourd. The genus belongs to the natural order *Passifloraceæ*. The flowers are diœcious. The male flowers are very small, in clusters concealed by an involucre; the perianth is bell-shaped, six-cleft, in two rows, the three outer hairy, shorter than the inner, which are petal-like. Within this are three scales adherent at the base to the inner divisions, and similar to them, but shorter and divided into two teeth or lobes at the apex. Stamens three, inserted near the throat of the perianth; filaments short, bearing the anthers, which are two-celled, introrse; ovary abortive; stigma three-toothed. The female flowers and fruit are not known. The plant is a climber, and a native of Abyssinia. [M. T. M.]

BOTRYOTHALLUS, a name applied to one or two acrostichaceous ferns included in *Polybotrya* and *Soromanes*.

BOTRYPUS. A synonyme of *Botrychium*.

BOTRYS. The term applied in Greek compounds to the raceme. A bunch.

BOTRYA. (Fr.) *Chenopodium Botrys*.

BOTRYTIS. A genus of filamentous moulds first proposed by Michelli, but now so divided that the original genus is almost swamped. Amongst those best known is the parasite which plays so important a part in the virulous potato murrain under the name of *B. infestans*; as, however, there are strong reasons for separating this and a host of allied plants, we must refer for their consideration to the article PERONOSPORA. The disease in silkworms called muscardine is produced by a mould called *B. Bassiana*, but this also in all probability will ere long find its place in some other genus, perhaps in *Botryosporum*. A few of the spores rubbed upon the skin of the caterpillar, or inserted carefully with a lancet, are sufficient to inoculate the animal. The spores soon germinate, and their threads prey upon the fatty tissue, till the caterpillar becomes mummified and resembles certain pastilles, from whence the name of the disease has been borrowed. In the silkworm houses the malady most commonly commences in the large intestine, as if from the germination of swallowed spores. The prevention of the disease consists in the most perfect cleanliness, and every precaution which may destroy the spores or prevent their access. [M. J. B.]

BOTULIFORM. Sausage-shaped.

BOUCAGE. (Fr.) *Pimpinella*; also *Œnanthe pimpinelloides*.

BOUCHEA. A genus of *Verbenaceæ*, containing fourteen species of herbs or undershrubs, natives of America, Africa, and Asia. They have sub-sessile flowers in a spicate raceme, which is either terminal or in the forking of two branches. The calyx is elongate tubular, with five ribs produced into small teeth, and five alternate furrows, and truncate between the teeth. The corolla is funnel-shaped. The four included didynamous stamens are inserted in the throat of the corolla. The ovary is two-celled, with a single anatropal ovule in each cell; the style is as long as the stamens. The capsule is surrounded by the persistent calyx; it is diœcious and has numerous seeds. [W. C.]

BOUGAINVILLÆA. A genus of the natural order *Nyctaginaceæ*, characterised by the flowers being almost concealed by large membranous or leafy bracts, which grow in triplets, and form magnificent masses of paniculate inflorescence. The perianth is tubular with a short limb; the stamens are seven or eight in number; the style lateral; the stigma thickened. *B. spectabilis* is a climbing shrub or small tree, with alternate leaves and small spines; the bracts are large and of rich rose colour; hence the pendent inflorescence is singularly handsome. The colour of the bracts varies. The plant is a native of tropical South America. [M. T. M.]

At least two other species of this gorgeous genus are grown in our gardens, *B. speciosa*, which has hairy leaves and stems, the latter furnished with strong short recurved thorns, and dense panicles of large soft rosy-tinted bracts; and *B. glabra*, which is of more slender habit, with smaller leaves, both these and the stems being nearly smooth, and bearing its showy bracts, which are of a lighter rose and rather smaller than in *B. speciosa*, in more open panicles. [T. M.]

BOUGUERIA. A genus of *Plantaginaceæ*, containing a single species, a native of Peru. It is a small perennial fleshy-rooted herb, growing in tufts, and having white linear leaves, and axillary peduncles, bearing compact heads, which blacken in drying. The flowers are polygamous, both sexes occurring on the same head. This genus occupies a position between *Plantago* and *Littorella*. [W. C.]

BOUILLARD. (Fr.) *Betula alba*.

BOUILLON-BLANC. (Fr.) *Verbascum thapsiforme*.

BOULE DE NEIGE. (Fr.) *Viburnum Opulus*, with double flowers.

BOULEAU COMMUN. (Fr.) *Betula alba*. —, ODORANT. *Betula lenta*.

BOULETTE AZUREE. (Fr.) *Echinops Ritro*.

BOUNCING BET. An American name for *Saponaria officinalis*.

BOUQUET PARFAIT. (Fr.) *Dianthus barbatus*, the common Sweetwilliam.

BOURBONNAISE. (Fr.) *Lychnis Viscaria*.

BOURDENE, or BOURGENE. (Fr.) *Rhamnus Frangula*.

BOURGOGNE. (Fr.) *Hedysarum Onobrychis*.

BOURRACHE. (Fr.) *Borago officinalis*. —, PETITE. *Cynoglossum Omphalodes*, or *Omphalodes verna*.

BOURREAU DES ARBRES. (Fr.) *Celastrus scandens*. — DU LIN. *Cuscuta epilinum*.

BOURSETTE. (Fr.) *Valerianella olitoria*.

BOUSSINGAULTIA. A name given in honour of a French philosopher, and applied to a genus of *Basellaceæ*. The plants have a perianth of six to eight pieces, and two small bracts on the outside; six stamens opposite the sepals; ovary elliptical; style thread-shaped, thickened at the base; stigmas three, club-shaped. Fruit roundish, compressed, membranous, one-seeded, indehiscent, crowned with the persistent style; seed kidney-shaped, smooth, sessile. *B. baselloides*, a native of the Andes, is an elegant climbing shrub, with alternate entire fleshy leaves, long clusters of fragrant whitish flowers, and thick fleshy roots. It is well adapted to grow in a stove, in a hanging basket, or to trail over trelliswork. [M. T. M.]

BOUTEILLEAU. (Fr.) A kind of olive.

BOUTINIANE. (Fr.) A kind of olive.

BOUTON D'ARGENT. (Fr.) *Ranunculus platanifolius*, with double flowers; also *Achillea Ptarmica*. —, D'OR. *Ranunculus acris*, with double flowers. —, ROUGE. *Cercis canadensis*.

BOUVARDIA. One of the genera of *Cinchonaceæ*, named in honour of Dr. Bouvard, a former superintendent of the Jardin du Roi, at Paris. It is distinguished by a calyx with a sub-globose tube, a limb of four linear awl-shaped lobes, occasionally with little teeth intermediate between the lobes; a funnel-shaped corolla, naked at the throat, and with a four-parted spreading limb; filaments adherent for some distance to the tube of the corolla; anthers linear, included; stigma divided into two lamellæ projecting beyond the tube of the corolla; capsule membranous, globose, compressed, with two compart-ments opening through the backs of the carpels by two valves; seeds numerous, winged. The plants are Mexican shrubs, with handsome flowers in terminal corymbs. Most of the species have red flowers, but in *B. longiflora* they are large white and fragrant; and in *B. flava* they are yellow. *B. triphylla* has three leaves with stipules between their petioles, thus presenting an approximation to the structure of the *Galiaceæ* or *Stellatæ*. [M. T. M.]

BOVA. A kind of vanilla.

BOVISTA. The small smooth nearly globose Puffballs which are so common in our fields and in large exposed pastures, distinguished by their having an outer coat which easily separates from the thin inner covering, belong to this genus, which contains also a few tropical and subtropical species. The smaller of these, *B. plumbea*, is one of our lesser puffballs, and is easily known by its leaden hue when dry; the larger, *B. nigrescens*, by the far firmer and darker inner coat. Both are eatable when young, but our own experience is not in favour of their use, as they are apt to have an unpleasant taste, if they have reached their full growth. Some of the foreign species have violet or russet spores, instead of the more sober hue of our own natives. In all the species it should seem the spores are seated on a short stalk, but this is not without example in *Lycoperdon*. [M. J. B.]

BOWDICHIA. A genus which belongs to the pea-flowered tribe of the leguminose family. The species are found in South America, and chiefly confined to Brazil. They are trees with alternate unequally-pinnate leaves, with from five to fourteen pairs of leaflets, which vary from half an inch to two inches long, and are often pubescent underneath. The flowers are very numerous, disposed in terminal panicles, and violet in colour. The pods are stalked, thin and papery in texture, containing six to eight seeds.

B. nitida, a Brazilian species, is a tree about fifty feet in height with a diameter of one foot; the wood is exceedingly hard, and the corollas bright blue with a slight purple tinge. Another Brazilian species, *B. virgilioides*, is one of the commonest trees of the Campos, and a great ornament to them, the upper part being clad with flowers of the finest amethystine blue, while the leaves are confined to the lower branches, the upper having fallen off. The bark is of a reddish-brown colour, and is known as Alcornoco bark. It is astringent, slightly bitter, and gives to the saliva a yellow colour. It was once recommended in pulmonary consumption, but its use is now obsolete. All the parts of *B. major* are said to have tonic qualities. The genus bears the name of J. E. Bowdich, who travelled in West Africa. [A. A. B.]

BOWMAN'S ROOT. *Tenardia alternifolia*; also applied in America to *Gillenia trifoliata*.

BOWRINGIA. A genus of leguminous plants, allied to *Baphia*, and consisting of a single species, which forms a smooth scandent shrub, with unifoliate leaves, the leaflet of which is ovate or oval-oblong and acuminate; it bears short axillary or subterminal racemes of from two to five small white pea-shaped flowers. The plant is abundant in the island of Hong Kong. The name has also been given to a genus of ferns, now referred to BRAINEA (which see). [T. M.]

BOW-WOOD. An American name for *Maclura aurantiaca*.

BOX. The common name for *Buxus*. —, BASTARD. *Polygala Chamæbuxus*. —, GREY, of Victoria. *Eucalyptus dealbata*. —, QUEENSLAND. *Lophostemon macrophyllus*. —, RED, of N. S. Wales. *Lophostemon australis*. —, SPURIOUS, of Victoria. *Eucalyptus leucoxylon*. —, TASMANIAN. *Bursaria spinosa*.

BOX ELDER. *Negundo fraxinifolium*.

BOX-THORN. The common name for *Lycium*.

BOXWOOD, AMERICAN. *Cornus florida*. —, JAMAICA. *Tecoma pentaphylla*.

BOYKINIA. A genus of perennial North American herbs, belonging to the natural order *Saxifragaceæ*, with alternate stalked palmately five or seven-lobed or cut leaves; flowers white, in cymes. It differs from *Saxifraga* by having the calyx (which adheres to the ovary) contracted at the top, and by having only five stamens. It also differs from *Sullivantia* by the calyx adhering completely to the ovary, and from *Heuchera* and *Tiarella* by the ovary being two-celled. [J. T. S.]

BRABEJUM. A genus of *Proteaceæ*, with apetalous flowers of four sepals, and four anthers on short filaments, attached to the base of the sepals. The flowers are borne on axillary spikes of about four inches in length. The seed-vessel is an elliptical nut, containing a single seed. The leaves are verticillate, about four inches in length, and one inch broad, remotely serrated. The plant, which attains the height of from six to eight feet, is a native of South Africa. The Cape settlers roast the seeds, which they call Wild Chestnuts, previously to eating them. [B. H.]

BRACHIALIS. An ell long; twenty-four inches long.

BRACHIATE. When branches spread, at nearly right angles, alternately in opposite directions.

BRACHIUM. An ell, or two feet.

BRACHYPODOUS. Having a short foot or stalk.

BRACHYS, in words of Greek origin, signifies short.

BRACHTIA. A genus of South American orchids, related to *Brassia*, of which it has the organs of vegetation. The flowers are, however, very different. They are small, secund, half hidden by bracts, and densely arranged. In front of the ovary, and forming part of it, is a hollow tumour like a goitre, from the superior edge of which rises a simplex lamellate lip. There is no tendency to the tail-like extension of the sepals and petals, so characteristic of *Brassia*. Three species are known. The genus has also been called *Oncodia*.

BRACHYCARPÆA. A genus of *Cruciferæ*, allied to *Scorbiera*. It consists of undershrubs from the Cape of Good Hope, with oblong or linear entire mucronate leaves, and elongated racemes of large yellow or purple flowers; pouch two-celled, constricted in the line of junction of the two portions, sub-compressed, tuberculate, indehiscent, each end with one seed. [J. T. S.]

BRACHYCHITON. A genus of tropical, or sub-tropical Australian trees, belonging to the Sterculiaceous family, with alternate entire or variously lobed leaves, which are either smooth or covered with starry pubescence. The flowers are sometimes produced from the old wood two or three together, but more generally in terminal panicles, and have a tubular coloured calyx, without corolla. They are male and female on the same plant, the males having a great number of stamens, the stalks of which are more or less united, and the anthers packed in a round mass. The fruit is composed of five woody follicles, clothed inside with starry hairs, as also are the seeds, which are numerous. *B. acerifolium* is called the Flame Tree about Illawarra, on account of its bright red flowers, which make the tree a conspicuous object at a distance. It attains a height of from 60 to 120 feet, and a diameter of two to three feet. The bark is used by the aborigines for making fishing-nets, and the wood is soft and spongy. *B. populneum* is found in Eastern tropical Australia, and grows to a height of thirty or fifty feet, with a diameter of from eighteen to thirty-six inches. Its leaves are smooth, on long stalks, generally ovate and long pointed, but sometimes trilobed. The wood is soft, and contains gum mucilage. The tap roots of the young trees, as well as the younger roots of the large trees, are used by the natives as an article of food. The seeds are eaten, and the bark is put to the same uses as that of the maple-leaved species. *B. Bidwillii*, a native of the Wide Bay district, was sent to England in 1851 by Mr. Bidwill. Its leaves are stalked, heart-shaped, entire or three-lobed, and covered with a soft pubescence. The flowers are of a bright red colour, and are arranged in axillary bunches. The stems of this species show a tendency to become gouty, like those of the nearly-related 'gouty stemmed tree' of Australia (*Delabechei*). Five species of the genus are known. [A. A. B.]

BRACHYCOME. An Australian genus of composites, belonging to the *Bellis* section of the order, and comprising several neat annuals, of dwarf habit. Of these the most interesting is the *B. iberidifolia*, or Swan River Daisy, an elegant little plant, of branched diffuse habit, having the segments linear, and loose terminal corymbs of cineraria-like blossoms, each nearly an inch across. The colour of the ray florets varies from violet-blue to white, the disk or centre being in all cases of a purplish brown. *B. glabra*, of more recent introduction, has solitary flower-heads or long foot-stalks, about as large as those of *iberidifolia*, with a white ray of numerous linear florets, tinged with violet beneath, and a yellow disk; its foliage is pinnatifid, with linear segments variously cut, the uppermost ones being nearly entire, and all rather fleshy, and ciliated. The only other species in cultivation is *B. diversifolia*, with yellow flower-heads, rather smaller than in either of the preceding species, and foliage variously cut and lobed, as the specific name implies. The genus is characterised by a slightly conical, pitted, and naked receptacle: a cup-shaped involucre, the scales of which are membranous at the margins; and laterally compressed fruit, crowned with a pappus of very short bristle-like hairs. [W. T.]

BRACHYGLOTTIS. The plants which composed this genus have been shown to differ in no way from the Groundsels (*Senecio*), and are now generally referred to that genus; yet, although the characters of the flower indicate this structural affinity, they have little resemblance to any of the species of *Senecio* found in Europe, for they are trees or small shrubs, with woody stems, which are covered, as well as the under surface of the thin leathery leaves, with long or short dense woolly hairs. *B. Forsteri* has large broad deeply-toothed leaves, and terminal panicles of numerous small yellow flower-heads. It is a native of New Zealand, as are all the species, and is there known as Puka-Puka by the natives, who use the leaves as paper, whence the same native name came to be applied by them to English paper. [A. A. B.]

BRACHYLÆNA. A genus of South African evergreen shrubs, numbering six species, and belonging to the composite family. Their leaves are stalked, alternate, entire or toothed, generally smooth above, and covered with short white pubescence underneath. The flower heads are arranged in terminal panicles or racemes — those on one plant containing female florets only, the males being on another. The florets are yellow in colour. The genus is nearly allied to the American genus *Baccharis*, but is readily distinguished from that by having tails to the anthers. [A. A. B.]

BRACHYLEPIS. A genus of *Asclepiadaceæ*, containing a single species, a native of India. It is a climbing hairy shrub, with opposite acuminate leaves, and many small purple flowers in tomentose cymes, with numerous imbricated scales, on interpetiolar peduncles. The calyx and corolla are five-parted. The exserted stamens, with short broad filaments, are inserted along with the five scales of the staminal corona in the throat of the corolla. The anthers adhere to the lower margin of the stigma; their oval pollen masses united throughout the whole length of their inner surface by a flat membrane. The stigma is five-sided. The hairy follicles are widely separated, oblong and obtuse. [W. C.]

BRACHYPODIUM. A genus of grasses, belonging to the tribe *Hordeaceæ*, or barley grasses. The genus is chiefly distinguished from *Triticum* by the glumes being unequal, a circumstance which some authors do not consider of sufficient importance as a generic distinction; hence the species are referred either to *Triticum* or *Festuca*. Two are natives of Britain — the False Brome Grass, *B. sylvaticum*, and the Heath False Brome Grass, *B. pinnatum*. The former is a very common kind, which generally grows in shady woods, or on dry hedge banks; but the latter is rare, and only found wild in England. They are not grasses of agricultural importance, though useful species in their natural localities. [D. M.]

BRACHYPTERYS. A name indicative of the short wing borne by the fruits of this genus of *Malpighiads*. The species are natives of Brazil and Guinea, of climbing habit, with yellow flowers disposed in umbels. The calyx is five-parted, glandular; the petals unequal, longer than the calyx; the stamens ten, with a more or less enlarged glandular connective; the styles three, dilated at the apex into a rather large foliaceous recurved or hook-like and compressed mass. Fruit of three distended carpels (fewer by abortion), having at the apex a short compressed wing. [M. T. M.]

BRACHYSEMA. A genus belonging to the pea-flowered section of the leguminous family, and chiefly natives of West Australia. Seven species are enumerated, the greater portion of them scandent shrubs, but some erect. The upper petal or standard being short compared with the others, gives rise to the generic name. The pods are stalked, ventricose, and many-seeded. *B. aphyllum* is, as its name implies, without leaves, the branches being singularly compressed and winged, so as to perform the functions of leaves. Here and there on the branches small brown stipules are found, and from the axils of these the flowers grow; they are single, large, of a bright blood-red colour, and curiously reversed, the keel, or boat-shaped petal, usually lowest, being uppermost. Another leafless species, *B. pungens*, seldom grows more than a foot high, has innumerable spiny branches, and a dense mass of scarlet flowers, produced just above the ground, at the base of the stems. *B. lanceolatum* is a very handsome species, and a great ornament to greenhouses, flowering as it does in the winter or spring months. Its leaves are generally opposite, ovate or lanceolate in form, with a glossy upper surface, and covered with silvery adpressed pubescence underneath. The flowers are in axillary clusters, large, and rich scarlet. [A. A. B.]

BRACHYSORUS. A name proposed for a fern which proves to be *Asplenium sylvaticum*. [T. M.]

BRACHYSTELMA. A genus of South African *Asclepiadaceæ*, containing several species of under shrubs, with erect annual stems and large perennial tuberous roots. The calyx consists of five sepals; the corolla is campanulate and five parted. The staminal corona consists of five trilobed leaves attached to the middle of the gynostegium, which is included; while the anthers are simple and without a membrane, and the pollen masses roundish, and capped by a pellucid margin, at the base of which the two masses are attached to a slender corpuscle by two short processes. The two follicles are long and slender with numerous comose seeds. The roots are edible, those of some species being much esteemed as a preserve by the Dutch inhabitants of S. Africa. [W. C.]

BRACKEN or BRAKE. A common English name of *Pteris aquilina*.

BRACTEÆ or BRACTS. The leaves placed immediately below a calyx, if they are at all altered from their usual form.

BRACTEATE. Having bracts.

BRACTEOLÆ, BRACTEOLES or BRACTLETS. Bracts of a second order, usually smaller and more changed than the true bracts; also any small bracts.

BRADBURIA. The name given to a Texan herb which belongs to the composite family. It is an annual plant with slender straight stems about three feet high, and altogether sparingly covered with hairs, which gives rise to the specific name *hirtella*. The leaves are numerous, linear, very narrow, and about an inch long; the flowerheads solitary at the ends of the branches, and the florets yellow. The genus bears the name of Mr. J. Bradbury, who travelled in America in 1809, and published some interesting notes on the botany of the Missouri country. [A. A. B.]

BRAGANTIA. A genus of *Aristolochiaceæ*, consisting of undershrubs with decumbent wavy branches, thick leaves with prominent nerves, a regular flower with a thread-shaped calyx-tube adherent to the ovary, and a bell-shaped three-cleft limb; stamens six or nine, inserted on a shallow disc, surrounding the upper part of the ovary, and adherent to the base of the four connate styles; capsule pod-like, quadrangular, four-celled, four-valved, many-seeded. These plants are also remarkable for the structure of their wood, which differs considerably from the ordinary wood of Exogens. They are natives of the tropical parts of Asia, and possess in some degree the properties of the Aristolochias. *B. tomentosa* is very bitter, and, according to Dr. Horsfield, is used medicinally in Java. Major Drury in his work on the useful plants of India, says that the natives of the western coasts of India use the leaves and roots of *B. Wallichii* rubbed up with lime juice, as a cure for snake bites; the whole plant mixed with oil in the form of an ointment is used in the treatment of inveterate ulcers. It used to be considered as an antidote to poison. A Malabar proverb says, as soon as the Alpam root, that is the root of this species, enters the body, poison leaves it. [M. T. M.]

BRAHEA. Certain fan-leaved palms, inhabiting Peru, the Andes, &c., have been collected by Martius into a genus with the above name. They are trees of moderate height, with fan-like leaves and spiny leafstalks; flowers hermaphrodite, greenish, with a calyx of three sepals overlapping at the margins; six stamens, connate in a sort of cup around the base of the ovary. [M. T. M.]

BRAHMIN'S BEADS. An Indian name for the corrugated seeds of *Elæocarpus*, which are used by the Brahmins, and also made into necklaces, &c.

BRAINEA. A genus of polypodiaceous ferns, now included in the group *Hemionitideæ*, in which it is distinguished by its primary veins anastomosing in an arcuate manner, so as to form a series of areoles next the costa, while the venules, which are parallel and oblique, are quite distinct to their apices. It has naked or non-indusiate sori continuous along the course of the transverse curved veins which unite to form the costal areoles, and often extended more or less along the parallel oblique free venules, becoming at length irregularly confluent.

B. insignis is the only species known. This is a native of Hong Kong, and forms a very handsome dwarf tree fern with a stem of three to four feet in height. The fronds are three feet long or more, pinnate, the pinnæ sometimes becoming pinnatifid; they are rigid and subcoriaceous in texture, and serrated along the margin. It is a very elegant and interesting plant. The genus has some points of resemblance to *Sadleria*, a genus of *Lomarieæ*, but differs in having naked instead of indusiate sori, and in some other particulars. We had formerly regarded it as presenting a connecting link between the *Meniscieæ* and the *Lomarieæ*, through the *Woodwardieæ*, and had placed it in the former group in consequence of its short transverse naked sori: but now that more perfect specimens in the fresh state have been examined, we are quite ready to adopt the suggestion of Sir W. J. Hooker, that it may be referred to the *Hemionitideæ*, the sori not proving to be short and lunate, but continuous along the arcuate veins. It is, however, even here, somewhat anomalous, the fructifications being merely branched and not truly reticulated. 'We have here,' Sir W. J. Hooker observes, 'a very remarkable, and, if I may so say, a new form among the Filices. In its arborescent caudex it reminds one of some of the cyatheaceous group of tree ferns, though not of one of the loftiest character; in its foliage it resembles several species of *Lomaria*; in its venation a *Woodwardia*, and in the more fully developed fructification an *Acrostichum*.' (T. M.)

BRAKES. A common English name of *Pteris aquilina* and the related species or varieties. —, ROCK. *Allosorus crispus*.

BRAMBLE. The common name for *Rubus fruticosus* and the allied plants. —, DOG. *Ribes Cynosbati*. —, MOUNTAIN. *Rubus Chamæmorus*, the Cloudberry.

BRAMIA. The title of a section of the genus *Herpestes* (Scrophulariaceæ) characterised by having the upper lip of the corolla deeply bifid. [W. C.]

BRANC-URSINE. (Fr.) *Acanthus mollis*; also *Heracleum Sphondylium*.

BRANDESIA. A section of the Amaranthaceous genus *Telanthera*, which consists of tropical plants (chiefly American) allied to the globe amaranth of the gardens. *Brandesia* is distinguished from the other sections by not having the calyx distinctly jointed to the extremely short pedicel, and its segments being nearly equal. The flowers are each accompanied by three bracts, and are in long-stalked globular or ovoid heads; the stamens are united into a tube by the adherence of the filaments. *B. porrigens* has the heads of flowers deep purple, resembling those of *Sanguisorba officinalis*, but dry like those of the 'everlasting' flowers. (J. T. S.)

BRANDY BOTTLES. A local English name for the flowers of *Nuphar lutea*.

BRANK. *Fagopyrum esculentum*.

BRASAVOLA. In the tropics of America, and in no other part of the world, occur many species of orchids with slender fleshy stems, solitary succulent usually pandaniform leaves, and large greenish flowers, with narrow acuminate or long-tailed petals, and a similar entire sometimes very broad lip. They have also a column with a pair of great falcate ears on each side of the front, and eight pollen masses. To these the name of *Brasavola* has been given. A few species have been added, in which the appendages or ears of the column are small and toothed. The most remarkable are *B. glauca*, with glaucous flat fleshy leaves, and very large flowers, from Mexico; and *B. Digbyana*, which differs in little, except having the margin of the lip broken up into long hair-like fringes.

BRASSICACEÆ. The Cabbage family, a natural order of Thalamifloral Exogens, to which the name of CRUCIFERÆ (which see) is usually given.

BRASSIA. An extensive genus of tropical American orchids, very nearly related to *Oncidium*, from which they are easily known in most cases by the lateral sepals being very much longer than the other parts of the flower. Since, however, this is also the case in *Oncidium phymatochilum* and some others, the distinction fails, and botanists are obliged to combine with long tail-like sepals, a short earless column, and a pair of vertical plates on the lip. In attempting to define what a *Brassia* is, Lindley enumerates seventeen species and many varieties, all pseudobulbous and bearing flowers more or less yellow, in simple racemes. They are chiefly handsome enough to deserve the gardener's care.

BRASSICA. A remarkable group of plants, of the order *Cruciferæ*. As constituted by Bentham, and characterised by its conduplicate cotyledons, and its siliquose beaked pods, this genus is made to include the mustards (*Sinapis*), an allocation to which we incline, both from experiment and observation. We shall, however, confine our remarks to the genus as constituted by Linnæus, of which the following are species :— *B. oleracea*, Wild Cabbage; *B. campestris*, Wild Navew, including *B. Rapa*, the Turnip; *B. Napus*, Rape or Coleseed. Of these, the first is in all probability the initiative. It occurs wild on rocks and cliffs by the sea shore; and we have now in cultivation some curious examples, derived from seed gathered from the rocky coast of Llandudno, North Wales, which already give indications of sports in several directions. Some have the short petioles and the close hearting condition of cabbages, of which form we have both green and red varieties, the tendency being much increased by repeated transplantation. Others, with longer petioles and lyrate leaves, seem to take on that looser method of growth which constitutes the 'Greens' and Kale of the garden; whilst some present that peculiar glaucous hue which belongs more particularly to rape. We should not, therefore, be surprised if experiment should ultimately establish the position that the *B. oleracea* is the only true species of the three above enumerated, and that the *B. campestris* and *B. Napus* are but agrarian forms derived from the cultivated varieties of this. This opinion is countenanced by the fact that nowhere are the two latter truly wild, but both track cultivation throughout Europe, Asia, and America. The protean forms induced from the *B. oleracea* are well known, such as many varieties—which are yearly increasing—of Cabbage, Broccoli, Cauliflower, Kale, and Kohl-rabi; whilst the no less numerous varieties of the common Turnip are all referred to *B. campestris*, with which, indeed, Bentham classes *B. Napus*.

As regards the Swedish Turnip, we are in the position to state that the seeding of rape and common turnips in mixed rows has resulted in the production of a small percentage of malformed Swedes, which, however, improved very much by careful cultivation; and our field observations have enabled us to detect in degenerate

Swedes a disposition to a negation of bulbs, and the production of monstrous rape plants—a tendency which is at once observable when this crop assumes a necky top, or many heads, which shows an inclination to 'run,' or when it forms a branchy finger-and-toe root growth, which indicates a breaking up of the bulb into ordinary roots. It may be remarked, as throwing some light on the nature of the changes by which the cultivated varieties of this genus have been attained, that experiments with seeds of plants showing any particular tendency, and especially if repeatedly grown in the same soil, will ever result in an increase of the peculiarity. [J. B.]

This genus comprises some of the most ancient and useful of our culinary vegetables, most of them possessing high antiscorbutic powers, which are believed to depend upon a certain acrid volatile oily principle—the chemical nature of which is imperfectly known. In common with the rest of the cruciferous order, they also possess a greater share of azote than any other tribe of plants, as is apparent in their fetid smell when fermented.

The Cabbage, *B. oleracea*, in its wild state, is a native of various parts of Europe, as well as of several places near the sea in England. It is a biennial, with fleshy lobed leaves, undulated at the margin, and covered with bloom; altogether, so different in form and appearance from the Cabbage of our gardens, that few would believe it could possibly have been the parent of so varied a progeny as are comprised in the Savoy, Brussels Sprouts, Cauliflower, Broccoli, and their varieties. A more wonderful instance of a species producing so many distinct forms of vegetation for the use of man is scarcely to be met with throughout the range of the vegetable kingdom.

The Common or cultivated Cabbage, *B. oleracea capitata*, is well known, and from a very early period has been a favourite culinary vegetable, in almost daily use throughout the civilized world. The ancients considered it light of digestion, when properly dressed, and very wholesome if moderately eaten. For the introduction of our garden variety of Cabbage, we are indebted to the Romans, who are also believed to have disseminated it in other countries. It is said to have been scarcely known in Scotland until the time of the Commonwealth, when it was carried there from England by some of Cromwell's soldiers; but it now holds a prominent place in every garden throughout the United kingdom. In general, cabbages are preferred when of a large size, thoroughly hearted and blanched within; they are not, however, then by any means so digestible and wholesome as when cut and eaten in a young state—that is to say, before the heart has become firm and hard. It is a remarkable fact, that all the varieties are sweeter and better flavoured after being touched with frost.

In Germany, salted cabbage, or *sauer kraut*, is much esteemed, and forms a kind of food, of which large quantities are prepared for winter use. It is made by cutting the cabbages into small shreds, and afterwards packing them in barrels, in layers three or four inches thick. Over each layer is thrown a certain quantity of salt and unground pepper, with a few cloves; and the whole is then well mixed, and pressed as hard as possible. Other layers are put in, and treated in the same way, until the barrel is full. A board is then placed on the top, on which heavy weights are put, and in this state it remains for ten or fifteen days, when it partially ferments, and a great deal of water rises to the surface. It is then placed in the cellar, and continues in excellent condition for use until late in the spring.

The Red Cabbage, *B. oleracea rubra*, is a very distinct variety, remarkable for the peculiar purple, or brownish-red colour of its leaves. It is chiefly used for preserving as a pickle, for which purpose it is greatly esteemed, and by proper management makes one of the most beautiful pickles that can be presented at table.

The Borecole, *B. oleracea acephala*, has every appearance of being one of the early removes from the original species. It is distinguished from the other sorts of cabbages by its leaves being beautifully cut and curled, of a green or purple colour, or variegated with red, green, and yellow, never closing, so as to form a head, nor producing eatable parts like the Cauliflower. Several sub-varieties of Borecole are well known under the names of German greens, Buda Kale, Scotch Curlies, or Kale—all of which are so hardy as to be able to endure severe frost, and continue green and fresh throughout the winter. The part which is used is the crown, or centre of the plant, cut so as to include the young and most succulent leaves. When properly dressed, they are tender, sweet, and delicate, more particularly after being exposed to frost.

The Large-ribbed Cabbage, or Couve Tronchuda, *B. oleracea costata*, is a variety peculiar to Trauxula, in Portugal, from whence it was introduced in 1821. It represents a singular race of the cabbage family, and is characterised by its leaves having very large midribs, which, when divested of their green parts, and thoroughly boiled, make an excellent vegetable for serving up in the manner of sea kale. The heart, or middle part of the plant, has likewise been found very delicate, tender, and agreeably flavoured.

The Savoy Cabbage, *B. oleracea bullata*, differs but little from the other kinds of heading or hearting cabbages, and is chiefly distinguished by its leaves being wrinkled in such a manner as to have a netted appearance. It has been cultivated in our gardens for three centuries. When fully headed, it forms an excellent hardy winter vegetable, for using in the same way as other cabbages, but it is not so delicately flavoured.

The Brussels Sprouts, or Bud-bearing Cabbage, *B. oleracea bullata minor*, originated in Belgium, and has been cultivated around Brussels from time immemorial; although it is only within the last twenty years that it has become generally known in this country. It is very hardy, and forms a head somewhat like a savoy, of which it is considered to be a sub-variety, differing in the remarkable manner in which it produces at the axils of the leaves, along the whole length of its stem, a number of small sprouts, resembling miniature cabbages, of one or two inches in diameter. These are peculiarly well-flavoured, and, as a winter vegetable, are more highly esteemed than any other kind of cabbage in cultivation.

The Cauliflower, *B. oleracea botrytis cauliflora*, is of great antiquity, but its origin is unknown, although it has been usually ascribed to Italy. It is mentioned by Gerarde, and must therefore have been in this country previous to 1597. It differs in a remarkable manner from all the other varieties of the cabbage tribe, whose leaves and stalks are alone used for culinary purposes. Instead of these being taken, the flower-buds and fleshy flower-stalks form themselves in a close firm cluster or head, varying from four to eight inches or more in diameter, and become one of the greatest of vegetable delicacies. It is not valued so much for its large size, as for its fine creamy white colour, its compactness, and regular form, without being warty, which features constitute the properties of a fine Cauliflower. The uses of Cauliflower are well known. When dressed it is served up at table, either plain boiled, or with white sauce. It forms an excellent addition to vegetable soups, and is often used for pickling. It may also be preserved for a considerable time, when pickled like saur kraut.

The Broccoli, *B. oleracea botrytis asparagoides*, is similar in form and appearance to the cauliflower, from which it is supposed to have originated. It was first brought into notice at the beginning of the last century. Two kinds, the white and purple, are mentioned by Miller (1724) as coming from Italy, and from these have arisen all the varieties that are now in cultivation. Broccoli is more robust and far more hardy than cauliflower, for which it becomes a valuable substitute during the winter and spring months, when the latter cannot be obtained. The heads vary in colour, from a brownish purple to a pure creamy white, in which state they are scarcely to be distinguished from cauliflowers. They are used for the same purposes, but are not so delicate in flavour.

The Turnip, *B. Rapa depressa*, is a hardy biennial, and, in its wild state, is found in corn fields in various parts of England. The change it has undergone by cultivation is no less remarkable than that of the Cabbage; but in this instance it is the root which has been transformed from a comparatively hard woody substance into the large fleshy bulb, which constitutes one of our most nutritious vegetables. The ancient Greeks and Romans were well acquainted with the Turnip; and, in the fifteenth century, we find it had become known to the Flemings, and formed one of their principal crops. The first Turnips that were introduced into this country are believed to have come from Holland in 1550; and, among all the varieties now cultivated for culinary purposes, the Early Dutch continues to be generally esteemed. The Turnip forms an ingredient in broths, soups, and stews; it is likewise cut into various figures for garnishing. In spring, when the plant is pushing up for flower, the points of the shoots are dressed as greens, and are acknowledged to be valuable as an antiscorbutic.

Rape, *B. Napus*, is a hardy biennial, indigenous to Britain. It is chiefly grown for cutting when quite young, and mixing with mustard, as a small salad. It is also sometimes cultivated in cottage gardens, for spring greens—the tops being cut first, and afterwards the side shoots. The Teltow Turnip, or 'Navet de Berlin petit' of the French (*B. Napus var.*), is very different from any of our cultivated varieties of Turnip, its root being long and spindle-shaped, somewhat resembling a carrot. Its culture in this country dates from 1790; but it was well known in 1671, and is noticed by Caspar Bauhin in his *Pinax*. It is much more delicate in flavour than our Common Turnip. In France and Germany it is extensively cultivated, and few great dinners are served up without it in one shape or other. It enriches all soups by the peculiar flavour contained in the outer rind, which is thin, and must not be cut away, but scraped. Stewed in gravy, it forms a most excellent dish, and, being white, is very ornamental when mixed and served with carrots. [W. B. B.]

BRAVAISIA. The *Onychacanthus Cumingii* of Nees von Esenbeck. [B. S.]

BRAVOA. A genus of Amaryllids, containing a single species, *B. geminiflora*, native of Mexico. This is a bulbous or rather a tuberous plant, with a tuft of radical linear keeled leaves, and an erect flower-stem, a foot or more in height, supporting a raceme of nodding flowers, which grow in pairs, and are of a rich orange-red outside and yellowish within. The tuber is somewhat elongated, tunicated, sending down several thick fleshy roots. The perianth, which is rather over an inch long, consists of a funnel-shaped curved tube widening at the throat, and having a very short six-cleft somewhat spreading limb. There are six stamens, and an inferior three-celled ovary, with a long exserted filiform style terminated by a dilated stigma. It is a very graceful plant. [T. M.]

BRAYERA. A genus of *Rosaceæ* named after a French physician, Dr. Brayer, who observed the valuable medicinal properties of the only species of this genus, and sent a specimen of the plant to Kunth. The plant is known by its top-shaped calyx, the limb of which is divided into ten segments, five exterior to the remainder, which are much smaller and of a different shape; it has two small bracts at the base. Petals five, small, linear, scale-like; stamens fifteen to twenty, inserted with the petals into the throat of the calyx, the filaments unequal in length. Carpels two at the bottom of the calyx, one to two-seeded; style terminal; stigma peltate.

B. anthelmintica, the only known species, is an Abyssinian tree with alternate pinnated leaves and diœcious flowers; in the true female flowers the petals and stamens

Brayera anthelmintica.

are entirely wanting. The flowers of this tree have been long used by the natives as a vermifuge, and have proved very efficacious in the removal of tape-worm in this country. The cause of its peculiar effects is not well understood. [M. T. M.]

BRAYETTE. (Fr.) *Primula (officinalis) veris*.

BRAZILETTO. The common name for *Cæsalpinia*; also specially, the colonial name of *C. braziliensis*, the timber of which is used for cabinet-work.

BRAZIL WOOD. A dye wood obtained from *Cæsalpinia echinata*, and other species.

BRAZORIA. A genus of *Labiatæ*, natives of Texas, and containing two species of erect branching herbs with the lower leaves petiolate and obovate-oblong, the upper serrate and lanceolate, and the flowers in simple terminal spikes. The calyx is campanulate and two-lipped, with the upper lip bilobed and the lower trilobed. The corolla tube is considerably exserted, the throat inflated, and the limb bilabiate, with the upper lip erect, slightly bilobed or entire, and the lower deeply trifid, with its roundish lobes spreading or recurved. [W. C.]

BREAD-FRUIT. *Artocarpus incisa*.

BREAD-NUT. The seed of *Brosimum Alicastrum*.

BREAD-ROOT. *Psoralea esculenta*.

BREAD, TARTAR. The fleshy root of *Crambe tatarica*.

BREAD-TREE, of N. Australia. *Gardenia, or Alibertia edulis*.

BREAK-YOUR-SPECTACLES. A vulgar name for the Blue-bottle, or Corn Bottle, *Centaurea Cyanus*.

BREATHING-PORES. See Stomates.

BREDES. (Fr.) *Solanum nigrum*. — D'ANGOLE. *Basella rubra*. — GLACIALE. *Mesembryanthemum crystallinum*.

BREDEMEYRA. A plant which is imperfectly known and referred doubtfully to the milkwort family (*Polygalaceæ*). It is described as being a shrub with alternate lanceolate entire leaves, which are stalked and smooth, and numerous yellow flowers which are disposed in terminal much-branched panicles. The calyx is five-leaved, two of the leaves petal-like; the petals are three in number, the intermediate one keeled; the stamens eight, united at their base; and the fruit an ovate two-celled drupe. The plant is said to be a native of Venezuela. [A. A. B.]

BREHMIA. A genus of *Loganiaceæ*, containing but one species, *B. spinosa*, which is, as its name implies, a spiny shrub. It grows about ten feet high, and is furnished with opposite stalked three or five-nerved entire leaves, elliptical in form, and small green flowers arranged in dense cymes at the apex of the branches. The hard-shelled ripe fruit resembles an orange in size and appearance, and contains many seeds immersed in a copious pulp. The genus differs chiefly from the nux-vomicas (*Strychnos*) in having a one-celled ovary. The plant is found in Madagascar, where it is called 'Voiva Vountaca,' in eastern Africa, and also on the west coast. The pulp of the fruit is commonly eaten by the natives wherever it grows; it is somewhat acid, and said to be delicious; but probably the seeds, from its near relationship to the nux-vomicas, are poisonous. It is, however, remarkable that the pulp of many species of *Strychnos*, whose seeds are a deadly poison, is perfectly harmless. [A. A. B.]

BREJEUBA. A kind of cocoa-nut, the wood of which is used by the Brazilian Indians in making their best bows.

BRESINE. (Fr.) *Zinnia multiflora*.

BREWERIA. A genus of herbs or undershrubs, natives of New Holland, tropical Asia, and Madagascar, belonging to the order *Convolvulaceæ*. They have alternate entire leaves, and solitary axillary flowers. The calyx consists of five sub-equal sepals; the corolla is campanulate and plaited. There are five included stamens. The ovary is two-celled, with two ovules in each cell, and bearing a style which is divided nearly half its length, and has a capitate stigma on each division. The capsule is two-celled and contains four seeds. [W. C.]

BREXIA. A genus of small trees belonging to *Brexiaceæ*, an order whose affinities are doubtful. They are natives of Madagascar, and have alternate petiolate leathery leaves, entire or furnished with spiny teeth. The flowers are in axillary umbels, of leathery texture and greenish hue; calyx with five short segments; petals also five; stamens five, arising from a toothed disk surrounding the base of the ovary, and adnate with it; ovary five-celled. Fruit drupaceous, five-ribbed, slightly papillose, about the size of an orange; seeds numerous. In cultivation they are handsome hothouse plants, usually called *Theophrastas*. [J. T. S.]

BREXIACEÆ. The genera *Brexia*, *Ixerba*, *Argophyllum*, and *Roussea*, each consisting of only one or two species, have been proposed as a small family allied to *Saxifragaceæ*. They are, however, not all very closely connected with each other, and neither form a natural group nor are they united by any well-marked common character. It is probable that when the very varied forms now provisionally grouped round the *Saxifragaceæ* shall have been thoroughly revised, the *Brexiaceæ* will be broken up, and the genera distributed into new combinations better defined than the present group.

BRIDGESIA. A genus of the soapwort family (*Sapindaceæ*) indigenous in the province of Coquimbo in Chili. The cut-leaved *B. incisifolia* is the only known species, and is a shrub three to five feet high with alternate stalked simple lobed leaves, and with the flower-stalks axillary and single, some of the flowers, which are small, bearing stamens only, others with both stamens and pistil. The fruit is a three celled bladdery capsule, each of the cells prolonged into a wing on the back, and containing a single seed. The genus may be distinguished from any in the family by its leaves alone. It bears the name of Mr. Bridges, a most extensive collector of Chilian plants. The same name has been also given to a group of Phytolaccals, now included in *Ercilla*; and to a group of composites, now included in *Polyachyrus*. [A. A. B.]

BRIER, or BRIER ROSE. The common larger-growing British species of Rose, especially the Dog Rose, *Rosa canina*; sometimes written Briar. —, SWEET. The Eglantine, *Rosa rubiginosa*.

BRIGALOW. *Acacia excelsa*.

BRIGNOLES. The dried fruits known as Provence prunes or French plums.

BRILLANTAISIA. A genus of *Acanthaceæ*, containing one or two species, natives of Guinea. They are erect branching herbs with ovate-cordate leaves on long petioles, and large purple flowers in terminal panicles, with small linear bracts. The calyx is five-parted, with unequal linear segments, the upper being the longer. The ringent corolla has the upper lip falcate and overarching with a trifid apex, and the inferior large, spreading, and shortly trifid. There are two fertile stamens inserted at the top of the tube, and having long linear bilocular anthers; the two barren stamens are represented by short filaments. The ovary is oblong, hairy, and surrounded by a disc, and bears a style of the same length as the corolla, terminated by an unequally bifid stigma. The capsular fruit is straight, narrow, tetragonous, and two-celled, with six to eight seeds in each cell. [W. C.]

BRIMSTONE, VEGETABLE. The inflammable spores of *Lycopodium clavatum* and *Selago*, employed on the continent in the manufacture of fireworks.

BRINJAL. The fruit of the egg-plant, *Solanum Melongena*. Mr. Bentham writes it Bringall.

BRINVILLIERA. (Fr.) *Spigelia Anthelmia*.

BRISTLEWORTS. A name applied by Lindley to the *Desvauxiaceæ*.

BRISTLY. Covered with stiff sharp hairs, or bristles.

BRITTLEWORTS. A name given by Lindley to the *Diatonaceæ*.

BRIZA. A genus of grasses belonging to the tribe *Festuceæ*, distinguished chiefly by the inflorescence being in panicles, the spikelets of which contain from five to twelve imbricated flowers; and in the two glumes being nearly equal, and like the pales membranous, with scarious margins. The Quaking Grasses are all handsome plants, so much so, that *B. maxima* and *B. minor* are frequently cultivated in gardens as ornamental annuals. Steudel describes thirty species, which are mostly natives of South America: Brazil, Chili, and Peru, being the principal countries which produce them. Two are British plants, *B. media* and *B. minor*; the former a very common species, on light limestone soils, &c., the latter confined to a few localities in England. They are not of agricultural importance, though *B. media* is a prevailing grass on some good permanent pastures. [D. M.]

BRIZOPYRUM. A genus of grasses belonging to the tribe *Festuceæ*. Eleven species are described, which are mostly natives of the Southern Hemisphere. One is, however, from Nootka Sound, *B. borea'e*, and another curious species found by Drummond in Australia, *B. scirpoides*, has leafless culms rising to the height of four feet. [D. M.]

BROAD SEED. The common name for *Clospermum*.

BROCCOLI. A cultivated variety of the Cabbage, *Brassica oleracea*, in which the young inflorescence is condensed into a depressed fleshy edible head.

BRODIÆA. A small genus of *Liliaceæ*, consisting of bulbous plants from Western North America, with linear leaves and naked scapes terminated by an umbel of rather large blue flowers. The base of the umbel is surrounded by an involucre of small scarious bracts; perianth funnel shaped, six-cleft; stamens three, attached to the perianth, alternating with three scales (abortive stamens). The ovary is surrounded by a fleshy three-lobed hypogynous disk. The bulbs are small, enveloped in a dark rough coat. [J. T. S.]

BROKEN, when applied to a whorl, signifies that the parts thereof are not all on the same plane. In fact, they form a part of an extremely short spiral, as may be seen in the calyx of any species of *Hypericum*.

BROME, FALSE. A common name for *Brachypodium*.

BROMELIACEÆ (*Bromeliæ, Tillandsieæ, Bromeliads, Bromeliworts*, the *Pine-Apple family*). A natural order of epignous monocotyledons included in Lindley's narcissal alliance. Short-stemmed plants with rigid channelled often scurfy and spiny leaves, and showy flowers. Outer perianth (calyx) three-parted, persistent; inner (corolla) of three withering petals; stamens six, inserted in the tube of the perianth; anthers opening on the side next the pistil; style single. Fruit either a dry capsule or succulent, three-celled, many-seeded; embryo very small, at the base of mealy albumen. Natives of the American continent and islands, whence they have been distributed to Africa and the East Indies. *Ananassa sativa*, the Pine-apple or Ananas, is one of the most important plants of the order. Its fruit is composed of the pistils and bracts of several flowers united into a succulent mass and crowned by a series of green leaves. It is *par excellence* the fruit of the Eastern Islands. The fibres of the plant are used in manufacture. *Bromelia Pinguin* is a remedy for worms in the West Indies. Some of the Bromeliads grow attached to the branches of trees, and are called Air-plants. One of these is *Tillandsia usneoides*, the Tree-beard of South America, which consists of a mass of black fibres. These are employed for stuffing cushions, under the name of Spanish Moss, Black Moss, or Long Moss. There are twenty-eight known genera and 174 species. Illustrative genera: *Ananassa, Bromelia, Æchmes, Bilbergia, Tillandsia, Bonapartea*. [J. H. B.]

BROMELIA. The natural order *Bromeliaceæ* takes its name from this genus, which consists of plants with short stems, and densely-packed rigid leaves, generally lance-shaped, with spiny margins, and channelled on the upper surface. The calyx is three-parted, much shorter than the corolla, which consists of three petals, convolute, erect or spreading at the top. The stigmas are three — short, fleshy, and erect. The fruit is succulent. The fruit of *B. Pinguin* yields in the West Indies a cooling juice, much used in fevers, etc. Many of them supply valuable fibre for textile purposes, and which might also be employed in the manufacture of paper. Several species are cultivated in stoves for their ornamental flowers. [M. T. M.]

BROMHEADIA *palustris*. In the Malay Archipelago there grows in bogs the orchidaceous plant to which this name has been given. It has the habit of such New World species as *Epidendrum elongatum*, the stems being erect, and clothed with leathery distichous leaves. The flowers, which are large and white, with a purple and yellow lip, are placed close together on a stiff zigzag rachis, which is in some cases branched. The lip, which is cucullate, and firmly fixed so as to be parallel with the column, has the unusual character of bearing a long woolly ridge in the middle. Mr. Finlayson first detected it near Singapore.

BROMUS. A genus of grasses, belonging to the tribe *Festuceæ*, distinguished chiefly by the inflorescence being in lax panicles, very rarely crowded; glumes unequal, containing from three to many flowers, the spikelets lanceolate and compressed; ovules two, the lower with a long awn attached nearly at the tip; styles below the top of the fruit. Steudel describes 141 species in his *Synopsis*. They have a very extensive geographical range. The greater number are, however, natives of temperate climates, and those that approach tropical limits generally grow at considerable elevations on mountains. About eight species are natives of Britain, along with some which have been introduced, and are now enumerated in British floras. They are not considered first-class agricultural grasses, though the Soft Brome Grass, *B. mollis*, constitutes a large portion frequently of good meadows, but being of annual duration only, it is not so common on good permanent pastures. *B. erectus* is a strong growing perennial species, which is rather abundant in some districts, and scarce in others. The Tall Brome Grass, *B. asper*, is one of the most beautiful of grasses, although a coarse kind, of little agricultural importance. [D. M.]

BRONGNIARTIA. A genus of the pea-flowered tribe of the leguminous family, numbering eight species, all of them natives of Mexico or Texas. They are shrubs, with unequally pinnate leaves, and many pairs of ovate or elliptical leaflets, which are generally about half-an-inch in length. The flower-stalks are twin, in the axils of the upper leaves, and the flowers flesh-coloured, or violet, the keeled petal yellow. The pods are stalked, thin, and in form like the blade of a table knife, but pointed, and contain six to eight seeds. None of the species are in cultivation. The genus is named in honour of Adolphe Brongniart, a famous French botanist. [A. A. B.]

BRONTESIS. A name given to express the injury done to plants by lightning. This is generally clear enough from the outward effects, the branches being broken, and the trunk shivered. The injury, however, may be more insidious, and, though no external damage may appear, or none which immediately excites attention, the connection of the component parts of the trunk may be dissolved more or less completely, by the sudden generation of gas, or the expansion of the sap, from the intense heat of the lightning. The whole vegetative power of a tree may also be at once arrested. But many of the cases of sudden death which are commonly attributed to lightning are the results of the spawn of some fungus attacking the roots, vegetation being kept up by a slight thread of sound tissue, as in the condition called gumming; and when this at last gives way, the plant at once perishes. [M. J. B.]

BROOK-BEAN. *Menyanthes trifoliata*.

BROOKLIME. *Veronica Beccabunga*.

BROOKWEED. The common name for *Samolus*.

BROOM. *Cytisus*, or *Sarothamnus scoparius*; also applied to *Lygeum Spartum*. —, AFRICAN. A common name for *Aspalathus*. —, DYER'S. *Genista tinctoria*. —, RUSH. A common name for *Viminaria*; also applied to *Spartium junceum*. —, SPANISH. *Spartium junceum*.

BROOM CORN. *Sorghum vulgare*, the branched panicles of which are made into carpet brooms and clothes-brushes. Also *Sorghum saccharatum*.

BROOM RAPE. The common name for *Orobanche*. —, NAKED. An American name for *Aphyllon*.

BROOMEIA. A most remarkable genus of puffballs, which has at present occurred only in South Africa. The inner sac, or peridium, is precisely like that of *Geaster*, but completely exposed, the outer sac being represented by a thick corky stratum, in which a multitude of individuals are half sunk, like jewels in a matrix. Some approach to it is made by a fine compound species of starry puffball, found in Ceylon and Cuba, though in that case the inner peridium is not at all exposed. [M. J. B.]

BROSIMUM. A genus of the order of artocarpads (*Artocarpaceæ*), containing six or seven species, natives of tropical South America. They are large trees, abounding in milky juice, and having entire leaves. Their male and female flowers are generally congregated into a globular head, but are sometimes borne on separate trees: they have neither calyx nor corolla, the males consisting of single stamens, separated from each other by shield-like scales, and the females of a solitary style, terminating in two stigmas. The fruit is a small one-seeded berry.

B. Alicastrum, the Bread-nut tree of Jamaica, has a tall straight trunk, and smooth shining deep-green elliptical lance-shaped leaves. Its pale-yellow heads of flowers are succeeded by round yellow fruits, about an inch in diameter, and containing a single seed, called Bread-nut in Jamaica. These so-called nuts are eatable, and are said to form an agreeable and nourishing article of food: when boiled or roasted, they taste like hazel-nuts. The young branches and shoots, also, are an excellent fodder for horses and cattle; and the wood, which bears some resemblance to mahogany, is used by West Indian cabinet-makers.

B. Aubletii, a native of British Guiana and Trinidad, also forms a large tree, often sixty or seventy feet high, and two or three feet thick. The leaves are of an oblong form, with their top end broader than the bottom; and they are covered with a whitish down on the under surface. The heart wood of this tree is exceedingly beautiful, being of a rich brown colour, and mottled with irregularly-shaped dark spots, on which account it is called Letter-wood, Snake-wood, or Leopard-wood. Unfortunately, however, it is only procurable in narrow pieces, and is therefore chiefly used for veneering small articles of furniture, and for making walking-sticks, which, however, are very liable to split.

B. Galactodendron, which is the Cow-tree of South America, yields a milk of as good quality as that from the cow. It forms large forests on the mountains near the town of Cariaco, and elsewhere along the sea-coast of Venezuela — growing to upwards of 100 feet high, with a smooth trunk six or eight feet in diameter, and without branches for the first sixty or seventy feet of its height. The leaves are of a leathery texture, strongly veined, and of a deep shining green colour. They are about a foot long, and three or four inches broad, of a somewhat elliptical form, terminating in a sharp point. In South America the Cow tree is called Palo de Vaca, or Arbol de Leche. Its milk, which is obtained by making incisions in the trunk, so closely resembles the milk of the cow, both in appearance and quality, that it is commonly used as an article of food by the inhabitants of the places where the tree is abundant. Unlike many other vegetable milks, it is perfectly wholesome, and very nourishing, possessing an agreeable taste, like that of sweet cream, and a pleasant balsamic odour; its only unpleasant quality being a slight amount of stickiness. The chemical analysis of this milk has shown it to possess a composition closely resembling some animal substances; and, like animal milk, it quickly forms a yellow cheesy scum upon its surface, and, after a few days' exposure to the atmosphere, turns sour and putrifies. It contains upwards of thirty per cent. of a resinous substance, called *galactin* by chemists. [A. S.]

BROSSÆA. An imperfectly known genus of *Vacciniaceæ*, comprising a West Indian shrub, bearing solitary axillary or a few termina flowers, whose stalks have two bracts, a conoid corolla, five included stamens. The capsule has five many-seeded compartments, and is covered by the enlarged limb of the calyx; seeds very small. *B. coccinea* is in cultivation, and is described as a cistus-like shrub, with scarlet flowers half an inch long. [M. T. M.]

BROUALLE E'LEVE'E. (Fr.) *Browallia elata.*

BROUGHTONIA *sanguinea* is a handsome West Indian epiphytal pseudobulbous orchid with oblong coriaceous leaves, and a short spike of deep crimson flowers. It has a spur completely immersed beneath the surface of the ovary. It is common in Cuba on bushes, but more usually comes from Jamaica. Its nearest affinity is with *Lælia* and *Cattleya.*

BROUSSONETIA. This genus is allied to the mulberry, and belongs to the same order of morads (*Moraceæ*). Three species have been defined, but they may probably be all referred to one, namely, *B. papyrifera*, the Paper Mulberry, which is so called on account of its fibrous inner-bark being used by the Japanese and Chinese for making paper. It grows wild in China and Japan, and also in many of the islands of the Pacific Ocean, where the natives manufacture a large part of their clothing from its bark. It forms a small tree, attaining about twenty or thirty feet high, with a trunk seldom more than a foot in diameter, and generally branching at a short distance from the ground. The young branches are covered with short soft hairs. The leaves are deciduous, and vary very much in shape, those of young trees being frequently divided into three or five sharp-pointed irregular lobes, while those of older trees are mostly entire and of a somewhat egg-shaped outline; they are very rough upon the upper surface, and slightly hairy beneath. It has distinct male and female flowers produced upon separate trees; the males being in cylindrical drooping catkins, each flower growing from the base of a small bract, and having a four-parted calyx and four stamens; while the females are congregated into round heads or balls about the size of marbles, and each have a tubular three or four-toothed calyx, a single style produced from the side of the ovary, and a tapering stigma. They are succeeded by deep-scarlet pulpy fruits, resembling a mulberry in structure, and of a sweetish flavour, but rather insipid.

The Japanese cultivate this plant very much in the same way that we grow osiers, and they use only the young shoots for the manufacture of paper; these are cut into conveniently sized pieces, and boiled until the bark separates readily from the wood, when it is peeled off and dried for future use. To convert this bark into paper, they proceed in the following manner:—The dried bark is first moistened by soaking for a few hours in water; all superfluous matter is then removed by scraping with a knife, after which the bark is boiled in a ley of wood-ashes until its fibres are thoroughly separated, when it is reduced to a pulp by beating with wooden batons; this pulp is then mixed with mucilage and spread upon frames made of rushes. The paper thus made is of a whity-brown colour, and very strong; it is in common use in Japan. Instead of paper, the natives of the South Sea Islands manufacture from this bark an exceedingly tough cloth, called tapa or kapa cloth, which they commonly use for clothing, either plain or printed, and dyed of various colours. This cloth is principally made by the women, who adopt the following method of manufacture:—The bark is first softened by being soaked in water for a considerable length of time; it is then placed upon a log of wood and beaten out with a baton

Broussonetia papyrifera.

until it is of the requisite degree of fineness; the baton is made of very hard wood, and has four flat sides, each of which is sharply ribbed. Two or four women usually work together, and as they keep time in beating, the noise they make is loud and musical. In some islands, however, another and inferior method is adopted, the bark being placed upon a flat board, and scraped with different kinds of sharp-edged shells while kept constantly wet. By employing mucilage obtained from the arrow-root, the natives join pieces of the cloth together, and Admiral Sir Everard Home states that the King of Tongataboo (one of the Friendly Islands) had a piece made which was two miles long and 120 feet wide. [A. S.]

BROWALLIA. The name of certain plants belonging to the order of linariads, characterised thus: calyx-teeth unequal; corolla salver-shaped with a border divided into five parts, all of a roundish outline and slightly notched at the tip, one piece broader than the others; end of the style or appendage of the seed-vessel four-lobed. The genus was named by Linnæus in honour of John Browallius, bishop of Abo, who strenuously supported the system of that great botanist. The plants of this genus are natives of tropical America, usually of erect habit, smooth, or hairy and viscid; the leaves alternate, stalked, ovate in outline; the flowers violet or blue, more rarely white. Their handsome flowers and easy cultivation render them favourite objects of culture. *B. elata*, an upright-growing species, and *B. demissa*, of more spreading habit, have been long in cultivation; the latter is a native of Panama, and has the leaves oblong-ovate, oblique at the base, the branches and flower-stalks downy, the corolla pure pale blue, tending to purple or red; sometimes all three colours are associated on the same plant. [G. D.]

BROWNEA. A genus of small evergreen trees belonging to the *Leguminosæ* and to that section having regular corollas. The species are peculiar to Venezuela, New Grenada, and some portions of central America, one of them being also found in Trinidad. The leaves are alternate, equally pinnate, and from one to one and a-half foot long, with from four to twelve pairs of entire leaflets. The flowers are rose-coloured or crimson, and disposed in dense terminal or axillary sessile heads. The pods are compressed cimiter-shaped, often covered with rusty pubescence, and contain many seeds. It would be difficult to point out a more beautiful genus of stove-plants than this, and few tropical plant-houses of any pretensions are without some of them. *B. grandiceps* has long pinnate leaves with about twelve pairs of leaflets and axillary or terminal flower-heads eight inches in diameter; the flowers are pink, very numerous, and arranged in tiers as it were round a conical axis, the outer ones expanding first, followed by the others until all are open, when the flower-head is not unlike that of a *Rhododendron*. The leaves droop during the day so as almost to hide the flowers from view; but they have been seen to rise up in the evening and remain erect all the night; the flowers are thus exposed to the falling dew, but the leaves drooping again during the day, protect the flowers from the heat of the sun. This species is a native of Venezuela, where it is called Rosa del Monte or Palo de Cruz, and was introduced to England in 1828. Altogether there are six species in cultivation, some of them with bright scarlet flowers, as in *B. coccinea*, which was the first known in our gardens. The genus is named in honour of Patrick Brown, who wrote a history of Jamaica. [A. A. B.]

BROWNIAN MOTION. A phenomenon sometimes called molecular motion, which occurs in minute particles, both of vegetable and mineral origin, consisting in a rapid whirling motion, the nature of which is obscure, but is certainly independent of evaporation or other appreciable external causes which produce motion in minute bodies. It may be seen admirably in a weak solution of gamboge, with a power of 250 linear. It is frequently observed in the minute anatomy of vegetables, especially when the tissues are diseased. [M. J. B.]

BROWNLOWIA. A genus of the lime-tree family. *B. elata*, a native of Chittagong in Burmah, is the only known species, and attains a great size, full grown trees being about fifteen feet in circumference at four feet from the ground; the branches are numerous and spreading, forming a large ovate shady head, and the leaves, like those of the lime-tree in form, are entire, five to seven-nerved, and often a foot in length and eight inches broad. The flowers are in terminal panicles, very numerous and showy, white or pale yellow in colour. The fruit is made up of five baccate carpels, each containing one seed. [A. A. B.]

BROWN RED. Dull red, with a slight mixture of brown.

BRUCEA. A genus of *Simarubaceæ*, so called in honour of the famous Abyssinian traveller. It consists of shrubs with compound leaves; flowers in heads, unisexual or sometimes hermaphrodite; parts of the flower in fours; stamens attached to a central gland-like four-lobed gynophore or stalk supporting the four drupes. The stamens are sterile in the female flowers. The species are natives of Abyssinia, China, &c., and some of them possess bitter properties similar to quassia, a drug furnished by a tree of the same natural order. Some of the species are cultivated as stove shrubs. [M. T. M.]

BRUEA. A genus of *Artocarpaceæ*, comprising a shrub with alternate somewhat heart-shaped serrated woolly leaves, and having leafy bracts, and terminal stalked dioecious flowers. Calyx tubular, irregularly four-toothed; ovary oblique; stigma lateral, sessile, very long, fringed; fruit hairy. Native of Bengal. [M. T. M.]

BRUGMANSIA. The name of a genus of *Solanaceæ*, or of one which was formerly included in that order, but which has been separated by Miers, under the name *Atropaceæ*. The species were formerly comprised under the genus *Datura*, as there is a close resemblance in the flowers; but these plants are shrubs, and their fruit is smooth, not spiny, and contains but two cells. *B. suaveolens* is a well-known ornament of our greenhouses, with its large fragrant tubular white blossoms, which are sometimes produced in great profusion; it is perhaps better known under its old name of *Datura arborea*. Other species with orange and red flowers are in cultivation. All are natives of Peru and the adjacent districts of South America. Their seeds are dangerous stimulating narcotics.

The name *Brugmansia* is also applied to a genus of plants parasitical, in Java, on the roots of certain species of *Cissus*. They consist of little else but flowers, which are of the size of the fist, hermaphrodite, with a whitish perianth, which is two or three-cleft and internally scaly or hairy. These plants are nearly allied to the curious and gigantic *Rafflesias*. [M. T. M.]

BRUGUEIRA. One of the genera of the mangrove family (*Rhizophoraceæ*), and known by having a top-shaped calyx adherent to the ovary below, and having a persistent five to thirteen-lobed limb; five to thirteen oblong petals, cleft into two segments, leathery, woolly at the margin, and so folded that each petal conceals two stamens, whose filaments are not of equal length, but all shorter than the petals opposite which they are placed in pairs; their anthers are linear, or arrow-shaped. The ovary has two to four compartments, each containing two ovules; stigma two or four-toothed on the end of a style, which is about the length of the stamens. The fruit is crowned by the persistent calyx, and the seed within it germinates before it has fallen from the branch, as in the true mangroves. The trees are natives of the East Indies, where the bark is used as an astringent, for tanning purposes, and for dyeing black. [M. T. M.]

BRUMAILLE. (Fr.) *Erica scoparia*.

BRUNELLIA. A genus of *Xanthoxylaceæ*, consisting of trees with simple or compound leaves, and unisexual flowers in axillary or terminal panicles. The calyx is four or five-parted. There are no petals. The eight or ten stamens of the male flowers arise from a depressed hairy disc. In the female flowers the stamens are absent, but there are four or five ovaries, each terminated by a short style. The fruit consists of four or five two-seeded capsules, which open inwardly. The species are natives of tropical America, and the Sandwich Islands. [M. T. M.]

BRUNFELSIA. A name given to a genus of *Scrophulariaceæ* in honour of Otto Brunfels of Metz, who lived about the middle of the sixteenth century, and contributed to the revival of botanical science. The genus is known by the possession of a five-cleft calyx; a corolla with a long tube, very slightly dilated at the top, and a flat limb, five-cleft with rounded lobes, bisluate in æstivation; four fertile stamens with anthers which are confluent at the top; and a style which is bent inwards at the top, where it is divided into two stigmatic lobes. The capsule is leathery or fleshy, more rarely indehiscent and drupe-like; seeds several, rather large, imbedded in pulp. The species are shrubs or small trees natives of South America and the West Indies, and have handsome fragrant flowers of a blue or white colour. Some of the species are in cultivation. [M. T. M.]

BRUNIACEÆ. A small family not separated by any positive character from *Hamamelideæ*, although very different in habit. They are mostly much-branched heath-like shrubs from South Africa or Madagascar. The leaves are usually small, crowded and entire, without stipules. The flowers in terminal heads, with an inferior or half-inferior one to three-celled ovary, having one to two pendulous ovules in each cell; a five-cleft calyx; five petals alternating with the calyx-lobes; five stamens alternating with the petals; and a simple or two and three-cleft style. The fruit is dry and indehiscent, or separates into indehiscent cocci. There are about sixty species known, distributed into fifteen genera, including *Grubbia* and *Ophira*, of which some botanists form a distinct family, still more nearly allied to *Hamamelideæ* in habit as well as in character. The *Bruniaceæ* will indeed probably hereafter be entirely included in *Hamamelideæ*, notwithstanding their want of stipules, which is now supposed to be the only constant differential character.

BRUNIA. A genus of epigynous exogenous plants, typical of the group *Bruniaceæ*, distinguished chiefly by having the flowers aggregate in little heads; calyx superior, five-parted; filaments of the stamens inserted into the claws of the petals; stigmas cleft, with small two-celled ovaries. The species are all natives of South Africa, and, consequently, require the protection of a greenhouse in England. *B. nodiflora* is the species which is most generally cultivated, and when well grown, it forms a very handsome plant. [D. M.]

BRUNNEUS. Deep brown; not much different from chestnut-brown.

BRUNNICHIA. A genus of *Polygonaceæ*, containing a single species, *B. cirrhosa*, a native of the warmer regions of North America. The stem is shrubby, twining, with alternate shortly-stalked smooth ovate-acuminate entire leaves; the leafstalks are dilated at the base, and half-clasping, a hairy line completing the circle round the stem; peduncles axillary and terminal often ending in tendrils; bracteles small with several flowers from the axil of each, the whole so arranged that the flowers are racemose on the peduncles; perianth herbaceous, very small, bell-shaped, five-parted; stamens eight or ten; styles three; nut three angled. [J. T. S.]

BRUNONIA, BRUNONIACEÆ. The genus *Brunonia* consists of two Australian herbs with capitate blue flowers, giving them the aspect of a *Scabiosa* or of a *Globularia*; whilst in their structure, and especially in their stigma, enclosed in a two-valved cup, they are more nearly allied to *Goodeniaceæ*. Robert Brown, in whose honour the genus was named, considered it as a section or anomalous genus of the latter family; whilst others have thought that the completely free ovary and exalbuminous seeds, combined with the inflorescence, are sufficient to mark it as a distinct family under the name of *Brunoniaceæ*.

The *Brunonias* grow up with tufts of entire spathulate radical leaves, and naked scapes terminated by the compact head of small blue flowers, which are surrounded by bracts. The five-cleft calyx has three bracts at the base; the corolla is five-parted, the two upper segments separate from the others; the five stamens are hypogynous, with the anthers slightly cohering; the ovary is free, one-celled and one-ovuled, with a simple style; and the fruit is a membranous utricle enclosed in the hardened tube of the calyx. [J. T. S.]

BRUNSVIGIA. A genus of *Amaryllidaceæ*, distinguished by broad recumbent biennial leaves, an autumnal precocious flower-scape, a very short-tubed recurved perianth, recurved style and filaments, the filaments not adhering beyond the tube, and a triangularly turbinate capsule. The *Brunsvigias* are rather remarkable bulbous plants of South Africa, closely related to *Amaryllis* itself. The typical species, *B. multiflora*, has a globose bulb as large as an infant's head, and produces distichous obtuse striated linguliform leaves seven or eight inches long, and a fleshy compressed scape, a span or more in height, supporting an umbel of from twenty to sixty purple flowers, which have lance-shaped segments spreading or revolute at the tips. There are but few other species referred to the genus. [T. M.]

BRUSE. (Fr.) *Ulex europæus*.

BRUSH-APPLE. The native Australian wood of *Achras australis*.

BRUSH-CHERRY. The native Australian wood of *Trochocarpa laurina*.

BRUSH-SHAPED. See Aspergilliform.

BRUSSELS SPROUTS. A cultivated variety of the Cabbage, *Brassica oleracea*, having the leaves blistered, and the stems covered by little close heads or hearts.

BRUYERE. (Fr.) *Calluna vulgaris*. — DU CAP. *Phylica ericoides*.

BRYACEÆ. A large group of acrocarpous mosses distinguished by the capsules having a double row of teeth, the inner of which are united at the base by a common plicate membrane. Very rarely there is a single row only, or the teeth are obsolete. The capsule is almost always pendulous. The stem is at first simple, but at length branched by means of new shoots, called innovations, given off near the tip, or the base, sometimes from subterranean creeping shoots. The leaves have a central nerve, and consist of large reticulations, and are mostly serrated at the margin and thickened. Very rarely the fruit is lateral as in *Mielichoferia*. Many of the species of *Mnium*, as *M. punctatum*, *rostratum*, *undulatum*, &c., are a great ornament to woods and rocks from their large leaves and handsome capsules, while various species of *Bryum* attract notice on walls, gravel-walks, and marshes, by their tufted habit and abundant pendulous capsules. Amongst these *Bryum argenteum* is peculiarly conspicuous from the silvery white of its leaves. *Bartramia* and one or two closely allied genera are remarkable for their nearly spherical capsules, which are almost always more or less streaked or furrowed, especially when dry. Our more common species, as *Bartramia pomiformis*, are subalpine, or occurs in bogs. Sometimes the term *Bryaceæ* is applied to the whole of the true mosses, as in Lindley's *Vegetable Kingdom*. [M. J. B.]

BRYA. A genus of leguminous plants (*Fabaceæ*: *Papilionaceæ*) consisting of three species, small trees or large shrubs, natives of tropical America. They have a five-toothed, somewhat two-lipped calyx; a papilionaceous corolla; and stamens united into a tube, which is split down one side. Their fruit is a flattened two-jointed pod, the upper half of which is generally imperfect, the lower containing a solitary seed. The leaves are solitary, or in clusters, or pinnate.

B. Ebenus, the Jamaica or West Indian Ebony-tree, is a large shrub or small tree, growing twenty or thirty or even forty feet high, with a trunk seldom exceeding four inches in diameter; it has long slender tough and flexible branches, which are armed with short sharp spines, and bear numerous small evergreen leaves, resembling those of the common Box, but rather broader at the top end. The flowers are of a bright orange-yellow colour, produced in great abundance upon the young branches, and have a very sweet odour. Although the wood of this tree is known in Jamaica by the name of Ebony, it is not the true ebony-wood, that being produced by a totally different tree. The Jamaica Ebony is of a greenish-brown colour, very hard, and so heavy that it sinks in water; it takes a good polish, and is used in Jamaica for making

Brya Ebenus.

various small wares. Part of the wood known in commerce as Green Ebony, and which is much used by turners and dyers, is probably obtained from this tree. The tough twiggy branches are used in Jamaica as riding-whips, and it is said that in former days they were kept at all the wharfs about Kingston to scourge the refractory slaves. [A. S.]

BRYANTHUS. A genus of *Ericaceæ* containing a single species, a native of Siberia and Kamschatka, so nearly related to *Menziesia* that it is generally considered as belonging to that genus. It differs chiefly in having a pentamerous arrangement of the flower, although Ledebour, and apparently also Swartz, have seen specimens in fruit with four divisions of the calyx and capsule. The divisions also are deeper than in *Menziesia*. [W. C.]

BRYOBIUM. A supposed genus of small unimportant orchidaceous epiphytes from India, not distinct from *Mycaranthus*, and like it now merged in *Eria*.

BRYOLOGY. The part of botany which treats of urn-mosses.

BRYONIA. The technical name of the genus to which the common bryony of the hedges belongs. Among the *Cucurbitaceæ* this genus may be known by the stamens and pistils being on the same plant, but in different flowers; by the calyx having five small teeth; the corolla five-lobes; stamens five in three parcels, the anthers sinuous; style three-lobed, with capitate stigmas; and fruit globular, succulent. *B. dioica*, the Common Bryony, has a thick tuberous rootstock of considerable length, yellowish-brown, and wrinkled transversely on the outer surface. The stems that spring from this are annual, and rough. They climb by tendrils, and, what is very unusual, the direction of the spiral is now and then altered, so that after proceeding in one course for some distance, the tendril suddenly changes to an opposite direction. The leaves are angular, three to seven-lobed, the terminal or middle lobe being the longest; they are rough like the stem. The male flowers are in clusters, bell-shaped, greenish-yellow, and veined; the female blossoms are smaller, disposed in a corymb or umbel, and have a globular ovary which ripens into a scarlet berry, containing several flattened seeds. The male and female flowers are sometimes on different plants, hence the name *dioica*, but this is not always the case. The plant has a fetid odor, and possesses acrid, emetic, and purgative properties, and from its elegant appearance, especially in autumn when it adorns the hedges with its brilliantly coloured fruit, accidents are likely to occur to children and others incautiously tasting the fruit. The root is used as an application to bruises, and occasionally as a purgative; but it is unsafe from its uncertain and sometimes violent action, whence the French call it Devil's-turnip. Its acridity is due to a chemical substance called bryonin. The writer of this notice was once called on to ascertain what vegetable substance had been administered to a farmer, his family, and his cattle, by a 'wise man,' who purported to be able to remove the spell of witchcraft, under which he said they were all suffering. The man succeeded in obtaining considerable sums of money at different times from the credulous farmer, whose suspicions were at length awakened by the dangerous illness of some of the members of his family. It was not distinctly proved that the man had administered bryony, but the symptoms complained of corresponded with those which would be produced by that root, a quantity of which was found in the man's house, and also a powder which was found to consist of the leaves of the hart's-tongue (*Scolopendrium vulgare*). When the mandrake was more esteemed than it is now, this root was frequently sold for it, as it occasionally branches in a similar manner, and, indeed, was forced to do so, by being grown in moulds. Even now it is occasionally to be met with in herbalists' shops as mandrake. The young shoots of bryony may be used as a vegetable with impunity, and are said, when boiled, to resemble asparagus in flavour. This plant must not be mistaken for the black bryony (*Tamus communis*), also a climbing plant, but whose leaves are heart-shaped, smooth, and shining.

Bryonia alba, a central European species has similar properties to the English Bryony, as also have *B. americana* and *B. africana*. The root of *B. abyssinica*, when cooked, is said to be eaten with impunity. The seeds of *B. callosa* are used in India as a vermifuge, and yield an oil used for lamps. *B. laciniosa*, *B. rostrata*, and *B. scabrella* are all used for medicinal purposes in India, while the leaves of some are boiled and eaten as greens. *B. epigæa* was at one time supposed to furnish calumba root, which it resembles both in appearance and properties. It is used in India as an external application and for other medicinal purposes. [M. T. M.]

BRYONY. The common name for *Bryonia*. —, BLACK. *Tamus communis*. —, RED. *Bryonia dioica*.

BRYOPHYLLUM. A name expressive of the peculiarity that the leaves have, under certain circumstances, of producing small buds on their margins. The genus to which the name applies, belongs to the house-leek family (*Crassulaceæ*), and is known by its bell-shaped distended calyx, which is four-cleft; the tube of the corolla somewhat quadrangular, the lobes of its limb, ovate or somewhat triangular; a number of gland-like compressed scales at the base of the carpels; and carpels on very short stalks. The leaves are unequally pinnate and fleshy. *B. calycinum*, when in flower, has loose panicles of drooping greenish-purple blossoms, which are very elegant. It is of particular interest from the formation of small buds at the notches on the margin of its leaves; sometimes these buds are produced naturally, but the plant may be made to form them by pegging a detached leaf close down to the soil, when the buds will root into the ground, and form new plants. The species is a native of the Moluccas, Madagascar, the Mauritius, &c., and grows in dry situations in the clefts of the rocks. In the Mauritius it is used as a fomentation or poultice in intestinal complaints. [M. T. M.]

BRYUM. A large genus of acrocarpous mosses, now subdivided, but formerly almost equivalent to the natural family BRYACEÆ, which see. [M. J. B.]

BUBANIA. A little known genus of *Plumbaginaceæ*, having the habit of *Goniolimon*, but possessing five clavate and not capitate stigmas. It differs from that genus, as well as from *Statice*, in having the styles united through a considerable extent of their length, and in the filaments being papillose at the base. The genus is founded on a single species from Algeria, which has not yet been satisfactorily described. [W. C.]

BUBON. A genus of *Umbelliferæ*, which has an obsolete calyx, and obovate entire petals, with the points bent inwards. The fruit is compressed and has a dilated flattened edge; while each half of it has on its outer surface four ridges, the central ones filiform, the lateral ones passing into the flattened margins of the fruit. In the channels between the ridges, in the interior of the fruit, are canals containing volatile oil, while on the inner face of the two halves are two such canals. The species are natives of the Cape of Good Hope, and have yellowish flowers. *B. Galbanum* secretes a resinous juice somewhat like galbanum. [M. T. M.]

BUCAIL. (Fr.) *Fagopyrum esculentum*.

BUCCÆ. The lateral sepals or wings of the flower of an aconite; seldom used.

BUCHANANIA. A genus of *Anacardiaceæ*, named in compliment to Dr. Buchanan Hamilton, a distinguished investigator of Indian botany. The genus consists of Indian trees with simple leathery leaves, hermaphrodite flowers in axillary panicles, with a five, or more rarely a three or four-cleft calyx; five petals rolled backwards; ten stamens shorter than the petals; and a ten-lobed disc wrapping round the ovaries, which are five in number, but only one perfect, the remaining four being represented only by the styles. The fruit is a drupe with one seed, borne on a little stalk within it. The seeds of *B. latifolia* are eaten by the natives as almonds, and they furnish an oil known as the cheroonjee oil; the fruits also supply a black varnish. The unripe fruits of *B. lancifolia*, according to Major Drury, are eaten by the natives in their curries. [M. T. M.]

BUCHNERA. A large genus of *Scrophulariaceæ*, generally distributed over the tropical and subtropical regions of the world. They are stiff scarcely-branched herbaceous plants, with the lower leaves opposite and the upper alternate, and with flowers in terminal spikes. The calyx is tubular with five short teeth; the corolla tube is straight and slender, and the limb has five nearly equal spreading lobes, the two upper ones inside in the bud. The two pairs of stamens are included in the tube; they have obtuse one-celled anthers. The style is club-shaped. The capsule is straight, opening loculicidally in two entire valves. [W. C.]

BUCHU. The same as Bucku.

BUCIDA. A genus of trees belonging to *Combretaceæ*, native of tropical America and the West Indies, with alternate wedge-shaped entire leaves, smooth or hairy on the margins, and axillary peduncles bearing rather small, spicate or capitate flowers. Calyx tubular, adhering to the ovary, above which it is bell-shaped and five-toothed at the margin; corolla none; stamens ten with long filaments; style simple, subulate; drupe one-seeded. The ends of the peduncles sometimes grow into spiny horn-like excrescences, from which the genus takes its name: (*bous*) ox. *B. Buceras*, the Olive-bark, or Black Olive of Jamaica, produces wood which is valuable on account of its not being liable to the attacks of insects; the bark is also used for tanning purposes. [J. T. S.]

BUCKBEAN. *Menyanthes trifoliata*.

BUCK-EYE. An American name for the species of *Pavia* and *Æsculus*, especially *Æ. ohiotensis*.

BUCKLANDIA. The name of a genus belonging to the order of witch hazels, having stamens and pistils in the same flower, or in different flowers on the same plant; or some plants have stamens only, while others have only pistils. The calyx is almost bell-shaped, adherent below to the seed vessel; the anthers are supported on awl-shaped filaments. The flowers are in head-like groups, each subdivision of which consists of eight flowers. The name *Bucklandia*, which has also been employed to designate certain fossil species of plants, was given in honour of the late Dr. Buckland, well known as a geologist. The only species is an Indian tree with the general aspect of a poplar; its leaves are alternate, stalked, and variable in outline. [G. D.]

BUCKLER-SHAPED. Having the form of a small round shield, like a Highland target.

BUCKTHORN. The common name for *Rhamnus*. —, DYER'S. *Rhamnus infectorius*. —, SEA. *Hippophäe rhamnoides*.

BUCKWHEAT. *Fagopyrum esculentum*.

BUCKWHEAT TREE. *Mylocaryum ligustrinum*.

BUCKU. A name applied in South Africa to several species of *Barosma*, especially *B. crenata*, *crenulata*, and *serratifolia*.

BUD. The young undeveloped branch or flower.

BUDDLEIA. A large genus of *Scrophu*-

lariaceæ, containing nearly eighty species from America, India, and South Africa. They are trees, shrubs, or herbs, generally tomentose or woolly, especially on the young branches, the under surface of the leaves, the peduncles and calyx, and sometimes even on the corolla. They have opposite leaves, and many-flowered peduncles, axillary or frequently in a terminal thyrse or panicle. The short campanulate calyx is divided into four equal teeth. The corolla is campanulate or tubular, with the limb spreading and divided into four equal teeth. There are four included stamens inserted either in the throat on very short filaments, or in the middle of the tube. The ovary is two-celled, and bears a simple style with a capitate stigma. The capsule dehisces septicidally with two valves; it contains numerous small seeds. [W. C.]

BUENA. One of the genera of *Cinchonaceæ* consisting of shrubs closely resembling the *Cinchona* itself, but distinguished by their solitary terminal flowers, and by the limb of the calyx being deciduous, so that the ripe fruit is not crowned by the calyx as in *Cinchona*. The species are natives of Peru and Western tropical America. [M. T. M.]

BUFFALO BERRY. *Shepherdia argentea.*

BUFFELHORN. The South African name of the wood of *Burchellia capensis.*

BUFFELSBALL. The South African name of the wood of *Gardenia Thunbergia.*

BUFFONIA. A genus of the alsineous group of *Caryophyllaceæ*, containing small herbs or undershrubs, natives of central Europe, the Mediterranean region, and temperate Asia. They have stiff slender stems, often paniculately branched, and somewhat resembling the toad-rush (*Juncus bufonius*) in habit; leaves awn-shaped, closely applied to the stem; flowers small cymose, arranged in a spicate, racemose, or paniculate manner; calyx four-parted, scarious, compressed; petals four, white; stamens four to eight; styles two; capsule two-valved; seeds two. One species, *B. annua*, is said to have been found in Britain in Plukenet's and Dillenius's time, but has not occurred since, and it is not improbable that some other plant may have been mistaken for it. [J. T. S.]

BUGBANE. An American name for *Cimicifuga.*

BUGLE. The common name for *Ajuga.*

BUGLE-WEED. The American name for *Lycopus virginicus.*

BUGLOSS. The common name for *Anchusa*. —, SMALL. *Lycopsis* or *Anchusa arvensis*. —, VIPER'S. The common name for *Echium*. —, WILD. The common name for *Lycopsis.*

BUGLOSSE. (Fr.) *Anchusa officinalis* —, PETITE. *Lycopsis arvensis.*

BUGRANE COMMUNE. (Fr.) *Ononis procurrens.*

BUGWORT. The common name for *Cimicifuga.*

BUIS. (Fr.) *Buxus sempervirens.* — DE MAHON. *Buxus balearica.*

BUISSON ARDENT. (Fr.) *Crataegus Pyracantha.*

BUKKUM WOOD. The wood of *Cæsalpinia Sappan*, used as a dye stuff.

BUKUL. *Mimusops Elengi.*

BULB. A leaf-bud, the scales of which are fleshy, and which propagates an individual. —, NAKED. A bulb whose scales are loose and almost separate, as in the crown imperial. —, SOLID. A corm, which see. —, TUNICATED. A bulb whose outer scales are thin and membranous.

BULBIL. An axillary bulb with fleshy scales, falling off its parent spontaneously, and propagating it.

BULBILLARIA. A genus of *Liliaceæ*, scarcely distinct from *Gagea*, which the only species, *B. gageoides*, from Mount Libanus, closely resembles, differing only by having the ovary on a conspicuous club-shaped stalk within the perianth; there are no radical and only one cauline leaf, which is linear. The plant is remarkable for the small bulbs which occur in the axils of the leaf-like bracts. [J. T. S.]

BULBINE. A section of the liliaceous genus *Anthericum*, containing several plants natives of South Africa. They have the segments of the perianth spreading and yellow; the filaments, or at least the alternate ones, bearded with short hairs; leaves somewhat fleshy, like those of the onion; root fasciculate; stem short. Several species are cultivated as greenhouse plants, and are not only pretty, but often fragrant. [J. T. S.]

BULBOCAPNOS. A section of the fumariaceous genus *Corydalis*, containing the species which have a large tuberous rootstock, a persistent style, and a digitate process at the base of the seed, which has an embryo, of which the two cotyledons are united into one. Stem usually succulent, with few thin glaucous twice-ternate leaves, having cut leaflets, and a terminal raceme of purple flowers, with paler markings. Several species occur in Europe and temperate Asia, but none are truly native in Britain, though one species, *Corydalis solida*, often found in gardens, flowering in spring, is naturalised in several places. This plant has solid tubers, a sheathing scale below the leaves, leaf-like bracts digitately cut, and rather large flowers. [J. T. S.]

BULBOCASTANUM. *Bunium Bulbocastanum.*

BULBOCHÆTE. A genus amongst the

confervaceous articulated *Algæ*, remarkable for its hyaline bristle-like branches, which are bulbous at the base. The fruit consists of globose capsules, with a green and then a dark red endochrome. The mode of impregnation in this genus, as also in *Œdogonium*, is very curious. Some of the cells produce little bodies, which are furnished with flagelliform appendages, by means of which they swim about till they fix themselves on or near the swollen joints, which are to produce the spores. These bodies become clavate, with one or two joints, and just when the contents of the swollen cells are ready for impregnation, a lid comes off, and makes way for the exit of one or more globose spermatozoids, which are admitted to the endochrome of the female cells by means of a little aperture. After impregnation, the endochrome acquires a membrane, and after a time becomes free. The spore, when liberated, elongates—in a few hours attaining twice its original length. The endochrome, by successive division, gives rise to four distinct bodies, which acquire a nearly globular form, and are furnished at one extremity with two sets of ciliary processes, by means of which they move about, and thus appear in the condition of zoospores, which ultimately reproduce the species. *B. setigera* is our most common species, but others occasionally occur in this country. [M. J. B.]

BULBOCODIUM. Bulbous plants with the habit of the *Colchicum*, and members of the same family, *Melanthaceæ*. The perianth consists of six coloured segments, with long taper claws or stalks, which form a slender tube; the upper portion of each segment is elliptical, and prolonged at the base into two small acute processes, so that the perianth may be described as consisting of six sagittate stalked segments. The stamens are six, attached to the segments of the perianth, and of unequal lengths; the style three-cleft, with simple stigmas. Ovary three-celled; ovules indefinite. Fruit a capsule, dividing when ripe into its component carpels. The species are pretty bulbous plants, natives of Europe, the Levant, &c. [M. T. M.]

BULBODIUM. The solid bulb of old botanists; the same as a corm.

BULBONAC. (Fr.) *Lunaria biennis*, and *rediviva*.

BULBOSI PILI. Hairs that proceed from a swollen base.

BULBOSPERMUM. A genus of *Liliaceæ*, containing a small fibrous rooted herb, from Java (*B. javanicum*), which has the stem somewhat bulbous at the base, with two or three long-stalked lanceolate radical leaves, which are membranous and many-nerved, but the sheaths and peduncles are frequently without any blade; scape short erect, with racemose flowers at the top; undermost bracts larger than the others, and frequently empty; flowers on long pedicels, with a six-parted greenish perianth; stamens six, monadelphous; ovary three-lobed, opening at the top when ripe, and showing the seeds, which are as large as peas, three or four in each of the three cells of the capsule, and with a soft thick seed-coat. [J. T. S.]

BULBOSUS. Having the structure of a bulb; having bulbs.

BULB-TUBER. A corm, which see.

BULL, or BULLET GRAPE. *Vitis rotundifolia*.

BULLACE. *Prunus insititia*.

BULLACE PLUM, JAMAICA. The fruit of *Melicocca bijuga*.

BULLATE. Blistered; puckered. When the parenchyma of a leaf is larger than the area within which it is formed.

BULL-HOOF. *Maruenja ocellata*.

BULLOCK'S HEART. A name given to the fruit of *Anona reticulata*, a kind of custard apple.

BULL-RUSH, or BULRUSH. *Scirpus lacustris*; and sometimes *Typha*.

BULLY, or BULLET TREE. A name given in Guiana to a species of *Mimusops*. —, BASTARD. *Bumelia retusa*. —, BLACK. *Bumelia nigens*. —, JAMAICA. *Lucuma mammosa*.

BULRUSHWORTS. A name given by Lindley to the *Typhaceæ*.

BUMELIA. A Greek name for the common ash, but applied in modern times to a genus of *Sapotaceæ*, having a corolla with a short tube, and a five-parted limb, at the base of each segment of which are two small scales. There are five fertile stamens attached to the tube of the corolla, opposite its lobes, and alternating with five petaloid barren stamens. The ovary has five one-seeded compartments, some of which, however, become arrested in their growth, so that the berry-like fruit frequently contains but one cavity and seed. The species consist of trees or shrubs, with a milky juice, a spiny stem, simple alternate leaves, and small white or greenish flowers. Some of them are sufficiently hardy to bear our climate, if protected by a wall, while others are grown in hothouses. [M. T. M.]

BUNCHOSIA. Tropical American trees or shrubs, belonging to the order *Malpighiaceæ*, and nearly allied to the genus *Malpighia*, but having the racemes of flowers axillary. Styles separate, or fused together; fruit fleshy, indehiscent, externally smooth, without angles and containing two or three seeds, which are convex on the back. The flowers are for the most part yellow. Several kinds are in cultivation as stove shrubs. The seeds of one species, *B. armeniaca*, a Peruvian tree, are reputed to be poisonous. [M.T.M.]

BUNGEA. A genus of *Scrophulariaceæ*, containing a single species, a native of America. It is a small herb, growing in densely leafy tufts. The leaves are linear, and deeply trifid. The flowers are on short pedicels, with two bracts. The calyx is tubular at the base, and has four long leafy divisions of the limb. The upper lip of the corolla is acuminate. The stamens are didynamous, hid under the upper lip, and have two equal mucronate cells. The style has a capitate stigmatose apex. The ovoid capsule dehisces loculicidally, and contains few largish seeds. This genus is very near to *Cymbaria*, from which, however, it is separated by its four leafy segments of the calyx, and its acuminate galea. From *Rhinanthus*, to which it was formerly referred, it differs in possessing two bracteoles under the calyx. [W. C.]

BUNIAS. A genus of *Cruciferæ*; herbs from central Europe, the Mediterranean region, and temperate Asia, having erect branched stems, entire or pinnatifid, often runcinate leaves, and elongated racemes of rather small yellow flowers, on short spreading pedicels. Pouch resembling a small four-sided ovoid pyramidal nut, often tuberculated or muricated, indehiscent two-celled; cells two-seeded; embryo with the cotyledons rolled up on themselves, which distinguishes the genus from all except *Erucaria*, which has a jointed pod breaking across into two segments. [J. T. S.]

BUNIUM. The five and twenty species composing this genus of tuberous-rooted umbelliferous plants (*Apiaceæ*) are chiefly inhabitants of Southern Europe and Western Asia. They are small herbaceous plants, seldom more than two feet high, and have very finely-cut leaves. Their flowers are white, and borne in compound umbels, generally destitute of an involucre, but occasionally with a few small bracts. The technical characters of this genus and its allies are derived from the fruit: in the present it is slightly flattened on two sides, and drawn in at the top, terminating in two straight styles; each half of the fruit having five indistinctly marked longitudinal ribs, with several oil cells between them.

B. flexuosum is a native of Western Europe, but is found wild in Britain. This grows erect about a foot or more high, with a few branches towards the top. Its leaves are very few in number, and very finely divided and sub-divided into numerous slender narrow divisions — those on the upper part of the stem having much finer divisions than the lower ones. The round tuberous roots of this plant have a sweetish aromatic taste, mingled with a considerable amount of acridity, which renders them unpleasant eating while raw, although they are often eaten in that state by children; but when boiled or roasted, they are very palatable, much resembling the chestnut in taste — hence one of the common names for them is Earth-chestnuts; they are also called Pig-nuts, Arnuts, Jur-nuts, Ynr-nuts, Kipper-nuts, &c.

B. ferulæfolium, which grows in the islands of Cyprus and Candia, produces tubers as large as filberts, which are eaten by the Greeks under the name of Topana. It has branching stems about a foot in height, and leaves primarily divided into three divisions, each of which are then subdivided into three leaflets. [A. S.]

BUN-OCHRO. An Indian name for *Urena lobata*.

BUNT. The common name of *Tilletia caries*, a parasitic fungus belonging to the section *Coniomycetes*. *Tilletia* differs from other genera of the group *Ustilaginei* in the perfectly globose spores having a cellular outer coat. These are at first developed from the ultimate branchlets of a very delicate web which at length completely vanishes, so that the inside of the seed in which they grow contains nothing but a mass of spores. These are held together for a long time in consequence of the toughness of the outer coat of the seed in which they grow, and accordingly the bunted grains are carried home with the rest of the produce, so that when the grain is threshed the spores of the bunt are dispersed, and many of them adhere to the seedcorn, ready to germinate when the seed is sown. The first thread protruded by the spores is thick and coarse, so that it cannot penetrate the tissue of the sprouting grain; but a tuft of far more delicate threads soon crowns its apex, and after becoming united with each other by means of little lateral processes, they produce secondary spores, which in their turn germinate. As the wheat crop often suffers seriously from bunt, many measures are adopted by the farmer to kill the bunt-spores. Arsenic and corrosive sublimate are ineligible because the grain, if not sown at once, is apt to lose its power of vegetating; sulphate of copper has not the same inconvenience, and is much used, as is also quicklime slacked with boiling water. The best practise is perhaps that pursued in some parts of France. The wheat is thoroughly wetted with a strong solution of Glauber's salts (sulphate of soda), and then dusted with quicklime. The effect of this is to set the caustic alkali free, while the sulphur and lime combine to form gypsum. Bunt scarcely occurs in barley, but it has been found in Aigierson *Hordeum murale*. The only other species of *Tilletia* occurs on *Sorghum*. [M. J. B.]

BUPHANE. A small group of amaryllids, remarkable in having precocious flower-scapes, supporting from 100 to 200 or more flowers in a single head. The flowers have a straight cylindrical tube and a regular six-parted expanded limb, their filaments being erect and distinct from the tube. The capsule which succeeds them is turbinate and dry, three-valved, with numerous distinct ovules. Only four species are referred to the genus by Herbert, and these are all South African. The peduncles, which are at first crowded and suberect, diverge so as to form a spherical head, the flowers of which are smaller than in the closely-allied *Ammocharis*. *B. toxicaria* is called the Poison Bulb, and is said to be fatal to cattle. The bulbs of *B. disticha* are met with as large as a man's head. The former of these produces crowded umbels of flesh-coloured flowers, the segments of which are linear-lanceolate, and its leaves are elongately lorate. [T. M.]

BUPHTHALMUM. A family of compound flowers deriving their name (equivalent to Ox-eye) from the broad open disk of their flowers. Among the plants of this family most frequently cultivated in English gardens are *B. grandiflorum*, a herbaceous perennial growing about a foot and a-half high, with narrow smooth leaves and large yellow flowers; and *B. cordifolium*, also a herbaceous perennial, forming a large tuft; the root-leaves are heart-shaped, the upper ones smaller, eggshaped and sessile; the flowers large, bright yellow with long rays. Both are natives of central Europe. [C. A. J.]

BUPLEURUM. Hare's-ear, Thorow-wax, or Thorow-leaf. The only common English species of this strongly-marked family of umbelliferous plants is *B. rotundifolium*, which occurs in corn-fields on a chalky soil, especially about Swaffham and in Cambridgeshire. It may be known by its roundish-oval leaves, which are alternate, and so extended at the base that 'every branch doth grow thorowe everie leafe, making them like hollowe cups or sawcers' (*Gerarde*). The flowers are small and of a greenish-yellow hue, and far less conspicuous than the large bracts at the base of the partial umbels. *B. fruticosum* is a shrubby species, a native of the South of Europe, with purplish branches and sea-green leaves. Several other species are cultivated, all of which are more or less remarkable for the unusual development of the floral bracts (involucre), and are of easy cultivation. French, *Buplèvre*, *Oreille de lièvre*; German, *Hasenöhrchen*. [C. A. J.]

BUPLEVER. An English name adapted from the French, proposed by Bentham for *Bupleurum*.

BUR-BARK. The fibrous bark of *Triumfetta semitriloba*.

BURCHARDIA. An Australian genus of the colchicum family (*Melanthaceæ*). The perianth is coloured, of six slightly-stalked segments, each having a nectariferous pore near the base; stamens six, inserted on the very base of the segments of the perianth; anthers peltate, opening outwardly; ovary triangular, containing three compartments, each with several ovules in two rows; styles three. Fruit a capsule, opening by the separation of its constituent carpels. *B. umbellata* is in cultivation; it is a herbaceous plant with thick rootlets, linear sheathing leaves; flowers white in umbels. [M. T. M.]

BURCHELLIA. A name given in honour of Mr. Burchell, an African traveller, and used to denote a genus of *Cinchonaceæ*. The characteristics of this genus are the flowers closely packed in a head, surrounded by a few bracts; corolla funnel-shaped, swollen above the middle; limb five-cleft, small, naked at the throat; stamens inserted above the middle of the tube of the corolla; anthers on very short filaments; stigma club-shaped. The fruit is succulent, two-celled, many-seeded, crowned by the deeply five-cleft calyx. The species are shrubs with handsome flowers, and are natives of S. Africa. *B. capensis* and *B. parviflora* are grown in greenhouses for the sake of their clustered handsome scarlet flowers. [M. T. M.]

BURDEE. An Arabic name for *Papyrus antiquorum*.

BURDOCK. The common name for *Arctium Lappa*; also applied to *Centotheca lappacea*. —, PRAIRIE. An American name for *Silphium terebinthinaceum*.

BURKEA. A genus of the pea family, and belonging to the section with regular flowers. *B. africana* is the only species of the genus, and is a shrub or sometimes a small tree, thirty feet high, with twice pinnate leaves, and very numerous oval leaflets from one to three inches long, and when young covered with silvery hairs. The flowers are small, white, and fragrant, disposed in panicles made up of long slender branching spikes. The pods are stalked, thin, and about one and a-half inches long, with one or two seeds. This is one of the many plants which are common to the eastern and western sides of tropical Africa. The genus is named in compliment to Mr. J. Burke, a plant collector, who made extensive collections of S. African and N. American plants. (A. A. B.)

BURLINGTONIA. A genus of epiphytal orchids inhabiting the tropics of Brazil. The species have large and often fragrant white yellow or pink flowers attached to a weak drooping or pendulous spike. All that are known are in cultivation in this country.

BURMANNIACEÆ. A family of monocotyledons, allied to orchids in their inferior ovary, either three-celled or with three parietal placentas, in their trimerous flowers, and especially in their minute seeds, with a loosely netted testa enclosing an apparently homogeneous nucleus or embryo; but differing in their perfectly regular flowers, with three to six distinct stamens and a central simple or three-cleft style. They are all herbaceous, with blue or white flowers, inhabiting marshy or shady places. In some genera the annual slender stems have no leaves except small colourless scales, which led former botanists to suppose them to be root parasites; but it has now been ascertained that they grow on rotten leaves and other decayed vegetable substances, and not on living plants. There are scarcely more than thirty species of *Burmanniaceæ* known, all tropical, except one North American *Burmannia*. They are distributed into ten or eleven genera, including *Tacca*, which some botanists treat as a distinct family under the name of *Taccaceæ*.

BURMANNIA. A genus of *Burmanniaceæ*, the principal one of the family, although consisting of only six or seven species. It is distinguished by the three-winged or three-angled ovary and capsule, completely divided into three cells, with numerous seeds attached to the inner angle of each cell. They are mostly marsh plants, with short flat sedge-like leaves, forming radical tufts or crowded at the base of the stem, and terminal blue flowers in short simple or two or three-branched spikes. One species, however, *B. capitata*, is a slender almost colourless plant, without other leaves than minute scales, and with very small capitate flowers. They are natives of the tropical regions of Asia, Africa, and America, one species extending northward as far as Virginia.

BURNET, GARDEN. *Poterium Sanguisorba*. —, GREAT. *Sanguisorba officinalis*. —, LESSER. The common name for *Poterium*. —, SALAD. *Poterium Sanguisorba*.

BURNING BUSH. An American name for *Euonymus atropurpureus*, and *E. americanus*; also sometimes applied in gardens to the Artillery plant, *Pilea serpyllifolia*.

BURR. The Burdock, *Arctium Lappa*.

BURSARIA. A genus of South Australian and Tasmanian shrubs belonging to *Pittosporaceæ*. Branches not unfrequently spiny; leaves alternate, subsessile, obovate wedge-shaped retuse and entire, or oblong-linear and toothed; peduncles terminal, ternate, or panicled, the flowers small, white, sometimes tinged with pink outside; sepals, petals, and stamens, five each. Ovary free; style thread-like. Capsule obcordate, compound, extremely like that of the shepherd's purse (*Capsella Bursa-Pastoris*), incompletely two-celled, two-valved at apex; seeds one or two in each cell. [J. T. S.]

BURSERA. One of the genera of *Amyridaceæ*, consisting of trees with alternate compound leaves, flowers in axillary clusters, a small three to five-parted calyx, a corolla of three to five petals, larger than the segments of the calyx, inserted with the six to ten stamens beneath an entire circular disc. Ovary sessile, with three compartments, each containing two suspended inverted ovules, placed side by side. Fruit globose or somewhat angular, with a leathery outer rind bursting into three pieces, and an inner hard shell, containing three bony seeds, surrounded by a small quantity of pulp, or a single seed, by the abortion of the rest. *B. paniculata*, called Bois de Colophane in the Isle of Bourbon, contains an abundance of oil, like turpentine, which exudes when the bark is pierced, and speedily congeals, till it acquires a buttery consistence; others of the species furnish a resinous substance. The shrubs are natives of the West Indies. [M. T. M.]

BURSICULA (adj. BURSICULATE). A small purse. A pouch-like expansion of the stigma, into which the caudicle of some orchids is inserted.

BURSINOPETALUM. A genus of *Olacaceæ*, containing an Indian tree (*B. arboreum*), which has ovate acuminate leathery leaves and small panicled white flowers, remarkable for the form of the petals, which have an inflexed lobe at the point, and terminate in two small sharp teeth. Calyx superior, with a five-cleft limb; stamens five, connivent. Fruit a one-celled drupe, with a groove down one side of the hard endocarp. [J. T. S.]

BURTONIA. A genus of dwarf heath-like shrubs belonging to the pea-flowered section of the leguminous family, all of them natives of West Australia. They have simple or trifoliate sessile leaves, which are usually awl-shaped. The flowers are axillary and often thickly-gathered on the ends of the branches; the corollas rich purple, the keel generally of a deeper colour, and the standard having sometimes a yellow blotch at its base. The pod is small, ovate, and sessile, with two seeds. The species are very pretty objects when in flower, and are often to be met with in greenhouse collections. *B. scabra* was introduced in 1803, but there are now five species in cultivation and nine species known. The genus bears the name of Mr. D. Burton, who collected plants in W. Australia for the Kew Gardens. [A. A. B.]

BURWEED. The common name for *Xanthium*.

BUSBECKIA. The name of a genus of *Capparidaceæ* characterised by a calyx of two sepals, valvate in the bud, deciduous; petals seven, inserted at the base of the hemispherical receptacle, unequal, imbricate in the bud; stamens several, inserted on the torus; ovary on a long stalk, one-celled, with two or more parietal placentæ, bearing several curved ovules; stigma sessile, round; berry globose, leathery, rough on the outer surface; seeds kidney-shaped, imbedded in pulp, and with a leathery coat. A Norfolk Island shrub of climbing habit, with alternate leaves furnished with spiny stipules, and solitary axillary stalked flowers. The fruit is of the size of a large orange. The *Busbeckia* of Martius is now included in the genus *Sulpichroma*. [M. T. M.]

BUSH SYRUP. A saccharine fluid obtained from the flowers of *Protea mellifera*, in the Cape Colony.

BUSSEROLE. (Fr.) *Arctostaphylos Uva-ursi*.

BUSSU. A S. American name for *Manicaria saccifera*.

BUTCHER'S BROOM. *Ruscus aculeatus*; also a common name for *Ruscus*.

BUTEA. The three or four species constituting this genus of leguminous plants (*Fabaceæ Papilionaceæ*) form either small trees or large climbing shrubs, and are all natives of India. Their flowers are produced in racemes consisting of numerous flowers arranged in threes. The calyx has two small bracts near its base, and is usually covered with black velvety down; it is bell-shaped and two-lipped, the upper lip being nearly whole, and the lower one three cut; the corolla is papilionaceous; the stamens are ten in number, nine of them being united into a tube, and the tenth separate. The fruit is a stalked flattened thin and membranaceous pod, containing one seed placed near the apex.

B. frondosa, the Dhak or Palas of India, is a fine tree growing to about thirty or forty feet high, common in the jungles of Bengal. Its leaves are composed of three roundish leaflets, covered with silky hairs, somewhat resembling the pile of velvet; the young branches likewise are hairy. The racemes of flowers are produced early in spring, before the leaves have made their appearance; each individual flower being about two inches long and of a very bright orange-red colour. Dr. Hooker states that when in full flower the Dhak tree is a gorgeous sight, the masses of flowers resembling sheets of flame, their 'bright orange-red petals contrasting brilliantly against the jet-black velvety calyx.' The Dhak tree supplies the natives of India with several articles of a useful nature. The most important of these is the red astringent juice which exudes from wounds in the bark, and which, when hardened by evaporation, forms one of the brittle ruby-coloured substances called kino, this particular variety being termed butea kino or gum butea. Sometimes, however, it goes under the name of Bengal kino; but it must not be confounded with East Indian kino, which is produced by *Pterocarpus Marsupium*. This substance is procurable in large quantities, but it has not yet come much into use. The natives employ it for tanning leather, and it has been tried in this country for the same purpose, but the dark colour which it communicates to the leather is considered objectionable; it might probably be turned to account by the dyer. The flowers are called teesoo or keesoo in India, and afford either a beautiful bright yellow, or a deep orange-red dye; but unfortunately these tints are not permanent. A coarse fibrous material obtained from the bark of the stems and roots is used in India for caulking the seams of boats as a substitute for oakum. The lac insect (*Coccus*) likewise frequents the Dhak tree, and by its punctures in the young twigs causes the formation of the substance known as stick-lac, which is used in the manufacture of sealing-wax and in dyeing. And, finally, the seeds yield a small quantity of oil, called moodooga oil, which the native doctors consider to possess anthelmintic properties.

B. superba is a large climbing shrub with leaves resembling those of the last species; its flowers, also, are of a similar bright orange-red, but rather larger, so that when in full flower the plant presents a very gaudy appearance. Its products are similar to those of the dhak; the flowers yielding a colouring matter, and the juice hardening into kino.

B. parviflora is a shrubby climber resembling the last in general appearance, but having very much smaller flowers than either of the preceding. The gum of this species is given, dissolved in arrack, in hysteria and colic. [A. S.]

BUTOMACE.E. (*Butomeæ*; the Flowering-rush family.) A natural order of hypogynous monocotyledons belonging to Lindley's alismal alliance. Aquatic plants, often milky, with very cellular leaves, and umbellate showy flowers. Perianth of six pieces, the three inner (corolla) being coloured like petals. Stamens either below or above twenty in number, hypogynous. Ovaries three to six or more, either separate or united; ovules numerous. Fruit consisting of achenes or follicles, separate or united. Seeds numerous, attached to a net-like placenta, which is spread over the whole inner surface of the fruit; no albumen. Natives of the marshes of Europe and Siberia, the north-western provinces of India, and equinoctial America. The flowering rush, *Butomus umbellatus*, is an ornament of our lakes; its underground stem is roasted and eaten in Asia. There are four genera and seven species. Illustrative genera: *Butomus*, *Limnocharis*. [J. H. B.]

BUTOMUS. The Flowering Rush: one of the stateliest and most elegant of English aquatics, improperly called a rush, though the similarity of its long smooth knotless flower-stalk to the stalk of the bulrush (*Scirpus*) sufficiently accounts for the name having been given. Gerarde, who suggests the name of Lillie-grasse, calls it the Water Gladiole or Grassie Rush, and says, that 'Of all others it is the fairest and most pleasant to behold, and serveth very well for the decking and trimming up of houses, because of the beautie and braverie thereof; consisting of sundry small flowers, compact of sixe small leaves, of a white colour mixed with carnation, growing at the top of a bare and naked stalk, five or sixe foote long, and sometime more.' The leaves are narrow, triangular, and very cellular, shorter than the flower stalks, but they, nevertheless, greatly exceed two feet, the dimensions assigned to them in botanical works, as the plant generally grows in water at least two or three feet deep. The bottom of the main stalk as well as the partial flower stalks are frequently tinged with purple. The flowers are large, of six sepals and contain each nine stamens and six styles. The seeds and root were formerly employed medicinally, and in the north of Asia, the latter is roasted and eaten. A variety is cultivated which has striped leaves. (French, *Butome*, German, *Blumenbinse*.) [C. A. J.]

BUTTER & EGGS. The double-flowered variety of *Narcissus* (*Queltia*) *aurantius*.

BUTTER AND TALLOW TREE. *Pentadesma butyracea*.

BUTTER-BUR. The common name of *Petasites*, a group of the *Tussilago* family.

BUTTERCUP. The popular name for *Ranunculus acris* and its near allies, *R. repens* and *bulbosus*.

BUTTER OF CACAO. A pleasant concrete-oil, obtained from the seeds of *Theobroma Cacao*. —, OF CANARA. Piney tallow, a solid oil obtained from the fruits of *Valeria indica*.

BUTTER TREE, INDIAN. *Bassia butyracea*. —, AFRICAN. The Shea tree, *Bassia Parkii*.

BUTTERFLY-PLANT. *Oncidium Papilio*. —, INDIAN. *Phalænopsis amabilis*.

BUTTERFLY-SHAPED.—See Papilionaceous.

BUTTERFLY WEED. *Asclepias tuberosa*.

BUTTERWEED. *Erigeron canadense*.

BUTTERWORT. The common name for *Pinguicula*.

BUTTON-BUSH. An American name for *Cephalanthus*.

BUTTON-FLOWER. The common name of *Gomphia*.

BUTTON-TREE. The common name of *Conocarpus*.

BUTTON-WEED. The common name of *Spermacoce*. Also an American name for *Diodia*.

BUTTON-WOOD. *Cephalanthus occidentalis*. Also an American name for *Platanus*.

BUTUA. The Brazilian name for the roots of *Botryopsis platyphylla* and *B. cinerea*. According to Pereira, Butua root is the root of *Cissampelos Pereira* or pereira brava of commerce. — DO CURVO. The Brazilian name of the roots of *Cochlospermum insigne*.

BUXBAUMIA: BUXBAUMIACEÆ. A most singular genus and division of mosses, in which the capsule bears an extraordinary proportion to the vegetative part, which is sometimes all but obsolete. It has a double peristome, of which the outer one is either nearly obsolete, or consists of a triple or quadruple circle of teeth, and the inner forms a truncate cone. The species are few in number. Two are found in this country, *Diphyscium foliosum*, which has a nearly sessile ovato-conical capsule, and occurs on the ground and on rocks in sub-Alpine districts; the other, *Buxbaumia aphylla*, remarkable for its long stalked capsule, being flat on one side, and convex on the other, like the roses of certain watering-pots, and the leaves being quite rudimentary. It is found, but rarely, on heaths or in heathy woods. [M. J. B.]

BUXUS. A small but important genus of spurgeworts (*Euphorbiaceæ*), one species of which is the well-known Box-tree of our gardens. They are shrubs or small trees, with opposite entire evergreen leaves, and their flowers being produced in clusters from the angles of the leaves, each cluster consisting of several male flowers, surmounted by one or two females. They have a calyx, consisting of four minute sepals, the males having four stamens, and the females three styles. The fruit is three-celled, containing two shining black seeds in each cell, and splitting open when ripe.

The Common or Evergreen Box-tree (*B. sempervirens*) is a native of both Europe and Asia. In Europe it extends as far north as the fifty-second parallel of latitude, and is found plentifully on the coast of the Black Sea, also in Spain, Italy, and the southern and eastern provinces of France. In this country the only place where it is really indigenous is Boxhill in Surrey. In Asia it is found in Persia, Northern India, China and Japan. It varies considerably in height, some varieties growing as high as twenty or thirty feet, with a trunk eight or ten inches in diameter; while others never exceed three or four feet, and have very small stems. As commonly seen in this country it is either a shrub eight or ten feet high, or artificially dwarfed and only a few inches high. Its leaves vary from half an inch to an inch long, and from an egg-shaped to an elliptical form; they are of a shining deep-green colour, and of a thick leathery texture. The wood of the Box-tree has long been celebrated for its hardness and closeness of grain; it is mentioned by Theophrastus, and also by Pliny, the latter asserting that it is as hard to burn as iron. Other early authors also mention it as being used for musical instruments, carving, turnery, &c. Its chief characteristics are excessive hardness, great weight, evenness and closeness of grain, light colour, and being susceptible of a fine polish. These are the qualities that render it so valuable to the wood engraver, the turner, the mathematical and musical instrument makers, and others. Between 2,000 and 3,000 tons are annually imported; in 1854, the imports amounted to 2,704 tons, valued at £8,276l. The finest quality, and the best suited for the engraver, comes from Odessa, Constantinople, and Smyrna, being grown in the vicinity of the Black Sea; it is generally in logs about four feet long, and seldom more than eight or ten inches in diameter. For the use of the engraver these logs are cut across the grain into slices about an inch thick. In the early days of wood engraving, these slices were cut lengthways with the grain, and it was not till the middle of the last century that the present method was adopted. For the turner and other manufacturers of small wares, wood of an inferior description from smaller trees is suitable, and large quantities of box-wood articles are consequently made in different parts of France, where the tree abounds, though it does not attain a great size. The Box-tree is greatly employed in ornamental gardening, particularly for the formation of geometric designs. The kind commonly used for the edges of flower beds is merely a dwarf variety of the common species.

The Minorca Box, *B. balearica*, is a native of several of the Mediterranean islands and of Asia Minor. It is a larger tree than the last, growing sometimes as high as sixty or eighty feet, with a straight smooth trunk; the leaves also are of a much paler green than those of the common box, and much larger, being about three inches long and of an elliptical shape. The wood much resembles that of the common box, but is said to be of a coarser grain; it no doubt forms part of the wood exported from Constantinople and Smyrna. [A. S.]

BYBLIS. A genus of *Droseraceæ* containing Australian herbs resembling sundews (*Drosera*). They have very short stems, and tufts of linear leaves, with revolute margins. The peduncles are axillary, one-flowered; sepals and petals five each, the latter blue; stamens five; style simple. The capsule, which is obcordate, ventricose and two-celled, contains but few seeds. [J. T. S.]

BYRSANTHES. One of the genera of the order *Lobeliaceæ*. Calyx tube adnate to the ovary, its limb five-parted; corolla funnel-shaped, leathery (hence the name Leather-flower), its limb five-parted, with erect equal segments; stamens five, inserted with the corolla on to the tube of the calyx, the anthers coherent in a tube, some or all of them hairy at the top. Ovary two-celled, containing several ovules, adhering to the two-lobed placentæ; style not projecting from the flower; stigma two-lobed, the lobes spreading, roundish, hairy. Shrubs inhabiting the Andes, covered with snow-white hairs; flowers stalked. [M. T. M.]

BYRSONIMA. An extensive genus of plants belonging to the order of malpighiads (*Malpighiaceæ*), and containing about eighty species, inhabitants of Tropical America. They form shrubs or small trees, seldom exceeding thirty or forty feet in height, and have opposite entire leaves, destitute of the glands common to those of allied genera. Their flowers are produced in racemes at the ends of the branches, and are generally of a yellow colour; the calyx has ten glands or wart-like swellings, two at the base of each sepal; their ten stamens are connected together by a ring at the base, and they have three distinct styles terminated by pointed stigmas. Their fruit has a fleshy pulp surrounding a hard three-celled stone, containing three seeds.

B. Cumingiana is common in New Grenada, Panama, and Veraguas, forming a small tree about twenty-five or thirty feet high. Its leaves are lance-shaped, about three inches long and an inch and a half wide, the widest part being at the top end; both their upper and under surfaces, but particularly the latter, are covered with a thick coating of light brown woolly scurf, which gives them a rather dull appearance. The racemes of flowers are about six inches long, and being of a deep yellow colour, give the tree a fine effect when in full flower. In Panama it is called Nanci, and the inhabitants consider the bark an efficacious remedy in certain skin diseases common in that country; they likewise use the wood for building purposes, and eat the small acid berries.

B. crassifolia is a native of the West Indies and the Northern part of South America, where it forms a small tree about fifteen feet high. It has oval leaves about four inches long and two broad, smooth upon the upper surface, and covered with brownish silky down underneath. The bark possesses astringent properties, and is used for tanning leather; it is also said to be useful as a medicine, a decoction being employed as an antidote to the bite of the rattlesnake; and in Cayenne it is employed as a febrifuge. The Carib Indians call the plant Moulae-le, and use its bark for painting their paddles and arrow-heads.

B. spicata is a tree thirty or forty feet high, growing in some of the West India Islands and in Brazil. It has lance-shaped blunt-pointed leaves about four or five inches long and an inch and a half broad, of a shining green upon the upper side and a dull rusty brown colour beneath. The bark of this, as indeed of all the species of the genus, is very astringent, and is commonly used by the Brazilian tanners, under the name of Muruxi bark; it also contains a colouring matter, and is used by the Indians for dyeing their garments red. The yellow acid berries of this plant are very good eating when ripe, but rather astringent; they are considered to act beneficially in cases of dysentery.

B. verbascifolia is a small shrub with a short thick knotty stem, the wood of which is of a bright red colour. Its leaves are about ten inches long and of an obovate shape, i.e. having the top half broader than the bottom; they are generally woolly on both sides, and a microscopical examination of the hairs of this and other species will show them to be centrally attached. In Brazil and Guiana, where this plant grows, a decoction of the roots and branches is used for washing ulcers, and is considered to possess healing properties. [A.S.]

BYSSACEOUS. Composed of fine entangled threads.

BYSSI. A name which formerly included a heterogeneous mass of perfect and imperfect plants of various affinities, but is now exploded, the term byssoid alone being retained to express a peculiar fringed structure in which the threads or fascicles of threads are of unequal lengths. [M.J.B.]

BYSSUS. The stipe of certain fungals.

BYTTNERIACEÆ. (*Buttneriæ*; *Byttneriads*, the *Chocolate* family.) A natural order of thalamifloral dicotyledons belonging to Lindley's malval alliance. Trees, shrubs, or undershrubs with simple leaves and deciduous stipules. Calyx four to five-lobed, valvate in bud. Corolla consisting of four to five petals twisted in æstivation (flower-bud), sometimes wanting. Stamens hypogynous (inserted below the ovary) united into a tube; anthers opening inwards, two-celled, generally splitting lengthwise. Ovary composed of four to ten carpels, arranged round a central column; ovules two in each carpel; styles united below, but branching into four to ten stigmas. Fruit usually a capsule, splitting through the cells or resolving itself into its original carpels by splitting at the partitions; seeds albuminous. Chiefly tropical or subtropical plants. Lindley gives the following distribution of the tribes into which the order is divided:—*Lasiopetaleæ* in Australia; *Hermannieæ* in South Africa; *Dombeyeæ* in Asia and Africa; *Eriolæneæ* in Asia; *Philippodendreæ* in New Zealand; *Byttnereæ* in Asia and America. These plants have mucilaginous qualities. Chocolate and Cocoa are prepared from the seeds (Cacao beans) of *Theobroma Cacao*, a small tree found in the forests of Demerara. The seeds contain a tonic substance called theobromine, allied to theine, and a fatty oil is expressed from them called the butter of cacao. From the pulp of the fruit a kind of spirit is distilled. Several of the plants yield fibres which are used for cordage. There are fifty known genera and about 420 species. Illustrative genera:—*Lasiopetalum, Byttneria, Theobroma, Hermannia, Dombeya, Astrapæa, Eriolæna*. (J. H. B.)

BYTTNERIA. This genus gives its name to the natural family to which it belongs. The species are upwards of fifty, and are widely distributed, being found in India, Java, and Madagascar, in the Old World, and in America as far north as Texas, reaching south to Buenos Ayres, and attaining their greatest number in Brazil. They are very diverse in appearance, some being small erect herbs about one foot high; others tall straight bushes with winged or angled stems and very narrow leaves; *B. catalpifolia* grows to a tree thirty feet high, with long-stalked heart-shaped leaves; but the greater number are scandent prickly bushes, scrambling over other plants as the brambles do in our hedges. The leaves in all the species are simple, and in the greater part more or less heart-shaped in form, with entire or notched margins. The flowers are small, generally dark purple in colour, and arranged in axillary simple or compound umbels. The petals are curiously hooded at the apex, and from the outer surface of the hood grow one, two, or three strap-like appendages. The fruit is a five-celled woody capsule, spherical in form, from half an inch to two inches in diameter, and armed with long or short rigid bristles. Each cell contains one seed. *B. heterophylla*, a native of Madagascar, is often to be found with entire or three-lobed leaves on the same plant; it is an extensive climber, scrambling over the tops of the highest forest trees, and is said to cover nearly the whole slope of the sides of the mountain called Tantinanarivo, and to occur nowhere else in the island. The genus is named in honour of D. S. A. Byttner, once professor of botany at Gottingen. (A. A. B.)

CAA-APIA. A Brazilian name for *Dorstenia brasiliensis*.

CAA-ATAICA. *Vandellia diffusa*.

CAA-CUA. A Brazilian name for some scrophulariaceous plant.

CAA'-TIGUA'. A Brazilian name for *Moschoxylon Catigua*, a plant which imparts a bright yellow stain to leather.

CAAPE'BA. The Brazilian name for the Pareira brava, *Cissampelos Pareira*.

CAAPIM DE ANGOLA. *Panicum spectabile*, a fodder grass of Brazil.

CAAPOMONGA. *Plumbago scandens*.

CABALLINE ALOES. Horse Aloes, *Aloe caballina*.

CABARET. (Fr.) *Asarum europæum*. —, DES OISEAUX. *Dipsacus sylvestris*.

CABBAGE. The common name for *Brassica*; specially applied to the plane-leaved hearting garden varieties of *Brassica oleracea*. —, DOG'S. *Thelygonum Cynocrambe*. —, ST. PATRICK'S. *Saxifraga umbrosa*. —, SKUNK. The fetid antispasmodic *Symplocarpus fœtidus*.

CABBAGE PALM. *Areca oleracea*.

CABBAGE-BARK TREE. The Worm Bark, *Andira inermis*.

CABBAGE-TREE. A common name for the genus *Areca*; also a garden name for *Kleinia neriifolia*. —, AUSTRALIAN. *Corypha australis*, the leaves of which are made into plait for hats, baskets, &c. —, BASTARD. *Andira inermis*.

CABBAGE WOOD. *Eriodendron anfractuosum*.

CABEZA DE NEGRO. Negro's head, the Columbian name for the fruit of *Phytelephas macrocarpa*.

CABOMBACE.Æ. (*Cabombeæ; Hydropeltideæ; Water-shields*.) A natural order of thalamifloral dicotyledons belonging to Lindley's nymphal alliance. Aquatic plants with floating shield-like leaves; sepals and petals three or four, alternating; stamens six to thirty-six. Carpels distinct, two to eighteen; seeds not numerous; embryo in a membranous bag, outside abundant fleshy albumen. The plants are obviously allied to the Water-lilies. They are found in America, from Cayenne to New Jersey, as well as in New Holland. There are two genera, *Cabomba* and *Hydropeltis*, which comprise three species. (J. H. B.)

CABOMBA. A genus of aquatic herbs giving its name to the small order of *Cabombaceæ*. The species are small water plants with shield-shaped entire floating leaves, and finely-cut submerged ones, like those of the common water ranunculus. The flowers have three sepals, four or five petals, six stamens with ovate four-cornered anthers, and two ovaries. They are natives of North America. [M. T. M.]

CABOTZ. *Brayera anthelmintica.*

CACALIA. The generic name of plants belonging to the composite order, distinguished by the flowers being all tubular, and having both stamens and pistils; the heads of flowers surrounded by a single row of leaf-like bodies varying from five to thirty in number. The appendage on the top of the fruit or seed is in the form of a short cone, hairy at the base; the fruit is oblong and smooth. The species are perennial herbs with the leaves alternate, toothed or lobed, and the flowers varying in colour. Most of them are plants of peculiar aspect, owing to their clumsy fleshy stems, and the dingy colour of their leaves. They are natives of the warmer parts of America, middle Asia, and Eastern Africa. The Chinese employ as food the leaves of *C. procumbens*, and those of *C. ficoides*, a native of the Cape, are also wholesome. (G. D.)

CACALIE E'CARLATE. (Fr.) *Emilia sonchifolia.*

CACAO or COCOA. The seeds of *Theobroma Cacao*, which form, or should form, the chief ingredient in chocolate.

CACHIBOU RESIN. A gum-resin obtained from *Bursera gummifera.*

CACHRYS. One of the genera of *Umbelliferæ* (*Apiaceæ*), deriving its name, it is said, from a Greek word indicative of the hot or carminative properties of the fruit. The prominent characteristics of the genus are: the absence of an involucre; the margin of the calyx five-toothed or wanting; entire petals bent inwards at the point; the stylopoda or thickened base of the styles not very distinct; the fruit thick and spongy, each half with five thick ribs, and containing several oil channels in its rind. The species are natives of Southern Europe, Siberia, &c. The Cossacks are said to chew the seeds of *C. odontalgica* as a remedy for toothache, the effects being due to the salivation they induce. Several species are in cultivation, but are of no particular interest. [M. T. M.]

CACTACEÆ. (*Cacti*; *Carteæ*; *Opuntiaceæ*; *Nopaleæ*; *Indian Figs.*) The cactus family, a natural order of calycifloral dicotyledons. They consist of succulent shrubs with remarkable spines clustered on the stems, which are angular two-edged or leafy, and have their woody matter often arranged in a wedge-like manner. Calyx of numerous sepals, combined and epigynous; petals numerous; stamens numerous, with long filaments. Ovary one-celled with parietal placentas; style slender; stigmas several. Fruit succulent; seeds without albumen. They are natives of America, whence they have been transported to various quarters of the globe. The fruit of many of the Indian Figs is subacid and refreshing; in some instances it is sweetish and insipid. The stems of some of the species are eaten by cattle. These stems present very varied forms; some are spherical, others jointed, others have the form of a tall upright polygonal column. Their succulent character enables them to thrive in arid climates, and some of them have been called vegetable fountains in the desert. A South American species, *Cereus peruvianus*, has stems thirty to fifty feet high, and one to two feet in diameter; *C. Thurberi* has a stem ten to fifteen feet high, and *C. Schottii* has one eight to ten feet in height. The spines and bristles on a specimen of *Echinocactus platycerus* were reckoned at 51,000, those of a *Pilocereus senilis* at 72,000. *Opuntia vulgaris*, the common Prickly Pear, has an edible fruit, and *O. cochinellifera*, the Nopal plant, supplies food to the cochineal insect (*Coccus Cacti*). The number of known genera is eighteen, and of species about 800. Illustrative genera: *Cactus* or *Cereus*, *Melocactus*, *Mammillaria*, *Opuntia*, *Pereskia*. (J. H. B.)

CACTUS. This name includes in popular estimation all the various species referred by botanists to *Cereus*, *Epiphyllum*, *Echinocactus*, *Echinopsis*, *Mammillaria*, and *Melocactus*; under which genera their several peculiarities will be noticed. It is this old familiar name, sometimes still used under the plural form of *Cacti*, which has given the title of *Cactaceæ* to the family to which these plants belong. [T. M.]

CACTUS, HEDGEHOG. *Echinocactus.*—LEAF. *Epiphyllum.*—MELON-THISTLE. *Melocactus.*—, NIPPLE. *Mammillaria.*

CADABA. A name applied to a genus of *Capparidaceæ*, characterised by a calyx of four sepals, distinct or coherent at the base only; petals sometimes wanting; stamens more or less united below; fruit berry-like, stalked, subtended by a strap-shaped nectary. The plants are natives of Africa, India, and Australia. The root of *C. indica* is said to be aperient and anthelmintic. [M. T. M.]

CADE. (Fr.) *Juniperus Oxycedrus.*

CADEN. An Indian name for *Phœnix sylvestris.*

CADETIA. A little-known genus of one-leaved epiphytal orchids with the habit of *Pleurothallis*. Five species are described from the Moluccas and New Guinea. The genus is hardly distinct from *Dendrobium*.

CADJII GUM. A South American gum obtained from *Anacardium occidentale.*

CADUCOUS. Dropping off.

CÆNOPTERIS. A name which has been sometimes adopted for the *Darea* section of *Asplenium*: a group of species usually distinguishable by the uniseriferous ultimate segments of their fronds. The name has also been given to another fern, now referred to *Onychium*. [T. M.]

CÆOMACEI. A term applied to those species of truly parasitic *Fungi* known familiarly under the name of Rust and Mildew, which have naked spores free from dissepiments. They are, however, so closely connected with those with septate spores that it is far more natural to unite them. We accordingly refer for further information to the article *Puccinieæ*. [M. J. B.]

C.ÆRULEUS or CŒRULEUS. Blue; a pale indigo colour.

CÆSALPINIA. A genus of leguminous plants typical of the section *Cæsalpinieæ*, containing about fifty species, most of which are small trees or large shrubs, inhabiting tropical countries. Their leaves are compound, being what is termed bipinnatifid. Their flowers are produced in racemes, and have a top-shaped calyx, divided at the end into five parts, the lowest of which is larger than the others; five unequal stalked petals, the upper one shorter than the rest; ten stamens, and a long slender style.

C. coriaria is a small tree twenty or thirty feet high, native of several of the West Indian Islands, Mexico, Venezuela, and North Brazil. The primary divisions of its leaves vary from nine to fifteen, each bearing from sixteen to twenty-four narrow oblong blunt leaflets, marked with black dots on the under surface. It has branched racemes of white flowers, which produce curiously flattened pods, about two inches long by three-fourths broad, and curved so as to bear some resemblance to the letter S. The large per centage of tannin in these pods renders them exceedingly valuable for tanning purposes: they are known in commerce under the names of Divi-divi, Libi-divi, or Libi-dihi, and are chiefly imported from Maracaibo, Paraiba, and St. Domingo.

C. crista, a native of the West Indian Islands, grows about twenty feet high, and has smooth prickly branches, and leaves with eight primary divisions, each having from three to five pairs of leaflets, which are generally notched at the top, and of an oblong shape, rather broader at the top end. The flowers are yellowish-red, and produce scimitar-shaped pods about three inches long, containing eight or ten seeds. *C. echinata* is a Brazilian tree with prickly branches, elliptical blunt-pointed leaflets and yellow flowers producing spiny pods. The woods known in commerce as Brazil, Pernambuco, Nicaragua, Lima, and Peach-woods, are said to be produced by this genus, but nothing certain is known upon the subject. They are generally attributed to the two last-named species, and to another called *C. brasiliensis* (the correct name of which is *Peltophorum Linnæri*), but which is not a native of Brazil. They are all exceedingly valuable to the dyer, producing various tints of red, orange, and peach-colour. The imports of Brazil wood in 1858 amounted to 1,052 tons, and of Nicaragua wood to 4,767 tons, the aggregate value of which was 133,627l.

C. Sappan, an East Indian tree growing about thirty or forty feet high, has prickly branches, the primary divisions of the leaves varying from twenty to twenty-four, and having ten or twelve pairs of obliquely oval-shaped leaflets, notched at the tip, with minute dots on the under surface. The brownish-red wood of this tree furnishes the Sappan wood of commerce, the Bukkum or Wukkum of India, from which dyers obtain a red colour, principally used for dyeing cotton goods. Its root also affords an orange-yellow dye. In 1858, 4,116 tons of sappan wood were imported into this country.

Of other useful species *C. Pipai* produces pods which possess some astringency, and are called Pipi pods; the seeds of *C. digyna*, an East Indian climber, yield an oil used for burning in lamps in India; the roots of *C. Nuga* are diuretic; and in China the pods of several species are called Soap pods from their being commonly employed for producing a lather as a substitute for manufactured soap.
[A. S.]

CÆSAREA. A genus of *Vivianiaceæ*, containing but few species, all of them natives of Southern Brazil. They are slender herbs one to three feet high, with opposite, or, towards the base of the stem, verticillate leaves, having serrate margins, generally smooth above, and covered with white down underneath. The flowers are axillary towards the ends of the branchlets, white, yellow, or of a reddish colour. None of the species are in cultivation, although they would be pretty greenhouse plants. The genus bears the name of César de S. Hilaire, a captain in the French navy, who first introduced the Mocha coffee to Bourbon. [A. A. B.]

CÆSIA. A genus of *Liliaceæ*, containing herbs from Australia and Tasmania, with fasciculate roots often with thickened tuberous fibres, grass-like radical leaves, and rather small white or blue flowers in simple or compound racemes. Perianth six-parted, the segments petaloid and spirally twisted after flowering; stamens six, with glabrous filaments; ovary three-celled, with two ovules in each cell; capsule sometimes one-celled; seeds with an appendage at the base. [J. T. S.]

CÆSIUS. Lavender colour.

CÆSPITOSE. Growing in tufts or patches.

CÆSULIA. The only species of this genus, *C. axillaris*, which belongs to the composite family, grows in moist places in many parts of India, and is a small weed with alternate linear toothed leaves, and what appear to be single sessile flower-heads, but which are in reality a number of flower-heads enclosed in a common involucre, each of them containing only one floret and provided with a two-leaved involucre, the lower part of which at length unites with, and forms part of the achene, the upper portions remaining free and giving the achene an eared appearance. The florets are purple or white. [A. A. B.]

CAFE' FRANÇAIS. (Fr.) *Cicer arietinum.* — MARRON. The wild Bourbon Coffee *mauritiana.*

CAFE'IER or **CAFFE'YER.** The Coffee tree.

CAFFER-BREAD. A South African name applied to various species of *Encephalartos.*

CAHINCA or **CAINCA.** A Brazilian drug obtained from *Chiococca densifolia.*

CAHOUN NUTS. The fruits of *Attalea Cohoun,* which yield an oil equal to that of the cocoa-nut.

CAIANNE. (Fr.) A kind of olive.

CAILLEBOTTE. (Fr.) *Viburnum Opulus.*

CAILLELAIT. (Fr.) *Galium verum.*

CAIOPHORA. A genus of loasads, distinguished from its congeners by having on the calyx ten spirally-arranged ribs; the divisions of the corolla notched at the tip or with three teeth; style or appendage on the ovary single, bifid at the end, the two pieces approximate. The name appears to be derived from the Greek verbs signifying 'I burn,' and 'I bear,' in allusion to the numerous stinging hairs which produce a burning sensation when they pierce the skin. The species are herbaceous plants, natives of Peru and Chili, of branched and climbing habit, armed with sharp stings. The leaves are opposite, lobed or deeply cut; the flowers solitary from the angles of the leaves or at the ends of the branches.

The plants have the general aspect of *Loasa,* a genus in which some of them were formerly included. One of the most notable is *C. coronata,* discovered by Dr. Gillies on the sides of the Cordillera, between Mendoza and Chile, at an elevation of 8,000 to 11,000 feet. Dr. Gillies observes of it: 'the general aspect of the plant is very peculiar, and on examining its whole economy we are struck with the care taken to protect the flower, and insure impregnation. It forms a large convex mass, rising one or two feet from the ground; the upper part is composed entirely of a great abundance of dark green leaves, along the margins of which, and protected by them, are arranged the large whitish flowers, forming one or two or more circlets or fillets, giving the whole a very singular and elegant appearance. The corolla, which is contracted towards its mouth, is of considerable size; the transverse section, at the widest part, being in some cases as large as that of a hen's egg. When the capsules are ripe, they are generally prostrate on the ground, the stalk being too weak to support them. [G. D.]

CAJANUS (from *Catjang,* the Malayan name for one of the species) is a small genus of leguminous plants of the section *Papilionaceæ,* forming shrubs, with leaves composed of three stalked leaflets, and flowers produced in racemes from the angles between the leaf stalks and stems. Their calyx is bell-shaped and cut half-way down into four divisions, the upper of which has two small teeth; their corolla is papilionaceous; and they have ten stamens, nine of which are united together, and the tenth free. The fruit is a pea-like pod, containing many seeds, and having its husk or shell constricted between each seed.

C. indicus is a native of the East Indies, but is now naturalised and cultivated in the West Indies, in tropical America and Africa, and in some islands of the Pacific Ocean. There are two varieties, differing only in height and in the colour of their flowers. It is a perennial shrub growing from three to ten feet high, but in places where it is cultivated, it is generally treated as an annual, the stems being pulled up and used for firewood as soon as the crop of seeds has been gathered. All parts of the plant are more or less covered with soft silky or velvety hairs. The leaves are composed of three oval-lance-shaped stalked leaflets. The variety *bicolor* generally grows from three to six feet high, and has yellow flowers marked with crimson streaks on the outside; its pods are spotted or marbled with dark lines. It is called the Congo pea in Jamaica. The variety *flavus* is a larger kind, forming bushes twenty feet in circumference, and varying from five to ten feet high; it has pure yellow flowers and uniformly-coloured pods. In Jamaica it is called the No-eye pea. Both of these varieties are cultivated in various parts of the tropics for the sake of their seeds or pulse. In India the pulse is called Dhal or Dhol, or Urhur, and it forms a large part of the food of all classes of natives, being ranked as third in value among the pulses. In the West Indies they are called Pigeon peas, being commonly used for feeding pigeons and other birds; besides which they are highly esteemed as an article of human food, the variety called No-eye pea being considered to be little inferior in a green state to our English peas, and, when dried and split, quite as good. The Congo pea is harder and coarser, and is only used by negroes, requiring a great deal of boiling. Pea meal of very good quality is prepared from both the varieties in Jamaica. Horses and cattle of various kind are very fond of the young branches and leaves, either in a fresh or dried state. The late Dr. McFadyen, speaking of this plant, says: 'There are few tropical plants so valuable. It is to be found round every cottage in the island (Jamaica), growing luxuriantly in the parched savannah and mountain declivity, as well as in the more fertile and seasonable districts.' [A. S.]

CAJU'PUTI. An old synonyme of *Melaleuca,* one species of which, *M. Cajuputi,* yields the stimulant oil of cajeput, used in medicine. [T. M.]

CAKILE. A sea-side herbaceous plant belonging to the *Cruciferæ,* easily distinguished by its oblong deeply-lobed fleshy leaves, which are smooth and of a glaucous hue; by its lilac flowers; and by its succulent pod, which when matured is found to be divided by a horizontal partition into two cells, the upper containing a single erect seed, the lower a pendulous one. It is known to sea-side visitors by the name of Sea Rocket, but has nothing to recommend it to notice but the singular structure of its seed-pods. Closely allied species inhabit the shores of the Mediterranean and the West Indian Islands. They are all annuals, and grow among the shingle or sand a short distance above high-water mark. French, *Caquille*. [C. A. J.]

CALABA TREE. *Calophyllum Calaba.*

CALABASH. *Crescentia Cujete*, a tropical tree bearing great gourd-like fruits. —, SWEET. *Passiflora maliformis.*

CALABASH NUTMEG. *Monodora Myristica.*

CALABUR TREE. *Muntingia Calabura.*

CALADENIA. A genus of exquisitely beautiful little terrestrial orchids inhabiting Australasia. They generally produce one grassy leaf, from within which arises a scape bearing a few ringent flowers, covered in various places in a very remarkable manner with glandular hairs, which have suggested the name. In many species the sepals or petals or both are prolonged into long slender tails; in others they have the usual oval outline. The genus is admirably illustrated in Hooker's *Flora Tasmanica.*

CALADIUM. The generic name of certain plants of the *Arum* family, having a hood-like spathe rolled round at the base; a spadix whose upper portion is entirely covered with stamens, but ultimately becomes bare at the extreme top, provided with blunt glands or sterile stamens in the middle, and ovaries beneath; the anthers shield-shaped and one-celled; the ovaries numerous, two-celled, with from two to four ascending ovules in each cell; the fruit a one or two-celled berry, with few seeds. These plants partake of the acrid properties which pervade the *Araceæ*, but, nevertheless, the rootstocks or rhizomes of some of the species are eaten as food in the West Indies, the Sandwich Islands, &c., in consequence of the abundance of starch contained in them, the process of cooking depriving them of their noxious qualities. It is stated that the rootstocks or tubers of *C. petiolatum* were on one occasion mistaken for potatoes on board ship, and were given to some animals with fatal results. The leaves of *C. sagittifolium* are boiled and eaten in the West Indies as a vegetable. The species are natives of the warmer regions of the globe, where they are cultivated in abundance for the above-named purposes. Several are also grown in hothouses in this country, latterly several varieties with beautifully variegated foliage have been introduced. See Plate 2, fig. 5. [M. T. M.]

CALAMAGROSTIS. A genus of grasses belonging to the tribe *Arundineæ*, distinguished chiefly by the inflorescence being in branched panicles, and only one flower in the spikelets, or within the glumes, which has long silky hairs at its base; sometimes the rudiment of a second floret is present; glumes nearly equal, keeled and pointed; pales two, small. There are eighty-six species described in Steudel's *Synopsis*. They have an extensive range over the globe. The greater number are, however, natives of rather temperate climates, and some reach the Arctic circle. Three species only are natives of the British Islands, and neither of these is of common occurrence; indeed, *C. stricta* is one of the rarest British grasses, and only grows sparingly in a few localities. They are not valuable for agricultural purposes, though very ornamental. [D. M.]

CALAMBAC. The commercial name of Aloes-wood, Eagle-wood, or Lign Aloes, which is produced by *Aloexylum Agallochum.*

CALAMINT. (Fr.) *Calamintha officinalis.*

CALAMINTHA. A genus of labiate flowers which as at present constituted, comprises several plants described in less recent works under the names of *Thymus, Acinos, Melissa,* and *Clinopodium.* The essential generic characters of *Calamintha* are: "calyx two-lipped; stamens diverging; upper lip of the corolla nearly flat, tube straight.' *C. Acinos*, or Basil-thyme (formerly called *Thymus Acinos* and *Acinos vulgaris*), is a low somewhat shrubby plant with stems from four to six inches high, small leaves, and rather showy violet-purple flowers, which grow in whorls of six together. The whole plant is fragrant and aromatic, and well deserves its name (from the Greek *basilicon*, royal), if, as Gerarde tells us, 'the seede cureth the infirmities of the hart, taketh away sorrowfulnesse which commeth of melancholie, and maketh a man merrie and glad.' It is most frequently found in chalky or gravelly pastures. *C. officinalis* (*Melissa Calamintha*), *C. Nepeta*, and *C. sylvatica*, the Calamints or 'Excellent Mints,' as their name imports, are herbaceous aromatic herbs to which great medicinal virtues were anciently ascribed. They bear their flowers in stalked tufts which proceed from the axils of the opposite leaves, and are only to be distinguished from one another by a minute comparison of characters. They all possess a strong aromatic odour resembling that of penny-royal, and are employed to make herb-tea. *C. Clinopodium*, the Wild Basil, formerly called *Clinopodium vulgare*, is a straggling hedge plant with hairy stems from one to two feet long, bearing its rather large purple flowers in dense whorls in the axils of the hairy ovate distant leaves, and having numerous bristly bracts at their base. The odour is aromatic, but not so agreeable as in the other species. [C. A. J.]

VEGETATION OF THE CAROLINE ISLANDS, ON THE OUTSKIRTS OF WOOD.

a Artocarpus incisa.
b Caladiums
c Pandanus odoratissimus in fruit.

CALAMOSAGUS. The four species constituting this genus of palms do not possess any individual features of interest beyond their technical characters. They are all natives of the forests of the Malayan peninsula, and have climbing whip-like stems, growing to a great length, and supporting themselves by means of their hooked spines. The footstalks of their pinnate leaves are likewise armed with prickles and hooked spines, and terminate in a long whip-like tail; the leaflets are of a green colour on the upper surface, and covered with a bluish bloom underneath; their top half is broad and very much jagged, the lower half being entire and wedge-shaped. One of the chief characteristics of the genus is the presence of a broad leafy expansion called the ligule, near to and partly surrounding the base of the footstalk of the leaf. They have perfect flowers, arranged in branching spikes resembling bunches of catkins, each flower being half buried in a dense mass of wool, and having a three-toothed calyx, a three-parted corolla, six stamens, and a three-celled ovary covered with scales, and crowned by a three-toothed awl-shaped style. None of the species are known to possess any useful properties, but as their stems bear a close resemblance to some of the species of *Calamus*, they are probably used for similar purposes. One species, *C. hariniæfolius*, is called Rotang Simote by the Malayans, and another, *C. ochriger*, Rotang Donam. [A. S.]

CALAMPELIS. *Eccremocarpus.*

CALAMUS. The stems of several species of this genus of palms are well known in this country under the names of Rattans or Canes. Upwards of eighty species are described, nearly all natives of Asia, abounding in the Malayan Peninsula and islands, also in the eastern and north-eastern provinces of India; two are found in Australia, and one in Africa. They have reed-like stems, seldom more than an inch or two in thickness, but often much less, generally growing to a great length, climbing over and amongst the branches of trees, and supporting themselves by means of hooked spines attached to their leafstalks; a few, however, form low bushes or small trees. Their leaves are pinnate, and in many of the species, the leafstalk is prolonged beyond the termination of the leaflets into a whip-like tail. The flowers are small, generally of a rose or greenish colour, and arranged very close together upon long branching spikes, the ultimate branches somewhat resembling catkins. They have a three-toothed calyx, and a three-parted or three-petaled corolla; the males having six stamens joined together at the base, and the females imperfect stamens, and a three-celled ovary, more or less covered with scales, and bearing three stigmas (no style). The spikes are surrounded by numerous bracts or spathes, which, however, do not completely enclose them, and each branch of the spike has a separate bract at its base. The fruits are covered with smooth shining scales, which are fixed by their upper edges, and overlap each other from the top downwards, like plates of mail; they generally contain a single seed, surrounded by an eatable pulp.

C. Rotang, C. rudentum, C. verus, C. viminalis, and probably several other species, furnish the canes or rattans so commonly employed in this country for the bottoms of chairs, couches, sides of carriages, and similar purposes; and of which no fewer than 18,625,368 were imported in 1858, and valued at 38,000*l*. In the countries where these palms abound, the inhabitants make use of them for a great variety of purposes, baskets of all kinds, mats, hats, and other useful articles being commonly made of them. Their most important use, however, is for the manufacture of the ropes and cables usually employed by junks and other coasting vessels. In the Himalayas they are used in the formation of suspension bridges across rivers, the construction of which Dr. Hooker thus describes: 'Two parallel canes, on the same horizontal plane, were stretched across the stream; from these others hung in loops, and along the loops were laid one or two bamboo stems for flooring; cross pieces below this flooring hung from the two upper canes, which they thus served to keep apart. The traveller grasps one of the canes in either hand, and walks along the loose bamboos laid on the swinging loops.'

C. Scipionum, the stems of which are much thicker than those of the preceding, furnishes the well-known Malacca canes so much prized for walking-sticks. They are imported from Singapore and Malacca, but are chiefly produced in Sumatra. Some are of a uniform rich brown colour, whilst others are variously mottled or clouded as it is called; the colour, however, is said to be artificially imparted to them by smoking. *C. Draco,* the species yielding the red resinous substance called dragon's-blood, is now placed in the genus *Dæmonorops,* as also are several other *Calami.* [A. S.]

CALAMUS. A fistular stem without an articulation.

CALAMUS AROMATICUS. *Acorus Calamus.* —**ODORATUS.** *Andropogon Schœnanthus.*

CALANDRINIA. A genus of purslanes consisting of smooth fleshy plants of annual or perennial duration, with entire leaves, and, in the case of the species in cultivation, showy purple or rose-coloured flowers expanding only in sunny weather. It is well distinguished among the other genera of the order by its two persistent sepals, which close over the seed-vessel after the petals have fallen; three to five petals, mostly the latter number, numerous distinct stamens, single style with its stigma three-lobed, and oblong one-celled fruit, splitting when ripe into three pieces or valves, and containing numerous seeds adhering to a central placenta. A few only of the species are introduced, but they include probably the most interesting members of the genus. As a type of one section, reference may be made to the *C. discolor*, which has large oblong or lanceolate pointed glaucous leaves, mostly radical, green on their upper surface and purplish beneath, whence its specific name; and flowers one and a half inch across, of a bright rose colour, produced in a long distant raceme on a sort of scape, the footstalks being deflexed before and after flowering, and furnished with one or two ovate bracts at their base; a calyx of two broad concave pieces spotted with black, and petals inversely heart-shaped. The *C. grandiflora* closely resembles it, but has, notwithstanding its name, rather smaller flowers, with leaves which are more tapering at the point and base, and green on both surfaces. Both of these species, though usually treated as annuals, are perennial in warmer latitudes than our own, and are, as well as the following plant, natives of Chili. *C. umbellata* differs very considerably in habit from the two preceding, forming a small spreading tuft with shrubby shoots thickly set with linear foliage, fringed at the margins. The flowers are produced in terminal umbels, more or less compound, according to the strength of the plant, each blossom being about half an inch in diameter, and of a rich purple-crimson colour. Possessing a hardier constitution than the foregoing plants, it frequently endures our winters in dry soils, though often treated as an annual. *C. speciosa*, a Californian annual, is of procumbent habit with numerous branched stems radiating from the crown of the root, thickly clothed with narrow spathulate glossy leaves, and producing singly from its axils a profusion of crimson purple flowers rather larger than in *umbellata*. The seeds of this species are lenticular in form, and of a glossy black colour, by which they are readily distinguishable from those of the three previously-named. There is a variety of this with flowers of a coppery-red colour. [W. T.]

CALANTHE. A large genus of terrestrial stemless vandeous orchids with broad many-ribbed leaves, and long spikes of flowers, the lip of which is calcarate and adherent to the column, while the waxy pollen masses are eight, adhering to a separable gland. Some thirty species are known, chiefly from tropical and extratropical Asia; a few are American. The flowers, which are white, or lilac, or purple, or copper-coloured, are ornamental, wherefore several species are in gardens. Of these *C. vestita* is one of the handsomest.

CALATHEA. A genus of *Marantaceæ*, deriving its name from its cup-shaped stigma. These plants have large leaves springing from the contracted stem near the root, from which they appear directly to emerge. The flowers are in terminal spikes and protected by bracts; they have a calyx of three segments; a corolla of six pieces, the external ones lance-shaped, the internal ones blunt and irregular in shape; three petal-like stamens, one of which bears a linear one-celled anther, attached to its edge, while the rest are sterile; and a petal-like style, the stigma hooded, angular. The species are natives of tropical America, and some of them are in cultivation for the sake of their handsome foliage, especially *C. zebrina*, the leaves of which have alternate dark-coloured and green stripes. The leaves of some of the South American kinds are used for making baskets. [M. T. M.]

CALATHIAN VIOLET. *Gentiana pneumonanthe*.

CALATHIDA, CALATHUS, CALATHIDIUM. The head of flowers borne by composites.

CALATHIFORM. Cup-shaped, or almost hemispherical.

CALATHODES. A genus of *Ranunculaceæ* containing one species, *C. palmata*, from Sikkim, growing at an altitude of about 10,000 feet. A perennial herb with the habit of *Trollius*, having palmately cleft leaves, a simple stem one and a-half foot to two feet high; flowers large terminal and solitary, with five ovate acute petaloid sepals; petals none; ovaries ten or more, oblong, gibbous externally, beaked; ovules eight or ten; style bent down outwards after flowering. [J. T. S.]

CALAVANCE. A name for several kinds of pulse, including *Dolichos barbadensis* and *D. sinensis*.

CALBOA. A synonyme of *Quamoclit*.

CALCAR (adj. *Calcaratus*). A spur; a hollow process of some part of a flower.

CALCAREUS. Dead-white, like chalk. Also growing in chalky places, or having the substance of chalk.

CALCARIFORM. Shaped like a calcar or spur.

CALCEARIA. *Coryanthes*.

CALCEOLARIA. A beautiful genus of *Scrophulariaceæ*, distinguished chiefly by the peculiar form of the corolla, which has two lips, the lower of which is inflated, somewhat elongated and turned downwards, having some resemblance to a shoe; the stamens are two in number. The name is derived from the Latin word 'Calceolarius,' a shoemaker. The peculiar form of the corolla, above described, is nevertheless not invariably a character of the genus; the plant known in collections as *C. violacea*, a native of Chili, has the corolla in the form of two equal gaping lips; it was formerly placed in the genus *Jovellana*, but is now considered by the best authorities as a true *Calceolaria*. The numerous species of this favourite and well-known genus are either herbaceous or shrubby in habit, with leaves in pairs or three together, rarely alternate, either entire toothed or deeply cut, often more or less hairy, the flowers variously grouped and distributed, the prevailing colours, yellow white or purple. They are natives of South America, confined either to the western side of the Andes, or to the southern extremity of the mainland and the adjacent islands. Some are found only near the level of the sea, and others are inhabitants of the higher parts of the Cordilleras; hence it is that, among the numerous introduced species in our collections, a few are more or less hardy, others require protection. *C. floribunda*, for instance, is a native of the vicinity of the city of Quito, at an elevation of 11,000 feet above the level of the sea, and several occur at low altitudes in the Falkland Islands, &c., forming a prominent feature of the native vegetation. In the *Flora Antarctica*, Dr. Hooker thus alludes to the *C. Fothergilli* of Port Famine: 'Though inferior in stature and beauty to many of its congeners, this is among the prettiest of the wild flowers of the Falklands, and the attention of the voyager who is familiar with the genus *Calceolaria* only in the conservatories of Britain, must be attracted by its appearance on the exposed shores of these inhospitable islands.' Many of the original pure species have been modified by hybridising, and are not now so common in collections. The hybrids are numerous and some of them greatly prized; not only is the size of the flower modified but the colour as well, the shades of yellow and purple being highly varied, as also the characters of the spots on the slipper-like portion. The handsome aspect of different species and crosses has always recommended them to the attention of cultivators, and acted as a stimulus to the exercise of ingenuity in discovering the proper method of treatment. The results have been such, that on plants attaining a height of two feet or little more, the flowers may sometimes be counted by hundreds, expanded about the same time. [G. D.]

CALCEOLATE. Having the form of a slipper or round-toed shoe.

CALCEUS. Dead-white, like chalk.

CALCITRAPA. The Star Thistle, *Centaurea Calcitrapa*.

CALDASIA. A genus of *Polemoniaceæ* containing annual herbs from Mexico with glandular hairs, branched stems, and alternate crenate-serrate leaves. Peduncles axillary, in pairs, one-flowered; calyx five-sided, five-toothed at apex, scarious in fruit; corolla violet blue, funnel-shaped, with a five-parted limb, the lobes notched at the apex, and two of them apart from the other three; stamens five, protruding, bent down; capsule three-celled; seeds in each cell with a spongy coat. [J. T. S.]

CALDCLUVIA. A genus of *Cunoniaceæ*, containing a small tree from Chili, with opposite simple lanceolate serrated leathery leaves, glaucous below; stipules lanceolate, deciduous; flowers in axillary panicles; calyx deciduous, four or five parted; petals four or five, inserted on a disk which has as many glandular notched lobes as there are petals; stamens eight or ten, inserted within the disk; ovary free, two or three-celled; ovules numerous; styles two or three, becoming reflexed. [J. T. S.]

CALEA. The species of this genus, which belongs to the composite family, are natives of tropical America, extending from Mexico to South Brazil. They are herbs or small shrubs with opposite or whorled entire or toothed leaves, generally three-nerved and very rough on the surface, many of them resembling those of the common nettle in form. In one group the species are dwarf and unbranched, bearing a long-stalked terminal flower-head, about an inch in diameter, containing both strap-shaped and tubular florets, the former having pistil only, the latter both stamens and pistil. In another group the plants are larger, the flower-heads small and numerous, disposed in corymbs at the ends of the branches, and bearing tubular florets only. The flowers of most of them are yellow and the pappus is made up of from five to twenty lanceolate pointed scales. Upwards of thirty species are known. *C. Zacatechichi*, a Mexican species with nettle-like leaves and small flower-heads, is known there by the name of 'Juralillo,' and is said to contain, in a fresh state, a considerable quantity of camphor, and to be employed against fevers, and the powdered leaves for healing wounds. The leaves of *C. jamaicensis* are said to be powerfully bitter, and steeped in wine or brandy are used as a stomachic in the West Indies; but this account is thought to apply rather to *Neurolæna lobata*. [A. A. B.]

CALEANA. A few brown-flowered terrestrial orchids confined to New Holland bear this name. They have simple filiform roots terminated by a small tubercle, solitary radical leaves, and a slender few-flowered scape. The column is broad, thin, and concave; the sepals and petals narrow and reflexed; the lip posticous, petiolate, unguiculate, and highly irritable. In fine weather or when undisturbed, this lip bends back and leaves the column uncovered; but if it rains or the plant is jarred, down goes the lip over the column, which it securely boxes up. See DRAKÆA and SPICULÆA, in which a similar phenomenon occurs.

CALEBASSE. (Fr.) The Bottle Gourd, *Lagenaria vulgaris*.

CALEBASSIER. (Fr.) The Calabash tree, *Crescentia Cujete*.

CALECTASIA. A genus of *Juncaceæ* containing a small branched shrubby plant from South Australia, with needle-shaped leaves sheathing at the base, and solitary flowers on short terminal branches, having a silver-shaped perianth with a six-parted limb of petaloid blue segments spreading like a star, the three outer pubescent; stamens six. The ovary is one-celled with three ovules. [J. T. S.]

CALEE KUSTOOREE. An Indian name for the Musk Ochro, *Abelmoschus moschatus*.

CALELYNA. A section of *Evelyna*.

CALENDULA. The name of a genus belonging to the composite order, having numerous flowers grouped on a nearly flat surface, those at the circumference strap-like, in two or three rows and with pistils only, those in the centre tubular with stamens only, both kinds hairy at the base, the whole surrounded on the outside by a series of scale-like leaflets. The name *Calendula* is founded on the circumstance that species may be in flower on the calends of every month. They are annual or perennial, chiefly natives of the Mediterranean borders, with yellow or orange-yellow flowers, usually of a powerful, not pleasant odour. One of them, *C. officinalis*, the Pot Marigold, formerly enjoyed repute as a domestic remedy, being used in forming a distilled water or vinegar. [G. D.]

CALF'S SNOUT. *Antirrhinum Orontium*.

CALICATE. Furnished with a calyx.

CALICINAL. When a flower becomes double by an increase in the number of lobes of the calyx or sepals.

CALICIUM, CALICIEI. A genus and family of lichens known at once by the sporidia forming ultimately a dusty stratum over a little orbicular disc which is either nearly sessile or supported upon a short stalk so as to look like a little nail more or less completely driven home. The sporidia, as in other lichens, are at first contained in asci, which soon, however, disappear. One of the most familiar species is *C. inquinans*, which is common upon gate-posts, and attracts notice from leaving the print of its discs upon the finger when touched. The crust is sometimes very obscure or almost obsolete. All the species of *Calicium* are, we believe, found in Europe, though several of them occur elsewhere. [M. J. B.]

CALICO BUSH. *Kalmia latifolia*.

CALICULAR. A term of æstivation, when the outer bracts of an involucre are much shorter than the inner.

CALIMERIS. The generic name of plants belonging to the composite order, having the flowers in heads, those at the circumference in one row, strap-like, the heads surrounded externally by two to four rows of nearly equal scale-like leaves. The surface supporting the flowers has numerous four-cornered pits or depressions toothed at the angles. The fruit is flat and hairy. The name *Calimeris* is of Greek derivation, and indicates general beauty of parts. The species are perennial herbs, natives of middle and Northern Asia, with the leaves entire or toothed and cut at the margin; the heads of flowers yellow in the middle, and white or blue at the circumference. [G. D.]

CALIPHRURIA. A genus of amaryllids, forming a link between *Eurycles* and *Griffinia*, and having, except in the inflorescence, much the appearance of *Eucrosia*. The species, *C. Hartwegiana*, has ovate bulbs, petiolate depressed perennial oval acuminate somewhat plaited leaves, with a blade six inches long or more, a glaucous scape a foot high bearing an umbel of about seven subdeclinate flowers, having a green tube and white limb. The tube of the perianth is narrowly funnel-shaped and nearly straight, the limb regular with the segments turned back in the form of a star, the sepaline ones rather the broader. The filaments of the six stamens are inserted at the base of the segments of the perianth, and have a white bristle on each side, and they are associated with a straight style terminated by a somewhat recurved three-lobed stigma. *C. Hartwegiana* is a native of New Grenada. [T. M.]

CALLA. A genus of *Orontiaceæ*, consisting of herbaceous marsh plants with creeping or floating stems, heart-shaped entire leaves, the stalks of which emerge from a sheath. The flowers cover a spadix, which is protected by a flat spathe, the flowers themselves having neither calyx nor corolla. The upper flowers are female, consisting of a one-celled ovary, from the base of which arise the ovules; the lower flowers are hermaphrodite with numerous thread-shaped stamens, flattened and dilated at the top, and springing from below the ovary. The species are natives of Northern Europe and North America, and possess acrid caustic properties. The rootstocks of *C. palustris* yield eatable starch, prepared by drying and grinding them, and then heating the powder till the acrid properties are dissipated. [M. T. M.]

CALLA D'ETHIOPIE. (Fr.) *Richardia æthiopica*.

CALLCEDRA-WOOD. The timber of *Flindersia australis*.

CALLERYA. The name formerly given to a plant of the leguminous family, but now found to be a species of *Millettia*, and perhaps the same as *M. nitida*, which is, like this, a native of NE. China. It is a small tree, with alternate unequally pinnate leaves, about a foot long, with two pairs of ovate leaflets from one to three inches in length; numerous flowers in terminal panicles; and two-valved pods one to three inches long, containing one to five seeds, covered externally with a velvety pubescence. [A. A. B.]

CALLIANDRA. A beautiful genus of leguminous plants peculiar to America, found as far north as California, and extending southwards to Buenos Ayres. A few are herbs not more than a foot high, but the greater portion shrubs or small trees, most frequently met with on river banks. The leaves of all are bipinnate, the leaflets varying much in size and number. In one section the leaves have one to four pairs of pinnæ, with few but large leaflets (one to eight inches long), the ultimate ones always the largest; while in another there are many pairs of pinnæ, the leaflets scarcely half an inch long, linear in form and almost numberless. The flowers are usually borne on stalked globose heads, but sometimes in terminal racemes; the corollas small and hidden by the very numerous long filaments of the stamens, which are almost always of a beautiful red colour. From this latter circumstance the genus is named *Calliandra*, signifying 'beautiful stamened.' It differs from all allied genera in the valves of its compressed pod rolling backwards in a remarkable manner from apex to base when the seeds are ripe. Many of the species are in cultivation in plant-stoves, and almost all of them produce bright red balls of flowers, which stand erect from amongst the ferny foliage of some of the species in great profusion. In *C. diademata* the stamens are beautifully curved backwards and pink in colour; the leaves twice pinnate with eight or nine pinnæ which have each from thirty to forty leaflets, so that

Calliandra Tweedii.

each leaf is made up of no fewer than six or eight hundred leaflets. This is a native of Brazil, and in cultivation. *C. hæmatocephala*, a lately introduced species, has binate leaves, each portion or pinna with about ten pairs of leaflets half an inch long, and its round balls of flowers are of a rich red colour. The Peruvian women decorate their hair with the flowers of *C. trinervia*, calling them seda-sisa or silk flower. More than sixty species are enumerated, all of them more or less ornamental. (A. A. B.)

CALLICARPA. A considerable genus of *Verbenaceæ*, chiefly from the tropical and subtropical districts of Asia, but found also, though more sparingly, in similar districts in Africa and America. They are shrubs, more or less woolly with stellate hairs, nearly glabrous, and often with numerous resinous glandular dots, especially on the under surface of the leaves. The flowers are small in axillary cymes. The calyx is truncate or four-toothed; the corolla tube is short, and the limb has four nearly regular lobes. There are four exserted stamens, a four-celled ovary, with a single ovule in each cell; and the fruit is a small juicy berry or drupe, with four distinct seed-like nuts or kernels. [W. C.]

CALLICOMA. A genus of *Cunoniaceæ*, containing small trees or shrubs from South Australia, with opposite simple lanceolate leaves, white beneath, furnished with elliptical membranous caducous stipules. The peduncles are long, axillary, with a dense globular head of small yellow flowers, which are sessile on a woolly receptacle, and surrounded by a four-leaved reflexed involucre. Flowers with four or six membranous bracts forming an involucel; calyx-tube very short, scarcely adhering to the ovary at the base; the limb four or five parted, persistent; corolla absent; stamens eight or ten; ovary woolly, two-celled, many-ovuled, the styles two, diverging. [J. T. S.]

CALLIGLOSSA. *C. Douglasii* is a pretty little yellow-flowered Californian annual of the composite family, with few strap-shaped leaves, toothed at the apex, about half an inch long and very narrow. The yellow flower-heads are single at the ends of the branchlets. Being a very free flowerer, it is often used as a bedding plant in flower gardens, and, like many of our best annuals, was introduced by Mr. Douglas. The genus does not differ from *Callichroa*, and the plant is therefore generally called *Callichroa Douglasii*. (A. A. B.)

CALLIGONUM. A genus of shrubs belonging to *Polygonaceæ*, natives of the Eastern Mediterranean region, and Central Asia. They are leafless plants with dichotomous jointed branches, each joint with a small membranous sheath at the base. Flowers small, on short-jointed pedicels springing from the axils of the sheaths; perianth coloured red, five-parted, reflexed in fruit, the two outer segments larger. The fruit is a large four-cornered nut with the corners expanded into double longitudinal spinous wings, the sides between the wings being covered with long branched shaggy filaments. [J. T. S.]

CALLILEPIS. A small genus of SE. African plants belonging to the composite family. They are herbs, about a foot high, branching from the base, or simple, with lanceolate entire or slightly serrated leaves, which are opposite on the lower part of the stem, and alternate above. The flower-heads single and terminal, nearly an inch in diameter, with strap-shaped ray florets bearing a pistil only, and tubular disc florets having both stamens and pistil. The pappus is made up of three unequal scales. The flowers are yellow. [A. A. B.]

CALLIOPSIS. A genus of plants belonging to the composite order, distinguished from their allies by the involucre or covering which surrounds the heads of the flowers being formed of two rows of scales, the outer short and spreading, the inner larger erect and united at the base. The receptacle or part supporting the flowers is flat, having on it narrow scales which fall early and are shorter than the flowers. The fruit is truncated, incurved, destitute of appendages. The name is from two Greek words which together signify 'beautiful eye, aspect, or appearance,' in allusion to the general elegance of the species or the eye-like spot on the flowers. The genus comprehends a number of interesting herbaceous plants, natives of North America, several of which were, and indeed still are, referred by some authorities to the genus *Coreopsis*. They are usually free from hairs, the leaves opposite, more or less divided; the flowers at the circumference of the heads yellow, with a dark purple or rose-coloured spot at the base, those in the centre yellow or purple. The elegance of the flowers, so marked in these plants, renders them desirable in flower-beds. The more hardy species, whether annual or perennial, are generally of easy cultivation. *C. rosea* has been long known, and may be specially alluded to as an example of the genus; having the stem smooth, leaves opposite, long and narrow, the heads of flowers small on short stalks. [G. D.]

CALLIPELTIS. An annual erect much-branched slender herb belonging to the order *Rubiaceæ*. The flowers grow in whorls of six, and are whitish four-parted and bell-shaped. The fruit, which is one-seeded by abortion, is partially enveloped by a large hollow membranous bract. The leaves and leaf-like stipules form whorls of four. *C. cucullaria*, the only species, a native of the Levant, is an unimportant plant growing from six to twelve inches high. [C. A. J.]

CALLIPHYSA. A genus of *Polygonaceæ*, differing from *Calligonum* only by having the nut not winged at the angles but rounded, and covered with bristles, and expanded at the apex into a bladder-like envelope to the nut. [J. T. S.]

CALLIPRORA. A genus of liliaceous plants, found in California, and consisting of dwarf bulbous herbs with small radical linear-ensiform leaves, and bearing the flowers in umbels at the top of a scape. The perianth is bell-shaped, six-parted, with equal-spreading segments; it is furnished with six stamens, all perfect, with petaloid bilobed filaments, the alternate ones shorter, and the anthers sessile between the lobes; and it has a stalked three-celled ovary containing many ovules, and surmounted by a three-lobed stigma. *C. lutea*, the only species, is a dwarf plant, producing umbels of yellow star-shaped flowers resembling those of an *Ornithogalum*. [T. M.]

CALLIPSYCHE. A genus of *Amaryllidaceæ*, founded on a Mexican species allied to *Eucrosia*, and named *C. eucrosioides*. The plant has roundish bulbs, furnished with a few green tessellated and pitted leaves, a foot long and four inches wide, and produces at a different season, before the leaves are developed, a glaucous scape upwards of two feet high, bearing an umbel of declinate flowers, which are stalked. The perianth consists of a short green tube, full of honey, and an erect regular limb nearly an inch long of bright red segments, the sepaline of which are boat-shaped, and the petaline obtuse. The six stamens are pale green, and with the style are about four times as long as the perianth. The filaments are free, inserted in the mouth of the tube, and are tuberculate at the base. In our gardens the leaves die away in the autumn, and the flower-scapes appear in spring before they are again developed. [T. M.]

CALLIPTERIS. A genus of polypodiaceous ferns of the group *Asplenieæ*. They belong to the diplazioid series, having the sori more or less abundantly and constantly placed in pairs back to back on the same vein; and are specially distinguished in the typal group by having the veins joined together in a connivent manner, that is, the main veins that spring out from the midrib are parallel, and the veinlets which branch out from them set off at an angle and meet the opposite ones in the centre, and so form a series of acute angles one above the other. In one group the junction of the veins is less regular and frequent. The species, ten or twelve in number, are almost all found in the eastern tropics, but one or two occur in the W. Indies and S. America. They are generally large growing plants with coarse pinnated or twice or thrice-pinnated fronds, the rachis sometimes proliferous. [T. M.]

CALLIRHOË. A genus of beautiful North American mallow-worts, comprising several perennial herbaceous species, sometimes known by the name of *Nuttallia*, which, however, belongs to a genus of *Rosaceæ*. They are very nearly allied to *Malva* itself, from which they differ in certain slight technical peculiarities of the fruit; and also in some of the species, by wanting the involucel or whorl of bracts which, is found exterior to the calyx in many of the genera of this order. The involucel, when present, consists of from one to three bracts, which are sometimes remote from the flowers. The calyx is five-cleft; the corolla five-petaled, the petals truncately wedge-shaped, and often erosely-toothed at the tip. The filaments of the stamens are united into a columnar tube which bears a tuft of many stamens at the end. The carpels are numerous, united by a short beak, and are one-seeded. About half-a-dozen species are recorded, and some of them are known in cultivation. *C. digitata*, which is one of the original typical species, has no involucel beneath the flowers. It is a herbaceous perennial, with palmately five-parted root-leaves, having lobed or toothed segments, and a smooth slender branching stem two to two and a half feet high, producing a few leaves towards the base, but leafless above, and producing the flowers in corymbosely racemose heads. The flowers are five-petaled, a couple of inches across, the petals fimbriately toothed at the truncate apex, and bearded at the base, of a rich dark crimson-purple, and very handsome. This plant is sometimes called *Nuttaltia grandiflora*. *C. Papaver*, another species of the genus, a good deal resembles *digitata*, but this is furnished with a three-leaved involucel. It has five-lobed leaves with lobate segments and large solitary long-stalked flowers from the upper axils, these being of a rich bright rosy-lake colour, and very showy. The name *Callirhöe* has also been given by Link, to a group synonymous with *Amaryllis*. [T. M.]

CALLISIA. A genus of the order of spiderworts, distinguished by three stamens having their filaments or supports bearded, and in the form of a flat circular surface at the top; the style or appendage on the top of the fruit thread-like and ending in three points. The name is derived from the Greek, and indicates the beautiful or handsome aspect of the species, which are natives of the warmer parts of America, having stems trailing at the base, the leaves sheathing the stem, their general outline lance-shaped, often with hard projections at the margin. *C. repens*, a native of the West Indies, is one long known in cultivation: its graceful habit, and brilliant leaves with purple edges are sufficient recommendations. [G. D.]

CALLISTACHYS. A genus of pretty Australian plants belonging to the leguminous family, and having alternate stalked entire smooth or silky leaves, and long racemes of yellow or purple flowers. The stalked pods are divided when young into as many partitions as there are seeds, but these divisions are obliterated as the pod ripens. The generic name is derived from the Greek, and signifies 'beautiful spike.' A number of the species are in cultivation in greenhouses. *C. lanceolata* has racemes of golden yellow flowers, nearly as large as those of the broom, and the stems and leaves are covered with beautiful silky hairs. *C. linearis* has dull purple flowers, while *C. longifolia* has racemes of yellow flowers with a purple keel. The species with one exception, *C. sparsa*, which is found in N.S. Wales, are all natives of the Swan River colony. [A. A. B.]

CALLISTEMON. A name indicative of the beauty of the stamens in the genus of *Myrtaceæ* to which it is given. The calyx tube is hemispherical, while the limb is divided into five obtuse lobes; petals five; stamens numerous, of considerable length, and not united together; style thread-like; capsule with three many-seeded compartments, included within the hardened tube of the calyx. These handsome flowering trees or shrubs are natives of Australia. *C. salignum* has much the appearance of the common weeping willow. The young foliage of some of the kinds is of a pink colour, so that the trees when putting forth their leaves appear from a distance to be in blossom. The outer bark of some of the kinds, according to Dr. Bennett, peels off in layers, hence the trees are called Paper Bark trees. Many of the kinds are grown in this country for their handsome flowers. [M. T. M.]

CALLISTEPHUS. The generic name of plants belonging to the composite order, the distinguishing characters of which are the following: the involucre or part surrounding the heads of flowers consists of three or four series of spreading scales fringed at the edge; the receptacle or surface which supports the flowers is somewhat convex and slightly pitted; the fruit compressed, thickest above, its pappus or crown in two rows, the outer of partially united bristles, the inner of longer rough hairs. The name is derived from two Greek words, which together signify 'beautiful crown,' in allusion to the appendages on the ripe fruit. The genus was originally founded on the characters of a plant long known as *Aster sinensis*. The species are annuals, chiefly natives of China; they have erect branched stems, with stalkless alternate and toothed leaves, the branches with single heads of flowers. The one already alluded to as *Aster sinensis*, and a very general favourite with cultivators, has the individual florets either strap-shaped or tubular, and presenting various tints of rose, violet and white; it is the 'Reine Marguerite' of gardeners. It has these recommendations: it is hardy, of easy cultivation, and flowers freely for weeks in succession; it is therefore a desirable plant in flower-beds. [G. D.]

CALLISTHENE. A genus of the *Vochisia* family, found in Brazil, and composed of a few somewhat resinous opposite-leaved trees, which differ from the others in the family in the following combined characters: the five-parted unequal calyx, the upper and larger segment of which is prolonged behind into a spur; the single inversely heart-shaped and stalked petal; and the solitary stamen whose anther is four-celled. The leaves are either smooth or downy, and have entire margins; in one species they are oval and about two inches long, while in another they are linear and scarcely half an inch in length. The yellow flowers (about the size of those of a pea) are either single or numerous in the axils of the leaves, and, like all the others of the family, are remarkable for the unsymmetrical arrangement of their parts. We have first an irregularly five-parted calyx, one of whose divisions is prolonged into a spur somewhat like that in a balsam flower; their cones a single yellow striped petal, instead of five, as would generally be the case; next a solitary stamen; and lastly a three-celled ovary, which, when mature, becomes a woody capsule about the size of a hazel-nut, containing a number of seeds, and splitting into three portions. [A. A. B.]

CALLITHAMNION. A beautiful genus belonging to the division *Ceramieceæ* of the rose-spored *Algæ*, to which it bears nearly the same relation as *Cladophora* does to the chlorosperms. The frond is generally more or less branched, and often most beautifully pinnate, consisting of jointed threads, the stem alone being occasionally slightly compound from decurrent branchlets, as in *Batrachospermum*. The tetraspores and capsules often occur on different plants, the latter containing irregularly distributed spores. Antheridia again are mostly produced on distinct plants. The species are extremely numerous and occur in most parts of the world on other algæ, and on almost any object which is washed by the waves. One or two species are found on rocks only occasionally immersed. From their beautiful ramification these plants are the delight of wanderers on the sea-shore, and afford great gratification to those who possess only imperfect microscopes. [M. J. B.]

CALLITHAUMA. A genus of Peruvian *Amaryllideæ* related to *Pancratium*, and remarkable for the large size of the staminal cup or coronet of its perianth, which is equal to that of the limb. *C. viridiflorum* has large oblong-cylindrical bulbs, long flat ensiform suberect leaves, and a flower scape, which is said sometimes to reach six feet in height in its native country, supporting four or five emerald green flowers, which have a horizontal slender tube two inches long, acuminate spreading limb segments, and a large cup or coronet. Mr. Mathews found this with scapes three feet high growing in dryish exposed situations. The other species, *C. angustifolium*, is similar in character, but rather smaller. [T. M.]

CALLITRICHE, CALLITRICHACEÆ. A small aquatic plant with simple entire opposite leaves and minute unisexual axillary flowers, so reduced in structure as to afford little indication of its real affinities, and to have induced botanists to propose it as a distinct family, under the plural name of *Callitrichaceæ*. The male flowers consist of a single stamen, between two small bracts; the females have a six-lobed four-celled ovary and fruit, crowned by two styles without any perianth, each cell enclosing one pendulous ovule and seed. The genus has been most frequently associated, with other minute-flowered aquatic plants, under *Haloragæ*, but, more recently, it has been proposed, upon more plausible grounds, to consider it as a much-reduced aquatic *Euphorbiacea*.

C. aquatica is common in our ponds and still waters, often floating over them in large masses, and it is found in most parts of the world. It varies much in its leaves, either all narrow and submerged, or more frequently the upper floating ones, oblong or obovate, in the size and form of the fruits, the erect or recurved styles, &c.; and it has been, therefore, variously divided into from two to twenty supposed species, which are now more generally admitted to be varieties of a single one.

This apetalous genus, which is so singular in its structure, consists of small herbaceous plants, natives of Europe and North America, growing in ponds and streamlets, usually immersed, but becoming more luxuriant in habit and producing much more seed when growing out of the water. The most common form in the British Islands is that called *C. verna*. The axillary flowers are usually unisexual, the males and females growing on the same plants; but not unfrequently they become hermaphrodite, apparently from the male flowers producing ovaries. The male flower consists of but one stamen without a calyx, its only envelope being two lateral bracts, which are in some species wanting; and the anther is two-celled, or more commonly one-celled, from the two cells having become confluent. The female consists of a four-celled ovary having but two stigmas, and is elevated on a short stalk, and enveloped by two lateral bracts as in the male. The cells contain one ovule each, suspended from the side, and the seed is albuminous. Mr. Babington states that at its first formation the ovary is only two-celled, and that the four-celled condition is produced by the midrib of each carpel extending inwardly between the two ovules to the centre of the ovary to which it becomes adherent. Very numerous flat glands have been observed on the young stems by Dr. Lankester and others (Linn. Proc. ii. 94). These give a glistening appearance to the plant when growing out of the water, something like that of the *Tetragoniaceæ*, which is also owing to the presence of minute glands. [B. C.]

CALLITRIS. A genus of conifers allied to *Thuya*, but differing from it in having the cones with four to six woody scales, which separate one from the other like the valves of a capsule: and three to six winged seeds to each scale. *C. quadrivalvis* is a large tree with straggling jointed furrowed branches, having rings of small scales at the joints. It is a native of Barbary, but is cultivated in this country in sheltered situations. The resin of this tree is used in varnish-making under the name of gum sandarach; while powdered it forms pounce, formerly used for the same purpose as blotting-paper now is. The timber also, according to Dr. Lindley, is durable, very hard, fragrant, and of a mahogany colour, for which reason it is largely used in the construction of mosques and similar buildings in the N. of Africa. [M. T. M.]

CALLIXENE. A genus of *Liliaceæ*, containing branched under-shrubs from extra-tropical South America, with the base knotted, scaly, and leafless, the upper part with alternate half-clasping elliptical leathery leaves with thickened margins, and terminal or axillary flowers on short peduncles. The perianth is six-parted, coloured red, the three inner segments with two glands at the base; stamens six; style thick; berry small, three-celled, with two or three seeds in each cell. [J. T. S.]

CALLOGRAMMA. A name given by Professor Fée to *Syngramma alismæfolia*.

CALLOSO-SERRATE. When serratures are callosities.

CALLUNA. The true 'Heather' of Scotland, called also Ling and Common Heath. A low much-branching tufted shrub, distinguished from *Erica* by having a calyx of four coloured leaves concealing a bell-shaped corolla, and accompanied by four bracts resembling an outer calyx, the true heaths having a calyx of four green leaves. *Calluna* derives its name from the Greek *calluno*, to 'cleanse or adorn,' an appropriate name, whether taken in reference to the use to which heather-brooms are applied, or to the exquisite beauty of its flowers. By this plant much of the moorland scenery of Great Britain is redeemed from utter sterility; for being indifferent to soil and capable of enduring a low temperature and the most parching winds, it everywhere finds itself a home, and when it has attained a moderate size hospitably affords shelter to other plants somewhat less hardy than itself. To red and black grouse it affords not only shelter but food, since both these birds are in the habit of concealing themselves among its branches and of feeding on its tender shoots; and it is no less serviceable to the mountain hare (*Lepus variabilis*). The moorland sportsman is therefore indebted to this plant for no small portion of his amusement. It is also much employed as fuel, for thatching houses, weaving into fences, covering underground drains; and a thick layer forms a by no means despicable bed. The flowers abound in honey, and are much frequented by bees. In various parts of Scotland and the north of England, bee-hives are carried, in the beginning of August, from the cultivated to the heathy districts, for the sake of the flowers, where they are allowed to remain two or three months, and are brought back in the autumn. Heather is a plant of slow growth, but very durable on this account; and because it is patient of any amount of clipping it is not unfrequently used as an edging in gardens instead of box. In the common form of the plant the flowers are purplish red, but varieties are cultivated in which this colour is replaced by crimson or white. Another variety with double flowers is well worthy of cultivation. The tint of the foliage varies considerably, being pale green, purplish, or hoary with down. In all the varieties the flowers retain their form and position long after they have ceased to perform their functions. Ling is abundant in all the moorland countries of almost the whole of Europe. It is the badge of the clan McDonell. French, *Bruyère commune*; German, *Heide*. [C. A. J.]

CALLUS (adj. CALLOSUS). A hardened part; anything which has acquired unusual hardness and toughness; also used in the sense of verruca; also the hymenium of certain fungals.

CALOCEPHALUS. A genus of the composite family found in Australia and Tasmania. The three known species are slender herbs one to three feet high, with opposite linear entire leaves, one to three inches long, and covered, like all parts of the plant, with white appressed down. The flower-heads are in dense round clusters, at the ends of the branches, each head containing three florets. In *C. lacteus* the flower-heads are white, and in *C. citreus* they are lemon-coloured. [A. A. B.]

CALOCHILUS *campestris* is a slender leafy-stemmed Australian orchid with testiculate roots and nearly closed greenish flowers, the tip of which is deeply clothed with long delicate hairs. *C. herbaceus*, supposed to be a second species, is regarded by Hooker as a mere form of the other.

CALOCHORTUS. A genus of beautiful bulbous plants belonging to the *Liliaceæ*, and closely allied to *Cyclobothra*, from which it differs in being destitute of a honey-pit on the segments of the perianth, and in having flat smooth instead of roundish angular seeds. They have tunicated bulbs, and produce rigid ensiform leaves, and an erect scape supporting a few large showy flowers which are racemosely arranged and remain open for several

Calochortus venustus.

days. The perianth is deciduous, six-leaved, the three outer or calycine divisions linear and beardless, the three inner petaloid, very much larger and broader than the outer, and bearded on the inside; the flowers, therefore, appear to consist of three large spreading petals, and three narrow sepals. There are six stamens adherent to the base of the perianth, and a three-celled ovary crowned by three subsessile stigmas. The few known species, which are found in Mexico, California, and N.W. America, are all plants of gorgeous beauty, but found to be exceedingly difficult of cultivation. *C. venustus* is one of the handsomest; it grows about two feet high, and produces large flowers, upwards of three inches across, with narrow green sepals, and broad roundish wedge-shaped petals which form a cup, and are white above, yellowish towards the base, each of them marked with a wedge-shaped deep crimson stain, terminating in a yellow spot, and above this, in the same line, with a deep red spot bordered with yellow, and a spot of lighter red. *C. macrocarpus* is another very fine species, growing nearly two feet high; this has three narrowish sepals very much longer than the petals, which are broad cuneately-obovate, forming a cup, and of a rich rosy-purple, paler towards the base, and beautifully bearded with yellow hairs. [T. M.]

CALODENDRON. A genus of *Rutaceæ*, so named from the beauty of the flowers and foliage. The flowers are regular, consisting of a five-parted calyx, five narrow spreading petals, hairy on the outside, five fertile stamens, alternating with and shorter than five petal-like sterile ones, which are tipped with a gland and placed on the outside of a shallow tubular disc; style long; fruit a stalked capsule with five angles, and five two-seeded cavities opening by as many valves. *C. capense* is a very ornamental tree, native of the Cape of Good Hope. [M. T. M.]

CALODRACON. A genus of liliaceous plants, which includes several species formerly referred to *Dracæna* and *Cordyline*. The species are natives of the Malayan and Australasian Islands and of China and Japan, and are handsome shrubs with slender cylindrical stems, crowded with leaves at top, the leaves lanceolate oblong, smooth, often beautifully coloured, and having channelled stalks. The flowers grow in large terminal panicles, and are white or rosy violet. The perianth is deciduous, tubulosely campanulate, six-cleft, with the segments somewhat unequal and imbricating; stamens six, with subulate filaments; style subulate with a trifid stigma. This genus, of which *Dracæna ferrea* is the type, agrees with *Cordyline* in having a tubular-campanulate perianth, and with *Dracænopsis* in having numerous ovules in each cell of the ovary. *C. Jacquinii*, the *D. ferrea* above referred to, is well known under the latter name, and that of *D. terminalis*, in the hot houses of this country, where it is prized for its highly-coloured red leaves, which render it gay at all seasons. *C. nobilis* is another species with the leaves richly variegated with red; and in *C. Sieboldii* they are deep green with paler blotches. The flowers being small, it is for their foliage and erect palm-like habit alone, that these plants are prized by cultivators. The same name has been given to a section of the genus *Dracocephalum*. [T. M.]

CALODRYUM. A genus of *Meliaceæ* inhabiting the islands of Madagascar, Mauritius, and Bourbon. The calyx is five-cleft; the petals five, more or less adherent; anthers projecting from the tube formed by the united filaments of the stamens; style thread-shaped; ovary five-celled with pendulous ovules. [M. T. M.]

CALOGYNE. A name expressive of the peculiarity and beauty of the stigma in the genus of *Goodeniaceæ*, to which it is applied. The genus consists of herbaceous plants with irregular flowers, and a style with three branches, each branch terminated by a kind of cup. The fruit is a two-celled capsule with several seeds. The plants are natives of the coast of tropical Australia, and one has lately been discovered in the neighbourhood of Amoy, in China. The flowers of this latter species are said by Bentham to have an odour like that of hay. [M. T. M.]

CALONYCTION. A genus of *Convolvulaceæ*, containing fifteen species, natives of the intertropical regions of Asia and America. They are twining herbaceous plants with alternate cordate leaves and very large showy flowers on axillary one to three-flowered peduncles. The calyx consists of five sepals; the corolla is funnel-shaped with a long tube, and large spreading limb. There are five exserted stamens, with filaments dilated at the base. The ovary is two-celled with two ovules in each cell; sometimes the rudiment of a secondary dissepiment makes it incompletely four-celled. The four-valved capsule contains four seeds. [W. C.]

CALOOSE. The Sumatran name for *Urtica tenacissima* and *Bohmeria nivea*, or their fibre.

CALOPAPPUS. The name applied to a Chilian genus of plants found on the Cordillera, and belonging to that section of the composite family with two-lipped corollas. They are low heath-like bushes with needle-shaped leaves set thickly on the stems, and single terminal flower-heads which are stalked or sessile, containing five florets, each having a pappus of about fifteen long needle-pointed awns. Two species are known. [A. A. B.]

CALOPHACA. A deciduous shrub allied to *Cytisus*, from which it may be distinguished by its not having all the stamens united into a tube, and by its pinnate leaves. It is a native of desert places near the rivers Don and Volga (hence its specific name *volgarica*). Being hardy and very pretty it is a desirable plant to have in gardens and shrubberies; but is less known than it ought to be in consequence of its being difficult of propagation except by grafting or from seed. The flowers are yellow, in clusters in the axils of the leaves, and are succeeded by reddish pods.   [G. A. J.]

CALOPHANES. A genus of *Acanthaceæ*, containing nearly thirty species of herbs or under-shrubs, natives of America. They are mint-like plants, more or less pubescent, and nearly related in structure to *Dyschoriste*. They have axillary opposite generally cymose flowers, with a blue corolla and spotted throat; the calyx is deeply five-cleft with setaceous divisions; the corolla is infundibuliform with a five-cleft limb; the filaments are united in pairs at the base, and have anthers with two parallel cells spurred at the base or rarely muticous. The capsule is lanceolate, with four seeds in the middle.   [W. C.]

CALOPHYLLUM. This genus of guttifers (*Clusiaceæ*) contains about twenty-five species, the majority of which are natives of the Eastern hemisphere, only four or five being found in America. They are large trees with shining leaves marked by numerous parallel transverse veins, and having racemes of flowers, some of which are of only one sex. Their calyx consists of two or four sepals; their corolla of four petals; the stamens are indefinite in number, their anthers bursting on the inner side; and the ovary is one-celled, the style being crowned with a shield-like lobed stigma. The fruit contains one seed. *C. Calaba*, a native of the West Indies and Brazil, is a tree about sixty feet high, having long elliptical oblong leaves, sometimes notched at the top. It has short racemes of white sweet-scented flowers, producing round green fruits about an inch in diameter, and containing a single seed. This tree is called Calaba in the West Indies, and an oil, fit for burning in lamps, is expressed from its seeds. *C. inophyllum*, an East Indian and Malayan tree, with a trunk about ten or twelve feet in diameter, and from eighty to 100 feet high, has the leaves elliptical and usually notched at the top, and it has white flowers resembling those of the last. The seeds of this tree yield a thick dark green strong-scented oil, employed in India for burning and also medicinally. Its timber is used for building purposes, and for masts and spars; and a greenish coloured resin which exudes from the trunk forms one of the kinds of East Indian Tacamahac. Other species likewise yield resin, such as *C. Tacamahaca* in Bourbon and Madagascar; and *C. brasiliense* in Brazil. The fruits of *C. edule* and *C. Mariruano* are eaten in South America; as also are those of *C. spurium* in Malabar. In Ceylon the timber of *C. tomentosum* is valued for building purposes, and an oil is expressed from its seeds.   [A. S.]

CALOPHYSA. A genus of *Melastomaceæ* containing a Brazilian shrub with opposite petiolate cordate acute seven-nerved toothed leaves, and short axillary crowded cymes of flowers; calyx-tube adhering to base of ovary; limb with four short lobes; petals four, obovate; stamens eight, without any appendage to the anthers; berry four-celled, with many seeds; whole plant more or less bispid.   [J. T. S.]

CALOPOGON. A small genus of tuberous orchids, inhabiting wet prairies or the edges of pine woods in all parts of the United States. They have grassy radical leaves and naked scapes bearing a small number of purple flowers at the summit. Four species are described: *C. pulchellus*, *multiflorus*, *parviflorus*, and *pallidus*. The generic name has been given in allusion to a handsome beard or tuft of hairs growing from the lip.

CALOPSIS. A genus of *Restiaceæ* from the Cape of Good Hope. Sedge-like herbs with deciduous glumiferous flowers in spikelets, arranged in spikes or panicles; stems branched, with split leafless sheaths. It is distinguished from *Restio* by having three stigmas and an indehiscent nut covered with a tough membrane.   [J. T. S.]

CALOSACME. *Chirita*.   [W. C.]

CALOSANTHES. A genus of *Bignoniaceæ*, consisting of a single species, a native of India. It is a very tall slender smooth tree with large opposite bipinnated leaves, the leaflets shortly petiolate subcordate ovate and acuminate. The racemes are terminal and erect; the flowers large, whitish within, exteriorly streaked with red, and having a fetid smell. The calyx is coriaceous, tubular and truncate; the corolla tube is short and campanulate; its limb sub-bilabiate, the upper lip with two, and the under with three lobes. The five fertile scarcely exserted stamens have the anthers pendulous from the apex of the filaments. The stigma consists of two roundish lobes. The pod-shaped capsule is very long, compressed and two-valved, containing numerous seeds which are surrounded with a large membranaceous wing. The wood is soft, spongy, and of no economic value.   [W. C.]

CALOSCORDUM. A genus of small-growing lilyworts, found in China. They are allied to *Allium*, from which they are distinguished by a few technical characteristics. *C. neriniflorum* has small bulbs and linear leaves which are thick and rounded behind, and the flowers which are small starry and rose-coloured form an umbel at the top of a scape. One or two other species are known. The plants have none of the onion-like odour which pervades the *Allium* family.   [T. M.]

CALOSERIS. The name given to a plant of the composite family which is found in Venezuela. It has much the habit and appearance of some of the coltsfoots, but belongs to a different section of the family, namely, that with two-lipped corollas. It has been described twice, under different names, and *Caloseris* being the last published, must give place to the first, *Isotypus*.   [A. A. B.]

CALOSTEMMA. A genus of *Amaryllidaceæ* consisting of bulbous herbs with linear lorate leaves, and bearing at the top of the scape a many-flowered umbel of pedunculated flowers. These flowers consist of a cylindrical tube, a funnel-shaped limb, and a coronet or crown, uniting the stamens into a cup, which is sometimes split. The filaments are short and erect; the stigma small and simple; and the ovary usually two-seeded. There are four or five recorded species, all natives of New Holland. *C. purpureum*, with purple flowers, has twelve triangular teeth placed between the filaments on the edge of the staminal cup. *C. luteum* has narrower leaves, and yellow green-ribbed flowers, with six purple spots at the base of the cup, which is toothed as in the former. *C. album* has white flowers and linear teeth to the cup; and *C. carneum* has pretty pale rose flowers and is without the teeth to the staminal cup, the spaces between the filaments being either emarginate or merely rounded. *C. candidum* is said to be fragrant, and *C. luteum* to have a strong smell of mint. [T. M.]

CALOSTIGMA. A genus of *Asclepiadaceæ*, containing three species of climbing shrubs, natives of Brazil. They have opposite elliptical or oblong leaves, and lateral interpetiolar peduncles with many flowers. The calyx is five-parted; the corolla bell-shaped with a five-cleft limb, the divisions being long, linear and spreading. The staminal corona is composed of five fleshy leaves, and adheres to the tube of the corolla, above which it projects. The gynostegium is short; the anthers terminate in a short membrane; and the elongated projecting stigma has a prominent dilated apex. The pollen masses are connected by a kneed and, in the upper portion, by a winged process to a linear corpuscle. [W. C.]

CALOTHAMNUS. One of the beautiful genera of *Myrtaceæ*, in which Australia abounds. The calyx limb has four to five teeth; the petals are four to five; the stamens are arranged in four to five bundles opposite the petals, some sterile or more or less joined to the neighbouring parcel, the anthers attached by the base; the many-seeded capsule is enclosed within the base of the hardened hemispherical calyx tube. The plants are shrubs with scattered needle-shaped leaves. The name indicates that the branches become covered with the beautiful flowers. [M. T. M.]

CALOTIS. A genus of simple or branched small Australian herbs of the composite family. The leaves are alternate, varying much in form, but most generally oblong and toothed. The flower-heads are terminal and solitary; the strap-shaped rayflorets, lilac, and rolled backwards spirally after expansion, the disc florets tubular and yellow. The seed crown (pappus) consists of two dilated ear-shaped scales, and a few long needle-shaped awns furnished with reflexed bristles. The genus is near that of the daisy, but differs in the pappus. It receives its name from the two ear-shaped scales of the pappus. *C. cuneifolia* is a slender herb about a foot high with small flower-heads. The awns of the pappus being furnished with very minute reflexed points get entangled in the wool of the sheep, and it is almost impossible to rid them of it. There are about twenty species known. [A. A. B.]

CALOTROPIS. A genus of asclepiads, consisting of three species, which form shrubs or small trees, and are natives of the tropics of Asia and Africa. Their flowers have a somewhat bell-shaped corolla, expanding into five divisions, the tube being composed of five angular swellings. The coronet of the stamens is composed of five narrow leaflets, which are united to the central column, but free and recurved at the base, with their edges rolled inwards. The fruits are produced in pairs resembling the horns of an animal, each being swollen or bulged out on the inside; they contain numerous seeds surmounted by tufts of beautiful silky hairs. *C. gigantea*, the largest of the genus, forms a branching shrub or small tree about fifteen feet high, with a short trunk four or five inches in diameter. Its leaves are about six inches long by two or three broad, and egg-shaped, covered on the under-surface with soft silky down, and they are arranged on the stem in pairs, each pair being at right angles with that above and below; its flowers are of a pretty rose-purple colour, and have the segments of the corolla bent downwards. This plant is called Mudar or Ak in Northern, and Yercum in Southern India. The inner bark of its young branches yields a valuable fibre, capable of bearing a greater strain than Russian hemp. All parts abound in a very acrid milky juice, which hardens into a substance resembling gutta percha; but in a fresh state it is a valuable remedy in cutaneous diseases. The bark of the root also possesses similar medical qualities; and its tincture yields mudarine, a substance possessing the property of gelatinizing upon the application of heat, and returning to its fluid state when cool. Attempts have been made to spin the silky down of the seeds, but its fibre is too short; a soft kind of cloth is, however, made by mixing it with cotton; paper has also been made from it. Another species, *C. procera*, a native of India, Arabia, Persia, and various parts of Africa, possesses similar qualities. It is a much smaller plant, and has white flowers with straight segments. [A. S.]

CALPANDRIA. *Camellia.* [B. S.]

CALTHA. A family of herbaceous plants belonging to the *Ranunculaceæ*, distinguished from *Ranunculus* by the absence of a green calyx, and from *Helleborus* by the absence of tubular petals (nectaries). *C. palustris*, the Marsh Marigold, is a stout herbaceous plant with hollow stems, large glossy roundish notched leaves, heart-shaped at the base, and conspicuous bright yellow flowers, each of which is composed of five roundish petals or sepals. It flowers freely from May to August, and is a native almost throughout the whole of Europe, as well as of Western Asia and North America, in marshy meadows and about the margins of ponds, rivers, and brooks. One of its rustic names is May-Blobs. The flowers, if gathered before they expand, are said to be a good substitute for capers. The juice of the petals boiled with alum stains paper yellow. A double-flowered variety is commonly cultivated in gardens, and the wild plant is liable to several variations, dependent on soil and situation. Several foreign species are enumerated by botanists, all of which are natives of marshes or shallow water, and more or less approach *C. palustris* in habit. The *Caltha* of the Latin poets is considered to be the common garden marigold. French, *Populage*; German, *Sumpf-dotter-blume.* [C. A. J.]

CALTROPS. The common name for *Tribulus*. —, WATER. That of *Trapa*.

CALUMBA, CALOMBA, or COLOMBO. The root of *Cocculus palmatus*, now called *Jateorhiza palmata*. —, FALSE or AMERICAN. The root of *Frasera Walteri*.

CALVUS. Quite naked; bald; having no hairs, or other such processes.

CALYBIO, CALYBIUM. A hard one-celled inferior dry fruit, seated in a cupule; as an acorn, or a hazel-nut.

CALYCANTHACEÆ (*Calycanths*). The Carolina Allspice family, a natural order of calyciforal dicotyledons belonging to Lindley's rosal alliance. Shrubs with square stems having four woody axes surrounding the central one, opposite entire leaves without stipules, and solitary lurid flowers. Calyx of numerous coloured sepals compounded with the petals, and all united below with a fleshy tube bearing numerous stamens on its rim; outer stamens opening outwardly, inner ones barren. Ovaries several, one-celled, adherent to the calycine tube; ovules one to two. Fruit consisting of achenes inclosed by the calyx; seeds without albumen. Natives of North America and Japan. Their flowers have an aromatic fragrance, and their bark is sometimes used as a carminative against flatulence. The bark of *Calycanthus floridus*, Carolina Allspice, is used as a substitute for cinnamon. There are two known genera, viz., *Calycanthus* of America, and *Chimonanthus* of Japan, comprising six species. [J. H. B.]

CALYCANTHUS. A genus giving its name to the family *Calycanthaceæ*, and composed of N. American shrubs with opposite oval or ovate lanceolate entire leaves, generally rough on the surface; axillary or terminal solitary stalked flowers made up of a great number of lurid purple-coloured narrow sepals and petals; and very numerous stamens, inserted on the mouth of the calyx-tube, which bears on its inner hollow surface numerous achenes, each with one or two seeds. *C. floridus* is a native of many parts of the United States, where it is called Carolina Allspice, or Sweet-scented shrub. Its wood and roots have a camphoric smell, and the aromatic bark is said to render it useful as a substitute for cinnamon in the United States. The flowers and leaves have a scent resembling that of the quince. This species and the following are often to be met with in English gardens. Some of its varieties are scentless, and it varies much in the form and pubescence of the leaves as well as in the colour of the flowers. These varieties have by some authors been considered as species. *C. occidentalis*, the only other species, is a native of California. It differs chiefly from the Carolina Allspice in its long flower-stalks, and the cordate base of the leaves. Its flowers are more than three inches across when fully expanded. [A. A. B.]

CALYCERACEÆ (*Boopideæ*). The *Calycera* family, a natural order of gamopetalous calyciforal dicotyledons included in Lindley's campanal alliance. Herbs with alternate leaves without stipules, and with flowers collected in heads. Calyx superior, of five unequal divisions; corolla regular, funnel-shaped, with a five-divided limb; stamens five, their filaments united, as well as the lower part of the anthers. Ovary one-celled; style smooth; stigma capitate. Fruit an achene, usually crowned by the rigid spiny segments of the calyx. The order occupies an intermediate place between *Compositæ* and *Dipsacaceæ*, differing from the former in their seed, which is pendulous and albuminous as in *Dipsacaceæ*, and from the latter in their anthers being united around the style as in *compositæ*. There are about twenty species, distributed into six or eight genera. They are natives of South America, found chiefly on the Andes of Chili; two species extend to the Cordillera of Peru; three are found near the straits of Magalhaens; seven in the eastern part of S. America, near the Rio Plata; and one from Rio Janeiro, as far as Bahia. The plants do not possess any marked qualities. Illustrative genera: *Boopis, Calycera, Acicarpha*. [J. H. B.]

CALYCERA. This genus gives the name to the family to which it belongs. It is confined to South America, and the species are mostly found on the Cordillera of Chili. They are small annual or perennial herbs, four to eight inches high; the leaves alternate oblong toothed or pinnatifid, and generally smooth; the flower-heads single terminal and shortly stalked. The genus differs from the others in the family by the presence of two sorts of flowers in the same head, the one set with the calycine teeth flattened and produced into spinous points after flowering, the other not so. The achenes are free and seated on a broad depressed receptacle. The few species are only interesting to the botanist. [A. A. B.]

CALYCIFLORÆ. A sub-class of exogenous or dicotyledonous plants characterised by having both calyx and corolla, petals separate and stamens attached to the calyx. [J. H. B.]

CALYCINAL. Of or belonging to the calyx.

CALYCINE. Of or belonging to a calyx; also a calyx of unusual size; or having the texture of a calyx.

CALYCOIDEOUS. Resembling a calyx.

CALYCOMIS. A genus of *Cunoniaceæ*, described by Don, and eight years afterwards renamed by Bentham, *Acrophyllum*. The latter name has been generally adopted, but contrary to the received laws of botanical nomenclature. [J. T. S.]

CALYCOPHYLLUM. A genus of *Cinchonaceæ*, remarkable for one of the five segments of the calyx being much larger than the rest and petal-like, a peculiarity observable also in an allied genus, *Mussænda*. The corolla is bell-shaped, the stamens inserted into its throat; the stigmas are two, reflexed; the fruit is an oblong capsule, opening at the top to allow of the escape of the numerous slightly-winged seeds. The plants are natives of the West Indies and Brazil. [M. T. M.]

CALYCOSERIS. The generic name of a little annual herb of the composite family, found by Mr. Wright in New Mexico, and named after its discoverer *C. Wrightii*. The plant has pinnatifid leaves with linear segments, and yellow flower-heads; and altogether it bears much resemblance to *Crepis virens*, a plant which is very often met with in dry pasture lands throughout Britain. The achenes being furnished with a double pappus, the outer small and cup-shaped, the inner of long soft white hairs, and the receptacle being furnished with numerous capillary bristles, are the most marked characters of the genus. [A. A. B.]

CALYCOTOME. A genus of the leguminous family, distinguished from that of the broom by the teeth of the calyx falling away early and leaving a notched membranous tube. The species are all thorny shrubs. *C. spinosa* is a stiff spiny bush with trifoliate leaves and numerous yellow flowers, in also like those of the laburnum, but single in the axils of the leaves. It is a native of Southern Europe and North Africa, as are all the species, and is well adapted for growing in shrubberies. It is in cultivation in England. The pods of *C. lanigera* are covered with long rusty hairs. All the parts of *C. intermedia*, an Algerian species, are covered with white silvery hairs. [A. A. B.]

CALYCULUS. A partial involucre, containing but one, or perhaps two flowers. Also the external bracts of a capitulum, when they form a distinct ring or rings.

CALYDOREA. The name of an iridaceous genus separated from *Sisyrinchium*, and of which *S. speciosum* is taken as the type. This plant, now called *Calydorea speciosa*, is a beautiful bulbous herb with a few narrow linear leaves, and a slender subramose scape, five to six inches high, bearing deep blue purple flowers with a yellow centre, the segments of which are spreading, the three inner ones smaller than the outer. It is distinguished from *Sisyrinchium* by its unequal instead of regular perianth, the petaline divisions of which are reflexed and much smaller than the sepaline, its subulate free filaments, and its trifid spreading style, with emarginate-spathulate fimbriated stigmas. The species is a native of Chili. [T. M.]

CALYMELLA. *Gleichenia.*

CALYMENIA, CALYXHYMENIA. These names occasionally met with in gardens, refer to some inconspicuous plants now referred to *Oxybaphus*. [M. T. M.]

CALYMMODON. A small genus of polypodiaceous ferns belonging to the *Gymnogrammeæ*. There are three or four species, found in Java and other eastern islands, and consisting of small plants with fasciculate thin pinnatifid fronds, growing from a short erect stem, the fertile lobes folded longitudinally so as to partially cover the sori, which, though elongated, has a tendency to the polypodioid structure. The veins are simple and the sori oblong, seated at the tip of the simple vein which occupies each lobe. [T. M.]

CALYPSO *borealis* is the most beautiful of northern orchids, being found all over the continent of Europe, America, and Asia in high latitudes, growing in woods, especially of fir, and appearing as soon as snow has melted. It is a tuberous terrestrial plant, with one leaf and one flower only. The leaf is thin, many-nerved, and either ovate or cordate. The rose-coloured flower appears at the end of a slender sheathed stem, and has something the appearance of a *Cypripedium*, owing to its forming a large pouch. The genus appears to be nearly related to *Calogyne* and especially to the section *Pleione*.

CALYPTRA (adj. **CALYPTRATE**). The hood of an urn-moss.

CALYPTRANTHES. This name of Lid-flower has been applied to a genus of *Myrtaceæ*, in allusion to a lid which the upper part of the calyx forms, and which falls off as the flower expands. These flowers have five very small petals, which are sometimes absent; stamens numerous, distinct; berry one-celled, one to four-seeded. They are American and West Indian shrubs, some of which are in cultivation. The flower buds of *C. aromatica* might according to Lindley, be used in the place of cloves. [M. T. M.]

CALYPTRIDIUM. A genus of the purslane order, chiefly distinguished from its allies by having the corolla composed of three pieces joined together so as to form a conical tube, three-toothed at the top, and covering the seed-vessel like a hood; the name indicates this, being derived from two Greek words signifying 'hood-like.' The only known species is a low succulent plant, a native of California, with alternate leaves and small flowers of a pale rose colour. [G. D.]

CALYPTRIFORMIS. Like an extinguisher, as the calyx of *Eucalyptus*.

CALYPTROSTIGMA. This name is sometimes given to a plant of the honeysuckle family, a N. Asian shrub with opposite leaves, between ovate and lanceolate in form, and having serrated margins. The flowers are yellow, six to eight in a cluster at the ends of the branches, and in size and form, much like those of the fox-glove. The stigma is more or less lobed, and sits like a cap on the top of the style, whence the generic name. By many the genus is not considered different from the well-known *Weigela* or *Diervilla*; and the plant is now in cultivation under the name of *Diervilla Middendorffiana*. [A. A. B.]

CALYSACCION. *C. longifolium* is the only species of this genus of guttifers (*Clusiaceæ*). It is a handsome large tree, found in abundance in South Western India, and also in China. Its leaves are opposite, and of a long narrow lance-like form, and thick leathery texture. Some of its flowers are perfect, while others are of distinct sexes, and sometimes borne on different trees. Their calyx, which is globular in the bud, bursts into two pieces; and their corolla consists of four, or rarely five, small concave petals of a yellowish tint streaked with red; the stamens are numerous, arranged in several rows, and either quite free or slightly connected at the base; while the two-celled fleshy ovary is terminated by a short style, and a broad very-fleshy flat-topped stigma. The fruit is unknown. This tree has several local Indian names, such as Suringee and Soorgee, and is interesting on account of the uses made of its flower-buds. These are known by the name of Nag-Kassar or Nagssar; but the same name is applied to the buds of a nearly allied plant, *Mesua ferrea*, with which they have been confounded. They are on long stalks and about the size of peppercorns, of an orange-brown or cinnamon colour, and very fragrant, possessing an odour like that of violets or orris-root. In India they are greatly esteemed on account of their fragrance, and are commonly sold in the bazaars; they are also used for dyeing silk, yielding a yellow, or, with sub-carbonate of potash, a deep-orange colour. A quantity of them were imported into London some years ago, but they did not receive the attention they deserved. [A.S.]

CALYSTEGIA. A genus of *Convolvulaceæ*, containing about twelve or fourteen species, widely diffused in extratropical regions all over the world. They are climbing or prostrate smooth herbs with milky juice. The leaves are alternate without stipules, and the large and beautiful flowers are solitary, axillary and pedunculed. The calyx of five sepals is enclosed in two leafy bracts. The corolla is bell-shaped, plaited and five-lobed. The ovary is semi-bilocular with four ovules, and bears a simple style and a stigma consisting of two obtuse lobes. The capsule has only a single cell. This is a very distinct genus, easily separated from *Convolvulus* and the allied genera, by the leafy bracts at the base of the calyx, and by the one-celled capsule. It includes the Common Bindweed, *C. sepium*. [W. C.]

CALYTHRIX or CALYCOTHRIX. A genus of *Chamælauciaceæ* from Australia. Small shrubs with short cylindrical sheath-like leaves, often on short footstalks, and small stipules; flowers axillary, nearly sessile, frequently clustered near the extremities of the branches; calyx with a long tube, adhering to the ovary at the base, and a five-lobed limb, each lobe terminating in a bristle from which the genus takes its name; petals five, purplish yellow or white; stamens ten or more; ovary inferior, one-celled, two-ovuled; capsule with five ribs, indehiscent. [J. T. S.]

CALYX. The most external of the floral envelopes; it is called adherent or superior when it is not separable from the ovary, free or inferior when it is separate from that part, and calyculate when it is surrounded at the base by bracts in a ring. Also the receptacle of some kinds of fungals. — COMMUNIS. The old name of the involucre of composites, &c.

CAMARA. A carpel. Also the name of a hard durable timber obtained in Guiana from *Dipteryx odorata*.

CAMARIDIUM. Under this name have been collected many species of orchids from tropical America, with the structure nearly of *Cymbidium*, but with distichous leaves and often proliferous stems. Some of them have been referred to *Isochilus*, a wholly different genus. About a dozen species are known, of little interest. The genus differs but little from *Ornithidium*.

CAMAROTIS. A small genus of scandent Orchids, with narrow hard leaves and lateral racemes of delicate yellowish, rosy or purple flowers. They are remarkable for having a long slender rostil, and a fleshy lip hollowed out into the form of a slipper. By means of very long hard roots they cling to the bark of trees in India, the Philippines, and New Guinea. *Micropera* is the same genus, with lemon-coloured blossoms.

CAMASSIA. The Quamash of the North American Indians is the only plant belonging to this genus of lilyworts (*Liliaceæ*). It is the *Camassia esculenta* of botanists, a small bulbous plant resembling the common blue hyacinth, but larger, its leaves being about a foot long, very narrow and grooved down the inside; and its flower-stalks growing a foot or a foot and a half high, and bearing from twelve to twenty blue or white flowers. The principal character of the flower consists in its having a calyx of six sepals slightly connected at the base, and spread out horizontally but not equally, the five upper ones being closer together and inclined upwards, whilst the lower or sixth stands by itself and is bent downwards, each petal having three prominent nerves on its outside, and a stamen attached to its base on the inside, and they do not fall off, but wither and remain till the fruit is ripe. The ovary is nearly round, and is divided into three cells, each of which contains numerous ovules attached to the centre in two rows. This plant grows in great abundance in swampy plains on the north-west coast of America and Vancouver's Island, and its bulbs form the greater part of the vegetable food of the Indians, the different tribes visiting the plains for the purpose of collecting them, immediately after the plant has flowered. The digging of Quamash is a time of feasting and rejoicing amongst the Indians; the entire labour, however, devolves upon the

Camassia esculenta.

women; and the unmarried females endeavour to excel each other in the quantity of the roots they collect, their fame as future good wives depending upon their activity in the Quamash plains. The roots are cooked by digging a hole in the ground and paving it with large stones, upon which a fire is lighted and kept up until they are red-hot, when they are covered with alternate layers of branches and roots till the hole is full; it is then covered with earth and a fire kept burning upon it for twenty-four hours, when the roots are taken out and dried, or pounded into cakes for future use. [A. S.]

CAMBESSEDESIA. A genus of *Melastomaceæ*, consisting of erect or ascending dichotomously-branched Brazilian shrubs, with the leaves at the apex of the branches, sessile, opposite, or verticillate, ovate, oblong, or linear, generally three nerved; flowers handsome, terminal and axillary in paniculate cymes; calyx bell-shaped, with a five-lobed limb; petals five, obovate, scarlet; stamens ten; ovary free, three-celled; capsule ovate-globose. [J. T. S.]

CAMBIUM. The viscid fluid which appears between the bark and wood of Exogens, when the new wood is forming. Also the mucus of vegetation out of which all new organs are produced.

CAMBOGIA. A genus of tropical shrubs belonging to the *Clusiaceæ*, and containing one of the plants which yields the well-known pigment gamboge. They have leathery simple leaves; the male and female flowers on different trees, and the petals white with a pink tinge towards the base. The name *Cambogia* is given from the circumstance of the drug being produced in greatest quantity in that part of Siam called Cambodja. Linnæus strangely confounded two Ceylon plants under the name *Cambogia Gutta*, the one having stalked and furrowed fruit, which is the true Ceylon gamboge, and has been called by subsequent authors *Garcinia Cambogia*, and *Hebradendron cambogioides*; the other with sessile fruit, not furrowed, which does not yield gamboge, is now called *Garcinia Morella*.

Two kinds of gamboge are known, the Ceylon gamboge and the Siam gamboge, both of them gummy-resinous exudations, obtained from the wounded stems of the trees or by breaking of the leaves and young twigs, and receiving the yellow juice as it drops in suitable vessels. That of Ceylon is sold in the bazaars on the Coromandel coast, and is said to be as good as the Siamese, but the process it goes through in preparation does not purify it sufficiently, and, therefore, it is not sold so readily as that from Siam. By far the greater portion of the gamboge so extensively used in the arts, as a water-colour, and as a varnish for lacquer work, as well as in medicine, is sent from Siam, and is supposed to be the produce of a species of *Garcinia*, but the plant is not known to botanists. It is said to form part of the tribute paid to the kings of Siam, and is sent to England from Singapore in boxes or bags, of from one to two hundred weight each, the amount annually imported being about 800 cwt. Gamboge is known in commerce in three distinct forms: in rolls or solid cylinders, in pipes or hollow cylinders, and in cakes. The two former are collected in the same manner, the juice when in a liquid state being run into hollow bamboos, about twenty inches long and one and a half in diameter, and allowed to harden. In this form it is known as pipe gamboge. The cake or lump gamboge occurs in round or square lumps, or masses several pounds in weight, and is generally inferior in quality to the former, which is an excellent and powerful purgative in doses of three, five, or seldom more than seven grains; on the other hand, it is a dangerous poison in large doses, causing death by violent inflammation of the bowels. Dr. Christison thinks that the fatal effects which sometimes follow the use of Morrison's Pills arise from the large amount of gamboge in their composition. A detailed account of the gamboge is given by Drs. Christison and Graham, in *Hooker's Companion to the Botanical Magazine* (ii. 193, 233). *Cambogia* and *Hebradendron* are now generally referred to the genus GARCINIA: which see. [A. A. B.]

CAMBON. An Indian name for the grain of *Pennisetum typhoideum*, the *Holcus spicatus* of Linnæus.

CAMBUY. The fruit of a species of *Eugenia*.

CAMEL'S HAY. *Andropogon Schœnanthus*.

CAMEL'S THORN. *Alhagi Camelorum*.

CAMELEE. (Fr.) *Daphne Cneorum*.

CAMELINA. A small genus of cruciferous plants (*Brassicaceæ*), containing two or three European and North American species. They are dwarf annual or perennial herbaceous plants, with stem-clasping leaves, and terminal racemes of yellow flowers. The fruit or pod is somewhat egg-shaped, with the broad end upwards, and has a broad partition parting it in two, each half being very convex, distinctly marked by a central rib or nerve, and having its edges flattened so as to form a narrow border round the pod. The seeds are numerous, and have their radicle, or rudimentary root, folded over upon the back of one of the cotyledons, or rudimentary leaves.

The most interesting species is the *Camelina sativa*. This plant is found growing in cultivated and waste places in Central and Southern Europe, and the temperate parts of Russian Asia; it is generally enumerated amongst the indigenous plants of the British Isles, but it is a very questionable native, being found only in corn and flax fields in England and Ireland, having most probably been introduced along with foreign seeds. It is an annual plant, growing about two feet in height and having a somewhat branching stem; its leaves are lance-shaped, and about two inches long, with their margins entire or slightly toothed, the lower ones having stalks, whilst those higher up have their bases shaped like those of arrow heads and clasp round the stem. The flowers are in long loose racemes, and produce pear-shaped pods, about a quarter of an inch long, containing numerous small seeds. The English name of the plant is Gold of Pleasure, but why it is so called is unknown. It is cultivated in some parts of the Continent, both on account of the fibre of its stems and the oil obtainable from its seeds, and it has been recommended for cultivation in this country, but it is not likely to prove a profitable crop. The seeds are sometimes imported under the name of Dodder seed, but they have nothing to do with the true dodder, which belongs to a widely different natural order. By pressure they yield a clear yellow-coloured oil, smelling something like common linseed oil; and the residual cake has been recommended as a food for cattle, but it is of too acrid a nature to be applied to such a purpose. The stems contain a considerable proportion of fibre, and are commonly used for making brooms in many parts of Europe. [A. S.]

CAMELLIA. A well-known genus belonging to the tea family (*Ternstrœmiaceæ*), and so nearly related to the teas (*Thea*) as to be with difficulty distinguished from them. The differences that do exist consist in the number of the parts and in the position of the flower. In *Camellia* the calyx leaves are numerous and fall early, the interior stamens twice the number of the petals, the styles generally five, and the flowers sessile and erect; while in *Thea* the calyx leaves are five in number, the interior stamens equal in number to the petals, and the flowers are stalked and drooping. These are generic distinctions as given by Dr. Seemann, and they involve the removal to the teas of a number of plants which have been known as species of *Camellia*.

Camellias are found in the eastern portion of the Himalaya, Cochin China, a great portion of China Proper, and Japan; two species, moreover, are found, the one in Java, the other in Borneo and Sumatra. The genus is named in honour of George Joseph Kamel, a Jesuit, who travelled in the East, the name being Latinised into Camellus. The first species cultivated in European gardens was the Japanese Camellia, *C. japonica*. It is said to have been introduced in 1739, by Robert James, Lord Petre; this was the single red flowered or normal form of the species. It was not until 1792 that any of the double-flowered varieties were brought to this country: then the double white and the striped were introduced, both from China; they were shortly followed by the double red. Many more were subsequently introduced, and with these introductions, and the varieties produced from them, through the exertions of cultivators, we have now an endless variety of forms of this beautiful plant. The most marked among them are the double white, the fringed white, which is the only variety with fringed petals, and the anemone flowered or Waratah Camellia, which has a margin of broad petals and a raised centre of smaller ones, somewhat like the flower of a double hollyhock.

The net-veined Camellia, *C. reticulata*, a native of Hong-Kong, is the largest-flowered of the species. The flowers are sometimes six inches or more in diameter, and not unlike those of a *Pæonia*. The petals are not so closely set as in the other species, but it is highly probable that cultivators will be able to do as much for this species as they have done for the Japanese one, although it is said to be difficult of propagation.

C. Sasanqua (Sasanqua is the Japanese

name of the plant) is found in many parts of China and Japan; it has small white scentless flowers, and is cultivated in English gardens. An oil is obtained from the seeds in China by crushing them to a coarse powder, afterwards boiling them, and finally subjecting them to pressure. The oil has an agreeable odour, and is used for many domestic purposes. The leaves are used in decoction by Japanese women to anoint the hair, and also in a dried state to mix with tea, on account of the pleasant odour contained in them. *C. drupifera* is nearly allied to *C. Sasanqua*, but differs in having a very long point to its ovate-lanceolate leaves; like it the flowers are small and white, but odoriferous; it is also in cultivation, and its seeds yield an oil used in medicine in Cochin China. This grows in great abundance on the eastern portions of the Himalayas. The lance-leaved Camellia is found in Sumatra and Borneo; and the only other species, *C. quinosaura*, is said to be a native of Java. The pink-flowered plant known sometimes in gardens as *Camellia Sasanqua*, as well as the plants usually called *C. rosæflora* and *C. maliflora*, are now referred to the genus THEA: which see. The present genus includes *Calpandria*. [A. A. B.]

CAMERARIA. A genus of the dogbane family (*Apocynaceæ*), having a small five-cleft calyx; a funnel-shaped corolla with a long tube inflated at each end, and a flat limb with five lance-shaped oblique segments; the connective of the anthers prolonged into a thread; and the two follicles swollen at the base on each side so as to appear three-lobed. The seeds are compressed and slightly winged at the top. Some of the species being shrubs with white or orange flowers, are cultivated in our stoves. [M. T. M.]

CAMÉRISIER. (Fr.) *Lonicera Xylosteum*.

CAMMOCK. The Rest Harrow, *Ononis arvensis*.

CAMOMILE. The common name for *Anthemis*; more frequently written Chamomile.

CAMOMILLE DES CHIENS. (Fr.) *Anthemis* or *Maruta Cotula*. — FAUSSE. *Anthemis arvensis*. — ROMAINE. *Anthemis nobilis*.

CAMOTE. A Spanish name for the Sweet Potato, *Batatas edulis*.

CAMPANILLE. (Fr.) *Wahlenbergia*.

CAMPANULACEÆ. (*Campanulæ*, Bellworts, Hare-bell family.) A natural order of calycifloral gamopetalous dicotyledons, characterising Lindley's campanal alliance. Milky herbs or undershrubs with alternate leaves having no stipules, and usually with showy blue or white flowers. Calyx above the ovary (superior), commonly five-cleft, persistent; corolla regular, bell-shaped, usually five-lobed, withering; stamens five, distinct; style with hairs. Fruit one or two-celled or many-celled; capsule opening by slits at the sides or by valves at the apex; seeds numerous, albuminous, attached to a central placenta. Chiefly natives of the north of Asia, Europe, and North America, and scarcely known in hot regions. In our hemisphere the greatest number of species are found between 36° and 47° of north latitude. The chains of the Alps, Italy, Greece, Caucasus, and the Altai are their true country. Several are found at the Cape of Good Hope. The species opening with internal slits in the seed-vessels are chiefly natives of the Northern hemisphere; those opening by valves at the top of their seed-vessels belong to the Southern hemisphere. The plants have a milky acrid juice; but the roots and young shoots are often cultivated as articles of food, as in the case of the Rampion *Campanula Rapunculus*. There are twenty-nine known genera and 540 species. Illustrative genera: *Jasione*, *Phyteuma*, *Campanula*, *Cyphia*. [J. H. B.]

CAMPANULA or Bell-flower. An extensive genus of herbaceous plants giving its name to the order *Campanulaceæ*. No less than 200 species of this family have been described, of which upwards of eighty are said to be either indigenous or cultivated in Great Britain. They are chiefly natives of the north of Asia, Europe, and North America, and are scarcely known in the hot regions of the world. In the meadows, fields, and forests of the countries they inhabit, they constitute the most striking ornament. Many abound in milky juice, which is rather acrid: but, nevertheless, the roots and young shoots of some species are occasionally eaten. *C. Rapunculus* (a diminutive of *rapa*, a turnip, whence the English name Rampion) is much cultivated in France and Italy, and sometimes in Britain, for the roots, which are boiled tender, and eaten hot with sauce, or cold with vinegar and pepper; *C. persicifolia* and *C. rapunculoides* may also be cultivated for the same purpose. Of the British species, *C. latifolia* is the finest and most stately; the flowers are very large, blue, or (in the Scottish woods) sometimes white. *C. Trachelium*, the nettle-leaved Bell-flower, formerly considered a specific for sore throat (Greek *trachelos*, a neck), is remarkable for the resemblance borne by its leaves to the common plant after which it is named. *C. glomerata* is a handsome plant with large erect flowers crowded into a kind of head. The more edible species, mentioned above, are sometimes also found apparently wild; but it is doubtful whether they have not escaped from cultivation, having been grown commonly in gardens before the time of Gerarde. The best-known species is *C. rotundifolia*, Hare-bell, or more correctly Hair-bell, the Blue-bell of Scotland, an elegant plant about a foot high, with a branched wiry stem and graceful drooping pale blue, sometimes white, flowers. The stem-leaves of this plant are exceedingly narrow, and seem to belie the name *rotun-*

difolia, but the root-leaves, which for the most part wither away early in the season, justify the appellation. It has been said that Linnæus gave this plant its name from having just seen the round leaves on the steps of the university of Upsal. This, however, may hardly be, as it is figured and described under the same name by Gerarde (1597). *C. hederacea* is an exquisite little plant, very abundant by the side of streams in the extreme west of England, generally growing with *Anagallis tenella*. The ivy-shaped leaves are of a remarkably fine texture, and delicate green hue; the flowers of a pale blue, sometimes slightly drooping, and supported on long stalks scarcely thicker than a hair.

Of the cultivated species *C. pyramidalis* was a very fashionable plant thirty years ago, and is still cultivated in Holland as an ornament to halls and staircases, and for being placed before fire-places in the summer season. It is still, too, a great favourite in cottage windows in England. In the shade it will continue in flower for several months. *C. Illifolia* is so called from its having at the summit of its stem a tuft of leaves resembling a double flower, which disperse as the stem elongates. 'All the species are elegant and handsome when in blossom, and are well adapted for decorating flower-borders. Some of the smaller perennial kinds answer well for decorating rockwork, or to be grown in pots, among other Alpine plants.' None are more worthy of being cultivated than the white variety of *C. rotundifolia*. French, *Campanule*; German, *Glockenblume*. [C. A. J.]

CAMPANULATE, CAMPANIFORM. Shaped like a bell.

CAMPANUMŒA. A genus of *Campanulaceæ* containing herbs from Java and India, with tuberous roots and milky juice. The leaves are opposite stalked ovate-cordate or oblong-linear, glaucous beneath. The flowers are solitary or subcorymbose; the calyx with a hemispherical tube, surrounded by a five-parted involucre, its limb truncate; the corolla five-parted; stamens five; ovary inferior, three-celled, capsule globose, five-angled. [J. T. S.]

CAMPEACHY or CAMPECHE-WOOD. The red dye-wood, better known as Logwood, obtained from *Hæmatoxylon Campechianum*.

CAMPELEPIS. An asclepiadeous genus belonging to the division *Periplocæ*, containing a single species, a native of Lower Bactria. It is an erect branching almost leafless shrub, the remote deciduous leaves being like scales, and the small coriaceous flowers in few-flowered cymes. The calyx is five-parted; the corolla rotate and five-cleft, its throat crowned with five short trilobed scales alternating with the segments; the five stamens have distinct filaments inserted in the throat of the corolla below the scales, and sagittate anthers, with the pollen-masses solitary and granular; the stigma is dilated; the follicles are slender, cylindrical and spreading, with numerous comose seeds. [W. C.]

CAMPELIA. The name of a genus belonging to the order of spiderworts, having three petals which remain attached after flowering, and form a cover to the fruit; the style or appendage on the top of the seed-vessel being smooth, bent down, and ending in a round bend which has three slight subdivisions. The species, natives of America and the warmer parts of Asia, are perennial herbs with erect stems, the leaves broadly lance-shaped and hairy on the lower surface. *C. Zanonia*, a native of the West Indies, &c., cultivated since 1759 under the name of *Tradescantia Zanonia*, is an interesting species. [G. D.]

CAMPHOR. A well-known stimulant drug, a kind of stearoptine, obtained from *Camphora officinarum*. —, BORNEO or SUMATRA. The drug produced by *Dryobalanops aromatica*, sometimes called *D. Camphora*.

CAMPHORA. The tree which furnishes camphor, *C. officinarum*, was referred by Linnæus to the genus *Laurus*, but subsequently it has been removed into a new genus of *Lauraceæ*, with a more significant appellation. This separated genus differs from *Laurus* in its ribbed leaves, the lesser number of its fertile stamens (nine), and its four-celled anthers. From *Cinnamomum* it differs in having its leaf-buds protected

Camphora officinarum.

by scales, and by the calyx being membranous instead of leathery. Camphor is prepared from the wood of the tree by boiling the chopped branches in water, when, after some time, the camphor becomes deposited, and is purified by sublimation.

CAMP] The Treasury of Botany. 210

It is produced principally in the island of Formosa, and is imported from Singapore, &c. Another kind of camphor is imported from the Dutch settlement of Batavia. What is known as Borneo camphor is the product of a tree of a different family; see DRYOBALANOPS. Camphor has acrid stimulant properties, and in large quantities is poisonous. There is a very prevalent but erroneous notion that camphor acts as a preventative in infectious diseases. It is, however, much used to prevent the ravages of insects in clothes, and in cabinets of natural history. The wood of the tree is occasionally imported to make cabinets for entomologists. [M. T. M.]

CAMPHOROSMA. A genus of *Chenopodiaceæ* consisting of small shrubs or herbs chiefly natives of the saline steppes of Central Asia, though one species occurs in the Mediterranean region. Leaves small, linear or awl-shaped, often downy, scattered or fasciculate; flowers very small, axillary, crowded; calyx tubular, compressed, four-toothed, two of the teeth larger and keeled; stamens four; style two or three-cleft. Fruit a membranous utricle contained in the unchanged calyx tube. The seeds are vertical with a membranous seed-coat and an annular embryo with green cotyledons; they contain a pungent volatile matter. [J. T. S.]

CAMPHRE'E. (Fr.) *Camphorosma monspeliaca*.

CAMPHUSIA. A genus of *Goodeniaceæ* characterised by a superior calyx; an irregular corolla having a curved tube and a three-cleft limb with narrow segments; anthers distinct; ovary with two cavities, each containing one ovule; style flattened, glabrous, wavy; stigma large, round, its cup ciliated. The genus has been separated from *Scævola*, and consists of one species, *C. glabra*, a tree inhabiting the island of Oahu, and bearing entire tufted leaves and large yellow solitary flowers. [M.T.M.]

CAMPION. *Cucubalus bacciferus*. —, BLADDER. *Silene inflata*. —, CORN. *Agrostemma Githago*. —, MEADOW. *Lychnis Flos-cuculi*. —, MOSS. *Silene acaulis*. — OF CONSTANTINOPLE. *Lychnis chalcedonica*. —, ROSE. *Lychnis coronaria* and *L. Flos Jovis*. —, RED. *Lychnis diurna*. —, WHITE. *Lychnis vespertina*.

CAMPIUM. A synonyme of *Pœcilopteris*.

CAMPSIDIUM *chilense*. The southernmost representative of the order *Bignoniaceæ*, and the only known species of the genus to which it belongs. It is a very handsome climber, with dark shining pinnate leaves, and flowers having a regular five-cleft calyx, a tubular almost regular corolla, of a rich orange colour; and five stamens, one of which is sterile, the anthers placed parallel (a peculiarity shared with only two other bignoniads, *Bignonia venusta* and *Millingtonia hortensis*). The plant grows in woods, ascending the trees to the height of forty or fifty feet. It is found in Chili and the adjacent islands, from latitude 40° to 44° south; the Isle of Huafo, where it was found by Eights, being the southernmost station at present known. According to Mr. Bridges, the inhabitants of Chiloe term it 'pilpil boqui.' [B. S.]

CAMPSIS. A genus of *Bignoniaceæ*, consisting of half-a-dozen species distributed over the Eastern Archipelago, China, Japan, and North America, and distinguished from all other members of the order by the branches being climbing and rooting like ivy, eminently qualifying these plants for covering walls and rocks, for which purpose two species, *C. adrepens* (*Bignonia*, or *Tecoma grandiflora*, of some writers) and *C. radicans* (*Bignonia*, or *Tecoma radicans* of botanists, the jasmin-trompette of the French, or Trumpet-flower as we call it) are already used in our gardens. The calyx is regular, with five acute lobes, valvate in æstivation; the corolla funnel-shaped, large; the stamens five in number, one of them being sterile, and the four fertile ones of unequal length. The capsular fruit is of oblong shape, two-celled, and the partition runs contrary to the direction of its valves, whilst the winged seeds are arranged in several rows at each side of the partition. The branches are slender; the leaves impari-pinnate, with the leaflets either entire or serrated; and the flowers arranged in terminal bunches, and either pink or of a rich orange colour. [B. S.]

CAMPTERIA. A genus of polypodiaceous ferns, of the group *Pterideæ*, distinguished by having the lowermost pairs of veins united, so as to form a series of arcs next the main costa or midrib. The sori are linear continuous and marginal, exactly as in *Pteris*. This group comprises eight or ten species, principally eastern. One of them, however, *C. biaurita*, has a very extended range not only through India, and the Eastern Islands to China, but is found also in the Mascaren Islands, South America, the West Indies, Tropical West Africa, and South Africa. [T. M.]

CAMPTOCARPUS. A genus of twining glabrous shrubs belonging to the order *Asclepiadaceæ*, natives of Madagascar, and the Isle of Bourbon. They have opposite leaves, and axillary few-flowered cymes. The small calyx consists of five sepals; the corolla is five-cleft and reflexed; the five-lobed staminal corona is inserted in the throat of the corolla, and the gynostemium is adnate to its base; the stamens have broad membranaceous filaments, and sagittate glabrous anthers attached to the margin of the stigma; the pollen mass is granular. The two long slender follicles contain many comose seeds. [W. C.]

CAMPTOSEMA. A genus of scandent or erect shrubs, belonging to the pea family, peculiar to South America, and for the most part found in Brazil. Their leaves are either simple or trifoliolate, the leaflets being oblong or elliptical in form, entire, and either smooth or tomentose. The flowers are disposed in axillary racemes; the calyces tubular, four-parted, and coloured or green; and the corolla from two to three inches long, either yellow or bright red. The pods are stalked, linear, compressed, and contain from three to six seeds. *C. rubicundum*, a native of South Brazil, is a climbing shrub of great length. The leaves are few with long stalks, their leaflets oblong or elliptical, smooth above, and pea-green beneath; the flowers are bright red, in long drooping racemes, like those of a laburnum. It is a beautiful object when in flower, and has long been in cultivation in English gardens. *C. grandiflorum*, also a Brazilian species, has yellow flowers, two to three inches long, disposed in axillary racemes. [A. A. B.]

CAMPTOSORUS. A genus of polypodiaceous ferns, of the group *Asplenieæ*, and of the scolopendrioid series, in which the sori are produced in pairs, set face to face on contiguous veins—the reverse of what occurs in the diplazioid series, in which they are set back to back in pairs on the same vein. The present is a small genus consisting of one North American and one Siberian species, both dwarf plants with simple spreading fronds, which are extended into a long narrow tail-like point, where is produced a young plant. The veins join to form a few angular unequal areoles near the midrib, and send out branches towards the margin. The sori, which are linear, and covered by linear indusia, are usually connivent in irregular unequal pairs, but are sometimes more scattered, owing to the irregularity of the venation. The variously directed irregularly-disposed yet generally opposite pairs of sori form the peculiar features of the genus. [T. M.]

CAMPTOTROPAL. An orthotropal ovule, curved downwards like a horse shoe, with the sides adherent.

CAMPYLANTHUS. A small genus, native of the Canary Isles, Tropical Africa, and India, consisting of branching undershrubs, growing chiefly in the fissures of rocks, having fleshy linear sessile leaves, and small jasmine-like flowers in loose terminal racemes. The calyx is deeply cleft into five linear-lanceolate divisions; the corolla tube is long, cylindrical, and slightly kneed near the middle, its limb deeply five-lobed. Two stamens on very short filaments rise from the curved portion of the corolla tube, and bear divaricate anthers. The capsule is compressed internally, and dehisces septicidally and septifragally, leaving the placentiferous column free; there are numerous roundish seeds. Webb seems to have satisfactorily referred this singular genus to *Scrophularineæ*, but so different is it from the other genera of the order, that he has been forced to make for its reception a new tribe which he calls *Campylantheæ*. [W. C.]

CAMPYLOBOTRYS. A genus of *Cinchonaceæ*, consisting of low-growing Brazilian shrubs, remarkable for their beautiful glossy foliage. They bear flowers with an obovate calyx-tube, having four small linear segments to its limb, and two or three small glands between them; a salver-shaped corolla; four short stamens, with anthers projecting from the short tube of the corolla; a four-cornered ovary, with two many-seeded compartments, and surmounted by a fleshy disc. *C. regalis* has elliptic leaves with a satiny lustre, and a bronzy-green colour, except the main rib and the larger side ones. *C. bicolor* and other species are cultivated in stoves for the beauty of their foliage. They are, however, now regarded as belonging to *Higginsia*. [M. T. M.]

CAMPYLONEURUM. A genus of simple-fronded polypodiaceous ferns of the group *Polypodieæ*. They have round naked sori as in the other genera of this group, from which they are distinguished by having the principal veins branching from the costa nearly parallel, and united by transverse curved venules, while from the outer side of these are produced two or three short straight veinlets on the middle or point of which the sori are placed. There are about a score of species, all West Indian and South American, and with two exceptions simple-fronded. One of these exceptions is *C. magnificum*, a splendid pinnate Venezuelan fern, of which the pinnæ measure eighteen inches long and four inches broad, and bear four rows of sori between the veins. *C. repens* is a well-known illustration of the simple-fronded series. [T. M.]

CAMPYLOSPERMOUS. When a seed or seed-like fruit is so rolled up as to have a furrow in the longer diameter of one side.

CAMPYLOSTACHYS. A genus of *Stilbaceæ* confined to South Africa. The only species known, *C. cernua*, is a heath-like bush, about one foot high, with closely set linear pointed leaves, about half an inch long, and terminal roundish spikes of flowers, which are reflexed when the seeds become mature. The flowers are very small, and have a long tube with a four-cleft border. The name *Campylostachys* has reference to the curved spike. [A. A. B.]

CAMPYLOTROPAL. An ovule, one of whose sides grows much faster than the other, so that while the chalaza remains at the hilum, the foramen is brought nearly into contact with it.

CAMPYNEMA. A genus of doubtful amaryllids found in Tasmania. It has been associated with *Anigozanthus* by Herbert, and has been regarded by Brown as intermediate between amaryllids and asphodels, coming near to *Melanthaceæ*. The only species, *C. lineare*, is a slender herb, about a foot high, with fasciculate fusiform roots; tufted grassy leaves; and one to four terminal inconspicuous yellowish-green flowers. It has a six-leaved perianth of persistent spreading equal elliptic-lanceolate segments; six stamens; and an inferior three-celled ovary, containing numerous ovules, and crowned by three recurved styles, terminating in simple stigmas. The name is sometimes written *Campylonema*. [T. M.]

CAMBUC, CAMRUNGA. *Averrhoa Carambola*.

CAMWOOD. A West African red dye wood produced by *Baphia nitida*.

CANADA BALSAM FIR. *Abies balsamea*. Canada Balsam is an oleo-resin obtained from this tree, and is extensively used in medicine and manufactures.

CANAGONG. The fruit of *Mesembryanthemum œquilaterale*.

CANALICULATE. Channelled, like the petioles of many leaves.

CANARINA. A genus of *Campanulaceæ*, containing a glaucous herb from the Canary Islands, which has a tuberous root with milky juice, and a branched stem thickened at the joints, the leaves opposite (rarely in a whorl of three), stalked, hastate-heart-shaped, irregularly toothed, shining above. The flowers are large nodding yellowish (a remarkable feature, as purple, blue, or lilac flowers are usually found in this natural order), solitary at the apex of short leafy axillary branches; calyx-limb six-cleft, reflexed; corolla bell-shaped, six-toothed; stamens six; ovary inferior, six-celled; style with six stigmas; capsule somewhat fleshy, and as well as the roots and young shoots said to be edible. [J. T. S.]

CANARIUM. A genus of *Amyridaceæ*, consisting of trees with compound leaves; the flowers panicled, diœcious, having a bell-shaped calyx, with three unequal lobes; three oblong concave petals; six stamens inserted beneath a cup-shaped disc; and a sessile globular ovary, with very short style, and three-lobed stigma. The fruit is a triangular drupe, with three, or, by abortion, one cavity, containing one seed. *C. commune* is cultivated in the Moluccas for its fruits, which are also eaten in Java, and from them an oil is expressed which is used at table when fresh, and for burning in lamps. A gum exudes from the bark which is said to resemble in its properties Balsam of Copaiba. *C. strictum*, according to Dr. Wight, is known in Malabar as the black Dammar tree, in contradistinction to the white Dammar (*Valeria indica*). The resin of Dammar is of a brownish or amber colour. [M. T. M.]

CANARY CREEPER. A garden name for *Tropæolum aduncum*, commonly but wrongly called *T. canariense*.

CANARY SEED. The grain of *Phalaris canariensis*, much used as a food for small domesticated birds.

CANARY WOOD. The timber of *Persea indica*, and *P. canariensis*.

CANAVALIA. About eighteen species of this genus of *Leguminosæ* are known. They are mostly shrubby climbing plants, with slender twining branches, and leaves composed of three leaflets, and are found inhabiting the tropical regions of both hemispheres. The flowers are in racemes produced from the axils of the leaves; their calyx is bell-shaped, two-lipped, with the upper lip largest, and either entire or cut into two lobes, while the lower is three-cut or entire; their corolla is papilionaceous; and their stamens are united into a column, one of their number being separated for the greater part of its length. The pods are large, with their sides swelled out, and having three elevated ribs or ridges along the upper edge; they contain numerous seeds, which are separated from each other by a quantity of cellular tissue.

C. gladiata is commonly found growing in woods in the East Indies, tropical Africa, Mexico, Brazil, the West Indies, &c. The leaves consist of three roundish or egg-shaped leaflets, terminating abruptly in a short point, and varying in size from two to six inches long. The flowers are dark-purple, and succeeded by scimitar-shaped pods, about a foot long, containing numerous red or white seeds, resembling large beans. According to Dr. McFadyen, this plant is called the 'Overlook' by the negroes in Jamaica, who plant it along their provision grounds from a superstitious notion that it 'fulfils the part of a watchman, and, from some dreaded power ascribed to it, protects the property from plunder. Even the better informed adopt the practice, although they themselves may not place confidence in any particular influence which this humble plant can exercise, either in preventing theft, or in punishing it when committed.' [A. S.]

CANCELLATE. Composed of veins only, all the parenchyma or intervening web being absent.

CANCER-ROOT. An American name for *Epiphegus* and *Conopholis*; also for *Aphyllon uniflorum*, sometimes called *Orobanche uniflora*.

CANCHE. (Fr.) *Aira*.

CANDIDUS. Pure white, but not so white as snow.

CANDLEBERRY MYRTLE. The common name for *Myrica*.

CANDLEBERRY TREE. *Aleurites triloba*, the nuts of which are commercially called Candle nuts.

CANDLE TREE. *Parmentiera cerifera*.

CANDLEWOOD, of Jamaica. *Gomphia guianensis*.

CANDOLLEA. A genus of Australian shrubs belonging to *Dilleniaceæ*, with obovate or wedge-shaped leaves, and handsome yellow flowers, which are solitary at the tips of the branches; sepals five, oval, mucronate; petals obovate or obcordate; stamens polyadelphous; style thread-like; carpels two to five, ovate. [J. T. S.]

The name *Candollea* was also given by Mirbel to a group of Polypodium-like Ferns, now included in *Niphobolus*.

CANDYTUFT. Any species of *Iberis*.

CANE. A common commercial name for the stems of various grasses, palms, &c. —, BAMBOO. *Bambusa arundinacea*. —, DRAGON. A kind of Rattan Cane. —, DUMB. *Dieffenbachia seguina*. —, GREAT RATTAN. *Calamus rudentum*. —, GROUND RATTAN. *Rhapis flabelliformis*. —, MALACCA. The stem of *Calamus scipionum*, imported for making walking-sticks. —, RATTAN. *Calamus Rotang* and its forms, now called *C. Royleanus*, *C. Roxburghii*, &c. —, REED. The stem of some grass often forty feet long, from New Orleans, largely imported for making weavers' shuttles. —, SWEET. *Andropogon Calamus aromaticus*. —, SUGAR. *Saccharum officinarum*. —, TOBAGO. The stem of *Bactris minor*, imported for walking-sticks.

CANE-BRAKE. The common name for *Arundinaria*.

CANELLACEÆ. Two or perhaps three, West Indian or tropical American aromatic shrubs, constituting the two genera *Canella* and *Cinnamodendron*, differ in so many respects from the several orders with which they have been compared, that it has been proposed to class them as a distinct family under the name of *Canellaceæ*. Their aromatic properties and the structure of their seeds have induced an approximation to *Wintereæ* (a tribe of *Magnoliaceæ*), from which, however, their flowers and ovary widely remove them. The stamens, united in a column, with the anthers sessile on the outside, have suggested an affinity with *Guttiferæ*, *Ternstromiaceæ*, or even *Sterculiaceæ*; but, upon the whole, it is probably with *Bixaceæ* and their allies that *Canellaceæ* have the nearest connection. They agree with them in their one-celled ovary, with parietal placentas, and they show no marked discrepancy in their foliage, flowers, fruit, or seed, except that the albumen is firmer, with a smaller embryo.

CANELLA. The tree yielding Canella bark has been placed in various natural groups by different writers. The characters of the genus, in brief, are the presence of three overlapping sepals; five petals; twenty stamens united below, and having narrow anthers; a one-celled ovary, with two or three pendulous ovules. The tree is a native of the West Indies, and furnishes a pale orange coloured bark, with an aromatic odour, which is used as a tonic. The negroes of the West Indies use it as a spice. The plant is frequently grown in botanic gardens. [M. T. M.]

CANELLA DE CHEIRO. The volatile oil of *Oreodaphne opifera*.

CANESCENS. Greyish-white; hoary. A term applied to hairy surfaces.

CANI. The sun-dried tubers of the Oca, *Oxalis tuberosa*.

CANKER. A disease resulting in the slow decay of trees, or other plants attacked by it. See CARCINODES. [M.J.B.]

CANKRIENIA. A genus of *Primulaceæ*, containing a single species from Java, a very beautiful Alpine plant, with erect radical leaves, often half a foot in diameter, verticillate nodding flowers, and erect fruit. The calyx is five-toothed and cup-shaped; the corolla is funnel-shaped, with a five-lobed limb; the five stamens, with short filaments, are inserted in the throat of the corolla opposite to its divisions; the ovary is globose with a rayed apex; the included style remains on the fruit, which is a globular capsule, containing numerous angular seeds. [W. C.]

CANNABINACEÆ. (*Cannabineæ*, Hemp-worts, the Hemp family.) A natural order of monochlamydeous dicotyledons, belonging to Lindley's urtical alliance. Rough-stemmed herbs with watery sap, alternate and lobed leaves having stipules, and small inconspicuous flowers. The plants have some flowers with stamens without pistils, and others with pistils without stamens. The staminate flowers are in clusters called racemes or panicles; calyx herbaceous and scaly; stamens few, opposite the sepals; filaments filiform. Pistillate flowers in spikes or cones, with a single sepal; a one-celled ovary containing a solitary pendulous ovule; stigmas two. Fruit a single-seeded nut; embryo hooked or spiral, without albumen. The plants are natives of the temperate parts of the northern hemisphere in the Old World. They possess narcotic qualities and yield valuable fibres. Hemp is the produce of *Cannabis sativa*. It is imported in large quantities from Russia. The plant grows in the cooler parts of India, and there develops narcotic qualities. These properties seem to reside in the Churrus or resin which covers the leaves. The names of Bhang, Gunjah, and Haschisch are given to the dried plant in different states. What are called Hemp seeds, used for the food of birds, are in reality Hemp fruits, each containing a single seed. *Humulus Lupulus*, the Hop, another important plant of the order, possesses both tonic and hypnotic properties, i.e. a power of inducing sleep. The scales of the hop-heads are covered with resinous matter, which has an aromatic odour. There are two genera in the order, viz., *Cannabis* and *Humulus*, and two species. [J. H. B.]

CANNABINE. A narcotic gum-resin obtained from the Hemp, *Cannabis sativa*.

CANNABIS. The Hemp-plant, *C. sativa*, which is the solitary species of the genus, is the type of the *Cannabinaceæ*. It is a native of India and Persia, and is generally cultivated, although it is only in hot dry climates that it forms the resin which gives it such value in the estimation of the natives, apart from its fibre-producing qualities. The dried plant, or portions of it, are sold in the bazaars of India under the name of Gunjah and Bhang, while the resin itself is known as Churrus. This resin is collected during the hot season in the following singular manner:—Men clad in leathern dresses run through the hemp fields, brushing through the plants with all possible violence; the soft resin adheres to the leather, and is subsequently scraped off and kneaded into balls. In Nepal, according to Dr. McKinnon, the leathern attire is dispensed with, and 'the resin is gathered on the skin of the naked coolies!' Gunjah is smoked like tobacco; Bhang is not smoked, but pounded with water into a pulp, so as to make a drink; both are stimulant and intoxicating; but the Churrus or resin possesses much more powerful properties. In small quantities it produces pleasant excitement, which passes into delirium and catalepsy if the quantity be increased; if still continued a peculiar form of insanity is produced. Many of

Cannabis sativa.

the Asiatics are passionately addicted to the use of this means of intoxication, as the names given to the hemp show — 'leaf of delusion,' 'increaser of pleasure,' 'cementer of friendship,' &c. &c. A recent traveller in East Africa, Capt. Burton, describes this plant as 'growing before every cottage door.' The Arabs smoke the sun-dried leaf with, and the Africans without, tobacco, in huge pipes. 'It produces a violent cough ending in a kind of scream after a few long puffs, when the smoke is inhaled, and if one man sets the example the others are sure to follow it. These grotesque sounds are probably not wholly natural. Even the boys may be heard practising them as an announcement to the public that the fast youths are smoking Bhang.' [M. T. M.]

The Hemp plant is an annual, growing in ordinary situations from four to ten feet high, but in Italy under very favourable circumstances it sometimes grows as high as twenty feet. The stem is grooved or angular, and, in plants growing singly, frequently much-branched, but when cultivated in masses for the sake of the fibre, it is generally straight and unbranched. It consists of a central pith surrounded by a layer of loose woody and cellular tissue, and enclosed in a thin bark containing the fibre which renders the plant so valuable. Its leaves have long stalks with minute awl-shaped stipules at their bases, and are composed of from five to seven long lance-shaped sharp-pointed leaflets, radiating from the top of the stalk, each leaflet having its margin cut into sharp saw-like teeth. The whole plant has a rough harsh feel from the presence of numerous minute asperities. The flowers are of separate sexes on different plants, the males being produced in racemes and generally crowded together towards the top of the plant or ends of the branches, having a five-parted calyx and five stamens; the females are in short spikes, their calyx consisting merely of a single sepal, rolled round the ovary, but open on one side, and they have two hairy stigmas. The fruit (commonly known as 'hemp seed') is a small greyish-coloured smooth shining nut, containing a single oily seed. Of whatever country Hemp is a native, it is certain that it was known in Europe in very early times, for Herodotus, writing upwards of 2000 years ago, mentions it as being cultivated by the Scythians, who used its fibre for making their garments. At the present day it is cultivated in most parts of Europe; in Arabia, Persia, India, China, and other Asiatic countries; in Egypt, and various other parts of the African continent; and in the United States. Russia and Poland, however, are the two great hemp-producing countries, and it is from them that our supply is mainly derived; but the best quality is produced in Italy; the United States and India likewise send hemp to this country, but the quality is inferior to the Russian. For the production of good fibre the seed is sown close, so as to produce straight stems without branches. The harvesting takes place at two periods; the male being pulled up as soon as it has done flowering, and the female not until the seeds are ripe. After pulling, the leaves are struck off with a wooden sword; the stems are then tied in bundles and steeped in water, or water retted as it is technically termed (two other processes, dew-retting and snow-retting, are sometimes substituted), the object being to loosen the fibre; they are then spread out to dry and bleach; this is called 'grassing,' after which the fibre is detached, either by pulling it off by manual labour, or by breaking the stems in a machine, and afterwards scutching them in a similar manner to that employed for the preparation of flax.

The uses of Hemp for the manufacture of cordage, canvas, &c., are too well known to require more than a passing allusion. The seeds are used for feeding caged birds, and an oil is expressed from them. The imports of Hemp in 1858 amounted to 739,339 cwts., the computed real value of which was 1,034,277*l.*; and of Hemp seed, 11,090 quarters; value 24,074*l.*   [A. S.]

CANNACEÆ. The Indian-shot family, a natural order of epigynous monocotyledons belonging to Lindley's amomal alliance. The name of *Marantaceæ* is also given to the order, and under that its characters and properties are stated.   [J. H. B.]

CANNA. The name of a genus of *Marantaceæ* distinguished by the flowers being in panicles; having a calyx of three sepals, a corolla of six pieces, five of which are erect, the other reflexed; these may be considered rather as abortive stamens than as petals; the one fertile stamen is petal-like, with an anther on the margin; the style is also petal-like with a linear stigma, and the fruit consists of a capsule covered with rough tubercles externally, and internally divided into three compartments, each of which contains a number of horizontally placed seeds; when ripe the fruit bursts into three divisions. The seeds of most of the species are round, hard, and black, hence the name of Indian Shot, which is applied to the plants.

Many of the species have brightly-coloured flowers—yellow, red or orange. The foliage, too, is highly ornamental and characteristic; hence they are favourite plants in cultivation, and produce a striking effect when grouped in beds out of doors during the summer months. The beauty of these plants is not their only feature of interest, as some of them are also of importance from their fleshy underground stems, containing an abundance of starch. *Tous les mois*, a superior kind of arrowroot, the grains of which are very large, is the produce of one of the West Indian species, probably *C. edulis*. The tubers of other species are eaten as a vegetable, while some have slight medicinal properties. In the Brazils the leaves are used for packing purposes, hence the French call these plants *Bansier*, from a Spanish word signifying cover. The seeds are also made use of as beads.   [M. T. M.]

CANNE A' SUCRE. (Fr.) *Saccharum officinarum*. —, DINDE. *Canna indica*. —, DE JONC. *Typha latifolia*. —, DE PROVENCE. *Arundo Donax*.

CANNEBERGE. (Fr.) *Oxycoccus palustris*.

CANNELLIER. (Fr.) The Cinnamon tree.

CANNILE'E. (Fr.) *Lemna minor*.

CANNOMOIS. A genus of *Restiaceæ*, differing from *Restio* in the fruit, which is a hard indehiscent nut; and from *Willdenowia* by having two distinct styles. *C. cephalotes*, the original species, has a rigid stem with numerous short barren stems at the base; flowers in a large ovate terminal head, with ovate acute imbricated bracts. This and another species are from the Cape of Good Hope.   [J. T. S.]

CANNON-BALL TREE. *Couroupita guianensis*.

CANOE BIRCH. *Betula papyracea*.

CANOE WOOD. *Liriodendron tulipifera*.

CANTERBURY BELL. *Campanula Medium*.

CANTHARELLUS. The scientific name of the Chantarelle.

CANTHIUM. A genus of *Cinchonaceæ* consisting of spiny rigid plants with solitary fragrant white flowers, having the stamens inserted near the throat of the corolla, and a thread-shaped protruding style terminated by a thick globular or mitre-shaped stigma. The fruit is a two-celled berry. *C. parviflorum*, an Indian plant, makes good fences, while the leaves are occasionally added to curries by the natives; but they have also medicinal properties. One or two species are in cultivation.   [M. T. M.]

CANTUA. A genus of *Polemoniaceæ*, containing six or eight species, natives of Peru. They are trees or shrubs with alternate fleshy entire or sinuate-dentate leaves, and large showy flowers in corymbs at the termination of the branches, rarely solitary and axillary. The calyx is tubular and five-cleft; the corolla is funnel-shaped with the spreading limb split into five obovate lobes; the five stamens are inserted at the base of the tube, and are more or less exserted; the ovary is three-celled with numerous ovules, and bears a simple style with a trifid stigma; the capsule is coriaceous and three-valved; the seeds have their apex produced into a wing. This genus is nearly related by its capsule and seeds to *Cobæa*, though in habit and inflorescence some of its species approach *Polemonium*.   [W. C.]

CANUS. Grey-white or hoary. A term applied to hairy surfaces.

CAOUTCHOUC. The elastic gummy substance known as India rubber, which is the inspissated juice of various plants growing in tropical climates in different parts of the world; *e. g. Ficus elastica* and other species of moraceous plants, *Castilloa elastica* and other artocarpads, *Siphonia elastica* and other euphorbiaceous plants, *Urceola elastica* and other apocynaceous plants, &c. The name is also given by the Popayans to the milky juice of *Siphocampulus Caoutchouc*, an elastic gum, very different from the caoutchouc of commerce.   [T. M.]

CAP. The convex part of an agaric or other similar fungal.

CAPANEA. A genus of *Gesneraceae* of the tribe *Besleriex*, consisting of dwarf herbs with subshrubby stems, and opposite oval stalked hairy leaves, from the axils of which spring the flowers two or three together from a common peduncle. The calyx is free, nearly regular, and five-parted. The corolla is irregularly bell-shaped, scarcely curved, somewhat ventricose beneath, with a short limb. There are four didynamous stamens, the filaments of which carry heart-shaped anthers, which are firmly joined together, and form in the mouth of the tuber a pale yellow star, with which the stigma is in contact. The ovary is free, surrounded by a disk of five obtuse fleshy lobes. The only species, *C. grandiflora*, a native of New Grenada, grows nearly a foot high, with moderate-sized oval-acuminate leaves, and large showy long-stalked flowers, seated in a tuft at the end of an axillary or terminal peduncle; these flowers are nodding gloxinia-like, with a limb of five broad spreading emarginate lobes, pubescent outside, white, elegantly painted on the inner face of the limb, or less frequently on the tube, with numerous crimson dots arranged in contiguous lines. Dr. Lindley writes the name of this genus *Campanea* in *Paxton's Flower Garden*, i. 91. [T. M.]

CAPE WEED. *Roccella tinctoria*, a dye lichen, obtained from the Cape de Verd Islands.

CAPER. *Capparis spinosa*, the flower buds of which, and of some allied species or varieties, form the well-known condiment of this name, for which the flowers of *Zygophyllum Fabago* are sometimes substituted.

CAPERONNIER. (Fr.) *Fragaria elatior*.

CAPER SPURGE. *Euphorbia Lathyris*, sometimes called Caper bush.

CAPER TREE, of New South Wales, *Busbeckia arborea*.

CAPILLACEOUS, CAPILLARY. Having the form of a thread.

CAPILLAIRE. A syrup prepared with *Adiantum Capillus-veneris*.

CAPILLAIRE. (Fr.) *Asplenium Trichomanes*. — DE MONTPELLIER. *Adiantum Capillus-veneris*. — DU CANADA. *Adiantum pedatum*. — NOIR. *Asplenium Adiantum-nigrum*.

CAPILLITIUM. Entangled filamentary matter in fungals, bearing sporidia.

CAPILLUS (adj. **CAPILLARIS**). The breadth of a hair; the twelfth part of a line.

CAPITÃO DO MATTO. A common Brazilian name for *Lantana pseudo-thea*.

CAPITATE. Pin-headed, as the stigma of a primrose, or as certain hairs. Also growing in heads, or terminal close clusters, as the flowers of composites, &c.

CAPITULUM. A close head of sessile flowers. Also a term vaguely applied among fungals to the receptacle, pileus, or peridium.

CAPNITES. A section of the genus *Corydalis*. Decandolle employs it in a sense synonymous with *Bulbocapnos*, but Endlicher used it to designate a part of Decandolle's section *Capnoides*, which includes the species of *Corydalis* without tuberous rootstocks. In this way it is equivalent to *Corydalis* of Bernhardi, and differs from *Capnoides*, as restricted by that author, by having the stem simple and branched, and the style persistent. There is, however, no natural division, and it is better to consider all the species of *Corydalis* without tuberous rootstocks, with two separate cotyledons, and with a cup-shaped appendage at the base of the seed, as belonging to the section *Capnoides*. The only British species is the small climbing Fumitory *Corydalis claviculata*, which has long branched trailing stems, and yellowish flowers in racemes. *C. lutea*, often cultivated, and naturalised in several localities, is easily known by its short stems and large bright yellow flowers. [J. T. S.]

CAPNODIUM. A curious genus of *Fungi* established by Dr. Montagne to receive a portion of the black smutty parasites which infest the leaves and twigs of shrubs in damp warm climates. It belongs to the division *Physomycetes*, and is characterised by the abundant creeping black threads which run over the several parts of the plants which it attacks. Shoots from these threads either intimately invest the fruit or are combined to form it. The fruit consists of irregular often elongated and branched cysts, which in the same species contain naked spores and sporidia, enclosed in asci. One species only, *C. elongatum*, has been found in the extreme south-west of this country on pear trees; others are the plague of coffee, lemons, olives, and other important plants. In a young state these plants are not distinguishable from *Antennaria*. The breathing pores or stomates of the plants which they attack are completely smothered, and direct light almost excluded, so that the functions of the leaves are greatly impeded. No remedy is known when the parasite is once developed. If any is applied, it must be directed to the destruction of the different species of coccus on whose dung or excretions these *Fungi* seem mostly to be developed. Lemons frequently arrive in this country in an unsaleable condition, incrusted more or less completely with a jet black felt, in consequence of the growth either of an *Antennaria* or the spawn of *Capnodium Citri*, which seems to increase greatly after the fruit is packed up for the market. [M. J. B.]

CAPPARIDACEÆ (*Capparids*.) A natural order of thalamifloral dicotyledons placed in Lindley's cistal alliance. Herbs, shrubs, or trees with alternate leaves and solitary or clustered flowers; sepals four, imbricate or valvate; petals four, arranged crosswise, sometimes eight; stamens usually numerous, and a multiple of four placed at the top of a stalk-like receptacle; disk much developed. Ovary usually supported on a stalk and one-celled, with parietal placentas. Fruit either pod-like and opening, or berried; seeds often kidney-shaped, without albumen. The order is divided into two suborders: 1. *Cleomeæ*, with dry dehiscent (splitting) fruit. 2. *Cappareæ*, fruit a berry. The plants are chiefly tropical. They abound in Africa and India. Some are found in Europe and in Canada. They have pungent and stimulant qualities, and have been recommended in scurvy. In their properties they resemble crucifers. The flower-buds of *Capparis spinosa* constitute capers. *C. ægyptiaca* is considered by some as the hyssop of Scripture. There are thirty-three known genera, and 355 species. Illustrative genera: *Cleome, Polanisia, Capparis, Cratæva.* [J. H. B.]

CAPPARIS. The genus so called gives its name to the natural order *Capparidaceæ*. It consists of shrubs having simple leaves, frequently with two little spines at their base, and showy flowers with a four-parted calyx, four petals, and numerous stamens, succeeded by a berry elevated on a long slender stalk. The most generally known plant of this genus is the common Caper, *C. spinosa*, which grows on walls, etc., in the South of Europe and Mediterranean regions. In its mode of growth it resembles the common bramble. The flower-buds, and in some parts of Italy, the unripe fruits, are pickled in vinegar, and form what are commonly known as capers. They are chiefly imported from Sicily, though the plant is also largely cultivated in some parts of France. All the species

Capparis spinosa.

contain, in greater or less quantity, an acrid principle, so that the bark of the root of some of them acts as a blister when applied to the skin; and the fruits of some of the Brazilian species are reported to be very poisonous. *C. Sodada* is described by Dr. Barth as forming one of the characteristic features in the vegetation of Africa from the desert to the Niger; the small berries have a pungent pepper-like taste, and when dried constitute an important article of food, whilst the roots, when burned, yield no small quantity of salt. Several species are in cultivation in this country, principally natives of warm and tropical climates. [M. T. M.]

CAPREOLUS. A tendril.

CAPRIER COMMUN. (Fr.) *Capparis spinosa*.

CAPRIFICATION. A fertilisation of flowers by the aid of insects, as that of the garden fig by a small fly.

CAPRIFICUS. The Wild Fig. This, according to Theophrastus and Pliny, is a tree of a wild kind which never ripens its fruit, but has the power of conferring on other trees a virtue which it has not in itself. Since, in accordance with the laws of nature, life springs from putrefaction, from the abortive fruit of the Wild Fig are generated certain winged flies, which, failing to find food in the corruption which gave them birth, fly to a tree of an allied species, and penetrating the fruit of the true fig, make a way for the admission of the heat of the sun and genial air, consume the immature juices, and help the fruit to ripen. To promote this end, the *Caprificus* is planted among fertile fig trees, or cut branches of the one are tied to growing boughs of the other. Fig trees growing in a poor soil exposed to the winds, and especially dust, do not, they say, need this assistance, as the fruit under these circumstances dries up of itself sufficiently to ripen. See *Pliny Nat. Hist.*, lib. xiv. cap. xix., and *Theophrastus de plantis*, cap. of lib. ii. This last passage is curious as containing an early recognition of the presence of sexes in plants. [C. A. J.]

CAPRIFOLIACEÆ. (*Lonicereæ, Caprifoils*, the Honeysuckle family.) A natural order of gamopetalous calyciﬂoral dicotyledons belonging to Lindley's cinchonal alliance. Shrubs or herbs, often twining, with opposite leaves which have no stipules; calyx adherent to the ovary, its limb four to five-cleft, usually with small leaves (bracts) at its base; corolla superior, regular or irregular; stamens four or five, alternate with the lobes of the corolla. Ovary usually three to five-celled; stigmas three or five. Fruit generally a berry, with one or more cavities, and crowned by the calyx-lobes; albumen fleshy. Natives of the northern parts of Europe, Asia, and America, found sparingly in Northern Africa, and unknown in the Southern hemisphere. Some of the plants are astringent; others have emetic and purgative qualities. Many have showy and fragrant flowers. The common honeysuckle or woodbine (*Lonicera Periclymenum*), one of the plants of this order, twines round the branches of trees, and often causes groovings in them. The elder (*Sambucus nigra*), the Guelder rose (*Viburnum Opulus*), the laurustinus (*Viburnum Tinus*), the snowberry (*Symphoricarpus racemosus*), as well as the *Linnæa borealis*, belong to the order. The black berries of the species of *Viburnum*, found on the Himalaya, are eatable and agreeable. There are sixteen genera and 230 species. Illustrative genera: *Linnæa, Lonicera, Viburnum, Sambucus*. [J. H. B.]

CAPRIFOLIUM. A family of well-known twining shrubs giving name to the order *Caprifoliaceæ*. No British shrub claims our favourable notice so early in the season as the Honeysuckle (*C. Periclymenum*); for even before the frosts of January have attained their greatest intensity, we may discover in the sheltered wood or hedge-bank its wiry stem throwing out tufts of tender green leaves from the extremity of every twig. Later in the season it engages our attention by its twisting stems clinging for support to some lustier neighbour till it has reached air and light, when it asserts its independence, loses a good deal of its twining character, and displays its numerous clusters of trumpet-shaped cream-coloured flowers, tinged with crimson, and shedding a perfume which in sweetness is surpassed by no other British plant. As the coils made by the honey-suckle in its effort to reach the summit of a tree never enlarge, but on the contrary, rather contract as the diameter of its stem increases, it is mischievous to any growing tree round which it twines; it should, therefore, be discouraged in young plantations; but trained against a wall or allowed to twine round a pole or the bole of a full-grown tree, it is harmless and always beautiful. The scarlet berries are clammy to the touch, glutinous and sweet to the taste, but mawkish. In October the woodbine endeavours to impart a grace to the fading year by producing a new crop of flowers, which, though not so luxuriant nor so numerous as the first, are quite as fragrant. Clusters of flowers and of ripe berries may then be found on the same twig, uniting autumn with summer as the early foliage united winter with spring. A variety with leaves sinuated like the oak is not of uncommon occurrence; and another variety, called Dutch Honeysuckle, is valued as a garden plant on account of its extreme fragrance (especially in the evening) and its early flowering.

The Perfoliate Honeysuckle (*C. Italicum*, sometimes called *Lonicera Caprifolium*) resembles the last in habit. It is a native of the middle and South of Europe, and is said to be naturalised in some parts of England. It may be distinguished from the common kind by having its upper leaves united at the base so as to form a kind of cup, and it bears whorls of flowers in the axils of these leaves as well as at the extremity of the shoot.

Among the other cultivated species, *C. flavum*, a native of America, has very fragrant yellow flowers, which as they fade become orange-coloured. The Trumpet Honey-suckle (*C. sempervirens*) is an evergreen twining shrub, the upper leaves of which are united at the base (connate), and the flowers, which are scarlet outside, and yellow within, are arranged in several terminal whorls; this is also a native of America, but thrives well in Great Britain in a dry open situation, bearing a profusion of beautiful but scentless flowers from May till August. *C. etruscum* approaches *C. Italicum* in habit, but the leaves are more obtuse and downy, and it flowers during a greater portion of the year. In France this species is more frequently cultivated than any other. [C. A. J.]

CAPSELLA. A common weed belonging to the cruciferous order, well marked by its heart-shaped pods, which when ripe separate into two boat-shaped valves, each enclosing numerous yellow seeds. There is but one species, *C. Bursa-pastoris*, Shepherd's Purse, so called from the resemblance of the pods to some ancient form of purse. A native of Europe, it has accompanied Europeans in all their migrations, and established itself wherever they have settled to till the soil. It is a troublesome weed, not refusing to grow and leave seed even in the poorest soil, but luxuriating in the richest. Hence its utilitarian popular name, 'Pickpocket,' is more appropriate perhaps than the sentimental one 'Shepherd's Purse.' When not in flower, it may be distinguished by its radiating leaves, of which the outer lie close pressed to the ground. It is less acrid than most of the cruciferous tribe, but was formerly used as a potherb, as is said to be still the custom in some parts of North America. French, *Bourse de Pasteur*; German, *Hirtentasche*. [C. A. J.]

CAPSICUM. One of the genera of *Solanaceæ*, deriving its name from the Greek word signifying 'to bite,' in allusion to the hot pungent properties possessed by the fruits and seeds. The genus consists of annual or biennial plants, frequently with a somewhat woody and bushy stem; a wheel-shaped corolla; five stamens protruding from the corolla, their anthers converging at their points, and opening by longitudinal slits; and a two to four-celled ovary, becoming, when ripe, a membranous pod containing several seeds. The shape of the fruit varies very much in the different species of the genus.

C. annuum, a native originally of South America, but introduced into India and elsewhere, furnishes the fruits known as Chillies; these, as well as the fruits of *C. frutescens*, and several other species or varieties, are used to form Cayenne pepper. For this purpose the ripe fruits are dried in the sun or in an oven, and then ground to powder, which is mixed with a large quantity of wheat flour. The mixed powder is then made into cakes with leaven, these are baked till they are as hard as biscuit, and are then ground and sifted. The Cayenne pepper of the shops is, however, usually largely adulterated with red lead and other less objectionable substances. The hot taste seems to be due to a peculiar acrid fluid called *capsicin*, which is so pungent that half a grain of it volatilised in a large room, causes all who respire the contained air to cough and sneeze. It is remarkable that the narcotic properties, which are possessed by most of the *Solanaceæ* to a greater or less extent, are not present in *Capsicum*—though this is open to some doubt, as it is said that some of the American species have narcotic properties residing in the pulpy matter in which the seeds are imbedded, this pulp being absent in those kinds which are used for their pungent properties.

Capsicum fruits are used medicinally, in powder or as a tincture, as an external application, or as a gargle in certain cases of sore throat, particularly those of a malignant character, and internally as a stimulant in cases of impaired digestion, &c. Several kinds are cultivated in this country, as objects of curiosity, and for the sake of their fruits. [M. T. M.]

The species of *Capsicum* are chiefly natives of the East and West Indies, China, Brazil, and Egypt, where they are much esteemed for their pungent fruit and seeds, which, under the name of Cayenne Pepper, or Chillies, form an indispensable condiment, which Nature herself appears to have pointed out to persons resident within the tropics. According to Sir R. Schomburgk, the natives in Guiana eat the fruit of these plants in such abundance as would not be credited by an European unless he were to see it. *Jour. Hort. Soc.* ii. 153. In Jamaica the species most esteemed is the Bonnet Pepper (*C. tetragonum*), the fruits of which are very fleshy, and have a depressed form, like a Scotch bonnet. The shrubby Capsicum, or Spur Pepper (*C. frutescens*), is a native of the East Indies, and has been in our gardens since 1656. It forms a dwarf bushy shrub, with white flowers, and bears numerous small oblong obtuse pods, which are very pungent, and in their green and ripe state are used for pickling, as well as for making Chilli vinegar. This is done by merely putting a handful of pods into a bottle, and afterwards filling it with best vinegar, which in several weeks will be fit for use. But the chief purpose for which this species is cultivated is for making Cayenne pepper, which is often prepared by drying the pods on a hot plate, or in a slow oven, and then pounding them in a mortar, and passing them through a handmill until the whole is reduced to the finest possible state. After this has been done, the powder is to be sifted through a thin muslin sieve, and preserved in well-corked glass bottles for use. The common annual Capsicum, or Guinea Pepper (*C. annuum*), was introduced into Europe by the Spaniards. It was cultivated in England in 1548, and is sufficiently hardy to thrive in summer against a south wall in the open air, and mature its fruit. The colour, direction, and figure of the latter is very variable—some being yellow, others red, and others black. In a green state they are used for pickling, and when ripe are mixed with tomatos, &c., to form sauces. They are also dried and ground for use like Cayenne pepper. The Berry-bearing Capsicum, or Bird Pepper (*C. baccatum*), is indigenous to both the East and West Indies, and has been grown in this country since 1731. Its pods are erect, roundish, egg-shaped, very pungent, and when ripe are dried and used for the same purposes as those of other kinds of *Capsicum*. They also form one of the chief ingredients in the preparation known in the West Indies as *man-dram*, which is usually resorted to by those affected with loss of appetite or weak digestion, and consists of cucumbers sliced very thin, shallots or onions chopped very fine, a little lime juice and Madeira wine, to which is added a handful of the pods of this pepper, and the whole are then mashed together, and mixed with as much liquid as may be thought necessary. Besides the three species noticed as being the kinds most generally cultivated, there are many other species and varieties occasionally grown for the sake of their pods, all of which yield a warm acrid oil, which acts powerfully on the stomach, and is thought to correct flatulency, and assist digestion. [W. B. B.]

CAPSOMANIA. An unnatural developement of pistils, which may consist either of an excessive multiplication of such a derangement as impedes their functions. In the first case the unusual demands for nutritive matter cannot be met, and the fruit becomes small and abortive; in the latter, as in green-centred roses, bladder plums, &c., the ovules being imperfect do not come to perfection. [M. J. B.]

CAPSULE. Any dry dehiscent seed-vessel. A spurious capsule is any dry seed-vessel that is not dehiscent. Also employed among fungals, to denote certain kinds of perithecia, or receptacles.

CAPUCHON. (Fr.) *Arisarum vulgare*.

CAPUCINE. (Fr.) *Tropæolum*.

CAPUT. The peridium of certain fungals. —, RADICIS. The crown of a root. The very short stem, or rather bud, which terminates the roots of herbaceous plants.

CAQUILLIER. (Fr.) *Cakile*.

CARABIN. (Fr.) *Fagopyrum esculentum*.

CARACHICHU. A Brazilian name for *Solanum nigrum*.

CARAGANA. (Fr.) *Caragana arborescens*. —, ARGENTÉ. *Halimodendron argenteum*. —, DE LA CHINE. *Caragana Chamlagu*. —, DE SIBÉRIE. *Caragana frutescens*.

CARAGANA. The Siberian Pea Tree. Trees or shrubs belonging to the leguminous order, natives of Siberia and the East, with pinnate leaves of which the midrib terminates in a bristle or spine instead of a leaflet, and axillary flowers, either solitary or crowded, but always single on thin stalks, of a pale *yellow* colour, with the exception of one species, *C. jubata*, in which they are white tinged with red. They are all ornamental or curious. Some of them being natives of Siberia, vegetate like most other Siberian plants, early in the spring, and their delicate pinnate foliage, of a yellowish green, independently altogether of their flowers, makes a fine appearance about the middle of April, or, in mild seasons, as early as the middle of March. The flowers, which are of a bright yellow, appear about the end of April, in the earliest Siberian species, and those which flower latest, are also latest in coming into leaf. Thus in a group consisting of the different species of this genus, in the climate of London, some plants may be seen, in the month of May, covered with leaves and flowers, and others in which the buds have just begun to expand. The yellow colour prevails in every part of the plants of this genus, even to the roots; and were it not that this colour is so abundant in common productions of the vegetable kingdom, there can be no doubt that the Caraganas would be used to afford a yellow dye. *C. arborescens* is a small tree with hard wood and a tough bark, which may be used as a substitute for ropes or cords, as the twigs are for withs. The seeds are good food for poultry, and the leaves are said to contain a blue colouring matter like indigo. *C. spinosa* is a thorny shrub plentiful in China about Pekin, where branches of it are stuck in clay upon the tops of the walls, in order that its spines may prevent people from getting over them. For other species see *London's Arboretum Britannicum*. [C. A. J.]

CARAGEEN or CARRAGEEN. A name given in Ireland to *Chondrus crispus*, and some other allied *Algæ* when dried and bleached. Vast quantities are collected for sale and supply a useful article for feeding cattle or making jelly for invalids. Its unequivocal sea taste and odour are against its being a perfect substitute for isinglass. There is no doubt, however, that in the sick chamber it is a far better substitute than gelatine, so that has very small, if any, nutritive qualities, a fact perhaps not sufficiently known. [M. J. B.]

CARAIPA. A genus of *Ternstræmiaceæ*, distinguished among the group having the petals contorted, and the capsule septicidally dehiscent, by its leaves being alternate, its stamens usually free, with the anthers glanduliferous at the apex, and fixed near the base, and by its having two or three pendulous ovules in each of the three cells of its ovary. The species, about eight in number, grow in Tropical America, and are trees bearing white sweet-scented flowers. The celebrated Balsam of Tamacoari is obtained from *C. fasciculata*, or a closely allied species. This substance, which is of the colour of old port wine, and the consistency of olive oil, is, according to Mr. Spruce (*Journ. Lin. Soc.* v. 63) of great use in the cure of the itch, a single application curing the most inveterate cases in twenty-four hours. [T. M.]

CARAJURA. A red colouring matter obtained from *Bignonia Chica*.

CARALLINE. (Fr.) *Ranunculus glacialis*.

CARALLUMA. A genus of *Asclepiadaceæ*, containing a few species of fleshy leafless herbaceous plants, natives of India and Arabia. The stems are sparingly branched, erect and four-sided, with teeth at the angles; towards the summit the stem becomes rounded, and from the teeth rise the peduncles bearing at their summits one or more drooping flowers. The calyx is five-parted. The rotate corolla is deeply five-cleft. The gynostegium is slightly exserted, and the bi- or trifid leaves of the staminal corona alternate with the stamens. The roundish pollen masses are capped by a pellucid membrane. The follicles are long and slender, with comose seeds. [W. C.]

CARAMBOLA TREE. *Averrhoa Carambola*.

CARANA PALM. A South American name for *Mauritia Carana*.

CARANA RESIN. A gum resin produced by *Bursera acuminata*, or, according to others, by *Icica Carana* or *Cedrela longifolia*.

CARAPA. A small genus of trees with abruptly-pinnate leaves, belonging to the order of melinds (*Meliaceæ*), and native of Tropical America, the West Indies, and Guinea. Their flowers have a calyx of four or sometimes five distinct sepals, and a corolla of the same number of oblong egg-shaped spreading petals; their stamens are united into a tube, the apex of which is divided into eight or ten rounded teeth, bearing the anthers on the inside, between the teeth; and the ovary is four or five-celled, each cell containing four ovules in pairs. The fruit is large and contains numerous oily seeds, and eventually splits into five pieces. *C. guianensis* is a large tree, sixty or eighty feet high, growing plentifully in the forests of Guiana where it is called Carapa and Androba. Its leaves are composed of from eight to ten pairs of elliptical lance-shaped leathery shining leaflets; and its fruit is nearly round, and about four inches in diameter. The bark of this tree possesses febrifugal properties, and is also used for tanning. Its timber, called Crab-wood, is obtainable in sticks, fifty feet long by fifteen inches square, and is used in Demerara for making articles of furniture, for shingles, and for the masts and spars of vessels; it is light, having a specific gravity of 0·603, and takes a good polish. By pressure the seeds yield a liquid oil, called Carap oil or Crab oil, suitable for burning in lamps, and which the natives use for anointing their hair; but in this country it hardens into a solid fat. *C. guinensis* is a native of Senegal, and scarcely differs from the last. Its seeds yield Tallicoonah or Coondi oil, which, besides being used for the same purposes as Crab oil, is employed as a purgative and anthelmintic. [A. S.]

CARAPIXO DA CALCADA. A Brazilian name for some species of *Triumfetta*.

CARATOE. A West Indian name for *Agave americana*.

CARAVELLA. An Indian name for the small black aromatic stimulant seeds of *Cleome pentaphylla*.

CARAVERU. A red pigment, so called by the Indians of Guiana, obtained from *Bignonia Chica*.

CARAWAY. *Carum Carui*, which yields the well-known carminative fruits called Caraway seeds.

CARBERRY. A local name for the Gooseberry, *Ribes Grossularia*.

CARCERULE. An indehiscent many-celled superior fruit, such as that of the linden. Also employed among fungals to denote their spore-case.

CARCINODES. A term applied to what is commonly called Canker in trees, which may in general be characterised as a slow decay inducing deformity. The appearances are very different in different plants, and the causes different. The same plant, as the apple, may even exhibit three or four different kinds of Canker. One form arises from the attack of the woolly aphis; a second from the developement of bundles of adventitious roots, whose tips decay and harbour moisture, and contaminate the subjacent tissues; a third exhibits itself without any apparent cause in the form of broad dark, or even black, patches, spreading in every direction; while a fourth shows pale depressed streaks which soon become confluent, and eventually kill, first the bark, and then, as a necessary consequence, the underlying wood. The only remedy is to cut out completely the affected parts, and that is not always efficacious. The canker of the plum and apricot is brought on by gumming. In many cases Canker arises doubtless from the roots penetrating into some uugenial soil, which vitiates the juices and induces death to the weaker cells, from which it spreads to surrounding tissue. The rugged appearance is generally due to a struggle between the vital powers of the plant and the diseased action. [M. J. B]

CARCINOMA. A disease in trees when the bark separates, an acrid sap exuding and ulcerating the surrounding parts.

CARCITHIUM. The mycelium of certain fungals.

CARCYTES. The same as Mycelium.

CARDAMINE. An extensive genus of herbaceous cruciferous plants, distinguished by the nerveless valves of the flat narrow pod, which, when the seeds are ripe, curl up with an elastic spring from the base upwards, thus scattering the seed. The Cuckoo-flower or Lady's-smock (*C. pratensis*) is a common and very pretty meadow plant, with large lilac flowers. 'They come with the cuckoo,' says Sir J. E. Smith, whence one of their English as well as Latin names (*Flos Cuculi*); and they cover the meadows as with linen bleaching, which is supposed to be the origin of the other. They are associated with pleasant ideas of spring, and join with the white saxifrage, the cowslip, primrose and harebell, to compose many a rustic nosegay. A double variety is sometimes found wild, which is remarkably proliferous, the leaflets producing new plants where they come in contact with the ground, and the flowers, as they wither, sending up a stalked flower-bud from their centres. This species is a native of the whole of Europe, Northern Asia, and Arctic America. The flowers and leaves are agreeably pungent, and may be eaten with other herbs in a salad.

C. hirsuta is a common weed everywhere, varying in size according to soil and situation, from six to eighteen inches in height. In dry localities it ripens its seeds in March and April, and withers away; but in damper places continues in flower all the summer. The leaves and flowers of this species also form an agreeable salad. This species, and it is said several others, produce young plants from the leaves. All that is necessary is to place them on a moist grassy or mossy surface. Two other British species are less common. The foreign kinds are less ornamental as garden plants than the double variety of *C. pratensis*. French, *Cresson*; German, *Gauchblume*. [C. A. J.]

CARDAMOM. The name applied to the aromatic tonic seeds of various zingiberaceous plants, as *Elettaria Cardamomum*, and *Amomum Cardamomum*, which, besides their medicinal use, form an ingredient in curries, sauces, &c. —, BASTARD. *Alpinia Cardamomum*.

CARDAMOMUM. The plants formerly so called are now included in AMOMUM and ELETTARIA; which see. [M. T. M.]

CARDE. (Fr.) *Cynara Cardunculus*.

CARD'ERE. (Fr.) *Dipsacus fullonum*.

CARDIANDRA. A genus of Hydrangeaceæ, containing an under shrub from Japan. It has alternate stalked leaves which are oblong-acute, serrated, and without stipules; and corymbose flowers, those at the margins of the corymb barren and radiant, with a large three-partite petaloid calyx. The fertile flowers have the calyx-tube adhering to the ovary, the limb five-toothed; petals five; stamens numerous, the anthers heart-shaped, from which the genus takes its name; styles three; capsule imperfectly three-celled, opening between the styles. [J. T. S.]

CARDIAQUE. (Fr.) *Leonurus Cardiaca*.

CARDINAL-FLOWER. *Lobelia cardinalis*; also *Cleome cardinalis*.

CARDIOCHLÆNA. A name proposed for a group of large-growing aspidium-like ferns, now referred to *Sagenia*. [T. M.]

CARDIOMANES. An unnecessary name under which it has been proposed to separate *Trichomanes reniforme* from the rest of the genus. [T. M.]

CARDIONEMA. A genus of *Illecebraceæ* containing a small perennial herb from Mexico, with numerous stems, opposite crowded linear leaves, and small sessile axillary greenish-white flowers, the calyx of which is five-parted, surrounded by an involucre of bracts, five of which are larger than the rest, serrulate, terminating in conical points; the petals absent; the stamens five, two sterile, the anthers subrotund; the ovary one-celled with a single ovule, and two revolute styles; the fruit an oblong-ovate utricle. [J. T. S.]

CARDIOSPERMUM. A genus of the soap-wort family (*Sapindaceæ*), composed of a number of scandent or climbing shrubs, or herbs having tendrils like the vine. The leaves are twice ternate or very compound, and the leaflets vary much in form; and the flowers, generally small, white or green, and disposed in short axillary racemes, which are furnished below the flowers with two tendrils. The fruit is a three-celled bladdery capsule, with few round seeds. The name of the genus is derived from the Greek, and signifies heart seed, in allusion to the prominent white heart-shaped scars on the seed, which indicate its point of attachment. The common Heartseed (*C. Halicacabum*), sometimes called also Winter cherry, or Heart-pea, is a widely distributed plant, found in all tropical countries. Its leaves are twice ternate, the leaflets lanceolate and coarsely toothed. In the Moluccas they are cooked and eaten as a vegetable, and on the Malabar coast are used with castor-oil, and taken internally for lumbago, &c. The root is laxative, diuretic, and demulcent. It is mucilaginous, but has a slightly nauseous taste, and is used in rheumatism. There are upwards of a dozen species known, the greater portion of them natives of South America, but there is no tropical country in which some of the species are not found. [A. A. B.]

CARD-LEAF TREE. A West Indian name for *Clusia*.

CARDON or CARDONETTE. (Fr.) *Cynara Carduncuculus*.

CARDOON. *Cynara Cardunculus*.

CARDOPATIUM. A genus of perennial thistle-like plants of the composite family, natives of the Mediterranean region, and also very common in Algeria. They vary in height from six inches to one and a half foot. The leaves are pinnatifid with much cut and spinous segments, and have considerable resemblance to those of the common wayside thistle (*Carduus*). The flower-heads are small, and disposed in dense corymbs at the ends of the branches; the outer scales of the involucre are pinnatifid and spinous, the inner entire and pointed; the florets are of a fine blue colour, all of them tubular, with a five-parted limb, and containing both stamens and pistil. The achenes are covered with villous hairs. According to Gibourt, the *C. corymbosum* is the true black Chamæleon of the Ancients; its roots contain an acrid caustic juice, and resemble those of the white chamæleon (*Carlina gummifera*), but differ in their caustic properties. [A. A. B.]

CARDO SANTO. A Brazilian name for *Argemone mexicana*.

CARDUNCELLUS. A genus of the thistle group of the composite family, and closely related to the saffron thistle (*Carthamus tinctorus*), but the achenes, instead of being naked, are crowned with a pappus consisting of numerous bearded hairs of unequal length united at the base into a ring. The stamens also have a tuft of hair on the middle of the filament. There are about nine known species distributed over the Mediterranean region. Some are stemless herbs, with toothed or pinnatifid spiny-pointed leaves lying close to the ground, and sitting in their midst is a large thistle-like flower-head, one to two inches across, containing numerous tubular florets of a blue colour, surrounded by an involucre of many scales, the outer row of which are often leafy, and have spinous teeth. Others have elongated simple or branched stems, one to two feet high, each branch terminating in a flower head. Some of the species are cultivated in botanic gardens. [A. A. B.]

CARDUUS. A genus of compound or composite flowers, distinguished among the thistle-like plants by having the perfectly smooth fruit crowned by a stalkless tuft of simple deciduous hair. *C. nutans*, a common English species, is distinguished by having the upper part of its stalk almost bare of leaves, and by its large solitary drooping rich purple flowers, which have a strong odour, thought by some to resemble that of the substance from which it derives its name, Musk-Thistle. This is sometimes called, but incorrectly, the Scottish Thistle (see ONOPORDUM). The Holy Thistle (*C. Marianus*) is well marked by the white veins on its large shiny leaves, fabled to have been produced by a portion of the milk of the Virgin Mary having fallen on them. The other British species are uninteresting weeds. Of the hundred species which the genus comprises, some are cultivated, and are considered ornamental plants. Care, however, should be taken how they are introduced into small gardens, many of the perennial species being exceedingly difficult to eradicate when they have once taken possession of the soil, and all having great facilities of dissemination by means of their downy seeds. The seeds of the thistle tribe are the favourite food of many of the hard-billed small birds, especially the goldfinch, which derives its name (*Carduelis elegans*) from the plant. The common statement that this bird lines its nest with thistle-down is scarcely accurate; the substance being, in most cases, the down of colt's foot (*Tussilago*), or the cotton from the willow, both of which are procurable at the building season, whereas thistle-down is at that time immature.

C. lanceolatus is the emblem of Scotland; the same plant, commonly called Spear Thistle, also forms the badge of the clan Stewart. [C. A. J.]

CAREILLADE. (Fr.) *Hyoscyamus albus*.

CAREYA. A genus of the myrtle family, and belonging to that section called *Barringtonieæ*, a group which differs from the true myrtles in having alternate leaves without transparent dots. The plants of this genus are for the most part trees, and are found in India, one species also occurring in North Australia. The leaves are stalked, serrate, and obovate. The flowers are large, red or greenish yellow, sessile, and forming a short head or spike, or stalked and somewhat corymbose; the calyx four-lobed; the petals, four; the stamens very numerous, their filaments united by their base into a ring; they generally fall in one piece when the flower withers, and have the appearance of a painter's brush. The fruit is a berry, crowned with the remaining calyx-lobes, and in *C. sphærica* is of the size and form of an orange, yellowish green in colour, and contains few seeds, embedded in pulp. This species is a native of the Malayan peninsula, where it attains a large size. The bark is ash-like, fibrous, and fit for cordage. The wood of *C. arborea* is used for various purposes, as making boxes, hoops, &c. It is, however, not a valuable timber, as it is liable to split when exposed to the sun, and is not impervious to wet; formerly it was employed for making the drums of the Sepoy corps, being flexible; it takes a good polish, and the colour resembles that of mahogany. The bark is made into a rough cordage; and prepared in a peculiar way, is said to be used in some parts of India as a slow match for firelocks. The fleshy calyx leaves are said to be used for curing colds in Scind. The genus is named in honour of Dr. W. Carey, an Indian botanist, who edited one of the editions of *Roxburgh's Flora Indica*. [A. A. B.]

CARGILLIA. A genus of the ebony family (*Ebenaceæ*), peculiar to Eastern tropical Australia. The two known species are trees, with alternate leathery oblong obtuse entire leaves. The flowers are small and white, collected in dense clusters in the axils of the leaves, the males and females on the same tree, the former containing eight stamens, surrounded by four petals and a four-parted calyx, and the latter like the males, but having only a few abortive stamens, and a four-celled ovary, which, when ripe, is a roundish drupe containing few seeds. The genus differs from its allies in the quaternary arrangement of parts of the flower. The Black Plum of Illawarra (*C. australis*) is a slender tree, from twenty to forty feet in height, and ten to fourteen inches in diameter, the wood of which is close-grained and useful; the fruits are the size of a large plum, and of a dark purple colour. The Grey Plum (*C. arborea*) grows to a height of fifty or a hundred feet, with a diameter of twelve to fourteen inches; its wood is tough and close-grained, but of no beauty. The fruits, which are produced in great abundance, are eaten by the aborigines. [A. A. B.]

CARIACA. A small variety of maize, much esteemed in British Guiana.

CARIBÆAN BARK. The bark of *Exostema floribundum*.

CARICA. This genus is the type of the order of papayads (*Papayaceæ*). It contains about ten species, natives of tropical America, forming small trees generally without branches, and having large variously-lobed leaves, resembling those of some kinds of palm; all parts exudine an acrid milky juice when wounded. Their flowers are borne in racemes, proceeding from the bases of the leaf-stalks, the males and females being usually on different trees. The males have a funnel-shaped corolla, into the throat of which the ten stamens are inserted in two rows, one above the other; and the females a corolla of five distinct petals. The fruit is fleshy, and does not split open when ripe.

The most remarkable species is *C. Papaya*, called the Papaw-tree. This is now generally acknowledged to be a native of tropical South America, but it is commonly cultivated in most tropical countries, and was at one time supposed to be indigenous to the East Indies. It is a small tree, seldom exceeding twenty feet in height, with a stem about a foot in diameter, tapering gradually to about four or five inches at the summit, and composed of soft spongy wood, mostly hollow in the centre. The leaves are frequently as much as two feet in diameter, and deeply cut into seven broad lobes terminating in sharp points, and having their margins irregularly waved or gashed; their foot-stalks are about two feet long, and diverge almost horizontally from the stem. The fruit, for which this tree is celebrated, is of a dingy orange-yellow colour, generally of an oblong form, about ten inches long by three or four broad, but sometimes shaped like a melon, with projecting angles; it has a thick fleshy rind, like that of a gourd, and contains numerous small black wrinkled seeds, arranged in five lines along the whole length of the central cavity. Throughout most of the West India islands the juice of this tree, or an infusion of its fruit or leaves, is reputed to possess the remarkable property of causing a separation of the muscular fibre of animal flesh, and thus rendering the toughest meat tender. It is asserted,

Indeed, that merely hanging the meat amongst the leaves of the tree will produce the same effect; but in this case it is probable that the result is rather attributable to the high temperature, than to any specific influence exerted by the tree. It is also said that if old hogs or poultry be fed upon the fruits and leaves, their flesh will not fail to be tender. The ripe fruit is seldom eaten raw, although, with the addition of pepper and sugar, it is said to be agreeable. It is generally made into sauce, or preserved in sugar, in the West Indies, and the unripe fruit is either pickled, or boiled and eaten like turnips. Its juice is used by the ladies as a cosmetic, to remove freckles; it is also a powerful vermifuge. And, according to the analysis of Vauquelin, it contains *fibrine*, a substance at one time supposed to be confined to the animal kingdom, but now known to exist in several vegetables. The leaves are employed as a substitute for soap. *C. spinosa* is a branching tree, about twenty feet high, with a spiny stem and branches; native of Guiana and Brazil, where it is called Chamburu. Its leaves are deeply cut into seven lobes, like those of *C. Papaya*, but the lobes are quite entire. The juice of this tree is of an exceedingly acrid nature, causing blisters and itching

Carica Papaya.

if applied to the skin. The fruits are insipid and are eaten only by a species of ant, neither birds nor other animals touching them; and the flowers have a disgustingly fetid odour. The fruits of some other species, such as *C. citriformis* and *C. pyriformis*, are eatable, but insipid. [A. S.]

CARIE. (Fr.) *Uredo Caries.*

CARIES. This word is used in vegetable pathology to denote decay of the walls of the cells and vessels, whether attended by a greater or less degree of moisture. Life is necessarily limited in all organic structure, and therefore the time must come when the oldest parts of trees must submit to decomposition; and as soon as this commences, it acts as a putrefactive ferment, and involves neighbouring sound tissues in plants of shorter duration, decay takes place from various causes, sometimes from mere constitutional peculiarities, sometimes from a cessation of vital functions, sometimes again from atmospheric or other outward agents, and sometimes from parasitic fungi. The rapidity with which the mischief spreads when once set up is exemplified by the potato murrain and the black spot of orchids; a few days in either case being sometimes sufficient to induce complete decomposition. The decay of fruit, though not due, as is sometimes supposed, to minute fungi, is certainly promoted by their presence, the mere contact of the tissues and parasite being sufficient to set up putrefactive action. [M. J. B.]

CARILLON. (Fr.) *Campanula Medium.*

CARIM-GOLA. An Indian name for the root of *Monochoria vaginalis.*

CARINA (adj. CARINATE). A keel. The two anterior petals of a papilionaceous flower, the three anterior in a milkwort, or any such. Also the thin sharp back of certain parts, as that of a glume of *Phalaris*, &c.

CARINATO-PLICATE. So plaited that each fold is like a keel, as in the peristome of some urn-mosses.

CARIOPSIS. A one-celled one-seeded superior fruit, whose pericarp is membranous and united to the seed, as in wheat, maize, and other kinds of corn.

CARISSA. A genus of apocynaceous plants consisting of shrubs with milky juice, and having axillary flower-stalks, some of which bear no flowers, but are reduced to the condition of spines. The corolla is funnel or salver-shaped, sometimes provided with hairs at its throat. Fruit a two-celled berry with few seeds. The species are natives of Asia and tropical Australia.

C. Carandas, a common Indian shrub, is used for fence-making, for which its thorny character renders it well adapted. Its fruits are also eaten by the natives as a conserve, &c. Some of the species have medicinal properties, being as bitter as gentian. The bark of *C. Xylopicron*, a native of Mauritius and Bourbon, is used by the Creoles in diseases of the urinary organs, while its wood, there called Bois amère, has a like reputation. Small cups are made of it in which water or wine is allowed to stand till it acquires the flavour of the wood, as in the bitter cups now so frequently used in this country. [M. T. M.]

CARLEMANNIA. A name applied by Bentham to a genus of cinchonaceous plants, in honour of Dr. Charles Leman, whose herbarium is now in the possession of the University of Cambridge. The plant is a native of Khasia and the Himalaya, and has leaves with saw-toothed margins, and minute stipules, while the four-parted flower has only two stamens, a circumstance which distinguishes the genus from all its allies. [M. T. M.]

CARLINA. A genus of prickly herbaceous plants distinguished among the thistle-like group of compound flowers by having the inner leaves of the calyx or involucre coloured, and of the texture usually called everlasting (scariose). The species, which closely resemble each other in habit are natives of most parts of Europe, growing on dry commons and sea cliffs. *C. vulgaris*, the only English species, is a common weed about a foot high, on dry heaths and soil which has been long undisturbed, less conspicuous from its dull purple disk than from the radiating straw-coloured involucre, which expands horizontally in dry weather, and becomes erect during rain. This portion of the flower is very durable, retaining its form long after the spiny leaves have been reduced to a skeleton. It preserves its hygrometric properties for a long period, and is sometimes gathered and suspended in the house to serve as a natural weather-gage. Olivier de Serres says that this plant received its name after the famous Charlemagne, whose army was cured of the plague by using it medicinally. Linnæus ascribes the name to the Emperor Charles V., whose army was relieved in Barbary from the same disease by a similar remedy. Several of the species, especially *C. gummifera*, contain an acrid resin in which the medicinal virtue of the plant is supposed to reside. The tender roots of some species are said to be eatable, and of others the flowers furnish a substitute for artichokes. French, *Carline*; German, *Eberwurz*. [C. A. J.]

CARLINE THISTLE. The common name for *Carlina*.

CARLUDOVICA. A small genus of screw-pines (*Pandanaceæ*) confined to tropical South America. Some of them have long climbing stems, sending out aerial roots which fasten upon the trunks of trees, or hang down like ropes, whilst others have no stems, and form dense thickets. They have large stiff plaited leaves, deeply cut into from two to five divisions. Their flowers are of separate sexes, and disposed in squares arranged very close together in a spiral manner, and forming cylindrical spikes, which, while young, are enclosed within four leafy bracts (spathes). Each square consists of a female flower surrounded by four males, giving the spikes a tessellated appearance. The males have a calyx cut into numerous lobes, and an indefinite number of stamens; and the females a calyx of four sepals, four barren stamens, and a square-sided ovary surmounted by a cross-like stigma, eventually producing a square-sided berry with numerous seeds.

C. palmata is a stemless species, common in shady places all over Panama and along the coasts of New Grenada and Ecuador. Its leaves are shaped and plaited like a fan, and are borne on three-cornered stalks from six to fourteen feet high; they are about four feet in diameter and deeply cut into four or five divisions, each of which is again cut. The Panama hats commonly worn in America, and now becoming common in this country, are manufactured from these leaves. Those of the best quality are plaited from a single leaf without any joinings, and, as the process sometimes occupies two or three months, their price is very high, a single hat often costing 150 dollars, and clear-cases of the same material 6d. each. The leaves are cut whilst young, and the stiff parallel veins removed, after which they are slit into shreds, but not separated at the stalk end, and immersed in boiling water for a short time, and then bleached in the sun. [A. S.]

CARMEL. The Arab name for *Zygophyllum simplex*.

CARMICHAELIA. A genus of New Zealand shrubs belonging to the pea-flowered group of the leguminous family. The branches are sometimes round, but more commonly flattened and tape-like. The plants when in a seedling condition are furnished with unequally-pinnate leaves, but after they are a few weeks old no more leaves are produced. The flowers are small, very numerous, pink or lilac in colour, and disposed in short racemes. The pods are roundish, slightly turgid, about half an inch long, and contain two or four seeds. They are remarkable in the family because of their having a thin partition (replum) between the valves of the pod, which remains after the valves have fallen; to this partition the seeds are attached. The genus is named in honour of Captain Carmichael, who published an account of the plants of the island Tristan d'Acunha. [A. A. B.]

CARNATION. A garden variety of *Dianthus Caryophyllus*. —, SPANISH. *Poinciana pulcherrima*.

CARNATION TREE. A garden name for *Kleinia neriifolia*.

CARNAÜBA. A Brazilian palm, *Copernicia cerifera*, the leaves of which yield a wax, which is used for making candles.

CARNEUS. Flesh-colour; the pale red of roses.

CARNILLET. (Fr.) *Silene inflata*.

CARO. The fleshy part of fruit. The flesh or tissue of which fungals consist.

CAROB TREE. The Algaroba Bean, *Ceratonia Siliqua*.

CAROLINEA. The designation given to a genus of *Bombaceæ* by the younger Linnæus in honour of the Princess Sophia Caroline of Baden, a name which he says will always be cherished by botanists. The plants are familiar in our hothouses under this name; but the inexorable law of priority has led botanists generally to adopt that of PACHIRA: which see. [T. M.]

CAROUBE A' SILIQUES or CAROUGE. (Fr.) *Ceratonia Siliqua*. —, A' MIEL. *Gleditschia triacanthos*.

CARPADELIUM. An inferior indehiscent two or more celled fruit with solitary seeds, and carpels which, when ripe, separate from a common axis, as in umbellifers.

CARPANTHUS. A synonyme of *Azolla*.

CARPEL (adj. CARPELLARIS). One of the rolled-up leaves of which the pistil is composed, whether they are combined or distinct.

CARPENTERIA. The name of a Californian shrub belonging to the order *Philadelphaceæ*, and having cymes of large white flowers, with a five or six-parted calyx; five to six petals; numerous thread-shaped stamens; and five to seven styles consolidated into one, and terminated by five to seven linear stigmas; capsules attached by their base to the calyx, five to seven celled, many-seeded. [M. T. M.]

CARPESIUM. A genus of the composite family, remarkable for its distribution only, two of the species being found in South Europe and the Caucasus, and appearing again in the Himalayan Mountains, where the greater portion of the species are found. They are smooth or pubescent erect branching herbs, with ovate or lanceolate toothed leaves. In one section of the genus the flower-heads are small and either solitary or two or three together, in the axils of the leaves, while in the other section they are much larger, single at the ends of the branches, and the outer scales of the involucre are leaf-like and reflexed. The florets in all are dull yellow, tubular, the central ones having both stamens and pistil, and those of the circumference, pistil only. The achenes are beaked, have slender furrows, and are destitute of pappus. [A. A. B.]

CARPET-WEED. A common name for *Mollugo*.

CARPOCERAS. The name of a group of *Pedaliaceæ*, now included in *Rogeria*. The same title, given by De Candolle to a section of *Thlaspi*, has been adopted as a generic name by Boissier, the section being raised to the position of a genus, and distinguished from the true *Thlaspi* by the absence of a wing around the pod. [W. C.]

CARPOCHÆTE. A genus of the composite family, comprising a few slender under shrubs, all of them natives of New Mexico. Their leaves are opposite, sessile, entire, very narrow, and furnished with glandular dots. The flower-heads purple or white, in loose terminal corymbs; each head with from six to eight florets, all fertile and about an inch in length. The pappus is composed of five to fourteen linear-lanceolate toothed scales, and the achenes have ten slender furrows. Three species are known. [A. A. B.]

CARPOCLONIUM. A free case or receptacle of spores found in certain algals.

CARPODETES. A small genus of *Amaryllidaceæ*, allied to *Coburgia*. It has oblong bulbs, ensiform leaves ten inches long and half an inch wide, and a short flower-scape with a large purple spathe, and bearing from one to three flowers; these are purplish-yellow, drooping, with a slender cylindrical curved tube, a limb of six short regular segments, and a short cup-shaped coronet bearing the six stamens on its margin. The species, *C. recurvata*, is a native of Peru. [T. M.]

CARPODETUS. A genus of New Zealand shrubs belonging to the order *Escalloniaceæ*. *C. serratus* has much the appearance of a *Rhamnus*, but in its fruit is more closely allied to *Escallonia*. The name of the genus is derived from two Greek words signifying fruit-bound, in allusion to the fruit being girt by the calyx. The principal characteristics are the presence of five petals, touching only by their margins, not overlapping as in allied genera; a viscid stigma; and a leathery succulent fruit, tightly girt with the margin of the calyx, and having four or five compartments containing several ovules. [M. T. M.]

CARPODINUS. Climbing shrubs with tendrils, natives of Sierra Leone, and belonging to the order *Apocynaceæ*. They have a funnel-shaped downy corolla, with oblique lance-shaped reflexed segments; five sagittate anthers; a globular stigma; and an orange-shaped fruit containing several seeds embedded in pulp. [M. T. M.]

CARPODONTOS. A genus of the St. John's-wort family, now generally referred to EUCRYPHIA; which see. [A. A. B.]

CARPOLOBIA. A genus of the milk-wort family (*Polygalaceæ*). The two known species are natives of West Tropical Africa. They are shrubs or small trees, with alternate ovate acuminate leaves, and short axillary racemes of yellow or white flowers. The calyx is five-leaved; the petals five, one of them keeled and crested at the apex; the stamens eight in number, their filaments united at the base, five of them bearing anthers, the others sterile. The ovary is two-celled with one ovule in each cell, and becomes when ripe a small fleshy somewhat three-angled fruit, containing one seed, which is covered with long silky hairs. [A. A. B.]

CARPOLOGY. That part of Botany which treats of the structure of fruits and seeds.

CARPOLYZA. A genus of South African amaryllids, the only species of which, *C. spiralis*, is a neat little plant, having ovate bulbs, short linear filiform leaves, which are twisted or recurved; a scape two to five inches high, singularly twisted in a spiral manner in the lower part, and bearing at the top an umbel of from one to four flowers. These flowers are white, the sepals reddish outside tipped with green; they have a short funnel-shaped tube, and a regular somewhat spreading limb the filaments are adnate to the tube, the three alternate ones shorter, and all bearing oblong anthers affixed by the base; the style is thick furrowed, triangular, more slender upwards, terminated by a trifid recurved fimbriated stigma. [T. M.]

CARPOMANIA. This affection, sometimes called Phytolithes, is scarcely a disease, for the grittiness of pears, medlars, quinces, &c. which the term has in view, is a condition which always exists, and the efforts of the gardener to reduce it as much as possible, are rather efforts to create a disease than to cure one. Grittiness depends upon the deposit of layer after layer of new matter within certain cells, till they become hard like stone. Cultivation has a tendency to make the fruit more juicy, but seldom if ever wholly prevents the formation of these stony cells. In the warm climate of Italy quinces are often so full of them as to become uneatable. A variety is said to exist in Chili completely free from grittiness, but this requires confirmation. [M. J. B.]

CARPOMORPHA. Those parts in cryptogamic plants which resemble true fruits without being such receive this name. The spores of lichens.

CARPOPHORUM. The stalk of the pistil above or beyond the stamens.

CARPOPHYLLUM. The same as Carpel.

CARPOPODIUM. A fruit-stalk.

CARPOPTOSIS. After the fruit is well-formed and impregnation has taken place, its progress is often suddenly arrested and after a short time it falls off. This frequently depends upon the fact that more fruit is set than the tree is equal to nourish, and the failure of the crop is in consequence either total or partial. If again the supply of nourishment is too great, from want of root-pruning or from any other cause, the demands of the young shoots are often such that the sap is diverted from the fruit, which consequently perishes. In Italy the rice crops are often somewhat similarly affected. In this case, however, the grain acquires a certain degree of maturity, though not its perfect condition, and is so slightly attached to the mother-plant that the slightest breeze shakes it off. It is not a mere case of over-ripeness, which, as in our own corn crops, may be avoided by early reaping. [M. J. B.]

CARPOSTOMIUM. The opening into the spore-case of algals.

CARRADORIA. A genus of *Globulariaceæ* containing a single species, a native of the Italian mountains. It is a glabrous herbaceous plant, with small scattered leaves. The flowers grow in a terminal head; the calyx is subequal; the upper lip of the corolla is simple and linear, and shorter than the lower lip; there is no nectary; the stigma is simple and the scales and paleæ of the involucre are persistent. In other respects it resembles *Globularia*, from which it has been but recently separated. [W. C.]

CARRAGEEN. *Chondrus crispus*; also written Carageen, under which name its properties are noticed.

CARRIA. The name sometimes given to a beautiful Ceylon tree, of the tea family (*Ternstræmiaceæ*). It attains a height of forty to fifty feet, and has entire sessile leaves, which are smooth, of a leathery texture, and elliptical in form; they vary from three to four inches in length, and one to two and a half in breadth. The fine large blood-coloured flowers proceed from the axils of the upper leaves, and are a good deal like those of some single-flowered camellias. The plant is now generally known as *Gordonia speciosa*. [A. A. B.]

CARRION-FLOWER. A common garden name for *Stapelia*. Also an American name for *Smilax herbacea*.

CARROT. *Daucus Carota*, the garden form of which furnishes the well-known esculent root. —, CANDY or CRETAN. *Athamanta cretensis*. —, DEADLY. A common name for *Thapsia*. —, NATIVE. A name given in Tasmania to the tubers of *Geranium parviflorum*.

CARROT TREE. *Monizia edulis*.

CARTHAGINIAN APPLE. *Punica Granatum*.

CARTHAME MACULE.' (Fr.) *Silybum Marianum*.

CARTHAMUS. A small genus of composites, containing two annual species whose flowers grow in heads at the ends of the branches, and are surrounded by numerous leafy bracts (involucre) in numerous rows, the outermost row being broad and spreading out flat, with their edges spiny, the middle ones more upright, of an oval form, and surmounted by an egg-shaped appendage with spiny edges, and the innermost much narrower, quite upright, with their edges entire, but terminated by a sharp spiny point. Each flower is perfect, and has an orange or yellow corolla longer than the involucre, their lower part being imbedded in a dense mass of fringed scales and hairs, but the chief characteristic consists in the absence of the bristles, technically termed pappus. The Safflower plant, or Bastard Saffron (*C. tinctorius*), the Koosumbha of India and Hoang-tchi of China, is extensively cultivated in India, China, and other parts of Asia, also in Egypt and Southern Europe; but its native country is unknown. It grows about two or three feet high, with a stiff upright whitish stem, branching near the top; and has oval, spiny, sharp-pointed leaves, scattered upon, and their bases half-clasping, the stem. Its fruits are about the size of barleycorns, somewhat four-sided, white and shining, like little shells. Under the name of Safflower, 11,954 cwts. of the flowers of this plant, made up into flat circular cakes about the size of half-crowns, were imported to this country, principally from India, and valued at 105,673*l*. Safflower contains two colouring matters, yellow and red, the latter being that for which it is most valued. It is chiefly used for dyeing silk, affording various shades of pink, rose, crimson and scarlet. Mixed with finely-powdered talc it forms the well-known substance called rouge. Another common use of safflower is for adulterating saffron, a more expensive dye stuff. The seeds yield an oil much used in India for burning and for culinary purposes. [A. S.]

CARTILAGINOUS. Hard and tough, like the skin of an apple-seed, or a piece of parchment.

CARTONEMA. The generic name of one of the spiderworts, characterised by having the filaments of the stamens without any hairs, but somewhat rough; the style or appendage on the seed-vessel thread-like, and bearded at the end. The name is from the Greek, and indicates the bare or shorn stamens. The only known species is *C. spicatum*, a native of New Holland, a plant covered with scattered hairs, the stem slightly branched, the leaves long and narrow, the flowers blue, arranged in spikes. [G. D.]

CARUM. A genus of *Apiaceæ* or *Umbelliferæ*, of some importance as producing the Caraway fruits, or seeds as they are improperly termed. The plants have finely cut leaves, and compound umbels, which in the true Caraway have but few bracts surrounding them, or sometimes none at all; petals broad, with a point bent inwards; fruit oval, curved, with five ribs, and one or more channels for volatile oil under each furrow. The Caraway, *C. Carui*, is cultivated in Essex and elsewhere, and may occasionally be found in a half wild condition. The fruits are used for flavouring as they contain an aromatic volatile oil. [M. T. M.]

CARUNCULA (adj. **CARUNCULATE, CARUNCULAR**). A wart or protuberance round or near the hilum of a seed.

CARUNCULARIA. A generic name given to a few plants from the Cape of Good Hope, separated by Haworth from *Stapelia*, but with characteristics scarcely sufficient to establish a new genus. It is consequently used to characterise that section of the genus *Stapelia* which is distinguished by having the staminal corona consisting of five spreading emarginate leaflets, with five bifid fleshy clavate appendages in the interior. [W. C.]

CARUTO. The lana dye, a permanent bluish-black obtained in British Guiana from the fruits of *Genipa americana*.

CARVA. *Billbergia variegata*.

CARVI. (Fr.) *Carum Carui*.

CARYA. The generic name of the Hickory trees of America, a genus belonging to the order *Juglandaceæ*, and at one time included with the walnuts under the name of *Juglans*, from which it is distinguished by having the male catkins produced in threes from a single stalk, each flower having a three-parted calyx, and not more than six stamens; and by the female flowers being destitute of a corolla, and having their four-lobed stigmas sessile upon the ovary. The husk of the fruit, also, splits into four equal-sized pieces, instead of irregularly as in *Juglans*. There are about a dozen species, all of them natives of North America, forming large forest trees. Their timber is coarse-grained, of great strength and toughness, and very heavy; but as it does not bear exposure to the weather, and is extremely liable to the attacks of insects, it is not suitable for building or similar purposes. It is, however, much used where toughness and elasticity are required, such as for barrel-hoops, press-screws, axe-handles, handspikes, &c., and common descriptions of furniture are also made of it. The nuts of some species are eatable, and resemble but do not equal our walnuts.

The Shell-bark, Scaly-bark, or Shag-bark Hickory, *C. alba*, is so called in consequence of its rough shaggy bark peeling off in long narrow strips. It is common throughout the Alleghany mountains from Carolina to New Hampshire, forming a tree eighty or ninety feet in height, with a trunk about two feet in diameter. Its leaves are about twenty inches long, and are composed of five or seven oblong sharp-pointed leaflets, which are hairy beneath, and have sharply saw-toothed edges. The fruit is nearly round, and has an excessively thick rind, enclosing a small white hard-shelled nut, slightly flattened upon two sides, and marked by four elevated angular ridges. These nuts stand second in point of flavour among the hickories, and small quantities of them are sometimes sent to this country. The Bitter-nut or Swamp Hickory, *C. amara*, produces small and somewhat egg-shaped fruits, having a thin fleshy rind, which never becomes hard and woody like that of the others; the nut is nearly round, flat-topped, and tipped with a short sharp point; its kernel is extremely bitter, and is not eaten by any kind of animal. The Pecan or Illinois-nut Hickory, *C. olivæformis*, is a common tree on the banks of the Ohio and Mississippi, attaining a height of sixty or seventy feet; having leaves from a foot to eighteen inches in length, composed of six or seven pairs of leaflets with an odd one, each leaflet being about three inches long, egg-shaped and tapering to a point, and having its edge finely serrated. The nuts of this species are enclosed in a thin woody husk, and are of a light-brown colour, shaped like an olive, and indistinctly marked by four slightly raised longitudinal ridges. They are much superior in flavour to those of the rest of the genus, and are occasionally to be met with in English fruit-shops. A very palatable oil is obtained from them by pressure. The Pig or Hog-nut, or Broom Hickory, *C. porcina*, is a noble tree seventy or eighty feet high, with a trunk upwards of a yard in diameter. Its wood is considered superior to that of the other species. The leaflets are seven in number, each about four inches long, lance-shaped, and tapering to a fine point, their edges being very regularly cut like the teeth of a saw. The fruit is pear-shaped, and has a thin husk which splits open only at the top end. The nut has a very thick hard shell, and is without the ridges common in other hickory-nuts; its kernel is small and sweet, and is eaten by pigs, squirrels, and other animals. [A. S.]

CARYOCAR. One of the two genera forming the order of Rhizobols (*Rhizobolaceæ*), and distinguished by its flowers having five petals, and only four styles, and by its leaves being always opposite; the other genus, *Anthodiscus*, having cohering petals, numerous styles, and often alternate leaves. There are about eight species of *Caryocar*, all large hardwooded trees, growing in the tropical regions of South America. The most interesting is *C. nuciferum*, which produces the Souari or Butter-nuts, occasionally met with in English fruit-shops. These nuts are shaped something like a kidney flattened upon two sides, having an exceedingly hard woody shell, of a rich reddish-brown colour, covered all over with round wart-like protuberances, and enclosing a large white kernel, which has a very pleasant nutty taste, and yields a bland oil by pressure. It is a lofty tree, frequently as much as 100 feet in height, inhabiting the forests of British Guiana, particularly the banks of the rivers Essequibo and Berbice, where its timber, which is very durable, is employed for ship-building. Its leaves are

Caryocar tomentosum.

composed of three broadly lance-shaped or elliptical taper-pointed leaflets, each about six inches long. Its flowers are of great size, and both calyx and corolla are of a deep purplish-brown colour. The fruit is nearly spherical, and about the size of a child's head, containing, when perfect, four of the above-mentioned nuts or seeds; but they are more frequently imperfect and contain only two or three.

Another species, *C. butyrosum*, also a native of Guiana, has white flowers, and leaves composed of five oval-pointed leaflets radiating from a central stalk. It is called Pekea by the natives, and its nuts resemble those of the last; its timber, also, is valuable for ship-building, mill-work, &c. [A. S.]

CARYODAPHNE. Under this name are included certain Javanese trees of the laurel family, possessed of scaly leaf-buds, three-nerved leaves, a funnel-shaped six-cleft perianth, and twelve stamens in four rows, the nine outer ones fertile. Of these stamens the three innermost have a stalked gland on each side of their base, and all have anthers opening by two valves, inwardly in those of the first and second row, outwardly in those of the third row. The three innermost sterile stamens are stalked, with a long pointed head. The drupe is one-seeded, adherent to the persistent tube of the perianth. *C. densiflora* has a bitter-tasting bark; its leaves are aromatic, and used in spasms of the bowels, &c. [M. T. M.]

CARYOLOPHA. A section of the genus *Anchusa*, one of the *Boragineæ*, containing *A. sempervirens*, which has a salver-shaped corolla with a very short straight tube, and the ring at the base of the nuts prolonged on the inner side into an appendage, in which it differs from the other sections of the genus. [J. T. S.]

CARYOPHYLLACEÆ (*Sileneæ, Alsineæ, Queriaceæ, Minuartieæ, Mollugineæ, Steudelia, Silenads, Cloveworts*, the *Chickweed* family). A natural order of thalamifloral dicotyledons belonging to Lindley's silenal alliance. Herbs with stems swollen at the joints, entire and opposite leaves, and a definite (cymose) inflorescence; sepals four to five, separate or cohering; petals four to five, with narrow claws, sometimes wanting; stamens usually as many or twice as many as the petals. Ovary often supported on a stalk(gynophore), usually one-celled with a free central placenta; styles two to five, with papillæ on their inner surface. Fruit a capsule, opening by two to five valves, or by teeth at the apex, which are twice as many as the stigmas; seeds usually indefinite; embryo curved round mealy albumen. There are three suborders: 1, *Sileneæ*, the pink tribe, with united sepals opposite the stamens, when the latter are of the same number. 2. *Alsineæ*, the chickweed tribe, with separate sepals, bearing the same relation to the stamens as in *Sileneæ*. 3. *Mollugineæ*, the carpet-weed tribe, in which the petals are wanting, and the stamens are alternate with the sepals when of the same number. Natives principally of temperate and cold regions. They inhabit mountains, bridges, rocks, and waste places. Humboldt says that cloveworts constitute $\frac{1}{35}$ of the flowering plants of France, $\frac{1}{37}$ of those of Germany, $\frac{1}{14}$ of Lapland, and $\frac{1}{35}$ of North America. The order has no very marked properties. Some say that the principle, called *saponine*, which is found in some of the plants, has poisonous qualities. There are some showy flowers in the order, such as pinks and carnations; but the greater number are mere weeds. The clove pink, *Dianthus*

Caryophyllus, is the origin of all the cultivated varieties of carnations, as picotees, bizarres, and flakes. The common chickweed (*Stellaria media*), and spurrey (*Spergula arvensis*) used as fodder for sheep, are other examples. There are about sixty genera and 1,100 species. Illustrative genera: *Dianthus, Saponaria, Silene, Lychnis, Alsine, Arenaria, Stellaria, Cerastium, Mollugo*. [J. H. B.]

CARYOPHYLLACEOUS, CARYOPHYLLATUS. A corolla whose petals have long distinct claws, as in the clove pink.

CARYOPHYLLATA. (Fr.) *Geum urbanum*.

CARYOPHYLLUS. One of the genera of *Myrtaceæ*, characterised by a long cylindrical calyx, whose limb is four-cleft; four petals adherent at their points; stamens numerous in four parcels; berry oblong, one or two-celled, and as many seeded.

The tree producing the well-known spice called Cloves (*C. aromaticus*) is a handsome evergreen, rising to from fifteen to thirty feet, with large elliptic leaves and purplish flowers arranged in corymbs on short-jointed stalks. The Cloves of commerce

Caryophyllus aromaticus.

are the unexpanded flower-buds, and derive their name from the French word *clou*, a nail, in allusion to the shape of the bud with its long calyx tube, and the round knob or head of petals at the top. These buds are collected by hand, or by beating the tree with sticks, when the buds, from the jointed character of their stalks, readily fall, and are received on sheets spread for the purpose. The Cloves are then dried by the sun. For many years the Dutch exercised a strict monopoly in the growth of this spice, by restricting its cultivation to the island of Amboyna, and even there extirpating all but a limited number of the trees; but they are now extensively grown in the West Indies and elsewhere. All parts of the plant are aromatic, from the presence of a volatile oil, but especially the flower-buds, hence its use for culinary purposes. The oil is occasionally used in toothache with the effect of lulling the pain, and as a carminative in medicine. [M. T. M.]

CARYOTA. A genus of very elegant lofty palms (*Palmaceæ*) with graceful twice-pinnate leaves, the leaflets of which differ very much from those of other plants of this order. In general the leaflets of pinnate-leaved palms are long, narrow, and tapering upwards to a point; but those of *Caryota*, on the contrary, are comparatively short, tapering to the base, very broad at their top end, where they are jagged as though gnawed by an animal. Nine species of this genus are known, all of them natives of India and the Indian Islands. They have flowers of separate sexes, borne upon the same spike, or sometimes on distinct spikes. The calyx is of three distinct sepals, and the corolla is three-parted; the male flowers have numerous stamens connected together at the base and forming a cup; and the females a one or two-celled ovary, with as many stigmas, and three barren stamens. The fruits are nearly round, somewhat fleshy, and generally of a purplish colour, containing one or two seeds.

C. urens is a beautiful tree with a trunk about a foot in diameter, growing to the height of fifty or sixty feet, and surmounted by an elegant crown of gracefully curved leaves. These leaves are eighteen or twenty feet long, and ten or twelve broad, and have a very strong central stalk, the base of which widens out so as to form a kind of sheath round the stem, and leaves a circular mark or scar when it falls away; they have, also, a curious black fibrous material at their base. The leaflets are shaped somewhat like a scalene triangle, one side being very sharply and irregularly jagged. The flower spikes are ten or twelve feet long, and issue from the trunk at the base of the leaves, hanging down like the tail of a horse; they are not produced until the tree has arrived at its full period of growth, and the manner in which the numerous spikes succeed each other is rather singular. The first spike issues from the top of the tree, and after it has done flowering another comes out below it, and so on, a flower-spike being produced from the angle of each leaf-stalk, or from the circular scar left by leaves that have fallen away from the trunk, until the process of flowering reaches the ground, when the tree is exhausted and dies. The fruits are reddish berries about the size of nutmegs, and have a thin, yellow, acrid rind. The tree is a native of Ceylon and many parts of India, particularly Malabar, Bengal and Assam; and it supplies the natives of those countries with several important articles. From its flower-spikes a large quantity of the juice called toddy, or palm wine, is obtained, and this, when boiled, yields very good jaggery or palm sugar, and also excellent sugar-candy. The whole of the sugar used in Ceylon is obtained from the present and two other palms (*Cocos nucifera* and *Borassus flabelliformis*, and a particular caste of natives are called *jaggeraros*, on account of their being solely employed in the preparation of this article. Another valuable substance supplied by this tree is

Caryota urens.

sago; it is prepared from the central or pithy part of the trunk, and is considered to be quite as good and nutritious as ordinary sago. When made into bread or gruel it forms a large part of the food of the natives. The fibre obtained from the leaf-stalks, called kittul or kitool fibre, possesses great strength, and is used for making ropes, brushes, brooms, baskets, &c.; and a woolly kind of scurf scraped off the leaf-stalks is used for caulking boats. The outside part of the stem furnishes a small quantity of hard wood. [A. S.]

CASCA D'ANTA. The Brazilian name for the aromatic bark of *Drimys granatensis*.

CASCA DE LARANGEIRA DA TERRA. The Brazilian name for a bark supposed to be that of *Esenbeckia febrifuga*.

CASCA PRECIOSA. The Portuguese name for *Mespilodaphne preciosa*.

CASCARA DE LINGUE. A Mexican tree bark.

CASCARA DE PINGUE. An astringent Mexican drug, supposed to be obtained from a species of *Curcuma*.

CASCARILLA. The aromatic bark of *Croton Eleutheria*.

CASCARILLA. A name applied by Weddell and other botanists to a genus of *Cinchonaceæ*, closely allied to the genus *Cinchona*, but distinguished from it by the fruit splitting into two halves from above downwards, instead of in the reverse manner, as in *Cinchona*, and—which is of more practical importance—by its not containing any of those chemical ingredients which render *Cinchona* so valuable. The shrubs are natives of Peru and Brazil. See also CROTON. [M. T. M.]

CASEARIA. A large genus of *Samydaceæ*, the species of which are found more or less abundant in all tropical countries, but principally in South America. They are small trees or shrubs, with alternate entire or serrated leaves, which in the greater number of the species are furnished with a mixture of round or linear pellucid dots, which can be seen with the aid of a lens, by holding the leaf between the eye and a good light, and serve to distinguish the plants of this genus from those of any other family with which they are likely to be confounded. The flowers are small, white, green, or rose-coloured, generally arranged in little umbels or corymbs, but sometimes acestic. The calyx is of four or five divisions; the petals wanting; the stamens are two, three, or four times as many as the calyx segments, and often the alternate ones are without anthers, and have commonly a tuft of hair in their place. The fruit is a one-celled fleshy capsule, containing few or many seeds.

C. ulmifolia, a native of Brazil, is used in that country as a remedy against snake bites. The Brazilians make a drink from the juice of the leaves, and apply the leaves themselves to the wounds. M. St. Hilaire asserts that this remedy has been employed with success against the bites of the most venomous serpents. *C. resinifera* has the young flowers enveloped in tears of a greenish resin, which, according to Spruce, is much used for killing cats and dogs; while another species, the Pao de rato of the Portuguese, is said to be poisonous to cattle. According to the same authority, *C. javitensis* is a constant constituent of all forests of recent growth, from the Amazon's mouth to the Orinoco; its habit is more or less corymbose, and the smooth glossy leaves in size and form somewhat like those of the Spanish chestnut.

C. esculenta, a native of the Circar Mountains of India, has purgative roots, which are used by the hill people, who also eat the leaves in stews. The bark of *C. astringens* is used in Brazil for poultices in cases of imperfectly healed ulcers, and is said to be wonderfully efficacious as a cleanser and stimulant of the raw flesh. The leaves of *C. Lingua*, a Brazilian species, are used in decoction in cases of fever or internal inflammation, while those of *C. consoria*, an Indian species, are used in medicated baths, and all the parts of the tree have a bitter taste. Nearly 100 species are enumerated. [A. A. B.]

CASHAW. *Prosopis juliflora.*

CASHEW NUT. The seed of *Anacardium occidentale.*

CASIMIROA. A Mexican genus belonging to the *Aurantiaceæ*, among which it is remarkable for its green-coloured flowers, which are borne in racemes; and by its five distinct stamens, whose filaments are dilated at the base. The fruit is of the size of a large apple. *C. edulis* is a tree, native of, and cultivated in, Mexico. Its fruit, when eaten, has an agreeable taste, but induces sleep, and is unwholesome. The seeds are poisonous. The bark of the tree is bitter, and it, as well as the leaves, and also the seeds, when burnt and reduced to powder, are used medicinally in Mexico. (*Seemann.*) [M. T. M.]

CASPARYA. A genus of begoniads, consisting of scandent (climbing) plants growing in Peru. The staminate flowers have four, and the pistillate six, sepals; anthers oblong, obtuse, the filaments very short, not united; style deciduous, tripartite, its branches papillose not tortuous. Seedvessel triangular, with three mucronate horns of a cartilaginous-corky consistence, attenuated at the apex into a short beak; placentas having two lamullæ. There are three known species, viz., *C. Airia, C. columnaris*, and *C. coccinea*. These species were formerly included in *Begonia*. The genus is named after Dr. Caspary, an eminent botanist of Bonn. [J. H. B.]

CASSAREEP. The inspissated juice of the cassava, which is highly antiseptic, and forms the basis of the West Indian pepper-pot.

CASSAVA. The purified fecula of the roots of the mandioc plant, *Janipha Manihot* (also called *Manihot utilissima* and *Jatropha Manihot*, and *J. Loflingii*. The Cassava juice, though at first poisonous, is rendered harmless by inspissation. In this state it is called cassareep, and is mixed with molasses to form an intoxicating liquor; it also forms a delicious sauce.

CASSE DU LEVANT. (Fr.) *Assaia Farnesiana*.

CASSE-LUNETTE. (Fr.) *Centaurea Cyanus*; and also *Euphrasia officinalis*.

CASSE-PIERRE. (Fr.) *Saxifraga granulata*.

CASSEBEERA. A genus of polypodiaceous ferns, belonging to the *Cheilanthea*, and distinguished by having the sori slightly within the margin, though terminal on the veins, and generally combined in pairs or three together on the emarginate lobes, and covered by one indusium. The veins are free but not readily seen. The fronds are coriaceous, three parted, pinnate, or bi-pinnate. There are three or four species, found in Brazil and Buenos Ayres. [T. M.]

CASSELIA. A limited genus of small shrubs or herbs from Brazil, belonging to the order *Verbenaceæ*. They have membranaceous opposite entire or serrated leaves, and small flowers in lax fewflowered axillary racemes. The calyx is tubular; the corolla funnel-shaped, with a short cylindrical tube, and a five-cleft limb; there are four didynamous stamens hidden in the lower part of the tube, and having very short filaments and two-celled anthers, which open longitudinally; and the ovary is two-celled, with a single ovule in each cell, the style equalling in length the shorter stamens. The drupe has two stones, and is covered by the persistent calyx. This genus is separated from *Tamonea* by the fruit, which in the latter has a single four-celled stone. [W. C.]

CASSIA. This genus is of much importance in a medical point of view, from its producing the well-known drug called senna. It is a member of the leguminous family (*Fabaceæ*); and is known by its five unequal sepals, its five petals of a yellow colour, not papilionaceous, and its ten stamens, three of which are long, four short, and three sterile or abortive, the anthers opening by pores at the top. The species are very numerous, and consist of trees, shrubs, or herbs, with compound pinnated leaves.

The leaflets of several species constitute what are known in medicine as senna leaves. These are of various shapes, and derived from various sources. Alexandrian senna consists of the lance-shaped leaflets of *C. acutifolia*, and the obovate ones of *C. obovata*, united with the leaves of other plants, which latter are readily detected, as the true *Cassia* leaflets, whatever their form, are unequal at the base, from the larger size of one side of the base of the leaflet as compared with the other. The pods of the two species of *Cassia* are also mixed with the leaves, and possess similar properties. East Indian or Tinivelly senna is a very fine kind, and consists of the large lance-shaped leaflets of *C. elongata*. Aleppo senna is the produce of *C. obovata*, a native of Northern Africa, but cultivated in the East Indies and elsewhere. There are other kinds of senna native to and grown in India, Northern Africa, the West Indies, &c. &c., but they are of less importance and value than those above mentioned. The leaves of a North American species, *C. marylandica*, possess similar properties. The heavy nauseous taste and smell of senna are due to a volatile oil, while the purgative effects seem to be due to a chemical substance known as cathartin.

The bark and roots of several of the Indian species are used as applications to ulcers and various skin diseases, as well as internally in diabetes and other disorders; they are likewise used for similar purposes in the Mauritius and the West Indies. The seeds of *C. Absus*, a native of Egypt and of India, are bitter, aromatic, and slightly mucilaginous. They are used in Egypt as a remedy for ophthalmia, as are the seeds of *C. auriculata* in India, where also the bark of this shrub is employed by the natives in tanning leather. *C. occidentalis*, a native of both the Indies, is now naturalised in the Mauritius, where the natives use the roasted seeds as a substitute for coffee, and with good effect in certain cases of asthma. It is related that Dr. Livingstone brought the seeds of a plant, which he found cultivated in the nterior of Africa, to the Botanic Garden at the Mauritius, without knowing what the plant was from which they were derived, but stating that the natives prepared and used them as coffee. On investigation the

Cassia lanceolata.

seeds turned out to be those of this species used for a like purpose in the Mauritius. *C. Fistula*, called the Pudding Pipe Tree from its peculiar pods, is a very handsome tree, with the foliage of the ash, and the inflorescence of the laburnum. It is a native of India, but has been introduced into the West Indies, Northern Africa, &c., whence its pods, called cassia pods, are imported. These pods are very unlike those of the other species, being cylindrical, black, woody, one to two feet long, not splitting, but marked by three long furrows, divided in the interior into a number of compartments by means of transverse partitions, which project from the placentæ. Each compartment of the fruit contains a single seed, imbedded in pulp. From this peculiarity of the fruit the plant is occasionally placed in a separate genus *Cathartocarpus*. The pulp surrounding the seeds is used as a mild laxative.

Several kinds of this extensive genus are in cultivation, most of them having handsome foliage and conspicuous yellow flowers. [M. T. M.]

CASSIA BUDS. A commercial name for the flower-buds of *Cinnamomum aromaticum*.

CASSIA, CLOVE. The bark of *Dicypellium caryophyllatum*. —, POETS. *Osyris*. —, PURGING. *Cassia* or *Cathartocarpus Fistula*.

CASSIA PODS. The black cylindrical woody pods of *Cassia* or *Cathartocarpus Fistula*.

CASSIDEOUS. Having the form of a helmet; as the upper sepal in the flower of an aconite.

CASSINE. A genus of South African plants belonging to the spindle-tree family, *Celastraceæ*. They are smooth, erect or climbing shrubs, with four-angled twigs, and opposite leathery entire or toothed leaves. The flowers are small and white, disposed in cymes; the calyx four or five-parted; the petals and stamens of a like number. The fruit is a fleshy drupe containing one or two seeds enclosed in a stony shell (putamen) and destitute of an aril. The Lagelhout or Ladlewood of the Cape, *C. Colpoon*, furnishes a useful and handsome wood for cabinet-work and other fancy purposes; it is hard and tough, and when polished, the veining has an exceedingly beautiful appearance; it grows to a height of ten feet, with a diameter of eight to twelve inches. The Hottentot Cherry, *C. maurocenia*, is a bush of like dimensions. The wood takes a good polish, and is particularly adapted for the manufacture of musical instruments. It is sometimes placed in a separate genus called *Maurocenia*. Seven species of *Cassine* are enumerated. [A. A. B.]

CASSINIA. A genus of the composite family comprising a number of elegant evergreen shrubs, natives of New Holland, Tasmania, New Zealand, and the Auckland Islands. The leaves are small, mostly linear, with the margins rolled backwards. The flower-heads are very numerous and small, white, pink, or yellow in colour, and disposed in terminal corymbs or panicles; the florets all tubular, having both stamens and pistil, or with a few slender female ones near the circumference. The receptacle is furnished with linear scales, like the inner ones of the involucre, and the presence of these scales serves to distinguish the genus (which is named in honour of M. Henri Cassini, an eminent French botanist) from *Ozothamnus*, to which it is nearly allied. *C. aurea*, a species with golden yellow flowers and linear leaves, is in cultivation. One species (*C. aculeata*) is found in Tasmania; and three in New Zealand, one of them (*C. Vauvilliersii*) occurring also in the Auckland Islands. The remainder are chiefly natives of the eastern portion of Australia. More than thirty species are known. [A. A. D.]

CASSIOBERRY BUSH. *Viburnum lævigatum*.

CASSIOPE. A genus of *Ericaceæ*, consisting of small Arctic or Alpine evergreen plants, resembling lycopods or heaths, with solitary flowers nodding on slender erect peduncles of a white or rose colour. The calyx consists of four or five nearly distinct ovate sepals, and is without bracts; the corolla is campanulate and deeply four to five-cleft; and there are eight to ten stamens, the anthers of which are fixed by their apex, and have ovoid cells, each opening by a large terminal pore, and bearing a long recurved awn behind. The ovoid capsule has four to five cells and as many valves, with a four to five-lobed placenta pendulous from the summit of the columella, and contains many smooth wingless seeds. [W. C.]

CASSIPOUREA. A genus belonging to that section of the mangrove family called *Legnotideæ*, containing three species, natives of the West Indies, Central America, Venezuela and Guiana. They are trees with opposite entire or serrated leaves, ovate or elliptical in form, and smooth and leathery in texture. The flowers are small, in axillary clusters, and sessile or shortly-stalked; the calyx four or five-lobed; the petals four or five, clawed and fringed like those of a Pink. The fruit is ovoid, about the size of a pea, somewhat fleshy, and containing few seeds. [A. A. B.]

CASSIS. (Fr.) *Ribes nigrum.*

CASSOLETTE. (Fr.) *Hesperis matronalis.*

CASSOUMBA. A pigment made by the Amboyniana of the burnt capsules of *Sereuba Bolonghas.*

CASSUVIUM. The plants formerly so-called are now considered to belong to ANACARDIUM; which see. [M. T. M.]

CASSYTHA. A curious genus of semi-parasitical leafless thread-like plants, usually considered as a section of the *Lauraceæ*. They grow sometimes in, and receive their entire nourishment from, the soil; but when they come in contact with other plants, they twine round them with their wire-like branches, and, at the place of contact, emit root-like tubercles, by which they derive their future nourishment from the plant to which they are fixed, the roots in the soil dying away. The flowers are small and white, disposed in short spikes which arise from the axils of small scales. The calyx is six-parted. The stamens are petal-like, twelve in number, arranged in four rows; the two external rows perfect, the anthers opening inwards with two recurved lids, the next row smaller and having a pair of glands at the base of each stamen, the anthers opening outwards, while the fourth row is scale-like and abortive. The fruit is about the size of a pea, enclosed in a berried calyx, and contains one seed. The plants of this genus are much like dodders in appearance, and are often called Dodder-laurels. They only differ from true laurels in the absence of leaves and the berried calyx. Some of the Australian species are called Scrub-vines; they grow so thickly in some places as to be almost impenetrable. The white drupes of *C. muscœiformis*, a N. Australian species, are eatable. *C. filiformis*, a common Indian species, is said to be reduced to a powder, mixed with sesamum oil, and used as a head-wash for strengthening the hair; it is also used by the Brahmins of S. India for seasoning their butter-milk; and in medicine as a remedy for cleansing in veterate ulcers, for which it is prepared by mixing the powdered plant with ginger and butter. The juice mixed with sugar is considered a specific in inflamed eyes. The species are found, more or less, in all tropical countries. [A. A. B.]

CASSYTHACEÆ. The genus *Cassytha*, consisting of leafless parasitical twiners, resembling the dodders in habit, is so very different in this respect from the trees or shrubs which constitute the *Lauraceæ*, that it has been proposed to establish it as a distinct family under the name of *Cassythaceæ*. The structure of the flower and fruit presents, however, no difference whatever; the number of parts, and the peculiar anthers are precisely the same, and *Cassytha* is more generally retained as an anomalous genus or tribe of *Lauraceæ*. There are five or six species, natives of the tropical regions both of the New and Old World, where their thread-like or wiry stems attach themselves to herbs or shrubs precisely like our dodders, only on a somewhat larger scale.

CASTANEA. The Chestnut. This, the most magnificent tree which reaches perfection in Europe, belongs to the *Corylaceæ*, and is so well known that any statement of its distinctive characters is superfluous. Up to a recent period, it appears to have been an almost generally received opinion that the Chestnut was an indigenous tree in Great Britain. This belief was founded mainly on the supposed fact that Chestnut timber existed in large quantities in old buildings. Evelyn says, 'It hath formerly built a good part of our ancient houses in the city of London, as does yet appear; I had once a very large barn near the city, framed entirely of this timber; and certainly the trees grew not far off, probably in some woods near the town, for in that description of London written by Fitz-Stephen, in the reign of Henry II., he speaks of a very noble and large forest, which grew in the boreal part of it,' &c. Other writers, equally deserving of credit, make mention of Chestnut timber being found in old buildings; and, among them, Hasted went so far as to broach a theory that a traffic was anciently carried on between Normandy and England, the latter supplying Chestnut timber in exchange for stone.

That this wood should be found in ancient buildings in very large quantities would carry great weight; but it has recently been discovered that the timber supposed to be Chestnut is in reality a kind of Oak (*Quercus sessiliflora*) or Denmark Oak, differing from common oak timber in those very characters which had been fixed on as distinctive of Chestnut. Besides this, Chestnut timber of large dimensions is neither in Great Britain nor the South of Europe found to possess the qualities, strength and durability, which were supposed to have recommended it to the notice of ancient builders. Evelyn's quotation from Fitz-Stephen is a very unhappy one, and the citation of the same passage from

Evelyn, by subsequent writers, is still more unfortunate, for the tree in question is neither described nor even mentioned by name. Evelyn honestly cited the passage as evidence that there formerly existed a great forest near London, in which he thought it probable that Chestnut timber, among other kinds, might grow; and the authors who followed him, perhaps not taking the pains to refer to the original work, and mistaking the drift of his remarks, took it for granted that the tree was mentioned, and considered the evidence conclusive, as well they might. Arguments founded on the facts that trees are individually mentioned as being in existence at periods more or less remote, and that there are in England several places which have long borne a name taken from these trees, e.g. Chesteney, Cheshunt, Cheston, Shesterhunte, Chastuners, &c., and consequently that the trees must have grown there in considerable abundance before such names were given — are far from conclusive; for when it is recollected that the Sycamore was, in the time of Gerard, a 'rare exotic,' and 250 years afterwards as common a hedge tree as the elm, we cannot deny that there was abundance of time between the Roman period and the earliest notice of Chestnut trees in our histories, for these trees to have propagated themselves to any extent. On the whole, then, rather than set aside the positive statement of ancient authors that the Chestnut was first introduced from Asia into Europe by the Greeks, and transported thence into Italy by the Romans, it may with reason be concluded that this tree, though long naturalised in England, is not an aboriginal native, but was introduced by the Romans at a very early period, and in process of time propagated itself so widely as to have raised a doubt whether it was not a really native tree.

Its history may be briefly told as follows:— It was first introduced into Europe from Sardis in Asia Minor, whence it was called the Sardian Nut, and at a later period Jupiter's Nut, and Husked Nut, from its being enclosed in a husk or rind instead of a shell. Several modern authorities, misquoting a passage in Pliny, attribute its introduction into Italy to Tiberius Cæsar, a palpable error, for it is evident from the writings of Virgil that Chestnuts were abundant in Italy long before the time of that emperor. By the Romans it was called Castanea from Castanum, a town of Magnesia in Thessaly, where it grew in great abundance, and from which it is said that they first brought it. From Italy and Greece it appears to have spread itself over the greater part of temperate Europe, ripening its fruit and sowing itself wherever the vine flourishes. In France, Italy, and Spain it attains a great size. On the Alps and Pyrenees it flourishes at an elevation of between 2,500 and 2,800 feet, the nuts having, perhaps, been carried to those lofty situations by the animals which lay up stores of winter food. It is still more abundant in Asia Minor, Armenia, and the Caucasus; and it is also found in America as far north as latitude 44°. It ripens its fruit in the warmer parts of Scotland, but rarely, if at all, in Ireland.

The Chestnut blossoms in July, and soon the upper part of the spike bearing the barren flowers withers and drops off, leaving the lower part of the spike still supporting the fertile flowers, with the embryo of the future nuts attached. Towards the end of September the latter begin to ripen, and in October fall to the ground, where they open with valves and expose the ripe nuts. Each case contains from two to five nuts, two or more of which are often mere empty rinds; but all, whether solid or otherwise, have the remains of the flower, in the shape of a few dry bristles, on their points. The Chestnut tree retains its leaves until late in the autumn, when they become of a rich golden hue. Owing to the tufted, and consequently weighty, character of the foliage, and the brittleness of the timber, the tree is liable to be injured by autumnal storms; but the leaves are rarely attacked by insects. The timber of young trees is applied to many useful purposes, but when matured is of little value, being brittle and apt to crack and fly into splinters. In the hop countries the growth of chestnut coppice is much encouraged, poles from this tree and the oak being preferred to all others. French, *Chataigner*; German, *Kastanienbaum*. [C. A. J.]

Theophrastus called it the Euboean nut, from Euboea, now Newrogont, where it was very abundant; and that being the case, the fruit may have been thence imported into Italy, although the tree, in a wild state, may have previously existed there. Professor Targioni observes that not only have the extensive woods in the Apian Alps, and other parts of the Apennines, every appearance of being really indigenous, but further evidence that woods of this tree existed in Tuscany from very remote times may be found in the number of places which have derived their names from them, such as Castagna, Castagneta, &c. He therefore concludes that we may safely give, as the native country of the wild Chestnut, the South of Europe, from Spain to the Caucasus. Some have even asserted that the tree is a native of Britain; but from the fact of its never being found here in such quantities as to form natural forests, whilst its seed only ripens in warm seasons or favourable localities, this seems very unlikely. It was probably introduced into this country by the Romans for the sake of its fruit. Gregor, in *Morton's Cyclopedia*, says the oldest Chestnut tree in England is supposed to be that at Tortworth, the seat of Earl Ducie, in Gloucestershire. Evelyn states it to have been remarkable for its magnitude in the reign of King Stephen (1135). It was then called ' the great Chestnut of Tortworth,' from which it may reasonably be presumed to have existed before the Conquest. It bore fruit abundantly in 1768. In 1820 its measurement, five feet from the ground, was fifty-two feet in circumference, so that the diameter, twelve feet, is equal to the width of a moderate-sized room. But these dimensions are small compared with the great Chestnut tree on Mount Etna, which measured 204 feet in circumference. When visited by M. Houel it was undergoing treatment by no means favourable to its prolonged existence. A house was formed in the interior, in which some country people were living; and they had an oven, in which, according to the custom of the country, they dried chestnuts, filberts, and other fruits which they wished to preserve for winter use, using for fuel, when they could find no other, pieces cut with a hatchet from the interior of the tree.

It has been said that the timber in the roof of Westminster Abbey is Spanish Chestnut; but Dr. Lindley has decided that such is not the case, and that Oak, *Quercus sessiliflora*, has been mistaken for it, in this and other old buildings. The timber employed in the construction of the old Louvre at Paris was also supposed to be Chestnut, but on examination by M. Daubenton it was found to be Oak. In this country, where it is certain that very fine oak trees abounded in natural forests, it is not likely that the Spanish Chestnut, requiring to be reared artificially, would be much employed for building purposes. The tree, doubtless, had been originally introduced and grown for the sake of its fruit. It is now, however, cultivated for posts, hop-poles, and hoops.

The fruit is enclosed in a round spiny husk, the inside of which is lined with soft silky pubescence; there are generally three chestnuts in each husk, occasionally more, but sometimes only one. There are many varieties. Some of a very large size are grown in Madeira, but they are not suited for the climate of England. The same remark applies to many of the French varieties, with the exception of the *Marron* corns. The Devonshire, Prolific, and Downton are amongst the best adapted for ripening in this climate. The Downton is remarkable for its short-spined husk. Chestnuts, after having been well-dried in the sun, may be kept amongst dry sand in casks. [R. T.]

CASTANHA DO JOBATA'. *Antsosperma Passiflora*.

CASTANOSPERMUM. A genus of plants so named in consequence of the supposed resemblance of the seeds to the sweet chestnuts of Europe. It belongs to the papilionaceous section of leguminous plants, and contains only one species, remarkable for its large woody long-stalked pods. This plant, *C. australe*, is a native of Moreton Bay, in Queensland, Australia, where it forms a tree forty or fifty feet in height. Its leaves are about a foot in length, pinnate, with an odd leaflet, the leaflets being smooth and of an elliptical form. Its pea-like flowers are produced in racemes, and are of a bright yellow colour; they have a two-lipped, short-tubed calyx, the upper lip having two, and the lower one three, divisions, and ten free stamens. The fruit is a pendulous cylindrical pod, of a bright brown colour, six or eight inches long, and tapering to both ends; it generally contains four seeds, which are rather larger than chestnuts, and of a roundish shape, but flattened on one side. The continent of Australia is remarkable for the paucity and inferior quality of its indigenous fruits or other esculents, the so-called apples and pears of the colonists being hard, woody, unestable productions; and the seeds of this tree, called Moreton Bay Chestnuts, are no exception to the rule, for, although they have been extolled, and placed upon an equality with our chestnuts, they are in reality not much superior to acorns, and have an astringent taste; they are improved by roasting, and no doubt proved acceptable to the travellers who first visited Moreton Bay. [A. S.]

CASTELA. A genus of tropical shrubs, belonging to the *Simarubaceæ*, having foliage like that of the olive, and small unisexual flowers arranged in axillary tufts. The male flowers have eight stamens, inserted beneath the margin of a fleshy eight-lobed disc, those opposite the petals shorter than the rest; the filaments adherent at their base internally to small hairy scales. The female flowers have four ovaries, on a short stalk; the four styles are detached at their origin, but are joined together in the middle for a short distance, and then again detached and recurved. The fruit consists of four fleshy bitter drupes. [M. T. M.]

CASTELNAVIA. One of several genera of most curious Brazilian plants, looking like mosses or *Hepaticæ*, belonging to the order *Podostemaceæ*, and which have been described with the greatest care and ability by M. Tulasne. The present genus consists of plants growing in the rapids, possessing no true leaves but a leaf-like stem or frond, dividing into forked lobes, and cut up at the margins into fringe-like segments. The flowers are either immersed in the substance of the frond, or placed on its margins. Some kinds have linear creeping branched stems, bearing a few linear leaves. The flowers have no calyx or corolla, but a tubular spathe or involucre divided at the margins into several thick thread-like segments; the stamens are two, slightly united one to the other; the fruit consists of a one-celled capsule, with two very unequal valves, surmounted by very long stigmas. [M. T. M.]

CASTILLEJA. A genus of *Scrophularineæ*, natives of America and Asia, containing about forty species of herbaceous plants, with alternate entire or cut-lobed leaves. The pale yellow or purplish flowers are in terminal spikes, with large coloured bracts usually more showy than the flowers. The calyx is tubular, flattened, cleft on the anterior side, and usually on the posterior also; the divisions are entire or two-lobed. The corolla-tube is included in the calyx; the upper lip is long and narrow, arched, keeled and flattened laterally, and incloses the stamens; the lower lip is short and three-lobed. There are four stamens with oblong-linear unequal anther cells, the outer attached by the middle, the inner pendulous. The pod contains numerous seeds. [W. C.]

CASTILLIER. (Fr.) *Ribes rubrum*.

CASTILLOA. A Mexican tree belonging to the *Artocarpaceæ*, and having male and female flowers alternating one with the other, on the same branch. The male flowers have several stamens, inserted into a hemispherical perianth, consisting of several united scales. The female flowers consist of numerous ovaries in a similar cup. The tree contains a milky juice, yielding caoutchouc. [M. T. M.]

CASTOR-OIL PLANT. *Ricinus communis*.

CASTRATUS. When an important part is missing, as in the case of filaments which have no anthers.

CASUARINACEÆ. A group of about a score of species of jointed leafless trees or shrubs, which, in their striated internodes and toothed-ribbed sheaths, have some resemblance to *Equisetum*, whilst in other respects they are allied in some measure to *Ephedra* and the *Coniferæ*, under which they were formerly classed, and still more with *Myriceæ* and other amentaceous groups, near to which they are now placed as a small distinct family. Their flowers are unisexual, the males in distinct whorls forming a cylindrical spike; each stamen is enclosed in four scale-like leaflets, the two outer ones considered as bracts, persistent at the base of the stamen, while the two inner ones or sepals, firmly cohering at the tips, are carried upwards by the anthers as the filament is produced. The female flowers are in dense axillary heads without any perianth. The ovaries, sessile within the bracts of the head, are one-celled, with a single ascending ovule, and bear two styles united at the base; the winged nuts are collected in a cone hidden under the thickened bracts. The *Casuarinas* are natives of Australia, of New Caledonia, or of the Indian Archipelago. They are too tender for this climate, but one species is occasionally planted in Southern Europe for its elegant drooping habit.

CASUARINA. A group of curious trees constituting of themselves a distinct family, *Casuarinaceæ*. They have very much the appearance of gigantic horse tails (*Equisetaceæ*), being trees with thread-like jointed furrowed pendent branches, without leaves, but with small toothed sheaths at the joints. The male flowers are in spikes with two bracts, and two sepals, which adhere at their points and are carried up like a hood by the anther of the single stamen. The female flowers are on the same plant, and are collected in dense heads; they have no calyx, but a one-celled ovary with one ascending ovule, and two styles: this ripens into a cone of woody bracts enclosing the seed-vessels, which are winged; the seeds are coated densely with spiral vessels.

These singular plants are met with most abundantly in tropical Australia, less frequently in the Indian islands, New Caledonia, &c. In Australia they are said by Dr. Bennett to be called Oaks. Their sombre appearance causes them to be planted in cemeteries, where ' their branches give out a mournful sighing sound, as the breeze passes over them, waving at the same time their gloomy hearse-like plumes.' The wood is used for fires, as it burns readily, and the ashes retain the heat for a long time. It is much valued for steam-engines, ovens, &c. The timber that is furnished by these trees is valuable for its extreme hardness. From its red colour, it is called in the colonies Beef-wood. The wood of *C. suberosa* is made use of for shingles to cover houses, and for other purposes where lightness, toughness, and durability are required. For further particulars of the Australian species, see Bennett's *Gatherings of a Naturalist* in *Australia*.

C. muricata is a native of Southern India, where it is valued for its showy wood, whose weight, however, forms an objection to its use. The bark furnishes a brown dye. The young branches of some of the species have a grateful acid flavour, much relished by cattle. *C. equisetifolia* is found in the South Sea Islands, the Indian Archipelago, and India. Its bark is astringent, and was formerly used by the South Sea Islanders to dye their cloth. The ashes of the tree yield a quantity of alkali, which is now used in the manufacture of a coarse soap. The wood furnished by it is called Iron-wood, from its colour, hardness, and durability. The natives avail themselves of these properties to make clubs, &c., of it. In Australia this species is called the Swamp Oak, though all the species thrive best in damp localities. Dr. Berthold Seemann mentions in a letter to the *Athenæum*, that the Fiji Islanders, or rather those among them that are cannibals, eat human flesh with forks made of the hard wood of a *Casuarina*, while they eat every other kind of food with their fingers. ' Every one of these forks is known by its particular often obscene name, and they are handed down as heirlooms from generation to generation.' So highly are they valued that it was difficult to obtain one. Several species of *Casuarina* are grown in greenhouses for the sake of their singular appearance. [M. T. M.]

CASSUMUNAR. The roots of *Zingiber Cassumunar*.

CAT. (Fr.) *Celastrus edulis*.

CATABROSA. A genus of grasses belonging to the tribe *Festuceæ*. The genus scarcely differs from *Glyceria*, except in the circumstance that there are only two florets in each spikelet. The British species, *C. aquatica*, is a handsome grass, but not of much agricultural importance. It is not uncommon in shallow ditches and the furrows of wet fields. [D. M.]

CATACLESIUM. A one-celled, one-seeded fruit, inclosed within a hardened calyx, as in *Mirabilis*.

CATALEPTIQUE. (Fr.) *Phyteuma*.

CATALPA. A genus of *Bignoniaceæ*, containing four or five species of trees, natives of the West Indies, North America, Japan, and China. They have large simple petiolate and opposite or terno-verticillate leaves, and flowers in terminal panicles. The calyx is deeply two-lipped; the corolla is bell-shaped, with a swollen tube and an undulate five-lobed spreading limb, irregular, and two-lipped. There are two or sometimes four fertile stamens, the one to three others being sterile and rudimentary; the anthers consist of two vertically diverging cells. The ovary is free, bearing a long slender style with a two-lipped stigma. The capsule is very long and slender, nearly cylindrical and two-celled, with the partition contrary to the valves. The seeds are numerous, broadly winged on each side, the wings being cut at their extremities into a fringe. On account of the beautiful and showy panicles of this genus, the species are cultivated in the various countries where they are found, as ornamental trees. They have been introduced into Europe; they thrive in France and Germany, and when planted in protected situations do well in the south of England, though they are very liable to be cut off by frosts or north-east winds. They grow rapidly. The wood is remarkably light of a grayish-white colour, and fine in texture, capable of receiving a brilliant polish, and when properly seasoned is very durable. The bark is said to be tonic, stimulant, and antiseptic; and the honey from its flowers poisonous. [W. C.]

CATANANCHE. A genus belonging to the chicoraceous tribe of compound flowers distinguished by its scarious involucre and the awned chaffy scales which crown its fruit. *C. cerulea* is a perennial herbaceous plant with slender stalks, long narrow leaves which are somewhat toothed at the base, and large heads of sky-blue flowers the scaly involucre of which is silvery-white tipped with reddish-brown. It is a native of the south of Europe, and as a border plant flourishes best in a light dry soil in a sheltered situation. Varieties are also cultivated with white or double flowers. *C. lutea* is an annual species with yellow flowers, a native of Candia. French, *Cupidone*; German, *Rasselblume*. (G. A. J.)

CATAPETALOUS. Having the petals slightly united by their inner edge near the base, as in the mallow. A form of polypetalous.

CATAPODIUM. A genus of grasses belonging to the tribe *Festuceæ*. The species which were included in this genus are described by Steudel under *Festuca*, in the *Synopsis Graminearum*.

CATAPUCE. (Fr.) *Euphorbia Lathyris*.

CATASETUM. A numerous genus of fleshy-stemmed terrestrial orchids from the tropical parts of the New World, where they form masses of considerable extent on decayed leaves, twigs, or other fragments of vegetation. The leaves are plaited and membranous. The flowers, always more or less green, spring in erect or drooping racemes from the base of great oblong fleshy stems, marked by circular scars, showing the places whence leaves have dropped away. The sepals and petals are of a firm leathery texture, sometimes converging into the form of a hood, sometimes spreading backwards. The lip is a fleshy body, not at all jointed with the column; sometimes it assumes the form of a casque, in other cases it is flat, lobed, and broken up into fleshy fringes; the first being characteristic of the original *Catasetum*, the second of what has been called *Myanthus*. The column is an erect fleshy body, terminating in a horn, and bearing about its middle a pair of long deflexed feelers or tendrils, except in a few instances, when the species without feelers have been called *Monachanthus*. In all cases the two fleshy pollen masses are ejected with considerable force by the sudden contraction of a glutinous gland, by which they adhere to surrounding objects. Among the most singular circumstances connected with this genus is the manner in which, upon the same spike, flowers of extremely

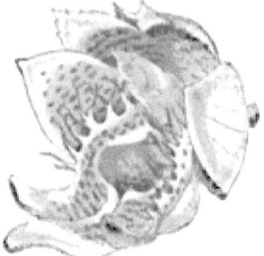

Catasetum Naso (flower).

different structure are produced. This was first noticed in Demerara by Sir R. Schomburgk, who published in the *Linn. Soc. Transactions* xxvii. 551 an account of the production of the three supposed genera, *Monachanthus*, *Myanthus*, and *Catasetum*, upon the same spike; and he expressed his opinion that the *Catasetum* was the female of these, because he found it producing seeds abundantly, while *Monachanthus* was uniformly sterile. Afterwards a similar specimen made its appearance in the garden of his Grace the Duke of Devonshire at Chatsworth. Mr. Darwin confirms that opinion. It has been well observed that 'such cases shake to the foundation all our ideas of the stability of genera and species, and prepare the mind for more startling discoveries than could have been otherwise anticipated.' For, according to the principles employed in botanical classification, no one could have doubted the distinctions between *Monochanthus*, *Myanthus*, and *Catasetum* being real, until the appearance of all their forms upon the same plant effectually dispelled the illusion.

CATAYA. A Brazilian name for a pungent species of *Polygonum*.

CATBRIER. An American name for *Smilax*.

CAT-CHOP. *Mesembryanthemum felinum*.

CATCHFLY. The common name for *Silene*.

CATCHWEED. The Cleavers, *Galium Aparine*.

CATECHU. The inspissated juice of the Areca palm (*Areca Catechu*), and of *Acacia Catechu*.

CATENULATE. Formed of parts united end to end like the links of a chain.

CATERPILLAR. A garden name for *Scorpiurus*.

CATERPILLAR FUNGUS. See COMPICEPS. [M. J. B.]

CATESBÆA. A genus of West Indian shrubs of the order *Cinchonaceæ*, with small spines above the leaves, and large trumpet-shaped whitish flowers, which are dilated in the throat, and have a four-parted limb; the stamens are four, with their anthers projecting from the corolla; the berry is of the size of a small egg. *C. spinosa* is a dwarf shrub with handsome flowers. Its fruit is yellow, pulpy, and of an agreeable taste. [M. T. M.]

CATHA. A genus belonging to the order *Celastraceæ*, separated from *Celastrus* on account of some not very well-marked technical characters. Its limits and the number of its species are not strictly defined. They are mostly natives of Africa, forming small shrubs, sometimes with spiny branches, and having simple leaves, from the bases of which small branching heads of flowers are produced. These flowers have a flat five-lobed calyx, five stalkless petals (those of *Celastrus* having short stalks), five stamens, a three-celled ovary half buried in the large disk which fills the bottom of the flower, and a very short style crowned by three pointed stigmas. The fruit or capsule is three-sided and three-celled, each cell containing a single seed, and splitting open when the fruit is ripe.

C. edulis, formerly *Celastrus edulis*, is a native of Arabia, and is by far the most interesting species of the genus, its leaves being used by the Arabs in the preparation of a beverage possessing properties analogous to those of tea or coffee. It is a shrub without spines, growing about ten feet in height, and is cultivated by the Arabs in the same ground as coffee. Its leaves are opposite on some branches, and alternate on others, smooth, and about two inches or more in length by an inch in width, of an elliptical form, and having their margins cut into blunt saw-like teeth. The flowers are very small and white. Under the name of Kât or Cafta, the leaves of this shrub, or, rather, twigs of it with the leaves attached, form a considerable article of commerce amongst the Arabs, large quantities of them being annually brought to Aden from the interior of Arabia, where the plant is cultivated. For the purposes of commerce they are made up into neat closely-pressed bundles of different sizes, according to quality, the best kind being in bundles a foot or fifteen inches long by three inches wide, and consisting of about forty slender twigs, tied together with strips of fibrous bark; each bundle selling at Aden for about two annas (or threepence sterling). The effects produced by a decoction of these leaves are described as resembling those of strong green tea, only more pleasing and agreeable. They are also chewed, either in a green state or when dried, and are said to have the effect of inducing great hilarity of spirits, and an agreeable state of wakefulness, so much so, indeed, that the Arabs who chew them are able to stand sentry all night long without feeling drowsy. The use of Kât in Arabia is said to be of great antiquity, and to have preceded that of coffee. Its stimulating effects induced some Arabs to class it with intoxicating substances, the use of which is forbidden by the Koran, but a synod of learned Mussulmans decreed, that, as it did not impair the health or impede the observance of religious duties, but only increased hilarity and good humour, it was lawful to use it. In addition to its powers as a stimulant, the Arabs have a superstitious idea that a twig of it worn in the bosom protects a person from the danger of infection; and that the shrub itself is a preventive against the plague. [A. S.]

CATHARANTHUS. A name sometimes met with in gardens, and given to certain shrubby plants belonging to *Vinca*. [T. M.]

CATHARTOCARPUS. The name under which certain species of *Cassia* are sometimes known. [T. M.]

CATHCARTIA. Dr. Hooker detected this beautiful plant of the poppy family in the eastern part of the Himalayan mountains, and named it in honour of Mr. Cathcart, an Indian judge who investigated the botany of the Sikkim Himalayas. It is a herb covered with soft yellow hairs, having lobed leaves, and golden drooping flowers, with a hemispherical sessile four-lobed radiating stigma, and an erect cylindrical capsule, bursting from above downwards into five valves. [M. T. M.]

CATHEDRA. A genus of Brazilian trees belonging to *Olacaceae*, having alternate shortly stalked elliptical leathery leaves, and small axillary clusters of nearly sessile flowers, with the floral coverings fleshy and green; calyx cup-shaped, petals six, stamens six, disk conspicuous. [J. T. S.]

CATINGA. A genus doubtfully referred by its author to the myrtle family. The two known species are described as trees of French Guiana; their leaves opposite or alternate, stalked, entire, ovate-oblong, with long points, and pellucid dots. The flowers are not known. The fruits are borne in axillary racemes; that of *C. moschata* resembling an orange in size, colour, and form; it is crowned with the four remaining calyx-lobes, and contains one seed, while the outer fleshy part of the fruit is covered with little bladders, containing an essential aromatic oil of a musky odour. *C. fragrans* has a fruit, in size and form like a citron, and has an odour like that of basil. [A. A. B.]

CATJANG. The native name of *Cajanus indicus*, a wholesome and much-used kind of pulse.

CATKIN. A deciduous spike, consisting of unisexual apetalous flowers. An amentum.

CATMINT. The common name for *Nepeta*; especially applied to *N. Cataria*. —, MALABAR. *Anisomeles malabarica*.

CATNEP, or CATNIP. *Nepeta Cataria*.

CATOBLASTUS. Two species of palms, formerly included in the genus *Iriartea*, have recently been separated under this name. They are both natives of New Grenada, where they grow together in masses, having trunks from thirty to fifty feet high, distantly marked with circular scars, supported a short distance above the level of the ground upon a tuft of aerial roots, and bearing a crown of pinnate leaves. They differ from *Iriartea* in the male and female flowers being borne on separate spikes, the males having a small rudimentary ovary in addition to the nine to fifteen stamens, whilst the females have scarcely any rudimentary stamens. The seed, also, has markings like a nutmeg (ruminate), and its embryo is placed upright at the base. [A. S.]

CATOCOMA. A genus of scandent or climbing shrubs, natives of the tropical parts of S. America, and belonging to the milkwort family. The leaves are alternate, entire, ovate or oblong, and leathery in texture. The flowers are numerous, disposed in terminal panicles, yellow or greenish white, often with a purple spot on the keel; the calyx five-leaved, the three exterior leaves small; the petals five, one of them large keeled and three-lobed, the two lateral ones small and scale-like, the other two oblong; the stamens eight in number, united into a tube which is cleft above; the anthers one-celled, opening by a terminal pore. The fruit is compressed, wedge-shape, fleshy, and two-celled, each cell with one seed, which is furnished with long silky hairs. *C. floribunda* is an extensive Brazilian climber, scrambling over the tops of the highest trees, and covering them with a crown of yellowish flowers. Its roots are used against snake-bites. Upwards of a dozen species are known, differing from each other chiefly in the form and pubescence of the leaves. [A. A. B.]

CATOSTEMMA. A genus of the tea family (*Ternstroemiaceae*) found in British Guiana. The only species known, *C. fragrans*, is a tree fifty feet high. The leaves are alternate, entire, obovate in form, with a little recurved point at the apex, two to four inches long, and one to two broad. The flowers are numerous and fragrant, disposed in fascicles in the axils of the upper leaves, about half an inch across; the calyx cup-shaped, with a two-lobed limb. The plant may be recognised from any other in the family by this character. The petals are five in number, inserted into the upper portion of the calyx tube, and falling away with it after withering; the stamens very numerous, in five parcels. The fruit not known. [A. A. B.]

CAT'S-CLAW, *Dolichos filiformis*. Also *Inga unguis-Cati*.

CAT'S-EAR. The common name for *Hypochoeris*, especially *H. radicata*; applied also to *Gnaphalium dioicum*.

CAT'S-MILK. *Euphorbia helioscopia*.

CAT'S-TAIL. The common name for *Typha*. —, GRASS. The common name for *Phleum*.

CATSUP or KETCHUP. A sauce prepared from mushrooms, walnuts, and other vegetable productions.

CATTEMUNDOO. A gum-elastic yielded by *Euphorbia antiquorum*. Sometimes called Callemundoo.

CATTERIDGE TREE. *Cornus sanguinea*.

CAT-THYME. *Teucrium Marum*.

CATTLEYA. An extensive genus of orchids inhabiting Central America and Brazil, where they are found on the bark of trees, and on rocks. The species all form pseudo-bulbs bearing one, or at the most two, fleshy leaves, from the axil of which rise two or more flowers for the most part rose-coloured, but occasionally yellow, or some tint of that colour. These flowers are often among the largest in the orchidaceous order, some being as much as seven inches across from tip to tip of the petals. The finest of all these grand species is *C. Warscewiczii* from the Amazons, whose flowers grow seven together on a raceme eighteen inches long. Next to it stand *C. Mossiae*, *labiata*, *crispa*, and *Skinneri*. The species called *C. Schilleriana* *guttata*, and *grandísea* have thick leathery flowers with crimson spots on a yellowish-green ground. *C. luteola* and *citrina* are wholly yellow. Many of the so-called species in gardens are mere varieties of others. The genus differs from *Lelia* in having four pollen masses instead of eight.

CATKUS. A catkin, or amentum, such as is borne by the hazel.

CATURUS. The name formerly given to a nettle-like plant of the spurgewort family, with long rat's-tail-like spikes of small green flowers, which are said to be used in the East Indian Islands either in a conserve or decoction, as a remedy for diarrhœa. The plant is now placed in the genus ACALYPHA; which see. [A. A. B.]

CAUCALIS. A genus of umbelliferous plants distinguished by its oblong fruit, ribbed with four rows of hooked prickles, with rough interstices. All the species are herbaceous, natives of Europe and the temperate parts of Asia and Africa. The Bur Parsley, *C. daucoides*, is a British plant, growing in corn-fields in a chalky soil, and is neither attractive in appearance, nor otherwise interesting. *C. latifolia* was formerly abundant in Cambridgeshire, but is now extinct. The leaves are broader and less divided than is generally the case with the umbelliferous tribe; the flowers are large and rose-coloured. The foreign species are equally unattractive. French, *Caucalide*; German, *Haftdolde*. [C. A. J.]

CAUDA (adj. CAUDATUS). Any long soft narrow terminal appendage, as that of the corolline lobes of *Strophanthus*, or the lateral sepals of *Cypripedium caudatum*.

CAUDEX. The axis of a plant, consisting of stem and root. — REPENS. A creeping stem; what is now called a rhizome. — DESCENDENS. The root.

CAUDICULA. The cartilaginous strap which connects certain kinds of pollen masses to the stigma, as in *Maxillaria*.

CAULERPA. A very beautiful genus of green-seeded *Algæ*, abounding in species, and assuming very different forms. The species are almost exclusively natives of warm climates, and occur on sand, on shaded rocks, or in deep water. All have a more or less decided green herbaceous hue, and however complicated may be their growth, or whatever size they may attain, they are formed of a single cell without any transverse divisions, branched and anastomosing in every part of the plant, amidst which a green chlorophyll is produced which ultimately gives rise to minute zoospores. The species are greedily eaten by turtles, of which they form the principal food. The nearest approach which is made to the genus on our coasts, is seen in *Codium*. [M. J. B.]

CAULET. (Fr.) A kind of cabbage.

CAULICULUS. A small stem produced at the neck of a root without the previous production of a leaf. Also the imaginary space between the radicle and cotyledons of an embryo. Also the stipe of certain fungals.

CAULIFLOWER. A garden variety of *Brassica oleracea*, in which the inflorescence while young is condensed into a depressed fleshy esculent head.

CAULIGENUS. Arising from a stem.

CAULINE. Of or belonging to the stem. — STIPULES. Such as adhere to the stem as much as to the petiole or leaf.

CAULINIA. A section of the genus *Naias*, sometimes considered as distinct on account of the anther being elliptical, one-celled, and without valves; while in *Naias* it is four-sided, four-celled, and opening by four valves, rolling inwards. The only British species of *Naias* (*N. flexilis*, found in lakes in Galway) belongs to *Caulinia*. [J. T. S.]

CAULIS. The stem or ascending axis; a name only given to the part in its customary state, growing in the air. — DELIQUESCENS. A stem which at a distance above the earth breaks into irregular ramifications, as in the oak. — EXCURRENS. A stem which shoots straight from the ground to the summit, having branches on the sides, as in *Abies*.

CAULOCARPOUS. A stem which lives many years, repeatedly bearing flowers and fruit; as a shrub or tree.

CAULOMA. The stem of a palm-tree. The stem-like portion of the thallus of such algals as some *Fuci*.

CAULON. In Greek compounds = stem.

CAULOPHYLLUM. A perennial herbaceous plant with tuberous roots, belonging to the order *Berberideæ*, remarkable for bearing only one leaf on each stem, directly under the cluster of flowers, and terminating the stem, which, consequently, has the appearance of being no more than a leaf-stalk. The true leaf-stalk is divided to the base into three parts, each part having three ovate deeply-cut leaflets. The foliage bears a resemblance to that of *Thalictrum* (hence the specific name *thalictroides*) or *Aquilegia*. The stems are about a foot high; the flowers small, yellowish-green, with six sepals and as many petals and stamens, and are succeeded by deep-blue globose berries, contracted below so as to approach pear-shaped. These berries are called by the Indians Cohosh, and the plant is esteemed medicinal. It is a native of North America, but is not of common occurrence. Two other species of similar habit are natives of the Altai mountains, and the country about Odessa. [C. A. J.]

CAUSTIC. Biting in taste, like Cayenne pepper.

CAVA or KAWA. An intoxicating beverage prepared from *Macropiper methysticum*.

CAVERN FERN. A name given in some books to *Antrophyum*.

CAVERNULI. The pores of certain fungals.

CAVITAS. The perithecium of certain fungals.

CAVUS. The peridium of certain fungals. — **SUPERUS.** The hymenium of certain fungals.

CAXAPORA DO GENTIO. A Brazilian name for *Terminalia argentea*.

CAYENNE PEPPER. The dried powdered fruits of various species of *Capsicum*. Miller calls it Cayan Pepper.

CEANOTHUS. Red root. A genus of shrubby plants belonging to the order *Rhamnaceæ*, allied to *Euonymus*, with which it agrees in having a three-celled three-seeded pericarp, but the seeds are not enclosed in a membrane (arillus). *C. americanus*, the species most commonly cultivated in English gardens, is a native of N. America, a shrub from two to four feet high, with downy leaves and stems, and small white flowers, which, being produced in great numbers, are very ornamental. They appear in June and July, and are succeeded by bluntly triangular seed-vessels. In America it is commonly known by the name of New Jersey tea, the leaves having been formerly used for the same purpose as those of the Chinese tea-plant, and for which it formed a general substitute during the War of Independence. In Canada it is used for dyeing wool of a nankin or cinnamon colour. Many other species are cultivated, and some of remarkable beauty, which have been introduced of late years, are amongst the finest of half-hardy shrubs. The English name has reference to their large red roots. [C. A. J.]

CEBADILLA, CEVADILLA. The seeds of *Asagræa officinalis*, from which veratria is obtained.

CECROPIA. A genus of large-leaved soft-wooded milky trees, native of tropical South America, and belonging to the order of *artocarpads*. The flowers are extremely minute, and are arranged upon short cylindrical spikes, several of which (sometimes as many as sixty) are enclosed within a large bract, which, however, soon falls away; their calyx is tubular, and rather thicker in the females than the males, the males containing two stamens, and the females a free ovary and short style terminated by a brush-like stigma. The latter are succeeded by short spikes of small fleshy one-seeded fruits. Upwards of twenty-five species are described.

C. peltata, the Trumpet-tree of the West Indies and tropical South America, so called because its hollow branches are used for musical instruments, is a very rapid growing tree, having a whitish trunk about a foot in diameter, and attaining a height of upwards of fifty feet, its branches growing out at right-angles, so as to form a large spreading head; the trunk itself being marked at regular distances by circular scars which indicate the places where leaves once grew, and the hollow inside having transverse partitions corresponding in number and position with them. The leaves are nearly circular, often more than a foot in diameter, and attached to their stalk from a point near the centre, their margins being deeply cut into nine oblong lobes, each of which is tipped with a short point; they are very rough upon the upper side, and thickly covered on the under side with snowy white down. The spikes of fruit are in clusters of from five to fifteen.

The Uaupé Indians, who inhabit the Rio Uaupes, a tributary of the Rio Negro, convert the hollow stems of this tree into a very curious kind of musical instrument, a species of drum, called by them Amboobas. They select a trunk four or five inches in diameter, and cut off a piece about four feet long, removing the partitions and rendering the inside smooth by means of fire; they then close up the lower end with leaves beaten down into a hard mass with a pestle, and cut two holes towards the top end, so as to form a handle. These rude instruments are commonly used in the native dances, the performer, holding by the handle, beats the lower end upon the ground, and moves his feet in unison with the sounds thus produced.

The inner bark of the young branches yields a very tough fibre, which is twisted into coarse ropes; and the old bark is employed medicinally as an astringent. The young buds are moreover eaten as a potherb, while the leaves are the common food of the sloth, and the milky juice hardens into caoutchouc. The wood is very light, and is commonly used in the West Indies for making floats for fishing nets; razor-strops are likewise made of it, and when dry the Indians use it for producing fire by means of friction. [A. S.]

CEDAR. The common name of various trees, but more especially applied to the Cedar of Lebanon, mentioned below. — BARBADOS. *Juniperus barbadensis*, — BASTARD BARBADOS. *Cedrela odorata*, also called the Sweet-scented Barbados Cedar. — BASTARD. *Guarana umifolia*; also a common name for *Cedrela*. — BERMUDA. *Juniperus bermudiana*. —, GUIANA. *Icica altissima*. —, HONDURAS. *Cedrela odorata*. —, INDIAN. *Abies* (or *Cedrus*) *Deodara*. — JAPAN. *Cryptomeria japonica*. — MOUNT ATLAS. *Abies atlantica*. — OF GOA. *Cupressus lusitanica*. — OF LEBANON. *Abies Cedrus*, often called *Cedrus Libani*. — OF N. S. WALES. *Cedrela australis*. — PRICKLY. *Cyathodes Oxycedrus*. — RED. *Juniperus virginiana*; of Australia; *Cedrela australis*. — SHARP. *Juniperus Oxycedrus*. — STINKING. *Torreya taxifolia*. — VIRGINIAN. *Juniperus virginiana*. — WHITE. *Cupressus thyoides*; of Australia; *Melia australis*; of it Guiana; *Icica altissima*; of Dominica; *Bignonia leucoxylon*.

CEDAR APPLES. The Pennsylvanian name of the curious excrescences on *Juniperus virginiana*, caused by the fungus called *Podisoma macropus*. [M. J. B.]

CEDRAT. A variety of the Citron, *Citrus medica*.

CE'DRE BLANC. (Fr.) *Cupressus thyoides*. — DES BERMUDES. *Juniperus bermudiana*. — d' ESPAGNE. *Juniperus thurifera*. — DE VIRGINIE or ROUGE. *Juniperus virginiana*. — PIQUANT. *Juniperus Oxycedrus*.

CEDRELACEÆ. (*Cedrelads*, the Mahogany family.) A natural order of thalamifloral dicotyledons, belonging to Lindley's rutal alliance. Trees with alternate pinnate leaves, without stipules. Flowers in panicles; calyx four to five-cleft; petals four to five; stamens eight to ten, inserted on a disk. Ovary three to five-celled. Fruit a capsule opening by valves, which separate from a thick axis; seeds numerous, flat, winged, and anatropal, i. e., with the opening near the hilum, and the chalaza at the opposite end. There are two suborders; 1. *Swieteniea*: filaments of stamens united. 2. *Cedreleæ*: filaments not united. Natives of the tropics of America and India, very rare in Africa. The plants of this order are generally fragrant, aromatic, and tonic. Many supply compact beautifully-veined timber, such as the mahogany of tropical America (*Swietenia Mahogoni*), satin-wood of India (*Chloroxylon Swietenia*), yellow-wood of New South Wales (*Oxleya xanthoxyla*), red-wood of Coromandel (*Soymida febrifuga*), and the toon of India or Simal-Kun of the Lepchas (*Cedrela Toona*). A kind of oil is procured from satin-wood; and the barks of *Cedrela febrifuga*, the mahogany tree and others, are used as remedies in intermittent fevers, as well as in dyspeptic complaints. There are nine known genera, and twenty-five species. *Swietenia*, *Soymida*, *Flindersia*, and *Cedrela*, serve as illustrations, of the group. [J. M. B.]

CEDRELA. A genus of large trees, giving its name to the order *Cedrelaceæ*. They bear compound leaves, regular flowers, five fertile stamens adherent to the stalk which supports the five-celled ovary, and five sterile stamens, which are very small, or altogether absent. The fruit is a capsule bursting by five pieces to liberate the seeds, which are winged. The trees are natives of the tropical parts of Asia and America, and are remarkable for their fine timber, sometimes called Cedar-wood. The trunk of *C. odorata*, a West Indian tree, is sufficiently large to be hollowed out into canoes; this, which is of a brown colour, and has a fragrant odour, is imported under the name of Jamaica or Honduras Cedar. *C. Toona*, a native of Bengal and other parts of India, furnishes timber much like mahogany in appearance but lighter. It is in great request, and is said to be one of the woods known as Chittagong wood. The bark is very astringent, and has been found valuable in fevers, dysentery, &c. The flowers are used in some parts of India for producing a red dye. The Red Cedar of Australia, *C. australis*, is now becoming scarce in that colony, the trees having been cut down for the sake of their timber, which was commonly used in the construction of houses. [M. T. M.]

CEDRINO. The small Italian Citron.

CEDRONELLA. A small genus of *Labiatæ*, natives of North America and the Canary Isles. They are sweet-scented perennial herbs, or rarely shrubs, with pale purplish flowers, in spikes or terminal racemes, and having the floral leaves bract-like, and the bracts themselves small and setaceous. The calyx is rather obliquely five-toothed, and many nerved. The corolla is very large, much expanded at the throat, and two-lipped, with the upper lip flattish or concave, and two-lobed, and the lower three-cleft, spreading, the middle lobe being largest. There are four ascending stamens, the lower pair are shorter than the others; the anthers have two parallel cells. The apex of the style is subequally bifid, with subulate lobes. The nucule is smooth. [W. C.]

CEDRUS. The name under which the Cedar of Lebanon, the Deodar or the Indian Cedar, and the Mt. Atlas Cedar, are sometimes separated from other coniferous trees. The characters mainly relied on to distinguish the genus are the evergreen leaves disposed many together in bundles or fascicles, and the erect cones with their carpels separating from the axis. The cedars are now generally included in *Abies*. [T. M.]

CEINBRA. (Fr.) *Pinus Cembra*.

CELANDINE. The common name for *Chelidonium*. — LESSER. *Ranunculus Ficaria*. — TREE. *Bocconia frutescens*.

CELASTRACEÆ. (*Celastrineæ*: Spindle-trees.) A natural order of calycifloral polypetalous dicotyledons belonging to Lindley's rhamnal alliance. Shrubs or small trees with alternate rarely opposite simple leaves, having stipules which fall off. Flowers in axillary cymes, small, green white or purple; sepals and petals four to five, imbricate, the petals sometimes wanting; stamens four to five, inserted on a large disk, which surrounds the ovary and encloses it. Fruit two to five-celled, capsular or drupaceous (cherry-like); seeds usually with an aril, albuminous, with a large straight embryo. Natives of the warmer parts of Europe, North America, and Asia, far more abundant beyond the tropics than within them. Many inhabit the Cape of Good Hope, some occur in South America, and a few in New Holland. There are two suborders; 1. *Euonymeæ*: fruit dry and capsular. 2. *Elæodendreæ*: fruit drupaceous or cherry-like. The plants of the order are more or less acrid in their properties some yield oils. The spindle-trees have a beautiful scarlet aril, which is derived from the sides of the opening in the seed. The species of *Euonymus* in America, from their crimson capsules and arils, are called burning-bush. *Celastrus scandens* from its aspect is denominated Wax-work in North America. The stimulating substance called by the Arabs khât is procured from *Catha edulis*. The wood of the European spindle-tree is used for cannon gunpowder in France. There are thirty-five known genera, and 280 species. Illustrative genera: *Euonymus*, *Celastrus*, *Elæodendron*. [J. R. R.]

CELASTRUS. A genus which gives its name to the family to which it belongs. It is on the one hand allied to *Euonymus*, from which it differs in its alternate leaves, and on the other to *Catha*, which embraces spiny shrubs, whose seeds are furnished with a small aril; whilst *Celastrus* is composed of small unarmed scandent shrubs or trees, having a large aril to their seeds. Their leaves are alternate, entire or serrated with minute stipules. The flowers are small, green or white, and disposed in terminal racemes or panicles. The name of the genus is derived from the Greek, signifying the latter season. The ancients considered the holly, the genista, and the celastrus, the trees which ripened their fruit latest in the season. The celastrus of the ancients is, however, supposed to have been a kind of *Euonymus*. *C. scandens* is a climbing North American shrub, popularly known as bitter-sweet or Wax-work. The capsules are orange-coloured when mature, and the seeds reddish-brown, coated with a bright orange or scarlet aril. It is sometimes planted as an ornamental climber because of its showy fruit. The seeds are said to possess narcotic and stimulating qualities, while the bark is purgative and emetic. The scarlet-coated seeds of *C. paniculatus*, a common Brazilian species, yield an oil which is sometimes used for burning in lamps, and is in repute among native doctors. The seeds have a hot biting taste owing to a resinous matter contained in them. The plants comprised in this genus, commonly called Staff-trees, are found in the temperate regions of tropical countries, and appear in greatest number in the Himalayas. [A. A. B.]

CELERIAC. A turnip-rooted variety of the garden celery.

CELERY. *Apium graveolens.*

CELLA. A name sometimes given to a form of the perithecium among fungals.

CELLS, CELLULES. Cavities in the interior of a plant. The cells of tissue are those which form the interior of the elementary vesicles. Cells of the stem, air-cells, &c., are spaces organically formed by a peculiar building up of tissue, for various vital purposes.

CELLULAR SYSTEM. That part of the plant which consists of cells or elementary vesicles.

CELLULARES. A name given to cryptogams, from a notion that they consist entirely of cells. A more accurate acquaintance, however, with their anatomy has shown that vascular tissue exists in many of the higher forms, and that even in *Fungi* there are genera which possess true spiral vessels, while in one or two higher *Algæ* the stem contains vascular threads, while the contents of the cells or endochrome are sometimes disposed in one or more spiral bands. In both, the cell-walls themselves have occasionally a spiral structure. *Podaxon* amongst *Fungi*, and *Conferva Melipesium* amongst *Algæ*, afford excellent examples. [M. J. B.]

CELLULOSE. The primitive membrane, free from all deposits of sedimentary or other matter. Its composition, according to the latest analysis, is C 24 H 20 O 10.

CELOSIA. A genus of amaranthads, consisting, with a few exceptions, of tropical annual plants, closely allied in their structure to *Amaranthus*, with which they agree in having the flowers three-bracted, a perianth of five-coloured scarious pieces, two-celled anthers, and an utricular seed-vessel splitting horizontally round when ripe; but differ in their five stamens being united at the base into a cup, in having a more or less elongated style, and in the utricle containing several seeds, instead of but one only. It is important to remark that the form of the *C. cristata* or Cockscomb usually found in cultivation, conveys a very incorrect idea of the inflorescence of this genus, the broad flattened stem with its terminal crest being a monstrosity, resulting from the lateral adhesion of the stems and branches by a process termed by botanists fasciation. In its normal phase the *C. cristata* is of erect habit, growing one to two feet high, with roundish striated stems pyramidally branched nearly to the base, alternate leaves of a lanceolate or ovate-lanceolate form, and flowers in either loose pyramidal panicles or compact spikes. In the beautiful, but now little known *C. aurea* of gardens, which is regarded by botanists as but a form of *cristata*, only a few of the flowers at the base of the panicle are perfect, those of the summit being abortive, and putting on the appearance of glossy yellow spirally-twisted scales, which gives the inflorescence a tassel-like form. There is a red-flowered variety of taller growth, with the blossoms in compact conical spikes. There are several other species agreeing with these in habit, but they are less ornamental, and possess little general interest. The flowers of the Cockscomb are reputed to be astringent, and are employed in India in diarrhœa and other maladies. [W. T.]

CELSIA. A small genus of linariads distinguished by a wheel-shaped five-lobed corolla, and didynamous bearded stamens. It is closely allied to *Verbascum*, which differs from it chiefly in having five perfect stamens. The species are annual or biennial plants, in the latter case sometimes of shrubby habit, with entire or pinnatifid foliage, and spikes of bright yellow mullein-like flowers. The biennial, *C. cretica*, found both in Candia and Northern Africa, is at the same time the best known and by far the showiest of the species. As cultivated in gardens, it attains a height of four or five feet, having the root-leaves of a lyrate form and the upper ones oblong, with a long terminal spike of large yellow blossoms, each of which arises from the axil of a small leaf or bract. The corollas have two brownish spots on the upper side near the centre, the two shortest stamens have their filaments bearded, and the segments of the calyx are sharply serrated. This plant affords a good example of what is termed by botanists a declinate style, this organ, as well as the two longer stamens, being very much bent upwards. *C. Arcturus*, a dwarf half-shrubby species, is sometimes met with in cottage-windows, and has, like the preceding, spikes of yellow flowers, but the calyx segments are all entire, and the filaments all bearded. [W. T.]

CELTIS. Nettle-tree. Handsome much-branched deciduous trees belonging to the *Ulmaceæ*, distinguished at once from the true elms by their bearing instead of a membranous fruit a hard fleshy drupe, which is edible, and, though small, is remarkably sweet and said to be very wholesome. Several species have been introduced into Great Britain. The European Nettle-tree, *C. australis*, is a tree from thirty to forty feet in height, with a straight trunk and a branched head. The branches are long, slender and flexible, with a grey bark spotted with white, and covered with a slight down at the extremities. The bark of the trunk is rich brown. The leaves are dark green, marked strongly with the nerves on the lower side, and, when young, covered with a yellowish down. They are oval-lanceolate, terminating in a point at the summit, and at the base having one side prolonged down the petiole. The flowers are small greenish and inconspicuous, and are produced at the same time as the leaves. The fruit, which, when ripe, is blackish and resembles a very small withered wild cherry, is said not to become edible until the first frosts, and it hangs on until the following spring. It is remarkably sweet, and is supposed to have been the Lotus of the ancients, the food of the Lotophagi, which Herodotus, Dioscorides, and Theophrastus describe as sweet, pleasant, and wholesome, and which Homer says was so delicious as to make those who ate it forget their native country. The berries are still eaten in Spain, and Dr. Walsh says that the modern Greeks are very fond of them. According to Dr. Sibthorpe, they are called in modern Greece Honey-berries. The tree grows rapidly, more especially when once established and afterwards cut down, sometimes producing shoots, in the climate of London, six feet or eight feet in length. *C. australis* is found on both the shores of the Mediterranean, throughout the whole of the south of France, Italy, and Spain. It is peculiarly abundant in Provence; and there is a celebrated tree at Aix, under which it is said that the ancient sovereigns of Prussia delivered their edicts to the people. It is much used in the south of Italy and the south of France for planting squares and public walks, when it is frequently found from forty to fifty feet high, with a trunk from one and a-half to three feet in circumference. The wood is extremely compact, ranking between that of the live-oak and box for hardness and density. The wood of the branches is elastic and supple, its compactness renders it susceptible of a high polish, and when it is cut obliquely across the fibres it very much resembles satin wood. It is used for furniture and carving, and the branches are extensively employed in making hay-forks, coach-whips, ramrods, and walking-sticks (Loudon). The North American Nettle-tree, *C. occidentalis*, differs from the European species in having longer leaves, which are of a lighter green, and in having the wood of a lighter colour in winter. The American, *C. crassifolia*, Huckberry is a very distinct species, and one of the finest trees which compose the dusky forests of the Ohio. The leaves are larger, more acuminated, of a thick texture with a rough surface. The fruit is round, and about the size of a pea. The Huckberry is found in the greatest abundance in the western states of America. The timber is of little value. *C. orientalis* and *C. acuminata* are low-spreading trees of inferior interest. French, *Micocoulier*; German, *Lotusbaum*. [C. A. J.]

CENARRHENES. A genus of *Proteaceæ*, found in Tasmania. Its flowers, which are apetalous, have four sepals with the points attenuated; four stamens with free filaments, inserted at the base of the sepals; and a filiform style with a simple stigma. These flowers are borne on spikes, rather shorter than the leaves. The fruit is a single-seeded berry. *C. nitida*, the only species, is a small tree about twenty-five feet in height, with shining coriaceous spathulate leaves, attenuated at the base, and remotely dentate, with a grooved petiole; they are from four to six inches in length and about one inch in width. [B. H.]

CENCHRUS. A genus of grasses belonging to the tribe *Paniceæ*, and scarcely differing from *Pennisetum*, except in the involucral scales being more hardened, broader, and more or less connate at the base. Steudel describes thirty species, which are chiefly inhabitants of rather warm and dry countries, consequently they require the protection of a conservatory when cultivated in Britain. [D. M.]

CENIA. A genus of the composite order, having the flowers at the circumference of the heads either strap-shaped or with two lips; those in the centre tubular and four-toothed; the receptacle or part supporting the flowers inflated or hollow; and the fruit two-ribbed, without any crown-like appendage. The name of the genus is from the

Greek word signifying hollow or void, in allusion to the hollow receptacle. The species are natives of the Cape of Good Hope, and have alternate leaves, which are twice pinnate, the divisions being long and narrow. *C. turbinata* has been long known, having been introduced about the beginning of the last century. [G. D.]

CENOBIUM. (adj. CENOBIONAR, CENOBIONEUR.) Such fruits as those of labiates, borageworts, &c., which consist of several distinct lobes, not terminated by a style or stigma.

CENOLOPHIUM. A genus of *Umbelliferæ*, nearly related to *Cnidium*, but differing in the mature seeds being enclosed in a loose pericarp, as well as in the ribs of the carpels being hollowed interiorly. *C. Fischeri*, the only species, is a tall smooth perennial weed, common throughout Russia, and sometimes cultivated in botanic gardens. It has many times ternate leaves, the segments narrow, lance-shaped, and nearly an inch in length; small white hemlock-like flowers, disposed in many-rayed umbels, with a general involucre of one bract, and numerous narrow bracts to the partial involucres; ovate-oblong nearly cylindrical fruits, each carpel having five sharp ribs, with an oil tube in each furrow, and two on the inner face. [A.A.B.]

CENTAUREA. An extensive and varied genus of *Compositæ*, comprising both annual and perennial herbaceous or half-shrubby plants, some of them common weeds, e.g., *C. nigra*, the Knapweed of our pastures, while a certain number are esteemed border flowers. They are distinguished by a globose or ovate involucre of many imbricated scales or leaflets which are either fringed at the tip or furnished with appendages varying in form and character; by a bristly receptacle; by the florets being all tubular, the outer row usually much the larger, spreading and sterile; and by a compressed fruit, with or without pappus of simple bristles, and a lateral depression or hilum near the base. The species present great diversity of habit and foliage, some being of prostrate growth, others quite erect; while the foliage varies from entire to pinnatifid or bipinnatifid, and the flowers from white to blue, yellow, and purple.

Of the perennial species, one of the most common in gardens is *C. montana*, which grows one and a half foot high, and bears entire lanceolate downy leaves, and large capitules, the outer florets of which are pale violet blue, and the central ones deep purple. *C. macrocephala*, an erect growing species, of stiff habit, with entire leaves, stalked at the root but decurrent on the stem, has large solitary flower-heads of a fine yellow colour. In *C. dealbata*, with reddish purple flowers, the twice-pinnatifid foliage is whitened on the under side, a circumstance to which the name is due. *C. candidissima*, a native of the Levant, has the lyrately pinnatifid leaves clothed on both surfaces with a white silky tomentum, which gives it a striking aspect; and *C. Ragusina*, a Dalmatian species, has similar foliage; both these latter have yellow flower-heads.

Of the annual species one of the most remarkable is *C. americana* or *Plectocephalus americanus* of some authors, which has a stout erect stem four to five feet or more high, oblong lance-shaped leaves, and very large capitules of a lilac-purple tint. *C. depressa* is a pretty dwarf plant from the Caucasus, of somewhat procumbent habit, with entire lanceolate leaves, and flowers of a fine blue. Better known than any of the preceding is the common Corn blue-bottle, *C. Cyanus*, an indigenous species of tall slender growth, the foliage greyish, and the flowers, in their wild state mostly of a light blue colour, but in gardens found varying from white to every shade of blue and purple. [W. T.]

CENTAURE'E DU NIL. (Fr.) *Centaurea Crocodilium*. — ODORANTE. *Amberboa odorata*. — PETITE. *Erythræa Centaurium*.

CENTAURELLA. A North American genus of herbaceous plants, belonging to the gentian family. It has terminal four-parted funnel-shaped flowers, and a one-celled ovary, surmounted by a two-lobed stigma. [M. T. M.]

CENTAURIDIUM. A genus of *Compositæ*. The only species, *C. Drummondi*, a Texan plant, has great resemblance to some of the knapweeds, but belongs to a different section of the family. The plant is biennial, a foot and a half high, much branched, with linear smooth jointed leaves, and single terminal yellow flower-heads. The ray florets are strap-shaped and female; those of the disc tubular and perfect. [A. A. B.]

CENTAURY. *Erythræa Centaurium*. — AMERICAN. A common name for *Sabbatia*.

CENTENILLE. (Fr.) *Centunculus*.

CENTINODE. (Fr.) *Polygonum aviculare*.

CENTRADENIA. A genus of *Melastomads*, containing under-shrubs from Mexico and Central America, with four-sided branches, and opposite leaves generally unequal in size and unequal-sided, which are ovate or lanceolate, entire, membranous, and three-nerved. The racemes are few-flowered, axillary; the flowers pink or white. Calyx tube four-sided, its limb four-parted; petals four; stamens eight; the two larger anthers spurred, the others with a glandular appendage to the connective; ovary free, four-celled, with a ring of hairs at the top. [J. T. S.]

CENTRANTHERA. A small genus of *Scrophulariaceæ*, natives of tropical Asia and Australia. They are scabrous herbaceous plants, with generally opposite leaves, and almost sessile axillary flowers. The calyx is compressed, and split down the inner margin, entire or two to five-toothed. The corolla tube is curved and dilated upwards; its limb has five broad lobes, the two upper being innermost in the bud. There are two pairs of included stamens, having transverse two-celled anthers, with mucronate cells. The capsule is obtuse. [W. C.]

CENTRANTHUS. A small genus of valerian-worts, consisting of smooth annual or perennial European plants, with mostly entire opposite leaves, and small red or white flowers in terminal corymbose panicles, the flowers arranged unilaterally along the branches of the panicle. A slender tubular spurred corolla with a five-lobed limb, one of the lobes standing apart from the rest; a single stamen; a superior calyx of feathery pappus-like appendages rolled inwards before the corolla falls, and only expanded as the fruit matures; and a one-celled, one-seeded fruit are the principal features of this genus. It differs from *Valeriana*, in having a spurred corolla and but one stamen. The pappose calyx is a pretty object under a lens. The Red Valerian, *C. ruber*, formerly known as *Valeriana rubra*, offers a good example of the genus. It is said to be eaten as a salad in Southern Italy, and its sweet-scented roots probably partake in some degree of the antispasmodic and tonic properties occurring in the true valerians. *C. macrosiphon* is a very pretty annual species from Spain, with smooth hollow stem, broadly ovate sessile leaves, entire, or pinnatifid, and very large corymbs of rose-coloured flowers. [W. T.]

CENTRIFUGAL. A term applied to those kinds of inflorescence which, like the cyme, flower first at the point or centre, and last at the base or circumference.

CENTRIPETAL. A term applied to those kinds of inflorescence which, like the spike or capitulum, flower first at the base or circumference, and last at the point or centre.

CENTROCARPHA. A group of the composite family, differing in no way from *Rudbeckia*. The species referred to it are N. American perennial herbs very frequently met with in gardens. Their leaves are alternate, entire or lobed, and generally scabrous. The flower-heads are large and yellow, terminating the stem or branches. In *C. grandiflora* (otherwise *Rudbeckia grandiflora*), the flower-heads are sometimes more than six inches in diameter, and much like those of the sunflower, but smaller. [A. A. B.]

CENTROCLINIUM. A genus of the composite family, belonging to that section of the order which has two-lipped corollas. The four known species are herbs or small shrubs found in the Peruvian Andes at an elevation of 6,000 to 8,000 feet. Their leaves are alternate, stalked, toothed or entire, and covered beneath, as well as the stems, with a white tomentum. The purple flower-heads are axillary and single, on long stalks, and about an inch in diameter; the ray florets few and female, those of the disc numerous, and containing both stamens and pistil. The achenes are five-angled, crowned with a pappus of numerous unequal rough hairs, and seated on a flat receptacle furnished with short bristles. *C. adpressum* and *C. reflexum* have been in cultivation, but their rose-coloured flowers, which smell of Hawthorn, are very sparingly produced. [A. A. B.]

CENTROLEPIS. A genus of *Desvauxiaceæ*, containing a few small tufted sedge-like herbs from Australia and Tasmania. Leaves setaceous, all radical; scapes short, terminated by a simple spike contained in a spathe formed by two slightly unequal bracts (glumes of some authors); glumes (pales of those who consider the spathe-bracts as glumes) two, membranous, stamen one; ovaries two to twelve, becoming utricles in fruit. [J. T. S.]

CENTROLOBIUM. A genus of leguminous trees found in Brazil, Guiana, and Venezuela. Their leaves are a foot or more in length, and unequally pinnate, the leaflets three to four inches in length, and, as well as all the young parts, clad with a rusty pubescence. The flowers are disposed in terminal panicles. The pod is the most remarkable part of the plant; it is like the fruit of the common Maple (*Acer*) in form, and about nine inches in length, the lower or seed-bearing portion globular, and clad with long straight prickles, the upper of winged portion thin, papery in texture, about two and a half inches in breadth, and bearing on its back near the base a long straight spurred spine, which is the hardened style. *C. paraense* furnishes one of the most esteemed timbers of the Orinoco; its colour is bright orange when fresh, but it fades to a brown after exposure; it is very strong, dense and durable. The name of the genus is derived from the spur-like hardened style which remains on the pod. [A. A. B.]

CENTRON, or CENTRUM. In Greek compounds = *calcar*, a spur.

CENTROPAPPUS. A genus of the composite family, found in Tasmania, nearly allied to *Senecio*, and differing chiefly in habit. The only known species, *C. Brunonis*, is found about the upper limits of the forest on Mount Wellington, at an elevation of 3,000 to 4,000 feet. It is a smooth shrub, seven to ten feet high. The leaves are sessile, gathered together towards the ends of the branches, three to four inches long, and one-quarter of an inch broad. The flower-heads are in terminal corymbs, and in form and appearance bear great resemblance to those of the common yellow ragwort. [A. A. B.]

CENTROPETALUM *distichum* and *C. Warscewiczi* are two small epiphytal orchids from the mountains of tropical America, with fleshy distichous leaves, and brownish solitary flowers, with a broad lip adherent to a hooded column. The pollen masses are four, free, attached in pairs to two filiform curved radicles which adhere to a common gland.

CENTROPOGON. A genus of *Lobeliaceæ*, consisting of undershrubs with irregular flowers on long axillary stalks. The five stamens are united into a tube, and spring from between the corolla, and a ring-like fleshy five-lobed disc, surrounding the inferior two-celled ovary. The two lower anthers are terminated by an ovate triangular cartilaginous point. The plants are natives of tropical America. [M. T. M.]

CENTROSEMA. A genus of prostrate or twining perennial plants belonging to the *Leguminosæ*, and distinguished from its nearest allies by its having on the back and near the base of the standard a short spur, from which circumstance the genus receives its name. The species are almost entirely American, and the greater number are found in Brazil. The leaves are made up of three leaflets, rarely of five or seven, the leaflets opposite and the terminal one rather distant; in two species they are digitately arranged and from three to five in number, while in a few others but one leaflet is present. The large and elegant pea-like flowers are single or in axillary racemes, and white violet rose or blue in colour. The pods are very narrow, compressed, thickened at both sides, and terminating in a long point; in some of the species they are eight inches long. The leaves of *C. macrocarpum* are eaten in Guiana. *C. virginianum* is found in Brazil and West Africa, as well as in the United States. Upwards of twenty species are known. [A. A. B.]

CENTROSIS. *Corysobis*.

CENTROSOLENIA. A genus of *Gesneraceæ* from British Guiana, founded on a single plant, which has a short creeping stem, subcordate petiolate leaves, and solitary axillary peduncles, sometimes bearing many pedicels. The calyx is five-parted with serrate segments; the tube of the corolla has a spur at its base, and the limb is slightly expanded into five small broad lobes. The four included didynamous stamens, with the rudimentary fifth, are inserted in the base of the tube. The ovary is oblong-conical and hairy. This genus is evidently allied to *Nematanthus*, but the spur of the flower, coupled with the habit and the toothed segments of the calyx, distinguish it. [W. C.]

CENTROSPERMUM. The name sometimes given to an annual cornfield weed of Spain and Algeria, very near to the genus *Chrysanthemum*, and very like our own corn-marigold, *Chrysanthemum segetum*. The achenes in the last-named plant are naked at top, but in this, *Ch. viscosum*, those of the ray florets have a pappus of three, and those of the disc of one awn, while the stems are smooth and not clammy. [A. A. B.]

CENTROSTEMMA. A genus of *Asclepiadaceæ*, containing five species, natives of the Indian Archipelago, the Moluccas, and the Philippine Islands. They are twining shrubs, with opposite coriaceous leaves, and umbels on interpetiolar and terminal peduncles, composed of many large yellowish flowers. The calyx is five-parted. The limb of the corolla is deeply five-cleft and reflexed; a hairy ring exists in the throat of the corolla around the base of the gynostegium, which is exserted. The staminal corona consists of five fleshy leaves inserted on the summit of the gynostegium and surpassing the stigma. The anthers are surrounded by a spreading membrane which attaches them to the stigma; the pollen masses being oblong, with a pellucid interior margin, and attached by short processes. The pentagonal stigma is lengthened out into a cone. The follicles are solitary, long and cylindrical, and contain numerous comose seeds. The hairy ring in the throat of the corolla separates this genus from *Hoya*, to which otherwise it is very nearly related. *C. multiflorum* is a well-known handsome hothouse shrub, often called *Cyrtoceras reflexum*. [W. C.]

CENTUNCULUS. Bastard Pimpernel. A minute herbaceous plant belonging to the *Primulaceæ*, and closely allied to *Anagallis*, from which it may at once be distinguished by its four-parted flowers and four stamens, which are glabrous. The whole plant consists of a small fibrous root, a simple or slightly-branched stem, which rarely exceeds an inch and a half in height, from a dozen to twenty, or less, ovate-pointed sessile leaves, and a few solitary sessile flowers of a pinkish hue and of very short duration. The seed-vessels resemble those of pimpernel, for a starved specimen of which the plant might be mistaken. It grows in many parts of Great Britain in sandy or gravelly places, especially where water has stood during the winter, and not unfrequently in company with another minute plant, *Radiola Millegrana*. French, *Centenille bassette*; German, *Centunkel*. [C. A. J.]

CEPHAELIS. The plant producing the true Ipecacuanha belongs to this genus of *Cinchonaceæ*, which is characterised by its flowers being collected together in heads surrounded by a leafy involucre; the limb of the calyx very small and five-toothed; the corolla funnel-shaped with five small lobes; the anthers inclosed within the corolla; and the fruit succulent with two compartments, each containing a single seed, striated on the outer side. The Ipecacuanha plant is a native of Brazil. Its root, the part used in medicine, is flexuose but little branched, and the rind is marked by a number of circular projecting knots or rings which are very characteristic. The stem is creeping and herbaceous, with oblong obovate leaves and drooping heads of flowers. The emetic properties of the root are due to a chemical principle called *emetin*.

Ipecacuanha is largely employed in medicine as a safe emetic, and in smaller quantities it acts on the skin, but especially on the bronchial passages. Some persons are so susceptible to the influence of this drug that they cannot remain in a room where there is ipecacuanha without severe suffering. It is likewise highly esteemed in dysentery, though not so much so now as formerly. Louis XIV. paid 1000 Louis d'or to a physician named Helvetius for the purchase of a remedy for dysentery, under which the Dauphin was then suffering. This remedy was Ipecacuanha. Helvetius derived his knowledge of it from a merchant, who from gratitude for attention paid him during illness, by Helvetius, gave the latter some of the root as a remedy for dysentery. [M. T. M.]

CEPHALANDRA. A diœcious climbing cucurbitaceous plant, native of the Cape of Good Hope, with thickened branches, simple tendrils, and large orange-yellow flowers with a five-toothed calyx. The five stamens grow in three parcels, inserted into the base of the corolla, and are adherent at the top into a globose head bearing the anthers—hence the name of the genus. The fruit is of the size of a pigeon's egg, and of a purple colour. [M. T. M.]

CEPHALANTHERA. A genus of orchids cut off by Richard from *Epipactis*, which the species entirely resemble in their tough fibrous roots and broad ribbed leaves, not only clothing the whole stem, but passing gradually into bracts. It differs from *Epipactis* in its anthers being terminal, as in *Arethusa*, not dorsal. The species have nearly regular white or red half-closed flowers with a saccate hypochil, and do not occur in the New World or the southern hemisphere. In the Old they are found from Western Europe to the extremest East of Asia, in the Japanese Archipelago. *C. pallens*, *ensifolia*, and *rubra* are wild in woods in this country.

CEPHALANTHIUM. The capitulum or flower-head of composites.

CEPHALANTHUS. A name expressive of the aggregation of the flowers into heads, and applied to a genus of cinchonaceous plants called in North America Button-wood. The calyx is tubular with an angular four-toothed limb; the corolla tubular, with a four-toothed limb; the stamens four in number, scarcely protruding from the corolla; the style protruded for a considerable distance from the throat of the corolla; and the stigma capitate. The fruit is inversely pyramidal in shape, crowned by the limb of the calyx, two to four-celled, each cell or compartment containing one seed, or sometimes two of the seeds are absent. The seeds are terminated by a small thickened knob at one end. *C. occidentalis* is a bushy shrub with leaves opposite, or sometimes three in a whorl, and yellowish white flowers in round heads of the size of a marble. [M. T. M.]

CEPHALARIA. A genus belonging to the teazelworts, characterised by having the leaves, which surround the heads of flowers, shorter than the appendages which are attached to the surface supporting the flowers. The covering, technically called involucel, which surrounds each flower, is four-sided, with eight grooves, and four to eight teeth at the margin. The name of the genus is derived from the Greek word signifying a 'head,' indicating the form assumed by the groups of flowers. There are about twenty species known, some of which are natives of Middle Europe, others occur in N. Asia and at the Cape; they are mostly perennial herbs, a few being annual, with opposite leaves, which are either toothed or deeply divided; the flowers white, yellow, or lilac. [G. D.]

CEPHALELYNA. A section of *Evelyna*.

CEPHALIUM. A peculiar woolly enlargement of the apex of the stem of *Melocactus*, among whose hairs the flowers appear.

CEPHALODIUM. A knob-like shield, such as occurs in the genus *Scyphophorus*. Also the capitulum of composites.

CEPHALOMANES. A name under which it has been proposed to separate a few species of *Trichomanes*, typified by *T. javanicum*. It is not generally adopted.

CEPHALOPHORUM. A term employed among fungals, sometimes to denote their receptacle, sometimes their stipe.

CEPHALOTACEÆ. The Australian Pitcher-plant, *Cephalotus follicularis*, a curious herb, with radical leaves mingled with pitchers, is a plant of very doubtful affinity. It has been considered provisionally as a distinct family, bearing the name of *Cephalotaceæ*. It has been compared with *Rosaceæ*, *Crassulaceæ*, and *Ranunculaceæ*; but it will probably ultimately find its place amongst or in the immediate neighbourhood of the *Saxifragaceæ*.

CEPHALOTAXUS. A genus of *Coniferæ* of the tribe or family of *Taxaceæ*, nearly allied to the yew (*Taxus*), in general habit, foliage, and essential characters; but the male flowers are in small heads, consisting of several closely-clustered catkins, and the fleshy disk, instead of forming an open cup round the base of the seed, completely closes over it into an entire pericarp, two or three of these fruits being collected into a drupe-like head. There are four or five species known, all from Japan or North China, one of which, *C. Fortuni*, is now frequently planted in our collections of conifers.

CEPHALOTUS. A genus of very singular dwarf pitcher-plants. *C. follicularis*, the only species, is a native of swampy places in King George's Sound, and may frequently be met with in gardens. It has a very short or contracted stem, with spoon-shaped stalked leaves, among which are mingled small pitcher-like bodies, placed on short stout stalks, and closed at the top with lids like the true pitcher-plants (*Nepenthes*). These pitchers are of a green colour, spotted with purple or brown, and provided with hairs; the mouth furnished with a thickened and regularly notched rim. The flowers are borne on a long spike, and have a coloured six-parted calyx, without a corolla; twelve stamens, six longer than the rest, inserted into a disc, the anthers provided with a large connective. There are six distinct carpels, each bearing a single seed. Dr. Hooker in a valuable paper on *Nepenthes* in the *Transactions of the Linnæan Society*, says

Cephalotus follicularis.

that there are no intermediate stages between the ordinary leaves and the pitchers of *Cephalotus*, but that the transition from one to the other is as sudden and abrupt as from the cotyledons to the pitchers in the seedling *Nepenthes* described by him. The writer of this notice, however, has on more than one occasion observed intermediate stages between the leaves and pitchers of this Australian Pitcher-plant, in the shape of leafstalks dilated and hollowed out at the point in the form of a horn, or of the mouth of a trumpet. [M. T. M.]

CEPHALOXYS. A section of the rush genus (*Juncus*) containing such species as have the capsule perfectly three-celled, the valves breaking away from the partitions, which remain attached to the central columella. The *J. repens* of the southern states of North America is the type of this section. [J. T. S.]

CEPHALUM. In Greek compounds = the head, or terminal mass, or thickened end of anything.

CERA DE PALMA. The Peruvian name for the waxy resinous matter secreted by the wax-palm, *Ceroxylon andicola*.

CERACEUS, CEREUS. Having the consistence or appearance of wax.

CERADIA. A genus of the composite family found on the south west coast of Africa. The only known species, *C. furcata*, is a shrub with fleshy horned and forking stems, bearing on their apex a number of bright green succulent veinless leaves, which are entire and spathulate in form. From the axils of these the flower-heads proceed; they are solitary, of a pale yellow colour, and placed on stalks hardly so long as the leaves. The name *Ceradia* has allusion to the horned appearance of the branches. From the wounded stems of the plant exude small tears of a gum resin, which in burning has a smell resembling that of myrrh, and has been called African Bdellium. [A. A. B.]

CE'RAISTE COMMUNE. (Fr.) Any wild *Cerastium*.

CERAMIACEÆ. A division of rose-spored *Algæ* distinguished amongst those which have their spores collected without order within a hyaline sac (*Gongylosperm*), by the capsular fruit being either naked or surrounded by a whorl of threads. The external walls of the capsule vary in character, and are sometimes membranous (*favella*), as if formed of a transformed mother cell. The frond is either compound or simple and filamentous. [M. J. B.]

CERAMIDIA. A name given to the globose ovate or conical capsules of rose-spored *Algæ*, mostly opening by a terminal pore, and quite distinct from the frond. They are, however, sometimes difficult to distinguish from coccidia. Examples are afforded by *Laurencia*. [M. J. B.]

CERAMIUM. A genus of articulated rose-spored *Algæ* known at once by its central thread being covered at intervals with a layer of cells which give it a knotted appearance. Sometimes the sepals project so as to give the frond somewhat the appearance of the stem of *Equisetum*. The tetraspores are sunk in the frond. Capsular fruit, consisting of a hyaline cell containing many angular spores. Several species occur on the coast, one or two of which are amongst the most ordinary parasites upon larger sea-weeds. *C. rubrum* is one of the sea-weeds most commonly collected by summer visitors of our coasts, abounding in company with the more delicate *C. gelatinosum* in almost every little pool amongst the rocks. [M. J. B.]

The name is also a synonym of *Bostrychiana*, a peculiar genus of South American *Algæ*. [T. M.]

CERANAIBA. The Brazilian name of a Palm called *Copernicia cerifera*.

CERASTIUM. A rather extensive genus of *Caryophyllaceæ*, containing small white-flowered plants, generally called Mouse-ear Chickweeds. Many of them are annuals, and are more or less hairy or glandular. They are distinguished from other genera of *Alsineæ*, by their cylindrical capsule opening by twice as many teeth as there are styles, the latter being usually five. The petals are generally bifid. The number of sepals, petals, and stamens varies; it is generally five in the two former, and ten in the staminal whorl. Several species occur in Britain. *C. trigynum* is an Alpine decumbent plant with only three styles, while in all the other British species there are five. *C. alpinum* and *C. latifolium* are Alpine plants with erect flowering stems, and petals much longer than the calyx; the former has soft, the latter short rigid pubescence. *C. arvense* is a common English plant, somewhat resembling the last two, but with much narrower hairs, and the bracts and sepals membranous at the edges. The other species have the petals scarcely exceeding the calyx, and often shorter than it. [J. T. S.]

CERASUS. A genus of *Drupaceæ*, frequently combined with *Prunus*, but distinguishable by having the following characters. The young leaves are folded in halves; the flowers are arranged in umbel-like tufts, appearing before the leaves or in terminal racemes which are produced with the leaves; the fruit is nearly globular in shape, destitute of the mealy bloom of the plum, or the down of the apricot, and having a roundish smooth stone. There are many species of this genus distributed over the temperate regions of both hemispheres; but as they are very subject to variations in habit and appearance, their discrimination is a matter of great difficulty. *C. Avium*, the Wild-cherry or Gean, is a native of Britain; it is a tree producing no suckers, its flower-buds are destitute of leafy scales, and the flesh of the fruit adheres to the stone, so as not to be readily separated from it. *C. vulgaris* is also a native of Britain; it is a shrubby plant, throwing up numerous suckers from its roots, the flower-buds have leafy scales, and the flesh of the fruit is readily separable from the stone. The wood of these trees is in great request in France, where mahogany is less common than with us; it is employed by cabinet-makers and musical-instrument makers. The bark also affords a yellow dye, while the leaves are said to be used to mix with tea. The fruits of *C. Avium* are employed in Switzerland and various parts of Germany in the distillation of a cheap spirit known as kirschewasser. Maraschino, ratafia and other liqueurs are made in part from the fruits of this tree or some of its varieties. The stalks of the fruits are said to be employed in France as a diuretic. A kind of gum, analogous to tragacanth, exudes in great abundance from these and also from other species of this genus. It is employed by hat-makers and others. A double-flowered variety of *C. vulgaris* is in cultivation; its flowers are very showy and interesting botanically from the fact that the pistil is replaced by two small green leaves. *C. Padus*, the Bird-cherry, is also a native of the British Isles; in Scotland it is known as the Hagberry. It differs from the foregoing in the flowers being arranged in terminal clusters or racemes. The fruit is small, black, and nauseous to the taste. In the north of Europe it enters into the formation of a palatable liqueur; the juice is also expressed and drunk with milk, while the residue of the fruit is kneaded up into cakes. *C. Mahaleb*, a native of the middle and south of Europe, is remarkable for the fragrance of its flowers, which, as well as the leaves are used by perfumers. A decoction of the leaves is also used in the manufacture of tobacco in France. The wood is prized by cabinet-makers, and in Austria the small branches are used for pipe-stems. *C. virginiana*, an American tree, frequently cultivated in this country, affords valuable wood for cabinet makers. Its bark is astringent and is esteemed for its febrifugal properties. From the fruits a liqueur is made, and when dried they are mixed with pemmican. *C. Capollin*, a native of Mexico, has also febrifugal properties. The rind of the root is used in cases of dysentery, and by tanners. The leaves and kernels of this, and indeed of most of the species, contain a greater or less proportion of prussic acid; thus the leaves of *C. virginiana* are dangerous on this account. *C. Capricida* derives its specific name from its fatal effects when eaten by goats in Nepal. It is this generally minute quantity of hydrocyanic or prussic acid that renders so many of these fruits useful for flavouring liqueurs; among others the kernels of *C. occidentalis* are used for flavouring noyeau.

The species heretofore mentioned have all deciduous leaves, but there are two well-known species, that have evergreen leaves. One is *C. lusitanica*, commonly called the Portugal laurel (though it has no botanical affinity with the true laurels: see *Laurus*), which is one of the commonest of evergreen shrubs, very hardy and very ornamental, especially when in flower. The leaves are dark green with reddish stalks; the flowers white, in clusters; and the fruits small, dark purple. These latter are much relished by birds. One of the largest bushes of this species is in the Duke of Marlborough's park at Blenheim. The other common evergreen species is *C. Lauroceraus*, the Cherry-laurel, or Common laurel as it is usually called. This has widely lance-shaped remotely serrate leaves of a bright shining green colour above, dull on the lower surface. The leaves, bark, and fruit, as well as the oil obtained from them, are more or less poisonous. The vapour of the bruised leaves is sufficient to destroy small insects. Cherry-laurel water is a watery solution of the volatile oil of this plant; it contains prussic acid, and its effects, medicinal and poisonous, are similar to those of that acid. Sweetmeats, custards, &c., flavoured with the leaves of this plant have occasionally proved fatal; hence it is better to discard the use of these leaves altogether for these purposes, and to employ the leaves of the Sweet Bay, *Laurus nobilis*, instead, as these are equally agreeable in flavour, and harmless. The Cherry-laurel was introduced into this country from the Levant in the sixteenth century. [M.T.M.]

The numerous varieties of cultivated Cherries have in all probability originated from *C. Avium* and *C. vulgaris*. Those belonging to *C. Avium*, of which the Bigarresu and the Black Heart may be instanced as typical of the better kinds, have generally large pendent leaves, waved on the margin, with sharp prominent veins beneath, coarsely serrated, of thinner texture, and of a more yellowish-green colour than those of the C. vulgaris; buds pointed; flowers large, proceeding from wood of not less than two years old; petals loosely set; stamens slender, irregular in length, some being longer and others shorter than the style. From C. vulgaris are derived such varieties as the May Duke, Kentish, and Morello. The leaves are generally smaller than those of the preceding species, and have their margins plain, with the veins beneath as they approach the margin scarcely at all prominent, the parenchyma or fleshy substance of the leaves being much thicker than in the former; their colour is deep green; petioles comparatively short and thick, supporting the leaves nearly erect; petals roundish, forming a regular cup-shaped flower, with strong stamens, generally shorter than the style. Fruit round, roundish heart-shaped, or oblate, with aqueous flesh; colour red, dark red, or nearly black, none being white, nor white and red.

Both these species appear to be natives of Europe, although Pliny states that there were no Cherries in Italy before the victory obtained over Mithridates by Lucullus, who was, according to the above author, the first who brought them to Rome, about sixty-eight years before the Christian era. It is also stated by the same authority that, "in less than 120 years after, other lands had Cherries, even as far as Britain beyond the ocean." Pliny's statements, Professor Targioni observes, gave rise to the tale, so generally received as a fact, that Cherries came originally from Cerasonte, now Zefano, and were therefore called Cerasus by the Latins. It may be here observed that nearly all the names of the Cherry in the south of Europe and Germanic languages are derived from the Κέρασος of the Greeks. Now, Decandolle says that the Cherry tree is decidedly wild in Europe, and especially in Greece, where it had existed from a very early period, for it is mentioned by Theophrastus B.C. 300, more than two centuries before its reputed first introduction to Rome by Lucullus from Cerasonte. Some authors are therefore of opinion that the name of that city had been derived from the tree, previously known as Cerasus in the south of Europe, and not that of the tree in question from the city. In the gardens of the latter, and in the surrounding country, Cherry-trees may have been so remarkably abundant as to occasion its being distinguished by their name.

When the Rev. Dr. Walsh visited Turkey in 1824, amongst other plants of which he gave an account (Trans. Hort. Soc. vi. 32), he mentions the abundance of Cherry-trees as follows: "Prunus Cerasus, two varieties. The first of these varieties is a Cherry of enormous size, that grows along the northern coast of Asia Minor, from whence the original Cherry was brought to Europe. It is cultivated in gardens always as a standard, and by a graft. The gardens consist wholly of cherry-trees, and each garden occupies several acres of ground. You are permitted to enter these, and eat as much fruit as you please, without payment; but if you wish to take any with you, you pay about a halfpenny per pound. The second variety is an amber-coloured transparent Cherry of a delicious flavour. It grows in the woods in the interior of Asia Minor, particularly on the banks of the Sakari, the ancient Sangarius. The trees attain a gigantic size; they are ascended by perpendicular ladders suspended from the lowest branches. The trunk of one which I measured was five feet in circumference; and the height where the first branches issued forty feet; from the summit of the highest branch was from ninety to 100 feet; and this immense tree was loaded with fruit.'

From a country naturally so favourable to the growth of the Cherry, it is probable that Lucullus may have brought some varieties different from any known at Rome; but, being indigenous to Italy, Cherries must have been familiarly recognized by a name common to them in the south of Europe long before the Romans extended their conquests as far as Asia Minor. In consequence of Pliny's statement, the existence of the Cherry as a native of Britain has been questioned; but Mr. Knight was of opinion that Pliny 'must have meant a cultivated variety of the Cherry, of which the Romans had many in his time; for the small black Cherry which abounds in our woods has much too permanent habits to have been derived from any cultivated variety.' The species to which Mr. Knight alludes is the C. Avium or C. sylvestris, commonly to be met with in the woods of this and other countries of Europe. Some of its varieties are occasionally found almost equal in size and quality to the cultivated sorts. Among these may be mentioned the Couronne Cherry, so called from its being as black as a crow, which reproduces itself from seed, and is very abundant in several parts of England, and particularly in Hertfordshire. C. vulgaris does not appear to be in general so plentiful as C. Avium; yet there is a variety of it which grows wild, abundantly, by the sides of the Como Lake in Italy, and which proves to be a sort of Morello, but smaller and more round than the common. Varieties resembling it, and evidently belonging to the same species, have also been found wild in Britain.

With regard to the present race of cultivated Cherries, doubtless many of them were introduced from Holland and Belgium. Evelyn says, 'It was owing to the plain industry of one Richard Haines, a printer to King Henry VIII., that the fields and environs of about thirty towns, in Kent only, were planted with fruit trees from Flanders, to the unusual benefit and general improvement of that county to this day.' The Kentish, sometimes called the Flemish, had probably been introduced at the above period, and likewise the Bigarreau: the former is the Cerise de Montmorency, and had most likely been obtained by the Dutch from France; but it would appear from Knoop that the Bigarreau tribe of Cherries had been introduced to the Continent from Spain; for, he says, in Germany and the Netherlands these are called Spanish Cherries (*Spansche Kersen*).

The cultivated varieties are now very numerous in this country. The following rank among the best: May Duke, Knight's Early Black, Elton, Bigarreau, Florence, Kentish and Morello. The last two are not properly dessert kinds, but are otherwise very useful. The Kentish is chiefly used for pies; its stalk is so strongly attached to the stone that the latter may be withdrawn from the fruit by it, so as to leave the cherry apparently whole, and in this state the fruit is laid on hair sieves and exposed to the sun, where it dries like a sultana raisin, becomes a delicious sweet-meat, and will keep thus for twelve months. The Morello is the sort chiefly employed for preserving in brandy.

Several highly-esteemed liqueurs are prepared from Cherries. The German Kirschwasser is made by distilling the fermented juice of the pulp with which the stones and kernels are ground and mixed. Maraschino, the most celebrated liqueur of Italy, is also obtained by the distillation of a small black Cherry, with which, while fermenting, honey, some cherry leaves, and the kernels of the fruit, are mixed. The celebrated Ratafia of Grenoble is prepared from pounded Cherries, to which brandy, sugar, and spices are added, the mixture being then placed in the sun or near a fire. The gum of the Cherry tree closely resembles gum Arabic in its nature and properties. The wood is hard and tough, and is used by the cabinet-makers. It has been occasionally employed for rifle stocks instead of walnut. [R. T.]

CERATANDRA. Under this name are collected several species of terrestrial Orchids, inhabiting the Cape of Good Hope. They have grassy leaves, covering the scape, and closely packed green or yellow flowers, turning black in drying. The anther is a great inverted horseshoe-shaped body; the lip, which is heart-shaped or angular, and bears some kind of process in its middle, is attached to the face of the column by a narrow unguis. The species grow in sand, into which they introduce long succulent hairy fibres; they seem to be uncultivable.

CERATIOLA. A small heath-like evergreen shrub, belonging to the *Empetraceæ*, among which it is distinguished by its two-leaved membranaceous calyx, with four scales at the base, two petals, and two stamens. *C. ericoides*, the only species, is an upright much-branched shrub, greatly resembling a heath, and varying from two to eight feet high; the branches are erect, somewhat whorled, and marked with the scars of the fallen leaves; the leaves are in whorls of four, very narrow and spreading. Flowers brownish, and very small, solitary in the axils of the upper leaves. 'A native of South Carolina, on the Edisto River, where it covers a space 300 or 400 yards in width, and two or three miles long, which appears to have been a sandbank formed by some of the ancient freshets of the river. According to Pursh, it is also found in the gravelly dry soil of Georgia and Florida; and in great plenty on the islands at the mouth of St. Mary's River.' (*London*.) [C. A. J.]

CERATITES. A name applied by Link to the long ragged species of *Aridium* which grow on the leaves of the mountain ash and whitethorn, sometimes attacking the fruit of the latter and distorting it. They are now placed in the genus *Rastelia*, to which we shall have occasion to refer hereafter. [M. J. B.]

CERATIUM. This is usually called a capsula siliquiformis. A long slender horn-like one-celled superior fruit, as in *Hypecoum*.

CERATOCALYX. A genus of *Orobanchaceæ*, containing a single species, parasitic upon the roots of other plants, a native of mountains in Spain. It has a simple scaly stem, and solitary sessile flowers in the axils of the bracts, like *Orobanche*, from which it scarcely differs, except in the structure of the calyx, which is gamosepalous, with a campanulate tube lengthened out laterally into two acute narrow lobes, truncate before and behind, and exhibiting no traces of the union of the sepals. [W. C.]

CERATOCAPNOS. A genus of *Fumariaceæ*, the four petals of which are spurred at the base and two-lobed at the apex; stamens six, united into two bundles; style simple, deciduous. Fruit either a one-seeded nut, marked with five ribs, and terminated by a long beak or a lance-shaped pointed capsule, two-valved and two-seeded, the valves marked with five ribs. The plants are scrambling shrubs, natives of Syria and Algeria. [M. T. M.]

CERATOCEPHALUS. A small genus of *Ranunculaceæ*, natives of Central and Southern Europe. They are small annuals covered with cottony hairs, having many-cleft radical leaves, and numerous short one-flowered scapes; calyx with five sepals; petals five, small, yellow; stamens five to fifteen; ovaries numerous. Achenes in an oblong spike on the receptacle; they have two protuberances and two empty cells at the base, and terminate in sword-shaped beaks, about half an inch long when mature. This beak which characterises the genus, is curved upwards in the commonest species, *C. falcatus*, but is nearly straight in *C. orthoceras*. [J. T. S.]

CERATOCHILUS. Under this name stand three very little known diminutive orchids with simple stems, fleshy distichous leaves, and minute axillary flowers. They inhabit trees in Java, where they live among mosses which partly conceal them. The supposed genus *Omea* is one of the species. The *Ceratochilus* of Loddiges' Botanical Cabinet is *Stanhopea*.

CERATOCHLOA. A genus of grasses belonging to the *Festuceæ*; only one species has been described, namely, *C. pendula*, which is *Bromus Schraderi*, a native of Carolina. [D. M.]

CERATODACTYLIS. A synonymn of *Lisæa cordifolia*, a beautiful Mexican fern. [T. M.]

CERATOGONUM. A genus of *Polygonaceæ* founded on a plant cultivated in the Calcutta Botanic Garden. The leaves are stalked, ovate-triangular or hastate, with ochreate stipules, ciliated at the apex, and extra-axillary lax filiform flower-spikes. The flowers are monœciously polygamous, the males with a five-parted coloured calyx, while that of the perfect flowers consists of six segments in two rows, the three inner ones petaloid, the three outer leathery, inserted into a tube; stamens eight; and adhering to the tube of perianth. [J. T. S.]

CERATONIA. A genus of leguminous plants remarkable on account of its flowers being destitute of a corolla, having only a small five-parted calyx, five stamens, and a pistil with a sessile stigma. The male and female organs are occasionally produced in distinct flowers on different trees.

C. Siliqua, the only species, is a native of the European, African, and Asiatic countries bordering on the Mediterranean, where it forms a small branching tree about thirty feet in height, having wood of a pretty pinkish hue. Its pinnate leaves are composed of two or three pairs of oval blunt-topped leaflets, of a leathery texture, and a shining dark-green colour. The flowers are in small red racemes; and are succeeded by flat pods, from six inches to a foot in length, an inch or rather more in width, and scarcely a quarter of an inch in thickness, of a shining dark purplish-brown colour; they do not split open like many other pods, and contain numerous small seeds arranged in a line along the centre of the pod, each seed being contained in a separate cell formed by the fleshy pulp of the pod. The tree is extensively cultivated in many of the above-mentioned countries, especially in such as suffer from periodical drought, its long roots penetrating to a great depth in search of water. It is called Algaroba by the Spaniards, and Kharoub by the Arabs, whence comes our English name Carob or Caroub, the pods being called carob-peas, or carob-beans, or sometimes sugar pods. These pods contain a large quantity of agreeably flavoured mucilaginous and saccharine matter, and are commonly employed in the south of Europe for feeding horses, mules, pigs, &c., and occasionally, in times of scarcity, for human food. During the last few years considerable quantities of them have been imported into this country and used for feeding cattle; but although they form an agreeable article of food, they do not possess much real nutritive property, the saccharine matter belonging to the class of foods termed carbonaceous or heat-givers, the seeds alone possessing nitrogenous or flesh-forming materials, and these are so small and hard that they are apt to escape mastication. They form one of the ingredients in the much-vaunted cattle-foods at present so extensively advertised, the green tint of these foods arising from this admixture. Some years ago they were sold by chemists at a high price, and were used by singers who imagined that they softened and cleared the voice. By fermentation and distillation they yield a spirit which retains the agreeable flavour of the pod.

Besides the name of Carob-beans, these pods are also commonly called Locust-pods, or St. John's Bread, in consequence of its having once been supposed that they formed the food of St. John in the wilderness, but it is now more generally admitted that the locusts of St. John were the animals so called, and which are at the present day used as food in Eastern countries. There is more reason, however, for entertaining the belief that these pods were the husks mentioned in the parable of the prodigal son. The small seeds are said to have been the original carat weight used by jewellers. [A. S.]

CERATOPETALUM. A genus of Australian shrubs or small trees, belonging to *Cunoniaceæ*. The leaves are opposite, ternate, with the leaflets coriaceous, serrated, the stipules somewhat leaf-like, caducous. The flowers are small yellow in terminal panicles; the calyx tube is adherent to the ovary, and the limb is five-parted; petals five, cut into a fringe of linear segments; stamens ten, the anthers beaked; capsule one-seeded, gaping at the apex, and crowned by the calyx limb. They have a gummy secretion. [J. T. S.]

CERATOPHYLLUM. CERATOPHYLLACEÆ. An aquatic floating herb, with numerous verticillate linear-filiform leaves several times forked; and minute sessile unisexual flowers of the most simple construction. There is no real perianth, but each flower is surrounded by a whorl of minute bracts; the males consist of twelve to twenty oblong sessile anthers; the females of a small ovary with a simple style, and containing a single pendulous ovule. The fruit is a small nut, smooth or more or less armed with prickly appendages, the seed has no albumen, and the embryo is remarkable for a highly developed placenta. The plant has some general resemblance to the aquatic *Haloragæ* or the *Callitriche*, but there is nothing in its nature to indicate any immediate affinity with the various families to which it has been appended, and it stands at present as an isolated genus or family. *C. demersum*, the only species known, is common in pools or slow streams over a great part of the world.

It varies much in the shape and excrescences of the fruit, and has been accordingly divided by some botanists into six or more supposed species, more generally considered as varieties.

CERATOPSIS. *Epipogum.*

CERATOPTERIDINE.E. One of the primary subdivisions or tribes of the polypodiaceous ferns, distinguished by the broad incomplete or rudimentary condition of the ring of the sessile globose spore-cases, the latter containing few large spores, concentrically striated on their three faces. [T. M.]

CERATOPTERIS. A peculiar genus of tropical aquatic ferns, constituting the group *Ceratopteridineæ*, or the *Parkerieæ* of some authors. They have sometimes been associated with the *Pterideæ*, or even the *Polypodieæ*, but seem to be more correctly regarded as a distinct group, characterised by having the ring of the spore-cases very broad, incomplete, or merely rudimentary and obliquely vertical, the spore-cases being sessile, or nearly so, and the spores few, comparatively large, obtusely trigonal, each of the faces being beautifully marked with concentric lines. The only species, *C. thalictroides*, is found scattered through the tropical and sub-tropical regions of Asia, Africa, America, and Australasia, either floating or attached to the soil in shallow still or slightly moving waters. The fronds are much divided, membranaceous, and succulent in the fresh state: the sterile ones more foliaceous and less divided, with evident reticulated veins; the fertile ones taller and more erect, and divided into linear somewhat slitpose segments, everywhere soriferous beneath the recurved indusium-like margin, and with the veins distinctly anastomosing. Both forms of frond, especially the sterile ones, are proliferous, often freely so. The succulent foliage of this fern is boiled and eaten as a vegetable by the poorer classes in the Indian Archipelago. [T. M.]

CERATOSTACHYS. A genus usually referred to *Combretaceæ*, containing a tree from Japan, with oblong entire smooth leaves, glaucous beneath, and axillary solitary spikes of flowers, forming dense heads. [J. T. S.]

CERATOSTEMMA. A genus of vacciniaceous plants, consisting of Peruvian shrubs, with superior five-toothed calyx; a tubular corolla with a five-toothed limb; ten stamens included within the corolla, the filaments united below into a cup, and the anthers opening by pores; and a five-celled ovary with several seeds, ripening into a kind of berry surmounted by the limb of the calyx. [M. T. M.]

CERATOSTYLIS. A small and unimportant genus of terrestrial orchids inhabiting tropical Asia. It contains two sections, one made up of species with long terete one-leaved simple stems with a dense cluster of minute flowers in the axil; the other with a branched stem like that of caulescent *Maxillarias*. These last constitute the spurious genus *Trigonanthus.*

CERATOTHECA. A genus of *Sesameæ*, containing a single species from tropical Africa. It is a herbaceous plant, with an erect tetragonous stem, opposite petiolate and dentate leaves, and single flowers on short axillary peduncles, with two glanduliferous bracteoles at their base. The persistent calyx is deeply divided into five acuminate lobes; the corolla tube is short and campanulate, and the limb bilabiate and five-cleft. There are four didynamous stamens, and no trace whatever of the fifth. The style is simple and deciduous, with a bilamellate stigma. The membranaceous truncate capsule has the corners of the apex produced into two or generally four horns. The free central placenta bears many flat obovate seeds. [W. C.]

CERATOZAMIA. The name of this genus of *Cycadaceæ* refers to its most prominent distinguishing feature: the presence of two horns on the scales of its zamia-like fruit. The stem is short and globular, giving off numerous pinnate leaves. The flowers are diœcious; the males in cones, whose scales are provided with two little teeth at the point, and with numerous anthers on their under surface; the females consisting of numerous scales with a thickened hexagonal disc-like top provided with two diverging horns, each scale concealing two seeds. The plants are natives of Mexico. [M. T. M.]

CERBERA. This name is intended to imply that the plants to which it belongs, are as dangerous as Cerberus; and some of them indeed are poisonous. Botanically, it is applied to a genus of *Apocynaceæ*, consisting of trees, natives of tropical Asia, with terminal flowers disposed in corymbs. The corolla is funnel-shaped, with the limb divided into five oblique lobes, and the throat provided with five teeth. The stamens are five, included within the corolla, their anthers tipped with a distinct spine. The ovary is two-lobed, with two compartments, having two to four seeds in each. The stigma is discoid, with a wavy margin. The fruit consists of two separate drupes, one of which is usually abortive. The inner shell of the drupe is fibrous, partly divided, when ripe, into two divisions, and, when seen in the dried state, much resembling a ball of string. These plants possess a milky juice of a poisonous character, though some of the species are said to be destitute of the venomous qualities possessed by the rest. The seeds of *C. Ahovai* are very poisonous, and the wood of this tree has an abominable odour. The seeds of *C. Manghas* are emetic and poisonous. *C. Odollam*, a Malabar tree, is cited by Lindley, as being innocuous, but this character applies probably to the fleshy drupe, the nut in the interior being narcotic and even poisonous. The bark is purgative: the unripe fruit, moreover, is dangerous, and is said to be used by the natives of Travancore to destroy dogs; the teeth of the unfortunate animals being, as is reported, loosened so as to fall out after mastication it. See Plate 6, fig. c. [M. T. M.]

CERCIDIUM. The mycelium or spawn of certain fungals.

CERCIFIX. (Fr.) *Tragopogon porrifolius.*

CERCIS. Judas Tree. This tree divides with the Elder the ignominy of being that on which the arch-traitor hanged himself, neither legend being worth the trouble of sifting. It is a native of the south of Europe and several countries of Asia from Syria to Japan, and is a handsome low tree with a spreading head, easily distinguished among the leguminous order by its simple gibbous kidney-shaped leaves, and by its purple flowers, which are produced abundantly in May before the leaves, not only from the young twigs, but from the mainned branches, and even the main trunk. The flowers are succeeded by flat thin brown pods, nearly six inches in length, which remain on the tree all the year. These are not generally produced in this country, unless the tree be planted against a wall, but in a warmer climate they perfect themselves in abundance and afford a ready means of propagation. The leaves are remarkable for their unusual shape, for the pale bluish green of their upper surface, and for their sea-green hue beneath. The flowers have an agreeable acid taste, and are sometimes mixed with salads or made into fritters with batter, and the flower buds are pickled in vinegar. This species is known as *C. Siliquastrum,* from the conspicuous appearance of its seed-pods.

C. canadensis (French *Bouton Rouge*), bears a general resemblance to the preceding, but is smaller and more slender. It may at once be distinguished by its leaves being heart-shaped and pointed. It is a native of North America, from Canada to Virginia, along the banks of rivers. The flowers are less numerous and of a paler rose colour; these are used by the French Canadians in salads and pickles, and the young branches to dye wool of a nankeen colour. The wood of both species is hard and variously marked with black, green, and yellow, on a grey ground. A new species, *C. chinensis,* which has been recently introduced from China, has sessile flowers, of which the standard is striped. French, *Guinier, Arbre de Judée;* German, *Judasbaum.* [C. A. J.]

CEREUS. An extensive genus of Cactaceæ, the species of which are remarkable for their singularity of form, and for the beauty of their flowers. Their stems are fleshy while young, but many of them harden and even become woody in course of time; they vary very much in form, some species having cylindrical and ribbed or fluted stems, whilst others have them nearly square or angular; some grow erect, others creep along the ground or up trees, and send out roots from their sides; many are unbranched, while others have numerous branches, and some are jointed. The majority are armed with spines, which radiate from little cushion-like tufts, placed at regular intervals along the ridges or angles of the stems. Their flowers are distinguished by the tube being somewhat funnel-shaped and generally armed with small spines, by the numerous stamens being nearly as long as the petals, and by the slender thread-like style scarcely exceeding the stamens in length.

C. giganteus, the Suwarrow or Saguaro of the Mexicans, is the largest and most striking species of the genus. It is a native of the hot, arid, and almost desert regions of New Mexico, extending from Sonora, in lat. 30° N., to Williams river, in lat. 35° N., and found growing in rocky valleys and upon mountain sides, often springing out from mere crevices in the hard rock, and imparting a singular aspect to the scenery of the country, its tall stems with upright branches looking like telegraphic posts for signalling from point to point of the rocky mountains. While young the stems are of a globular form, gradually becoming club-shaped, and ultimately almost cylindrical, and from fifty

Cereus giganteus.

to sixty feet in height, with a diameter of about two feet at middle height, and gradually tapering both upwards and downwards to about one foot. They are most frequently unbranched, but some of the older ones have branches, which issue at right angles from the stem and then curve upwards and grow parallel with it. The stems are regularly ribbed or fluted, the ribs varying in number from twelve to twenty, and have, at intervals of about an inch, thick yellow cushions bearing five or six large and many smaller spines. The flowers are produced near the summit of the stems and branches, and are about four or five inches long by three or four in diameter, having light cream-coloured petals. The fruits are about two or three

CACTUS VEGETATION OF NEW MEXICO
AFTER MOLLHAUSEN

a Cereus giganteus

inches long, of a green colour and oval form, having a broad scar at the top caused by the flowers falling away; when ripe they burst into three or four pieces, which curve back so as to resemble a flower. Inside they contain numerous little black seeds imbedded in a crimson-coloured pulp which the Pimos and Papagos Indians make into an excellent preserve; and they also eat the ripe fruit as an article of food, gathering it by means of a forked stick tied to the end of a long pole.

C. MacDonaldiæ is one of the night-flowering kinds, and is of great beauty, its flowers when fully expanded being as much as fourteen inches in diameter, having numerous radiating red and bright orange sepals and delicately white petals. The stems are cylindrical, creeping, and branched, not much thicker than the little finger, and having here and there small swellings with a spine in the centre. It is a native of Honduras. The most common night-flowering kind is the *C. grandiflorus*, a native of the West Indies. [A. S.]

CERFEUIL. (Fr.) *Scandix Cerefolium.* — A' AIGUILLETTES. *Scandix Pecten-Veneris.* — CULTIVÉ. *Anthriscus Cerefolium.* — DES FOUS. *Anthriscus vulgaris.* — MUSQUÉ. *Myrrhis odorata.*

CERINTHE. A small genus of borageworts, consisting, with one exception, of annual plants, with oval glaucous stem-clasping leaves, and tubular flowers in one-sided drooping leafy racemes. The species are mostly European, and are more remarkable for their singularly glaucous aspect than for beauty. Two species, *C. major* and *C. minor*, have been long cultivated in gardens under the name of Honeywort, an appellation due to the abundance of honey secreted by their blossoms, which are much resorted to by bees. *C. major* grows about a foot high, with a branched stem, oval stem-clasping leaves, minutely toothed at the margin, set with rough white dots, and covered with a bluish white bloom. The crook-like racemes of flowers have on each side a row of imbricated oval leaves, the purplish corolla being about an inch long, contracted at the mouth, with a narrow five-toothed spreading margin, and a fruit of two conical black nuts. *C. minor* has smaller yellow flowers, the segments of which are conniveut and not reflexed. In *C. retorta* the tube of the corolla is curved, and the leaves are blotched with silvery-white. [W. T.]

CERINUS. The colour of yellow wax.

CERIOPS. Trees distinguished from the neighbouring genus *Rhizophora*, by their small five-parted flowers, the petals of which are hairy at the points. The ten stamens are placed in pairs before the petals. The lower part of the ovary has three compartments and six ovules, while the upper part is solid, and ends in a style which is longer than the stamens. Like the rest of the mangrove family the seed has the curious habit of germinating and protruding from the fruit while still attached to the bough. The trees are natives of the shores of tropical Asia and Australia. [M. T. M.]

CERISETTE. (Fr.) *Solanum pseudocapsicum.*

CERISIER A' BOUQUETS. (Fr.) *Cerasus vulgaris.* — D'AMOUR. *Solanum pseudocapsicum.* — DE LA TOUSSAINT. *Cerasus semperflorens.* — NAIN. *Lonicera tatarica*, and also *Cerasus Chamæcerasus.* — PETIT DES HOTTENTOTS. *Celastrus lucidus.*

CERIUM, CERIO. Same as *Caryopsis*.

CERNUE. (Fr.) *Agrostis stolonifera.*

CERNUOUS. Inclining a little from the perpendicular; generally applied to drooping flowers.

CEROCHILUS. *Rhamphidia.*

CEROPEGIA. A genus of *Asclepiadaceæ*, containing more than fifty species of perennial herbaceous plants or undershrubs, natives of India and Africa. They have a bulbous root, and short erect or slender twining stems, with opposite leaves and interpetiolar umbels of few or many flowers. The calyx is five-parted. The corolla tube is slender in the middle, expanding more or less below as well as above, where the limb divides into five generally slender portions, which being united at their points form a globose head. The staminal corona consists of five, ten, or fifteen ligulate lobes in one or two series. The gynostegium is included. The anthers have no membrane. The pollen masses are rounded, have a pellucid interior margin, and are connected by short processes. The slender follicles are cylindrical with comose seeds. Several species are employed for food; in some cases the whole plant is eaten as a salad, in others the fleshy leaves, stems and tubers, are used as pot vegetables. [W. C.]

CEROPTERIS. A name formerly proposed but not adopted for the species of *Gymnogramme*, which have the surface covered by a coloured powdery secretion, and which are familiarly known as Gold Ferns and Silver Ferns, from the colour of this substance which is of a waxy nature, whence the name. [T. M.]

CEROXYLON. This genus of palms is by some botanists combined with the genus *Iriartea*, from which, however, it is distinguished by the spathe or bract which covers the young flower spikes being entire (in *Iriartea* it is divided), by some of its flowers being perfect, while those of *Iriartea* are all imperfect, and also by a slight difference in the position of the embryo in the seed. Both calyx and corolla are three-parted, the calyx being very minute; the stamens are generally twelve in number, but occasionally vary from nine to fifteen; and the females have a three-celled ovary and three stigmas. The fruit is a small round berry containing one seed. Three species of this genus are known, two of which are noble trees of great height.

C. andicola, the Wax Palm of New Grenada, was first made known and described by the celebrated travellers Humboldt and Bonpland, who found it growing in great abundance in very elevated regions on the chain of mountains separating the courses of the rivers Magdalena and Cauca, in New Grenada, extending almost as high as the lower limit of perpetual snow, which is a remarkable fact when it is remembered that the generality of the palm tribe inaugurate in tropical climates. It has a straight trunk of great height and about a foot in diameter, cylindrical for the first half of its height, after which it swells out, but again contracts to its original dimension at the summit; but the most singular feature connected with the trunk is the circumstance of its being covered with a thin coating of a whitish waxy substance which gives it a curious marble-like appearance. It is surmounted by a tuft consisting of from six to eight

Ceroxylon andicola.

handsome pinnate leaves, each of which is about twenty feet long, and has a strong thick footstalk, the base of which spreads out and clasps round the trunk, leaving a circular scar when it falls away; the leaflets are densely covered on the under side with a beautiful silvery scurf, while the upper side is of a deep green colour. The waxy substance of the trunk forms an article of commerce amongst the inhabitants of New Grenada. It is obtained by cutting down the tree and scraping it with a blunt implement, each tree yielding about twenty-five pounds. According to the analysis of Vauquelin, it consists of two parts of resin and one of wax, and is therefore of too inflammable a nature to be used by itself; but by mixing it with one-third part of tallow, very good candles for ordinary purposes are manufactured from it. The candles used by the inhabitants for offerings to the Saints and Virgin are, however, made without any such mixture; but on account of their resinous nature the priests will not allow them to be used for the high ceremonies of the Romish Church. The wood is very hard towards the exterior, and is commonly employed for building purposes; and the leaves are used for thatching. [A. S.]

CERVANTESIA. A genus belonging to the order of santalworts, characterised by the disk, or part intervening between stamens and pistil, being five-cleft, shorter than the flowers, and adherent to it below, the style or appendage on the seed-vessel thick and slightly notched at the end. The name was given in honour of Cervantes. The species are trees or shrubs, natives of Peru, having scattered entire simple leaves. The fruit of C. tomentosa is used as food in Peru. [G. D.]

CERVINE. Deep tawny, such as the dark part of a lion's hide.

CESTREAU A' BAIES NOIRES. (Fr.) Cestrum Parqui.

CESTRUM. A genus of solanaceous shrubs, of which several are in cultivation in this country, though of no great beauty. They have a funnel-shaped yellowish fragrant corolla concealing the stamens, whose anthers open longitudinally. The fruit is a dark-coloured berry, enclosed within the calyx, with two compartments (or from the union of the placentæ and breaking down of the partition, one only) with few seeds, and a straight embryo. The plants are natives of Brazil. Some of them possess a bitter principle like quinine, while others are used as diuretics, and for other medicinal purposes. [M. T. M.]

CETERACH. A genus of polypodiaceous ferns of the group Asplenieæ, distinguished by having distinct simple sori, reticulated veins of which the marginal veinlets are free, and fronds clothed thickly with scales, amongst which the sori are hidden. One species is a common native fern called Miltwaste or Scale Fern, and another of twice the stature is found in the Canary Islands, both being alike coriaceous, and clothed on the under surface with a thick covering of imbricated tawny scales, by which peculiarity the British species may be readily known from all other native ferns. To this plant was formerly attributed a marvellous influence over the spleen, and Vitruvius states that it had the effect of destroying that organ in certain Cretan swine which fed upon it. So Gerarde writes:— 'There be empiricks or blinde practitioners of this age who teach that with this herbe, not only the hardness and swelling of the spleen, but all infirmities of the liver, may be effectually, and in a very short time removed. But this is to be reckoned amongst the old wives' fables, and that also which Dioscorides telleth of touching the gathering of Spleenewort in the night, and other most vaine things which are found here and there scattered in old books.' It is said, however, to be still usefully employed as a bait for rock-cod fishing on the coast of Wales. The genus is a somewhat anomalous one as to classification, the indusium, which is one of the characteristics of the *Aspleniæ*, being here either wholly wanting or merely rudimentary. The sori are nevertheless unilateral, and something like an indusium has been detected, so that it is now generally associated with the *Aspleniums* as it was by Linnæus. The name *Ceterach* has also been used by Presl to distinguish a section of *Gymnogramma*. [T. M.]

CETRARIA. A genus of lecidineous lichens distinguished by the fructification being fixed laterally to the borders of the thallus, and consequently margined by it. It is not, however, peltate. It deserves notice here as containing *C. islandica*, or the well-known Iceland Moss, which affords at once a nutritious article of food, and a doubtful medicine. Before using it requires to be steeped for several hours to get rid of a bitter principle. It is sometimes boiled to form a jelly, which is mixed with milk or wine; sometimes it is reduced to powder and used as an ingredient in cakes or bread. It is esteemed by many useful in pulmonary complaints or as a restorative, but after the bitter principle has been extracted it seems to possess no active qualities. [M. J. B.]

CEVADILLA. The seeds of *Asagræa officinalis*.

CHA DE FRADE. A Brazilian name for a decoction of *Casearia lingua*. — DE PEDRESTE. A Brazilian name for *Lantana pseudo-thea*.

CHABRÆA. The generic name of plants belonging to the composite order, having the flowers uniform, smooth, two-lipped, the lips bent down, the outer largest and three-toothed, the inner with two teeth; the fruit narrow below, covered with short projections, and crowned with feathery appendages. The name was given in honour of Chabre, a botanist of Geneva. The species of this genus are natives of Chili and of the Straits of Magellan; they have alternate leaves, those below mostly twice pinnate; and the heads of flowers are purplish. Dr. Hooker, in the *Flora Antarctica*, alludes in these terms to *C. suaveolens*: 'The odour of this plant, which is a great ornament to the grassy hills of the Falkland Islands, is decidedly that of benzoin.' [G. D.]

CHACA, or CHOCO. *Sechium edule*.

CHÆNANTHE. *Diadenium*.

CHÆNESTHES. A genus of trees or large shrubs, belonging to the solanaceous family, and having long crimson or orange-coloured flowers of much beauty, like those of *Dunalia*, an allied genus; but the stamens, in the present instance, are destitute of the lateral appendages which characterise *Dunalia*. There is another distinguishing feature in the peculiarity of the tubular calyx, which splits open by the growth of the fruit. The trees grow in the valleys of the Andes. [M. T. M.]

CHÆNOSTOMA. A considerable genus of herbs or undershrubs, belonging to *Scrophulariaceæ*, natives of South Africa. They have opposite dentate rarely entire leaves, and axillary or racemose pedicellate flowers, which do not blacken in drying. The calyx is five-parted, the deciduous corolla is funnel-shaped, sometimes with a short tube, and its limb is five-cleft. There are four didynamous stamens the length of the corolla or slightly exserted. The style is simple, and the stigma subclavate. The capsule is membranaceous and two-celled. [W. C.]

CHÆRADOPLECTRON. *Gisewla*.

CHÆTA. A bristle. The slender stalk of the spore-case of mosses, also called Seta.

CHÆTACHME. A small spiny S. African genus, belonging to the *Ulmaceæ*. It differs from the elm in not having winged fruits, and from *Sponia* or *Celtis* in its natural habit more than in anything else. The leaves are smooth or downy, oval or elliptical in form, with entire or toothed margins, and from one to two inches long; they are generally terminated by a bristle, and accompanied at the base of the stalks by two short spines. The flowers are small and green, male and female on the same plant; the males are numerous in the axils of the leaves, and have a five-parted calyx with five stamens opposite its division; the females are single in the axils of the leaves, with a similar but smaller calyx, enclosing a one-celled ovary, which is crowned with two reflexed stigmas. The fruit is a little oval nut, about the size of a pea, with one seed. In some works the name has been spelt *Chatachyne* and *Chetnchme* by mistake. [A. A. R.]

CHÆTOGASTRA. A genus of *Melastomaceæ*, natives of tropical America, allied to *Arthrostemma*, but with the parts of the flower in fives. Like that genus the present is an unnatural one, the species having only trifling technical characters in common. [J. T. S.]

CHÆTOSTOMA. A genus of small dry heath-like Brazilian shrubs, belonging to *Melastomaceæ*. Stems leafless at the base; flowers solitary, rather small, purple with yellow anthers; parts of the flowers in fours or fives, the stamens being twice as many as the petals; capsule free, cylindrical, three-celled. [J. T. S.]

CHÆTURUS. A genus of grasses belonging to the tribe *Agrostideæ*. The only species described, *C. fasciculatus*, is a small annual grass, a native of Spain. [D. M.]

CHAFF, CHAFFY. The same as paleaceous.

CHAFF-FLOWER *Alternanthera Achyrantha*.

CHAFF-SEED. An American name for *Schwalbea*.

CHAFF-WEED. *Centunculus minimus*.

CHAGAS DA MINDA. A Portuguese name for *Chytnocarpus*.

CHAILLETIA. A genus which gives its name to the family of *Chailletiaceæ*. It is found more or less in most tropical countries, but represented in greatest numbers in Brazil. The species are small erect trees or shrubs, but sometimes (as in *C. pedunculata*, a Guiana species) extensive climbers, reaching the tops of the highest trees. The leaves are shortly-stalked, alternate, entire, and generally oval in form. The flowers are small white, often odoriferous and disposed in axillary cymes and racemes: the calyx five-leaved; the corolla of five cleft petals; the stamens five; and the ovary two or three-celled, crowned with a like number of styles, and becoming when ripe a somewhat dry drupe, with one or two seeds. The only extra-tropical species is *C. cymosa*, which is a native of South Africa, and has oblong obtuse leaves; and the only species whose uses are recorded is *C. toxicaria*, a native of Sierra Leone, where the seeds of this plant are said to be used by the colonists for poisoning rats, and by them called Ratsbane. Upwards of thirty species are known. [A. A. B.]

CHAILLETIACEÆ. A family of dicotyledons, nearly allied to *Celastraceæ*, but differing in their usually notched petals, in the five distinct glands which take the place of the perigynous disk of the latter order, and generally in the want of albumen to the seeds. They are remarkable also by the great tendency of the peduncles to combine with the petioles, so that the flowers, which are really axillary, appear to spring from the leaf itself at the summit of the petiole. They are all trees or shrubs, with alternate stipulate entire leaves, often white underneath. The flowers are small, in paniculate cymes or compact clusters. There are usually five sepals, petals, and stamens, regularly alternating with each other; but these numbers are, in one genus, *Tapura*, irregularly reduced. The ovary is superior with two or three cells, and two pendulous ovules in each cell; the style is simple; the fruit a rather dry drupe with one to three seeds. There are nearly twenty species, natives of tropical regions, and dispersed over both the New and the Old World. They have been distributed into four or five genera, of which the principal are *Chailletia*, *Moacurra*, and *Tapura*.

CHALAZA (adj. CHALAZINES). That part of the seed where the nucleus joins the integuments: it represents the base of the nucleus, and is invariably opposite the end of the cotyledons.

CHALEF. (Fr.) *Elæagnus*.

CHALK WHITE. Dull white, with a dash of grey.

CHAMARAS. (Fr.) *Teucrium Scordium*.

CHAMŒCERISIER DES HAIES. (Fr.) *Lonicera Xylosteum*. — ROSE. *Lonicera tatarica*.

CHAMÆBATIA. *C. foliolosa*, the only representative of this genus, which belongs to the rose family, is a beautiful Californian shrub, about three feet high. All the young parts of the plant are covered with small glands, which secrete a resinous fluid, having a pleasant balsamic odour. The leaves are unlike those of any other plant in the family, and bear great resemblance to those of the milfoil (*Achillea*), but are of a much harsher texture, and generally from two to three inches long. The flowers are in terminal cymes, and in size and colour very much like those of the hawthorn. The plant is in cultivation, having been introduced in 1850, and will prove a great acquisition to our gardens. [A. A. B.]

CHAMÆBUXUS. *Polygala Chamæbuxus*.

CHAMÆCERASUS. *Cerasus Chamæcerasus*; also *Lonicera Ledebouri*.

CHAMÆCISTUS. *Rhododendron Chamæcistus*.

CHAMÆCYPARIS. A little group of Conifers forming a section of the genus *Cupressus*, from which it is separated by some botanists, and characterised by the seeds being two only under each scale. It is sometimes restricted to the American species, sometimes extended to those Japanese ones, which have been separated under the name of *Retinospora*.

CHAMÆCYPARISSUS. *Santolina Chamæcyparissus*.

CHAMÆDOREA. A genus of Palms, containing between thirty and forty species. They have reed-like stems marked by rings of scars, and seldom more than fifteen or twenty feet high, and one or two inches thick, and surmounted by tufts of leaves, which are either pinnate or nearly entire. All of them are natives of tropical America, inhabiting forests and forming dense masses of underwood. Their flowers are of separate sexes, borne on distinct plants, very small, and produced in great quantities on long branching spikes: the males having a cup-shaped three-lobed calyx, a corolla of three roundish sepals, and containing six stamens and a rudimentary or barren ovary; and the females, a three-parted cup-shaped calyx, a corolla like the males, and a three-celled ovary crowned by three short stigmas, and without any rudimentary stamens. The fruit is a small roundish berry containing a single long seed. The stems of most of the species serve for walking-sticks and similar purposes; and their young unexpanded flower-spikes are used by the Mexicans as a culinary vegetable, under the name of Tepejilote.

C. Ernesti-Augustii is a small species, native of New Grenada, having a stem about four or five feet high. Its leaves are two feet long, wedge-shaped at the base and almost entire, being merely divided, for about half-way down the centre, into two spreading sharp-pointed plaited lobes; the foot-stalks of the leaves widen out at their bases and clasp round the stem, giving it a swollen appearance. The female flower spikes of this species are cylindrical, about a foot long and undivided, and form a very striking object, being at first of a dark green colour and studded with red bead like flowers; but when these latter fall away, the spike becomes a bright coral-red colour. [A. S.]

CHAMÆDRYS. An old herbalist's word, literally signifying 'dwarf oak,' applied to both *Teucrium Chamædrys* and *Veronica Chamædrys*.

CHAMÆJASME. *Androsace Chamæjasme*; also *Stellaria Chamæjasme*.

CHAMÆLAUCIACEÆ. A tribe of *Myrtaceæ*, sometimes considered as a separate family. They are distinguished by their heath-like habit and foliage, their one-celled ovary with few ovules, their stamens partially reduced to staminodia, and by their sepals often extended into bristles or broken up into fringes. The latter character is, however, evanescent in some genera, and the others may be more or less traced through *Baeckea* or its allies into the true *Myrtaceæ*. There is a considerable number of species, all Australian, and distributed under fourteen or fifteen genera, of which the principal ones are *Calytrix*, *Lhotskya*, *Verticordia*, *Chamælaucium*, *Genetyllis*, &c.

CHAMÆLAUCIUM. A genus of *Chamælauciaceæ*, containing small heath-leaved shrubs, from South Australia. The leaves are opposite, crowded, semi-cylindrical or three-edged, with dots formed by small cavities containing essential oil. The flowers are white, shortly-stalked, axillary or terminal. The calyx has two concave mucronate bracts at the base, which enclose the bud; the calyx tube adheres to the short ovary at the base; the limb is five-cleft and subpetaloid; the petals are five; stamens ten, the alternate ones abortive, strap-shaped; capsule one-celled, indehiscent, few-seeded. [J. T. S.]

CHAMÆLEON, BLACK. *Cardopatium corymbosum*. —, WHITE. *Carlina gummifera*.

CHAMÆMELES. A genus of apple-worts, having the free border of the calyx truncate, and obscurely five-toothed; petals five, small, irregularly-toothed; stamens ten to fifteen; style, or appendage on the seed-vessel, simple, slightly notched at the tip; fruit one-celled; cotyledons convolute. The name means literally 'pigmy apple,' to indicate the general nature of the fruit, and the low habit of the plant. The genus was founded by Dr. Lindley, to comprehend a dwarf shrub very like Box, a native of the sea cliffs in Madeira, having simple shining evergreen mostly entire leaves, and flowers growing in clusters, which are leafy at the base. [G. D.]

CHAMÆMESPILUS. *Pyrus Chamæmespilus*.

CHAMÆMORUS. *Rubus Chamæmorus*.

CHAMÆNERION. A subdivision of the genus *Epilobium*, comprising those species which have regular erect flowers (though in some cases drooping while in bud), and either club-shaped or four-cleft, not cruciform stigmas. [C. A. J.]

CHAMÆPEUCE. A genus of the composite family, allied, on the one hand, to plume-thistles (*Cirsium*), and, on the other, to true thistles (*Carduus*). From the first of these it differs in the covering of the achenes being hardened, not membranaceous; and from the second, in the pappus being feathery, not simple-haired. A few of the species have narrow entire leaves with recurved margins, but the greater portion of them are hostile-looking thistle-like plants, from one to six feet high : the leaves generally lanceolate in form, smooth above, but as well as the stems covered underneath with a white cottony substance, and their margins furnished with numerous long spiny teeth ; the flower heads one to two inches in diameter, arranged in corymbs or long leafy racemes ; the corolla purple or white, and enclosed by an involucre made up of many spiny-pointed scales. The fifteen known species are natives of the Mediterranean region, and extend eastwards to the Caucasus. The name also belongs to *Stæhelina Chamæpeuce*. [A. A. B.]

CHAMÆPITYS. *Ajuga Chamæpitys*.

CHAMÆRHODOS. A genus of the rose family, allied, on the one hand, to *Potentilla*, from which it differs in having a definite number of stamens and carpels; and on the other to *Sibbaldia*, which has a double calyx composed of ten segments in two rows, while the calyx in this genus is of five segments in one row. The species are small perennial plants, seldom attaining more than a foot in height, and generally having decumbent stems which are furnished with alternate three or many-parted leaves, about half an inch long, their segments narrow and covered with greyish pubescence. The flowers are small, white or purple in colour, either simple in the axils of the leaves, or numerous and arranged in leafy panicles. These plants are found in Siberia, N. China, and Thibet (where *C. sabulosa* grows at an elevation of 15,000 feet), and also in the Rocky Mountains of N. America. [A. A. B.]

CHAMÆROPS. This is the most northern genus of palms. It contains about ten or twelve species, inhabitants of Northern Asia, Africa, and America, and Southern Europe. They are mostly of dwarf habit, but sometimes grow as high as thirty feet.

Their leaves are shaped and plaited like a fan, having the margin deeply cut into numerous sharp-pointed divisions; and the bases of their long and generally prickly footstalks are inserted into a mass of coarse fibrous matter. Their flowers are produced in panicles from among the bases of the leaves, and are either perfect or of separate sexes, and consist of a three-parted calyx, and a corolla of three petals with from six or nine stamens attached to their bases; the fertile ones having, in addition, three distinct ovaries tapering into awl-shaped styles. The fruit is a berry about the size of an olive, containing one seed.

C. humilis, the only European species of the palm tribe, does not extend farther north than Nice. It is generally very dwarf, not more than three or four feet high, sending up numerous suckers from its creeping roots, and thus forming dense tufts, which, in Sicily and North Africa, take the place of our furze bushes; but if these suckers are not allowed to grow, the plant forms a trunk twenty or thirty feet high. The leaves of this Palm are commonly used in the south of Europe for making hats, brooms, baskets, &c., and for thatching houses; they also yield a large quantity of fibre, from which the French manufacture a material resembling horse-hair—for which it is substituted. The coarse fibre from the bases of the leaves is used by the Arabs for mixing with camel's hair to make their tent covers.

C. Fortuni grows to about twelve or twenty feet in height, and is a native of the north of China, but is perfectly hardy in the southern parts of England, a plant having attained ten feet in height in Her Majesty's garden at Osborne. The Chinese agricultural labourers use the coarse brown fibre, obtained from the bases of the leaves, for making hats and also the garment called *So-e*, worn in wet weather. [A. S.]

CHAM.ESPH.ERION. The name given to a plant of the composite family found in W. Australia. The whole plant is about the size of a large pea, and consists of a globular dense cluster of white flower-heads, surrounded by a rosette of narrow leaves a quarter of an inch in length. The genus differs from its nearest allies in the crown-shaped lacerated pappus and its few flower-heads. The generic name has reference to the appearance of the plant, and is derived from two Greek words signifying 'on the ground' and 'a little sphere.' [A. A. B.]

CHAMBURU, *Carica digitata*.

CHAMISSOA. A genus of tropical herbs belonging to *Amaranthaceae* with alternate leaves and flowers in axillary or terminal spikes or globular heads; differing from *Amaranthus* by having the seeds furnished with a small white axil at the hilum, and the radicle of the embryo superior. [J. T. S.]

CHAMOMILE. *Anthemis nobilis*, sometimes written Camomile. —, WILD. *Matricaria Chamomilla*.

CHAMP. The timber of *Michelia excelsa*.

CHAMPIGNON. The French name for mushrooms in general, but applied in this country only to *Agaricus* (*Marasmius*) *Oreades* or by mistake to very different and often dangerous species. In some parts of the country it is known under the name of Scotch Bonnets. The Champignon grows in fairy rings, generally of a few feet only in diameter. It seems to luxuriate most in a sandy soil, but occurs everywhere in exposed pastures. The pileus when moist is of a dull fawn colour, when dry of a creamy white; the stem is tough with a villous bark, the gills broad, cream-coloured, free from any attachment to the stem, and very distant. The only species with which it can be fairly confounded is *A.* (*Marasmius*) *urens*, which has narrow browner gills, and leaves a burning sensation in the throat, while the true champignon is the mildest and most sapid of fungi. It is excellent as a fricassee, or stewed like common mushrooms, and it has the great merit of drying admirably. Few comparatively are acquainted with its excellent qualities, but those who are, gladly avail themselves of it as a most welcome article for the table. The Champignon cultivé of the French is *Agaricus campestris*. [M. J. B.]

CHAMPIONIA. A genus named after the late Lieut.-Col. Champion, who was mortally wounded at Inkerman, containing a single species, an undershrub from Ceylon, belonging to the cyrtandreous division of *Gesneraceae*, which is characterised as having the seeds without albumen, and the fruit wholly free. The plants of this genus have opposite oblong leaves, and short axillary trichotomous peduncles. The calyx is hairy and cut into five equal linear-subulate lobes; the white glabrous and rotate corolla is longer than the calyx, and has a very short tube, and a four-parted limb, the lobes of which are equal and oblong-lanceolate. There are four equal stamens and no hypogynous glands. The ovary is one-celled with two parietal placentae; and the style filiform with a capitate stigma. The oblong capsular fruit exceeds the persistent calyx; it is one-celled and contains many ovate seeds with a reticulated testa. [W. C.]

CHANDELIER TREE. *Pandanus Candelabrum*.

CHANNELLED. Hollowed out like a gutter, like many leaf-stalks.

CHANTARELLE. The French name for *Cantharellus ciborius*, adopted in this country. The genus *Cantharellus* is distinguished from *Agaricus* by the gills of the latter being replaced by veins which are frequently branched, and if they ever approach the appearance of gills, they are distinguished by their very obtuse edge.

the shorter ones not being distinct as in mushrooms, but connected with the longer as if immediately given off by them. The Chantarelle is a common though seldom an abundant inhabitant of our woods. The rich yolk-of-egg yellow and fragrant fruity smell at once distinguish it. It is rather acrid when eaten raw, but makes an excellent fricassee if steeped before dressing in boiling milk, and then stewed very gently. It is, however, of far less frequent use in this country than on the continent, where it is highly esteemed. We are not aware that there is any deleterious fungus with which it can be confounded. [M. J. B.]

CHANVRE. (Fr.) *Cannabis*. — D'EAU. *Bidens tripartita*. — SAUVAGE. *Galeopsis Tetrahit*.

CHANVRINE. (Fr.) *Eupatorium cannabinum*.

CHAPEAU D'EVÊQUE. (Fr.) *Epimedium alpinum*.

CHARACEÆ. A small natural order of *acrogens* consisting of two or at most three genera. The species are all aquatic, and are found in almost all parts of the world, but are most common in temperate countries. In the genus *Nitella*, the structure of the plant has much resemblance to that of *Cladophora*, which circumstance, combined with the aquatic habit, has caused these plants to be associated with *Algæ*. In *Chara*, however, the axis is coated with tubes, and a large quantity of calcareous matter is deposited upon them. The branches are given off in whorls, those of the fruit-bearing branchlets, however, being imperfect on the outer side. The species are either monoecious or dioecious, the two kinds of fruit being often seated close to each other. The female fruit consists of an ovate nucleus extensively coated with spirally-arranged tubes, the tips of which are free and look like so many stigmas, and secondly with a firm spirally-ribbed integument, the cells of which abound in starch granules. The male fruit is globose and brick-red, the surface being divided into eight equal asca consisting of tubes radiating from a common centre. From each of these a short tube is given off within the eight tubes, meeting in the centre, and joined to a cellular mass, which is supported by a ninth bell-shaped process which is fixed by the broader end to the plant, and keeps the globule from falling prematurely. At this point of junction a number of jointed threads are attached, each cell of which contains a spiral spermatozoid with two long slender thong-shaped processes at one end, by means of which they move about. Wallroth asserts that he has seen the globules vegetate, a circumstance which is not impossible after they have performed their function. A more common mode of reproduction is by means of little tuberiform bodies attached to the creeping roots. Each articulation in these plants has a distinct system of circulation which seems to be connected with the manner in which the grains of chlorophyll are arranged on the walls of tubes, a free longitudinal colourless space being left round which the juices circulate from the base upwards at the rate of about two lines in a second. An ordinary microscope is amply sufficient to show this interesting phenomenon. A little alcohol, as also many other chemical substances, at once arrests the motion, as is also the case when the distribution of the chlorophyll is disturbed. We know of no use to which these plants can be applied. The smell which they emit resembles that of sulphuretted hydrogen, and it is to this cause probably that they have an evil report as productive of fevers. Their nucules, known to mineralogists under the name of *Gyrogonites*, are found for the first time in the lower freshwater formations. [M. J. B.]

CHARACTER. A short phrase expressing the essential marks by which a given plant or group of plants is distinguished from others. A specific character distinguishes one species from other species; and so on.

CHARAGNE. (Fr.) *Chara*.

CHARBON. (Fr.) *Uredo Carbo*.

CHARDINIA. A genus of the composite family with a single species, *C. xeranthemoides*, which is a pretty little annual herb, a few inches high, found in Asia Minor and Persia. It has alternate lance-shaped entire leaves, nearly an inch long, covered with white pubescence; and twigs terminating in solitary silvery flower-heads, which when mature are nearly an inch across, and owe their beauty to the shining chaffy lance-shaped pappus scales which crown the cylindrical striate achenes, and are nearly half an inch in length. In the closely-related genus *Xeranthemum* the inner scales of the involucre are much longer than the others, bent out at the top, and often of a bright pink colour, so that they look like ray florets; here, however, the inner scales are erect like the outer, not much longer, and of the same silvery hue. [A. A. B.]

CHARDON. (Fr.) *Carduus*. — A' BONNETIER. *Dipsacus fullonum*. — ARGENTÉ. *Silybum Marianum*. — ETOILÉ. *Centaurea Calcitrapa*. — HEMORRHOÏDAL. *Carduus arvensis*. — MARIE. *Sylybum Marianum*. — ROLAND. *Eryngium campestre*.

CHARDS. The late summer blanched leaves of the Artichoke, *Cynara Scolymus*.

CHARIANTHUS. A genus of *Melastomaceæ* from the West Indies. Erect shrubs with opposite stalked five-nerved leaves, generally entire. Flowers purple in a trichotomous corymbose cyme; calyx-tube adhering to the ovary, its limb slightly four-lobed; petals four; stamens eight; fruit a globose berry depressed in the centre, with four cells and numerous seeds. [J. T. S.]

CHARIEIS. A genus of the composite order, having the heads of flowers surrounded by a covering of scales in two rows, forming an involucre, those of the outer row being plane, those of the inner keeled; the receptacle or part supporting the flowers is pitted, the pits slightly toothed at the margin; the fruit is broadest at the upper part, having a border composed of one row of hairs. The name is derived from the Greek word signifying 'graceful' or 'elegant.' The only species, *C. heterophylla*, is an annual, a native of the Cape of Good Hope, having the stem erect, striated and hairy, all the leaves stalkless, the lower ones, opposite, those at the upper part of the stem, alternate, narrow lance-shaped, and the heads of flowers yellow in the centre, and violet at the circumference. [G. D.]

CHARLES' SCEPTRE. *Pedicularis Sceptrum Carolinum.*

CHARLOCK. *Sinapis arvensis.* —, JOINTED. The wild Radish, *Raphanus Raphanistrum.*

CHARLWOODIA. A genus of *Liliaceæ*, closely allied to *Cordyline* and *Dracæna*, with the former of which it is, indeed, often associated. Dr. Planchon, in his recent revision of the genera of this group of plants, considers it a well-marked genus, sufficiently distinguished from *Dracæna* by the numerous ovules in each cell, and from both *Dracæna* and *Cordyline* by the persistent perianth, and by the remarkably biserial insertion of its lobes, which are very much imbricated in their æstivation. Their general habit is that of the *Cordylines*. The type of the genus is *C. congesta*, an Australasian species, of elegant habit, with elongate nervosely-striate leaves, and crowded many-flowered panicles. Three or four other species are associated with it. [T. M.]

CHARME COMMUN. (Fr.) *Carpinus Betulus.* —, HOUBLON, or D'ITALIE. *Ostrya vulgaris.*

CHARRAH. The Arabian name for the Trumpet-Gourd, *Lagenaria vulgaris clavata.*

CHARTACEOUS. Having the texture of writing-paper.

CHARTOLOMA. A genus of *Cruciferæ*, allied to *Isatis*, but with the radicle of the embryo bent over the edges of the cotyledons, not over the back of one of them. The only species, *C. platycarpum*, is an annual, with oblong sinuate-toothed leaves, yellow flowers, and large deflexed pods, which are eight or ten lines long, by six or eight broad, and are indehiscent, one-celled, and one-seeded. [J. T. S.]

CHASSE-BOSSE. (Fr.) *Lysimachia vulgaris.*

CHASSE-RAGE. *Lepidium graminifolium.*

CHASTE TREE. *Vitex Agnus-castus.*

CHÂTAIGNE D'EAU. (Fr.) *Trapa natans.*

CHÂTAIGNIER. (Fr.) *Castanea vulgaris.*

CHATE. The hairy Cucumber, *Cucumis Chate.*

CHAUBARDIA. An obscure genus of orchids, apparently allied to *Maxillaria*. It is said to have altogether the habit of *Kefersteinia.*

CHAUDRON. (Fr.) *Narcissus Pseudo-Narcissus.*

CHAULMOOGRA. The seeds of *Gynocardia odorata.*

CHAUSSE-TRAPPE. (Fr.) *Centranthus Calcitrapa.*

CHAVICA. A genus of *Piperaceæ*, producing two important plants, namely, the Long Pepper and the Betel Pepper. The genus is distinguished from the true peppers (*Piper*) by its perfectly unisexual flowers, which are sessile on spikes placed opposite the leaves, each flower being protected by a stalked quadrangular peltate bract. *C. Roxburghii* is largely distributed in India, where it is cultivated to furnish the Long Pepper of the shops, which consists of the spikes of flowers which, while yet immature are gathered and dried in the sun. The natives employ them for various medicinal purposes, as also the roots, and the stem cut into small pieces. In chemical composition and qualities, Long Pepper resembles ordinary black pepper, like which

Chavica Betel.

it contains *piperin.* The Long Pepper which is imported by the Dutch is said to be produced by an allied species, *C. officinarum. C. Betel*, and *C. Siriboa* furnish the betel already mentioned under ARECA; which see. The betel leaf is chewed with lime, and a slice of the Areca nut. The saliva is tinged of a bright red in consequence. It acts as a powerful stimulant to the digestive organs and salivary glands, and causes, when swallowed, giddiness and other unpleasant symptoms in persons unaccustomed to its use. [M. T. M.]

CHAW-STICK. *Gouania domingensis.*

CHAY-ROOT. *Oldenlandia umbellata.*

CHEAT or CHESS. An American name for *Bromus secalinus.*

CHEESE RENNET. *Galium verum.*

CHEESEBOOM. The common name in some parts of the country for *Agaricus arvensis,* or Horse Mushroom. This fungus grows in large rings, often many yards in diameter, and in some years, as in the wet summer of 1860, occurs in extraordinary abundance. It is known from true mushrooms by its large size, paler gills, generally thick rings, which are double at the edge, but especially by their turning yellow when bruised. It constitutes the greater part of the mushroom baskets in the Covent Garden market, and is consumed in large quantities in Leeds and other important towns in the north. When properly dressed and eaten in moderate quantities with plenty of bread, to insure mastication, these horse mushrooms are an excellent article of food, though they occasionally prove unwholesome, partly from over-indulgence, and partly from their having undergone decomposition before use. The term is sometimes applied to species of *Boletus,* several of which are highly dangerous. [M. J. B.]

CHEILANTHEÆ. A section of polypodineous ferns, in which the sori are punctiform at the apices of the veins, and covered by indusia, which,—sometimes short and rounded, sometimes elongated continuous and therefore pteroid—consisting of portions of the margin inflected over them, are therefore necessarily transverse to the margin of the frond or of its segments. [T. M.]

CHEILANTHES. A genus of polypodineous ferns of the group of *Cheilantheæ,* which it typifies. The species, which are numerous and scattered over the tropical and temperate regions both of the Old and New World, generally inhabiting dry rocky situations, are much varied in aspect, and for the most part are dwarf plants of tufted habit, with more or less compound fronds, the under surface in some cases being covered with silvery or gold-coloured powder, as in *Gymnogramma.* The distinguishing features of the genus consist in its producing small punctiform sori at the ends of the veins close to the margin of the frond, the margin itself becoming membranaceous, and bent over them to form the indusia, which are either linear and continuous, or take the shape of roundish lobes. The veins are free. *Cheilanthes* has a considerable affinity with *Nothochlæna,* the species of which possess a similar habit, but have naked or non-indusiate sori. Owing, however, to the different degrees in which the margin becomes attenuated and reflexed, it is sometimes not easy to decide between the two. *C. argentea,* a pretty dwarf tripartite silvery species, is found in Siberia; *C. fragrans,* a dwarf bipinnate species, whose fronds have a grateful anthoxanthoid or new-hay-like odour, occurs throughout the region of the Mediterranean, and reaches as far north as Switzerland; whilst Arabia, Abyssinia, South Africa, India, the Eastern Islands, Australasia, North and South America, and the West Indies, yield a variety of species, some of which, like *C. tenuifolia,* are distributed over a very wide area. One of the most beautiful species, and one which is familiar in gardens, is *C. farinosa,* a fine bipinnatifidly-pinnate plant, with tallish fronds silvered beneath and having black stalks. A peculiar group of the species has sometimes been separated under the name of *Myriopteris:* in this, the segments are small, roundish, pouch-shaped, the indusium entire and almost closing over the back of the segment, which, when reversed, looks not unlike a small roundish watch-pocket. The difference is hardly important enough to warrant their separation. [T. M.]

CHEIRADENIA *cuspidata* is a small Demerara orchid with the aspect of some equitant *Oncidium.* It has the lateral sepals adnate to the prolonged foot of the column, a pair of solid pollen masses, and a round lip bearing five processes near the margin, arranged like the fingers of an expanded hand: a circumstance alluded to in the name of the genus.

CHEIRANTHERA. A genus of *Pittosporaceæ,* containing an Australian undershrub with erect stems, and narrowly linear acute leaves, which have fascicled leaves in the axils; peduncles terminal, with small blue corymbose flowers; calyx of five sepals; the petals and stamens five each, the latter all bending to one side; fruit dry, scarcely berry-like, two-celled. [J. T. S.]

CHEIRANTHUS. A genus of cruciferous flowers, all so nearly resembling in habit and characters the common species as to be easily distinguished. *C. Cheiri,* the Wallflower, is a native of all Southern Europe, growing on old walls, in quarries, and on sea-cliffs. In its wild state the flowers are always single and of a bright yellow colour, but the varieties obtained by cultivation are of various tints, many of them beautiful, and all fragrant, especially in the evening. Seeds of numerous beautiful varieties are annually imported from Germany; and small gardens, in which the supply of ornamental early summer flowers is limited, may be made very gay by planting them liberally with these German wallflowers. The wallflower has long been a favourite cottage-garden flower, and has been praised in many a rustic lay; it is supposed by many to be the *Viola* of the Latin poets. Its French names are *Giroflée jaune, Violier, Ravenelle, Rameau d'or, Baton d'or,* &c.; German, *Leucoje.* Several other species are also worthy of cultivation. Among these the

Giroflée de Delile of the French, perhaps a variety of *C. linifolius*, forms small tufts, the extremities of which are covered with flowers which, during expansion, pass through several shades of purple; it continues in bloom during a great part of the year. *C. Marshallii*, a low tufted plant with bright evergreen leaves and numerous large orange-coloured flowers, blooms early in the year. All these, and several other species, are well suited for adorning rock-work. [C. A. J.]

CHEIROGLOSSA. A name under which *Ophioglossum palmatum* was proposed to be separated from the other species of this genus of ferns. [T. M.]

CHEIROPLEURIA. A synonyme of *Anapausia*, applied to *A. Vespertilio* and *A. bicuspis*, two ferns which are remarkable in bearing fronds of a form resembling bats' wings. [T. M.]

CHEIROSTEMON. The Hand-flower tree, or *Macpalxochitlquahuitl* of the Mexicans, is the sole species of this genus of sterculiads. The plant, *C. platanoides*, is a tree growing thirty or more feet in height, and having plane-like leaves of a deep green colour on the upper surface, but covered underneath with a rust-coloured scurf composed of star-like hairs, each leaf being about six inches long by five broad, deeply indented at the base and divided at the margin into from three to seven blunt, rounded lobes. Its flowers are two inches

Cheirostemon platanoides (flower).

long by as much broad, and are destitute of a corolla, but have a leathery, rusty-red, cup-shaped calyx, deeply cut into five broad, sharp-pointed divisions, the bottom of the cup having five bright yellow cavities which secrete a quantity of sweet fluid. The arrangement of the stamens is most remarkable; they are of a bright-red, and united together for about one-third of their length (four inches), when they separate into five curved claw-like rays, and thus bear some resemblance to the human hand. The club-shaped style emerges from the centre of the stamens, and is terminated by a pointed stigma. The fruit is five-cornered, and splits open in five places when ripe, allowing the escape of the numerous seeds.

A solitary specimen of this tree was first discovered growing near the town of Toluca in Mexico. It was of great age, and an object of veneration among the Indians, both on account of the remarkable structure of its flowers, and because they supposed that no other tree of the kind existed elsewhere; but forests of it have since been discovered near the city of Guatemala, from whence it is probable that the Indians of Toluca had transported it in very early times. [A. S.]

CHEIROSTYLIS. A genus of terrestrial orchids, consisting of little plants with the habit of *Anœctochilus*, to which it is nearly allied. Its most distinguishing character is having the three sepals united into a short tube, from the front of which hangs down a lip divided into narrow lobes. The column, moreover, has four arms, half its own, and half belonging to the stigma.

CHELIDOINE PETITE. (Fr.) *Ficaria verna*, also known as *Ranunculus Ficaria*.

CHELIDONIUM. The Greater or Common Celandine, a plant frequently found in this country in the neighbourhood of villages or old ruins, is the only species of this genus of the poppy family, and is not to be confounded with the lesser celandine (*Ficaria verna*). The Greater Celandine is a glaucous hairy annual plant, with pinnately-lobed leaves, small yellow flowers in a loose umbel, and a fruit, consisting of a long pod, bursting from below upwards by two valves, and containing a number of seeds with a small crest on them, near to the place where they are attached to the interior of the pod. The whole plant is full of a yellow juice which is of an acrid poisonous nature, and has been used in certain diseases of the eye, and as a caustic to destroy warts, &c. [M. T. M.]

CHELIDOSPERMUM. A section of the genus *Pittosporum*, containing a few species from New Guinea, with the calyx deeply five-parted, valvate, and the seeds with long seedstalks. The leaves are oblong; the flowers grow in a pedunculated terminal umbel; and the capsule is two-seeded, with leathery valves. [J. T. S.]

CHELONANTHERA. *Pholidota*.

CHELONE. A small genus of linariads, very closely allied to the *Pentstemon*, from which, however, it is easily distinguished by its indehiscent winged seeds, by its sterile fifth stamen being shorter than the other four, and by its flowers being arranged in short dense bracted spikes. The form of the corolla in this genus is also very distinct, the broad keeled upper lip and scarcely open mouth giving it some resemblance to the head of a tortoise or turtle, to which feature is due both its scientific appellation and the popular American name of Turtle-head. The best-known representative of this genus is the *C. obliqua*, a perennial with creeping roots, erect smooth bluntly four-angled stems, opposite serrated lanceolate leaves varying considerably in breadth and acuteness, and flowers in terminal spikes, with corollas mostly of a rose-purple colour. The so-called *C. glabra* is now regarded as but one of the forms of *C. obliqua*. *C. Lyonii*, with the same habit, has smaller flowers and longer and thinner leaves. *C. nemorosa* seems to be intermediate between *Chelone* and *Pentstemon*, having the winged seeds of the former genus, with the inflorescence and habit of the latter. It has ovate serrated leaves, and dull purple pentstemon-like flowers produced from the upper axils. It is proper to note that several popular border flowers pass for *Chelones* which are in fact true *Pentstemons*; as examples may be cited the *Pentstemon barbatus*, *P. campanulatus*, and *P. centranthifolius*, all of which have been improperly classed under the present genus, though they possess none of its distinguishing features. [W. T.]

CHEMISE DE NOTRE-DAME. (Fr.) *Convolvulus* or *Calystegia sepium*.

CHENA, or CHAINA. An inferior kind of Indian Millet, *Panicum pilosum*; also sometimes applied to *Panicum miliaceum*.

CHÊNE. (Fr.) *Quercus.* — A' GRAPPES, *Quercus pedunculata.* — ANGOUMOIS, *Quercus Toza.* — A' TROCHETS, *Quercus sessiliflora.* — AU KERMES, *Quercus coccifera.* — BROSSE, *Quercus Toza.* — COMMUN, *Quercus pedunculata.* — CYPRÈS, *Quercus fastigiata.* — DES PYRÉNÉES, *Quercus fastigiata.* — GREC, *Quercus Ægilops.* — NOIR D' AMÉRIQUE, *Catalpa longissima.* — PETIT, *Teucrium Chamædrys*; and also *Veronica Chamædrys.* — QUERCITRON, *Quercus tinctoria.* — ROURE, *Quercus sessiliflora.* — VELANI, *Quercus Ægilops.* — VERT or VEUSE, *Quercus Ilex.* — TAUZIN or TOZA, *Quercus Toza.* — ZANG or ZEEN, *Quercus Mirbeckii*.

CHENILLETTE. (Fr.) *Scorpiurus*.

CHENOPODIACEÆ. (*Chenopods*, the Goose-foot family.) A natural order of monochlamydeous dicotyledons, characterising Lindley's chenopodal alliance. Herbs or undershrubs with alternate sometimes opposite leaves without stipules, and small flowers which are sometimes unisexual, i.e. have stamens and pistils in separate flowers. Perianth (calyx) deeply-divided, sometimes tubular at the base, persistent; stamens inserted into the base of the perianth and opposite to its divisions. Ovary free, one-celled, with a single ovule attached to its base. Fruit an utricle (inflated) or an achene, sometimes succulent; embryo curved round mealy albumen, or spirally curved without albumen. Inconspicuous plants found in waste places in all parts of the world, but abounding in extra-tropical regions. Many of them, as species of *Salicornia* and *Salsola*, inhabit salt-marshes in the northern part of Europe and Asia. Some of them are used as potherbs, for instance, spinach, (*Spinacia oleracea*), orach (*Atriplex hortensis*), beet (*Beta vulgaris*), English mercury (*Chenopodium Bonus Henricus*), Australian spinach (*Chenopodium crosum*). The mangold-wurzel is a variety of beet used for the food of cattle. The beet is much cultivated in France for its sugar. Some of the plants yield soda, others supply essential oils which render them useful in cases of worms and in spasmodic diseases. The seeds of *Chenopodium Quinoa* are used as food in Peru. They abound in starch, but have a bitterish taste. The seeds of *Chenopodium Bonus Henricus* are used in the manufacture of shagreen. There are seventy-four known genera and 533 species. Illustrative genera: *Salicornia*, *Atriplex*, *Spinacia*, *Beta*, *Blitum*, *Salsola*, *Chenopodium*. [J. H. B.]

CHENOPODIUM. A genus of annual and perennial herbs giving its name to the natural order of chenopods, and chiefly remarkable for the weedy character of the species composing it, of which the Common Goosefoot, a plant found everywhere in waste places, with triangular leaves covered with a whitish mealiness, and numerous small flowers in terminal clusters, is an example. It includes, however, a few species interesting for their utility, and one which has some merit as an ornamental plant. The latter is *C. Atriplicis*, a tall branched annual of erect pyramidal habit, growing four to five feet high, with reddish stems, rhomboidly-ovate and often sinuate leaves, covered while young with a glittering purple meal, and numerous small flowers in terminal compound spikes of a dark-purple colour, and also clothed with purple meal. *C. ambrosioides*, or Mexican Tea, the *Ambrina ambrosioides* of some botanists, a tropical species, contains an essential oil to which it owes tonic and antispasmodic properties; and *C. anthelminticum*, a species differing from the preceding, of which it is perhaps but a variety, chiefly in its leaves being more deeply cleft, and the flower-spike mostly leafless, yields the wormseed oil, a popular vermifuge in the United States. The species to which the greatest interest attaches is, however, *C. Quinoa*, indigenous to the Pacific slopes of the Andes, where it is largely cultivated in Peru and Chili for the sake of its seeds, which are extensively used as an article of food. They are prepared either by boiling in water like rice or oatmeal, a kind of gruel being the result, which is seasoned with the Chili pepper and other condiments; or the grains are slightly roasted like coffee, boiled in water and strained, the brown-coloured broth thus prepared being seasoned as in the first process. This second preparation is called 'carapulque,' and is said to be a favourite dish with the ladies of Lima. However prepared, the Quinoa is unpalatable to strangers, though it is probably a nutritious article of food from the amount of albumen it contains. Two varieties are cultivated, one producing very pale seeds called the White, which is that employed as food, and a dark-red fruited one called the Red Quinoa. A sweetened decoction of the seeds of the latter is used medicinally, as an application to sores and bruises, and cataplasms are also made from it. This species attains a height of four to five feet, and has a stout furrowed branched stem, large triangular-ovate deeply-sinuate leaves, on long footstalks, and densely-clustered small green flowers, produced in axillary and terminal panicles. Botanically, the genus *Chenopodium* is distinguished by a five-parted perianth, five stamens, two styles crowning the ovary which contains a single round flattened seed. [W. T.]

CHERAMELLA. An Indian name for the subacid fruits of *Cicca disticha*.

CHERIMOYER. *Anona Cherimolia*, a delicious Peruvian fruit.

CHERMESINE. A kind of crimson.

CHE-ROOT. *Oldenlandia umbellata*.

CHERRIS. An Indian name for the resinous exudation of the Hemp, *Cannabis sativa*.

CHERRY. A well-known fruit produced from cultivated varieties of the Wild Cherries, *Cerasus avium* and *C. vulgaris*. —, BARBADOS. *Malpighia glabra*. —, BASTARD. *Cerasus Pseudo-cerasus*. —, BEECH or BRUSH. *Trochocarpa laurina*. —, BIRCH. *Betula lenta*. —, BIRD. *Cerasus Padus*. —, CHOKE. *Cerasus virginiana*; also *C. serotina*, *hiemalis*, and *borealis*. —, CLAMMY. *Cordia Collococca*. —, COWHAGE. *Malpighia urens*. —, CORNELIAN. *Cornus mascula*. —, GROUND. *Cerasus Chamaecerasus*; also an American name for *Physalis*. —, HOTTENTOT. *Cassine Maurocenia*. —, NATIVE, of Australia. *Exocarpus cupressiformis*; —, of N. S. Wales. *Nelitris ingens*. —, WINTER. *Physalis Alkekengi*; also sometimes applied to *Physalis angulata* and *Cardiospermum Halicacabum*.

CHERRY-CRAB. A variety of the Siberian Crab, *Pyrus Malus baccata*.

CHERRY-LAUREL. *Cerasus Lauro-cerasus*.

CHERRY-PEPPER. *Capsicum cerasiforme*.

CHERRY-PIE. A garden and popular name for the Heliotrope.

CHERUI. (Fr.) *Sium Sisarum*.

CHERVIL. A garden potherb, *Chaerophyllum sativum*, also called *Anthriscus Cerefolium*. The name Chervil is also applied generally to the plants referred to *Chaerophyllum*. —, GREAT. *Myrrhis odorata*. —, NEEDLE. *Scandix Pecten veneris*. —, PARSNIP. *Chaerophyllum bulbosum*, or *Anthriscus bulbosus*. —, SWEET. *Myrrhis odorata*. —, WILD. *Chaerophyllum sylvestre*.

CHERVIS. (Fr.) *Sium Sisarum*.

CHESNEYA. A genus of dwarf woody plants, belonging to the pea-flowered Leguminosae, nearly related to *Calophaca* and *Colutea*, from both of which it differs, in having the spaces between the seeds in the pod occupied by a spongy pith-like substance. The leaves are alternate, unequally pinnate, with from three to nine pairs of wedge-shaped leaflets, about half an inch long, and downy. The flower-stalks are axillary, bearing on their apex one to three yellow or violet-coloured flowers, whose tubular calyces are curiously swollen above at the base. The pods are from one to two inches long, roundish, or somewhat flattened, containing four to six seeds. There are about eight species known: one of them, *C. cuneata*, found in Tibet at an elevation of eight to twelve thousand feet, but the greater portion in W. Asia, and chiefly in Persia. [A. A. B.]

CHESTNUT. The common name for *Castanea*. —, HORSE. *Aesculus Hippocastanum*. —, MORETON BAY, or NEW HOLLAND. The large fleshy seeds of *Castanospermum australe*. —, SPANISH. *Castanea vesca*, the fruits of which are known as Sweet Chestnuts. —, TAHITI. *Inocarpus edulis*. —, WATER. *Trapa natans*. —, WILD. A name given by the settlers at the Cape to the seeds of *Brabejum*. —, YELLOW. *Quercus Castanea*.

CHESTNUT OAK. *Quercus Castanea*; also sometimes applied to the timber of the sessile-fruited English oak, *Quercus sessiliflora*.

CHEVEUX DE VENUS. (Fr.) *Adiantum Capillus-Veneris*; also applied to *Cuscuta major*, and *Nigella damascena*. —, DU DIABLE. *Cuscuta major*.

CHEVREFEUILLE. (Fr.) *Lonicera*. — DES BOIS. *Lonicera Periclymenum*. — D'ITALIE. *Lonicera etrusca*. — DE VIRGINIE. *Lonicera sempervirens*.

CHEVRILLE. (Fr.) *Lactuca perennis*.

CHEYNIA. A handsome-flowered genus of the myrtle family, native of the Swan River territory. The plant is a small much-branched shrub, with fine heath-like leaves arranged in four rows, and bears handsome scarlet flowers with a long calyx tube, and a five-parted limb. The five petals are inserted into a thick rim lining the throat of the calyx, as also are the numerous stamens, which are separate from each other, and of unequal lengths, the connective of the anther being slightly swollen; the ovary is five-celled and many-seeded. [M. T. M.]

CHIAZOSPERMUM. A genus containing an annual herb from temperate Asia, allied to *Hypecoum*, and like it forming a connecting link between the orders *Papaveraceae* and *Fumariaceae*. It differs from *Hypecoum* by having the seeds somewhat four-sided, each side with a cross-marked elevation. [J. T. S.]

CHIBOU RESIN. A product of *Bursera gummifera*.

CHICASAW PLUM. *Cerasus Chicasa*.

CHICHA. *Sterculia Chicha*, the seeds of which are eaten as nuts by the Brazilians; also a colouring-matter obtained from the leaves of *Bignonia Chica*.

CHICHE. (Fr.) *Lathyrus Cicera*.

CHICKEN-WEED. A name under which *Roccella tinctoria* has been sometimes imported.

CHICHOW. The seeds of *Cassia Absus*, an Egyptian remedy for ophthalmia.

CHICKRASSIA. A latinised version of the Bengalee name of a lofty Indian tree, belonging to the order *Cedrelaceæ*. The leaves are pinnated; the flowers large, in terminal panicles with ten stamens united by their filaments into a tube. Ovary three-celled, placed on a broad disc, with pendulous ovules, arranged in two rows. The fruit is a capsule opening from above downwards by three valves, leaving a central column. The seeds are winged. The wood of *C. tabularis* is close-grained, light-coloured, and elegantly veined; hence it is in much request by cabinet makers, who call it chittagong wood, though there are other woods with a similar appellation. The bark of this tree is astringent but not bitter. [M. T. M.]

CHICKWEED. The common name for *Alsine*. The well-known weed of this name is *Alsine*, or *Stellaria media*. —, BASTARD, *Buffonia tenuifolia*. —, FORKED, *Anychia dichotoma*. —, INDIAN. An American name for *Mollugo*. —, MOUSE-EAR. The common name for *Cerastium*; also specially *C. vulgatum*. —, SEA. *Arenaria peploides*. —, SILVER. *Paronychia argyrocoma*. —, WATER. *Montia fontana*; also sometimes applied to *Malachium aquaticum*, and *Callitriche verna*.

CHICO. A kind of beer, made in Chili from the Indian corn, *Zea Mays*.

CHICON. (Fr.) *Lactuca sativa*.

CHICOREE. (Fr.) Succory, *Cichorium Intybus*. —, FRISEE. Curled Endive, a variety of *Cichorium Endivia*.

CHICORIA DE LA TIERRA CALIENTE. A South American name for *Achyrophorus scandiflorus*.

CHICORY. *Cichorium Intybus*, or Succory.

CHICOT, or CHICHOT DU CANADA. (Fr.) *Gymnocladus canadensis*. The term Chicot is also applied to the seeds of *Moringa pterygosperma*.

CHIENDENT. (Fr.) *Cynodon Dactylon*. A' BALAIS, *Andropogon Ischæmum*. — A' CHAPELET, *Avena bulbosa*. — DES BOUTIQUES, *Triticum repens*.

CHILLI. The fruit of *Capsicum annuum*, and other allied species.

CHILOCARPUS. An imperfectly known genus of climbing shrubs, natives of Java, with a salver-shaped corolla, capitate stigma, and a capsular fruit. The genus is referred to the *Apocynaceæ*. [M. T. M.]

CHILODIA. A genus of *Labiatæ*, containing a single species, a native of New Holland. It is a branched glabrous or slightly pubescent shrub, with small entire linear-sessile leaves and single flowered axillary peduncles, with two small subulate bracts below the calyx. The calyx is campanulate with a short striated tube, and a bilabiate limb, the upper lip being entire and the lower emarginate or bidentate. The corolla is campanulate and faintly two-lipped. There are four stamens shorter than the tube; the anthers have two smooth parallel cells, without appendages. The apex of the style is slightly bifid with sub-equal lobes. In habit and structure this genus is very near *Prostanthera*, differing only in having no appendages to the anther-cell. [W. C.]

CHILOGLOTTIS. Under this name stand a small number of terrestrial Australasian orchids, bearing radical leaves in pairs and solitary galeate reddish flowers at the end of a short naked scape. Like *Caladenia* its lip is marked by prominent glands; nor, indeed, does it differ much from that genus, except in having a very broad arched dorsal sepal.

CHILOPSIS. A genus of *Bignoniaceæ*, consisting of a single species of erect branching shrubs from Mexico. It has long linear entire alternate leaves, and beautiful flowers in terminal dense spicate racemes, on short bibracteolate pedicles. The bilabiate calyx is membranaceous, inflated, and deeply-cleft in front; the corolla-tube is dilated upwards, and the two-lipped limb is five-lobed. The four stamens are didynamous, the sterile fifth being very minute. The style is filiform, and the stigma bilobed. The pod-like capsule is two-celled, with the partition bearing the placentæ contrary to the valves. The seeds are transversely winged. [W. C.]

CHILOSCHISTA *usneoides* is a leafless Indian epiphyte of the orchidaceous order, with narrow, flat, green roots, which cling to the branches of trees and appear to serve the purpose of leaves, as also happens in the leafless *Angræceum*.

CHIMAPHILA. A small genus of *Pyrolaceæ*, natives of Europe, Siberia, and North America, differing from *Pyrola* by the hairy filaments, very short style, and capsule splitting from the apex downwards with the edges of the valves not woolly. The plants, called Winter Greens in America, have woody subterranean shoots, and a short stem with a tuft of thick shining evergreen leaves, oblong, wedge-shaped, or lanceolate—in the latter case variegated with white. The scape is corymbosely or umbellately branched at the apex, the pedicels one-flowered, bearing handsome, bell-shaped, white flowers, tinged with purplish-red, and very sweet-scented. The leaves contain a bitter extractive matter, on which account they have been used in medicine, in North America. [J. T. S.]

CHIMNEY PLANT. *Campanula pyramidalis*.

CHIMONANTHUS. The Japan Allspice, *C. fragrans*, is the only representative of this genus of the Calycanthus family, and it is well-known in gardens for its early flowering and the sweet scent of its blossoms. It was introduced from China in 1766, and for a long while was known under the name of *Calycanthus præcox*, until it was shown to differ from that genus in having but ten stamens arranged in two rows; while in *Calycanthus* they are very numerous, and arranged in four rows. The Japan Allspice is a much-branched shrub, and generally treated as a wall-plant in gardens; its leaves are opposite, stalked, between oval and lanceolate in form, and very rough on the surface; they generally fall late in the autumn, but sometimes a few remain till the spring. The flowers are sessile on the branches, about an inch in diameter, and made up of a large number of pale yellow waxy petals, arranged in several rows; the inner series in one variety chocolate-coloured, and in another mottled with red. These flowers in mild winters often appear about Christmas, and last for a long time. [A. A. B.]

CHINA ASTER. *Callistephus chinensis*, also called *Callistemma hortense*.

CHINA BARK. The bark of *Buena hexandra*, an indifferent febrifuge.

CHINA GRASS. The fibre of *Bœhmeria nivea*, the Rheea, or Ramee.

CHINA ROOT. The tuberous rhizome of *Smilax China*.

CHINCAPIN. (Fr.) *Castanea pumila*.

CHINCHIN. A Chilian name for *Polygala thesioides*.

CHIN-CHON. A gummy or glutinous matter, much used as a glue or varnish in China and Japan, and supposed to be the produce of *Pistacia terebinthus*.

CHINESE SWALLOWS' NESTS. These curious productions, which sell at such a high price in China, though they have no especial points of recommendation beyond many other gelatinous ingredients in soups, were formerly supposed to be made of some species of the rose-spored Algæ, as *Sphærococcus lichenoides*; but this is now ascertained to be a mistake, and it is known that they are formed of a secretion from the mouth of the bird itself. [M. J. B.]

CHINESE TREE. *Pæonia Moutan*.

CHINESE VARNISH. *Rhus vernicifera*.

CHINKWORT. The popular name in some districts for the different species of *Opegrapha* and their allies, which grow on the trunks of trees. These lichens are also sometimes called Letter-lichens, or Scripture-worts. [M. J. B.]

CHINQUAPIN. An American name for *Quercus prinoides*; also for *Castanea pumila*.

CHIOCOCCA. A genus of the Cinchonaceous family, consisting of small shrubs, with a funnel-shaped yellowish corolla, concealing the five stamens, which are provided with hairs. Ovary two-celled, with two inverted ovules. Fruit a berry with two seeds. The species are remarkable for the violent emetic and cathartic properties possessed by the roots, which are administered in Brazil as a certain remedy for snake bites, though their intense action would seem to be, from the account of Von Martius, almost as dangerous as the wound they are intended to cure. The name is derived from two Greek words, signifying 'snow-berry,' in allusion to the white fruit. [M. T. M.]

CHIONANTHUS. The Snowdrop tree of North America, or the Snow-flower, as the name implies, belongs to a genus of *Oleaceæ*, and is distinguished by its deciduous leaves, and the long narrow ribbon-like segments of the corolla. The fruit is a drupe like that of the olive. *C. virginica* is a deciduous shrub or small tree, with large smooth leaves like those of a *Magnolia*, and bearing flowers in terminal panicles. It blossoms in this country in June, and is highly ornamental. [M. T. M.]

CHIONOPHILA. A genus of Scrophulariaceæ, nearly allied to *Pentstemon*, but differing from that genus in its five-toothed (not five-cleft) calyx, as well as in habit. *C. Jamesoni*, the only known species, found in the Rocky Mountains near the snow limit, is a small unbranched herb about two inches high, with a few smooth linear leaves which are enveloped near the base by a number of membranaceous scales. The tubular flowers grow one or two on the apex of a short scape. The fruit is not known. [A. A. B.]

CHIP. A material used for plaiting into various articles of ornament and use, and obtained from the leaves of the palm called *Thrinax argentea*.

CHIQUICHIQUI. The Venezuelan name for *Attalea funifera*, which yields the Piassaba fibre of commerce.

CHIRATA. An Indian tonic, *Agathotes Chirayta*; also called Chiretta or Chiretta.

CHIRITA. A small genus of Gesneraceæ, natives of tropical Asia. They are herbaceous plants with a short stock or a simple leafy stem, the leaves opposite, and the flowers solitary or umbellate, on axillary or radical peduncles. The calyx is five-lobed; the corolla tubular, the limb two-lipped. Of the four stamens the two upper are small and sterile, and the fertile pair have divaricate anther-cells cohering laterally. The stigma is flattened and emarginate or two-lobed. The linear capsule contains many minute seeds without appendages. [W. C.]

CHIRONIA. A genus of the gentian family, somewhat singularly named after Chiron, one of the reputed fathers of medicine, inasmuch as the species inhabit a district unknown in those days, to wit, the Cape of Good Hope. The genus consists of herbs or small shrubs with narrow ribbed leaves, and a corolla with a short tube, and a five-cleft bell-shaped limb with deciduous segments. The five stamens are short, inserted on the throat of the corolla and bent downwards, and the anthers open by two pores at the top. The ovary is partly two-celled, by the bending inwards of the placentas, bearing the numerous seeds; the style terminal, curved at the top, and directed away from the stamens. The capsule has a somewhat fleshy external rind, and an inner membranous one. Several kinds are in cultivation. They have for the most part pretty pink flowers. [M. T. M.]

CHIRONIS. (Fr.) *Sium Sisarum.*

CHIROPETALUM. A genus of *Euphorbiaceæ,* allied to *Croton,* but differing in the stamens being united into a column, not free, and also to *Ditaxis,* which, however, has ten stamens in two tiers, instead of five in one tier. The plants are herbs or small shrubs confined to the temperate parts of South America, some of them having all their parts covered with little simple hairs. The leaves are alternate entire or serrate, generally lanceolate in form and three-nerved. The small green flowers are disposed in axillary or terminal racemes, the upper portion of the raceme being occupied by the males, which are the most numerous, the lower by the females. The calyx is five-parted, and the petals, of a like number, are three or seven-lobed. The ovary is crowned with three styles, each forked at the summit in the form of a Y, and bent back on the fruit which is three-lobed and contains three seeds. The leaves of some of the species are of a reddish-brown colour owing to the presence of colouring matter. [A. A. B.]

CHITONIA. A genus of West Indian shrubs of the family *Melastomaceæ,* some species of which are grown in this country as ornamental stove-plants. They form shrubs or small trees, and have opposite ovate acute five-nerved leaves, and terminal panicles with three-flowered branches. The limb of the calyx is described as being in two rows, the outer consisting of awl-shaped teeth, the inner of short very blunt membranous processes, adherent to the base of the outer teeth; the anthers open by one pore only; the ovary is enclosed within the tube of the calyx, and has six compartments. [M. T. M.]

CHITTA-EITA. An Indian name for *Phœnix farinifera.*

CHITTAGONG WOOD. The timber of several Indian trees, especially of *Cedrela Toona,* and *Chickrassia tabularis.*

CHITTAH-PAT. The Assam name for *Licuala peltata.*

CHIVES or CIVES. *Allium Schœnoprasum,* a garden esculent.

CHLÆNACEÆ. A small family consisting of only four genera of one or two species each, all from the island of Madagascar, and as yet but very imperfectly known. They are trees or shrubs with the habit, alternate leaves, stipules, and terminal inflorescence of some *Sterculiaceæ,* of which they have also the free petals, monadelphous stamens and anthers; and the structure of the ovary fruit and seed is the same as in some genera of that family; but the calyx is said to be always three-cleft or composed of three sepals, and enclosed in a five-toothed involucre, an anomaly which has prevented the absolute union of *Chlænaceæ* with *Sterculiaceæ.*

CHLAMYDANTHUS. A name now applied to a section of the genus *Thymelæa,* in which the tubular calyx remains attached after withering and encloses the nut. The plants embraced in this section are low woody-stemmed bushes, chiefly natives of the Mediterranean regions. Their bark is very tough as in all the plants of the family to which they belong (*Thymelaceæ*). Their leaves are seldom more than half an inch long, and generally linear in form; and the flowers are small and inconspicuous in the axils of the leaves. [A. A. B.]

CHLIDANTHUS. A genus of South American amaryllids having truncated bulbs, linear-lorate leaves sheathing at the base, developed after the flowers, and a scape, one and a half foot high, supporting an umbel of a few large fragrant flowers. The perianth has an erect cylindrical tube widened at the mouth, and a nearly-equal somewhat spreading limb of six segments. The filaments of the six stamens are inserted in the points of the alternately unequal teeth of a thin membrane adhering completely to the tube and base of the petals, but partible. This membrane Dr. Herbert regarded as an incipient manifestation of the staminiferous cup of his paneratiform section of amaryllids, with which *Chlidanthus* thus becomes a connecting link. *C. fragrans,* the only species, has glaucous erect leaves about a quarter of an inch wide; its flowers are yellow, fragrant, sub-sessile, with the tube two to four inches long, and the limb one inch and a half. [T. M.]

CHLOANTHES. A genus of *Verbenaceæ* from extra-tropical New Holland, consisting of undershrubs thickly covered with opposite or ternate sessile linear and revolute leaves, and having solitary axillary flowers with short peduncles. The calyx is campanulate, five-cleft, and spreading. The tube of the corolla has a woolly ring on its interior above the apex of the ovary, and the ringent limb has the upper lip bifid, and the flower tripartite, the middle lobe being the longest. The four didynamous stamens are inserted in the corolla-tube, and the ovary is four-celled, with a single ovule in each cell; the slender style, as long as the stamens, has a bifid stigma. The capsule is dicoccous, each coccus being two-celled. (W. C.)

CHLOIDIA. Among the terrestrial orchids with the habit of small bamboos are two species referred to *Neottia* by Swartz: *N. flava* and *N. polystachya*, the first found in swamps in Jamaica and Brazil, the second inhabiting barrens on the highest mountains of Jamaica. Both look like a *Corymbis* or *Cnemidia*. They seem to fill the same position among *Neottiea* as *Evelyna* among *Epidendreæ*. *C. decumbens* is six feet high; but *C. versalis* is not more than a foot.

CHLORA. An annual herbaceous plant, well marked among the *Gentianaceæ* by its eight-cleft flowers and eight stamens. *C. perfoliata*, called Yellowwort, the only British example, is a singularly erect slender plant, about a foot high, with but few root-leaves, opposite stem-leaves which are united at the base to onesixt, and stems which are forked towards the extremity, having a single flower in each fork and others crowded at the extremity. The whole plant is perfectly smooth and of a decided glaucous hue. The flowers, which are rather large, and of a delicate clear yellow, expand only during the sunshine, like the genus *Erythræa*, to which *Chlora* is allied. The whole plant is intensely bitter, and may be employed with advantage as a tonic; it also dyes yellow. It is of tolerably common occurrence in chalky pastures, especially near the sea. Two other species resembling *C. perfoliata* in habit occur on the European continent, one a native of Germany and Hungary, the other of Southern Europe. French, *Chlore*; German, *Bitterkraut*. (C. A. J.)

CHLORÆA. An extensive genus of terrestrial orchids exclusively found in the southern districts of South America. Botanically they are allied to *Arethusa*, although very different in habit from that genus. Their roots are coarse fascicled glutinous fibres. The leavesare all radical. The scape is clothed with thin herbaceous sheaths. The flowers grow in spikes or racemes in the manner of the genus *Orchis*, are greenish, whitish, or yellow, occasionally marked by deep brown specks. Some thirty or forty species are known, none of which are in cultivation, although they have been occasionally introduced, among which is the plant called, in the *Botanical Magazine* (t. 7856), *Chlorea grandiflora*, the native country of which is unknown.

CHLORANTHACEÆ. A small family of dicotyledons with flowers of a very simple structure, allied to those of *Piperaceæ* and *Saururaceæ*. They are trees, shrubs, or rarely herbs, with opposite leaves connected by sheathing stipules. The minute flowers are in simple or branched terminal spikes, often articulate, as in *Gnetum*. There is no perianth. One or more stamens are adnate to the ovary when the flowers are hermaphrodite. The ovary contains a single pendulous ovule, and is crowned by a thickened sessile stigma. The fruit is a small drupe, the embryo very minute in the top of a fleshy albumen. There are but very few species, all tropical and contained in two genera: *Chloranthus* in Asia, and *Hedyosmum* in America.

CHLORANTHUS. A genus of tropical *Chloranthaceæ*, the only floral envelope of which is a very small calyx, consisting of one scale adhering to the side of the ovary. It consists of small evergreen shrubs, having jointed stems with tumid articulations, and opposite simple leaves with minute intervening stipules. The apparently single stamen, which is the most remarkable part of its structure, consists of three, the central one of which has a perfect two-celled anther, and the other two, one on each side of it, have only half an anther, so that they are only one-celled; or the two lateral half anthers may be deficient, leaving a single perfect stamen. They are attached to the side of the ovary immediately above the calyx. The three stamens grow together except at their points, so as to become monadelphous, which has given rise to different opinions as to their structure. The ovary is one-celled, consisting of a single carpel with one pendulous ovule; and the seed has a large quantity of albumen, the embryo being very minute.

Chloranthus inconspicuus.

The roots of *C. officinalis*, a native of Java, occasionally seen in our hot-houses, are an aromatic stimulant, which, Dr. Blume states, has proved of the greatest service in a typhus fever of that island, accompanied with symptoms of extreme debility, languid pulse, and stupor. It was also employed most beneficially in malignant intermittent fever; and he adds there can be no doubt that it is one of the most valuable stimulants in such cases. It was given in infusion, and was usually combined with a decoction of *Cedrela Toona*. The roots are also employed there with the greatest success, mixed with carminatives, as anise, in the malignant small-pox in children. *C. brachystachys* has similar properties. [D. C.]

The detached flowers of *C. inconspicuus*, which are fragrant, are used in China under the name of Chu-lan, for scenting some of the perfumed teas. They are placed with the prepared leaves in alternate layers under pressure, and thus impart their fragrance to the leaves. [T. M.]

CHLORETTE. (Fr.) *Chlora perfoliata.*

CHLORIS. A genus of grasses, typical of the tribe *Chlorideæ*, distinguished chiefly by the spikes of inflorescence being in finger-like fascicles, rarely two, or only one. Flowers polygamous; glumes two, containing from two to six florets; lower flowers one to three, hermaphrodite; male flowers often stalked; pales with terminal awns; stamens three; styles two. Sixty-nine species are described in *Steudel's Synopsis*, and these are mostly natives of warm, dry countries, and consequently require the protection of a conservatory in Britain. *C. radiata* is a pretty annual grass, frequently cultivated in greenhouses, in consequence of its ornamental and curious appearance. Many of the other species are handsome also. [D. M.]

CHLORO. In Greek compounds = green.

CHLOROCHROUS. Having a green skin.

CHLOROPHYLL. The green resinous granular colouring matter secreted below the surface of plants.

CHLOROSA *latifolia*, is an insignificant Javanese orchid, allied to *Neottia* and *Cryptostylis*, from the latter of which it differs in the pollen, which is strictly powdery, and in the anther, which is terminal. It has small insignificant green flowers.

CHLOROSIS. One of the most formidable diseases to which plants are subject, and often admitting of no remedy, especially where it is constitutional. It consists in a pallid condition of the plant, in which the tissues are weak and unable to contend against severe changes, and the cells are more or less destitute of chlorophyll. It is distinct from blanching, because it may exist in plants exposed to direct light on a south border, but is often produced or aggravated by cold ungenial weather and bad drainage. Plants may, however, be affected by this disease as soon as the cotyledons make their appearance, and the seedlings of chlorotine plants partake often of the weak constitution of the parent. The best culture will not always restore such plants to health. The most promising remedy is watering them with a very weak solution of sulphate of iron. Many forms exist, of which those of clover, onions, cucumbers, and melons, are perhaps the best known. Melons have become so subject to chlorosis, from some unknown cause, that their cultivation is daily becoming more difficult; and cucumbers are still more generally affected, the fruit even partaking of the malady, and not only losing its brilliant green, but becoming distorted from gumming and partial decay. [M. J. B.]

CHLOROSPERMEÆ. One of the three great divisions of *Algæ* characterised by the green colour of the spores. To this there are occasional exceptions, and in some of those the spores are originally green. The species are in general far less compound than in the two other orders, though in some instances the phenomena of fructification are more striking. The green powdery or gelatinous productions, which are so common upon damp walls or rocks; the curious microscopic few-celled productions which abound in our pools or infest other *Algæ*; the green floating masses which form a scum upon our pools, or the shrubby tufts of the same colour in running streams or on sea rock; the flat slimy membranes which occur both in fresh and salt water, are so many members of the division; to which may be added, the spongy *Codiums* and the herbaceous tinted *Caulerpæ*, which often assume the more solid appearance of the more perfect *Algæ*. In a few genera large quantities of carbonate of lime are deposited, so as to give them a coral-like appearance. To avoid repetition the peculiar features of each group will be stated in its proper place. In *Diatomaceæ* the spores, which are however rarely produced, their multiplication being chiefly effected by repeated cell divisions, are of a yellow brown, and in an artificial system they might be referred to the *Melanospermæ*. Though, however, they are in some respects so peculiar as to stand apart from other *Algæ*, they are so closely connected with *Desmidiaceæ*, that they can scarcely be separated from true Chlorospermæ.

The spores of most members of this great division when they are first liberated are endowed with active motion, which is produced by long thorny-like appendages and by short cilia. In most cases they are very minute. Such spores are called, from their resemblance to Infusoria, Zoosperms. In some instances, as in *Conjugate*, the admixture of the contents of two contiguous cells, either in the same or different individuals, is requisite for the production of the perfect spore. In the latter case, short lateral tubes are thrown out, by means of which different threads are united, or they become adherent without any distinct connecting thread. Male organs have been found in many of the divisions. [M. J. B.]

CHLOROXYLON. A genus of *Cedrelaceæ*, generically distinguished by its fruit having only three cells, and splitting into three parts instead of five.

The satin-wood tree of India, *C. Swietenia*,

forms a fine tree fifty or sixty feet in height. It is a native of Ceylon, and the Coromandel coast, and also of other parts of India. Its leaves are pinnate, consisting of numerous pale-coloured leaflets, of a somewhat egg-shaped outline, but with the two sides unequal. These leaflets are readily distinguishable from those of all the allied genera, with the exception of *Flindersia*, an Australian genus, by their substance being dotted with minute pellucid glands or oil cells. The small whitish flowers of this tree are borne in large branching panicles, growing at the ends of the young branches. They have a small five-parted calyx; five spreading petals with short stalks; ten awl-shaped spreading stamens, all of which are distinct and fertile; and a three-celled ovary, which is half buried in the disk from which the stamens rise. The fruit contains four seeds in each cell, and the seeds are prolonged at one end into a thin wing or membranous expansion.

This tree yields the satin-wood of India, a handsome light-coloured hard wood, with a satin-like lustre, and sometimes beautifully mottled or curled in the grain, bearing some resemblance to box-wood, but rather deeper in colour. The best kind of satin-wood, however, comes from the West Indies, and is the produce of a different but unknown tree. In 1858 the imports of this wood amounted to 248 tons, valued at 2,467*l.*; the Indian wood being in circular logs of nine to thirty inches in diameter, and that from the West Indies (St. Domingo and New Providence) in square logs or planks varying from nine to twenty inches across. The principal use of satin-wood is for making the backs of clothes- and hair-brushes, and for articles of turnery ware; the finest mottled pieces, however, are cut into veneers and used for cabinet-making and similar purposes. [A. S.]

CHNOOPHORA. A name sometimes given to certain ferns usually referred to *Alsophila*. [T. M.]

CHOCO. *Sechium edule*, a tropical esculent of the cucurbitaceous order. Not used in this country.

CHOCOLATE ROOT. *Geum canadense*.
—, INDIAN. *Geum rivale*.

CHOCOLATE TREE. *Theobroma Cacao*. The Chocolate-nut is the seed of this tree, and the chocolate of the shops a preparation of these seeds.

CHŒRADODIA. A genus referred by Herbert to the alstrœmeriform amaryllids. It has fibrous roots, numerous radical linear acute erect glabrous leaves, and a scape five to six feet high, bearing three or four smaller alternate clasping leaves, and supporting a corymb of flowers, of which the sepals and petals are very unequal in size, the one white the other tipped with red. It is a little-known plant of Chili, where it is called *Theiui*. A cold infusion of its leaves is purgative and diuretic. [T. M.]

CHOHO. An Abyssinian name for *Indigofera argentea*.

CHOIN. (Fr.) *Schœnus*.

CHOISYA. A Mexican rutaceous shrub, with ternate leaves, a panicled inflorescence, with large deciduous bracts beneath the flower-stalks; white flowers sprinkled with glandular dots; the five petals and ten stamens inserted on a short stalk supporting the ovary, which consists of five carpels fused into one. The style is short with five furrows, hairy like the ovary; stigma capitate. The fruit is a capsule with five furrows. [M. T. M.]

CHOKE-BERRY. An American name for *Pyrus arbutifolia*.

CHOKE, BLACK. *Cerasus hiemalis*.

CHOLA. An Indian name for Gram, *Cicer arietinum*.

CHOLLU. An Indian name for the grain of *Eleusine coracana*.

CHOLUM. The great Millet, *Sorghum vulgare*.

CHOMORO. *Podocarpus cupressinus*, one of the best timber trees of Java.

CHONDRILLA. A genus of the composite family, nearly allied to the lettuce (*Lactuca*), which has the achenes prolonged into a beak and smooth; while those of *Chondrilla* are often rough and furnished at the base of the beak with five small scales, arranged in the manner of a little calyx. The plants are herbs, with generally pinnatifid root-leaves, having a large terminal lobe and small lateral ones; those of the stem, few small and entire. The yellow flower-heads are solitary and terminating the branches, or in corymbs or leafy spikes. *C. juncea*, a native of the south of Europe, a straggling much-branched plant, is almost destitute of leaves when in flower; a narcotic gum is said to be obtained from it in the island of Lemnos. About twenty species are enumerated, all of them weedy plants, natives of South Europe, the East, and Siberia. [A. A. B.]

CHONDRODENDRUM. A genus of climbing shrubs belonging to the *Menispermaceæ*, and closely allied to *Cocculus*, from which it is distinguished by the stigmas, which are ovate and simple; by the globose fruit, which consists of one drupe, owing to the suppression of the others; and by the flat orbicular seeds with a striated margin. *C. convolvulaceum* is called by the Peruvians the Wild Grape, on account of the form of the fruits, and their acid and not unpleasant flavour. The bark is esteemed as a febrifuge. [M. T. M.]

CHONDRORHYNCHA rosea is a terrestrial orchid related to *Cymbidium*, inhabiting Central America. It has long ribbed broad grassy leaves, and large dirty purple radical flowers, with the upper sepal united to the back of the column, and the lip in the form of a large boat-shaped body. The pollen masses are four, secured in pairs to a long soft gland attached to a hard cartilaginous rostel.

CHONDROSPERMUM. A genus of climbing evergreen shrubs, natives of India, with opposite petiolate and three-nerved leaves; the flowers are in very short pedunculate panicles with small lanceolate bracts. The calyx and corolla consist each of four parts; the corolla has a long tube, and spreading limb, cleft into four linear clavate lobes; there are two scarcely exserted stamens, a two-celled ovary with a single erect ovule in each cell, and two very short styles or stigmas. The yellow flowers and climbing stems, together with the erect ovules, have caused this genus to be referred to *Jasminaceae*. The whole structure of the flower seems, however, to unite it more closely to *Oleaceae*. [W. C.]

CHONDRUS. A small genus of rose-spored *Algae*, with a forked fan-shaped frond, and the capsules, which contain several masses of spores, immersed in the frond without any definite border. The type of the genus is *C. crispus*, the true *Carrageen*. It is very common on our

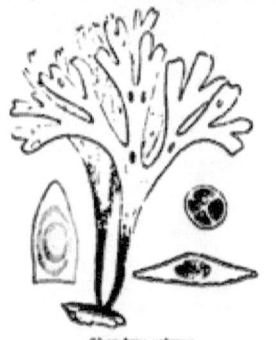

Chondrus crispus.

coasts, as it is also on the Atlantic coast, from the shores of British America to those of Long Island. The colour varies from a dull livid purple to greenish and yellowish. [M. J. B.]

CHONEMORPHA. A genus of *Apocynaceae*, closely allied to *Echites*, and differing from it principally in the funnel-shaped corolla. The species are, moreover, Indian not American. The root and leaves of *C. malabarica*, a plant of Malabar, are used medicinally by the natives. [M. T. M.]

CHOOA or CHOUA. An Indian name for *Amaranthus frumentaceus*, and *oleraceus*.

CHOOPA. *Pierardia dulcis*, a Malacca fruit.

CHORDA. A genus of dark-seeded *Algae*, with a simple cylindrical tubular frond, in the surface of which are imbedded a number of obconical spores. In *C. filum*, the frond is slimy, perfectly cylindrical, and sometimes twenty, or even forty feet in length. It is occasionally used instead of fishing lines, for which, however, it must be a poor substitute. 'It is in quiet land-locked bays,' says Dr. Harvey, 'with a sandy or somewhat muddy bottom, and in from three to six fathoms of water, that it reaches its greatest size. In such places it frequently forms extensive submarine meadows so dense as seriously to affect the passage of boats, and to endanger the life of the unfortunate swimmer who may chance to become entangled in its slimy cords, which when growing have considerable tenacity.' [M. J. B.]

CHORDA PISTILLARIS. A line of tissue reaching from the stigma down to the cavity of the ovary.

CHORDARIEAE. A natural order of dark-spored *Algae*, distinguished by their compound gelatinous frond consisting of vertical and horizontal threads variously interlaced, the cysts being contained in the substance of the frond, and not external as in *Ectocarpeae*. Some are as gelatinous as *Batrachospermum*; while *Leathesia* presents an irregular firm but hollow truffle-like mass; *Ralfsia* forms an adherent crust. They are principally inhabitants of colder regions, though species are found at Port Natal and amongst the Philippines. The spore-like cysts are often of two kinds, both producing zoospores. The tips of the terminal threads are often swollen so that they appear to be fruit. [M. J. B.]

CHORETIS. A genus of Mexican and Texan *Amaryllidaceae*, with the habit of *Ismene*, and the flowers of *Hymenocallis*. The perianth has a long slender nearly straight tube, a reflexed limb of long narrow segments, and a large rotate cornet incerted at the margin, the long filaments being spreading-connivent. *C. glauca* is a beautiful species with black-coated bulbs, erect glaucous leaves, a foot and a half long and two and a half inches wide, and a scape upwards of a foot high, supporting three or four sessile flowers. The tube of the perianth is six inches long, quite slender, green, the limb three and a half inches or more, linear white-ribbed with green, the cup, or coronet above an inch long, white, rotate, with large teeth between the stamens, which have long locurvo-connivent filaments attached to the upper part of the anther in a prominent callosity. *C. galvestonensis* is a smaller-flowered species from Texas, producing four-flowered umbels, the tube of the perianth and the limb each about two and a half inches long. [T. M.]

CHORETRUM. A genus belonging to

the order of sandalworts. The flowers have both stamens and pistils; the divisions of the calyx vaulted and covering the five stamens; the style, or appendage on the top of the ovary, is very short, ending in a star-like surface. The name is derived from the Greek word signifying a rustic, probably in allusion to the plain aspect of the species. They are natives of New Holland, having the form of shrubs resembling our native broom. The leaves are very small and scattered, confined mostly to the vicinity of the flowers, which are also small, white, and subtended by four leaflets or bracts. (G. D.)

CHORION. A carpel; also the pulpy matter which fills the interior of a young seed before impregnation.

CHORIPETALUM. A genus of scandent shrubs or trees of the ardisiad family, distinguished among its allies in its petals being four in number and free, not united, as well as in its racemed flowers. The leaves are alternate, shortly-stalked, and entire, generally about five inches long, between ovate and elliptical in form, and furnished with glandular dots. The small white or yellowish flowers are borne in little axillary racemes, and the berries when ripe are scarlet in colour and contain few seeds. Those of C. undulatum are, according to Dr. Hooker, eaten in Sikkim as well as the leaves, which are sour to the taste. This species grows to a straggling tree of sixty feet, and, along with a few others, is a common plant in the temperate regions of the Himalayas. One species is found in Ceylon; another, with small leaves, rounded at the apex and narrowed towards the base, is found in Hong Kong, the eastern limit of the genus, Java being the southern, and Bombay the western. [A. A. B.]

CHORISIA. A genus of small prickly-stemmed trees of the sterculiad family, peculiar to South America. Their leaves are alternate stalked and digitate, made up of five oblong, or elliptical smooth leaflets, each from three to six inches in length. The flowers (one to three in the axils of the upper leaves) are large, rose-coloured, and composed of a bell-shaped three or five-lobed calyx; five narrow petals from one to three inches long, either entire or with crisped margins, and covered with silky hairs; a double staminal tube, the outer one short and bearing on its apex ten barren stamens, the inner much longer and bearing ten fertile stamens. The number and arrangement of the stamens serve to distinguish the genus from its allies. The fruit is a one-celled pear-shaped capsule, containing many seeds, which are covered with silky or cottony hairs. The tough bark of C. crispiflora is used in Brazil for making cordage; and the white cottony hairs of the seeds of C. speciosa are used by the Brazilians for stuffing pillows and cushions. The tree is known by the name of Arvore de Paina in Brazil. M. de St. Hilaire

remarks, that the cottony hairs which cover the seeds and line the walls of the fruit of many of the plants of this *Bombax* family, are used wherever they grow—whether in India, Africa, or America, for precisely similar purposes. (A. A. B.)

CHORISMA. The only species of this genus, C. repens, a little plant of the composite family, grows on the sandy sea-shores, from Loo Choo and Hong Kong to Kamtschatka. The stems, about the thickness of a straw, creep along the sand and emit roots where the leaves are given off; these leaves have stalks about three inches long, and are generally three or five-lobed, but sometimes three-parted with three-lobed segments, the blades much shorter than their stalks and quite smooth. The yellow flower-heads are from one to three, supported on a stalk a little longer than the leaves. This plant is nearly related to the sow-thistles, but the peculiar habit and form of its leaves readily serve to distinguish it. It has been called *Chorisis repens*, and is now placed in the genus *Ixeris*. (A. A. M.)

CHORISPORA. A genus of *Cruciferæ*, allied to *Cakile*, but differing in the longer cylindrical pod, which breaks across into one-seeded portions. Annual plants, natives of Siberia and the Altai, with purple, white or yellow flowers. (J. T. S.)

CHORISTER. A genus of much-branched Mexican shrubs, belonging to the *Cinchonaceæ*. The flowers are few in number, with a top-shaped calyx, having four persistent, short divisions; a somewhat bell-shaped corolla; four stamens; a two-celled ovary, becoming a capsule, dividing from the top into two divisions, to liberate the numerous seeds. [M. T. M.]

CHORISTOPHYLLOUS. Separate-leaved.

CHORISTYLIS. A South African genus of *Escalloniaceæ*, represented by a shrub with panicles of small green flowers, having five awl-shaped silky three-nerved petals, valvate in the bud, and four stamens with a fleshy connective between the lobes of the anther. The ovary is inferior, with two compartments, and surmounted by two styles, united partially at first, but ultimately becoming disjoined. The fruit is a capsule bursting into two pieces to liberate the many seeds it contains. [M. T. M.]

CHORIZANDRA. A genus of cyperaceous plants, belonging to the tribe *Scirineæ*. The spikes are many-flowered, and androgynous; the exterior flowers male and monandrous, the central ones solitary, female, with two to three-cleft styles. Three species are described of these little-known plants, which are all natives of Australia. [D. M.]

CHORIZANTHE. A genus of *Polygonaceæ*, of the tribe *Eriogoneæ*. Herbs, natives of California, or under-shrubs from Chili. Leaves alternate, crowded at the base of the stem, woolly; inflorescence cymose, lax, or contracted into heads; involucres one-flowered, tubular, three-sided, six-toothed; perianth herbaceous, tubular, with a six-lobed limb, the lobes in two rows; stamens nine, the filaments cohering at the base; styles three; fruit a three-sided nut. [J. T. S.]

CHORIZOPTERIS. A name proposed for one or two Acrostichoid ferns, now classed with *Neurocallis*, which are remarkable in having the parts of the fronds articulated, so that they readily fall to pieces when dry. [T. M.]

CHOROZEMA. A genus of pretty West Australian bushes belonging to the pea-flowered *Leguminosae*. It is nearly allied to *Callistachys*, but differs in having the keeled petal shorter than the wings, as well as in the inside of the pod being destitute of any pithy substance. The plants are very often to be met with in greenhouses, upwards of a dozen species being in cultivation, the greater portion of them producing their graceful elegant flowers in the spring months. The leaves are simple, either entire or with spinous teeth, generally smooth, and varying much in form. In the greater portion the flowers are in racemes, but in a few they are axillary and solitary; the pods are generally oval in form, turgid, and about half an inch long, containing a number of seeds.

The first species of the genus, *C. ilicifolium*, was found by Labillardière in West Australia. This botanist was attached to the expedition sent by the French government in search of the lost La Pérouse, and on one of his excursions suffered much, with his party, for want of water; at last they met with an ample supply, and near it with this plant, which he named *Chorozema*, a name said to be derived from *choris* a dance, and *zema* a drink, in allusion to the joyful feelings of the party on meeting with a supply of water.

Amongst the most beautiful of the Chorozemas known in cultivation are:—*C. Henchmanni*, with long terminal leafy racemes, of a beautiful red colour, the standard having a green spot at its base; the leaves are awl-shaped, about half an inch long, and generally disposed in clusters of three. *C. spectabile*, a twiner of great beauty, producing long drooping racemes of orange-coloured flowers, which appear in the winter months; its leaves are oblong-lanceolate, with a little point at the apex. *C. cordatum*, a plant very common in gardens, and having ovate short-stalked leaves, heart-shaped at the base, the margins armed with prickly teeth; the flowers, in loose racemes, are red, the standard spotted with yellow at the base. *C. Dicksoni*, a handsome plant with larger flowers than the others; the leaves are entire and lance-shaped, generally having on both surfaces a few long spreading hairs. There are upwards of twenty species known. [A. A. B.]

CHOU. (Fr.) *Brassica*. —, CABAÏBE *Caladium sagittifolium*, and *Colocasia esculenta*. — DE CHIEN. *Mercurialis perennis*. — FLEUR. *Brassica oleracea botrytis*. —, MARIN. *Crambe maritima* —, PALMISTE. *Areca oleracea*.

CHRISTISONIA. A genus of Orobanchaceæ containing ten or twelve species, natives of India. They are parasitic plants, with fleshy stems, scattered or imbricated scaly leaves, and the flowers terminal or in the axils of the upper leaves; the calyx is tubular and five-toothed; the corolla infundibuliform and somewhat two-lipped; the anthers two-celled, with one of the cells barren and subulate; and the ovary imperfectly two-celled, the inflexed septa only partially meeting in the axis, and the placentiferous margins remaining free, and hence reflexed form two loose placentæ in each cell. [W. C.]

CHRISTMAS ROSE. *Helleborus niger*.

CHRISTOPHER, HERB. *Actæa spicata*; also *Osmunda regalis*.

CHRIST'S EYE. *Inula Oculus Christi*.

CHRIST'S THORN. *Paliurus aculeatus*.

CHRISTYA. A Cape shrub forming a genus of *Apocynaceæ*, and having erect rod-like branches and large handsome flowers, with a calyx divided into five lance-shaped divisions, each with a cleft gland at its base; a somewhat bell-shaped corolla with a row of cleft fleshy scales at its mouth, alternating with the linear divisions of its limb; five anthers, hairy on their back, and cohering with the velvety cushion-like stigma. The two ovaries contain several seeds. [M. T. M.]

CHROMATIDIUM. The colouring matter of plants.

CHROMISM. Preternatural colouring of plants, as that of leaves when they become red, &c.

CHROMULE. The fluid colouring matter of vegetation.

CHROOLEPUS. A curious genus of *Algæ*, referred to the green-spored division on account of its clear natural affinities, but exhibiting, when fresh, orange not green tints. The species grow on damp walls, rocks, evergreen leaves, bark, &c., and when fresh often emit a scent like that of violets. The minute zoospores are contained in lateral cysts, and by these the genus is at once distinguished from *Callithamnion*, which in some respects it resembles. The black productions, commonly referred to this genus are evidently fungi and not alcæ. *Chroolepus* sometimes occurs in a rudimentary state on exposed stones, and then obtains the name of the sweet-scented *Byssus*. [M. J. B.]

CHRYSALOIDEUS. Rolled up and folded up at the same time.

CHRYSANTHEME DES INDES OU POMPON. *Pyrethrum indicum*.

CHRYSANTHEMUM. A genus of herbaceous or slightly shrubby plants belonging to the corymbiferous group of the order *Compositæ*, distinguished by their hemispherical involucre composed of imbricated scales which are membranous at the edges, large naked receptacles, and by the absence of a pappus from the fruit. The family is represented in Britain by the familiar Ox-eye Daisy, *C. Leucanthemum*, and Corn Marigold, *C. segetum* : the former a common weed in hay-fields, where its flowers, which are white with a yellow disk, are conspicuous; and the latter a handsome but mischievous weed in corn-fields, where it is sometimes so abundant as to be more conspicuous with its large golden-yellow flowers than the crop which it tends to impoverish. Many species have been introduced from various countries, and are cultivated in our gardens, of which *C. grandiflorum* from the Canaries, *C. pinnatifidum* from Madeira, and others, are of a shrubby habit and flower during a large portion of the year, but require protection in winter; white *C. coronarium* from the Levant, and *C. carinatum* (called also *C. tricolor*) from Barbary, are ornamental border annuals. The species, however, which holds so high a rank, and with reason, among florists' flowers is *C. sinense*, a plant which has long been familiar to us from its frequent appearance in Chinese drawings, but has of late years been improved to such an extent as to be prized for its intrinsic beauty, and not simply from its valuable property of blooming in November and December. This plant, popularly known as *the* Chrysanthemum, is more generally referred by botanists to the genus *Pyrethrum*, as *P. sinense*. Chrysanthemums are classed by growers into Large-flowered, Anemone-flowered, Pompons, and Anemone-flowered Pompons. [C. A. J.]

CHRYSANTHUS. Yellow-flowered.

CHRYSEIS. A name sometimes given to the species of ESCHSCHOLTZIA; which see.

CHRYSIPHIALA. A synonyme of *Stenomesson*, adopted in some systematic books.

CHRYSO. In Greek compounds = golden yellow.

CHRYSOBACTRON. A genus of *Liliaceæ*, near *Anthericum*, from the Auckland and Campbell Islands, New Zealand. It has linear leaves and racemose flowers (which are occasionally dioecious) of a bright yellow colour. The perianth is six-lobed; the anthers connected (almost in the female flowers). The ovary has three furrows; style thick; stigma capitate, three or six-lobed; capsule ovoid, three-celled, the cells usually two-seeded. *C. Hookeri* is a pretty little bog plant, which has been introduced into this country, where it requires the protection of a greenhouse. [J. T. S.]

CHRYSOBALANACEÆ. A family of dicotyledons closely allied to *Rosaceæ*, or more generally considered as a tribe of that order taken in its most extended sense. They are distinguished from the other tribes by a frequent irregularity in the stamens, and more especially by their solitary carpels, with the style always proceeding from the base, and containing two ascending ovules. The fruit is free, either drupaceous or capsular. They are all trees or shrubs with alternate stipulate leaves and several of them produce edible fruits. There are nearly one hundred species, more or less known (including several as yet unpublished), dispersed over the tropical regions both of the Old and New World, although much more abundantly so in the latter. They are distributed into about twelve genera, of which the principal ones are *Chrysobalanus*, *Hirtella*, *Couepia*, *Parinarium* and *Prinsepia*.

CHRYSOBALANUS. This genus, the type of the family to which it belongs, is composed of shrubs and small trees, natives of the tropical parts of Africa and America, one species being found in Florida. It differs from others of the family in having its stamens, in number about twenty, arranged in a regular whorl, not inserted on one side of the flower, as well as in the nut of the fruit being one-celled only. The leaves are alternate, stalked, entire, and obovate in form, having both the surfaces smooth. The flowers, borne in short panicles or racemes, are small, white, and made up of a bell-shaped five-cleft calyx; five petals; about twenty stamens; and an ovary the style of which arises from one of its sides near the base, which latter is one of the principal characters of the family. The Cocoa-plum, *C. Icaco*, is one of the commonest species. The fruits are about the size of a plum, and vary much in colour, being either white, yellow, red, or purple. The pulp is sweet, a little austere, and not disagreeable. The shell of the kernel is hard and six-grooved. In the West Indies, according to McFadyen, the fruits prepared with sugar form a favourite conserve with the Spanish colonists, and large quantities are annually exported from Cuba. The kernels yield a fixed oil, and an emulsion made with them it is said to be used in dysentery. An astringent bath recommended in leucorrhœa and blennorrhœa is prepared from the leaves and roots. Four species are known. [A. A. B.]

CHRYSOCHROUS. Having a yellow skin.

CHRYSOCOMA. A genus of South African shrubs or undershrubs of the composite family, nearly allied to *Linosyris*, from which it differs in the hairs of its pappus being in a single series. About fifteen species are enumerated. Their leaves in most cases are linear in form and entire. The yellow nearly spherical flower-heads are about the size of a pea, and single on the ends of the branches; the florets all tubular and perfect. The achenes are laterally compressed, somewhat bi-pid, and seated on a naked honey-combed receptacle. *C. Coma aurea* is in cultivation, and is said to be a very common species about Cape Town; its leaves are linear, and about half an inch long. [A. A. B.]

CHRYSOCORYNE. A curious genus of small annual Australian plants belonging to the composite family. They are branched from the base, and seldom exceed three inches in height. The leaves are small, linear, and covered with loose white wool; but the most marked feature in the plants is the arrangement of the flower-heads; these are disposed in short yellow club-shaped spikes, and each flower head is almost hidden by a yellow bract, and contains but two florets. The florets are tubular and bi- or tridentate, an unusual circumstance in this family. The achenes are covered with wart-like glands, and are destitute of pappus. Five species are known; they are chiefly found in the western and southern parts of Australia. [A. A. B.]

CHRYSOCYCNIS *Schlimii* is a New Grenada epiphytal orchid, with a thick creeping rhizome, from which arise at long intervals small flat pseudo-bulbs, each bearing one stalked oblong leathery leaf. From the base of the pseudo-bulbs rise numerous short one-flowered peduncles clustered in the manner of some *Maxillarias*. The flowers are furnished with a sagittate stalked lip, and are rose-coloured or dull purple, by no means yellow, as shown in Reichenbach's *Xenia*, t. 55, and as the name implies. The genus, if it be one, is near *Cumaridium*.

CHRYSODIUM. A synonyme of the typical species of *Acrostichum*, represented by *A. aureum*. [T. M.]

CHRYSOGLOSSUM. Under this name Blume has a genus of two species only, *C. ornatum* and *villosum*, inhabiting the mountainous parts of Java. Both are terrestrial one-leaved orchids, allied to *Liparis*, and have fleshy subterranean rhizomes. In *C. villosum*, a shaggy plant, and the best known of the two species, the leaf is large, plaited and ovate, while the scape is about two feet high, bearing orange and yellow flowers the size of a wild pansy.

CHRYSOGONUM. This genus of the composite family differs from its allies in its achenes being crowned with a two or three-toothed pappus. Its only representative is *C. virginianum*, a dwarf perennial herb found in many parts of the United States. All its parts when young are covered with hairy tomentum. It is nearly stemless when it begins to flower, but soon sends out several stems, some of which are erect and flower-bearing, while others take the form of runners and creep along the ground. The leaves are opposite on long stalks, ovate, with notched margins. The flower-heads, stalked, solitary, and terminating the branches, are made up of numerous bright yellow florets, those of the ray few and strap-shaped, and those of the disc numerous and tubular. [A. A. B.]

CHRYSOMA. A genus of North American plants of the composite family, considered by the authors of the *Flora of North America* to be the same as that of the golden rod (*Solidago*). The species are perennial plants, with alternate lance-shaped entire or serrated leaves, sometimes furnished with pellucid dots, and they bear terminal corymbs of yellow flower-heads, each of which contains from six to eight florets, one to three of them being strap-shaped. [A. A. B.]

CHRYSOPHYLLUM. A name expressive of the golden colour on the underside of their leaves, which the trees of this genus possess. It is a group of *Sapotaceæ*, and consists of trees with a milky juice, alternate leaves with numerous transverse closely aggregated ribs, and golden hairs on the under surface. The corolla is somewhat campanulate; its tube bears five fertile stamens and no sterile ones; the ovary is five to ten-celled with a short style; the fruit is a berry with ten cells, or one only from the suppression of the rest. Some of the species are grown in this country for the sake of their handsome foliage; while in the West Indies the fruit of *C. Cainito* is esteemed a delicacy under the name of the Star apple, inasmuch as it is of the size of a large apple, while the interior, when cut across, reveals ten cells, and as many seeds disposed regularly around the centre. [M. T. M.]

CHRYSOPSIS. A genus of annual or perennial North American plants of the composite family, the greater portion of the species having all their parts covered with villous or silky hairs. The oblong or linear leaves are usually entire and sessile. The showy yellow flower-heads, usually terminating the branches and often corymbose, have an involucre of many linear scales enclosing numerous florets; those of the ray strap-shaped and bearing pistil only, those of the disc tubular and perfect. The genus differs from its allies in having the pappus of the ray and disc florets similar and double, the exterior short and scale-like, the inner copious and capillary. *C. villosa*, a plant with oblong hairy leaves about an inch and a half long, and numerous yellow flower-heads, half an inch in diameter, is said to be one of the commonest plants on the prairies of the Saskatchawan. *C. graminifolia* extends southwards to Mexico; its leaves are clad with beautiful close-pressed silvery hairs. [A. A. B.]

CHRYSOPTERIS. A synonyme of *Phlebodium*, a genus of ferns which includes Linnæus's *Polypodium aureum*, the specific appellation seeming to have suggested this generic name. [T. M.]

CHRYSORRHOE. A genus of *Chamælauciaceæ*, consisting of a rigid shrub from the Swan river, with narrow terete leaves, and terminal corymbs of bright yellow flowers. The sepals are five in number, and cut into many pilose segments; the petals five, serrate; the stamens five, ten fertile and ten imperfect and shorter than the others; the ovary completely covered by the disc. The genus is closely allied to *Vericordia*, but that has monadelphous stamens, and the sterile ones longer than the fertile. [J. T. S.]

CHRYSOSPLENIUM. Golden Saxifrage. A small genus of unimportant herbaceous plants belonging to the *Saxifrageæ*, among which they are discriminated by their one-celled seed-vessel, and by being destitute of petals. Two species are indigenous to Britain, and scarcely differ from one another, except that one has the leaves opposite, the other alternate. They grow on the margins of streams, forming extensive patches, and in hilly districts often betray, by a line of bright green, the course of a mountain-spring which has worn a narrow way for itself down the slope. The roots are intermatted and send up numerous delicate green very succulent stems, to the height of three or four inches. The leaves are roundish, somewhat fleshy, and sprinkled with longish hairs. The flowers, which are bright yellowish-green, appear in April and May, growing in flat tufts at the summit of the stems. *C. oppositifolium* is the commonest species. *C. alternifolium* is more abundant in the north. The genus is represented in various parts of the world by plants of similar habit, none of which are worthy of cultivation. In the Vosges, the species are used as a salad under the name of *Cresson de Roche*; French, *Dorine*; German, *Goldmilz*. [C. A. J.]

CHRYSOSTEMMA. Under the name of *C. tripteris* is sometimes cultivated in gardens a tall smooth North American herb of the composite family, with opposite leaves, those on the lower part of the stem pinnately five-parted, the upper ones three-parted, with lance-shaped segments one to four inches long, and having the yellow-rayed flower-heads arranged in a corymbose manner at the ends of the twigs, each head one to two inches across. The plant is placed by modern authors in the genus *Coreopsis*, with the same specific name, and may be recognised from others in that genus by the achenes being narrowly-winged, with a toothed fringe at the summit of the wing. [A. A. B.]

CHRYSOXYLON. The name of a South Bolivian tree, now referred to HOWARDIA, which see. It derived its name from the yellow colour of its wood.

CHRYSURUS. A genus of grasses belonging to the tribe *Festuceæ*. Only one species is described, *C. cynosuroides*, which is the *Lamarkia aurea* of some authors. This handsome dwarf-habited annual grass is a native of the south of Europe and north of Africa, and is occasionally cultivated in botanical gardens, where it makes a pretty appearance during the summer months. [D. M.]

CHU'CHUNCHULLI. The root of *Ionidium microphyllum*.

CHU'-LAN. *Chloranthus inconspicuus*, a tea-scent used in China.

CHUMBELEE. *Jasminum grandiflorum*.

CHURN-STAFF. *Euphorbia helioscopia*.

CHURRAS. The Nepalese name of the resinous exudation of the Hemp, *Cannabis sativa*.

CHUSSALONGO. The vulnerary, Matico, *Eupatorum glutinosum*.

CHYMOCARPUS. A genus of scandent herbs belonging to the *Tropæoleæ*. The flowers consist of a coloured calyx, prolonged behind into a hollow spur, and divided at the margin, in a somewhat two-lipped manner, into five nearly equal lobes; a corolla of two petals inserted in the mouth of the tube-like spur; and eight hypogynous stamens. The sessile three-lobed three-celled ovary grows into a three-lobed sweet fleshy edible berry, which remains attached to the front of the persistent calyx. This black juicy berry, which is not unlike in appearance and flavour to the Zante or currant grape, is the most remarkable peculiarity of the genus, which was founded on *C. pentaphyllus*, a plant of Buenos Ayres, long cultivated in our gardens. This is a handsome species, with a thick fleshy fusiform tuber, and smooth filiform stems, climbing several feet high, and furnished with alternate stalked five-parted leaves, having oblong-elliptic leaflets. The flowers are solitary in the axils of the leaves, the spur of the calyx funnel-shaped, above an inch long, bright orange-red, the limb green, marked with dark red spots. The two petals are very small, purple. [T. M.]

CHYSIS. Under this name are collected about four species of orchids from tropical America, with fleshy stems covered with sheaths, thin-ribbed leaves, and lateral spikes of large handsome white or yellow flowers of the consistence of wax. The finest species is *Chysis bractescens*. They all have eight pollen masses attached to a broad yellow pulverulent somewhat rectangular plate.

CIBOTIUM. A genus of polypodiaceous ferns, belonging to the *Dicksonieæ*, among which it is distinguished by having the indusia or involucres two-valved, both the valves being coriaceous, the outer one larger and cucullate, the inner one operculiform. They are large growing and very handsome ferns, in some cases arborescent, the fronds bipinnate and often glaucous beneath. The fructification is remarkably pretty. *C. Barometz*, sometimes called *C. glaucescens*, is believed to be the Barometz, *Agnus Scythicus*, or Tartarian Lamb, about which travellers have told so wondrous a tale. This 'Lamb' consists merely of the decumbent shaggy caudex of a kind of fern, which is no doubt the species just referred to. When inverted, the basal part of the stipes of four of the fronds suitably placed, having been retained as legs, and the rest cast away, these caudices present an appearance which may be taken as a rude representation of some small woolly animal. The 'traveller's tale' is that on an elevated uncultivated salt-plain of vast extent, west of the Volga, grows a wonderful plant, with the appearance of a lamb (Baran in Russian), having feet, head, and tail distinctly formed, and its skin covered with soft down. The 'lamb' grows upon a stalk about three feet high, the part by which it is sustained being a kind of navel; it turns about and bends to the herbage, which serves for its food, and pines away when the grass dries up and fails. The fact on which this tale is based appears to be, that the caudex of this plant may be made to present a rude appearance of an animal covered with silky hair-like scales, and if cut into is found to have a soft inside with a reddish flesh-coloured appearance. When the herbage of its native haunts fails through drought, its leaves no doubt droop and die, but both perish from the same cause, and independently of each other. 'Thus it is,' observes Dr. Lindley, 'that "simple people have been persuaded that there existed, in the deserts of Scythia, creatures half animal, half plant."' This condition of the rootstock of some ferns,' writes Sir W. J. Hooker, 'being engaged the attention of early writers of the marvellous, and many strange figures were published of it; but Dr. Beyne, of Dantzig, in 1725, declared that the pretended *Agnus Scythicus* was nothing more than the root of a large fern covered with its natural villus or yellow down, and accompanied by some of the stems, &c., in order, when placed in an inverted position, the better to represent the appearance of the legs and horns of a quadruped.' He also adds, 'that the down or villus is the *pow tempie*, or "golden moss," so much esteemed by the Chinese for the purpose of stopping hæmorrhage,'—the very use to which it has been found to be applied elsewhere in modern times. A substance called Pulu, consisting of silky fibrous hairs, used for stuffing mattresses, &c., is obtained from three species of this genus, *C. glaucum*, *Chamissoi*, and *Menziesii*, natives of the Sandwich Islands, whence this article has become a regular export, to the extent of some thousands of pounds annually. This Pulu consists of the hair-like scales found on the crown of the stem and about the base of the frond-stalks of the ferns; only a small quantity, about two or three ounces, is found on each plant, and it takes about four years for the plants to reproduce this amount. The ferns which produce the Pulu grow on all the high lands of the Sandwich Islands at an elevation of about 1000 ft. The silken golden-coloured hairs of *Dicksonia Culcita* are employed in the same way in Madeira and the adjacent isles. A similar fibrous substance, used medicinally as a styptic, is derived in the islands of the Eastern Archipelago from the caudex and stipes of *C. Baromets*; and also from *Dicksonia chrysotricha*, of which latter a plantation belonging to the Dutch government exists in the interior of Java, and the produce of this plantation has been exported to Holland for public sale. This substance is called Penghawar Djambi. Its styptic properties seem attributable to

Cibotium Baromets (caudex, in a natural state, and formed into a Tartarian Lamb).

the rapidity with which its filaments, acting by capillary attraction, absorb the aqueous particles of the blood, and thus cause its immediate coagulation. *C. Menziesii*, one of the species said to furnish Pulu, has large thick coriaceous bipinnate fronds, the large oblong acuminate sinuato-pinnatifid pinnules with rounded lobes, bearing several large corneous opaque involucres. This may be taken as a fair representative of the other species, one of which, besides those already mentioned, is found in Assam, and another of very graceful habit in Mexico. [T. M.]

CIBOULE. (Fr.) *Allium ascalonicum*. —COMMUNE. The Welsh Onion, *Allium fistulosum*.

CIBOULETTE. (Fr.) *Allium Schœnoprasum*.

CICATRICULE. The scar formed by the separation of a leaf from its stem.

CICATRISATE, CICATRICOSE. Marked with scars.

CICATRIX. Any kind of scar formed by the separation of one part from another.

CICCA. A genus of *Euphorbiaceæ*, comprising a number of small trees or shrubs, natives of the tropical parts of India, Africa, and America. The leaves, stalked, entire, and generally oval, are furnished with minute stipules; the small green flowers are shortly stalked, generally four to five in the axils of the leaves, but sometimes in long-bracted racemes; the males and females being on the same, or on different plants. The males have a calyx of four divisions, no petals, and four free stamens inserted on a disc, which arrangement of the parts in fours serves to distinguish the genus from its allies. The calyx of the females is like that of the male. The ovary is three or five-celled, crowned with a like number of styles, each divided at the apex in the form of a V, and each cell contains two seeds. *C. disticha*, sometimes called the Otaheite Gooseberry, is an elegant small tree and a native of India, where it is cultivated, as well as in many other countries, for the sake of its fruits. The slender leaf-bearing branches, about a foot long, are furnished with numerous oval smooth leaves, and might by a superficial observer be taken for pinnate leaves. The racemes of small flowers proceed from the old wood, and are provided with a number of small scales, each of which bears in its axil six to eight stalked flowers. The fruits in size like those of a gooseberry are green, three or five-furrowed, and somewhat acid and cooling. In India they are used as an article of food, either in a raw state or cooked in various ways. Europeans pickle or make preserves of them, and also use them in tarts. In Java they are brought to the markets and sold for preserving at threepence per gallon. A decoction of the leaves is used to cause perspiration, and the roots are emetic, but too violently so to be used.

C. indica, sometimes called *Prosoris indicus*, is a tree of thirty or forty feet high, found in the Bombay district and also in Ceylon; its ovate-lanceolate entire leaves are pale green underneath, and the flowers are in axillary fascicles. The bright blue seeds are contained in a dry capsule, and according to Mr. Thwaites are a favourite food of the green pigeon. Its wood is white, tough, and used for building purposes in Ceylon. [A. A. B.]

CICELY, SWEET. *Myrrhis odorata*; also an American name for *Osmorrhiza*.

CICER. A genus of leguminous plants, which, in combination with five or six others, closely allied, forms the vetch tribe of that order. About a dozen or fifteen species, natives of Southern or Eastern Europe, Western Asia, and Abyssinia, are described. They are annuals or perennial herbaceous plants or undershrubs, clothed with glandular hairs, and having pinnate leaves, consisting either of a definite number of leaflets in pairs with the leaf-stalk terminating in a tendril, or of several pairs of leaflets with an odd one at the end, the leaflets being conspicuously marked by veins. The generic character consists in the tube of the five-lobed calyx being puffed out on the upper side, and two or three of the lobes being pressed upon the upper petal of the pea-like corolla. The pods have their sides swollen out, and contain only a few (seldom more than two or three) seeds, which bear some resemblance to peas, but are of an irregular shape.

C. arietinum is the Chick-pea, or Egyptian pea of the English, the 'Cece' of the Italians, the 'Garbanzos' of the Spaniards, and the 'Gram' of India. It is an annual plant, growing about a foot or more in height, and is a native of the south of Europe, and also of India. Its leaves consist of from three to seven pairs of leaflets with an odd one at the end, the leaflets being egg-shaped, and having their edges cut into very sharp teeth. Both leaves and stems are covered with glandular hairs containing oxalic acid, which exudes from them in hot weather and hangs in drops, ultimately forming crystals. The flowers are either white or rose-coloured, and are produced singly upon stalks growing from the bases of the leaves. The pods are from an inch to an inch and a half long, of a rhomboidal form, with puffed-out sides, and generally contain two seeds, but sometimes only one. These seeds vary in size and colour in different varieties, the finest kinds being nearly a quarter of an inch in diameter, slightly pointed, and of a pale-yellow colour, with their skins netted in consequence of inward shrivelling, and having two swellings on one side; the peculiar form of these peas has given rise to the specific name of the plant *arietinum*, which alludes to their supposed resemblance to a ram's head.

This plant is extensively cultivated in India and other eastern countries, and likewise in the south of Europe. In India the seeds form one of the pulses known under the name of 'Grams,' and are greatly used as an article of food by the natives, being ground into meal, and either eaten in puddings or made into cakes. They are also toasted or parched, and in this state are commonly carried for food on long journeys; rolled in sugar-candy, these toasted peas form a rough sort of comfits, and gram-flour made up with sesamum oil and sugar-candy is an Indian sweetment. Small quantities of these peas come to this country from Turkey, and are used for grinding into pea-meal. Attempts have been made to employ them as a substitute for coffee. In *Paris* they are greatly used in soups. [A. S.]

In Mysore the natives collect the dew from the 'Gram' plants by means of muslin cloths, which become saturated with it. The liquid thus obtained, which is very acid, is preserved in bottles for use, and is regarded as a sure medicine in cases of indigestion, being administered in water. It is stated that the boots of a person walking through a dewy gram field will be entirely destroyed by the pungency of this acid given out by the leaves. [T. M.]

CICHE. (Fr.) *Astragalus Cicer*; also *Cicer arietinum*.

CICHORACEÆ. (Chicory family.) A suborder of the natural order *Compositæ* or *Asteraceæ*, under which its full characters are given. The plants have numerous florets (small flowers) on a common head, and all of them are irregular, having a ligulate form in consequence of the corolline tube being split down on one side, and a tongue or strap-like process formed by the united petals projecting on the other side. The suborder sometimes receives the name of *Liguliflora* from the form of the flowers. The stamens are united, and their anthers are as other syngenesious or composite plants. The fruit is an achene adherent to the calycine tube, and furnished with pappus or a hairy calycine limb at the top. The plants abound in a milky juice, and they have bitter and sometimes narcotic qualities. Some of them, as the dandelion, act on the kidneys and the liver. Some of them are esculent vegetables and salads. They abound in cold regions. Their heads of flowers have usually the property of opening under the influence of light and closing in darkness. Chicory or wild succory (*Cichorium Intybus*) is much cultivated in France and Germany, its roots being used as an addition to coffee. The admixture, without the due indication of it, is forbidden in Britain. *Cichorium Endivia* supplies the salad called endive. *Lactuca virosa* furnished lactucarium, a drug employed in place of opium to procure sleep. Common lettuce is the produce of *Lactuca sativa*, skirret is the root of *Scorzonera hispanica*, while salsafy is obtained from *Tragopogon porrifolius*. The root of dandelion (*Leontodon Taraxacum*) is sometimes used as coffee. See ASTERACEÆ. [J. H. B.]

CICHORIUM. A genus of composites which includes the chicory and the endive, and belongs to the division characterised by the presence of ligulate or strap-shaped florets only in the heads of flowers, and by the presence of a milky juice. It consists of perennial plants, with stiff branching stems, and sessile heads of blue flowers, surrounded by an involucre consisting of two rows of bracts, the outer of which are reflexed and shorter than the inner. The fruits are crowned by two rows of minute scales, constituting the limb of the calyx.

The Wild Chicory or Succory, *C. Intybus*, is a perennial plant found in this country by roadsides and in dry, especially chalky, soil. It has a long tap root, and a rigid slightly hairy branched stem, with a few sessile clasping leaves. The lower leaves spread on the ground, and are pinnately lobed and coarsely toothed, while the upper ones are scanty and embrace the stem by the two pointed lobes at their base. The heads of flowers are few, sessile, of the size of a penny-piece, and of a brilliant light blue colour. The leaves of chicory are blanched and used as a salad under the name of Barbe du Capucine. The root roasted was largely used to mix with and adulterate coffee, but within the last few years grocers mixing chicory with coffee are bound to affix a label on the outside of the package announcing the admixture, so that purchasers can now have pure coffee, or coffee mixed with chicory, as they prefer—for there are some who like the mixture. It need hardly be said that chicory is entirely destitute of those properties which render coffee an agreeable and nutritive beverage, while on the other hand it possesses medicinal properties closely like those of dandelion, and which therefore render it unwholesome for constant use. Moreover, the chicory used to mix with coffee is very often largely adulterated with carrot, mangold-wurzel, oak-bark, tan, mahogany saw-dust, baked horse liver, Venetian red, &c., &c. The detection of these several materials is easily accomplished by the aid of the microscope and the test tube as shown in Dr. Hassall's work on the adulteration of food. Chicory is readily cultivated in this country. That grown at Canterbury was acknowledged to be finer than that imported from abroad, and would have been a very profitable crop, but that the buyers arbitrarily fixed a lower price upon the English than upon the imported. The herbage forms good food for cattle. [M. T. M.]

The Endive, *C. Endivia*, is a hardy annual indigenous to the northern provinces of China, and other parts of Asia, and, according to the *Hortus Kewensis*, was cultivated in this country in 1548. Macintosh in his *Book of the Garden*, believes it is also a native of Egypt, and that it was carried from thence to Italy, and afterwards into Britain. Be this as it may, there is no doubt of its having been used as an esculent from a very early period by the Egyptians, who probably communicated it to the ancient Greeks and Romans, along with their manner of using it. Endive, radishes, and succory are mentioned by Ovid as forming part of a garden salad; and Pliny states that endive in his time was eaten both as a salad and potherb. As such it has been used in this country for three centuries, and it is a singular fact that the manner in which it was prepared for winter use, as described by Gerarde in 1597, differs but little from the mode that is often practised at the present day. The plant has numerous large sinuate smooth toothed, or in some varieties much undulated and finely-curled deep-green leaves. The flower-stem rises about two feet high and produces numerous pale-blue flowers. It is cultivated solely for the stocky head of leaves, which after being blanched to diminish their bitterness, are used in salads and stews during winter and spring. The different varieties of endive are arranged in two classes, namely: 1, the *Batavian* (Scaroles of the French), which comprises all with large broad leaves, slightly ragged or torn; and 2, the *Curled* or *Chicorées* of the French, being all those with crisp and finely-frizzled leaves. [W. B. B.]

CICONIUM. A section of the genus *Pelargonium*, comprising the species with the petals all the same colour, the two upper ones shorter and narrower than the rest; stamens short and erect, the two lowest shorter, with the anthers nearly sessile. The stems are somewhat shrubby and fleshy. [J. T. S.]

CICUTA. A densely genus of *Umbelliferæ* or *Apiaceæ*, known by their diæresipedicates, by their compound umbels without any general involucre, but with partial involucres consisting of several awl-shaped bracts, and by the teeth of the calyx projecting above the fruit, which is roundish, compressed at the side, and marked with ten scarcely prominent ridges, five to each half of the fruit, while beneath each furrow in the rind of the fruit there runs a single channel filled with volatile oil.

C. virosa, the Cowbane or Water Hemlock, is a plant occasionally found wild in this country by the side of ponds and ditches. The rootstock is large, white, and fleshy, covered externally with fibres, and internally hollow and divided into several compartments, by transverse partitions, filled with a yellowish milky juice. The stem is erect, hollow, striated, somewhat branched, and attains a height of three or four feet. The leaves are twice or thrice-pinnate, with narrow lance-shaped segments, one to one and a half inches long, and slightly toothed at the margin. The umbel consists of from ten to fifteen principal rays, unprovided with any involucre or with only a few small bracts. The flowers are whitish. This plant is dangerously poisonous, having qualities like those of *Conium*; indeed, it is called Water Hemlock. It produces tetanic convulsions, and is fatal to cattle eating the herbage. In April 1857, two farmer's sons were found lying paralysed and speechless close to a ditch where they had been working. Assistance was soon rendered but the poor fellows shortly expired.

Cicuta virosa.

A quantity of the Water Hemlock grew in the ditch where they had been employed. A piece of the root was subsequently found with the marks of teeth in it, near to where the men lay, and another piece of the same root was discovered in the pocket of one of them, so that there can be no doubt that they were poisoned by eating the root of this plant in mistake for some other. The root of the American *C. maculata* is even more virulent. [M. T. M.]

CICUTAIRE. (Fr.) *Cicuta*.

CIERGE. (Fr.) *Cereus*.

CIGUE AQUATIQUE. (Fr.) *Cicuta virosa*. —, D'EAU. *Œnanthe Phellandrium*.

—, GRANDE. *Conium maculatum*.
—, PETITE. *Æthusa Cynapium*.

CILIÆ (adj. CILIATED). Marginal hairs forming a fringe.

CILIATO-DENTATE. When the teeth of anything are finely serrated as if fringed.

CILIATO-SERRATE. When the serratures of anything end in a hair.

CIMICIFUGA. Bugbane. A genus of *Ranunculaceœ*, allied to *Actœa*, but differing by having several carpels, which are follicules, not berries. The species are natives of Eastern Europe, Siberia, and North America.

The European species, *C. fœtida*, which has twice-ternate leaves and racemes of inconspicuous flowers arranged in a terminal panicle, is extremely fœtid, and has been used to drive away vermin, whence the generic name. [J. T. S.]

CIMICINE. Smelling of bugs, as *Coriander*.

CINCHONA.* This important genus gives its name to the order of which it is a member. The genus consists of evergreen trees or shrubs growing in the tropical valleys of the Andes. The flowers are of a white or pinkish colour, very fragrant, arranged in panicles. The corolla is salver-shaped, and nearly, if not quite, conceals the five stamens. The ovary is crowned with a fleshy disc; the style is simple; the stigma two-cleft. The fruit is an ovate capsule, grooved on both sides, crowned by the limb of the calyx, and dividing from below upwards, in order to allow of the escape of the numerous winged seeds.

There are, according to Weddell, twenty-one species of this genus, but only some of them yield commercial Cinchona, or Peruvian bark. Of this there are several varieties, the most esteemed of which are the Calisaya or yellow bark, the produce of *C. Calisaya*; grey, or Huanuco bark, the produce of *C. micrantha* and *C. nitida*; Loxa, or crown bark, the produce of *C. Condaminea*; red bark, &c. The great value of these barks as tonics and remedies for fevers, depends upon the presence of certain alkaloid substances called quina, cinchonia, and quinidina, which exist in the bark, especially in the liber or inner bark, in combination with kinic and tannic acids. It is found that certain of the barks contain more of one principle than of another; hence their greater or less value commercially, and the skill and complex knowledge required by the manufacturer to distinguish the different varieties of bark one from the other. Quina is the most useful of the alkaloids, and this is found in greatest abundance in Calisaya bark; cinchonia occurs most abundantly in the best grey and red barks; while Loxa bark furnishes the largest amount of quinidia. The several alkaloids have all similar properties, but varying in degree. Quina, in its combination with sulphuric acid, is the most gener-

ally used under the name of sulphate of quinine—next to opium and calomel, probably the most important of all drugs. The alkaloids extracted from the barks are recognised by their distinctive chemical characteristics, while the barks producing them are likewise distinguished by a careful scrutiny of their external appearance, the lichens, &c., growing on them, the way in which they break, their taste, odour, &c., as well as by their microscopical and chemical characteristics. All these varied points require long practical experience for their due acquirement. The way in which the barks break, or the fracture, as it is termed, depends on their anatomical structure, that is to say, on the size and arrangement of their cellular and woody portions. Where the former preponderates, the fracture is smooth and even, and such barks are said to yield the greatest quantity of quinidin. When there is less cellular tissue, or the constituent cells are smaller, then a fibrous or stringy fracture is observable, and a short stringy fibrous fracture is considered to be an indication of the presence of quinine. Withal there is still much uncertainty as to the trees producing the various kinds of bark. No doubt the same tree, in varying circumstances, may, nay does, produce different sorts of bark. Similar-looking barks too may be produced by very different species, and the same package may contain the produce of more than one species of *Cinchona*. These difficulties are enhanced by the jealousies and restrictions of the various governments, and of the merchants.

The *Cinchona* trees grow in the forests of Bolivia, Peru, &c., in groups or clusters. The *cascarilleros*, or bark collectors, encamp in these forests, and ascertain where the trees are to be found, a process in which the sagacity and endurance of the Indians are put to a severe test. They are reported to be able to tell the trees at a distance by a peculiar movement of the leaves, and by the colour of the masses of bloom. When the position of the trees has been ascertained, there is frequently much difficulty in getting to the spot; this done, however, the trees are felled; no light labour, for the intertwining climbing plants sustain the trunks when cut through. When the trees are at length felled, the bark is stripped off all round, and cut into pieces of a convenient size for carriage; and particular care is taken to secure the bark near the root, as it is there thicker and more valuable. The bark from the small branches rolls up when stripped into cylindrical pieces or quills, while the larger pieces are placed in stacks to dry, with a heavy weight on the top. The carriage of the packages of the bark to the place of encampment, by a route which is traversed with much difficulty by the unembarrassed Indian, is a work of great hardship and labour.

In the process above described, the trees are necessarily destroyed, and hence the supply of this valuable drug is likely to be greatly diminished, if means be not taken to secure the growth of these trees. Chemists, however, tell us we need not despair of finding a substitute. Thanks, nevertheless, to the labours of Messrs. Markham, Spruce, and others in South America, as well as to those of Mr. McIvor and other cultivators in India, there are now established in many of the hilly districts of the latter country large plantations of the most valuable kinds of cinchona. Mr. Wilson also has met with tolerable success in the cultivation of these plants on the higher mountains of Jamaica. Mr. Howard's reports on the analysis of bark derived from these sources are very encouraging.

Bark was first employed in Europe in the middle of the seventeenth century. The discovery of its medicinal value is a matter of fable and conjecture. The name Cinchona is derived from the wife of a Viceroy of Peru, who is said to have brought the drug from South Ame-

Cinchona Calisaya.

ries in 1630. Afterwards the Jesuits used it; and it became generally known when Louis XIV. purchased of Sir R. Talbot, an Englishman, his heretofore secret remedy for the cure of intermittent fever, and made it public. For full information on the subject of *Cinchona* and its barks, the reader should consult the magnificent works of Weddell and Howard, the account of Mr. Markham's travels, Mr. Spruce's report of his explorations, or the valuable epitome contained in Pereira's *Materia Medica*, from which sources the greater part of this notice has been gleaned. The aspect of a *Cinchona* forest is shown in Plate 12. The name is now sometimes written *Chinchona*. (M. T. M.)

CINCHONACEÆ. (*Rubiaceæ*, *Cinchonads*, the Peruvian bark family.) A natural order of gamopetalous calyciflorai dicotyledons, characterising Lindley's cinchonal alliance. The order is sometimes considered as a sub-order of the natural family of *Rubiaceæ*, or Madderworts. Trees, shrubs, or herbs, with simple opposite leaves, having glandular stipules placed between the bases of the leaf-talks (interpetiolar), and flowers arranged in panicles or corymbs. Calyx adherent, entire, or toothed; corolla regular; stamens attached to the corolla. Ovary two-celled; style one. Fruit inferior, either dry or succulent, splitting into two or not opening; seeds either definite in number, or numerous, containing a small embryo in horny albumen. Chiefly found in tropical regions, where they constitute 1/25 of the flowering plants. In northern regions the order is represented by *Galiaceæ*, which some regard as a sub-order of *Rubiaceæ*. The order furnishes many important products. The plants supply remedies for intermittent fevers; some are emetics and purgatives, others act in strengthening the tone of the stomach. The various medicinal barks are yielded by species of *Cinchona*, which grow in the Andes between 3,000 and 9,000 feet of elevation above the level of the sea. *Coffea arabica* supplies coffee, which is the hard albumen of the seeds. *Cephaelis Ipecacuanha* yields the well-known Ipecacuan root which is used commonly as an emetic. A dye called soorangee is procured from the root of *Morinda citrifolia*. White gambier, a kind of catechu, is the product of *Uncaria Gambir*. Gardenias have showy as well as fragrant flowers, and *G. Rothmannia* yields an edible fruit. There are upwards of 300 genera and 2,000 species in the order. Illustrative genera: *Spermacoce, Cephaelis, Coffea, Ixora, Hedyotis, Pentas, Cinchona, Nauclea, Gardenia, Musscenda*. [J. H. B.]

CINCINALIS. This name as originally employed by Gleditsch is a synonymy of *Pteris aquilina*; as however used by subsequent writers it is synonymous with *Nothochlæna*, a genus of ferns. [T. M.]

CINCLIDIUM. A fine genus of acrocarpous mosses belonging to *Bryaceæ*, and closely allied to *Mnium*, agreeing with it in the characters of the stem and large leaves, but differing in having the inner peristome cup-shaped with sixteen short outer teeth. It occurs in spongy bogs and is rare in Great Britain. It resembles in general appearance *Mnium punctatum*, but the stems are more densely matted together with the purple rootlets. Only one other species is known, *C. arcticum*, which has been found in Norway. [M. J. B.]

CINCLIDOTUS. A genus of aquatic acrocarpous mosses belonging partly to that division which has been called *Cladocarpi*, because in the majority of species the fruit terminates in short lateral branches. It is named from the lattice-like structure of the peristome, which consists of thirty-two teeth anastomosing at the base. This structure obtains in *C. fontinaloides*, which grows in large tufts on rocks and stones in rivulets and on the borders of lakes, especially in hilly limestone districts, and also in *C. riparius*; but in *C. aquaticus* the peristome is quite rudimentary. [M. J. B.]

CINCTUS. A term applied to albumen when surrounded by an annular embryo.

CINENCHYMA. That kind of tissue in which latex, or the proper juice of plants, is supposed by some to be conveyed from place to place. Probably a form of the intercellular passages.

CINERACEOUS. Ash-greyish.

CINERAIRE A' FLEURS BLEUES. (Fr.) *Agathæa amelloides*.

CINERARIA. A family of compound flowers, difficult of discrimination, and containing many species which are referred by some botanists to the genus *Senecio*, &c. As at present constituted, *Cineraria* does not contain any native examples, but is well known as an ornament of the conservatory and window garden. Some of the species are half-shrubby, but the majority are herbaceous and of easy cultivation; and some may be so managed as to be made to bloom almost at any season. *C. cruenta*, a native of Teneriffe, has heart-shaped leaves, variously toothed at the edge, tinged with red or purple, or of unmixed green; the upper leaves clasp the stem and are auricled at the base. In the wild state of the plant, the flowers have a deep purple disk with bright purple rays; but since it has been taken up as a florist's flower, a countless number of varieties have been raised from seed, with flowers in which white, purple, rose-colour, crimson, violet, azure, &c., are combined in ever-varying proportions. 'The early flowering of this plant,' says *Le bon jardinier*, 'its long duration, which allows it to be an ornament of the conservatory and window during several months, have given some importance to its culture.' [C. A. J.]

CINEREUS. Ash-grey; a mixture of white and black.

CINNABAR, CINNABARINUS. Scarlet touched with orange.

CINNAMODENDRON. A genus allied to *Canella*, and like it belonging to *Pittosporaceæ*, or, in the opinion of some authors, forming a separate order *Canellaceæ*. The *C. axillare* is a Brazilian tree with smooth whitish bark cracking transversely; leaves alternate, stalked, elliptical, leathery, smooth and entire; peduncles axillary, three-flowered; calyx of three sepals; petals five, alternate with as many scales; stamens ten, forming a tube round the ovary. The bark is aromatic, and used as a tonic and antiscorbutic. [J. T. S.]

CINNAMOMEUS. The colour of cinnamon.

CINNAMOMUM. The trees furnishing cinnamon and cassia barks belong to a genus of *Lauraceæ*, or true laurels, characterised by the presence of ribbed leaves, leaf-buds not provided with scales, a six-cleft leathery calyx, nine fertile stamens in three rows, with four-celled anthers which open inwardly, except those of the third or innermost row, which open towards the outside of the flower. The stamens of this third row are moreover provided with two sessile glands, one on each side of their base, and within them is a fourth row of abortive stamens. The fruit is berry-like, one-seeded, in a cup-like calyx.

C. zeylanicum is largely cultivated in Ceylon, for its bark, which furnishes the best Cinnamon. The bark is stripped off the branches, when it rolls up into quills, the smaller of which are introduced within the larger and then dried in the sun. The thinner the bark is as a rule, the finer its quality. Cinnamon is largely used as a condiment for its agreeable flavour, while its astringent and cordial properties give it a medicinal value. It is said to possess the special property of restraining uterine hæmorrhage.

C. Cassia furnishes Cassia bark, which is much like cinnamon, but thicker, coarser, stronger, less delicate in flavour, and cheaper; hence it is frequently used to adulterate cinnamon. Its admixture, however, can be readily detected, even in a powdered state, according to Dr. Hassall. Cassia is grown in China, Java, &c. The German and Russian chocolate-makers prefer cassia to cinnamon, as affording a stronger flavour. The same, or some closely-allied trees, furnish Cassia buds, which are something like cloves, and, like them, consist of the unexpanded flower-buds; but they possess properties similar to those of the bark.

Other species of this genus afford aromatic barks; such as *C. Culilawan*, a native of Amboyna, whose bark has a flavour of cloves. *C. iners*, a native of Malabar, is employed medicinally in fevers and dysentery; the seeds are the parts used; the bark is likewise employed as a condiment. The leaves of *C. nitidum*, dried, are said to have furnished the aromatic leaves called "folia Malabathri". Indeed, it is surprising that the leaves of the cinnamon are not more often imported, as they, like the inner bark, though to a less extent, contain the volatile oil on which the fragrant aromatic properties depend.　　　　[M. T. M.]

CINNAMON. *Cinnamomum zeylanicum*, a tree cultivated in the tropics for its aromatic bark. —, BASTARD, *Cinnamomum Cassia*. —, BLACK. *Pimenta acris*. —, ISLE OF FRANCE. *Oreodaphne cupularis*. —, SANTA FE. *Nectandra cinnamomoides*. —, WILD. *Canella alba*; also *Myrcia acris*.

CINQUEFOIL. The common name for *Potentilla*. —, MARSH. *Comarum palustre*.

CIONIDIUM. A small genus of Australasian polypodiaceous ferns belonging to the *Dicksonieæ*, distinguished by having the indusium cup-shaped and standing out beyond the margin of the frond, and having the veins reticulated. The only species known, *C. Moorei*, has a short decumbent rhizome, and pedately bipinnato-pinnatifid fronds of membranaceo-herbaceous texture, which are studded with soft around the margin. The fructification of *Cionidium* is that of *Deparia*, the distinction between these two consisting in the reticulated venation of the former, and the free venation of the latter.　　[T. M.]

CIPURA. A small genus of Iridaceous plants closely allied to *Morica*, consisting of bulbous herbs with ensiform leaves and terminal heads of flowers. The species, which are but few in number, are found in tropical and subtropical America. The perianth has a very short tube and a six-parted limb, of which the inner or petaloid divisions are much the smaller; there are three stamens with distinct filaments inserted in the tube of the perianth, and a three-celled ovary containing numerous ovules, and surmounted by a very short style, and three petaloid undivided styles alternating with the stamens. *C. paludosa*, a native of humid meadows in Cayenne, has conico-globose bulbs, radical linear-lanceolate plaited leaves from three to five inches long, the scape shorter than the leaves, and bearing a short densely-imbricated distichous terminal spike of bluish flowers.　　　[T. M.]

CIRCÆA. A plant with a name so ominous as Enchanter's or Enchantress-Nightshade might well be supposed to be gifted with the most potent properties. It is, however, a humble herbaceous plant, belonging to the *Onagraceæ*, growing to the height of about a foot and a half, with delicate egg-shaped leaves which taper to a point, and small white flowers tinged with pink, which are succeeded by small roundish seed-vessels thickly covered with hooked bristles. *C. Lutetiana*, the common species, is abundant in shady woods, where it frequently covers a large space of ground. It often too finds its way into shrubberies, where it is a pretty but troublesome weed, creeping extensively, and very difficult to eradicate. It has no affinity with any of the true nightshades, and is conjectured to have received its name from the tenacity with which its prickly seed-vessels attach themselves to the person and clothes of passengers, and from its habit of lurking in obscure places. *C. alpina*, a closely-allied species, scarcely differs from the preceding except that it is smaller and of more delicate habit; it is not unfrequent in Scotland and the north of England. French, *Circée*; German, *Hexenkraut*.　　　[C. A. J.]

CIRCINALIS, CIRCINATE. Bent like the head of a crosier, as is the young leaf of a fern when it begins to grow.

CIRCUMPOSITIO. A layer; that is to say, a branch laid into the ground or layered in order that it may strike root.

CIRCUMSCISSILE, CIRCUMSCISSUS.

cut circularly round the sides, as the seed-vessel of *Anagallis*.

CIRCUMSCRIPTIO. The outline of anything.

CIRCUMSEPIENTIA FOLIA. Leaves which rise up like a funnel and surround the stem as if to protect the young shoots, as in the marvel of Peru. Such is De Candolle's definition, but the term is very rarely used.

CIRIER or CIRIER DE LA CAROLINE. (Fr.) *Myrica cerifera*.

CIRRHÆA. A genus of pseudobulbous orchids from tropical America, with solitary ribbed leaves, and drooping racemes of flowers, yellowish, greenish, or spotted with purple. They are remarkable for their long column, which bears a one-celled anther at the back of the upper extremity, curving gracefully over a deeply three-lobed lip, the middle division of which turns back from the side ones. The perceolla or stigmatic point is extended into a slender tendril-like thread, whence the name.

CIRRHIFEROUS. Bearing a tendril.

CIRRHIFORM. Shaped like a tendril.

CIRRHOPETALUM. An extensive genus of small epiphytal orchids, with solitary fleshy leaves proceeding from the top of roundish pseudobulbs. Their flowers are remarkable for having the lateral sepals prolonged into narrow streamers, by which the species are readily distinguished from *Bolbophyllum*. Between thirty and forty species are known, all from tropical Asia except *C. Thouarsii*, which inhabits the Mascaren and South Sea Islands. The singularly-formed flowers have made a few favourite objects of cultivation. The best are *C. fimbriatum*, *refractum*, *chinense*, and *Cumingii*.

CIRRHOSITAS. The production of tendrils.

CIRRHUS (adj. CIRRHOSE). A tendril. A slender twining organ by which a plant climbs.

CIRSIUM. A genus of compound flowers belonging to the thistle group, distinguished from *Carduus* by having the receptacle covered with chaffy bristles, and the achenes crowned with a soft feathery pappus. Several British species are described by English botanists as belonging to the genus *Cnicus*. Numerous others occur in various parts of the continent, some having purple, and others yellow flowers, but none of sufficient interest to require further notice. (C. A. J.)

CISSAMPELOPSIS. The name given to a number of trailing shrubby plants of the composite family which are found in India and the adjacent islands, as well as in S. Africa. They differ in little except habit from groundsels (*Senecio*), in which genus indeed they are placed by some authors. Most of them have heart-shaped stalked leaves with toothed margins, and their under-surface is invariably covered with short close-pressed white hairs. The yellow flower-heads, arranged in terminal or axillary panicles or corymbs, have an involucre of eight or ten scales, enclosing about a dozen florets, all of them tubular. The achenes have no beak, are somewhat angular in form, and crowned with a pappus of many rough hairs, arranged in a single series. [A. A. B.]

CISSAMPELOS. The plants so named have the climbing character of the ivy —*kissos* of the Greeks, and the clustered fruit of the vine—*ampelos*. Their flowers are diœcious. The male flowers have four sepals and four petals combined into a cup; the female flowers have two sepals fused into a somewhat fleshy two-serrated scale, frequently notched at the margin, and having externally a small bract, formerly considered as a sepal: the ovary is solitary. In drawing up the differential characteristics of this genus, the explanations of Hooker and Thomson as to the structure of these flowers have been adopted as being probably correct, though at variance with the account given by other writers. The most important plant of the genus is the Velvet Leaf, *C. Pareira*, a native of the West Indies, Central America, and India. It is an exceedingly variable plant with a climbing stem, the leaves of variable rounded shape, and dotted with velvety pubescence; male flowers in stalked hairy cymes, and female flowers in clusters, with large rounded bracts, and succeeded by sub-globose hairy scarlet drupes. The root of this plant furnishes the 'Pareira brava' of the druggists, which is used with much benefit in diseases of the bladder and urinary organs. Many other species are used as tonics and diuretics, while *C. globerrima* and *C. ebracteata* are used as remedies for serpent bites. The root of *C. obtecta* is used in the manufacture of an intoxicating drink. [M. T. M.]

CISSAROBRYON. A genus of *Vivianieæ* found in the Andes of Chili, and differing from its congeners in having a five-parted calyx, five petals, and three conspicuous slender styles. The only species known, *C. elegans*, is a little prostrate branching plant with slender woody stems and opposite roundish leaves an inch long; their stalks as long as the blade, which has three to seven deep notches, is slightly hairy above, and is covered underneath with a hoary down. The flowers are blue, in size and form made like those of the wood sorrel, and single from the axils of the leaves, supported on long slender stalks. [A. A. B.]

CISSUS. A genus of *Vitaceæ* scarcely differing from the vine (*Vitis*). The petals, however, usually separate before they fall, instead of remaining united at the tips as in *Vitis*, and are usually four instead of five; the disk is more conspicuous. The leaves are often more deeply divided. Most of the species are found within the

Tropics, especially in Asia; a few occur in North America. [J. T. S.]

CISTACEÆ. (*Rock-rose* family.) A natural order of thalamifloral dicotyledons, characterising Lindley's cistal alliance. Shrubs or herbs, often viscid, with simple entire leaves and showy flowers. Sepals three to five, persistent, unequal, the three inner twisted in the bud. Petals five, rarely three, falling off, often crumpled, twisted in an opposite direction from the sepals. Stamens numerous, not united. Fruit a one-celled capsule with parietal placentas, or imperfectly three to five-celled with central placentas. Seeds with mealy albumen; embryo curved or spiral. The plants are found chiefly in the south of Europe and north of Africa. They are very rare in North America, still more uncommon in South America, and scarcely known in Asia. They are usually resinous, and have a balsamic fragrance. The resin called *Ladanum* is procured from several species of *Cistus*. *Helianthemum vulgare*, the common rock-rose of Britain, has remarkably irritable stamens, which in sunny weather move on being touched. There are eight known genera, and about two species. Illustrative genera: *Fumana*, *Cistus*, *Helianthemum*. [J. B. B.]

CISTELLA, CISTULA. A cell-like shield found among lichens in the genus *Sphærophoron*.

CISTOME. A membranous sac which according to some, penetrates stomates, and reaches the bottom of the subjacent chamber. If this statement is correct the cistome must be a fold of the cuticle.

CISTOPHORUM. The stipe of certain fungals.

CISTOPTERIS. A mode of spelling which is sometimes adopted instead of *Cystopteris*. [T. M.]

CISTUS. A genus of the rock-rose family, to which it gives the name, composed of handsome shrubs, many of them in cultivation, natives of Southern and Western Europe, North Africa, and the Canary Islands. They are commonly known as Rock-Roses and Gum Cistus, but the latter name is the better of the two, as the former is equally applied to *Helianthemum*, from which this genus differs in having an imperfectly five or ten-celled capsule, while in *Helianthemum* the capsule is imperfectly three-celled. The greater portion of the species are elegant erect bushes, with opposite entire or sometimes toothed leaves, generally oblong or lance-shaped, and axillary or terminal flower-stalks bearing one or many flowers. These are made up of a five-leaved calyx, five large petals, numerous stamens, and an ovary crowned with a simple style. The flowers of all are handsome, and many of them in size and appearance resemble those of the dog-rose; they seldom last more than a few hours after expansion, and do not open in dull weather when there is no sunshine. In one group of species the petals are white and furnished with a yellow or purple mark at their base; while in a second the petals are rose-coloured, each with a yellow spot at its base. None of them have yellow flowers, a colour so common among the *Helianthemums*.

The Ladanum or Labdanum of Crete is a well-known gum, which exudes from the leaves and branches of *C. creticus*, and some other allied species. This plant is a handsome shrub, with oblong obtuse rough leaves with waved margins and about an inch in length. The flowers are terminal and single or twin, the petals purple with a pale yellow spot at the base. The gum is collected in Crete by means of a kind of rake, 'with a double row of long leathern straps, employed in the heat of the day when not a breath of wind is stirring. Seven or eight country fellows, in their shirts and drawers, whip the plants with these straps, which, by rub-

Cistus creticus.

bing against the leaves, lick off a sort of odoriferous glue sticking to the foliage.' Formerly it was said to be gathered from the beards of goats, which are fond of browsing on the foliage of the plant. The gum, by gently rubbing in the hands, emits a very pleasant balsamic odour, from the presence of a volatile oil. It was once used in time of the plague as a stimulant and expectorant, and as a constituent in plaisters. About fifty hundredweight of it is annually sent from Crete to Turkey, where it is used as a perfume, and as a fumigation. The Ladanum of Spain and Portugal is derived from *C. ladaniferus*. This is one of the most beautiful of the genus, and is very frequently to be met in gardens. In Portugal it is said to cover leagues of country. Its leaves are lance-shaped, entire, and three-nerved, the upper surface covered with a clammy gluten, and the under surface prominently three-nerved and covered with a dense white tomentum. The large white flowers are sometimes more than three inches across in one variety, the petals having a deep purple blotch at the base. The gum is said to be obtained by boiling in water the summits of the branches. It has an odour similar to that of the former, but is not in much repute.

A much more common plant in gardens is C. cyprius, which is often confounded with the former, but has large and always solitary flowers, while this has three or four flowers on a common stalk; the leaves are also stalked, while in C. ladaniferus they are sessile.

One of the most beautiful of the rosy-flowered species is C. roseus, a native of Teneriffe. Its hairy leaves are lance-shaped, three-nerved, and dilated towards the base, while the splendid large rose-coloured flowers are very numerous and in terminal panicles. The petals are crumpled and have wavy margins, bent inwards, with a yellow spot at their base. A large number are in cultivation, and upwards of thirty coloured figures of these plants are given in Sweet's Cistineae. [A. A. B.]

CISTUS, GROUND. *Rhododendron Chamaecistus.* —, GUM. *Cistus ladaniferus*, and *C. Ledon*.

CISTUSRAPES. A name given by Lindley to the group of Cytinaceous parasites.

CITHAREXYLON. A considerable genus of trees or shrubs, belonging to the order *Verbenaceae*, natives of tropical and sub-tropical America. They have tetragonous sometimes spiny branches, opposite or verticillate leaves, and small racemose flowers each with a minute bracteole. The persistent calyx is cup-shaped or tubular; the limb of the corolla is sub-equally five-parted. The included stamens are inserted below the throat of the corolla on short filaments; they are sub-equal or the fifth is shorter than the others, sometimes rudimentary and sterile, or altogether abortive. The ovary is four-celled, with one ovule in each cell. The juicy drupe is surrounded by the large cup-shaped calyx, and is two-stoned, each stone being two-celled. [W. C.]

CITREOUS, CITRINOUS. Lemon-coloured.

CITRIOBATUS. A genus of small thorny Australian trees or shrubs belonging to *Pittosporaceae*. Leaves alternate, shortly stalked, obovate, leathery, entire. Flowers small, solitary, sessile, axillary, with five sepals bracteated at the base; five petals united at the base; and five stamens. Fruit an orange berry with a leathery skin, sub-globular, about one inch and a half in diameter, eaten by the natives; seeds large. The plants are called the Native Orange and Orange Thorn by the Australian colonists. [J. T. S.]

CITRIOSMA. A genus of opposite-leaved bushes or small trees belonging to the *Monimiaceae*, confined to the tropical parts of South America, and numbering upwards of fifty species. A large proportion of them have their parts, especially the leaves, covered with glands which secrete an oily substance of a strong citron colour. Some of the species are known on this account by the name of Limoncillo, and the genus also derives its name from this circumstance. The leaves, sometimes three or four in a whorl, are either entire or toothed, and very often covered with rusty hairs, but sometimes smooth. The small green or yellow flowers without petals are numerous, disposed in axillary cymes, and either male and female on the same or on different plants. They are made up of a three or six-lobed cup-shaped calyx, which in the male encloses few or many stamens, and in the female a number of one-celled and one-seeded ovaries, each with a simple style; these are at length entirely enveloped by the fleshy tube of the calyx. This latter circumstance serves to distinguish the genus from its allies. The fruit is about the size of a pea when ripe. The name of the genus was formerly written *Citrosma*. [A. A. B.]

CITRON. *Citrus medica.* —, FINGERED. *Sarcodactylis*.

CITRONELLA. *Andropogon citratum*, which yields an essential oil used in perfumery.

CITRONELLE. (Fr.) *Artemisia Abrotanum*.

CITRONNELLE. (Fr.) *Melissa officinalis*.

CITRONNIER. (Fr.) *Citrus medica*.

CITRONWORTS. A name given by Lindley to the family of aurantiaceous plants to which the orange and citron belong.

CITROUILLE. (Fr.) A race of large oblong Gourds derived from *Cucurbita Pepo*. —, PASTEQUE. *Cucumis Citrullus*.

CITRUL. The Water Melon, *Cucumis Citrullus*.

CITRULLUS. The bitter Cucumber or Colocynth, which furnishes a well-known cathartic drug, belongs to this genus of the gourd family, *Cucurbitaceae*, and is known by its unisexual flowers, which have a persistent five-parted calyx and corolla. In the male flowers, the stamens are five, united into three bundles, and the anthers are sinuous. The female flowers have an inferior three to six-celled ovary, a cylindrical three-cleft style, and kidney-shaped stigmas. The fruit is a many-seeded gourd. *C. Colocynthis* was originally a native of the warmer parts of Asia, but has now become widely diffused. The drug known as Colocynth consists of the round fruits or gourds, which are imported either with the rind on or peeled, from Spain, the Levant, &c. The pulp in the interior of the fruit is light and spongy, and very bitter; from it a watery extract is made, which is much employed as a purgative in the form of pills. Some discrepancy exists as to the seeds of this plant, which some describe as bland and nutritious, while others say that they are bitter and purgative. Certainly the dried dark-coloured seeds met with in this country are so; but it is stated that

Citrullus Colocynthis.

the seeds are used as food at the Cape of Good Hope. An oil is also extracted from them for burning in lamps. [M. T. M.]

CITRUS. The Orange, Lemon, Citron, and other well-known fruits of a similar kind, are included in this genus of *Aurantiaceæ*. Its distinguishing characteristics are: the presence of a cup-like calyx, numerous stamens irregularly united by their filaments into several bundles, a cylindrical style, and a pulpy fruit with a spongy rind. The leaves of these trees are also remarkable inasmuch as they consist of one leaflet, separated from the leaf-like stalk supporting it, by a distinct joint.

The most important species, in a medical or pharmaceutical point of view, are the Citron, *C. medica*, which furnishes two distinct kinds of oil, used by perfumers, the essential oil of citron and the essential oil of cedra. The Lemon, *C. Limonum*, is employed in medicine for the sake of its aromatic bitter rind; its odour is due to the volatile oil in which it abounds. The juice of the Lemon is used as a refreshing beverage in fevers and scorbutic affections, and as effervescing lemonade to check sickness and nausea. As it is apt to decompose, crystallized citric acid is usually employed in its place as an antiscorbutic, and with the best effects. Lately it has been recommended in acute rheumatism. Lime juice is employed for similar purposes. The Seville or Bitter Orange, *C. Bigaradia*, is used for the sake of its rind and its flowers, which possess a stronger flavour and odour than the sweet orange. The rind is used as a stomachic and tonic, while the flowers yield by distillation orange-flower water. [M. T. M.]

The Citron, Orange, Lemon, Shaddock, and Lime have been referred to various species of *Citrus*, with regard to which botanists, however, are not agreed. It is even doubtful whether all of them, with their very numerous varieties, have not originated from *C. medica*. On this point the following observations by Dr. Lindley in the *Journal of the Horticultural Society* (ix. 171), are important. He states that the above-mentioned fruits 'are all of Eastern origin, and mostly introduced into Europe in comparatively modern days, but of very ancient and general cultivation in Asia. The varieties known are very numerous and difficult to reduce according to their species, on the limits of which botanists are much divided in opinion. Those who have bestowed the most pains in the investigation of Indian botany, and in whose judgment we should place the most confidence, have come to the conclusion that the Citron, the Orange, the Lemon, the Lime, and their numerous varieties now in circulation, are all derived from one botanical species, *C. medica*, indigenous to, and still found wild in, the mountains of East India. Others, it is true, tell us that the Citron, the Orange, and the Lime are to be found as distinct types in different valleys, even in the wild state; but these observations do not appear to have been made with that accuracy and critical caution which would be necessary in the case of trees so long and so generally cultivated.'

The Citron, *C. medica*, is described by Theophrastus as abundant in Media, that is to say, in the north of Persia. Its fruit, according to Professor Decandolle, was carried to Rome in the beginning of the Christian era, or perhaps at an earlier period. The first attempts at its cultivation in Italy proved unsuccessful, and according to Gallesio, it was not established there till about the third or fourth century. The Jews cultivated the Citron at the time they were under subjection to the Romans, and used the fruit then, as at the present day, in the Feast of Tabernacles; but there is no proof of their having known this tree in the time of Moses. It is likely they found it at Babylon during their captivity, and brought it to Palestine on their return. Whatever may have been the diffusion of the species in Western Asia at that remote date, there is no evidence of its having been indigenous to Media, nor have modern travellers found it wild in Persia; but Dr. Royle found the species in the forests of Northern India. The Citron is cultivated in Cochin China, and in China, but Thunberg does not mention it as existing in Japan. Taking all the above facts into consideration, it is evident that the species is originally from the north of India, and as the habitat of every one of the Orange tribe is naturally rather limited, Professor Decandolle does not think that this extended in the case of the Citron, as far as the north of Persia. Probably the Citron was carried in that direction, and also into China at a very early period. In many countries they are easily naturalised. The seeds sow themselves in several of the colonies; for instance, in Jamaica. In its wild state the Citron grows erect with spiny branches. The flowers are purple on the outside and white inside. The fruit is large, oblong or ovate, sometimes six inches long, the skin covered with protuberances, and of the well-known citron-yellow colour when the fruit is ripe. Of the cultivated varieties, some are oval, others round, and that called the Madras Citron has the form of an oblate sphere. In China there is a variety with its lobes separating into finger-like divisions, and hence called the Fingered Citron.

The Lemon, *C. Limonum*, of some botanists, *C. medica Limonum* of others, is, according to Dr. Royle, who found it growing wild in the North of India, named in Hindostanee Neemoo, Leemoo, Leeboo; in Arabic Limoun; and in Italian, Limone. Professor Decandolle states that it was unknown to the ancient Greeks and Romans; and that its culture only extended into the west with the conquests of the Arabs. On their spreading over the vast regions of Asia and Africa, they carried with them everywhere the Orange and Lemon. The latter was brought by them in the tenth century from the gardens of Oman into Palestine and Egypt. Jacques de Vitry, writing in the thirteenth century, very well describes the Lemon, which he had seen in Palestine; and doubtless it was by the crusaders first brought into Italy, but at a date which cannot be exactly ascertained. From the north of India it appears to have passed eastward into Cochin China and China, and westward into Europe, and it has naturalised itself in the West Indies and various parts of America. Fruit oval or ovate, terminated by a small blunt nipple-like point; skin smooth, rind much thinner than that of the Citron. The varieties are numerous. Lemons are chiefly imported for their agreeably acid juice and essential oil, and also for the manufacture of citric acid.

The Orange, *C. Aurantium* of those botanists who do not consider it to be probably only a variety of *C. medica*, is associated with the latter as a native of the north of India. According to Gallesio, instead of being found in the north of Africa, Syria, or even in Media, it was not at the time of Alexander the Great in that part of India which he penetrated; for it is not mentioned by Nearchus among the productions of the country which is watered by the Indus. But the Arabs, carrying their conquests further into India than Alexander, found the Orange more in the interior; and according to Professor Targioni it was brought by them into Arabia in the ninth century. Oranges were unknown in Europe, or at all events in Italy, in the eleventh century, but were shortly afterwards carried westward by the Moors. They were in cultivation at Seville towards the end of the twelfth century, and at Palermo in the thirteenth, for it is said that St. Dominic planted an orange for the convent of St. Sabina in Rome, in the year 1200. In the course of the same thirteenth century, the crusaders found Citrons, Oranges, and Lemons, very abundant in Palestine; and in the following or fourteenth century, both Oranges and Lemons became plentiful in several parts of Italy. It appears, however, that the original importation of Oranges from India into Arabia and Syria occurred about a century earlier than that of Lemons. Gallesio states that Oranges were brought by the Arabs from India by two routes; the sweet ones through Persia to Syria, and thence to the shores of Italy and the south of France; and the bitter, called in commerce Seville Oranges, by Arabia, Egypt, and the north of Africa to Spain. Of the numerous varieties of this esteemed fruit, our limits will only admit of our noticing some of the more important.

The Sweet Orange has the leaves ovate-oblong, acute, somewhat serrate, with the stalk more or less winged. The flowers are white. The fruit is well known. There are many varieties; that called the China Orange is the common Orange of the markets. The Blood-Red or Malta Orange has the fruit round, rough red or reddish yellow outside, with a pulp irregularly mottled with crimson. The Saint Michael's Orange has the fruit rather small, pale yellow and seedless, with a thin rind and very sweet pulp; it is one of the most delicious and productive varieties. The Noble, or Mandarin Orange is small flattened and deep orange, with a thin rind which separates so obviously from the pulp, so that when quite ripe the latter may be shaken about inside; it is exceedingly rich and sweet. In China, where this delicious variety has been raised, the fruit is chiefly consumed in presents to the Mandarins, hence its name. It is now, however, very successfully cultivated in Malta and in the Azores. The Sweet-Skinned Orange is the Pomme d'Adam, or "Forbidden fruit" of the shops of Paris, but not of London; its skin is smooth, deep yellow, with a thick sweet soft rind. The above are some of the principal sorts of sweet oranges; but there are many other varieties, many of which possess, however, but little merit.

The Common Seville, or Bitter Orange, or Bigarade, *C. Bigaradia*, has a round dark fruit with an uneven, rugged, extremely bitter rind. This sort is largely imported for the manufacture of bitter tincture, and the preservation of the candied orange-peel. To this section are referred the various kinds of Bigarades, among which may be named the Horned, Curled-leaved, Purple, Double-flowered, and Myrtle-leaved. These, especially the Horned and Curled-leaved, are cultivated chiefly for their flowers, which are powerfully fragrant.

Of the Bergamot Orange, *C. Bergamia*, both flowers and fruit possess a peculiar fragrance; and from each of them an essence of a delicious quality is extracted.

The Lime, *C. Limetta*, bears ovate or roundish pale yellow fruit with a boss at the point; its juice is acid and slightly bitter. There are varieties differing in form and in the thickness of their rind. Among these is one called by the Italians Pomo d'Adamo, because they fancy the depressions on its surface appear as if it still bore the marks of Adam's teeth.

The Shaddock, *C. decumana*, derives its common name from Captain Shaddock, by whom it was first carried from China to the West Indies, early in the eighteenth century. The shoots are pubescent; the leaves large with a winged stalk; the fruit very large, weighing sometimes ten to twenty pounds, roundish, with a smooth, pale yellow skin, and white or reddish sub-acid pulp. When the fruits attain their largest size they are called Pompoleons, or Pompelmouses; those of the smallest size form the 'Forbidden fruit' of all the English markets.

The Orange tribe cannot be grown in this country without protection in winter. In some parts of Devonshire, however, they require but very little, as for example, at Combe Royal, near Kingsbridge, where very fine specimens of Oranges, Citrons, and Lemons, &c., have been for many years obtained from trees planted against a wall, and protected only with a movable wooden shelter in winter. The first Oranges, it is stated, were imported into England by Sir Walter Raleigh, and reared by his relative Sir Francis Carew at Beddington in Surrey. These trees are mentioned by Bishop Gibson, in his additions to *Camden's Britannia*, as having existed for a hundred years previous to 1695; but finally they were entirely killed by the great frost in 1739-40, after they had attained the height of eighteen feet, with stems nine inches in diameter. Trees of the Orange tribe naturally live to a very great age in a soil and climate which suit them. Even under artificial treatment there are some remarkable instances of their longevity. There may be seen, in the orangery at Versailles, a tree which was sown in 1421. It is growing with its roots in a large box, and appeared very healthy when we saw it lately. The Orange tree at the convent of St. Sabina at Rome is thirty-one feet high, and said to be upwards of 600 years old. At Nice, where the tree may be considered naturalised, growing quite in the open air, there was in 1789, according to Risso, a tree which generally bore 5,000 or 6,000 oranges, and which was more than fifty feet high with a trunk which required two men to embrace it. In Cordova, the noted seat of Moorish grandeur and luxury in Spain, there are Orange trees still remaining, which are considered to be 600 or 700 years old.

Under favourable circumstances, the productiveness of the Orange is astonishing. In an account of the gardens and orange-grounds of St. Michael's in the Azores, by Mr. Wallace (*Journal of the Hort. Society*, vii. 236), we are informed by the author, who resided at St. Michael's for several years, that the orange grounds vary from one to sixty acres in extent, and are surrounded with high walls and tall-growing trees as shelter, not from the cold but from the sea-breeze. The grounds are rarely occupied wholly by Orange trees, for Limes, Citrons, Lemons, Guavas, &c., are scattered about in them. Orange trees were first introduced to the Azores by the Portuguese. There are only two kinds of oranges cultivated at St. Michael's, viz., the Portugal and the Mandarin; many varieties of the former exist, and they are greatly improved by the genial climate of St. Michael's. The Mandarin Orange has not been many years in the island, nevertheless there are some trees of it fourteen feet high. This capital little orange has lately been exported to England, where it realises a higher price than the common St. Michael's. The largest orange tree which Mr. Wallace measured was thirty feet high, the stem being seven feet in circumference at the base; but many larger trees, destroyed by the coccus, had been cut down. The produce of the trees is almost incredible; props are always used to prevent the weight of the fruit from breaking down the branches. An orange tree in the quinta, or orange garden of the Barão das Laranjeiras produced twenty large boxes of oranges, each box containing upwards of 1,000 fruit—in all 20,000 oranges from one tree. Two hundred ship-loads of oranges are annually exported from St. Michael's, being nearly 200,000 boxes. [R. T.]

CIVETTE. (Fr.) *Allium Schœnoprasum*.

CLADENCHYMA. Branched parenchyma.

CLADIUM. A genus of cyperaceous plants belonging to the tribe *Rhynchosporeæ*. The spikelets are one to two-flowered; glumes five or six; bristles wanting; nut with a thick fleshy coat, tipped with the conical base of the jointless style. Twenty-one species are mentioned in Steudel's *Plantæ Cyperaceæ*; these have an extensive geographical distribution, the majority being natives of New Holland. *C. Mariscus* is a native of Britain, and the most northerly of the species. It is a handsome aquatic plant, not of frequent occurrence, though plentiful in some districts. [D. M.]

CLADODIUM. An obsolete name of *Scaphyglottis*.

CLADOCARPI. A small section of mosses containing those anomalous genera in which the fruit is not truly lateral but terminates short lateral branchlets. The British genera belonging to this section are *Sphagnum*, *Mielichoferia*, *Fissidens*, and *Cinclidotus*; but the two latter contain species which are not truly cladocarpous. [M. J. B.]

CLADOCAULON. A Brazilian eriocaulaceous plant, an undershrub with much-branched leafy stems and flowers in heads, the male flowers being in the centre, the females at the circumference of the head. These latter present the distinguishing feature of the genus, that is to say, a double perianth, each row of three linear oblong segments adherent at the base, the outer segments reduced, and ultimately deciduous, the inner ones shorter, thinner, and persistent. In other particulars it does not differ from the other genera of the order. [M. T. M.]

CLADODYSTROPHIA. An affection to which oaks and other trees are subject in light soils, or when past maturity. The upper branches are less perfectly nourished than the lower, and therefore more rapidly decay. It has also been supposed that trees become stag-headed in consequence of decay of the tap root, possibly from the attacks of fungi, the terminal branches having an especial reference to it. This, however, is mere matter of speculation, though the main branches and roots have in many cases direct communication with each, some particular root more especially supplying some particular branch, as is indicated by the buttress-like spurs which connect the two. Where the tree is without leading shoots, the tips of the several branches sometimes assume a similar condition. [M. J. B.]

CLADONIA. A genus of lecidineous lichens which is characterised by its ultimately globose or button-shaped fruit growing at the tips of vertical hollow shrub-like or cup-shaped processes, arising from a foliaceous or crust-like thallus, to which they bear an inverse proportion. The fructifying disc is often of the brightest scarlet, but sometimes assumes other tints as pinkish-brown or black. The species are numerous even when reduced within reasonable limits, and extend into the coldest regions, while some are cosmopolites. *C. rangiferina* extends almost to the extreme limits of vegetation, and affords an abundant supply of excellent food to the reindeer, without which the inhospitable northern parts of our continent could scarcely be inhabited. *C. pyxidata*, a common species in woods and hedge-banks, is supposed to afford a good medicine in the whooping cough; while *C. sanguinea*, rubbed down with sugar and water, is successfully applied in the thrush of infants in Brazil. [M. J. B.] *C. rangiferina*, the reindeer moss, is the badge of the clan McKenzie.

CLADOPHORA. A genus of chlorosperms closely allied to *Confervæ*, and distinguished by its branched habit. The species are numerous, and grow in various situations, but the most characteristic, as *C. glomerata* and *rupestris*, occur either in rivulets or on sea-rocks. A few of the species of warm countries attain a considerable size. *C. mirabilis* was once celebrated as affording a supposed instance of transformation from a green-spored into a rose-spored alga; but it has since been ascertained that the cladophore serves merely as a matrix to the rhodosperm which surrounds the threads with its dilated base. The reproductive bodies are minute zoospores with thong-like appendages contained in the articulations. The species are found in most parts of the globe. [M. J. B.]

CLADOPTORIA. A name given to a singular affection to which several of our forest trees, as the oak and willow, are subject, in which the small branches snap off with a regular circular fissure, leaving a cup-shaped scar, somewhat similar to that which takes place when a leaf or fruit separates at the stalk. The branch was of course previously dead, and the separation seems to depend upon a vital process by means of which the dead are thrown off from the living tissues. After pears have fallen, a repeated separation into joint-like portions takes place, in a somewhat similar manner, between the component parts and the branch which gave them birth. After cold summers, vine branches are apt to fall off, a process which is facilitated by the peculiar formation of the stem, there being a transverse layer of cellular tissue at each bud. This is sometimes called l'brisanoptosis. Larger branches occasionally fall off in a like manner in the elm, but more generally, though the line of demarcation is well-marked, the branch does not fall till it is tightly pressed by the new layers of bark destined after its disappearance to close up the cavity. [M. J. B.]

CLADOR. In Greek compounds = a branch.

CLADOSPORIUM. A genus of naked-spored moulds of which one species, *C. herbarum*, is found in all habitable parts of the world on decaying substances, whether animal or vegetable, covering them with olive patches which when in fruit are shot with green. It consists of short brown-jointed waved threads, which bear on their sides oblong or elliptic spores with one or two transverse divisions. In damp seasons, wheat is often discoloured at the tip with this fungus, and it is then said to be taxed. Another species, *C. dendriticum*, is common on apple leaves, and when attacking the fruit causes the orbicular dark spots which are so common on apples, and which have been named by Fries *Spiloma*. The same species, or a mere variety of it, attacks pears in a similar way, and sometimes materially affects the health of the tree by infesting the leaves and young shoots. [M. J. B.]

CLADOSTACHYS. A genus of *Amarantaceæ* allied to *Celosia*, but having the stamens free, not cohering as in that genus, and the three stigmas linear and revolute, not capitate. *C. muricata* is a much-branched Indian shrub with alternate stalked ovate-acute smooth leaves, and elongated paniculately-arranged spikes of small white flowers. [J. T. S.]

CLADOTHAMNUS. A genus of *Pyrolaceæ* consisting of a shrub from Western Arctic America with much-branched stems, sessile wedge-shaped oblong glabrous leaves, glaucous below, and solitary axillary shortly-stalked flowers, with a five-parted calyx, five petals, ten stamens, a thread-like style, incurved at the apex, with a globular five-lobed stigma; capsule sub-globose, five-celled. [J. T. S.]

CLANDESTINA. A genus of *Orobanchaceæ* containing a single species, a parasite on roots in damp woods in the South of Europe. It is a small plant with a short branching scaly subterraneous stem, the bluish-violet flowers being seen in clusters rising from the apex of the stem as if from the earth. The four-cleft calyx is bell-shaped; the upper lip of the corolla helmet-shaped, the lower short and bifid. The ovary is surrounded at the base by a half-moon-shaped hypogynous gland. The capsular fruit contains four or five seeds attached to two linear parietal placentæ. This genus is nearly related to *Lathræa*, in which it was formerly included; it differs chiefly in having erect flowers, and a definite number of seeds on small placentæ. [W. C.]

CLAOXYLON. A genus of *Euphorbiaceæ* composed of trees or shrubs, natives of the tropical portions of the eastern hemisphere. They are nearly allied to *Mercurialis*, but differ in their arborescent habit as well as in the petal-like disc of the female flowers. The leaves are long-stalked, oval or lanceolate, and entire or toothed at the margins. The nerves of the leaves and the various parts of the flowers of many species are deeply tinged with a dark red colouring matter which is said to be used as a dye. The inconspicuous generally green flowers are arranged in slender racemes furnished with bracts, each bearing in its axil a cluster of flowers; these in the male are made up of a calyx with three or four deep divisions enclosing numerous stamens; and in the female of a similar calyx enclosing a three-lobed ovary, crowned with a three-branched style, and seated on a disc formed by three dark red petal-like glands. The capsular fruits are three-celled, about the size of a pea, and each cell contains one seed. [A. A. B.]

CLARKIA. A small genus of onagrads, indigenous to California and North Western America, contributing to our gardens two of the best known and most esteemed of popular annuals. The genus is well characterised by its clawed petals, eight stamens, of which the alternate four are shortest and sterile, four-lobed stigma with broad roundish spreading lobes, and cylindrical four-furrowed four-celled seed-vessel, opening when ripe by four valves. The species are all erect branching plants, with entire or toothed foliage, and showy reddish-purple flowers produced singly from the axils of the leaves. *C. pulchella* has the largest flowers, and is remarkable in its typical form for its petals being three-lobed with a tooth on each side of the claw, though in the variety *integripetala* of garden origin the lobes are obliterated. The leaves of this species are long and narrowly lance-shaped, quite entire, and the stem and branches are drooping at the summit before the expansion of the flowers. *C. elegans* is a taller plant with slender twiggy shoots, quite erect in all stages of their growth, ovate toothed foliage, the flower-buds drooping before expansion, and the bluntly rhomboidal petals quite undivided. [W. T.]

CLARY. *Salvia Sclarea*. Hormimum Clary is *Salvia Horminum*, and Vervain Clary, *Salvia Verbenaca*.

CLATHRUS. A genus of gasteromycetous *Fungi* belonging to the phalloid group, remarkable at once for the beauty of their colour and cleanness of form, combined with the most abominable odour. The receptacle to which the deliquescent fruit-bearing cells are attached, forms a scarlet net-work, which bursts forth from a gelatinous volva. In *C. crispus*, which occurs in warmer climates, the edge of the meshes is beautifully crisped. The closely-allied *Ileodictyon cibarium* is known to the New Zealanders by a name implying Thunder-dirt, and forms a coarse article of food. *C. cancellatus* is common in the south of Europe, and occurs occasionally in the southern parts of England, as at Torquay and the Isle of Wight, also in Ireland. [M. J. B.]

CLATHRATUS (adj. CLATHRATUS). A lattice; a membrane pierced with holes and forming a kind of grating, as in the *Ouvirandra fenestralis*.

CLAUDEA. The most beautiful genus of rose-spored *Algæ*. It is named after Claude Lamouroux, a distinguished French algologist. Three species only are known, of which two occur on the coasts of Australia and Tasmania, and the third on those of Ceylon. The frond proceeds from a thread-shaped stem, which is continued into the marginal rib of a flat unilateral open net-work formed of several series of anastomosing slender mid-ribbed leaflets. Each net-work, when fully formed, is ten or twelve inches long, and about an inch broad, and is elegantly recurved like a scimetar. The capsules are in the mid-rib of metamorphosed primary and secondary leaflets, and contain at the base a dense tuft of pedicellate pyriform spores. The tetraspermes are contained in the swollen bars of the second series of net-work in transverse rows. *C. elegans* sometimes grows at the mouth of rivers where the saltness is much modified, and then assumes a large size with increased delicacy. The above account is taken from Dr. Harvey's *Phycologia Australasica*, a work which ought to be in the hands of every lover of seaweeds. [M. J. B.]

CLAUDINETTE. (Fr.) *Narcissus poeticus*.

CLAUSILE. A name given by Richard to his macropodal embryo, when its radicle is united by the edges and entirely incloses all the rest of it.

CLAVALIER. (Fr.) *Xanthoxylon Clava Herculis*.

CLAVARIA. A genus of the clavate division, *Clavariei*, of hymenomycetous

Fungi, distinguished by their fleshy substance and confluent stem. The species are either simple or branched, and are extremely numerous, and from the great difference of form, colour, and division assumed under different circumstances, often extremely difficult to recognise. The surface is mostly smooth, but sometimes wrinkled longitudinally. Many of them afford excellent articles of food, but they are not much used in this country, probably from the scarceness of the larger species. *C. vermicularis*, which comes up frequently on our lawns, looking like little bundles of candles, is sometimes very abundant, and extremely delicate when dressed. [M. J. B.]

CLAVARIEI. A natural order of hymenomycetous *Fungi* distinguished by their vertical growth and superior hymenium, which extends to the very apex, and is distributed equally on all sides (amphigenous). The species generally grow on the ground amongst leaves, a few appear on rotten wood, and some of the lower kind on decaying herbaceous stems. We believe all the species which produce white spores are wholesome; some, moreover, with yellow spores are esculent, though one or two are doubtful. [M. J. B.]

CLAVATUS, CLAVIFORMIS. Gradually thickening upwards, from a very taper base; as the appendages of the flower of *Schievenckia*, or the spadix of *Arum maculatum*.

CLAVICULA. A tendril.

CLAVIGERA. The name applied by the elder De Candolle to three Mexican plants of the composite family, since shown by Dr. Asa Gray to differ in no way from *Brickellia*, and therefore placed in that genus which numbers about thirty species, mostly Mexican, and is distinguished from *Eupatorium* by the many-striate instead of five-angled achenes. [A. A. B.]

CLAVIJA. A genus of the myrsine family, comprising a number of shrubs or small trees, confined to the tropical parts of South America. Their unbranched rodlike stems are furnished at the top with a crown of large alternate coriaceous leaves, often two feet in length, quite smooth, oblong in form, and entire or spinously-toothed at the margin. The waxy white or orange-coloured flowers are small and disposed in erect or drooping racemes which are shorter than the leaves, and either proceed from their axils or from the bare stem where the leaves have fallen. The tube of the corolla being very short, and the five stamens having five roundish fleshy scales alternating with them, are characters which distinguish the genus from the others in the family. The fruits are fleshy and contain numerous seeds embedded in a pulp which is said to be eatable. In size they vary, but are seldom larger than a pigeon's egg. The genus bears the name of J. Clavijo Fazardo, a Spanish naturalist.

C. ornata, a native of Brazil and Guiana, is frequently to be met with in plant stoves, where it is always a prominent object from its straight unbranched stems, bearing on their apex a cluster of large handsome leaves often a foot or more in length. Its starry wax-like flowers, of a bright orange colour, are produced in great abundance, and are disposed in erect racemes. The root of some of the species is said to be emetic. [A. A. B.]

CLAVULA. The receptacle, or sporecase of certain fungals.

CLAVUS The disease which produces ergot in grasses; so called because it causes the young grain to grow into the form of a nail or club.

CLAW The long narrow base of the petals of some flowers; the analogue of the petiole.

CLAYTONIA. A genus of purslanes, chiefly North American, consisting of dwarf annual or tuberous-rooted perennial plants with entire leaves, and small white or flesh-coloured flowers in terminal racemes. Generically they are distinguished by a calyx of two oval permanent sepals, five petals usually with short claws cohering at the base, five stamens inserted on the claws, one style with its apex three-cleft, and an ovary ripening into a one-celled capsule, opening by three valves, and containing from three to six seeds. Of the annual section, *C. perfoliata*, one of the best known, is a weedy little species with fibrous roots, broadly ovate veinless radical leaves on long foot-stalks, and numerous simple naked flower-stems, bearing at the summit a roundish leafy bract formed by the cohesion of two opposite leaves, from which arise one or more short racemes of small white flowers with notched petals. The leaves of this plant are used like those of the common purslane, *Portulaca oleracea*. The perennial *Claytonias* have for the most part small tuberous or spindle-shaped roots, from which arise a few simple stems a foot high, bearing about the middle a single pair of opposite linear or lanceolate leaves, and being terminated by a loose drooping raceme of pink flowers veined with red. The species are rare in cultivation, but *C. virginica* is sometimes met with. They are popularly known in America by the name of Spring Beauty, from the early season at which they flower. [W. T.]

CLEARING NUT. An Indian name for the nut of *Strychnos potatorum*.

CLEARWEED. An American name for *Pilea pumila*.

CLEAVERS. *Galium Aparine*.

CLEGHORNIA. A Cingalese and Indian genus of *Apocynaceæ*, the plants of which have small white flowers with a calyx of five lobes alternating with five glands; a salver-shaped corolla with oblique lobes, and without scales in its throat; included anthers, arrow-shaped and sharply-pointed at the top; two ovaries, with a short style and large stigma; the fruit consisting of two large follicles. [M. T. M.]

CLEISOSTOMA. A genus of caulescent orchids, with leathery narrow distichous leaves, and long tough roots by which they cling to the bark of trees in various parts of the East Indies. They have the pouched lip and fleshy flowers of *Sarcochilum* and *Sarcanthus*, differing from the former in having the orifice of the pouch closed by a large projecting tooth, and from the latter in the pouch being absolutely one-celled. Sixteen or seventeen species are known, all having small flowers of little beauty.

CLEISTES. A genus of terrestrial leafy-stemmed orchids inhabiting tropical America. In habit they resemble *Arethusa*. The flowers are terminal and nearly solitary, of some purple tint. *C. rosea*, with large nodding flowers, is one of the finest; *C. paludosa* is quite insignificant.

CLEMATIS. An extensive genus of twining shrubs with variously-cut opposite leaves, belonging to the *Ranunculaceæ*, among which they are distinguished by their single perianth (a coloured calyx but no petals), and by the long feathery tail attached to their one-seeded carpels. The only English species, *C. Vitalba*, Virgin's-Bower, is so called on account of its being used for covering bowers; another name, Traveller's Joy, was probably given to it because of its being, in winter, among the most conspicuous and ornamental of wayside plants, often covering hedges for a considerable distance with its feathery seed-vessels, from the resemblance of which to grey hair the plant is sometimes called Old-Man's Beard. The flowers are greenish-white, and destitute of perfume. French *Herbe aux gueux*, from its ragged appearance. *C. Flammula* is the sweet-scented species common in gardens, a native of Southern Europe and Northern Africa; a variety of this, *C. rubella*, has larger flowers tinged with rose-colour, expanding in October. Other ornamental species are *C. florida*, of which a variety with large double white flowers is to be preferred as being the handsomest and remaining the longest in bloom; *C. Viticella*, of which there are several varieties with single or double flowers, blue, purple, or red; *C. austriaca*, bearing in June and July solitary large blue flowers with numerous abortive stamens simulating petals; and *C. azurea* and *C. lanuginosa*, magnificent blue-flowered Japanese species. *C. tubulosa* is a showy perennial with large blue flowers. [G. A. J.]

CLEMATITE COMMUNE. (Fr.) *Clematis Vitalba*. — ODORANTE. *Clematis Flammula*.

CLEMATITIS. *Aristolochia Clematitis*.

CLEOME. A genus of capparids chiefly found in the tropical regions of the New World, and presenting, in common with the other genera of the order, some interesting features. It is distinguished by the possession of a calyx of four pieces; a corolla of four erect petals, usually with long claws; six stamens having long distinct filaments; and a many-seeded pod-like fruit borne on a stipe or stalk of varying length. Most of the species are annual plants of erect habit, with digitate leaves of from three to seven lanceolate leaflets, and flowers in terminal bracted corymbs lengthening into racemes. One of the commonest species is *C. pungens*, a robust clammy plant, attaining a height of four or five feet, with spiny stipules, foot-stalks as well as under side of midribs armed with sharp prickles, and racemes of rosy-purple flowers; the anthers of this species are yellow, by which it may be known from *C. spinosa*. *C. rosea* resembles *pungens* in general habit, but is quite free from prickles, is less robust, and its leaves consist of but five leaflets, the uppermost and lowest of three only. *C. speciosissima* has handsome rose-coloured flowers, leaves with five to seven leaflets, petals as long as the flower-stalk, and a pod on a stipe longer than itself. The species are chiefly remarkable for their beauty, but are reputed to possess a pungent taste like that of mustard. [W. T.]

CLEOMELLA. A small genus of annual *Capparidaceæ*, the leaves of which are trifoliate, and the flowers have four somewhat spathulate petals with short claws and six separate stamens attached to the stalk supporting the ovary, which latter is gourd-shaped and one-celled, becoming a pod-like capsule. The plant is a native of Mexico and N. America. [M.T.M.]

CLERODENDRON. A considerable genus of *Verbenaceæ*, natives of tropical districts chiefly in Asia, but found also in Africa and America. They are shrubs or trees with opposite or ternate simple leaves, and loosely cymose or capitato flowers in terminal panicles or thyrses, more rarely axillary. The calyx is campanulate or inflated, and five-toothed or five-lobed. The corolla-tube is slender, the limb spreading and nearly equally five-lobed. There are four stamens inserted in the tube of the corolla, and usually much exserted; the anthers have two parallel cells, opening longitudinally. The ovary is four-celled, with a single pendulous or laterally attached ovule in each cell. The slender exserted style has two acute stigmatic lobes. The fruit is a drupe surrounded by the calyx, its kernel usually large, separating into two two-celled or four one-celled nuts. This genus is nearly related to *Volkameria* and *Ægiphila*, but is separated from the former by its fruit, and from the latter by its pentamerous flower. Nearly eighty species have been described. They have been arranged under two sections:— 1. *Euclerodendron*, in which the corolla is salver-shaped with a short tube scarcely longer than the calyx; and 2. *Siphonanthus*, in which the corolla is funnel-shaped with a very long tube. The plants have slightly bitter sub-astringent properties, and on this account some of them are used in Indian medicine. [W. C.]

CLESTINES. Large cells of parenchyma, in which raphides are often deposited.

CLETHRA. A genus of *Ericaceæ*, consisting of shrubs or trees, with alternate serrate deciduous leaves, and bearing white flowers in terminal hoary racemes. They are natives of North and tropical America. The calyx is five-parted; the corolla has five distinct obovate-oblong petals. There are ten hypogynous stamens, with inversely arrow-shaped anthers, which open by terminal pores or short slits. The ovary is three-celled with many ovules in each cell. The style is slender with a three-cleft stigma. The capsule is three-celled, with many seeds in each cell, three-valved, and enclosed in the calyx. [W. C.]

CLEYERA. A genus of *Ternstrœmiaceæ*, comprising a few Indian and Japanese evergreen bushes with camellia-like leaves, and small axillary white or yellowish flowers, sometimes sweet-scented. These flowers are stalked, and have a calyx of five leaves, five petals, numerous stamens in two or three series, and an ovary surmounted by a style which is three-parted at the top. The five free petals, and the numerous stamens slightly adhering to their base, are the chief distinguishing features of the genus. [A. A. B.]

CLIANTHUS. A genus of *Leguminosæ* found in New Zealand, Norfolk Island, and New Holland. It is nearly related to *Sutherlandia*, a Cape genus which has bladdery pods, while the pods in the present are coriaceous. The plants are herbaceous or woody branching shrubs, with unequally-pinnate leaves made up of eight to sixteen pairs of linear or elliptical leaflets half an inch long. The large handsome flowers are in terminal or axillary racemes. The calyx is bell-shaped and five-toothed. The upper petal or standard is oval, pointed, and bent backwards, much larger than the wings and shorter than the keel, which is skiff-shaped. The pod is stalked, somewhat woolly inside, and contains a number of seeds. The name of the genus is derived from the Greek, and signifies Glory Flower, a name peculiarly applicable to the plants. The best known species is *C. puniceus*, called Parrot's-Bill-in New Zealand, from the resemblance of the keeled petal to the bill of that bird. This plant was introduced in 1831, and is often to be met with in greenhouses, or on open walls with a southern aspect, where it flowers freely if protected in winter. It seldom attains more than six feet in height, although in Ireland, where the climate seems to suit it better, it is sometimes to be seen covering on a wall a surface of twelve or fourteen feet square. The pinnate-leaves are about six inches long, and the leaflets, about half an inch in length, are smooth above and slightly pubescent underneath. The flowers grow in oval clusters hanging from the leaf-axils, each flower more than three inches from the tip of the standard to the tip of the keel, and of a deep blood colour.

C. Dampieri is a native of the desert regions of Australia, and is also in cultivation. In habit it is much like the former, but it does not grow to such dimensions. The whole plant is of a pale green colour, and is thickly covered with long white hairs. The peduncles proceed from the axils of the leaves, and bear on their apex four or five scarlet flowers, larger and of a much brighter colour than those of the former, the standard having also a large black-purple base at its base. This plant has the most beautiful flowers in the genus, but is unfortunately difficult of cultivation. The only other known species is *C. carneus*, a native of Norfolk Island; it has flesh-coloured flowers, and although a pretty plant, is not to be compared with the others, the flowers being much smaller. [A. A. B.]

CLIDEMIA. A genus of *Melastomaceæ* from tropical America, containing hairy branched shrubs with opposite stalked leaves, generally unequal in size, with three to seven ribs, and white or rose-coloured flowers, often silky. The calyx is adherent to the ovary at the base; petals five or six, rarely four; stamens twice as many as the petals; ovary hairy, with as many cells as there are petals; berry fleshy, often edible. [J. T. S.]

CLIFFORTIA. A genus of small apetalous South African bushes, belonging to the rosaceous family, whose principal distinction lies in the three-toothed calyx and very numerous stamens. The leaves are alternate, small, and composed of two or three leaflets; when the latter is the case, the two lateral ones are small, and more or less united to the central one, so that the leaves appear to be simple. The flowers are small, and seated in the axils of the leaves. In the males the calyx tube is contracted at the top, and bears about thirty stamens. In the females the calyx is similar to that of the male, and encloses one or two achenes, each furnished with a lateral bearded or feathery style. The holly-like leaves of *C. ilicifolia* are used by the Boers as an emollient and expectorant in coughs. *C. crenata* is remarkable for the form and arrangement of the leaves, which are composed of two orbicular leaflets with notched margins, and are so closely set on the stems that they lap over each other in the manner of the scales of a fish. [A. A. B.]

CLIFFORTIACEÆ. A name given sometimes to *Rosaceæ* proper, including *Sanguisorbeæ*, as distinguished from *Amygdaleæ* and *Pomaceæ*.

CLINANDRIUM. The bed of the anther of orchids; an excavation of the top of the column, in or on which the anther lies.

CLINANTHIUM. A flat or broad space, on which flowers are packed closely; the receptacle of composites; a shortened widened axis.

CLINANTHUS. A name given to a group of Peruvian amaryllids; subsequently changed to *Clitanthes*, and now merged in *Coburgia*. [T. M.]

CLINIUM. In Greek compounds = receptacle. Also an accessory part of certain fungals, consisting of very small long simple or branched cells, bearing a spore at their end.

CLINOPODIUM. One of the names of the Wild Basil, now referred to *Calamintha*.

CLINTONIA. A small genus of lobeliads, consisting of dwarf annuals with the aspect and habit of *Lobelia*, but differing from that genus in the corolla being without a tube, and in the character of the seed-vessel, which, instead of being a half egg-shaped two-celled capsule opening when ripe by pores at the summit, is a long slender three-angled pod of one cell only, with seeds attached to two parietal placentas, and splitting when ripe into three narrow thong-like valves. Of the several species composing the genus, but two are known in British gardens, *C. pulchella* and *C. elegans*, both natives of California. The former is an elegant little plant, with slender prostrate branched stems, sparingly clothed with linear blunt foliage, and producing from its upper axils numerous flowers, with the upper lip of two spreading deep blue segments, and the lower lip very broadly wedge-shaped, three-lobed, blue at the margin, the centre being white and yellow with several deep purple spots. The pod is so long and slender that it presents the appearance of a foot-stalk rather than that of a seed-vessel. *C. elegans* is distinguished by its leaves being ovate instead of linear, and its flowers of a pale blue colour. The name of this genus has also been applied by Rafinesque to a small group of plants belonging to the lily tribe. [W. T.]

CLIOCARPUS. A genus of Brazilian shrubs of the family *Atropaceæ*, remarkable for being densely covered with small star-shaped hairs. The flowers are stalked, and are set in the axils of the leaves. The calyx is hairy, platter-shaped at its base, with five small pouches near its junction with the flower-stalk, the upper portion divided into five lance-shaped spreading segments, which increase in size as the fruit ripens, and become erect; their margins also are everted and touch those of the adjacent segments, so that a kind of tube is formed. The corolla is wheel-shaped, hairy, its divisions with a prominent nerve. The five stamens arise from a thickened rim at the base of the corolla, and have short wavy filaments, and large four-celled anthers. The fruit is a many-seeded berry included within the calyx. [M. T. M.]

CLIOCOCCA. A genus of *Linaceæ* from South Australia, scarcely distinct from *Linum*, the only differential characters being the imbricated, not contorted, æstiva-tion of the corolla, and the capsule splitting into ten cocci. [J. T. S.]

CLITANTHES. A name proposed for a group of amaryllids, since referred to *Coburgia*. [T. M.]

CLITOCYBE. A sub-genus of white-spored agarics with strongly decurrent, or acutely-adnate gills, the stem elastic with a fibrous outer coat, and the pileus convex when young, though depressed when old. It contains a great many species, some of which are excellent articles of food. *Agaricus nebularis*, for example, which occurs in woods with a compact obtuse pileus, clouded with grey, is one of the most delicate of mushrooms; and *A. geotropus*, especially the form called *sub-involutus*, is not to be despised. [M. J. B.]

CLITOPILUS. A sub-genus of rose-spored mushrooms with decurrent gills, and the pileus confluent with the fleshy or fibrous stem. *Agaricus prunulus*, which is a frequent inhabitant of our woods, and readily recognised by its primrose-whitish depressed pileus, narrow rose-coloured decurrent gills, and mealy scent, belongs to this sub-genus, and is excellent either stewed or pickled. It must not be confounded with *A. gambosus* (see TRICHOLOMA) which sometimes bears the same name. Most of the species are too small to be of much value. [M. J. B.]

CLITORIA. A large genus of pea-flowered plants belonging to the leguminous family, and nearly related to *Centrosema*, but differing in the standard having no spur-like appendage near its base. The genus is widely distributed, being found in tropical Asia, Africa, and America; in the latter country in the greatest numbers, and almost exclusively on the eastern side of the Andes. The greater portion of the species are large climbers, scrambling over trees to a great height; some few are erect, and several are twiners among bushes. The alternate pinnate leaves are made up of one or many pairs of opposite leaflets, and a terminal odd one. The peduncles arise from the axils of the leaves, and bear one or many large purple blue white or red flowers, often two to three inches long. The tubular five-toothed calyx is furnished with two bracts at its base; the standard is large and oval, notched or bifid at the apex, and narrowed into a claw at the base, the wings are much smaller than the standard, and the keel smaller than the wings and sometimes almost hidden by them. The straight pod is sometimes winged, and contains a number of seeds.

C. Ternatea, so called because the seeds were first brought from the island of Ternate, one of the Moluccas, is a very common plant in most tropical countries, and has long been in cultivation in England. In habit it is much like the common pea. Its leaves have two to four pairs of oval leaflets and a terminal odd one. The large handsome flowers grow in the axils of the leaves, and are of a beautiful blue colour, the standard with a white or yellow blotch at its base. They sometimes occur double, and a variety with white flowers also exists. The corollas of the blue variety are said to afford a blue dye in Cochin China, but it is not permanent, and Rumphius says that they are used for colouring boiled rice in Amboyna. The root is reputed to be as powerfully purgative as jalap; and in India, where it is sold in the bazaars in pieces about the thickness of two quills, it is given to children to promote sickness and vomiting. The Butterfly Pea, *C. Mariana*, has a curious distribution, being found in the Southern American States and Mexico, and appearing again in the Khasin Mountains in India without being found in any intervening place. It is a slender twining plant with leaves made up of three oval or lanceolate thin leaflets, about two inches long, and axillary peduncles bearing one or three flowers of a light blue colour. *C. arborescens*, a native of the West Indies and the adjoining mainland, is the only one of the numerous large scandent species peculiar to South America, which are in cultivation. Its leaves are pinnate and more than a foot in length, and the leaflets sometimes eight inches long and four broad. The large pale-blue flowers are numerous and in racemes, which are shorter than the leaves. Some of the species were formerly known under the names *Neurocarpum* and *Ternatea*. [A. A. B.]

CLIVERS, or CLEAVERS. *Galium Aparine*.

CLIVIA. A beautiful genus of amaryllids, to which the name *Imantophyllum* has also been applied. The latter, however, corrected to *Imantophyllum*, Sir W. J. Hooker now proposes to apply to a distinct though allied plant of South Africa called *I. miniatum*. The *Clivias* consist of herbs with fasciculate fleshy roots, and distichous lorate radical persistent dark-green leaves, from among which springs a plano-convex scape, bearing at top a crowded umbel of drooping flowers. These are formed of a six-leaved cylindrically funnel-like perianth curved on the upper side, the divisions having fourfold diversity, and being connivent into the form of a tube, overlapping and partially united at the base; the three exterior ones are the shortest; there are six equal slightly protruded stamens affixed to the base of the segments, a three-lobed stigma, and an inferior three-celled ovary containing many ovules and seeds, the cells, according to Herbert, being three-seeded. The species are of South African origin. *C. nobilis* is a very handsome plant, often seen in greenhouses, remarkable for its sturdy-looking harsh evergreen retuse two-ranked leaves, and producing a large head of numerous (forty to fifty) pendulous, club-shaped, orange-scarlet flowers tipped with green. *C. Gardeni* is a similar plant from Natal. *C. cyrtanthiflora*, a plant raised in the Belgian gardens, and known under the name of *Imantophyllum cyrtanthiflorum*, is said to be a hybrid between *C. nobilis* and the *Imantophyllum miniatum* above referred to. It has distichous lorate leaves, and a erect flower-scape, bearing numerous drooping slender funnel-shaped flowers, of a pale flame-colour with green tips. [T. M.]

CLOCHETTE DES CHAMPS. (Fr.) *Convolvulus arvensis*. — D'HIVER. *Galanthus nivalis*.

CLOSTERANDRA. A poppy-like papaveraceous plant imperfectly known. The filaments of the stamens are dilated in the middle. The ovary is obovate, one-celled, surmounted by five radiating stigmas which fall off when the capsular fruit is ripe. [M. T. M.]

CLOT-BURR. A North American name for *Xanthium*.

CLOUDBERRY. *Rubus Chamæmorus*.

CLOUDED. When colours are unequally blended together.

CLOVE BARK. The bark of *Cinnamomum Culilawan*.

CLOVE CASSIA. The bark of *Dicypellium Caryophyllatum*.

CLOVE GILLIFLOWER. The aromatic-scented double-flowered whole-coloured varieties of *Dianthus Caryophyllus*.

CLOVE NUTMEG. The fruit of *Agathophyllum aromaticum*.

CLOVE TREE. *Caryophyllus aromaticus*. The cloves of commerce are the dried aromatic flower-buds. —, WILD. *Eugenia acris*.

CLOVER. The common name for *Trifolium*, especially applied to the sorts cultivated for fodder. —, BOKHARA. *Melilotus leucantha*, a fodder plant, very grateful to bees. —, BUSH. An American name for *Lespedeza*. —, PRAIRIE. An American name for *Petalostemon*. —, SOOLA. *Hedysarum coronarium*. —, SWEET. An American name for *Melilotus*.

CLOVES. The small bulbs formed within the mother-bulb of certain plants; such as garlic.

CLOVEWORTS. A name sometimes used for the caryophyllaceous family to which the clove gilliflower belongs.

CLOWESIA rosea. A very rare orchid, with the habit of *Catasetum*, said to be a native of Brazil. It has erect racemes of concave white flowers delicately edged with rose-colour, broad fringed petals, and a saccate three-lobed lip the edge of which is broken up into innumerable thread-shaped glands. The anther lies at the bottom of an upright toothed hood. The caudicle resembles an hour-glass slit at the back.

CLOWN'S ALLHEAL. *Stachys palustris*.

CLUBBING. A peculiar condition or hypertrophia affecting the roots of cabbages and other allied parts, in which the whole force of vegetation is carried downwards to the destruction of the leaf and stem. The main root is mostly affected, but the disease sometimes affects the laterals. The structure of the root is much altered, so that on division it looks marbled like a truffle, and many of the cells gorged with highly nitrogenous matter. The disease is local, or where not local, capricious, and probably depends upon peculiar chemical conditions of the soil. In districts which are subject to it, the most effectual remedy appears to consist in putting a small quantity of wood-ashes, which contain several salts of potash, into the hole in which the root of each plant is placed. [M. J. B.]

CLUB-GRASS. A common name for *Corynephorus*.

CLUB-MOSS. A common name for *Lycopodium*.

CLUB-RUSH. A common name for *Scirpus*.

CLUB-SHAPED. The same as Clavate.

CLU′SIA. A large genus taken as the type of the *Clusiaceæ* or *Guttiferæ*, the latter name referring to the fact that the greater portion of the plants secrete in more or less quantity a milk-like or yellow resin. Clusia is chiefly distinguished by its capsular five or ten-celled fruit, which splits when ripe, each cell having many seeds; and by the numerous stamens, whose anthers open along their whole length, and not by a small pore or slit at the apex. All are trees or shrubs peculiar to Tropical America, and grow in very humid hot places. A great portion of them are parasitical on other trees, and a few send down stout root-supports from their thick branches similar to those of the banyan tree. The leaves are opposite, entire, very leathery in texture, mostly obovate in form, and furnished with numerous parallel nerves which are very evident in dried specimens, but almost imperceptible in the living plants. The greater portion have roseate flowers, but in a few they are white or yellow; in the larger-flowered species there are seldom more than two or three together in the axils of the upper leaves, but in the smaller-flowered ones they are numerous and disposed in a sort of panicle. In the males the calyx is of four to six leaves, the petals four to eight, and the stamens very numerous. In the females, which have a calyx and corolla like the male, a few abortive stamens surround the ovary, which is crowned by a flat radiating stigma. The fruit is a dry or fleshy capsule splitting up when ripe into five or ten portions.

The genus bears the name of Charles de l'Écluse or Clusius, a celebrated botanist of the sixteenth century. The leaves vary little in form throughout the genus; those of *C. grandiflora*, a native of Surinam, are from seven inches to a foot long, and its beautiful white flowers from five to six inches in diameter. Nearly allied to this, but smaller in all its parts, is *C. insignis*, a Brazilian plant, whose flowers 'weep a considerable quantity of resin from the disc and stamens, so much so indeed, that Von Martius says he obtained an ounce from two flowers'; this resin rubbed down with the butter of the chocolate-nut, the Brazilian women employ to alleviate the pain of a sore breast.' Other large flowered species, such as *C. alba*, *C. rosea*, and *C. flava* in the West Indies, yield an abundant tenacious resin from their stems, which is largely used for the same purposes as pitch; it is at first of a green colour, but when exposed to the air assumes a brown or reddish tint. The Caribs use it for painting the bottoms of their boats.

Among the smaller-flowered species the most interesting is the *C. Galactodendron*, a native of Venezuela. This plant, according to M. Devaux, is one of the Palo de Vaca or Cow-trees of South America. Its leaves are about three inches long, obovial in form, and narrowed towards the base. The bark is thick, covered with rough tubercles, and its internal tissue becomes red when exposed to the air. In extracting the milk from this tree the inhabitants make incisions through the bark till they reach the wood, these incisions are said to be made only before the moon is full, as they imagine the milk flows more freely then than at any other time. One tree is said to yield a quart in an hour. When the inhabitants find themselves at a distance from their homes, they make use of the milk for themselves and their children; its use is accompanied by a sensation of astringence in the lips and palate, which is said to be characteristic of all edible vegetable milks.

C. Deca yields a resin known in Columbia by the name of Ituca, and burnt for the sake of its pleasant odour. Upwards of thirty species are enumerated. [A. A. B.]

CLUSIA′CEÆ. The mangosteen family, a natural order belonging to the thalamifloral dicotyledons, usually called GUTTIFERÆ; which see. [J. H. B.]

CLU′STERED. Collected in parcels, each of which has a roundish figure; as the flowers of *Cuscuta*.

CLU′TIA. A genus of *Euphorbiaceæ* composed of numerous diœcious bushes, confined to Africa and found in the greatest number at the Cape. The double disc of the male flowers readily serves to distinguish the genus from its allies. The alternate stalked leaves are destitute of stipules and vary in form from oval to linear; in some they are evergreen, but in others they fade in the autumn, whilst a few are charmed, as well as the young branches, with glandular dots. The small, generally green flowers are in cymes in the axils of the leaves, numerous in the males, and few or single in the females. In the former they are made up of a five-leaved calyx, five petals, and five stamens supported on a central column and arranged like the branches of a chandelier; the base of the column is surrounded by two rows of glands, five of them large and two- or three-lobed, and five smaller, each of them entire or two-lobed. In the female flower the calyx and corolla is the same as in the male, but the disc is made up of five blind glands only, and the three-lobed ovary is crowned by a three-branched style, each branch bifid at the point and bent back on the ovary. The fruit is a three-celled capsule with three seeds. The only reported useful species is *C. lanceolata*, a native of Abyssinia, where it is said to be used for stopping dysentery in cattle. [A. A. B.]

CLYPEA. A name which has been given to certain *Menispermaceæ*, now referred to *Stephania*. [M. T. M.]

CLYPEATE. Having the form of an ancient buckler; the same as Scutate.

CLYPEOLA. A genus of small annual herbs, belonging to *Cruciferæ*, natives of Southern Europe and temperate Asia. They have the habit of the annual species of *Alyssum*, but differ in having an indehiscent orbicular, flattened and margined pouch containing a single seed. [J. T. S.]

CNEMIDIA. If to the flowers of a large *Physurus* are added the foliage and habit of some herbaceous-leaved *Cypripedium*, the reader will form some idea of this singular genus of orchids. The few species known are all Indian. *C. angulosa* has also been called *Govindoria nervosa*, and *Dressiana angulosa*.

CNEMIDOSTACHYS. A genus of *Euphorbiaceæ*, known also as *Microstachys*, and composed of herbs seldom more than two feet high, with twiggy branches, and alternate, linear, entire or serrate leaves. The inconspicuous flowers are male and female on the same plant; the males in slender spikes, have a three-parted calyx and three free stamens; the females, single in the axils of the leaves, have a calyx like the males, and a three-lobed ovary, crowned with a three-parted style. The capsule, about the size of a pea when ripe, is either smooth or covered with rough points, and is three-celled, each cell with a single seed. The greater portion of the species are Brazilian. One (*C. Chamælea*) is common to India and Africa, and another, which has been called *Elachocroton asperiocccum*, is found in Tropical Australia. [A. A. B.]

CNEORUM. A genus of uncertain position, but closely allied to the *Rutaceæ*. It consists of small shrubs inhabiting the Mediterranean region, the Canary Isles, etc. They have narrow, entire leaves; yellow flowers with three or four sessile equal petals, larger than the sepals, inserted beneath the disc; three or four stamens attached to the stalk bearing the three or four-lobed ovary, which has two ovules in each of its three or four compartments. The fruit when ripe consists of three or four segments, which separate one from the other, and are fleshy externally, bony internally, and divided into two cavities by a spurious transverse partition. The species will grow in the south of England in sheltered situations. [M. T. M.]

CNEORUM. (Fr.) *Daphne Cneorum.*

CNESTIDIUM. Dr. Planchon has described under this name a Central American tree of the order *Connaraceæ*. It has compound leaves covered with thick red down; clustered flowers, which have a calyx consisting of five parts, adhering together for a time, but at length breaking irregularly into two or three divisions; ten stamens, five of which are shorter than the rest, and confluent in a ring at their base; and five ovaries with as many thread-shaped styles. The fruit consists of a single follicle from the suppression of the remaining four, covered with red down, and containing a single seed. [M. T. M.]

CNESTIS. A name derived from the Greek word signifying to scratch, in allusion to the hairs on the fruit, which irritate the skin. It is applied to a genus of *Connaraceæ* consisting of shrubs frequently of climbing habit, with alternate compound thick leaves, and clusters of five-parted flowers; which bear ten stamens, five shorter than the remainder, and five sessile ovaries, with two ascending ovules. The fruit consists of five or fewer follicles, covered with stinging hairs, and containing but one seed. Two or three species are in cultivation, natives of Guinea, the Mauritius, etc. [M. T. M.]

CNICUS. A thistle-like genus of *Compositæ*, known by the following characters: — Bracts of the involucre leathery, extended into a long hard pinnated spine; fruits furrowed, marked with a broad scar on one side; pappus in three rows, the outer horny, short, the next composed of ten long bristles, the third of ten short bristles. The English plume thistles, formerly included in a genus of the same name, but differing from the above, are now referred to *Carduus*. Of the true genus *Cnicus* the most remarkable is *C. benedictus*, a native of the Levant and Persia, but now widely distributed. The plant was formerly esteemed as a tonic, diaphoretic, etc., but is now little used. [M. T. M.]

CNIDOSCOLUS. A genus of *Euphorbiaceæ*, composed of a few shrubs or herbaceous plants, all of them confined to tropical America. On the one hand they are nearly related to *Jatropha*, on the other to *Manihot*, but differ from the former in having no petals, and from the latter in the filaments of their stamens being united into a central column, not free. Their stems are often fleshy and gouty, and are furnished with stalked leaves, which in most cases are armed with straight hairs, which sting most virulently; the blades are sometimes entire, but mostly palmately-lobed. The small white flowers are arranged in terminal or axillary cymes, the females few and occupying the central portion of the cyme; the males more numerous and occupying the internal parts. In both males and females the calyx is tubular with a five-lobed limb, and encloses in the former ten stamens united into a column and arranged in two tiers; and in the latter a three-lobed ovary crowned with three stigmas torn at the apex. The three-celled capsular-fruit is about the size of a large pea, and covered with sharp hairs, each cell containing but one seed.

C. stimulans is a plant of the Southern American states, and has palmately-lobed leaves from four to eight inches long. The lacerated segments are covered with spreading hairs, which sting fearfully the bare feet of the negroes when they tread on them; it is sometimes called on this account 'Tread Softly.' Its tuberous roots are said to be eatable like those of the cassava or manihot. *C. quinquelobus* has been in cultivation, but it stings so terribly that few people care to keep it. The effects of the sting are various on different constitutions. Some on being stung fall down and are quite unconscious for a length of time; but others are not so affected. In both cases an excruciating pain is felt, which lasts for some days, and the parts swell and sometimes continue swollen, accompanied with an itching sensation for months. [A. A. B.]

COACERVATE. The same as Clustered.

COADNATE, COADUNATE. The same as Connate.

COALITIO. The growing of one thing to another; as that of petals, which produces a monopetalous corolla, &c.

COARCTATE. Contracted; drawn close together.

COARCTURE. The neck of a plant. See Collum.

COREA. This small genus of phloxworts consists of climbing tendrilled plants, with pinnate foliage, and large bell-shaped flowers produced singly from the leaf axils. Although at first sight they appear to have little in common with the other plants of this order, and really differ essentially in habit, they yet agree with them in their most important structural features. The genus is distinguished by its large leafy permanent five-parted calyx; declinate stamens and style; three-celled ovary surrounded at its base by a fleshy annular disc; and large flat winged seeds, imbricated in a double row. *C. scandens*, the most interesting species, is a well-known summer climber of very rapid growth. Its leaves are composed of three pairs of elliptic leaflets, the midrib being terminated by a branched tendril; it has large bell-shaped flowers, which are at first green, but ultimately assume a deep violet hue. *C. macrostema* has smaller yellowish-green flowers, with stamens twice as long as the corolla, and the segments of the calyx lanceolate. [W. T.]

COB-NUT. A variety of the Hazel, *Corylus Avellana*. —, JAMAICA. The seeds of *Omphalea triandra*.

COBURGIA. A genus of ornamental *Amaryllidaceæ*, having tunicated bulbs, loratety linear glaucescent leaves, and a two-edged scape supporting a terminal umbel of few showy flowers. The perianth is funnel-shaped, with an elongated angular incurved tube, swollen towards the top, a regular six-parted imbricated somewhat spreading limb, and a short campanulate cup, bearing on its margin the six stamens and six intermediate bidentate lobes; the ovary is three-celled with numerous ovules. There are eight or ten species known, and these are natives of Peru. The type of the genus, *C. incarnata*, is a very handsome plant, with bulbs like those of the Jacobæa lily, five or six oblong linear hluntish, slightly glaucous leaves, and a scape two and a half feet high, supporting a four-flowered umbel of pendent flowers, about five inches in length, of a brilliant salmon-orange colour, the tube of which is bluntly three-cornered, very slender at the base, widened upwards and dividing into a moderately-spreading limb of six ovate-elliptic segments an inch long, lighter in colour and more pinky than the tube, and with a green central stripe. The crown is short and erect, with six green bifid lobes between the stamens, which about equal the limb in length and are shorter than the style. *C. trichroma*, a species with a five-flowered umbel and flowers three inches long, the tube light red, the limb white within, green without, and with green-tipped teeth to the cup, is said to be cultivated in pots with great care in Mexico, where it flowers at various seasons. *C. variegata* has four-flowered umbels, the tube of the flowers yellow and red, and the limb yellow outside, white within, margined with rose, and tipped with green. *C. lutea*, formerly named *Clinanthus*, and subsequently *Chlanthes*, has a two-flowered scape, the flowers yellow and about two inches long. The genus was named in honour of the Prince of Saxe Coburg, now king of the Belgians, who, when resident at Claremont, was a great patron of horticultural and botanical science. The name has also been applied to another group of amaryllids, now merged in *Hippeastrum*. It is written *Coburghia* by Dr. Herbert. [T. M.]

COBWEBBED. Covered with loose, white, entangled, thin hairs, resembling the web of a spider.

COCA. *Erythroxylon Coca*, the leaves of which are used as stimulants by the Peruvian Indians.

COCALLERA. A Brazilian name for a decoction of *Croton perdicipes*.

COCARDEAU. (Fr.) *Mathiola fenestralis*.

COCCIDIA. A name applied to that form of the conceptacles in the rose-spored *Algæ*, which consists of globular tubercles with a cellular wall continued from the substance of the frond, whether partly confluent with it or free, and not opening in general by a terminal pore. Examples are afforded by Rhodymenia and Grevillaria. The elongated processes in such alsæ as Gigartina mamillosa are simply called tubercles. In this species, at least, there is a pore for the exit of the spores. [M. J. B.]

COCCIGROLE. (Fr.) *Fritillaria Meleagris*.

COCCINEUS. Pure carmine colour, slightly tinged with yellow.

COCCINIA. A climbing shrub of the gourd family, common in the hedges of India, where it grows like our bryony. *C. indica*, the only species, has large white dioecious flowers, with five stamens united together by their filaments into a column bearing three parcels of wavy anthers. The female flower has three sterile stamens, united in three parcels; and the style is short and trifid. The fruit is oblong, marked with ten white lines; when ripe it is of a red colour, bursting irregularly, and having several seeds provided with a gelatinous covering. The ripe fruit is used by the natives in their curries; the leaves and other portions are also used medicinally. [M. T. M.]

COCCOBRYON. A South African climbing shrub of the pepper family has been made the type of a genus with the above name. The flowers are perfect, in densely crowded stalked spikes placed opposite to the leaves, and each flower protected by stalked peltate roundish bracts; there are two stamens, and sometimes a third; and the ovary is sessile, the style short. The fruit is a berry, crowned by the persistent style. *C. capense* possesses stomachic properties. [M. T. M.]

COCCOCYPSELUM. A genus of *Cinchonaceæ* with a four-parted calyx; a funnel-shaped imitated corolla; a two-celled, many-seeded berry; and a style partly divided into two. The name refers to the vase-like form of the fruit. *C. repens*, a West Indian creeping-plant, is in cultivation, and is interesting from its blue-purple berries. [M. T. M.]

COCCODES. Resembling pills; consisting of spheroidal granulations.

COCCOLOBA. A genus of polygonaceous plants, one of which, *C. uvifera*, is known in the West Indies as the Seaside Grape, from the peculiarity of the perianth, which becomes pulpy, and of a violet colour, and surrounds the ripe fruit. By this character also the genus is distinguished amongst its fellows. The pulpy perianth has an agreeable acid flavour. An extract is prepared from the plant, which is so astringent as to rival kino in its effects. [M. T. M.]

COCCULUS. This name is liable to mislead the general reader, who might suppose it to apply to the plant producing the poisonous berries called *Cocculus Indicus*. These, however, are the produce of an allied genus *Anamirta*; which see. *Cocculus* belongs to the same family, *Menispermaceæ*, and consists of climbing shrubs, with unisexual flowers having six sepals, six petals, and six stamens; the female flowers have three ovaries placed on a short stalk, the styles erect, cylindrical. The fruit is a drupe with a bony shell, containing a curved seed. *C. laurifolius* forms an exception to the general rule in this genus, inasmuch as its stems are erect, not climbing. The plant producing Columba root was formerly referred to this genus, but is now included in *Jateorhiza*. The root of *C. villosus*, an Indian species, is used in decoction in cases of rheumatism, &c., while the fruits furnish a kind of ink. See also *Tinospora*. [M. T. M.]

COCCUS. A shell; a carpel, which separates with elasticity from an axis common to itself and others.

COCE DOLCE. The Italian name for the seeds of Sweet Fennel, *Fœniculum dulce*.

COCHÊNE. (Fr.) *Pyrus Aucuparia*.

COCHINEAL-FIG. *Opuntia cochinillifera*.

COCHLEAR. A term used in describing æstivation; when one piece, being larger than the others, and hollowed like a helmet or bowl, covers all the others; as in *Aconitum*.

COCHLEARIA. A genus of *Cruciferæ*, represented in this country by the dissimilar-looking Horse-radish and the Scurvy-grass, in the essential parts of whose flowers, however, the correspondence is close. The points of distinction between this genus and its allies are the entire white petals, the stamens not being toothed, and especially the roundish pod or silicula, the valves of which are very convex, the partition between them very broad. The embryo is so folded up that the young root or radicle lies along the edge of the two flat cotyledons or seed leaves.

C. Armoracia is the common Horse-radish whose large coarsely-toothed rough leaves, and tall stem bearing a profusion of white flowers, are well known. The lowest leaves are frequently deeply and irregularly divided like the teeth of a comb, the upper ones become smaller and narrower. The root-stock is the part used for culinary purposes for its pungent taste. Dreadful accidents have occurred from mistaking the root of aconite for Horse-radish, as mentioned under *Aconitum*. The plant very rarely perfects its fruit in this country. *C. officinalis*, the Scurvy-grass, is a small low growing plant, with thick egg-shaped-cordate leaves, the upper of which clasp the stem; unlike the preceding species, the pods have a very prominent rib in the centre of each valve. This plant was formerly used as an antiscorbutic, and is still used in salads, as watercress is. It is common in some parts of Scotland. *C. danica* and *C. anglica* are probably only varieties of this species. [M. T. M.]

COCHLEARIFORM. Spoon-shaped.

COCHLEATE. Twisted in a short spire, so as to resemble the convolutions of a snail-shell; as the pod of *Medicago cochleata*, or the seed of *Salicornia*.

COCHLIA *violacea*. A small orchidaceous epiphyte from Java, with fleshy leaves, and small purple flowers growing in heads.

COCHLIDIOSPERMATE. Seeds which are convex on one side, and concave on the other, owing to unequal growth, or anomalous structure, as in *Veronica*.

COCHLIDIUM. A synonym of *Monogramma*, a genus of curious small tropical ferns. [T.M.]

COCHLIODA *densiflora*. A handsome Peruvian epiphytal orchid, with thin pseudobulbs and parchment-like leaves. The flowers appear in dense spikes, and have the lip adnate to the column as in *Epidendrum*; but the pollen apparatus is that of the *Vander*. A second unpublished species of the genus has been sent from the Quitinian Alps by Dr. Jameson.

COCHLOSPERMUM. A genus of small trees or shrubs, natives of Tropical India, Africa, and America, as well as in North Australia. They are placed by some among the *Cistaceæ*, and by others among the *Ternstrœmiaceæ*, but are easily recognised from any genus in either of these families by their palmately-lobed leaves. These are alternate, furnished with long stalks, and bear much resemblance to those of some of the maples. The large yellow flowers are in terminal panicles, and generally open and wither before the leaves make their appearance. They are composed of a five-divided calyx, five large nearly round petals, and very numerous stamens surrounding a one-celled ovary crowned by a single unbranched style. The capsular fruit when ripe is in size and form like a pear, and opens with three or five valves. The seeds are small, very numerous, and covered with a cottony down.

C. Gossypium is a shrub or small tree found in the peninsula of India. Its five-lobed leaves are smooth above and downy underneath, and, including the stalk, more than a foot long. The numerous yellow flowers in terminal panicles are about four inches across. From the stem of this plant a gum called Kuteera is obtained, and it is used as a substitute for gum tragacanth because of its viscidity. The cottony substance which adheres to the seeds is sometimes used for stuffing pillows and cushions. Much like this is *C. insigne*, a native of Brazil, but its leaves are smaller and have serrate lobes. The Brazilians make use of a decoction of the roots of this plant against internal pains, and principally against those which are the result of falls and other accidents; they also affirm that this decoction cures abscesses which have already formed.

C. Planchoni, a native of Western Africa, is a shrub about five feet high, with alternate three or five-lobed leaves which are pubescent underneath. According to Mr. Barter, who gathered the plant, 'each shoot rises from a stool, is unbranched, and bears on the apex a cluster of yellow flowers three to four inches across. The roots are large and succulent, and yield the only yellow dye with which the people are acquainted. It is a common plant on the river Quorra.' Another species, *C. tinctorum*, a native of Senegambia, is said to have a thick tuberous root-stock, which furnishes a yellow dye, known to the natives as Fayar, and used for dying cotton stuffs, as well as in medicine in cases of amenorrhœa. The flowers of this only are known, and very likely it is not different from the last-mentioned species. The woolly covering of the seeds gives rise to the name of the genus. [A.A.B.]

COCKLE-BURR. An American name for *Xanthium*.

COCKSCOMB. *Celosia cristata*.

COCK'S-HEAD. *Onobrychis Caput-galli*.

COCK'S-SPUR THORN. *Cratægus Crus-galli*.

COCOA or CACAO. The seeds of *Theobroma Cacao*.

COCOA-NUT. The nut of *Cocos nucifera*. —, DOUBLE or SEA. The nut of *Lodoicea seychellarum*.

COCOA-PLUM. The fruit of *Chrysobalanus Icaco*.

COCOA-ROOT or COCO. The root of *Colocasia antiquorum*, used as an esculent in tropical countries.

COCO, LE PETIT. *Theophrasta Jussiæi*.

COCOS. The well-known Cocoa-nut tree is the type of this genus of palms, to which, in addition, about a dozen other species belong. They mostly form tall graceful trees, and the majority of them are natives of the tropical regions of America, one only, the common Cocoa-nut, being found in Asia or Africa. Their leaves are very large and pinnate. Their flowers are of separate sexes produced on the same spike, both having a calyx consisting of three sepals, and a corolla of three petals, the males containing six stamens united at the base, and the females an egg-shaped ovary, with a short style and three stigmas, and sometimes six barren stamens. The fruit is either elliptical, or egg-shaped and three-sided, and contains a single seed enclosed in a hard bony shell, which has three round holes at its base, and is surrounded by a dry fibrous husk.

The Cocoa-nut Palm, *C. nucifera*, is now so extensively cultivated throughout the tropics, that it is impossible to ascertain its native country; there can be no doubt, however, that it is indigenous to some part of Asia, probably Southern India. It exists in vast quantities on the Malabar and Coromandel coasts, and adjacent islands, growing in the greatest luxuriance upon sandy or rocky sea-shores, and evidently preferring the vicinity of the sea, although it sometimes occurs a considerable distance inland. It is also common in Africa, America, and the West Indies. Its extensive geographical distribution is accounted for by the fact of the tree growing in such close proximity to the sea, that the ripe fruits, falling on the beach, are washed away by the waves, and afterwards cast upon some far-distant shores, where they readily vegetate. It is in this way that the coral islands of the Indian Ocean have become covered with these palms. It is also worthy of remark, that the triangular form of the fruit facilitates its progress through the waves.

The Cocoa-nut Palm has a cylindrical trunk, sometimes as much as two feet in diameter, and rising to the height of sixty or one hundred feet, its outside being marked with scars, indicating the places from which leaves have fallen away. It is surmounted by a crown of gracefully curved feathery or pinnate leaves, each of which is from eighteen to twenty feet in length, and composed of a strong tough central footstalk, with numerous narrow long and sharp-pointed leaflets arranged along both sides of it, giving the entire leaf the appearance of a gigantic feather; the base of the stalk spreads out so as to clasp the stem, and is surrounded by a kind of fibrous network of a light-brown colour. The flowers are arranged on branching spikes five or six feet long, and enclosed in a strong tough pointed sheath (spathe), which splits open on the under side, displaying the delicately white but inconspicuous flowers. They are succeeded by bunches containing from twelve to twenty fruits, each of which is about a foot long by six or eight inches wide, of a three-sided form, and covered by a thick fibrous rind or husk, enclosing a single seed contained in a hard shell, which is what is commonly called the Cocoa-nut in this country.

The uses of this palm are so numerous that space will only allow us to give a brief outline of them. In this country we know comparatively little of its value. It is true that we are indebted to it for several very useful articles, such as cocoa-nut fibre, cocoa-nut oil, and the cocoa-nuts themselves; but they are all articles that we might contrive to do without. In tropical countries, however, such as Southern India and the adjacent islands, the case is very different; there the Cocoa-nut Palm furnishes the chief necessaries of life, and its culture and the preparation of its various products afford employment to a large part of the population. Every part of the tree is put to some useful purpose. The outside rind or husk of the fruit yields the fibre from which the well-known cocoa-nut matting is manufactured. In order to obtain it the husks are soaked in salt water for six or twelve months, when the fibre is easily separated by beating, and is made up into a coarse kind of yarn called coir. In 1855 we imported 81,138cwts. of this fibre.

Besides its use for matting, it is extensively employed in the manufacture of cordage, being greatly valued for ships' cables, and although these cables are rough to handle and not so neat-looking as those made of hemp, their greater elasticity renders them superior for some purposes. Other articles of minor importance are now made of this fibre, such as clothes- and other brushes, brooms, hats, &c.; and when curled and dyed it is used for stuffing cushions, mattresses, &c., as a substitute for horse-hair.

The next important product of the fruit is the oil procured by boiling and pressing the white kernel of the nut (all unmen). It is liquid at the ordinary temperature in tropical countries, and while fresh is used in cookery; but in this country it is semi-solid, and has generally a somewhat rancid smell and taste. By pressure it is separated into two parts—one, called stearine, is solid, and is used in the manufacture of stearine candles; the other, being liquid, is burned in lamps. As an article of food the kernel is of great importance to the inhabitants of the tropics. In the Laccadives it forms the chief food, each person consuming four nuts per day, and the fluid, commonly called milk, which it contains, affords them an agreeable beverage. While young they yield a delicious substance resembling blanc-mange. The hard shells of the nut are made into spoons, drinking cups, lamps, &c.; reduced to charcoal and pulverised they afford an excellent tooth-powder, and very good lamp-black is made from them.

Amongst other products of this palm may be mentioned 'toddy,' which is obtained by the same process as that described under Borassus flabelliformis. When fermented it is intoxicating, and strong arrack is distilled from it, besides which it yields vinegar and 'jaggery' or sugar.

The leaves are greatly used for thatching houses, for platting into mats, baskets, hats, and similar articles; and from strips of the hard footstalk very neat combs for the hair are made. The unexpanded leaves cut out of the heart of the tree are used in the same way that we use cabbages. The brown fibrous network from the base of the leaves is substituted for sieves, and also made into fishermen's garments. And the extremely hard wood obtained from the outer portion of the trunk is used in the construction of both houses and their furniture. In this country, under the name of Porcupine wood, it is made into work-boxes, and other fancy articles. Finally, we may mention that the natives attribute various medicinal qualities to this palm. The flowers they employ as an astringent, the roots as a febrifuge, the milk in opthalmia, &c.

Few of the other species of this genus present particular features of interest. C. butyracea, a native of New Grenada, yields toddy, but the manner of extracting it is very different to the process employed in Eastern countries. The tree is cut down, and a long cavity excavated in its trunk near the top; in three days' time this cavity is found to be full of toddy, which, it must be borne in mind, is the sap of the tree. Its seeds yield a semi-solid oil. *C. coronata*, a small Brazilian species not more than thirty feet high, has a pithy substance in the interior of its stem, which is used as food; its seeds also yield oil. The Cocoa-nut Palm is represented in Plate 7, fig. d. (A. S.

COCOTIER. (Fr.) *Cocos nucifera*.

COCRISTE. (Fr.) *Rhinanthus major*.

CODACANTHUS. A small genus of Indian herbaceous plants, belonging to *Acanthaceæ*, and having the habit of *Campanula ranunculoides*. The drooping blue flowers are in compound one-sided racemes at the ends of the stem or branches; they are furnished with small bracts and bracteoles. The calyx is equally five-parted; the corolla has a short campanulate tube, and a five-cleft limb; there are only two included stamens owing to the non-development of the other pair; the style is free. The racemose inflorescence of this genus obviously separates it from the allied genera *Phlebophyllum* and *Endopogon*, which have their flowers in spikes. (W. C.)

CODAZZIA. A name given by Karsten and Triana to *Delostoma integrifolium*.

CODDA-PANNA. The Talipot Palm, *Corypha umbraculifera*.

CODESO DEL PICO. A name applied in Teneriffe to *Adenocarpus frankenioides*.

CODIA. A synonyme of *Pisonia*.

CODIÆUM. A genus of the spurgewort family found in the Moluccas, and the islands to the north of Australia. It is composed of shrubs which have much the appearance of *Aucuba*. They differ from *Croton*, to which they are most nearly allied, in having very numerous stamens in the male flowers, and in the females being destitute of petals. Their beautiful painted leaves, which are shortly stalked and collected principally at the apex of the branches, vary much in form in the same species, being either linear or broadly oval, generally about six inches long, and quite smooth with entire margins. The green inconspicuous flowers are male and female on different racemes on the same plant; the males with a calyx of five divisions, five small petals, and very numerous stamens, and the females with a similar but smaller calyx, no petals, and a three-lobed ovary crowned with a trifid style. The fruit is a three-celled capsule about the size of a pea; each cell with a single seed.

C. pictum is a shrub often met with in stoves, where it is cultivated for the sake of its beautiful leaves, which are of a deep-red colour, or sometimes yellow mottled and variegated with green. In the Moluccas, its native country, it is cultivated about the houses, and used for fences. The inhabitants also decorate their triumphal arches with its leaves, and strew them about on occasions of festivity. The bark and root excite a burning sensation in the mouth when chewed. This is the plant so often found in gardens under the names of *Croton variegatum* and *Croton pictum*. The two other known species are plants of very similar appearance. (A.A.B.)

CODIUM. The most highly organised of the siphonaceous division of green-spored *Algæ* which occurs upon our coasts. The species resemble sponges. The frond is composed of branching filaments without any partitions, having on their lateral branchlets little cysts containing numberless minute zoospores. *C. tomentosum* has a more or less cylindrical or compressed forked green frond, and is found from the equator almost to the polar basin, but is scarcely found on the eastern coasts of North America, though common on the north. It extends also southward to Cape Horn, Australia, &c., without any essential change. (M. J. B.)

CODLIN. A variety of the Apple, *Pyrus Malus*.

CODLINS AND CREAM. *Epilobium hirsutum*.

CODON. A genus containing a single species from the Cape of Good Hope. It is an annual herb, covered over with white spines, and having alternate petiolate leaves and large flowers in terminal racemes. The calyx is ten to twelve-parted; the corolla is campanulate with as many lobes as the calyx, and like the sepals long and short alternately; there are ten to twelve stamens inserted at the base of the corolla tube; the ovary is sub-two-celled, free, and ovoid-acute with two parietal placentæ, to which are attached numerous ovules. The capsule is surrounded by the persistent calyx, and surmounted by the style, and contains numerous angular tuberculated seeds; it dehisces loculicidally.

This genus has a very uncertain position. It has been most generally referred to *Solanaceæ* or to *Hydrophyllaceæ*, though by some to *Scrophulariaceæ* and even to *Boraginaceæ*. Its one-celled multilocular ovary and parietal placentæ separate it from *Boraginaceæ*. Its ten to twelve-lobed regular corolla, ten to twelve equal stamens, and one-celled ovary separate it from *Scrophulariaceæ*. Its habit and structure approach nearer to *Solanaceæ*, but it can scarcely be united to this order on account of its one-celled ovary and loculicidal dehiscence. In most characters, and in its whole habit, it is more nearly related to *Hydrophyllaceæ*, though differing remarkably from any other genus of the order. (W. C.)

CODONANTHEMUM. A genus of *Ericaceæ*, consisting of several species of heath-like plants, with ternate whorled or scattered leaves, and the flowers crowded together at the end of very short branches. It has a four-toothed calyx, and a hypogynous persistent corolla, both campanulate; the four stamens are inserted below the hypogynous disc, and have lateral exserted anthers; the ovary is one-celled with a single pendulous ovule, and the stig-

CODONANTHUS. The name formerly given to a West African plant of the convolvulus family, but now generally placed in the genus *Prevostea*, and called *Prevostea africana*. It is a branching tree of middling stature, with alternate oblong leaves narrowed at both ends, having entire margins and about six inches in length. Three or four white flowers grow in the axils of the leaves; the two exterior calyx leaves are large and heart-shaped, the others small and narrow; the corolla, which is bell-shaped, with a slightly recurved five-toothed margin, encloses five stamens, and an ovary surmounted by a bifid style, each of whose branches is furnished with a shield-shaped stigma. [A. A. B.]

CODONOCALYX. Small Brazilian hairy plants with dioecious flowers, constituting a genus of *Euphorbiaceæ*. The male flowers have a calyx with five deep divisions, a corolla of five overlapping segments, a disc of five free glands alternating with the petals, and ten stamens longer than the rest. [M. T. M.]

CODONOCARPUS. A genus of *Gyrostemoneæ* containing small shrubs from South Western Australia, with branched stems, alternate linear subulate leaves, and solitary axillary stalked flowers, which are dioecious, with a six or seven-lobed calyx and no petals. The male flowers have numerous sessile anthers; and the female flowers numerous carpels combined around a central column into a many-celled ovary; styles short recurved. The fruit is obovate, depressed, separating into numerous one-seeded cocci. [J. T. S.]

CODONOPSIS. A genus of *Campanulaceæ*, natives of the mountains of Northern India. They are glabrous herbs, often twining, with stalked crenate leaves whitish below, and axillary or terminal stalked flowers, which are yellow, bluish, or purple. Calyx-limb five-lobed; corolla slightly fleshy, bell-shaped, five-lobed at the apex; stamens five; style with three stigmas; capsule hemispherical, three-celled, three-valved at the apex. [J. T. S.]

CODONORCHIS, literally Bell-orchis, in allusion to its campanulate flowers, is a small terrestrial genus, occurring in the southernmost parts of South America. The best known form, *C. Lessonii*, is a simple-stemmed plant with two three or four verticillate leaves near the base of a scape from four to six inches high, terminated by a single rose-coloured spotted flower, the upper surface of whose lip is covered with sessile or stalked glands. A supposed second form, called *C. Poeppigii*, is regarded by Hooker, fil., as a mere variety.

CODONOSTIGMA. A genus of *Ericeæ*, containing a single species from South Africa, a heath-like shrub with ternate verticillate leaves, and flowers in terminal buds. It has a four-toothed campanulate calyx; a persistent globular cup-shaped corolla; four exserted stamens, attached below the hypogynous disc, with hairy anthers; and a one-celled ovary containing a single pendulous ovule, and surmounted by a small cup-shaped stigma. Its one-celled ovary allies it to the genus *Omphalocaryon*, but it has the calyx of *Codostigma*. [W. C.]

CŒLANTHIUM. A genus of *Caryophyllaceæ*, of the tribe *Mollugineæ*, consisting of glabrous annuals from the Cape of Good Hope, with obovate stalked radical leaves in rosettes, while those of the stem are thread-like and verticillate, with fringed stipules. The stems are forked at the top; the flowers racemose with a funnel-shaped five-cleft calyx, having petaloid lobes; petals none; stamens five; stigmas three; capsule three-valved. [J. T. S.]

CŒLEBOGYNE. A genus of *Euphorbiaceæ*, found in the eastern tropical portion of New Holland, and represented by *C. ilicifolia*, a bush which in everything but its flowers is very like the common holly, or still more like the Japanese Osmanthus. The inconspicuous green flowers are male and female on different plants. The males, in the axils of the leaves, are arranged in short-bracted spikes, each bract toothed and supporting a number of

Cœlebogyne ilicifolia.

flowers, which have a calyx of four divisions enclosing from four to eight stamens. In the female plant the flowers are arranged in a similar manner, or in little cymes at the ends of the branches; the calyx is of four or six divisions, and often accompanied with one or two lateral glands near its base; the ovary is crowned with a three-lobed stigma, whose branches are large and lie flat on its summit. The fruit is a three-lobed capsule, about the size of a pea, with three cells, each of which contains one seed. The genus is nearly allied to *Omoriroba*, and differs only in the number of the calyx divisions.

This plant has excited much interest because it is said to ripen its seeds without the aid of pollen. Female plants (and females only) were sent to Kew by Allan Cunningham in 1829, where they flowered and perfected their seeds apparently without the aid of pollen. The circumstance was noticed by Mr. Smith, who made it the subject of a communication to the Linnæan Society. This led to careful examinations by Kiotzsch, Radlkofer, and A. Braun, besides other continental botanists. The former of these demonstrated from the formation of the seed that it contained no embryo but a bud; while the other two came to the opposite conclusion; and A. Braun made "a most important observation, still unexplained by him, namely, that he found a pollen grain on the stigma of *Cœlebogyne*." Naudin and Decaisne, in France, made experiments on *Hemp*, *Mercurialis*, and *Bryony*, as well as some other plants, and came to the conclusion that female plants of any of these, when sufficiently guarded against the accidental influence of pollen from the male flowers, produce perfectly ripe seeds. More lately, Henri in Russia has made extensive experiments on these plants, and affirms that no plant with evident sexual organs can produce perfect seeds without the aid of pollen. This is the opinion held by most botanists. [A. A. B.]

CŒLESTINA. A genus of erect annual Mexican plants of the composite family, seldom more than two feet in height. Their leaves are opposite, shortly-stalked, and generally heart-shaped in form with notched margins, and often clad with short rough hairs. The blue flower-heads are about the size of a pea, and disposed in terminal corymbs. The florets are all tubular and perfect, and their pappus is cup-shaped and slightly toothed. In this latter character only does the genus differ from *Ageratum*, which has a pappus of from five to ten awned scales. The species are plants of little beauty. [A. A. B.]

CŒLIA. A genus of terrestrial orchids with long grassy leaves, and dense spikes of rather small flowers supported by linear acuminate bracts. Three or four species are said to be known, but the genus has been little examined. *C. Baueriana*, with fragrant white flowers, from the West Indies, is that on which the genus was founded. It has a spurless lip, a three-winged ovary, and eight pollen masses without a gland.

CŒLOGLOSSUM. An obscure genus of Indian terrestrial orchids with the habit and general structure of *Platanthera* or *Peristylus*, but with a concave lip, and a pair of adnate processes arising from the orifice of the spur. All have small green flowers.

CŒLOGYNE. There occurs in the tropical and sub-tropical regions of Asia a race of pseudobulbous orchids, conspicuous for large coloured membranous flowers, with converging and slightly-spreading sepals, petals of like nature but narrower, a great cucullate lip usually bearing fringes on its veins, and a broad membranous column. The pollen masses are four in number, waxy, and cohering by a granular substance; the stigma is prominent, deeply hollowed out (whence the name), and two-lipped. Most of the species are beautiful objects, and therefore favourites in cultivation. Some have tough persistent leaves and loose racemes of flowers; others have flowers peeping up from the soil in the absence of the leaves, in the same way as the crocuses of Europe; to the latter the name of *Pleione* has been given. Between forty and fifty species are known, the finest of which are *C. cristata*, with ivory-white flowers, whose veins are fringed with yellow; *C. odoratissima*, unsurpassed for fragrance; and *C. præcox* (a *Pleione*), an Alpine plant, ornamenting with its large rich rose-coloured flowers the branches of oaks, at the elevation of 7,500 feet above the sea in lat. 30° N.

CŒLOSPERMOUS. Hollow-seeded; when the seed, or seed-like fruit, is hemispherical, and excavated on the flat side, as in coriander.

CŒNANTHIUM. The receptacle of flowers in the inflorescence called a Capitulum; same as Clinanthium.

CŒNOBIO. The same as Carcerulus.

CŒNOCLADIA. A name applied to the natural grafting which is so common in the beech in our own country and in many tropical trees. Both branches and roots, when growing so close together that there is no room for their proper developement, become intimately united, and form a sort of network. Amongst herbaceous plants, as in *Asparagus*, *Hyacinths*, &c., union often takes place between two contiguous stems, which in this case are generally flatter than usual. Some cases of wide-flattened stems arise from this cause, but others apparently from the attack of insects. If two or more buds concur in the formation of such a stem, and they have different rates of growth, we have curled fasciated branches such as not uncommonly appear on the ash. Similar branches are produced in the elder by a species of *Æcidium*. The roots of contiguous firs sometimes unite, so that when one of the trees is cut down, the stump still increases in diameter, in consequence of receiving nutriment from the tree with which it is united. [M. J. B.]

CŒNTRILHO. A Brazilian name for *Xanthoxylum hiemale*.

CERULEUS. Blue; a pale indigo colour.

CESIUS. Lavender colour.

COFFEA. A genus of *Rubiaceæ* or *Cinchonaceæ*, composed of between fifty and sixty species, one of which yields the well-known article coffee. All are shrubs or small trees, seldom more than twenty feet high, and inhabit the tropics of both hemispheres, the greatest number, however, being found in the Western. Their flowers have a small cup-shaped globular or top-shaped calyx, divided at the summit into four or five short teeth, and a tubular corolla, shaped like a funnel, with four or five spreading divisions; the stamens agreeing in number with the divisions of the corolla, and being either fixed to the top of its tube and protruded beyond it, or about half-way down on its inside, and entirely included within it. The fruit is a small fleshy berry, sometimes crowned by the remains of the calyx, and contains two seeds enclosed in a thin parchment-like shell, each seed being convex on the outside, but flat and marked by a longitudinal furrow on the inside.

The most interesting species is the Coffee shrub, C. arabica. This, when allowed to grow freely, will attain a height of twenty feet, with a stem three or four inches thick, but in a cultivated state it is seldom permitted to grow higher than ten or twelve feet, in order to facilitate the gathering of the berries. Its leaves are smooth and shining, and of a dark green on the upper surface,

Coffea arabica.

but paler beneath, about six inches long by two and a half wide, and of an oblong somewhat oval shape, with wavy edges, and terminated by a long narrow point. The flowers are produced in dense clusters at the bases of the leaves, and, being of a snowy white colour, they give the shrub a beautiful appearance, but are of ephemeral duration; their corolla is cut into five divisions, bearing the stamens fixed round the top of the tube, and protruded beyond it. They are succeeded by numerous little red fleshy berries resembling small cherries, each of which contains two of the seeds commonly called coffee.

At the present day the Coffee shrub is cultivated throughout the tropics, but its native country is the mountainous regions at the extreme south-west point of Abyssinia, the word Coffee being derived from Caffa, the name of one of the provinces of that country. From Abyssinia the Coffee shrub was first introduced into Arabia by the Arabs, and cultivated in Yemen, or Arabia Felix as it was anciently called, and for upwards of two centuries Arabia supplied all the coffee then used. Towards the end of the seventeenth century, however, the Dutch succeeded in transporting it to Batavia, and from thence a plant was sent to the Botanic Garden at Amsterdam, where it was propagated, and in 1714 one was presented to Louis XIV. The credit of introducing the Coffee shrub into the Western Hemisphere is a disputed point. One story asserts that the French introduced it into Martinique in 1717; while, on the other hand, the Dutch are said to have previously taken it to Surinam. In either case, it is certain that we are indebted to the progeny of a single plant for all the coffee now imported from Brazil and the West Indies.

The early history of the use of coffee is enveloped in obscurity, and consequently there are many fables regarding its origin. According to the best accounts, the custom of drinking coffee originated with the Abyssinians, by whom the plant has been cultivated from time immemorial; and it was not introduced into Arabia until the early part of the fifteenth century, when a learned and pious Schiekh, named Djemaleddin-Ebn-Abou-Alfagger, returning from Abyssinia, brought a quantity of coffee with him to Aden, where it soon superseded the beverage made from the leaves of the kât (Catha edulis), and its use gradually spread over the rest of Arabia. It, however, met with great opposition from the priests, who classed it among the intoxicating beverages forbidden by the Koran, and therefore prohibited its use, but the most learned physicians having declared it to be harmless, the prohibition was removed. The European use of coffee dates from the middle of the sixteenth century, when it was introduced into Constantinople; and a century later, namely, in 1652, the first coffee-shop was established in London. Since then its use has become so general, that the consumption of this article in Europe and the United States is now estimated to be not far short of nine hundred million pounds, nearly half of which is the produce of Brazil. Ceylon, however, supplies the greatest portion of that consumed in this country. In 1858 the total quantity imported into the United Kingdom was 60,697,708 lbs, of which 35,208,093 lbs, was retained for home consumption, and the remainder re-exported. The import duty, being three pence per pound on raw, and four pence on roasted, yielded a revenue of £440,475.

When ripe, the coffee berries are gathered, and the soft outer pulp removed by a machine called the pulper; they are then steeped in water for twenty-four hours to remove all mucilaginous matter, after which they are carefully dried, and the parchment-like covering of the seeds removed by means of a mill and a winnowing machine. In Brazil, however, the berries as gathered are simply dried in the sun, and afterwards passed through a mill which crushes the shells and allows the separation of the seeds.

Before being used for the preparation of the well-known beverage, coffee undergoes the process of roasting. By this means it gains nearly one half in bulk, and loses about a fifth in weight; besides which its essential qualities are greatly changed, the heat causing the developement of the volatile oil and peculiar acid to which the aroma and flavour are due. Coffee acts upon the brain as a stimulant, inciting it to increased activity, and producing sleeplessness; hence it is of great value as an antidote to narcotic poisons. It is also said to exert a soothing action upon the vascular system, preventing the too rapid waste in the tissues of the body, and by that means enabling it to support life upon a smaller quantity of food than would be otherwise required. These effects are due to the volatile oil, and also to the presence of a peculiar crystallisable nitrogenous principle termed *caffeine*; and it is not a little remarkable that closely allied, if not identical, principles exist in many similar beverages used by mankind, such for instance as tea, cocoa, Paraguay tea, and others. The leaves of the Coffee shrub likewise contain caffeine, and in the island of Sumatra the natives prefer an infusion of them to that of the berries. A patent has been taken out for the introduction of Coffee-tea into this country, but it has not been successful. A Javanese Coffee-plantation is shown in Plate 3. [A. S.]

COFFEE. *Coffea arabica*, the roasted seeds of which form the Coffee of the shops. —, SWEDISH. The seeds of *Astragalus baeticus*.

COFFEE-BEAN TREE. *Gymnocladus canadensis*.

COGWOOD. *Ceanothus Chloroxylon*. —, JAMAICA. *Hernandia sonora*.

COHESION. The union or superficial incorporation of one organ with another.

COHNIA. An obscure genus of orchids, related to *Oncidium*, whose terete-leaved species it resembles in habit. The only knowledge of it is derived from a solitary specimen from Guatemala in the Vienna Herbarium, and from Reichenbach's description and figure in his *Xenia Orchidacea*.

COHOSH. An American name for *Actæa* and *Leontice*. —, BLUE. *Leontice thalictroides*.

COHUNE OIL. An oil obtained from the fruit of *Attalea Cohune*.

COIGNASSIER. (Fr.) *Cydonia vulgaris*.

COIGNASSIER DU JAPON. (Fr.) *Cydonia japonica*.

COILOSTIGMA. A genus of Cape *Ericaceæ*, containing several heath-like shrubs, with ternate verticillate leaves, and flowers clustered at the end of the branches. The calyx has four divisions, generally equal, though sometimes with one larger than the others; the persistent corolla is small and ovate; the four stamens are inserted below the hypogynous disc, and have hairy anthers; the ovary has from two to four cells with a single ovule in each, and a cyathiform stigma. The members of this genus have the habit of *Simocheilus*. They are separated from the allied genus *Codonanthemum* by the shape of the stigma, and from *Codonostigma* by the several-celled ovary. [W. C.]

COIR. Cocoa-nut fibre.

COIX. A genus of grasses belonging to the tribe *Phalarideæ*. The flowers are monoecious. The males grow in lax spikes; glumes two, membranaceous; paleæ two; stamens three. The females grow in two-flowered spikelets, the inferior flower being neuter with one pale, while the perfect flower has two fleshy paleæ, of which the superior is two-nerved. The best known species is *C. Lachryma*, commonly called Job's Tears, a native of the East Indies and Japan. This is frequently cultivated, but requires the shelter of a conservatory. The large round shining fruit have, when young, some resemblance to heavy drops of tears, hence the fanciful specific name. Its medicinal qualities are said to be strengthening and diuretic, and for these qualities it is sometimes used in the countries where it grows. [D. M.]

COLA. A genus of *Sterculiaceæ*, consisting of two species only. They are middle-sized trees, with smooth entire leaves, and inhabit western tropical Africa. Their flowers are destitute of a corolla, but have a coloured five-cut calyx with the segments spreading like the spokes of a wheel; the stamens are united into a very short column, which bears the anthers in a single row, the cells of the anthers spreading apart (in allied genera they are parallel). The ovaries are five in number and cohere together, each having a slender stigma, but no style. The fruit consists of two (sometimes more) separate pods (follicles), which split open on the inner side, and contain several seeds about the size of horse chestnuts.

C. acuminata grows about forty feet high, and bears pale yellow flowers spotted with purple; its leaves are about six or eight inches long, and pointed at both ends. Under the name of Cola, or Kolla, or Goora nuts, the seeds of this tree are extensively used as a sort of condiment by the natives of western and central tropical Africa; and likewise by the negroes in the West Indies and Brazil, by whom the tree has been introduced into those countries. In Western Africa the trees grow mostly in the vicinity of the coast, and an extensive trade is carried on in Cola nuts with the natives of the interior; the practice of eating Cola extending as far as Fezzan and Tripoli. A small piece of one of these seeds is chewed before each meal as a promoter of digestion; it is also supposed to improve the flavour of anything eaten after it, and even to render half-putrid water drinkable. There are several varieties of Cola nuts; the common kind has an astringent taste, whilst another, called bitter Cola, is intensely bitter, and is thought to possess febrifugal properties. Powdered Cola is applied to cuts. [A. S.]

COLAX. A small genus of epiphytal orchids, near *Maxillaria*, under which name some have been published. It is especially remarkable for the condition of its caudicle, which seems to have no distinct gland, but consists of a thin wavy membrane gradually narrowing to the point where a gland usually occurs. *Maxillaria viridis* and *placanthera* are the best known species.

COLBERTIA. A genus of *Dilleniaceæ*, the type of which is a tree from tropical Asia with oblong or obovate shortly stalked serrated leaves, and large yellow flowers on one-flowered peduncles, several of which arise from the same scaly bud. It differs from *Dillenia* by the greater separation of its ovaries, which are generally fewer than in that genus. [J. T. S.]

COLCHICUM. The well-known Meadow Saffron, or, as it is erroneously called, Autumn Crocus. The genus appertains to *Melanthaceæ*, and is known by its bell-shaped coloured perianth, with a long tube; six stamens inserted into the upper part of the tube; a three-celled ovary placed at the bottom of the tube, and surmounted by three long thread-shaped styles; and a three-celled capsule which bursts by as many openings. The appearance of the flower is so like that of the crocus, that it is frequently mistaken for it; but in the crocus there are three stamens only, and the ovary is placed below the tube of the perianth, not within it, as in the *Colchicum*; or, more correctly speaking, in the latter the ovary is free, while in the former it is united to the lower part of the tube of the perianth. *C. autumnale*, the Meadow Saffron, found wild in some parts of England, has a subterranean bulb-like stem, called a corm, from which in autumn the light purplish mottled flowers arise. The leaves do not appear till afterwards; they are fully developed in the following spring, in the shape of loose green sword-shaped blades, among which the ripened fruit may be found raised from below the surface of the ground by the lengthening of the flower-stalk.

The *Colchicum* is valued not only for its appearance, but more particularly for its medicinal properties. The dried corms and the seeds are the parts employed, the former have much of the appearance of tulip bulbs, but are not scaly like them, but solid in the interior. The active principle is said to be an alkaline substance of a very poisonous nature called *colchicine*. Colchicum is principally used in medicine for the alleviation or cure of gout. In some cases its use is very beneficial, but, like other remedies, it has no claim to be considered infallible. It is acrid, sedative, and acts upon all the secreting organs, particularly the bowels and the kidneys. It is apt to cause undue depression, and in large doses acts as an irritant poison. Dr. Lindley relates the case of a woman who was poisoned by the sprouts of *Colchicum*, which had been thrown away in Covent Garden Market, and which she mistook for onions.

The Hermodactyls of the Arabians, formerly celebrated for soothing pains in the joints, are said by Dr. Royle to belong to *C. variegatum*. [M. T. M.]

COLDENIA. A genus of *Ehretiaceæ*, consisting of herbs from India and Ceylon, with wedge-shaped stalked plicate serrated leaves, which are often more developed on one side of the mid-rib than on the other. Flowers small, white, axillary, solitary; calyx five-parted; corolla funnel-shaped. The nuts have a somewhat fleshy covering, and are rugose. *C. procumbens* is used in India for promoting suppuration, for which purpose it is dried and powdered, and mixed with the seeds of the fenugreek. [J. T. S.]

COLD-SEEDS. In the old materia medica the seeds of the cucumber, gourd, pumpkin, &c.

COLEA. A genus of *Bignoniaceæ*, natives of Madagascar, Mauritius, and the neighbouring islands. They consist of glabrous shrubs or small trees, with impari-pinnate bi- or many-jugate leaves. The calyx is sub-campanulate and five-toothed; the corolla is funnel-shaped, and the limb is cleft into five spreading lobes. The four didynamous stamens are inserted on the corolla, and have two-celled anthers. The fruit is oblong, fleshy, and indehiscent, with two cells containing many imbricated wingless seeds. [W. C.]

COLEBROOKIA. An East Indian genus of shrubs, belonging to the family of labiates. They are covered with reddish down. The flowers are clustered, of a white colour, with a bell-shaped equally five-parted calyx, the segments of which are feathery, and whose tube becomes confluent with the ripe fruits; a short-limbed corolla divided into four nearly equal divisions, the upper lobe notched, and four stamens equidistant one from the other; the anthers with two parallel cells. *C. oppositifolia* and *C. ternifolia* are in cultivation as greenhouse shrubs. [M. T. M.]

COLEONEMA. A beautiful genus of Cape *Rutaceæ*, related to *Diosma*, and consisting of evergreen shrubs with sharp linear leaves, and white flowers, consisting of five petals attached to the base of a five-lobed disc, which is adherent to the tube of the calyx, and having a broad stalk or claw which is furrowed longitudinally. There are ten stamens, five sterile, concealed in the furrows of the claws of the petals and adherent to their base, and five fertile, opposite to the lobes of the disc, longer than the sterile ones, and having their anthers tipped with a minute sessile gland. The fruit is a capsule of five carpels, each provided with a small horn-like process at the top and opening by two valves. *C. album* is the best known species.   [M. T. M.]

COLEOPHORA. The name given to a tree of the daphne family found in Brazil, the leaves of which are not known. From the little scaly buds, which are scattered over the trunk of the tree, the flowers proceed. They are small, yellow and brown, and borne on short racemes. The tubular calyx has a four or five-toothed border fringed with hairs, and inside of it, and surrounding the stalked ovary, is a little four-toothed petal-like cup, about half the length of the calyx tube. The stalked ovary, surrounded by the peculiar cup, serves to distinguish the genus.   [A. A. B.]

COLEOPHYLL, or COLEOPTILE. The first leaf which follows the cotyledon in endogens, and ensheaths the succeeding leaves.

COLEORHIZA. The sheath formed at the base of an endogenous embryo, where it is pierced by the true radicle.

COLEOSTYLIS. Herbaceous plants covered with glandular hairs, natives of New Holland, and closely allied to *Stylidium*, but distinguishable from it by the following characters :—The limb of the corolla is divided into five segments, four of like form, stalked; the fifth or lip is unlike, jointed to the tube of the corolla, stalked, its blade boat-shaped, notched at the point or prolonged; and the column, which consists of the stamens and style united together, is shorter than the lip, erect, and passing at its base through a kind of sheath, whence the name.   [M. T. M.]

COLESEED, or COLLARD. The Rape, *Brassica Napus*.

COLESULA. The small membranous bag which contains the spore-case of liverworts.

COLEWORT, or COLLET. The Cabbage, *Brassica oleracea*.

COLEUS. A considerable genus of *Labiatæ*, found in Asia and Africa. It consists of annual herbs, sometimes with perennial stocks, rarely shrubs. The flowers are in loose or dense six or many-flowered verticillasters. The calyx is ovate-campanulate, bending back when in fruit, and the limb is five-toothed or bilabiate. The corolla-tube is longer than the calyx, and the limb is bilabiate with the upper lip obtusely three to four-cleft, and the lower entire, lengthened, and concave, often curved and enclosing the four stamens. The style is bifid with subulate lobes. The nucule is compressed and smooth.   [W. C.]

COLICODENDRON. A genus of *Capparidaceæ*, consisting of tropical American trees or shrubs, covered with small star-shaped hairs, and having clusters of flowers, with a cup-shaped calyx, divided into four or five segments, provided internally and at their base with a petaloid scale. The four or five petals are inserted on to the calyx; the stamens are from eight to twenty, inserted on a stalk, and united at the base into a shallow cup; the ovary is also on a long stalk. The fruit is a roundish or elongated berry, knotted and containing several kidney-shaped seeds. The genus possesses an acrid principle, which, according to Martius, is so potent in *C. Yco* as to be dangerous to mules and horses.   [M. T. M.]

COLIC ROOT. *Aletris farinosa*.

COLIGNONIA. A genus of Peruvian herbs or undershrubs belonging to the order *Nyctaginaceæ*, and having flowers arranged in an umbellate manner, surrounded by deciduous bracts. The perianth is coloured, bell-shaped, with a five-cleft limb from which the five stamens protrude; the style is simple; the stigma is fringed. The fruit is hardened, pentangular, crowned by the upper part of the perianth.   [M. T. M.]

COLLABIUM nebulosum. A terrestrial orchid, with a slender creeping rhizome, known only by a brief description of Blume, who says it has distant stalked membranous radical leaves clouded with purple, and small nodding spiked flowers, whose sepals are reflexed.

COLLANIA. A genus of amaryllids allied to *Alstrœmeria*, from which it differs in having a pulpy fruit, and in 'the great prominence of the operculum of the germen, making it at least half superior instead of inferior.' The species are natives of Peru, and are very ornamental plants. They have rigid erect stems curved at the summit, bearing simple rigid leaves, and a pendulous umbel of flowers, of which the six-leaved perianth is tube-formed and not at all spreading. *C. dulcis* has stems about a foot high, erect with a little tortuosity but not prehensile, clothed with oblong obtuse glaucous leaves, which are narrowed at the base, and terminating in a four-flowered umbel of cylindraceous purple flowers tipped with green, the three petaline segments longer and bright green. This plant is called 'Campanillas coloradas' in its native country, and the fruit is sweet and agreeable to the taste, and much sought by children, the seeds being enveloped in a reddish gelatinous substance. *C. andimarcana* is a much larger plant, with a stem terminating in a fine umbel of leafy racemes of large pendulous sub-cylindraceous flowers, upwards of two inches long, of which the sepals are orange red tipped with black, and the petals yellow tipped with green. The name *Collania* has also been applied to another genus of the same order, now called *Urceolina*. [T. M.]

COLLAR. The ring upon the stipe of an agaric; see also *Coltum*.

COLLARE. The ligule, or transverse membrane that stands in grasses at the junction of the blade and sheath of the leaf.

COLLATERAL. Standing side by side.

COLLECTORS. The hairs found on the style of such plants as the *Campanula*, and which collect or brush out the pollen from the anthers.

COLLEMACEÆ. A natural order of lichens, distinguished principally by their gelatinous substance and the green globules or gonidia, which are so distinctive a mark of lichens in general, forming necklace-like threads. They are found in various parts of the world, and though in general attracting little notice when dry, a few hours rain swells them out and exhibits often extremely beautiful forms. One of the most curious genera is *Myriangium*, which occurs in the southern part of England, Algeria, Australia, and the United States, on the trunks of living trees, and is remarkable for the high development of the sacs or asci in which the sporidia are contained. These plants have been considered as a distinct group from lichens, but such a notion is at present received with little favour. The resemblance of the young plant to *Nostoc* is so striking, both in appearance and structure, that the one has been supposed to be the infant state of the other, but without sufficient grounds. The species grow on trees, rocks, and the bare ground, and, if *Lichina* be included, in situations exposed to frequent immersion in the sea. One at least of the species has a very fetid smell. We are not aware that they have ever been applied economically. [M. J. B.]

COLLENCHYMA. The cellular matter in which the pollen is generated; usually absorbed, but remaining and assuming a definite form in some plants, as in orchids, or delicate threads, as in *Œnothera*.

COLLETIA. A genus of American Rhamnaceæ inhabiting Chili, Peru, and Mexico. They are much-branched shrubs, scantily furnished with minute leaves, and having spines which stand at right angles with the stem in alternate pairs. The flowers, which are yellowish or white, are either solitary or in tufts in the axils of the leaves, or rise from beneath the base of the spines. Two or three species are known in gardens. [C. A. J.]

COLLINSIA. A genus of dwarf annuals belonging to *Scrophulariaceæ*, all indigenous to North America, chiefly of the north-western regions, and including several showy border plants. Its most important features are a deeply five-cleft calyx; a two-lipped irregular corolla, with the tube bulging at the base on the upper side, the upper lip two-cleft with its lobes erect, the lower lip three-cleft, the middle lobe forming a pouch-like cavity in which the stamens and style are enclosed; and a gibbose two-celled many-seeded capsule. All the species are of branching habit, and furnished with opposite leaves, and flowers in erect whorled racemes. *C. bicolor*, one of the best known, grows twelve to eighteen inches high, and has sessile ovate-lanceolate toothed leaves, either opposite or in threes, and strongly nerved, the flowers, which are in whorls of five or six blossoms, having their upper lip very pale lilac or whitish, and the lower one deep lilac-purple. *C. heterophylla* has rather larger and deeper coloured flowers, with the calyx clothed with coarse hairs, and the lower leaves three-lobed and stalked. In *C. multicolor* the upper lip has a broad white central spot speckled with crimson, and the leaves beneath the whorls are tinged with purple. These characters are, however, somewhat inconstant under cultivation, and it is doubtful whether this plant, as well as *C. heterophylla*, may not be a mere variety of *C. bicolor*. *C. verna* is a very pretty little species scarcely known in this country, though the first discovered; it differs from the preceding in its flowers having longer pedicels, as well as in their colour, which is pure white in the upper lip, and blue, of variable intensity, in the lower one. *C. grandiflora*, a species common in our gardens, is sometimes confounded with *verna*, but has shorter pedicels, and the upper lip of flower is lilac. [W. T.]

COLLINSONIA. A genus of *Labiatæ*, containing a few species of strong-scented perennial herbs, natives of North America. They have large ovate leaves, and yellowish flowers on slender pedicels in loose and panicled terminal racemes. The calyx is ovate and two-lipped, with the upper lip truncate and three-toothed, and the lower two-toothed; it is declined in fruit. The corolla is elongated, expanded at the throat, and somewhat two-lipped, with the four upper lobes nearly equal, but the lower larger, longer and pendent, toothed or lacerate-fringed. There are two, sometimes four, much exserted diverging stamens, with divergent anther cells; the apex of the style is subequally bifid; and the nucule is smooth. [W. C.]

COLLINUS. Growing on low hills.

COLLOMIA. A small genus of phloxworts, having a five-cleft campanulate calyx, a corolla with salver-shaped limb and slender tube, five stamens inserted in the middle of the tube, and a three-celled capsule, each cell containing one or two seeds. It is nearly related to *Gilia*, from which it differs chiefly in habit, colour of flowers, and form of corolla. The species are all dwarf annuals with red or buff-coloured flowers, natives of the Western Hemisphere, and chiefly of California. With one or two exceptions they are quite devoid of interest as ornamental plants, their flowers being small and without effect. *C.*

coccinea or *Caravilleeii* grows nearly a foot high, with a branched hairy stem, alternate linear-lanceolate leaves incised near the extremity, and terminal clusters of brick-red flowers each nearly half an inch across, the tube and underside of corolla being a buff-yellow, and the calyx glandular. *C. grandiflora*, the only other species worth cultivating, is of taller growth and more robust habit, with shining lanceolate leaves coarsely serrated, and buff or nankeen-coloured flowers larger than those of *coccinea*. The species have no known properties, but their seeds are remarkable for the quantity of mucus existing in their testa or outer covering—whence the name of the genus from *kolla*, glue,—which gives rise, under certain conditions, to a singular and interesting phenomenon. When these seeds are thrown into water the mucous matter is dissolved and forms a cloud around them. This cloud, Dr. Lindley tells us, 'depends upon the presence of an infinite multitude of exceedingly delicate and minute spiral vessels lying coiled up, spire within spire, on the outside of the testa, and the instant water is applied, they dart forward at right angles with the testa, each carrying with it a sheath of mucus, in which it for a long time remains enveloped as in a membranous case.' [W. T.]

COLLOPHORA. The name of a little-known Brazilian tree, mentioned by Von Martius, as abounding in a milky juice furnishing caoutchouc. The genus belongs to the *Apocynaceae*. It has a salver-shaped corolla, without scales in the throat of its tube; and the stigma is cylindrical. The fruit is a berry containing several seeds embedded in pulp. [M. T. M.]

COLLUM. The point of junction between the radicle and plumule; the point of departure of the ascending and descending axes, that is to say, of the root and stem, which is often called the collar. Also the lengthened orifice of the ostiolum of a lichen; *Colliform* is sometimes applied to an ostiolum, whose orifice is lengthened into a neck.

COLLYBIA. A sub-genus of white-spored *Agarics* with the outer coat of the stem cartilaginous, the margin of the pileus at first involute, and the gills not decurrent. *Agaricus fusipes*, which is not uncommon at the foot of old oaks, growing in dense tufts of a more or less decided rufous tint, though too tough for stewing, is excellent when pickled. *A. esculentus* also, which, though small, is brought abundantly into the German markets under the name of *Nagelschwamme*, belongs to the same sub-genus. One of the best known species of the group is *A. visitipes*, which grows on almost every decayed tree, conspicuous for its velvety stem and rich yellow shining pileus. Few plants are more patient of cold than this, for the severe Christmas frost of 1860 did not destroy it, specimens after the thaw being as vigorous as ever. [M. J. B.]

COLOCASIA. A genus of *Araceae*, very closely allied to *Caladium*, but differing from it in the spadix having a club-shaped or pointed top destitute of stamens. The middle portion of the spadix is provided with stamens, above and below, which latter are rudimentary organs. The anthers are two-celled, opening by pores, and having a broad wedge-shaped connective. The ovaries, at the base of the spadix, are one-celled, with six erect ovules. The plants are Indian herbs, with tuberous or stem-like rootstocks, and peltate leaves. *C. antiquorum*, the *Arum Colocasia* of Linnæus, is cultivated in most tropical countries, Egypt, India, &c., for the sake of its leaves, which when uncooked are acrid, but by boiling, the water being changed, lose their acridity, and may be eaten as spinach. *C. indica* is cultivated in Brazil for its esculent stems and small pendulous tubers. *C. esculenta*, *C. macrorhiza*, and many varieties of these species, are cultivated in the Sandwich Islands under the name of Taro; and their rootstocks being filled with starch, furnish a staple article of diet among the natives. The leaves are likewise used as a vegetable. [M. T. M.]

COLOCYNTH. *Citrullus* or *Cucumis Colocynthis.* — HIMALAYAN. *Citrullus* or *Cucumis Pseudo-colocynthis.*

COLOMBINE PLUMEUSE. (Fr.) *Thalictrum aquilegifolium.*

COLOQUINTE. (Fr.) *Citrullus Colocynthis.*

COLOMBO or CALOMBA. The Calumba root, *Jateorhiza* (formerly *Cocculus*) *palmata.* —. AMERICAN. *Frasera Walteri.*

COLOUR (adj.) COLOURED, COLORATUS. Any colour except green. In technical botany white is regarded as a colour, and green is not.

COLPENCHYMA. Sinuous cellular tissue

COLPOON TREE. *Cassine Colpoon.*

COLQUHOUNIA. A genus of *Labiatae*, containing three species of climbing or erect shrubs, natives of India, with petiolate ovate acuminate leaves, and scarlet flowers scattered in axillary verticillasters or crowded in a terminal spike, and having small bracts. The calyx is tubular, campanulate, ten-nerved, and unequally five-toothed; the corolla tube is longer than the calyx, its throat dilated, and its limb bilabiate, with the upper lip entire, and the lower with three small ovate lobes; the stamens are covered by the galea or helmet; the apex of the style is subequally bifid, with subulate lobes; the nucule is oblong, dry and smooth, with a membranaceous apex. [W. C.]

COLTSFOOT. The common name for *Tussilago.* —. SWEET. An American name for *Nardosmia.*

COLUBRINA. Snake-wood, so called from the twisted wood of one species, which inhabits the forests of Martinique. A family of plants belonging to the order

Rhamnacea, comprising small trees and shrubs, some of which are climbing, natives of South America and the warmer regions of Asia and Africa. They are closely allied to *Ceanothus*, but possess no properties which render them worthy of cultivation. (C. A. J.)

COLUM. An obsolete term for the placenta.

COLUMBINE. The common name for *Aquilegia*.

COLUMELLA. A little column; the firm centre of the spore-case of an urn-moss, from which the spores separate. The long axis round which the parts of a fruit are united: in reality, the ripened growing point. A slender axis, over which the spore-cases of such ferns as *Trichomanes* are arranged.

COLUMELLIA. A genus of epigynous exogens having a monopetalous corolla, the structure of which, and especially of the anthers, is so remarkable that it has been separated as a distinct order under the name of *Columelliaceæ*. It consists of a few evergreen shrubs or trees, natives of Mexico and Peru, having opposite entire or slightly serrated leaves, and small yellow flowers. The calyx is superior, five-cleft; the corolla five-lobed; the stamens two only, attached to the tube of the corolla, and the anthers are as usual only two-celled, but each cell is elongated, more so than in any other plants comparatively with the size of the anther, but being doubled and redoubled on themselves they form a globular mass. The anther of *C. oblonga* has the shortest cells of any of the species. The ovary is two-celled, each cell containing numerous ovules, and the seed has a large quantity of albumen. The station of *Columelliaceæ* in the natural system is near *Stylidiaceæ*, the stamens in the latter being only two, although differently attached. (See *Ann. and Mag. Nat. Hist*, ser. 2, l. 109.) (B. C.)

COLUMELLIACEÆ. *Columellia*, which consists of two or perhaps three species from the Andes of America, having no immediate affinity with any of the orders with which it has been compared, has therefore been considered as forming a family of itself. It consists of evergreen shrubs or small trees with opposite serrate leaves without stipules; a superior five-cleft calyx; a five-lobed spreading corolla bearing in its short tube two stamens, each with three waved anthers. The ovary is inferior, two-celled, with numerous ovules; the fruit capsular; the seeds numerous, with the embryo in the axis of a fleshy albumen. These characters, as well as the habit, remove the genus from the generality of *Monopetalæ*, and indicate several points of connection with *Saxifrageæ* and their allies, amongst which *Columellia* may possibly take its place as a gamopetalous form.

COLUMN. The combined stamens and styles forming a solid central body, as in orchids.

COLUMNARIS. Having the form of a column, as the stamens of a mallow.

COLUMNEA. A genus of erect or climbing slender herbaceous plants or undershrubs, with opposite fleshy and hairy leaves, and solitary or crowded axillary peduncles bearing scarlet flowers. They are natives of tropical America, and belong to the order of *Gesneraceæ*. The calyx is free and five-parted. The corolla is tubular, with the limb two-lipped, the upper one entire, erect or overarching, the lower tri- and patent; the four didynamous stamens are inserted in the tube of the corolla, and with them the rudiment of a fifth; the ovate anthers have two cells. The one-celled ovary is free, surrounded by a five-lobed disc, and contains two two-lobed parietal placentæ with anatropal ovules. The fruit is a berry containing many obovate seeds, and two fleshy placentæ. The genus is near to *Besleria*, differing chiefly in the form of the corolla. [W. C.]

COLURIA. A genus of the rose family, very nearly allied to *Geum*, but differing in the styles being jointed and falling from the achenes when mature, while in *Geum* they remain attached and become feathery. *C. geoides*, the only species of the genus, is a plant about six inches high with pinnatifid leaves having cut segments, and a peduncle bearing one to three little yellow flowers. Altogether it bears much resemblance to the silver-weed, *Potentilla anserina*, but its leaves although pubescent are not clothed with silvery hairs. It is found on the less elevated mountains in Siberia, growing in rocky places. [A. A. B.]

COLUTEA. The technical name of a genus of *Leguminosæ* consisting of certain shrubs, indigenous to the south of Europe and the Mediterranean region in general, and especially characterised by having, with the ordinary papilionaceous flowers, membranous bladder-like pods. The leaflets of *C. arborescens* and other species have purgative properties like those of senna, and are sometimes mixed with senna leaves. The distended pods, when pressed suddenly, burst with a loud noise; hence the common name, Bladder-senna. Two or three species are cultivated as deciduous plants in this country, but they seem to be more abundantly used on the Continent than with us. *C. arborescens* is said to grow on the crater of Vesuvius, where there is little other vegetation. [M. T. M.]

COLVILLEA. The name given to a tree of Madagascar belonging to the leguminous family. The genus is nearly related to *Cæsalpinia*, but is readily distinguished by the form of its calyx, which is two-lipped, the upper lip convex and four-toothed, and the lower linear in form and entire. *C. racemosa* is a beautiful tree, which attains a height of forty or fifty feet, and is furnished with elegant fern-like twice-pinnate leaves about three feet long; these are made up of from twenty to thirty pairs of pinnæ, each pinna with a like number of opposite leaflets, which are nearly linear in form, and about half an inch long. The beautiful scarlet flowers are in dense racemes, which arise from the axils of the upper leaves, and are either simple or branched, and about a foot and a half in length. The calyx, like the petals, is of a scarlet colour; the petals are five in number, the standard the smallest and nearly bidden by the others, the two oblong wings next in size, and the two free petals, which form the keel, the largest; the ten free stamens are of unequal length. The pod is straight, about six inches long, containing a number of seeds. The genus bears the name of Sir Charles Colville, once governor of the Mauritius. [A. A. B.]

COLZA. (Fr.) *Brassica Napus oleifera.*

COMA (adj. COMOSE). The hairs at the end of some seeds; the empty leaves or bracts at the end of the spike of such flowers as those of the pine-apple.

COMANDRA. The generic name of plants belonging to the sandalwort order; having the calyx adherent to the seed-vessel, its upper part with an adherent disk whose border is five-lobed, on which the stamens are inserted between its lobes and opposite those of the calyx, the anthers being connected with the calyx by a tuft of hair-like threads. The fruit is nut-like and filled with the globular seed. The name is derived from the Greek words signifying 'hair' and 'stamen,' indicating a character above mentioned. The plants are low perennials with herbaceous stems springing from a woody base; the leaves alternate, stalkless, oblong; flowers greenish-white in small clusters. One species, *C. umbellata*, is common in North America, and attaches itself as a parasite to the roots of trees. [G. D.]

COMAROSTAPHYLIS. A genus of Ericaceæ, containing fourteen species of small trees or shrubs, with the habit of *Arbutus*, natives of Mexico and Guatemala. They have coriaceous oblong evergreen leaves, and flowers in terminal bracteate racemes or panicles. The hypogynous calyx is five-parted; the corolla is inserted on the calyx, is campanulate, enlarged below, and has the limb five-lobed. The ten stamens are inserted on the base of the calyx; the filaments are short, and the anthers oval and compressed. The ovary is placed on a ten-angled hypogynous disc, and has five cells, with a single ovule in each. The style is simple, and the stigma obscurely five-toothed. The drupe is globose and fleshy, containing a single stone (pyrena), which has five rarely more cells, with a single seed in each. [W. C.]

COMARUM. A herbaceous marsh plant with a stout creeping stem, rather large and handsome leaves composed of seven, five, or three deeply-serrated leaflets, which are slightly-branched stalked and dingy-purple flowers. The fruit somewhat resembles that of the strawberry, but is spongy instead of juicy, and does not fall off when ripe. It is of common occurrence in marshes and boggy meadows in most parts of England, and extends over a great part of Europe and North America. The roots and stems have been used to dye wool of a dirty-red colour, and are sufficiently astringent to be employed in tanning. In some parts of Scotland the fruits are called Cow-berries, on account, it is said, of their being used to rub the inside of milk-pails for the purpose of thickening milk. The Marsh Cinquefoil, *C. palustre*, is rarely cultivated, though Gerarde says with some pride that he brought some plants from Bourne Pinne, half a mile from Colchester, for his garden, and that they there flourished and prospered well. French Comaret; German Fünffblatt. [C. A. J.]

COMBESIA. Abyssinian herbs belonging to the *Crassulaceæ*, and having five-parted flowers with petals united for a short distance at their base; five stamens with anthers opening inwardly; five small scales at the base of the five sessile two-seeded ovaries, which ripen into follicles, bursting by a long slit towards the centre of the flower. [M. T. M.]

COMBINATE-VENOSE. When the lateral veins of a leaf unite before they reach the margin.

COMBRETACEÆ (*Myrobalans*). A natural order of polypetalous calycifloral dicotyledons, belonging to Lindley's myrtal alliance. Trees or shrubs, with alternate or opposite entire leaves having no stipules. Sometimes the flowers are imperfect, some having stamens only, others pistils only, and occasionally the petals are wanting. Calyx adherent, its limb four to five-lobed, falling off. Petals arising from the orifice of the calyx, alternate with the lobes. Stamens often eight or ten. Ovary one-celled, with two to four suspended ovules. Fruit succulent or dry, one-celled, and one-seeded. Seeds without albumen; cotyledons of the embryo rolled up. Natives of the tropical parts of Asia, Africa, and America. The plants of the order have astringent qualities; some are cultivated for ornament, others yield timber. The astringent fruit, known by the name of Myrobalan, is produced by *Terminalia Bellerica* and *T. Chebula*. The bark of *Bucida Buceras* is used for tanning. There are twenty-three known genera, and upwards of 200 species. Illustrative genera: *Terminalia, Combretum, Gynocarpus.* [J. H. B.]

COMBRETUM. The typical genus of *Combretaceæ*, inhabiting tropical regions of both hemispheres, and consisting of trees or shrubs, often trailing or climbing by the indurated leaf-stalks, which are persistent and act as hooks to support the plant. The leaves are opposite, rarely alternate, entire, exstipulate. The flowers, which grow in spikes, axillary or sometimes terminal, solitary or arranged in a panicle, are polygamous; calyx-tube adhering to the ovary, above which it is constricted, the limb bell-shaped; petals four; stamens eight. The fruit is leathery, with four membranous wings, indehiscent and one-seeded by the abortion of several of the ovules. Many of the species are very handsome. (J. T. S.)

COMB-SHAPED. The same as Pectinate.

COMESPERMA. A genus of *Polygalaceæ*, consisting of erect or twining plants found in Australia and Tasmania, and numbering about twenty species. It is nearly related to the South American genus *Cotocoma*, but differs in the corolla being composed of three united petals, instead of five. The stems, which are not much thicker than a crow-quill, are furnished with alternate leaves, mostly linear in form. The flowers, disposed in axillary or terminal racemes, either yellow, white, blue or purple, the three latter colours sometimes found in the same species, as they are in our own common milkwort, *Polygala vulgaris*, to whose flowers those of the plants of this genus bear much resemblance, but are generally larger. The calyx is five-lobed; the corolla three-lobed, the middle lobe largest; the stamens eight; the ovary two-celled, crowned with a curved style. The fruit is a wedge-shaped capsule with two seeds, each furnished with a tuft of silky hairs. This latter circumstance gives rise to the name of the genus.

C. volubilis, the Blue-creeper of Tasmania, is a graceful little plant, twining among other bushes and covering them with its great abundance of beautiful blue flowers. Its thin twining stems are furnished with leaves which are between linear and lance-shaped in form. This plant grows in various parts of Australia, as well as in Tasmania, and is universally admired. It has been in cultivation in England under the name *Comesperma gracilis*. (A. A. B.)

COMFREY. *Symphytum officinale*.

COMIN. (Fr.) *Ervum Ervilia*.

COMMELYNACEÆ. (*Spiderworts*.) A natural order of hypogynous monocotyledons, belonging to Lindley's xyridal alliance. Herbs with flat leaves, usually sheathing at the base. Outer perianth (calyx) of three parts, herbaceous; inner (corolla) also of three, coloured; stamens six or three, the anthers opening on the side next the pistil. Ovary three-celled with a central placenta; style one. Fruit a two to three-celled capsule, opening by two or three valves, which bear the dissepiments (partitions) on the middle; seeds with a linear hilum; embryo pulley-shaped. Natives of the East and West Indies, New Holland, and Africa. A few are found in North America, but none in Northern Asia or Europe. The underground stems of many of the plants yield starch, and are used for food. The filaments of the *Tradescantias* have jointed hairs, in which a circular movement is seen under the microscope. There are sixteen known genera, and a species. The best known are *Commelyna*, *Tradescantia*, and *Cyanotis*. (J. B. B.)

COMMELYNA. The typical genus of the order of spiderworts, distinguished by having usually three petals, dropping one of the three different in form from the others, or wanting; the filaments or stalks of the anthers smooth and naked; the style or appendage on the seed thread-like, and entire at the end. The species are herbs, natives of tropical and Northern America, East India, and New Holland, having ovate or lance-shaped leaves, and the flowers in groups, either issuing from an involucre or sheath-like bracty, or destitute of such covering, the former constituting *Commelygna* proper, the latter *Aneilema*. The genus was named in honour of J. and G. Commelyn, well-known Dutch botanists. The species are numerous, and several have been long known in our collections. They require various modes of treatment, some being hardy, others requiring a high temperature. *C. cœlestis*, notable for the delicate blue of the flower, has oblong lanceolate leaves, and the sheaths ciliated; it is a half-hardy species, which under proper treatment displays a succession of azure flowers from July to September. One of more recent introduction is *C. scabra*, a half hardy perennial from Northern Mexico, having straggling reddish stems, the leaves lance-shaped, waved and hard at the margins, and the flowers of a dull purple brown. Dr. Lindley, in his *Vegetable Kingdom*, states that 'the fleshy rhizomes of *C. cœlestis*, *tuberosa*, *angustifolia*, and *striata*, contain a good deal of starch mixed with mucilage, and are therefore fit for food when cooked. The Chinese employ those of *C. medica* as a remedy in cough,' &c. (G.D.)

COMMIA. The name given to a plant of Cochin-China, which forms a genus of *Euphorbiaceæ*. This plant (not well known to botanists) has been described by Loureiro as a tree from which a white tenacious gum exudes, said to be of a purgative and emetic nature, and valuable in dropsy, but requiring careful administration. The leaves are stalked, lance-shaped, entire and smooth. The inconspicuous flowers are male and female on different plants; the males in short axillary bracted spikes, having neither calyx nor corolla, and the stamens united into a column which bears on its summit a number of anthers; the females in terminal racemes having a three-leaved calyx enclosing a three-lobed ovary, crowned by three short recurved styles. The fruit a three-celled capsule with three seeds. (A. A. B.)

COMMISSURE. The face by which two carpels come together or cohere, as in umbellifers.

COMMON PETIOLE. The first and principal leaf-stalk in compound leaves; the secondary petioles are called partial.

COMOLIA. A genus of *Melastomaceæ*, consisting of Brazilian trees or shrubs, with four-sided branches and obovate three-nerved leaves clothed with adpressed hairs; flowers axillary, solitary, sessile, white; tube of calyx adhering to the base of ovary, its limb four-lobed; petals four; anthers one-celled curved; ovary glabrous; capsule two-celled. [J. T. S.]

COMPAGNON BLANC. (Fr.) *Lychnis dioica*.

COMPARETTIA. A small genus of epiphytal orchids inhabiting tropical America. Four species are known, all with small pseudobulbs, coriaceous leaves, and gracefully bending racemes of long spurred rose coloured purple or scarlet flowers.

COMPASS PLANT. *Silphium laciniatum*, which is said to present the faces of its leaves uniformly north and south.

COMPLEXUS. Tissue: *C. cellulosus*, cellular tissue; *C. membranaceus*, the thin membrane, which is the foundation of all tissue—elementary membrane; *C. tubularis* tubular tissue, or woody fibre; *C. utricularis*, angular cellular tissue; *C. vascularis*, spiral vessels, properly so-called; often, however, extended to all sorts of tubes with markings on the side, thus losing precision, and with it its value as a scientific term.

COMPLICATE. Folded up upon itself.

COMPOSITÆ. The more familiar name of the *Asteraceæ*, a large natural order of gamopetalous calyciforal dicotyledons belonging to Lindley's campanal alliance, consisting of herbs and shrubs with alternate or opposite leaves having no stipules; the stamens and pistils either in the same or in separate flowers, which are collected into a head on a common receptacle (hence the name Compositæ or compound flowers), and surrounded by a set of floral leaves or bracts, called an involucre. The fruit is single-seeded, crowned with the limb of the calyx. The plants are found in all parts of the world, in warm countries sometimes assuming arborescent forms. They were included by Linnæus in his class *Syngenesia*. The properties of the order are various; but bitterness seems to prevail in it, and this is accompanied with tonic, stimulant, aromatic, and sometimes even narcotic qualities. *Lactuca sativa*, the common lettuce, and *L. virosa*, supply lactucarium, a substance used like opium. [J. H. B.]

COMPOSITION. The arrangement of organs, or their order of development, or their manner of branching, &c.

COMPOUND, COMPOSITE. When formed of several parts united in one common whole; as pinnated leaves, and all kinds of inflorescence beyond that of the solitary flower.

COMPRESSED. Flattened lengthwise; as the pod of a pea.

COMPTONIA. A deciduous bushy shrub belonging to the order *Myricaceæ*, bearing both male and female flowers in catkins, and on the same plant. A native of North America in moist peaty soils. The leaves are long and narrow, alternately arranged and cut on each side into rounded and numerous lobes, so as to resemble the fronds of *Osmunda* (hence the name *C. asplenifolia*), downy and sprinkled with golden resinous transparent dots, which, as well as the rest of the plant, have an aromatic scent. It was introduced in 1714 by the Duchess of Beaufort, and was named in honour of Henry Compton, Bishop of London, the introducer and cultivator of many curious exotic plants, and a great patron of botany and horticulture. It is hardy, but requires a peat soil and shade. In America it is called the Sweet Fern Bush; in France *Comptonie*, or *Liquidambar à feuilles de Osmunda*; but it must not be confounded with *Liquidambar Styraciflua*, Sweet Gum. [C. A. J.]

CONANTHERA. A genus of *Liliaceæ* containing a few small Chilian bulbous plants, with linear leaves, and a scape supporting paniculate blue flowers. Perianth six-parted, adhering to the base of the ovary, and breaking away by a transverse split as the fruit ripens; stamens six, united into a cone. [J. T. S.]

CONCEPTACLE. A term sometimes applied to the capsular fruit of red-spored *Algæ*, in contradistinction to the fruit in which the reproductive mass is ultimately divided into four bodies, and hence called tetrasperms. Modifications have received the names of ceramidia, cystocarps, favillæ, nuclei, &c. The explanation of these terms belongs rather to a treatise on Algæ than to the present work. [M. J. B.]
Also, a special organ, developed in some fungals on the surface, or in the interior of a receptacle, and containing the organs of reproduction as well as their accessories; it differs from a spore-case in the latter being itself one of the accessories, and only containing spores.

CONCHIDIUM. *Eria*.

CONCHIFORM. Shaped like one valve of a common bivalve shell.

CONCHOCHILUS. *Appendicula*.

CONCOLOR. Of the same colour as some other thing compared with it.

CONCOMBRE D'ÂNE, or SAUVAGE. (Fr.) *Ecbalium agreste*, the *Momordica Elaterium* of some.

CONDAMINEA. The name of a genus of *Cinchonaceæ*, consisting of Peruvian shrubs, some of which have similar tonic properties to those contained in the true *Cinchona*, while others are used for dyeing purposes. The genus is known by the cup-shaped tube of the calyx, whose limb is five-toothed, and separates from the tube by a circular line. The corolla is tubular, concealing the stamens, which are attached near to the middle of its tube

The fruit is a top-shaped truncated capsule, opening by two valves, and containing several wingless seeds. [M. T. M.]

CONDUPLICANT. Doubling up; as when the leaflets of a compound leaf rise up and apply themselves to each other's faces.

CONDUPLICATE, CONDUPLICATION. A term of æstivation; when the sides of an organ are applied to each other by their faces.

CONDYLIUM. The antherid of a *Chara*.

CONE. The strobilus or conical arrangement of scales in the fruit of a *Pinus* or fir-tree.

CONE-FLOWER. An American name for *Rudbeckia*. —, PURPLE. An American name for *Echinacea*.

CONE-HEAD. A garden name for *Strobilanthes*.

CONENCHYMA. The conical cells which constitute hairs.

CONFERRUMINATE. Glued together.

CONFERTUS. When parts are pressed closely round about each other; packed close.

CONFERVA. The typical genus of *Confervaceæ*, the species of which are either attached to various bodies or float in dense masses on ponds swollen up with bubbles of gas, from whence the genus takes its name. The branched *Confervæ* are now separated under the name of *Cladophora*. *Confervæ*, when dried, were once used as a packing instead of tow, to support fractured limbs, a use which is now quite obsolete. The name was also applied to a vast heterogeneous mass of plants, as may be seen in *Dillwyn's History of British Confervæ*. [M. J. B.]

CONFERVACEÆ. A division of the green-spored *Algæ* characterised by their simple or branched articulated threads, diffused endochrome, and small zoospores. The articulations are mostly as long as, or longer than, their diameter, which forms one of the main technical distinctions between them and the *Oscillatoriæ*, which have, however, a very distinct habit. The genera are numerous, and in some cases, as in *Chætophora*, the threads are compacted into a solid mass by means of gelatine. The zoospores sometimes occur in the ordinary cells of the threads, but sometimes in distinct cysts, and sometimes privileged cells are multiplied by cell-division for their production, as in *Stygeoclonium*. They are found in all parts of the world, but are most plentiful in temperate regions. They are sometimes so abundant that, after floods, they form a thick coat, like paper, on the ground, to which the name of meteoric paper has been given. *Chroolepus* differs from the rest in its being developed in the air and not in water, and in its golden colour when fresh, but the species when dry become green. The fruit, however, like that of *Callithamnion*, externally contains zoospores. [M. J. B.]

CONFLUENT. The fastening together of homogeneous parts. Gradually uniting organically.

CONFORM. Of the same form as some other thing.

CONGELATION. In countries where frost is severe, most forest trees exhibit marks of serious injury, either in formidable fissures caused by differences of temperature in the different parts of the trunk, or in the death of portions of the bark and wood. Trees thus become accurate registers of severe winters. While, however, some plants give way at once under a slight degree of frost, others may be turned into a solid mass of ice without losing their vital powers, especially if the mass is thawed gradually, and in the dark. The outward parts of plants sometimes escape, when more delicate and protected parts are destroyed. Pear blossom, for instance, may be apparently unaffected by frost and expand as usual, when the pistils are completely destroyed. The effect of frost on plants depends greatly upon the condition of soil. The wetter the soil, and the more saturated the plants with moisture, the more destructive is it. A degree of cold, which is quite harmless when the cells are comparatively empty, is positively destructive under other circumstances. [M. J. B.]

CONGESTED. Crowded very closely.

CONGLOBATE. Collected into a ball, as the florets of *Echinops*.

CONGLOMERATE. The same as Clustered.

CONGLUTINATE. Glued together, not organically united.

CONIDIA. Many *Fungi*, besides their true fruit, produce little reproductive cells in different parts, especially on the spawn or mycelium which are known by the name of conidia. The substance called ergot is a good example, the conidia appearing some months before the perfect fungus. When these conidia are contained in distinct cysts or perithecia, they are called stylospores. In some cases undoubtedly, as in *Erysiphe*, the conidia are reproductive, but in others it is possible that they may perform the functions of male organs. The subject of impregnation, in *Fungi*, is so imperfectly known that it is not possible to speak with certainty about it. [M. J. B.]

CONIDIUM. The gonidium of a lichen.

CONIFERÆ. (*Conaceæ*, *Pinaceæ*, *Conifers*, the *Pine* family.) A considerable and important family, constituting with the smaller groups of *Cycadeæ* and *Gnetaceæ* the sub-class gymnosperms of dicotyledons. It consists of trees or shrubs, mostly with resinous secretions. The leaves are stiff, sometimes linear or needle-like, sometimes short and scale-like, or more rarely broad, lobed, or divided. The flowers are unisexual, either in cylindrical or short catkins with closely packed scales, or the females are solitary. There is no perianth. The stamens in the males are either inserted on the axis of the catkin under the scales, or the anther-cells are sessile on the inside of the scales themselves, which then form part of the stamens. The ovules and seeds are naked, that is, without ovary, style, or pericarp, although sometimes more or less enclosed in two bracts, or in a fleshy or hardened disk. The seeds are albuminous, with one or sometimes several embryos in the centre, each embryo having sometimes more than two cotyledons. There are probably nearly two hundred species known, dispersed over a great part of the globe, several of them forming large forests in temperate climates, or more rarely within the tropics; while some of them extend almost to the utmost limits of woody vegetation in high latitudes, or at great elevations. They are distributed into about twenty-five genera, forming three tribes or sub-orders: 1. *Abietineæ*, with the fruits collected in cones, and inverted ovules; of this the principal genera are *Pinus* (including *Abies*), *Araucaria*, *Cunninghamia*, *Sequoia*, &c. 2. *Cupressineæ*, with the fruits collected in cones, and erect ovules; including *Juniperus*, *Callitris*, *Thuja*, *Cupressus*, *Taxodium*, *Cryptomeria*, &c. 3. *Taxineæ*, sometimes considered as a distinct family, with the fruits solitary or loosely spiked, including *Podocarpus*, *Dacrydium*, *Phyllocladus*, *Salisburia*, *Taxus*, &c.

The woody tissue of the trees of this family is seen to be marked with peculiar circular dots or punctations when examined under the microscope. The ovules at the base of each cone-scale are generally held to be naked, each ovule having a large opening at its apex, to which the pollen from the stamens is applied directly. But some say that the ovules are not naked, but are contained in a proper ovary which is closely applied to the seed; that the outer membranous scales are modified leaves; and that the bard scales are altered branches bearing the pistillate flowers.

Some botanists look upon Conifers as the highest type of true dicotyledons. They are most abundant in temperate regions, both in the northern and southern hemispheres. In Europe, Siberia, and China, and in the temperate parts of North America, we meet with species of pine, spruce, larch, cedar, and juniper. In the southern hemisphere they are replaced by *Araucaria*, *Eutassa*, *Dammara*, *Podocarpus*, and *Dacrydium*.

Conifers are of great importance to mankind. They supply valuable timber, and yield resin, oil, pitch, and turpentine. Some attain a great size. Thus *Wellingtonia gigantea* has been known to attain in Oregon, a height of 450 feet, with a circumference of 110 feet at the base. *Taxodium sempervirens* also attains an enormous size. The various species of *Pinus* have their leaves in clusters of two, three, four, five, or six, surrounded by a membranous sheath at the base. *Pinus sylvestris* is the common Scotch fir, which abounds in cold climates, and which supplies timber, turpentine, and pitch, as well as a hemp-like fibre from its leaves, which is used for stuffing pillows and cushions under the name of pine wool. *Pinus Pinaster*, the Bordeaux pine, thrives well on the sea-shore. *Abies* includes different species of fir and spruce, in all of which the leaves come off from the stem and branches singly. *Abies excelsa* is the Norway spruce; *A. balsamea* the balm of Gilead fir; *A. canadensis* the hemlock spruce; and *A. pectinata* the silver fir. *Cedrus* comprises those cedars which have clustered persistent leaves. *Cedrus Libani* is the cedar of Lebanon, the tree of the Bible; while *Cedrus Deodara*, a local variety, is the sacred cedar of India. *Larix* includes the species of larch, which have clustered deciduous leaves. *Larix europea* is the common larch; *L. Griffithii* the Himalayan larch. The *Araucarias* have single-seeded scales, with adherent seeds and many-celled anthers; *Araucaria imbricata* is a Chilian species; *A. Bidwillii* is from Moreton Bay; both have edible seeds. *Eutassa excelsa* is the Norfolk Island pine, which yields valuable wood. *Cryptomeria japonica* is the Japan cedar. *Cupressus sempervirens* is the common cypress. The Junipers have a peculiar succulent fruit. *Juniperus Bermudiana* furnishes the cedar for pencils. The species of *Thuja* are known by the name of Arbor Vitæ. [J. H. B.]

CONIMA. The fragrant gum resin of the incense tree, *Icica heptaphylla*.

CONIOCYSTR. Closed spore-cases resembling tubercles, and containing a mass of spores.

CONIOGRAMMA. A name given by Fée to the species of *Gymnogramma* represented by *G. javanica*. [T. M.]

CONIOMYCETES. A family of Fungi distinguished by the predominance of the spores over the receptacle. The spores are simple or articulated, solitary or chained together, and sometimes fasciculate, naked or enclosed in a distinct cyst. The plants, however, in which this last structure obtains are probably, for the most part, mere conditions of *Sphæriæ* and other pyrenomycetous Fungi. The most important members of the family, are the numerous parasitic species which affect the living organs of plants, and cause great mischief, especially amongst our corn crops, by exhausting the energies of the mother plant, and preventing the full development of the seed. One or two of these, which grow on different species of Juniper, approach *Tremella* in outward appearance, as all do in some peculiarities of structure. In most of these the spores exhibit bright colours, while in other divisions they are as generally black. The dark soot-like patches which are so common on old rails and dead wood are formed mostly by these dingy *Coniomycetes*, which notwithstanding their unpromising appearance, are often full of interest when closely examined. [M. J. B.]

CONIOSELINUM. A genus of four species belonging to the *Umbelliferæ*, and found in mountain districts in Central Europe, Siberia and North-west America. *C. Fischeri*, the best known species, and one which has long been cultivated in botanic gardens, is a biennial herb one to three feet high, with much the appearance of the hemlock, and having its small white flowers disposed in many-rayed umbels, without common, but with partial involucres of five to seven narrow linear leaves, which are equal in length to the flower-stalks.

The principal distinguishing features of the genus are found in the fruit, which is elliptical, dorsally compressed and about one-third of an inch long; each of the carpels has five winged ribs, the marginal ones twice the breadth of the others; and in the lateral furrows, there are three oil tubes (vittæ), seen in the form of dots when the fruit is cut across, usually two in the dorsal furrows, and four to eight on the inner face. [A. A. B.]

CONIOTHECÆ. The cells of an anther.

CONIUM. The genus to which belongs the well-known Hemlock. The botanical name has been given under the supposition that this is the plant mentioned by the Greeks, under the same name, and which was administered, as a judicial means of execution, to Socrates and Phocion. The distinguishing characters reside in the fruit, which is somewhat globular in shape, and each half is marked with five wavy ridges. There are no vittæ or channels for oil, and the albumen is deeply furrowed on its inner surface. Such are the botanical characteristics, but the poisonous nature of *C. maculatum*, the common Hemlock, and its frequent growth in hedges and by roadsides in this country, demand a more full description.

The Hemlock is an erect branching biennial plant, with tap-shaped root, a smooth shining hollow stem, two to five feet in height, frequently marked with purple spots, though these vary very much in number and intensity of colour. The leaves are much divided, with numerous small egg-shaped or lance-shaped deeply cut segments; the upper leaves are smaller. When bruised they emit a peculiarly nauseous odour, not at all aromatic, as is usually the case in our native umbelliferous plants. The inflorescence is a compound umbel, with ten or more rays, surrounded by a general involucre of three to seven leaflets; the partial umbels or umbellules, have at their base a small involucre of three bracts, which are all turned to one side, and do not surround the umbel, as in the case of the general involucre. The flowers are white or greenish white, and the fruits have the important characteristics before mentioned. In attempting the discrimination of this plant, all the above points must be attended to, as there are many plants possessing some of the characteristics of the true Hemlock, and which are in consequence frequently mistaken for it.

The active principle of Hemlock is a peculiar oily-looking fluid, lighter than water, and called *conia*. It exists in all parts of the plant, but especially in the fruits. It acts first as an irritant poison, but speedily causes paralysis of all the muscles, convulsions and death. The plant is of course much less dangerous than its extract, but in poisonous doses it produces similar symptoms, and sometimes coma, and other effects like those produced by opium. Medicinally *Conium* has been used for promoting the absorption of tumors, and glandular swellings, and as an antispasmodic and anodyne. [M. T. M.]

Conium maculatum.

CONJUGATÆ. A tribe of green-spored *Algæ* distinguished from *Confervaceæ* by their endochrome being spiral, stellate, or otherwise disposed, and not equally diffused, or simply denser in the centre; and by the large zoospores formed by the union of the endochromes of two contiguous cells, or one divided into two for the purposes of fructification in the same or in two different plants. In a few, impregnation is effected, in the manner described under *Bulbochæte*, by means of free antheridia, which ultimately fix themselves near the spore-bearing cell. In some instances the bodies perfected by impregnation undergo cell-division, and the component parts become so many zoospores. The species are either attached or float freely in the water. Almost all are fresh-water plants, and are found in various parts of the globe, but especially in temperate regions.

The term *Conjugatæ* does not strictly apply to all. In *Œdogonium* there is no conjugation, but fructification takes place by the division of a cell, one of the two divisions only proving fertile. In this genus, as also in some others, the spores are often of a brilliant scarlet or vermilion. The same spore, however, may be, in different stages of growth, either green or red, a phenomenon not very uncommon amongst the green-spored *Algæ*. Conjugation takes place also in *Desmidiaceæ* and *Diatomaceæ*, and also amongst moulds, as in *Syzygites*, so common on decaying toadstools. [M. J. B.]

CONJUGATE. Paired; when the petiole of a leaf bears one pair only of leaflets.

CONJUGATO-PALMATE. When a leaf divides into two arms, each of which is palmate.

CONJUGATO-PINNATE. When a leaf divides into two arms, each of which is pinnate.

CONJUNCTORIUM. The operculum of the spore-case of an urn-moss.

CONJURER OF CHALGRAVE'S FERN. A name assigned by Relhan in his Flora of Cambridgeshire to *Puccinia anemones*; but whether a popular name or not we are unable to say. It is derived from the external resemblance of its little heaps of protospores to the fructification of Ferns. *P. anemones* is 'filix lobata, globulis pulverulentis undique aspersa' of Ray's Synopsis, where it is figured, and named after a specimen in Bobart's Herbarium marked by his own hand—'This capillary was gathered by the Conjurer of Chalgrave.' The elder Bobart, it may be observed, died in 1680, and his son, who succeeded him, in 1719. [M. J. B.]

CONNARACEÆ. (*Connarads.*) A family of calycifloral dicotyledons, closely allied on the one hand to *Xanthoxylen*, and on the other to *Leguminosæ*, differing from the former chiefly in the more completely apocarpous ovary, and from the latter in the perfectly regular flowers, and in the seed in which the radicle is always at a distance from the hilum. They are trees or shrubs, sometimes climbing, with alternate usually pinnate leaves; the stipules either small and deciduous, or wanting; the flowers small, in terminal or axillary racemes or panicles. There are five sepals and petals, ten stamens, and one to five carpels, with two ovules in each, and distinct terminal styles. There are about forty species, natives of the tropics both of the New and the Old World. They are distributed in six or seven genera, of which the principal ones are *Rourea*, *Connarus*, and *Cnestis*. The aril in some species of *Omphalobium* is entire. Zebra-wood is obtained from *Omphalobium Lamberti*.

CONNARUS. A genus of shrubs or trees, forming the type of the order *Connaraceæ*. The leaves are compound, without stipules; the flowers regular, with ten stamens united by their filaments at the base, the five which are opposite to the petals shorter than the rest. Of the five ovaries, four are generally abortive, and reduced to the condition of styles, while the fifth contains two ascending ovules; the stigma is dilated. The fruit is a kind of pod, but it does not open, and contains only one seed, the other being suppressed. The trees are natives of India, and tropical South America. [M. T. M.]

CONNATE. When the bases of two opposite leaves are united together. Also when any parts, originally distinct, become united in after-growth.

CONNECTIVAL. Of or belonging to the connective.

CONNECTIVE. The part which intervenes between the two lobes of an anther and holds them together; it is subject to great diversity of form. It appears to be analogous to the midrib of a leaf, and is only absent when an anther is strictly one-celled; that is to say, when the whole of the interior of the end of the stamen is converted into pollen.

CONNEMON. The fruit of *Cucumis Conomon*, cultivated everywhere in Japan.

CONNIVENT. Having a gradually inward direction, as many petals; converging.

CONOCARP. A fruit consisting of a collection of carpels arranged upon a conical centre, as the strawberry.

CONOCARPUS. A genus of *Combretaceæ*, consisting of trees and shrubs from tropical America (one species extending northwards as far as Florida) and Western Africa, with alternate leathery entire leaves, and densely aggregated stalked heads of flowers on globular or oblong receptacles. Calyx about the length of the ovary to which it adheres; petals none; stamens five to ten exserted; ovary compressed, two-ovuled. The fruit is leathery, scale-like, forming imbricated cone-like heads. The Indian species, which were formerly placed in this genus, are now, separated under the name of *Anogeissus*, having the calyx tube prolonged upwards far above the ovary. They produce very valuable timber, nearly as durable as teak, if kept dry. [J. T. S.]

CONOCLINIUM. The name given to a genus of the composite family, composed of a number of tropical American weeds, rarely exceeding three feet in height. It is characterised by a setose pappus, conical naked receptacle, and bell-shaped involucre, made up of two or three series of linear scales. The species, of which about ten are enumerated, bear much resemblance to each other. *C. cœlestinum* is a common plant in thickets and waste places in the Southern and Western United States. It is a smooth or slightly hairy herb with opposite stalked leaves, which are oval in form, with notched margins. The flower-heads, in terminal corymbs, are very numerous, and about the size of a pea; the florets, of a bright blue or purple colour, are all tubular, and have a fragrant odour. The genus differs from *Eupatorium* only in the conical receptacle. [A. A. B.]

CONOIDAL. Resembling a conical figure, but not truly one, as the calyx of *Silene conoidea*.

CONOMORPHA. A genus of small evergreen trees of the *Myrsine* family, found in the tropical parts of South America. The species have alternate stalked entire leaves of a leathery texture and full of dots, oblong or elliptical in form, and varying from three to seven inches in length. The small white or green flowers are borne on short stiff racemes, and have a calyx of four divisions; a funnel-shaped corolla with a four-parted border, enclosing four stamens; and a one-celled ovary, which is crowned with a short style. The fruit is a berry about the size of a pea, and contains few seeds. [A. A. B.]

CONOPHOLIS. A genus of *Orobanchaceæ*, containing a single species, a native of South America. It is a singular plant growing in clusters among fallen leaves, in oak woods. The stem is crowded with scales, which are at first fleshy, then dry and hard. The upper scales form bracts to the flowers, the lower are closely and regularly imbricated. The flowers are in a thick scaly spike, and have an unequally four to five cleft calyx, a bilabiate slightly curved corolla swollen at the base, protruded stamens, and a depressed stigma. The fruit is an ovoid pod, with four placentæ approximated in the middle of each valve. The genus is nearly related to *Orobanche*, differing chiefly in having a bibracteolate calyx, and exserted stamens. [W. C.]

CONOPSIDIUM. *Platanthera*.

CONOSPERMUM. A genus of *Proteaceæ*, containing about forty species. It is distinguished by having a tubular four-cleft calyx, one of the segments of which is occasionally much larger than the others. There are four stamens on short filaments (one of which is sterile), inserted at the base of the calyx segments; the three anthers cohere together; style filiform, with a free oblique stigma. The fruit is a nut, containing a single silky seed. The inflorescence is either in spikes or panicles. The habit of the different species varies considerably; some are tall erect shrubs, while others are of much humbler growth. The foliage is very variable: in *C. imbricatum* the leaves are oval, scarcely a quarter of an inch in length; in *C. filiforme* and *C. ericifolium* they are narrow and sharp-pointed; in *C. cæruleum* they are spathulate, on very long footstalks; in *C. longifolium* and *C. flexuosum* they are nearly a foot in length, and not more than a quarter of an inch in width; while in *C. teretifolium* and *C. tenuifolium* they are filiform, and a foot in length. A few species, as *C. ephedroides*, *C. polycephalum*, are nearly leafless. This genus is confined to the extra-tropical portions of Australia: one species (*C. taxifolium*) is likewise found in Tasmania. [R. H.]

CONOSTEPHIUM. A genus of *Epacridaceæ*, containing a single New Holland species, a branched erect shrub, with scattered leaves, and solitary recurved axillary flowers. The calyx is five-parted, and is surrounded with four or more bracts; the corolla is five-toothed; the oblong anthers are included; the ovary is five-celled, with a single pendulous ovule in each cell. The hard indehiscent drupe is one-celled from the abortion of the other cells. [W. C.]

CONOSTYLIS. A genus of New Holland *Hæmodoraceæ*, consisting of perennial herbs with distichous cuniform radical leaves, partially sheathing and equitant at the base, and corymbose or subspicate heads of flowers at the top of a simple scape. The perianth is lanately woolly outside, its tube connate with the ovary, and the limb regular, persistent, and half expanded in a bell-shaped form; it has six stamens with short erect filaments, and a conically dilated hollow persistent tripartible style. There are about half-a-dozen described species. [T. M.]

CONOTHAMNUS. A myrtaceous shrub, native of the Swan River Colony, having linear lance-shaped leaves, and flowers in heads, surrounded by ovate hairy bracts. The calyx is hairy and four-toothed at the margin; the stamens are numerous, united into five parcels, opposite to the petals; ovary three-celled; fruit a capsule included within the tube of the calyx united at the base with the branch, and containing one seed in each of its three compartments. [M. T. M.]

CONRADIA. A genus of *Gesneraceæ*, containing several species of shrubs or herbaceous plants, natives of the West Indies, and reaching into the southern districts of North America. They are shrubs, or rarely herbs, with petiolate generally dentate leaves, and axillary peduncles with a single flower or sometimes with many-flowered cymes. The calyx tube is adherent to the ovary, the limb five-cleft, or more or less deeply five-parted; the corolla is tubular or campanulate, and its limb nearly equally five-cleft. There are four didynamous stamens, with the rudiment of a fifth, but neither hypogynous disc nor gland. The capsule is two-valved, and has two-parietal placentæ with numerous minute seeds. This genus can readily be separated from its allies, by the absence of disc or glands around the ovary. [W. C.]

CONSOLEA. A name proposed for a genus of *Cactaceæ* in honour of M. Michel-Angelo Console, assistant-director of the Botanic Garden at Palermo, by whom the peculiar feature which serves to distinguish it from *Opuntia* was first observed in 1860-61. This peculiarity consists in the presence of a cupuliform disk at the summit of the ovary, within which the substipitate base of the style is inserted. The species, which include both unarmed and prickly plants, are shrubby, with tall simple continuous and inarticulated stems, bearing a few lateral-apical branches, which fall off as the stem increases in height. The flowers resemble those of *Opuntia*, and are produced near the apices of the branches; they are succeeded by oblong compressed berries. *C. rubescens* is an example of the unarmed species; and *C. ferox* and *spinosissima* of the aculeate series. [T. M.]

CONSOLIDA. A section of the ranunculaceous genus *Delphinium*, containing annual species with only one carpel. *D. orientale* and *D. Ajacis*, the rocket larkspurs, are often cultivated; and the blue variety of the latter occurs in Cambridgeshire as a corn-field weed, though it is usually considered as *D. Consolida* by British authors. The true *D. Consolida* differs by having glabrous carpels, and a corymbose inflorescence, not racemose or paniculate, as in *D. Ajacis*. It has been found in Jersey, but perhaps not truly wild. [J. T. S.]

CONSOUDE GRANDE. (Fr.) *Symphytum officinale.* — HE'RISSE'E. *Symphytum echinatum.* — MOYENNE. *Ajuga reptans.*

CONTINUOUS. The reverse of articulated. A stem is said to be continuous which has no joints.

CONTORTED. An arrangement of petals or corolline lobes, when each piece, being oblique in figure, and overlapping its neighbour by one margin, has its other margin in like manner overlapped by that which stands next it, as in oleander.

CONTORTUPLICATUS. Twisted back upon itself.

CONTRAYERVA. *Dorstenia Contrayerva.*

CONULEUM. The name given to a West African bush of the elæagnus family, with opposite entire leaves, which are oboval in form and pointed, while both surfaces are covered with scurfy scales. The small flowers are not known. The females, arranged in forked racemes, have a calyx with a cylindrical tube and a conical limb, and are provided with a little opening at the top through which the style protrudes. The fruit is not known. [A. A. B.]

CONVALLARIA. The Lily of the Valley is a plant so well known, and one which is so universally a favourite, that little need be said by way of description. A slender irregular stalk, a few inches high and slightly curved, bears from eight to twelve small bell-shaped milk-white flowers, arranged one above another, each on a stalk of its own, all bending towards the ground, symmetrically elegant in form, and of a delicate perfume. This stalk rises from the base of a pair of broadly-lanceolate leaves, tapering towards each extremity of a somewhat glaucous hue, clasped together at the base by sheathing scales, and scarcely unfolded by the time the flowers are in perfection. Without poetical or fanciful conventionalities, the Lily of the Valley is as perfect an emblem of purity, modesty, and humility, as the floral world can afford. It may seem idle to observe that a flower of this description cannot be that referred to in the Sermon on the Mount; but as that opinion is frequently broached in popular works, it may simply be observed, that it never grows in the open fields, and that there is nothing in its array to which the term 'glory' is applicable. Not a little unprofitable commentary might have been spared if the same general meaning had been attached to the term 'Lilies of the Field,' which has by common consent been ascribed to the parallel phrase, 'Fowls of the Air,' while the passage itself would have gained in force and dignity by being kept clear from botanical disquisitions. The Lily of the Valley is an inhabitant of the woods in many parts of England, and has long been admitted into every garden. A variety with double flowers, and another of a reddish hue, are also cultivated; but these are far inferior to the wild form of the plant. Notwithstanding the fragrance of the flowers, they have a narcotic odour when dried, and if reduced to powder excite sneezing. An extract prepared from the flowers or roots partakes of the properties of aloes. A beautiful and durable green colour may be prepared from the leaves with lime. The genus belongs to the *Liliaceæ*. *C. majalis* is the only species retained, some others which were formerly included being now referred to *Polygonatum.* French, *Muguet de Mai, Lis de Mai,* or *des Vallées*; German, *Mayblume.* [C. A. J.]

CONVERGENTI-NERVOSE. When simple veins diverge from the midrib of a leaf, and converge towards the margin.

CONVERGINERVED. When the ribs of a leaf describe a curve and meet at the point, as in *Plantago lanceolata.*

CONVOLUTE, CONVOLUTIVE. When one part is wholly rolled up in another, as in the petals of the wallflower.

CONVOLVULACEÆ. (*Bindweeds.*) A natural order of corolliflora dicotyledons, included in Lindley's solanal alliance. Herbs or shrubs, usually twining, and with a milky juice, having alternate leaves without stipules, and regular flowers; the flower-stalks (peduncles) bear one or many flowers. Calyx five-divided, imbricated, persistent; corolla plaited; stamens five, alternate with the corolline lobes; ovary free, two to four-celled; ovules one to two in each cell; styles united, often divided at the top. Fruit a two to four-celled capsule, rarely one-celled, valves breaking off and leaving the dissepiments and placentas in the middle of the fruit; seeds large with mucilaginous albumen; embryo curved. Abundant in tropical countries, and rare in cold climates; they twine around other plants and creep among weeds on the sea-shore. The plants are characterised chiefly by their purgative qualities, and many of them are used medicinally. Jalap is procured from the root, or rather underground stem of *Exogonium* (*Ipomœa*) *Purga*, while the gum-resin called scammony is produced by *Convolvulus Scammonia.* *Ipomœa Bona nox* is the moonflower of Ceylon and other warm countries. *Batatas edulis*, sweet potato or Batatas, is cultivated in Carolina, Japan, and China, and succeeds within an annual isotherm of 59°. It is cultivated also in Spain and Portugal. In the Philippine islands the batatas or camotes are used for making soup, as well as roasted. There are forty-six known genera, and nearly 700 species. Illustrative genera: *Calystegia*, *Convolvulus*, *Exogonium*, *Ipomæa*, *Batatas*, *Pharbitis*. [J. H. B.]

CONVOLVULUS. An extensive and widely-distributed genus, typical of the order of bindweeds, consisting of twining or trailing annual and perennial plants, mostly with showy flowers expanding during the early part of the day. Among the allied genera of the order it is distinguished by its naked bractless calyx, funnel-shaped corolla, two linear often revolute stigmas, and two-celled capsule, each cell containing two seeds. The species share largely in the medicinal properties found in some other genera of the family; qualities which depend on the presence of a peculiar resin with purgative properties. *C. Scammonia* furnishes the scammony of the druggist; and in most of the perennial species, including the indigenous *C. arvensis* and *C. Soldanella*, the same principle occurs. *C. dissectus* abounds in hydrocyanic acid, and is said to be one of the plants from which the liqueur noyau is prepared. Some of the species are popular ornaments of the flower-garden, and with one at least everybody is familiar; viz. *C. tricolor* or Minor Convolvulus of the seedsman, a dwarf Mediterranean species with large flowers of a beautiful violet blue, the centre white and yellow. Of the perennial climbing species, *C. althæoides* with silky deeply-cleft ovate foliage and rose-coloured flowers is an example. *C. bryoniæfolius*, *C. italicus*, and *C. Sibthorpii* are closely related to it, and not easily distinguished. *C. lineatus* with very narrow entire foliage, and flesh-coloured flowers, is occasionally met with in gardens; it is dwarfer and less showy than the preceding. A very distinct species is the *C. Cneorum*, indigenous to the south of Europe, of shrubby habit with persistent lanceolate foliage clothed with silvery hairs, and whitish flowers produced in terminal bunches in spring. *C. mauritanicus* is a pretty dwarf trailing species with oval wavy foliage, and numerous axillary flowers of a pleasing violet colour. [W. T.]

CONYZA. A genus of herbaceous or shrubby plants belonging to the radiate group of compound flowers, among which it is discriminated by its naked receptacle, its three-cleft outer-florets, and the rough pappus which crowns its fruit. The species possess no properties to render them attractive. They were formerly supposed to have the power, when suspended in a room, of driving away fleas; hence the English name Flea-bane, given also to an allied genus. *C. camphorata* and *maritandica* give out a strong smell of camphor. *C. carolinensis* is an evergreen shrub, a native of Carolina, growing to the height of five feet, and producing purple flowers from July to October. *Baccharis halimifolia*, a shrubby species with insignificant white flowers is by some authors placed in this genus. French, *Herbe aux Puces*; German, *Durrwurz*. [C. A. J.]

COOKIA. A genus of *Aurantiaceæ*, named in honour of the famous circumnavigator. It consists of small trees with compound leaves; whose leaflets are unequal at the base. The flowers have four to five concave petals; eight to ten stamens, distinct one from the other; the ovary on a very short stalk, four to five-celled, with two ovules in each compartment; and the style short and surmounted by a four to five-toothed stigma. The fruit is a globular berry, with five, or by suppression, fewer compartments, filled with juice. The fruit of one species, *C. punctata*, is esteemed in China and the Indian Archipelago, where it is known under the name of Wampee. [M. T. M.]

COONDA OIL. The oil of *Carapa guianensis*.

COOPERIA. A genus of *Amaryllidaceæ*, allied to *Zephyranthes*. They are bulbous plants with linear tortuous leaves, and one-flowered scapes. The perianth consists of a long erect slender cylindrical tube widened at the mouth, and a stellate limb of six regular equal segments; the filaments are nearly equal, erect, inserted in the mouth of the tube; the style erect with a three-lobed fimbriated stigma. The species, of which but few are known, are natives of Texas. The typical one, *C. Drummondiana*, has narrow tortuous leaves, twelve to eighteen inches long, and a scape of six inches to a foot high, bearing at the end a single flower, of which the tube is four and a-half inches long, greenish, often fading red, and the limb, rather over an inch long, and white. The flower always expands in the evening, and is not usually perfect after the first night, the limb becoming less stellate, and its margins curled, but it lasts three or four days in that state. 'The nocturnal flowering of this plant is an anomaly in the order, and the more remarkable because its nearest kin, *Zephyranthes*, requires a powerful sun to make it expand. The flower is fragrant, smelling like a primrose.'—*Herbert*. *C. pedunculata*, called also *Sceptranthus*, is also a nocturnal-blooming plant, with pure white primrose-scented flowers. [T. M.]

COOPER'S WOOD. *Alphitonia excelsa*.

COPAIVA TREE. *Copaifera officinalis*, which, with other species of *Copaifera*, yields Copaivi balsam.

COPAI YE'. The wood of *Vochya guianensis*.

COPAL. A name applied to a gum-resinous product of various tropical trees. —, BRAZILIAN, obtained from several species of *Hymenæa*, and from *Trachylobium Martianum*. —, INDIAN, produced by *Vateria indica*. —, MADAGASCAR, produced by *Hymenæa verrucosa*. —, MEXICAN, supposed to be the produce of some *Hymenæa*.

COPALCHE PLANT. *Strychnos pseudo-quina*, which furnishes the Brazilian copalche bark; also *Croton pseudo-china*, the bark of which is called copalche bark in Mexico.

COPALM BALSAM, or COPALME D'AMÉRIQUE. (Fr.) A liquid balsam obtained from *Liquidambar styraciflua*.

COPERNICIA. A genus of palms named in honour of the celebrated Copernicus. It comprises six species, inhabiting tropical America, but three of them are almost unknown. They grow twenty, thirty, rarely forty feet high, their trunks being covered by the remains of leaf-stalks, and surmounted by tufts of fan-shaped leaves, from amongst which the branching spikes of small greenish flowers are produced, each spike having several sheathing bracts scattered along its stalk. The flowers are either perfect or imperfect, and have a cup-shaped calyx with three small teeth, a bell-shaped corolla with the upper part cut into three divisions, six stamens fixed to the inside of the corolla, and three ovaries more or less cohering together. The fruit is yellowish, of an elliptical form, and contains a single seed.

The Carnaüba or Wax-Palm of Brazil, *C. cerifera*, grows about forty feet high, and has a trunk six or eight inches thick, composed of very hard wood, which is commonly employed in Brazil for building and other purposes, and is sometimes sent to this country and used for veneering. The upper part of the young stems, however, is soft, and yields a kind of sago; and the bitter fruits are eaten by the Indians. The young leaves are coated with wax, called carnaüba wax, which is detached by shaking them, and then melted and run into cakes; it is harder than bees' wax, and has been used by Price and Co. for making candles, but as no process of bleaching has been discovered, they retain the lemon-coloured tint of the raw wax. The leaves are also used for thatching, making hats, &c., and while young as fodder for horses. [A. S.]

COPPERY. Brownish red, with a metallic lustre.

COPRINUS. A genus of gill-bearing Fungi remarkable for their dark spores and deliquescent pileus. The gills moreover adhere together in consequence of the great projection of the transparent processes supposed to be antheridia. The species are numerous, of extremely rapid growth, and are developed for the most part on dung hotbeds or very rich manured soil. They have even been found on the dressings of fractured limbs. A few hours is often sufficient for their complete development and decay. *C. atramentarius* yields a very dark juice which has sometimes been used for ink, and both that and some other species are mixed with other fungi to make ketchup. *C. comatus* is sometimes eaten when young and is said to be both delicate and wholesome. *C. bolbi-tius* is distinguished by its salmon-coloured spores. [M. J. B.]

COPROSMA. A genus of cinchonaceous shrubs, owing their name to their fetid smell. The flowers are polygamous, each whorl of from four to nine divisions; the stamens project from the somewhat bell-shaped corolla; the ovary has two to three compartments, and is surmounted by an epigynous disc. The fruit is a berry with two or three seeds. The leaves of *C. foetidissima* are used by the New Zealand priests to discover the will of the gods. The leaves are attached with a cord of flax to sticks, which are laid on the ground, each stick representing a separate party. The priests retire to pray, and after a time the chiefs are summoned to examine the sticks, which are found to have been moved, and some have disappeared entirely; this is considered a certain sign that one of the party will be destroyed. Others are found turned over. If the leaf be turned down, the omen is bad, but if the reverse should occur, it is a sign that the party represented by the stick will prosper in their undertakings. See *Bennett's Gatherings of a Naturalist in Australia*. [M. T. M.]

COPTIS, Gold Thread. A genus of *Ranunculaceæ* containing a few North American and North-east Asian herbs (one of which extends into Russia) with creeping rootstocks and trifoliate or biternate radical leaves and simple or branched scapes with small white flowers, with five or six petaloid deciduous sepals and as many petals; fifteen to twenty-five stamens, and five to ten follicular stalked carpels diverging in the form of a star, with four to eight seeds in each. The bitter rhizomes are used in America as a tonic, and also yield a yellow dye. [J. T. S.]

COPTOPHYLLUM. The name of a section or group of *Anemia* in which the caudex produces distinct sterile and fertile fronds. It contains one or two beautiful dwarf species, as *A. bunifolia* and *A. millefolia*. [T. M.]

COQUARDEAU. (Fr.) *Cheiranthus fenestralis*.

COQUE. (Fr.) *Cocculus*.

COQUELICOT. (Fr.) *Papaver Rhœas*.

COQUELOURDE. (Fr.) *Anemone Pulsatilla*; also *Lychnis coronaria*.

COQUELUCHIOLE. (Fr.) *Cornucopiæ*.

COQUERELLE. (Fr.) *Anemone Pulsatilla*.

COQUERET. (Fr.) *Physalis Alkekengi*. —, COMESTIBLE. *Physalis peruviana*.

COQUILLA NUTS. The seeds of *Attalea funifera*.

COQUITO. The Chilian name of the palm *Jubæa spectabilis*.

CORACAN. (Fr.) *Eleusine Coracana*.

CORACINUS. Raven-black.

CORAÇOA DE JESU. *Mikania officinalis.*

CORAIL DES JARDINS. (Fr.) *Capsicum annuum.*

CORAL BERRY. An American name for *Symphoricarpus vulgaris.*

CORALLIFORM, CORALLOID. Resembling coral in general appearance.

CORALLINA, CORALLINEÆ. A genus and division of rose-spored *Algæ*, the latter characterised by their calcareous rigid fronds, which when fresh are purple, fading to cream; white. Some are shrubby and jointed, others are crustaceous, and often adhere closely to their matrix, as pebbles, shells, seaweed, &c., while others present clavate or nodular forms, and are at length free from any attachment. When treated with hydrochloric acid their structure becomes visible under the microscope, and in some, as in *Corallina*, tetraspores have been discovered. The whole group, however, requires further investigation, and when the fruit is discovered in all the genera, it is probable that they will be absorbed into other groups. From the great quantity of carbonate of lime which they contain, some of the species, but especially *Corallina officinalis*, which is very common on our coasts, have been employed in medicine. They have, however, no specific properties beyond common chalk, which is a much more convenient substance. Crabs' eyes, crabs' claws, and red coral may be considered as belonging to the same pharmaceutical category. [M. J. B.]

CORALLINES. See *Corallineæ.*

CORALLORHIZA. A genus of orchids consisting of a small number of brown or yellowish terrestrial parasitical herbs, natives of moist woods and shady places in Europe, North America, and Northern Asia. Their leaves are reduced to small scales of the colour of the stems; their flowers are small in a loose terminal spike, the sepals and petals nearly alike, the lip larger and often white, the column short, with a terminal lid-like anther, and two pairs of globular pollen masses attached laterally. *C. innata*, the only European species, occasionally occurs in some parts of Scotland. It is a slender plant of six to nine inches high, of a pale colour, and remarkable for its rootstock, formed of a number of short thick whitish fleshy fibres, repeatedly divided into short blunt branches, and densely interwoven, which, from their resemblance to coral, have given the name to the genus.

Two species are found in Mexico, of which one, *C. bulbosa*, has its stem distended into a kind of corm at the base. The largest flowered species inhabits North-west America. *C. indica* was found by Dr. T. Thomson in the North-western Himalays. A supposed species called *C. foliosa*, because it bears a true leaf, now forms part of the genus *Oreorchis.*

CORAL-ROOT. The common name for *Corallorhiza*; also sometimes applied to *Dentaria bulbifera.*

CORAL-TREE. The common name for *Erythrina.*

CORALWORT. *Dentaria bulbifera.*

CORBEILLE D'ARGENT. (Fr.) *Iberis sempervirens.* — **D'OR.** (Fr.) *Alyssum saxatile.*

CORBULARIA. A genus of amaryllids, commonly called Hoop-petticoats. It is a small group sub-divided from *Narcissus*, and its chief peculiarities are a funnel-shaped tube to the perianth, an inconspicuous limb with small narrow spreading segments, and a large funnel-shaped cup, which is longer than the tube itself; the filaments and style are declinate and recurved, the sepaline filament inserted at the base, and the petaline near the base of the tube. The species are pretty dwarf hardy bulbs with very narrow half-terete leaves, and comparatively large showy flowers, one to three together on the scape. The species are found in the middle and south of Europe. *C. Bulbocodium*, the common Hoop-petticoat and the type of the genus, is a small plant, with conical bulbs as large as a nut, three or more leaves from four to eight inches long, and a one-flowered scape four to six inches long; the flower an inch long, pale yellow, with narrow linear lanceolate segments, the cup or coronet prominent truncate deep yellow. The few species vary chiefly in size and colour. [T. M.]

CORCHORUS. This genus of *Tiliaceæ* contains between forty and fifty species of herbaceous plants or small shrubs, with simple leaves, inhabitants of both hemispheres, but seldom found far beyond the tropics. Their flowers are produced either singly or in clusters opposite the leaves. They have a calyx of five deciduous sepals, and a corolla of five petals, with numerous stamens, a very short tubular style, and from two to five stigmas. Their fruit is long and pod-like or roundish, and splits when ripe into five divisions, each of which has numerous seeds arranged in rows on either side of a longitudinal partition.

C. capsularis is an annual Asiatic plant, growing about ten or twelve feet high, and having a straight cylindrical stem as thick as the little finger, and seldom branching till near the top. Its leaves are about six inches long by one and a half or two broad towards the base, but tapering upwards into a long sharp point, and having their edges cut into saw-like teeth, the two teeth next the stalk being prolonged into bristle-like points. The flowers are yellow, and produced in clusters of two or three together; they are succeeded by a small almost globular but flat-topped fruit. This species, as well as *C. olitorius*, yields the exceedingly valuable fibre known under the name of Jute. Only twenty years ago, Jute was hardly heard of out of India, where it had long been in use amongst the natives for making cordage and cloth, but it now forms a very important article of commerce; no less than 738,085 cwt., valued at £19,68sl., having been imported to this country alone in 1858. The plant is largely cultivated in India; also by the Malays and Chinese. The fibre is separated by the ordinary process of steeping in water. It is frequently as much as twelve feet in length, very soft, silky, and separable into fine filaments, which are easily spun. Jute is much used in the manufacture of carpets, and some kinds of cloth; but is not suitable for cordage, as it will not bear exposure to wet. Its most important use, however, is for the manufacture of the gunny-bags, so extensively used for packing cotton, rice, and other dry goods, enormous quantities of them being exported from India to the United States for that purpose. Very good paper is made from the refuse fibre, and also from worn-out gunny-bags; and a kind of whisky, resembling corn-spirit, has been distilled from the waste ends of the stems.

C. olitorius, is a native of India, but is now naturalised in all parts of the tropics, and extends as far north as the shores of the Mediterranean. It is an annual plant much resembling *C. capsularis*, the principal difference existing in the fruit, which in this species is two inches long, almost cylindrical, and about the thickness of a quill. The young shoots of this species are commonly used as a pot-herb in tropical countries, as are those of *C. capsularis*; it is much grown for this purpose in Egypt and Syria, and being used by the Jews, it has obtained the name of Jews' Mallow. It yields part of the Jute of commerce.

C. siliquosus, a common species in the West Indies and tropical America, is an herbaceous plant about two or three feet high; its leaves differ from those of the two last in not having bristles on the two bottom teeth, and there is usually a line of minute hairs along the stem. The negroes in the West Indies use it for making besoms, and the inhabitants of Panama drink an infusion of the leaves as a substitute for tea; hence they call it tea.   [A. S.]

CORCULUM. The embryo; and also, the small axis of growth in such dicotyledonous embryos as the walnut.

CORDATE. A plane body, having two round lobes at the base; the whole resembling the heart in a pack of cards.

CORDATO-HASTATE. Between cordate and hastate.

CORDATO-OVATE. Between cordate and ovate.

CORDATO-SAGITTATE. Between cordate and sagittate.

CORDELYSTYLIS. A little known genus of Indian orchids described by Falconer in the *Journal of Botany* (iv. 75). It seems to be related to *Spiranthes*.

CORDIA (including *Myxa*, *Pilicordia*, *Rhabdocalyx*, and *Sebestena*). A genus of *Boraginaceæ*, containing nearly 300 species of plants scattered over the tropical and sub-tropical regions of the world. They are trees or shrubs with alternate rarely subopposite petiolate and entire or sudden-tate leaves, and flowers variously arranged, sometimes polygamous, or monœcious from the abortion of parts. The calyx is tubular with four or five teeth; the corolla is funnel-shaped with the limb four to five-parted, rarely six to twelve-lobed; the stamens are as numerous as the divisions of the corolla, and are inserted in the tube; the ovary is four-celled, and bears a doubly-bifid style, with a stigmatic surface on each division. The drupaceous fruit is ovate or globose, pulpy, generally surrounded by the persistent calyx, and four-celled or one to three-celled from the abortion of one or more cells; there is a single seed in each cell. This large unwieldy genus has been divided into the following sections from characters obtained from the calyx; and it would be well if these sections were raised into genera:—

Gerascanthus, having a cylindrical ten-grooved calyx.

Pilicordia, with an oblong or cylindrical ten-striate calyx.

Physoclada, having a membranaceous calyx, hispid at the apex with setæ, and at length irregularly torn.

Sebestenoides, having a cylindrical or ovate smooth three to ten-toothed calyx.

Myxa, the calyx not grooved, four to five-toothed, the teeth short or rarely awn-shaped.

Cordiopsis, with an obovate or oblong calyx terminating in five setaceous divisions.

The fruit of some species is eaten, as of *C. latifolia* and *C. Myxa*, two Indian species which have succulent mucilaginous and emollient fruits. From their mucilaginous qualities, combined with some astringency, they have been employed as pectoral medicines, under the name of Sebestens. The fruit of *C. abyssinica* is used in the same way in Abyssinia. The bark of *C. Myxa* is a mild tonic, and is used in India for astringent gargles. Some species supply useful and ornamental timber; the wood of *C. Rumphii* is brown, beautifully veined with black, and smells of musk. *C. Gerascanthus* yields a timber of importance in the West Indies. The wood of *C. Myxa* is soft, and of little use except for fuel. It is reckoned one of the best kinds for kindling fire by friction, and it is said to be the wood which was used by the Egyptians in constructing their mummy-cases. See Plate 6, c, and Plate 10, e.   [W. C.]

CORDIACEÆ. A tribe or suborder of *Boraginaceæ*, often considered as forming a distinct family. They differ from true *Boragineæ*, but agree with *Ehretiaceæ* in their concrete entire ovary internally divided into four or more cells; and are distinguished from both of those suborders by their branching style, and most frequently by their plaited cotyledons. They are trees, shrubs, or rarely herbs, with alternate rough leaves; their flowers are in terminal cymes, sometimes gyrate as in true *Bora-*

gineæ, or rarely solitary. The fruit is usually more or less drupaceous. There are above 150 species, natives of the tropical or subtropical regions both of the New and the Old World, and have been distributed into about twelve genera, most of which have, however, been since reduced to *Cordia* itself.

CORDICEPS. A fine genus of Sphæriaceous *Fungi* distinguished by its fleshy texture, vertical stipitate stroma and filiform articulate spores, which separate at the articulations. The species are the most remarkable amongst the very important group to which they belong. A few grow upon dead leaves, decaying branches, or ergoted grains, the rest upon pupæ or larvæ of insects. The New Zealand *C. Robertsii* occurs on the caterpillar of a species of *Hepialus*, and is frequently brought home as an object of curiosity. We have two or three fine species in this country, of which *C. militaris* is remarkable for its brilliant scarlet hue. *C. alutacea*, which is of a pale tan, grows upon pine leaves, and a form of it, or distinct species, on *Ulex europæus*. There is no doubt that, in many cases, the fungus-bearing insects are attacked during their lifetime; and there is one species of *Cordiceps* which occurs on wasps in the West Indies, which is considerably developed before the insect dies. The wasps so attacked are known by the name of Goëpes vegétantes. The peculiarities of the species which grow on ergot will be noticed under that head. *C. sinensis* is supposed by the Chinese to have healing properties, and is sold as a drug in little bundles. [M. J. B.]

CORDIFORM. When a solid has the form of cordate.

CORDLEAFS. A name given by Lindley to the group of restiaceous plants.

CORDON DE CARDINAL. (Fr.) *Polygonum orientale*.

CORDYLANTHUS. A genus of *Homalineæ* from Java. It is allied to *Blackwellia*, but with an elongate club-shaped perianth tube, adhering to the ovary, the limb ten or twelve-parted, and the segments in two rows, the inner longer and petaloid; leaves alternate, shortly-stalked, leathery, elliptical, toothed; flowers white, racemose, axillary; peduncles one to three-flowered; stamens fifteen or twelve; styles three to five; ovary one-celled. [J. T. S.]

CORDYLINE. A genus of erect-stemmed shrubby palm-like *Liliaceæ*, bearing spreading and very ornamental heads of narrow elongate striated leaves, and terminal panicles of numerous small flowers. The perianth is deciduous, tubulosely bell-shaped, with a six-cleft or six-parted spreading limb of linear segments, inserted in two rows; and there are six stamens with linear filaments inserted in the mouth of the tube. The ovary is three-celled with one ovule in each cell; and the style is filiform with a capitate three-lobed stigma. The fruit is a globose three-celled berry, often by abortion one or two-seeded. The species are found in tropical Africa, in Madagascar, and the Mascarene Islands, and in the Malayan Archipelago. The typical species, *C. reflexa*, a native of the Mauritius, St. Helena, and Madagascar, has a naked simple stem, bearing a crowded head of numerous ensiform striated leaves, six or seven inches long, and scarcely half an inch wide. The flowers are fragrant, numerous, yellowish green, in a branched terminal raceme. *C. fragrans*, a West African species, has a tall stem with a terminal head of lanceolate leaves, two to three feet long, and two to three inches broad, and divaricately-branched panicles of fragrant white flowers, collected into dense umbellate heads. *C. Sieboldii* is a compact growing species with oblong leaves, four to six inches long, deep green, ornamentally blotched with paler green, and producing short terminal panicles of greenish-white flowers. Some very ornamental species formerly included in *Cordyline*, are now referred to the genera *Caladracon* and *Dracænopsis*; and others less striking in their appearance to *Charlwoodia*. [T. M.]

CORDYLOBLASTE. The name of a Javanese tree doubtfully referred to *Meliaceæ*. It has elliptic entire pointed leaves; flowers in whorls of five; stamens united into a tube, the upper edge of which has six anthers on it and ten or twelve teeth, while numerous other anthers are attached to the inner surface of the tube of the stamens; style simple; ovary seated on a fleshy disc, which is adherent to the base of the calyx, with one many-seeded compartment. [M. T. M.]

CORDYLOGYNE. A genus of *Asclepiadaceæ*, consisting of a single herbaceous plant growing at a height of 4,000 feet on the mountains of Southern Africa. The plant has many erect slender stems about a foot high, long linear leaves, and pale green flowers clustered in many-flowered long peduncles. The calyx consists of five small hairy sepals; the corolla is five-parted, the divisions erect, and at length spreading; the staminal crown consists of five oblong leaves with angular processes on their lateral margins; the anthers are terminated by a triangular opaque apex, adpressed to the base of the oblong fleshy stigma; the pollen masses are attached by slender-kneed processes to a small simple corpuscle; the follicle is solitary, slender, and erect, with comose seed. [W. C.]

COREMA. Portugal Crakeberry. An erect much-branched low shrub of rigid habit, closely allied to *Empetrum*, from which it is distinguished by having no scales at the base of its calyx, and by its white three-seeded globose berries. The branches are slightly downy; the leaves obtuse, small, and narrow, with revolute edges, and sprinkled with resinous dots; flowers white, growing in terminal groups very like those of *Empetrum*, but larger. It is a native of Portugal and other countries of Southern Europe, and is described by some authors under the name of *Empetrum lusitanicum*. [C. A. J.]

COREOPSIS. A genus of American herbaceous composite plants remarkable for the singular shape of its seeds, which are flat on one side, convex on the other, membranous at the edge, and having the pappus furnished with two horns not unlike the antennæ of an insect. Hence its name, which in Greek signifies 'bearing resemblance to a bug.' Many species are cultivated, among which *C. diversifolia* is a perennial with branching stems, small three to five-lobed leaves and large terminal flowers, the disk of which is purple, and the rays yellow, marked with a purple stain at the base. Several beautiful annual species, as *C. tinctoria*, *C. coronata*, *C. Atkinsoniana*, and *C. Drummondii*, are now referred to *Calliopsis*. *C. verticillata* is a handsome shrubby perennial, continuing long in flower; its flowers are used in North America to dye cloth red. [C. A. J.]

CORESES. Dark red, broad, discoid bodies, found beneath the epicarp of grapes.

CORETTE POTAGE'RE. (Fr.) *Corchorus olitorius*.

CORETHROSTYLIS. A genus of W. Australian bushes belonging to the byttneriads, remarkable for the form of the style, which is elongated and furnished with numerous tufts of recurved hairs, giving it the appearance of a bottle-brush. This curious appearance has suggested the name. About seven species are known, all of them having their parts more or less covered with rusty-coloured starry hairs. Their leaves are alternate, mostly heart-shaped, and either entire or notched. The flowers are in branched racemes, which arise from opposite the leaves, each flower supported by a bract, and consisting of a five-parted petal-like calyx covered with soft hairs; no petals; five stamens with short stalks, and anthers opening at the apex by a small pore, surrounding a three-lobed ovary, which, when ripe, becomes a three-celled capsule with three seeds. *C. bracteata* is a pretty bush sometimes seen in greenhouses: it has heart-shaped entire leaves about an inch in length, covered like all parts of the plant with rusty hairs. The pink starry flowers, with pink bracts, appear in great profusion. [A. A. B.]

CORIACEOUS. Having the consistence of leather.

CORIANDER. *Coriandrum sativum*.

CORIANDRUM. A genus of *Umbelliferæ* producing the fruits erroneously called Coriander seeds. There is but one species, *C. sativum*, a native of Southern Europe, the Levant, &c., and cultivated even in this country, where it is also sometimes met with in a half wild condition. It has a branching annual stem, one to one and a half foot high, with the lower leaves pinnately divided into broad or wedge-shaped deeply-cut segments, while the upper leaves are more finely cut. The umbels have five to eight rays without a general involucre, and the partial ones consist of only a few small bracts; the flowers are whitish or pink. The most characteristic feature, however, is the globular fruit, which is crowned by the teeth of the calyx, and has no oil channels on the outer surface, but two on the inner face of each half of the fruit; the ridges are five and rather indistinct. The two carpels of which the fruit is composed do not readily separate one from the other. Coriander fruits or seeds are carminative and aromatic, and are hence used for flavouring purposes in curries, &c., &c. The odour and taste depend upon a volatile oil. The fresh plant has a strong smell of bugs. [M. T. M.]

CORIARIA. A genus of shrubs of uncertain position, by some made to constitute a distinct family under the name of *Coriarieæ*. The leaves are opposite, simple, ribbed, and entire. The flowers are in clusters, either hermaphrodite, monœcious, or diœcious, calyx five parted, bell-shaped, petals five, fleshy, with a prominent ridge internally; stamens ten, arising from beneath the ovary, which consists of five carpels arranged obliquely upon a thickened receptacle; stigmas five; ovules solitary, pendulous, inverted. Fruit of five crustaceous indehiscent one-seeded carpels, concealed by the membranous sepals and fleshy petals. These shrubs are natives of Southern Europe, the Mediterranean, Peru, Nepaul, and New Zealand. *C. myrtifolia*, the European species, is a low deciduous shrub with myrtle-like leaves. Its fruit is poisonous, and is said to have proved fatal to some French soldiers in Catalonia. The leaves have also been used to adulterate senna—a dangerous fraud, as they are stated to have caused tetanic convulsions, and subsequent coma. *C. myrtifolia* is also used in dyeing black. *C. sarmentosa*, the Wineberry shrub of the settlers in New Zealand, has pendulous branches, greenish white flowers in long slender clusters, and shining-black berry-like fruits, full of a dark red juice of sweet taste, and free from any deleterious properties, but the seeds if eaten are poisonous; the natives therefore having expressed the juice from the fruits, strain it before they drink it, or soak their baked fern root in it. The 'missionaries at the Bay of Islands,' says Dr. Bennett, from whose *Wanderings in Australia* this notice is taken, 'make an agreeable wine from the berries of the shrub, which tastes like that made from elderberries.' The effects that result from eating the seeds are convulsions and delirium, which continue for several hours, and frequently terminate fatally. The fruit of *C. nepalensis* is also eaten in Northern India. [M. T. M.]

CORINDE. (Fr.) *Cardiospermum Halicacabum*.

CORINTHS. The berries of the Corinthian grape, the Currants of the shops.

CORIS. A genus of *Primulaceæ*, containing a single species, a native of the western coasts of the Mediterranean. It is a lowly branching herbaceous plant, with alternate linear coriaceous leaves, and flowers in dense terminal spicate racemes. The calyx is campanulate with a double limb, the outer ray sublabiate, with the upper lip six-toothed and the lower five-toothed, the inner portion being cleft into five triangular lobes, of which the upper two are the largest; the corolla is tubular, with the limb bilabiate and cleft into five emarginate lobes, the two upper of which are the smallest; the stamens are scarcely exserted; the slender filaments have glands at their base on the corolla tube; the ovary is obovate, and has a subglobose placenta; the globose capsule has five valves and five seeds. [W. C.]

CORISPERMUM. A genus of *Chenopodiaceæ*, containing wiry-stemmed hairy annual herbs from Eastern Europe and temperate Asia. Leaves narrow, sessile; flowers very small, solitary in the axils of the leaves, forming spikes; perianth of a variable number of small scales, rarely of one, or absent; stamens one to five, but generally three, the lateral ones often sterile; ovary compressed with short style and two stigmas; fruit compressed often margined. Abundant in the marshy steppes of Southern Russia. [J. T. S.]

CORK, KORKER. The name in the Scotch Highlands of *Lecanora tartarea*, where, Dr. Lindsay informs us, it is made into a domestic dye by macerating the powdered lichen for some weeks in putrid urine, with the addition of kelp or salt, and when the requisite crimson or purple tint is obtained, forming the paste into balls or lumps with lime or burnt shells, and hanging it in bags to dry. When used it is powdered, and then boiled in water with a little alum. In the island of Shetland both the dye and the lichen are called Korkalett. [M. J. B.]

CORK-TREE. *Quercus Suber*, the bark of which is cork.

CORK-WOOD. *Anona palustris.* — NEW SOUTH WALES. *Duboisia myoporoides.* — WEST INDIAN. *Ochroma Lagopus*.

CORM. A fleshy underground stem, having the appearance of a bulb, from which it is distinguished by not being scaly.

CORMAU or **CORNIAU.** (Fr.) A kind of olive.

CORMIER. (Fr.) *Sorbus domestica*.

CORMOPHYLLUM. A name given by Newman to a genus of Ferns having an erect caudex 'eventuating in fronds,' and in which he proposed to unite the species usually referred to *Cyathea*, *Hemitelia*, and *Alsophila*. [T. M.]

CORN. A general term applied to the cereal or grain-producing grasses. — BROOM. *Sorghum Dora*, the panicles of which are made into brooms, and the grain used for poultry food. — KAFFIR. A species of *Sorghum*, probably *S. saccharatum*. — GOOSE. *Juncus squarrosus*. — GUINEA. *Sorghum vulgare*; also applied in the West Indies to several grain-bearing species of *Panicum*, as *P. pyramidale*, *scabrum*, &c. — INDIAN. The maize, *Zea Mays*.

CORNACEÆ. An inconsiderable natural order of polypetalous calycifloral dicotyledons, belonging to Lindley's umbellal alliance. Trees or shrubs usually with opposite leaves having no stipules; flowers in cymose clusters or in heads surrounded by an involucre; calyx adherent, its limb four-toothed; petals four, valvate in bud; stamens four, alternate with the petals; styles united into one; ovary two-celled; ovules solitary, pendulous. Fruit a two-celled drupe (like a cherry). Natives of the temperate parts of Europe, Asia, and America. The plants of this order are used as tonics and in ague. *Cornus mascula* is the akenia of the Greeks, and the kizziljick or red-wood of the Turks. From the wood of this plant the Turks obtain the dye for their red fez. The fruit stewed and mixed with water forms a good drink in hot weather, and from its astringency it is useful in bowel-complaints. Various species of *Cornus* or dogwood are used in America as substitutes for Peruvian bark. There are nine known genera and forty species. Illustrative genera: — *Cornus*, *Benthamia*, *Aucuba*. [J. H. B.]

CORNARET. (Fr.) *Martynia*.

CORN CAMPION, CORN COCKLE. *Agrostemma Githago*.

CORNE-DE-CERF. (Fr.) *Coronopus vulgaris*.

CORNEILLE. (Fr.) *Lysimachia vulgaris*.

CORNEL. (CORNOUILLER, Fr.) The Cornelian cherry, *Cornus mascula*. — WILD or FEMALE. The dogwood, *Cornus sanguinea*.

CORNEOUS. Horny; hard and very close in texture, but capable of being cut without difficulty, the parts cut off not being brittle; as the albumen of many plants.

CORN-FLAG. The common name for *Gladiolus*.

CORN-FLOWER. *Centaurea Cyanus*.

CORNICULATE. Terminating in a process resembling a horn; as the fruit of *Trapa bicornis*. If there are two horns the word *bicornus* is used, if three *tricornis*, and so on.

CORNIDIA. A genus of trees and shrubs from Peru and Chili belonging to *Hydrangeaceæ*. They have opposite ovate or obovate stalked leaves, which are leathery and generally serrated, and bear their flowers in a terminal corymbose cyme of many rays; calyx-tube adhering to the ovary, the limb four or five-toothed; petals four or five on an epigynous ring; stamens eight or ten; styles two to four; capsule with two or four imperfect partitions. [J.T.S.]

CORN-SALAD. The Lamb's lettuce, *Valerianella olitoria.*

CORNU (adj. CORNUTUS). A horn-like process, commonly solid, and usually a metamorphosed state of some other organ. Also employed in the sense of Calcar.

CORNUCOPIÆ. A genus of grasses belonging to the tribe *Phalarideæ.* It is distinguished chiefly by the involucre being large, one-leaved, cup-shaped or funnel-shaped, many-flowered; glumes two, united at the base, mitre-formed, and equal; pales one, bladder-shaped, split on one side, with an awn below the middle; stigmas long. Only one species is described, *C. cucullata,* the Horn of Plenty grass, a native of Greece and Asia Minor, which is frequently cultivated in gardens amongst curious annuals. [D. M.]

CORNUELLE. (Fr.) *Trapa natans.*

CORNUS. The typical genus of the order of cornels, consisting of twenty or thirty species distributed throughout temperate Europe, Asia, and America, generally forming small trees or shrubs, some, however, being humble herbs only a few inches high. Their leaves are undivided and generally opposite; their flowers have a calyx composed of four minute teeth, and a corolla of four yellow or white petals; and their fruit contains a hard two-celled stone with two seeds, and is marked at the top with a scar from the remains of the calyx.

C. florida, a deciduous tree about thirty feet high, is common in the woods in various parts of North America. It has shining branches, and egg-shaped sharp-pointed leaves, clothed with closely-pressed hairs on both sides; and its heads of yellowish flowers are surrounded by four large white bracts. In the United States the bark of this tree is substituted for Peruvian bark in intermittent fevers. Mixed with sulphate of iron it makes a good black ink; and the bark of the root dyes a scarlet colour. Its wood is hard, heavy, and close-grained, but being of small size it is only used for handles of tools, &c.; the young branches stripped of their bark are used for whitening the teeth.

The Cornelian Cherry, *C. mascula,* is a native of many parts of Europe and Northern Asia, forming a large shrub or small tree about fifteen or twenty feet in height, having smooth branches with oval sharp-pointed leaves, and producing its heads of small yellow flowers early in spring, before the appearance of the leaves. Its pulpy fruits resemble a cornelian in colour, and are about the size and shape of olives, for which they are sometimes substituted. The ripe fruits have a harsh acid taste, and are scarcely eatable, but they are sold in the markets in some parts of Germany, and eaten by children, or made into sweetmeats and tarts. The Turks use the flowers in diarrhœa, and the fruits against cholera, or for flavouring sherbet. The wood is exceedingly hard and durable, and also tough and flexible; in central Europe it is used for making forks and other implements, ladder-spokes, &c., and the young branches for butcher's skewers. *C. sanguinea,* which grows wild in England, is known under the names of Dogwood, Dogberry tree, or Hound's tree, in consequence of a decoction of its bark having formerly been used for washing mangy dogs. It is a shrub about six feet high with dark red branches and broadly egg-shaped pointed leaves, which are hairy when young; and bearing heads of dull white flowers without bracts, producing globular, nearly black, and very bitter fruits, which yield an oil fit for lamps. Its hard wood is used like that of the other species, and its young branches for skewers. *C. suecica* is a humble little plant not more than six inches high, native of Britain, Northern Europe, Asia, and America. Its creeping roots produce annual stems having a few stalkless egg-shaped leaves, and terminated by a head of very minute purple flowers, surrounded by four large petal-like white bracts. The little red berries of this plant form part of the winter stock of food collected by the Esquimaux; and in the Scotch highlands they are a reputed tonic, and are supposed to increase the appetite, the plant being called insa-chrosis, or Plant of gluttony. [A. S.]

Chemical analysis shows that the bark of the root, stem, and branches of *C. florida,* which are bitter, astringent and aromatic, contain, in different proportions, the same substances as are found in *Cinchona,* except that there is more gum, mucilage, gallic acid, and extractive matter, and less resin, quinine, and tannin. The principle obtained from it is called *cornine,* and its salts have, according to Dr. Blackie, all the properties of those of quinine, though not so strongly marked; the principle is also difficult to obtain in any quantity. The extract of Dogwood, though inferior and less astringent than the best cinchona, is said to be better than the inferior kinds; this extract contains all the tonic properties, while the simple resin is merely a stimulant. In cases of debility, Dogwood is a valuable corroborant. Country people often use it as a decoction, or chew the twigs as a prophylactic against fevers. Drunkards sometimes employ a tincture of the berries to restore the tone of the stomach, and combat the pains of dyspepsia. The powdered bark of the plant makes one of the best tooth powders, as it preserves the gums hard and sound, and at the same time renders the teeth extremely white. Rubbing the fresh twigs on the teeth has this effect, and the Creoles of the West Indies, the pearly whiteness of whose teeth is universally acknowledged, use another species in this way. [T. M.]

CORNWEED. *Biserrula Pelecinus.*

COROLLA (adj. COROLLARIS, COROLLINE). That part of a flower which intervenes between the calyx and stamens; its parts are called petals.

COROLLIFLORÆ. A subclass of dicotyledons or Exogens, characterised by the petals being united so as to form a monopetalous corolla, inserted below the ovary, and by the stamens being usually attached to the corolla, but sometimes inserted separately below the ovary. Such orders as the heath family, the gentians, and the labiates, may serve as illustrations. [J. H. B.]

CORONA. A coronet. Any appendage that intervenes between the corolla and stamens, as the cup of a daffodil, or the rays of a passion-flower. — **STAMINEA.** A coronet formed from transformed stamens.

CORONANS. Situated on the top or crown of anything. Thus, the limb of the calyx may crown an ovary; a gland at the apex of the filament may crown a stamen.

CORONARIA. A section of the caryophyllaceous genus *Lychnis*. The type of the group is the Rose Campion, *Lychnis coronaria*, a native of S. Europe, commonly cultivated for its beauty. The leaves of this plant are elliptical, white with soft wool, as are the stems and calices; corolla with the petals nearly entire, red or white, with a firm scale at the base of the limb of each; these scales form the crown. The most natural group to combine with the Rose Campion are the remaining species of the discarded genus *Agrostemma*, which have not the deeply-bifid petals of *Lychnis*; this is the arrangement of Fries. [J. T. S.]

CORONATE. Furnished with a coronet. Also used in the sense of Coronans.

CORONILLA. A genus of pretty annual or perennial plants of the pea family, characterised by the flowers being borne on stalked umbels, as well as by the articulated, round, and nearly straight pod. The plants of this genus are found in Europe, Asia Minor, and North Africa, and in greatest abundance in the countries bordering on the Mediterranean Sea. Between twenty and thirty species are enumerated. The Scorpion Senna, *C. Emerus*, a plant not unfrequently seen in gardens is a much-branched pretty bush, about five feet high. Its leaves are alternate, pinnate, from one to three inches long, and composed of three or four pairs of small wedge-shaped leaflets of a pea-green colour; these are said to produce a dye like indigo by proper fermentation, and are also reported as laxative. The yellow flowers, in their form and arrangement, are a good deal like those of the bird's-foot trefoil (*Lotus corniculatus*), and are produced in great abundance, making their appearance in May or June, and continuing in succession till the frost appears. The slender-jointed pod has been compared to a scorpion's tail. *C. varia* is a perennial plant with creeping roots, and slender angular stems, from one to three feet long. The leaves, from two to three inches in length, have numerous oblong leaflets, and the flowers vary much in colour, being either white, rose, or violet. It grows in various parts of Southern Europe, and has been recommended as a forage plant, but its leaves are too bitter, and are even said to be poisonous. The plants of this genus bear much resemblance to each other, and in almost all, the foliage is of a peculiar pea-green colour. The yellow flowers of many emit a strong odour. [A. A. B.]

CORONULE. The small calyx-like body which crowns the nucule of *Chara*.

CORPUS. The mass of anything; thus, *C. ligneum*, or *lignosum*, signifies the mass of the woody tissue of a plant, and *C. medullare* the mass of its cellular tissue in the pith.

CORPUSCULES. The spore-cases of certain fungals. — **VERMIFORM.** Spiral vessels in a contracted, strangled, distorted condition.

CORREA. The pretty greenhouse shrubs so named are now familiar to most persons. They belong to a genus of *Rutaceæ*, and have simple dotted leaves, covered more or less with down. The handsome reddish or greenish flowers have a cup-shaped nearly entire calyx; a corolla of four petals united into a tube; eight stamens attached beneath the ovary; and four one-celled ovaries placed on a small eight-lobed disc, and covered with dense star-like hairs, the styles confluent into one. The fruit consists of four follicles bursting each by two valves, and one-seeded by abortion. These shrubs are natives of the Southern and Eastern parts of Australia, where they are sometimes called Native Fuchsias, from the slight resemblance of the blossoms to those of the fuchsia. It is said too that the leaves of some of the species are used as tea. [M. T. M.]

CORRIGIOLA. A genus of *Illecebraceæ*, small herbs growing in Europe (especially the Mediterranean region), and at the Cape of Good Hope. They have numerous slender slightly-branched procumbent stems, bearing linear and oblong fleshy glaucous leaves; stipules scarious, small; flowers small, green and white striped, forming compound corymbs or racemes at the end of the stem and branches; calyx five-parted, herbaceous, with a petaloid margin; petals five, very small; stamens five; style very short, three-cleft; fruit, a hard nut enclosing a single seed. *C. littoralis*, found in the extreme south-west of England, is a small annual with narrow leaves extending to the tips of the stems, and there intermixed with clusters of small flowers which arise from the axils, and also from a small terminal corymb. [J. T. S.]

CORROYÈRE. (Fr.) *Coriaria myrtifolia*.

CORRUGATED, CORRUGATIVE. When

the parts are crumpled up irregularly, as the petals of the poppy, or the skin of some seeds.

COR SEMINIS. An old name for the embryo.

CORTEX. The bark. Also the peridium of certain fungals.

CORTICAL INTEGUMENT. The bark, or false bark of endogens.

CORTICAL STRATUM. The superficial layer of tissue in the thallus of a lichen.

CORTICATE. Harder externally than internally; having a rind, as the orange.

CORTINA. The filamentous ring of certain agarics.

CORTINARIUS. A large genus of *Fungi*, separated from *Agaricus* more from habit than from any striking characters. The spider-like veil, and bright red-brown spores resembling in tint that of peroxide of iron, are the most easily recognised characters. In the woods of Sweden they form by far the larger part of the mass of Fungi, and in our own country are sometimes very abundant. There is scarcely a single species which is received into European cookery, but in Bhotan one or two are eaten. Many of the species are extremely beautiful in point of colour, especially when young. They alter wonderfully in this respect in dry weather or as they pass maturity. [M. J. B.]

CORTINATE, CORTINARIOUS. Having a cobweb-like texture.

CORTISIA. A genus referred to *Ehretiaceæ*, consisting of a much-branched shrub from the Pampas of South America, having alternate sessile wedge-shaped leaves trifid at the apex, and small white tubercles on both surfaces, from which tubercles spring white hairs. Flowers solitary, sessile, generally terminal, with a tubular calyx having ten small teeth; corolla yellowish-white tubular, with a five-lobed spreading limb; stamens five exserted; style thread-like, cleft at the apex; fruit, an ovate drupe with two seeds. [J. T. S.]

CORTUSA. A genus of *Primulaceæ*, containing a single species, a native of alpine and boreal districts in the Old World. It is a herb with the radical leaves on long petioles, and with simple scapes bearing pedicellate flowers in umbels. The calyx is five-parted; the corolla has a very short tube, and a campanulate limb; the five included stamens are inserted at the base of the limb with very short filaments and obcordate anthers; the capsule is five-valved, and many-seeded, and dehisces from the apex. [W. C.]

CORTUSALES. A name given by Lindley to a group of perigynous exogens, containing among others the primrose and the thrift families.

CORYANTHES. Under this name, formed in allusion to the resemblance of a part of the flower to a helmet, is collected a set of epiphytal orchids inhabiting tropical America, which are the strangest of all the strange forms of that extraordinary order. From one or two-leaved pseudobulbs hang down few-flowered racemes of flowers varying in length from two to five inches long, with the following singular structure. For the sepals there are two large membranous plates folding like a bat's wings, with a smaller interposed. In front hangs down a fleshy lip, bucket-like at the base, and expanding into a great helmet-shaped terminal lobe, whose weight keeps it always downwards, the cavity being turned upwards. The column is a long twisted recurved body with a vertical anther, containing a pair of excavated pollen masses. At the foot of the column are two fleshy feet, from whose toe perpetually distils a clear honey-like fluid, which drops into the hollow of the helmet. The meaning of so strange an apparatus is at present unexplained. Six species are known, of which *C. Fieldingi* has the largest flowers, five inches long and three wide when closed, and *C. distillatoria* the smallest.

CORYCIUM. A remarkable genus of orchids related to *Ceratandra*, and, like it, turning black in drying; the most marked difference between the two consisting in the petals of *Corycium* being saccate, and the lateral sepals connate, so as to form a narrow concave lower lip. Nine or ten species have been described, all inhabiting the Cape of Good Hope, and having close spikes of purplish or greenish flowers. One of them, *Corycium orobanchoides*, has been in cultivation.

CORYDALIS. A genus of *Fumariaceæ*, containing succulent-stemmed herbs, natives of the Northern Hemisphere. They have ternate or twice ternate leaves, and racemose flowers, which are very irregular. Calyx of two lateral sepals; corolla of four petals, the upper one spurred or gibbous at its base; stamens six, in two bundles of three each, the filaments forming a ribbon which is three-cleft at the end, the middle lobe with a two-celled, the others each with a one-celled anther; a spur-like process projects backwards from the upper ribbon, and is received into the hollow space of the upper petal. The capsule is a two-valved one-celled pod with numerous seeds, which have an appendage at the hilum. The genus is divided into several sections:—*Capnites* with no thickened tuberous rootstock, *Bulbocapnos* with a roundish or ovoid enlarged rootstock and alternate leaves, *Cryptoceras* with a fusiform rootstock and opposite leaves. [J. T. S.]

CORYDALIS, CLIMBING. An American name for *Adlumia*.

CORYDANDRA. *Eulophia*.

CORYLACEÆ. (*Cupuliferæ, Castaneæ, Quercineæ, Mastworts.*) A natural order of monochlamydeous dicotyledons, belonging to Lindley's querual alliance. Trees or shrubs bearing catkins, with simple, alternate, stipulate, often feather-veined leaves, and frequently staminate and pistillate (monœcious) flowers. Barren flowers (staminate) in catkins; stamens five to twenty, inserted in the base of scales, or of a membranous valvate perianth. Fertile flowers (pistillate) aggregate, or in a spike. Ovary with several cells, enclosed in an involucre or cup (cupule); ovules in pairs or solitary; stigmas, several. Fruit a nut with a husk or cup; seed solitary, without albumen. The plants abound in the forests of temperate regions in the form of oaks, hazels, beeches, and chestnuts. They afford valuable timber and edible seeds, and their bark is astringent. *Quercus* includes the various species of oak, which are well characterised by their acorns. Liebman says that there are 230 species of oaks known, belonging chiefly to the Northern Hemisphere. To the south of the Line they occur in the Sunda Islands. They are not met with in the temperate zone of the Southern Hemisphere. *Quercus pedunculata* or *Robur* is the common British oak, which has usually stalked acorns. *Q. sessiliflora* is the Durmast with sessile fruit, which by some is reckoned only a variety of the former. The Durmast furnishes the best timber. In the common oak the medullary rays are large and the wood is easily rent; in the Durmast the rays are small and the wood not easily rent. Common oak taken from a ship broke under an average weight of 931 lbs., only bending 4½ inches; while Durmast from the same ship broke with an average weight of 1,032 lbs., and deflected 5½ inches before breaking. Durmast grows faster than common oak, and it was used in many ancient buildings, as in Glasgow Cathedral. The cups of *Quercus Ægilops* are used by dyers under the name of valonia. The outer bark of *Quercus Suber* supplies cork. *Corylus Avellana*, the common hazel, yields excellent charcoal for drawing. *Fagus sylvatica*, the beech, and *Castanea vulgaris*, the Spanish chestnut, are cultivated for timber. *Castanea chrysophylla* is the golden chestnut from Oregon. There are eight or nine known genera, and about 280 species. Illustrative genera:— *Corylus, Carpinus, Fagus, Castanea, Quercus.* (J. H. B.)

CORYLOPSIS. The name of a genus belonging to the order of Witch-hazels, characterised by the calyx being adherent to the ovary, and divided above into five unequal pieces; corolla of five pieces, broadest upwards; filaments or stalks of the stamens awl-shaped and free; five short scales in the spaces between the stamens; styles or appendages on the ovary two in number, each thickest at the base, and ending in a round head or stigma. The name is derived from the Greek, and means 'Hazel-like,' indicating the general habit of the species, which are shrubs, natives of Japan, with alternate stalked leaves, heart-shaped or entire at the base and of short duration; the flowers are yellow. (G. D.)

CORYLUS. A small tree belonging to the *Corylaceæ*, and under the name of Hazel too well known to need any statement of the characters by which it may be identified. The usual form of the Hazel in its wild state is a straggling bush consisting of a number of long flexible stems from the same root. The bark on the young branches is a h-coloured and hairy, that on the old stem mottled with bright brown and gray. *C Avellana* includes not only the hazel, but all the European varieties of filbert and cobnut. Among the wild animals which feed on these nuts the most destructive are the squirrel, which carries them off for a winter hoard, or demolishes them on the spot, splitting the shell into two halves; the dormouse, which climbs the trees, and nibbles a round even hole, extracting the contents piece-meal; and the nuthatch, a bird not much bigger than a sparrow, belonging to the tribe Scansores, which carries them off singly, and fixing them in the crevice of an oak or some other rough-barked tree takes his position above, and, head downwards, hammers away with his strong beak until he has made an irregular angular hole. Many nuts are also rendered worthless by a beautiful little beetle (*Balaninus nucum*), which in early summer lays within the tender shell of a nut a single egg, which when the kernel is approaching maturity is hatched into a small grub. This, when the period of transformation to the pupa state is approaching, eats its way through the shell, and falling to the ground buries itself and constructs a cell from which it comes forth in the following season as a perfect insect. The hazel is rarely found of sufficient size to supply building materials, but the young rods being tough and flexible are much used for hoops, walking-sticks, fishing-rods, &c.; and from their smoothness and pleasing colour they are well adapted for making rustic seats and tables for summer-houses; they are also good fire-wood. The charcoal crayons used by artists for drawing outline are also prepared from hazel-wood. A purple-leaved variety to be obtained at the nurseries is a great ornament to shrubberies. Other species occasionally cultivated in England are *C. tubulosa* from Europe, *C. americana* and *rostrata* from America, and *C. Colurna* from Turkey. French, *Noisetier*; German, *Haselstande*. (C. A. J.)

The name of *Avellana* is said by Pliny, according to Prof. Targioni, to be derived from Abellina in Asia, supposed to be the Valley of Damascus, its native country. He adds that it had been brought into Greece from Pontus, hence it was also called *Nux pontica*. The nuts were called, by Theophrastus, Heracleotic nuts, from Heraclea, now Ponderachi, on the Asiatic shores of the Black Sea. Others admit that a variety of hazel nut or filbert was brought from Pontus to Abella, a town in Campania, and hence the name of Avellana was applied to these trees. In France, at the present day, the better varieties are called *Avelines*. But the above indications of an Eastern origin can only refer to particular kinds, for the species, it is well known, is common enough in Italy, as well as in other parts of Europe. It is also found over a great part of Asia in a wild indigenous state. It bears the common names of Hazel, Hazle, or Hasel, not only in this country, but also in Germany, Holland, Sweden, and Denmark. The plant is indigenous to all these countries. Its habitat extends from the extreme south of Europe to the most northern parts of Britain. According to De Candolle it is found wild in the mountains of the Island of Sardinia; and he is not certain whether its growth may not be natural in some ravines near Algiers. It is said to be not now found in Shetland; but formerly it had existed there, for the shells of its nuts are found plentifully in bogs, as they are likewise in similar places throughout Scotland. The ancient nut-shells are often met with in fragments, but many are found quite entire, at various depths below the surface; some of them are larger than those of the Wild Hazel, growing near the same localities at the present day.

The Hazel generally forms large bushes, from its great disposition to produce suckers; but if grown with a single stem it assumes the form of a low tree. One at Gordon Castle, North Britain, measured thirty feet in height, with a trunk three feet in circumference. The plants often form a sort of jungle on precipitous banks of rivers and streams, and may frequently be seen growing out of crevices and fissures of rocks, sometimes much confined for root-space, yet in that case roots will extend far downwards, naked along the face of the rock, till they reach soil below. The wood when two years old and upwards is tough and elastic, and it is well adapted for hurdles, crates, hoops, walking-sticks, &c. Its charcoal is esteemed for making gunpowder.

Nut leaves are large, roundish cordate, and somewhat pointed. The same tree bears male and female flowers, distinct from each other, proceeding from different buds. The male flowers begin to make their appearance in autumn, and acquire their full developement early in spring; they are at first compact cylindrical bodies of a greyish colour, afterwards they become long pendulous catkins of a yellow colour, giving the trees, then destitute of leaves, a conspicuous and rather ornamental appearance. The female flowers do not appear till spring. They exhibit a few crimson thread-like styles issuing from the apex of a bud. This bud elongates, and forms a small branchlet, at the extremity of which the cluster of nuts is borne. Until the nuts are nearly full-sized their yet soft green shell is filled with a milky juice, but this does not constitute the kernel. The latter may be observed at the same time not larger than the head of a pin. As it grows the milky substance is absorbed, all except the fibrous portion, which is deposited on the inside of the shell, forming a soft lining for the kernel. The calyx or husk has a fleshy base, to which the lower part of the nut is strongly attached until fully ripe, when the husk dries up and permits the nut to drop out, except in the case of some varieties, more especially those called filberts, which have long tubular husks contracted beyond the apex of the nuts. These were formerly called Full-beards, whilst those with short husks were simply termed Nuts or Hazel-nuts.

There are numerous varieties, differing in the form of the nuts, and in the relative length of their husks. The Red Filbert and White Filbert are similar in external appearance, but in the former the thin pellicle which forms the immediate coating of the kernel is red or crimson, that of the latter white or pale-brown. Both these are esteemed because they admit of being kept fresh in the husks. Short roundish nuts with a strong thick shell are called Cob nuts. Of this description are most of those imported from Spain. The Cosford nut is of an oblong form with a comparatively thin tender shell, finely striated longitudinally. The sorts above-named, together with the Downton large square nut, and the large Spanish, are amongst the best sorts for cultivation.

In this country, the neighbourhood of Maidstone in Kent is the most celebrated for the cultivation of filberts. The foreign supply is chiefly from Spain. Phillips states that from a single wood near Recus, 60,000 bushels have been gathered in one year, and shipped from Barcelona, whence they are called Barcelona nuts. 'In the neighbourhood of Avelino in Italy,' says Swinburn, 'the whole face of the neighbouring valley is covered with nut trees, and in good years they yield a profit of 60,000 ducats.' According to French authors the nuts of Provence and Italy are preferable to those of Spain and the Levant. [R. T.]

The common Hazel, *C. Avellana*, is the badge of the clan Colquhoun.

CORYMB (adj. CORYMBOSE). A raceme, whose pedicles are gradually shorter as they approach the summit, so that the result is a flat-headed inflorescence, as in candy-tuft. — COMPOUND. A branched corymb, each of whose divisions is corymbose.

CORYMBIFERÆ. Corymb-bearing composite plants, a sub-order of the natural order *Compositæ* or *Asteraceæ*, containing plants with numerous flowers on a common receptacle, forming a head surrounded by a set of floral leaves or bracts called an involucre. The heads of flowers are either placed singly on stalks; or there are several stalked heads supported on a common axis, and so arranged as to have collectively the form of a corymb, the lower stalks being longer so as to bring the heads to nearly the same level. The flowers in the circumference of the heads are usually ligulate and bear pistils only, while those of the centre are tubular and have both stamens and pistils. The style of the perfect flowers is not swollen below the stigma. Such plants as chamomile, the daisy, the ox-eye, the dahlia, everlasting, sunflower, cineraria, ragwort, and groundsel belong to this sub-order. The plants have bitter qualities; some of them induce sleep, and they usually contain more or less of a volatile oil. [J. H. B.]

CORYMBIS. Under this name, and those of *Corymborchis, Centrosis, Rhyncanthera, Macrostylis*, and *Hysteria*, botanists have described a singular herbaceous plant found in the tropical parts of Africa and Asia, with the habit of a small bamboo, and long slender white flowers. After flowering the column grows to a great length, with the remains of the other organs at the base. Only one species, *C. disticha*, is well known; two others very like it are described.

CORYMBIUM. A genus of S. African plants of the composite family, some of them common on the flats about Cape Town, and on Table Mountain. They are perennial plants about two feet high, with grassy root leaves, which have parallel nerves, and are furnished at the base with a tuft of woolly hairs. The stem bears a number of small linear leaves, and terminates in a dense corymb of flowerheads, each of which contains but a single floret —a circumstance unusual in the family, by far the greater portion having many florets collected in one flower-head. The achene is clothed with long soft hairs, and crowned by a pappus of short scales. [A. A. B.]

CORYNEPHORUS. A genus of grasses belonging to the tribe *Aveneæ*. It is not considered essentially distinct from *Aira*, under which it is described in Steudel's Synopsis, as the *Aira canescens* of Linnæus. It is a rare grass in Britain, and is not found wild out of England. [D. M.]

CORYNEUM. One of the most remarkable genera amongst the coniomycetous *Fungi*, distinguished by the dark naked elongated articulated spores, radiating in every direction from a little raised cushion-like receptacle. It is distinguished from *Hendersonia*, which has somewhat similar spores, by the absence of any surrounding cyst or perithecium. From *Bactridium* it is separated by its more developed receptacle and dark, not coloured, spores. The species grow on dead twigs. *C. Kunzei*, which is not uncommon on oak, affords a pretty microscopical object. [M. J. B.]

CORYNIDIA. Processes sunk into the margin of the germinating leaf of ferns, and containing spiral threads.

CORYNOCARPUS. A New Zealand genus of handsome trees belonging to the order *Myrsinaceæ*. The leaves are entire and smooth; and the flowers small, white, in terminal clusters. The sepals and petals are five in number, the latter provided with a narrow claw; alternating with the petals are five ascending scales, each with a small globular gland attached to it; there are five stamens; the ovary is globular. The fruit is club-shaped, hence the name of the genus; it contains but one seed. *C. lævigatus* is in cultivation in this country. The tree, according to Dr. Bennett, is valued in New Zealand for the sake of its fruit and seeds; the former is of the size of a plum, pulpy in the interior and sweet. The seeds are used in times of scarcity, and contain a tasteless farinaceous substance. The raw seeds, however, are poisonous, and produce spasmodic pains, giddiness, and partial paralysis, to obviate which effects they are steamed for twenty-four hours, and then either buried in the ground, or allowed to soak in water for some days. [M. T. M.]

CORYNOSTYLIS. Tropical American climbing shrubs of the violet family, with entire saw-toothed leaves, deciduous stipules, and large handsome flowers. The sepals are nearly equal; the five petals very irregular, the anterior ones the smallest, the lateral ones erect, the hinder one very large and prolonged at the base into a spur; the five stamens have short filaments prolonged into a hairy appendage at the base, the anthers surmounted by a membranous crest; the ovary is somewhat globular, three-celled; the style terminal, club-shaped; the stigma ciliated, lateral; the fruit a capsule with many seeds. [M. T. M.]

CORYPHA. A genus of fan-leaved palms composed of about five species, all natives of tropical Asia, and mostly forming tall trees. All their flowers are perfect, and produced on branching spikes, which are surrounded at the base by numerous leafy bracts. They have a cup-shaped calyx, the rim of which is cut into three teeth; a three-petaled corolla; six stamens, whose bases are dilated so as to join one another; and three ovaries, which cohere, and have their awl-shaped styles united together and crowned by a simple stigma. The fruit is a one-seeded berry.

The Talipot palm, *C. umbraculifera*, is a native of Ceylon and the Malabar coast, where it grows to sixty or seventy feet high, with a straight cylindrical trunk, marked by rings, and surmounted by a crown of gigantic fan-like leaves. These leaves have prickly stalks six or seven feet long, and when fully expanded form a nearly complete circle of thirteen feet in diameter, and composed of from ninety to a hundred radiating segments, joined together and plaited like a fan till near the extremity, where they separate and form a fringe of double points. Large fans made of these leaves are carried before people of rank among the Cinghalese; they are also commonly used as umbrellas, and tents are made by neatly joining them together; besides which they are used by the natives as a substitute for paper, being written upon with a style. Some of the sacred books of the Cinghalese are composed of strips of them. The hard seeds are suitable for turnery purposes. *C. Taliera* is a native of India, and is closely allied to the preceding, but does not grow more than thirty feet high. Its leaves are used for the same purposes. *C. Gebanga* is called Gebang in Java, where it is a native. The leaves are used for thatching, plaiting into baskets, hats, and similar articles. From the interior of the trunk a kind of sago is obtained; and the sliced root is said to be an efficacious remedy for diarrhœa. [A. S.]

CORYSANTHES. Curious little swamp orchids, inhabiting Australia and Java, have received this name in allusion to their large dorsal sepal having the form of a helmet. They have thin roundish solitary leaves, from the axil of which rises a single purple and green flower. One species, named *Calceolaria* by Blume, grows among damp moss on the summit of Mount Salak in Java.

COSCINIUM. A remarkable genus of *Menispermaceæ*, characterised by its large petals, the irregularly-mottled albumen, and the structure of its embryo, which has its radicle superior, pointing towards the apex of the drupe-like fruit, while the cotyledons are rounded, widely-spreading, either perforated with holes, or, according to Miers, deeply-gashed; but they are so thin as not readily to be taken from the albumen on which they lie. *C. fenestratum*, formerly called *Menispermum fenestratum*, is considered in Ceylon to be a valuable stomachic and tonic. The wood, which has a peculiar structure, described in Hooker and Thomson's *Flora Indica*, is of a yellow colour, and yields a yellow dye. Medicinally the wood, bark, and root, are used as tonics. [M. T. M.]

COSMANTHUS. A small genus of annual hydrophylls, closely allied to *Eutoca* and *Phacelia*, so closely in fact, that by some botanists both *Eutoca* and *Cosmanthus* are regarded as only sections of the genus *Phacelia*. It scarcely differs from the latter, but in its fringed corolla and procumbent habit; from *Eutoca* it is distinguished by the former character, and by its fewer and larger seeds. The only species at all known in this country, *C. fimbriatus*, and which may serve as a type of the genus, is a neat procumbent plant, with rather succulent branched angular spreading stems, pinnatifid leaves, those of the stem stalkless and stem-clasping, those at the root on long stalks, all with ovate entire lobes; it has very pale lilac purple flowers produced in a curled or crook-like raceme, the corolla wheel-shaped, and having at the base of each lobe a scale rolled into a tubular form; five linear calyx segments, five stamens with hairy filaments, a single style with a circle of hairs at its base, and a four-seeded pod, complete the description. The species are all natives of North America, and appear to be destitute of any marked properties. [W. T.]

COSMELIA. A genus of *Epacridaceæ*, consisting of two species of erect marshy plants, with glossy leaves sheathing the stem, and solitary reddish flowers at the termination of the short branches. The foliaceous calyx is surrounded with many imbricated bracts; the corolla is five-parted; the anthers are exserted; there are five hypogynous scales; the capsule is five-valved and many-seeded. The species are natives of New Holland. [W. C.]

COSMIDIUM. A genus of composites, recently separated from *Coreopsis*, from which it differs in having an elongated obscurely four-angled, and minutely mammillated fruit, crowned by two short thick horns, and partially enveloped in a membranous scale, which remains attached when the fruit separates from the receptacle. In general habit and aspect, the species approach very closely to *Calliopsis*, having, like that, smooth erect branched stems, opposite leaves, pinnatifidly cut into distinct thread-like segments, and flower-heads an inch and half in diameter, with a ray of about eight broadly wedge-shaped florets, and a double involucre surrounding the capitule, each series consisting of eight leaflets, the innermost broad and erect, the outer narrow, spur-like and spreading. *C. filifolium*, till recently the only species generally known or cultivated, has the ray florets yellow, and the disk or centre crimson-brown. The beautiful *C. Burridgeanum*, of gardens, which is perhaps but a variety of the preceding, has larger flower-heads, with the ray florets of a deep purple brown at their base, the tip only being orange yellow. The fruit of this plant is considerably shorter and thicker than in *C. filifolium*. [W. T.]

COSMOPHYLLUM. The name given to a genus of the composite family found in Guatemala. *C. cacaliæfolium* is the only known species; it is described as a shrub or small tree, with leaves one to two feet in length, oval in contour, with seven triangular lobes, and having their surface clothed with short white down. The flower heads have some resemblance to those of the chamomile, and are disposed in terminal corymbs; the outer florets are white, strap-shaped and contain a pistil only; the inner are yellow, tubular, and perfect. The four-sided achenes bear on their angles rough points, and are crowned with a hard short pappus composed of a number of unequal-cut scales. [A. A. B.]

COSMOS. A small genus of composites allied to *Bidens*, with large showy reddish-purple or yellow flower-heads, and finely divided or pinnate foliage. They are better known in gardens by Willdenow's name of *Cosmea*, but *Cosmos* has priority in its favour. The genus has a double involucre, as in *Coreopsis*, each series composed of from eight to ten ovate leaflets, the outer ones spreading, the inner ones erect; the receptacle is flat and set with membranous coloured scales, drawn out to a thread-like point; and the fruit is four-angled, tapering to both ends, and crowned with from two to four deciduous barbed awns. *C. bipinnatus* is a handsome annual, attaining in moist soil a height of four or five feet, with a smoothish erect furrowed stem, spreadingly-branched; opposite bipinnate leaves, the segments of which are linear, pointed, and somewhat curled; and flower heads two inches or more in diameter, on long peduncles, the ray florets about eight in number, of a bright red purple, the disk being composed of yellow florets tubular,

The fruit of the species is smooth and usually furnished with three awns; but that of *C. tenuifolius*, a dwarfer species with more finely-divided foliage, and darker flowers, is rough, and more frequently has but a single awn. All the species are natives of Mexico. Under the name of *Dahlia Zimapani*, a new species, *C. diversifolius atropurpureus*, has recently been introduced, with pinnate dahlia-like foliage, and flower heads varying from blackish-purple to red-purple, on very long peduncles, the disk being of the same colour as the ray florets. [W. T.]

COSMOSTIGMA. A genus of *Asclepiadaceæ* containing a single species, a branched twining shrub, that climbs over trees of great height in India. It has opposite leaves with conic glands at their base, and many small flowers in racemes on interpetiolar peduncles. The small calyx is five-parted; the corolla is rotate and five-parted; the staminal corona consists of five bifid divisions, which are irregularly toothed on their upper and inner margins; the anthers are terminated by a broad membrane; the oval pollen masses attached by long-kneed processes to a small bifurcate corpuscle; the follicles are large, linear, oblong, and smooth, with ovate comose seeds. [W. C.]

COSSIGNIA. A genus of *Sapindaceæ*, differing from the others in the family in having flowers with petals, together with a capsular but not bladdery fruit, which is three-celled, each cell containing two or three small black seeds. The two known plants of this genus are natives of the Mauritius, where they are known as Bois de Judas. They are small trees with pinnate leaves made up of one or three pairs of oblong or oboval entire leaflets and an odd one; these are about two inches long, smooth above, and covered underneath (as are all the young parts) with a short white down. The small white flowers, disposed in terminal panicles, have a five-parted calyx, four or five oval petals larger than the calyx, and a like number of stamens inserted on a disc. The three-lobed ovary is crowned with a single style. [A. A. B.]

COSTA. The midrib of a leaf; that part which is a direct extension of the petiole, and whence the veins arise; a leaf may have many costæ.

COSTATE. When there is only one rib, as in most leaves. Also the mere adj. of costa.

COSTATO-VENOSE. When the parallel side-veins of a feather-veined leaf are much stouter than those which intervene.

COSTMARY. *Pyrethrum Tanacetum*, sometimes called *Balsamita vulgaris*.

COSTUS. A genus of tropical herbs belonging to the *Zingiberaceæ*, and having tuberous roots, somewhat fleshy leaves, and flowers in spikes with overlapping bracts. The calyx is tubular and three-cleft; the tube of the corolla is funnel-shaped, the outer segments of the limb equal, the inner lateral ones (sterile stamens) wanting, while the innermost or middle segment, called the lip or labellum, is large, bell-shaped, cleft at the back; the filaments are petaloid, prolonged beyond the anther on all sides. Ovary with three compartments; the style thread-like, passing between the cells of the anther; the stigma two-cleft, with two small horns at the base. Many of the species are highly ornamental as stove plants, such as *C. speciosus*, the roots of which are used by the natives in a kind of preserve. [M. T. M.]

COSTUS. The roots of an Arabian plant, supposed to be allied to *Cardopatium corymbosum*. The Costus of the ancients has, however, been ascertained to be the root of *Aucklandia Costus*, now *Aplotaxis Lappa*.

COTONEASTER. A family of small trees or trailing shrubs belonging to the order *Rosaceæ*, and allied to *Mespilus*, inhabiting the northern parts of Europe and the mountains of India. The leaves are small and entire at the edge, downy beneath; in some species evergreen; the flowers, which are white or pinkish, grow either in lateral clusters, like those of the hawthorn, or singly, and are succeeded by scarlet, or less commonly black, berry-like fruit. 'The species are very desirable from the beauty of their foliage, flowers, and fruit; the fruits of *C. frigida* and *C. affinis*, in particular, being produced in great abundance, and being of an intense scarlet colour, have a very splendid appearance, and remain on the trees the greater part of the winter. Though the greater part are natives of Asia, yet in Britain they are found to be as hardy as if they were indigenous to the north of Europe, especially such of them as are true evergreens. *C. vulgaris*, a species with deciduous leaves, has been in cultivation in British gardens since 1656, and was always considered a foreign plant, till it was found in a wild state at Orme's Head in Carnarvonshire.'—(*Loudon*). *C. microphylla* is a yet more valuable plant. In this species the branches are trailing, the leaves small and evergreen. It is perfectly hardy and, wherever it grows, ornamental. 'Its deep glossy foliage, which no cold will impair, is, when the plant is in blossom, strewed with snow-white flowers, which, reposing on a rich couch of green, have so brilliant an appearance, that a poet would compare them to diamonds lying on a bed of emeralds.'—(*Lindley*). *C. marginata*, *rotundifolia* and *buxifolia*, are of similar habit. The last species were introduced from the hills of Hindostan in 1824 and 1825. [C. A. J.]

COTONNIER. (Fr.) *Gossypium*.

COTTON. This well-known valuable textile commodity is the hairy covering of the seeds of *Gossypium herbaceum* and other species of *Gossypium*, especially of *G. religiosum*, *barbadense*, *indicum*, and *arboreum*. — COREWOOD. A name given in Trinidad to the down of *Ochroma Lago-*

pus. —, NATAL. A textile material resembling true cotton, obtained from the pods of a species of *Datura*. —, SILK. A common name for *Bombax*. *B. jaxiandrum* is called the Indian cotton-tree.

COTTONIA macrostachya is an orchideous epiphyte from the Madras presidency and Ceylon, with a few greenish purple-lipped flowers at the end of a long lateral slender peduncle. The foliage is that of a *Saccolabium* or *Vanda*, so that the name of *V. pedunculuris* has been applied to the plant. Another species, *C. Championi*, found on both Victoria Peak Hong Kong, and the Khasya mountains, has smaller dirty lemon-coloured flowers in racemes little longer than the distichous leaves, which are mucronate and even serrate at the point.

COTTON-GRASS. The common name for *Eriophorum*.

COTTON-ROSE. A common name for *Filago*.

COTTON-RUSH. A name sometimes given to *Eriophorum*.

COTTON-SEDGE. A name given by Bentham to *Eriophorum*.

COTTON-THISTLE. A common name for *Onopordum*.

COTTON-WEED. *Diotis maritima*.

COTTON-WOOD. An American name for *Populus monilifera* and *P. angulata*.

COTULA. A genus of weedy compound flowers allied to *Anthemis*, from which it is distinguished by its hemispherical naked receptacle, four-cleft florets of the disk, and by the ray being almost wanting. There are numerous species, of which one only, *C. coronopifolia*, is found in Europe. There is no British example. *Cotula* is a diminutive of *Cota*, the old name of some species of *Anthemis*. [C. A. J.]

COTYLEDON. A genus of shrubs and herbaceous plants belonging to the Crassulaceæ, among which they are distinguished by their five sepals, tubular five-cleft corolla bearing ten stamens, and a scale at the base of each of the five carpels. The only British species, *C. Umbilicus*, Navelwort, or Penny-wort, is a common weed in the west of England and some parts of Wales and elsewhere, growing on the sides or in the crevices of damp rocks and walls, where it is conspicuous during the winter and spring months by its orbicular concave peltate exceedingly succulent leaves, called by children Penny-pies. In summer it sends up a stalk, the lower portion of which bears succulent leaves, which gradually lose their peltate form and pass into bracts. The stalk, when the plant grows in a dry situation, is from four to six inches long, and bears a simple spike of drooping green flowers; but in a more genial situation grows to the height of a foot or more and is branched. After the seeds have ripened, the stems wither and turn brownish red, but retain their form during a great part of the winter. Of the foreign species several are natives of the Cape of Good Hope; these are evergreen under-shrubs, and are sometimes found in the green-houses of the curious. *C. orbiculata*, which is the one most frequently cultivated, has thick and succulent leaves tinged at the edge with purple. The flowers are large drooping, and have the divisions revolute and of a reddish hue: they last from June to September. *C. lutea* is by some authors enumerated among British plants, but without due grounds. It is a native of Portugal. [C. A. J.]

COTYLEDONS. The seed-lobes; the primordial leaves in the rudimentary plant or embryo.

COTYLIFORM. Dished. Resembling rotate, but with an erect limb.

COUCH-GRASS. *Triticum repens*.

COUCOU. (Fr.) *Primula officinalis*.

COUCOURZELLE. (Fr.) A kind of gourd.

COUDRIER. (Fr.) *Corylus Avellana*. — DU LEVANT. *Corylus Colurna*.

COUEPIA. A genus of the chrysobalan family, whose distinguishing characters are its one-celled ovary, which adheres to the calyx tube, and its numerous stamens (twenty to forty or more), which arise from one side only of the mouth of the calyx, or are disposed round it in a perfect ring. The genus comprises upwards of a dozen species, all of them trees of South America, generally small, but sometimes attaining a height of fifty feet. Their leaves are entire, usually oblong, and very often covered with short white down underneath. The flowers, numerous and seldom more than half an inch in diameter, are either white or cream-coloured, and when in bud have a shape exactly like that of a clove; they are disposed in terminal or axillary panicles or racemes, and are composed of a tubular calyx with a five-parted border, five petals, numerous stamens, and an ovary with a simple style arising from near its base. The oval stoned fruits of a number of species are eaten. *C. chrysocalyx* is a beautiful tree of a pyramidal form, branching to the base, and attaining a height of thirty feet. According to Mr. Spruce, it grows plentifully all along the Amazon river from the Barra upwards. The Indians plant it also near their houses for the sake of its edible fruits, and a large puebla on the Marmon of Cucama Indians derives its name 'Parinari' from the abundance of this tree, so called. Its oblong pointed or blunt leaves have a smooth upper surface, and are covered underneath with short white down. The flowers, about an inch in length, have a calyx covered with yellow down, and are borne in axillary racemes much shorter than the leaves. *C. guianensis* is, according to Aublet, a tree of sixty feet high, with grey shining bark, and dark red-coloured wood, which is durable and heavy. The leaves are oval, acute, and stalked. The Indians make use of the bark in the manufacture of their pottery. The Caribbean name of the tree is Couepi, whence the origin of the generic name. *C. bracteata*, a Brazilian tree forty feet high with leaves half a foot long, and panicles of flowers furnished with large bracteas, is remarkable in the family, according to Mr. Spruce, for the fetid odour of its cream-coloured flowers. [A. A. B.]

COUGOURDETTE. (Fr.) *Cucurbita ovifera*.

COULEUVRE'E. (Fr.) *Bryonia dioica*.

COUMARIN. The fragrant principle of the Tonka bean, *Dipterix odorata*, and also of *Melilotus cœrulea*, the latter of which gives its peculiar odour to Chapsiger cheese.

COUNTRYMAN'S TREACLE. An old name for *Ruta graveolens*.

COURGE. (Fr.) *Cucurbita maxima*. — DE SAINT-JEAN. *Cucurbita Pepo*.

COURONNE DES BLE'S. (Fr.) *Agrostemma Githago*. — IMPE'RIALE. *Fritillaria Imperialis*, and also *Cucurbita Melopepo*.

COUROUPITA. A genus of trees belonging to the order *Lecythidaceæ*, and natives of tropical America. The clusters of flowers spring from the trunk and branches. The flowers are large whitish or rose-coloured, with a top-shaped calyx-tube, adherent to the ovary, its limb having six deciduous segments; the corolla consists of six petals inserted into a disc, which surrounds the top of the ovary; the cup formed by the union of the filaments of some of the stamens is inserted with the petals: on one side it is very short, on the other it is prolonged into a petal-like hood overlapping the style, and bearing anthers at its top; the stamens at the base of the cup are minute and barren, those at the apex of the petal-like hood are fertile; ovary with six compartments; stigma sessile, hexagonal. The fruit is large, globular, and woody, marked with a circular scar indicating the point of detachment of the limb of the calyx; the seeds are numerous and imbedded in pulp. The fruit of *C. guianensis* is called from its appearance the Cannon-ball fruit, its shell is used as a drinking vessel, and its pulp when fresh is of an agreeable flavour. [M. T. M.]

COURY. A kind of Catechu, obtained by evaporating a decoction of the nuts of *Areca Catechu*.

COUSINIA. A genus of prickly-leaved thistle-like plants of the composite family, found in Western Asia, occurring as far east as Kunawur in the Himalaya, having their western limit in Asia Minor and found in greatest numbers in Persia. They are nearly allied to *Carlina*, but differ in having a simple-haired, not feathery, pappus. Upwards of thirty species are enumerated, some of which are annual, others perennial; some dwarf and prostrate, others tall and erect. The root leaves of many are pinnately-parted, with spiny segments, and covered, especially underneath, with a loose white cottony substance; those of the stem, similarly cut and spiny, often have their bases decurrent, which gives the stem a winged appearance. Others have leaves, which in size and form are not unlike those of the holly. The flower-heads are either large and few on the ends of the branches, or numerous and small; their involucres, made up of many spiny-pointed scales, enclose a great number of yellow or pink florets. The achenes are smooth, or have rough points, or longitudinal furrows, and in some cases they are compressed and angled. The pappus is composed of two or three series of short and unequal rough hairs. [A. A. B.]

COUSSAPOA. A genus of tropical American trees, abounding in a milky juice, and belonging to the family *Artocarpaceæ*. The trees are described as being at first mere climbing shrubs, but after reaching the summit of the tree upon which they are supported, they send down branches into the earth, these branches becoming fused together so as to encircle completely the tree which originally sustained them, and cause its death. The branches are spongy in texture, and hollow in the interior. The flowers are diœcious and clustered in heads, the male flower encircled by three or four small bracts, and consisting of a tubular perianth, from whose base two conjoined stamens arise; the female flower without bracteles surrounding its perianth, and consisting of four leaflets in close approximation. The one-celled ovary becomes succulent when mature, as also does the investing perianth, so that a mulberry-like fruit is produced. [M. T. M.]

COUSSINET. (Fr.) *Oxycoccus palustris*.

COUTAREA. A genus of cinchonaceous trees inhabiting Guiana, &c., and having large whitish flowers. The corolla is funnel-shaped, its tube short, so that the six stamens project from it, and its limb six-parted. The fruit is a leathery capsule, bursting by two valves, and containing several kidney-shaped seeds. *C. speciosa* is a very handsome stove plant; its bark is used in Guiana as a substitute for cinchona. It is also known by the name of *Portlandia hexandra*. [M. T. M.]

COVENTRY BELLS. *Campanula Medium*, also called Canterbury Bells.

COWAGE. The Cow-itch, *Mucuna pruriens*.

COWANIA. A genus of the rosewort family, distinguished from its congeners by the ten-cleft calyx; corolla of five petals; seed vessels five to ten, closely covered with fine down, and when ripe, each crowned with a feathery appendage, consisting of the enlarged persistent style. The genus was named by David Don in honour of Mr. Cowan, who, in the course of visits to Mexico and Peru, introduced many plants of those countries into Britain. *C. plicata* or *mexicana*, the only species, is an interesting shrub, about two feet high when mature, with alternate small narrow leaves, the edges turned down, covered with glands on the upper surface, and on the lower, white with fine down. The flowers are numerous and of a yellow colour, very much resembling those of certain species of *Potentilla*. [G. D.]

COWBANE. *Cicuta virosa*; also an American name for *Archemora*.

COWBERRY. *Vaccinium Vitis idæa*. The name Cowberry is also applied in some parts of Scotland to the fruits of *Comarum palustre*.

COW-GRASS. *Trifolium medium*.

COWHAGE-CHERRY. The fruit of *Malpighia urens*.

COW-HERB. *Saponaria Vaccaria*.

COWITCH, COWAGE, or COWHAGE. The hairs of the pods of *Mucuna pruriens*, which are used as a mechanical anthelmintic.

COW-PARSLEY. *Heracleum Panaces*; also commonly applied to *Chærophyllum sylvestre*.

COW-PARSNIP. A common name for any *Heracleum*.

COW-PLANT, CEYLON. *Gymnema lactiferum*.

COW-QUAKES. *Briza media*.

COWRIE PINE. *Dammara australis*.

COWSLIP. *Primula veris*. — AMERICAN. The common name for *Dodecatheon*. — VIRGINIAN. *Mertensia* or *Pulmonaria virginica*.

COW-TREE. The Palo de Vaca of South America, *Brosimum Galactodendron*, sometimes called *Galactodendron utile*; also the Hya Hya of the same continent, *Tabernæmontana utilis*. The name has besides been given to *Ficus Saussureana*, and other species of figs; and is, according to M. Desvaux, applied to *Clusia Galactodendron*.

COW-WEED. *Chærophyllum sylvestre*.

COW-WHEAT. A common name for *Melampyrum*.

CRAB. *Pyrus Malus*. — QUEENSLAND. *Petalostigma quadrilocularis*. — SIBERIAN. *Pyrus baccata* and *P. prunifolia*.

CRAB OIL. The oil obtained from *Carapa guianensis*.

CRAB'S EYE LICHEN. *Lecanora pallescens*, which was formerly gathered under this name in the north of England for the dyers. [M. J. B.]

CRAB'S-EYES. The seeds of *Abrus precatorius*.

CRAB-WOOD. The timber of *Carapa guianensis*.

CRACCA. The name given to a few slender perennial herbs or small bushes of the pea family, which were at one time placed in the genus *Tephrosia*, from which they differ in having no cup-shaped disc round the ovary. Six species are enumerated, all of them confined to tropical America. Their leaves are unequally pinnate, with four to twelve pairs of small opposite leaflets mostly elliptical in form, and the flowers (about the size of those of a vetch) are arranged in axillary racemes. The straight narrow pods are thin, smooth, and contain a number of seeds. [A. A. B.]

CRAKEBERRY. *Empetrum nigrum*. — PORTUGAL. *Corema alba*.

CRAM DES ANGLAIS. (Fr.) *Cochlearia Armoracia*.

CRAMBE. A genus of *Cruciferæ*, consisting of several species, of which two are edible, namely, *C. maritima* and *C. tatarica*. The former is our well-known Sea Kale. The latter is the Tatar Kenyer or Tartarian bread of the Hungarians, of which an interesting account is given in Loudon's *Encyclopædia of Plants*, p. 557; but we are not aware of any attempt having been made to cultivate it in this country, although the plant is stated to have been introduced in 1789.

The Sea Kale, *C. maritima*, is a hardy native perennial, found on various parts of the coast, growing among sand and shingle. It is easily recognised by its broad wavy toothed gray-coloured leaves, which, with the stem, have a peculiar appearance, from being glaucous, or covered with a very fine bloom. The flowers are white and have a strong smell of honey. It appears to have been known to the Romans, who gathered it in its wild state, and preserved it in barrels for use during long voyages. From a remote period it has also been used in this country by residents near the sea, but its introduction into our gardens is comparatively of recent date, although it is recorded that bundles of it were exposed for sale in Chichester market in 1753. It was not known about London until 1767, when Dr. Lettsom cultivated it at Camberwell, and was the first to bring it into general notice. It has now become a common vegetable, and when blanched, the young shoots and leaves, before their complete development, are cut and tied up in small bundles for boiling. When thoroughly dressed they are served like Asparagus, and are esteemed exceedingly choice and delicate. [W. B. B.]

CRANBERRY. The fruit of *Oxycoccus palustris*, also sometimes applied, according to Lindley, to those of *Vaccinium Vitis idæa*. — AMERICAN. *Oxycoccus macrocarpus*. — TASMANIAN. *Astroloma humifusum*.

CRANE'S BILL. The common name for *Geranium*.

CRANICHIS. A rather numerous genus of American orchids, mostly tropical, with the habit of *Spiranthes*, but with a dorsal concave not convolute lip. The flowers are insignificant, and the species scarcely more than weeds.

CRANIOLARIA. A genus of Pedaliads, distinguished from its congeners by the somewhat bell-shaped calyx, which is cleft or five-toothed, and by the tube of the corolla widening toward the upper part, where it is bell-shaped and two-lipped, the upper lip of two pieces, the lower of three, the middle piece of the latter longer than the other two. The name of the genus was given in allusion to some resemblance which the ripe fruit has to the skull, in Latin 'cranium.' The species are herbaceous, natives of the tropical parts of America, usually very hairy and viscid; the leaves are opposite angled or five-lobed, the flowers from the axils of the leaves or terminal, the corolla being generally pale, with the throat variegated. The genus was originally formed to comprehend a plant known as the *Martynia Craniolaria*, first introduced in 1733, and which is now *Craniolaria annua* : a handsome greenhouse plant, easily cultivated, attaining a height of two feet, with leaves somewhat heart-shaped, five-lobed and toothed, the tube of the corolla longer than the calyx, which has at the base two leaflets or bracts. Dr. Lindley states 'that its fleshy and sweet root is preserved in sugar by the Creoles as a delicacy. In the dry state it is said to be a bitter cooling medicine.' Another species is the *C. unibracteata*, which is perennial; the tube of the corolla is as long as the calyx, which has one bract. The flowers are in clusters, sulphur yellow, with purple dots. [G. D.]

CRANIOSPERMUM. A small genus of Siberian herbs belonging to *Boraginaceæ*. They are hairy with obovate or linear leaves, and rather small rose-coloured flowers with a five-parted calyx, a tubular corolla, five-cleft at the mouth, the segments erect, the throat without scales; stigma capitate; nuts four, obliquely depressed at the apex, affixed to a four-sided pyramidal central column, the disk subconcave with a narrow margin. [J. T. S.]

CRANSON. (Fr.) *Cochlearia officinalis.* — RUSTIQUE. *Cochlearia Armoracia.*

CRAPAUDINE. (Fr.) Any *Sideritis.*

CRAQUELIN. (Fr.) *Fragaria collina.*

CRASPEDARIA. A name given by Link and others to various polypodiaceous ferns, now referred to the genera *Goniophlebium*, *Niphobolus*, &c. [T. M.]

CRASS. Something thicker than usual. Leaves are generally papery in texture; the leaves of cotyledons, which are much more fleshy, have been called crass.

CRASSULACEÆ. (*Sempervivum, Succulentæ, House-leeks, Stonecrop* family.) A natural order of polypetalous calyciflorai dicotyledons, included in Lindley's violal alliance. Succulent herbs or shrubs with exstipulate (no stipules) leaves, and clustered flowers, which are often turned towards one side; sepals three to twenty, more or less combined; petals three to twenty, separate or united; stamens equal in number to the petals or twice as many; ovary composed of numerous one-celled carpels, having scales at their base; fruit consisting of follicles. Natives of dry places in all parts of the world. They are found on naked rocks, old walls, or hot sandy plains, alternately exposed to the heaviest dews of night, and the fiercest rays of the noon-day sun. Acridity prevails in many plants of this order. Some species are cooling in their properties, others are astringent. *Sedum acre* is very acrid, and is hence called Wall-pepper; it is abundant on sandy shores. *Sempervivum tectorum*, the Houseleek, is so called from being grown on the tops of houses. *Bryophyllum calycinum* has the property of producing leaf-buds along the margin of its leaves. There are about 470 species, distributed among twenty-four genera, of which *Crassula*, *Bryophyllum*, *Sedum*, *Sempervivum*, and *Penthorum* are examples. [J. H. B.]

CRASSULA. A well-known genus giving its name to the order *Crassulaceæ*. It consists of herbs or shrubs, with, for the most part, more or less fleshy leaves and stems, and white or pink flowers in loose cymes or compact heads. The form and disposition of the leaves vary in the different species; frequently the two opposite leaves are conjoined at the base, as in *C. perfoliata*.

Crassula perfoliata (stem and leaves).

The sepals are five, shorter than the five petals; the stamens are five, perigynous; there are also five hypogynous scales; the ovaries are five, distinct one from the other, and ripening into as many few or many-seeded follicles. Some of them are found in the Mediterranean region, but the head-quarters are at the Cape of Good Hope. Numerous species are in cultivation, some of them frequently produce little leaf-buds in place of flowers on their inflorescence. [M. T. M.]

CRATÆGUS. A well-known family of moderate-sized trees, commonly called thorns, belonging to the sub-order *Pomeæ*, of rosaceous plants, closely allied to the medlar, *Mespilus*, from which it is distinguished by the small (not leaf-like) segments of the calyx, and by the different form of the fruit. The thorns are natives of Europe, North America, and the temperate regions of Asia and Africa, bearing for the most part a great resemblance to one another in habit of growth, and agreeing generally in having cut leaves, white fragrant flowers, and scarlet berries, though there are exceptions to all these characters. All are hardy and ripen their fruit in the climate of Great Britain, and being very ornamental, both when in flower and fruit, are highly prized by the landscape gardener. *C. Oxyacantha*, the Hawthorn, is to be met with on a dry soil in most parts of Europe, in the North of Africa, and in Western Asia, varying greatly in size according to soil and climate, and presenting in the shape, size, and surface of its leaves, and in the colour of its berries, numberless shades of difference. The leaves vary also in their amount of pubescence; and the flowers, though generally white and fragrant, sometimes have an unpleasant fishy smell; they are either tinged with red, or, in some cultivated varieties, are of a full pink or crimson. The fruit or 'haw,' too, varies greatly in size, shape, and colour, being sometimes oblong, sometimes nearly globular, sometimes downy, at other times smooth and polished. Varieties have been observed in which it changes its usual crimson hue for black, orange, golden-yellow, or white. In some districts, each haw contains a single nut, in others they more frequently contain two. In spite, however, of all these liabilities to variation, a hawthorn tree can be distinguished at any season of the year without recourse being had to botanical characters; and a mere cursory examination of almost any other species of crataegus will suffice to assign it to its proper genus. Most of the cultivated species blossom in the month which has given to the Common Hawthorn the name of Maytree; but no one of them is more worthy of the title than that which has so long held it. Collections of thorns exist in various places in Europe, some containing from fifty to eighty sorts, including varieties: for a full account of which the reader should consult *London's Arboretum Britannicum*. French, *Aubépine*; German, *Hagedorn*. The hawthorn is the badge of the Ogilvies. [C. A. J.]

CRATERA. The cup-shaped receptacle of certain fungals.

CRATERIFORM. Concave, hemispherical, a little contracted at the base.

CRAT.EVA. A genus of the caper family consisting of shrubs or trees, natives of tropical regions, whose flowers have a four-parted calyx, a corolla of four stalked petals inserted on the margin of a hemispherical fleshy receptacle, and eight to twenty stamens inserted with the petals; ovary on a long stalk; stigma sessile; berry globular, one or two-celled, containing pulpy matter, in which the seeds are imbedded. *C. Nurvala*, a native of Malabar and the Society Isles, is a sacred tree in the latter islands, and is planted in graveyards. Its leaves are aromatic, bitter, and stomachic, and other parts of the tree are likewise used medicinally. The bark of the root of *C. gynandra*, the Garlic Pear, blisters like cantharides. Some of the species have a strong smell of garlic. [M. T. M.]

CRATOXYLON. A genus of oppositeleaved bushes or small trees of the St. John's wort family, found in the Malayan peninsula, China, Java, and the adjacent islands. Its chief distinguishing characters are the winged seeds, contained in a three-celled capsule, which when ripe is surrounded by the withered calyx. The leaves are stalked, or sessile and entire, generally lance-shaped or elliptical in form, but sometimes obovat. The flowers are white, chocolate, or rose-coloured, arranged in terminal panicles, or arising from the axils of the leaves. They have a five-leaved calyx, five roundish petals, and three or five parcels of stamens surrounding an ovary crowned with three styles. About ten species are known. *C. Hornschuchii*, a Javanese species, is said to be slightly astringent and diuretic. [A. A. B.]

CRAWFURDIA. A genus of Nepalese gentianaceous herbs with twining stems and large axillary flowers. They have a bell-shaped corolla whose limb is five-cleft, or ten-cleft, with five of the divisions smaller than the rest; filaments of the five stamens dilated; ovary one-celled, style straight; stigma two-cleft with oblong recurved lobes; disc hypogynous, five-lobed; capsule stalked, one-celled, many-seeded. [M. T. M.]

CREAM-COLOUR. White, verging to yellow, with little lustre.

CREAM FRUIT. *Bourpellia grata*.

CREAM OF TARTAR TREE. *Adansonia Gregorii*.

CREEPER, TRUMPET. An American name for *Tecoma radicans*.

CREMANIUM. A genus of tropical American shrubs or small trees belonging to *Melastomaceæ*. They have terminal panicles of small white flowers with the parts in fours or fives; the stamens twice as many as the petals; the berry globose, depressed at the apex, blue or violet, adhering to the circumscessile calyx, with three to five cells, and numerous seeds. *C. reclinatum* and *tinctorium* yield a yellow dye. [J. T. S.]

CREMASTRA. A little-known genus of terrestrial orchids from India and Japan, with broad ribbed leaves, and radical scapes bearing each a spike of dull-red tubular flowers. Two species are known. *Hyacinthorchis* of Blume is the same genus.

CRÈME D'ABSINTHE. A bitter aromatic liqueur prepared from *Artemisia Mutellina* and *A. spicata*.

CREMOCARP. Such fruits as that of umbellifers, consisting of two or more indehiscent inferior one-seeded carpels adhering round a distinct and separable axis.

CREMOLOBUS. A genus of *Cruciferæ* from Peru and Chili, consisting of herbs or undershrubs, with oblong or ovate leaves, and elongated racemes of numerous yellow flowers; filaments not toothed; pouch stalked, laterally compressed, constricted at the partition as in *Biscutella*, with orbicular valves winged on the back, tipt by the persistent style; seed solitary in each valve. [J.T.S.]

CREMOSTACHYS. M. Tulasne's name for a genus of *Nyctaginaceæ*, which had previously, unknown to him, received that of GALEARIA: which see. [A.A.B.]

CRENA, CRENATURE, CRENEL. A round or convex flat tooth.

CRENATE, CRENELLED. Having convex flat teeth. When these teeth are themselves crenated, bicrenate is the term which is used.

CRENATO-DENTATE. Divided at the edge into triangular notches.

CRENATO-SERRATE. When serratures are convex, and not straight.

CRENULATE. Having the edge divided into small crencls.

CRE'ON. (Fr.) *Pinus Pumilio.*

CREPIS. A genus of herbaceous plants, known as Hawk's-beards, belonging to the chicory tribe of compound flowers, and distinguished among its congeners by the soft whitish deciduous pappus which crowns the cylindrical achenes, which are destitute of a beak, or furnished with but a very short one. The species are common hedge plants throughout Europe, and are uninteresting. The most frequent British species is *C. virens*, a branched herb from twelve inches to two feet high or more, with leaves not unlike those of the dandelion (*Leontodon*), and numerous small yellow flowers. Its most favourite habitat is the moss of thatched cottages, but it grows also in dry hedges and in waste ground. *C. paludosa* is a much larger plant, not uncommon in moist woods, where it grows to the height of six feet or more. [C.A.J.]

CRESCENTI-PINNATISECT. When the lobes of a pinnated leaf become gradually larger as they approach the end.

CRESCENTIA. The typical genus of *Crescentiaceæ*. Its four species are inhabitants of the forests of tropical America, and are either small trees or large shrubs, having simple or trifoliate leaves arranged alternately or in clusters upon the stem. The flowers are produced upon the stem or old branches, and are distinguished by having a two-lipped calyx, with the lips undivided; the corolla being somewhat bell-shaped, and having a long tube puffed out on one side. Their fruits have a hard woody shell or rind, and contain numerous seeds nestling in pulp. *C. alata* is a native of Western Mexico, growing mostly in the vicinity of the sea-coast; but it is cultivated in the Philippine and Ladrone or Marianne Islands. It is called Tecomate in Mexico, and forms a tree about thirty feet high, with its leaves growing in clusters of three; the two outer ones being undivided and stalkless, while the central one is composed of three distinct leaflets, with a long winged stalk, and is compared to a cross by the inhabitants of the Philippines, the tree being called Hojacruz, and a decoction of its leaves used as a remedy against spitting of blood. The fruit is about the size and of the same colour as an orange, and contains a sourish-bitter pulp, which the Mexicans boil with sugar and administer internally as a cure for chest complaints; while the shells are converted into drinking cups.

C. cucurbitina, the Calabazo de playa of the Panamians, is a shrub about twelve or fifteen feet high, found growing very commonly on the coasts of Central America, the West Indian, and some of the Pacific Islands, and cultivated in Java. Its leaves are placed singly and alternately upon the stem, and vary very much in shape. Its fruit is either round, egg-shaped, or elliptical, and has a very brittle shell. This shrub has been reported to possess poisonous properties, but as the rest of the plants belonging to the order are of a harmless character, probably some mistake has occurred.

C. Cujete, commonly called the Calabash-tree, from the Spanish word Calabazo, which means a gourd or pumpkin, and alludes to the resemblance of the fruits, is a tree about thirty feet high, and is found growing either wild or cultivated in various parts of tropical America, and the West Indies. Its flowers are variegated with green, purple, red and yellow; and its leaves are arranged in clusters of five, all of them undivided, and of a narrowly elliptic form, the upper half being broader and terminated by a short point, while the lower tapers gradually to the base. The fruits are generally of a globular form, or sometimes slightly oval, and have a very hard woody shell, which is made to serve many useful purposes in the domestic economy of the inhabitants of the above-mentioned countries—basins, cups, spoons, water-bottles, pails, and even kettles being made of them: the latter, it is said, standing the fire several successive times before they are destroyed. In fact they in great measure take the place of pottery-ware, and many of them are carved and polished or stained in various quaint devices. The pulp is esteemed as a medicine, acting as a purgative, and considered to be beneficial in diseases of the chest; it is also roasted and used as a poultice for bruises and inflammations. The wood of the Calabash-tree is light, tough, and pliant, but is only obtainable in planks six or eight inches broad. [A.S.]

CRESCENTIACEÆ. (*Crescentiads.*) A small family of corolliflorai dicotyledons, closely allied to *Bignoniaceæ*, and often associated with them as a tribe or suborder, but differing in their one-celled ovaries with parietal placentas, and in their large succulent fruits, with almond-like wingless seeds. They are usually trees with alternate or rarely opposite leaves, and rather large flowers growing out of the old stems or branches. Calyx at first undivided, but at length splitting into irregular pieces. Corolla gamopetalous, irregular, somewhat two-lipped; stamens four, inserted in the corolla, two long and two short, often with the rudiment of a fifth; ovary free, one-celled, with two or four parietal placentas; ovules numerous. Fruit woody, not splitting, and containing large seeds immersed in pulp; embryo without albumen. They are tropical and subtropical plants, extending from 30° S. to 30° N.; they abound in Madagascar, the Mauritius, the Seychelles, and other islands of Eastern Africa. In America they are represented by ten species, in Asia by two only, and they are not found in Europe, nor on the continent of Australia. Some, as *Kigelia pinnata*, yield timber, which is used for canoes and for pillars. *Crescentia Cujete* is the Calabash tree, whose gourd-like fruits have been seen two feet in diameter in the west of Africa. A large Calabash can support two men in crossing a river. *Parmentiera cerifera* yields wax, and is called the Candle-tree in Panama. The fruit of *P. edulis* is the Quexhilote of Mexico, and is edible. The fruit of *Tanæcium Jaroinum* and of *Colea Telfairiæ* is eaten. There are eleven known genera and thirty-four species. Illustrative genera:— *Crescentia*, *Parmentiera*, *Colea*, *Kigelia*, and *Tanæcium*. [J. H. B.]

CRESS, AMERICAN. *Barbarea præcox*. —, AMERICAN WATER. *Cardamine rotundifolia*. —, AUSTRALIAN. The Golden Cress, a broad yellowish-leaved variety of *Lepidium sativum*. —, BASTARD. The common name for *Thlaspi*. —, BELLEISLE. *Barbarea præcox*. —, BITTER. A common name for *Cardamine*. —, GARDEN. *Lepidium sativum*. —, GOLDEN. A variety of *Lepidium sativum*. —, INDIAN. *Tropæolum majus*; the name of Indian Cresses is also given to the order *Tropæolaceæ*. —, LAND. *Barbarea vulgaris*. —, MEADOW. *Cardamine pratensis*. —, MOUSE-EAR. *Arabis Thaliana*. —, PARA'. *Spilanthes oleracea*. —, PENNY. *Thlaspi arvense*. —, PETER'S. An old name for *Crithmum maritimum*. —, ROCK. A common name for Arabis; also an old name for *Crithmum maritimum*. —, SPANISH. *Lepidium Cardamines*. —, SPRING. *Cardamine rhomboidea*. —, SWINE'S. *Senebiera Coronopus*. —, THALE. *Arabis Thaliana*. —, TOOTH. A common name for *Dentaria*. —, TOWEL. *Arabis Turrita*. —, VIOLET. *Ionopsidium acaule*. —, WALL. *Arabis Thaliana*; also a common name for Arabis. —, WART. *Senebiera Coronopus*; also a common name for *Senebiera*. —, WATER. *Nasturtium officinale*. —, WINTER. *Barbarea vulgaris*; also a common name for *Barbarea*. —, YELLOW. *Nasturtium palustre*, and *N. amphibium*.

CRESSA. A genus of *Convolvulaceæ*, containing probably a single species, though very variable from the different conditions under which the plant grows, as it is a common sublittoral undershrub in tropical and sub-tropical regions all over the world. It has scattered entire leaves, and crowded flowers in the axils of the uppermost leaves. The calyx consists of four sepals; the corolla is funnel-shaped and five-cleft; the ovary is two-celled with two ovules in each cell; the capsule contains from one to four seeds. [W.C.]

CRESSON ALE'NOIS. (Fr.) *Lepidium sativum*. —, AMER. *Cardamine amara*. — D'EAU or DE FONTAINE. *Nasturtium officinale*. — DE PARA'. *Spilanthes oleracea*. — DES JARDINS. *Lepidium sativum*. — DES PRE'S. *Cardamine pratensis*. — DU BRE'SIL. *Spilanthes fusca*. — DU PE'ROU. *Tropæolum majus*. —, DE ROCHE. *Chrysosplenium*; also *Cardamine petræa*.

CRESS-ROCKET. *Vella Pseudo-cytisus*.

CRESTED. Having an elevated, irregular or notched ridge, resembling the crest of a helmet. This term is chiefly applied to seeds, and to the appendages of anthers; it also belongs to bracts which form with their edges an appearance like that of a crest, as in *Melampyrum*.

CRETACEOUS. Very dull white, with a little touch of grey; chalky.

CRETE DE COQ. (Fr.) *Celosia cristata*; also *Erythrina Crista-galli*; also *Rhinanthus major*. —, MARINE. *Crithmum maritimum*.

CRE'TELLE COMMUNE. (Fr.) *Cynosurus cristatus*.

CREVE-CHIEN. (Fr.) *Solanum nigrum*.

CREYAT. The Indian name for *Justicia paniculata*.

CRIBRARIA. One of the most elegant genera of myxogastrous Fungi. The lower half of the spore-case or peridium is permanent, but the upper half partially shells off, and leaves behind a complicated network. The species are confined to the northern temperate regions. Two species have been found in this country.
[M. J. B.]

CRIBROSE. Pierced (like a sieve) with numerous close small apertures.

CRINITE. Having tufts of long weak hairs, growing from different parts of the surface.

CRINODENDRON. The name of a small Chilian tree of the lime-tree family, having opposite or alternate shortly-stalked and smooth leaves, with their margins toothed near the points. The flower-stalks, which are single from the axils of the leaves, and longer than them, are thickened towards the apex, and bear a rose-coloured flower, which has a two-lobed five-toothed calyx; five pyramidal fleshy petals hollowed at their base; about twelve stamens with anthers as long as their stalks, and a globular ovary crowned with a single style. The fruit is a four or five-celled capsule about the size of a cherry, and containing numerous seeds. Chequehue is the name given by the Chilians to this plant, which is known to botanists as *Crinodendron Hookeri* or *C. Patagua*. [A. A. B.]

CRINUM. A genus of remarkably handsome amaryllidaceous plants, well-known in gardens. They are tropical or sub-tropical herbs, generally of large size, with columnar or spherical bulbs, lorate-lanceolate leaves, and a solid scape bearing a many-flowered umbel. The perianth has a long slender tube scarcely enlarged at the mouth, and a six-parted limb of nearly equal segments, which are erect, spreading or reflexed. The six stamens are inserted in the mouth of the tube. The ovary is three-celled, containing many ovules; the style filiform and inclined, and the stigma obtuse or obsoletely three-lobed. There are numerous species of Asiatic, Australasian, and South American origin, while one or two are met with in Western Africa, and some of a hardier character in South Africa. Many very fine cross-bred varieties have also been obtained in gardens. One of the best known species is *C. amabile*, which Dr. Herbert regards as a spontaneous cross, probably between *C. procerum* and *C. zeylanicum*, also stating that it is cultivated for its beauty in Sumatra. This plant has thick pyramidal bulbs, and sheathing strap-lance-shaped erect leaves, three to six feet long and three to six inches wide in the centre. The scape is much compressed, three to four feet high, and bears an umbel of from twenty to thirty large rosy fragrant flowers, having a tube five or six inches long, and a limb of lanceolate-linear lobes as long as the tube, and pale flesh-coloured within. The South African, *C. capense*, is sufficiently hardy to grow in a protected border out of doors in warm situations. This has roundish ovate bulbs, and lanceolate-linear glaucescent leaves, two to three feet long, ending in long narrow points; the flowers are pleasantly scented, flesh-coloured, and about six inches long. It is sometimes called *C. longifolia*. Among the interesting hybrids is one called *C. Mitchamiæ*, raised between *australe* and *capense*; this is a very handsome plant, perfectly hardy in favourable positions, and produces a succession of flower-scapes till the winter. Another is *C. Herbertii*, raised between *scabrum* and *capense*, a plant of great beauty, bearing about a dozen flowers on a scape three feet high, the tube four inches long, the limb three and a half inches, the colour blush with deep-red stripes. See Plate 2, fig. d.
[T. M.]

CRISPATURE (adj. CRISPUS). When the edge is excessively and irregularly divided and puckered; also when the surface is much puckered and crumpled. Good examples are afforded by 'curled' endive, 'curled' kale, and the like. Also a diminutive of Bullate.

CRISTALLINE. (Fr.) *Mesembryanthemum crystallinum*.

CRISTATE. The same as Crested.

CRISTATO-RUGOSE. When the wrinkles of a surface are deep and sharp-edged.

CRISTE MARINE. (Fr.) *Crithmum maritimum*.

CRITHMUM. The Samphire, an umbelliferous plant, easily distinguished from all others of the same order by its glaucous twice-ternate leaves, the divisions of which are very succulent and taper towards either extremity. The flowers are greenish-yellow and inconspicuous, except from the contrast between their general hue and the blue tinge of the foliage. The whole plant is ' of a spicie taste with a certaine saltnesse,' on which account it has been long held in great repute as an ingredient in salads, and was declared by Gerarde to be ' the pleasantest sauce, most familiar, and best agreeing with man's bodie for digestion of meates.' For this purpose it is now nearly gone out of use, but it is still so much valued as a pickle that other succulent marine herbs are not unfrequently offered for sale under the name of Samphire, for example *Salicornia* and *Sueda*. All these substitutes, which are worthless for the purpose of pickling, may be infallibly detected on a simple examination of the leaf. Samphire is exclusively confined to the rocky sea-shore, and, like many other marine plants, has an extensive geographical range, being found on most of the shores of Europe, from the Crimea to the Land's End, and extends even to the Canaries. The best pickled Samphire is made from leaves which have been gathered in May, before the appearance of the flower-stalk; otherwise it is apt to be tough and stringy. The etymology of the name *Samphire* is somewhat curious; it was formerly written *Sampier*, a corruption of *Saint Pierre*; and, more anciently still, it was called by the French *Perce-pierre*; by the Italians, *Herba di San Pietro*, and in Latin, *Petrus crescentius*. Thus a herb properly enough called Rock-cress from its growing in the crevices of rocks, came to be known as Peter's cress (the name Peter meaning a rock). The change to Saint Peter's Herb was an easy one; the postfix 'herb' being dropped, San Pietro became Sampier and that Samphire. French, *Bacille*; German, *Meerfenchel*. [C. A. J.]

CRITHO. A genus of grasses belonging to the tribe *Hordeæ*, not considered by modern authors to be distinct from *Hordeum*, under which it is described by Steudel. The only species is the curious Nepal Barley, *C. ægiceras*, which is cultivated at great elevations on the Himalayas and Thibet. The grain has been frequently sent to Europe from those countries, recommended as a very hardy kind, arriving at maturity within an unusually short period after sowing. It has not, however, been found of much value in Britain, where it is chiefly cultivated in botanical gardens. [D. M.]

CRITHOPSIS. A genus of grasses belonging to the tribe *Hordeæ*, scarcely distinct from *Elymus*. One species is described, namely, *C. rachitrichus*, a native of Syria and Persia. [D. M.]

CROCEOUS, CROCATUS. Saffron-coloured.

CROCOSMIA. A beautiful genus of *Iridaceæ*, separated from *Tritonia*, and consisting of one species, *C. aurea*, a native of South Africa. It is a perennial Ixia-like herb, with fleshy corms, slender erect compressed stems terminating in a branched flower-spike, the leaves narrowly sword-shaped, and the flowers, sessile on the branches, large, deep orange-coloured, and not inaptly compared to large crocus blossoms. The perianth has a longish curved slender tube, and a nearly regular six-parted limb of oblong segments spreading in a star-like form, which causes the long filaments and style, which are fully as long as the segments, to stand out very prominently. The ovary is oblong with about ten or twelve ovules in each of its three cells; this grows into a three-lobed subglobose capsule, having about three seeds in each cell. In this particular, and in not having the throat of the perianth enlarged, this plant differs from *Tritonia*, with which it had been associated. [T. M.]

CROCUS. A well-known genus of *Iridaceæ*, very much prized in gardens as affording some of the earliest of spring flowers. The species and varieties are numerous, and exceedingly beautiful, best known as early spring bloomers, a large proportion of them flowering at that season; but also including several which are very handsome autumn-flowering kinds. They are all dwarf herbs, with fleshy corms and grassy leaves, the latter not fully developed till after the flowers have faded. The perianth is funnel-shaped with an elongated tube, and a six-parted limb of concave petaloid segments, of which the inner are rather smaller than the outer series: these segments are erect and closed in cloudy weather and at night, but expand under the influence of sunshine. There are three erect included stamens inserted in the throat of the tube, and an elongated style terminated by three dilated wedge-shaped fleshy cleft or fimbriated stigmas. The ovary is three-celled, containing numerous seeds. The species are mostly found in the southern and eastern parts of Europe, and in Asia Minor. A few species extend to central Europe, and one or two, long cultivated for ornamental purposes, have become established in some localities in England. *C. vernus*, one of these latter, is a handsome plant producing in very early spring its large bluish-purple flowers with orange-coloured stigmas. Another of them is *C. pyrenæus* or *nudiflorus*, an autumnal-blooming species, producing light purple flowers. *C. vernus* and *versicolor* have yielded many of the fine garden spring-flowering sorts, other favourite ones blooming at that early season being, *C. Imperatinus*, *nivalis*, *reticulatus*, *annulatus*, *lagenæflorus*, with its variety *luteus*. Of the autumnal-blooming species some of the most beautiful are *C. speciosus*, *pulchellus*, *Visianicus-Cartwrightianus*, *cancellatus*, *medius*, *Biory, anus*, *byzantinus*, and *odorus*. *C. sativus*, which is a light-purple autumnal-flowering species, formerly cultivated about Saffron Walden, and partially naturalised, yields the saffron of the shops, which consists of the deep orange-coloured stigmas of the flowers gathered with part of the style, and carefully dried. According to Dr. Pereira, a grain of good commercial saffron contains the stigmas and styles of nine flowers, and consequently 4,320 flowers are required to yield one ounce of saffron. English grown saffron is now rarely, if ever, met with in commerce. The best comes from Spain, while that imported from France is usually considered of second-rate quality. The quantity imported varies between 5,000 and 20,000 lbs. weight per annum. Saffron has a bitter taste, and a penetrating aromatic odour, and was formerly considered to possess stimulant, emmenagogue, cordial, and antispasmodic properties, but when administered in large quantities it is narcotic. It is employed, especially on the continent, as a flavouring and colouring ingredient in culinary preparations, liqueurs, &c., and in modern medicine is only applied for similar purposes, except when included in the domestic pharmacopœia. Saffron gives to water and alcohol three-fourths of its weight of an orange-red extract, which is largely employed in painting and dyeing. Another colouring agent of the same deep orange colour, called safflowers, is quite different from saffron, and consists of the florets of *Carthamus tinctorius*. [T. M.]

CROISETTE. (Fr.) *Gentiana cruciata*.—**VELUE.** *Galium cruciatum*.

CROIX DE JE'RUSALEM or **DE MALTE.** (Fr.) *Lychnis chalcedonica*. The name Croix de Malte is also applied to *Tribulus terrestris*. —, **DE ST. JACQUES.** *Sprekelia formosissima*.

CROSSANDRA. A genus of Indian *Acanthaceæ*, consisting of shrubs or herbs with subentire verticillate leaves, and large red flowers in terminal four-cornered spikes, with broad bracts, and narrow membranaceous bracteoles. The calyx is five-parted, with broad lobes, the inner ones being smallest; the corolla has a long tube, and a flat five-cleft limb; the four didynamous stamens are included in the tube; the one-celled anthers are hairy and ciliated at the margin; the capsule is compressed and two-celled, with four ovate seeds at the base. The genus is nearly related to *Stenandrium*, which has, however, a more prostrate habit, and more slender anthers. [W. C.]

The same name is given to a little known genus of terrestrial orchids, near *Gastrodia*.

CROSSOSTEMA. A climbing shrub of the passion-flower family, found in Sierra Leone. The calyx and corolla each consist of five segments; those of the corolla are larger than the sepals, more deeply coloured, and three to five-nerved; within them is a 'crown' consisting of one row of filaments. The ovary is placed on a short stalk which is expanded into a disc-like mass, with five short acute teeth at the margin, alternating with the five stamens which arise from the same place; it is terminated by a slender style with a dilated stigma, and is internally one-celled with several ovules attached to the walls of the ovary. [M. T. M.]

CROSSOSTYLIS. A genus of trees placed by Lindley and others among *Lecythidaceæ*, but by Bentham referred to *Rhizophoraceæ*. The trees are natives of the Society and Feejee Islands. They have opposite entire leaves; flower-stalks arranged somewhat umbel fashion, jointed in the middle; flowers greenish with four or five segments to the calyx, and as many shortly-stalked petals; stamens about twenty on a short disc alternating with an equal number of sterile stamens; ovary superior, with five to twelve compartments, in each of which are two ovules; fruit fleshy, but ultimately opening by two valves. [M. T. M.]

CROSSOTOMA. The name of an Australian shrub, of the order *Goodeniaceæ*, separated by Don from *Scævola*, but by others ranked with the latter, from which it differs in the calyx being imperfectly developed or obsolete, and in the segments of the corolla being fringed. [M. T. M.]

CROSS-SPINE. *Stauracanthus aphyllus.*

CROSSWORT. The common name for *Cruc anella*; also applied to *Galium* or *Vaillantia cruciata*, and to *Eupatorium perfoliatum*. It is further sometimes applied to the cruciferous family.

CROTALARIA. A very extensive genus of papilionaceous leguminous plants, containing between 250 and 300 species, natives of the tropics and sub-tropics of both hemispheres. They are either herbs or small shrubs, some having simple and others compound leaves. Their flowers are produced in racemes, either opposite the leaves or at the ends of the branches, and are usually of a yellow colour. They have a somewhat two-lipped calyx; a papilionaceous corolla, the upper petal or standard being heart-shaped, and the lower or keel sickle-shaped; and the stamens united into a column which is split down one side. The legume or pod is curved inwards, and of an oblong form, with its sides puffed or swollen out.

C. Burhia is a small shrub with numerous spreading stiff branches, slightly armed with spines, growing in arid sandy places in Sindh. Its leaves are of an oblong form and wide apart on the branches; and the whole plant is covered with silky hairs. The tough twiggy branches are used in Sindh for twisting into tough ropes.

C. Espadilla, a harsh shrubby plant about a foot high, growing in sandy places in Venezuela, has bluntly lance-shape leaves, covered with stiff, close-pressed, shining hairs, and when young of a fine golden colour. This plant is a common domestic medicine in Venezuela; a decoction of it is a sudorific, and it is used in fevers.

C. juncea, the Sunn-hemp of India, is a shrubby plant growing from eight to twelve feet high, with a branching stem marked with longitudinal furrows; when cultivated, however, it is sown close, so as to prevent branching as much as possible. Its leaves are on short stalks, and are either bluntly lance-shaped, or very narrow and sharp-pointed, from two to six inches long, thickly covered with shining silky white hairs, which give them a silvery appearance. The flowers are of a beautiful bright-yellow colour, resembling those of the common broom; they are produced in long racemes at the ends of the branches, and are succeeded by club-shaped stalkless pods about two inches long, containing numerous kidney-shaped seeds. This plant is extensively cultivated in different parts of Southern Asia, particularly in India, on account of the valuable fibre yielded by its inner bark; and which is known by the names of Sunn-hemp, Bombay-hemp, Madras-hemp, Brown-hemp, &c. The stems after being cut are steeped in water for two or three days in order to loosen the bark, they are then taken out in handfuls and bent so as to break the interior wood without injuring the fibre; the operator then beats them upon the surface of the water until the fibrous part is entirely separated, when it is washed and hung upon bamboo poles to dry, and afterwards combed to separate the filaments from each other. The fibre thus obtained is very strong, and is considered to be equal if not superior to some kinds of Russian hemp: it is employed for cordage, canvas, and all the ordinary purposes of hemp. A variety produced at Jubbulpore in Malwah, and called Jubbulpore-hemp, has been supposed to be the produce of a different species, *C. tenuifolia*, but that species is now united with the present. Besides its use as a fibrous plant, it is grown in the Madras territories as a food for milch cows, and is said to be very nourishing.

C. retusa, a native of the East Indies, but naturalised in the West Indies and Brazil, is an annual plant with smooth branching stems, from four to six feet high, and oblong wedge-shaped leaves notched at the top, smooth upon the upper surface, but covered with short silky hairs underneath. This is cultivated for its fibre in the Madras territory. [A. S.]

CROTON. An important genus of *Euphorbiaceæ*, among which it may be known by the flowers being monœcious, with a five-parted calyx. The male flowers have five petals, and ten stamens, and the female flowers are destitute of petals, but have three styles, divided into two or more branches. The fruit consists of three carpels separating one from the other, and each containing one seed. The species are numerous and vary very much in general appearance, some being herbs, others trees, and some having entire, others divided leaves.

C. *Tiglium* is the most important tree of this genus in a medicinal point of view, as it produces the seeds whence croton oil is extracted. The tree is a native of Coromandel, the Indian Archipelago, &c., and has oblong-pointed leaves covered with stellate hairs, when young. One seed is sufficient to act as a purgative, but the oil expressed from the seeds is yet more powerful, though sometimes uncertain in its action; one drop is usually sufficient, hence the great value of this drug in cases where smallness of dose, speediness of action, and powerful effects are required, as in mania, apoplexy, dropsy, &c. It is so acrid that it is exhibited usually in

Croton Tiglium.

pills in order to avoid the burning heat it occasions in the throat if swallowed by itself; on this account it is not used in any case where there is inflammation of the bowels. In large doses it acts as a frightful poison, producing symptoms like those of cholera. Externally it has been used as a counter-irritant. It is obtained by submitting the seeds to pressure, an operation which affects the men engaged in it with irritation of the eyes, and air passages, and purging. Dr. Pereira gives the case of a workman who suffered very severely from inhaling the dust of the seeds, he having been occupied for some time in emptying packages of them. The seeds of *C. Pavana* and *C. polyandrum*, Indian shrubs, are also used as purgatives.

Many of the species have aromatic properties. Of these the most important are *C. Eleutheria*, the tree yielding Cascarilla bark, which is chiefly collected on the island of Eleuthera, one of the Bahamas. This bark is esteemed in this country as an aromatic bitter tonic, without astringency, in cases of simple indigestion. It has a fragrant smell when burnt, on which account it is said to have been at one time mixed with tobacco for smoking. *C. pseudo-China*, called in Mexico Copalche, yields a bark having similar properties with the above, and which is used in Mexico in place of cinchona. *C. balsamiferum*, a West Indian shrub, furnishes a spirituous liquor called Eau de Mantes, which is used in irregular menstruation; whilst others are employed in the West Indies, the Cape, &c., for their aromatic, fragrant, and balsamic qualities. *C. lacciferum* in Ceylon, and *C. Draco* in Mexico, yield resin used for varnish-making, &c. The plants known in cultivation as *C. pictum*, &c., are referred to *Codiæum*. [M. T. M.]

CROTONOPSIS. A North American herb of the euphorbiaceous family, scattered over with bran-like scales; the fruit and calyx with stellate hairs. The flowers are monœcious, the males having a five-parted calyx, with five petals, and as many stamens; the females likewise have a five-parted calyx, two of the segments of which are frequently suppressed, with five petaloid scales opposite the sepals; the ovary has three two-lobed stigmas. The fruit is one-seeded. [M. T. M.]

CROTTLES. A name given by the lichen gatherers in Scotland to various species, which they distinguish under the names black, brown, dark, light, white, stone crottles, &c. In Scotland the name is applied indifferently, but the merchants and dyers distinguish all the species with an erect or pendulous habit by the name of weeds, while the flat imbricated species, as *Parmelia saxatilis*, are called mosses. The word Crottles is not confined to Scotland, but is used in some parts of England. [M. J. B.]

CROWBERRY. *Empetrum nigrum*. —, BROOM. An American name for *Corema*.

CROWEA. Pretty greenhouse shrubs with simple dotted leaves, and purple flowers, constituting a genus of *Rutaceæ*, and natives of New Holland. The whorls of the flower are in fives; there are ten stamens with hairy filaments, five of which, opposite the petals, are shorter than the remainder; the anthers have an awl-shaped hairy appendage prolonged from the summit; the carpels are five on a five-lobed disc, with five styles fused into one. The fruit consists of five dry segments, which burst into two pieces, each containing one seed. [M. T. M.]

CROWFOOT. The common name for *Ranunculus*.

CROW GARLIC. *Allium vineale*.

CROWNBEARD. An American name for *Verbesina*.

CROWN IMPERIAL. *Fritillaria Imperialis*.

CROWNWORTS. A name given by Lindley to the group *Malesherbiaceæ*.

CROW'S-FOOT. *Echinochloa crus-corvi*.

CROWSILK. A name sometimes given to the *Conferva* and other delicate green-spored *Algæ*. [M. J. B.]

CROZOPHORA. A genus of *Euphorbiaceæ* found in tropical and northern Africa, and extending eastwards as far as India. It consists of annual or perennial low-growing plants, having all their parts densely clothed with starry hairs or shield-shaped scales. The stalked leaves have an oval or heart-shaped blade with either entire lobed or curled margins. The minute green flowers are borne on terminal or axillary bracted racemes, the lower portion of which is occupied by the females, the upper by the males. The latter have a calyx of five divisions, five petals and a central column of five to ten stamens, but most commonly eight, these being arranged in two whorls, the outer one of five short stamens, the inner of three longer, and all of them opposite the calyx leaves. The number and disposition of these stamens afford the chief distinguishing character of the genus. The ripe capsule is about the size of a pea, and covered with shield-shaped scales. It contains three seeds.

C. tinctoria, which grows wild in the countries bordering the Mediterranean, is cultivated in the South of France for the sake of a dye which is obtained from it. This dye is called Turnsole, and is obtained by grinding the plants, little herbs seldom more than a foot high, to a pulp in a mill, when they yield about half their weight of a dark green coloured juice, which becomes purple by exposure to the air or under the influence of ammonia. It is chiefly exported to Holland, and is prepared for exportation by soaking coarse linen rags or sacking with it, the rags being previously washed clean. After soaking they are allowed to dry, and are exposed to the influence of ammonia by being suspended over heaps of stable manure. They are then packed in sacks, and ready for shipping to Holland. Not much is known of the uses the Dutch put the dye to, but it is supposed to be chiefly employed as a colouring matter for cheese, and perhaps confectionary, wine, &c. This dye has been confounded by some authors with the litmus of our chemists. [A. A. B.]

CRUCIANELLA. A genus of herbaceous plants, called Crosswort and Petty Madder, and belonging to the *Rubiaceæ*. The corolla is funnel-shaped with an exceedingly slender tube and narrow inflected lobes; the seeds are in pairs, linear not crowned with the calyx. They are found in the southern parts of Europe and Asia, and are of humble growth, bearing thin leaves inserted in opposite pairs, and having stipules at their base so arranged as to simulate a whorled form of growth. The species are rarely cultivated except in botanical gardens, with the exception of *C. stylosa*, a native of Persia and the Caucasus: this is a low tufted herb with rose-coloured flowers, which bloom during the greater part of the summer; it is well adapted for rockeries. French, *Croisette*; German, *Kreuzblatt*. [C. A. J.]

CRUCIATE, CRUCIFORM. Having the form of a flowers of ... oss, with equal arms, as the disk or wallflower.

CRUCIBU...M. A genus of gasteromycetous Fu..., belonging to the natural order *Nidu...*ries. It is distinguished from *Cyathus* by ... peridium being homogeneous and not co... posed of distinct strata, and by the sp... angia being supported by a cord endin... above in a globular swelling sunk in a ... t of the sporangium, and including a... elastic complicated thread. There is bu... one species which is common all over Eur... pe, and occurs in the north of Africa, and ... ew Zealand. It is especially fond of the ... old fronds of ferns, but occurs also on ... sticks, old ropes, and various other veget... ble substances. [M. J. B.]

CRUCIFERÆ. (*Brassicaceæ*, *Cruciferæ*, the *Cruciferous* family.) A natural order of thalamifl...ral dicotyledons, belonging to Lindley's cr... al alliance. Herbs with alternate leaves ... aving no stipules, and flowers, usually yell... w or white, arranged in racemes or co... ymbs without bracts; sepals four, falling off; petals four, arranged like a cross; stamens six, of which four are long and two short. Fruit, a siliqua or silicula, that is, a long or short pod opening by two valves, with a partition (septum) in the centre; seeds without albumen; embryo with its radicle folded on the cotyledons. The plants of this very natural order were included by Linnæus in his class Tetradynamia. They are generally distributed, but most abound in cold and temperate regions, especially in Europe. This order has been divided into suborders and tribes according to the nature of the fruit or the embryo. Considering the fruit we have these six divisions:— 1. *Siliquosæ*, a siliqua or long pod opening by two valves from below upwards; 2. *Siliculosæ latiseptæ*, a silicula or short pod opening with two flat or convex valves, the septum (partition) being in the broadest diameter; 3. *Siliculosæ angustiseptæ*, a silicula with folded or keeled valves, the septum in the narrow diameter; 4. *Nucamentaceæ*, a silicula whose valves do not open, one-celled, having no septum; 5. *Septulatæ*, valves with transverse partitions on their inside; 6. *Lomentaceæ*, a pod dividing transversely into single-seeded portions, the beak sometimes containing one or two seeds, while the true pod is abortive. The nature of the embryo gives origin to five subdivisions, namely:—1. *Pleurorhizeæ*, the radicle folded on the edge of the cotyledons; 2. *Notorhizeæ*, the radicle folded on the back of the cotyledons; 3. *Orthoploceæ*, the cotyledons folded on the radicle; 4. *Spirolobeæ*, cotyledons twice-folded; 5. *Diplecolobeæ*, cotyledons thrice-folded. Cruciferæ are pungent, and occasionally acrid in their properties. None of them are poisonous; many are culinary vegetables. From containing much nitrogen and sulphur in their composition they give out a fetid odour when decaying. Among the common cruciferous garden flowers may be enumerated wallflower, stock, rocket, honesty. *Brassica oleracea* is the origin of the cabbage, cauliflower, broccoli, savoy and curled kale. *Brassica Rapa* is the origin of the turnip. The Swede or Swedish turnip is by some said to be a variety of *Brassica campestris*, by others a hybrid between *B. Rapa*, the turnip, and *B. Napus*, the wild navew, rape or coleseed. *Crambe maritima* supplies sea-kale, which is subjected to the process of blanching in order to fit it for the table. Among the pungent plants of the order are *Sinapis nigra*, the black seeds of which supply the best mustard; *S. alba*, or white mustard, which is less pungent; *Lepidium sativum*, common cress; *Nasturtium officinale*, water-cress; *Cochlearia Armoracia*, horse-radish; and *Raphanus sativus*, the radish. *Isatis tinctoria*, woad, yields a blue dye; and *I. indigotica* is used as indigo in China. Many of the species grow on the sea shore, and have been used as fresh vegetables by the crews of ships affected with scurvy. Hence, *Cochlearia officinalis* receives the name of scurvy-grass. Oil is procured from the seeds of many of the plants; thus we have rape oil, and oil of mustard, and camelina oil. After pressing out the oil from rape-seeds the cake is used as food for cattle. There are 208 known genera, and about 1730 species. Illustrative genera:—*Cheiranthus, Arabis, Lunaria, Draba, Thlaspi, Teesdalia, Hesperis, Erysimum, Capsella, Isatis, Brassica, Bunias, Senebiera*, and *Schizopetalum*. [J. H. B.]

CRUICKSHANKIA. The name of certain Chilian herbs, constituting a genus of *Cinchonaceæ*. The plants have branching wavy stems, and yellow flowers in terminal heads. Their calyx tube is globular, its limb with four stalked roundish netted segments, having two stipules at the base of each; the corolla is salver-shaped. The fruit is a membranous two-celled and two-valved capsule. The most remarkable feature in the genus is the curious condition of the calyx before mentioned. [M. T. M.]

CRUSTA. The upper surface of lichens.

CRUSTACEOUS. Hard, thin, and brittle; as the seed-skin of asparagus, and the thallus of many lichens.

CRUSTOLLE. (Fr.) *Ruellia*.

CRYBE *rosea* is a small tuberous orchid with grassy leaves, from Guatemala. It has the habit of *Bletia*, but its pollen is that of an Arethuseon.

CRYPSIS. A genus of grasses belonging to the tribe *Agrostideæ*. The inflorescence is generally between a thyrse and a capitule; spikelets one-flowered; glumes two, compressed and carinate; pales two, lanceolate, the inferior one nerved; stamens two to three; styles two. Thirteen species are described, mostly annuals, and little known in a cultivated state. [D. M.]

CRYPTA. The sunken glands or cysts which occur in dotted leaves. The same as Cyst.

CRYPTADENIA. A genus of *Thymelaceæ*, composed of a few heath-like dwarf bushes, natives of S. Africa. They differ from their allies in their tubular calyx bearing on its inner surface near the apex of the ovary eight anther-like glands. Their minute linear leaves are numerous, opposite and smooth; the pink flowers, single or in pairs at the apex of the twigs, or from the axils of the upper leaves, consist of a coloured tubular calyx with a four-parted border, covered outside with short silky hairs, and bearing on its tube eight stamens, four of which are short and included, the others longer and slightly protruding beyond the mouth of the tube. *C. uniflora* is a slender pretty bush with pink flowers at the ends of the branches, and is sometimes seen in greenhouses. Five species are known. The name of the genus has reference to the eight hidden glands of the calyx tube. [A. A. B.]

CRYPTANDRA. A genus of heath-like under-shrubs, belonging to the order *Rhamnaceæ*, natives of New Holland. They are erect branching plants, with alternate entire glabrous leaves, and flowers aggregated at the summits of the branches, or sometimes solitary. The coloured calyx has a campanulate occasionally cylindrical tube attached below to the ovary, but free above, and having a five-cleft limb cut into acute segments. The small hooded petals are inserted in the throat of the calyx, and cover the stamens, which have short filaments, and two-celled anthers opening longitudinally. The three-celled ovary is semi-inferior, each cell containing a single erect ovule; the style is simple, with a three-lobed stigma; the capsule is covered with the persistent calyx. There are upwards of seventy species. [W. C.]

CRYPTANTHUS. A Brazilian epiphyte belonging to the *Bromeliaceæ*. Its leaves are lanceolate, and conceal the flowers; hence the name. The flowers have the arrangement and structure common to the order, with six stamens inserted on a fleshy epigynous disc, three of them moreover are united to the base of the inner petal-like segments of the perianth. The stigmas are three in number, twisted and hairy. [M. T. M.]

CRYPTARRHENA. A very singular genus of tiny stemless epiphytal orchids with spikes of minute yellowish flowers, living in the forests of Surinam and Mexico. They have a lip divided into attenuated segments, and a column furnished at the upper extremity with a hood, under which the anther lies. A plant called *Orchidofunkia pallidiflora* belongs to the genus.

CRYPTOCARYA. A genus of *Lauraceæ*, consisting of trees natives of the tropics of both hemispheres, and of Australia. The leaf buds are scaly. The flowers are hermaphrodite, with a somewhat funnel-shaped six-cleft perianth; stamens twelve in four rows, the nine outer ones fertile, the three inner sterile; the innermost row of the fertile stamens has stalked glands at each side of each stamen, and the anthers of this row open outwardly, while those of the two outer rows open inwardly, in either case by two valves; the one-celled ovary is immersed in the calyx tube, which becomes succulent as the fruit ripens, concealing the latter, hence the name of the genus. Brazilian Nutmegs are the produce of *C. moschata*. [M. T. M.]

CRYPTOCERAS. A section of the fumariaceous genus *Corydalis*, containing a few species from the warmer parts of temperate Asia. They have enlarged fusiform rootstocks, simple stems with two opposite leaves, which are ternate with imbricated segments, and very large flowers. [J. T. S.]

CRYPTOCHILUS *sanguinea*. A terrestrial orchid from the cooler parts of India, with leathery lanceolate leaves, and scapes bearing spikes of crimson tubular flowers. There is another species from the same country, the flowers of which are smaller and yellow.

CRYPTOGAMS. Many names have been applied to the vast class of plants comprehended under this name, as Asexual, or Flowerless Plants, Acrogens, Agamæ, Anandræ, Acotyledons, Cryptogams, Cryptophyta, Cellulares, Exembryonata, &c. Some of these have been given to them by authors collectively, while others have been appropriated to one of the two great sections into which Cryptogams are divisible. Of these we have chosen the term CRYPTOGAMS as liable to fewer objections than most others, and predicating little that is exposed in the present state of our knowledge to much contradiction. We have already stated the objections to which some are subject, as Asexual Plants, Acotyledons, Anandræ, and Cellulares; others will be mentioned hereafter. The great distinctive point of Cryptogams does not consist in the absence of decided male and female organs, nor in their minuteness, for in the greater part their presence has been ascertained beyond all doubt, and the analogous organs in phænogams often require the assistance of the lens to make out even their external form clearly. The main point is that the reproductive organs are not true seeds containing an embryo, but mere cells consisting of one or two membranes inclosing a granular matter. These bodies, whether called spores or sporidia, produce by germination a thread or mass of threads, a membrane, a cellular body, &c., as the case may be, which either at once gives rise to the fruit or to a plant producing fruit. Indeed the differences are so great that these spores seem rather to be relatives, or what is technically termed homologues, of pollen grains, than of true seeds.

The Cryptogams are divided into two great classes, THALLOGENS and ACROGENS, whose distinctive characters will be found under those heads. It is scarcely possible to give any general character of the whole except that which we have indicated above, as these two divisions are as distinct from each other as Cryptogams themselves are from phænogams. Many of them indeed consist entirely of cells, but so do some more perfect plants, and vascular tissue exists in many Cryptogams. The greater part increase from the tips of the threads, but cell division takes place occasionally in other parts; while even in exogens, the main growth of the cells of which the wood and bark are composed is similar. Again, if they have no true pistils and anthers, they have their analogues, while in several an embryo is at length produced, and in *Selaginella* something even like cotyledons. Both the embryo and cotyledons are, however, aftergrowths, and not derived immediately from the spore. The consideration of the relations between the reproductive organs of phænogams and Cryptogams is one of the most interesting which is to be found in Botany, but it is also one of the most abstruse and difficult, and can be followed out only by those who have an intimate knowledge of the structure and functions in either branch of the vegetable kingdom. Such considerations would be wholly out of place in a work like the present. [M. J. B.]

CRYPTOGLOTTIS *serpyllifolia* is a little trailing moss-like orchid growing on trees in the Malayan archipelago. Its flowers are very minute. It is the same as the *Hexameria* of Brown, and notwithstanding its diminutive dimensions, is nearly related to the showy *Angræcums*.

CRYPTOGRAMMA. A small genus of polypodiaceous ferns of the group *Platylomeæ*. They are very closely related to our native *Allosorus*, with which they are indeed sometimes, and, perhaps rightly, united. The typical species of the present genus, *C. acrostichoides*, has, however, the spore-cases continued in lines along the course of the veins from the margin a short distance inwards, so as to be unmistakeably oblong or linear-oblong, and hence has the distinguishing characteristic of the *Platylomeæ*; while in *Allosorus*, as now restricted, the sori are normally punctiform, and therefore polypodioid. They simulate the *Pterideæ*, in consequence of the reflexed herbaceous margin resembling an indusium. The aspect of the plants is quite that of *Allosorus crispus*, being of dwarf and tufted habit, with dimorphous fronds, and having the fertile pinnules formed like a silicle or short pod. There are three species, *C. acrostichoides*, found in Arctic America, *C. sitkensis*, found in Sitka, and *C. Brunoniana*, found in India. [T. M.]

CRYPTOMERIA. A lofty evergreen tree, forming a genus of *Coniferæ* of the tribe or suborder *Cupressineæ*. The leaves are shortly linear, falcate, rigid and acute, crowded but spreading. The flowers are monœcious, the males in axillary catkins, the peltate scales bearing five anther-cells at their base. The fruits are in small terminal globular cones, with palmately-lobed imbricate scales, each one covering four to six winged seeds. *C. japonica*, the only species known, is a native of North China and Japan, and being hardy enough to sustain our climate without injury, is now very generally planted in collections of Conifers. It is not, however, suited to heavy soil.

CRYPTONEMATA. Small cellular threads produced by cryptostomata.

CRYPTONEMIACEÆ. One of the largest natural orders amongst the rose-spored *Algæ*, belonging to the section *Gongylospermeæ*, in which the inarticulate cartilaginous frond consists of a number of jointed threads compacted by gelatine. In the membranous species it is sometimes formed of many-sided cells, decreasing in size towards the surface. The capsules are immersed and are sometimes compound, and the spores are congregated without order. These arise either from several congregated fertile cells, which at length enlarge their endochrome, giving rise to a multitude of spores, or from a single cell, according as they are compound or simple; in the former case all trace of the original structure is frequently lost when the fruit is perfected. The genera and species are numerous, and occur in all climates. *Chondrus crispus* with several species of *Iridæa* and *Gigartina* belonging to this order, abound in gelatine, and in consequence are useful for many domestic purposes. [M. J. B.]

CRYPTOPHYTES. A synonym of cryptogams. [M. J. B.]

CRYPTOPUS *elata* (*Beclardia* of Rich.) is a handsome epiphytal orchid from the Isle of Bourbon. It has the habit of *Epidendrum elongatum*, the double gland and caudicle of an *Angræcum*, and flowers with deeply-lobed petals and lip; their colour is white dotted with purple.

CRYPTOS. In Greek compounds=concealed; thus Cryptogams are plants with concealed sexes.

CRYPTOSANUS. *Leochilus*.

CRYPTOSEMA. A name sometimes given to a West Australian bush of the pea family, also called *Jansonia*: which see. [A. A. B.]

CRYPTOSORUS. A very appropriate name proposed for a few species of small-growing Ferns, having sunken punctiform nonindusiate sori, but which are not generally considered sufficiently distinct from *Polypodium*. [T. M.]

CRYPTOSTEGIA. A genus of twining shrubs, belonging to the natural order *Asclepiadaceæ*, and containing a single species from India and another from Madagascar. They have opposite leaves, and large reddish-white flowers in terminal cymes. The calyx consists of five lanceolate sepals; in the tube of the corolla there are five linear bipartite scales; the stamens are included, and have very short filaments inserted at the base of the tube, and the oval pollen masses are solitary and attached to the five glandular points on the globose stigma. The large three-sided follicles are widely divaricate, with an incurved apex and comose seeds.

The plants of this genus abound in milky juice, which when exposed for a short time to the sun is converted into pure caoutchouc. [W. C.]

CRYPTOSTOMATA. Little circular nuclei found on the surface of some algals.

CRYPTOSTYLIS. A small genus of brown-flowered terrestrial orchids inhabiting New Holland, Java, and Ceylon. The main character consists in its having a great dorsal lip hollowed out at the base to receive the column." The abolished genus *Zosterostylis* is one of the species.

CRYPTOTÆNIA. A genus of *Umbelliferæ*. The Honewort, *C. canadensis*, is the only species, and is one of a goodly number of plants common to North America and Japan. It is a smooth perennial erect herb, one to two feet high, having ternate stalked leaves with ovate coarsely-toothed leaflets, and numerous umbels of small white flowers, curiously disposed in an almost panicled manner, which is very unusual in the family. The fruit is linear-oblong, contracted at both sides, each of the carpels having five equal obtuse ribs, with an oil tube (vitta) in each furrow, and one under each rib. [A. A. B.]

CRYPTOTHECA. A genus of *Lythraceæ*, containing bog herbs or undershrubs from Japan with angular stems, opposite shortly stalked lanceolate or linear-lanceolate leaves, and axillary many-flowered peduncles. The calyx is funnel-shaped, four-cleft; corolla of four small petals or absent; stamens two, with roundish anthers; style lateral; capsule one-celled, irregularly circumscissile, inclosed in the calyx tube. [J. T. S.]

CRYPTOTHECII. A small group of mosses, represented by *Spiridens*.

CRYSTALWORTS. A name given by Lindley to the *Ricciaceæ*.

CTENOMERIA. A genus of slender twiners of the spurgewort family, found in South Africa. The slender cobwebby filaments of the male flowers, together with the pectinately-toothed calyx leaves of those of the females, serve to distinguish it from its allies. The wiry stems are furnished with distant nettle-like heart-shaped leaves, and the small green flowers are disposed in racemes which arise from opposite the leaves. [A. A. B.]

CTENOPTERIS. A name originally proposed as a sectional division of *Polypodium* by Blume, a Dutch botanist, and subsequently adopted as a genus, with various modifications by modern pteridologists. It is, however, synonymous with the true or typal species of *Polypodium*. [T. M.]

CUBEBA. A genus of *Piperaceæ*, the distinguishing features of which are, the diœcious flowers partially covered by sessile bracts and the fruits elevated on a sort of stalk, formed from the contraction of the base of the fruit itself, so that they are not really but only apparently stalked. They are shrubs frequently of climbing habit, indigenous in the tropics of Asia and Africa. *C. officinalis*, a native of Java, furnishes the cubeb fruits of com-

Cubeba canina.

merce, which are like black pepper but stalked. They have an acrid hot aromatic taste, and are specially useful in diseases of the bladder and urinary passages. In large doses they give rise to symptoms of irritant poisoning. *C. canina* is also said to furnish some portion of the commercial cubebs. [M. T. M.]

CUBEBS. The fruits of various species of *Cubeba*, as *C. officinalis*, *C. canina*, and others.

CUCHUNCHULLY or CUICHUNCHULLI. *Ionidium microphyllum*.

CUCKOLD TREE. *Acacia cornigera*.

CUCKOO-FLOWER. *Cardamine pratensis*; also *Lychnis Flos-cuculi*.

CUCKOO-PINT. *Arum maculatum*.

CUCUBALUS. A genus of *Caryophyllaceæ*, of the tribe *Sileneæ*, containing a single European herb which has been found in the Isle of Dogs, but doubtless introduced. It has trailing stems, opposite ovate leaves, and shortly stalked drooping whitish flowers in dichotomous cymes. The calyx is bell-shaped; the petals deeply cleft; stamens ten; styles three; fruit a globular berry, at first reddish, but black when ripe; seeds numerous. [J. T. S.]

CUCULLATE. When the apex or sides of anything are curved inwards, so as to resemble the point of a slipper, or a hood as in the lip of *Cypripedium* and *Calypso*.

CUCULLUS. A hood or terminal hollow.

CUCUMBER. *Cucumis sativa*. —, BITTER. *Citrullus* or *Cucumis Colocynthis*, commonly called Colocynth. —, INDIAN. *Medeola virginica*. —, ONE-SEEDED STAR. An American name for *Sicyos*. —, SINGLE-SEEDED. A common name for *Sicyos*. —, SNAKE. *Trichosanthes colubrina*; also *Cucumis flexuosus*. —, SPIRTING or SQUIRTING. *Ecbalium agreste*, formerly called *Momordica Elaterium*.

CUCUMBER-ROOT. An American name for *Medeola*.

CUCUMBER-TREE. An American name for *Magnolia acuminata* and *M. Frazeri*.

CUCUMBERTS. A name proposed for the *Cucurbitaceæ*.

CUCUMIS. A genus of *Cucurbitaceæ*, comprising a number of species, among which the most remarkable are the Cucumber, *C. sativa*, so well known as one of our most ancient table esculents, and the Melon, *C. Melo*, equally familiar to us as one of our most ancient and luscious fruits. Some of the species possess valuable medicinal properties. Nearly all are annuals and natives of the warmer parts of Asia, Africa, and America. It is worthy of note that the tender tops of all the edible species of *Cucurbitaceæ*, boiled as greens or spinach, are even a more delicate vegetable than the fruit.

The Cucumber, *C. sativa*, is a tender annual, having rough trailing stems, with large angular leaves, and yellow male and female flowers borne in the axils of the leaf stalks. It is a native of Asia and Egypt, where it has been cultivated for more than 3,000 years. It is mentioned as one of the things for which the Israelites longed while in the wilderness, and complained to Moses (Numbers xi. 5). At a very early period it was grown by the Greeks and Romans, and according to Pliny, the Emperor Tiberius had Cucumbers at his table every day in the year. They were known in England in the time of Edward III. (1327), but during the wars of the Houses of York and Lancaster their cultivation was neglected, and the plant lost until the reign of Henry VIII., when it was again introduced. Since then it has gradually increased in public favour until it has now become of such importance as to be an object of rivalry with gardeners to produce fruit for the great and wealthy at all seasons. In summer such is the demand for this esculent that in order to obtain a sufficient supply, it is grown extensively in forcing frames, and in the counties near the metropolis, whole fields are devoted to cucumbers as a crop. Although cold and watery, and by some considered unwholesome, still the fruits are generally much esteemed as forming a most grateful salad when cut into very thin slices, and dressed with vinegar, &c. In a young state when small they are called Gherkins, and are in great request for preserving in vinegar, or for pickling with other vegetables. It is recorded that the village of Sandy in Bedfordshire has been known to furnish for the London market, 10,000 bushels for this purpose in one week! [W. B. B.]

The Melon, *C. Melo*, is the Pepon of Dioscorides, the Melopepon of Galen, and the Melo of Pliny. In Greece at the present day it is named Peponia. In Italy in 1539, the names of Pepone, Melone, and Mellone were applied to it. In Sardinia, where, it is remarked by De Candolle, Roman traditions are well preserved, it is called Meloni. From the Spaniards in the beginning of the sixteenth century, it received the name of Melon, which it retains in France, England, and with but slight modifications in other countries throughout Europe, where indeed the uniformity of name seems to indicate an introduction not very remote. De Candolle is of opinion that the species was originally confined to the valleys in the south of the Caucasus, and chiefly to the southern coasts of the Caspian. But its cultivation in the open air has long been extensively practised over a great part of Asia. It even appears to have been introduced into Italy early in the first century, if not before, as it is mentioned by Pliny, who died from suffocation caused by the great eruption of Vesuvius in A.D. 79. In his works he describes the modes by which melons were grown or forced, so as to be obtained for the Emperor Tiberius at all times of the year. Their cultivation, however, appears to have been very limited in Europe till within the last three centuries. According to M. Jacquin, *Monographie complète du Melon*, the Cantaloup variety derives its name from Cantaluppi, a seat belonging to the Pope, near Rome, where this sort, brought from Armenia by the missionaries, was first cultivated. He states further that it was received into France from Florence; that from France it passed into Spain, and thence into England, where, according to some authors, it has been cultivated since 1570; but the precise time of its introduction is uncertain. Probably the cultivation of Melons had been attempted much earlier. Till lately they were called in this country Musk Melons to distinguish them from water melons, which belong to a different species. Persia is noted for the excellence of its Melons, and the extensive scale on which their cultivation is carried on. Some nobles and wealthy individuals keep, it is said, from 10,000 to 20,000 pigeons, chiefly for manuring their melon beds, pigeon's dung being there considered the best manure for these plants. A collection of seeds of the best Persian varieties was sent in 1824 to the Horticultural Society by Sir Henry Willock, ambassador at the court of Persia; and some of the kinds when grown under particular treatment in this country proved excellent, but they are apt to degenerate. The melons of Bokhara are of the highest excellence, although in our climate they are liable to the same objection with regard to degeneration as those of Persia. Burnes in his *Travels* says 'The Melon is the choicest fruit of Bokhara. The Emperor Baber tells us that he shed tears over a melon of Turkistan which he cut up in India after his conquest: its flavour brought his native country and other dear associations to memory. There are two distinct species of Melons which the people class into hot and cold; the first ripens in June, and is the common Musk or Scented Melon of India; the other ripens in July, and is the true Melon of Turkistan: in appearance it is not unlike a water melon, and comes to maturity after being seven months in the ground. It is much larger than the common sort and generally of an oval shape, exceeding two and three feet in circumference. Some are much larger, and those which ripen in the autumn have exceeded four feet. One has a notion that what is large cannot be delicate or high-flavoured; but no fruit can be more luscious than the Melon of Bokhara, nor do I believe their flavour will be credited by any one who has not tasted them. The Melons of India, Cabool, and even Persia, bear no comparison with them—not even the celebrated fruit of Ispahan itself. There are various kinds: the best is named Kokechu, and has a green and yellow-coloured skin; another is called Ak nnlat, which means white sugar candy; it is yellow and exceedingly rich. The Winter Melon is of a dark green colour, called Kara koshuk, and said to surpass all the others. Bokhara appears to be the native country of the Melon, having a dry climate, sandy soil, and great facilities for irrigation.' (*Burnes' Travels in Bokhara.*)

Provided the soil is moist below, the Melon succeeds in all countries where the summer is sufficiently hot, even although the winters are cold, as is the case at Cabul, where severe winters are succeeded by very hot summers. There, Melons are produced in great abundance. Being an annual, its vegetation only commences naturally when the soil and air are warm; the fruit ripens in summer or before winter; and the plant then dies off before cold weather sets in. In the middle and southern states of America, Downing informs us, Melons are raised as field crops by market gardeners, the seeds being sown in the open air in May, and ripe fruit is obtained in August. In Australia likewise Melons are produced with the greatest ease in extraordinary abundance. There are many varieties of Melons, differing in size, form, and colour. Some are round or oblate, others oblong or oval; the surface of some is smooth, of others ribbed, netted, or warted. The flesh is either white, greenish, salmon-coloured, or red. The green-fleshed varieties are now generally preferred.

The Water Melon, *C. Citrullus*, is supposed to be of more ancient introduction to Europe than the foregoing. Rauwolf, in 1574, found it in abundance in the gardens of Tripoli, Rama, and Aleppo, under the name of Bathieca, the root of which word is from the Hebrew Abhattichim, one of the fruits of Egypt which the Jews regretted in the wilderness. It still forms chiefly the food and drink of the inhabitants of Egypt for several months in the year. It is very much cultivated in India, China, Cochin-China, Japan, the Indian archipelago, in America, and in short in most dry hot parts of the world, on account of its abundant refreshing juice, which, however, is not so rich and sugary as that of the common Melon. It is not esteemed in this country, where it is rarely grown. [R. T.]

CUCURBITACEÆ. (*Nhandiroba, Cucurbits, the Cucumber and Gourd family.*) A natural order of polypetalous and gamopetalous calycifloral dicotyledons, characterising Lindley's cucurbital alliance. Succulent climbing plants with tendrils in place of stipules, alternate palmately-veined rough leaves, and staminate and pistillate flowers. Calyx adherent, its limb five-toothed, or obsolete. Petals four to five, usually united (gamopetalous), reticulated. Stamens generally five, distinct or combined; anthers long and wavy. Ovary one-celled, inferior, with three parietal placentas, which often send processes into the cavity so as to reach the centre, and there unite; stigmas thick. Fruit succulent, a pepo (gourd); seeds flat, without albumen; cotyledons of the embryo leafy. Natives chiefly of hot countries; they abound in India and South America, a few are found in the North of Europe and North America; some are also met with at the Cape of Good Hope and in Australia.

The plants of this order possess generally a certain amount of acridity. Many of them are powerful purgatives, such as the melon, cucumber, vegetable marrow, gourd, pumpkin and squash; while of others the fruits are edible when cultivated. The seeds are usually harmless. The pulp of the fruit of *Citrullus Colocynthis*, the coloquintida, or bitter apple, is the colocynth of the shops; this is supposed to be the wild gourd of Scripture. *Ecbalium purgans* or *agreste* (*Momordica Elaterium*) is called squirting cucumber on account of the elastic force with which its seeds are scattered; the deposit from the fluid of the fruit constitutes the powerful purgative called elaterium. *Cucumis sativus* is the common cucumber, *C. Melo* the melon, and *C. Citrullus*, the water-melon. *Cucurbita Pepo*, the gourd, is a scrambling plant, to which belong the vegetable marrows, which are edible, the orange gourds, which are bitter, the egg-gourds, giraumons, crooknecks, Turks' caps, and warted gourds. *C. maxima*, the pumpkin, bears immense fruit; and *C. Melopepo*, the Squash, forms a bush about 3 ft. high, and may be had in the shops under the names of Pâtisson, Elector's Cap, and Jerusalem Artichoke Gourd. The seeds of *Hodgsonia* are eaten in India. *Lagenaria vulgaris* supplies fruit, which after the pulp is removed is used for carrying water, under the name of bottle-gourd. The fruit of *Luffa Ægyptiaca* is cut up when dry and used as a flesh brush, under the name of towel-gourd. *Sechium edule* yields an edible fruit called chocho or chacha. The species of *Bryonia* are purgative. There are three divisions of this order; 1. *Nhandiroba*, anthers not wavy, placentas adhering in the axis of the fruit, seeds numerous; 2. *Cucurbiteæ*, anthers wavy, placentas and seeds as in the first; 3. *Sicyeæ*, placentas not projecting into the cavy, seeds solitary from the top of the cell. There are about seventy genera, and 340 species. *Bryonia, Citrullus, Momordica, Luffa, Cucumis, Cucurbita, Coccinia, Trichosanthes, Telfairia, Feuillea*, and *Sicyos* are examples. [J. H. B.]

CUCURBITA. The typical genus of the *Cucurbitaceæ*, and composed of herbaceous mostly climbing plants, that are natives of hot countries in both hemispheres, chiefly within the tropics. A few are found in the north of Europe and North America, but India appears to be their head quarters. Those which are annuals readily submit to the climate of northern latitudes during summer. Although we best know the cucurbits by the use of the melon, cucumber, vegetable marrow and similar plants, yet it must be borne in mind that acrimony and a drastic tendency pervade many species, the fruits of some of which afford cathartics of remarkable power. Such being the predominant quality of the family it is well to be cautious in the use of even the best known species. (*Lindl. Veg. King.* p. 313.)

The Pompion or Pumpkin Gourd, *C. Pepo*, of which there are many varieties, is a tender or half-hardy annual, a native of Astrachan, and is stated to have been cultivated in England since 1570. It has large rough heart-shaped five-lobed leaves, and hispid branching tendrilled stems, which in good soil will grow rapidly and cover a large space in the course of a season; the flowers are large deep yellow. The fruit is oblong egg-shaped, varying both in form and size, and is used for soups or stews, but more frequently in this country it is mixed with sliced apples, to which a little sugar and spice are added, and after being baked is eaten with butter under the name of pumpkin pie. Until 1815, according to London, this was the principal kind of gourd cultivated in British gardens—in those of the rich chiefly for ornament, and in those of the poor, in some parts of England, as a culinary vegetable.

The Egg-shaped or Succade Gourd, or Vegetable Marrow, *C. ovifera succada*, sometimes regarded as a variety of *C. Pepo*, is believed to have been originally brought from Persia, but the date of its introduction is not exactly known. It is one of the most valuable sorts of gourd for culinary purposes that we possess. The plant is similar in habit and appearance to the other kinds of trailing gourds; and the leaves are rough, middle-sized, and deeply-lobed. The fruit is of an uniform pale greenish yellow, of an elongated oval-shape, slightly ribbed and about nine inches long. It is used in every stage of its growth, and is peculiarly tender and sweet; when very young it is good if fried in batter, but it is in the intermediate or half-grown state that it deserves the name of Vegetable Marrow. It is then excellent when plain boiled and served with rich sauces. For many years this valuable esculent was only to be met with in the gardens of the wealthy, but it is now extensively cultivated, and during the latter part of summer and autumn it forms one of our common vegetables.

The Melon Pumpkin, *C. maxima*, is one of the largest examples of the gourd tribe. It is a native of the Levant, and is recorded to have been introduced in 1547. The stems are angular, rough and trailing, with large heart-shaped five-lobed tooth-letted rough leaves. The flowers are large bell-shaped deep-orange. The fruit is roundish, often flattened at top and bottom, slightly ribbed, of a pale buff or salmon colour, and thickly netted over its surface with narrow vermicular processes. When dressed it has a peculiar flavour not unpleasant to the taste, and forms an excellent substitute for carrots or turnips. It is the *Potiron* of the French, who use it largely in soups, as well as mashed in the manner of potatoes. In North America it is extensively cultivated as an article of food, and as it keeps well it affords a supply through a great part of the winter. The fruit often attains a large size. One grown at Luscombe in Devonshire is mentioned in the *Gardener's Magazine* (vii. 102), as having weighed 245 lbs. Another, grown at Lord Rodney's in 1834, weighed 212 lbs., and was 8 ft. round. Yellow, green, and grey varieties are cultivated.

Besides the gourds just noticed as being the sorts that have been longest cultivated and best known in this country, there are many other sorts well deserving of attention. Among these we would particularly mention the Custard Marrow Squash, and the Improved Custard Marrow or Bush Squash, both of which are prolific and highly esteemed for their superior excellence, as well as for the peculiar form of their fruit, which for culinary purposes are remarkably handsome and in great request. Many kinds of gourds are also exceedingly ornamental. [W. B. B.]

CUDBEAR. A name given in Scotland to a crimson dye prepared from *Lecanora tartarea* and some other lichens, by treating them with alkaline substances. The collection of the lichen formerly employed a great number of hands, but it is now much neglected. A person so employed could earn fourteen shillings a week, the lichen being sold at about three halfpence a pound. It is now principally procured from Sweden and Norway, the manufacture being chiefly in the hands of the English. The name was derived from Dr. Cuthbert Gordon who first introduced the manufacture in Glasgow. [M. J. B.]

CUDRANIA. Climbing spiny shrubs, belonging to the *Artocarpaceæ*; they are natives of the Moluccas, Philippines, and India, and have entire diœcious flowers, the females in globose or oblong heads, each with a four-leaved perianth, and a pendulous ovule. [M. T. M.]

CUDWEED. The common name for *Gnaphalium*.

CUICHUNCHULLI. *Ionidium microphyllum*.

CUITLAUZINA. *Odontoglossum*.

CUJUMARY BEANS. The fruits of *Aydendron Cujumary*.

CULANTRILLO. The Chilian name for *Tetilla*, an astringent plant.

CULCASIA. A little known genus of *Araceæ*, comprising a tropical African species, with entire stalked leaves, and a brownish spathe enclosing a spadix bearing male and female flowers, and intermediate rudimentary organs. Ovaries crowded, each with one ovule. [M. T. M.]

CULCITA. *Dicksonia Culcita*. The name has sometimes been used generically to separate this species from the rest of the genus *Dicksonia*. [T. M.]

CULCITIUM. A genus of *Compositæ*, composed of woolly herbs or small bushes found in the Andes of Peru and Columbia near the snow limit at an elevation of 14,000 or 15,000 feet above the level of the sea. The name derived from *Culcita*, a cushion, is given, because all parts of the plants, except the upper surface of the leaves of a few, are covered with dense white or rusty coloured woolly hairs, which serve as beds for those travellers who may be forced to spend the night in the open air at this great elevation. The manner of making the bed is, by first amassing a quantity of the plants, and after taking the soft woolly pappus from the flowers, laying the branches, with the leaves attached, on the ground. On this first layer the soft warm pappus hairs are scattered, then a third layer is placed of leaves only, and, lastly, another layer of pappus hairs. On this couch the traveller reposes after the toils of the day without fear of frozen limbs. The genus *Espeletia* also belongs to this family, and growing on the high Andes, bears much resemblance to this in the woolly clothing of the leaves and stems, but the present is easily distinguished from it, the florets being all tubular, while in *Espeletia* there is an outer row of strap-shaped florets in the flower-head. Their nearest relationship is to the groundsels, *Senecio*, from which they may be at once recognised by their appearance. About a dozen species are known, some attaining a height of five or six feet, and having lance-shaped root leaves from six inches to a foot in length clasping the stem with their sheathing bases; these are sometimes called Lion's ear. [A. A. B.]

CULEN. A Chilian name for *Psoralea glandulosa*.

CULILAWAN BARK. The bark of *Cinnamomum Culilawan*, or Clove Bark.

CULLUMIA. A genus of little Cape bushes belonging to the composite family, and distinguished from its allies by the achenes being destitute of pappus, as well as by the curiously spinous margins of the leaves. These are seldom more than an inch long (generally much shorter), oblong in form, sessile, and often closely pressed to the stem. In a great many the margins are bordered with a single row of slender bristles about an eighth of an inch in length and in a few there is a double row of these bristles, one set pointing upwards, the other directed downwards. In all cases the leaves are terminated by a bristle. The yellow flower-heads are single on the ends of the branches, and half an inch or more in diameter. The scales of the involucre, in many rows, are furnished with bristles like the leaves. The florets of the outer row are strap-shaped and barren, of the inner tubular and fertile. About twenty species are enumerated. [A. A. B.]

CULM. The straw of corn; a kind of hollow stem.

CULMIFEROUS. Producing culms.

CULVER'S ROOT or CULVER'S PHYSIC. American names for *Veronica virginica*.

CUMIN or CUMMIN. *Cuminum Cyminum*. —, BLACK. The pungent seeds of *Nigella sativa*. —, SWEET. The Anise, *Pimpinella Anisum*. —, WILD. *Lagoecia cuminoides*.

CUMIN CORNU. (Fr.) *Hypecoum procumbens*. —, NOIR. *Nigella sativa*.

CUMINUM. Fennel-like plants, belonging to the *Umbelliferæ*, and botanically characterised by the presence of both general and partial involucres, the latter one-sided; by the calyx having five lance-shaped teeth; and by the elongated fruits, slightly contracted at the side, and each half provided with five thread-like ridges, and four intermediate ones more prominent and slightly prickly, beneath each of which there is an oil channel or vitta. The cumin seeds or fruits are the produce of *C. Cyminum*. They are much like those of caraway, but larger and of lighter colour, and with nine in place of five ridges on each half of the fruit. They are but little used, as caraways are more agreeable and more efficacious. The seeds of cumin smoked were considered by the ancients to produce pallor of the countenance. [M. T. M.]

CUMINGIA. A genus of *Liliaceæ*, consisting of bulbous Chilian herbs, with linear-lanceolate nervose leaves, and branched scapes bearing panicles of nodding blue flowers. The perianth is bell-shaped, the tube adhering to the base of the ovary, the limb six-parted with spreading segments. The six stamens are inserted in the tube, and have short compressed filaments; the ovary is three-celled with many ovules, the style subulate and the stigma simple. The genus is near *Conanthera*, but differs in having a less divided perianth, in the same way as *Hyacinthus* differs from *Scilla*. *C. campanulata* is a very interesting plant, with linear-channelled leaves, and a stem from a span to a foot high, bearing a racemose panicle at top, the flowers violet, paler in the throat around which they are spotted with blackish purple. [T. M.]

CUNEATE, CUNEIFORM. Wedge-shaped, inversely triangular, with rounded angles.

CUNICULATE. Traversed by a long passage, open at one end, as the peduncle of *Tropæolum*.

CUNILA. A genus of *Labiatæ*, containing several species of perennial herbs or undershrubs, natives of N. America. They have small white or purplish flowers, in corymbed cymes or close clusters. The calyx is ovate-tubular, equally five-toothed, and hairy in the throat; the corolla is two-lipped, with the upper lip erect, flattish, mostly notched, and the lower somewhat equally three-cleft; the two inferior stamens, which alone are fertile, are erect, exserted, and distant, and there are no traces of the superior stamens; the apex of the style is shortly bifid with subulate lobes. The nucule is smooth. [W. C.]

CUNIX. The separable space which intervenes between the wood and bark of exogens; an obsolete word.

CUNNINGHAMIA. A lofty evergreen tree, forming a genus of *Coniferæ* of the suborder or tribe *Abietineæ*. The linear falcate or lanceolate stiffly-pointed leaves are nearly those of the American *Araucarias*, but of a brighter green and less rigid. In the flowers and cones, the genus is nearly related to *Pinus*, but there are three or four anther-cells instead of two to each scale of the male catkins, and three instead of two ovules or seeds to each scale of the females. *C. sinensis*, the only species known, is a native of South China, and too tender for our climate without protection; but it is occasionally to be seen in our conservatories, where, from the elegance of its habit, it is a welcome inmate when there is room for its development.

CUNONIA. A genus of *Cunoniaceæ*, consisting of a small tree from the Cape of Good Hope, where it is called Rood Else by the Dutch colonists. It has reddish twigs, and opposite pinnate leaves with oblong coriaceous serrated leaflets, and ovate caducous stipules. The dense racemes of small white flowers are axillary and opposite, with the pedicels fascicled; calyx five-parted, deciduous; corolla of five oblong petals; stamens ten; ovary free, with two diverging styles; capsule conical, two-celled, separable into two many-celled carpels. [J. T. S.]

CUNONIACEÆ. (*Ochranthaceæ*, *Cunoniadæ*.) A family of dicotyledons, closely allied to *Saxifrageæ*, and very generally considered as a tribe only of that family, differing more in their habit than in the structure of their flowers or fruit. They are shrubs or trees with opposite leaves, simple or compound, and have stipules between the leaf-stalks. The calyx is half-superior or nearly inferior, the petals and stamens perigynous, the latter definite or more rarely indefinite. The ovary is two-celled, with two or more ovules in each cell; the styles usually distinct; the fruit capsular or indehiscent. They are natives chiefly of tropical regions or of the southern hemisphere, and especially of Australia. There are above a hundred species, distributed into about twenty genera, among which may be cited as the most generally known, *Weinmannia, Callicoma, Acrophyllum, Ceratopetalum, Cunonia, Cuidelaria, Belangera*, &c.

CUPANIA. A large genus of trees or shrubs belonging to the *Sapindaceæ*, numbering upwards of fifty species, more or less frequent in all tropical countries, but found in greatest numbers in South America. They are distinguished from their near allies by having a dry capsular fruit, which bursts when ripe; those genera more immediately related to them having more or less fleshy fruits which do not burst when ripe. In all cases the leaves are pinnate, varying in length from six inches to two feet, and composed of few or many leaflets. The flowers are small, generally green or white, and arranged in terminal or axillary racemes or panicles; some of them contain stamens only, others both stamens and pistil. The calyx is five-parted; the petals five, with or without a little scale-like appendage; and surrounding the ovary is a fleshy ring, inside of which the stamens (eight to ten in number) are inserted. The ovary is crowned with a simple style, generally trifid at the top, and becomes when ripe a two or three-lobed capsule, woody or thin in texture, with two or three cells, each containing one seed; the latter in all the species are furnished with a large or small fleshy cup-shaped aril, which is frequently of a bright yellow colour, while the outer coating of the seed is generally black and polished. *C. edulis*, the Akee Tree, is sometimes called *Blighia sapida*; which see. The Tulip Wood of eastern tropical Australia is furnished by the *Cupania* or *Harpulia pendula*, a tree of lofty growth, with a stem varying from eighteen to twenty inches in diameter. The light coloured wood is interspersed with darker mahogany-coloured patches, and is susceptible of a high polish; it bears much resemblance to that of the Tamarind tree. A very curious circumstance has been noticed by Mr. Spruce in connection with the seeds of *C. cinerea*, a Peruvian tree with pinnate leaves, and wedge-shaped leaflets covered underneath with a white down. He says, 'The embryos fall out of the seeds, while the outer coating or husk of the seeds with their aril contained in the burst capsules still remain on the tree.' Loblolly-wood is the name given in Jamaica to the wood of a number of trees of this genus. [A. A. B.]

CUP FLOWER. *Scyphanthus elegans*.

CUP GOLDILOCKS. *Trichomanes radicans*.

CUPHEA. A genus of *Lythraceæ*, consisting of herbs or undershrubs, often viscid, natives of Tropical America, one species extending northwards as far as New York. The leaves are opposite, rarely verticillate, entire; flowers solitary, on short often-curved stalks, and not unfrequently arranged in a racemose manner, purple, red or white; calyx tubular, inflated below, and gibbous or spurred at the base on the upper side, strongly nerved, the limb plaited and six-toothed, often with six smaller intermediate teeth, the whole coloured and often forming the most conspicuous part of the flower; petals six, rarely absent, unequal, the two uppermost generally much larger than the others; stamens about twelve, unequal, in two sets; ovary free, one or two-celled, few ovuled, with a slender style and two-lobed stigma. Capsule oblong, usually ruptured before the seeds are ripe, in which case the placentas with the seeds attached, protrude. [J. T. S.]

CUPIDONE. (Fr.) *Catananche cærulea*.

CUP-PLANT. An American name for *Silphium perfoliatum*.

CUPRESSUS. A genus of evergreen trees and shrubs, giving its name to the tribe *Cupressineæ*, of the family of conifers. Their foliage is not often to be distinguished from that of some species of juniper, consisting, as in that genus, of either small scale-like closely-appressed leaves, or of longer linear spreading ones, acute or acuminate, always opposite, and both forms occurring sometimes in different parts of the same tree or shrub. The fruit or cone is, however, very different from that of *Juniperus*, being much larger, with peltate woody scales opening to let out the seeds when ripe, and not at all succulent; and the seeds are winged. There are about ten species natives of the northern hemisphere, all extratropical or penetrating into the tropics only in mountain regions. They may be readily distributed into two sections, considered sometimes as distinct genera: *Cupressus* proper, with several seeds under each scale of the cone; and *Chamæcyparis*, with two seeds only to each scale. But the species themselves are very difficult to mark out, being distinguished rather by general habit than by any very positive botanical character.

C. sempervirens of Linnæus, the common Cypress, is a native of Persia and the Levant, but so generally planted in the East that the precise limits of its indigenous area have not been well ascertained. It has two very remarkable forms. One, *C. fastigiata*, with erect closely-appressed branches, is the well-known tall Cypress, celebrated by Oriental poets for its elegant slender pyramidal form, and extensively planted in Southern Europe and Western Asia, especially in Mahommedan and Armenian burial grounds. It will there reach a height of above sixty feet, densely clothed with leafy compact branches to within four or five feet of its base, the trunk below the branches attaining twelve to fifteen feet in circumference. In our country, however, it is only in a few favoured spots that it will rise much above a bush of ten to fifteen feet, for it is of very slow growth, and much liable to injury from wind and severe frost. The second variety, *C. horizontalis*, or spreading Cypress, with all its branches more or less spreading or quite horizontal, is so different in aspect that it would be difficult to conceive it to belong to the same species, were it not that it will frequently spring from the seed of *C. fastigiata*. In the south of Europe it readily grows to a tree, having much the form of a Cedar, but it is seldom planted in England.

C. torulosa, from the Himalaya, is one of the most elegant of modern introductions to our pinetums and shrubberies; the branches are erect or ascending, but less compact than in the common tall cypress, and the colour is not so dark. It is hardly enough to bear well the climate of some parts of England, but in others suffers much in severe winters. *C. glauca*, another East Indian species, is much more tender, and will seldom outlive our winters without protection, but it is much planted in Portugal, and has thence acquired the name of *C. lusitanica*. *C. funebris*, from North China, with its long branches, said to droop like those of a Weeping Willow, promises to be a valuable addition to our hardy evergreens. To these must be added *C. macrocarpa* and *Goveniana*, both Californian. The first a noble tree with the habit of *C. sempervirens*, the second of much smaller dimensions and with a less compact habit.

Of the section *Chamæcyparis*, two species, *C. thyoides* and *C. nutkaensis* (*Thujopsis nutkaensis* of our garden catalogues), from North America, and *C. squarrosa* (*Retinospora squarrosa* of our garden catalogues) from Japan, are to be met with in our plantations of conifers.

CUPULE. The cup or husk of the acorn, Spanish chestnut, &c.; a collection of bracts; a sort of involucre; a cup-like body found in such fungals as *Peziza*.

CUPULA-SHAPED. Slightly concave, with a nearly entire margin; as the calyx of *Citrus*, or the cup of an acorn.

CURAGE. (Fr.) *Polygonum Hydropiper*.

CURANA WOOD. The wood of *Icica altissima*.

CURATELLA. A genus of small trees from Tropical America, belonging to *Dilleniaceæ*, with alternate ovate rough leaves often with winged leaf stalks; flowers small, white, racemose; calyx of four unequal roundish sepals; petals four or five; stamens numerous; ovaries two, subglobose, united at the base, with subulate styles; capsules leathery, bispid, one-celled, two-seeded; seeds with a membranous aril. The rough leaves of *C. americana* are used in Guiana for polishing. [J. T. S.]

CURCAS. A genus of *Euphorbiaceæ* formed for the reception of the Physic-nut tree, *C. purgans*, or, as it was formerly called, *Jatropha Curcas*. It differs from *Jatropha* merely in having a bell-shaped corolla, while the latter has a corolla of five distinct petals. It forms a large bush or sometimes a tree of twenty feet high, with soft spongy wood and smooth bark, and is indigenous in Tropical America, but is very generally found in all tropical countries, being cultivated for the purgative oil of the seeds. Its leaves, generally crowded at the apex of the branches, are smooth, entire, and heart-shaped, or more commonly three or five-lobed, and including the stalks, from six to eight inches in length. The small green flowers are supported on stalked cymes about the length of the leaves; the males occupy the extremities of the ramifications, and the females the forks. The former have a calyx of five leaves; a bell-shaped corolla with a five-lobed border; and a double stamen-tube of ten stamens, the five inner longer than the others. The females have a similar calyx and corolla, and a three-lobed ovary crowned with a tripartite style, each branch forked at the apex.

Dr. Bennett in his *Gatherings of a Naturalist*, states that this tree 'contains a milky acrid glutinous juice, which when dropped on white linen produces an indelible stain, at first of a light blue colour, but after being washed, changing to a permanent brown: it might therefore form a very excellent marking ink. The fruit is globular and fleshy, about the size of a filbert, and contains three seeds in distinct cells. When immature, it is of a green colour, and when ripe black. On removing the husk from the oblong seeds, a white kernel remains, which contains much oil, and has an agreeable almond-like taste. The seeds are collected by the natives of the Philippine Islands for the purpose of expressing the oil, which they use for

Curcas purgans.

burning in their lamps, as well as for medicinal purposes. The leaves are employed for fomentations, and the juice of the young buds or other parts of the tree as a beneficial application to the ulcerated surface of wounds.' The seeds are employed by the native doctors of the Philippine Islands, and are considered excellent and mild purgatives, in doses of from one to four seeds. The effects which result from an overdose are vomiting, purging, a burning sensation in the stomach and bowels, with a determination of blood to the head. The only antidote used by native practitioners is cold water; warm water they affirm would be injurious. The kernels are administered entire, or are pounded in a mortar with water, and after being strained given as a draught. Dr. Bennett has himself administered these seeds to Europeans, but has found their effects very irregular, and occasioning in all cases a burning sensation in the bowels, followed with nausea and vomiting.

The oil is said to be sometimes boiled with oxide of iron, and used by the Chinese as a varnish. It is of a light colour, and has been imported into England and used as a substitute for linseed oil, as well as for dressing cloth, burning in lamps, &c. Its qualities differ little from those of castor oil according to Dr. Christison, who says that twelve or fifteen drops of it are equal to an ounce of castor oil. The white milky juice in which the plant abounds is reported as having healing properties, and a decoction of the leaves is used in the Cape de Verd islands to excite secretion of milk in women.

The only other species of the genus is *C. spathulata*, sometimes called *Moringa spathulata*, a low bush found in Mexico, with stout succulent stems, having olive-coloured bark, and furnished with numerous warty excrescences from which the leaves and flowers arise. The former are small and spathulate, and the latter inconspicuous. [A. A. B.]

CURCULIGO. A genus of hypoxids found in extratropical South Africa, in tropical New Holland, and in India. They are herbs with grassy ribbed leaves, and short scaparous spikes or fascicles of small inconspicuous flowers, which have a cylindrical tube adhering to the style, a regular spreading six-parted limb, and six stamens inserted in the mouth of the tube. The roots of *C. orchioides* are bitter and aromatic, and are used medicinally in India; while those of *C. stans* are eaten in the Marianne Islands. [T. M.]

CURCUMA. A genus of *Zingiberaceæ*, consisting of plants with perennial rootstocks and annual stems. The flowers are in spikes with concave bracts; they have a tubular three-toothed calyx; the tube of the corolla is dilated above, five of its lobes are equal, but the middle one of the inner row of the lip is larger and spreading; the filament is petaloid, three-lobed at the top, with a two-spurred anther on the middle lobe. The substance called Turmeric consists of the old tubers of *C. longa*, and perhaps some other species. The powder is used as a mild aromatic, and for other medicinal purposes in India. It enters into the composition of curry powder, and is used as a chemical test for the presence of alkalies, which change its yellow colour to a reddish brown. The young colourless tubers of this plant furnish a sort of arrowroot; another species, however, *C. angustifolia*, furnishes East Indian arrowroot, which is prepared by bruising and powdering the tubers, and throwing the powder into water, which is frequently changed till the starch loses its originally bitter taste. *C. rubescens* and *C. leucorhiza* also furnish starch. *C. aromatica* and *C. Zedoaria* furnish Zedoary tubers, which are used by the natives of India as aromatic tonics, and as a perfume. Several species with yellow or reddish flowers are cultivated in hot-houses. [M. T. M.]

CURL. A formidable disease in potatoes, referrible to Chlorosis, in which the tubers produce deformed curled shoots of a pallid tint, which are never perfectly developed, and give rise to minute tubers. It is supposed to arise from the tubers being overripe. It is, however, a local disease, and is quite unknown in many districts. It must not be confounded with a curled state of the foliage, which arises from the presence of aphides. [M. J. B.]

CURLS, BLUE. An American name for *Trichostema*.

CURRANT. The common name for *Ribes*, but especially applied to *Ribes rubrum*, the red, and *R. nigrum*, the black currant of the gardens. The currants of the shops are the dried berries of the Corinthian grape. —, AUSTRALIAN. *Leucopogon Richei*. —, INDIAN. An American name for *Symphoricarpus vulgaris*. —, NATIVE, of New South Wales. *Leucopogon Richei*. —, NATIVE, of Tasmania. A name applied to some species of *Coprosma*.

CURRANTWORTS. A name given by Lindley to the *Grossulariaceæ*.

CURRA-TOW. *Ananassa Sagenaria*.

CURRORIA. A genus of *Asclepiadaceæ*, containing a single species from Western Tropical Africa. It has a five-parted calyx, with ovate-lanceolate sepals; the corolla tube is short and subglobose, the divisions of the limb are linear-lanceolate, and have a twisted æstivation; there are five linear scales in the throat of the corolla; the gynostegium is included; the pollen masses are slightly stalked and erect; and the stigma is short. [W. C.]

CURRY-LEAF TREE. *Bergera Kœnigii*.

CURTISIA. A genus belonging to the order of cornels, having a four-parted calyx, four blunt petals, four stamens alternate with them, and a stone fruit, the hard part of which is four or five-celled. The name was given in honour of Mr. Curtis, a well-known English Botanist. The only species is a large and fine tree, a native of the Cape, with opposite shining broad or toothed leaves, of a rusty colour beneath; the flowers small and numerous. The natives of the region where it abounds employ it to form shafts for their javelins or assagays; hence the common name Assagay Tree. [G. D.]

CURVATIVE. When the margins are slightly turned up or down, without any sensible bending inwards.

CURVE-RIBBED. When the ribs of a leaf describe curves, and meet at the point; as in *Plantago lanceolata*.

CURVINERVED, CURVE-VEINED. The same as Convergenti-nervose.

CUSCO BARK. A kind of cinchona bark.

CUSCUTACEÆ. (*Dodders.*) A natural order of corolliflorial dicotyledons, belonging to Lindley's solanal alliance. The plants are included by some in a suborder of *Convolvulaceæ*. Leafless parasitic twining herbs, with flowers in dense clusters, Calyx inferior, four to five-parted; corolla persistent, four to five-cleft; scales alternating with the segments of the corolla, and adhering to them; stamens four to five; ovary two-celled, with two ovules in each cavity; styles two or wanting; fruit two-celled, either capsular or succulent; seeds with fleshy albumen; embryo spiral, filiform, having no cotyledons. The seeds germinate in the soil in the usual way, and afterwards become true parasites by attaching themselves to plants in their vicinity, and growing at their expense. Some of them destroy flax, clover, and other crops. Dodder, or scaldweed, is also the pest of beans and hops in some places. These parasites are found in the temperate regions of both hemispheres. They seem also to possess acrid and purgative qualities. The farmer requires to take care that dodder seeds are not mixed with those of his crops. They may be separated by careful sifting. There are upwards of fifty species included in four genera, of which *Cuscuta*, *Lepidanche*, and *Epilinella* are examples. [J. H. B.]

CUSCUTA. The Dodders, a genus of annual leafless parasitic plants, the stems of which consist of small wire-like tendrils that twine round the plant destined to be the foster parent, and into the texture of which they send out aerial roots at the points of contact, and through these imbibe the sap of the attacked plant. Our native flora contains two species: *C. europæa*, a plant which is described by Sir J. Smith as climbing 'two or more feet high upon thistles, oats, and any plants that are crowded together and will afford it nourishment;' and *C. Epithymum*, a smaller plant which grows on heath, thyme, &c. Besides these are now recognised *C. Epilinum*, the Flax Dodder, and *C. Trifolii*, the Clover Dodder, species, or probably varieties, which it would appear have been introduced with foreign seeds of their respective crops in the cultivation of which they are so gradually becoming most serious impediments. This is so much the case that we were induced to experiment largely on their mode of growth, with a view if possible to obviate the evil. The extent of the mischief may be judged from the fact that it was reported in the *Agricultural Gazette* for 1850 that one grower of flax had separated no

Cuscuta Epilinum.

less than seventy bushels of Dodder seed from his flax crop. With some of this seed we carried out the following experiments:—

Exp. 1.—On sowing some seeds in a saucer with fine mould, the following appearances presented themselves. In four days, the radicle was extended. In five days the germ was elevated above the soil, bearing the seed-covering on its apex. In six days the young thread-like plant was as it were on the look-out for a foster parent, and by the eighth day, not finding a foster parent, it emerged from the soil and died. (See diagrams in *Agric. Gaz.* 1850, 746.) Thus, then, all the plants which germinated freely died within a few days, the thread-like germs gradually becoming elevated out of the soil, and then withering away. However, on planting young examples of flax, chickweed, tomato, and others among them, the later germinated seeds immediately directed their threads towards them, and commenced that parasitic mode of growth which was so fully shown in the next case.

Exp. 2.—A saucer was sown with a mixture of flax and Dodder. In a few days both germinated, and the Dodder threads were attracted to the stems of the young flax, their history and progress being as follows. In seven days the Dodder had just clasped a flax plant. In nine days, both Dodder and flax having grown, the elevation of the flax stem had lifted the firmly attached Dodder out of the soil. In eleven days the Dodder was throwing out buds for new shoots, and the lower unattached part was dying away. (See diagram already referred to.) This explains the method by which the Dodder first becomes attached to the plant upon which it grows. It makes one or two tight coils around its future support, and during the time these coils are progressing, the foster-parent is increasing in size, the compression of the former around the latter becomes tighter, thus causing the bark of the foster-parent to be more delicate, while the parasite is preparing a series of aërial roots to penetrate it; it having done this, its position is firmly established, its own natural root dies quite away, and thenceforward its true parasitic growth is astonishingly rapid. Experiments 3 and 4 were repeated during the present summer, 1860, as follows:—

Exp. 3.—A plot of pure flax seed was sown in the botanical garden of the Cirencester Royal Agricultural College; this came up well, and afforded a good crop of fine flax.

Exp. 4.—A plot of flax seed and Dodder seed intermixed. In this the flax and Dodder came up simultaneously, and the thread-like germ of the latter soon twisted round the flax stems, and in time sent out branches in every direction, which in turn twined about fresh flax stems until the whole plot was borne down by the parasite, and both it and the crop went through the processes of flowering and seeding simultaneously: so that in harvesting the crop both would be gathered together, and of course, unless carefully separated, such flax seed would perpetuate the evil.

The same remark applies equally to the clover crops as to those of flax. If crops are to be free from the Dodder pests, the farmer must take care not to sow them with the seed for the crop, for it is now evident that this is their mode of propagation. *C. Trifolii* grows precisely in the same way, but the whole plant is smaller; the seeds on this account are not so readily detected, so that it is much on the increase. [J. B.]

CUSPIDARIA. A genus of *Bignoniaceæ*, natives of Brazil, containing several species, forming erect or subscandent glabrous shrubs. The leaves are opposite, petiolate, and simple or trifoliate, with petiolulate ovate acuminate and ciliate leaflets. The flowers are in terminal panicles. The cup-shaped calyx is cut into five long cuspidate teeth; the corolla tube is ventricose-campanulate, and the limb is five-lobed; one of the five stamens is sterile; the stigma is bilamellate, with long acute lobes; the four angles of the capsular fruit are produced into wings; the seeds also are winged. This genus is nearly related to *Bignonia* and *Lundia*. It is separated from the former by its ciliated anthers, from the latter by its awn-like sepals, and from both by its tetrapterous fruit. [W. C.]

This name has also been applied to a genus of ferns, which have since been called *Dicranoglossum*. [T. M.]

CUSPIDATE. Tapering gradually into a rigid point; also abruptly acuminate, as the leaflets of many *Rubi*.

CUSSO. The Abyssinian *Brayera anthelmintica*.

CUSSONIA. The name of a genus belonging to the order of Ivyworts, distinguished by the top-shaped calyx, which is adherent to the seed vessel, its border having from five to seven short teeth; the petals five to seven, adhering to a conical disk on the upper part of the seed vessel; stamens five to seven, adherent to the petals; fruit almost round, with little juice, two to three-celled, one seed in each cell. The genus was named in honour of Cusson, a botanist of Montpelier. The species are shrubs, natives of the Cape or of New Zealand, having a soft stem, with leaves alternate, smooth, stalked, in three to seven large lobes: the flowers are greenish.

Two species have been known in our collections since the end of the last century; they are chiefly interesting on account of their peculiar aspect. *C. thyrsiflora* has the leaflets sessile, wedge-shaped, truncate, and three-toothed at the end. *C. spicata* has the leaflets wedge-shaped, acuminate, and serrated at the end, the flowers in spikes. *C. triptera* is by some considered to be a hybrid, having numerous leaflets, like those of *C. spicata*, but without stalks, as in *C. thyrsiflora*. [G. D.]

CUSTARD-APPLE. The common name for *Anona*.

CUTICLE. The external homogeneous skin of a plant, consisting of a tough membrane overlying the epidermis. The word is also used for the skin of anything, including the epidermis.

CUTIS. The peridium of certain fungals.

CUVY. The name of the large common form of *Laminaria digitata* in Orkney, where the narrow plant with a smooth stem (*Laminaria flexicaulis*) is distinguished by the name of tangle. The situations in which the two plants grow, are, according to Mr. Clouston, very different: 'the Cuvy growing so far out in the sea that the highest limit can only be approached at the lowest stream tides, and from this it runs out into the ocean as far as the eye can penetrate, and probably much farther; while the tangle may be approached at ordinary tides, and forms a belt between the Cuvy and the beach. The general aspect also differs: the stems of the Cuvy stand up like a parcel of sticks, and the leaves wave from them like little flags; while the tangle lies prostrate on the rocks, the leaves mingle together and form a darker belt round the shore. Six or eight feet is reckoned a good length for a Cuvy, while tangles may be found from twelve to twenty feet.' [M. J. B.]

CYAMIUM. A kind of follicle, resembling a legume.

CYANANTHUS. A genus of *Polemoniaceæ*, containing a few species of annual procumbent or erect herbs, found on lofty situations on the Himalayas. They have alternate entire or lobed leaves, and few solitary and generally terminal showy blue flowers. The calyx is inferior, tubular-campanulate, and five-cleft; the corolla is funnel-shaped, with a large five-cleft limb; the five stamens are inserted at the base of the corolla, alternate with its lobes, the anthers being adpressed to or connate with the ovary. The ovary is free and five-celled, with many ovules in each cell, and bears a simple style, and a five-lobed stigma. The capsule is oblong-conical, dehiscing loculicidally. Some botanists, overlooking the superior ovary, have referred this genus to *Campanulaceæ*, because of its five-celled ovary and five-lobed stigma, but in every other respect it appears more nearly connected with *Polemoniaceæ*. [W. C.]

CYANELLA. A genus of herbs from the Cape of Good Hope, belonging to *Liliaceæ*, and having lanceolate-elliptical or linear radical leaves sheathing at the base, and racemose blue or yellow flowers. Perianth coloured, six-parted; stamens six, with glabrous filaments; the lower perianth segments, the lowest stamen, and the style declinate; capsule three-celled, with numerous seeds. They are pretty greenhouse plants. [J. T. S.]

CYANEOUS, CYANÆUS, CYALINUS. In composition *Cyano*. A clear bright blue.

CYANOCHROUS. Having a blue skin.

CYANOPHYLLUM. A genus of *Melastomaceæ*, containing one or two undershrubs with large five-nerved leaves having a metallic lustre. The flowers are small, with five petals and ten stamens. *C. metallicum*, which grows at an altitude of 6,000 or 7,000 feet, has a blue metallic lustre on the under-surface of its leaves. *C. magnificum*, one of the grandest of what are now commonly cultivated in hothouses as ornamental-leaved plants, has its very large opposite leaves of a rich shaded green above, and purple beneath. They are from Tropical America. *C. ———*, a similar plant, is of eastern origin. [J. T. S.]

CYANOSTEGIA. A genus of *Verbenaceæ*, found in West Australia, and composed of small upright bushes with narrow lance-shaped or linear entire leaves, often covered with a cottony substance. The blue flowers, in terminal branching racemes, are numerous and remarkable for their frill-like papery calyx, with a five-lobed border, which increases in size after the flowers have withered, and when mature is about half an inch in diameter. The somewhat irregular corollas are small and tubular. The great profusion of the blue flowers with their remarkably enlarged calyces of a paler colour, together with their neat bushy habit, would no doubt render them favourite greenhouse plants were they in cultivation. Three species are enumerated. [A. A. B.]

CYANOTIS. The generic name of plants belonging to the spiderwort order, characterised by having the calyx in three divisions joined into a tube at the lower part, and persistent; the three petals also joined to form a tube, but soon falling off; the style thickened upwards, ending in a point—the stigma—which is hollow and covered with hairs. The name of the genus is derived from the Greek words signifying 'blue' and 'ear,' in allusion to the colour of the flowers. The species are showy plants, natives of Tropical Asia, annual or perennial, hairy or woolly, seldom naked; the stems trailing below, sometimes erect above. [G. D.]

CYATHEA. An extensive genus of arborescent ferns representative of the *Cyatheaceæ*. The genus belongs to that series or subgroup which has an indusium or involucre placed in the form of a cup beneath or so as to contain the spore-cases, the fructification being seated on the under surface of the fronds. The species are numerous, and rank amongst the most striking features of tropical scenery. They are most abundant in South America and in the West Indies, in India, the Eastern Islands, and the Pacific Islands; a few are met with in New Zealand and South Africa. In some the trunk is short, but in others it reaches a height of forty or fifty feet or even more, and is crowned with a magnificent head of fronds, which, in many cases, are of gigantic size, and are always large. The greater number have the fronds bipinnate, with the pinnæ deeply pinnatifid; but in one, *C. Brunonis*, found in Malacca and Penang, they are pinnate, the fronds being two to three feet long, and the pinnæ six or eight inches; and in another, *C. sinuata*, found in Ceylon, they are simple and lanceolate, with a sinuated margin. This latter has a slender trunk, about an inch in diameter, on which the elegant crown of simple wavy fronds is upborne. *C. medullaris*, a fine bipinnated or tripinnated species of New Zealand and the Pacific

Cyathea medullaris

Isles, and known in gardens as a hardy tree-fern of comparatively hardy character, forms in its native country a common article of food with the natives. The part eaten is the soft pulpy medullary substance, which occupies the centre of the trunk, and which has some resemblance to sago. *C. dealbata*, another beautiful species of New Zealand, is said to be eaten in a similar way. This has a trunk of from ten to fifteen feet high, crowned with a noble tuft of fronds, which are white beneath with a silvery powder. [T. M.]

CYATHEINE.Æ, CYATHE.Æ. The former is a principal sub-division or tribe of the polypodiaceous ferns, in which the receptacles are elevated and the sessile or subsessile spore-cases are oblique-laterally compressed, and burst horizontally, the ring or annulus being narrow, nearly complete, and more or less obliquely vertical. The latter is a section of this group, in which the sori have involucres or inferior indusia, the fructification being borne on the back of the fronds. [T. M.]

CYATHIFORM. The same as Cup-shaped.

CYATHOCALYX. A genus of *Anonaceæ*, characterised by having their petals hollow and constricted at the base, but expanding above into a flat blade; the stamens numerous; and the ovary solitary, embedded in a hollow receptacle, with several ovules attached to the line of union of the margins of the carpels. The genus includes a Cingalese tree, with flowers opposite to the smooth shining leaves. [M. T. M.]

CYATHOCNEMIS. A genus of begoniads, consisting of succulent Peruvian plants. The staminate and pistillate flowers have each two sepals: the anthers elongated with slightly united filaments; the style persistent, its branches furnished with a continuous papillose band, making two spiral turns; the seed-vessel margined with three equal wings. The peduncles at their dichotomous divisions are surrounded by a large cup-like bract. There is one known species, viz., *C. obliqua*, found on rocks in the Andes of Peru. It was formerly a *Begonia*. [J. H. B.]

CYATHODES. A genus of *Epacridaceæ*, consisting of fifteen species, natives not only of Australia, but, like very few other genera of this order, found also in New Zealand and the Pacific Islands. They are small branching woody heath-like shrubs, with small axillary white or yellow flowers. The pedicles are covered with imbricated bracts, which are gradually larger upwards, and appear to pass into the sepals; the corolla is funnel-shaped, with a naked or bearded limb, and a smooth tube; the stamens are included or exserted; the drupe is more or less fleshy, with a bony five to ten-celled and five to ten-seeded nut, seated on a fleshy cup-shaped disc. [W. C.]

CYATHOGLOTTIS. An obscure genus of terrestrial orchids, with the ribbed foliage of an *Evelyna*, to which genus it is probably more nearly allied than to *Sobralia*, with which it has been usually compared. Two Andine species are mentioned: one with white, the other with yellow flowers.

CYATHUS. The cup-like body which contains propagula, or the reproductive bodies of *Marchantia*.

CYATHUS. One of the genera to which the curious *Fungi* belong which are commonly named Bird's Nest Peziæ. It is distinguished from *Nidularia* by the more complicated structure of the walls, and the stouter peduncle of the sporangia. We have two species generally distributed throughout England: *C. striatus*, which has a bright-brown shaggy cup, deeply grooved within, and *C. vernicosus*, which is mouse-grey, with the outer surface tomentose, and the inner polished. [M. J. B.]

CYBELE. *Peristylus*.

CYBISTAX (including *Yangua*). A genus of *Bignoniaceæ* confined to Peru, Bolivia, and Brazil, and easily distinguished from its allies by its lax plicate calyx, and broad pods traversed by twelve deep furrows on the surface. There seems to be only one species, *C. antisyphilitica* (*Yangua tinctoria* of Spruce), which forms a bush or small tree, and has when young duplicato-pinnate, when old digitate leaves. The bark of the younger branches is considered, in Brazil, one of the most powerful remedies against syphilitic swellings of a malignant character. The decoction is chiefly used, and also the bark dried and powdered and applied externally. In the Peruvian Andes, the tree is termed Yangua or Atunyangua, and the inhabitants dye the cotton cloths of their own manufacture a permanent blue by simply boiling them along with its leaves. About every three months all the leaves that can be got at are stripped off, and the trees seem not to suffer from being thus denuded; but they rarely put forth flowers till they grow beyond the reach of spoliating hands. The panicles are small, the calyx whitish, and the tubular corolla and the fruit of a greenish colour. In Brazil and Peru the plant is cultivated; it was also, at one time, an inmate of our gardens. [B. S.]

CYCADEACE.Æ. (*Cycads*.) A natural order of achlamydeous dicotyledons belonging to the gymnospermous (naked-seeded) alliance. Small palm-like trees or shrubs, with unbranched stems, occasionally dividing into two, marked with leaf-scars, and having large rays in the wood along with punctated ligneous tubes. Leaves pinnate, and usually rolled up like a crozier while in bud. Flowers staminate or pistillate, and without any envelope (achlamydeous); staminate flowers in cones, the scales bearing one-celled anthers on their lower surface; pistillate flowers consisting only of ovules on the edge of altered leaves, or placed below, or at the base of scales. Seeds either hard, or with a soft spongy covering; embryo hanging by a long cord in a cavity of the albumen; cotyledons unequal. Natives chiefly of the tropical and temperate regions of America and Asia. They are found also in southern Africa, and in Australia. Cycads are mucilaginous and starchy. *Cycas revoluta*, a native of Japan, supplies a kind of starch which is used as sago; and a similar kind of false sagois supplied by *Cycas circinalis* in the Moluccas

Caffre-head is made from the starch of a Cape species of *Encephalartos*, many species of which genus exist in Australia. In the West Indies some species of *Zamia* yield a kind of arrowroot. Cycads occur in a fossil state after the coal epoch. There are seven known genera, and about fifty species. Examples: *Cycas, Dion, Encephalartos,* and *Zamia.* [J. H. B.]

CYCAS. A remarkable genus giving its name to the order *Cycadaceæ*. It consists of trees of no great height, with cylindrical usually unbranched stems, terminated at the top by a crown of handsome deeply-cut pinnate leaves of thick texture. The male flowers grow in cones, consisting of scales bearing anthers on their under surface. The female plants bear in the centre of the crown of leaves surmounting the stem, a tuft of woolly pinnately-cleft leaves, in the notches of whose margins the naked or uncovered ovules are placed. The species are natives of the tropical regions of Australia, Polynesia, and Asia.

C. circinalis furnishes in Malabar a sort of sago, which is prepared from the seeds, which are dried and powdered; medicinal properties are attributed to the seeds, but these are of little importance. The plant is said to be singularly tenacious of life. The pith in the interior of the stem of *C. revoluta* abounds in starch, which is highly esteemed in Japan. A clear gum exudes from the trunks of these trees, which is said to be employed by the natives of India in promoting speedy suppuration. These elegant species are great ornaments in our plant-houses. A fine group of them and of the allied genera may be seen at one end of the large palm-house at Kew. They are popularly but erroneously called Sago-palms, as they furnish none of the sago of commerce. See Plate 6, figs. b and d, the latter showing an old branched stem. [M. T. M.]

CYCLADENIA. A genus of apocynaceous plants, natives of California, and allied to the genus *Vinca*, but abundantly distinguished from it, says Mr. Bentham, by the cymose inflorescence, the funnel or bell-shaped corolla, and the ring-like disc at the base of the stamens, from which latter the genus derives its name. *C. humilis* is described as being a most beautiful plant, resembling *Villarsia pumila*. [M. T. M.]

CYCLAMEN. A strongly marked genus of plants belonging to the order *Primulaceæ*. In all the species the leaves and flowers spring direct from a solid tuberous rootstock, which is shaped like an orange; the leaves are deeply heart shaped at the base, toothed or crenate at the edge, and in outline more or less orbicular. The flowers are of one petal, deeply divided into five oblong segments, which being erect while the mouth of the tube is turned downwards, present something of the appearance of a turban. After flowering, the flower-stalk (scape) coils itself up into a spiral form, having the seed-vessel in the centre, and bends itself towards the ground, in which position the seeds are ripened. The fleshy rootstocks, though of a highly acrid nature, are in Italy and Sicily greedily sought after by swine; hence the name *Pane porcino*, from which the English name Sowbread is adopted, the plants not being found in Great Britain in any situation to which swine have access. One species is, indeed, sometimes included in the British Flora, but is in all probability an outcast from a garden. Most of the species are hardy, and as they flower early are much prized as border flowers or for pot cultivation. *C. persicum*, the handsomest of all, requires artificial heat during the winter, but with care may be made to flower freely in the window garden. French, *Pain de Porceau*: German *Erdscheibe*. [C. A. J.]

CYCLANTHACEÆ. A name sometimes given to the family of *Pandanaceæ*, of which the *Cyclantheæ* are a tribe.

CYCLANTHERA. A Mexican climbing herb, belonging to the *Cucurbitaceæ*. It has bifid tendrils, and small green flowers. The female flowers are sessile, arising from the same point as the males. The latter are stalked, and their peculiarities have given rise to the name of the genus, and to the means of distinguishing it from its allies: the stamens, that is to say, are combined below into a short column which expands above into a round disc, bearing the anthers at its circumference. [M. T. M.]

CYCLANTHUS. A remarkable genus of Tropical American plants, referred to *Pandanaceæ*. From a contracted stem they throw up leaves, which are fan-shaped and cleft into two divisions. The flowers are unisexual, and arranged in spiral bands around the spadix, the bands consisting alternately of male and female flowers. The former have many stamens with four-celled anthers; the latter several ovaries, which become blended into a fleshy many-seeded fruit. The spadix is protected by a spathe which consists of four overlapping bracts. [M. T. M.]

CYCLE. A term employed in the theory of spiral leaf-arrangement to express a complete turn of the spire which is assumed to exist.

CYCLICAL. Rolled up circularly, as many embryos.

CYCLOBOTHRA. A genus of Liliaceous plants, allied to *Calochortus*, from which they are distinguished by having all the divisions of the perianth bearded within, and furnished with a honey-pit in the centre, forming a bump or gibbosity on the outside. The species are found in Mexico and California, and are very singular and handsome objects. They have tunicated bulbs, and erect leafy stems, the leaves linear and acuminated, and the stems bearing at the top the nodding flowers, which are sometimes arranged in the form of an umbel, and have considerable general resemblance to those of certain species of *Fritillaria*.

FOREST ON GUAHAN, ONE OF THE MARIANNE ISLANDS

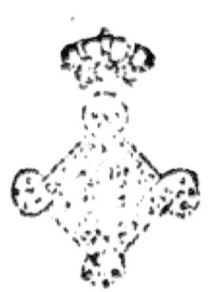

The perianth is bell-shaped or ventricose, with the three outer segments about half as large as the inner ones; and there are six stamens inserted in the base of the perianth, and a three-cornered ovary surmounted by three sessile stigmas. *C. luteus* grows about a foot high, and has oblong bulbs, long linear leek-like leaves, and two or three terminal yellow flowers, the divisions of which are scarcely connected at the base, the three exterior ones, which are smaller, greenish on the outer side, the three broader inner ones being bearded with purple hairs; the stem is bulbiferous in the leaf axils. *C. purpurea* is a more showy plant, growing two feet high, with the lower leaves elongately-linear, and the upper ones broader and more lanceolate; these also bear numerous little bulbs in their axils. The stem produces about three flowers, two from the point, and one from a side branch; these are large, with the sepals green marked with purple outside and yellow within, and the petaline segments brownish-purple outside, yellow within, and ciliated; the nectariferous pit is spade-shaped. [T. M.]

CYCLODIUM. A genus of polypodiaceous ferns belonging to the *Aspidieæ*. They are distinguished by having pinnate veins from a central costa, these producing connivently anastomosing venules, which form arcuate or angulate areoles. The sori, which are globose, are covered by peltate indusia. The species are thick-fronded robust pinnate ferns, with the fertile fronds more or less contracted, and sometimes so much so that the sori almost cover the surface. There are two or three closely allied species found in South America, and one in the Philippine Islands. [T. M.]

CYCLOGYNE. A genus of *Leguminosæ* found in Western Australia, the chief distinguishing character of which is found in the style, which is coiled inwards at the point, and much bearded. *C. canescens*, the only species known, is an astragalus-like bush with unequally pinnate leaves and leafy stipules; the leaflets (thirteen or fifteen in number) being small, obval, smooth above, and clad underneath with white hairs. The white appearance these hairs give to the plant, together with the profusion of purple flowers, render it an attractive object. The flowers are disposed in erect racemes, and have a bell-shaped five-toothed calyx; a nearly round standard, with a green blotch at its base, two short wing petals, and a keel longer than the wings. The inflated oblong pods contain a number of seeds. [A. B.]

CYCLOMYCES. A curious genus of hymenomycetous *Fungi*, allied to *Polyporus*, in which the walls of the pores form gills concentric with the stem or with the border of the pileus. The species originally described was gathered in Mauritius, but another occurs in North America, and a third has been found in the Sikkim Himalaya. [M. J. B.]

CYCLOPELTIS. A name applied to two pinnate aspidiaceous ferns, with peltate indusia and free veins, now referred to *Polystichum*. [T. M.]

CYCLOPHORUS. A name given by Desvaux to a group of polypodiaceous ferns, subsequently called *Niphobolus* by Kaulfuss, on the ground that Desvaux's name was in use among conchologists. *Niphobolus* has been generally adopted. [T. M.]

CYCLOPIA. A genus of dark-coloured South African bushes of the leguminous order, belonging to that group of the family in which the filaments of the stamens are quite free. They may be recognised from their allies by having a circular depression at the base of the calyx, round the pedicel. The leaves are sessile and made up of three generally linear smooth or pubescent leaflets, which often have their margins rolled backwards. Their bright yellow broom-like flowers are generally produced in great profusion from the axils of the upper leaves, and have their stalks always furnished with two boat-shaped bracts. The pods are oblong, compressed, and contain a number of seeds. *C. genistoides*, a plant sometimes seen in greenhouses, has smooth linear leaflets strongly-recurved at the margin, and pretty yellow broom-like flowers. The leaflets of this plant are used at the Cape in infusion or decoction for promoting expectoration in chronic catarrh and consumption. It is called Bush Tea, and has an agreeable tea-like smell, with a sweet astringent taste. [A. A. B.]

CYCLOSIS. A supposed motion of fluids, occurring in the kind of tissue called cinenchyma.

CYCNOCHES. A singular genus of orchids, with the habit of *Catasetum*. The name, which signifies Swan-neck, was suggested by the long curved column which in the original species rose gracefully from a broad convex lip. The character of the column is proper to all the species, but the lip varies from a broad solid oval plate to a stalked disk whose margin is broken up into numerous glandular rays. And, what is most strange, the same stalk bears flowers of both kinds, with others intermediate between the two. Here, therefore, we have a repetition of the singular variations already described in *Catasetum*. Upon this the Editor of *Paxton's Flower Garden* makes the following remarks: 'In Mr. Bateman's magnificent work we are told how the long-spiked small purple-flowered *C. Egertonianum* is only the short-spiked large green-flowered *C. ventricosum*; how the same plant at one time bears one sort of flowers, and at another time another sort; and we have ourselves shown how the same plant, nay the same spike, is sometimes both the one, the other, and neither. *C. Egertonianum* is then a 'sport,' as gardeners say, of *C. ventricosum*. But what, again, is *C. ventricosum*? Who knows that it is not another 'sport' of *C. Loddigesii*, which has indeed been caught in the very act of showing a false countenance, something wonderfully suspicious, all things considered, and justifying the idea that it is itself a mere Janus, whose face is green and short on one side, and spotted and long on the other? Then, if such apparently honest species as *C. Egertonianum*, *ventricosum*, and *Loddigesii* are but counterfeits, what warrant have we for regarding the other so-called species as not being further examples of plants in masquerade? For ourselves we cannot answer the question; nor should we be astonished at finding some day a *Cycnoches* no longer a *Cycnoches*, but something else: perhaps a *Catasetum*. If one could accept the doctrine of the author of the *Vestiges*, it might be said that in this place we have found plants actually undergoing the changes which he assumes to be in progress throughout nature, and that they are thus subject to the most startling conditions only because their new forms have not yet acquired stability.'

The principal species of this curious genus are *C. Loddigesii*, *ventricosum*, and *chlorochilon*, which have a sessile perfectly entire fleshy lip; and *C. pentadactylon*, *aureum*, *maculatum*, and *Egertonianum*, with a stalked flat lip, whose edges are broken up into numerous finger-like rays. These plants are all from Tropical America, and chiefly from the central states. See LUDDEMANNIA.

CYDONIA. A genus of the pomaceous division of the *Rosaceæ*, allied to *Pyrus*, from which it is distinguished by its leafy calyx-lobes, and the many-seeded cells of its fruit, those of *Pyrus* being dispermous. It comprises a few species, one of which is the well-known Quince; and another, *C. japonica*, one of the most ornamental deciduous shrubs in our gardens. The latter reaches some five or six feet in height, and is clothed in summer with oval crenately serrated leaves having kidney-shaped serrated stipules, and in spring with a multitude of glowing red flowers, to which it owes its beauty. [T. M.]

The common Quince tree is called *C. vulgaris*. The name of *Cydonia* was given to this by the ancients from its growing abundantly near Cydon, in the isle of Crete, now Candia. It is stated by some authors to have been introduced from Greece to Italy; but this can only refer to a particular variety, for Pliny in his fifteenth book says, 'There are many kinds of this fruit in Italy, some growing wild in the hedge-rows, others so large that they weigh the boughs down to the ground.' Sir Joseph Banks (*Trans. Hort. Soc.* i. 153), referring to Martial (xiii.24), states that the Romans had three sorts of Quinces, one of which was called Chrysomela from its yellow colour; they boiled them with honey as we make marmalade. According to the best modern botanists, the species grows spontaneously on the hills and in the woods of Italy, in the south of France, in Spain, Sicily, Sardinia, Algeria, Constantinople, the Crimea, and in the south of the Caucasus; it also grows abundantly on the banks of the Danube. It is found in Cashmere, and even in the north of India, according to Drs. Roxburgh and Royle. De Candolle thinks its native country extends perhaps as far as Hindoo-Coosh; but it is not cultivated in the north of China. In Imiretta, a region in the interior of Mingrelia, a variety is said to have been found with fruit as big as the head of a child. It appears from the above that the Quince is indigenous over a great extent of Europe and Asia, and that it is likewise found in the north of Africa. Phillips says in his *Historical and Botanical Account of Fruits known in Great Britain*, 'The learned Goropius maintains that Quinces were the golden apples of the Hesperides, and not oranges, as some commentators pretend. In support of his argument, he states that it was a fruit much revered by the ancients, and he assures us that there has been discovered at Rome a statue of Hercules that held in its hand three Quinces; this,' he says, 'agrees with the fable which states that Hercules stole the golden apples from the gardens of the Hesperides.' Galesio, in his treatise on the Orange, has shown that the orange tree was unknown to the Greeks, and that it did not naturally grow in those parts where the gardens of the Hesperides were placed by them. The Quince tree, according to the *Hortus Kewensis*, was introduced into this country in 1573; but Gerarde, who was alive at that date, says it was often planted in hedges and fences to gardens in his time, and from this it may be concluded the tree was common long before the period above mentioned.

The Quince is a hardy deciduous tree, fifteen to twenty feet high, with numerous crooked branches, forming a bushy spreading head; the leaves are roundish or ovate. The flower-buds push early in spring, and elongate into a branch, with five or six leaves, and at the extremity a single flower, white or pale red and of large size, is produced as late as May or June. The fruit is large, roundish, turbinate, pear-shaped, or irregularly oval, according to the variety. On approaching maturity it assumes a fine golden yellow colour, giving the tree a very ornamental appearance. The Portugal Quince is considered the best, but it does not bear so abundantly as the more common apple and pear-shaped varieties. All the varieties have a strong odour, with an austere flavour, so that they are unfit for eating raw; but the fruit is much esteemed along with apples in pies and tarts, and in confectionary it forms an excellent marmalade and syrup. Indeed, the name of marmalade is said to be derived from Marmelo, the Portugese name of the Quince. The plants are much used as stocks for pear trees, especially those intended to be kept dwarf. [R. T.]

CYLINDRENCHYMA. Cylindrical cellular tissue, such as that of *Confervæ*, of many hairs, &c.

CYLINDROLOBUS. *Eria*.

CYLISTA. A genus of *Leguminosæ*, found in the Bombay districts of India, and only represented by a single species, *C. scariosa*, which is a perennial twiner growing among bushes, with ternate leaves, having oval, pointed, and entire leaflets with short white pubescence. The yellow flowers, borne on erect bracted racemes, are remarkable for their large papery calyx, which is much more conspicuous than the corollas, and is deeply four-cleft; the upper segment being two-lobed, the lateral ones much smaller, and the lowest very large, all of them beautifully veined. The little oval one-seeded pod is completely enveloped in the peculiar calyx, which affords the most marked character in the genus. [A. A. B.]

CYMBALAIRE. (Fr.) *Linaria Cymbalaria*.

CYMBELLÆ. Reproductive locomotive bodies, of an elliptical form, found in some algals.

CYMBIDIUM. A name given by Swartz to a large group of tropical orchids, growing in the ground, with simple fleshy hairy roots, throwing up tufts of sword-shaped leaves, and producing radical spikes of flowers, which are erect or pendulous, many-flowered or few-flowered, and conspicuous for their beauty, or quite inconspicuous. All have a pair of curved ridges on the lower part of the lip: an essential character. Many plants in which this character is absent, and which have been erroneously referred to the genus, are now eliminated; nevertheless some twenty or thirty legitimate species remain. Of them the most important are *C. sinense*, a strong Chinese species with erect spikes of brown flowers emitting the most delicious fragrance; *C. giganteum*, an Indian plant with racemes of very large brown tessellated flowers; *C. eburneum* from India, with large radical ivory-white flowers smelling like lilacs; and *C. elegans*, also Indian, with great massive pendulous spikes of yellowish flowers. There are also many yellow Cape species not yet known in cultivation.

CYMBIFORM. Having the figure of a boat in miniature: that is to say, concave, tapering to each end, with a keel externally, as the glumes of *Phalaris canariensis*.

CYMBOCARPA. A genus of *Burmanniaceæ*, consisting of a single slender leafless annual, closely allied to *Dictyostegia*.

CYME. A kind of inflorescence, produced by the rays of an umbel forming one terminal flower, and then producing secondary pedicels from below it, in the centrifugal manner, as in the laurustinus.

CYMINOSMA. Small trees with opposite or alternate entire dotted leaves on a jointed stalk. They are of uncertain position, but are generally placed in *Rutaceæ*, and by some authorities are included in *Aeronychia*. The flowers are white or greenish, in axillary or terminal corymbs, and have a four-parted calyx and corolla;

eight stamens with flattened filaments, four longer than the others; an ovary with four two-ovuled cells, placed on a fleshy disc; a short style; and a berry-like fruit. The species are natives of China, the East Indies, and Australia. [M. T. M.]

CYMODOCEA. A genus of *Zosteraceæ*, containing a diœcious plant resembling *Zostera*, and found in the Mediterranean Sea. It has creeping branched rhizomes, and ribbon-like leaves faintly serrulated towards the apex: the flowers have no perianth, and consist of a pair of male flowers each reduced to a single two-celled stamen, or a pair of female flowers reduced to a single ovary, with a short style and two stigmas, which are long and thread-like. The fruit is produced in pairs. [J. T. S.]

CYNANCHUM. A genus of South European and Mediterranean herbs, belonging to the order *Asclepiadaceæ*, and characterised by its wheel-shaped corolla, and by the coronet of the stamens being tubular, with from five to ten divisions at its upper margin, and with five inner segments exterior to, and parallel with, the anthers. The fruit consists of two cylindrical follicles. The *Arghel*, the leaves of which are used to adulterate Alexandrian Senna, was formerly considered to belong to this genus, but is now included in *Solenostemma*: see also *Vincetoxicum*. [M. T. M.]

CYNARA. A genus of *Compositæ*, of which many of the species are prickly troublesome weeds, some are handsome, but scarcely any are useful besides the two familiarly known as the Artichoke and Cardoon.

The Cardoon or Chardoon, *C. Cardunculus*, very much resembles the artichoke. It is a hardy perennial, a native of the south of Europe and the northern parts of Africa. The earliest writer on gardening who has noticed it is Parkinson, who calls it *Carduus esculentus* in his *Paradisus Terrestris*, published in 1629. Its introduction into this country is stated to have been in 1656, and according to Dr. Neill, it was even cultivated in Holyrood Palace Garden so early as 1683; but it has never been considered a vegetable of much excellence, and at the present day it is only to be met with in a few of our best gardens. On the continent, however, the Cardoon is regarded as a wholesome esculent, which in the hands of a skilful cook forms an excellent dish. The parts which are used are the stalks of the inner leaves, rendered white, crisp, and tender by blanching. These stalks are either stewed, or form an ingredient in soups and salads during autumn and winter. When permitted to flower, the plant has a fine appearance, and attains a greater height than the artichoke. The flowers have also the property of coagulating milk, for which purpose they are frequently used by the French, after being gathered and dried in the shade.

The Artichoke, *C. Scolymus*, is a hardy perennial, a native of Barbary and the south of Europe. Although it is mentioned by Pliny as being a vegetable that was much esteemed by the Romans, it does not appear to have been known in this country until introduced from Italy in 1548. The plant has some resemblance to a large thistle. The leaves are numerous, ample, pinnatifid, somewhat spiny, from three to four feet long, and covered with an ash-coloured cottony down. The flower stems grow erect, and attain the height of from four to six feet. They are each terminated by a large globular head of imbricated oval spiny scales of a purplish-green colour, which envelope a mass of flowers in the centre. These flower-heads in an immature state contain the parts that are eatable; which comprise the fleshy receptacle usually called the 'bottom,' freed from the bristles and seed-down, commonly called the 'choke,' and the thick lower part of the imbricated scales or leaves of the involucre. Although Artichokes are a common vegetable, they are not so much in request with us as on the continent, where by various modes of cooking they are made to form favourite dishes. In France, the bottoms are often fried in paste, and enter largely into ragouts. They are occasionally used for pickling, but for this purpose the smaller heads which are formed on the lateral shoots that spring in succession from the main stem, are generally preferred when about the size of a large egg. The Chard of Artichokes, or the tender central leaf-stalk blanched, is by some considered to be equal to the cardoon. The flowers are very handsome, and are stated to possess the property of coagulating milk. [W. B. B.]

CYNAROCEPHALÆ. The artichoke-headed composites, a suborder of the natural order *Compositæ* or *Asteraceæ*, having numerous flowers collected in a common receptacle, and surrounded by a series of leaves or scales so as to form a compact head. The flowers are all tubular, and either have stamens and pistils, or those of the circumference (the ray) are abortive; the style is swollen below the stigma. Among the plants of this suborder are the artichoke, the cardoon, the burdock, the safflower, and thistles. They are usually bitter and tonic; some are esculent. See COMPOSITÆ. [J. H. B.]

CYNARRHODON. Such a fruit as that of the rose, in which many bony achenia are enclosed in a fleshy hollow enlargement of the apex of the flower-stalk.

CYNOCTONUM. A genus of *Asclepiadaceæ*, containing more than thirty species of perennial herbaceous plants or twining shrubs, natives of Africa, India, and tropical America. They have cordate leaves, and lateral peduncles springing from between the petioles, and bearing many-flowered umbels. The calyx and corolla are five-parted; the staminal corona is tubular and simple, with five or ten lobes, and without any appendages in the interior; the gynostegium short; the anthers surmounted by membranaceous appendages; and the projecting stigma is bilobed or with a bifid linear apex. [W. C.]

CYNODON. A genus of grasses belonging to the tribe *Chlorideæ*, distinguished chiefly by the spikes of inflorescence being in short spreading finger-like heads. The spikelets one-flowered, awnless; the glumes nearly equal, spreading; pales equal; stamens three; and styles three. Fourteen species are described, only one of which is a native of Britain, *C. Dactylon*, which inhabits the southern coasts of England. The creeping roots of this and some other grasses are said to possess some of the medicinal properties of sarsaparilla. [D. M.]

CYNOGLOSSUM. Houndstongue. A genus of *Boragineæ*, consisting of herbs from the temperate zones, especially of the northern hemisphere. Leaves often covered with silky-white hairs; flowers in scorpioid racemes, often bracteous, dull-red or blue; calyx five-parted; corolla salver-shaped with the throat closed by five obtuse scales, and the limb five-lobed; stamens five, included; nuts four, muricated depressed externally. Two species occur in Britain, *C. officinale*, with leaves covered with soft white hairs, dull-red flowers, and strongly-margined nuts; and *C. montanum*, a much more local plant, with green roughish leaves without soft hairs, blue-veined flowers, and nuts without a prominent margin. [J. T. S.]

CYNOMORIUM. One of the genera of the singular family *Balanophoraceæ*. It is represented by a fleshy red herbaceous plant, about a foot in height, covered with scales, the flowers of which are unisexual, the males and females mixed in the same heads, and surrounded by numerous scales; occasionally the flowers are hermaphrodite. The perianth in either case consists of six divisions.

C. coccineum, the *fungus melitensis* of old writers, was formerly valued as a styptic and astringent. The plant is not confined to Malta, but extends also to the Levant, Northern Africa, and the Canary Islands, in which, according to Mr. Webb, it is esteemed good to eat. It was formerly used to procure abortion in Malta, and was so highly valued as a remedy for dysentery that the place where it grew was guarded with the utmost vigilance; and even up to a recent date the plant was gathered, and its growth secured by a person specially appointed to the office by the English Government. [M. T. M.]

CYNORCHIS. A Mascaren genus of terrestrial orchids, differing from *Habenaria* in little except the lip being connate with the face of the column. The species have testiculate roots, like the orchids of Europe. One, *C. fastigiata*, has been in cultivation (see *Bot. Register*, t. 1948). Blume's genus, *Mitostigma*, is a synonym.

CYNOSURUS. A genus of grasses belonging to the tribe *Festuceæ*, and distinguished chiefly by the inflorescence being in crowded close thyrsoid panicles, with flowers pointing to one side; glumes nearly equal, scarious, and strongly keeled, two or more-flowered; each spikelet with a pectinated bract at its base. The genus comprises five species, only one of which, the Dog's Tail grass, *C. cristatus*, is truly a native of Britain. This is considered an excellent species for permanent sheep-pasture. The roots penetrate deep into the earth, which enables the plant to withstand droughts better than many of the other pasture-grasses; hence it may often be seen looking quite fresh when they are partially withered up. *C. echinatus* is an annual species, which is occasionally cultivated in British collections of grasses. It is a southern plant, but extends as far north as the Channel Islands. [D. M.]

CYPELLA. A genus of beautiful *Iridaceæ*, consisting of a single species, *C. Herberti*. The perianth is six-parted, concave at the base, the outer segments larger and spreading, the inner ones small convolute and reflexed at top. There are three erect stamens, united at the base of the filaments; a slender style; and a three-lobed stigma with trifid segments, which are appendiculate on both sides at the base. The chief distinction of the genus consists in its spreading not reflexed sepaline segments, and in their being deeply indented or hollowed out, as it were, at the base, so as to form a kind of bowl or cup. *C. Herberti* is a very slender plant, with fleshy corms, long lanceolate acute plaited glaucescent leaves, and a slender stem 1½ to 2 feet high, branched at top and producing in succession many flowers which last for several days, unlike those of some allied plants which are very fugacious. The flowers are bright orange yellow, the three outer segments with a central dark purple stripe, the three inner whitish in the centre, spotted with purple. It is a native of Buenos Ayres. [T. M.]

CYPERACEÆ. (*Cyperoideæ*, Sedges.) A natural order of glume-bearing monocotyledons belonging to Lindley's glumal alliance. Grass-like tufted plants, having solid, usually jointed, and frequently angular stems; leaves with their sheaths entire (not split, as in grasses); and flowers either perfect or incomplete (staminate and pistillate), each borne on a solitary bract or scale, and all united in an imbricated manner so as to form a spike. In the pistillate flowers there is often a membranaceous covering within the scale. Stamens hypogynous, varying from one to twelve, usually three; anthers attached at their base to the filament. Ovary superior, often surrounded at the base by bristles; ovule one; style two to three-cleft. Fruit a crustaceous or bony achene; embryo lens-shaped, and lying at the base of fleshy or mealy albumen. The plants are generally distributed over the world, and abound in moist situations. Some of the sedges are demulcent, others are bitter and astringent. Some by means of their creeping underground stems bind together the loose sands of the sea-shore. Their cellular tissue is sometimes used for paper. The underground stems of several species of *Cyperus* are used as food. *Carices* abound in moist temperate and cold regions. *Carex arenaria* is one of the sandy-shore plants; its underground stems are used for sarsaparilla. The species of *Eriophorum*, or cotton grass, have long white silky hairs surrounding the fruit. *Papyrus antiquorum* appears to be one of the plants called bulrush in scripture. It formerly grew abundantly at the mouth of the Nile, which was hence called by Ovid papyriferous, but it is now gone. The cellular tissue of its stems was used in place of paper. *Scirpus lacustris*, the bulrush, is used for making mats, baskets, and the bottoms of chairs. In South America it is used for making balsas or boats; a similar use is referred to in Isaiah (xviii. 1, 2). There are 120 known genera, and upwards of 2,000 species. The genera *Carex*, *Cladium*, *Scirpus*, *Eleocharis*, *Eriophorum*, *Cyperus*, and *Papyrus* afford examples. A plant of this family is shown in Plate 10, fig. a. [J. H. B.]

CYPERORCHIS. A name proposed by Blume for the *Cymbidium elegans* of Lindley, on account of its having a prominent stigma and pyriform pollen masses.

CYPERUS. A genus of plants giving its name to the sedge family, *Cyperaceæ*. It is distinguished chiefly by the stem being triangular, and leafy at the base; spikelets distichous, imbricated, in clusters or heads, with a leaf-like involucre under them; glumes several in each spikelet, with one flower in each glume; seed without bristles. According to Steudel's untrustworthy *Synopsis*, the genus contains 573 species, widely distributed over the warmer parts of the earth, and gradually disappearing as the extremes of north and south are reached. Two species only are natives of Britain, both of which are rare and not found out of England. Dr. Lindley states that the roots of these plants are succulent, and filled with an agreeable and nutritive mucilage. The English species, *C. longus*, contains also a bitter principle, which gives its roots a tonic and stomachic quality. The tubers of *C. hexastachys* are said to be successfully used by Hindoo practitioners in cases of cholera, who call the plant Mootha. Those of *C. pertenuis*, or Nagur Mootha, are, when dried and pulverised, used by the Indian ladies for scouring and perfuming their hair. The root of *C. odoratus* has a warm aromatic taste, and is given in India in infusions as a stomachic. The roots of some of the species are also used as an article of diet. *C. esculentus* yields tubers which are called by the French *Souchet comestible* or *Amande de terre*, and are used as food in the south of Europe. According to Dr. Royle, they have been proposed as a substitute for coffee and cocoa when roasted. The tubers of another species, *C. bulbosus*, are said to taste like potatoes when roasted, and would be valuable for food only they are so small. Some are also useful for textile purposes, *C. textilis* being employed in making ropes and mats for covering the floors of houses; others are valuable for covering the sand and loose soil on the borders of rivers and streams; thus, *C. inundatus* helps to bind the banks of the Ganges, protecting them from the rapidity of the stream, and the force of the tides. (See *Lindley, Veg. King.* 118.) [D. M.]

CYPHEL. *Cherleria sedoides.*

CYPHELIA. Collections of gonidia in the form of cups; a term only used in speaking of lichens.

CYPHELLÆ. Pale wart-like spots, found on the under-surface of the thallus of some lichens.

CYPHIA. A genus of three or four South African species, intermediate, as it were, between *Campanuleæ* and *Lobelieæ*, and therefore, when these two tribes are considered as independent families, *Cyphia* is raised to the same rank under the name of *Cyphiaceæ*. The species are all slender herbaceous twiners, with small nearly regular bell-shaped flowers, and united anthers. They possess no peculiar interest, except that the Hottentots are said to eat the tuberous roots of at least one species.

CYPHOCARPUS. A genus of *Campanulaceæ*, containing a rigid scabrous pilose herb from Chili, with erect stems, and oblong spinosely dentate radical leaves. Bracts three together, spinosely dentate; calyx tubular-curved, contracted at the mouth, with a five-parted limb, having spinous-toothed segments; corolla very irregular, two-lipped; capsule one-celled, resembling a follicle. [J. T. S.]

CYPHONEMA. A genus of cyrtanthiform *Amaryllidaceæ*. The only species, *C. Loddigesianum*, produces scapes with about two erect flowers, which measure an inch and three quarters, and are whitish, striped with green. The perianth has a straight slender cylindrical tube, campanulate above, with a regular reflexed limb. The plant had been supposed to have been imported from Valparaiso, but Dr. Herbert, by whom it was described, suspected it to be South African. [T. M.]

CYPRES. (Fr.) *Cupressus sempervirens.* —, CHAUVE or DE LA LOUISIANE. *Taxodium distichum.*

CYPRESS. The common name for *Cupressus*, especially applied to *C. sempervirens.* —, BALD. An American name for *Taxodium.* —, BROOM. *Kochia scoparia.* —, DECIDUOUS. *Taxodium distichum.* —, GROUND. *Santolina Chamæcyparissus.* —, SUMMER. *Kochia scoparia.*

CYPRESS KNEES. See Exostosis.

CYPRIPEDIUM. In the north of England the eye of the botanist has been now and then delighted by the discovery of one of the rarest of native plants, *C. Calceolus*, once called *Calceolus Marianus* or the Slipper of our Lady. It has a branching fibrous root; single stems, a foot or more high, bearing three or four broad ovate rather downy ribbed leaves, clasping the stem at the base, and one or two large flowers. These consist of two lanceolate brown purple sepals, and a pair of somewhat narrow wavy petals crossing each other at right angles (decussating); from the midst of these projects a great yellow pouch or bag, within which lurks the column, for the plant is an orchid. From other orchids it differs, however, in having two lateral anthers instead of one that is dorsal, the latter being represented by a great broad angular plate, in front of which projects a stalked three-lobed stigma. This curious deviation from the ordinary state of an orchid flower is characteristic of the genus *Cypripedium* (that is to say, shoe of Venus). Great numbers of species of the same genus occur in both the Old and New World, in the ice-bound woods of Canada and Siberia, the warm glades of Mexico and Nepal, and in the torrid regions of Central India and Continental (not Insular) America. Some of them have yellow flowers, and they are the most frequent; others are white and pink; many are more or less purple; and one, *C guttatum*, a Russian plant, is richly bloodstained. Two principal forms are to be distinguished, one having thin ribbed leaves, and the other narrow carinate veinless ones. The latter, which are all from warm countries, are easily cultivated, and are common in gardens under the names of *C. venustum, insigne, purpuratum, Lowei, Dayanum, Fairieanum, villosum*, &c. The others, though often introduced, live for only a short time and disappear. In addition to these, another race, exclusively found in Tropical America, distinguished by having a three-celled ovary, might be added. We prefer, however, to notice it under the name of *Selenipedium*. The curious *Cypripedium caudatum* belongs to that race.

CYPSELA. The dry one-celled one-seeded inferior fruit of composites.

CYPSELIA. A genus of *Tetragoniaceæ*, consisting of a small fleshy annual herb, resembling *Montia*, from St. Domingo. Stems prostrate, with small oval or obovate stalked alternate or opposite leaves, and fringed stipules; flowers small, solitary, shortly stalked, with a free five-parted calyx and no corolla, the two inner segments of the calyx broader and membranous; stamens one to three; ovary one-celled; capsule bursting transversely; seeds numerous. [J. T. S.]

CYRILLACEÆ. A small family of Dicotyledons, most nearly related perhaps to *Ericaceæ*, although differing in their free petals and anthers opening in slits; or to some of the groups connected with *Saxifragaceæ*. They have also been compared with *Olacaceæ* and with *Aquifoliaceæ*, both of which are much farther removed. They are shrubs or small trees, with alternate evergreen undivided leaves without stipules, the flowers usually in racemes. There are four or five calyx lobes and petals, and as many or twice as many slightly perigynous stamens. The ovary is two, three, or four-celled, with one pendulous ovule in each cell, and bears as many stigmas as cells. The fruit is usually succulent; the seeds albuminous with an axile embryo. There are about six species known from North or Tropical America, constituting four genera, *Cyrilla*, *Mylocaryum*, *Elliottia*, and *Purdiæa*.

CYRILLA. A genus of *Cyrillaceæ*, consisting of plants from the warmer parts of North America, with the habit of some of the larger shrubby species of *Andromeda*. Leaves alternate, wedge-shaped; racemes lateral, elongated, aggregated; flowers small, white, with a five-cleft calyx, five petals, and five stamens; ovary two-celled; capsule fleshy, two-valved, two-seeded. *C. caroliniana* is a handsome greenhouse shrub. [J. T. S.]

CYRTANDRACEÆ or CYRTANDREÆ. A tribe of *Gesneraceæ*, formerly considered as a separate family, including all the Asiatic genera which have no albumen in the seed, whilst the American genera were all believed to be possessed of albumen. These and some other slight distinctions have, however, all proved less constant than had been supposed, and the two groups are now acknowledged to be tribes of one family. The *Cyrtandreæ* proper consist of above thirty genera, including *Ramondia* and *Haberlea* from Europe, *Æschynanthus*, *Chirita*, *Didymocarpus*, and many others from Asia, *Streptocarpus* from South Africa, *Klugia* from Eastern Tropical Asia and Mexico, and perhaps *Napeanthus* from Brazil.

CYRTANDRA. A genus of Cyrtandrous *Gesneraceæ*, containing a considerable number of caulescent undershrubs or herbs, natives of the Moluccas. They have opposite leaves, equal or frequently with one side dwarfed or aborted. Their flowers are in axillary fascicles or heads, seldom solitary. The calyx is tubular, with five more or less deeply-cut lobes; the corolla funnel-shaped, with the limb spreading and cleft into five obtuse lobes; there are four to five stamens, two of which only are fertile; the ovary is cylindrical, containing many ovules attached to two two-lobed revolute parietal placentæ; the stigma is obtuse or emarginate. The fruit is a many-seeded ovate berry. [W. C.]

CYRTANTHERA. A genus of *Acanthaceæ*, natives of Tropical America, consisting of some eight species of caulescent shrubs, with broad petiolate leaves, and large scarlet flowers like those of *Aphelandra*, arranged in a beautiful dense terminal thyrse, except in one species, in which they are in axillary cymes. The calyx is cleft into five equal coloured parts; the ringent corolla has a long tube, and the limb is divided into two lips, the lower of which is trifid; there are two stamens inserted at the base of the tube and adherent to it beyond the middle; the anthers are two-celled; the stigma is obtuse and undivided. [W. C.]

CYRTANTHUS. A genus of handsome *Amaryllidaceæ*, consisting of bulbous herbs, with two-ranked narrow elongate leaves, and many-flowered umbels of flowers. The perianth has a curved narrow funnel-shaped tube, which is often a little ventricose, and a limb of six short subequal segments; the filaments of the six stamens straight, decurrent, inserted in the upper portion of the tube. They are South African plants, the type of the genus being *C. obliquus*. This has globose bulbs as large as a man's fist, persistent lanceolate entire leaves an inch wide, and an erect scape supporting a loose umbel of numerous pendulous flowers, orange-colour mixed with yellow and green, the tube sensibly widened upwards, an inch and a half long, and the limb spreading, nearly as long as the tube. In another group of the species the leaves are deciduous. One of them, *C. striatus*, has subacute leaves a foot long, and half an inch wide, and an umbel of three or four pendulous narrow funnel-shaped flowers two and a half inches long, red, striped with yellow. *C. odorus* has fragrant crimson flowers; whilst in *C. collinus* they are poppy scarlet. [T. M.]

CYRTOCERAS. *Centrostema*.

CYRTOGONIUM. *Pœcilopteris*.

CYRTOGYNE. A genus of succulent-leaved undershrubs, belonging to the order *Crassulaceæ*, having white flowers in cymes, with a five-parted corolla whose segments are much longer than those of the calyx. The stamens are inserted into the base of the corolla, with whose lobes they alternate, and within them are five small hypogynous scales. The ovary consists of five oblong carpels, gibbous at the top, and ending in long styles. *C. albiflora*, a native of the Cape of Good Hope, is in cultivation. [M. T. M.]

CYRTOLEPIS. A genus of *Compositæ*, composed of a few small annual herbs, found in northern Africa and Asia Minor. They have much resemblance to the chamomile (*Anthemis*), and are nearly related to that genus, differing only in having winged achenes, the wings toothed, those of *Anthemis* not being winged. They have alternate pinnatisect leaves with linear segments, yellow flower-heads with an involucre of one series of roundish scales, which enclose a large number of tubular five-toothed florets. [A. A. B.]

CYRTOMIUM. A genus of polypodiaceous ferns, belonging to that series of the *Aspidieæ* which have reticulated veins and peltate indusia. The characteristics of *Cyrtomium*, as shown in the more typical plants, consist in the veins being pinnato-furcate from a central costa, the lower anterior venules being free, and the rest angularly and irregularly anastomosing, forming unequal subhexagonal areoles, within which two or three excurrent veinlets are produced. Sometimes only the upper venules are anastomosed. The species are robust evergreen pinnate ferns, of very ornamental character, the pinnæ being of a

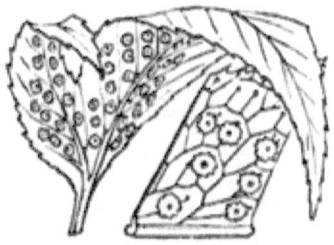

Cyrtomium caryotideum.

deep shining green, broad, and of a singular rhomb-lanceolate form, sometimes approaching to hastate. The species are few, the typical ones natives of India, China, and Japan, one or two, somewhat anomalous, occurring in South America. [T. M.]

CYRTONEMA. Herbaceous plants of the gourd family, *Cucurbitaceæ*, with tuberous rhizomes, simple tendrils, and monœcious greenish flowers, the males in clusters, the females solitary. The stamens are five, in three bundles, with straight anther-lobes, and curved filaments thickened at the top. The fruit is a spindle-shaped beaked gourd with three compartments and few seeds. The plants are natives of the Cape of Good Hope. [M. T. M.]

CYRTOPERA. A genus of tuberous tropical orchids with a tall radical inflorescence and membranous plaited leaves, sometimes not appearing along with the flowers. Some are American, some Asiatic, a few are South African, and one, *C. plantaginea*, is from Madagascar. *C. Woodfordi*, a plant with dense upright racemes of greenish-purple flowers, is the only one in cultivation. The genus is perhaps hardly distinct from *Eulophia*.

CYRTOPHLEBIUM. A name proposed for the species of *Campyloneurum*, a genus of polypodiaceous mostly simple-fronded ferns. [T. M.]

CYRTOPODIUM. Under this genus are collected some species of terrestrial orchids from Tropical America, with long fusiform fleshy stems, bearing membranous plaited leaves, and long racemes or panicles of showy yellow or spotted flowers, whose scape rises from the base of the stem. They are very fine objects in cultivation, well grown specimens measuring as much as five feet in height. The handsomest species are *C. Andersoni* and *cardiochilum*, brilliant yellow, and *C. punctatum*, yellow and brown. *Tylochilus flavus* of the Germans is the same as *Cyrtopodium Andersoni*.

CYRTOSIA. A remarkable genus of leafless and usually climbing orchids, allied to *Vanilla*, but producing a more or less dry capsule instead of a fleshy aromatic fruit. Few species are known, among which are the *Dendrobium cassythoides* of New Holland, or *Ledgeria aphylla* of Mueller; and the *Erythrorchis* or *Hæmatorchis* of Blume. The best known species is *Cyrtosia Lindleyana* of Sikkim, admirably represented by Dr. J. D. Hooker. This is a stout erect plant, with a great woody rootstock, a strong reddish brown stem, and panicles of bright yellow flowers, succeeded by velvety brown flat pods which after a long time open into flat valves. The seeds are surrounded by a thin jagged wing, which forms a pretty microscopical object.

CYRTOSPERMUM. A name applied by Mr. Bentham to a tree growing in tropical South America, forming a genus of *Anacardiaceæ*, and yielding a yellow-coloured resin. The calyx is five-parted, and there are ten stamens arising from beneath a hypogynous disc; but the chief peculiarity resides in the drupe-like fruit, whose bony inner shell is divided into two compartments by a hard curved partition: one of these compartments is small and empty, the other forms a horse shoe-shaped cavity, containing a simple seed of the same form attached to the apex of the cavity. The same name has been applied to a genus of *Umbelliferæ*, now known as *Lereschia*. [M. T. M.]

CYRTOSTYLIS. Under this name are collected a few little Australasian terrestrial orchids, with thin solitary roundish flat leaves, a slender naked scape, and two or three greenish distant flowers. They inhabit dry rocky spots on the edge of ravines.

CYST. The spore-case of certain fungals; also the hollow spaces in parenchyma in which oily matter collects, as in the rind of the orange.

CYSTANTHE. A genus of *Epacridaceæ*, containing a few species, natives of the mountains of Tasmania. They are small glabrous bushes with erect naked stems, marked, as are also the lower portions of the branches, with annular scars, where the leaves, which are sheathing at the base, were inserted. They have a subfoliaceous bracteate calyx; the corolla is a closed conical calyptra, which dehisces transversely a little above its base, the upper part falling away, and the lower being very persistent; the stamens are hypogynous and persistent; the ovary is five-celled, each cell containing many ovules, attached to a pendulous placenta. The plants resemble *Sprengelia* and *Pilitis* in stature and habit; from the first they are separated by the structure of the corolla, and from the second by the absence of hypogynous scales. [W. C.]

CYSTEA. A fanciful alteration, which has not met with acceptance, of *Cystopteris*, the name of a genus of ferns. [T. M.]

CYSTIDIA. Salient cells, accompanying the basids or asci of fungals; by some regarded as antheridia. [M. J. B.]

CYSTIDIANTHUS. A genus of *Asclepiadaceæ*, containing a few species of climbing shrubs, natives of the Indian Archipelago, with opposite leaves, and numerous pedicellate flowers in interpetiolar and terminal umbels. The calyx is five-parted; the corolla bell-shaped, five-toothed, and spreading; the staminal corona consists of five fleshy leaves attached to the short gynostegium; the anthers are terminated by a membrane adpressed to the stigma, which is convex, pentagonal, and smooth; the follicles are solitary, long and slender, with numerous comose seeds. This genus has the habit of *Centrostemma*. [W. C.]

CYSTOCAPNOS. A genus of *Fumariaceæ*, containing a glabrous climbing branched herb from the Cape of Good Hope, with stalked twice-pinnate leaves, having three-lobed segments, and small white racemose flowers. It differs from the other genera of the order in the capsule, which is inflated and bladdery, containing several seeds. [J. T. S.]

CYSTOCARPIUM. A case including a great many spores; a term confined to algals.

CYSTOPTERIDEÆ. A section of polypodineous ferns, in which the sori are punctiform or dot-like, and covered by cucullate or fornicate indusia, which, being attached behind them, are inflected over them in the earlier stages. [T. M.]

CYSTOPTERIS. A genus of dwarf polypodiaceous ferns, typical of the group *Cystopterideæ*. In that group, it is distinguished at once by its sori being medial on the veins, that is, placed some distance below the apex. The species, numbering about a dozen, are small membranaceous plants with a tufted or creeping caudex, and twice or thrice-pinnated annual fronds; they are furnished with punctiform sori, covered by roundish ovate indusia, which are fornicate or subhemispherical, affixed by their broad base, and sometimes lacerate or acuminate at the apex. *C. fragilis*, which has lance-shaped fronds, is a widely distributed British species; *C. montana*, with a creeping caudex and triangular fronds, has been gathered in a few Scottish habitats. The genus is scattered from the poles to the tropics. [T. M.]

CYSTOPUS. Under this name Blume has collected a few little white-flowered Java orchids near *Goodyera*.

CYSTORCHIS. A genus of terrestrial orchids, allied to *Goodyera*. Blume mentions three species with small pink or yellow flowers. They are especially known by having the glands found inside the lip in so many of these little plants enclosed in a pair of cysts or pockets, whence the name has been formed.

CYTHERIS. *Nepholaphyllum*.

CYTINACEÆ. The *Cytinus Hypocistis*, either alone or in conjunction with two African root-parasites, *Hydnora* and *Hypolepis*, has been considered as constituting an independent family of very uncertain affinities. It is a native of the Mediterranean region, growing on the roots chiefly of *Cistus monspeliensis*, and rises to a few inches above ground in the form of a tuft of succulent stems covered with imbricated scales, and terminating in a head of flowers, the whole plant of a rich yellow or orange-red colour. The flowers are polygamous, with a tubular four-lobed perianth, and four two-celled anthers, sessile on a central column attached to the perianth-tube. The ovary is inferior, one-celled, with several parietal placentas, and numerous ovules. The plants contain gallic acid, and have been used in consequence as astringents and styptics. [M. T. M.]

CYTISE A' GRAPPES or AUBOURS or DE VIRGILE. (Fr.) *Cytisus Laburnum*. —, PETIT. *Cytisus sessilifolius*.

CYTISOPSIS. A genus of *Leguminosæ*, containing but one species, *C. dorycniifolia*, a small prostrate perennial plant found in the mountains of Syria and Cilicia. This has sessile leaves made up of from three to seven small oblong leaflets, which are covered with silvery hairs; and the axillary yellow flowers have a tubular calyx nearly an inch long, and a corolla of five nearly equal clawed petals, a little longer than the calyx. The pods are narrow, thick, elongated, and contain a number of seeds. The genus is nearly allied to *Anthyllis*, but differs in the sessile digitate leaves, and in the calyx and corolla falling after withering. [A. A. B.]

CYTISUS. An extensive and well-known genus of trees and shrubs belonging to the *Leguminosæ*. *C. Laburnum*, with which all are familiar under the name of Laburnum, is a native of the mountains of France, Switzerland, and Southern Germany, where it attains the height of twenty feet and upwards. It was introduced into England previously to 1597, at which time Gerarde appears to have had it in his garden under the names of *Anagyris*, *Laburnum*, and Bean Trefoil. This and the lilac are the commonest ornamental trees in suburban gardens; but the Laburnum is seen to the greatest advantage when planted in front of loftier trees in a park or extensive shrubbery. The heart wood is of a dark colour, and, though of a coarse grain, it is very hard and durable; it will take a polish, and may be stained to resemble ebony. It is much in demand among turners, and is wrought into a variety of articles which require strength and smoothness. The seeds, it should be remembered, act so violently as an emetic that they are justly deemed poisonous.

C. purpureus is an elegant procumbent shrub, a native of Carniola. It seldom exceeds a foot in height, and is either used for ornamenting rockwork, or is grafted on the Laburnum. *C. purpurascens* (Fr. *C.*

d'*Adam*), the purple Laburnum, is a hybrid between the two preceding. It was originated in Paris in 1828, by M. Adam, and has since been much cultivated in England. A peculiarity of this tree has often been noticed, which is interesting to the physiological botanist as showing the influence exercised by the stock on the scion. 'This purple Laburnum is a hybrid between the common yellow Laburnum and *C. purpureus*. The branches below the graft produce the ordinary yellow Laburnum flowers of larger size; those above often exhibit a small purple Laburnum flower, as well as reddish flowers, intermediate between the two in size and colour. Occasionally the same cluster has some flowers yellow and some purple' (*Balfour*). *C. alpinus* differs principally in having its leaves rounded at the base, and in having the pods smooth and few-seeded; whereas *C. Laburnum* has the leaves white with down beneath, and the seed-pods many-seeded and downy. The *Cytisus racemosus* and *rhodopnœus*, so generally to be observed among the plants offered for sale in spring in the streets of London, are referrible to *Genista*. [C. A. J.]

CYTOBLAST. That elementary spherule, derived from organic mucus, which produces a cell from its side, according to Schleiden. It is the nucleus of R. Brown and others.

CYTTARIA. A curious genus of ascomycetous *Fungi*, consisting of a subglobose cartilaginous receptacle in which are sunk a number of ovate pits lined with the hymenium. The mouth of these at last becomes open, and the whole plant has then the appearance of a little waxlike wasps' nest, from whence the genus takes its name. There are three or four species, all of which grow parasitically upon the living branches of evergreen beeches, and one in Tierra del Fuego for several months affords the staple food of the inhabitants. It is, however, almost tasteless, and has been compared to cow-heel. The species are confined to a portion of the southern hemisphere. The individual plants are sometimes solitary, but frequently they form dense clusters, and it is probable that the same branch yields more than one crop from the same spawn. [M. J. B.]

CZACKIA. The name of a group now generally regarded as a subdivision of the genus *Anthericum*, from which it is, however, still sometimes separated. It is distinguished by having the segments of the flower brought together, or connivent into a kind of bell-shaped form, and by having the stamens glabrous. [T. M.]

DABŒCIA. A section of *Menziesia*, or as it is sometimes considered a separate genus of *Ericaceæ*, distinguished chiefly by its tetramerous instead of pentamerous structure, the calyx being four-cleft, the corolla limb four-toothed, the stamens eight, and the capsule four-celled, with four valves. The plant called *D. polifolia* is the St. Dabœc's Heath. [T. M.]

DACHA. A Hottentot name for *Cannabis sativa*.

DACRYDIUM. A genus of *Taxaceæ*, consisting of a few evergreen trees inhabiting the East Indies and New Zealand. The distinguishing characteristics reside in the female flower, which consists of a boat-shaped bract, bearing an ovule which at first lies on the scale, but ultimately becomes erect, and when fully developed has a short outer fleshy integument, from which the inner bony investment of the seed protrudes. *D. cupressinum*, which has pendulous feathery branches and slender needle-like leaves, is a very graceful lofty tree. *D. Franklinii* is the Huon Pine. *D. taxifolium* is said to acquire a height of 200 feet in New Zealand; its shoots may be made into a beverage having the same antiscorbutic properties as spruce beer. *D. laxifolium*, also a native of New Zealand, is a low growing shrub not unlike *Empetrum nigrum*. [M. T. M.]

DACRYMYCES. A genus of tremelloid *Fungi*, of which *D. stillatus* is almost universal in the form of small bright-orange gelatinous tear-like masses on decayed pine or larch rails, accompanied sometimes by a larger species, *D. deliquescens*. The former plant has often been supposed to consist of a mass of branched threads, terminated by chains of oblong spores. These, however, are merely conidia, the perfect fruit being developed in the same way as in *Tremella*, and consisting of slightly-curved septate spores, from the edge of which minute secondary spores are given off. The scarlet gelatinous fungus so common on dead nettle stems is now believed to be a condition of *Peziza fusarioides*. [M. J. B.]

DACTYLÆNA. The name of a herb of the capparidaceous family, whose native country is not known. It has a calyx of four sepals, the anterior one longer than the rest; four petals, two longer than the others; six stamens, inserted into a hemispherical receptacle, which is provided with a gland at the back; four of these stamens are antherless, while the two remaining ones are completely joined together, so that the anther appears four-lobed. [M. T. M.]

DACTYLANTHUS. A genus of *Balanophoraceæ*, founded on a root-parasite from New Zealand. It is attached to the roots of beeches and *Pittosporï* by a thick tuberous rhizome, the stems rising in clusters two or three inches from the ground, covered with imbricated scales. The flowers are diœcious, small and numerous, in dense spadices, of which several are clustered together at the summit of the stems surrounded by the upper scales.

DACTYLICAPNOS. A genus of *Fumariaceæ*, distinguished by having the two outer petals bulging out at the base, the fruit berry-like, and the seeds crested. It is considered by Drs. Hooker and Thomson as merely a section of *Dicentra*, from which it differs only in having the walls of the fruit fleshy. It contains two Indian herbs with weak stems climbing by means of tendrils, compound triternate leaves, and racemes opposite the leaves. [J. T. S.]

DACTYLIS. A genus of grasses belonging to the tribe *Festuceæ*, and distinguished by the flowers being in very crowded panicles, and subsecund, i.e., pointing nearly all to one side. The glumes are unequal and many-flowered, acute and herbaceous, with terminal setæ. *D. glomerata*, the Cock's-foot Grass, is the only British species, and one of the best known of our native grasses. It is also the strongest grower among the superior kinds, and derives its English name from the fancied resemblance the three-branched panicles of flowers bear to the foot of a fowl. It forms a portion of most good pastures, particularly where the soil is loamy or chalky. It is also suitable for sowing alone on boggy land which is in the course of being reclaimed, for, although it does not grow on this sort of soil naturally in great quantities, it produces a good crop when cultivated on it artificially Steudel describes twenty-nine species, which have a wide range of habitats over the globe. [D. M.]

DACTYLIUM. A genus of filamentous moulds, of which the genuine species have hyaline threads bearing at their tips clusters of septate spores. *D. roseum*, which was formerly referred to *Trichothecium* from an insufficient knowledge of its structure, belongs essentially to this genus, and is remarkable not only as being one of the most widely-diffused species, distinguished by its delicate pink hue, but as occurring not unfrequently in the closed cavities of nuts. The spawn of these delicate moulds will, however, soon penetrate the firmest vegetable tissues if there be proper conditions of moisture. Another species, *D. oogenum*, occurs in the inside of eggs, where its presence is difficult to account for without having recourse to the wild and unphilosophical notion of equivocal generation. In *D. roseum*, besides the common large spores, there are conidia of a small size, which may have greater power of penetration than the larger. The function, however, of these bodies is uncertain, and they may be spermatia rather than conidia. [M. J. B.]

DACTYLOCTENIUM. A genus of grasses belonging to the tribe *Chlorideæ*, distinguished by the inflorescence being in finger-like spikes, the flowers on the spikelets pointing to one side; the glumes two, compressed, keeled, and subherbaceous, the exterior one cuspidate; stamens three; ovary smooth; styles two; stigma hairy and branched. There are only seven species described, all natives of Africa, with one exception, *D. radulans*, which is a New Holland grass. They are mostly annuals, and little known in cultivation. [D. M.]

DACTYLORHIZA. An affection of some agricultural plants, as turnips and carrots, in which the root divides and becomes hard and worthless. It is commonly called Fingers and Toes, and must be distinguished from anbury, which arises from the attacks of insects. It is in fact not properly a disease, but a tendency to a reversion to the wild state, which can only be remedied by a careful selection of seed. It is sometimes thought that it arises from an unequal distribution of manure, but this is probably a mistake. [M. J. B.]

DACTYLOSTEMON. A genus of the spurge-wort family, composed of a number of trees or shrubs found in the tropical parts of South America, and chiefly distinguished by their flowers being destitute of a true calyx, the males containing three, or more generally from four to seven stamens. The leaves are lance-shaped, entire, and glossy; either alternate or whorled, and varying from two to eight inches in length. In their axils the little green flowers are arranged in short catkins, the males towards the apex, and the females near the base, the former entirely naked or accompanied with one or more little scales which represent the calyx, the latter also naked or having a calyx of three small divisions. The fruit is a brown polished three-lobed woody capsule, about the size of a pea, and contains three seeds. The name *Actinostemma* is sometimes given to the plants of this genus. [A.A.B.]

DÆDALEA. A genus belonging to the spore-bearing section of the higher *Fungi*. In this genus the cavities, instead of being circular or only slightly distorted, are sinuous and intricate from the partial breaking-up of the cell-walls. *D. quercina*, a fungus of a hard corky texture, is not uncommon upon oak stumps or rails, and sometimes makes its appearance in buildings or conservatories, where the wood has been impregnated with its spawn before being felled. [M. J. B.]

DÆDALEUS. When a point has a large circuit, but is truncated and ragged. Or, wavy and irregularly plaited as the hymenium of some agarics.

DÆMONOROPS. A genus of palms closely allied to *Calamus*, in which the greater number of the forty species referred to it were formerly placed. Its distinguishing peculiarities consist in the flowers being loosely scattered along the branches of the flower spikes, not collected into catkins as in *Calamus*, and also in the spathes or bracts being complete, i.e., entirely enclosing the young spikes. All the species are natives of the eastern hemisphere, principally of the Malayan Peninsula and Islands; they have long thin flexible stems, furnished with pinnate leaves, the prickly stalks of which are frequently prolonged into whip-like tails.

D. Draco (formerly *Calamus Draco*) is a native of Sumatra and other islands of the Indian Archipelago, and is called the Dragon's Blood Palm, in consequence of its fruits yielding a portion of the substance known in the arts as dragon's blood. The fruits are about the size of cherries, and, when ripe, are covered with a reddish resinous substance, which is separated by shaking them in a coarse canvass bag. The resin thus obtained forms the best kind of dragon's blood, while inferior sorts are obtained by boiling the fruits after they have undergone the shaking process. Several varieties of dragon's blood (sticks, reeds, tears, and lumps) are known in commerce, but some are yielded by plants belonging to widely different natural orders. It is chiefly used for colouring varnishes, for dyeing horn in imitation of tortoise shell, and in the composition of tooth-powders and various tinctures. [A. S.]

DAFFODIL. *Narcissus Pseudo-Narcissus*, also called Daffy-down-dilly. —, PERUVIAN. *Ismene Amancaes*. —, SEA. *Ismene calathina*.

DAGGER-FLOWER. *Machæranthera*.

DAGGER PLANT. A name for *Yucca*.

DAHLIA. A well-known herbaceous plant belonging to the compound flowers, and distinguished by its chaffy receptacle, the absence of a pappus, and by the double involucre of which the outer is many-leaved, the inner of one leaf divided into eight segments. The Dahlia is named after Dr. Dahl, a pupil of Linnæus, but is also known, especially on the continent, by the name *Georgina*. Countless as are the varieties of this flower, there are, at the most, only two species in cultivation, *D. superflua*, of which the outer involucre is reflexed, and *D. frustranea*, in which it is spreading; while under the name *D. variabilis* both these are united. The Dahlia is a native of Mexico, where it grows in sandy meadows at an elevation of 5,000 feet above the sea, and from whence the first plants introduced to England were brought by way of Madrid in 1789, by the Marchioness of Bute. These having been lost, others were introduced, in 1804, by Lady Holland. These also having perished, a fresh importation was made from France, when the continent was thrown open by the peace of 1814. The first introduction into France had taken place about 1800; and the plant was cultivated there for the sake of its tubers, which were said to be eatable. Owing, however, to their acrid and medicinal flavour, they found no favour with the human species, and were rejected by cattle. The roots are large, spindle-shaped, and assembled into bundles from the centre of which rises the stem. The flowers, in the examples first introduced, were single, with a yellow disk and dull scarlet rays having a velvety surface. The seeds of these soon produced flowers of various tints, some double, others variegated. Flowers of a better colour and form were successively propagated; in some the petals, or rather florets (for in what is called the 'double Dahlia' the fulness of the flower is owing to the conversion of disk into ray florets), assumed the shape of a horn or funnel with singular regularity, in others the florets were arranged in the form of a perfect rose. Finally, in the course of years, horticulturists flatter themselves that they have brought the Dahlia to the highest point of beauty, though among the numerous seedlings raised every year, there are constantly occurring individuals which are considered as surpassing their predecessors in some point of floral excellence. A race of pompons with remarkably small flower-heads has been obtained. [C. A. J.]

DAIS. A genus of *Thymelaceæ* or *Daphnaceæ*. Its characters are: flowers surrounded by an involucre; calyx coloured, funnel-shaped, with a four or five-divided limb, and without scales in its throat; stamens eight to ten in two rows, included within the calyx; no hypogynous scales; ovary one-celled, with a single pendulous ovule. Fruit a drupe enclosed by the persistent calyx; albumen fleshy; embryo orthotropal. Shrubby plants found at the Cape of Good Hope and in the tropical and subtropical parts of Asia. There are seven known species. [J. H. B.]

DAISY. The common name for *Bellis*. —, AFRICAN. *Athanasia annua*. —, AUSTRALIAN. *Vittadenia triloba*. —, BLUE. *Globularia vulgaris*. —, CHRISTMAS. A popular name for some of the species of *Aster*. —, MICHAELMAS. A popular garden name for *Aster*, especially for *A. Tradescanti*. —, OXEYE. *Chrysanthemum Leucanthemum*. —, SWAN-RIVER. *Brachycome iberidifolia*.

DAISY-STAR. *Bellidiastrum*.

DALBERGIA. A large genus of leguminous forest trees and climbing shrubs principally inhabiting the tropics of the Eastern Hemisphere. Most of the species have pinnate leaves with numerous leaflets arranged alternately, but sometimes reduced to three leaflets only. The flowers are borne in axillary racemes, and have a bell-shaped calyx, the mouth of which is cut into five divisions, a papilionaceous corolla, and nine to ten stamens, either all joined together into a sheath, which is split along the upper side, or divided into two equal bundles of five each. The pods are thin, very much flattened, not winged, and either long and straight, or short and crescent-shaped, containing one or several flat seeds.

D. latifolia, the Black-wood or East Indian Rose-wood tree, and the Sit-sal of the Bengalese, is common on the Malabar and Coromandel coasts, and forms a magnificent tree, yielding one of the most valuable furniture woods. The timber is procurable in planks four feet broad, exclusive of the sap-wood, and is of a dark purplish colour, very heavy, close-grained, and susceptible of a fine polish. It comes to this country under the names of Blackwood and East Indian Rosewood, but it has not the agreeable perfume of the true rosewood, nor is it marked with the black lines of resinous matter which add so much to the beauty and value of the Brazilian wood. In India it is greatly used for making the most expensive descriptions of furniture. *D. sissoides*, is a smaller tree, but yields an equally valuable timber, which also goes by the names of Blackwood and Rosewood in Madras, where it is employed in the construction of gun-carriages.

D. Sissoo is an East Indian species, but found farther north than either of the preceding, abounding principally in Bengal and the provinces as far north as the Punjab. It is a large and very rapid-growing tree, yielding a strong tenacious compact timber of a dark brown colour, but not so fine-grained as the Blackwood. This wood is called Sissoo or Sissum, and being very durable it is included among those which are authorised to be employed for the sleepers of Indian lines of railway. In Bengal it is used in the construction of gun-carriages, and it also supplies the ship-builders of that presidency with crooked timbers and knees, besides which it is extensively employed for all the ordinary purposes connected with house building. [A. S.]

DALEA. A genus of sub-shrubby or herbaceous plants of the pea family found in America, appearing in greatest numbers in New Mexico, and having their northern limit in the United States, and their southern in Chili, very few being found in the north-eastern part of the continent. Its most marked features are the flowers in terminal spikes, and the pods small, one-seeded, and not longer than the calyx. Its nearest affinity is with *Petalostemon*, in which the stamens are five, while here they are generally ten, and never fewer than nine. In the great bulk of the species the leaves are unequally pinnate, and composed of numerous small wedge-shaped or oblong leaflets, which are often covered with small glandular dots like those seen in the St. John's-wort. The white, yellow, pink, or purple flowers are about the size of those of a vetch, and arranged in terminal spikes or heads; the calyx nearly equally five-toothed or cleft; and the keeled petal and wings united with the staminal tube and jointed to it, but the standard or upper petal quite free. The little pod is wholly enveloped in the calyx.

One of the most remarkable species is *D. spinosa*, which inhabits the desert regions of California, and has simple narrow leaves, and large deep violet flowers arranged in a spiked manner on the spiny-pointed branches. The plant attains a height of four or five feet. Like many desert plants the stems have a bleached appearance. *D. arborescens*, found in the Sierra Nevada mountains, is remarkable as being the only one which attains the dimensions of a small tree. *D. Jamesii* attains only a height of about six inches, and is altogether covered with silky hairs; it is also remarkable as being the only species with trifoliolate leaves. Upwards of sixty species are enumerated. The genus is named in honour of Thomas Dale, an English botanist of the last century. [A. A. B.]

DALECHAMPIA. A genus of spurgeworts found in the tropics of both hemispheres. Their slender stems are generally found twining among bushes, but sometimes scrambling to a great height amongst trees. The leaves are alternate, stalked, heart-shaped, entire or three to five-lobed, sometimes divided to the base. The small green flowers are borne on stalked heads which proceed from the axils of the leaves, a circumstance that at once serves to distinguish the genus. The heads contain a number of flowers of both sexes, and are enveloped by an involucre of two leafy, beautifully veined green or coloured bracts. The male flowers have a four or five-parted calyx, and very numerous stamens; the females a calyx of five or six divisions which are often fringed with hairs, and an ovary surmounted by a cylindrical or club-shaped style, which is entire, with a terminal or lateral stigmatic opening. The fruit is a three-celled three-lobed capsule about the size of a large pea. The names *Cremophyllum* and *Rhopalostylis* are given by some authors to plants of this genus. [A. A. B.]

DALHOUSIEA. A smooth simple-leaved shrub of the pea family found in Silhet, where it bears its white blossoms in May, and ripens its pods in the end of the season. Its beautifully veined glossy leaves are stalked, oval, and entire; the peduncles which arise from their axils are once or twice forked, and at the points of forking furnished with small round bracts; each flower is also supported by two similar bracts, which completely hide the five-toothed calyx. The upper petal or standard is deeply notched, and the ten stamens are quite free to the base. The dark brown polished pods are of a woody consistence, from three to four inches long, tapered at each end into a sharp point, and containing two or three flat seeds. The simple leaves, bracted peduncles, and free stamens, together with the nature of the pods, are its most marked features. [A. A. B.]

DALIBARDA. A genus of herbs or small shrubs with white or yellow flowers, belonging to the *Rosaceæ*, distinguished from the allied genus *Rubus* by having dry fruit, and terminal, not lateral, styles. The herbaceous species, which are hardy, have creeping stems and solitary flowers; they may be grown in a peaty soil, and are fit for ornamenting rock-work. The shrubby species have the flowers in panicles, and being natives of Java require to be grown in a hot-house. [C. A. J.]

DAMAR. A viscid resinous product of *Canarium microcarpum*.

DAMAS. (Fr.) *Hesperis matronalis*.

DAMASONIUM. A floating aquatic belonging to the *Alismaceæ*, better known under the name of *Actinocarpus Damasonium*. It is found, though somewhat rarely, in our ponds and ditches, and forms a tuft of radical floating long-stalked leaves, from amongst which issues the flower-stem bearing one or more whorls of white flowers. They are each succeeded by six or eight two-seeded carpels, arranged in the form of a star. [T. M.]

DAME D'ONZE HEURES. (Fr.) *Ornithogalum umbellatum*.

DAMIER. (Fr.) *Fritillaria Meleagris*.

DAMMARA. A genus of *Coniferæ* or *Pinaceæ*, the name of which is derived from the native one in Amboyna. Flowers diœcious, that is, some with stamens only, and

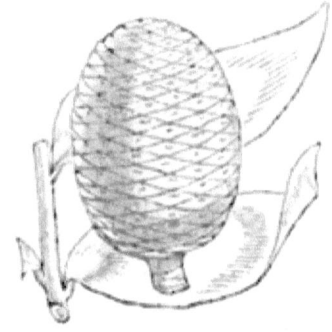

Dammara obtusa.

others with pistils only on separate plants: the staminate flowers in catkins, the numerous stamens having very short filaments, and eight to fifteen-celled anthers; the pistillate ones in ovate or globular cones with persistent scales without bracts. There is a single unequal-winged seed with two cotyledons under each scale. The species are large trees with scattered leathery leaves. They are found in the East Indian Islands, New Zealand and New Guinea.

D. australis, the Kauri Pine of New Zealand, is a tree from 150 to 200 feet in height, producing a hard brittle resin like copal. *D. macrophylla* is a large tree 100 feet high, found on Vanicolla, one of the Queen Charlotte Islands in the South Sea. *D. Moorii* is a tree forty feet high, found in New Caledonia. *D. obtusa* is a large timber tree used in ship-building, found in the New Hebrides. *D. orientalis*, the Amboyna Pine, is a tree of the Moluccas, 100 feet high, which yields the fine transparent resin called Dammar. [J. H. B.]

DAMMER TREE, BLACK. *Canarium strictum*. —, **WHITE.** *Vateria indica*, the resin of which is called Dammer pitch.

DAMOUCH. An Arab name for *Nitraria tridentata*, which is believed to be the Lotus tree of the ancients.

DAMPIERA. A genus of *Goodeniaceæ*, named after the celebrated navigator Captain W. Dampier. It is distinguished by having a calyx whose limb is short or obsolete; a monopetalous, two-lipped, blue or purple corolla, the segments of the upper lip of which are auricled on the inner margin; five stamens with coherent anthers; and a style with a stigma seated at the base of a cup, termed an indusium. The flowers are axillary or terminal; and the leaves alternate. The plants, which are shrubby or herbaceous, are natives of Australia and Tasmania. [R. H.]

DAMSON. A small austere variety of plum. —, **BITTER,** or **MOUNTAIN.** *Simaruba amara*.

DANÆA. A remarkable genus of ferns of the danæineous division of the *Marattiaceæ*. The species are not very numerous, and are all South American or West Indian. They have large woody rhizomes, and pinnate rarely simple fleshy coriaceous fronds, the pinnæ of which are usually articulated. The fertile fronds are more or less contracted. The sori are very remarkable; they are linear, occupying the whole length of the veins, and crowded so as to cover the whole under-surface of the divisions of the fertile fronds. The sporecases are consolidated into a fleshy mass, which represents an involucre, each fleshy case at length opening at the top by a small round pore, so that the contiguous fructiferous ridges appear to be each pierced by a double line of small apertures. In some species represented by *D. nodosa*, which has the joints of the fronds thickened, the sori are affixed to the veins by

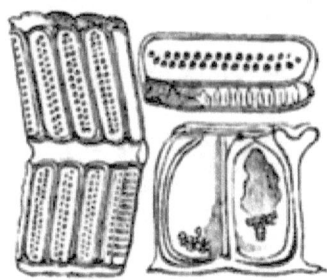

Danæa alata (fructification).

their whole length; in others, as in *D. stenophylla*, they are said to be affixed only by the centre. [T. M.]

DANÆACEÆ. The name of a natural order of ferns, also called **MARATTIACEÆ**; which see.

DANÆOPSIS. A doubtful genus of ferns, separated from *Danæa*, on the ground of its having anastomosing veins. Nothing definite, however, appears to be known of the plant. [T. M.]

DANCING GIRLS. *Mantisia saltatoria*.

DANDELION. *Taraxacum Dens Leonis*, the Dent de Lion.

DANEWORT. *Sambucus Ebulus*.

DANGLE-BERRY. An American name for *Gaylussacia frondosa*.

DANTHONIA. A genus of grasses belonging to the tribe *Aveneæ*, distinguished by the glumes being two or many-flowered; the outer pale smooth and coriaceous below, many or sometimes nine-nerved, emarginate at the apex, with an intermediate broad tooth, which sometimes terminates in a geniculate twisted awn; stamens three; styles two. Sixty species are described, nearly all natives of South Africa, and many of them useful there as pasture grasses. [D. M.]

DAPANIA. A genus considered by Korthals to belong to *Chrysobalanaceæ*, but referred to *Oxalidaceæ* by Planchon, who considers it closely allied to *Averrhoa*. The hypogynous flowers, and five-celled fruit with pendulous exalbuminous seeds, are sufficient to distinguish it from the *Chrysobalanaceæ*. [J. T. S.]

DAPHNACEÆ. A natural order of monochlamydeous dicotyledonous plants, synonymous with THYMELÆACEÆ or THYMELACEÆ: which see. [J. H. B.]

DAPHNADS. A name used by Lindley for the *Thymelaceæ*.

DAPHNE. A well-known thymelaceous genus of shrubs, the species of which are very widely distributed, being found in the temperate and tropical parts of Europe, Asia, America, and Australia. The flowers are hermaphrodite; the calyx tubular, its limb divided into four segments; petals none; stamens eight, inserted in two rows upon the inner surface of the tube; ovary one-celled, style short, stigma button-shaped; fruit a drupe. Many of the species are remarkable for the beauty and fragrance of their flowers; while all have a more or less stringy bark, and all are more or less acrid.

D. Mezereum, the Mezereon, a common shrub in cottage gardens in this country, where it is also said to grow wild, is remarkable for the appearance of its fragrant pink flowers in early spring before the leaves expand. It forms a dwarf bush with erect branches, along the sides of which the flowers are placed, while the leaves occur in tufts at the extremity of the twigs. The fruits are red and pulpy, of the size of a pea, and poisonous to human beings, though birds eat them with avidity. There is a variety with white flowers and yellowish fruits. The bark of this species, and especially that of its roots, has been used in medicine as a sudorific and alterative in scrofulous, venereal, and other diseases. It is extremely acrid to the taste, and is now rarely employed.

D. Laureola, the Spurge Laurel, occurs in woods in this country. It is a smaller plant than the preceding, and has bright green oblong evergreen leaves, and greenish flowers which are scentless; the fruits are oval and of a bluish-black colour. This species participates in the acrid properties of the mezereon, though it is not employed in medicine. It finds a place in shrubberies, on account of its evergreen character, and because it thrives well beneath the shade of trees.

The tough fibrous nature of the inner bark of these plants is made available for the manufacture of paper in various parts of India and China. In Nepal the bark of *D. cannabina*, *D. Edgeworthii*, and other species is thus employed. For this purpose it is scraped and boiled in water, with a small quantity of the ashes of the oak; after this it is washed and beaten to a pulp on a stone, and then spread out on a mould or frame made of bamboo mats. This paper is of various qualities; the best is strong and tough, is not liable to crack or break upon being folded, is not eaten by insects, and does not suffer from damp. From its durability it is used in many parts of India for deeds and records.

Several kinds are cultivated in this country as hardy shrubs or in greenhouses. Among them are *D. pontica*, which resembles the common spurge laurel, but is of larger growth, has more fragrant flowers, and grows well under the shade of trees; *D. alpina*, a low growing shrub with deciduous leaves and white fragrant flowers, well adapted for rock-work; as also is *D. Cneorum*, a charming plant with procumbent stems, lance-shaped evergreen leaves, and clusters of pink sweet-scented flowers. The foregoing are hardy. *D. odora*, *D. indica*, *D. chinensis* and others require the protection of a greenhouse. The Spurge Laurel is the badge of the Grahams. [M. T. M.]

DAPHNIDIUM. Indian trees constituting a genus of *Lauraceæ*, having unisexual flowers within an involucre of scaly bracts. The perianth is six-cleft; the male flower has nine stamens in three rows, the three innermost having glands at each side of their filaments. Fruit berry-like, one-seeded, partly enclosed within the persistent base of the perianth. [M. T. M.]

DAPHNIDOSTAPHYLIS. A small group of shrubby plants, separated by some from *Arctostaphylos*.

DAPHNOPSIS. A genus of *Thymelaceæ*, consisting of diœcious Brazilian plants. The male flowers have a four-cleft perianth, with eight stamens, and a rudimentary ovary. The perianth of the female flower is persistent at the base of the drupe, which is of a fibrous texture. [M. T. M.]

DARBYA. A North American tree or shrub, described by A. Gray as constituting a distinct genus of *Santalaceæ*, but reduced by A. De Candolle to a section of *Comandra*.

DAREA. A section of *Asplenium*, sometimes called *Cænopteris*, and characterised by the prevalence of uniseriferous segments to the fronds. [T. M.]

DARLINGTONIA. A remarkable genus of *Sarraceniaceæ* found in California. The only species, *D. californica*, known as the

Californian Side-saddle flower or Pitcher-plant, is a perennial herb growing in marshy places. Its leaves all rise from the root, the adult ones being from eighteen inches to a foot or more in length, the stalk or pitcher tubular, gradually tapering downwards and singularly twisted on the axis about half a turn, marked with strong veins and slender veinlets, and the summit vaulted and formed into a sac about the size of a hen's egg, on the under side of which is an oval orifice about half an inch in diameter opening into the cavity of the pitcher; the upper part of this tube is of a dull orange colour. The blade, which is borne on the end of the stalk or pitcher, is narrowed at the base and deeply divided into two spreading nearly lance-shaped lobes, which are curved downwards, and also often backwards, resembling the lop-ears of some varieties of rabbit. The pitcher inside the hood is furnished with short conical hairs which point downwards, and towards the base there are long slender hairs also pointing downwards; remains of insects are sometimes found at the bottom. Dr. Torrey writes 'The orifice of the pitcher being placed directly under

Darlingtonia californica.

the vaulted summit, cannot receive either rain-water or dew, and yet Mr. Bracken-ridge thinks he found some of the leaves containing water; still I cannot think the water was secreted by the hairs in the tube.' The flowers are single and nodding at the apex of a smooth stalk, which is furnished with straw-coloured scales, and varies from two to four feet in length. When fully expanded the flower is about two inches in diameter; the calyx consists of five straw-coloured acute sepals; the petals, of a like number, and pale purple in colour, are narrowed and concave at the apex and broad below; the twelve to fifteen stamens are nearly hidden by the projecting summit of the ovary, which is top-shaped, slightly five-angled, and crowned by a short style with a five-lobed stigma. The fruit is a five-celled capsule about an inch in length, with numerous seeds. The forked blade of the leaf and the form of the stigma distinguish the genus from Sarracenia, which has an umbrella-shaped stigma. The genus is named in honour of Dr. Darlington, of Pennsylvania.

This account is chiefly abridged from a paper of Dr. Torrey's in the *Smithsonian Contributions to Knowledge* (Washington 1853), where a full history of this interesting plant is given. It has been in cultivation in England. [A. A. B.]

DARNEL. *Lolium temulentum.*

DARWINIA. A small genus of *Chamœlauciaceœ*, named after Dr. Darwin. It has a five-cleft calyx, the lobes of which are roundish-cordate, concave, and full of pellucid dots; petals wanting; stamens from ten to fifteen, often joined in threes, inserted in the limb of the calyx, and having very short flat glabrous filaments, and anthers inserted by their bases; style tapering, bearded at the apex; stigma a pruinose dot; ovary one-celled, single-seeded; seeds pentagonal pitted on the surface. Heath-like shrubs of lowly growth found in the extra-tropical portions of Australia. The leaves are marked with pellucid dots. [R. H.]

DASYA. A lovely genus of rose-spored *Algœ*, allied to *Polysiphonia*, from which it differs in its more compound stem, with persistent coloured branchlets, which originate the pod-like receptacles of the tetraspores or stichidia. The species are far more common in the Southern Ocean. We have, however, a few fine species, of which *D. coccinea* is well known to most collectors of *Algœ* from its bright scarlet tint; and there are representatives in the Northern Hemisphere of four out of the five sub-genera into which Dr. Harvey has disposed the species in his *Nereis Australis*. In *Polysiphonia*, it may be observed, the tetraspores are imbedded in the branches themselves, and not in distinct organs. [M. J. B.]

DASYCLADE.Æ A small natural order of green-spored *Algœ*, which are either naked or coated with carbonate of lime, and have a one-celled simple or branched axis which is whorled either throughout its whole length or near the summit with jointed branchlets. The fruit is contained in free or laterally united sporangia. In *Acetabularia* the stem is filiform, and ends in a target shaped disc composed of spore-bearing cells; from the centre of this the stem is continued bearing whorls of forked fibres, and as the fruit cells fall off below, new discs are formed above. *Dasycladus*, the typical genus, has threads free from any crust, and the axis is clothed everywhere with whorls of jointed trifid branchlets. The thread-shaped forked distinctly jointed frond of *Cymopolia*, on the contrary, is densely incrusted, the crust being pierced with pores, and the nodes fringed with byssoid multifid fibres. We have no representative of this curious order on our coasts. Both *Dasycladeœ* and *Valonieœ* were first separated by Kutzing from *Siphoneœ*, and are adopted by Dr. Harvey in his admirable work on North

American *Algæ*, to which we have been largely indebted. [M. J. B.]

DASYLIRION. A genus of *Bromeliaceæ*, consisting of Mexican plants with short stems, and densely crowded linear leaves which droop gracefully, and generally have a little brush-like tuft of fibres at the point. From amid these leaves the flower-stalks rise to a considerable height, the upper portion being crowded with a dense panicle of flowers, which are diœcious. The perianth consists of six nearly equal segments in two rows; and there are six stamens with filaments thickened in the middle, and having a gland at the base. The female flowers differ in having antherless stamens, a superior six-seeded one-celled ovary, with membranous angles, the six ovules in pairs; a short style with a dilated three-lobed stigma; and a nut-like fruit, one-seeded by abortion. *D. acrotrichum* is a handsome kind grown in greenhouses. [M. T. M.]

DASYLOMA. The generic name of plants belonging to the umbelliferous order, characterized by having five ribs on each half of the fruit, three on the back smaller than the two at the sides, the latter being larger and thicker, a character indicated by the name, which is derived from two Greek words signifying 'thick border.' The species are natives of India, and are herbaceous plants, with hollow stems, and twice pinnate leaves, the leaflets of which are wedge-shaped, toothed at the end. [G. D.]

DASYMALLA. A genus of small West Australian bushes of the *Myoporum* family, having their leaves and stems covered with dense white wool. The forked style and bracted flower-stalks, together with the woolly nature of the stems and leaves, are its chief distinguishing features. The four-angled stems are furnished with opposite entire leaves, oblong obovate in form. In the axils of these the flowers are found in little bundles or cymes shorter than the leaves. The calyx is five-parted, and the purple tubular corollas are widened at the top and two-lipped, the upper lip two-lobed, the lower three-lobed. In the inside of the tube and near its base the four stamens (two long and two short) are inserted. The ovary is densely hairy and crowned with a filiform style forked at the top; when ripe it becomes a somewhat dry four-celled berry with one seed in each cell. Two species are known. [A. A. B.]

DASYNEMA. A name once given to a few South American trees of the lime-tree family; they, however, belong to SLOANEA; which see. [A. A. B.]

DASYPHYLLUM. *Flotovia*.

DASYPOGON. A genus referred to *Juncaceæ*, in which it is distinguished by its capsular one-celled fruit with basilar ovules, and longitudinally dehiscent incumbent anthers with filaments thickened at the apex. It comprises two undershrubs from South Australia, with simple leafy stems, and grass-like leaves rough at the margin; flowers sessile in a globular terminal head. [J. T. S.]

DASYSTEMON. A genus of Australian herbs, covered with scaly pimples, and having linear fleshy opposite leaves united at their bases; greenish yellow flowers; a calyx of three to seven leaf-like segments; corolla of three to seven petals rolled under at the point, and slightly united at the base; stamens three to seven inserted on to the calyx round the three to five ovaries. The filaments are thick, hence the name of the genus 'thick-stamen.' It is included among the *Crassulaceæ*. *D. calycinus* is occasionally grown as a greenhouse plant in this country. [M. T. M.]

DATE. *Phœnix dactylifera.* —, WILD. *P. sylvestris.*

DATISCACE Æ. (*Datiscads*.) A natural order of dicotyledonous plants included in the sub-class *Monochlamydeæ*, and referred by Lindley to the cucurbital alliance. Herbs or trees with alternate leaves having no stipules; some flowers have stamens only, others have pistils only; the corolla is wanting; the calyx or perianth adheres to the ovary, and is divided into three or four parts; stamens three to seven; ovary one-celled, with three or four many-seeded parietal placentas. Fruit a one-celled capsule, opening at the top; seeds having a reticulated skin, and a cup-like swelling at one end; there is no separate albumen. The plants consist of few species, which are scattered over North America, northern India, Siberia, the Indian Archipelago, and the south-eastern part of Europe. They have bitter and purgative qualities. It is said that the ovary of *Datisca cannabina* can produce perfect seeds without the application of pollen to the pistil. *Tetrameles Horsfieldii* is a large tree of the order. There are but three known genera, *Datisca*, *Tetrameles*, and *Tricerastes*, and these comprise but four species. [J. H. B.]

DATISCA. A genus of plants typifying the *Datiscaceæ*. The characters are: flowers diœcious; calyx five-parted in the staminate flowers, three to five-toothed in the pistillate flowers; no corolla; stamens five to fifteen, collected in the middle of the flower; ovary united with the calyx, inferior, one-celled, with three to five parietal placentas; styles three to five. Fruit a one-celled capsule opening by a round hole at the apex. Seeds numerous, striated, with a cup-like covering at the base. Annual herbaceous plants found in Nepal and in Asia Minor. They have unequally-pinnate alternate leaves, and racemose bracteated greenish flowers. There are two known species. [J. H. B.]

DATLIER COMMUN. (Fr) *Phœnix dactylifera.*

DATURA. A genus of *Solanaceæ*, or, according to Mr. Miers, of *Atropaceæ*, the species of which are eminently poisonous; while in small quantities they act as valuable remedial agents. They are known by their tubular calyx, the upper part of which falls off as the fruit ripens, while a small portion remains as a circular rim around the base of the fruit; the corolla is funnel-shaped and plaited. The fruit is a capsule with four compartments and four valves.

The best known plant of this genus is the common Thorn Apple, *D. Stramonium*, which springs up in a half wild state on the borders of cultivated fields, rubbish heaps, &c., in this country, and is found in similar situations in all the warmer parts of the globe. It is a coarse strong-smelling annual, growing one or two feet high, with widely-spreading forked branches, and large ovate leaves with irregularly-waved or sinuately-toothed margins. The flowers are large, placed on short stalks arising from the forks of the stem; the calyx is tubular and angular; the corolla is double the length of the calyx, funnel-shaped, with a large plaited five-toothed limb, generally of a pure white colour, but sometimes in hot climates pink or purple; the capsule is ovate, of the size of a walnut, somewhat four-celled, bursting by four valves, which are covered with stout triangular spines, whence the name Thorn Apple. The poisonous principle of this plant is an alkaline crystalline substance called *daturin*. The effects produced by medicinal or poisonous doses of Stramonium are similar to those induced by belladonna, but to this is added a certain degree of acridity and of anodyne power not possessed by the other plant. Stramonium has been found beneficial in neuralgia, epilepsy, mania, &c.; while in some cases of asthma relief has been experienced from smoking the leaves.

D. fastuosa, a common Indian plant, is possessed of properties similar to those of stramonium, and is employed by the native doctors for the relief of rheumatic and other painful affections. The seeds are used in India and China to stupefy or even poison an enemy. *D. alba* or *D. Metel*, also an Indian plant, produces similar effects. The Rajpoot mothers are said to smear their breasts with the juice of the leaves, so as to poison their newly-born female infants. It has been conjectured that the seeds of *D. Stramonium* were used by the priests of Apollo at Delphi to produce those frantic ravings which were called prophecies, a suggestion which derives some support from the fact 'that in the temple of the Sun, in the city of Sagomozo (Peru?), the seeds of the Floripondio, *D. sanguinea*, are used for a similar purpose.' The Peruvians also prepare an intoxicating beverage from the seeds, which induces stupefaction and furious delirium if partaken of in large quantities. The Arabs of central Africa are said by Lieut. Burton to dry the leaves, the flowers, and the rind of the rootlet, which is considered the strongest preparation, and smoke them in a common bowl, or in a water-pipe. It is esteemed by them a sovereign remedy for asthma and influenza, and although they do not use it like the Indian Datura poisoners, accidents nevertheless occur from its narcotic properties. See BRUGMANSIA. [M. T. M.]

DAUBENTONIA. A genus of bushy plants of the pea family, comprising three species found in Texas and Buenos Ayres. They are chiefly remarkable for their curious quadrangular pods, which are three to four inches long, stalked, pointed, and furnished with wings along the angles. The only other genus with four-angled pods nearly related to this is *Piscidia*, which has unequally pinnate leaves; while here there is no odd leaflet, but the leaves are made up of ten to twelve pairs of oblong leaflets, each about an inch in length. The red or yellow flowers, a good deal like those of the laburnum, are borne on axillary racemes shorter than the leaves. *D. punicea* is a common plant on the banks of the Uraguay, and in various parts of Banda Oriental and Rio Grande, where it grows into a large handsome shrub with leaves like those of the false acacia, and bears abundant racemes of brilliant red flowers, between cherry and orange-colour. The genus is named in honour of M. Daubenton, an eminent French naturalist and physician. [A. A. B.]

DAUBENYA. A genus of one or two species of bulbous *Liliaceæ* from the Cape of Good Hope. *D. aurea*, the typal species, has a pair of oblong leaves seated close to the earth, and in their sinus a sessile umbel of yellow flowers, whose perianth is tubulose with a two-lipped limb, both lips being three-toothed, the upper short, the lower one larger in the ray flowers and depauperated in those of the centre or disk. There are six stamens with unequal declinate filaments, somewhat joined at the base; and a filiform style with a capitate stigma. The genus is dedicated to Prof. Daubeny of Oxford. [T. M.]

DAUCOSMA. A North American genus of *Umbelliferæ*, represented by an annual herb, with the odour of the wild carrot, whence its name. Its distinguishing characters are its petals, which are bent inwards; its five-toothed calyx; and its distinct carpophore or stalk bearing the two halves of the fruit. The first of these characters separates it from *Cynosciadium*, the second from *Æthusa*, and the last from *Œnanthe*. [M. T. M.]

DAUCUS. A genus of *Umbelliferæ*, consisting of several species of dwarf weedy-looking plants, having thin deeply-cut pinnatifid leaves; and flower-stems rising from two to three feet high, and bearing in a terminal umbel a number of small white or rosy-coloured flowers. It is distinguished by the long prickles to its carpels, the prickles being long, flat, and straight. Of one of its species cultivated as a vegetable, there are many varieties.

The Carrot, *D. Carota*, is a biennial, a native of Britain, usually found, in its wild state, in light sandy soil. Notwithstanding the great difference between its dry sticky root, and that of the large succulent root of our garden Carrot, it is generally admitted to be the stock from which all the cultivated varieties have sprung; although Miller states that he in vain endeavoured to improve the quality of the wild plant by cultivation. As an esculent, the Carrot was known to the ancients; and Pliny says the best came to Rome from Candia. Gerarde, writing in 1597, tells us they do not grow in Candia only, but are found upon the mountains in Germany, and about Geneva. How or when they were first introduced into this country is unknown, but it is generally believed to have been by the Dutch during the reign of Queen Elizabeth (1558), and that they were first grown about Sandwich in Kent.

Scarcely any vegetable is better known, or in greater demand for culinary purposes than the Carrot. Its root contains a large portion of saccharine matter, and is used in soups and stews, as well as a vegetable dish during winter. In order to supply the demand for young carrots during the spring and summer, large quantities are grown by artificial heat. The various sorts of Carrots in cultivation are divided into two classes, known as Horn Carrots and Long Carrots: the former short and early; the latter becoming mature in autumn for winter use.

Parkinson, writing in 1629, says, that in his day ladies wore Carrot leaves in place of feathers; and London states (*Encycl. of Gard.*, p. 835) that in winter an elegant chimney ornament may be formed by cutting off a section from the head or thick end of a Carrot, containing the bud, and placing it in a shallow vessel of water. Young and delicate leaves unfold themselves, forming a radiated tuft of a very handsome appearance. [W. B. B.]

The Carrot yields two British species, *D. Carota* and *maritimus*; but we agree with Sir W. J. Hooker in deeming them 'scarcely permanently distinct.' The Carrot of the garden and farm is a well known derivation of one of these; we almost think of the latter, as our experiments in ennobling the common *D. Carota* have been unfortunate, though we have had reports of success in this experiment by those with whom the ennobling of the parsnip has not succeeded as it has with us. [J. B.]

DAUPHINELLE. (Fr.) *Delphinium.*

DAURADE. (Fr.) *Ceterach officinarum.*

DAVALLIA. A fine and extensive genus of polypodiaceous ferns, typical of the group *Davalliex*. They have scaly creeping rhizomes, which feature has given rise to the name of Hare's Foot Fern, applied to *D. canariensis*. The fronds are sometimes pinnate, but more frequently pinnately decompound, very elegantly cut into multitudes of small divisions, and bearing numerous fructifications, which form a series of cups or cysts at the margins of the segments. These cysts assume two somewhat different forms: the one, rather shallow cup-shaped, represented by *D. tenuifolia* and *D. aculeata*, the latter of which is quite scandent and bramble-like in habit; the other tubulose, represented by *D. elegans* and *D. solida*. The genus is well marked by natural features, and is one of the most elegant to be found in our gardens. Several offshoots have been separated from it, as *Acrophorus, Humata, Loxoscaphe,* and *Microlepia*. [T. M.]

DAVIESIA. A large genus of New Holland and Tasmanian bushes of the pea family, easily recognised among their allies with ten free stamens and two ovules, by the form of their pods, which are short, nearly triangular, with a straight upper and a much curved under edge. In some species the leaves are much like those of the juniper, and in a large number they take the form of spines like those seen on the furze, to which plant many of them bear a strong resemblance. In a few the leaves are heart-shaped and embrace the stem; in others they are oblong; and a few are entirely destitute of leaves, but in these the stems are usually flattened and perform leaf functions. The flowers are small, usually yellow, sometimes blue or purple, arranged in little tufts or racemes or stalked cymes arising from the axils of the leaves, or from those of little scales where no leaves exist.

A very common plant in greenhouses, and one of the most beautiful in the genus, is *D. latifolia*, a native of Tasmania and the south parts of New Holland. This plant has smooth oblong leaves, in the axils of which the pretty yellow flowers are found disposed in dense erect racemes. Another scarcely less beautiful species is *D. cordata*, the leaves of which, about the largest in the genus, are sessile, heart-shaped, acute, and embrace the stems at the base; they are quite smooth and beautifully veined. The flowers are in stalked corymbs, each supported by two leafy bracts which envelope a number of stalked flowers having a yellow standard and a purple keel. *D. epiphyllum*, a West Australian species, is remarkable for having white flattened and variously lobed stems without leaves, but having much the appearance of the antlers of a stag. The flowers are curiously placed on the middle of the flattened portion, and arise from the axils of little scales, two or more together. *D. juncea* has rush-like branches devoid of leaves, and furnished at distant intervals with bundles of yellow flowers; and an allied species has similar but much thicker stems, nearly half an inch in diameter, with soft pith-like wood. It would be difficult to point to a genus comprising more diversity of form among its species, of which there are upwards of fifty known. It bears the name of the Rev. Mr. Davies, a Welsh botanist. [A. A. B.]

DAVYA. A genus of small oppositeleaved trees or scandent bushes of the *Melastomaceæ*, found in various parts of tropical America, and numbering about a dozen species. They are chiefly characterised by the capsular (not berried) fruit, and the peculiar structure of their stamens, which are eight to ten in number, nearly equal in height and similar in form; the anthers linear or awl-shaped, curved outwards and opening at top by a little pore; the connective or point of junction of the anther with its stalk produced behind into an obtuse or acute horn, sometimes forked at the point and parallel to the anther. The leaves are stalked, lance-shaped, oval or elliptical, entire or toothed. The flowers are yellow, rose, or purple, disposed in terminal panicles or cymes, and having an entire or five-toothed calyx; five obovate petals, and a filiform style crowning an ovary which becomes, when ripe, a five-celled capsule with numerous seeds. The genus bears the name of Sir H. Davy, the eminent chemist. [A. A. B.]

DAY-FLOWER. An American name for *Commelyna*.

DEAL. The wood of various pine and fir trees.

DEALBATE. Covered with a very opaque white powder.

DECA. In Greek composition = ten.

DECAISNEA. A genus of plants named after Decaisne, a celebrated French botan-

Decaisnea insignis.

ist, by Drs. Hooker and Thomson. It belongs to the natural order *Lardizabalaceæ*, and is an erect shrub with large pith, pinnate leaves, racemose inflorescence, and greenish flowers; sepals six, linear and awl-shaped; petals none; flowers sometimes abortive or becoming staminate or pistillate; stamens six, free or united by their filaments; ovaries three with an oblique style; ovules very numerous, on two thread-like placentas. The fruit consists of follicles filled with pulp. The only known species is *D. insignis* found at Sikkim and Bhotan in the Himalayas at the height of 6,000 to 10,000 feet, flowering in May and fruiting in October. The fruit is very palatable, and is eaten by the Lepchas of Sikkim. The name is also a synonym of *Onemidia*. [J. H. B.]

DECAMALEE or DIKAMALI. A gum obtained in India from *Gardenia lucida*.

DECANEMA. A genus of *Asclepiadaceæ*, containing a single species from Madagascar. A leafless branched undershrub, remarkably like *Sarcostemma aphylla*, except in the structure of the flower. The flowers are small in terminal or lateral umbels; the calyx is five-parted; the corolla rotate and five-cleft; the staminal crown consists of two series of five lobes, the outer being opposite to, the inner alternating with, the lobes of the corolla, and its lobes are rounded and terminate in a long linear blade exceeding the corolla. The long round follicles contain comose seeds. [W. C.]

DECASPORA. A small Tasmanian genus of *Epacridaceæ*, having small ovate or lanceolate leaves, and flowers in terminal spikes of a reddish hue. The calyx with two bracts at the base; corolla campanulate, the limb slightly bearded; stamens exserted with five scales united at the base; a ten-celled ovary with a single seed in each cell. The fruit is a violet-coloured berry. [R. H.]

DECIDUOUS. Finally falling off; as the calyx and corolla of crucifers.

DECKERIA. A name recently proposed for a genus of palms, but the characters upon which it is founded not being of sufficient importance to warrant its adoption, other botanists have since referred the species to the older genus *Iriartea*, to which three of them originally belonged. They are natives of tropical South America, and are remarkable on account of the singular shape of their trunk, which,

Deckeria ventricosa.

though cylindrical throughout its entire height, like that of numerous other palms, while young, after attaining a certain age, swells suddenly out at a point about midway between the ground and its crown of leaves, to more than double its previous diameter, again contracting to its original size and cylindrical form at a short distance from the summit. This peculiarity is more particularly evident in the species called by the Indians on the Amazon Paxiuba barriguda (i.e. pot-bellied Paxiuba: Paxiuba being a general term applied to the *Irarteas*), the *Deckeria* or *Iriartea ventricosa* of botanists, a common palm in the forests bordering the Amazon and Rio Negro, where the natives take advantage of its swollen trunks in the construction of their canoes, its natural shape saving them much labour. They also use the hard black wood of the outer portion of the trunk to make harpoons for spearing the cow-fish. [A. S.]

DECLINATE. Bent downwards.

DECODON. A genus of *Lythraceæ*, nearly related to *Lythrum*, and differing chiefly in the calyx-tube being shortly bell-shaped, instead of cylindrical. *D. verticillata*, the Swamp Loosestrife, grows on the borders of swamps in the United States, and is the only known species. It is a pretty bush six to eight feet high, having slender recurved stems furnished with privet-like leaves, placed in whorls of three round the stem, and bearing in their axils clusters of stalked rose-coloured flowers much like those of *Lythrum Salicaria*. The plant is also called *Nesæa verticillata*. According to Torrey it is used as an emmenagogue. [A. A. B.]

DECOMPOUND, DECOMPOSITE. Having various compound divisions or ramifications.

DECUMARIA. A climbing shrub of the Southern States of North America, forming a genus of *Philadelpheæ*. The flowers are white, arranged in corymbs, sweet-scented, and in gardens are observed to be sometimes unisexual, though this has not been found to be the case in wild specimens. The calyx-tube is adherent to the ovary, and is marked by from seven to ten prominent nerves; the style is consolidated, expanded above into a stigma, with seven to ten rays. The capsule is divided into seven to ten compartments, and is crowned by the persistent style and limb of the calyx. It contains not only numerous seeds, each surrounded by an aril, but also, at least in dried specimens, a quantity of small crystals (raphides) interspersed among them. [M. T. M.]

DECUMBENT. Reclining upon the earth, and rising again from it.

DECURRENT. Prolonged below the point of insertion, as if running downwards.

DECURSIVELY PINNATE. When a petiole is winged by the elongation of the base of the leaflets; hardly different from pinnatifid.

DECUSSATE. Arranged in pairs that alternately cross each other.

DEDUPLICATION. The supposed unlining process which some botanists believe in when one organ in a flower is produced opposite another.

DEER BALLS. A synonym of Hart's Truffles, Lycoperdon Nuts, and *Elaphomyces*. [M. J. B.]

DEERBERRY. *Gaultheria procumbens* also an American name for *Vaccinium stamineum*.

DEERINGIA. A genus of *Amaranthaceæ*, distinguished by its fruit being a many-seeded berry. They are smooth weak-stemmed shrubs from India and Australia, with alternate leaves, and spikes of small flowers, having a five-leaved calyx, five stamens united below into a cup, a short style, three stigmas, and an inflated berry. *D. celosioides*, from New Holland, bears long spikes of red berries, about the size of currants. [J. T. S.]

DEFERENT. Conveying anything downwards.

DEFOLIATION. The casting off of leaves.

DEFORMATION. An alteration in the usual form of an organ by accident or otherwise.

DEGRADATION. A change consisting of an abstraction, loss, abortion, or nondevelopement of usual organs.

DEHAASIA. A genus of *Lauraceæ*, consisting of trees with hermaphrodite or monœcious flowers, the perianth of which is six-cleft, the three outer divisions being much smaller than the inner ones. Stamens nine or twelve in three or four rows, the inner row sterile; of the fertile stamens, the two outer rows have their anthers opening inwardly, while those of the inner row open outwardly, the filaments of this latter series having glands on each side at the base. The fruit is a one-seeded berry placed upon a thickened fleshy flower stalk. [M. T. M.]

DEHISCENCE. The act of splitting into regular parts, or in some manner dependent upon organic structure.

DELABECHEA. The Bottle-tree of Northeastern Australia, *D. rupestris*, is the only plant of this genus, which belongs to the *Sterculiaceæ*, and is very nearly related to *Brachychiton*. The Bottle tree is of middling stature, and is chiefly remarkable for the curious form of the trunk, which is bulged out in the middle in the form of a barrel. The stem abounds in a mucilaginous or resinous substance resembling gum tragacanth, which is wholesome and nutritious, and is said to be used as an article of food by the aborigines in cases of extreme need. Dr. Lindley, in describing the tree, says, 'the wood has a remarkably loose texture; it is soft and brittle, owing to the presence of an enormous quantity of very large tubes of pitted tissue, some of which measure a line and a half across; they form the whole inner face of each woody zone. When boiling water is poured on shavings of this wood, a clear jelly resembling tragacanth is formed, and becomes a thick viscid mass; iodine stains it brown, but no trace of starch is indicated in it.' Usually the leaves are from two to four inches long, entire, stalked, and lance-shaped; sometimes, however, they are digitate and composed of seven to nine sessile leaflets of the same form as the simple leaves. The digitate leaves are probably found only on young plants. The flowers are inconspicuous, and borne on short panicles arising

Delabechea rupestris.

from the axils of, and shorter than, the leaves; in the males the calyx is five-cleft, and the stamens numerous; the females are not known. The fruit is composed of five stalked smooth brown leathery follicles, covered internally with a thick fur of starry hairs; each of these contains about six seeds, which have their lower portion covered with similar hairs, and are smooth above. The genus is named in honour of the late eminent geologist, Sir H. T. De la Bêche. [A. A. B.]

DELAIREA. The name sometimes given to a trailing South African Groundsel (*Senecio mikanioides*), with stalked, smooth, and fleshy leaves, which are cordate at the base, and five to seven-lobed. The flower-heads are numerous, and disposed in axillary corymbs longer than the leaves. In gardens it is called German Ivy. [A. A. B.]

DELASTREA. A genus of *Sapotaceæ*, represented by a lofty tree native of Madagascar, distinguished from its allies by the lobes of its corolla, which are eighteen in number, twelve external, six internal, opposite to which latter are six stamens, all of them fertile; and by its ovary, which contains twelve compartments. [M. T. M.]

DELESSERIA. A genus of rose-spored *Algæ*, belonging to the section in which the spores form little necklaces (*Desmiospermeæ*) containing many of the most beautiful and delicate species which adorn our coasts, a great part of their beauty arising from the symmetry of the frond, and the contrast between the dark midrib and the membranous border. The capsules contain a placenta formed of branched threads bearing short chains of spores, the ultimate members of the chains being the first to ripen. The species are numerous, and many of them are widely dispersed. The beautiful ash-leaved seaweed formerly called *D. sanguinea*, has fruit of a different structure, and is now referred to a distinct genus, *Wormskioldia*. [M. J. B.]

DELIMA. A small genus of *Dilleniaceæ*, all, with the exception of one Asiatic species, natives of the tropics of the Western hemisphere. They have very small flowers disposed in loose panicles at the ends of the young branches: the calyx consisting of five permanent sepals, and the corolla of four or five white petals, which soon fall away. The ovary is solitary, nearly globular, and terminated by a curved tapering style; it ultimately becomes a small dry oval fruit, which splits open along the inner edge when ripe, exposing a solitary arillate seed.

D. sarmentosa is widely distributed throughout the eastern countries of tropical Asia, including Ceylon, Malaya, Ava, Silhet, Java, Southern China, the Philippine Islands, &c. Its leaves vary very much in shape, but are generally somewhat oval; their edges either entire or cut into teeth tipped with short hard points; the upper surface of these leaves is completely covered with little asperities, which are so hard and render the leaves so rough that they are commonly employed in most of the above-mentioned countries as a substitute for sand-paper, and are thus used for polishing various domestic utensils, and other articles made of either wood or metal. In Ceylon the plant is called Kora-sawel, and in the Philippine Islands, Bois de râpe. [A. S.]

DELIQUESCENT. Branched, but so divided that the principal axis is lost trace of in ramifications; as the head of an oak tree.

DELISSEA. A genus of shrubs, natives of the Sandwich Isles, and included in the order *Lobeliaceæ*. The main characteristics of the genus are a hemispherical calyx tube, which is united to the ovary, and is surmounted by a limb with five very short teeth; a tubular corolla with a two-lipped limb: filaments and anthers combined into a tube; fruit a somewhat globular berry, two-celled, crowned by the limb of the calyx. [M. T. M.]

DELOSTOMA (including *Codazzia*). A genus of *Bignoniaceæ*, remarkable for its double calyx, and flat oblong capsule divided into two cells by a partition placed contrary to the direction of the valves. There are four species, all confined to the Andes of South America, where they range from New Granada to Peru. They are small trees, with simple oblong leaves generally covered with hair, and terminal panicles bearing fine pink or purple blossoms. The outer calyx is five, the inner three-cleft; the corolla tubular, slightly curved; the stamens four in number; the capsule smooth, with the winged seeds arranged in several rows. One of the handsomest species is *D. integrifolium* (*Codazzia speciosa*), frequent in the Andes of Quindin. *D. latifolium* is identical with *Callichlamys riparia*, and *D. Steudobium* with *Stenolobium stans*. *Amphilophium* is the only other bignoniaceous genus which has a double calyx. [B. S.]

DELPHINIUM. A genus of *Ranunculaceæ*, commonly known by the name of Larkspur. The species are numerous, and widely distributed over the temperate regions of the Northern hemisphere. They are herbaceous plants, with erect branching stems, and finely cut or palmately-divided leaves. The flowers are in loose racemes towards the end of the branches; they have a calyx of five-coloured sepals, the upper one prolonged at its base into a long tapering spur, and four (or two) petals concealed partially within the spur of the calyx. The fruit consists of from one to five many-seeded follicles. The flowers resemble those of some species of *Aconite*, but they have a spurred, not a hooded calyx, and they have not the peculiar hammer-like petals of the aconite. Larkspurs partake largely of the acrid properties for which the order is in general so remarkable.

D. Staphisagria, or Stavesacre, was used medicinally by the Greeks, and still finds a place in the pharmacopœia, though now rarely used. The seeds contain the active principle in greatest abundance, and hence are ordered to be used in the form of ointment to destroy vermin. Delphinia is an extremely acrid bitter white powder prepared from the seeds, and used externally in cases of rheumatism and neuralgia. Numerous species and varieties of this genus are cultivated in gardens. *D. Consolida*, a common European plant, is occasionally found in a half-wild state on the borders of fields. Its name was given in reference to its power, real or imaginary, of healing or consolidating wounds. *D. Ajacis*, a common garden plant, derives its name from certain markings on the petals, presenting more or less resemblance to the letters A I A I; hence also it has been conjectured to be the 'hyacinth' of the ancients, described as possessing similar markings. Dr. Daubeny, the latest commentator on the plants mentioned in ancient Greek and Latin writers, concludes, 'that the term *Auakinthos* was in general applied to some plant of the lily tribe; but that the poets confounded with this the larkspur, which has upon it the markings alluded to; and that the name hyacinth was given, in the first instance, to the plant which most distinctly exhibited them.'

Some of the cultivated species, such as *D. grandiflorum*, *D. chinense*, *D. sibiricum*, &c., are called Bee Larkspurs, from the resemblance of the petals, which are studded with yellow hairs, to a humble bee whose head is buried in the recesses of the flower. One of the most beautiful species in cultivation is *D. formosum*, with large rich blue flowers; and *D. cardinale* is remarkable for its scarlet flowers. [M. T. M.]

DELTOID. A solid, the transverse section of which has a triangular outline, like the Greek Δ. Also applied to the outline of thin bodies.

DEMATIEI. A natural order of filamentous moulds, separated from the white or brightly-coloured species by the dark threads, which look as if they were smoke-dried or carbonised; and in the more typical species have an investing membrane. Some of our common moulds, as *Cladosporium herbarum*, belong here. [M. J. B.]

DEMERSED. Buried beneath water.

DEMIDOVIA. A genus of *Trilliaceæ*, founded on the *Paris incompleta* of Bieberstein. It differs from *Paris* by not having any inner series of perianth segments. The leaves are six to twelve, oblong or oblong-oblanceolate, acuminate; the perianth segments green, ovate acuminate, twice as long as the eight to twelve stamens; styles four, longer than the stamens. The only species, *D. polyphylla*, is a native of southern Russia. [J. T. S.]

DENDROBIUM. A well-known genus of epiphytal orchids, comprising more than 200 species, of which upwards of eighty have been cultivated in hothouses for the sake of their beautiful flowers. The great mass comes from India and its Archipelago; a few are found in East Australia and the Pacific Islands; and one in New Zealand. 'The genus varies extremely in the habit of its species, some being little larger than the mosses among which they grow; while others are surpassed in stature by few in the order. Like the *Oncidia* of the New World, there are some species of which the foliage is ancipitous, others having it terete, while in the majority it is in the usual flat condition. A few have no other stem than a wiry creeping rhizome; others have small conical pseudo-bulbs; many form clavate horny stems, leafy only at the summit; but the greater part produce long leafy branches. In the majority the colour of the flowers is some shade of purple; a few are destitute of all colour except green; and a rather considerable group is especially distinguishable by the rich yellow tint of their blossoms.'—*Lindley*. In arrangement, the flowers are either solitary, fascicled, or in racemes. According to Dr. Lindley, all agree in having a two-celled anther with four pollen masses, which have no caudicle or separate stigmatic gland, and are of uniform breadth at either end; the latter character separating them from *Eria*, which bears pear-shaped pollen masses; whilst, from the nearly-related genus *Bolbophyllum*, they may be recognised by the sessile and not unguiculate (clawed) lip.

Of cultivated species, with flowers in which purple predominates, we have *D. nobile*, perhaps the most beautiful in the genus. It has erect stems one to two feet high, bearing at intervals two or three-flowered peduncles, the flowers when expanded being two to three inches across. The petals and sepals are faintly rose-coloured at the base, and bright purple towards apex; the lip rolled up so as to be nearly trumpet-shaped, with a recurved border which is greenish-yellow at the edges and purple at the end, while the tube is of a deep blood-red colour. There are a number of fine varieties of this plant cultivated. *D. macranthum*, from Manilla, has rich rose-coloured flowers, sometimes five inches across; the ovate lip is margined with a delicate fringe of hairs, and marked at the base on either side with a deep purple blotch. *D. Macanthiæ*, called in Ceylon Wissak-mal, meaning rainy-month flower, has slender stems one to two feet long, and three to five-flowered racemes; the flowers of a pale purple, three inches wide. *D. Falconeri*, from Bhotan, is readily recognised by the markedly tumid joints of its slender stems; the beautiful large solitary flowers have pale rose-coloured petals and sepals tipped with dark purple, the lip having a deep purple blotch at the base bordered by a yellow ring. These are all lovely plants. In the yellow-flowered group we have *D. fimbriatum* from Nepal, with racemes of fine yellow flowers from near the apex of the naked stems; a variety of this occurs with a deep red spot at the base of the beautifully fringed lip. *D. densiflorum* has stout stems which end in a tuft of glossy leaves, setting off to great advantage the fine dense clusters of drooping golden-yellow flowers: this is one of the finest in the genus. Mr. Darwin, in his book on orchids, gives an account of the self-fertilization of *D. chrysanthum*, which belongs to this group. Amongst a host of species with drooping stems, we have *D. Pierardi*, with delicate pale lilac flowers; and the beautiful little *D. Devonianum*, named after the late Duke of Devonshire, the lovely flowers of which have a white ground colour, the sepals and petals tipped with pink, and the heart-shaped frilled lip marked with a pink blotch at the apex, and two yellow spots near the base. No collection should want the *D. Hillii*, of Australia, which is an improvement on the better known *D. speciosum*. Its stout stems bear a number of large glossy green leaves, and a profusion of dense flowered racemes, the creamy-white narrow-petaled flowers of which have a highly agreeable odour. The generic name is derived from the Greek, signifying tree and life, from the plants living on trees. [A. A. B.]

DENDROCHILUM. A genus of orchids found growing on branches or trunks of trees in the Malayan Archipelago. They have short and fleshy pseudo-bulbs, each with a single coriaceous leaf, and their small green, white, or yellowish flowers are arranged in slender, terminal, or lateral bracted spikes six to eight inches long; the bracts arranged in a two-ranked manner. The anther is two-celled, with four incumbent pollen masses; while the column has two short horns in front, and the lip is entire. About a dozen species are known; of which one is *D. glumaceum*, a very pretty Philippine Island plant, cultivated in orchid houses for the sake of its graceful drooping spikes of ivory-white flowers, the leaves resembling those of the lily of the valley; and another is the graceful little *D. filiforme*, in which the flowers are bright yellow. [A. A. B.]

DENDROID. Divided at the top into a number of branches, so as to resemble the head of a tree; only applied to small plants like mosses.

DENDROLOBIUM. A genus of small leguminous trees found in the tropical countries of the eastern hemisphere, but in greatest abundance in India. They only differ from *Desmodium* in their small jointed pods, about an inch in length, being somewhat rounded, and in the disposition of the flowers. The leaves are made up of three oblong or oval leaflets, usually downy or covered underneath with silvery hairs. The flowers, in little axillary fascicles or umbels, are white and inconspicuous. [A. A. B.]

DENDROMECON. A genus of shrubby *Papaveraceæ* found in California, and having two ovate caducous sepals, four petals, numerous stamens with filiform filaments and linear anthers, two short thick sessile stigmas, and a siliquæform one-celled two-valved pod, with a marginal placenta and numerous seeds. *Dendromecon*, literally Tree Poppy, is a most appropriate name, the plant having all the aspect and character of the poppy tribe, combined with a woody stem and branches. The species, *D. rigidum*, has lance-shaped glaucous leaves, and yellow flowers resembling those of *Meconopsis cambrica*. [T. M.]

DENDRON. In Greek compounds = a tree.

DENDROPEMON. A genus of *Loranthaceæ*, parasitic shrubs from the Antilles, with small white or purplish flowers in simple racemes, rarely paniculate or corymbose. It differs from *Loranthus* in having the alternate anthers abortive, the style filiform, and the flowers conspicuously bracteated. [J. T. S.]

DENDROPHTHOË. A genus of *Loranthaceæ*, natives of Australia, Asia and the Cape of Good Hope, distinguished from *Loranthus* and its near allies by having the petals united into a tube. They are parasitic shrubs, with long green yellowish or purple flowers, the peduncles several flowered, racemose or fasciculate. [J. T. S.]

DENDROSERIS. A few small trees peculiar to the island of Juan Fernandez make up this composite genus, which is nearly allied to the hawk-weeds, though the plants have more the appearance of gigantic sow-thistles, from which they are, however, easily recognised by their tawny pappus hairs, those of sow-thistles being silvery white. The stems seldom exceed twelve feet in height. The leaves vary much in form and size, some being entire and two or three inches long, while others are afoot or more in length, and pinnatifid. The flower-heads few and large in some, or numerous and small in others, are arranged in terminal panicles, and the numerous florets are either of a white or tawny-yellow colour. The achenes, compressed or triangular with winged angles, are crowned with a pappus of rough unequal hairs.

The most striking species of the genus is *D. macrantha*, whose lower leaves are stalked, oblong, coarsely toothed, and obtuse, while the upper ones are small and entire, and clasp the stem by their base. The flowerheads are more than an inch in diameter. Seven species are enumerated. The name *Rea* is sometimes given to these plants. [A. A. B.]

DENHAMIA. A genus of tropical Australian trees or shrubs of the spindle-tree family, chiefly distinguished by their bony capsules and numerous seeds. Their pale-green stalked leaves are oval or lance-shaped, and have entire or spiny margins. The flowers are small, green, arranged in terminal panicles, and have a five-cleft calyx, five petals, and five stamens, inserted on a minute fleshy ring. The fruits are imperfectly three or five-celled bony capsules, the seeds being enveloped in a beautiful red aril. *D. heterophylla* has some of its branches furnished with lance-shaped entire leaves, and others with oval leaves having spiny teeth like those of the holly. Five species are known. [A. A. B.]

DENNISONIA. The only species of this genus, which belongs to the vervain family, is *D. ternifolia*, a North Australian bush, with straight stems clad with glandular hairs, and a great abundance of mint-like leaves, which are sessile, oval and sharply toothed. The little rose-coloured flowers are single in the axils of the leaves and shortly stalked, the corollas being two-lipped. The genus bears the name of Sir W. T. Dennison, governor of New South Wales. [A. A. B.]

DENNSTÆDTIA. A genus of herbaceous ferns of the group *Dicksonieæ*, distinguished from *Dicksonia* itself, chiefly by having a cup-shaped instead of a two-valved indusium, this being reflexed so as to stand at a right angle to the plane of the frond. They have creeping rhizomes, and for the most part large herbaceous bipinnate or decompound fronds. *D. punctilobula*, *D. cicutaria*, *D. apiifolia*, &c., are familiar examples. The same group has been sometimes called *Sitobolium*, or by error *Sitolobium*. [T. M.]

DENS. A toothing; adj. DENTATE: having sharp teeth with concave edges. When these teeth are themselves toothed, the part is *duplicato-dentate*: not *bidentate*, which means two-toothed.

DENTARIA. A family of herbaceous perennials belonging to the *Cruciferæ*, and closely allied to *Cardamine*, from which it differs in having broad seed-stalks, and in its creeping roots being singularly toothed; hence the systematic name, and the English one of Toothwort. There are many species, which inhabit mostly the temperate regions of Europe and America, and are ornamental plants with terminal corymbs of light purple, sometimes white or yellow flowers. The roots of *D. diphylla* have a pungent mustard-like taste, and are used by the natives of the mountains of North America, from Pennsylvania to Canada, instead of mustard, under the name of Pepperwort. The genus is represented in England by *D. bulbifera*, a slender plant about eighteen inches high with pinnate leaves and a few pretty light purple flowers. In the axil of every stem-leaf is a small bulb of a purple hue, by which the plant, which rarely perfects seeds, is propagated. Though local it is very abundant in some of the woods of Hertfordshire, creeping extensively by means of its curiously toothed white roots, and forming dense patches. The root-leaves are all pinnate, those of the stem pinnatifid, the upper ones nearly simple. [C. A. J.]

DENTATO-CRENATE. The same as Crenato-dentate.

DENTATO-LACINIATE. When toothings are irregularly extended into long points.

DENTATO-SERRATE. When toothings are taper-pointed and directed forwards, like serratures.

DENT DE CHIEN. (Fr.) *Erythronium Dens Canis*.

DENT DE LION. (Fr.) *Taraxacum Dens Leonis*.

DENTELAIRE. (Fr.) *Plumbago europæa*.

DENTELLA. Little creeping annuals, natives of marshy places in India and the Indian Islands, constituting a genus of *Cinchonaceæ*. The flowers are small, white, on axillary flower-stalks, with a roundish hairy calyx-tube united to the ovary; the limb of the calyx is five-cleft; the corolla is funnel-shaped with a dilated throat, its limb five-cleft, each of the petals having on either side a small acute tooth-like process; the stamens are concealed within the corolla; and the fruit is a two-celled berry, surmounted by the lobes of the calyx. [M. T. M.]

DENTICULATE. Having very fine marginal teeth.

DENUDATE. When a surface which has once been hairy, downy, &c., becomes naked.

DEODAR. *Abies*, or *Cedrus*, *Deodara*.

DEOPERCULATE. A term used in describing mosses, when the operculum will not separate spontaneously from the spore-cases.

DEPAUPERATE. When some part is less perfectly developed than is usual in plants of the same family; thus, when the lower scales of the head of a cyperaceous plant produce no flowers, such scales are said to be depauperated, or starved.

DEPPEA. The name of a Mexican shrub of the cinchona family, the wood and bark of which are of a red colour. The flowers are yellow arranged in cymes; the limb of the calyx has four small teeth; the corolla is wheel-shaped; the filaments are very short, nevertheless, the anthers project from the corolla; the fruit is a capsule bursting by two valves. [M. T. M.]

DERMA. In Greek compounds = the bark or rind.

DERMIS. The skin of a plant.

DESCENDING. Having a direction gradually downwards.

DESERT ROD. *Eremostachys*.

DE'SESPOIR DES PEINTRES. (Fr.) *Saxifraga umbrosa*.

DESFONTAINEA. The name of a genus of Peruvian shrubs of doubtful affinity, but somewhat allied to *Solanaceæ* and *Gentianaceæ*. The leaves are thick with spiny margins like those of a holly; the flowers are axillary, stalked, five-parted; the corolla tubular, more than twice the length of the calyx, the lobes of its limb imbricated before expansion; stamens five, concealed within and attached to the corolla; anthers opening longitudinally; ovary one-celled with five parietal placentæ; style thread-like. The fruit is berry-like with numerous seeds. *D. spinosa*, with its deep green spiny leaves, and splendid scarlet flowers, is a most ornamental greenhouse plant. [M. T. M.]

DESICCATIO. In very hot countries, and in dry seasons in those which have a more temperate climate, not only is the duration of annual plants cut short, but many perennials fall a sacrifice. Trees which send their roots down deeply into the soil may stand the trial better, while those with more superficial roots suffer; but even in climates like our own, two years of annual drought like 1858 and 1859 will cause the death of many a deep-rooting tree, where the vitality was previously low. Where plants have suffered from want of water, a too liberal supply at once is apt to bring mischief; and in young trees which have been long kept out of the ground, the application of damp moss to the bark in a shady place is better than immediate planting. [M. J. B.]

DESMANTHUS. A genus of tropical and subtropical Indian and American herbs of the leguminous family. The stems seldom exceed three feet in height, and are furnished with twice-pinnate leaves composed of numerous small leaflets like those of the sensitive plant; the leaf-stalks are furnished with one or more glands, and at their base are two small setaceous stipules. The small green or white flowers are numerous, and borne in round stalked heads which arise from the axils of the leaves, and consist of a bell-shaped calyx, five petals, and five or ten stamens, though sometimes flowers are found in which there are neither stamens nor pistil. The pods are flat, smooth, membranaceous, several-seeded, and about an inch in length; when ripe they split into two portions, while in *Mimosa*, to which this genus is nearly allied, they break up into as many portions as there are seeds. The little brown polished seeds of *D. virgatus* are in Jamaica strung like beads, and used for making bracelets, work bags, &c. *D. brachylobus* is a Texan plant, sometimes known as *Darlingtonia*; but that name is now given to the Californian pitcher plant. [A. A. B.]

DESMIDIACEÆ. A natural order of green-spored *Algæ*, remarkable for their mode of reproduction, and for the eccentric and varied forms assumed by many of the species. The more typical species of the group, as the name implies, consist of a chain of connected joints, increasing by the continued addition of two new half-joints in the centre, so that the two extreme members of the chain are the oldest and the two in the centre the youngest. In the majority of instances, however, the disarticulation takes place on the formation of the first new half-joints, in such a manner that the two new individuals consist of half the old plant connected with half of the new, a mode of increase which obtains also in *Diatomaceæ*. Fructification takes place, though rarely, by the conjugation of two individuals by means of lateral tubes or simple contact, as in *Conjugatæ*, the spore affecting a variety of interesting forms, and being often strongly spinulose, the spines being occasionally complicated in structure. The new individual is produced from this by the formation of a vertical partition in the centre, and the subsequent formation of two new half-joints, so that the proper form of the species is not attained till the third generation, if so soon.

Desmidiaceæ differ from *Diatomaceæ* in their green colour, and the absence of silex. The general appearance of the plants, moreover, is totally different. They occur in pools, running streams, &c., and appear to be more frequent in Europe than elsewhere; though North America produces many species, and *Closteria* occur in the Himalayan collections. We are not aware that they are ever attached at any period of growth. In one or two instances the endochrome is spiral or not equally diffused. In general the joints are deeply constricted, but this is not always the case, and in *Closterium*, in which the plant consists of two elongated curved cones applied to each other by thin bases, there is not the slightest constriction. In this genus the joints are often as distinctly grooved or striate as in *Diatomaceæ*. Besides the increase of the species by means of cell division and spores, minute zoospores with lash-like appendages have been discovered in *Pediastrum*, a genus which belongs to a small group in which the cells remain united so as to form a little flat frond. In *Closterium* there is, moreover, an organ at the extremity of the frond consisting of a cell inclosing active molecules. This is probably the male apparatus. The armed spores are sometimes found in a fossil state enclosed in flints and other transparent minerals. Like *Diatomaceæ*, Ehrenberg has attempted to refer them to the animal kingdom, but all good authorities seem now convinced that the proper place is amongst the *Algæ*. Mr. Ralfs' beautiful work on *Desmidiaceæ* may be consulted by those who wish for fuller details. It does not appear that any individual of the order can be applied to any economical purpose. [M. J. B.]

DESMIOSPERMEÆ. One of the main divisions of rose-spored *Algæ*, in which the spores are not scattered or simple, but form distinct chains like little necklaces. These are attached to a placenta, which may either spring from the walls or their base, or may be strictly central. Far the larger portion of the more compound species belong to this section. [M. J. B.]

DESMOBRYA. A term proposed to designate that group of ferns in which the fronds are produced terminally, that is, from the apex of the caudex, and are adherent to it: see EREMOBRYA. [T. M.]

DESMOCLADUS. An Australian genus of *Restiaceæ*, a sedge-like plant, with the branches of the stem rigid, the barren ones awl-shaped, the flowering ones with a single ovate few-flowered spike. [J. T. S.]

DESMODIUM. An extensive genus of herbs, shrubs, or small trees, of the pea family, found more or less in all extra-European countries, but chiefly confined to the tropics. They are easily recognised by the form of their pods, which are flat, straight or curved, with two or many joints, each jointed portion enclosing one seed; in form, size, and thickness, they are much like the blade of a pen-knife, but the under edge is always notched, and occasionally the upper also. The leaves are commonly made up of three leaflets, but sometimes they are simple and lance-shaped or linear. The flowers are white, pink, purple, or blue, and usually disposed in terminal or leaf-opposed racemes or panicles; they have a bell-shaped four or five-toothed calyx, five narrow petals, and ten stamens, all inserted into a tube or one of them free. The most interesting, although by no means the most beautiful, plant in the genus is *D. gyrans*, the Moving plant, a native of India, and often found in cultivation in plant stoves; its leaves are made up of three oblong or lance-shaped smooth leaflets, the two lateral ones much the smallest. The flowers are violet, and arranged in terminal racemes. The singular rotatory motion of the leaflets of this plant renders it an object of great interest. In the trembling poplar, the leaf-stalk is so constructed that the least breath of wind causes the leaf to whirl; in the sensitive plant when the leaves are touched,

Desmodium gyrans.

they are perceptibly affected; but in this the motion in the leaves goes on if the air be quite still, and they are scarcely influenced by mechanical irritation. The leaflets move in nearly all conceivable ways, but do not fold on themselves; two of them may be at rest and the other revolving, or all three may be moving together. Sometimes one leaf or two on the plant only are affected, and at others the movement is nearly simultaneous in all the leaves. More commonly the lateral leaflets are seen to move up or down, either steadily or by jerks. The movements are most evident if the plant be in a close hothouse with a strong sun shining. It is said that by arresting the vital action going on in the leaflets, by giving them a coating of gum, and thus preventing transpiration and respiration, the movements are stopped, but that they recommence when the gum is removed by water. Upwards of 100 species are known, a great proportion natives of South America and India. [A. A. B.]

DESMONCUS. A genus of palms inhabiting the forests of tropical America. They have long slender flexible stems, and pinnate leaves with the leaf-stalks prolonged into whip-like tails, resembling in general appearance the calami of the Eastern Hemisphere, and like them also, they climb over and amongst the branches of trees, and support themselves by means of the hooked or recurved spines attached to all parts of their leaf-stalks. The flower spikes are simply branched, and have male flowers upon the upper, and females upon the lower part of the branches. The males have a thin three-cornered calyx, a corolla of three petals of thicker substance than the calyx, and six stamens with narrow erect anthers; and the females have a cup-shaped calyx with the rim entire or divided into three small teeth, a bell-shaped corolla with the mouth drawn in, and an ovary with one perfect and two imperfect cells, surmounted by a short style and three sharp stigmas. The fruit is small and nearly round.

D. macracanthos, the Jacitára of the Amazon and Rio Negro, grows fifty or sixty feet long, with a stem not thicker than an ordinary cane, and either climbs up trees or trails among the underwood, where it offers an annoying obstruction to persons wearing clothes, the sharp curved spines upon its leaves taking such firm hold of the garments that great care and patience are required to detach them. The Indians use strips of the stem for platting the tipitis or strainers used for squeezing out the poisonous juice of the mandioc root.   [A. S.]

DESMOPODIUM. A subgroup of *Polypodium*.

DESMOS. In Greek compounds=anything bound to another or brought into close contact with it.

DESMOSTACHYS. A genus of *Icacinaceæ*, founded on a climbing shrub from Madagascar, with alternate, ovate or lanceolate smooth leathery stalked leaves, and several slender spicate racemes growing out of each axil. The flowers are very small, bracteated, with a five-toothed calyx, five linear oblong thin petals, and five stamens.   [J. T. S.]

DESVAUXIACEÆ. (*Centrolepideæ, Bristleworts.*) A natural order of monocotyledonous plants with incomplete flowers, included in Lindley's glumal alliance. They are small tufted herbs with bristly leaves, and flowers enclosed in a spathe or sheath. Glumes one or two; pales either none or represented by one or two delicate scales; stamen one, rarely two; ovaries one to eighteen, attached to a common axis, distinct or united partially, one-celled, with a single stigma to each; ovules single orthotropal. Fruit consisting of one-seeded carpels, opening lengthwise; seed pendulous; embryo having a lens-like form. They are found in the South Sea Islands and in New Holland. There are about fifteen species described, and four genera, of which *Centrolepis* and *Aphelia* are examples.   [J. H. B.]

DETARIUM. A genus of West African *Leguminosæ*, of which two species are known. The four-lobed calyx, absence of petals, and rounded succulent fruit distinguish them from most genera; and from *Dialium*, to which they are most nearly allied, they are readily recognised by having ten stamens, five of which are longer than the others. *D. senegalense* is a tree of twenty to thirty feet high, with pinnate leaves, having oval entire leaflets, and numerous small white fragrant flowers arranged in axillary panicles shorter than the leaves. The fruits are between oval and orbicular, slightly compressed, and about the size of an apricot. Underneath the thin outer covering there is a quantity of green farinaceous edible pulp intermixed with stringy fibres that proceed from the inner and bony covering which encloses the single seed. According to M. Richard there are two varieties of this fruit, one bitter, the other sweet. The latter is sold in the markets and prized by the negroes as well as eagerly sought after by monkeys and other animals. The fruits of both are so similar that the negroes often mistake the one for the other, and do not find out their error until after having tasted them. The bitter variety they regard as a violent poison. The wood of the tree is hard and resembles mahogany in colour.   [A. A. B.]

DEUTZIA. A genus of *Philadelphaceæ*, consisting of shrubs, whose leaves are rough with star-shaped hairs. The flowers are handsome, arranged in panicles, with a bell-shaped calyx, five petals inserted beneath a disc which surrounds the top of the ovary, ten stamens inserted with the petals, the five between the petals longer than the others, the filaments flat, awl-shaped at the top or three-lobed, the middle lobe bearing the anther; ovary inferior, three to four-celled; styles three or four, thread-shaped, erect; stigmas club-shaped; capsule leathery, surmounted by the disc, bursting in the middle by three or four slits. *D. scabra* is a hardy shrub, whose clusters of white flowers give it a very ornamental character. Its leaves are used by polishers in Japan on account of their rigid star-shaped hairs; these latter, too, and especially those of *D. staminea*, are sought after by microscopists, as affording objects of great beauty. *D. gracilis* is a particularly elegant early-flowering green-house shrub.   [M. T. M.]

DEVERRA. The generic name of plants belonging to the umbelliferous order, characterised by the fruit being round or ovate and covered with scales or hairs. The species are natives of Africa, usually of small size and of a bare rigid aspect, broom-like; hence named Deverra, after the 'goddess of brooms.' In the earlier stages the plants usually have small linear leaves; at more advanced periods of growth few of these remain, hence the peculiar habit of the species.   [G. D.]

DEVILLEA. A genus of *Podostemaceæ*, comprising Brazilian herbaceous species, with hermaphrodite flowers unprotected by a bract; one stamen, whose anther opens inwardly; small globular stigmas; and smooth fruit, dividing by two unequal-sized valves. *D. flagelliformis* has in its leaves somewhat the appearance of *Ranunculus aquatilis*.   [M. T. M.]

DEVIL IN A BUSH. *Nigella*.

DEVIL'S APRON. The American name for the very broad form of *Laminaria sac-*

charina. Dr. Harvey says of the United States plant, 'numerous varieties, which perhaps demand future study, occur on the American coast. *L. Lamourouxii*, which has been sent me from Boston Harbour and from Newfoundland, looks almost like a species with its thick broadly-elliptical scarcely waved frond and its slightly-branching root.' The species, in fact, varies from one foot to six or ten feet in length, and from one to twelve inches in breadth. [M. J. B.]

DEVIL'S BIT. *Scabiosa succisa*; also *Chamælirium luteum*, sometimes called *Helonias dioica*.

DEVIL'S GUTS. A vulgar name for the species of *Cuscuta* or Dodder.

DEVIL'S LEAF. *Urtica urentissima*.

DEVIL'S MILK. *Euphorbia Peplus*.

DEVIL TREE. *Alstonia scholaris*.

DEWAZ. The Caspian name for the grape Vine.

DEWBERRY, *Rubus cæsius*, and *R. canadensis*.

DEWETA. A genus of the umbelliferous order, having five sharp tooth-like projections on the top of the fruit; the latter is oblong and oval, each half with five elevated ribs. The only species is perennial, herbaceous, a native of North America, with the leaves simply divided into pinnæ, the divisions large, ovate or heart-shaped, with numerous sharp teeth; the flowers pale-yellow. [G. D.]

DHAEK. The flowers of *Grislea tomentosa*, used in India, mixed with *Morinda*, for dyeing.

DHAK TREE. *Butea frondosa*, which yields Butea kino.

DHAL or DHOL. *Cajanus indicus*.

DHAMNOO. The timber of *Grewia elastica*.

DHAROOS. A Bengalee name for *Abelmoschus esculentus*.

DHENROOS. A Bengalee name for the fibre of *Abelmoschus esculentus*.

DROONA. The balsamic resin of *Shorea robusta*.

DHOONA-TIL. The Cinghalese name for the balsam obtained from *Dipterocarpus*.

DHOOP. *Vateria indica*.

DHOURIA. An Indian name for wormwood.

DHURRA DOURAH or DURRA. An Indian name for the grain-bearing *Sorghum vulgare*.

DI. In Greek compounds = two.

DIACHYMA. The green cellular matter of leaves.

DIACALPE. A beautiful eastern fern allied, on the one hand, to *Peranema* (the *Sphæropteris* of some authors), from which it is distinguished by having the globose involucres, which enclose the spore-cases, sessile instead of being stalked; and on the other to *Woodsia*, from which it may be known by the hard texture of the indusia, and by their irregular mode of bursting. *D. deparioides* is a herbaceous species, with finely divided decompound fronds, and is found in Java and some parts of India. Two other species have been described, one from Java, the other from Madagascar, but little is known respecting them. [T. M.]

DIADELPHOUS. Consisting of two parcels or fraternities of stamens.

DIADENIUM *micranthum* is a stemless orchid about a span high, found growing on trees in Peru. The leaves are oblong-lanceolate, seldom more than two or three in number, and the small rose-coloured flowers are arranged in a loose panicle. The anther is two-celled with two waxy pollen masses attached to the end of the caudicle which is dilated above, and furnished with two glands at the apex, whence the generic name. *Comparettia* is the most nearly related genus, but that has two instead of one caudicle to the pollen masses. [A. A. B.]

DIAGNOSIS. The short character or description by which one plant is distinguished from another.

DIALIUM. A genus of leguminous trees found in tropical India, Africa, and America, and numbering about seven species. The chief distinguishing features of the genus are found in the flowers having but two stamens (most leguminous plants have ten), and in the fruits being round or slightly compressed, and containing an edible pulp surrounding the seeds. All have unequally pinnate leaves, and terminal panicles of small white or rose-coloured flowers. These have a five-parted calyx, and are usually destitute of petals; some flowers, however, are found with a solitary petal. *D. acutifolium*, the Velvet Tamarind of Sierra Leone, is a tree of about twenty feet high with slender branches, and pinnate leaves of five to seven smooth oval entire leaflets; the flowers are pale rose-colour; and the pod, about the size and form of a filbert, is covered with a beautiful black velvet down, while the farinaceous pulp which surrounds the seeds has an agreeable acid taste, and is commonly eaten. The fruits of *D. ovoideum*, a Ceylon plant, are sold in the bazaars; they have also an agreeable acid flavour. The wood of this plant is said to be strong, durable, and suitable for ornamental furniture. *D. floribundum*, a Brazilian species, has round smooth fruits about the size of a marble, containing one or two seeds surrounded with a pulp which has a taste and smell like that of currants. The Tamarind Plum of the East Indies, *D. indum*, has a delicious pulp resembling that of the tamarind, but not quite so acid. [A. A. B.]

DIALYPETALÆ. Plants with distinct petals, in contradistinction to *Gamopetalæ*, which have the petals united into a single corolla. The term is a modern one proposed to be substituted for *Polypetalæ*, which is more generally used in the same sense, although it signifies literally plants with many petals.

DIALYPETALOUS. The same as Polypetalous.

DIALYPHYLLOUS. The same as Polysepalous.

DIAMORPHA. The name of a small crassulaceous North American herbaceous plant, with whorled branches, alternate cylindrical leaves, and small white flowers with four-parted whorls. The ovary consists of four carpels adherent at the base, but divergent at the top; the fruit is a four-celled capsule. [M. T. M.]

DIANELLA. A genus of *Liliaceæ*, containing herbs from Australia and Tropical Asia, distinguished by their fruit being berry-like, their stem leafy, the flowers perfect, the stamens inserted at the very bottom of the six-parted perianth, and the filaments incurved, thickened at the apex. They have fibrous roots, grass-like leaves, and paniculate blue flowers on drooping pedicels. The berries are blue, many-seeded. [J. T. S.]

DIANTHUS. The Pink. An extensive genus of *Caryophyllaceæ*, distinguished by having two styles, and a cylindrical calyx tube bracteated at the base. Most of the species are natives of Europe, temperate Asia, and the North of Africa. The leaves are often rigid, glaucous and grass-like; the flowers crimson or pink in more or less regular dichotomous cymes, sometimes reduced to fascicles or compact heads; in these latter the central flowers have no bracts at the base of the calyx tube, but in this case the lateral flowers, and in by far the greater number of species, all the flowers have two or more close-fitting scales or bracts, often like a small outer calyx. In Britain the following occur: *D. prolifer* and *D. Armeria*, both annuals with clustered flowers; and *D. plumarius*, or Pheasant's Eye; *D. Caryophyllus*, or Clove Pink; *D. cæsius* and *D. deltoides*, all which are perennials with separate or solitary flowers. *D. Caryophyllus* is the original of the garden Carnations. *D. barbatus*, which has fasciculate corymbose flowers and broad leaves, is often seen in cultivation under the name of Sweet William. [J. T. S.]

DIAPENSIACEÆ. (*Diapensiads.*) A natural order of dicotyledonous plants, belonging to the subclass *Corolliflora* and to Lindley's gentianal alliance. Prostrate undershrubs with crowded heath-like exstipulate leaves, and solitary terminal flowers. Calyx formed of five rather unequal sepals, surrounded by bracts; corolla gamopetalous and regular; stamens five, equal; filaments dilated and attached to the corolla; anthers two-celled, opening transversely; ovary superior, three-celled; ovules seven or numerous; style single. Fruit a membranous or papery capsule, surmounted by the persistent sepals, and terminated by the rigid style; seeds pitted, peltate; embryo very small, in fleshy albumen. The plants inhabit the northern parts of Europe and North America. There are two genera, *Diapensia* and *Pyxidanthera*, and but two or three species. [J. H. B.]

DIAPENSIA. Two beautiful little Alpine plants are the only representatives of this genus, which gives the name to its family. The best known is *D. lapponica*, originally discovered in Lapland, but since found in many parts of Northern Europe, Asia, and America, where it has been gathered as far south as the White Mountains in New Hampshire; it is also found in Japan. *D. himalaica* was found by Dr. Hooker in Sikkim growing on rocks and in moist places in the sub-Alpine valleys at an elevation of 8,000 to 10,000 feet. Both are evergreen, and grow in dense tufts scarcely rising more than an inch above the ground. The stems are clad with closely imbricated spathulate, and entire leaves, which in *D. lapponica* are nearly half an inch long, and in *D. himalaica* much smaller; in the former the stems are terminated by a peduncle about an inch long bearing a solitary white bell-shaped flower about half an inch across, surrounded by a five-leaved calyx; the border of the corolla has five rounded flat lobes, and alternating with these lobes are five stamens which have their filaments dilated upwards. The Himalayan species has much the habit of the procumbent *Azalea* of the Scotch mountains, and its purple flowers with short stalks call to mind those of the opposite-leaved saxifrage. The flower-stalks continue growing after the flower withers, and when the capsule is ripe are frequently more than two inches long. The only other genus in the family (*Pyxidanthera*) has got awned points to the anther-cells, while in *Diapensia* the anthers are awnless. [A. A. B.]

DIAPHANOUS. Transparent, or nearly so.

DIAPHYSIS. A præternatural extension of the centre of the flower, or of an inflorescence.

DIARRHENA. A genus of grasses belonging to the tribe *Festuceæ*, distinguished by the panicles of inflorescence being simple and contracted; the spikelets roundish, two to five-flowered; glumes two, unequal, acute, mucronate; stamens two or three; styles two, feathery. Only one species is described, *D. americana*, which has creeping stoloniferous roots, and erect simple stems, three to five feet high. [D. M.]

DIASCIA. A genus of *Scrophulariaceæ*, consisting of South African herbs, mostly annuals, very nearly allied to *Nemesia* and *Hemimeris*. They differ from the former in the corolla, which is flattened or concave, with two spurs or pouches at the base instead of one, and in the capsule, which is not flat; while from *Hemimeris* they are chiefly distinguished by their four stamens, all usually bearing anthers, the filaments of the lower ones curved round at the base so as to embrace the upper ones. There are about twenty species known.

DIASPASIS. A genus of *Goodeniaceæ*, containing a single species, *D. filifolia*, a native of the south-west coast of Australia. This has an adnate calyx with five short teeth, a nearly regular salver-shaped rose-coloured corolla with a five-parted limb, and free included stamens. The peduncles are axillary and single-flowered; the leaves alternate and nearly terete. [R. H.]

DIASTEMELLA. A genus of *Gesneraceæ*, containing a single species from Costa Rica. It is a slender hairy herbaceous plant, with ovate serrate and petiolate leaves, and flowers in axillary racemes. The corolla is slightly oblique and ringent, and the limb bilabiate, with the upper lip two-lobed, and the lower one trifid. The four stamens are included, and with the rudimentary fifth are inserted on the base of the corolla. The capsular fruit is membranaceous. [W. C.]

DIASTEMMA. A genus of *Gesneraceæ*, containing thirteen species natives of South America. They are perennial stoloniferous scaly herbs with opposite leaves, and small flowers in axillary corymbs. The calyx is adherent to the base of the ovary; the corolla is oblique, erect in the calyx, with a tube subcylindrical or increasing upwards, and a five-lobed spreading limb; the four stamens are included, the fifth rudimentary; the anthers are small and coherent. The ovary is surrounded by five elongate glands, and surmounted by a bilamellate stigma. [W. C.]

DIATOMACEÆ. A very distinct natural order of green-spored *Algæ*, remarkable for the enormous quantity of silex contained in their frond, and for their yellow-brown colour. The mode of increase so closely resembles that of *Desmidiaceæ*, that in this respect, we refer for information to that article. Their claims to a place amongst animals was even more strongly contested than in that order, but Mr. Ralfs' discovery of the formation of spores by conjugation in several genera has effectually put an end to controversy. The species are often attached by a slender peduncle when young, and in some genera this is repeatedly dichotomous. The joints often remain connected for a long time, separating in some instances alternately above and below so as to form a curious chain. When connected they form various shaped fronds, as linear, flabelliform, circular, &c.; but in a multitude of instances disarticulation takes place with the formation of each new individual. The separate joints which have received the name of frustules exhibit frequently a totally different outline when seen dorsally and laterally, and they are almost always adorned with delicate streaks and other markings. In *Coscinodiscus* they form a disk with circular apertures like a colander. In many cases the frustules have distinct external apertures in the siliceous coat, without which it is not easy to see how there could have been a proper communication with the surrounding medium from which they must derive their nourishment. In consequence of the large proportion of silex which they contain, the frustules are capable of retaining their form after all vegetable constituents have fled, and thus they are admirably adapted for preservation in a fossil state. Vast beds accordingly occur, many feet in thickness, consisting entirely of effete frustules, known under the name of Tripoli, and affording an admirable material for polishing, for which they are used extensively. 'The phonolite stones of the Rhine,' says Dr. Hooker, 'and the Tripoli stones, contain species identical with what are now contributing to form a sedimentary deposit, and perhaps at some future period a bed of rock extending in one continuous stratum for 400 measured miles. I allude to the shores of the Victoria barrier, along whose coasts the soundings examined were invariably charged with diatomaceous remains constituting a bank which stretches 200 miles north from the base of the Victoria barrier, while the average depth of water above it is 300 fathoms or 1800 feet. Vast quantities again occur in bed under the guise of a white powder, which is called mountain meal, and is actually mixed with flour in some parts of Sweden, though it is perfectly inert, and can serve merely to increase the bulk of the food, a circumstance of some importance where it is scarce.[*] The walls of the frustules are so thin, and the little cells of silex so light, that they are often wafted to great distances by the trade and other winds, so that species of remote regions may occasionally occur in a dead state in countries where they could not maintain their existence. *Diatomaceæ* form a large portion of the food of some of the lower mollusks, which in turn are preyed on by sea birds; and as the shells are capable of resisting digestion, they are found, frequently in great quantities, in the beds of manure which are collected for agricultural purposes under the name of guano. Many unique species have been obtained by travellers from the stomachs of fish, which sometimes afford an abundant harvest for the microscope. *Diatomaceæ* occur in all parts of the world, and abound amongst the ice and in the deep sea of polar regions. They probably are the plants above all others capable of enduring extreme degrees of cold without annihilation; while, on the contrary, several occur in springs of high temperature. The striæ on the walls

[*] Experiments in cattle-feeding show that the relative quantity of nutritious matter in food, independent of the bulk, is not the only point worthy of observation. The stomach must be properly filled, or, as it is termed in French, lesté, or the due effect of the nutriment will not be obtained.

are often so regular that the frustules form admirable tests for ascertaining the comparative merit of microscopes.

Though *Diatomaceæ* are for the most part free or only attached for a time, there are a few genera in which an enormous quantity of mucus is thrown out by the frustules, which accordingly, as in *Schizonema*, *Dickiea*, &c., form variously shaped filiform or alveoid fronds. In *Cymbelleæ*, a suborder, the quantity of silex is comparatively so small that the plants are more easily destructible than in the other sections. The peculiar motions in the genus *Bacillaria* have been noticed above. In many other genera motion has been observed, but it is now well known that even active motion is not incompatible with the nature of vegetables. For full information we refer to Mr. Smith's beautiful work on *Diatomaceæ*. [M. J. B.]

DIBLEMMA. The name of a Philippine Island fern, in which the sori are of two kinds: the first linear continuous, seated on a submarginal receptacle; the second roundish or oblong, and irregularly scattered. *D. samarensis* has simple fronds and uniformly reticulated venation, short free veinlets being included in the unequal areolæ. [T. M.]

DICALYX. The name given by Loureiro to a few Asiatic bushes which were described as belonging to the tea family. Modern authors have shown, however, that they are genuine species of SYMPLOCOS: which see. [A. A. B.]

DICELLA. A genus of Brazilian climbing shrubs belonging to the *Malpighiaceæ*. The calyx has five segments each provided with two glands at its base; the petals are stalked, unequal in size, and downy on the outside; the stamens are ten, united below into a tube, the anthers hairy; the ovary is two-celled, surmounted by two hook-like styles. Drupe woody, one-celled, one-seeded. [M. T. M.]

DICENTRA. A genus of *Fumariaceæ*, the *Dielytra* or by mistake *Diclytra* of some authors. They are known by the two outer petals being spurred or bulging at the base, the seeds crested, and the capsule with two dry valves. The species are natives of the Northern Hemisphere, and are generally stemless herbs with ternately compound leaves, and succulent stems terminating in a raceme of large nodding flowers, which are white, rose-coloured, or purplish. The section *Eucapnos* has the outer petals merely bulging at the base, and the racemes compressed; while *Cucullaria* has the outer petals produced backwards into two long spurs at the base, and its racemes are simple. The two most common American species, which belong to the second group, have white flowers. *D. Cucullaria* is known in the United States as Dutchman's Breeches, from the shape of the spurred flower, and *D. canadensis*, which is fragrant, as Squirrel Corn. A stemless species from Virginia and North Carolina, with rose-coloured flowers, *D.*

formosa, is often cultivated in gardens; but the best known and most beautiful is *D. spectabilis*, from Northern China, which has a leafy stem, and flowers nearly an inch long, of a beautiful rose colour, with the narrow constricted inner petals white; the leaves are like those of the Moutan peony in miniature. [J. T. S.]

DICERANDRA. The name of a genus belonging to the labiate order, chiefly distinguished from its congeners by the presence of two straight and pointed appendages on the upper part of each stamen, hence the name, derived from Greek words which together signify 'two-horned stamens.' *D. carolinensis* is a small shrub, a native of the United States, having erect stems and narrow entire leaves. (G. D.)

DICEROS. A name successively given by different authors to species of *Artanema*, *Limnophila*, and *Vandellia*.

DICHÆA. A genus of orchids found growing on tree stems in the West Indies and the adjoining mainland. They are small tufted plants having short erect or creeping stems, thickly clad with small ovate-oblong or linear leaves arranged in a two-ranked manner, and solitary inconspicuous axillary greenish flowers. About a dozen species are known. [A. A. B.]

DICHÆTA. A genus of small annual Californian composite herbs, of which two species are known. They seldom exceed six inches in height, and are found on the margins of pools or in wet places. The stems and leaves are covered when young with loose white wool. The lower leaves are generally pinnatifid with linear segments, and the upper entire; and the yellow flower-heads are single on the ends of the stems. The genus is nearly allied to *Burrielia*, but differs in the pappus being composed of from four to eight oblong-obtuse scales, with generally two which are awl-shaped and awned. [A.A.B.]

DICHASIUM. A name once given to an Indian fern which proves to be the same as the English *Lastrea Filix-mas paleacea*.

DICHERANTHUS. A genus of *Illecebraceæ* allied to *Pteranthus*. Small shrubs from the Canary Islands, with opposite or verticillate fleshy linear-cylindrical leaves, dilated and clasping at the base; and flowers in small dense compound corymbose cymes at the apex of the branches; calyx segments mucronate, hooked when in fruit; corolla none. [J. T. S.]

DICHILUS. A genus of slender erect or prostrate South African leguminous herbs, nearly related to *Argyrolobium*, but differing in the keeled petal being rather longer than the vexillum, and in the pods being swollen at intervals (torulose), not flat, and clad with silky hairs. The stalked leaves are made up of three narrow leaflets. The little yellow flowers are either solitary or racemed in the axils of the leaves, their calyx distinctly two-lipped, and the pod is smooth, narrow, an inch or more in length. Three species are known. (A. A. B.)

DICHLAMYDEOUS. Having both calyx and corolla.

DICHOGAMOUS. When the florets of an inflorescence are of two separate sexes.

DICHONDRA. A genus of *Convolvulaceæ* containing two species, one a native of the tropical and sub-tropical regions both of the Old and the New Worlds, the other found in tropical America. They are prostrate herbs with small flowers. The calyx five-parted; the corolla campanulate and deeply five-lobed; the ovary consisting of two distinct carpels with one ovule in each of the two cells. The two styles are distinct from the base, with thickened stigmas. [W. C.]

DICHORISANDRA. A genus of *Commelynaceæ* with the habit of *Tradescantia*, but with the filaments neither hairy nor dilated at the apex. They are Brazilian herbs, with lanceolate acuminate leaves, and racemose flowers, either terminal or produced from the base of the stem. (J. T. S.)

DICHOSEMA. There is a group of small leguminous West Australian bushes in which the stamens are ten in number and quite free, and the pods have their margins rolled inwards, so that they are imperfectly or altogether two-celled, and a cross section of them would be somewhat like the figure 8. To that group *Dichosema* belongs. It differs from the others in having a very broad vexillum which is bilobed at the apex, and much longer than the wings, these in their turn being a little longer than the keel. There are about half a dozen species, all of them little spiny bushes seldom more than two feet high. The slender stems are clad with minute linear or oblong leaves generally arranged in parcels of three, and accompanied by slender spines which often exceed them in length. The flowers are small, yellow, or purple, solitary in the leaf-axils or arranged in short racemes. [A. A. B.]

DICHOTOMIA (adj. **DICHOTOMOUS**). Having the divisions always in pairs; a term equally applied to branches, or veins, or forks.

DICHROCEPHALA. A genus of Asiatic, African, and Australian *Compositæ*, which differs from its near allies chiefly in the convex receptacle of the flower-heads. They are branching herbs, with oval coarsely toothed or lyrate sometimes pinnatifid leaves; the branches being terminated by panicles of nearly globular flower-heads, about the size of a small pea. The achenes are compressed, those of the outer florets without pappus, and those of the inner series with a pappus of one or two short hairs. Of the five species known, all are common weeds in the countries where they grow, and of no beauty. (A. A. B.)

DICHYNCHOSIA. A genus of *Cunoniaceæ* from Celebes. A tree with opposite pinnate leaves, the few leaflets of which are oblong ovate, coarsely serrated, with the under surface (as well as the branchlets and inflorescence) covered with stellate down. The stipules are large and kidney-shaped. The flowers grow in axillary panicles, which are much branched in a corymbose manner; the calyx five or six-parted, persistent; stigmas two, diverging; capsule two-beaked; seeds numerous, with a membranous wing. (J. T. S.)

DICKIEA. A curious genus of *Diatomaceæ*, in which the frond assumes an ovoid form, as it does a filiform in *Monema* and *Schizonema* and a globose in *Berkeleia*. When the gelatinous element in these genera is removed, the frustules are found to be of precisely the same nature as those in genera where the gelatinous element is extremely reduced, or where it only tends to keep a quantity of frustules together in an irregular stratum. [M. J. B.]

DICKSONIA. A genus of noble mostly arborescent ferns of the polypodiaceous group, and typical of the section *Dicksonieæ*. Their stems are often thick and trunk-like, but sometimes decumbent and criniferous. The fronds are large, generally decompound, and leathery, forming a noble tuft or crown; and the sori are globose or shortly oblong, transverse, and marginal, with a coriaceous indusium of two valves, of which the outer, formed of a lobule of the frond, is cucullate, and the inner usually smaller and less convex; the veins are free. *D. antarctica* is a very beautiful tree fern often seen in green-houses, having been freely imported from our Australasian colonies. Others occur in St. Helena, Brazil, Juan Fernandez, Columbia and Java. One pinnate species, *D. abrupta*, which is only found in Bourbon, has quite the aspect of a *Nephrolepis*. The sori are always more or less recurved from the plane of the frond. [T. M.]

DICLESIUM. A one-seeded indehiscent fruit enclosed within a hardened perianth, as in the marvel of Peru.

DICLIDANTHERA. A genus of dicotyledons, founded on two Brazilian shrubs which are in many respects allied to *Sapotaceæ*. Differing, however, as it does in a slight irregularity in the flowers, in the curious two-valved anthers, and in some measure in the structure of the ovary, the genus has been removed from that family, and eminent botanists have severally proposed associating it with *Polygalaceæ*, *Hamamelidaceæ*, or even *Byttneriaceæ*.

DICLIDIUM. A genus of plants belonging to the *Cyperaceæ*. Only one species is described, namely, *D. ferox*, a native of South America. [B. M.]

DICLIDOCARPUS. The name given to a genus of *Tiliaceæ*, remarkable for the form of its fruit, which is a somewhat woody compressed two-celled capsule, with numerous seeds. It is about an inch long, rather more in breadth, and nearly inversely heart-shaped in form. When ripe it splits into two portions, and has the appearance of some bivalve shell. The only known species, *D. Richii*, was found by Mr. Rich in the Feejee Islands, where it grows to a tree of forty feet high, with oval entire nearly smooth leaves, having two lateral ribs at the base parallel to the central one. The fertile flowers are unknown. The sterile ones are small and crowded in axillary cymes, which, as well as the flowers, are clothed with minute white down. [A. A. B.]

DICLIDOPTERIS. A genus of polypodiaceous ferns belonging to the *Pleurogrammeæ*, having linear continuous sori, sunk in a deep oblique furrow on each side and near to the costa, towards which the opening is directed. The veins are reduced to the costa, and the intermarginal receptacle parallel with it. The only species, *D. angustissima*, found in the Pacific Islands, is a very small plant, with narrow simple fronds. The genus is related closely to *Monogramma* and *Pleurogramma*. [T. M.]

DICLIDOSTIGMA. A cucurbitaceous plant of Cuba, with the aspect of *Bryonia*. Both calyx and corolla are five-cleft, the segments of the latter, in the male as well as in the female flowers, being rough and glandular: there are five stamens in three parcels with separate wavy anthers; in the female flowers there is a five-lobed glandular disk surrounding the base of the style, which latter is terminated by three stigmas, each of which is divided into two plates. The fruit contains six to nine seeds. [M. T. M.]

DICLINOUS. Having the stamens in one flower and the pistil in another.

DICLIPTERA. A considerable genus of *Acanthaceæ*, containing nearly seventy species, dispersed over the tropical and subtropical regions of the New and Old Worlds. They are herbs with entire leaves, and with flowers in axillary clusters and short cymes, usually surrounded by four bracts, of which the outer two are the larger. The calyx consists of five sepals; the corolla is two-lipped, and the tube is so twisted that the upper entire or two-toothed lip becomes the lower; there are two stamens whose anthers have each two similar cells, but with the one inserted much below the other. [W. C.]

DICLIS. A genus of *Scrophulariaceæ*, consisting of slender herbaceous creepers resembling in habit *Linaria Cymbalaria*, and with very similar corollas, but the anthers have only one cell, and the capsule is nearly globular, opening loculicidally in two valves. There are three species known, all from south-eastern Africa or Madagascar.

DICLISODON. A name proposed for a curious genus of ferns, in which the sori occupy small projecting marginal teeth, and have scale-like covers. Hence it has been regarded as having a two-valved indusium, and as associating with the *Dicksonieæ*, the outer valve being described as a small rounded herbaceous projecting lobe of the frond; and the inner a proper indusium, larger than the lobe, membranaceous, and distinctly reniform, affixed by the sinus. The sori, though not stalked, project from the margin so as to resemble those of *Deparia*, but instead of a marginal cup, as in that genus, the involucre consists of the two valves lying flat in the plane of the frond; the veins are free. Some writers, however, regard the plant as a *Lastrea* with ex-serted sori. *D. deparioides* is a very beautiful bipinnate fern, found in Ceylon. [T. M.]

DICLYTRA. An erroneous mode of spelling sometimes adopted for *Dielytra*, a synonyme of *Dicentra*.

DICOCCOUS. Splitting into two cocci.

DICOLORATIO. As petals are mere modifications of leaves, we need not be surprised if leaves, though not in a state of transmutation to petals, occasionally exhibit vivid colours, especially in variegated plants. It does not appear, however, that coloured varieties grafted on those which are not coloured, communicate their colour in the same way in which variegated grafts affect the stock. The change of colour in leaves as autumn advances, appears rather to be a chemical than a vital action, and is owing to some change in the chlorophyll on which the healthy green tint of the leaves depends. The contents of the cells, like the cell walls themselves, have performed their office, and are therefore, like other inert bodies, subject to chemical changes, which would not affect them while their vital powers were active. [M. J. B.]

DICORYNEA. A genus of large trees of Brazil and Guiana belonging to the leguminous family. Some attain a height of sixty, and a diameter of three to four feet. All have pinnate leaves a foot or more in length, made up of five or seven smooth leaflets. The branches are terminated by very large panicles of numerous white flowers, which are interspersed with fawn-coloured bracts. Each flower is about half an inch long, and composed of a calyx of three divisions; five unequal petals, the two exterior like the calyx leaves, the upper broadly orbicular at the point and narrowed below into a claw, and the two internal obliquely orbicular and shorter; two stamens with broad and thick filaments of unequal length; and an ovary crowned by a curved style. The pods are obliquely oval, thin, about one and a half inch long, and contain one or two seeds. Five species are known. [A. A. B.]

DICORYPHE. A genus belonging to the order of witch-hazels. The name indicates one of its obvious characters, viz., the presence of two horn-like appendages on the upper part of the fruit. *D. stipulata* is a native of Madagascar, having slender branches with oblong, entire, and shortly-stalked alternate leaves, and below each, a pair of unequal heart-shaped appendages, the stipules. [G. D.]

DICOTYLEDONOUS. Having two cotyledons.

DICOTYLEDONS, DICOTYLEDONEÆ. Plants having two seed-leaves or seed-lobes, which are called cotyledons. This is one of the primary divisions or classes of the vegetable kingdom, including about 7,000 known genera, and about 70,000 known species of flowering plants. The class also receives the name of *Exogens* or *Exogenæ*, from the structure of the stems. The plants in this great class have spiral vessels; their stems are formed by additions externally in the form of zones or rings; stomata or pores exist in the leaves, which have a reticulated or netted venation. The plants have stamens and pistils, either in the same or in different flowers. The symmetry of the flowers is represented by five or two, or multiples of these numbers. The ovules are contained in an ovary, or more rarely are naked; and the embryo has two, sometimes more, cotyledons.

In De Candolle's system this class of Dicotyledons is divided into four sub-classes:—1. *Thalamifloræ*, petals distinct; stamens hypogynous; 2. *Calyciflora*, petals distinct or united; stamens perigynous or epigynous; 3. *Corolliflora*, petals united; stamens usually attached to the corolla, which is hypogynous; 4. *Monochlamydeæ*, including *Gymnospermæ*, a calyx only, or no floral covering. Lindley divides the class into four subclasses: 1. *Diclinous*, those plants which have separate staminate and pistillate flowers. Those which have stamens and pistil in every flower are divided into—2. *Hypogynous*, stamens not adhering either to calyx or corolla; 3. *Perigynous*, stamens adhering to either calyx or corolla; and 4. *Epigynous*, stamens, calyx, and corolla, all adhering to the side of the ovary. *Gymnogens*, or plants with naked seeds, represent a separate class according to Lindley. The age of Dicotyledonous trees can be computed by counting the number of annual concentric rings of wood. [J. H. B.]

DICILEA. Herbaceous plants, natives of Madagascar, &c., constituting a genus of *Podostemaceæ*, characterised by hermaphrodite flowers unprotected by a bract; monadelphous stamens; and ribbed fruit opening by two equal valves. [M. T. M.]

DICRANODIUM. *Gymnogramma leptophylla*.

DICRANOGLOSSUM. A genus of polypodiaceous ferns of the group *Tæniteæ*, in which the sori are naked, linear, continuous, and submarginal as in *Tæniopsis*; but the veins, instead of being straight and free, or combined by the transverse receptacle, describe a series of simple elongated arcs, each one uniting with the next, and thus forming a continuous irregular curved sub-marginal receptacle to which the spore-cases are affixed. *D. sub-pinnatifidum*, a South American and West Indian plant, with furcately-lobed fronds, is the only species. [T. M.]

DICRANOLEPIS. A genus of thymelaceous plants, the flowers of which have a salver-shaped perianth with a five-parted limb, and ten scales inserted in its throat; stamens ten, attached to the perianth; ovary stalked, with a cup-like disk at the base, one-celled, containing a single pendulous ovule. There is only one species, *D. disticha*, which grows at Sierra Leone; a shrubby plant with distichous leaves, and solitary axillary flowers. [J. H. B.]

DICRANOPTERIS. A synonyme of *Gleichenia*; also applied by some writers to a section of *Polypodium*.

DICRANOSTIGMA. A genus of *Papaveraceæ*, represented by a plant indigenous in the Himalayan mountains. It has numerous radical pinnately-lobed leaves covered with short hairs; the stems are about a foot in height, and bear at the top two or three golden-coloured flowers, with a flask-shaped ovary, surmounted by thickened stigmas with two erect awl-shaped arms alternating with the placentas. Its nearest ally is *Chelidonium*, from which the form of the ovary and stigmas abundantly distinguish it. [M. T. M.]

DICRANUM. A large and important genus of acrocarpous mosses, distinguished by the unequal cernuous capsule, the hood-like calyptra, rostrate lid, and single peristome consisting of sixteen equidistant teeth which are confluent at the base, and split half way down or more into two unequal portions, the medial line being continued to the base, and occasionally perforated. *Leucobryum* is distinguished by the peculiar structure of the leaves, and their consequent pallid hue. The species, from the different habits which they assume, are distributed into several distinct sections. They grow variously on rocks, or on the ground, or more rarely on the trunks of trees. Some of them, as *D. scoparium*, are amongst the larger mosses, and remarkable for their long and often curved leaves, while others are minute. It is observed by Wilson, in his *Bryologia*, that in several of the larger species, which have the stem covered with a dense layer of radical fibres, the male plants appear to be replaced by minute bulbs, nestling among the fibres; and this is all that is known of the male inflorescence of certain species; but in *D. scoparium* the inflorescence may sometimes be traced from these radicular gemmæ up to the perfect development of male plants. A somewhat analogous process is observable also in a few species of *Hypnum*. [M. J. B.]

DICTAME BLANC. (Fr.) *Dictamnus albus*. — **DE CRÈTE.** *Origanum Dictamnus*.

DICTAMNUS. A small genus of *Rutaceæ*, found in southern Europe, Asia Minor, &c. *D. Fraxinella* and *D. albus* are both cultivated in gardens for their fragrant leaves, as well as for the handsome appearance of their flowers. They are perennial plants with unequally-pinnate leaves, the main stalk between the four or five pairs of leaflets being winged and leaf-like. The inflorescence, as well as the outer parts of the flowers themselves, is covered with glands secreting a resinous or oily matter, so volatile, that the air surrounding it becomes inflammable in hot weather. The calyx has five sepals, the two lowermost of which are longer than the rest; the five petals, which are stalked and inserted into the stalk bearing the ovary, are of unequal size, the four upper ones erect, and the lowest one bent downwards; the stamens are ten, bent downwards; the five ovaries are placed on a short stalk, each one-celled. The fruit consists of a capsule, the constituent carpels of which are confluent below but separate above, and when mature burst each into two pieces: they contain two or three seeds. [M. T. M.]

DICTYANTHUS. A genus of *Asclepiadaceæ*, containing twenty species, natives of Central America. They are twining undershrubs, with cordate membranaceous leaves on long petioles, and one or two-flowered peduncles. The corolla is campanulate, spreading, and five-cleft, and the staminal crown consists of five small lobes adnate to the tube; the stigma is fleshy with five prominent angles, and very small glandular corpuscles. [W. C.]

DICTYDIUM. A beautiful genus of *Fungi* allied to *Cribraria*, but distinguished by the outer coat of the peridium disappearing to the very base, and leaving behind a beautiful net-work. In *D. umbilicatum*, which is not uncommon on decayed fir stumps, the peridium is deeply, umbilicate, and looks like an elegant balloon. [M. J. B.]

DICTYMIA. *Dictyopteris.*

DICTYOCALYX. Creeping pubescent herbs, allied to *Nicotiana*, but constituting a distinct genus of *Solanaceæ* or *Atropaceæ*, characterised by the presence of a cylindrical five-lobed calyx, the tube of which becomes distended after the expansion of the corolla, and is marked by a network of prominent veins. The corolla is membranous and funnel-shaped. [M. T. M.]

DICTYOCLINE. A genus of interesting hemionitoid ferns, which grow in India and China. *D. Griffithii*, found in Assam and Khasya, is a coarse herbaceous pinnate fern, with three or four pairs of pinnæ, and having the sori reticulated between the primary pinnate veins, which transversely anastomose so as to form two or three series of roundish hexagonal areoles between them. The aspect of the plant approaches that of some of the larger species of *Aspidium*. [T. M.]

DICTYOGENS. (*Dictyogenæ*.) A sub-class of monocotyledons or Endogens according to Lindley. The plants are characterised by having net-veined in place of parallel-veined leaves, which usually disarticulate with the stem. The woody matter on the rhizomes of the plant is often disposed in a circular wedge-like manner. The name is derived from the Greek word *dictyon*, a net. This subclass includes *Dioscoreaceæ* or yams, *Smilaceæ* or sarsaparillas, and *Trilliaceæ*, *Roxburghiaceæ* and *Philesiaceæ*. Some *Araceæ* and *Liliaceæ* have, however, net-veined leaves. [J. H. B.]

DICTYOGLOSSUM. A genus of acrostichoid ferns, now called *Hymenodium*.

DICTYOGRAMMA. A genus of polypodineous ferns, found in Japan and the Feejees, and belonging to the group *Hemionitideæ*, with naked linear reticulated sori, among which *Dictyogramma* is distinguished by having the primary veins arcuate so as to form costal areoles, and the venules reticulated, except those of the margin, which are free. The sori are narrow, linear, and sub-parallel, the lines sparingly united towards either end. The fronds are pinnate and somewhat leathery, with a few large pinnæ. *D. japonica*, the typical species, is, as its name implies, found in Japan. *D. elongata*, from the Feejees, is the same fern which has been called *Syngramma pinnata*. The name has been used in place of *Selliguea*. [T. M.]

DICTYOLOMA. A genus of Brazilian trees belonging to the *Simarubaceæ*. The flowers are unisexual; calyx minute five-parted; petals five, sharply-pointed, or prolonged into a linear appendage; stamens five, attached below to a two-cleft scale. In the female flower there are five ovaries, five styles, and a five-toothed stigma. [M. T. M.]

DICTYOPTERIS. A genus of ferns belonging to the reticulated division of the *Polypodieæ*, and comprising a few species found in the East and in Australia. They have either simple or bipinnate fronds, sometimes of large size; and dot-like naked sori, which are seated at the confluence of several veinlets (compital). The areoles of the reticulated veins are without free included veinlets, which, together with their uniformly reticulated, not connivently-anastomosing venation, separates them from all other genera of ferns with netted veins, and naked dot-like sori. [T. M.]

DICTYOSTEGIA. A genus of *Burmanniaceæ*, consisting of a very few species from tropical America, all small slender leafless annuals, with very small flowers in a terminal cyme or head. They grow on rotten leaves in damp shady woods, and differ from *Burmannia* chiefly in their capsules opening by lateral pores.

DICTYOTA. A small genus of dark seeded *Algæ*, with thin flat ulva-like forked fronds, producing spores in little superficial disks. The species are of an olive-green, and are widely diffused in either hemisphere. *D. dichotoma* is one of the commonest *Algæ* on our coast, and assumes a great variety of forms as regards the length, breadth, and division of its fronds.

The development of the frond is curious, each division ending in a 'single cell by the constant division of which at its lower side the other cells of the frond are formed, the terminal cell being then continually pushed onwards.' This is the same mode of growth as that which obtains in exogenous stems. [M. J. B.]

DICTYOTE.Æ. An order of dark-seeded *Algæ* with superficial spores or cysts, disposed in definite spots or lines. The fronds are sometimes flat, sometimes thread-like, and occasionally branched and tubular. In *Hydroclathrus* it is pierced with large holes. Some beautiful *Algæ*, as *Padina, Zonaria, Haliseris*, belong to this order, which has representatives in every part of the world, but very few are found in high latitudes. *Padina pavonia*, the turkey feather layer, is common in warm countries, but extends to our own coasts as far as lat. 51°, though in North America it does not pass farther than lat. 25°. In *Cutleria* there is reason to believe that true spermatozoids are produced ; but in some other genera, as *Stelophora*, two kinds of fruit occur, the one of which produces large, the other small zoospores, both of which have lash-like appendages. The cysts, which produce the large zoospores, are called *Trichosporangia*; those which produce the smaller, *Oosporangia*. [M. J. B.]

DICTYOXIPHIUM. A genus of polypodiaceous ferns related to *Lindsæa*, from which it is distinguished in the first place by its compoundly-reticulated veins having free included veinlets in their areoles; and in the second, by its indusium exceeding and being inserted over the margin of the frond. The fronds are simple, narrower in the fertile parts, and the sori are linear continuous and marginal, with the indusium opening outwardly. There are only a couple of species, which are found in Panama and New Grenada. [T. M.]

DICYPELLIUM. The name of a Brazilian tree of the laurel family. The flowers are dioecious. The male flowers are not described, but the female ones have a six-parted perianth ; twelve barren stamens in four rows, the outermost petal-like, the innermost small and scale-like, the intermediate ones glandular. The fruit consists of a one-seeded berry, surrounded by the thickened fleshy perianth, which, with the sterile stamens, is persistent. The bark of *D. caryophyllatum* furnishes Clove Cassia. [M. T. M.]

DICYRTA. A genus of *Gesneraceæ*, containing a single species, a native of Guatemala. It is a perennial stoloniferous herb, with opposite leaves on long petioles, and solitary axillary flowers, the small corollas of which have a slightly-curved tube and an equally five-lobed limb. There are four didynamous stamens, with the rudiment of a fifth, inserted at the base of the tube. The disk is fleshy five-sided ; the stigma capitate, depressed. [W. C.]

DIDERMA. A genus of myxogastrous Fungi, characterised by a double peridium, of which the outer is quite smooth and crustaceous ; the inner delicate and attached to the straggling hairs amongst which the spores are seated. In some species the peridium bursts by regular radiating fissures, so as to look like a little flower, while in others it is ruptured irregularly. One of the most common species, *D. vernicosum*, is characterised by its obovate shining chestnut-coloured outer peridium. It is common in woods, on mosses, twigs, &c., and is often very conspicuous. The flower-like species are by no means common. The genus is found more or less frequently in all temperate regions. [M. J. B.]

DIDICLIS. *Selaginella.*

DIDISCUS. A genus of umbellifers, characterised by the fruit being very much flattened laterally, each half with five ridges, the middle ridge most prominent. The name of this genus is intended to indicate the double disk-like fruit. The species are herbaceous and natives of Australia. *D. cæruleus* is a showy plant, covered with hairs ; its leaves three-parted, each division again subdivided ; its flowers blue. The fruit when mature is covered with small tubercles. Another species, *D. albiflorus*, has no hairs, and the flowers are white. [G. D.]

DIDISMUS. A genus of *Cruciferæ*, with pods breaking across into joints which have one or two seeds in each, the uppermost joint ending in a striated beak, the lower one truncate at the apex. Flowers white or yellow. The species occur in Greece, Syria, and N. Africa. [J. T. S.]

DIDYMIUM. A genus of myxogastrous Fungi, distinguished by the outer coat of the peridium being scurfy, mealy, scaly, tomentose, &c., and bursting irregularly. The species are numerous and sometimes beautiful. One of the most common is *D. cinereum*, which occurs everywhere, and is easily known by its stemless cinereous peridium, and the snow-white flattish hairs amongst which the dark spores are dispersed. The genus belongs essentially to temperate climates. [M. J. B.]

DIDYMOCARPUS. A genus of *Cyrtandraceæ*, containing fully thirty species, natives of India. They are caulescent or stemless herbs or undershrubs, with the leaves serrate or crenate petiolate, those on the stem being opposite or rarely alternate ; the flowers blue or white, in cymes ; the calyx five-cleft ; and the corolla funnel-shaped and unequally five-lobed. There are four stamens, two of which only are generally fertile ; the long capsule bursts longitudinally, and contains many naked sessile pendulous seeds. [W. C.]

DIDYMOCHITON. A genus of *Meliaceæ*, consisting of trees or shrubs, natives of the Moluccas. They have soft compound leaves, and flowers in axillary spikes or heads. The corolla has five linear petals, attached below to the tube of the stamens, which is divided at the top into ten lobes, and contains within it ten anthers. The ovary is sessile, five-celled, surrounded by a gourd-shaped five-lobed disk; and the fruit is berry-like. [M. T. M.]

DIDYMOCHLÆNA. A genus of polypodiaceous ferns, having indusiate sori of an oblong form attached longitudinally along its centre to a crest-like elevation of the receptacle, and free all round the margin; besides which the veins are free. *D. lunulata* is a fine South American arborescent fern with bipinnate fronds, the articulated pinnules of which are dark green, coriaceous, and shining. *D. dimidiata*, a South African plant, differs in having ecostate pinnules. [T. M.]

DIDYMOGLOSSUM. A division of the genus *Trichomanes*, in which the funnel-shaped involucres are two-lipped instead of truncate at the mouth, which is, in fact, an approach towards the two-valved involucre of *Hymenophyllum*. The group is considered by some writers to form a distinct genus. [T. M.]

DIDYMOTHECA. A genus of *Phytolaccaceæ*, from Tasmania. A smooth-branched undershrub with scattered linear semicylindrical leaves and axillary divisions; flowers on short stalks; the perianth four-lobed, two of the lobes larger than the others. [J. T. S.]

DIDYMOUS. Double; growing in pairs, as the fruit of umbellifers.

DIDYNAMOUS. Having two stamens longer than the two others.

DIEFFENBACHIA. A genus of arads, consisting of about fifteen species, all inhabitants of tropical South America and the West Indian Islands, where they flourish in moist shady places in the woods. Their stems are fleshy, and vary from two to six or eight feet long, partly lying upon the ground and partly erect, the erect portion bearing the greatest number of leaves. The leaves have fleshy foot-stalks, the lower part of which expands and forms a sheath round the stem; they are generally of an oblong form; in most species they are green, but some are marked or variegated with white or yellowish irregularly shaped spots, and all have numerous veins diverging from the midrib, and running parallel with each other until near the margin, where they curve upwards and unite. The spadix or flower spike is enclosed in a green or yellowish spathe, which does not wither like that of some allied genera, but remains fresh until the fruit is ripe; the lower part of the spike bears female flowers only, each consisting merely of an ovary surmounted by a stalkless stigma, and surrounded by from two to four rudimentary or imperfect stamens; the upper part is free and thickly covered with male flowers only.

D. seguina has acquired the name of Dumb Cane in the West Indies, in consequence of its fleshy cane-like stems rendering speechless any person who may happen to bite them, the juice of the plant being so excessively acrid as to cause the mouth to swell, and thus to prevent articulation for several days. It is said that the West Indian planters were formerly in the habit of punishing their refractory slaves, by cruelly forcing them to bite a piece of this plant; and accidents have occasionally occurred with it in this country, where, however, it is only to be found growing in the hothouses of the curious. The negroes in the West Indies make an ointment for rubbing dropsical swellings, by boiling the juice of the plant in hog's lard; and a physician in the reign of Charles II. recommended the juice to be administered internally as a cure for dropsy, but it is so excessively acrid that it is almost impossible to swallow it. Notwithstanding the acridity, however, a wholesome starch has been obtained from the stem. The plant grows from six to eight feet long, and has a stem an inch and a half thick, bearing green leaves about ten inches long by four broad. When the leaves are pulled away the stem has a cane-like appearance. [A. S.]

DIELLIA. *Schizoloma*.

DIELYTRA. The name sometimes given to a very handsome genus of *Fumariaceæ* made familiar in gardens by the beautiful Chinese perennial called *D. spectabilis*. It is sometimes called *Diclytra*, but more correctly referred to *Dicentra*. [T. M.]

DIENIA. A small genus of terrestrial orchids, the species of which are found in the Himalayas, Siberia, and Mexico. They seldom exceed a foot in height; the stems in some being furnished with one leaf, in others with several. These are membranaceous, plaited, and usually ovate or ovate-lanceolate in form. The flowers are minute, green or yellowish, and disposed in slender erect spikes.

The four pollen masses are collateral (oooo), while in the nearly-related British genus *Malaxis* they are incumbent ($^{8}_{8}$). *Microstylis*, also nearly related, has the lip at right angles to the column instead of parallel with it, as in *Dienia*. [A. A. B.]

DIERVILLA. A genus of caprifolls, distinguished from the honeysuckle and others allied to it, by its funnel-shaped three-cleft corolla, and one-celled fruit. The name was assigned by Tournefort in compliment to Dierville, a Frenchman, who discovered a species in Acadia, and sent it to that botanist. The species are erect shrubs, natives of North America and of Japan. That best known in cultivation is *D. canadensis*, a shrub from three to four feet high, with the leaves shortly stalked, smooth, sharply ovate, the edges serrate; the flowers are yellow and appear in early summer. In its wild state it is widely distributed in Canada, and is found about Hudson's Bay and on part of the Rocky Mountains.

Those which are natives of Japan are reported by Siebold, in his account of the plants of that country, as notable on account of the beauty of their flowers. These showy eastern species are the *Weigelas* of our modern gardens. [G. D.]

DIFFUSE. Spreading widely.

DIGITALIFORM. Like campanulate, but longer and irregular, as the corolla of *Digitalis*.

DIGITALIS. A genus of *Scrophulariaceæ*, represented in this country by the well-known Foxglove; which is the badge of the Farquharsons. The genus consists of several species, which are biennials or perennials, with flowers having a calyx deeply divided into five unequal segments; an irregular tubular corolla, the tube of which is distended in the middle, the limb four or five-lobed, the lowest lobe the longest; four concealed stamens; and the fruit a capsule opening by two valves.

D. purpurea, the common foxglove, is a well-known ornament of woods and roadsides in this country and the central parts of Europe. It has an erect stem three to four feet high, marked with a few longitudinal ridges and covered with greyish down; the leaves are alternate, ovate-lanceolate or oblong, covered with down, especially on the under surface, their margins crenate or divided into small rounded lobes, and the base tapered gradually into the leafstalks. The raceme is at the extremity of the stem, and consists of a number of flowers each protected by a bract, and all drooping on one side of the stem; the corollas are irregularly bell-shaped, and upwards of an inch in length, and of a pinkish-purple colour, marked in the interior with circular dark spots, which are interspersed among a number of delicate light-coloured hairs. This plant from its stately beauty is cultivated in shrubberies and gardens, where likewise a variety with white flowers may be frequently observed. In cultivated plants there frequently occurs a malformation, whereby some one or two of the uppermost flowers become united together, and form an erect, regular, cup-shaped flower, through the centre of which the upper extremity of the stem is more or less prolonged. All parts of this plant possess powerful medicinal properties, which are due to an extremely poisonous substance called *digitalin*. In medicine the leaves are the parts used, in the form of tincture and infusion. The effects of this drug are various and remarkable; that most frequently observed is a lessening of the force and frequency of the pulse. This occasionally takes place to a dangerous degree, and more than one instance is recorded, of a patient under the influence of this medicine, having died immediately on making a sudden effort to change his posture. The heart, enfeebled by the drug, has been unequal to the fulfilment of its functions under the increased requirements made upon it by the change in position. Hence, although it may be, and is sometimes used in large doses with impunity, its action must always be watched with great care, the more particularly as occasionally when employed in small but frequently repeated doses dangerous symptoms accrue. Foxglove likewise acts as a diuretic, and in large doses causes vomiting, purging, and fainting. It is now most frequently employed in certain cases of dropsy and of heart disease with great benefit, though its use demands care and vigilance on the part of the practitioner. Lately it has been recommended in large doses in delirium tremens.

Several other species are grown in gardens, such as *D. grandiflora* and *D. lutea*, with yellow flowers, and *D. ferruginea* with brown flowers, but none rival our indigenous foxglove in beauty, though they may do so in their poisonous qualities. [M.T.M.]

DIGITARIA. A genus of grasses belonging to the tribe *Paniceæ*, distinguished by the inflorescence being in fingered spikes; spikelets in pairs, on one side of the flattened rachis, awnless, one-flowered with an inferior rudiment of a second; seed invested with the hardened pales. This genus is nearly allied to *Panicum*, under which all the species are described by Steudel. They are mostly natives of the middle and south of Europe, one, *D. humifusa*, reaching to the southern counties of England. [D. M.]

DIGITATE. When several distinct leaflets radiate from the point of a leaf-stalk.

DIGITINERVED. When the ribs of a leaf radiate from the top of the petiole.

DIGIT'S (adj. DIGITALIS). The length of the Index finger.

DIGLOTTIS. A name applied to a Brazilian shrub of the rue family, characterised by its bell-shaped calyx; its corolla of five partially united petals; its five stamens, three of which are sterile and adherent to the tube of the corolla, while the two fertile stamens have flattened filaments, hairy at the top, and anthers whose connectives are prolonged into acute hairy strap-like processes; ovaries five. [M. T. M.]

DIGRAMMARIA. A genus of polypodiaceous ferns proposed by Presl, and figured by him in his *Tentamen Pteridographiæ*, but somewhat doubtful as to its identity, no fern with indusia such as he describes being known to possess venation such as he figures. Some reward *Callipteris ambigua* as the plant intended by Presl; while others consider it to be the plant he afterwards named *Heterogonium*, which latter view we adopt. This fern has linear oblong naked sori, borne on the two branches of the forked veins, and looking like double lines of spore-cases united below: hence appropriate to the name. The veins too are arcuate, forming costal areoles, with free marginal venules as in Presl's figure. [T. M.]

DIGRAPHIS. A genus of grasses belonging to the tribe *Phalarideæ*, and now generally referred to *Phalaris*. [D. M.]

DILIVARIA. A small genus of *Acanthaceæ*, containing probably not more than three species, erect shrubs, natives of

India and Africa. They have entire or generally spinose and dentate leaves, and showy bracteate flowers in leafless spikes; the corolla consisting of a single three-lobed lip enclosing four didynamous stamens, with one-celled anthers, the margins of which are ciliated. [W.C.]

DILL. *Anethum graveolens.*

DILLENIACEÆ (*Dilleniads*). A natural order of thalamifloral dicotyledons included in Lindley's ranal alliance, consisting of trees, shrubs, or undershrubs with exstipulate alternate leaves; five persistent sepals in two rows; five deciduous imbricated petals; stamens more than twenty, often turned to one side. Fruit consisting of two or five distinct or united carpels; seeds surrounded by an aril; albumen homogeneous. There are about thirty known genera and 230 species. They are found chiefly in Australia, India, and Equinoctial America. They have astringent qualities. Some are large timber trees. *Dillenia speciosa* is an Indian tree with showy flowers and an edible acid fruit. There are two suborders: 1. *Dillineæ*, connective of the anthers equal or narrow at the point, found in Asia and Australia; 2. *Delimeæ*, connective of the anthers dilated at the point, found chiefly in America. Illustrative genera: *Dillenia, Candollea, Delima,* and *Tetracera.* [J.H.B.]

DILLENIA. The species of this genus of dilleniads are handsome lofty trees inhabiting dense forests in India and the Malayan Peninsula and Islands, one only reaching as far as the base of the Himalayan mountains. They have large alternate generally oval or oblong leaves, strongly marked with parallel veins running from the midrib to the margin, where they form the points of sharp teeth. The flowers, which are frequently large and showy, have five fleshy concave sepals, and five white or yellow petals, the sepals increasing in size after flowering and eventually closely covering the ripe fruit. The stamens are very numerous, and arranged in several series round the pistil, those composing the inner rows facing outwards, while the outer ones face inwards, the anthers opening by pores or holes at the top. The fruit consists of from five to twenty cells (or carpels) growing together round a fleshy centre, and surmounted by as many radiating styles; each cell containing numerous seeds, surrounded by a gelatinous pulp.

D. pentagyna is common throughout the peninsula of India, Birmah, and Malaya, and forms a handsome forest tree, with a broad spreading head. Its leaves are of extraordinary size, averaging from one to two feet long, but in young trees sometimes as much as four or five feet; they are pointed at the top, and gradually taper from the middle to the base, the edges being either toothed or waved. The flowers are yellow, about an inch in diameter, and produced in clusters upon the naked branches before the appearance of the leaves; they have only ten stamens and five styles. According to Dr. Cleghorn, it is probable that this tree, and not the *Calophyllum Inophyllum*, as generally supposed, furnishes the valuable poon spars used for Indian shipping.

D. speciosa is also a very handsome tree, growing about forty feet high, and commonly cultivated in India on account of its ornamental appearance. It is found in all parts of tropical India, as well as in the Malayan Islands and peninsula. The leaves are from ten inches to a foot in length seated upon broad foot-stalks; and the flowers are produced at the same time as the leaves, which are of great size and beauty, measuring, when fully expanded, as much as nine inches in diameter; the petals white and contrasting with the bright yellow stamens, which are extremely numerous, and form a dense globular mass in the centre of the flower, with the stigmas radiating like a white star upon the summit. The fruit is about three inches in diameter, enclosed in the swollen and fleshy calyx, which, as well as the fruit, is eatable but very acid, and is also said to be slightly laxative. The natives in India use it in their curries or for making jelly, and the acid juice sweetened with sugar forms a cooling fever drink. The hard tough wood, also, is used for making gun-stocks. The fruits and calices of another species, *D. scabrella*, are likewise used in the same way as those of the last species; and the Cinghalese employ a decoction of the leaves of *D. retusa*, for cleansing foul ulcers. [A.S.]

DILL-SEED. The name applied by Bentham to *Anethum graveolens*.

DILLWYNIA. A genus of pretty yellow-flowered juniper-leaved bushes of the leguminous family, numbering about a dozen species, three of which are found in Tasmania, and the remainder in the southern and western parts of Australia. The leaves have no stipules, the pedicels are furnished with little bracts, and the vexillum or upper petal is broad. These characters taken together serve to distinguish the genus from *Pultenæa*, to which it is most nearly allied. In a few the leaves are more like those of a heath than a juniper; they are either smooth or slightly rough, and sometimes covered with a grey pubescence. The little yellow pea-flowers, scarcely half an inch across, make up for their minuteness by their great profusion; they are arranged in axillary or terminal clusters which seldom exceed the leaves. The minute pods, unless carefully searched for, will be readily overlooked; they are nearly oval, ventricose, and one or two-seeded.

A number of these plants have long been in cultivation in greenhouses, and richly repay the attention of the cultivator in the great profusion of their yellow blossoms. One of the best is *D. ericæfolia*, which has solitary or twin bright yellow flowers, arranged so closely towards the ends of the branches as nearly to hide the leaves: this plant is sometimes called *D. floribunda* from the abundance of its flowers. In *D. parvifolia* the leaves are scarcely a quarter of an inch long, and the flowers are in terminal clusters of four or five, of a pale-yellow colour, with the standard marked at the base by a reddish tint. One of the most desirable species, from its flowering while not more than eight inches high, is *D. scabra*, which has linear leaves about half an inch long covered with minute tubercles; this is remarkable in having *stalked* corymbs of bright nearly scarlet flowers at the ends of the twigs. The genus is named after L. W. Dillwyn, Esq., an English botanist. [A. A. B.]

DILOPHIA. A genus of *Cruciferæ* from Thibet, a small annual with spathulate leaves, and the flowering racemes contracted into umbels; pouch tuberculated with a partition, having a wide opening through it. [J. T. S.]

DIMIDIATE. When one half of an organ is so much smaller than the other as to seem as if missing; hardly different from oblique except in degree; also slit half-way up.

DIMIDIATO-CORDATE. When the larger half of a dimidiate leaf is cordate.

DIMORPHANTHUS. A genus belonging to ivyworts. The name means 'two-formed flower,' indicating that there are some flowers which are in every respect perfect and produce fruit, and others in which no perfect seeds are formed. The flowers of the first kind have the calyx oblong and bell-shaped, or ovate and pentagonal; the styles or appendages on the upper part of the seed-vessel are more or less spreading; in the other flowers the tube of the calyx is in the form of a hemisphere and very short, while the styles approach each other. The species are shrubs or herbs, natives of China and Japan; some are prickly, others unarmed; the leaves are alternate, once or twice pinnate, the leaflets serrate. Dr. Lindley, in his *Vegetable Kingdom*, states that '*D. edulis* is employed in China for exciting the action of the skin and producing perspiration; its young shoots are a delicate article of food, and its root, which is bitter, aromatic, and pleasant to the taste, is employed by the Japanese in winter, as we use Scorzonera.' [G. D.]

DIMORPHOLEPIS. An Australian genus of the composite family represented by one species, *D. australis*, an annual branching herb one to three inches high, with linear nearly smooth leaves, and stems clothed with loose tawny hairs, and terminated by small yellow flower-heads, which have an involucre of two sorts of scales, while the florets are all tubular, and the few outer female ones three-toothed. [A. B.]

DIMORPHOTHECA. The Cape Marigold, known also under the names of *Calendula pluvialis* and *Meteorina gracilipes*. An annual herbaceous plant with narrow sinuated and toothed leaves, very slender weak stems, and pretty flowers which bloom from June to September. These have the florets of the ray white above, violet below, and those of the disk brown. *D. pluvialis* is so called because the florets of the ray fold together and close at the approach of rain. French, *Souci pluvial* or *hygromètre*. [C. A. J.]

DINEMANDRA. A genus of heath-like Peruvian shrubs, belonging to the *Malpighiaceæ*. They bear flowers in clusters, on small jointed stalks; the calyx in five divisions, each provided with one or more glands at its base; the stamens ten, united below, eight being sterile and short. The fruits consist of three-winged carpels. [M. T. M.]

DINKEL. (Fr.) *Triticum monococcum*.

DINOPHORA. A genus of the *Melastoma* family, nearly related to *Spennera*, from which it differs in its five, not three-celled ovary. It is represented by a single species, *D. spenneroides*, which is a smooth slender branching bush of three to five feet, found in moist places in Fernando Po, and bearing opposite stalked oval-acute leaves, the branches terminated by loose panicles of little pink flowers, which have a top-shaped calyx tube, five oval acute petals, and ten stamens. [A. A. B.]

DIODIA. A genus of *Cinchonaceæ*, consisting of herbaceous plants or small shrubs, natives of Tropical America and Africa. They have small white flowers, with a calyx divided into two or four equal or unequal segments; a funnel-shaped corolla, the tube of which is lined with hairs, and the limb divided into four lanceolate divisions; four stamens inserted into the throat of the corolla; and an ovary adherent to the calyx tube, surmounted by a fleshy disk, and internally divided into two compartments, each containing a single ovule. [M. T. M.]

DIŒCIA (adj. DIŒCIOUS, DIOICUS). When the sexes of a plant are borne in different flowers by distinct individuals, as in willows. Expressed by the signs ♂ ♀.

DIOICO-POLYGAMOUS. When some of the flowers of a diœcious plant produce hermaphrodite flowers.

DIOLENA. A genus of *Melastomaceæ* found in Venezuela, and nearly allied to *Sonerila*, but differing in having the parts of the flower in fives; and also to *Bertolonia*, from which it is recognised by the form of the anthers, which are ten in number, short obovate and open at top by two pores, while at the junction of the anther with its stalks there are two slender erect spur-like appendages. The only known species, *D. hygrophila*, is a dwarf unbranched herb, with opposite long-stalked oval pointed entire leaves, and terminal one-sided raceme of small white flowers, succeeded by three-celled and three-winged capsules. [A. A. B.]

DIONÉE ATTRAPE-MOUCHE. (Fr.) *Dionæa muscipula*.

[DION] **The Treasury of Botany.** 410

DION.ÆA. A singular plant referred to in most works on structural and physiological botany, as affording a striking instance of vegetable irritability. *D. muscipula*, Venus's Flytrap, the only species, belongs to the order *Droseraceæ*, and is an humble marsh plant bearing from the root, on a smooth leafless stalk a few inches high, a corymb of white flowers. The root is composed of scales almost like a bulb with a few fibres. From this proceed in a radiating manner a number of leaves on longish stalks, which are winged like those of the orange-tree. The lamina of the leaf itself is divided by the midrib into two nearly semicircular halves, each of which is fringed with stiff hairs, and furnished near the middle with three minute bristles arranged in a triangle, which bristles are extremely irritable, and when touched by a fly or other insect cause the two sides of the leaf to collapse with a sudden spring, imprisoning the intruder until it is either

Dionæa muscipula.

dead or ceases to move. Some time after all motion has ceased, they open again spontaneously. It is a native of the swamps of North Carolina, but is often cultivated in English stoves. *Dionæa* is derived from Dione, one of the Greek names of Venus; *muscipula* is in Latin 'a fly-trap.' As might be expected, the same result is produced by touching the irritable bristles with any fine-pointed substance, as a pin or bit of straw. French, *L'attrape-mouche*; German, *Venus die fliegenfängerin*. [C. A. J.]

DIONYSIA. A genus of *Primulaceæ*, closely allied to *Gregoria*, and including all the Oriental species previously described under the latter name, but distinguished by Boissier from the European *G. vitaliana* on account of some slight differences in the seeds and in the shape of the corolla. They are all small Alpine tufted plants, with flowers intermediate between those of a *Primula* and of an *Androsace*.

DION. A family of Mexican *Cycadeaceæ* with a simple *Zamia*-like stem clothed with woolly hairs, and bearing light-green pinnate leaves, whose leaflets are sword-shaped, very sharp, attached to the petiole by their whole base. The female cone is about the size of a child's head, and consists of flat lance-shaped scales, covered with wool, and two-lobed at the base; each scale bears two large seeds of the size of chestnuts. The seeds of *D. edule* yield a large quantity of starch which is used as arrowroot. [M. T. M.]

DIORYCTANDRA. This name, which has but slender claims to euphony, is applied to a shrub of the violet family, in allusion to the passage of the style through the anthers. The genus is closely allied to *Alsodeia* from which it differs in the greenish petals, which are stalked, not sessile; and in the stamens which have slender filaments as long as the stalks of the petals. [M. T. M.]

DIOSCOREACEÆ. (*Yams.*) A natural order of monocotyledonous or Endogenous plants belonging to the subclass of *Dictyogenæ*. Twining shrubs or herbs with tubers either above or below ground, usually alternate leaves with reticulated venation, and small staminate and pistillate flowers growing in spikes. Perianth six-cleft, in two rows, herbaceous and adherent; stamens six, inserted into the base of the perianth; ovary inferior, three-celled; ovules one or two, suspended; style three-cleft. Fruit compressed, three-celled, two cells often abortive; seeds albuminous; embryo in a cavity. They are found chiefly in tropical countries. *Tamus* is, however, a native of Europe and of the temperate parts of Asia.

Acridity prevails in the order, but it is often associated with a large amount of starch. Various *Dioscoreas* produce edible tubers, which are known as yams and are used like potatoes. *Tamus communis*, black bryony, has an acrid purgative and emetic tuber, and a berried fruit of a red colour. *Testudinaria Elephantopis* has a remarkably tuberculated stem, and has been called elephant's foot or the tortoise-plant of the Cape. The central part is eaten by the Hottentots. There are seven genera and 160 species. *Tamus*, *Testudinaria*, and *Dioscorea*, are examples. [J. H. B.]

DIOSCOREA. The typical genus of the order of yams. Upwards of 150 species are described, most of them being confined to tropical countries, principally in America and Asia, the majority, however, belonging to the former continent; about a dozen are found in Africa, and three or four in Australia. They are herbaceous perennials or undershrubs, with twining stems generally turning to the left hand; and fleshy tuberous roots; their leaves are usually produced alternately, but occasionally opposite, and, except in a few species where they are divided into several radiating lobes, they are always entire, and have several strongly-marked veins running throughout their entire length. The flowers, which are very small and inconspicuous, are produced in spikes from the bases of the leaves, and consist of a perianth of variable form, but usually either bell or funnel-shaped; the males have six stamens, and the females a three-celled ovary, surmounted by a style separating into three stigmas.

Under the name of Yams, the large fleshy tuberous roots of several species of this genus are extensively used for food in many tropical and subtropical countries, where they are largely cultivated, and take the place of our potatoes. Among the species most commonly employed for this purpose are: *D. sativa*, which is a native of Malabar, Java, and the Philippines; *D. alata* of the Moluccas and Java, and *D. aculeata* of Malabar, Cochin China, and Java, all of which are cultivated in various parts of tropical Asia, and likewise in the West Indies, where they have been introduced; besides which, *D. globosa*, *D. purpurea*, *D. rubella*, and *D. fasciculata* are cultivated in India, and other species elsewhere. Yams vary greatly in size and colour, according to the species or variety producing them; many attain a length of two or three feet, and weigh from 30 to 40 lbs.; some are white, others purplish throughout, while some have a purple skin with whitish flesh, and others are pink, or even black. Like potatoes they contain a large quantity of starch; and a nutritious meal, used for making cakes, puddings, &c., is prepared from them in the West Indies, where, also, they are commonly sliced and dried in the sun in order to preserve them.

One species, the Chinese or Japanese Yam, *D. Batatas*, has recently come into notice in this country, where it has been recommended for cultivation as a substitute for the potato; but although it succeeds very well when properly managed, it has not as yet found much favour among agriculturists. The chief drawback connected with it, is the great depth to which the roots penetrate into the earth, and the consequent difficulty of extracting them. It is extensively grown and used for food in China and Japan. [A. S.]

DIOSMA. A genus of heath-like plants, natives of the Cape of Good Hope, and belonging to *Rutaceæ*. It is nearly allied to *Barosma*, but differs in that the flowers have five fertile stamens, and no sterile ones, in the style being shorter than the stamens, in the more narrow leaves, and in other minor points. They possess a fragrance not unlike that of the kinds of Bucku (*Barosma*), and many of them are cultivated for their white or pinkish flowers, the most frequently met with being *D. capitata* and *ericoides*. [M. T. M.]

DIOSPYROS. Large hard-wooded trees, or rarely shrubs, belonging to the *Ebenaceæ*, which is so named in consequence of several species of this genus yielding the black wood called ebony. There are upwards of 100 species, the greater part of them natives of Asia and the Mauritius, only about a dozen being found on the American continent, and three or four in Africa; for the most part they are confined to the tropics, but a few extend as far north as latitude 44°. They have flowers of separate sexes on different trees, and borne in little clusters, or singly at the bases of the leaves; the calyx divided into from four to six lobes; and the corolla tubular or bell-shaped. The fruit is fleshy or pulpy, generally either globose or egg-shaped, and varies greatly in size.

Ebony wood is obtained from several species of this genus. The best and most costly kind, with the blackest and finest grain, is that imported from the Mauritius, which is yielded by *D. reticulata*. East Indian ebony is mostly procured from two species, *D. Melanoxylon* and *D. Ebenaster*; while the best kind of Ceylon ebony is obtained from *D. Ebenum*. It is only the inner part of the trunk or heart-wood, as it is called, that yields the black ebony, the outer portion or sap-wood being white and soft. The chief uses of ebony are for fancy cabinet-making, mosaic work, and turnery, and for making a vast number of small articles, such as knife handles, door knobs and plates, pianoforte keys, &c.

D. quæsita produces the beautiful wood called Calamander in Ceylon, and which the Cinghalese use for making the finest kinds of ornamental furniture. It is a very large tree, and the wood is so extremely hard that it is only worked with great difficulty. *D. Embryopteris* is a tree called Gaub by the Hindus. Its fruit is powerfully astringent, and is employed for tanning purposes. The juice of the unripe fruit is very viscid, and is used in India for paying the seams of boats; fishing nets are also coated with it to render them more durable.

The fruit of the Kaki or Chinese Date Plum, *D. Kaki*, is as large as an ordinary apple, of a bright red colour, and contains a yellow semi-transparent pulp resembling the flesh of a plum, both in appearance and flavour. The Chinese dry them in the sun, and make them into sweetmeats. The tree is a native of China and Japan, but is cultivated in India. *D. virginiana* is the Virginian Date Plum or Persimon, a native of the United States, where it attains a height of fifty or sixty feet, with a trunk about a foot and a half in diameter, the heart-wood of which is of a brown colour, hard and elastic, but liable to split. The fruit of the Persimon is an inch or more in diameter, nearly round, and of a yellowish orange colour, very austere and astringent even when quite ripe, but when bletted or softened by the action of frost it becomes eatable. In the Southern States, Persimons are pounded and made into cakes with bran, and by adding yeast and hops to an infusion of the cakes a kind of beer is brewed; or, by fermenting and distilling them they yield a spirituous liquor. The bark of the tree is very bitter, and possesses febrifugal properties; it has been successfully employed by American physicians in cases of cholera infantum and diarrhœa. [A. S.]

DIOTIS. A Siberian shrub belonging to the *Chenopodiaceæ*, deriving its name, signifying two-eared, from the calyx of the female flower which is of one piece, but deeply divided, and ending in two horns. *D. cernuoides*, the only species, is a dwarf bushy hoary plant, plentifully furnished with slender spreading branches. The leaves are narrow tapering towards each end and alternate. Neither male nor female flowers are showy, but the former from their number and the prominence of the stamens, render the flowering of the plant obvious. They have a slight scent of a honey-like sweetness. [C. A. J.]

DIPETALOUS. Consisting of two petals.

DIPHYLLEIA. A genus of *Berberidaceæ* containing a North American herb, with thick horizontal rhizomes, sending up a large roundish peltate umbrella-like leaf deeply-lobed, or a flowering stem with two alternate excentrically peltate deeply-cleft leaves with wedge-shaped segments and a terminal cyme of rather small white flowers, having six sepals, six petals, and six stamens; fruit a few-seeded berry. The only species, *D. cymosa*, a native of the southern United States, is there called the *Umbrella Leaf*. [J. T. S.]

DIPHYLLOUS. Two-leaved.

DIPHYSCIUM. A curious genus of mosses allied to *Buxbaumia*, with large oblique nearly sessile capsules, an obscure or obsolete outer peristome, the inner being formed of a conical membrane with sixteen folds thickened at the prominent angles, as though so many thread-shaped outer teeth were united with it. There is but one well-established species which occurs on shady banks and barren places in mountainous districts. The leaves are narrow and linear, and resemble somewhat those of *Polytrichum*. The plant looks like a monstrous *Phascum*. [M. J. B.]

DIPLACUS. A genus of *Scrophulariaceæ*, closely allied to *Mimulus*, from which it chiefly differs in a shrubby habit, and in the capsule which, on opening, carries away the seed-bearing placentas attached to the valves. There are three or four species known, all natives of Mexico or California. *D. glutinosus*, a native of Northern California, has long been cultivated in our gardens under the names of *Mimulus glutinosus*, *M. aurantiacus*, or *D. puniceus*. It is an erect branching plant, becoming more or less shrubby at the base, the young branches being often very viscid. The leaves are opposite, varying from broadly-oblong to narrow-lanceolate. The flowers are rather large, solitary in the upper axils, and vary from a pale yellow to a rich orange or scarlet.

DIPLADENIA. The generic name of plants belonging to the order of dogbanes, distinguished principally by the presence of two blunt glands at the base of the seed-vessel, each of which is apparently formed of two conjoined. The name *Dipladenia*, "double gland," appropriately indicates the chief character. The species are climbing shrubs or undershrubs, natives of Central America, having opposite entire leaves, and at their point of attachment often provided with glands or bristles; the flowers are handsome, springing from near the point of insertion of the leaves, or in terminal clusters. These plants are near allies of the well-known genus *Echites*, in which indeed some of them were formerly included.

The species of *Dipladenia* are divided into two sections: 1. Those in which the glands at the base of the seed-vessel are large and well developed, and the appendages at the insertion of the leaves, small or wanting; 2. Those having the glands small. There are some differences in the general outline of the corolla. In certain species it is almost salver-shaped, the tube slightly inflated at the top; in others the tubular part is cylindrical below, and funnel-shaped above. Some approach the herbaceous character, with narrow leaves; others are undershrubs with broader leaves. Several species must be ranked with the finest of our stove plants, and are among the more important and recent acquisitions of collectors.

The twining habit, the large and graceful flowers and general appearance of the foliage are sufficient recommendations. *D. crassinoda*, *nobilis*, *splendens*, and others, occupy a prominent place as stove climbers. The charming and finely-coloured convolvulus-like flowers of *D. splendens* succeed each other for weeks. [G. D.]

DIPLANDRA. A genus of onagrads, distinguished by having the calyx in four lanceolate divisions, two of which are often joined; the corolla has four divisions, one larger than the others, all attached to the calyx and shorter than it. The name *Diplandra* indicates another character, viz. the presence of two stamens only, opposite to two pieces of the calyx. The only species, *D. lopezoides*, a native of Mexico, is a branched hairy shrub, with opposite shortly-stalked leaves, which are oblong and narrow toward the end, almost entire, and hairy on both surfaces. The flowers are purple, forming clusters. [G. D.]

DIPLANTHERA. A scrophulariaceous tree, native of tropical Australia, with large four-lobed stalked leaves, which have two glands at their base, and terminal clusters of handsome flowers, with yellow two-lipped corolla, and four projecting stamens. [M. T. M.]

DIPLARCHE. A genus of *Ericaceæ*, consisting of evergreen, heath-like undershrubs, with prostrate stems, and small rose-coloured flowers arranged in terminal heads. It is botanically characterised by the presence of ten stamens in two rows, the upper placed upon the corolla (perigynous), the lower arising from beneath the ovary (hypogynous), a most unusual circumstance. These shrubs are natives of the Himalayan mountains. [M. T. M.]

DIPLASPIS. A genus of *Umbelliferæ*, consisting of two species, natives of the

South Eastern Alps of Australia and of Tasmania. They are small herbs with radical stalked cordate or reniform leaves, and simple scapes bearing a small simple umbel of flowers. They have thus the habit of *Hydrocotyle*, whilst the fruit is nearly that of a *Bolax*.

DIPLAX. A genus of grasses belonging to the tribe *Oryzeæ*, distinguished by the inflorescence being in panicles, the spikelets two-flowered; glumes two, unequal, the inferior one nerved, blunt and ovate, the superior much larger; lower floret sterile; stamens two or one; styles short and smooth. *D. arenacea*, the only species, is a native of New Zealand. [D. M.]

DIPLAZIUM. A genus of polypodiaceous ferns, belonging to that group of the *Asplenieæ* which have the indusia connate in pairs set back to back on the same vein, the veins in this case being free. The limit between *Diplazium* and *Asplenium* is not very definite, certain of the species bearing but few of the double sori of *Diplazium* amongst many of the single sori characteristic of *Asplenium*. On this account the two groups have been reunited by some modern botanists. It is, however, more convenient to keep them distinct. The species are rather numerous and very varied in size, form, and habit, some bearing simple fronds like *Scolopendrium*, others very large bipinnate or tripinnate fronds. There is a tendency in many of them to develope a short stem. [T. M.]

DIPLECOLOBEÆ. A subdivision of cruciferous plants, embracing those in which the cotyledons are twice folded, and the embryo, when cut across, presents this appearance, O ‖ ‖ ‖, in which O represents the cut radicle, which is placed on the back of the two cotyledons marked by lines ‖ to show that they are cut across three times. Among the genera in this section of *Cruciferæ* are placed *Senebiera*, *Brachycarpæa*, *Subularia*, *Heliophila*, *Schizopetalum*, and a few others. [J. H. B.]

DIPLESTHES. A name sometimes given to the Cape species of *Salacia*.

DIPLOCALYMNA. An imperfectly described genus, included by its author, Sprengel, among *Pentandria*, and subsequently referred to *Thunbergia* (*Acanthaceæ*), but incorrectly if Sprengel's description can be trusted; and also to *Convolvulaceæ*, with no genus of which, however, does it seem to be allied. The genus is founded on a twining plant, without a locality, having the appearance of a *Convolvulous*. It is described as possessing a double calyx, the outer two-valved and the inner ten-toothed; the corolla infundibuliform, and subplicate; the anthers sagittate and included; the stigma urceolate and subbilobed. [W. C.]

DIPLOCENTRUM. A genus of epiphytal orchids found growing on tree stems in the Madras presidency. The three known species, *D. recurvum, lancifolium,* and *congestum*, are furnished with strap-like channeled leaves notched at the apex, and axillary racemes, or panicles of small pink flowers with a crimson lip, or the petals are dull brown with a lilac lip. The lip has two instead of one short spur, whence the name, and this is the only character which separates the genus from the well-known *Vanda*. [A. A. B.]

DIPLOCLINIUM. A genus of begoniads, separated from *Begonia* by Lindley, but subsequently restricted by Klotzsch. It contains plants which are found in the East Indies and in Java. The staminate flowers have four, the pistillate three sepals; anthers oblong with narrow lateral fissures; filaments slightly united at the base; style persistent with two innate branches furnished with a continuous papillose band; placentas split lengthwise. There are five species. The name refers to the divided placenta. [J. H. B.]

DIPLOCLISIA. This genus of *Menispermaceæ*, proposed by Miers, has been referred by Drs. Hooker and Thomson to the genus *Cocculus*, from which it differs only in the elongated drupe, a character not of sufficient importance in the judgment of those authors to constitute a new genus. [M. T. M.]

DIPLOE. That part of the parenchyme of a leaf which intervenes between the two layers of epiderm.

DIPLOGENEA. A genus of *Melastomaceæ*, nearly related to *Medinilla*. The only known species, *D. viscoides*, is found in Madagascar, where it grows on trees, and has somewhat the appearance of mistletoe, but is not like that, a parasite. It has fleshy, smooth three-nerved leaves, between oval and elliptical in form, and small flowers arranged in axillary cymes. These have a bell-shaped calyx, with a nearly entire fleshy border, four oval petals, and eight equal stamens. Its leaves are said to be furnished with dots like those seen in myrtles. [A. A. B.]

DIPLOLÆNA. A genus of shrubs natives of New Holland, belonging to *Rutaceæ*. They have alternate stalked dotted leaves with stellate hairs on the upper surface, and thick white down on the lower. The flowers are borne within a many-parted involucre, the bracts of which are arranged in three rows, the outermost being woolly, the inner petaloid. [M. T. M.]

DIPLOLOMA. A genus of *Boraginaceæ* allied to *Cynoglossum* and more nearly to *Solenanthus*. It has a tubular corolla with five bosses at the throat and an erect five-cleft limb; stamens longer than the corolla; nuts adhering to a central column by their inner angle, crowned and margined by a ring. A native of the Altai. [J. T. S.]

DIPLOMORPHA. A name at one time given to a few plants of the daphnad family which are now generally known as species of *Wickstrœmia*. [A. A. B.]

DIPLOPAPPUS. A genus of perennial bushes or dwarf herbs of the composite family, very near *Aster*, and only differing in the nature of the pappus, which is double, the outer row of short stiff bristles, the inner of capillary bristles as long as the disk florets; whilst in *Aster* the pappus is single. About twenty species are known, some found in South Africa, others in China and the Himalayas, and the remainder chiefly in North America. The most of the Cape species are smooth bushes with small linear or oblong leaves, and solitary stalked flower-heads terminating the twigs. *D. asper*, of the same country, is an herb about a foot high, with sessile lance-shaped leaves, entire or toothed at the margin, and handsome flower-heads, which are solitary and supported on long naked stalks, and nearly two inches across, the ray florets being strap-shaped and purple, those of the disk tubular and yellow. A goodly number of those found in the Himalayas are handsome Alpine plants, with short unbranched stems, furnished with oblong toothed or entire leaves, and terminated by single flower-heads one to two inches in diameter, the outer florets strap-shaped and violet, the inner yellow and tubular. The North American species are mostly shrubby, with linear or lanceolate leaves, and terminal corymbs of flower-heads of which the ray florets are either blue, purple, or white. Almost the only species found in South America is *D. lavandulifolia*, a large handsome bush found on the Peruvian Andes at an elevation of 11,000 to 12,000 feet; its closely packed leaves are covered underneath with white down, and the numerous little twigs are each terminated by a purple-rayed flower-head. This plant and a few of the North American species are also known under the generic name of *Diplostephium*, and some of the former are known also by the name *Eucephalus*. [A. A. B.]

DIPLOPELTIS. A genus of *Sapindaceæ* composed of a few West Australian herbs from one to three feet high, with alternate wedge-shaped and toothed, or sometimes pinnatifid leaves, and terminal panicles of numerous pretty pink flowers, each about half an inch across. All the parts of the plant are usually covered with a short white glandular pubescence. The flowers are male and female on the same plant, the former with a calyx of five leaves, five oblong petals, and usually eight stamens; the latter with a similar calyx and corolla, and a three-lobed ovary crowned with a simple twisted style. The herbaceous nature of the plants is almost enough to distinguish them in the family, which is for the most part composed of bushes or trees with pinnate or trifoliolate evergreen leaves. [A. A. B.]

DIPLOPHYLLUM. A name at one time given to *Veronica crista galli*, a species much like *V. Buxbaumii*, which is indigenous to Britain. [A. A. B.]

DIPLOPOGON. A genus of grasses belonging to the tribe *Pappophoreæ*, distinguished by the inflorescence being in close head-like spikes; spikelets one-flowered; glumes two, lax, membranaceous, and awned; stamens three; ovary sessile; styles two, joined at the base; stigmas feathery. *D. setaceus*, the only species, is a native of New Holland. [D. M.]

DIPLOPTERYS. A genus of *Malpighiaceæ*, consisting of climbing shrubs natives of Guiana, with yellow flowers disposed in an umbellate manner, and surrounded by a series of bracts. The calyx has five segments, four of which are provided with two glands; stamens ten, slightly coherent at the base; ovary three-lobed, three-celled; styles three; fruit with five somewhat woody wings. [M. T. M.]

DIPLOSIPHON. A genus of *Hydrocharidaceæ*, an annual herb growing in rice fields in India, with radicle leaves arranged in a rosette; and axillary perfect flowers from a spathe split at the apex. The perianth tube is twice as long as the spathe, the three outer segments of the limb herbaceous, the three inner larger, petaloid, white; stamens three; style long, adhering to the perianth tube; fruit membranous, many-seeded. [J. T. S.]

DIPLOSPORA. A Chinese shrub of the cinchona family, but imperfectly known. The calyx tube is obovate, its limb somewhat bell-shaped, four-toothed; the corolla with a wide tube, hairy at the throat, and with a limb divided into four fleshy spreading segments; the anthers four, sessile, projecting. *D. dubia* or *Canthium dubium* is a shrub with axillary tufts of white flowers. [M. T. M.]

DIPLOSTEMONOUS. Having twice as many stamens as petals.

DIPLOSTEPHIUM. *Diplopappus.*

DIPLOTAXIS. A family of unimportant herbaceous plants, belonging to the *Cruciferæ* and allied to *Sinapis*, distinguished by having the seeds arranged in two rows in a long compressed pod. *D. tenuifolia* is a slender glabrous perennial plant with a branched stem shrubby at the base, bluntly divided leaves, and rather large light-yellow flowers. It grows in quarries, on rubbish and walls, near large towns. *D. muralis* is a smaller and much rarer species, an annual whose stems and leaves are rough with scattered hairs. [C. A. J.]

DIPLOTEGIA. An inferior capsule.

DIPLUSODON. A genus of *Lythraceæ*, consisting of Brazilian herbs and shrubs with opposite often four-angled branches, opposite or verticillate entire leaves, and solitary axillary nearly sessile flowers, arranged in a racemose or even capitate manner. They have a bell-shaped twelve-toothed calyx, with the teeth arranged in two rows; six petals, and from twelve to forty stamens. [J. T. S.]

DIPODIUM. A genus of terrestrial leafless orchids of Australia and New Caledonia, belonging to the tribe *Vandeæ*. They have thick branching roots, and stems one to two feet high furnished at intervals with brown scales, and terminating in large racemes of numerous rose-coloured nearly regular flowers about an inch across. The oblong clawed lip is two-eared at the base and slightly bearded at the apex. There are two pollen masses each with a separate candicle, whence the generic name signifying two feet. There are three known species. A beautiful figure of *D. punctatum* will be found among the illustrations to Dr. Hooker's *Flora of Tasmania*. [A. A. B.]

DIPSACACEÆ. (*Teazelworts*.) A natural order of gamopetalous calycifloral dicotyledons or Exogens, belonging to Lindley's campanal alliance, embracing herbs or undershrubs with opposite or whorled exstipulate leaves, and flowers in heads surrounded by an involucre; calyx adherent, membranous, surrounded by a separate covering or involucel; corolla tubular, with an oblique four to five-lobed limb; stamens four; anthers distinct; ovary one-celled; ovule pendulous. Fruit dry, not opening, crowned by the pappus-like calyx; seed albuminous. Natives chiefly of the south of Europe, Barbary, the Levant and the Cape of Good Hope. Astringent qualities reside in some of the species. Some are used in dressing cloth. *Dipsacus Fullonum* is the fuller's teazel, the dried heads of which, with their hooked spiny bracts, are used in fulling cloth. The opposite leaves of the wild teazel, *D. sylvestris*, unite at their bases so as to form a cavity in which water collects; hence the plant was called *Dipsacus* or thirsty. There are six known genera and about 170 species. *Morina, Dipsacus, Cephalaria,* and *Scabiosa* afford examples. [J. H. B.]

DIPSACUS. The Teazel family, typical of the order *Dipsacaceæ*. It forms a small genus of prickly biennial plants, natives of Europe and Northern Asia, having oblong or globular heads of flowers, surrounded by an involucre of several narrow bracts, the individual flowers separated by long prickly scales, and inserted into a small angular outer calyx (involucel). The true calyx has a small cup-shaped border surmounting the involucel, and the corolla is divided into four unequal lobes.

D. sylvestris, the common Teazel, is a native of the southern parts of England and Ireland, also of central and south Europe, and Russian Asia. It grows from four to six feet high, and is very prickly in all parts; the leaves long, lance-shaped, and stalked, those on the upper part of the stem growing together by their bases, and forming a cup, which is generally found full of clear water. The heads of flowers are cylindrical, and between two and three inches long, by one and a half broad, having an involucre of from eight to twelve stiff prickly bracts curved upwards, and the scales separating the flowers terminate in a long straight sharp point.

D. Fullonum, the Fuller's Teazel, is by most botanists supposed to be merely a variety of the preceding, from which it only differs in the scales of the flower-heads being hooked instead of straight, and the involucral bracts being shorter and spreading. The flower-heads of this plant, under the name of Teazels, form an article of considerable importance to the cloth manufacturer, who employs them for raising the nap on cloth, no machine having yet been invented to supplant them. For this purpose they are fixed in regular order upon cylinders, which are made to revolve in such a manner that the hooks of the Teazels come in contact with the surface of the cloth, and thus raise a nap, which is afterwards cut level. The plant is cultivated in some parts of this country, also in France, Austria, and other parts of Europe. In 1859 the enormous number of 18,907,120 teazel-heads were imported, all of which came from France, and were valued at five shillings per thousand. [A. S.]

DIPTERACANTHUS. A large genus of *Acanthaceæ*, containing nearly 100 described species, chiefly from Central and South America and Asia, with a few from Africa and Australia. They are creeping or erect herbs or rarely shrubs, with solitary or fasciculate flowers, collected at the ends of the stem and branches into racemes. The lower flowers have large leafy bracts, which become small and narrow in the crowded racemes; the calyx is more or less deeply five-cleft, and the corolla is funnel-shaped with a five-cleft limb; the four didynamous stamens are included, and the stigma is bilamellate. [W. C.]

DIPTERACEÆ. (*Dipterocarpeæ, Dipterads*.) A natural order of thalamifloral dicotyledons or Exogens, belonging to Lindley's guttiferal alliance, containing large trees with resinous juice; alternate involute leaves with convolute stipules; long unequal calyx lobes; twisted petals, and stamens above twenty, distinct or united in several bundles. Fruit leathery, one-celled, surrounded by the calyx, the enlarged divisions of which form winged appendages; seeds single, without albumen. Tropical Indian trees found especially in the islands of the Indian Archipelago. They yield a resinous balsamic juice. *Dipterocarpus lævis* or *turbinatus*, the gurjun of Chittagong, yields wood-oil which exudes from the trunk, and is used as pitch, varnish, and medicine. *Dryabalanops Camphora* or *aromatica*, a tree from 100 to 130 feet high, supplies the hard camphor of Sumatra, which exists in a solid state in the interior of the stem, sometimes in pieces weighing from 10 to 12 lbs. It also yields by incision a resinous oily fluid called the liquid camphor or camphor-oil of Borneo. Sometimes five gallons of the liquid are found in a cavity in the trunk. The wood of *Vateria* or *Shorea robusta* is used in India under the name of sál. Dhoona pitch is also procured from the plant. *Vateria indica* yields the piney resin or piney dammer of India, which is used as a varnish, and for lighting. There are seven known genera and forty-seven species, including *Dipterocarpus, Dryobalanops, Valeria,* and *Shorea.* [J. H. B.]

DIPTERIS. A beautiful genus of polypodiaceous ferns, sometimes referred to *Polypodium* itself, but differing in the netted venation, and in the binate digitato-palmately-lobed or repeatedly dichotomously-partite fan-like fronds. Two or three species only are known, and these are beautiful plants of India and the Archipelago, with tall slender rod-like stipes, and fan-shaped palm-like coriaceous fronds, which rise from a freely creeping woody rhizome. The sori are small, round, and very numerous in *D. conjugata* and *D. Wallichii,* in which the costa is dichotomously-branched in the ultimate segments of the frond; but uniserial in *D. Lobbiana,* in which there is a simple central costa in each of the narrow and more completely separated ultimate divisions. The reticulation of the veins is highly compound. [T. M.]

DIPTERIX. One of the few genera of leguminous plants bearing a single-seeded fruit, which does not open naturally at maturity; the pod which bears this is called drupaceous. There are eight species belonging to the genus, all of them large trees inhabiting the forests of Brazil, Guiana, and the Mosquito country, and having pinnate leaves and panicles of flowers. The flower is characterised by having a two-lipped calyx, the upper lip consisting of two large lobes spreading like wings, while the lower is very small and either of three teeth or only one: the stamens are eight or ten in number and united together into a sheath, which is split on the upper side.

D. odorata yields the fragrant seed called Tonquin, Tonka, or Tonga bean, used for scenting snuff. Perfumers also obtain an extract from it, which forms an ingredient in some bouquets, and the pulverised seed is employed in the preparation of sachet powders. The odour resembles that of new-mown hay, and is due to the presence of *coumarine.* The tree producing these seeds grows sixty or eighty feet high, and is a native of Cayenne. The fruit bears some resemblance to that of the almond tree, and the seed or bean is shaped like an almond, but much longer, and is covered with a shining black skin.

D. ebenensis, the Ebœ tree of the Mosquito shore, has a fruit and seed greatly resembling the preceding in appearance, but entirely destitute of the odoriferous principle. It, however, contains a large quantity of fatty oil, which the natives of the Mosquito country extract and use for anointing their hair, for which purpose it is said to be peculiarly suitable. It is a large tree, and produces an excessively heavy yellowish-tinted timber. (A. S.)

DIPTEROCARPUS. A name indicative of the two calycine wings, which surmount the fruit of these plants, which give their name to the order of *Dipteraceæ* or *Dipterocarpeæ.* The genus consists of lofty trees, abounding in resinous juice, with leathery leaves, covered in some instances with star-shaped hairs. The flowers are in clusters, large, white or pink, fragrant; the calyx divided into five unequal segments, two of them becoming very large and leaf-like; petals five; stamens numerous, with linear anthers; ovary with two ovules in each of its three compartments, included within the tube of the calyx. Fruit woody, one-celled, one-seeded by abortion, surmounted by the persistent and enlarged calyx. These trees are natives of the Indian islands, where the resin is made use of medicinally, and for burning in torches. *D. lævis* yields in Eastern India and the Malay Islands a thin liquid balsam called wood-oil, which is employed for painting ships and houses. The resinous fluid is collected by cutting

Dipterocarpus trinervis.

a deep notch in the trunk of the tree near the ground, where a fire is kept until the wood is charred, when the liquid begins to ooze out. This wood-oil is now imported into this country as a substitute for balsam of Copaiba, which it greatly resembles. By the application of heat it becomes concentrated and semi-solid. The resin mixed with dammer is valuable in preserving timber from the ravages of white ants, according to Dr. Wight. [M. T. M.]

DIPTEROUS. Having two wing-like processes.

DIPTERYGIUM. The name applied to an Arabian herbaceous plant, with rather thick leaves, and flowers in terminal clusters provided with bracts. The calyx and corolla four-parted; stamens six, four somewhat longer than the other two; ovary four-cornered, one-celled, with a cylindrical style and capitate stigma; pod indehiscent, compressed, provided with a membranous wing, one-celled, one-seeded. This plant seems to have nearly equal claims to be comprised among *Cruciferæ* and *Capparidaceæ.* [M. T. M.]

DIPYRENA. A genus of *Verbenaceæ,* found in Chili, and represented by *D. glabrescens,* an erect rigid bush, with narrow oblong somewhat fleshy entire leaves, alternate on the stems, and often arranged in bundles of four or five. The twigs terminated by a loose spike of tubular sweet-scented flowers resembling those of the *Verbenas* so commonly seen in our flower-beds. Indeed the plant would be a *Verbena* were it not that the fruit is composed of two little nuts or pyrenæ (whence the name) instead of four. A still closer relationship exists between this plant and *Priva*; the latter, however, has an herbaceous stem. [A. A. B.]

DIPYRENOUS. Containing two stones or pyrenæ.

DIRCA. A genus of *Thymelaceæ*, with hermaphrodite flowers, the perianth coloured, somewhat bell-shaped and oblique; the stamens eight, inserted in two rows in the tube of the perianth; the ovary one-celled, with a single pendulous ovule. The fruit is drupaceous. There is one species, *D. palustris*, a North American shrub called Leather-wood, Moose-wood, and Wicopy; the twigs are used as thongs; fruit poisonous; leaves alternate entire; flowers pale yellow. [J. H. B.]

DIRCÆA. A genus of *Gesneraceæ*, consisting of Brazilian herbs with tuberous rhizomes, and herbaceous stems bearing large opposite leaves, and long-tubed showy panicled flowers, often of a rich scarlet colour. The group is typified by the species formerly known as *Gesnera faucialis, bulbosa*, &c., and is distinguished by the great development of the upper lip of the corolla. [T. M.]

DIS. An Algerian name for the fibrous stems of *Festuca patula* and *Arundo tenax*, which are used for cordage.

DISA. A numerous genus of terrestrial orchids peculiar to South Africa and

Disa grandiflora.

Abyssinia. The species vary much in habit, but most agree in having the sepals usually much larger than the petals, and the posterior sepal instead of the labellum, as in *Habenaria* and other allied genera, is furnished with a more or less evident hood-like spur. *D. grandiflora* is perhaps the most beautiful of all terrestrial orchids, and is spoken of by Dr. Harvey as the pride of Table Mountain, where it grows in great profusion on the borders of streams and water pools which are dry in summer, producing its gorgeous flowers in February and March. The stems grow two and a half feet high, and are furnished with a number of broad grassy leaves, and terminated by from one to four splendid flowers, measuring from three to five inches across. The lateral sepals are of a bright crimson, the dorsal one paler on the outside, and blush-coloured and delicately veined with crimson within. Unfortunately this plant is very difficult to cultivate, and is therefore not so frequently seen in our gardens as it deserves to be. It is beautifully represented in Lindley's *Sertum Orchidaceum*, t. 49. *D. spathulata* is a most remarkable species from the long and slender stalk of the lip, which much exceeds the flower in length, and has a trowel-shaped more or less lacerated apex. Many species have rose-coloured flowers, but in a goodly number there is a charming mixture of blue, white, green, and purple, in the same flower. [A. A. B.]

DISANDRA. A trailing plant often seen in greenhouses, referred by some botanists to *Sibthorpia*.

DISCANTHUS. A palm-like plant from the Andes of Eastern Peru, forming a genus of *Cyclanthaceæ*. It has the long radical trifid leaves, and the inflorescence of a *Carludovica* or of a *Cyclanthus*, and most of the characters of the latter genus; but it differs chiefly in the perianth consisting of distinct disks embracing the spadix, and in the ovules being naked from their first appearance. The lobes of the leaves have also only one strong rib, and are not plicate as in *Cyclanthus*.

DISCARIA. A genus of *Rhamnaceæ*, nearly allied to *Colletia*, but differing in having no petals. One species, *D. australis*, is common to Tasmania, New Zealand, and Australia, and the others are found in extra-tropical South America. All of the six known species are spiny undershrubs of no beauty, some almost leafless, and others with minute oblong or spathulate smooth leaves. The small opposite secondary branches terminate in a sharp spine, and towards their base are found, in twos or threes, the little flowers, which have a short bell-shaped calyx tube, and from four to five small scale-like hooded petals. [A. A. B.]

DISCHISMA. A genus of *Selaginaceæ*, containing nine species from Southern Africa. They are herbs or herbaceous shrubs, with linear entire or dentate leaves, and flowers in more or less hairy terminal bracteate spikes, the corolla tube short, and the limb fissured in front, and consisting of a single four-lobed lip; there are four sub-sessile stamens with one-celled anthers. [W. C.]

DISCIFORM. Flat and circular; the same as Orbicular. Also a name given to the chambered pith of such plants as the walnut.

DISCIPLINE DE RELIGIEUSE. (Fr.) *Amaranthus caudatus.*

DISCOCACTUS. A genus of *Cactaceæ*, consisting of three or four species, natives of the West Indies and Brazil, remarkable for having very short flat fleshy stems, which are only about two inches in height, and from four to six broad, with eight or ten ridges bearing at intervals little bundles of stiff prickles. The flowers are produced from out of a mass of silky wool and slender spines with which the plant is crowned; they have a long narrow tube, the sepals spreading and coloured, the petals white and spreading out very flat, the stamens of different lengths closing up the tube of the flower, and the style thread-like, shorter than the stamens, and divided at the top into five radiating stigmas. The flowers of *D. insignis* have a very pleasant odour, somewhat resembling that of orange flowers; while that of *D. alteolens* is not so pleasant. [A. S.]

DISCOCAPNOS. A genus of *Fumariaceæ*, distinguished by having the fruit membranous, orbicular, flattened, and winged all round. The flowers are nearly as in *Corydalis*, but with the inner petals united. It is a Cape annual with bipinnate leaves made up of wedge-shaped segments, glaucous beneath, and climbing by the petioles; the flowers are in racemes opposite the leaves. [J. T. S.]

DISCOCARPIUM. A collection of fruits placed within a hollowed receptacle, as in many roseworts.

DISCOIDAL. Orbicular, with perceptible thickness, slightly convex, and a rounded border.

DISCOLOR. Parts having one surface of one colour, and the other of another colour. Also any green colour altered by a mixture of purple.

DISCOPHORA. A genus of *Icacinaceæ*, containing a shrub from Guiana, with large smooth leathery shortly-stalked leaves, and axillary racemes of small flowers articulated with the flower-stalks. [J. T. S.]

DISCOPODIUM. The foot or stalk on which some kinds of disks are elevated.

DISCOSTEGIA. A name proposed for a few marattiaceous ferns including *Marattia alata*. [T. M.]

DISCOSTIGMA. *Garcinia.*

DISEASES OF PLANTS. Plants like animals are subject to diseases both functional and organic. They arise from various causes, being often strictly constitutional and hereditary; and frequently, on the other hand, induced by bad food, imperfect nutriment, depraved atmosphere, defect of light, &c. A very important class again arises from the attacks of parasitic animals and *Fungi*, while others are the direct consequences of injury from external agents. Many of the objects of cultivation, in which some particular organ or element of the plant is preternaturally developed, are really in a diseased state, the peculiar condition being induced artificially, or, at least, encouraged to supply the wants of man, exactly as the livers of geese are compelled to put on a diseased action to afford materials for the pâté. The blanched stems and leafstalks of celery, the swollen stems of kohl-rabi, the enlarged roots of turnips and carrots, &c., are all so many instances of diseased action compelled to administer to our necessities.

The study of vegetable diseases is essential to good cultivation, for though little can be done towards arresting disease in any individual plant, much may be done, either rationally or empirically, in preventing the spread of those which are infectious or contagious, and more by guarding against those conditions which induce disease. The principal maladies to which plants are subject will be noticed briefly under their respective heads. [M. J. B.]

DISEMMA. A genus of *Passifloraceæ*, closely allied to *Passiflora*, but distinguished from it by the coronet, which consists of an outer row of thread-like processes, and an inner tube with longitudinal plaits. They are shrubs, natives of tropical Australia, and have entirely the appearance of passionflowers. [M. T. M.]

DISEPALUM. A Borneo tree forming a genus of *Anonaceæ*, remarkable for the sepals and the petals of each series being two only, instead of three, as in the rest of the order.

DISETTE. (Fr.) A kind of Beet.

DISK. An organ intervening between the stamens and ovary; it assumes many forms, the most common of which is a ring or scales; it is apparently composed of metamorphosed stamens. Also the receptacle of certain fungals, or the hymenium of others.

DISOCACTUS. A genus of *Cactaceæ*, of which only one species is known. This plant, *D. biformis*, is a native of Honduras, and forms a weak trailing shrub or bush, with stem and older branches nearly cylindrical, gradually tapering upwards, and woody; while the younger branches are broad and flat, with blunt teeth, resembling leaves in appearance, but of a succulent or fleshy nature. Like most plants of the order, it has no real leaves. The flowers are produced singly from one of the notches at the upper end of the young branches, and are characterised by having only four sepals and four petals, both of a deep pink colour, and about two inches in length, the sepals very narrow and bent backwards, and the petals broader and growing so close together for the greater part of their length as to form a tube. The fruit is of a beautiful shining deep crimson colour, shaped like a little florence-flask; it contains numerous seeds, imbedded in a soft pinkish pulp, which has a sweetish sub-acid taste. [A. S.]

DISOCARPUS. A genus of the spurgewort family, composed of a few tropical South American trees, with smooth oval entire leaves two or three inches long, a good deal like those of the Portugal laurel, and axillary bundles of small sessile flowers of which the male and female are on different plants. The males have a cup-shaped calyx of five unequal divisions, no petals, and five stamens; while the females have five petals, five rudimentary stamens, and a three-lobed ovary. Three species are known. The genus differs from its near allies in the absence of petals in the male flowers, and the presence of rudimentary stamens in the females. [A. A. B.]

DISOON. A genus of *Myoporaceæ*, represented by *D. floribundum*, a smooth slender graceful bush, six feet high, found in South-eastern Australia. It has alternate linear leaves, and a great profusion of little bell-shaped flowers arranged in axillary clusters, and having a five-toothed calyx which does not grow larger after the flower withers, a five-toothed border to the corollas, and four protruding stamens. The fruit is a little two-celled drupe with two seeds. The nature of the fruit, and the calyx not enlarging after the fading of the flower, are the most marked characters. [A. A. B.]

DISPHENIA. A small set of cyatheaceous ferns, now generally included in *Cyathea* itself, but separated by some authors on account of the elevated receptacle being split into two wedge-shaped divisions. [T. M.]

DISPORUM. A genus of *Melanthaceæ*, belonging to the group connecting that order with *Liliaceæ*, of which *Uvularia* is the type. The species which occur in India are herbs with subsessile leaves and few-flowered axillary peduncles, the perianth six-cleft, with each division keeled and bulging at the base, the whole forming an angular tube. [J. T. S.]

DISSECTED. Cut into many deep lobes.

DISSEPIMENTS. The partitions in a fruit caused by the adhesion of the sides of carpellary leaves. —, SPURIOUS. Any partitions in fruit which have not the origin just explained.

DISSOMERIA. A genus of *Homaliaceæ*, represented by a shrub native of Western tropical Africa, the parts of whose flowers are arranged in fours, the eight petals alternate with as many glands; the stamens numerous, in eight bundles opposite to the petals, the anther-lobes separated by a thick fleshy connective; ovary one-celled; styles four or three. Fruit indehiscent, seeds few by abortion. [M. T. M.]

DISSOTHRIX. A genus of the composite family found in Brazil. *D. Gardneri*, the only species, is a slender annual herb, a foot and a half high, with erect stems terminating in a loose panicle of small flower-heads, and furnished with stalked nearly oval leaves toothed at the margin, opposite on the lower part of the stem and alternate above. Each flower-head has from five to eight tubular five-toothed florets, enclosed in an involucre formed of two series of lance-shaped scales. The achenes are five-angled, and crowned with a pappus of numerous hairs of two sorts, the greater proportion capillary, but five longer than the rest, more rigid, and corresponding to the angles of the achenes. The nature of the pappus serves to distinguish the genus from *Stevia*, to which it is most nearly allied. [A. A. B.]

DISSOTIS. A genus of West African melastomaceous herbs, nearly allied to *Osbeckia*, from which it differs in having dissimilar stamens. The few known species are erect herbs one to three feet high, with opposite lance-shaped three to five-nerved leaves, which as well as the four-sided stems, are clothed with soft-spreading hairs. The rosy or purple flowers generally in threes at the ends of the twigs, and about an inch across, have the tube of the calyx beset with hairy tubercles, and its border five-toothed; five rounded petals; and ten stamens, the latter of two sorts, the five opposite the petals having their anthers joined to the filament by a long slender connective, while those opposite the calyx teeth have a very short or almost obsolete connective. *D. Irvingiana*, a pretty species found in Abbeokuta, is now cultivated in England. [A. A. B.]

DISTEGANTHUS. The name of a parasitical bromeliaceous plant, with yellow flowers, which have a six-parted perianth, the three inner divisions of which form a kind of spiral tube below, while above they are petal-like and somewhat concave; stamens six, thick, hidden by the scales of the inner divisions of the perianth; style twisted at the base, divided above into papillose convolute stigmas. [M. T. M.]

DISTEGOCARPUS. A name sometimes given to a few Japanese species of hornbeam, *Carpinus*, which differ from the others in having the bracts of the male catkins narrowed into a stalk. In other respects they are very like the common hornbeam of our shrubberies. [A. A. B.]

DISTEPHANUS. A genus of shrubs of the composite family from Mauritius and Madagascar, nearly related to *Vernonia*, and differing in having appendiculate apices to the scales of the involucre. Of the three known species, the most common is *D. populifolius*, a bush with stalked oval pointed leaves covered on both surfaces with soft white pubescence. The flower-heads, each about half an inch in diameter, are numerous, and disposed in terminal corymbs; the florets being numerous and all tubular. [A. A. B.]

DISTICHIA. A genus of *Juncaceæ*, from elevated table-land in Peru, forming small tufted plants with dichotomous stems, subulate distichous leaves sheathing at the base, and a six-parted perianth with three stamens. [J. T. S.]

DISTICHIS. A name at one time applied to a few terrestrial orchids of India and Mauritius, now shown by Dr. Lindley to belong to *Liparis*. [A. A. B.]

DISTICHOUS. When parts are arranged in two rows, the one opposite to the other, as the florets of many grasses.

DISTICTIS. A genus of *Bignoniaceæ*, containing a few species, natives of America and the West Indies. They are slender climbing shrubs, with opposite petiolate leaves, sometimes trifoliate, more generally doubly bifoliate; the apex of the petiole is commonly produced into a tendril. The white flowers are in terminal few-flowered racemose panicles, the corolla funnel-shaped, cut into five unequal roundish lobes, and enclosing four didynamous stamens, with the filaments kneed and hairy on the inner surface of the angle; the fifth is sterile. [W. C.]

DISTRACTILE. Divided into two parts as if torn asunder, like the connective of some anthers.

DISTYLIS. A genus of *Goodeniaceæ*, found on the West coast of Australia, and containing only a single species. It is distinguished by having a five-parted calyx adnate to the ovary; a five-parted spreading somewhat bilabiate corolla, the segments of which have winged margins and the tube cleft behind. There are five distinct stamens and a bipartite style. The fruit is a capsule, which is crowned by the permanent calyx. *D. Berardiana* is an annual plant, with alternate toothed leaves, and yellow axillary solitary flowers on long footstalks. [R. H.]

DISTYLIUM. An evergreen tree, native of Japan, belonging to the order of witch-hazels. The flowers are sometimes perfect, having stamens and pistils; while others have stamens only or pistils only. One marked character implied by the name, is the presence of two cylindrical erect appendages, the styles, which remain attached to the fruit. [G. D.]

DITASSA. A considerable genus of *Asclepiadaceæ*, containing nearly forty species of small twining or erect undershrubs, natives of tropical America. They have opposite coriaceous leaves, and small whitish interpetiolar flowers, either solitary or umbellate, with rotate five-cleft corollas; the staminal crown double, its outer whorl consisting of five linear or ovate-acuminate lobes; and its inner of five generally shorter leaflets opposite the outer lobes; the follicles are long, round, and smooth. [W. C.]

DITAXIS. A genus of *Euphorbiaceæ*, comprising about seven species, which are found in various parts of America, south of Mexico. They are white-barked shrubs, with alternate entire or finely-toothed lance-shaped or oboval leaves, and have small green flowers, either male and female on the same, or on different plants, and arranged in little axillary racemes or cymes. The males have a calyx of five deep divisions, five fringed petals, and ten stamens arranged in a candelabra-like manner in two tiers, their filaments united below into a column, round the base of which is a disk of five glands; the females are nearly similar, having calyx and corolla; and a three-lobed hairy or nearly smooth ovary, crowned with a three-forked style. A purplish colouring matter is found in the leaves and flowers of some species. The calyx-leaves do not overlap in the buds, this serves to distinguish the genus from *Jatropha* and other of its allies. [A. A. B.]

DITTANDER. *Lepidium latifolium*.

DITTANY. *Cunila mariana*. —, BASTARD. *Dictamnus Fraxinella*. —, OF CRETE. *Origanum Dictamnus*.

DIURNAL, DIURNUS. Enduring but for a day, as the flower of *Tigridia*.

DIURIS. A genus of terrestrial tuberous-rooted orchids found in Australia and Tasmania. They are slender herbs, having stems one to two feet high, furnished below with several grassy leaves, and terminating in a loose raceme of pretty flowers, which are usually of a rich yellow colour marked with purple spots; more rarely white or purple. The two lateral sepals are long and narrow, suggesting the generic name—from the Greek, signifying two tails. The lip is trilobed, and the column is furnished on either side with a short erect petal-like appendage. Four of the species are well represented in Dr. Hooker's *Flora of Tasmania*. [A. A. B.]

DIVARICATE, DIVARICATING. Straggling, spreading abruptly, and at an obtuse angle, such as 140.°

DIVERSIFLOROUS. When a plant or inflorescence bears flowers of two or more sorts.

DIVIDIVI. The astringent pods of *Cæsalpinia coriaria*.

DIVI LADNER. A Cinghalese tree, *Tabernæmontana dichotoma*.

D'JURNANG. A natural secretion of the fruit of *Calamus Draco*, commonly known as Dragon's-blood.

DOBERA. The latinised form of an Arabic name for a tree with opposite-stalked leaves, whose stalks are thickened, and of a yellow colour, and whose flowers grow in terminal panicles, and have a four-toothed calyx, four petals, and four stamens with the filaments combined below into a tube, and having four little scales between them and the petals; the ovary is superior, and becomes an ovate fleshy one-seeded edible warted fruit. The genus is referred to the *Salvadoraceæ*. [M. T. M.]

DOBINEA. An Eastern Himalayan bush of the maple family. It grows to about ten feet in height, and has opposite stalked lance-shaped or oval toothed leaves, and minute flowers, male and female on the same plant, disposed in long terminal panicles. The males have a four-toothed bell-shaped calyx, four oblong clawed petals, and eight stamens. The females are quite naked, and sit on the middle of a thin yellowish beautifully-veined bract, which is nearly round, and about half an inch in diameter. The circumstance of the female flower arising from near the middle of a veined bract is highly curious, and not paralleled in the family, nor is it met with in any family more nearly related than that of the lime tree. [A. A. B.]

DOCK. The common name for *Rumex*.—GROVE, *Rumex Nemolapathum*.—WATER, *Rumex Hydrolapathum*.

DODARTIA orientalis is an erect glabrous herb with stiff rush-like very spreading branches, and few small leaves, forming a genus of *Scrophulariaceæ*, with flowers much like those of the smaller *Antirrhinums*, but with a globular capsule opening in two short nearly equal valves. It is a native of the dry saline steppes of southern Russia.

DODDER. *Cuscuta*.

DODDER-CAKE. An oil cake made from the refuse of *Camelina sativa*.

DODDER-LAURELS. A name applied by Lindley to the *Cassythaceæ*.

DODECA. In Greek compounds=12.

DODECAS. A genus of *Lythraceæ* from Surinam. It consists of glabrous shrubs with four-angled branches, opposite oblong-obovate entire leaves, and axillary usually one-flowered peduncles; the calyx is urceolate with a four-cleft spreading limb, the petals four, small and round, and the stamens twelve. [J. T. S.]

DODECATHEON. A genus of *Primulaceæ*, known by the reflexed segments of the deeply-cleft corolla, and the cylindrical capsule opening at the apex by five teeth. They are smooth perennial herbs, with fibrous roots, and rosettes of oblong or obovate root leaves; the scape is simple, bearing an umbel of large nodding rose-purple or white flowers, with long reflexed segments, and five short monadelphous filaments with long anthers which are exserted and form a slender cone. The well-known American Cowslip, *D. Meadia*, grows in woods in the warmer parts of North America. In the Western States, where it is more common, it is called the Shooting Star. The name, signifying twelve divinities, is one of fanciful application. [J. T. S.]

DODONÆA. A genus of viscous shrubs of the order *Sapindaceæ*, comprising about ninety species, the greater proportion of which are found in extratropical Australia, and the remainder are thinly scattered over other tropical countries. Few of them exceed ten feet in height, and almost all have their leaves more or less covered with a clammy gum. In the most commonly diffused group these organs are lance-shaped or spathulate; in another they are linear; in a third they are wedge-shaped and toothed; while in a fourth they are pinnate, made up of numerous little wedge-shaped or linear leaflets. The apetalous flowers are unisexual or polygamous, arranged in axillary or terminal racemes or panicles. The fruits are membranous, with their angles produced into thin papery rounded wings. The leaves of *D. viscosa*, one of the most widely diffused species, have a somewhat sour and bitter taste, and the plant is from this circumstance, called in Jamaica, Switch Sorrel. According to Mr. Bennett, this plant is known in Tahiti as Apiri, and 'fillets of it were once used for binding round the heads and waists of victors after a battle, and during the pursuit of the vanquished.' The leaves of *D. Thunbergiana*, a native of South Africa, are said to be used against fevers, and as a purgative. The genus bears the name of Dodoens, a Belgian botanist and physician of the sixteenth century. [A. A. B.]

DODRANS (adj. DODRANTALIS). Nine inches, or the space between the thumb and the little finger separated as widely as possible.

DOGBANES. A name given by Lindley to the *Apocynaceæ*.

DOGBERRY-TREE. *Cornus sanguinea*.

DOG MERCURY. *Mercurialis*.

DOG-POISON. *Æthusa Cynapium*.

DOG'S-BANE. A common name for *Apocynum*; also *Aconitum Cynoctonum*.

DOG'S-CHOP. *Mesembryanthemum caninum*.

DOGWOOD. A common name for *Cornus*.—, AMERICAN. *Cornus florida*.—, BLACK. *Piscidia carthaginensis*.—, JAMAICA. *Piscidia Erythrina*.—, NEW SOUTH WALES. *Jacksonia scoparia*.—, TASMANIAN or VICTORIAN. *Bedfordia salicina*.—, WHITE. *Piscidia Erythrina*.

DOH. A Javanese name for the horse-hair-like fibres of the Gomuti palm, *Saguerus saccharifer*.

DOLABRIFORM. Fleshy, nearly straight, somewhat terete at the base, compressed towards the upper end; one border thick and straight, the other enlarged, convex, and thin.

DOLIA. A genus of *Nolanaceæ*, containing a few South American littoral plants with the habit of some of the smaller maritime *Chenopodiaceæ*. Heath-like branched shrubs with fleshy linear leaves and small flowers, with salver-shaped corollas, and eight or ten ovaries variously united. [J. T. S.]

DOLICHANDRA. A small genus of *Bignoniaceæ*, inhabiting extratropical parts of Brazil, and remarkable as the only known climber of the order having a capsule the partition of which runs in a contrary direction to that of the valves. In habit it much resembles *Macfadyena*, the branches being climbing, the leaves either trifoliate or conjugate, and furnished with tendrils, and the flowers in the axils of the leaves; the calyx is spathaceous; the corolla is long and tubular, whilst the stamens (four in number with the rudiment of a fifth), as well as the stigma, project beyond the corolla. The typical species is *D. cynanchoides*. [B. S.]

DOLICHANDRONE. A small genus of bignoniaceous trees, inhabiting tropical Asia and Australia. Their leaves are either simple or impari-pinnate, and the leaflets either ovate, lanceolate, or, in *D. filiformis* of New Holland, reduced to very narrow linear bodies. The flowers are white and arranged in panicles; the calyx is spathaceous, and the corolla has a tube twice or thrice the length of the calyx; the stamens are four in number, with the rudiment of a fifth; the fruit is a flat capsule opening at the margin, but being divided by a partition, which runs contrary to the direction of the valves. Some of the Asiatic species yield timber. [B. S.]

DOLICHOS. A genus of leguminous plants, consisting of herbaceous or shrubby plants, which for the most part have twining stems. Between sixty and seventy species are known, and are found equally distributed throughout the tropical and temperate regions of Asia, Africa, and America. The plants, formerly called *D. Lablab* (*Lablab vulgaris*), *D. sinensis* (*Vigna sinensis*), *D. bulbosus* (*Pachyrhizus angulatus*), *D. Catjang* (*Vigna Catjang*), all produce edible legumes and pulses. The species of *Dolichos* have trifoliate leaves; and their flowers are produced, either solitary or in racemes, from the bases of the leaves. The pods are generally more or less flattened, and neither winged nor prominently nerved.

D. sesquipedalis is a native of the West Indies and tropical South America, but is cultivated in warm sheltered places in France, and some parts of the south of Europe. The French call it *Dolic asperge*. It has smooth twining stems, six or eight feet in height, with large egg-shaped pointed leaflets, and yellowish-green flowers. Its pods are from a foot to a foot and a half long, cylindrical and pendulous, and of a shining light-green colour, containing from seven to ten kidney-shaped seeds. The young or green pods of this plant are cooked and used as a table vegetable, and, being without the tough parchment-like skin of the common pea-pod, they form an excellent dish. *D. tuberosus*, a native of Martinique, has a fleshy tuberous root, which the inhabitants cook as an article of food, and they also use the pulse for the same purpose. It has a shrubby stem, with twining branches, and leaves with roundish-pointed leaflets. *D. uniflorus* is an annual plant having an erect stem and twining branches, with leaves composed of three egg-shaped leaflets, and yellow flowers, which produce narrow flat pods curved something like a reaper's sickle, and covered with soft hairs. This plant is a native of the East Indies, where it is grown for food under the name of Horse Gram. [A. S.]

DOLIOCARPUS. A small genus of *Dilleniads*, consisting of about half a-dozen species, nearly all of which are climbing shrubs, inhabitants of tropical South America. It is closely allied to *Delima*, but the leaves are not rough, and the flowers are produced from the sides instead of the ends of the branches; besides which, the fruit is pulpy and does not burst open when ripe. *D. Calinea* is a climbing shrub with woody stems, having oblong pointed leaves, and small white flowers collected into dense heads, a portion only perfect, the rest being male or female. The fruit is a small fleshy shining berry. [A. S.]

DOLOMIÆA. A genus of *Compositæ*, nearly related to *Saussurea*, but differing in the pappus-hairs being rough instead of feathery. *D. macrocephala*, the only known species, is a perennial stemless herb found at elevations of 10,000 to 13,000 feet in N. W. India; it has pinnately parted much-lobed leaves clothed with white down beneath, while the centre of the plant is occupied by a cluster of shortly-stalked flower-heads, each an inch or more in length, and furnished with an involucre of numerous lance-shaped scales, which enclose many purplish tubulous florets. According to Royle, it is used by the inhabitants of the hills in their religious ceremonies, and is called by them Googlan. [A. A. B.]

DOMBA-OIL. A fragrant oil obtained from the seeds of *Calophyllum Inophyllum*.

DOMBEYACEÆ. A tribe of plants included in the natural order *Byttneriaceæ*. The petals are flat; stamens fifteen to forty, united at their base, usually some of them sterile. Ovary with five or many cells, having two or more ovules in each. Fruit a capsule; embryo within fleshy albumen. Trees or shrubs growing in tropical regions of the Old World. In this tribe are included the genera, *Pentapetes*, *Brotera*, *Dombeya*, *Melhania*, *Astrapæa*, and a few others. [J. H. B.]

DOMBEYA. A genus of handsome African shrubs or small trees of the *Byttneria* family, a goodly number of them cultivated in plant stoves for the sake of their handsome foliage and flowers. They are found in the greatest number in Madagascar and Mauritius, and extend as far north as Abyssinia. The leaves are often like those of the maple or the plane, but in some are much smaller, heart-shaped and nearly entire; while the flowers are borne in axillary cymes or umbels, each flower being supported by an involucre of three small leaves which fall early. It has a five-parted calyx, five petals, and fifteen to twenty stamens, accompanied by five filiform or strap-shaped sterile ones, all slightly united at the base into a ring. The fruits are little hairy five-celled capsules. Ropes and various sorts of cordage are made in Madagascar from the bark of *D. platanifolia*, as well as from some other of the species. *D. mollis* has large heart-shaped leaves, three-lobed at the apex, covered with a soft dense down, and its rose-coloured flowers with narrow petals, are disposed in dense stalked umbels, and smell like hawthorn. The genus bears the name of M. Dombey, a French botanist and traveller in S. America. [A. A. B.]

DOMPTE-VENIN. (Fr.) *Vincetoxicum officinale*.

DONALDIA. A genus of S. American begoniads whose staminate flowers have two, and pistillate five sepals; anthers elongated, with a dark-brown small connective, the filaments not united; the style is persistent, its branches furnished with a continuous papillose band, which makes three spiral turns; the placentas are split lengthwise. There are two species, viz., *D. ulmifolia* and *D. Ottonis*, both formerly included in Begonia. [J. H. B.]

DONATIA. A genus of *Saxifragaceæ* from the Straits of Magalhaens: small herbs resembling *Saxifraga groenlandica*, with tufted stems, and thick linear lanceolate obtuse glabrous leaves having wool in their axils; flowers, terminal, sessile, white, with the calyx tube adhering to the ovary, and the limb four or five-toothed, and having eight or ten petals. [J. T. S.]

DONDIA. *Hacquetia*.

DONDISIA. The name applied to an Indian shrub of the order *Cinchonaceæ*. The tube of the corolla is lined with rigid hooked hairs; its limb is divided into five acute lobes; stamens five, inserted into the throat of the corolla; style thread-like dilated in the middle; stigma ovate. [M. T. M.]

DONIA. The name sometimes applied to an American genus of yellow-flowered composite plants, better known as *Grindelia*. It has been also applied to *Clianthus*.

DONKLAERIA. A garden name sometimes applied to *Centradenia*.

DONZELLIA. A genus of polypetalous dicotyledons, established by Tenore on a shrub grown in the plant-houses in the Botanic Garden of Naples. It is, however, so imperfectly described, that it has not been recognised in our own collections.

DOOB or DOORBA. Indian names for *Cynodon Dactylon*, which is there a fodder grass.

DOODIA. A group of polypodiaceous ferns related to *Woodwardia*, with which they are incorporated by many modern botanists notwithstanding considerable differences of size, habit, and aspect. They differ from *Woodwardia* chiefly in having superficial instead of sunken sori, and in having the indusia less convex or vaulted, and more lunate. These differences seem rather to indicate sectional than generic distinction. [T. M.]

DOOGHAN. *Myristica spuria*.

DOOLOO. A kind of rhubarb.

DOONA *zeylanica* is a large resinous dipteraceous tree with rose-coloured flowers in panicles. Three of the five sepals of its flowers are larger than the other two, and increase in size after the fall of the corolla; the petals are united at the base; there are sixteen stamens in two rows with dilated filaments, and four-sided anthers with a club-shaped appendage; ovary three-celled, six-seeded. [M. T. M.]

DOOPADA. Indian Copal or Piney Varnish, a resin obtained from *Vateria indica*.

DOORA. *Sorghum vulgare*.

DOORNIA. A genus of *Pandanaceæ*, native of Bourbon or Madagascar, having the appearance of screw pines. The female flowers, which alone are known, are seated on a branched spadix, and consist of ovaries arranged in groups of three or four. The fruit consists of a number of fibrous or woody drupes arranged in groups, and separated from neighbouring parcels by a fibrous material. These collections of drupes form six-sided conical masses on a common stalk. [M. T. M.]

DOORWA. *Cynodon Dactylon*, a fodder grass of India.

DOOR-WEED. *Polygonum aviculare*.

DORADILLE. (Fr.) *Asplenium*.

DORATOMETRA. A genus of begoniads, consisting of East Indian undershrubs, whose staminate flowers have four, and whose pistillate flowers five sepals; the anthers are short, rounded on both sides, with united filaments; the style is persistent, its branches surrounded by a continuous papillose band which makes two spiral turns; the placentas are undivided and stalked, their transverse sections cordate-ovate acute. The seed-vessels have three equal wings, and are attenuated at the apex. There is only one species, *D. Wallichiana*, which has been separated from *Begonia*. [J. H. B.]

DORELLE. (Fr.) *Linosyris vulgaris*.

DOREMA. A genus of *Umbelliferæ* or *Apiaceæ*, comprising certain Persian herbs with branching proliferous umbels, and flowers imbedded in a woolly substance, but having no involucre; the calyx is slightly toothed at the margin. The fruit is compressed, surrounded by a broad border, and marked on the back by five ridges, the three central ones thread-like, and more prominent than the two lateral; oil channels four, on the inner surface of each half of the fruit. *D. ammoniacum* furnishes the drug now known as ammoniacum. It is a native of Persia, and abounds in a milky juice which exudes upon the slightest puncture being made, and dries upon the stem in little rounded lumps, or tears as they are called. This gum resin is used as a stimulant expectorant, and as an external application, but its powers are not great. The ammoniacum of the ancients is said to have been the produce of *Ferula tingitana*. [M. T. M.]

DORINE. (Fr.) *Chrysosplenium*.

DORITIS. A small genus of caulescent epiphytal orchids found in Cochin-china and New Guinea. They have ovate or oblong leaves, and axillary panicles of small white or purple flowers. The sepals are oblong, the lateral ones decurrent with the column; the petals, nearly equal and wedge-shaped; the lip trifid, with a long claw attached by an elastic joint to the produced foot of the column, and the two bilobed pollen masses are borne on the end of a long slender caudicle attached to an ovate gland. [A. A. B.]

DORONICUM. A family of herbaceous perennials belonging to the order of compound flowers. The florets of the ray are destitute of a pappus, while those of the disk have a hairy pappus. *D. Pardalianches*, though enumerated among British plants, is not generally considered to be indigenous to the soil. It is to be found in waste ground near houses in several parts of England, and yet more frequently in Scotland. Under the name of *Pardalianches*, or Leopard's-bane, it had the reputation of possessing 'virtues so ambiguous,' says Gerarde, 'and so doubtfull: yea, and so full of controversies, that I dare not to commit that to the world which I have red. It is reported and affirmed that it killeth panthers, swine, wolves, and all kindes of wilde beasts, being given them with flesh. Theophrastus saith that it killeth cattle, sheepe, oxen, and all fower-footed beasts within the compasse of a day: yet he writeth further, that the roote being drunke is a remedie against the stingings of scorpions, which sheweth that this herbe or the roote thereof is not deadly to man, but to divers beasts onely, which thing also is found out by triall and manifest experience: for Conradus Gesnerus, a man in our time singularly learned, and a most diligent searcher of many things, sheweth that he himself, in a certain epistle written to Adolphus Occo, hath oftentimes inwardly taken the roote hereof greene, drie, whole, preserved with honie, and also beaten to powder, and that even on the very same day in which he wrote these things, he had drunke, with warme water, two drams of the rootes made into fine powder, neither felt he any hurt thereby.' The fact appears to be that the leopards and other 'fower-footed beasts' were poisoned with aconite, one of the author's synonyms for *Pardalianches*; while the human experimentalist found the powdered root of the latter plant inert. Leopard's-bane is a robust plant, with large roughish leaves and conspicuous yellow flower-heads. There are several species natives of Europe or Asia, some of which are cultivated as ornamental plants. French, *Doronic*; German, *Gemsenwurz*. [C. A. J.]

DORSIFEROUS. Bearing something on the back.

DORSTENIA. A genus of moraceous plants named after Dorsten, a German author. It is associated with mulberries

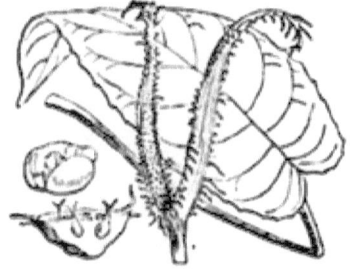

Dorstenia ceratosanthes.

and figs. The genus has a flat and somewhat concave receptacle bearing numerous flowers. The staminate flowers have no perianth, but two or more stamens. The pistillate flowers are also without a perianth; the ovary is one-celled with a lateral style and bifid stigma, containing one ovule. The fruit-bearing receptacle becomes somewhat succulent. There are thirty-six known species. They are herbaceous plants found in tropical America. They have radical leaves which are palmate or pinnatifid, and the receptacle terminating the scape is quadrangular or rounded, or occasionally linear and forked. *D. Contrayerva* and other species have a stimulant and tonic rhizome, which is used medicinally under the name of Contra-yerva-root. [J. H. B.]

DORSUM. The back of anything; in the parts of the flower, that surface which looks towards the outside.

DORYANTHES. A genus of New Holland *Amaryllidaceæ*, having what Herbert calls imperfect bulbs, a tall straight stem twenty feet high, springing from an aloe-like tuft of broadly ensiform-spreading basal leaves, the stem itself clothed with much smaller appressed ones, and terminated by a bulky compound flower-head composed of crimson flowers emerging from great half-sheathing crimson bracts. The perianth is six-parted and funnel-shaped, the segments nearly equal; the six stamens, having long erect anthers, are inserted into the base of the perianth segments; the style is three-furrowed, with a three-cornered stigma; and the ovary grows into a three-celled turbinately oval capsule.

Doryanthes excelsa.

It is a fine plant, sometimes met with in cultivation. [T. M.]

DORYCNIUM. A genus of *Leguminosæ*, comprising a few erect twiggy herbs, nearly related to *Lotus*, from which they may be recognised by the keeled petal being obtuse, not beaked. Their leaves are sessile, and made up of three to five linear leaflets about half an inch long. The minute pink or white flowers are collected into round stalked heads, a good deal like those of the white clover, but smaller. The pods are turgid, scarcely larger than the calyx, and contain two to four seeds. The species are confined to the countries bordering on the Mediterranean. [A. A. B.]

DORYCNOPSIS. A genus of *Leguminosæ*, with the habit of *Dorycnium*, but differing in the stamens being of equal instead of unequal length, as well as in the pod not bursting when ripe. The only known species is *D. Gerardi*, found in the south of Europe, a perennial branching herb one to two feet high, having slender stems furnished with unequally-pinnate vetch-like leaves, the twigs terminating in little clover-like heads of small rosy flowers. The minute one-seeded pod is quite hidden in the calyx. [A. A. B.]

DORYOPTERIS. A name proposed for a group of ferns belonging to the *Pterideæ*, and having uniformly reticulated venation, which is sunk in the substance of the frond, and is on that account generally obscure. The genus is not, however, materially different from *Litobrochia*, with which it is now frequently united. [T. M.]

DORYPHORA. The generic name of the Sassafras tree of New South Wales, which belongs to the *Atherospermaceæ*, and is somewhat nearly related to the Sassafras of Tasmania, *Atherospermum moschatum*. It differs, however, in having the anthers prolonged into a tail-like process. *D. Sassafras*, the only species of its genus, grows to a fine symmetrical pyramidal tree of sixty or one hundred feet high, with a diameter of two to three feet; and is furnished with opposite smooth lance-shaped or elliptical toothed leaves. The flowers are small, perfect, and three together, supported on axillary peduncles shorter than the leaves, and enveloped by two silky bracts, each with a calyx border of six divisions and twelve stamens, six fertile and six sterile, the fertile ones having the anthers prolonged into a tail. The ovaries are numerous and become one-seeded nuts, the styles remaining attached to the ripe fruits in the form of feathery awns. The leaves, bark, and wood emit an agreeable aromatic odour which, when fresh, is said to resemble fennel. The bark is also said to be used by the colonists as a tonic, and is much esteemed. The wood is of little value, being extremely soft and light. It is sometimes used for making packing cases and similar articles. [A. A. B.]

DORYSTIGMA. A genus of *Solanaceæ*, consisting of low-growing herbaceous plants, with solitary extra-axillary flower-stalks; the corolla is funnel-shaped, hairy within, the anthers green, concealed within the corolla; the ovary is two-celled. They are natives of the Andes. [M. T. M.]

DOSSINIA *marmorata* is the name of a beautiful little Bornean orchid cultivated in gardens for the sake of its olive-green velvet-like leaves, the nerves and nervelets of which are of a paler colour, thus giving them a marbled appearance, whence the specific name. The creeping stems have five or six ovate leaves, two or three inches in length by one or two broad; and the flower spike is about a foot high, bearing a number of small white flowers tinged with pink. The plant is sometimes called *Cheirostylis* or *Macodes marmorata*, as well as *Anœctochilus Lobbii*; it differs from *Anœctochilus* in the absence of a bearded fringe to the lower part of the lip, as well as in the boat-shaped process of the column. It is dedicated to E. P. Dossen, a Belgian botanist. [A. A. B.]

DOTHIDEA. A large genus of sphæriaceous *Fungi*, differing from *Sphæria* and its more immediate allies in not having the walls of the fruit-bearing nucleus so perfectly developed or so distinct in colour and structure from the stroma. *D. ribesia* is one of our commonest *Fungi*, forming little black spots on the dead stems of currants, &c.; the sporidia in *Dothidea* seldom acquire complicated forms like those which make *Sphæria* so abundant a source of objects for the microscopist. [M. J. B.]

DOTTED. Furnished with transparent receptacles of oil, looking like dots; marked with punctures.

DOUBLE-BEARING. Producing twice in the same season.

DOUBLY. Having a form or structure repeated; doubly-toothed = teeth themselves toothed, and so on.

DOUCE-AMÈRE. (Fr.) *Solanum Dulcamara.*

DOUCETTE. (Fr.) The common *Valerianellus*, which were called *Valeriana locusta*, by Linnæus.

DOUCIN. (Fr.) Certain varieties of *Pyrus Malus.*

DOUGLASIA. A genus of primworts, distinguished from its allies by the funnel-shaped corolla, the tube of which is partly dilated. The name was given by Dr. Lindley as an appropriate compliment to David Douglas, a well-known botanical collector, to whose energy and zeal we owe the introduction of many interesting plants. *D. nivalis*, which is the best known species, was discovered by Douglas not far from the sources of the Columbia river, near snow, at an elevation of 12,000 feet; another, *D. arctica*, was found by Sir J. Richardson, on the Arctic shore between the Mackenzie and Coppermine rivers. These plants have forked and closely tufted stems, linear leaves, and are covered with numerous short stiff hairs. [G. D.]

DOURA, or DURRA. The great Millet, *Sorghum vulgare.*

DOUVE, GRANDE. (Fr.) *Ranunculus Lingua.* —, PETITE. *R. Flammula.*

DOUX-GUILLAUME, — also **DOUX-JEAN.** (Fr.) *Dianthus barbatus.*

DOUZE DIEUX. (Fr.) *Dodecatheon Meadia.*

DOVEA. A genus of *Restiaceæ*, consisting of South African sedge-like plants, distinguished by their simple one-celled anthers, three-celled capsule opening at the angles, and three (rarely two) sessile stigmas. The rhizome is creeping, scaly; the stems wand-like with remote sheaths; the flowers diœcious. [J. T. S.]

DOVE-FLOWER. *Peristeria.*

DOVE'S-FOOT. *Geranium dissectum.*

DOWNY. Covered with very short weak close hairs.

DRABA. Whitlow Grass. An extensive genus of small annual or perennial herbaceous plants of the cruciferous order, among which they are distinguished by having the frond compressed, with the dissepiment in the broadest diameter, and numerous seeds in each cell. They are most numerous in the cold mountainous countries of Europe; a few are natives of America, and several of Great Britain. Of these last, *D. aizoides* grows on walls and rocks near Swansea, and is remarkable for its bright yellow flowers, and glossy leaves margined with hairs. It is a pretty plant, well adapted for rock-work, as, indeed, are several of the foreign species, being of humble growth, and tufted habit, and made conspicuous by their white or yellow flowers, which, though small, are numerous and bright. Of the other British species, *D. verna*, called also *Erophila vulgaris*, an humble little annual with scanty foliage and inconspicuous white flowers, is not without interest from its appearing very early in the year. It grows on wall-tops and dry banks. Fr. *Drave*; German, *Hungerblumchen.* [C. A. J.]

DRACÆNA. A genus of monocotyledons of the order *Liliaceæ*, remarkable for the elegant palm-like character assumed by the greater number of the species. The genus as formerly constituted was a rather extensive one, but it has lately been remodelled by Dr. Planchon, who removes from it all but the *Dracæna Draco*, or Dragon tree of Teneriffe, and refers the other species to *Dracænopsis, Cordyline, Calodracon, Charlwoodia,* and *Cohnia.* Thus limited, *Dracæna* is distinguished by having a bell-shaped perianth deeply separated into six equal segments, furnished with six stamens inserted at the base of the segments, and succeeded by a fleshy berry containing one, two, or rarely three seeds, the ovary, which is three-celled, with a single ovule in each cell, seldom perfecting all of them.

D. Draco has a tree-like stem, simple or divided at the top, and often when old becoming much branched, each branch terminated by a crowded head of lanceolate linear entire leaves of a glaucous green colour, which leaves embrace the stem by their base, and on falling off at maturity leave a ring-like cicatrix or scar. The flowers form a large terminal panicle, and are individually small and of a greenish-white colour. As seen in our stoves and greenhouses, the plant is usually unbranched, being in its 'first age' or infancy, which lasts in its native country from twenty-five to thirty years. The 'second age,' or period of maturity and reproduction, and the 'third age,' or period of decay, are of indefinite extent. During the former of these, the scars of the leaves disappear, and the thickness of the trunk is at length increased by the formation of branches, and the consequent deposit of new matter; while in the latter stage, aërial roots appear, and glandular excrescences are formed. It is only when of great age that it branches. This tree derives its common name from a resinous exudation known in commerce as dragon's-blood. The resin has been found in the sepulchral caves of the Guanches, and has hence been supposed to have been used by them in embalming their dead. It appears at one time to have formed a considerable branch of export from the Canaries, and has never wholly fallen into disuse. The colossal Dragon tree at the town of Orotava in Teneriffe is a giant amongst the plants of this type of vegetation, being according to Meyen seventy feet high, and forty-eight feet in circumference, with an antiquity which must at least be greater than that of the pyramids. The trunk of this tree is hollow, and may be ascended by a staircase in the interior up to the height at which it begins to branch. Near the ground Le Duc found it to be seventy-nine feet in circumference. As to its great age, Humboldt mentions that when he saw it, it had the same colossal size—a diameter of more than sixteen feet—which it had when the French adventurers, the Bethencourts, conquered these gardens of the Hesperides in the beginning of the fifteenth century, yet it still flourishes as if in perpetual youth, bearing flowers and fruit. A tree like this of slow growth, which four centuries have changed so little, may well be believed to possess great antiquity. [T. M.]

DRACÆNOPSIS. A genus of *Liliaceæ*, separated from *Dracæna* by Dr. Planchon, and consisting of plants agreeing in the following peculiarities: a six-parted marcescent campanulate perianth, with the segments biseriate; six stamens inserted at the base of the perianth segments; a three-celled ovary with many ovules in each cell; and a pea-shaped berry containing several seeds in each of its three cells. To this genus are referred *D. australis* and *D. indivisa*, two beautiful Australian arborescent species, with erect simple stems, and *Yucca*-like heads of crowded lanceolate-ensiform leaves. [T. M.]

DRACOCEPHALUM. This alarming name, literally Dragon's-head, has been given to a genus of from twenty-five to thirty species of herbaceous labiates, distinguished by having the throat of the corolla inflated, and the upper lip concave. They grow to the height of from six inches to three feet, and in habit somewhat resemble *Salvia*. *D. canariense* or *Cedronella triphylla* is better known as Balm of Gilead, a designation which it hardly merits, being a native of America and the Canaries, and having no healing properties, though the foliage is fragrant. It is distinguished by its plukish spiked flowers, and ternate leaves. *D. Moldavica* is an annual with reddish stems, oblong blunt leaves, and whorled purplish blue or white flowers, forming a leafy spike. *D. virginianum*, also called *Physostegia*, bears numerous large light bluish flowers, arranged in four ranks, of which, it is said, 'the position may be altered at pleasure, and as they are placed, so they will remain for several hours.' Fr. *Dracocéphale*; Ger. *Drachenkopf*. [C. A. J.]

DRACONTIUM. A genus of *Orontiaceæ*, comprising certain tropical species, with a thick fleshy rhizome, whence proceed a number of stalked pedate leaves, a sessile spadix with a hooded spathe, and very fetid flowers, which are hermaphrodite and have a five to eight-cleft perianth; stamens five to eight, the anthers with two transverse cells; ovary three-celled, each cell containing a single ovule; style awl-shaped; berries distinct, with one to three seeds. *D. polyphyllum*, a native of some parts of India, Japan, &c., possesses powerful stimulant properties. In Guiana it is considered as a remedy against the Labarri snake, which it resembles in the colour of its spotted leaf-stalks. [M. T. M.]

DRACOPHYLLUM. A genus of *Epacridaceæ*, which is distinguished by having a calyx of five coriaceous leaves; a broad-tubed glabrous corolla with five spreading lobes curved in at the point; and the stamens inserted on the corolla in the New Zealand species, hypogynous in those of Australia and New Caledonia; the ovary is five-celled with five glands at its base. They have narrow grassy leaves sheathing at the base, and white flowers forming a raceme, spicate, or paniculate inflorescence. Most of them are natives of New Zealand, where their peculiar habit gives a striking character to the scenery. Some few are tall trees, the others only shrubs. [B. H.]

DRACOPIS. A genus of annual composite plants consisting of one N. American species, *D. amplexicaulis*, which has oblong-cordate stem-clasping leaves, and conspicuous flower-heads with a yellow ray and prominent black disk. It is an old garden plant, and is allied to *Rudbeckia*. [T. M.]

DRACUNCULUS. A genus of *Araceæ*, consisting of certain South European plants, with tuberous rhizomes and pedate leaves, scarcely differing from *Arum*, except in the upper part of the spathe being flat not convolute. One species, *D. vulgaris*, the old *Arum Dracunculus*, is common in gardens, where its pedately-divided leaves and spotted stems render it very ornamental. [M. T. M.]

DRAGON. *Dracunculus vulgaris*; also applied to the orontiaceous genus *Dracontium*. —, GREEN. *Arisæma Dracontium*.

DRAGONNE. (Fr.) *Tulipa turcica*.

DRAGONNIER. (Fr.) *Dracæna Draco*.

DRAGON ROOT. *Arisæma atrorubens*; also an American name for *Arisæma Dracontium*.

DRAGON TREE. *Dracæna Draco*.

DRAGON'S-BLOOD. A dark-red astringent resinous secretion of the fruit of *Calamus Draco*; another kind is obtained from *Dracæna Draco*. *Ecastaphyllum monetaria* yields a similar resinous product.

DRAGON'S-EYE. *Nephelium Longanum*.

DRAGON'S-HEAD. A common name for *Dracocephalum*. —, FALSE. *Physostegia*.

DRAGON'S-MOUTH. *Epidendrum macrochilum*.

DRAKEA *elastica* is a curious terrestrial orchid of West Australia with woolly roots ending in fleshy tubercles; a single orbicular leaf three quarters of an inch across, growing quite close to the ground, and a slender erect smooth scape twelve to eighteen inches high, bearing at the apex a solitary dull-coloured flower three-quarters of an inch across. The shield-shaped labellum 'is placed on a long arm with a moveable joint in the middle, and is stated by Mr. Drummond to resemble an insect suspended in the air moving with every breeze.' This is the only species known. [A. A. B.]

DRAPETES. A genus of *Thymelaceæ* with hermaphrodite flowers, and a coloured funnel-shaped perianth with a four-cleft limb, and no scales in its throat; stamens four, inserted on the perianth; no hypogynous scales; ovary one-celled. The fruit is a single-seeded nut, included in the base of the persistent perianth. *D. muscoides*, the only species, is found at the Straits of Magalhaens. It is a shrubby plant with opposite decussate sessile leaves. [J. H. B.]

DRAVE. (Fr.) *Draba.*

DRAYTONIA. A genus nearly related to *Saurauja* (which is placed by some botanists with the dilleniads, and by others in the tea family), but differing in the styles being united to the apex. *D. rubicunda*, so called from the reddish hue of the leaves, is found in the Feejee Islands, and is the only species. It is an ornamental shrub, or sometimes tree, of forty to fifty feet high, with long alternate stalked papery oblong serrated leaves, and axillary stalked cymes of small red flowers, which have a calyx of five roundish sepals, five obovate petals, about forty stamens slightly united below, and an ovary crowned by a columnar style tipped with a three-lobed stigma. The fruit is a small capsule about the size of a pea, enclosing numerous seeds. The genus bears the name of Mr. J. Drayton, an American naturalist and artist. [A. A. B.]

DREGEA. A genus of *Asclepiadaceæ*, containing two species, natives of Africa and Arabia. They are shrubs with opposite membranaceous leaves, and small glabrous flowers in umbels on interpetiolar peduncles. The calyx consists of five sepals, and the rotate corolla is five-cleft, with faintly emarginate lobes, while the staminal crown consists of five small kidney-shaped leaflets attached to the gynostegium. The two follicles are four-winged and divaricate, and contain few comose seeds. In habit and structure this genus is very near to *Marsdenia*; it differs from it chiefly in the structure of the staminal crown and in the tetrapterous fruit. [W. C.]

DRIMIA. A genus of *Liliaceæ* from the Cape of Good Hope, containing bulbous herbs, with oblong orchis-like or linear root leaves, and scapes bearing a raceme of flowers, with a six-parted reflexed perianth, varying in colour in different species, being purple, yellow, white, or red, often tinged with green. The juice of the bulbs is said to be very acrid, causing blisters when applied to the skin. [J. T. S.]

DRIMIOPSIS. A genus of *Liliaceæ* from the Cape of Good Hope, containing bulbous herbs with radical leaves, and a scape with a raceme of greenish yellow flowers,

which are bell-shaped, and have six equal stamens inserted on the perianth segments. *D. maculata* is a greenhouse bulb with spotted leaves. [J. T. S.]

DRIMYS. A genus of *Magnoliaceæ*, consisting of trees natives of South America, New Zealand, &c. They have their carpels crowded, berry-like, and many-seeded, and the cells of the anther are separated by a thickened connective. *D. Winteri*, a native of Chili and the Straits of Magalhaens, furnishes the bark known as Winter's Bark, which both in appearance and properties is much like canella bark, but is of a darker colour internally. It is a stimulant aromatic tonic, but is seldom used. The bark was first brought to Europe by Capt. Winter in 1579, he having accompanied Sir Francis Drake to Magalhaens' Straits. In Brazil the bark of *D. granatensis* is used against colic. *D. piperita* is a native of Borneo. [M. T. M.]

DRIMYSPERMUM. A Malayan shrub with alternate leaves and umbellate flowers surrounded by an involucre. The perianth is coloured, tubular, with a four-parted limb; stamens eight, inserted into the throat of the perianth. The base of the ovary is surrounded by a membranous tube, the ovary itself being free, with one ovule in each of its two compartments, and crowned by a short style with a button-like stigma. Fruit berry-like, two-celled, two-seeded. It is included among the *Aquilariaceæ*. [M. T. M.]

DROGUE AME'RE. A bitter tincture, of which *Andrographis paniculata* is the basis; it possesses stomachic and tonic properties.

DROP-SEED. *Muhlenbergia diffusa.*

DROPWORT. *Spiræa Filipendula*; also *Potentilla Filipendula*. —, WATER. The common name for *Œnanthe*.

DROSERACEÆ. (*Sundews.*) A natural order of thalamifloral dicotyledonous or exogenous plants belonging to Lindley's berberal alliance. Herbs often covered with glandular hairs. They have alternate leaves with fringes at their base, and a circinate vernation; sepals five, persistent; petals five; stamens as many as the petals, or twice or three times as many; styles three to five. Fruit a one-celled three to five-valved capsule with loculicidal dehiscence. The plants are found inhabiting marshes in Europe, India, China, Cape of Good Hope, Madagascar, North and South America, and New Holland. They have acid and slightly acrid properties. Hooker thinks that the order should be placed near the *Saxifragaceæ*. Some of the Antarctic species are perigynous. The species of *Drosera* are remarkable for their glandular hairs, which are covered with drops of fluid in sunshine; hence the name of Rossolis, and of the Italian liquor Rossoli, in the preparation of which a species of *Drosera* is used. Some include *Parnassia* in this order. There are seven known genera, including

Drosera, Dionæa, Drosophyllum, and Aldrovanda, and about 100 species. [J. H. B.]

DROSERA. A genus of plants giving name to the order *Droseraceæ*, and distinguished by having five sepals, petals, and stamens, three to five-cleft styles, and a one-celled many-seeded capsule. Their most striking character, however, is connected with their leaves. These in the British species all spring from the root in a radiating manner, and in their early stage are rolled up in a circinate form like the fronds of a fern. When expanded they are somewhat concave, and are thickly set with red glandular hairs, those nearest the edge being the longest. Each hair is tipped, especially in bright weather, with a minute drop of viscid fluid, hence the name *Drosera* (from the Greek *drosos*, dew), and the English name Sundew. The hairs are not so decidedly irritable as in the allied genus *Dionæa*, but when any small fly or other insect alights on a leaf, it is held entangled, at first by the viscid fluid, and, subsequently, the hairs bend down over it until decomposition has taken place. And this is no unusual occurrence; on the contrary, one can scarcely ever examine a plant without finding the wings and legs of insects on one or more of the leaves. The viscid fluid with which the hairs are furnished, is said to be acrid and caustic, to curdle milk, and to remove warts, corns, freckles, and sunburns. It is also said to cause the rot in sheep. The sanitary virtues ascribed to it may be real or imaginary; but with respect to its mischievous effects on sheep, there can be no doubt that where Sundew grows, there flocks are not likely to fatten, for the herbage with which it is associated is mostly moss, rushes, cotton-grass, and other juiceless weeds. There are three species of Sundew indigenous to Britain, which differ in the shape and size of their leaves, and agree in having small inconspicuous flowers on a leafless wiry scape. Some of the foreign species have leafy stems. The hairs of *D. lunata* are said to close upon insects which alight upon them. French, *Rossolis*; German, *Sonnenthau*. [C. A. J.]

DROSOPHYLLUM. A singular half-shrubby plant belonging to the *Droseraceæ*, distinguished by its ten stamens, and one-celled capsule opening with five valves, which bend inwards so as almost to make the capsule five-celled. *D. lusitanicum*, the only species, a native of the sandy hills of Portugal, grows about six inches high, bearing narrow leaves thickly set with stalked glands, and having large sulphur-coloured flowers. [C. A. J.]

DROUILLIER (Fr.) *Pyrus Aria*.

DRUMMONDIA. A name formerly given to a group of N. American herbs of the saxifrageous order, now more commonly regarded as a section of *Mitellopsis*, and known by their stamens being opposite the pinnatifid petals, and by the bilobed condition of their stigmas. [T. M.]

DRUMMONDITA. A genus of heath-like rutaceous undershrubs, with yellow flowers, natives of South-western Australia. They may be known by their stamens, which are combined into a long hairy tube of a purple colour. Of the ten stamens which form this tube, five are fertile, and five sterile, the latter being feathery. Ovaries five, placed on a five-lobed fleshy disk; style thread-like, protruding; stigma button-like. [M. T. M.]

DRUMSTICK TREE. *Cathartocarpus conspicua*.

DRUPACE.E. (*Drupiferæ*, *Amygdaleæ*, *Almundworts*.) According to Lindley this is a distinct natural order, while other botanists regard it as a suborder of *Rosaceæ*. The order belongs to the class of dicotyledons, and the sub-class *Calyciflora Polypetalæ*, and to Lindley's rosal alliance. Trees and shrubs with simple alternate stipulate leaves. Flowers white or pink, in umbels or single; calyx five-toothed, lined with a disk, the fifth lobe superior or next the axis. Petals five, perigynous. Stamens about twenty, arising from the throat of the calyx. Ovary superior, one-celled; ovules two, suspended. Fruit a drupe, with a hard endocarp; seed usually solitary; no albumen. The plants are found in cold and temperate climates of the northern hemisphere. The leaves, flowers and seeds yield hydrocyanic or prussic acid. The bark is astringent, and yields gum. The fruit is in many cases edible. *Amygdalus communis*, the almond-tree, a native of Asia and Barbary, is cultivated in the South of Europe. There are two varieties, one producing sweet, the other bitter almonds. The kernels of the former contain a fixed oil and emulsin, while those of the latter contain also amygdalin, which by combination with emulsin produces prussic acid. *Cerasus communis* yields the common cherry. *C. Lauro-cerasus*, the cherry-laurel or bay-laurel, yields a hydrocyanated oil. The kernels of species of *Cerasus* impart flavour to noyeau, ratafia, cherry-brandy, and maraschino. *Prunus communis* furnishes the common plum, and *P. Armeniaca*, the apricot. *Amygdalus persica* supplies the peach, and a variety gives the nectarine. There are five known genera, and 110 species. [J. H. B.]

DRUPARIA. A Brazilian herbaceous plant of the gourd family, with a furrowed stem, and branching tendrils; female flowers in clusters. The fruit is four-celled, four-seeded. [M. T. M.]

DRUPE (adj. DRUPACEOUS). A fleshy or succulent fruit, with a bony putamen or lining, as a plum. —, SPURIOUS. Any fleshy body inclosing a stone.

DRUPEOLE. A little drupe.

DRYADANTHE. A genus of the rose family, nearly allied to *Sibbaldia*, but differing in the parts of the flower being arranged in fours. *D. Bungeana*, the only known species, is a little Alpine plant from two to four inches high, found in the Altai mountains; it grows in dense tufts, and all its parts are covered with silky hairs; the leaves are about a quarter of an inch long, and made up of three leaflets, the central one with three, the lateral ones with two teeth. The stems are terminated by one or two little flowers, each with a four-parted calyx border, four petals, and four stamens, or in the females a like number of ovaries. [A. A. B.]

DRYANDRA. A large proteaceous genus, named after Dr. Jonas Dryander, a celebrated botanist, who was librarian to Sir Joseph Banks. It is distinguished by having four-parted apetalous flowers, generally clothed on the exterior with reddish-brown wool; four linear nearly sessile anthers, inserted on the concave extremities of the segments of the flower, bursting longitudinally; a round occasionally furrowed style, slightly exserted, and a cylindrical or clavate stigma. The fruit is a woody follicle. The flowers grow in sessile terminal heads, with a closely imbricated involucre, clothed with dense reddish-brown wool, the outer bracts elliptical, acuminate, the inner ones subulate with a pencil of rufous wool at the point.

Dryandra, like its congener *Banksia*, is more remarkable for the variety and peculiar forms of its generally rigid foliage than for the beauty of its flowers. The leaves are either linear or oblong, and with very few exceptions coarsely serrated, lobed or pinnatifid (in *D. speciosa* they are entire), varying considerably in size, some being from a foot to a foot and a half in length, and not more than a third of an inch in breadth, as *D. longifolia, D. Brownii, D. tenuifolia*, &c.; whilst in others, as *D. præmorsa, D. cuneata, D. floribunda*, &c., they are only two inches long, and half an inch broad. The genus has only been found on the south and south-west parts of Australia, the larger number of the species having been discovered in the immediate vicinity of King George's Sound and Swan River. [R. H.]

DRYAS. A genus of herbaceous plants with shrubby stems, giving name to the suborder *Dryadeæ* of the *Rosaceæ*. The species are elegant little evergreen plants of humble growth, with rather large simple leaves which lie prostrate on the ground, and showy white or yellow flowers like the *Potentillas* and *Geums*, but well distinguished from both by having the seed-vessels furnished with a long unjointed feathery appendage or tail. They are found either in high latitudes, or in Alpine or sub-Alpine regions, in both hemispheres. *D. octopetala*, the only British species, well marked by its eight white petals, is not unfrequent in the mountainous parts of England, Ireland, and Scotland, the last especially. French, *Driade*: German, *Silberkraut*. [C. A. J.]

DRYMARIA. A genus of *Illecebraceæ* allied to *Spergula*, and like it rather to be referred to a section of *Caryophyllaceæ*. It consists of tropical or sub-tropical herbs with slender diffuse stems often rooting at the joints, opposite leaves varying from subrotund to linear, often with small caducous stipules, and white flowers in paniculate or corymbose cymes. [J. T. S.]

DRYMODA *picta*. The name of a curious minute epiphytal orchid, with pseudobulbs, and apparently no leaves, found growing in Birmah, and described and figured by Dr. Lindley in the *Sertum Orchidaceum*, t. 8. The flower is single, on the end of a short scape, and inverted, that is' the labellum is uppermost. 'The column with its two long petal-like arms is undermost, and the long foot of the column stands over it, bearing at the apex a pair of pink and white lateral sepals, between which hangs down the deep red, fleshy, and hairy labellum.' The other parts of the flower are yellow with brown spots. The four pollen masses without caudicles, attached to a large globose fleshy stigmatic gland, make this plant a link between *Epidendreæ* and *Vandeæ*. [A. A. B.]

DRYMOGLOSSUM. A genus of small creeping polypodiaceous ferns, with simple fronds, belonging to the group *Tæniitideæ*. The fronds are either of two forms, the fertile ones more or less revolute or contracted, or else the fertile apex of the frond is contracted. The sori form thickish lines at or near the margin on the lower surface. The veins are reticulated, and very frequently obscure; they are, however, uniform, and form roundish or oblong-hexagonal areolæ, which enclose a few free veinlets. The species are not very numerous, but are widely scattered, occurring in India, China, and Japan, extending to Norfolk Island, and again occurring in the West Indies. The lines of sori, which are not covered, are sometimes placed directly on the surface of the frond, sometimes sunk in a little groove or channel. In some species the sterile fronds are nearly round; in others they are subcordate, or elliptic, or spathulate, while the fertile are twice their length, and of a linear or linear-oblong outline. The common typical species is *D. piloselloides*, a wide-spread eastern plant. [T. M.]

DRYMONIA. A genus of South American shrubs, belonging to the *Gesneraceæ*. They are twiners upon trees in moist places, throwing out rootlets from any part of the stem, and they have opposite serrated petiolate leaves, and large flowers on solitary axillary peduncles, the corolla being campanulate-ringent, gibbous at the base on the posterior side, and with the upper lip two-lobed and the lower three-lobed. The four included didynamous stamens are inserted at the base of the corolla tube, without any trace of a fifth. Seven species have been described. [W. C.]

DRYMOPHILA. A genus of *Liliaceæ* from Tasmania, consisting of herbs with erect stems leafless below, but with two-ranked narrowly-lanceolate sessile acute leaves above, and axillary and terminal one-flowered peduncles, supporting white flowers with six spreading segments. The fruit is a pendulous blue sub-globose three-celled berry. [J. T. S.]

DRYMYRRHIZE.E. A synonyme of *Zingiberaceæ*, under which the characters of the plants are given. [J. H. B.]

DRYNARIA. A genus of polypodiaceous ferns, generally distinguishable by the production of two separate kinds of fronds: the one pinnate or pinnatifid in the usual way, and bearing sori; the other very short, always sterile, coarsely veined, and soon acquiring a harsh dried appearance, quite stalkless, and lobed at the edge so as to resemble the leaf of an oak, whence they are called querciform. The fronds have a very compound venation, two or three series of irregular quadrate areoles being formed within each other, and free veinlets being produced in the ultimate areoles. The fructification is that of *Polypodium*. The genus, which is very well marked, is therefore known by its polypodioid fructification, its compoundly anastomosing venation, and its dwarfed querciform sterile fronds. The segments or pinnæ of the larger fronds readily fall away, being articulated at their base. In *D. quercifolia*, which is the type of the genus, the sterile oak-leaf fronds are four to six inches long, and the larger fertile ones from one to two feet or more in length, dark shining green, with long segments bearing a row of sori on each side of their costa. The few species now retained in the genus are all eastern, being found in India, and in the islands of the Pacific, extending as far as Australia and the Feejee Islands. In one species, *D. coronans*, the two forms of frond become combined in one, the fronds of this species being sessile and querciform at the base, but elongated upwards so as to bear the fertile segments on the upper part. Though normally and usually round, as in *Polypodium*, the sori in *D. coronans* sometimes become confluent in lines between the primary veins, and in that state are very similar to those of *Selliguea*. [T. M.]

DRYOBALANOPS. A tree, native of the island of Sumatra, yielding a kind of camphor. It constitutes a genus of *Dipteraceæ*, characterised by the calyx having a cup-shaped tube, and a limb divided into five leafy erect segments. The fruit is a capsule, enclosed within the cup of the calyx, and bursting when ripe by three valves: according to Professor Oudemans, of Rotterdam, the most recent investigator of this plant, and who has enjoyed better opportunities for so doing than his predecessors. It appears from his description in the *Annales des Sciences Nat.* (4 ser. v. 100), that the valves of the fruit, in separating from each other, carry with them the investment of the seed, so that the embryo is left exposed in the cavity of the fruit. The fruit is usually described as containing but one seed, but this is not always the case, as in some instances two seeds have been found. Standing up in the centre of the fruit is a little stalk or columella, which is concealed in a furrow of the seed, where it divides into two wings concealed beneath the edges of one of the cotyledons, which is considerably larger than the other. The seeds have been observed to germinate in the ripe fruit after the dehiscence of its valves.

D. aromatica or *D. Camphora* furnishes a liquid called camphor oil, and a crystalline solid known as Borneo or Sumatra Camphor. Camphor oil, which is obtained by incising the tree, has a fragrant aromatic odour, and has been employed to scent soap. The solid camphor is found in the cracks of the wood, and is obtained by cutting down the tree, dividing it into blocks and small pieces, from the interstices of which the camphor is extracted. It is rarely seen in this country, but fetches a very high price. It differs from ordinary camphor by its six-sided crystals, and its greater hardness and brittleness. It does not so readily become condensed on the sides of the bottle wherein it is kept, as ordinary camphor does. This camphor is much sought after by the Chinese, who attribute many virtues to it. It seems to have been long known, as it is mentioned by Marco Polo in the thirteenth century, and Camoens, in 1571, also mentions it as 'the balsam of disease.' [M. T. M.]

DRYOMENIS. A curious and somewhat anomalous genus of ferns belonging to the group having naked sori, and having the sori small and oblong, but arranged transversely to the veins and parallel with the costa, thus indicating a technical relationship with *Meniscium*. It has a compound form of venation, the pinnate veins being first united by transverse venules, and then again once or twice united by zigzag veins forming irregular areoles, from which in the sterile fronds free included veinlets branch out. The receptacles are seated on the transverse veins which join the primary veins proceeding from the costa, so that the sori are placed parallel to the costa. The only species, *D. menisciicarpon* of the Philippine Islands, is a rather coarse-growing fern with broad fronds, becoming taller and contracted with a less copious venation when fertile. It is sometimes associated with the *Polypodieæ*. [T. M.]

DRYOPTERIS. A name originally given by Adanson to the common male fern now called *Lastrea*, and subsequently also applied to a group agreeing with this in general structure. It has not, however, been generally adopted, the name *Lastrea* being preferred by some, who separate the free and netted veined species, and that of *Nephrodium* by others, who, irrespective of venation, combine in one group all the aspidioid plants with reniform fructification. It has been applied by some writers to a section of *Polypodium*. [T. M.]

DRYOSTACHYUM. A small genus of

ferns, remarkable for the diversity of the different parts of its fronds. The species are generally referred to the *Polypodieæ* group of true ferns, but on account of the sori being seated on a broad receptacle, consisting not of a point on one vein, but of a crowded network of fine veinlets or little veins, they have been sometimes placed along with *Platycerium*, in a small group called *Platycerieæ*, in which the same feature occurs. The fronds are leathery in texture, with prominent veins, merely pinnatifid, with the parts broad at the base, but in the upper part deeply divided into narrow or contracted segments, which are fertile. The parts are all articulated so as to separate spontaneously from the mainrib or rachis. The sori are large and generally quadrangular, closely set along each side of the costa, each of them covering or seated on a fine network of veins. They are without indusia. The venation is very compound, the veins and venules in the sterile parts anastomosing freely in almost equal-sided areoles, and enclosing free veinlets in the ultimate spaces. There are only two species known, both of which are natives of the Philippine Islands. [T. M.]

DRYPETES. A genus of *Euphorbiaceæ*, comprising a few West Indian and South American trees or shrubs, which have alternate oval or elliptical pointed leathery leaves, and inconspicuous flowers arranged in axillary fascicles, the males and females on different plants. The males have a calyx of four to six divisions, no petals, and two to six free stamens; the females have an ovary of one or two cells seated in a fleshy disk. The fruit is a hard elliptical dry drupe. The genus is nearly allied to *Hemicyclia*, but the latter has numerous stamens. [A. A. B.]

DRYPIS. A genus of *Caryophyllaceæ*, distinguished by having a one-seeded utricular capsule, which breaks across transversely. *D. spinosa*, a Mediterranean herb, has branched rigid fragile stems, with opposite subulate leaves ending in spines, and small rose-coloured flowers in dense corymbose cymes, with only five stamens in each. [J. T. S.]

DRY ROT. We are concerned with this subject only so far as it may be the effect of *Fungi*, or as calling those *Fungi* into especial notice. As, however, Dry Rot may be the effect of slow chemical combustion as well as of *Fungi*, and the results are much the same in either case, it is well that any mycologist who may attempt the investigation of the subject should be aware of the fact.

Dry Rot may be produced by various species, as *Polyporus hybridus*, *Thelephora puteana*, *Merulius lacrymans*, &c. In oak it is generally due to the first, and in the wood of conifers to the last. Different kinds of timber, moreover, in tropical countries have their own enemies, but these at present have not been sufficiently investigated. The spores of the *Fungi* may be brought into the dock-yard with the wood, in which case they are mostly the result of some ancient malady, and may have remained dormant in the wood altogether, or may have existed in the shape of minute spawn. The foxy oak, which is grown on old stools, owes its colour to incipient decomposition accompanied by delicate spawn, and when exposed to circumstances favourable to fungal development, the perfect form of the fungus by which it was injured will soon make its appearance. Elm trees are often strongly impregnated with spawn before they are felled, and we have lately seen *Polyporus ulmarius* bursting forth from the cut surface of an elm tree which fell a sacrifice to the spring gales of 1860, the whole wood being evidently affected.

When the fungus attacks the surface, it soon runs over it, and its spawn penetrates the wood, destroying all before it. The best remedy against Dry Rot consists in careful selection of wood, perfect ventilation, and patient seasoning, added to the employment of such kinds of wood for particular purposes as may be most suitable to the situation they are intended to occupy. Mineral salts may also be employed, but the remedy on which authorities in the present day insist the most is creosote, which has the property of coagulating albumen, and making it enter into combinations unfit for vegetation. It is, however, found that where the proper conditions have been secured, it is quite as economical to do nothing, for even with bad materials Dry Rot is not universal, and with good, attended by proper precautions, there will be little or no Dry Rot except under accidental circumstances which are favourable to its progress. In damp situations with imperfect ventilation, even should no fungus be present, decomposition is sure to take place, destructive to the wood, and prejudicial to the health of those who are constantly in its neighbourhood. In cellars and domestic buildings where the fungus has not already committed too much mischief, it may be effectually checked by washing it with a strong solution of corrosive sublimate. Where fungus does not exist, the remedy is scarcely applicable when decay has commenced, though it may be useful in the first instance. [M. J. B.]

DUBOISIA. A name applied to an Australian shrub, placed by Miers in *Atropaceæ*, but by others referred to *Scrophulariaceæ*. Its flowers are in axillary clusters, white, with a two-lipped calyx; corolla funnel-shaped, the limb five-parted; stamens five, included within the corolla, four fertile (two long, two short), and one rudimentary; ovary with two many-ovuled compartments; fruit berry-like. [M. T. M.]

DUBYÆA. A genus of *Compositæ*, nearly related to *Hieracium*, but differing in having beaked achenes. Of the three known species, one with the habit of a sowthistle is found in Australia; another, somewhat like a dandelion, is found in Armenia; and a third, like a hawkweed, grows in the Himalayas from Sikkim westward to Kumaon, at elevations between 8,000 and 12,000 feet. Its upper leaves are hairy, oblong, entire or toothed, and embracing the stem, while the lower are nearly triangular, and narrowed into a winged stalk. The yellow flower-heads have their lance-shaped involucral scales beset with black hairs, and the compressed striated achenes are narrowed into a beak, and crowned with a pappus of numerous rough hairs. The genus bears the name of M. Duby, a French botanist. [A. A. B.]

DUC DE TOLE. (Fr.) *Tulipa suaveolens.*

DUCHARTREA. A genus of *Gesneraceae*, containing a single species, a native of the mountains of Cuba. It is an erect branching shrub, wrinkled with resinous warts, and having coriaceous toothed leaves and greenish flowers in few-flowered corymbs. The corolla campanulate, slightly constricted on the underside, and the limb cut into five unequal roundish lobes, furnished with awned teeth around the margins; the stamens are didynamous with a sterile fifth. The base of the style is surrounded by an erect pentagonal cup. The warty oval fruit is crowned by the persistent calyx. [W. C.]

DUCHASSAINGIA. *Erythrina.*

DUCHESNEA. The name sometimes applied to an East Indian strawberry, *Fragaria indica*, with insipid fruit and yellow flowers. [A. A. B.]

DUCK'S-FOOT. *Podophyllum.*

DUCKMEAT or DUCKWEED. The common names of the curious floating aquatics, which form the genus *Lemna.*

DUCTS. Tubular vessels marked by transverse lines or dots; apparently in some cases modifications of spiral vessels, when they are called *closed, annular, reticulated,* and *scalariform;* sometimes analogous to pitted tissue, when they are called *dotted,* and form bothrenchyma.

DUCU. The resin of *Clusia Ducu.*

DUDAIM. A biblical plant, regarded as the Mandrake, *Mandragora officinalis.*

DUFOUREA. A genus of *Convolvulaceae*, containing five species of South American twining undershrubs with alternate entire leaves, and numerous white flowers in panicles on axillary or terminal peduncles. The calyx consists of five sepals, the two outer of which are membranaceous, and coloured, very large, almost hiding the funnel-shaped corolla, within which are five included stamens, with short subulate filaments. The two-celled ovary is surmounted by two styles or a single one deeply-cleft, with capitate stigmas. [W. C.]

DUFRESNEA. A Persian annual of the order *Valerianaceae.* The leaves are entire; the flowers in close cymes, sometimes unisexual; the calyx limb has three unequal ovate acute netted segments, which increase in size as the fruit ripens; the corolla is tubular, regular, spurless, its limb five-lobed; stamens three. The fruit is membranous, very hairy, crowned by the calyx lobes, three-celled, with two of the cells empty and distended. [M. T. M.]

DUGUETIA. A genus of *Anonaceae*, consisting of Brazilian trees with scurfy branches. The flower is not described, but the receptacle bearing the fruits is divided transversely into two sections, the lower globular, woody, marked with the scars of the fallen stamens, the upper portion somewhat conical, spongy, pitted; the carpels inserted on this receptacle are numerous, ovate, angular, terminated by the persistent styles, woody and one-seeded. *D. quitarensis* furnishes the light elastic wood, called Lance-wood, imported from Cuba and Guiana, for the use of coach-builders principally. [M. T. M.]

DUK. The horsehair-like fibres of the Gomuti palm, *Saguerus sacchariferus.*

DULCIS. Any kind of taste which is not acrid.

DULSE. A name given in Scotland to several different kinds of rose-spored *Algae*, but especially to *Rhodymenia palmata* and *Iridaea edulis*, which are extensively eaten on the sea-coasts, and which occasionally make their appearance in the market. We have ourselves been thankful for this coarse and parchment-like food amongst the Western Isles, when it was impossible to procure any other kind of sustenance. *Laurencia pinnatifida* affords an inferior Dulse, known under the name of Pepper Dulse. These species are generally eaten raw. When cooked they have an unmistakeable sea-twang, which, in spite of all the pains of Soyer, forbids their entrance into any acceptable food, where more sapid articles are procurable. [M. J. B.]

DUMERILIA. A genus of perennial Mexican herbs, belonging to the lip-flowered group of *Compositae.* They are smooth plants from one to three feet high, with sessile oval rigid leaves, which embrace the stem by their base, and shortly-stalked flower-heads disposed in terminal corymbs; each capitule is about half an inch long, and contains from five to fifteen white florets, enclosed by an involucre of about three series of lance-shaped scales. The achenes are slightly beaked, dilated at the apex, and crowned by a pappus of one series of numerous white pilose bristles. In the nearly-allied *Perezia*, the hairs of the pappus are in two series; but according to Dr. A. Gray, this character is here of little importance, and he would unite the genus to *Perezia* along with *Clarionea* and *Homoianthus.* The roots of the two known species are stringy, and the stem at the base is furnished with a tuft of rusty hairs. From the roots of *D. Alamani*, a curious chemical production known as Pipitzahuac is prepared; it resemhies flakes of gold, and is said to be powerfully drastic, with an odour of valerian, and useful as a dye. The plant is also known as *Perezia fruticosa* and sometimes *Acourtia rigida*. [A. A. B.]

DUMUS (adj. DUMOSE). A low branching shrub.

DUNBARIA. A small genus of twining plants of the pea family, the species of which are found in India, Java, and the surrounding Islands. In foliage and habit they are somewhat like *Phaseolus*, but the leaves are smaller. The large flowers are generally bright-yellow, and disposed in loose axillary racemes. The calyx is four-cleft to the middle; the corolla remarkable for the large membranous standard, much longer than the calyx, which embraces and hides the other petals, and has two callosities at its base. In some species it is an inch long. The pod is flattened and hairy, strongly compressed between the seeds. The genus is nearly related to *Cylista*, which, however, has a large membranous calyx completely hiding the corolla. It is named in honour of Prof. Dunbar of Edinburgh. [A. A. B.]

DUNGAN. *Myristica spuria*.

DUODENI. Growing twelve together.

DUPERREYA. A genus of *Convolvulaceæ*, containing a single New Holland species, a twining under-shrub, with petiolate narrow leaves, and solitary axillary flowers at the ends of the branches, having a somewhat funnel-shaped corolla. The capsule contains a single seed. [W. C.]

DUPLICATE. Growing in pairs. In composition the word indicates the repetition of a character: thus *duplicato-crenate* is when each crenel is itself crenate; *duplicato-dentate*, when each toothing is itself toothed; *duplicato-pinnate*, when the leaflets of a pinnate leaf become themselves pinnate; *duplicato-serrate*, when each serrature is itself serrated; and so on.

DUPLO. Twice as much as, or twice as many as.

DUPONTIA. A genus of grasses belonging to the tribe *Aveneæ*, distinguished by the inflorescence being in contracted panicles; spikelets ovate, two-flowered, with the rudiment of a third floret; stamens three; ovary smooth; styles two, feathery. The species are from the extreme northern limits of phænogamous vegetation: *D. Fischeri* from Melville Island, and *D. psilosantha* from Russian North America. [D. M.]

DUPUISIA. A genus of *Anacardiaceæ*, consisting of trees natives of Senegal. The calyx is cup-shaped, persistent, slightly five-toothed; petals five, concave, longer than the sepals; stamens five, inserted with the petals into the calyx; ovary one-celled, one-seeded. [M. T. M.]

DURAMEN. The heart-wood, or that part of the timber of a tree which becomes hardened by matter deposited in it. It is next the centre in Exogens, and next the circumference in Endogens.

DURANTA. A genus of S. American bushes of the vervain family, easily distinguished by the racemed flowers, and by the nature of the fruits, which are composed of four nuts enclosed in the calyx tube, which is contracted at top; they are hard and about the size of a pea, each nut with two one-seeded cells. Some are spiny, others unarmed; but all are straggling bushes with four-angled grey twigs, and opposite or whorled stalked leaves, in some like those of the privet, in others toothed. The pretty blue flowers are borne in great profusion in racemes towards the ends of the branches, each about half an inch long, and having a tubular five-ribbed five-toothed calyx, and a corolla about three times the length of the calyx, with a flat border of five unequal rounded lobes, nearly half an inch across. Some of them are said to be poisonous, and the seeds are not eaten by birds. *D. Plumieri* is in cultivation, and may often be seen in plant-stoves. About six species are known. [A. A. B.]

DURELIN. (Fr.) *Quercus sessiliflora*.

DURIÆA. A genus of *Ricciaceæ*.

DURIAN. *Durio zibethinus*.

DURIO. The tree producing the celebrated Durian fruit of the Indian Archipelago, *D. zibethinus*, is the only species of this genus of *Sterculiaceæ*. It forms a large forest tree, attaining sixty or eighty feet in height, with somewhat the general appearance of an elm. The leaves are entire, oblong, rounded at the base and taper-

Durio zibethinus.

ing upwards into a long point, densely covered beneath with minute scales, which give them a silvery red appearance. The flowers are yellowish-green, produced in little clusters upon the trunk or main branches, each flower having two large concave bracts at its base; the calyx is tubular and five-toothed; the corolla has five petals, which are partly joined so as to form a short tube; the stamens are numerous collected into five bundles, and have twisted or uneven anthers; and the scaly ovary is surmounted by a long thread-like style, and a simple round stigma. The fruit varies in shape, being either globular or oval, and measures as much as ten inches in length; it has a thick hard rind, entirely covered with very strong sharp prickles, and is divided into five cells, each of which contains from one to four seeds rather larger than pigeons' eggs, and completely enveloped in a firm luscious-looking cream-coloured pulp, which is the eatable portion of the fruit.

This tree is very commonly cultivated throughout the Malayan Peninsula and Islands, where its fruit, during the period it is in season, forms the greatest part of the food of the natives. Considerable diversity of opinion exists among epicures as to the relative merits of several well-known tropical fruits, including the Durian, the mangosteen, the cherimoyer, and the pine-apple, any one of which is made to occupy the foremost place, according to individual taste. The flavour of the Durian, however, is said to be perfectly unique; and it is also quite certain that no other fruit, either of tropical or temperate climes, combines in itself such a delicious flavour with such an abominably offensive odour — an odour commonly compared either with putrid animal matter, or with rotten onions. It might be supposed that a fruit possessing such an odour could never become a favourite; but it is said that when once the repugnance has been overcome, the Durian is sure to find favour, and that Europeans invariably become extremely fond of it. Mr. A. Wallace observes that 'a rich custard highly flavoured with almonds gives the best general idea of it, but there are occasional wafts of flavour that call to mind cream-cheese, onion-sauce, sherry wine, and other incongruous dishes. Then there is a rich glutinous smoothness in the pulp which nothing else possesses, but which adds to its delicacy. It is neither acid, nor sweet, nor juicy; yet it wants none of these qualities, for it is in itself perfect. It produces no nausea or other bad effect, and the more you eat of it the less you feel inclined to stop. In fact, to eat Durians is a new sensation worth a voyage to the East to experience.' The unripe Durians are cooked as a vegetable, and the pulp of the ripe fruit is salted and preserved in jars; while the seeds are roasted and eaten like chestnuts. [A. S.]

DURMAST. *Quercus sessiliflora pubescens.*

DURRA. *Sorghum vulgare.*

DUST BRAND. *Ustilago.*

DUTCHMAN'S BREECHES. *Dicentra Cucullaria.*

DUTCHMAN'S LAUDANUM. A tincture of *Passiflora rubra*, or, according to some, of *Murucuja ocellata.*

DUTCHMAN'S PIPE. An American name for *Aristolochia Sipho.*

DUTTONIA. A name originally proposed by Dr. Mueller for an Australian composite plant, which proved to be the same as *Dimorpholepis*; and afterwards applied by him to a myoporaceous shrub from South Australia, which he published as a new genus, but which he has more recently reduced to *Eremophila.*

DUVALIA. A name given by Haworth to some species of *Stapelia.*

DUVAUA. A genus of *Anacardiaceæ*, consisting of trees or shrubs, natives of China and the Sandwich Isles. They are sometimes armed with axillary spines; the leaves are entire; the flowers are in clusters, each with a four or five-cleft persistent calyx, four to five petals inserted beneath an eight-lobed disk; eight to ten stamens, those alternate with the petals longer than the others; and a sessile one-celled ovary. The drupe is pea-shaped, having the smell of juniper. Some of the species are grown as evergreen wall shrubs, with white or greenish flowers. Dr. Lindley remarks ' that the leaves of *D. latifolia* expel their resin with such violence, when immersed in water, as to have the appearance of spontaneous motion in consequence of the recoil.' [M. T. M.]

DWALE. The Deadly Nightshade, *Atropa Belladonna.*

DYCKIA. Brazilian herbs, named in honour of Prince Salm-Dyck, an amateur and patron of science. They constitute a genus of *Bromeliaceæ*, having lance-shaped pointed leaves, and bearing flowers in panicles, with spiny bracts. The perianth is six-parted, the three outer segments calycine, the three inner ones petal-like, bell-shaped, rather fleshy; the six filaments of the stamens are united below into a tube adherent to the inner segments of the perianth; the ovary is free, three-lobed, with three spreading forked stigmas. *D. rariflora* is a very showy plant with orange-coloured flowers. [M. T. M.]

DYER'S-WEED. *Reseda Luteola*; also *Genista tinctoria*, and *Isatis tinctoria.*

DYNAMIS. A power. A figurative term employed by Linnæus to express the degrees of development of stamens. Thus his *Didynamia* signified stamens of two different lengths, or of two different degrees of development.

DYSOPHYLLA. A genus belonging to the labiate order, distinguished from its congeners by the corolla having a short tube, the border divided into four nearly equal pieces, the upper division entire or slightly notched. The few species belonging to it are herbs, natives of India and Java; the leaves opposite or in whorls; the flowers in more or less dense clusters. The name is derived from Greek words signifying 'fetid leaf,' and descriptive of the odour of the plants, in which property they differ from most of the species of the same order. [G. D.]

DYSOXYLON. Large Javanese trees forming a genus of *Meliaceæ*, with compound leaves, whose leaflets are oblique at the base; the flowers are in axillary panicles with four or five-parted whorls; the tube, formed by the union of the stamens, is eight to ten-toothed with as many anthers in the interior; ovary three or four-celled surrounded at the base by a small disk; capsule three-celled, the seeds solitary in each compartment. [M. T. M.]

DYSSOCHROMA. A climbing Brazilian solanaceous shrub. It has a calyx of five persistent segments; a fleshy funnel-shaped corolla with the limb divided into five acute revolute segments; five protruding stamens, the anthers opening longitudinally, and surmounted by a small point; an erect style thickened at the top; and a two-celled ovary placed on a large fleshy disk. [M. T. M.]

DYSSODIA. A genus of composite herbs, nearly related to *Tagetes*, but differing in the nature of the pappus, which is composed of a number of chaffy scales pinnately or palmately divided above, and entire below, so that they appear like a polyadelphous pappus. Of the eight known species, two are found in the United States and the others in Mexico. Some have linear or lance-shaped entire or toothed leaves; in others they are pinnatisect. The yellow flower-heads are disposed in loose corymbs, or panicles at the ends of the branches, and have an involucre of one series of scales, more or less united by their edges, and often surrounded by an outer series of bracts. In a few species the florets are all tubular and perfect, but in most of them the outer ones are strap-shaped and contain a pistil only. Most of these herbs emit an unpleasant odour from the presence of oily matter secreted by the glandular dots of the leaves. *D. chrysanthemoides*, a dwarf annual with pinnatisect leaves, grows in great profusion over the western prairies of Illinois, and in autumn exhales so unpleasant an odour as to sicken travellers. [A. A. B.]

E, EX. In composition = without; thus ex-albuminous signifies without albumen.

EAGLE-WOOD. The timber of *Aloexylon Agallochum*; and also of *Aquilaria ovata*, and *A. Agallocha*.

EARAIHAU. *Ascarina polystachya*.

EARCOCKLE. The name of a curious disease in wheat, in which the grain becomes blackened and contracted, and mealy within from the presence of myriads of worms belonging to the genus *Vibrio*. The little animals are extremely tenacious of life, and though apparently reduced to dust, when steeped in warm water for a short time, after being dry for many months, they recover their former activity. The disease not only impairs the value of the wheat, but the little worms are very annoying to the miller from filling up the pores of his bolting-cloths. The affection is local, and quite unknown in many parts of England. In some districts it is called Purples. [M. J. B.]

EARED. The same as Auriculate.

EARTH-GALL. *Ophiorrhiza Mungos*.

EARTH-NUT. *Arachis hypogæa*.

EARTH-STAR. *Geaster*.

EARTH-TONGUE. *Geoglossum*.

EAU D'ANGE. A perfume distilled from the flowers of *Myrtus communis*. — DE COLOGNE. A well-known alcoholic perfume, to which *Lavandula vera* and *Rosmarinus officinalis* contribute their fragrance. — DE CRE'OLE. A stomachic distilled from the Mammee apple, *Mammea americana*. — DE MANTES. A liqueur distilled from *Croton balsamiferum*. — ME'DICINALE. A gout medicine prepared from *Gratiola officinalis*. — D'OR. A liquid distilled from *Convallaria majalis*. — D'ORME. A liquid secreted in certain galls of the elm.

EBENACEÆ (*Ebenads*). A natural order of corollifloral dicotyledons, belonging to Lindley's gentianal alliance. Trees or shrubs, not milky, with alternate exstipulate leathery and entire leaves; flowers hermaphrodite, or staminate and pistillate; calyx three to seven-cleft, persistent; corolla three to seven-cleft; stamens usually twice or quadruple the number of the corolline segments. Ovary three or several-celled, with one or two pendulous ovules in each cell. Fruit a round or oval berry; seeds albuminous. Chiefly Indian and tropical. A few are found in Europe, North America, the Cape of Good Hope, and New Holland. The trees of this order yield hard and durable timber. The bark of some is astringent, and the fruit is sometimes eatable. The heartwood of different species of *Diospyros* constitute the ebony of commerce, of which there are many varieties, e. g. *D. Ebenum*, Mauritius ebony, *D. Melanoxylon*, the ebony of Coromandel, and *D. Ebenaster*, the bastard ebony of Ceylon. *Diospyros hirsuta* yields the variegated calamander wood of Ceylon and the coasts of India, which is shipped from Bombay and Madras. The keg-fig of Japan is the edible fruit of *Diospyros Kaki*; while the persimmon is the fruit of *D. virginiana*. There are fifteen known genera and about 180 species. Illustrative genera: *Diospyros, Muba, Cargillia, Holochilus*. [J. H. B.]

EBE'NIER. (Fr.) *Diospyros Ebenum*.

EBENUS. A genus of the pea family, numbering about a dozen species, nearly related to *Onobrychis*, but the pods are smaller, and not toothed or crested. They are elegant little shrubs or biennial plants, chiefly confined to the high mountainous regions of Eastern Europe and Asia Minor, though *E. pinnata* is found in Algeria, and another as far east as Beloochistan. All their parts are commonly crowned with silky hairs; and the leaves are usually unequally-pinnate, made up of three to five pairs of linear or lance-shaped leaflets, though in a few they are digitate or simple. The peduncles are axillary or terminal, and bear dense spikes or round heads of pink, or violet blossoms, in which the deeply-lipped calyx is conspicuous, and densely clothed with silky hairs. The keel of the corolla has the very minute triangular wings adhering to its claw near the base, and of the ten stamens, nine are united into a tube, and one is free. [A. A. B.]

EBOE TREE. *Dipterix ebœnsis.*

EBONY. The timber of various species of *Diospyros*, especially *D. Ebenum, Ebenaster,* and *Melanoxylon.* — AMERICAN. *Brya Ebenus.* — GREEN. *Excæcaria glandulosa,* and also *Jacaranda ovatifolia.* — JAMAICA, or WEST INDIA. *Brya Ebenus.* — MOUNTAIN. *Bauhinia variegata.*

EBRACTEATE. Having no bracts.

EBURNEUS. Of the colour of ivory.

ECALCARATE. Having no calcar, or spur.

ECASTAPHYLLUM. A small genus of leguminous shrubs found in S. America and W. Africa, nearly related to *Dalbergia,* which has long straight thin pods, while these have flat, nearly orbicular one-seeded pods, whose valves have a tendency to a corky thickening. The leaves are sometimes simple, but more generally pinnate, and made up of three to five pairs of oval leaflets, and an odd one. Their little white straw-coloured or reddish-purple pea flowers are disposed in short axillary cymes. The calyx is bell-shaped and five-toothed; the stamens eight or nine, but more usually ten, nine united and one free. *E. Monetaria,* a pinnate-leaved species with white flowers found in Surinam, has red wood, which is said to furnish a resin like dragon's blood, and the root when cut emits a purple juice. The name is sometimes written *Hecastophyllum.* [A. A. B.]

ECBALIUM. A genus of *Cucurbitaceæ* closely allied to *Momordica,* from which it differs in the absence of tendrils, and of rudimentary stamens in the female flowers, and by the peculiarity of the fruit, which when ripe separates from the stalk and expels, with considerable force, the brown seeds through the aperture made by the removal of the stalk. *E. agreste (Momordica Elaterium),* the Squirting Cucumber, a native of waste places in the south of Europe, is an annual plant with prostrate branching stems, and heart-shaped rough leaves. The flowerstalks are axillary; the male flowers in clusters with bell-shaped yellow green-veined corollas; the females solitary. The fruit is a small elliptical greenish gourd, covered with soft triangular prickles. These fruits forcibly eject their seeds, together with a mucilaginous juice, a phenomenon said by Dutrochet to be due to endosmosis.

The drug known as Elaterium is the dried precipitate that is deposited from the juice which flows from the fruit or rather from the pulp surrounding the seeds. So powerful is pure elaterium, that one eighth part of a grain is sufficient to produce strong cathartic effects; it is, however, rarely obtained pure. It is of great value in certain cases of dropsy and of cerebral disease, where an active remedy is required; but as its action is violent it

Ecballium agreste.

requires to be administered with great caution, and in cases where there exists no objection to its use. The active principle of elaterium is a crystalline substance called *elaterin.* The plant is grown for medicinal purposes at Mitcham and elsewhere. It is related of Dr. Dickson, who was formerly lecturer on botany at St. George's hospital, that he suffered severely from the effects of this plant, in consequence of having conveyed some specimens of it in his hat from the Jardin des Plantes to his lodgings in Paris. [M. T. M.]

ECBLASTESIS. The production of buds within flowers, in consequence of monstrous developement; or on inflorescences.

ECCREMOCARPUS. Handsome climbing plants with a somewhat shrubby stem, long succulent branches, much-divided leaves, terminating in a branched tendril, and tubular yellow or green flowers, which are divided into five equal lobes. The stamens are four, two longer than the others, with the rudiment of a fifth. The seeds are produced in a one-celled two-valved ovate capsule, and are surrounded by a membranous wing, on which account they are favourite objects for microscopes of low power. The genus belongs to the *Bignoniaceæ. E. longiflorus* has a red calyx and a very long corolla with a yellow tube and green limb. *E. scaber,* a handsome Chilian species with orange-coloured flowers, much cultivated as an ornamental creeper, is sometimes called *Calampelis scabra.* [C. A. J.]

ECHALOTTE. (Fr.) *Allium ascalonicum.*

E'CHARBOT. (Fr.) *Trapa natans.*

ECHEANDIA. A genus of *Liliaceæ,* nearly related to *Phalangium*, from which it differs in the club-shaped filaments of the stamens being furnished above with short obtuse recurved teeth. The six known species, which extend from Mexico southwards to Brazil, are perennial herbs, with roots consisting of fascicles of fleshy fibres, grassy root leaves six inches to a foot or more in length, and rising from the midst of these a branching flower-stem, six inches to four feet high, with narrow bracts at the forking points, and white or orange-yellow asphodel-like drooping flowers, disposed in racemes, the individual flowers seldom more than half an inch across. *E. terniflora*, a yellow-flowered Mexican species, has been cultivated in English gardens. [A. A. B.]

ECHEVERIA. A handsome genus of succulent often fruticose plants belonging to the *Crassulaceæ*, and chiefly natives of Mexico. The leaves, which are generally glaucous, and sometimes excessively so, are not uncommonly spathulate in form, sometimes disposed alternately along the stem, sometimes collected into rosulate tufts. The flowers are in racemes or cymes, often secund, and generally of a bright scarlet or yellow colour, and very ornamental; they have a five-parted calyx, a perigynous five-parted corolla, whose erect segments close up into a pitcher-like form, ten included stamens, with short hypogynous scales, and five free one-celled ovaries, which become many-seeded follicular capsules. Many of the species are in cultivation, and they are esteemed as including some of the most interesting and beautiful of greenhouse succulents. *E. secunda* and *glauca* are particularly ornamental dwarf herbaceous species, well adapted for indoor window gardens. [T. M.]

ECHIALES. One of Lindley's alliances, which includes the *Boraginaceæ*, *Labiatæ*, &c.

ECHINACANTHUS. A small genus of *Acanthaceæ*, containing four species, natives of India. They are herbs with denticulate leaves, and small flowers which grow in secund axillary cymes running into a terminal panicle, and furnished with narrow bracts and no bracteoles. The calyx is deeply five-cleft, the corolla funnel-shaped, the stamens four, included, didynamous, united in pairs at the base of the filaments, and the stigma simple. The round two-celled capsule bears many seeds. [W. C.]

ECHINAIS. A small genus of *Compositæ*, found in Armenia, Siberia, and N. W. India. The leaves and flower-heads are very like those of our own *Carduus arvensis*. The chief characters which separate these plants from *Carduus* are the thin and lacerated apices of the involucral scales, which end in short spiny points, and the short lacerated tails seen at the base of the anther lobes. [A. A. B.]

ECHINARIA. A genus of grasses belonging to the tribe *Pappophoreæ*. The inflorescence is in simple globose spikes; spikelets two to four-flowered, the superior flower stalked; glumes two, membranaceous and keeled, the lowest with two awns at the tip, shorter than the superior one, which has only one awn at the apex; paleæ or inner glumes two, the lowest five-nerved and cleft at the tip. Of this small genus only two species are described, namely, *E. capitata*, which is a native of Africa as well as Syria, and *E. pumila*, a native of Spain. [D. M.]

ECHINATE. Furnished with numerous rigid hairs, or straight prickles; as the fruit of *Castanea vesca*.

ECHINOCACTUS. The plants composing this genus of Indian figs, like many others belonging to the same natural order, assume most grotesque forms. The name is derived from two Greek words, *echinos*, a hedgehog, and *kaktos*, a prickly plant, in allusion to many of the species being globular and thickly beset with spines, resembling a rolled-up hedgehog. There are hosts of species enumerated in botanical works, more than half of them natives of Mexico, and the rest distributed throughout South America, extending as far south as Buenos Ayres and Mendoza.

Like the generality of the order, they delight in hot, dry, sandy, or stony places, exposed to the full power of the sun. They consist merely of a fleshy stem, without leaves, and are either of a globular form slightly flattened at the top, or oblong, or cylindrical, and only attain a large size when they are very old. Whatever their shape or size, the stems are always either more or less fluted and ribbed, or covered with tubercular swellings, the number of the ribs varying in the different species, being numerous and sharply defined in some, while in others they are fewer and merge into each other. Nearly all the species are armed with stiff sharp spines, arranged in clusters, and seated upon little woolly cushions placed at intervals along the edges of the ridges, or on the tips of the tubercles. The flowers are generally large and showy, and are produced at or near the top of the plant, growing from the upper side of the younger fascicles of spines; but in some species the top of the plant is densely covered with light brown wool, from out of which the flowers proceed. The calyx has a broad generally short tube, the lower or outermost sepals being of a scale-like character, and the upper ones more like petals, into which, in fact, they gradually pass, the inner petals spreading out and radiating. The numerous stamens are fixed to the inside of the calyx-tube, and are shorter than the petals. The style is columnar, and separates into from five to ten radiating stigmas, which project very slightly beyond the stamens. The fruit is generally scaly or prickly, and is crowned with the withered remains of the flowers.

E. Visnaga, which is perhaps the largest of the genus, is a native of San Luis de Potosi, in Mexico. Large plants of this have from forty to fifty sharp ridges, with the clusters of spines sunk into their edges at short intervals. The aggregate number of these spines upon a single plant is something enormous; a comparatively small plant in Kew Gardens was

Echinocactus Visnaga.

estimated to have 17,600, and a larger specimen, at the same place, could not have had less than 51,000. The Mexicans commonly use them for toothpicks, hence the specific name *Visnaga*, which means a toothpick. The flowers are bright yellow. Some years ago a plant of this species, weighing one ton, and measuring nine feet in height by three in diameter, was forwarded to Kew, where, however, it lived only a short time. [A. S.]

ECHINOCARPUS. A small genus of *Tiliaceæ*, found in India, Java, and Eastern Tropical Australia. They are most nearly related to *Sloanea*, but differ in having petals. All are large trees with alternate stalked oval oblong or lance-shaped leaves. The flowers are arranged in short axillary racemes or fascicles, seldom in terminal panicles; and each is about half an inch in diameter, with a five-parted calyx, five lacerated petals, numerous stamens with pointed anthers, and a five-celled ovary crowned with a simple style. In some species the fruits are beset with straight prickles, and resemble those of the Spanish chestnut; in others the outer covering of the fruit consists of short crisp closely-packed rigid hairs. In all they are five-celled with five seeds, and split when ripe into five woolly portions. This prickly covering of the fruit has suggested the name. [A. A. B.]

ECHINOCEREUS. A genus of *Cactaceæ*, sometimes combined with *Cereus*, but in the latter the tube of the flowers is very long, while in *Echinocereus* it is always short, besides which the fruit is crowned with the withered remains of the flower, and the seeds are always rough or warted, not smooth, as in true *Cereus*. The species number between twenty and thirty, and are all natives of the hot dry regions of Mexico and Texas. They seldom exceed a foot in height, the stems being simple or branched, and either divided into very numerous ridges, or with only from four to ten, all being formidably armed with sharp spines. *E. pectinatus* grows about eight inches high, and two inches thick, and has about twenty ridges bearing at short intervals dense clusters of very small yellowish and rose-coloured spines. The fruit, like that of several other species, is of a purplish colour, and very good eating, resembling a gooseberry. The Mexicans, who call the plant Cabeza del Viego, eat the fleshy part of the stem as a vegetable first carefully freeing it of the spines. [A. S.]

ECHINOCHLOA. A genus of grasses of the tribe *Paniceæ*. The species are now generally included under *Panicum*. [D. M.]

ECHINOCYSTIS. A North American cucurbitaceous annual with climbing stems, palmate leaves, branching tendrils, and small greenish flowers; the males in clusters, the females in juxtaposition, either solitary or in tufts upon a short stalk. The calyx segments and petals are six in number, the stamens three, in two parcels, with connate wavy anthers. In the female flower there are three abortive stamens. The two-celled fruit is somewhat globular, spiny, at first juicy, but subsequently dry and fibrous. [M. T. M.]

ECHINOLÆNA. A genus of grasses belonging to the tribe *Paniceæ*, now included in *Panicum*. [D. M.]

ECHINOPE DE RUSSIE. (Fr.) *Echinops sphærocephalus*.

ECHINOPHORA. A genus of umbellifers distinguished by the prickly character of the parts which surround the flowers and fruit. The species are perennial herbs, having generally a rigid habit, and, as the name implies, partly covered with spines. They are chiefly found on the borders of the Mediterranean, and are more of interest owing to their peculiar appearance than on account of any useful or economical property which they possess. *E. tenuifolia*, found on some parts of the Mediterranean shore, is, however, reported as acting moderately upon the kidneys. One species, *E. spinosa*, still holds a place in the British Flora, having been reported as found on the sandy sea-shores of Lancashire and Kent; but it is now extinct. [G. D.]

ECHINOPOGON. A small genus of curious grasses belonging to the tribe *Agrostideæ*, having the inflorescence in crowded ovate panicles; stamens three; styles two, with plumose stigmas; fruit oblong-lanceolate, awned. The species are all natives of New Holland. [D. M.]

ECHINOPS. A genus of the composite family, numbering upwards of thirty species, many of them known as Globe Thistles. They are remarkable for having the heads one-flowered and arranged in dense round clusters at the ends of the branches, so that each cluster of flower-heads has the appearance of a single head containing many florets. They are found as far eastward as Kumaon in the Himalayas, extend westward to Spain, and appear in greatest numbers in Asia Minor. Some are annuals, but most of them are biennial or perennial erect simple or branching herbs from two to six feet or more high, furnished with large thistle-like spiny leaves, once, twice, or thrice pinnately-parted, the lower surface usually covered, like the stems, with loose white wool. The flowers are white or pale blue, and the compound heads one to three inches in diameter, surrounded by a common involucre of narrow scales, while each separate single-flowered head has an involucre of numerous scales, the outer hair-like, the inner broader and spiny-pointed. The silky cylindrical achenes are crowned with a pappus of numerous short bristles. *E. strigosus*, an annual species, native of Spain, is said to yield the substance known as Spanish tinder. Three sorts of it are prepared, one from the pubescence of the flower-heads, another from that of the leaves, and a third from that of the stems. [A. A. B.]

ECHINOPSIS. A genus of Indian figs, formerly combined with *Echinocactus*, but now separated and placed with the *Cereideæ*, distinguished by the flowers being produced from the side of the stem, instead of at the top, as in the *Echinocactideæ*. They have fleshy stems of a flattened globular or cylindrical form, divided into numerous ridges, which either run uninterruptedly from the apex to the base and bear clusters of spines at intervals, or are waved or notched, and have the spines placed in the depressions. In some species the spines are of great length. The flowers are very large, and in many species exceedingly handsome, forming a striking contrast with the ungainly appearance of the plants themselves: they have a very long tube, more or less covered with bristly or hairy scales, which increase in size towards the upper end of the tube, and at length merge into sepals, the sepals in their turn passing into petals. The stamens are arranged in two series, the inner attached to the bottom of the tube, and the outer growing to the tube throughout its whole length, and becoming free at the orifice, forming a circle around it. The thread-like style, scarcely longer than the stamens, is surmounted by a many-rayed stigma. Between twenty and thirty species, natives of Bolivia, Chili, Mexico, Brazil, and Texas, are described. [A. S.]

ECHINOPTERYS. The name of a Mexican shrub, constituting a genus of *Malpighiaceæ*, with yellow flowers in terminal clusters, and which are jointed to the stalks supporting them. The calyx is without glands; the petals five, stalked, of unequal length; stamens ten, all fertile, the filaments united into a tube at the base, the anthers hairy; ovary three-lobed, densely hairy. The fruit consists of three indehiscent spiny carpels. [M. T. M.]

ECHINOSPERMUM. A genus of *Boragineæ*, distinguished by having a salver-shaped corolla, which has the throat closed by five small scales; and three-edged nuts, with the anterior face margined and often bordered with one or more rows of hooked prickles; and by the inner angle of each of the four carpels adhering by its whole extent to a central column. They are hairy herbs resembling *Myosotis*, with narrow leaves and small blue flowers, in bracteated scorpioid racemes. The species are most abundant in the temperate regions of the northern hemisphere. *E. Lappula*, which is one of the erect-fruited species, has been found in England at Southwold, but doubtless an accidental introduction. [J. T. S.]

ECHIUM. A genus of *Boragineæ*, distinguished by its tubular bell-shaped corolla, open at the throat (without scales or plaits), and with an irregular limb, bearing some resemblance to that of some of the labiates. They are bristly or hairy plants, generally distributed, especially abundant in the Mediterranean region, where most of the species are herbaceous, and in the Canaries, where the greater number are shrubby. The flowers are usually large, in small curled spikes, arranged in a compound spike or panicle. *E. vulgare*, the common Viper's Bugloss, is a very rough plant with strap-shaped leaves, narrow at the base, and bright blue flowers whose stamens exceed the corolla; this occurs throughout Britain. *E. violaceum* is not found in Britain proper, but is common in Jersey; its flowers are much larger, more purple, with shorter stamens, and the leaves clasp the stem by a broad base. [J. T. S.]

ECLAIRE, or ECLAIRE GRANDE. (Fr.) *Chelidonium majus*. — PETITE, or ECLAIRETTE. (Fr.) *Ranunculus Ficaria*.

ECLIPTA. A genus of erect or prostrate annual or biennial weeds of the composite family, approaching dahlias in the structure of their flowers, but widely different in habit, and pretty equally distributed over all tropical countries. The leaves are usually opposite and lance-shaped, with entire or toothed margins, and the white stalked flower-heads, growing one to three together, proceed from the axils of the leaves, and are about half an inch across. The receptacle is flat and furnished with bristle-like scales, between the florets. The achenes of the ray-florets are triangular, those of the disc compressed; and the pappus is either absent altogether, or when present reduced to a minute toothed border. [A. A. B.]

ECOSTATE. Not having a central or strongly-marked rib or costa.

ECTADIUM. A genus of South African *Asclepiadaceæ*, containing a single species, an undershrub with opposite coriaceous leaves, and small yellow flowers in sub-axillary racemes. The calyx is five-parted; the corolla salver-shaped with five oblong unequal lobes; the staminal crown of five lanceolate included scales; the stamens included, free, their anthers densely hairy at the back; and the pollen-masses adpressed to an oblong truncate corpuscle. The stigma is pentagonal and apiculate. The follicles are smooth, slender, obtuse, and divaricate, with comose seeds. [W. C.]

ECTOCARPEÆ. A natural order of dark-spored *Algæ*, consisting of olive-jointed threadlike seaweeds, whose spores are mostly external, attached to the branchlets or formed in a swelling of their substance. It differs principally from *Chordarieæ* in the less compound frond and external spores. The fructification is often of two kinds in the same species. They are most abundant in temperate regions, though several are found in warm seas. [M. J. B.]

ECTOCARPUS. A genus of dark-spored *Algæ*, with a branched threadlike jointed soft flaccid frond, and remarkable for the different aspects assumed by the fruit. The secondary form is disposed in podlike bodies, which are variously articulated. A good many species are found on our coasts, and are more easily distinguished by their fruit than by the character of the frond. The cells of the pods produce zoospores. It is not quite certain whether the endochrome of the so-called spores is ever resolved into zoospores. *Ectocarpus* is known from *Sphacelaria* by the less elegant branching, and the soft not rigid threads. These are sometimes collected in bundles by the action of the waves, but never essentially combined, as in *Mesogloea*. [M. J. B.]

ECTONEURA. *Polybotrya*.

ECTOZOMA. A genus of *Atropaceæ*, represented by a shrub, native of Ecuador, of somewhat climbing habit, and with flowers in terminal panicles. The calyx is thick, bell-shaped, with five triangular erect divisions; corolla fleshy, tubular, somewhat dilated in the middle, its lobes roundish, and overlapping before expansion; stamens five, with very short filaments attached to a thin hairy ring surrounding the ovary; stigma globular. The fruit is unknown. [M. T. M.]

ECUELLE D'EAU. (Fr.) *Hydrocotyle vulgaris*.

EDDOES. The tuberous stems of various araceous plants, as *Colocasia esculenta*, *antiquorum*, &c., *Caladium bicolor*, *violaceum*, and others.

EDDYA. A genus of *Boraginaceæ* from Texas and New Mexico, containing a small much-branched very hispid prostrate undershrub with crowded leaves and solitary axillary white flowers; corolla salver-shaped, naked at the throat; stamens inserted at the apex of the tube of corolla; nuts ovate, cohering by their internal angles, muricate. [J. T. S.]

EDENTATE. Not having teeth.

EDGED. When one colour is surrounded by a very narrow rim of another.

EDGWORTHIA. A genus of plants belonging to the *Thymelaceæ*, named by C. A. Meyer in honour of Mr. Edgworth, an Indian botanist. The flowers have a single perianth, the limb of which is divided into four ovate blunt lobes. There are no perigynous scales, but one emarginate hypogynous one; stamens eight, nearly sessile, arranged in two distinct lines, one above the other; ovary covered with hairs, one-celled, containing a single suspended ovule; style threadlike, ending in an elongated awl-shaped stigma. There are two species of the genus, *E. chrysantha*, found in Chusan by Mr. Fortune, a shrub with yellow flowers, and oblong-lanceolate leaves of a very dull green, covered with hairs closely pressed to the surface; and *E. Gardneri*, found in Nepal. [J. H. B.]

EDMONSTONIA. A genus of plants named after Thos. Edmonstone of Shetland, naturalist of the Herald. It belongs to the *Samydaceæ*, and has a coloured persistent four-cleft perianth; four stamens inserted into the bottom of the calyx, the filaments free, the anthers introrse; and a free one-celled ovary with three parietal placentas, and numerous ovules. There is one known species, *E. pacifica*, which is a shrub ten to twelve feet high, native of the promontory of Corrientes in Darien. [J. H. B.]

EDRAIANTHUS. The generic name of plants belonging to the order of bellworts, and characterised by the number five prevailing in the flower; the stamens free, their filaments broad at the base; the ovary with two or three cells; and the seeds ovate and plain. The name is defined from Greek words signifying 'sessile or stalkless flower.' The species are natives of Southern Europe, and usually in the form of small tufted herbs with narrow alternate leaves, which are often furnished with stiff hairs; the individual flowers are stalkless but grouped in heads. [G. D.]

EFFLORESCENT. The action of beginning to flower.

EFFOLIATION. The removal of leaves.

EFULCRATE. Said of buds from below which the customary leaf has fallen.

EGENOLFIA. *Polybotrya*.

EGERIA. A genus of *Hydrocharidaceæ* from South America, consisting of water plants with the habit of *Anacharis*, having dichotomous branches and verticillate linear leaves with finely serrated margins. The spathe of male flowers is axillary and sessile, the flowers themselves resembling those of *Hydrocharis*; female flowers unknown. [J. T. S.]

EGG-PLANT. *Solanum esculentum* (*Melongena*) and *ovigerum*.

EGG-SHAPED. The same as Ovate.

E'GILOPE. (Fr.) *Ægilops*.

E'GILOPS. (Fr.) *Quercus Ægilops*.

EGLANTIER. (Fr.) *Rosa Eglanteria*. — JAUNE. *Rosa lutea*. — ODORANT, or

ROUGE. *Rosa rubiginosa.* — SAUVAGE. *Rosa canina.*

EGLANDULOSE. Not having glands.

EGLANTINE. *Rosa Eglanteria*, and *Rubus Eglanteria*: also applied to *Rosa rubiginosa*, the Sweet Brier.

EGREVILLE. (Fr.) *Lactuca perennis.*

EHRETIACE.E (*Ehretiads*). A natural order of dicotyledonous plants belonging to De Candolle's subclass *Corolliflorae*, and to Lindley's cohial alliance of peryginous exogens. The plants are closely allied to the borageworts, differing in their terminal style, perfectly concrete four-celled ovary, and drupaceous fruit. Trees, shrubs or herbs covered with rough hairs; leaves alternate, simple, without stipules; inflorescence scorpioid; aestivation imbricate. Calyx inferior with five divisions; corolla gamopetalous and tubular; stamens five, alternate with the corolline segments. Ovary on a circular disk, two to four-celled, with a terminal style or two-lobed stigma. Fruit fleshy, with a single seed in each cell. Chiefly tropical plants, though some occur in the South of Europe, others in the Southern States of America. They have scarcely any important properties. A few are febrifugal, astringent, and alterative. The Peruvian heliotrope, cultivated since 1740, has a delightful odour. There are fifteen known genera and about 330 species in the order. Illustrative genera: *Ehretia, Tournefortia, Heliotropium.* [J. H. B.]

EHRETIA. A genus of *Ehretiaceae*, consisting of tropical trees or shrubs with paniculate or corymbose flowers which are usually white. The calyx is deeply five-parted, the corolla salver-shaped with a five-parted limb, the stamens five, the ovary four-celled, and the fruit a berry-like drupe with two or four stones, each containing a single seed. *E. buxifolia*, an Indian shrub with sessile wedge-shaped shining scabrous leaves and axillary few-flowered peduncles, is employed as an alterative, and is also regarded as an antidote to vegetable poisons. *E. serrata*, an Indian tree, with oblong serrated smooth leaves and fragrant flowers, yields tough light and durable wood. [J. T. S.]

EHRHARTA. A genus of grasses belonging to the tribe *Oryzeae*, distinguished by the inflorescence being in compressed spikelets, three-flowered; flowers nearly together; the two lower neutral, one-paled, thick and keeled, mucronate or with short awns; the terminal hermaphrodite, two-paled; stamens six; styles two, with feathery stigmas. The species belonging to this curious genus are mostly natives of the southern hemisphere, South Africa, and New Holland. [D. M.]

EICHORNIA. A genus of *Pontederaceae* from South America. Kunth restricts *Pontederia* to the species in which two of the cells of the ovary are abortive, while *Eichornia* has a three-celled three-valved many-seeded capsule. They are aquatic plants with roundish rhomboidal stalked radical leaves, and a scape with a single leaf or spathe like the root-leaves, and a spike of lilac or blue flowers. *E. speciosa*, widely spread on the continent of South America, is a very handsome plant with a ten or twelve-flowered spike and the petioles of the leaves curiously swollen, the enlargement consisting of very loose spongy tissue. It is often cultivated in stoves under the name of *Pontederia azurea*, or *crassipes*. [J. T. S.]

EICHWALDIA. A genus of the *Reaumuriaceae*, distinguished by its many-leaved calyx. The only known species, *E. oxiana*, found on the Oxus river, which flows into the Caspian Sea, is a scrubby little bush with white stems, alternate linear fleshy leaves, and few somewhat racemed flowers, almost half an inch across. Inside the calyx of numerous round bract-like leaves are five-clawed petals, numerous stamens, and an ovary crowned with five styles. The fruit is a little capsule opening by five valves, apparently one-celled at top, but distinctly five-celled below. [A. A. B.]

ELACHISTA. A small genus of parasitic *Algae* allied on the one hand to *Ectocarpus*, and on the other to *Choriaria*. In *E. scutulata* the threads are so intimately combined with the tissue of *Himanthalia lorea* that it is impossible to say where the one begins and the other ends. Indeed, did not the species produce distinct fruit, they might justly be reckoned as mere transformations of the cells of the mother plant. [M. J. B.]

EL.EAGIA. A genus of lofty cinchonaceous trees, natives of the Cordilleras. The flowers are arranged in terminal clusters; they have a cup-like calyx; a corolla with the short tube bulging at top, and a spreading limb; and stamens with very short filaments, and broadly ovate anthers. The globose capsule is ribbed, and bursts into two or four valves. These trees are remarkable for the quantity of green resinous or waxy matter which is secreted by the stipules, which invest the unexpanded buds. This resin is collected by the Indians, and is employed by them to varnish boxes and many other useful or ornamental objects. For this purpose it is purified by immersion in hot water; its fragility is then removed by chewing it till it becomes ductile; and after these processes it acquires a yellow tint, and is ready to receive the various colours imparted by adding colouring matter to it when melted. The resin when thus prepared and coloured is laid on in thin layers by the aid of heat and pressure, and by means of differently coloured layers placed one upon another and cut out into various shapes, a sort of design is produced. To procure a metallic lustre on the objects covered with the varnish, the Indians first coat the surface with a layer of silver foil. The natives speak of the tree producing this resin, *E. utilis*, as the Wax tree or

Varnish tree. M. Triana, to whose account in the *Bulletin de la Soc. Bot. de France*, 1858, p. 500, we are indebted for these particulars, dwells with justice on the importance of developing this manufacture, and of cultivating this and allied plants producing similar secretions in other localities. The temperature of the district where the *Elæagia* is chiefly found ranges from 54° to 74° F. Some better method of preparing the resin might no doubt be adopted. [M. T. M.]

ELÆAGNACEÆ (*Oleasters*). A natural order of monochlamydeous dicotyledons, belonging to Lindley's amental alliance of diclinous Exogens. Trees or shrubs usually covered with scales or scurf, having exstipulate entire leaves, and usually imperfect flowers. Staminate flowers in catkins, arising each from a scale-like bract; perianth of two to four leaves, sometimes united; stamens three, four, or eight. Pistillate and perfect flowers with a tubular perianth and a fleshy disk; ovary free, one-celled, with one ovule; fruit a crustaceous achene, enclosed within the succulent perianth. Chiefly natives of the northern hemisphere. Represented in Britain by *Hippophäe rhamnoides*, the sea buckthorn, a spiny shrub which grows well near the sea, and forms a good fence; it is covered with silvery scurf, which is a beautiful object under the microscope; its fruit is sometimes eaten. *Elæagnus parvifolia* bears clusters of red edible berries, mottled with scales. The fruit called in Persia zinzeyd is the produce of *Elæagnus orientalis*. Some of the plants of the order are said to possess narcotic qualities. There are four known genera and thirty species. Examples: *Shepherdia, Hippophäe, Elæagnus*. [J. H. B.]

ELÆAGNUS. The Oleaster or Wild Olive tree. A small tree native of the southern countries of Europe and several parts of Asia, which received its name from its resemblance to the true olive, from which, however, it differs in not bearing useful fruit. The two plants in reality belong to different orders, the present plant giving name to the order *Elæagnaceæ*. *E. hortensis*, the species most commonly grown in English gardens, attains the height of from fifteen to twenty feet. The leaves are long and narrow, covered, as well as the young shoots, with stars of hairs of a hoary colour. The branches are brown and smooth, more or less spiny. The flowers are of two kinds, some containing stamens and pistils, which are four-cleft, pale yellow within; the others, with stamens and an abortive pistil, are five to eight-cleft, and of a golden yellow within; all are axillary, two or three together on short stalks, and fragrant. It flowers in May, and ripens its fruit, which is of a red-brown colour, something like a small date, in August. The blossoms, which are produced in great abundance, perfume the air for a considerable distance round. For this reason it is a most desirable tree for a lawn or shrubbery. French, *Olivier de Bohème*; German, *Wilde Oelbaume*. [C. A. J.]

ELÆIS. A genus of palms comprising the Oil Palm of Africa, and another closely allied American species. They have thick trunks of no great height, indeed the American species creeps along the ground, and bears a tuft of large pinnate leaves, with strong prickly stalks. The male and female flowers are borne in distinct heads, generally upon different trees, but occasionally upon the same; each head consisting of numerous little branches of minute flowers, gathered together into a dense mass and enclosed while young in two complete spathes. The males are packed very close together, so that the branches resemble catkins; the females are spread farther apart. The fruit, which is yellow or bright red, is irregular in form, gener-

Elæis guineensis.

ally angular and somewhat three-sided, and larger at the bottom than the top. It consists of an outer coating of fibrous oily flesh, surrounding a hard nut. *E. guineensis*, the African Oil Palm, which yields the celebrated palm oil, is a native of tropical Western Africa, where it is found in great abundance; and from whence it has been introduced into the West Indies. It grows twenty or thirty feet high, the trunk being covered with the remains of the stalks of dead leaves. The fruits are borne in dense heads, measuring a foot and a half or two feet long, and from two to three feet in circumference, the individual fruits being about an inch and a half long, by an inch in diameter. The part yielding the palm oil is the outer fleshy coating of the fruit, but the seed, which is enclosed in a hard shell, likewise affords an oil, small quantities of which occasionally come to this country. Commercial palm oil is about the consistence of butter, of a bright orange-red colour, and has a rather pleasant violet-like odour when perfectly fresh. It is obtained by boiling the fruits in water and skimming off the oil as it rises to the surface; and as its production and preparation is carried on solely by the negro population, who bring it to the merchants in small quantities for sale, it is anticipated that ere long the Negro kings will find the trade in palm oil more profitable than that in human beings. In 1860 the imports of palm oil into the United Kingdom amounted to 804,338 cwts., representing a money value of 1,786,895*l*. The chief use to which this substance is applied is for the manufacture of candles, and it is the principal article used for that purpose in the extensive works of Price's Patent Candle Company; besides which it is greatly employed in soap-making, and likewise for greasing the wheels of railway carriages. In Africa it is eaten as butter, and a kind of soup is made by boiling the fruits. The hard black shell of the nut takes a fine polish, and is frequently made into rings and other ornamental articles by the negroes. [A. S.]

ELÆOCARPUS. A genus of *Tiliaceæ*, natives mostly of tropical parts, principally of India and Java, a few occurring in Australia and New Zealand. They either form trees, attaining sometimes the height of fifty or sixty feet, or large shrubs; they have simple leaves, and racemes of small flowers, with a calyx of five sepals, and five petals either toothed or beautifully fringed, the stamens indefinite, inserted upon a swollen lobed disk, and having long downy unequal-celled anthers usually terminating in a bristle. The ovary is from two to five-celled, and the fruit contains a very hard rough-shelled nut, divided into as many one-seeded cells as the ovary, or sometimes all but one cell imperfect. *E. Ganitrus*, a tree, growing forty or fifty feet high, is native of India and the Malay Islands, where the hard stones of the fruit are commonly used for stringing into rosaries, or for making necklaces, bracelets, buttons, beads of pins, and similar articles. *E. Hinau*, the Hinau of the New Zealanders, is a tree fifty or sixty feet high, with a trunk three or four feet thick, producing a very hard white timber, which, however, is not very valuable, being apt to split when exposed to wet or heat. The bark affords an excellent and permanent dye, varying from light brown, to puce, or deep black; it is greatly used by the natives for dyeing their garments. The pulp surrounding the stone of the fruit of this and other species is eatable; and in India the fruits of several are either used in curries or pickled like olives. [A. S.]

ELÆODENDRON. A genus of trees or shrubs belonging to the *Celastraceæ*, occurring in greatest numbers in S. Africa, but also represented in Australia, India, and the W. Indies. The leaves are opposite or alternate, elliptical or lanceolate and smooth; and the inconspicuous green or white flowers are disposed in axillary cymes, and have a four or five-parted calyx, a four or five petaled corolla, inserted under a fleshy ring and longer than the calyx, a like number of stamens inserted on the margin of the fleshy ring in which the ovary, crowned with a short style and a rounded stigma, is immersed. The fruits are green fleshy drupes, sometimes about the size of a hazel-nut, but often much smaller, with a thin fleshy outer covering, surrounding a hard three to five-celled nut. *E. australe* furnishes a close-grained firm wood, which is used in N. S. Wales for turning and cabinet work; this tree attains a height of thirty to forty feet, with a diameter of eight to fourteen inches. The drupes of *E. Kubu* are eaten at the Cape. The bark and roots of *E. Roxburghii*, an Indian species, are considered efficacious in all cases of swelling, and are used externally rubbed with water. The root is also said to be powerfully astringent and useful in snake bites. [A. A. B.]

ELÆOSELINUM. The generic name of plants belonging to the order of umbellifers, distinguished from their allies by having each half of the fruit with five principal and four secondary ribs, two of the latter being wing-like. *E. meoides* is a native of Sicily, and occurs also in Algiers; its leaves are twice pinnate, rough on the stalks and nerves, the leaflets numerous and very narrow. [G. D.]

ELAIO. In Greek compounds = olive colour, a mixture of green and brown.

ELAPHOGLOSSUM. A genus of polypodiaceous ferns of the tribe *Acrosticheæ*, distinguished by their simple fronds, and simple or parallel forked free veins, which are club-shaped at the apex. Thus defined the genus includes a large proportion of the species formerly referred to *Acrostichum*. In some of them the fronds are smooth and naked, but in others they are clothed with variously shaped and often strongly coloured scales which form pretty objects for microscopical examination. Upwards of 150 species are admitted, the larger proportion of them occurring in the West Indies and South America, a considerable number in India and the East, and others extending to the Mascaren Islands, Madagascar, the Cape, and Sierra Leone, and to Australia and the islands of the Pacific. The fertile fronds are distinct from the sterile ones, generally more or less often very much contracted, and not unfrequently elevated on longer stalks, their under surface being entirely covered with spore-cases. [T. M.]

ELAPHOMYCES. Underground *Fungi*, differing from truffles by reason of the contents of the thick peridium being ultimately resolved into a mass of dusty sporidia from the absorption of the asci. They were, in consequence, for a long time associated with puffballs. We have three wild species in this country, two of which are pretty generally diffused. The peridium is either smooth or rough with warts. Of our more common species, *E. granulatus* is far less rough externally, and has a thinner peridium, which is not variegated within like *E. variegatus*. All the species appear to be involved in an intricate mass of rootlets and spawn. They had once the reputation of being aphrodisiac, arising from a false notion as to their origin, and they are still kept by the herbalists in Covent Garden under the name of lycoperdon nuts. Several species have been found in France which at present have not rewarded the researches of British mycologists. *E. granulatus* gives rise occasionally to *Cordiceps capitatus*, as does *E. variegatus* to *C. ophioglossoides*. [M. J. B.]

ELAPHRIUM. The name of a genus of trees or shrubs belonging to the *Amyridaceæ*, abounding in resinous juice, and natives of tropical America. The characteristics reside in the hermaphrodite flowers, which have a persistent four-parted calyx; four petals inserted beneath an entire disc along with the eight stamens; a sessile ovary, with two ovules in each of its two compartments; and a short style with two stigmas. The fruit is a pea-shaped drupe with a thick rind, and an inner shell containing generally one seed surrounded with pulp. *E. elemiferum*, a native of Mexico, according to Dr. Royle furnishes Mexican Elemi, a greenish resin. *E. tomentosum* also yields a resin. [M. T. M.]

ELATERIUM. A drug prepared from the pulp of the fruits of *Ecbalium agreste*, formerly known as *Momordica Elaterium*.

ELATERS. Cells containing a double spiral which occur in the capsules of *Jungermanniaceæ* and *Marchantiaceæ* in company with the spores. The young capsules contain two sets of cells, the one narrow, the other broader; the endochrome of the narrower gives rise to a single or double spiral thread, while the broader cells by transverse and longitudinal cell-division give rise to the spores. The threads in the peridia of *Trichia* resemble elaters very closely, but there is still a controversy as to their real structure. [M. J. B.]

ELATINACEÆ (*Water-peppers*). A natural order of thalamifloral dicotyledons, belonging to Lindley's rutal alliance of hypogynous Exogens. Small annuals growing in marshes with opposite leaves, having interpetiolar stipules, and minute axillary flowers; sepals and petals three to five; stamens as many or double the number; styles three to five; stigmas capitate. Fruit a three to five-celled capsule, opening at the partitions. Seeds numerous, exalbuminous attached to a central placenta. The family is nearly allied to the chickweed order, but differs in the stigmas, the mode in which the fruit opens, and the straight, not curved, embryo. Lindley thinks that the affinities are chiefly with the rue family. The plants are generally distributed over the world. Some of them possess acridity, hence their English name. There are six known genera, and twenty-four species. Examples, *Elatine, Bergia, Anatropa*. [J. H. B.]

ELATINE. A genus of dwarf annual aquatics with rooting pipe-like stems and opposite leaves. There are two British species, called Waterworts, both of unusual occurrence. *E. hexandra* is a minute plant hardly an inch high, which grows on the margins of lakes, forming a moss-like mat sometimes extending under the water, and in dry seasons when it is left by the receding water assuming a crimson hue. The flowers, which are minute and flesh-coloured, grow in the axils of the opposite leaves. *E. Hydropiper* scarcely differs from the preceding except that the flowers are octandrous. They might be sown with advantage on the shelving banks of artificial water to conceal the unsightliness of mud. [C. A. J.]

ELCAIJA, ARABIAN. *Trichilia emetica*.

ELDER. The popular name for *Sambucus*.

ELECAMPANE. *Inula Helenium*.

ELECTRA. A genus of small Mexican bushes of the composite family, nearly allied to *Coreopsis*, but the latter has the ray florets barren, while in this genus they are female. The young twigs are four-sided and furnished with lance-shaped smooth leaves; and the yellow-rayed flower-heads, nearly an inch across, are disposed in terminal corymbs. Each head is surrounded by an involucre of two series of scales, the outer very narrow, the inner broader and membranaceous; the strap-shaped ray florets are female, the tubular ones of the disc five-toothed and perfect. The achenes compressed, nearly elliptical, crowned with two bristles, or quite naked. The receptacle is furnished with golden chaffy scales. Only two species are known. [A. A. B.]

ELEGIA. A genus of *Restiaceæ* allied to *Restio*, but differing in having the male flowers with the three inner glumes larger than the three outer. They are sedge-like plants from the Cape of Good Hope, with leafless stems, and paniculate or spicate flowers. [J. T. S.]

ELEMI. The name of certain stimulant gum-resins, derived from various plants. —AMERICAN or BRAZILIAN. The gum-resin of *Icica Icicariba*. —EASTERN or MANILLA. The gum-resin of *Canarium commune*. —, MEXICAN. The gum-resin of *Elaphrium elemiferum*.

ELEOCHARIS. A genus of cyperaceous plants, belonging to the tribe *Scirpeæ*, distinguished by one or two of the lowest glumes being larger than the others, and empty; bristles three to six, or wanting; nut compressed, crowned with the persistent dilated base of the jointed style. The species have a wide geographical range, some of them growing within the torrid zone and others reaching nearly to the arctic. In Steudel's *Synopsis Plantarum Cyperacearum*, 118 species are described, four of which are natives of Britain. The commonest is *E. palustris*, which in many places fills up bog holes and ditches with its long-matted entangled stems. The flowers are in small brown spikes. [D. M.]

ELEPHANTOPUS. A genus of erect annual or perennial hairy weeds belong ing to the composite family. About a dozen species are known, natives of America, one, however, being a common weed in most tropical countries. The alternate leaves are linear, or more generally oblong, narrowed below. The compound flower-heads, half an inch or more in diameter, are arranged in loose terminal corymbs, or in a spicate manner, each separate head having an involucre of narrow-pointed bracts, which enclose three to five white or purple tubular four-toothed florets deeply cleft on one side so as to appear unimate. The achenes are compressed, ribbed, and crowned with a pappus of numerous chaffy bristles. An infusion of the leaves of *E. Martii* is used by the Brazilians in pectoral affections, and is known as Erva Grasso; it grows from two to three feet high, and has narrow oblong root leaves, and compound flowerheads disposed in loose corymbs terminating the simple stems. The leaves of *E. scaber*, a plant very like the former in appearance, are used in Travancore, boiled and mixed with rice, for pains in the stomach, swellings in the body, &c. The twigs of *E. spicata* are used in Jamaica for making brooms. [A. A. B.]

ELEPHANTORHIZA. A genus composed of two Caffrarian bushes with tuberous woody root-stocks which are said to bear some resemblance to an elephant's foot. They belong to the *Mimosa* group of the leguminous family, and are most nearly allied to *Prosopis*, but differ in the nature of their pods, which are nearly straight, compressed, about six inches long, and an inch broad, and contain numerous seeds; when ripe the two valves fall away, and leave the entire rim of the pod behind. The pods in *Prosopis* are spirally twisted, and do not open when ripe. The leaves, like those of many species of *Acacia* and *Mimosa*, are bipinnate and alternate, and are made up of six to ten pairs of pinnæ, each of which has eighteen to thirty pairs of linear smooth leaflets. The numerous small flowers are arranged in simple or compound spikes. [A. A. B.]

ELEPHANT'S-EAR. The common name for *Begonia*.

ELEPHANT'S-FOOT. *Testudinaria Elephantipes*; also the common name for *Elephantopus*.

ELETTARIA. A genus of *Zingiberaceæ*, consisting of plants having much the appearance of *Amomum*, from which genus the present is distinguished by the elongated filiform tube of the corolla, by the presence of the internal lateral lobes in the shape of very small tooth-like processes, and by the filaments not being prolonged beyond the anther. They are natives of the tropical parts of India. *E. Cardamomum* furnishes the fruits known as the Small or Malabar Cardamoms of commerce. These are collected either in their wild state or from cultivated plants. In the forests of Travancore the Cardamom springs up spontaneously when the trees are felled. In four years' time the plant attains its full development, and produces its fruits, which are gathered in November and ripened in the sun. The plant continues to yield fruit till the seventh year, when the stem is cut down, new plants arising from the stumps (*Drury*). As imported the fruits are ovate triangular capsules of a dirty yellow colour, containing several angular seeds. Three principal varieties occur in commerce, called, according to their length, shorts, short-longs, and long-longs! The seeds are used medicinally in this country, for their cordial aromatic properties, which depend on the presence of a volatile oil. In India the fruits are chewed by the natives with their betel, and are also used in bowel complaints. Ceylon Cardamoms are said by Dr. Pereira to be the produce of *E. major*. [M. T. M.]

ELEUSINE. A genus of grasses belonging to the tribe *Chlorideæ*, distinguished by the inflorescence being in close finger-like spikelets at the apex of the rachis; glumes five to seven-flowered, the valves obtuse; pales obtuse, upper bifid-toothed; scales truncate, fimbriate; styles two, confluent at the base. The species are mostly natives of the warmer parts of the globe, where some of them are of considerable importance. *E. coracana* is cultivated in Japan as a corn crop for its large farinaceous seeds, and also on the Coromandel coast, where it is called Natchnee. [D. M.]

ELEUTHEROS. In Greek compounds—distinct, separate.

ELISENA. A genus of pancratiform amaryllidaceous bulbs, found in Peru. They produce erect linear-lorate leaves, and scapes supporting a few flowers, the short tube of which is cylindrical, the limb reflexed, its segments linear, two of them as well as the cup and filaments declinate; the cup is cylindrical with a repand recurved margin. One or two species are known. *E. longipetala* is a pretty plant with white flowers, sometimes met with in gardens. [T. M.]

ELIZABETHA. A genus composed of two beautiful leguminous trees found in British Guiana, nearly related to *Brownea*; but the stamens are nine in number, three of them longer than the others and bearing anthers, the remainder sterile; while in *Brownea* they are more numerous, and all fertile. The unequally pinnate leaves of *E. princeps* are made up of twenty to forty pairs of narrow leaflets, while those of *E. coccinea* have fewer and larger leaflets. The rose or scarlet flowers are arranged in dense terminal bracted spikes from two to four inches long. The tubular calyx is five-toothed, the upper tooth broader than the others; the five narrow petals are nearly of equal length; and the nine stamens are free or shortly united at the base. The broadly linear pod is compressed and thickened on the upper edge, from three to five inches long, velvety, and containing a number of seeds. The genus was dedicated by Schomburgk to the Princess Royal of Prussia. [A. A. B.]

ELLEANTHUS. *Evelyna.*

ELLEBORE BLANC. (Fr.) *Veratrum album.*

ELLERTONIA. A Malabar climber, forming a genus of *Apocynaceæ*, distinguished from *Alstonia* and *Rinberopus* chiefly in having peltate seeds expanded at each end into a broad membranous wing.

ELLIOTTIA. A genus of *Cyrillaceæ* containing a North American shrub with alternate entire leaves and terminal racemes. Flowers with a four-parted calyx, a very deeply six-parted corolla, eight stamens with glandular filaments, and a four-celled ovary and capsule. [J. T. S.]

ELLIPEIA *cuneifolia*, the only species of the genus, is a climbing shrub, native of Malacca, and belonging to the order *Anonaceæ*, in which it is distinguished by the following characteristics:—petals overlapping one another in the bud; carpels oblique, distinct one from another, and each having a single ovule attached to the ventral suture. [M. T. M.]

ELLIPSOIDAL. A solid with an elliptical figure.

ELLIPTIC. A flat body, which is oval and acute at each end.

ELLISIA. A genus of *Hydrophyllaceæ*, containing six species of North American branching annual herbs, with opposite or alternate oblong and pinnatisect leaves, and white flowers on solitary peduncles, opposite the leaves below, and in loose racemes above. The calyx is five-parted, without reflexed appendages to the sinuses; the corolla tubular, campanulate and caducous, with ten small scales in the tube; the stamens included; the nectary surrounding the ovary, and rising into five gland-like teeth; and the capsule ovoid-globose with four seeds. This genus scarcely differs from *Nemophila*, except in wanting the appendages to the calycine sinuses. [W. C.]

ELLOBOCARPUS. *Ceratopteris.*

ELM. The common name for *Ulmus*.—of New South Wales. *Epicarpurus orientalis.*—SPANISH. *Cordia Geraschanthus*, or *Geraschanthus vulgaris*; also said to be applied to *Hamelia ventricosa*.—WYCH or WITCH. *Ulmus montana.*

ELODEA. A genus of *Hypericaceæ*, differing from *Hypericum* only in having scale-like glands alternating with the bundles of stamens. *Hypericum Elodes*, which is found in various parts of Britain, is referred to this genus; and there are a few European and Western Asiatic species, and two found in the United States. Most of these are perennial smooth pea-green herbs, with opposite shortly stalked or sessile elliptical or lance-shaped leaves, furnished with transparent dots, and yellow or purplish flowers disposed in axillary or terminal few-flowered cymes. A stomachic tincture is said to be prepared from the leaves of *E. virginica*. [A. A. B.]

ELONGATE. Lengthened or stretched out, as it were.

ELS, ROOD. *Cunonia capensis.*—, WIT. *Weinmannia trifoliata.*

ELSHOLTZIA. A genus of labiate plants, distinguished by having the calyx ovate or bell-shaped (changing its form and becoming longer as the fruit ripens), and having five equal teeth; the tube of the corolla about as long as the calyx, rarely longer, its border two-lipped, the upper slightly notched, the lower three-lobed and spreading. The species are herbs or undershrubs of little interest, natives of Eastern India and Java, rare in Central Asia. The genus was named after Elsholtz, a Prussian botanist. [G. D.]

ELVASIA. A genus of Brazilian shrubs, belonging to the *Ochnaceæ*. They have small flowers in terminal clusters, a four-leaved calyx, four petals, eight stamens, and a four-celled ovary, with an ovule arising from the base of the inner angle of each compartment. [M. T. M.]

ELYME DES SABLES. (Fr.) *Elymus arenarius.*

ELYMUS. A genus of grasses belonging to the tribe *Hordeæ*, distinguished by the inflorescence being in simple spikes, very rarely branched; spikelets two to three together; glumes two, both on the same side of the spikelet, without awns, enclosing one to seven florets. In *Steudel's Synopsis* there are forty-nine species described. These have an extensive geographical range; nearly all are inhabitants of the temperate zones, but some extend even to the Arctic circle. Only two species are natives of Britain, *E. geniculatus* and *E. arenarius*; the latter, Sea Lyme-grass, is useful for binding with its long creeping roots the land where it grows. They are all coarse grasses, and of little importance for agricultural purposes. [D. M.]

ELYNA. A genus of cyperaceous plants belonging to the tribe *Cariceæ*, distinguished by the scales being imbricated or slightly lapping over each other by their edges, covering a spikelet of two flowers; lower floret fertile, upper barren. Small grass-like plants, having the habit of some carices. They are mostly natives of Alpine countries, and rarely met with. [D. M.]

ELYNANTHUS. A genus of cyperaceous plants, belonging to the tribe *Rhynchosporeæ*, and distinguished by the inflorescence being in close bundled spikes; flowers polygamous, the terminal one hermaphrodite; styles three-cleft, thickened and bulbous at the base; seeds triangular. The species are mostly natives of the Southern Hemisphere, South Africa, and New Holland. [D. M.]

ELYTRANTHE. A genus of *Loranthaceæ*, containing Indian parasitical shrubs with compact abbreviated spikes of few whitish purple or orange flowers, which are inserted in the rachis, and each furnished with three bracts; petals cohering in a tube at the base; stamens six. [J. T. S.]

ELYTRARIA. A genus of *Acanthaceæ*, containing a few species scattered over the tropical regions of America, Africa, and India. They are stemless herbs, with entire dentate or repand radical leaves and small flowers. The calyx is four or five-parted. The corolla is two-lipped or ringent; there are two fertile and two barren stamens, all included; and the capsule contains many pitted seeds in each cell, attached to the placenta without any retinacula. [W. C.]

EMARCID. Flaccid, wilted.

EMARGINATE. Having a notch at the end, as if a piece had been taken out.

EMBELIA. A genus of *Myrsinaceæ*, nearly allied to *Mæsa*, from which it differs in its free ovary, and from the other genera in the family with free petals in its slender racemes of flowers, which usually form a terminal panicle. It is composed of about twenty species of straggling shrubs, found in India and the Islands of the Indian Archipelago, and those to the east of Africa. The alternate stalked leaves are lance-shaped, elliptical, or oval, furnished with transparent dots. The minute green, white, or pink flowers are borne in great profusion, and arranged in simple or compound racemes towards the ends of the branches; they have a five-parted calyx, five free spreading petals, opposite to which are five stamens, and an ovary crowned with a short style and rounded stigma. The berries are minute, round, and either red or black when ripe. Those of *E. Ribes*, one of the most common Indian species, with ovate-lanceolate smooth leaves, are gathered and sold to traders, who use them for adulterating black pepper, which they somewhat resemble, and have, moreover, a slight pungency, owing to a resinous substance contained in them. They possess anthelmintic properties, and are sometimes given in infusion. *E. Basaal*, another Indian species with larger elliptical and more or less downy leaves, is useful in various ways. The young leaves in combination with ginger are used as a gargle in cases of sore throat; the dried bark of the root is a reputed remedy for the toothache; and the berries mixed with butter are used as an ointment, which is applied to the forehead as a specific for pleuritis. [A. A. B.]

EMBLICA. A genus of *Euphorbiaceæ*, differing only from *Phyllanthus* in the more deeply divided style, and in the nature of the fruit, which is about the size of a small gooseberry, with a fleshy outer covering, and a hard three-celled nut, which splits when ripe into six portions, and contains six seeds: the fruit of *Phyllanthus* being usually a dry membranous capsule. *E. officinalis*, or, as it is sometimes called, *Phyllanthus Emblica*, is the only species, and is found wild and cultivated in various parts of India and the Indian Archipelago. It is a tree sometimes of large growth, but more generally of twenty to thirty feet, with an abundance of simple alternate linear leaves, which are smooth, and arranged on slender branches in a distichous manner, so that they appear like leaflets of pinnate leaves; in their axils the little green flowers are found in cymes, the females mixed singly with the males. The latter have a six-parted calyx, no petals, six glands, and three to five stamens united into a short column. The females, with a similar calyx, have a cup-shaped disc, and an ovary crowned with a style which has three thick recurved two-lobed branches. In Borneo, the bark and young shoots are used to dye cotton black, for which purpose they are boiled with alum. The fruits are often made into a sweatmeat with sugar, or eaten raw as a condiment, but they are exceedingly acid. The wood is hard and valuable, as it resists damp well. In India the bark is used in tanning, and the root-bark mixed with honey is applied to inflammation of the mouth. The fruits also are used as a pickle, or preserved in sugar; when ripe and dry they are given in cholera, diarrhoea, &c., under the name Myrobalant Emblici. The seeds are employed in nausea and bilious affections, and given in infusion in fevers. An infusion of the young leaves mixed with sour milk is also used in dysentery. The natives of Travancore have a notion that the plant imparts a pleasant flavour to water, and therefore place branches of the tree in their wells, especially when the water is charged with an accumulation of impure vegetable matter. [A. A. B.]

EMBOLUS. A plug; a process which projects downwards from the upper part of the cavity of the ovary in *Armeria*, and closes up the foramen of the ovule.

EMBOTHRIUM. A small genus of *Proteaceæ*, having an elongated tubular calyx, bursting longitudinally, and a sub-globose four-cleft limb bearing the anthers, which are sessile, on the concave points of the segments. The fruit is a leathery many-seeded follicle. They are trees or shrubs with simple, oval or lanceolate entire leaves, greyish on the under-side, and red generally smooth flowers. They are found in the Western and Antarctic portions of South America. [R. H.]

EMBRACING. Clasping with the base. The same as Amplexicaul.

EMBRYO (adj. EMBRYONAL). The rudimentary plant, engendered within a seed by the action of pollen. — FIXED. A leaf bud.

EMBRYO-BUDS. Spheroidal solid bodies, of unknown origin, resembling woody nodules, formed in the bark of trees, and capable of extending into branches under favourable circumstances.

EMBRYOTEGIUM, EMBRYOTEGA. A little papilla, often separating as a lid, which covers over the radicle of some kinds of embryo. It is the hardened apex of the nucleus.

EMERICELLA. A most curious genus, connecting, apparently like *Coniocybe* and some others, the myxogastrous *Fungi* with *Calicei*. The stem consists of a spongy central column, giving off threads which have gonidia like those of *Paulia*, and resembling some species of *Palmella*, to which we shall have occasion to recur hereafter. These bodies become blue when treated with iodine. The spores are purplish, furnished with very long spines, seated in the same plane, and inclosed in a globose peridium. The only species has been found on decaying leaves of *Euphorbia neriifolia* at Secunderabad. A figure will be found in Berkeley's *Introduction to Cryptogamic Botany*, p. 341. [M. J. B.]

EMERUS (Fr.) *Coronilla Emerus*.

EMEX. A genus of *Polygonaceæ*, closely allied to *Rumex*, from which it is distinguished by the perianth segments being united at the base, and by the flowers being polygamous. *E. spinosus*, the only species, is a salt marsh annual, of the Mediterranean region, the Cape of Good Hope, and the Antilles. In habit it closely resembles *Rumex pulcher*, except that the leaves are broadest in the middle, and the perianth has much larger spines when the fruit has arrived at maturity. [J. T. S.]

EMILIA. A small group of composite plants, separated from *Cacalia*, and consisting mostly of annuals, represented by *E. sagittata*, the *Cacalia coccinea* of gardens. They are natives of India, China, and the South African islands. The flower-heads are subcorymbose, the florets being all tubulose, and in the common garden forms either orange, scarlet, or yellowish. The lobes of the florets are linear elongate, and the pentagonal achenes are ciliated at the angles, and crowned with a many-rowed pappus of filiform hairs. [T. M.]

EMMENANTHE. A genus of *Hydrophyllaceæ*, containing a single species from California. It is an elegant erect herb with alternate pinnatifid leaves, and pendulous flowers in erect racemes, the calyx being five-parted, the corolla campanulate and persistent, and the stamens included. The ovary is surrounded by a small disc, and the capsule is oblong, and two-celled from the meeting of the enlarged placentas in its centre. [W. C.]

EMPETRACEÆ (*Crowberries*). A natural order of monochlamydeous dicotyledons belonging to Lindley's euphorbial alliance of dielinous Exogens. Shrubs with heath-like evergreen exstipulate leaves, and small axillary flowers which are usually imperfect. Perianth of four to sixahypogynous persistent scales, the innermost sometimes petaloid and united. Stamens two to three, alternate, with an inner row of scales. Ovary free in a fleshy disk, two to nine-celled. Fruit fleshy, with two to nine nucules; seed solitary. Natives chiefly of the northern parts of Europe and America. A few are found in the South of Europe, and even at the Strait of Magalhaena. The order is represented in Britain by *Empetrum nigrum*, the black crowberry, the fruit of which is eaten in northern countries, and is used by the Greenlanders to prepare a fermented liquor. The leaves and fruit of some of the plants are somewhat acid. There are four known genera, and five species. Examples :— *Empetrum, Corema, Ceratiola*. [J. H. B.]

EMPETRUM. Crowberry or Crakeberry. Small evergreen heath-like plants of the order *Empetraceæ*, distinguished by the following characters: calyx of three leaves with six imbricated scales at the base; three petals, and as many stamens; berry depressed, containing from six to nine bony seeds. *E. nigrum*, the badge of the M'Leans, is a small procumbent much-branched shrub, with rough wiry branches and small narrow leaves, the edges of which are so much recurved as almost to form a tube. The flowers are of a dark red colour, small and situated in the axils of the upper leaves, and are succeeded by brownish-black berries, about the size of juniper berries, of a firm fleshy substance, and insipid in taste. A native of moors and the sides of boggy hills throughout the north of the Eastern continent, and the islands towards America. In Great Britain it is most abundant on the Scottish hills, where it affords abundant food to the moor-game, and is also found in the moorland districts of the north of England. Its berries are eaten by the Highlanders and Russian peasants, and are considered wholesome. Boiled in alum-water, they furnish a dingy purple dye, and Linnæus states that they are used by the Laplanders for dyeing otter and sable skins black. *E. rubrum*, a native of the extreme south of South America, has red berries, which are said to be pleasant to eat. It is most abundant along the sandy coast. Both species are easy of cultivation as bog plants, but are slow growers. French *Camarine*: German *Rauchbeere*. [C. A. J.]

EMPHYSEMATOSE. Bladdery, resembling a bladder.

EMPHYSOPUS. A name formerly applied to a little perennial herb of the composite family, which is common on pasture lands in Tasmania, and in habit and form of leaves a good deal resembles the common daisy. The flower-heads, however, supported on naked stalks, one to three inches long, are not more than a quarter of an inch in diameter, and the leaves are clothed with a soft down. The plant is now placed in LAGENOPHORA; *which see*. [A. A. B.]

EMPLEUROSMA. The name of a small shrub of the rue family, a native of Swan river, and having leathery linear leaves, rolled under at the margins, and unisexual flowers, the males with a four-parted calyx, and eight stamens, whose anthers are tipped with a gland. The plant is imperfectly known. [M. T. M.]

EMPLEURUM. A genus of *Rutaceæ* consisting of shrubs, natives of the Cape, with oblong glandular serrulated leaves, and axillary flowers, solitary, or in twos or threes. They have a four-cleft calyx, thickened at the base, no corolla, and four stamens, opposite to, and longer than the lobes of the calyx, with anthers having a gland at the top. The ovary is solitary, one-celled, terminating at top in a long horn, the style lateral, as long as the horn-like extremity of the ovary; ovules two. The horned capsule is one-seeded. *E. serrulatum* is a pretty greenhouse shrub with pinkish flowers. [M. T. M.]

ENARTHROCARPUS. A genus of *Cruciferæ* from the coasts of the Mediterranean; it is allied to *Raphanus* but with a different pod, which breaks across into only two parts, the lowermost of which is persistent, short, obconical with one to three seeds, the uppermost long and knotty, with numerous seeds. Rough annuals with lyrate-pinnatifid root leaves, toothed stem leaves, and elongate racemes of yellow or purplish flowers. [J. T. S.]

ENCALYPTA. A genus of mosses belonging to the natural order *Encalyptei*, distinguished by the large funnel-shaped persistent veil which covers the capsule. *E. vulgaris* is a remarkable moss, and occurs here and there on the tops of walls, though not so general as some other wall mosses. The capsule in this genus is either even or grooved; the peristome is either single, double, or altogether wanting. The genus is almost exclusively confined to Europe and North America, though there are traces of it in Chiloe, Peru, and Kumaon. It is the only genus of the order. [M. J. B.]

ENCEPHALARTOS. A genus of *Cycadeaceæ*, having tall cylindrical trunks, with a terminal tuft of pinnate thick spiny leaves. The male flowers, like those of *Cycas*, are collected into a terminal stalked cone, consisting of a number of oblong wedge-shaped scales, with anthers on their under-surface; while the female flowers are collected in terminal stalked cones, consisting of peltate stalked scales, on the under-surface of which the ovule is placed, as in *Zamia*. The interior of the trunk, and the centre of the ripe female cones, contain a spongy farinaceous pith, made use of by the Caffers as food, and hence the trees are called by the name of Caffer-bread. [M. T. M.]

ENCHOLIRIUM. A name applied to a genus of *Bromeliaceæ*, represented by a Brazilian herbaceous plant, with tufted spiny leaves, and clustered flowers, with a calyx of three short equal segments, a corolla of three petals, and six hypogynous stamens, with curved filaments dilated at the base. The ovary is free; the style triangular with three stigmas. [M. T. M.]

ENCKEA. Shrubs, or less frequently trees, forming a genus of *Piperaceæ*, characterised by the bracts of the inflorescence, which are hood-like and bent inwards. Stamens five to seven, placed round the ovary; filaments persistent, the anthers kidney-shaped, deciduous. Ovary sessile, sometimes prolonged at the top; stigmas three to five. Fruit aromatic, berry-like, with a thick rind. The roots of *E. unguiculata* and *E. glaucescens* are used medicinally in Brazil. [M. T. M.]

ENDECA. In Greek compounds=eleven.

ENDIVE. *Cichorium Endivia*.

ENDIVE PETITE. (Fr.) *Cichorium Endivia angustifolia*.

ENDOCARP. The lining of a carpel; the inner surface of a fruit, representing at that time the upper surface of a carpellary leaf. The stone of a cherry is its endocarp.

ENDOCARPEI. A natural order of lichens, in which the capsule-like fruit is constantly immersed in the foliaceous or crust-like frond. The walls of the fruit moreover are pale and never carbonised as in *Verrucariei*. The best known species are *Endocarpon miniatum*, which is so common about waterfalls, presenting a peltate leathery greenish frond tinged with red below; and *Pertusaria communis*, which is still more common on smooth-barked trees, especially the oak and the beech, and which by a peculiar degeneration produces the white patches which, according to their more or less mealy condition, are referred to the now exploded genera *Variolaria* and *Lepraria*. [M. J. B.]

ENDOCHROA. A supposed interior layer of the cuticle.

ENDOCHROME. The colouring matter of plants. A term applied to the contents of the cells, especially amongst *Algæ* and *Fungi*, though frequently applicable to the simple structures in phænogams. The colour of flowers, funguses, &c., depends generally upon the colour of the endochrome, the cell wall itself being hyaline. In *Algæ* and *Fungi* it frequently acts an important part, being either concentrated into a single spore or zoospore, or resolved into a definite or indefinite number of either, while at times it gives rise to spermatozoids. [M. J. B.]

ENDOGENS. A large class of plants to which the names of *Monocotyledones* and *Amphibryæ* are also given. They have a cellular and vascular a) stem—the latter exhibiting spiral vessels. Their stem is endogenous, that is, increases in diameter by addition of woody vessels towards its interior, the outer part being the oldest and densest (hence the name Endogens or inward-growers); bundles of woody, spiral and pitted vessels are scattered throughout the cellular tissue; there is no pith, no separable bark, no woody rings or zones, and no true medullary rays. The age of woody Endogens cannot be determined by counting concentric rings as in Exogens. The leaves are usually continuous with the stem, and do not fall off by articulations; when at length they separate their bases leave marks or scars at definite intervals on the stem, as seen in palms. The stems of endogens are often subterranean, in the form of corms, rhizomes, or bulbs. The leaves have stomates, and their venation is usually parallel, though in a few cases it is slightly reticulated. The flowers have stamens and pistil, and three-membered symmetry. The ovules are contained in an ovary, and the embryo has one cotyledon or seed-lobe, whence they are called monocotyledonous.

The class has been divided into two subclasses: 1. *Petaloideæ* or *Floridæ*, in which the flowers consist either of a coloured perianth or of scales arranged in a whorl; 2. *Glumiferæ*, in which the flowers, in place of sepals and petals, have imbricated bracts or scales called glumes. Lindley has added a third subclass called *Dictyogenæ*, on account of the net-veined leaves. Among the *Petaloideæ* there are three sections: 1. *Epigynæ*, having perfect flowers and a superior perianth, as orchids, gingers, irids, amaryllids, &c.; 2. *Hypogynæ*, having perfect flowers and an inferior perianth, as lilies, rushes, and palms; 3. *Incompletæ*, with imperfect flowers without a proper whorled perianth, as screw-pines and arums. Among *Glumiferæ* there are included the two orders of grasses and sedges.

Permanent endogenous stems are well illustrated by palms. In these the hardest part is on the outside, and the trunks are usually unbranched, and are limited as regards their increase in diameter. They increase principally by forming a crown of leaves, and if this growing point is destroyed they die. Some Endogens, as *Dracænas*, attain a great diameter of stem, and divide in a forked manner. [J. H. B.]

ENDOGONIUM. The contents of the nucule of a *Chara*.

ENDONEMA. A Cape of Good Hope shrub belonging to the *Penæaceæ*. Its leaves are flat and overlapping; the flowers axillary, solitary, with a coloured tubular perianth divided into four short lobes, and four stamens, inserted into the top of the perianth tube, alternately with its lobes. The ovary has four compartments, each containing four ovules, the upper pair ascending, the lower pendulous. [M. T. M.]

ENDOPHLŒUM. The liber of bark; the inner layer, containing woody tissue, lying next the wood.

ENDOPHYLLOUS. Formed from within a sheathing leaf; as the young leaves of endogenous plants.

ENDOPLEURA. The innermost skin of a seed-coat.

ENDOPTERA. A genus of *Compositæ*, nearly related to *Crepis*, but differing in the achenes of the ray florets having a wing on their inner face. There are but two species, *E. Dioscoridis* and *E. aspera*, both annual branching weeds of S. Europe and Asia Minor, and in appearance much like our species of *Crepis*. The generic name refers to the wings on the inner face of the achene. [A. A. B.]

ENDOPTILE. Said of an embryo whose plumule is rolled up by the cotyledon, as in endogens.

ENDORHIZAL. That kind of germination in which the original radicle forms a sheath round the first root which comes from within the former.

ENDORHIZEÆ. A name applied by Richard to endogenous or monocotyledonous plants, on account of the mode in which the young root is developed. The embryo of these plants, when it germinates or sprouts, usually sends out from a definite point a bundle of rootlets, which pierce through the integument, and are covered each by a sheath called coleorhiza. This is well seen in the sprouting of the grains of grasses. The embryo is hence called endorhizal, meaning root within. [J. H. B.]

ENDOS. In Greek composition = within, or in the inside of anything.

ENDOSMOSE. That force which causes a viscid fluid lying within a cavity to attract to itself a watery fluid through an organic membrane.

ENDOSPERM. The albumen of a seed.

ENDOSTOME. The aperture in the inner integument of an ovule.

ENDOTHECIUM. The lining of an anther.

ENEMION. An herbaceous perennial belonging to the *Ranunculaceæ*, with five petal-like deciduous sepals, and from two to six carpels, which when mature are arranged in a stellate manner, and contain two oval seeds. *E. biternatum*, the only species, grows to about the height of six inches, and bears flowers about the size of *Anemone quinquefolia*. It is a native of Kentucky. [C. A. J.]

ENERVIS. When there are no ribs or veins visible.

ENGELHARDTIA. A genus of *Juglandaceæ*, numbering about ten species, found in India, Java and the Philippine islands. They are trees with pinnate leaves a good deal like those of the walnut, and inconspicuous flowers disposed in drooping spicate panicles, the outer and shorter branches of which bear sterile flowers, the inner fertile. These are succeeded by the little fruits, which are about the size of a pea, each seated on the base of a three-lobed beautifully veined and coloured bract. The beautiful catkin-like spikes of these bracted fruits are often more than a foot long, and hang very gracefully among the foliage. [A. A. B.]

ENGELIA. A genus of *Acanthaceæ*, containing two species, natives of Columbia. They are climbing undershrubs, with one-flowered axillary peduncles. The calyx is reduced to a mere ring. The corolla tube is bent, and the limb fissured in front and parted into five roundish lobes; there are four didynamous stamens, with a rudimentary fifth. The ovary is one-celled, with one ovule in each cell; and the fruit is a fleshy one-celled drupe with a single seed. [W. C.]

ENGELMANNIA. A genus of *Euphorbiaceæ*, found in Texas and the neighbouring states. It is nearly allied to *Croton*, but differs in having fewer stamens, and in its little capsular fruit, about the size of an orange-pip, being composed of two not three cocci. *E. Nuttalliana*, the only species, is an erect branching herb one to two feet high, with alternate stalked oval leaves, silvery-white underneath. The minute flowers are disposed in little clusters in the forks of the branches, the males and females together. The genus bears the name of Dr. Engelmann, of St. Louis, an American botanist. A composite plant has also been dedicated to him, but that now bears the name *Angelandra*. [A. A. B.]

The name is also applied to a section of *Cuscuta*, elevated into a genus by Pfeiffer, containing those species which have a four or five-cleft monosepalous calyx, and a capitate stigma, and in which the capsule dehisces at the apex. [W. C.]

ENGLISH MERCURY. *Chenopodium Bonus-Henricus*.

ENGRAIN. (Fr.) *Triticum monococcum*.

ENHALUS. A genus of *Hydrocharidaceæ*, allied to *Stratiotes*, from which it differs chiefly in having the inner segments of the perianth linear. It is found in the estuaries of the rivers, in Ceylon and other Indian Islands. The leaves are radical, linear, serrated at the apex, and the spathe of the female flowers is two-leaved with a bearded keel. [J. T. S.]

ENKYANTHUS. An elegant glabrous shrub, with deciduous leaves and showy red flowers, often tipped with white. It is a native of South China, and has been introduced into our gardens. It forms a genus of *Ericaceæ* of the tribe *Andromedeæ*, distinguished by a campanulate five-lobed corolla, ten stamens having the anther-cells tipped with awn-like points and opening longitudinally to the base, and a free hard five-celled capsule opening loculicidally in as many valves. The flowers are terminal, pedicellate and drooping, issuing, several together, from a tuft of coloured bracts. *E. quinqueflorus* is probably the only species known, for *E. reticulatus* appears to be only a slight variety of it.

ENKYLIA. A genus of *Cucurbitaceæ*, consisting of Indian climbing herbaceous plants, with pedate downy leaves, having somewhat spiny margins, and small flowers arranged in panicles. The male flowers have a five-fold calyx and corolla, and five stamens completely united into one parcel; the female flowers have a similar calyx and corolla, and an inferior ovary, with a single pendulous ovule in each of the two or three compartments. Fruit berry-like, of the form of a pea. [M. T. M.]

ENNEA. In Greek compounds = nine.

ENSATÆ. A name given by Linnæus to a natural order of monocotyledonous or endogenous plants, including *Iris, Gladiolus, Antholyza, Ixia, Sisyrinchium, Commelyna, Xyris, Eriocaulon*, and *Aphyllanthes*. These plants are now distributed over five separate orders. [J. H. B.]

ENSIAO. *Sempervivum glutinosum*.

ENSIFORM, ENSATE. Quite straight, with the point acute, like the blade of a broad-sword, or the leaf of an *Iris*.

ENTADA. A genus of leguminous plants containing about half-a-dozen species of climbing tropical shrubs, which have twice-pinnated leaves, and flowers produced either in spikes at the bases of the leaves, or in bunches at the ends of the branches; these flowers have a bell-shaped calyx, five white or yellow petals, and ten stamens. The most remarkable feature of the genus is the extraordinary length of its pods, which are flat and woody, divided into numerous joints, each containing one large flat polished seed. In *E. scandens*, a native of the tropics of both hemispheres, the pods often measure six or eight feet in length. The seeds are about two inches across, by half an inch thick, and have a hard woody and beautifully polished shell, of a dark brown or purplish colour. In the tropics the natives convert these seeds into snuff-boxes, scent-bottles, spoons, &c., and in the Indian bazaars they are used as weights. Occasionally they are sent to this country and are hawked about the streets of London under the name of West Indian Filberts, but they are not eatable. Sometimes they are conveyed by the great oceanic currents to the shores of the west of Scotland and the Orkneys, and they are occasionally carried as far as the Loffoden Isles and the Norway coast. [A. S.]

ENTANGLED. Intermixed in so irregular a manner as not to be readily disentangled, such as the hairs, roots, and branches of many plants.

ENTELEA. A genus of *Tiliaceæ*, peculiar to New Zealand, and represented by a single species, *E. arborescens*, a small branching tree from five to ten feet high, with large alternate heart-shaped or three-lobed leaves, and white flowers, somewhat like those of a small dog-rose, disposed in little umbels which terminate the branches of an axillary or terminal panicle. They have a four or five-leaved calyx, a like number of somewhat crumpled petals, and numerous fertile stamens. The four to six-celled capsular fruits are about the size of a hazel nut. In New Zealand the light wood is used by the natives as floats for their nets. *Sparmannia*, to which the plant is most nearly allied, differs in having numerous sterile stamens intermixed with the fertile ones. [A. A. B.]

ENTEROMORPHA. A genus of green-spored *Algæ*, comprising those species of *Ulva* which have a tubular frond, whether simple or more or less branched. The most general species, *E. intestinalis*, known by its bullate crisped fronds, occurs in fresh as well as salt water, *E. compressa* being the more common species on tidal rocks, and having simple or branched narrower fronds, dilated above. The species run closely into each other, and are probably too much multiplied. [M. J. B.]

ENTIRE. Having no kind of marginal division.

ENTOPHYTE. A plant which grows from within others, as some rhizanths and fungals.

ENULA-CAMPANA. (Fr.) *Inula Helenium*.

EOUSE. (Fr.) *Quercus Ilex*.

EOUVE. (Fr.) *Pinus Cembra*.

EPACRIDACEÆ (*Epacrids*). A natural order of corollifloral dicotyledons, included in Lindley's erical alliance of hypogynous Exogens. Shrubby plants, with usually alternate simple leaves; flowers regular and perfect, in spikes or racemes; corolla gamopetalous; stamens five, equal in number to the lobes of the corolla; anthers one-celled, opening by a longitudinal slit. Ovary superior, five-celled, with five scales, distinct or combined at its base. Fruit either fleshy or capsular; embryo with albumen and very small cotyledons. There are two sections of the order: 1. *Epacreæ*, with a capsular many-seeded fruit; 2. *Styphelieæ*, with a drupaceous one-seeded fruit. The plants are natives of the Indian Archipelago and Australia, and represent the heaths in those countries; but they differ from true heaths (*Erica*) in their pentamerous symmetry, their anthers being one-celled without appendages, and in the attachment of the stamens and the corolla.

They are cultivated in greenhouses for the beauty of their flowers. Some yield edible fruits. The berries of *Leucopogon Richei*, called native currants, are said to have supported the French naturalist Riche, who was lost for three days on the south coast of New Holland. *Astroloma humifusum* is called the Tasmanian cranberry. There are 32 known genera, and 336 species. Examples:— *Epacris, Dracophyllum, Styphelia, Leucopogon*. [J. H. B.]

EPACRIS. A large genus typical of the *Epacridaceæ*, distinguished by having a coloured calyx with many bracts, a tubular corolla with a smooth limb, stamens affixed to the corolla, and a five-valved many-seeded capsule. They are branched shrubs with the leaves lanceolate or cordate, generally sharp-pointed, and the flowers axillary, white red, or purple, usually in leafy spikes. The species are distributed over the extra-tropical portions of Australia, Tasmania and New Zealand, and many of them, from the abundance and beauty of their flowers, are deservedly great favourites in the greenhouse. There is very much diversity in the habits of the plants. In *E. pulchella*, *E. rigida*, and *E. microphylla*, the leaves are very small, and the flowers white.

Epacris grandiflora.

In *E. grandiflora* the leaves are much larger, heart-shaped and sharp-pointed, the flowers nearly an inch in length, of a brilliant reddish purple at the base, and pure white at the apex. In *E. impressa*, *E. ruscifolia*, and *E. tomentosa* the flowers are of a deep rose-colour; while in *E. nivea*, *E. obtusifolia*, *E. heteronema*, and *E. paludosa* they are large and of a pure white, the plants having narrow lanceolate sharp-pointed leaves. The New Zealand species are rather inconspicuous in their flowers. [R. H.]

ÉPEAUTRE. (Fr.) *Triticum Spelta*.

EPERUA. The Wallaba, *E. falcata*, a very large timber tree, is the only member of this genus of leguminous plants. It has pinnate leaves composed of two or three pairs of leaflets; and its red flowers are borne in drooping long-stalked bunches. The calyx consists of four thick concave sepals with their bases connected, the upper sepal being broader than the others; the corolla is a solitary roundish fringed petal, inserted into the middle of the calyx; and there are ten stamens. The curiously curved flat pod bears some resemblance to a hatchet, and generally contains from three to four very flat seeds. The tree is abundant in the forests of British Guiana, where it attains a height of fifty feet, with a girth of about six feet. The timber is of a bright red-brown colour, marked with whitish streaks, hard and heavy, but rather coarse-grained. In consequence of the readiness with which it splits, it is commonly employed in Demerara for shingles, palings, &c., and being impregnated with a resinous oil, it is very durable. The bark of the tree is bitter, and the Indians employ a decoction of it as an emetic. They also use the gum as an application to cuts. [A. S.]

ÉPERVIÈRE. (Fr.) *Hieracium.* — ORANGÉE. *Hieracium aurantiacum.*

EPHEBE. A curious and anomalous genus proposed by Fries, which, after oscillating between lichens and sea-weeds, was for a time joined with *Byssoideæ*, but whose real affinities were quite unintelligible till the discovery of the fructification, which clearly places it in close conjunction with *Lichina* and its near allies. The frond consists of branched threads composed when young principally of large brownish cells more or less perfectly disposed in transverse rows, and often divided vertically or horizontally into four. In this condition it seems to indicate an intimate relation with the algal genus *Scytonema*. In older branches, however, there is a distinct cellular tissue both external and within the layer of large cells, and towards the extremities the branchlets swell and contain nuclei, like those of *Dothidea*, filled with a gelatinous mass consisting of fertile asci, each of which contains eight sporidia, while in other similar swellings nuclei are produced whose gelatinous contents produce myriads of granules, supposed to be the male fruit of the plant. The plant is therefore clearly a lichen, allied to *Collema* and *Lichina*, receding from the common type in the nature of its gonidia, which depart from the usual green tint, and seem to be propagated like such genera as *Hæmatococcus*. The species, which are not numerous, occur on irrigated rocks and stones. They are, as known at present, confined to the temperate regions of the northern hemisphere. [M. J. B.]

EPHEDRA. A genus of *Gnetaceæ*. The plants have stamens and pistils in separate flowers; the staminate flowers in catkins and with a membranaceous perianth; the pistillate flowers terminal on axillary stalks, within a two-leaved involucre. The fruit is a succulent cone, formed of two carpels, with a single seed in each. They are branching shrubs, natives of the sandy sea shores of temperate climates in both hemispheres. The branches are slender, erect or pendulous; leaves very small, scale-like, articulated and united into a sheath at the base. There are twenty-five known species. *E. distachya* abounds in the southern parts of Russia; its fruit is eaten by the peasants and by the wandering hordes of Great Tartary. The branches and flowers of some of the *Ephedras* have been used to stop bleedings and discharges. [J. H. B.]

ÉPHÉMÉRINE. (Fr.) *Tradescantia.*

EPHEMERUM. *Tradescantia.*

EPHEMERUS. Enduring but a day.

EPHIPPIUM. A name applied by Blume to certain epiphytal orchids now referred to various genera, such as *Bolbophyllum*, *Sarcopodium*, and *Cirrhopetalum*. The most notable species is *Sarcopodium grandiflorum* from New Guinea, which has creeping wiry stems with four-sided pseudo bulbs at intervals, and a solitary flower and leaf; the flower, borne on a stalk, is said to be eight inches across, and of a pale yellowish-green colour. [A. A. B.]

EPI. In Greek compounds = upon.

EPIAIRE. (Fr.) *Stachys.* —, GRANDE. *Stachys sylvatica.*

EPIBLAST. A small transverse plate (a second cotyledon), found on the embryo of some grasses.

EPIBLEMA. An epidermis consisting of thick-sided flattened cells.

EPIBLEMA grandiflorum is a terrestrial tuberous rooted orchid of W. Australia, with a slender erect stem eighteen inches high, bearing a single grassy leaf with a few sheathing bracts, and terminating in a raceme of from one to five pretty blue flowers, each about an inch across. According to Fr. Lindley, the genus, which belongs to the tribe *Neotteæ*, differs from *Thelymitra*, of which it has the habit, not only in the clawed lip with long slender processes at the base, but also in the anther bed not being cucullate. [A. A. B.]

EPICALYX. The involucellum, or external series of envelopes beyond the calyx, as in *Malva*.

EPICARP. The outermost layer of the pericarp, corresponding with the under side of the carpellary leaf.

EPICEA. (Fr.) *Abies excelsa.*

EPICHARIS. A genus of *Meliaceæ*, comprising certain trees, natives of the Molucca Islands, which, added to the general characteristics of the order to which they belong, present the following distinguishing features: corolla of four spreading or reflected petals; stamens eight, their filaments united so as to form a tube, the upper margin of which is divided into eight notched lobes, and encloses the anthers; ovary sessile, four-celled, enclosed within the tube of the stamens. Fruit a capsule bursting by two or four divisions, each compartment containing a single seed provided with a fleshy arillus. [M. T. M.]

EPICHILE. The upper half of the lip of an orchid, when that organ is once jointed or strangulated.

EPICLINAL. Placed upon the disk or receptacle of a flower.

EPIDENDRUM. A vast genus of South American orchids, numbering more than 300 species, and exhibiting great diversity of growth. They are mostly epiphytes on trees, whence the generic name, though not a few are terrestrial. The stems are elongated and leafy in some, and reduced to a pseudo-bulb in others; the leaves are leathery in texture and usually strap-shaped; and the flowers are either solitary or disposed in axillary or terminal spikes, racemes, or panicles. According to Dr. Lindley, the essential character of the genus resides in the lip being more or less united by a fleshy base to the edge of a column, which is bornless, and considerably elongated, but not petaloid and winged; in the pollen masses being four, equal, compressed, with as many pulverulent caudicles folded back on them; and finally, in the presence of a cuniculus more or less deep at the base of the lip. *E. nemorale*, often miscalled *verrucosum*, is one of the handsomest in cultivation. It is a Mexican plant, with ovate pseudo-bulbs bearing two glossy strap-shaped leaves, and panicles of handsome rosy flowers, each about one and a half inch across, the lip streaked with lines of a darker colour; it takes its name from the minute rough points on the branches of the panicle. Similar in size and colouring of flower is *E. Skinneri* from Guatemala, but it has elongated stems, the flowers arranged in drooping terminal racemes, and the lip with three yellow crests. Perhaps the most desirable species is the Mexican *E. vitellinum*, from its brilliant deep orange-coloured and long-enduring flowers: it has oblong pseudo-bulbs with two short leaves, and a flower scape six to twelve inches high, the individual flowers about an inch in length. Then we have *E. cuspidatum*, notable for its very large yellow flowers, with a curious tri-lobed lip, the central lobe linear, and the two lateral crescent-shaped with beautifully fringed borders. To the same group belong *E. ciliare* and *E. nocturnum*, the first smaller in all its parts, with greenish flowers, the latter destitute of the fringe to the lip, and emitting a very agreeable odour in the evening. The singular colouring of the flowers in *E. prismatocarpum* renders it attractive, the ground colour being yellow-green with many dark purple blotches across the sepals and petals, and the lip pink. It has ovate pseudo-bulbs with a leathery strap-shaped leaf a foot long, and the flowers are in many-flowered scapes. [A. A. B.]

EPI D'EAU. (Fr.) *Potamogeton*. —, DE LAIT or DE LA VIERGE. *Ornithogalum pyramidale*. — DE VENT. *Agrostis Spicaventi*. —, FLEURI. *Stachys*.

EPIDERMIS. The true skin of a plant below the cuticle.

EPIDERMOID. Of or belonging to the skin.

EPIGÆA. The generic name of shrubs of the heathwort order, characterised by having three leaflets on the outside of the five-parted calyx; and by the corolla being salver-shaped, five-cleft, with its tube hairy on the inside. The name, derived from Greek words signifying 'upon the earth,' is sufficiently expressive of the mode of growth or trailing habit of the species. One of them, *E. repens*, a native of North America, has been long known in cultivation; it is an ornamental procumbent shrub, with fragrant flowers, usually white with a reddish tinge. [G. D.]

EPIGENOUS. Growing upon the surface of a part, as many fungals on the surface of leaves.

EPIGEOUS. Growing close upon the earth.

EPIGONE. The membranous bag or flask which incloses the spore-case of a liverwort or scale-moss when young. Also the nucule of a *Chara*.

EPIGYNIUM. East Indian shrubs, so named, in consequence of the disc which surmounts the ovary. They constitute a genus of *Vacciniaceæ*, known by their five-parted flowers, bell-shaped or cup-shaped corolla, ten separate stamens, and five-celled ovary containing many ovules, and surmounted by a five-lobed disc, as well as

Epigynium leucobotrys.

by the limb of the calyx. The fruit is succulent. *E. acuminatum*, a greenhouse shrub, has racemes of richly-coloured red flowers. *E. leucobotrys*, another species in cultivation, has a tuberous root like a yam, and the berries are white and wax-like; hence the name. [M. T. M.]

EPIGYNOUS. Upon the ovary; a term applied when the outer whorls of the flower adhere to the ovary, so that their upper portions alone are free and appear to be seated on it, as in umbellifers, myrtals, campanals, &c.

EPILEPIS. A genus of the composite family allied to *Coreopsis*, but differing in having applied to the outer surface of each wingless achene a three-toothed chaffy scale. The only known species, *E. rudis*, is an erect hispid Mexican herb with opposite pinnatisect leaves, whose stems are terminated by a corymb of numerous yellow-rayed flower-heads, each about an inch across; these have an involucre of two series of scales, neuter ray-florets, those of the disc tubular and perfect. The compressed achenes crowned with two short awns.1 [A. A. B.]

EPILINELLA. A section of *Cuscuta*, containing those in which the calyx consists of five fleshy sepals, keeled on the back, and with membraneous margins united at the base. It has been raised to a generic position by Pfeiffer. [W. C.]

EPILITHES. The name of a small herbaceous plant, which covers the rocks in certain parts of the island of Java, and is described by Dr. Blume as belonging to *Nyctaginaceæ*, though his description seems rather to apply to a plant of some other order. The flowers are monœcious, four-parted, the females without petals; ovary inferior with one ovule; stigmas four, brush-like; fruit berry-like. [M. T. M.]

EPILOBIUM. A somewhat extensive genus of mostly perennial herbaceous plants belonging to the order *Onagraceæ*, among which they are distinguished by their flowers having eight stamens, and by bearing numerous cottony seeds in an elongated pod-like seed-vessel. They are found in all situations, by rivers, in woods, or on waste ground, and some are Alpine. In habit they are mostly erect and but little, if at all, branched; the leaves are narrow and opposite, frequently toothed at the base; and the flowers, which are either axillary or in terminal spikes, are generally of a purple hue, apparently stalked, but in reality supported on the slender rudimentary capsule. There are several British species, most of which are unpretending weeds; but *E. hirsutum*, a tall species growing from four to six feet high, is frequently ornamental to the banks of rivers and ponds. The flowers of this plant are large and of a delicate pale pink, with a conspicuous four-cleft white stigma. The whole plant is downy, soft and clammy, exhaling a peculiar acidulous scent, which has gained for it the popular name of Codlins and Cream. *E. angustifolium* is not often found truly wild, but is a common ornament of cottage gardens, when, if suffered to range at its will, it soon overpowers all other herbaceous vegetation. It is sometimes planted with advantage in shrubberies when luxuriant undergrowth is desired, but should not be admitted into a small garden, as it is most difficult of eradication. In this plant the leaves are scattered and destitute of all pubescence, and the flowers are irregular, large, rose-red, and grow in a terminal spike. French, *Épilobe, Laurier St. Antoine, Osier fleuri*; German, *Weiderich*. [C. A. J.]

EPIMEDIUM, Barrenwort. A genus of *Berberidaceæ*, known by having the parts of the flower in fours, there being four sepals, eight petals and four stamens. They are Alpine herbs, found in Europe, Middle Asia, and Japan. *E. alpinum*, the only European species, is a low herb with a creeping rhizome, and long-stalked ternate leaves, with large ovate-cordate serrated leaflets, and panicles opposite the leaves bearing rather small dull purplish flowers, with the inner petals bulging at the base; it has been stated to grow in Scotland and the north of England, but only where planted. [J. T. S.]

EPINE BLANCHE. (Fr.) *Cratægus Oxyacantha*. — DU CHRIST. *Paliurus aculeatus*. — NOIRE. *Prunus spinosa*.

ÉPINETTE ROUGE. (Fr.) *Larix americana*.

ÉPINE-VINETTE. (Fr.) *Berberis vulgaris*.

ÉPINARD. (Fr.) *Spinacia*. — BLANC DU MALABAR. *Basella alba*. — D'HIVER. *Spinacia spinosa*. — DE HOLLANDE. *Spinacia inermis*. — DU MALABAR. *Basella rubra*. — FRAISE. *Blitum virgatum*. — IMMORTEL. *Rumex Patientia*. — SAUVAGE. *Chenopodium Bonus-Henricus*.

EPIPACTIS. A genus of terrestrial orchids, consisting of erect herbs with fibrous roots, and a leafy stem, bearing a loose simple raceme of purplish-brown or whitish flowers occasionally tinged with red. The perianth is spreading, without any spur; the petals and sepals are nearly similar; the lip free from the column, thick and concave at the base, the terminal portion broad and petal-like, with two protuberances at the base; the column short with a terminal anther. There are but few species, natives of the temperate regions of the northern hemisphere. Two only are British: *E. latifolia*, not unfrequent in woods and shady places, but usually singly, attaining two feet in height or even more, the lower leaves ovate, the upper ones small and narrow, the flowers varying from green to a dingy brown, and hanging in a long loose one-sided raceme; and *E. palustris*, which is more local, although abundant in particular spots, and is not so tall, but a more showy plant, the leaves narrower, the racemes more compact, with larger slightly drooping flowers, the sepals pale greenish-purple, the petals and lip white, more or less streaked with pink.

EPIPETALOUS. Inserted or growing on a petal.

EPIPHEGUS. A genus of *Orobanchaceæ*, containing a single species from North America. It is a brownish fleshy herb, parasitic only upon the roots of the beech, and furnished with a branched stem, and small remote scales, from the axils of which spring root fibres as well as flowers. The flowers on the upper portion of the branches are hermaphrodite and have a large corolla, but are generally barren, while those on the lower parts of the branches are small, have a short corolla, and are always fructiferous. In the hermaphrodite flowers, the corolla is ringent, compressed and four-cleft with the lower lip flat; while in the female flowers the corolla is short, obsoletely four-toothed and deciduous. The capsule is small, roundish, imperfectly two-valved, with numerous ovate seeds. [W. C.]

EPIPHORA *pubescens* is a South African epiphytal orchid about a span high, with short ovate pseudo-bulbs bearing two or three oblong linear leaves a little oblique at the apex, and a terminal erect raceme of numerous fragrant bright yellow flowers streaked with red, and nearly half an inch across. It is a highly desirable plant, as it keeps on flowering for nine months of the year. The relationship of the genus is with *Polystachya* amongst the *Vandeæ*, and it differs chiefly from that genus in the four pollen masses being attached to a distinct though short caudicle. The inside of the little trident-shaped lip, which is uppermost in the flower, is bearded with long hairs. There is only one species known. [A. A. B.]

EPIPHLŒUM. The layer of bark immediately below the epidermis. The cellular integument of the bark.

EPIPHRAGM. A membrane drawn over the mouth of the spore-case in urn-mosses, and closing it up.

EPIPHYLLOUS. Inserted upon a leaf.

EPIPHYLLUM. A small genus of *Cacineæ*, commonly cultivated in conservatories in this country on account of the showy pink or crimson flowers. Only three species are known, all natives of Brazil, where they are generally found upon the trunks of trees. They grow two or three feet high, and have thin cylindrical stems, and branches composed of numerous short leaf-like joints growing out of one another, and resembling leaves joined together by their ends. The flowers are produced singly at the extremities of these branches, and are upright and regular in one species, but bent downwards and somewhat two-lipped in the others. The sepals and petals are numerous and coloured alike, so that they are scarcely distinguishable, though the innermost have their bases united into a tube; the stamens are numerous, arranged in two series. The fruit is a small very smooth berry, sometimes having angular ribs.

E. truncatum is the species most frequently cultivated in this country, and there are several garden varieties of it, distinguishable only by the size and colour of their flowers. It is a native of Brazil, particularly of the Organ Mountains, but is seldom found at a greater elevation than 4,500 feet. The flat joints of the branches are about two inches long, broad at top, but tapering towards the base, and the flowers, which are produced from the broad ends of the joints, are bent downwards, one side of the expanded part being larger than the other; they are pink, crimson, or orange-coloured, with white stamens. *E. Russellianum*, also Brazilian, is readily distinguishable from the last by its flowers being straight, and the petals expanding in a regular manner; the stamens, also, are of the same pink colour as the flower. [A. S.]

EPIPHYTE (adj. EPIPHYTAL). Plants which grow upon the surface of others, as many mosses and orchids.

EPIPODIUM. A form of disk consisting of glands upon the stipe of an ovary. Also the stalk of the disk itself.

EPIPOGIUM *aphyllum* is a curious leafless pale-coloured herb, forming a genus of terrestrial orchids. The rootstock has a number of short thick fleshy fibres like those of *Corallorhiza*. The stem, about six inches high, bears some small scale-like bracts, and three or four rather large pale yellowish flowers with narrow sepals and petals, and an ovate somewhat concave lip with a thick projecting spur underneath; the column is short, with a shortly stalked terminal anther. The species has a very wide range in Europe, and temperate Asia, but is generally very scarce, growing here and there among rotten leaves, in woods, and shady places. In Britain it has only been found in a single locality, near Tedstone Delamere in Herefordshire.

EPIPTEROUS. Having a wing at the summit.

EPIRHIZOUS. Growing on a root.

EPISCIA. A small genus of *Gesneraceæ*, containing six species, natives of America. They are fleshy, creeping, and rooting herbs, with opposite petiolate leaves, and solitary or aggregated axillary flowers, whose small calyx is free and five-parted, and the corolla erect within the calyx, then obliquely salver-shaped, with the limb five-lobed. The ovary is surrounded at the base by a disc, which swells behind into a gland. The capsule is membranaceous, two-celled, with numerous oblong seeds. [W. C.]

EPISCOPEA. *Themistoclesia*.

EPISPERM. The skin of a seed.

EPISPORANGIUM. The indusium of a fern when it overlies the spore-cases, as in *Aspidium*.

EPISPORE. A skin which covers some spores.

EPISTYLIUM. A genus of the spurgewort family peculiar to Jamaica, containing only a couple of species, one of which is a shrub, the other a tree of about twenty feet; both have smooth alternate laurel-like leaves, and minute yellowish-green or reddish flowers disposed in little clusters or racemes, which in *E. axillare* proceed from the axils of the leaves, and in *E. cauliflorum* from the bare stems. The sterile and fertile flowers are in the same cluster, the former with a four-parted, the latter with a five-parted calyx. The fruits are little oblong three-sided capsules, with three cells and one or two seeds in each. The genus is by some authors united with *Phyllanthus*, from which it chiefly differs in the four-lobed calyx of the male flowers. [A. A. B.]

EPITHELIUM. An epidermis consisting of young thin-sided cells, filled with homogeneous transparent colourless sap.

ÉPURGE. (Fr.) *Euphorbia Lathyris*.

EQUISETACEÆ, EQUISETUM. A natural order and genus of the higher cryptogams, remarkable for the external resemblance which they bear in habit to *Casuarina* or *Ephedra*, and as regards the heads of fructification to *Zamia*. All resemblance, however, ceases there, and the natural affinities of the plants are with ferns. The plants are often perennial, new shoots being thrown up from the creeping rhizomes. The spores germinate like those of ferns, and produce a sort of prothallus, which bears archegones and antherids. The latter yield large spiral fringed spermatozoids like those of ferns. The shoots are jointed, each articulation having a toothed membranous sheath, and are often repeatedly divided, with whorls of branches and branchlets. The fructification is produced in the form of terminal cones, consisting of a number of peltate scales, each of which produces a circle of spore-cases, perpendicular to the axis, and opening by a longitudinal fissure, the walls of which consist of very delicate spiral tissue. The spores have a spiral coat, which ultimately splits up into two bodies, each with two clavate ends, and attached by their centre so as to look like four stamens. These, however, are nothing more than the unrolled spiral of which the spore coats consist.

The structure of the rhizome and of the lower part of the stem is very curious, and quite different from anything in ferns. In an early stage there is a central column of cellular tissue in the rhizome, from which eight plates radiate, being connected with an external cylinder of the same nature, and leaving between them distinct cavities. At a later period new tissue grows from the walls of the plates, and ultimately obliterates the cavities. Opposite to each of the plates is a vascular bundle, consisting of distinct annular vessels passing into spirals. In ferns, on the contrary, the vessels are mostly scalariform. In the fruit-bearing stems the cavities are more abundant with various modifications.

Equiseta, or Horsetails, are found in most parts of the world, though they are wanting in Australia and New Zealand. In the temperate regions they are mostly inhabitants of fields and wet places, and sometimes of loose sands, which they tend to bind together by their delicate rootlets, and have stiff erect stems capable of supporting themselves. But in warmer regions, and even in Lisbon, as *E. debile* and *elongatum*, they require the support of bushes to which they cling. They sometimes attain a considerable size, as *E. giganteum*, though never reaching the dimensions of undoubted fossil *Equisetaceæ*. An immense quantity of silica, amounting sometimes to half their weight when consumed, is taken up into their substance; and, according to the observations of Brewster, the particles, each of which has a double axis of refraction, are disposed in rows parallel to the axis, and occasionally forming ovals connected together like the jewels of a necklace. In consequence of this abundance of silica, like Tripoli, some of the species are used for polishing various articles, and large quantities of *E. hyemale* are imported into this country under the name of polishing or Dutch rushes. Some of the species have been used in medicine,

Equisetum xylochætum.

but their virtues are doubtful. The rhizomes contain a considerable quantity of starch, and the starch cells sometimes exhibit a kind of circulation. [M. J. B.]

EQUITANT. When the two sides of a leaf are brought together and adhere except at the base, where they enclose an opposite leaf whose sides are in the same state: hence they look as if they rode on each other.

ERABLE. (Fr.) *Acer*. — DE NORVEGE *Acer platanoides*. — DURET. *Acer opulifolium*. — JASPÉ. *Acer pensylvanicum*, — NÉGUNDO. *Negundo*. — OBIER. *Acer opulifolium*.

ERAGROSTIS. A very extensive genus of grasses, belonging to the tribe *Festuceæ*, distinguished by having the inflorescence in more or less compound or decompound panicles; glumes four to ten-flowered; pales imbricated in two ranks, the upper reflexed with the edges turned back; stamens two or three; styles two, with feathery stigmas; seeds loose, two-horned, not furrowed. In Steudel's *Synopsis* there are 243 species described; these range more or less over the whole surface of the globe, Asia being the quarter where they mostly abound. Europe has only six species, all of which are natives of the southern portion only. The appellation is derived from two Greek words, signifying when combined Love-grass. Most of the kinds are handsome, and some of them are sufficiently hardy for being cultivated as ornamental grasses in Britain. [D. M.]

ERANTHEMUM. A considerable genus of *Acanthaceæ*, containing nearly fifty species, widely distributed over the tropical and subtropical regions of the Old and New Worlds, chiefly growing at a greater or less height, on mountains. They are shrubs or under-shrubs, with entire or serrated leaves, and showy often spicate flowers, whose corolla is salver-shaped, with a long slender tube and an unequally lobed limb. There are two fertile stamens. [W. C.]

ERANTHIS. A highly prized little herbaceous plant belonging to the *Ranunculaceæ* and allied to *Helleborus*, from which it may at once be distinguished by the more delicate texture of its leaves, and by having its solitary flowers surrounded by an involucre cleft into numerous segments. It is most commonly known by the name of Winter Aconite, because its foliage resembles that of the aconites, and its bright green involucre and pretty yellow flowers are in perfection when snowdrops bloom. Being a low-growing plant, but a few inches high, it is well adapted for the front of borders. It is perfectly hardy as to temperature, and will thrive in any soil. *E. hyemalis*, the species most generally cultivated, is a native of central and southern Europe, in moist shady places, and on hills. *E. sibiricus*, a native of Eastern Siberia, a plant of precisely similar habit, has five sepals; whereas *E. hyemalis* has six to eight. French, *Eranthis d'hiver, Helleborine*. [C. A. J.]

ERASMIA. A genus of *Piperaceæ*, comprising a low-growing Mexican herb, with lance-shaped leaves, and branching spikes bearing scattered persistent peltate bracts; the filaments of the stamens are rather thick, short; anthers globular; ovary sessile, cylindrical; stigma conical. The fruit is an elongated smooth berry. [M. T. M.]

EREMÆA. A genus of shrubs of the myrtle family, natives of Swan River, and nearly allied to *Melaleuca*, but distinguished from it by the stamens, which are either entirely detached, or more or less united into groups. The anthers are fixed moreover by their base, and not by their backs; and the flowers grow singly at the end of the branches, where they are covered with overlapping bracts. [M. T. M.]

EREMIA. The generic name of shrubs, natives of the Cape of Good Hope, belonging to the heathworts, having the calyx bell-shaped or somewhat globose, and the stamens more than four, usually six or eight, very rarely five. The name *Eremia* was assigned for the purpose of indicating another mark (not however confined to these plants), viz. one seed in each cell of the fruit. The species have the general aspect of heaths, with leaves three or four in a whorl, spreading or bent down, and having stiff hairs. [G. D.]

EREMOBRYA. A term proposed to designate that group of ferns in which the fronds are produced laterally on the rhizome, and articulated with it. See also *Desmobrya*. [T. M.]

EREMOCARPUS. A genus of *Euphorbiaceæ*, remarkable for having its little hairy fruit, about the size of an orange-seed, composed of a single carpel, not of three, which is the usual number in the family. *E. setigerus*, so named from the bristle-like hairs on the stems, is peculiar to California, and is the only species of the genus. It is a small, prostrate annual herb, having all its parts densely clothed with soft, white starry hairs. The stalked alternate leaves have broadly-oval obtuse blades, and the small green flowers come in dense clusters in the forks of the branches, males and females together, the females sessile. The whole plant has a strong disagreeable odour, even in a dried state. [A. A. B.]

EREMODENDRON. A genus of *Myoporaceæ* containing a single species from New Holland. It is a beautiful tree, with long narrow lanceolate leaves, and axillary flowers on the tops of the branches. The large coloured lobes of the five-parted calyx are oblong, obovate, narrow at the base, and not changing in fruiting, while the corolla has an incurved tube, and an unequally five-lobed limb. The ovary is ovoid-oblong, compressed, and two-celled. This genus is scarcely separable from *Eremophila*, except by the peculiar lobes of calyx. [W. C.]

EREMOLEPIS. A genus of plants included in the order *Loranthaceæ*. The flowers are diœcious, and have no petals; the staminate flowers have a tripartite calyx, with three stamens which are inserted opposite to the calycine segments; and the pistillate flowers have a tripartite calyx, an inferior ovary, a short style, and a simple stigma. They are parasitic shrubby plants, with alternate leaves destitute of a terminal scale, the staminate flowers in catkins, the pistillate in clusters. The genus is allied to the *Eubrachion* of Hooker. Two species, natives of South America, have been described, *E. punctulata* and *E. verrucosa*. [J. H. B.]

EREMOPHILA. A genus of *Myoporaceæ*, containing four species of broom-like shrubs, natives of New Holland. They have opposite or alternate leaves, and axillary crowded or solitary peduncles supporting flowers which have a five-parted scarious calyx, and a corolla with a large tube and bilabiate limb. [W. C.]

EREMOSTACHYS. A genus of labiate plants distinguished by the upper lip of the corolla being elongated and helmet-like, narrow below, and hairy on the outside, the lower lip with three spreading rounded lobes, the middle being broadest. The name is derived from two Greek words signifying 'solitary' and 'spike.' The species are hardy plants of little importance. One of them, *E. laciniata*, has been long known in cultivation as a hardy perennial, a native of dry hills in the eastern part of the Caucasian range; it has large spindle-shaped fleshy roots well adapted to resist the drought to which, in its native wilds, it is sometimes subjected. [G. D.]

EREMOSYNE. A genus of *Saxifragaceæ* from New Holland, with rosettes of obovate entire root leaves, and pectinate-pinnate stem leaves; the flowers small, white, in compact dichotomous cymes; with a hemispherical calyx-tube adhering to the ovary, and five linear petals; ovary two-celled, with solitary ovules. [J. T. S.]

EREMURUS. A genus of *Liliaceæ* closely allied to *Asphodelus*, but differing by having the filaments not dilated at the base and the seeds smooth. They are herbs from the Caucasus, Siberia, and Asia Minor, with fasciculate roots, linear radical leaves, and a naked scape terminating in an elongated raceme of yellow or white flowers with narrow spreading perianth segments, and exserted stamens. [J. T. S.]

EREMUS. A ripe carpel separating from its neighbours, and standing apart.

ERGOT. An affection of the seeds of different grasses in which the seed becomes black and elongated, so as to resemble in form a cock's spur, whence it derives its name. In an early stage the Ergot is partially covered with a thin crust producing abundant conidia. These conidia appear sometimes to grow like yeast globules, so as to assume the form of an *Oidium*, whence the fungus has received the name of *Oidium abortifaciens*. It is at least supposed that the grains of the conidia and of the *Oidium* are identical. If the Ergot, however, is kept moist, either by excluding the outward air, or by sowing it in damp soil, different species of *Cordiceps* invariably appear, which are supposed to be the perfect state of the fungus. Ergot is a destructive disease amongst corn, but especially in rye; but it derives its greatest notoriety from its peculiar properties in producing contraction of the uterus, properties of which the surgeon avails himself for the expulsion of the fœtus and preventing hæmorrhage. It is moreover combined with chloroform with a view to produce contraction without pain. Ergot is a valuable remedy in the hands of the regular practitioner, but a most formidable one in those of the quack, by whom it is often given to produce abortion. In this case a second quality comes into play, namely, that of causing dangerous gangrene, which it does where it forms a considerable portion in bread-corn, or is taken medicinally for a continuance. Instances are on record where the most frightful gangrene has ensued from its use, sometimes affecting a whole district. Ergot is often extremely abundant in our pastures, and causes sheep and cows to slip their young. No doubt many cases of gangrene in our flocks and herds are attributable to its prevalence. [M. J. B.]

ERGOT DE COQ. (Fr.) *Crataegus Crusgalli.*

ERIA. A genus of epiphytal orchids peculiar to India and the adjacent islands, and numbering about seventy species. Some are minute stemless herbs, consisting of a small pseudobulb with a solitary leaf and flower; others have creeping wiry stems furnished at intervals with flattish pseudobulbs and short spikes or racemes of small white or greenish flowers; whilst not a few have erect or drooping terete stems, with lance-shaped often plaited leaves, and axillary or terminal racemes or panicles. None of the species are remarkable for their beauty, though many have fragrant flowers. The genus is nearly related to *Dendrobium*, but has eight instead of four pollen masses. It takes its name from the Greek *erion*, wool, the flowers of many of the species being clad with soft white down. The lip is usually trilobed, with a crested disc, and jointed to the much produced base of the column. [A. A. B.]

ERIACHNE. A genus of grasses belonging to the tribe *Aveneæ*. The inflorescence of the species is panicled, the spikelets two-flowered, sessile or stalked, and hermaphrodite; glumes two, membranaceous, about equal to the short awns; stamens three; ovary smooth. There are twenty-three species described in Steudel's *Synopsis*, which are nearly all natives of the southern hemisphere, South Africa, and New Holland, where some of them are valuable as pasture grasses. [D. M.]

ERIANTHERA. A genus of *Acanthaceæ*, containing two species, natives of India. They are low undershrubs, with few leaves, and flowers without bracts, on one or two-flowered axillary peduncles; the calyx equally five-parted, and the corolla two-lipped, with the broad upper lip bifid, and the lower trifid. [W. C.]

ERIANTHUS. A genus of grasses belonging to the tribe *Andropogoneæ*, scarcely differing from *Saccharum*, under which the species are included by Steudel. [D. M.]

ERICACEÆ. (*Heathworts*). A natural order of corollifloral dicotyledons, typical of Lindley's erical alliance among hypogynous Exogens. Shrubs or undershrubs, with evergreen, rigid, entire, whorled or opposite, exstipulate leaves; calyx inferior, four to five-cleft; corolla four to five-cleft; stamens eight to ten or twice those numbers, hypogynous; anthers two-celled, with appendages, opening by pores. Ovary surrounded by a disk or scales. Fruit capsular, rarely berried; seeds numerous, albuminous. There are two sections of the order :— 1. *Ericeæ*, fruit opening loculicidally, rarely septicidally; buds naked; 2. *Rhododendreæ*, fruit capsular, septicidal; buds scaly, resembling cones.

The genus *Leiophyllum* is remarkable on account of its having a polypetalous corolla. The common heath (*Calluna*) is separated from the heaths (*Erica*) by its capsules having a septicidal and not a loculicidal dehiscence. The genus *Erica* reaches its maximum at the Cape of Good Hope. Some of the heathworts are astringent, others have edible fruit, and others, such as species of *Rhododendron*, *Kalmia*, and *Ledum*, are poisonous. *Arbutus Unedo* is the strawberry tree, common near the Lakes of Killarney. *Rhododendron arboreum*, and other species, in India, sometimes attain a height of forty feet; some species grow at the elevation of 16,000 to 18,000 feet in the Himalayas. *Rhododendron hirsutum* and *ferrugineum* grow on the Alps and Pyrenees at an elevation of 4,000 to 6,000 feet, and are called the Roses of the Alps. *Andromeda fastigiata* is called Himalayan heather. *Gaultheria Shallon* and other species yield edible baccate fruits. *Azalea procumbens* grows on the Scotch mountains, and is also a native of the Arctic regions, of the Alps, of Northern and Southern Europe, Siberia, and North America. There are about fifty known genera and nine hundred species. Examples: *Erica, Clethra, Arbutus, Azalea, Kalmia, Rhododendron, Bejaria, Ledum.* [J. H. B.]

ERICA. The generic name of shrubby plants belonging to the heathwort order, from which, indeed, the scientific designation of it, *Ericaceæ*, is derived. They are distinguished from their congeners by the four-leaved calyx, and four-lobed corolla, the lower part of which is either globular or tubular and dilated; the stamens have the lobes of the anthers distinct, sometimes with an awn-like appendage, and opening by an oblong pore; the fruit is dry, four or eight-celled, many-seeded, bursting loculicidally.

The genus *Erica* comprehends a great number of species of much interest and beauty, and therefore general favourites with horticulturists, especially since the best method of growing them has been found out, and in this much credit is due to the late Mr. M'Nab of Edinburgh. There is a marked tendency to repetition of the number four in the different parts of the flower, viz. calyx, corolla, stamens, and fruit; and this is true even of the grouping of the leaves and of the flowers. The usual absence of any odour is compensated for by elegance in the general aspect of the plants, as well in their foliage as flowers, which combine to render most of the species worthy of a place in collections. In the corolla especially, the beauty of form, delicacy of aspect, and variety of tint can scarcely be surpassed. The shapes of the flower, a study for the modeller, present considerable variety of modification, being long and tubular, straight or arched, in some very small and dilated, in others smooth and brilliant, or covered with clammy hairs. As to colour, we find the purest white, passing into very pale rose, purples of various hues, red, less frequently yellow, and sometimes green. In some instances the calyx rivals even the corolla in appearance. Plants of this genus are confined to the old world; in Africa especially they abound, and the Cape of Good-Hope is the main source whence we have derived those now so well known as ornaments of our horticultural collections, where, under skilful treatment, they even far surpass in luxuriance those which occur in the wild state. In Britain six species are usually counted as indigenous, only two of which are, however, widely diffused and cover immense tracts, viz. *E. Tetralix* and *E. cinerea*; the remaining four are more local and confined to the southern and western parts of the United Kingdom.

The true Heaths are of little importance in a medical point of view, none possessing any active property. In our own country the two more common species above mentioned are used for brooms and for bedding cattle; their buds and tender shoots constitute part of the food of some of our native birds; and they often contribute largely to the formation of peat. The Scotch Heath, *E. cinerea*, is the badge of the M'Alisters, and *E. Tetralix* that of the M'Donalds. [G. D.]

ERICAMERIA. A small genus of the composite family, found in Oregon and California, related to *Linosyris*, but differing in having rayed as well as tubular florets, and also in having smooth achenes. They are dwarf resinous shrubby plants, much branched and leafy, with the aspect of heaths, the leaves awl-shaped and numerous, and the small yellow flower heads in corymbs at the ends of the branches. The smooth achenes are crowned with a pappus consisting of numerous capillary unequal bristles. [A. A. B.]

ERICINELLA. A genus of heathworts, having the calyx in four divisions, one of them larger than the others; corolla bell-shaped, the border deeply divided into four; stamens four, rarely five, usually with awn-like appendages; style or appendage at top of the seed vessel, ending in a shield-like surface. The name *Ericinella* is the diminutive of *Erica*, the species having the general aspect of heaths; leaves three in a whorl, flowers small and terminal, without bracts or leaflets at their base. They are small shrubs, natives of Madagascar, Tropical Africa, or Caffraria. [G. D.]

ERIGERON. A genus of unpretending herbaceous plants of humble stature belonging to the *Compositæ*. The flowers are radiate, the florets of the ray in several rows, very narrow and of a different colour from those of the disk, which are fertile, with a hairy pappus; the involucre is imbricated with several rows of narrow scales. Two or three weedy species are natives of this country, and many foreign species are described by authors, all marked rather by the absence of bright colours than by any desirable qualities. The name *Erigeron* denotes 'soon becoming old,' and is most appropriate, for in many of the species the plant, even when in flower, has a worn-out appearance, giving the idea of a weed which has passed its prime. French, *Vergerette*; German, *Scharfe*. [C. A. J.]

ERINEUM. A name given to numerous productions which appear upon the leaves of trees and shrubs, and very rarely on those of herbaceous plants, which were formerly referred by authors to *Fungi*, but are now almost universally acknowledged to be merely diseased states of the cuticular cells. The spongy spots on the leaves of vines and lime trees afford a good example. The forms which these diseased cells assume are extremely various; and they are interesting to the physiologist, as showing the alteration to which the component cells of plants are subject when free from the pressure of neighbouring cells and subjected to new conditions. Illustrations will be found in the works of Corda and Greville, and a complete account in a work on the subject by Fee. [M. J. B.]

ERINOSMA. A genus of *Amaryllidaceæ* containing the plant sometimes called *Leucojum vernum*, an early spring-flowering herb, with ovate bulbs, linear-lorate leaves, and one-flowered scapes. The flowers are fragrant, and differ from those of the snowdrop in having petals like the sepals, white, with a yellowish-green spot outside; and from those of the snowflake in having a club-shaped style. [T. M.]

ERINUS. Low herbaceous Alpine plants belonging to the *Scrophulariaceæ*, distinguished by having a five-leaved calyx, a corolla with a five-cleft equal limb, and short reflexed upper lip, and a two-celled capsule. They are pretty little plants, with tufted foliage and simple racemes of purple or yellow flowers, and are therefore desirable for the decoration of rockwork or old walls, for which purpose no plants can be more fitted, as they produce their numerous blossoms during most of the summer months. The species most frequently cultivated, is *E. alpinus* (French, *Erine des Alpes*), a native of the European mountains. *E. hispanicus* is smaller than the preceding, and has downy leaves. [C. A. J.]

ERIOBOTRYA. The Loquat, or Japanese Medlar, *E. japonica* (*Mespilus japonica* of Linnæus), one of the *Pomaceæ*, is a native of Japan and the southern parts of China, and is cultivated as an edible fruit in many parts of India. It was first made known to us by Kaempfer, who saw it growing in Japan, which he visited in 1690. It was more fully described in 1712 by Thunberg, who met with it growing near Nagasaki, Yedo, and elsewhere commonly in Japan. In that country it is called Bywa and Kuskube, in China Lo-quat. It was brought to Europe by the French in 1784, and planted in the National Garden at Paris; and three years later it was imported from Canton to Kew.

The tree, according to Thunberg, attains a large size in its native country. The leaves are evergreen, large, oblong, rugose like those of the medlar, bright green above, somewhat downy beneath. The flowers are produced in October and November, in spikes at the ends of the branches; their petals are like those of the hawthorn, but larger and perfectly white. The fruit is oval, of the size of a small apple, pale orange with a faint blush of red, the flesh pale yellow, with a sharp subacid flavour resembling that of an apple. It ripens in spring, or early in summer. The tree is hardy enough to bear the cold of our ordinary winters, but it has been killed when exposed to frosts of unusual severity, such as that which occurred in 1814, 1838, and 1860. We are not aware of its having fruited in this country except under glass, and with the aid of artificial heat, it has not fruited at Paris in the open air; but it is successfully cultivated as a standard in the south of France, and its fruit is even common in the markets of Hyeres and Toulon. At Malta it succeeds admirably. Improved varieties, as regards the size and quality of the fruit, have been there raised, and introduced into England; but in consequence of the tree naturally producing its flowers at the commencement of winter, it is not adapted for bearing fruit in the open air in this country, the blossoms being either cut off by frost, or so much checked by cold that the growth of the embryo fruit cannot go on. There is, however, no difficulty in fruiting it under glass. This was done at Blithfield in Staffordshire in 1818; and an account of the means adopted is given by Lord Bagot in the *Transactions of the Horticultural Society* (iii. 299) accompanied by a coloured plate. The plant was fruited in a pot kept in a stove during winter, the fruit ripening in March or April, two months earlier than its period of ripening in its native country. We may therefore conclude, that the amount of heat to which the plant was subjected in the stove was greater than that which prevails between the time of flowering and the ripening of the fruit in Japan. It appears, however, that the fruit artificially produced at Blithfield was of excellent quality. [R. T.]

ERIOCAULACE.E. (*Pipeworts*). A natural order of incomplete monocotyledons included in Lindley's glumal alliance among the Endogens. Marsh plants, with narrow cellular spongy leaves, sheathing at the base, and a capitate inflorescence. The flowers are very minute, some having stamens, others pistils. Glumes two to three. Ovary superior, three, rarely two-celled, surrounded by a membranous tube; ovules solitary, orthotropal; style very short; stamens two or three. Capsule with loculicidal dehiscence; seeds solitary, pendulous, with a winged or hairy covering. The species abound in South America, and some plants of the order are found in North America and Australia.

There are ten known genera and upwards of two hundred and twenty species. Examples: *Eriocaulon, Lachnocaulon, Cladocaulon, Philodice*. [J. H. B.]

ERIOCAULON. The typical genus of *Eriocaulaceæ*. The name is derived from two Greek words, meaning 'wool' and 'stem,' on account of the woolly character of the stalks of some of the species. Flowers diœcious, in a compact scaly head, the staminate ones in the centre, and the pistillate ones in the circumference of the head. The species are found in the principal parts of Asia, America, and New Holland. They are rare in North America. One species, *E. septangulare*, occurs in Britain, being found in the Isle of Skye and in Galway. One hundred species have been described. Some of those found in Brazil attain a height of six feet. [J. H. B.]

ERIOCEPHALUS. A genus of S. African *Compositæ*, comprising nearly twenty species, which form much branched bushes, usually with linear, somewhat fleshy leaves, covered with silky hairs, but sometimes large and variously toothed, a good deal like those of some wormwoods, and like them with an aromatic odour. The white flower-heads, sometimes solitary but usually arranged in corymbs or umbels, are a good deal like those of the milfoils in size and appearance. They are remarkable for having the inner scales of the involucre clothed with long woolly hairs; these are not very perceptible when the plant is in flower, but after the flowers wither, and the anthers approach to ripeness, the heads are completely enveloped in the hairs, and look like little balls of cotton about the size of a pea. The hairs when fresh are white, but at length become rust-coloured, and are used by various birds for building their nests. [A. A. B.]

ERIOCHLOA. A genus of grasses belonging to the tribe *Paniceæ*, now referred to *Helopus*. [D. M.]

ERIOCHOSMA. *Nothochlæna*.

ERIOCNEMA. A genus of *Melastomaceæ*, nearly allied to *Sonerila*, but having the parts of the flower arranged in fives instead of in threes. The species are dwarf hairy Brazilian herbs, scarcely a foot high, with somewhat fleshy stems, bearing near their base a few oval leaves, heart-shaped at the base, and densely clothed with rusty hairs. The small white flowers are few, and arranged in little umbels, on the end of a naked stalk. *E. marmoratum* has the leaves beautifully variegated. [A. A. B.]

ERIOCOCCUS. The name given by some authors to a species of *Riedia*, whose capsules are clothed with soft short wool.

ERIOCOMA. The Silk Grass, *E. cuspidata*, is peculiar to North America, where it is found usually in barren spots from Lake Winipeg, west and south to New Mexico. It grows one to three feet high, has wiry leaves with the margins rolled inwards, and very lax panicles of flowers, each spikelet supported on a long slender stalk. Like the feather-grasses, to which this is nearly allied, the spikelets are one-flowered, the outer glumes are membranaceous, remarkably inflated below, and contracted suddenly at the apex into a short pointed beak. The inner glumes are very silky at the base, and end in a short awn. [A. A. B.]

ERIOCYCLA. A genus of *Umbelliferæ*, characterised chiefly by the fruit being clothed with wool-like hair. The only species is an inconspicuous herb, a native of the Himalayas, having the leaves thrice-pinnate; the secondary divisions of the umbels somewhat capitate. [G. D.]

ERIODENDRON. A genus of tropical trees, referred by some botanists to the *Sterculiaceæ*, and by others to the *Malvaceæ*. It is nearly allied to *Bombax*, from which it differs in the staminal column being five-cleft, each branch bearing two or three anthers, that of *Bombax* being divided at top into an indefinite number of filaments bearing single anthers. They have digitate leaves, and one-flowered axillary or subterminal peduncles, which are either solitary or fasciculate, the flowers being rather large, white or rose-coloured. The habit of *E. indicum* is represented in plate 13 b. [T. M.]

ERIOGLOSSUM. A genus of *Sapindaceæ*, nearly related to *Sapindus*, but differing in the nature of its fruits. These in *Sapindus* are made up of two or three one-seeded carpels, which are united their whole length, or slightly separate at top, so that they form one berry, while here the elliptical berried carpels, which are two or three in number, are quite free to the base. *E. edule* is a common tree in the Malayan peninsula and the neighbouring islands, extending to North Australia; it has alternate unequally-pinnate ash-like velvety leaves, about one foot long, and the small greenish-white flowers are disposed in branching panicles. The four petals are each furnished with a strap-shaped and bilobed woolly appendage on the inside near the base, the name of the genus, which signifies 'woolly tongue,' having reference to these. The wood is valuable, being strong and durable. In the only other species, *E. cauliflorum*, the racemes of flowers proceed from the old wood. [A. A. B.]

ERIOGONUM. A genus of *Polygonaceæ*, forming the type of a tribe distinguished by the absence of stipules, and the involucrate flowers. They are natives of western North America, rarely occurring in the Southern States, or on the east coast. Herbs or undershrubs, usually woolly, with radical leaves in tufts, and alternate or tufted stem leaves. The peduncles often form a compound umbel or head. The perianth is herbaceous, six-cleft, with the segments arranged in two rows. [J. T. S.]

ERIOLÆNA. This genus, known also as *Schillera* and *Microlæna*, belongs to the *Sterculiaceæ*, in which it is notable from having perfect flowers with petals which do not wither and remain attached, but fall early, together with a column of numerous stamens in many series, all the stamens perfect, and not as in many of the family having sterile stamens (staminodia), alternating with the perfect ones. There are seven known species, all East Indian trees or shrubs, with alternate stalked heart-shaped leaves resembling those of a lime tree in form and size, and axillary or terminal panicles of rather large mallow-like yellow flowers. [A. A. B.]

ERION. In Greek compounds = woolly.

ERIOPETALUM. A small genus of *Asclepiadaceæ*, the species of which are natives of India, and form erect branching herbs with scale-like adpressed leaves, and small flowers in lateral or terminal sessile umbels. The corolla is subcampanulate and five-cleft, with long linear segments, and the staminal crown gamophyllous and fifteen-lobed; the five inner lobes rest on the anthers, the others are erect, and adhere to the inner series. This genus agrees in habit with *Microstemma*, but differs in the form of the staminal crown; on the other hand, it resembles *Boucerosia* in the crown, but has a widely different habit. (W. C.)

ERIOPHORUM. A genus of cyperaceous plants belonging to the tribe *Scirpeæ*, distinguished by the inflorescence being either in single or compound spikes; glumes nearly equal, the lowest sometimes empty; bristles ultimately silky; nut, trigonous. The British species all grow on wet bogs or turfy moors, where they frequently form very conspicuous masses of vegetation, in consequence of the long showy silky bristles of the flowers. The English name Cotton Grass is very expressive, the flowers of some of the species appearing like tufts of cotton. [D. M.]

ERIOSOLENA. A genus of *Thymelaceæ*, or *Daphnaceæ*. Perigone coloured, villous externally, funnel-shaped, with a four-cleft limb, the alternate segments shorter, the throat naked. Stamens eight, inserted in two rows into the upper part of the tube of the perigone, the alternate ones longer. Ovary one-celled, with a single ovule; style short; stigma capitate. Fruit drupaceous, single-seeded. Shrubs from Java and India, with alternate oblong-lanceolate coriaceous leaves, which are glaucous below; flowers in solitary axillary heads with long peduncles, and a two to four-leaved involucre. There are three species. By De Candolle they are included under *Daphne*. [J. H. B.]

ERIOSORUS. *Gymnogramma.*

ERIOSPHÆRA. A genus of *Compositæ*, consisting of a few South African herbs, nearly related to *Helichrysum*, differing chiefly in their less numerous, and densely woolly involucral scales. Some are unbranched, erect, and about six inches high; others much branched, with slender prostrate stems; and all have their parts clothed with a short white wool. The leaves are oboval or spathulate, and entire, and the yellow spherical flower-heads are few or numerous, and disposed in dense clusters on the ends of the stem, each being about the size of a pea. [A. A. B.]

ERIOSTEMON. A genus of shrubby *Rutaceæ*, whose main characteristics are, a corolla of five petals which do not soon fall off, but remain on the plant for some time in a withered condition; ten hispid stamens,—hence the name of the genus, which signifies 'woolly stamen;' fruit of five carpels which separate and open by a long cleft to liberate the usually solitary seed. They are natives of New Holland, and have for the most part white or pinkish flowers. [M. T. M.]

ERIOSYNAPHE. A genus of umbellifers, the name of which is derived from two Greek words, signifying 'wool,' and 'junction,' or commissure, and points out a prominent character—the presence of a downy or wool-like covering near the line which indicates the junction of the two halves of the fruit. *E. longifolia* is a perennial shrub, with the divisions of the leaves long and narrow; the flowers yellow. It is a native of Siberia, along the course of the Volga. (G. D.)

ERISMA. A curious genus of tropical American trees, belonging to the *Vochyaceæ*, and remarkable for the enlarged calyx segments which crown the somewhat pear-shaped ripe fruit. The species are some of them upwards of 100 feet high, with smooth, opposite or whorled laurel-like leaves of a leathery texture; some are oval, pointed and entire, others oblong, attenuate below into a stalk, and notched at the apex. The pretty blue or yellow flowers, disposed in terminal panicles, smell like primroses, in some species. They are like the others in the family, very unsymmetrical, having a calyx of four or five teeth; a single nearly fan-shaped petal narrowed below into a claw; one fertile and four barren stamens; and a one-celled ovary crowned with a simple style.

The Japura of Brazil, *E. Japura*, is a tree of 80 to 120 feet, with stalked, whorled, oblong leaves, and panicles of yellow flowers. Mr. Spruce, its discoverer, thus speaks of it:—This noble tree, called by the Indians Japura, is frequent on the Upper Rio Negro, and on the Uaupés. It is said to be abundant on the Japura, and to have given the name to that river. As I came up the Rio Negro from the mouth of the Uaupés to San Carlos, in March 1853, the large heads of the Japura, clad with red fruits, were observed dotted everywhere about the forest. The kernels are pleasant eating both raw and boiled: they are also prepared in this way; having been boiled from morning till night, they are well covered up, and put into baskets in running water, where they remain two or three weeks. When at the end of this period they are opened out, they have a disagreeable stercoraceous odour. They are now beaten in a mortar until they have the appearance and consistence of pale butter. To receive this, a large cylindrical basket, three to five palms long by one in diameter, is made of strips of the trunk of the gravatana palm (*Iriartea prurience*), and lined with the leaves of a *Heliconia*. The basket is placed on a stage over the fire, where it is customary to put things that require to be kept dry, and there the butter will remain good for two or three years. Japura butter (as it may be called) is eaten along with fish and game, being melted in the gravy along with the fruits of various species of *Capsicum*, which is an essential ingredient in the moblo at every Brazilian table, whether the guests be red or white. People who can get over its vile smell, which is never lost, find it exceedingly savoury. The fruits call to mind those of the Indian *Dipterocarpus*. [A. A. B.]

ERISMA. The rachis or axis of grasses.

ERITHALIS. A name applied to a genus of West Indian shrubs, in consequence of their shining deep green leaves. They are included among the *Cinchonaceæ*, and have axillary panicles of white flowers, with five or ten parted whorls, a wheel-shaped corolla, an inferior ovary with from five to ten compartments, one ovule hanging from the summit of each of the cavities. The fruit is a berry crowned by the limb of the calyx. [M. T. M.]

ERITRICHIUM. A genus of *Boraginaceæ*, consisting of small woolly Alpine plants forming dense cushions; racemes short, bracteated, bearing a few small bright blue flowers, with a salver-shaped corolla closed at the throat by five small obtuse scales. [J. T. S.]

ERNESTIA. A genus of *Melastomaceæ*, represented by *E. tenella*, which grows in the mountain woods of New Granada, and is a slender suffruticose hairy herb, with opposite stalked oval leaves, and white flowers disposed in loose terminal panicles. The form of the stamens is that which chiefly distinguishes the genus from its allies; the anthers are awl-shaped, and their connective has two erect bristle-like appendages, about the length of the anthers, and is produced below into a short spur. The genus bears the name of Ernest Meyer, a Hanoverian botanist. [A. A. D.]

ERNODEA. A genus of low-growing cinchonaceous plants, with lance-shaped leaves; sheathing many-parted stipules; a salver-shaped corolla, with four to six linear segments rolled back; and an inferior two-celled ovary, surmounted by a fleshy disc. The fruit is a berry, crowned by the limb of the calyx, and contains two one-seeded stones. *E. montana*, a Sicilian plant, has dark red flowers. [M. T. M.]

ERODIUM. Stork's Bill. A genus of *Geraniaceæ*, known by having five of the ten stamens without anthers, and the tails of the carpels bearded on the inside; they coil up spirally when they split away from the central column. The species are generally distributed; a great many of them inhabit the Mediterranean region; and three occur in Britain, of which the most common is *E. cicutarium*, which has the leaflets of the pinnate leaves deeply pinnatifid, and the flowers pink or white. *E. moschatum* is much more rare, and has the leaflets of the pinnate leaves only deeply toothed, and the flowers are smaller. [J. T. S.]

EROPHILA. A section of the genus *Draba*, distinguished by having the petals bifid, and the seeds numerous in each cell of the pod. The common British *Draba verna*, or Whitlow Grass, belongs to this section; it is one of the earliest flowering plants we have, and is often scarcely an inch high. [J. T. S.]

EROSO-DENTATE. Toothed in a very irregular manner, as if bitten.

EROSTRATE. Not having a beak.

EROSE, ERODED. Having the margin irregularly toothed, as if bitten by an animal.

ERPETINA. A genus of *Melastomaceæ*, nearly allied to *Medinilla*, but differing in the structure of the anthers. These, in *Medinilla*, open at top by a little pore, but here they open by two slits along the inner face, from base to apex. The only species, *E. radicans*, is a smooth slender epiphytal plant, growing on the stems of trees in the Solomon islands. The stems, about the thickness of a crow-quill, are furnished with opposite stalked elliptical fleshy leaves, the little stalked flowers being produced singly in their axils. [A. A. B.]

ERPETION. *Viola*.

ERUCA. A genus of *Cruciferæ*, closely allied to *Brassica*, *Sinapis*, and *Diplotaxis*, but differing by having the beak of the fruit compressed, strap-shaped, and acute. The seeds are in two rows, as in *Diplotaxis*. Erect annuals, with lyrate-pinnatifid leaves, and rather large white or yellow flowers. The species occur in the Mediterranean region; the most common, *E. sativa*, which has large white flowers veined with purple, and very acrid leaves, is used in Southern Europe as a salad. [J. T. S.]

ERUCARIA. A genus of *Cruciferæ*, known by its pod breaking into two parts, the lower with two cells, the upper one-celled and ensiform. The species are annuals from South-east Europe and Western Asia; the leaves pinnatifid and smooth, and the racemes of purplish or white flowers, terminal or opposite the leaves. [J. T. S.]

ERS. (Fr.) *Ervum Ervilia*.

ERVA DE RATA. A Brazilian name for *Psychotria noxia*, and *Palicourea Marcgravii*. — MOIRA. A Brazilian name for *Solanum nigrum*.

ERVALENTA. The same as Revalenta, a meal prepared from the seeds of *Ervum Lens*.

ERVUM. A genus of leguminous plants, containing about twenty species of weak-stemmed annuals, with pinnate leaves generally terminating in tendrils. It is very closely related to *Vicia*, both in general appearance and botanical characters, the principal difference consisting in the calyx of *Ervum* having narrow sharp segments of nearly equal length, and almost as long as the papilionaceous corolla, while in *Vicia* they are broader, and the two upper ones are shorter than the others. The pods contain from two to four seeds.

E. Lens, the common Lentil, grows about a foot and a half high, and has a weak branching stem, leaves composed of from eight to twelve oblong leaflets, and pale lilac flowers borne in twos or threes. The pods are nearly as broad as long, smooth, and contain one or two seeds.

The Lentil was probably one of the first plants brought under cultivation by mankind for the purpose of affording food. It is several times mentioned in the Bible: for instance, in Genesis xxv. we read that Esau sold his birthright to his brother Jacob for a mess of red pottage, made of lentils. At the present day Lentils are still extensively cultivated throughout most parts of the East, including Egypt, Nubia, Syria, India, &c.; and likewise in most of the countries of Central and Southern Europe, but not to any extent in England. There are several different kinds, the most common being the French and Egyptian. The former is of an ash-grey colour, large and very flat, resembling a lens in shape: in fact, the lens derives its name from the resemblance it bears to the lentil seed; while the latter is much smaller and rounder, with a dark skin, and of an orange-red

Ervum Lens.

colour inside. On the Continent, and also in India and other eastern countries, Lentils are largely employed as an article of human food, but in this country their use is not so general, although considerable quantities are annually imported. Thus, in 1859, the imports into the United Kingdom amounted to 131,892 bushels, valued at $24,379l.$, or 4s. per bushel, nearly the whole of which came from Egypt. Their principal use with us is for the preparation of the so-called invalids' food, which under the names Ervalenta and Revalenta have attained no little celebrity. These articles are nothing more than lentil meal, sweetened with sugar or flavoured with salt; but under cover of their high-sounding names they are palmed off upon a credulous public at a price far above their real commercial value. As an article of food lentils rank first among the pulses, containing three per cent more flesh-forming or nutritive matter than the common pea, but like many other eatable leguminous seeds, they are very indigestible when not freed from the outer skin. [A. S.]

The generality of readers will wish to know if there is any real foundation in the widely extended belief that Lentil powder, and combinations of it with other vegetable ingredients, have the medicinal powers attributed to them. To this question the answer is in the affirmative, allowing, however, for some degree of exaggeration. Lentil powder, and the prepared foods alluded to, are reported to be a remedy for almost every variety of indigestion and bilious disorder, to relieve pains in the stomach, and to be so far aperient, as in most cases to obviate the necessity of habitually taking aperient medicine; and there is, in truth, no doubt that they act as a mild deobstruent on the entire of the digestive organs, producing an increased flow of gastric juice, bile, and other secretions. But it should be distinctly understood, that these beneficial effects can only be secured by selecting lentils of the best quality, and completely depriving them of the various extraneous substances and decayed and injured seeds which they always contain, as well as of their outer skin.

The proper mode of cooking Lentils as a remedy for indigestion, &c., is boiling them for twenty minutes, or till they are quite soft (but never more than half an hour), in soap or beef-tea, to which a small quantity of salt has been previously added. In this mode of cooking them, the peculiar vegetable principles on which the remedial powers depend, a great part of which are extracted by the liquid during the boiling, are eaten with the soup, beef-tea, or other convenient vehicle; and it is probable that Lentil-powder owes part of its reputation to its being taken entire, the direction given being to mix it with milk.

Peas possess in some degree the same qualities, and haricot beans in almost an equal degree, but this is for the most part destroyed by the length of time required in boiling them. As an article of diet Lentils are extensively used in various parts of the world, and are a favourite food in the East, where the Hindoo adds them to his rice, making doubtless a salubrious mixture. Like other leguminous seeds, they contain much caseine, and constitute one of the most nutritious of vegetable products, 100 parts by analysis yielding: Water, 14·0; caseine, 26·0; starch, 35·0; sugar, 2·0; gum, 7·0; fat, 2·0; woody fibre, 12·5; mineral matter, 1·5. [B. C.]

The Lentil is easily cultivated in England, and is worthy of attention, as being capable of yielding a large supply of a highly nutritious and wholesome food. Half a pint of seed drilled in rows a foot apart, would not badly occupy a portion of the cottager's potato garden, and the seeds ground into meal would make a pottage which would be of great value in rearing a family.

We have two native species, viz. *E. hirsutum*, the Hairy Tare, and *E. tetraspermum*, the Smooth Tare. These are readily distinguished by the hairy two-seeded pod of the former, and, the smooth four-seeded pod of the latter. These plants are of frequent occurrence about bushes, among which their slender stems climb for support. They are also common as agrarian weeds, especially in corn-fields, the hairy form being the most general, as being fond of all kinds of soil. Where it establishes itself amidst the wheat, it is a great pest, as it sometimes climbs about it to such an extent, as to bear it to the earth, to the great danger of the crop. The smooth form is less common, but it is not unfrequent in clays. [J. B.]

ERYCIBE. A genus of climbing shrubs, containing seven species, natives of tropical Asia. They have entire leaves, and flowers in terminal panicles; the calyx consisting of five sepals, the corolla deeply five-cleft, with large bifid lobes, having a triangular sericeous part on the middle of the back, the five stamens inserted on the tube of the corolla, and the ovary cylindrical-ovoid, glabrous, and one-celled, surmounted by a large fleshy ten-ribbed stigma. The fruit is a one-seeded berry. An order, *Erycibeæ*, has been established for the reception of this anomalous genus. Nearly approaching *Convolvulaceæ*, it differs from this order in having a sessile radiating stigma like a poppy. The sessile stigma exists in *Ebenaceæ*, but in most other respects *Erycibe* has no relation with that group. [W. C.]

ERYNGIUM. A well-marked genus of umbelliferous plants, distinguished by spiny leaves, and hemispherical or oblong heads of sessile flowers, the base of which is surrounded by a whorl of conspicuous bracts, most frequently rigid and spiny. *E. maritimum*, Sea Eryngo, or Sea Holly, is a common plant on most of the sandy shores of Great Britain, where it is conspicuous by the glaucous hue of its short rigid leaves and stems, and its thistle-like heads of blue flowers. It has extensively creeping cylindrical fleshy roots, the gathering of which, for the purpose of converting them into a sweetmeat, was formerly an occupation of some consequence to the sea-side population. Candied Eryngo-root is still to be obtained in some places, but its medical powers, which were at one time highly extolled, are now held in no repute. The venation of this plant, as well as of other species, being remarkably strong and durable, the leaves and flowers are frequently employed as fit subjects for skeleton bouquets. *E. campestre*, was formerly to be found in a few places in England, but has recently become extinct. Of the foreign species of *Eryngium*, which are numerous, the most worthy of notice are *E. amethystinum*, so called from the brilliant blue tint, not of its flowers only, but of the bracts and upper part of the stem; it is a native of Dalmatia and Croatia, but is frequently cultivated in English gardens. *E. alpinum*, a smaller plant of a still more brilliant colour, is a native of the Swiss Alps. French, *Panicaut*; German *Krausdistel*. [C. A. J.]

ERYNGO. *Eryngium maritimum* and *campestre*.

ERYSIMUM. A genus of *Cruciferæ*, distinguished from the other long-podded genera, which have the radicle of the seed bent round and lying on the back of one of the cotyledons, by having the pods four-angled and elongated. The species are usually biennials, found in Europe and temperate Asia, with narrow leaves often attenuated at the base, and terminal racemes (at first corymbs) of yellow, or very rarely white, flowers. *E. cheiranthoides*, with narrow-based leaves and small yellow flowers, is not uncommon in England; and *E. orientale*, with the stem leaves amplexicaul, and the flowers small and white, has occurred in some of the eastern counties, but scarcely even naturalised. *E. Perofskianum* and *E. arkansanum*, are handsome cultivated species, the former with rich orange-coloured flowers. [J. T. S.]

ERYSIPHE. A large assemblage of ascigerous *Fungi*, now broken up into a number of distinct genera. The mycelium is white, or in parts slightly tinged with brown, creeping over the green parts of plants, or more rarely, bursting through the stomates, and sending out here and there suckers which exhaust the juices of the matrix. The creeping threads send forth here and there perpendicular branches, which are articulated, and break up at the tips into large conidia, which either germinate immediately or produce a multitude of threads from the granular contents. Some of these joints occasionally become cellular and produce in their centre a multitude of minute conidia or spermatia. At different points in the creeping threads little swellings are formed, which ultimately become perithecia, and are fringed with curious appendages, which are sometimes straight and pointed with a bulb-like base, sometimes waved, sometimes hooked or incurved, sometimes repeatedly forked either with straight or divaricate branches, and sometimes end in a thick spongy body. The perithecia contain occasionally only a single ascus, as in *Sphærotheca*, while in other genera, the asci vary in number, but are generally few, and never so numerous as in *Sphæria*. Perithecia sometimes occur which are not distinguishable from the true, but which, instead of containing asci, yield a multitude of minute spores joined together with mucous matter. Five kinds of fructification, therefore, have been found in these plants.

In an early stage, the species, which are then described as *Oidia*, constitute the white mildew so destructive to various plants, as vines, hops, peaches, &c. In this state they are easily checked by the application of sublimed sulphur, which seems to combine with the nascent oxygen to form sulphurous acid. [M. J. B.]

ERYTHRÆA. Herbaceous plants, growing in many parts of the world, with simple or branched stems, and pink or pale yellow flowers in cymose panicles; they differ from those of the allied *Gentiana*, by their calyx being divided to the base, by their anthers, which become spirally twisted as they wither, and by the greater length of the style. *E. Centaurium* is a common English plant, in dry, sandy, or chalky soils especially; and found also throughout Europe and Central Asia. It is an annual, with erect square generally branched stems, broad egg-shaped leaves at the base, and flowers of a pale pink colour in a much branched cyme. This plant varies very much in the size of the flowers, the size of the leaves, and the degree of branching, so that it may be found as a simple stem half an inch high, with only a single flower, or one or two feet in height, with very numerous blossoms; hence some of the more marked varieties have been considered to form distinct species. The plant partakes of the bitter qualities of the order, and might be used in place of gentian. Besides the English species, others from the south of Europe, the Azores, &c., with yellow or pink flowers, are occasionally grown in gardens. [M. T. M.]

ERYTHRINA. A genus of handsome leguminous trees or shrubs, popularly known as Coral trees. They are pretty generally distributed through the tropics of both hemispheres. Some attain great dimensions, while others are dwarf bushes with woody rootstocks; a few have the stems and leaf-stalks beset with prickles. The leaves are trifoliate, with long stalks, the leaflets oval lanceolate elliptical or triangular. Many of the species are cultivated in hothouses for the sake of their beautiful large generally blood-red pea flowers, which are arranged in terminal racemes. In some species the tubular calyx is two-lipped or equally five-toothed, the petals all narrow, and nearly of equal length, while the keel is composed of two distinct petals. Some botanists consider that these alone should form the genus *Erythrina*. The name *Chirocalyx* is given by some authors to a few species in which the calyx is sheath-like, split above, and five-toothed at the apex; in a third group, called *Duchassaingia*, the keel is of one petal, bifid at the point, and is equal in length to the wings, which are about twice as long as the calyx, while the erect standard is broad, generally oval, and narrowed below into a claw; while in a fourth group, called *Micropteryx*, the keel is also of one petal beaked at the point, but the wings are small, generally scale-like, and included in the calyx. The pods in most species are long, narrow, round, and constricted between the seeds, which are often bright red with a black spot, and about the size of a pea. These hard red seeds are frequently strung into necklaces. The Amasisa of Peru, *E. Amasisa*, is the only species whose pods split when ripe. This plant is described by its discoverer, Mr. Spruce, as one of the most beautiful trees of the country, attaining a height of 100 feet, and clad in spring and autumn with large flame-coloured or vermilion flowers. [A. A. B.]

E. Caffra, the Kaffirboom of the Dutch, or Kaffir's tree, is a native of South Africa, where it forms a tree fifty or sixty feet in height. Its trunks are commonly hollowed out and made into water-troughs and canoes. The wood is soft, but is said to be durable when tarred; and it is so light that it is used as a substitute for cork for floating fishing nets. *E. indica*, a small tree, native of the East Indies, growing about thirty feet high, is commonly cultivated in India and the Malayan peninsula and islands, for supporting the weak stems of the pepper plant, for which purpose it is kept dwarf. It affords a very soft porous wood, greatly used in India for making toys, light boxes, and similar articles, which are usually overlaid with a thick coating of varnish or lacquer. In Ceylon the young tender leaves are eaten in curries. *E. umbrosa*, which attains a height of fifty or sixty feet, is a native of tropical South America, and is commonly cultivated there, as well as in some of the West India Islands, for the purpose of protecting the cocoa plantations from the effects of high winds, and at the same time to induce a proper degree of moisture in their neighbourhood. [A. S.]

ERYTHRINE. A colouring matter found in lichens.

ERYTHROCHITON. A small Brazilian rutaceous tree with long alternate simple fragrant leaves, and flowers placed on short jointed stalks arising from the leaf axils, in groups of two or more; the calyx is red, large, tubular; the corolla white, salver-shaped. [M.T. M.]

ERYTHROCOCCA. A genus of *Euphorbiaceæ*, composed of a single W. African species. *E. aculeata*, a smooth low shrub with stalked oval leaves, having short straight prickles in their axils (prickly plants are rare in the family), and little fascicles of minute green flowers, the males and females on different plants, both having a three-parted calyx. The ripe fruits, about the size of a peppercorn, are of an intense scarlet colour. [A. A. B.]

ERYTHROLÆNA. A genus of *Compositæ* found in Mexico, represented by a single species, *E. conspicua*, which was introduced to English gardens about 1838, and is commonly known as the Scarlet Mexican Thistle. It is a tall plant eight to ten feet high, with rigid leaves, somewhat like those of a common wayside thistle: those at the base of the stem pinnatifid, with cut and spiny-pointed segments, and about two feet long; the stem-leaves smaller, lanceolate, with spiny teeth, and all more or less downy underneath. The flower-heads, clustered at the ends of the branches, are about three inches long, and very handsome, because of their scarlet involucral scales. The florets are all tubular, yellow, and perfect; and the smooth achenes are crowned with a feathery pappus. [A. A. B.]

ERYTHRONIUM. A genus of *Liliaceæ*, consisting of nearly stemless herbs, with a long narrow solid scaled bulb, and two very smooth elliptical leaves usually spotted with purple. The scape is one-flowered; the flower large, nodding, lily-like, with the perianth of six separate portions, bell-shaped or recurved, the three inner segments furnished with a callous tooth on each side.

The common Dog's-tooth violet, *E. Dens canis*, has purple flowers; it is a native of

Southern Europe and temperate Asia, and is an exceedingly ornamental garden plant, as well as an early flowerer. The most common American species, *E. americanum*, has narrow perianth segments of a pale yellow colour. [J. T. S.]

ERYTHRO. In Greek compounds = any pure red.

ERYTHROPHYLL. The red colouring matter of plants.

ERYTHROPHYSA. A genus of *Sapindaceæ*, nearly related to *Cardiospermum*, but differing in its five-lobed bell-shaped and petal-like calyx. The only known species, *E. undulata*, is a smooth stunted South African bush, with rigid stems, furnished near the apex with unequally pinnate leaves, composed of four to six pairs of small elliptical leaflets, and an odd one; the flowers, which all seem to be yellow or scarlet, are disposed in little clusters at the ends of the twigs. The fruits are three-celled bladdery capsules, of a fine red colour, and suggest the generic name, which signifies 'red bag.' The plant is sometimes called *Erythrophita undulata*. [A. A. B.]

ERYTHROPOGON. Two neat little erect S. African bushes form this genus of *Compositæ*, which differs from its nearest ally, *Metalasia*, in having stalked flower-heads and sessile achenes. In *E. umbellata* the minute heath-like leaves are of a silvery white colour, rounded, linear, curiously spirally twisted, and disposed in numerous crowded bundles. In *E. imbricata* they are fewer and nearly lance-shaped. In both the top-shaped flower-heads, with white or purple florets, are few and disposed in little umbels on the ends of the twigs. The smooth beakless achenes are crowned with a pappus of one series of rough hairs, of an intense purple colour. [A. A. B.]

ERYTHRORCHIS. A remarkable genus of leafless terrestrial orchids found in the Birman empire and adjacent islands. They are perhaps the most gigantic plants in the family. The stems of *E. scandens* are from fifty to a hundred feet long, scrambling over trees in dense wet jungles. They are of a pale dull red, furnished with brown scales which supply the place of the leaves; and the flowers are disposed in panicles or racemes, the sepals and petals whitish-yellow, and the lip tinged with pale blue. It differs from *Vanilla* in the lip being free instead of connate with the column; and from *Cyrtosia* in the capsular not berried fruit, with winged seeds. [A. A. B.]

ERYTHROSPERMUM. A genus of bixads, composed of a few Mauritian and one Ceylon tree, and differing from the others in the family in having a definite number of stamens. They have smooth oval lance-shaped or oblong leaves, either alternate, opposite or whorled; and the white myrtle-like flowers are arranged in racemes or panicles. [A. A. B.]

ERYTHROSTIGMA. A Japanese tree, belonging to the *Anacardiaceæ*, and remarkable for being covered with red dots; the leaves are unequally pinnate; the five-parted flowers are arranged on a much-branched panicle; the five hair-like filaments are united together at their bases; the ovary is stalked, and contains a single ovule; and the fruit is a kidney-shaped drupe. [M. T. M.]

ERYTHROSTOMUM. Any aggregate fruit like that of a strawberry or *Ranunculus*.

ERYTHROXYLACEÆ. (*Erythroxyls*.) A natural order of thalamifloral dicotyledons belonging to Lindley's sapindal alliance of hypogynous Exogens. Shrubs or trees with alternate smooth stipulate leaves, and small whitish or greenish flowers on axillary peduncles, covered at the base with imbricated scaly bracts. Sepals five, united at the base, persistent; petals five, equal, with plaited scales at their broad bases; stamens ten, monadelphous; anthers innate, with longitudinal dehiscence. Ovary three-celled, with three styles and eight capitate stigmas; ovule anatropal, Fruit drupaceous, one-seeded. They are chiefly West Indian and South American plants. Some have stimulating qualities; others have a tonic bark. The bark of *E. suberosum* supplies a reddish-brown dye. There is only one genus, *Erythroxylon*, and above seventy species. [J. H. B.]

ERYTHROXYLON. This genus contains numerous species, the majority natives of tropical South America and the West Indian Islands, but some occurring in Madagascar and the Mauritius. They are mostly bushy shrubs, or occasionally they form small trees.

E. Coca is the most interesting of the species, on account of its being extensively cultivated, and its leaves largely employed as a masticatory, under the name of Coca, by the inhabitants of countries on the Pacific side of South America. It is a shrub of six or eight feet high, somewhat resembling a blackthorn bush. The Coca leaves are of a thin texture, but opaque, oval, tapering towards both extremities, their upper surface dark green, the lower paler and strongly marked with veins, of which two, in addition to the midrib, run parallel with the margin. Small white flowers are produced in little clusters upon the branches, in places where the leaves have fallen away, and stand upon little stalks about as long as themselves.

The use of Coca in Peru is a custom of very great antiquity, and is said to have originated with the Incas. At the present day it is common throughout the greater part of Peru, Quito, and New Grenada; and also on the banks of the Rio Negro, where it is known as Spadic. Coca forms an article of commerce among the Indians, and wherever they go they carry with them a bag of the carefully dried leaves, and also a little bottle-gourd filled with finely powdered lime, and having a wooden or metal needle attached to its stopper. Four times a day, whatever the nature of his occupation, whether employed in the mines, the fields, as a muleteer, or as domestic servant, the Indian resigns himself to the pleasures of Coca chewing, mixing the leaves with lime or the ashes of *Cecropia*. When used in moderation Coca exerts a pleasurable influence upon the imagination, and induces a forgetfulness of all care; it is also a powerful stimulant of the nervous system, and when under its influence Indians are able to perform long and rapid journeys, and carry heavy loads, without requiring any other sustenance. But when taken in excess it produces intoxication, of a character resembling that of opium rather than alcohol, but not so violent, although the consequences of its prolonged use are quite as injurious, and very few of those who become slaves to the habit attain an old age.

Erythroxylon Coca.

Spruce says that an Indian, with a chew of Spadic in his cheek, will go two to three days without food, and without feeling any desire to sleep. [A. S.]

ESCALLONIACEÆ. (*Carpodetea, Escalloniada.*) A natural order of calycifloral dicotyledons belonging to Lindley's grossal alliance of epigynous Exogens. Evergreen shrubs, often odoriferous, with alternate exstipulate leaves, and axillary conspicuous flowers. Calyx superior, five-toothed; corolla of five petals, alternate with the divisions of the calyx, æstivation imbricated; stamens five, attached to the calyx, and alternating with the petals. Ovary inferior, two to five-celled, with a large central placenta and numerous ovules; style simple, surrounded at the base by an epigynous disk; stigma two to five-lobed. Fruit a capsule or berry crowned by the persistent calyx and style; seeds minute with oily albumen. The order is allied to the gooseberry family, and some think that it has an affinity to saxifrages. The species are natives chiefly of South America; but some are found in the southern parts of Australia and New Zealand. On the mountains of South America they grow at an elevation varying from 6,500 to 14,700 feet, and form a marked region of vegetation. There are seven known genera and about sixty species. Examples: *Escallonia, Itea,* and *Carpodetus*. [J. H. B.]

ESCALLONIA. A genus of *Escalloniaceæ*, named in honour of a Spanish traveller, the companion and friend of the botanist Mutis. It consists of trees or shrubs, natives of South America, Chili, &c. They have simple leaves, covered with resinous dots; flowers variously arranged, white, pink, or red, with five-parted whorls; and petals and stamens attached to the margin of a cup-like disc which surmounts the ovary. The fruit is a capsule. Several species are in cultivation as greenhouse or half-hardy shrubs. *E. rubra* has tubular red flowers, and is very handsome when trained against a wall. *E. macrantha* is even more beautiful. [M. T. M.]

ESCARIOLE, or SCAROLE. (Fr.) *Cichorium Endivia latifolia.*

ESCENS. A termination equivalent to the English ish; thus, rubescens = reddish.

ESCHERIA. A synonym of *Salisia*, a genus of gesnerads, of which *Gloxinia maculata* is the type.

ESCHSCHOLTZIA. A Californian genus of herbaceous plants belonging to the Papaveraceæ, distinguished by its singular calyx, which, unlike that of the true poppies, is lifted off in one piece by the expanding petals instead of separating into two sepals. The petals are four in number, and the seed-vessel resembles the silique of the cruciferous order, being two-valved and bearing the seeds on the edges of the valves. There are several species or varieties, all from California. *E. californica*, the best known, is a large bushy herb with straggling branches, which, as well as the finely divided leaves, are very glaucous. The flowers are large, bright yellow, saffron-coloured in the centre, and expand only in the sunshine. It is a perennial, but in British gardens is mostly treated as an annual, as it flowers the first year and sows itself freely. *E. crocea*, with saffron-coloured flowers, and *E. compacta*, of a less straggling habit, are probably mere varieties. [C. A. J.]

ESCHWEILERA. A genus of Brazilian trees, belonging to the *Lecythidaceæ*, and only differing from *Lecythis* in the limb of the calyx being bent backwards so as to touch the tube. [M. T. M.]

ESCOBEDIA. A genus of *Scrophulariaceæ*, consisting of two South American or Mexican species, erect stiff nearly simple herbs, very rough to the touch, with opposite entire or toothed leaves, and large white flowers, nearly sessile in the upper axils. The calyx is long, tubular, and herbaceous; the corolla-tube very long, with a broad spreading limb; the capsule is two-valved, included in the persistent calyx. Neither of the species has been as yet introduced into our gardens, although both are said to be handsome. They may, however, possibly be parasitical on the roots of other plants, in which case their cultivation would be very difficult.

ESCORZONERA. A Chilian name for *Achyrophorus apargioides* and *A. Scorzonera*.

ESENBECKIA. A genus of arboreous *Rutaceæ* remarkable for their bark, which contains tonic properties. In one of the Brazilian species cinchonin has even been detected. The flowers have five-parted whorls, the five stamens ultimately bent downwards, and, like the petals, inserted at the base of a cup-shaped disc, in which the ovary is placed; the latter is warty on the surface, and five-lobed. [M. T. M.]

ESPADÆA. The generic name of a Cuban plant said to belong to the *Verbenaceæ*, and to have alternate leaves, and an ovary united half its length with the tube of the calyx. These are characters, however, quite at variance with those of the family. *E. amœna* is described by M. Richard as a much branched bush, with rusty down on its twigs, which are furnished with obovai and obtuse smooth leathery leaves, narrowed towards the base; the solitary flowers in the axils of the leaves are stalked, and have a bell-shaped calyx, a funnel-shaped arched corolla, with an oblique border of five erect unequal lobes, and four stamens, two long and two short. The fruits are globose drupes, with two cells, and one seed in each. [A. A. B.]

ESPAGNOLE. (Fr.) A kind of olive.

ESPARCETTE. (Fr.) *Onobrychis sativa*.

ESPATHATE. Not having a spathe.

ESPELETIA. A genus of remarkable *Compositæ*, found near the snow limit at elevations of 13,000 to 14,000 feet and upwards in the Andes of N. Grenada, and Equador. A few of them do not exceed a foot in height, and have grassy rigid root-leaves, quite white from a covering of silky hairs. The greater number, however, are taller, and furnished with long strap-shaped root-leaves wholly covered with dense white or rusty-coloured wool, which forms for them an admirable protection from the cold, their thick texture and warm woolly covering no doubt suggesting the name 'Lion's ear' which is sometimes given to them by the Spaniards. The stems terminate either in a single flower-head, or more commonly in a corymb of yellow flower-heads, some an inch or more across, and surrounded with an involucre, which, like all parts of the plant, is clothed with wool.

These plants bear much resemblance to *Culcitium*, which is found in the same regions, and the Spanish appellation 'Fraislejon' is common to both. They differ abundantly, however, in having strap-shaped ray florets, and achenes destitute of pappus. About seven species are known. A resinous substance is present in most of them, but is produced in greatest quantity by *E. grandiflora*; it is of a beautiful yellow colour, and is valued by the printers of Santa Fé de Bogota, who use it in the composition of their ink, and give to it the name of tremintina (terebinthine), though it has neither the odour nor the consistence of the turpentine of commerce. The genus was named by Mutis in compliment to M. Espeleta, who rendered him much service in his botanical labours about Santa Fé. [A. A. B.]

ESPRIT D'IVA. An aromatic liqueur of which *Ptarmica moschata* is the basis.

ESQUINANCIE. (Fr.) *Asperula cynanchica*.

ESTERHAZYA. A genus of *Scrophulariaceæ*, closely allied to *Gerardia*, and differing chiefly in the stamens projecting far beyond the corolla, with the anthers thickly clothed with long woolly hairs. There are two or three species, natives of Southern Brazil, erect branching shrubs or undershrubs, with opposite or scattered entire leaves, and large, very showy flowers of a rich red or pink colour, forming short terminal leafy racemes. Notwithstanding their beauty, they have not been introduced into our gardens, and perhaps, like the *Gerardias*, their cultivation may be very difficult.

ESTIVATION. The manner in which the parts are arranged in a flower-bud.

ESTRAGON. (Fr.) *Artemisia Dracunculus*.

ESULE. (Fr.) *Euphorbia Esula*. — RONDE. *Euphorbia Peplus*. — GRANDE. *Euphorbia Lathyris*. — PETITE. *Euphorbia exigua*.

ETÆRIO, ETAIRIUM. Such a kind of aggregate fruit as that of the *Ranunculus* or strawberry.

ETERNELLE. (Fr.) *Helichrysum orientale*.

ETERNUE. (Fr.) A kind of *Agrostis*.

ETHULIA. A genus of the *Compositæ*, distinguished by the four or five-angled achenes being surmounted by a minute and entire crown-like ring. It is made up of about seven species, all of them branching weeds of no beauty, found in various tropical and subtropical countries of the eastern hemisphere, extending as far west as Syria in Asia, and Senegambia in Africa. The little purple or white flower-heads are numerous, about the size of a small pea, disposed in a corymb at the end of the twigs. [A. A. B.]

ETIOLATED. Deprived of colour by being kept in the dark; blanched.

EUBOTRYS. A genus of deciduous ericaceous shrubs, better known under their former name of *Lyonia*. The main characters of the genus are: a five-parted calyx with two small bracts at the base, a more or less cylindrical corolla with a reflexed limb, ten stamens with short flattened filaments, a truncate stigma, and a five-celled five-valved capsule. The species are handsome North American shrubs, many of them cultivated in this country. The leaves of *E. arborea* have an acid flavour, whence the name of Sorrel-tree. Hunters in the mountains are said to use these leaves as a means of alleviating thirst. [M. T. M.]

EU'BRACHION. A small kind of leafless mistletoe, growing on myrtles on the river Paraguay, in South America, and constituting a genus of *Loranthaceæ*. It has the male and female flowers mixed in small catkin-like spikes on the terminal branches.

EUCALYPTUS. The gigantic Gum-trees, Stringy-barks, and other timbers of the Australian and Tasmanian forests, constitute this genus of *Myrtaceæ*, of which between 100 and 150 species are described, though, owing to the widely different appearances assumed by individual trees at different periods of growth, it is extremely difficult to arrive at a correct estimate of their number. Australia is the headquarters of the genus, numerous species being distributed throughout all parts of that continent; several are also found in Tasmania, where they form extensive forests; and a few extend as far north as Timor and the Molucca Islands. The majority of them are trees, some growing to an immense height and having proportionately thick trunks. Their leaves are of a thick leathery texture, always quite entire, very variable in shape. In young plants they are always opposite, but they generally become alternate as the plant gets older, and their stalks then acquire a peculiar twist, so that the leaves present their edges to the branches. The flowers grow from the angles between the leaves and stem, and are either solitary or in clusters; the calyx is hard and woody, and separates into two pieces, the upper of which resembles a lid or cover, and falls away in a single piece when the flower opens, carrying along with it the corolla, which is intimately combined with it, while the lower is persistent, and bears the very numerous stamens, which form a fringe round its summit. The fruit is closely enveloped in the woody calyx.

The Australian colonists distinguish many of the trees of this genus by characters derived from the bark; some having smooth, others rough or cracked bark; some are solid (Iron-bark), while others are fibrous (Stringy-bark); and, finally, in some species the bark scales off in flakes, either from the whole tree or from the upper part only. They are also called Gum-trees, in consequence of the quantity of gum that exudes from their trunks. The timber is exceedingly valuable, and is in common use in our Australian and Tasmanian colonies. In the latter, the three following species yield the best quality of timber, namely; *E. globulus*, the Blue Gum; *E. gigantea*, the Stringy-bark; and *E. amygdalina*, the Peppermint-tree. But of these the first-mentioned is considered the most valuable, although the Stringy-bark attains the largest size. Trees of the latter species have been felled, measuring upwards of 300 feet high, by 100 feet in girth at a yard from the ground. The blue-gum timber is greatly used by colonial ship-builders, also by mill-wrights, carpenters, and implement-makers, and by engineers in the construction of works requiring beams of great span; it is exceedingly strong and very durable. A plank

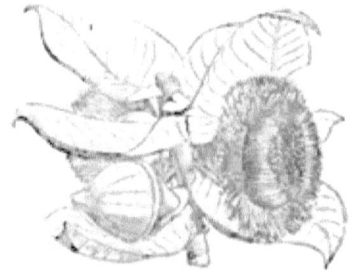

Eucalyptus macrocarpa.

of the swamp-gum, forwarded to the International Exhibition of 1862, measured 230 feet in length. For some particulars as to the strength of the timber of these Gum trees, see *Gardener's Chronicle*, 1862, 571.

Among other interesting products of this genus, we may briefly mention that many species yield a red resinous juice, which hardens into a substance resembling kino, and possessing powerful astringent qualities. *E. Gunnii*, the Tasmanian Cider-tree, yields a cool refreshing liquid, from wounds made in its bark during the spring. A saccharine substance, resembling officinal manna, exudes from *E. mannifera*, and other species; *E. piperita* yields an essential oil; and the large flakes of bark obtained from several of the species are used by the aborigines for making huts, canoes, &c. [A. S.]

EUCEPHALUS. *Diplopappus.*

EUCER&AE. A genus of *Samydaceæ*, nearly related to *Casearia*, from which it may be recognised by its minute greenish-white flowers being disposed in axillary panicles longer than the leaves, not in short axillary fascicles or cymes. *E. nitida*, a Brazilian bush, is the only species. Its spreading branches are furnished with alternate stalked entire leaves, between oval and lance-shaped in form. [A. A. B.]

EUCHARIDIUM. A pretty little annual introduced from N. America in 1840, belonging to the *Onagraceæ*, and closely allied to *Clarkia*. It is distinguished by the long and slender tube of the calyx terminating in four deciduous sepals, by its four three-cleft petals, and by its four-celled, four-valved capsules containing numerous seeds. It grows to the height of about a foot, with somewhat downy stems and foliage; the leaves are stalked, ovate, and undivided, bearing in their axils solitary purplish flowers. [C. A. J.]

EUCHARIS. A beautiful genus of broad-leaved evergreen pancratiform *Amaryllidaceæ*, having ovate bulbs, broadly elliptic or ovate long-stalked leaves, and tall scapes bearing several large white nodding fragrant flowers. The blossoms have a long slender tube, six broad ovate spreading perianth segments, and a large bell-shaped white coronet or cup bearing the six stamens on its margin, each having a lateral tooth at its base. The species are few in number, and are all natives of South America. *E. grandiflora* is one of the most beautiful of stove bulbs. [T. M.]

EUCHEUMA. A genus of rose-spored *Algæ* belonging to the natural order *Gelidiaceæ*, consisting of cartilagino-gelatinous tuberculated or spinous species, remarkable for thick-walled capsules, containing a central placenta, which at length becomes hollow in the middle and bears necklaces of spores. *E. speciosum* is the Jellyplant of Australia, and is one of the best species for making jelly, size, cement, &c. A very fine species, possessing doubtless similar properties, occurs in the United States. [M. J. B.]

EUCHILUS. A genus of much branched Australian under-shrubs of the leguminous family, and nearly allied to *Pultenæa*, but differing in having a disproportionately large upper lip to the calyx, instead of two nearly equal lips, and a stalked instead of a sessile pod. They have slender twiggy stems, separately or densely clothed with leaves, which are sometimes juniper-like, and sometimes small, round, or inversely heart-shaped; while the little yellow peaflowers are solitary in the axils of the leaves, and either few, distant, and stalked, or numerous at the ends of the twigs, and nearly sessile. About ten species are known. [A. A. B.]

EUCHRESTA. A genus of *Leguminosæ*, nearly allied to the W. Indian cabbage trees (*Andira*), but differing in having the base of the tubular five-toothed calyx slightly projecting above, and in the standard or upper petal being very narrow, not broad and rounded. *E. Horsfieldii*, the only known species, grows in mountain districts in Java and Formosa, and is an elegant smooth shrub with unequally pinnate leaves; the slender branches terminated by erect racemes of numerous white, waxy, vetch-like flowers, succeeded by stalked elliptical one-seeded pods. According to Dr. Horsfield, the whole plant is intensely bitter, and much sought after by the natives because of its medicinal properties. They employ the seeds against any poison that may have been taken into the stomach, exhibiting one of them triturated with water to counteract the effects. According to Leschenault, the powdered fruits mixed with food are regarded as having the power of preventing diseases and giving tone to the stomach.

Mixed with lemon juice, they are applied to wounds caused by any venomous animal. The fruits are sold for five or even ten sous, French money, each. This plant holds the first rank among the medicinal plants of the island in the opinion of the natives. [A. A. B.]

EUCLEA. A genus of simple-leaved African bushes, of the family *Ebenaceæ*, numbering about twenty species. One is found in Abyssinia, another on the west coast south of the line, and the remainder in the southern districts. Their nearest relationship is with *Royena*, from which they are readily distinguished by their flowers being disposed in racemes, not one or three together in the axils of the leaves, and by the males and females growing on different plants. The leaves are alternate or opposite, entire, oval lance-shaped or oblong, sometimes crisped or wavy; and the little white flowers, disposed in racemes shorter than the leaves, are a good deal like those of some whortleberries, having a bell-shaped corolla with five to seven teeth. The fruit is globular, fleshy, and juicy, sometimes as large as a cherry. Those of many of the species, known by the colonists as Guarry, are eaten, and are sweet and slightly astringent. The wood of *E. Pseudo-Ebenus*, a species with narrow lance-shaped leaves, is said to be extremely hard and black. [A. A. B.]

EUCLIDIUM. A genus of *Cruciferæ*, found in South-eastern Europe and Western Asia. It is known from the other genera in which the radicle of the seed is bent round and lies along the edges of the cotyledons, by having a smooth subglobular indehiscent pouch, with a complete partition, and a single seed in each cell. *E. syriacum*, the only European species, is a small annual with rigid branches, runcinate root-leaves, and lateral spikes of small white flowers. [J. T. S.]

EUCNIDA. An annual belonging to the order *Loasaceæ*, and often cultivated under the name of *Microsperma*. *E. bartonioides*, the only species, a native of Mexico, whence it was introduced in 1849, grows to the height of about a foot, with bristly stems, ovate, lobed, and denticulated leaves, and axillary, very large yellow flowers, consisting of five spreading petals, at the base of which is a distinct tuft of long golden stamens, exceeding the petals in length. The style is slender; the stigma five-furrowed, but not divided; the capsule one-celled, containing numerous minute seeds. [C. A. J.]

EUCODONIA. A Mexican gesnerad allied to *Achimenes*, forming a dwarf herb, with ovate obtuse pubescent crenated leaves, and lilac flowers, of which the calyx is deeply five-parted, and the corolla large, ventricose, campanulate with an oblique spreading limb; there are four included stamens, a membranaceous entire glandular ring, and a bilobed stigma. The plant, which is cultivated for the sake of its flowers, is sometimes called *Schœria lanata* and *Mandirola lanata*. [T. M.]

EUCOMIS. A genus of *Liliaceæ*, consisting of South African bulbs, with broad root-leaves, and a simple raceme of rather large usually greenish flowers, surmounted by a tuft of empty leaf-like bracts, called a coma. The perianth is six-parted and spreading, the stamens inserted in its segments, with the filaments dilated at the base. The capsule is three-winged, the few seeds with a hard black seed-coat. Several handsome half-hardy species are in cultivation. *E. bifolia* has only two leaves lying flat on the ground, and a short raceme of pale green flowers. [J. T. S.]

EUCOSIA *carnea* is a terrestrial orchid of Java, with a stem about a span high, bearing four to six ovate leaves, and terminating in a downy spike of about three small flesh-coloured flowers, each supported by a long narrow bract. The plant has the habit of some *Goodyeras*, but the internal structure of the flower is quite different. The remarkable thing is, that the anther is attached to a filament which grows out quite distinctly from the base of the column, whereas in most orchids the anther has no filament, but rests in a niche at the apex of the column. The plant is beautifully represented by Blume in his splendid folio work on the orchids of the Indian Archipelago. [A. A. B.]

EUCRINUM. A subgenus of *Fritillaria*, proposed by Nuttall to include a few species which approach to *Lilium*, but differ from both genera, in having an undivided stigma, and the cup formed by the perianth segments broadly funnel-shaped. The habit is that of *Fritillaria*. The *F. tulipifolia* from the Caucasus, and the American *F. pudica*, belong to this section. [J. T. S.]

EUCROSIA. A genus of stove bulbs of the order *Amaryllidaceæ*. *E. bicolor*, the only species, a native of S. America, has round bulbs, wide petiolated leaves, and a tapering scape, supporting an umbel of four or more vermilion-coloured flowers, which have an oblique perianth tube, compressed recurved limb segments, a declined cup, abbreviated and rostrate above, and shovel-formed and prolonged below. It comes near *Eusena*. [T. M.]

EUCRYPHIA. A genus of the St. John's-wort family, differing from all the other genera except *Cratoxylon*, in having winged seeds, and from the latter genus in its four to five-leaved calyx being thrown off in the form of a cap as the flower expands. The four to five petals are roundish; the stamens numerous and disposed in many series; and the ovary five to twelve-celled, surmounted with as many styles as there are cells. The fruits are little woody capsules. The four known species are oppositely-leaved trees or shrubs, two found in Tasmania, and two in Chili. *E. Billardieri* is one of the most beautiful trees of Tasmania, sometimes growing to sixty or a hundred feet high, and covered in February with an abundance of white cistus-like flowers, one to two inches across. These are solitary and stalked in the axils of the narrow obtuse leaves, which are gummy above, and white underneath. *E. pinnatifolia*, one of the Chilian species, is remarkable in the family, as well as in the genus, for having pinnate leaves, which are made up of two pairs of smooth serrated leaflets, and an odd one; the single white flowers themselves are a good deal like those of a wild dog-rose, and solitary near the ends of the branches. The name *Carpodontos* is sometimes given to the Tasmanian species, which differ from the Chilian ones in having five-celled fruits. [A. A. B.]

EUCYCLA. A genus of *Polygonaceæ*, allied to *Eriogonum*, having the plicate coloured perianth salver-shaped, the three outer divisions diverging and orbicular, and the three inner linear oblong, forming a cylinder. The flowers are yellow or purple. The species are natives of the Rocky Mountains. [J. T. S.]

EUDESMIA. A handsome Swan River myrtaceous shrub. The branches are four-cornered, with lance-shaped thick leaves; the flowers red, disposed in umbels, with four-parted stamens, united into four bundles; capsule four-celled. *E. tetragona* is in cultivation. [M. T. M.]

EUDIANTHE. A section of *Lychnis*, comprising those species which have the calyx tube contracted at the top. The pretty lilac-flowered *L. Cœli-rosa*, which is often cultivated as an ornamental annual, under the name of *Viscaria*, belongs to this section. [J. T. S.]

EUDOXIA. Peruvian herbs belonging to *Gentianaceæ*, with large handsome drooping flowers, in a terminal panicle. They have a bell-shaped membranous calyx; a bell-shaped five-cleft corolla, having the segments convolute before expansion; and the filaments of the stamens are channelled on the inner surface. Ovary two-celled, two-valved; stigma sessile, with two dilated revolute segments. [M. T. M.]

EUFRAGIA. A genus separated from *Bartsia* mainly on account of the structure of its seeds, which are 'slightly angular, very minute, crenate-ribbed, the hilum basal.' *E. viscosa*, an erect unbranched annual with viscid foliage and yellow flowers, not uncommon in marshy ground in the west of England and some parts of Ireland and Scotland, is the same as *Bartsia viscosa*. [C. A. J.]

EUFRAISE. (Fr.) *Euphrasia officinalis*.

EUGENIA. A genus of *Myrtaceæ*, comprising several trees or shrubs, for the most part natives of tropical America and the West Indies. The flowers are placed in the axils of the leaves, white, with a four-parted calyx, four petals, and numerous stamens. The berry is crowned by the calyx, one or two-celled, and contains one or two seeds.

The most important species is *E. Pimenta*,

which furnishes Allspice. This consists of the fruits gathered before they are quite ripe, and dried in the sun. The Allspice tree is cultivated in the West Indies and Jamaica, where the trees are planted in rows called pimento walks; the produce is sometimes very large. The Allspice or Pimento berries of commerce are of the size of a small pea, of a dark colour, and surmounted by the remains of the calyx. The odour and flavour are supposed to resemble a combination of those of cinnamon, cloves, and nutmeg, hence the name allspice; they are due to a volatile oil, which is obtained by distillation. Allspice is largely used for flavouring purposes, being cheap. The oil is occasionally employed as a carminative.

Many of the species yield agreeably tasting fruits, such as *E. cauliflora*, which furnishes the Jabuticaba fruits of Brazil, described as being of the size of a greengage, and very refreshing; it is cultivated in some parts of Brazil. The Rose Apples of the East are the produce of *E. malaccensis* and *E. Jambos*. *E. Ugni*, a native of Chili, has lately been introduced into English gardens, where it is at least as hardy as its near ally, the myrtle. Its fruit is highly esteemed in Chili. Those grown in this country are glossy black when ripe, and have an agreeable flavour and perfume. Numerous other species are grown either for their handsome foliage or for their flowers. *E. Luma* is one of the most beautiful of these. [M. T. M.]

EUKYLISTA *Spruceana*, the only species of the genus, is described as a tree attaining the height of fifty to seventy feet, with bark which scales off like that of the plane tree: its flowers show it to be one of the *Cinchonaceæ*. The flower buds are at first enclosed within membranous bracts, which ultimately fall off; the limb of the calyx is scarcely developed; the tube of the corolla is short, its limb divided into six to eight lobes, and its throat lined with dense hairs; stigmas two. Fruit a capsule, dividing into two pieces, the seeds winged. [M. T. M.]

EULOBUS. A Californian annual, with narrow leaves, and rather large white flowers often tinged with red, constituting a genus of *Onagraceæ*, distinguished from *Œnothera* chiefly by its long slender linear capsules incompletely divided into four cells.

EULOPHIA. A numerous genus of epiphytal or terrestrial orchids, natives of tropical Asia, Africa, and America, but occurring in greatest numbers at the Cape. They have either pseudobulbs with one or two leaves, or tuberous rhizomes of the size of potatos or larger, with the leaves and flower-scapes arising laterally from near the base. The leaves are grassy, or lance-shaped and plaited; and the flower-scapes either simple or branched, bearing few or many flowers, which seldom exceed an inch in diameter, the prevailing colour being yellow. The sepals and petals are nearly equal; the lip pouched or spurred, with an entire or trilobed limb, bearded or crested in the middle; the column with a terminal helmet-shaped anther-case, enclosing the two pollen masses with their very short caudicle, attached to a rather large diverging gland. A few of the species have been in cultivation. *Cyrtopera, Galeandra,* and *Zygopetalum,* have all been referred to this genus by Dr. Blume. [A. A. B.]

EUMORPHIA. The name of a pretty little South African bush of the composite family, nearly related to, and having flower-heads like those of the chamomile. It differs, however, in the achenes, which are four or five-angled, and destitute of pappus. The leaves also are very different, being minute, heath-like, and closely packed on the twigs, which are terminated by three white-rayed flower-heads. The plant was first gathered by Mr. Drege, a collector in South Africa, and is named after him *E. Dregeana*. [A. A. B.]

EUONYMUS. The Spindle-tree, a common hedge shrub or small tree, better known among mechanics by the names Dogwood, Pegwood, Skewerwood, and Prickwood. It may be discriminated in summer by its ovate lanceolate shining leaves, and by its small pale green flowers, each composed of four petals, issuing cross-wise from a whitish disk. These are borne two to five together on a stalk in the axils of the leaves, and are succeeded by top-shaped seed-vessels of three blunt lobes, and as many cells, each containing a solitary seed. Towards autumn these become more conspicuous among the leaves (now turning yellow) by their assuming a pink hue; and when the tree has entirely lost its foliage, they are highly ornamental. Each of the lobes of the capsule, which has by this time acquired a bright rose-coloured hue, opens at the projecting angle, and discloses the seed wrapped in an orange-coloured arillus. The foliage, flowers, and fruit of the Spindle-tree are poisonous, but the last are sometimes used as a dye. The wood, which is of a light yellow hue, being strong, compact, and easily worked, is applied to many useful purposes. 'Skewers, pegs for shoes, spindles, toothpicks,' readily suggest the derivation of its various names. The charcoal made from the young shoots is also much approved by artists for its smoothness, and the ease with which it may be erased. Among foreign species cultivated in British gardens, *E. latifolius* is the handsomest, from its broad shining leaves and its large red pendulous seed-vessels, with orange-coloured seeds, which, when the capsules open, are highly ornamental. *E. japonicus* is an evergreen species with rounded ovate-toothed leaves. French, *Fusain*; German, *Spindelbaum*.

EUPATOIRE D'AVICENNE. (Fr.) *Eupatorium cannabinum.* — DE MÉSUÉ. *Achillea Ageratum.*

EUPATORIUM. An extensive genus of *Compositæ*, consisting for the most part of herbaceous plants. The species agree in having all the florets tubular, perfect, and furnished with a long branched style, and in colour either purple, pink, or white, never yellow. They are mostly natives of America; but one species, *E. cannabinum*, Hemp Agrimony, a tall plant with unbranched stems, downy leaves, and terminal crowded corymbs of dull pale purplish flowers, inhabits watery places and damp hedges in Britain. *E. perfoliatum* has some reputation in America as a tonic stimulant, and is administered in the form of a decoction of the leaves; it is employed also as a remedy in intermittent fevers. French *Eupatoire*; German *Abkraut*. [C. A. J.]

The leaves of *E. glutinosum* have been considered to be the Matico of the Peruvians, a substance that is used as a styptic and for other medicinal purposes. It is possible that more than one plant bears the name Matico, but the leaves brought to this country under that name are those of *Artanthe elongata*. [M. T. M.]

EUPETALUM. A genus of begoniads, represented by undershrubs found in Peru. The staminate flowers have four, and the pistillate from five to eight sepals. There are four species. The name is derived from two Greek words, *eu* well or good, and *petalon* a petal, in allusion to the character of the sepals. The species were formerly included under *Begonia*. [J. H. B.]

EUPHORBIACEÆ. (*Pseudanthea, Tricoccæ, Spurgeworts.*) A natural order of monochlamydeous dicotyledons, typical of Lindley's euphorbial alliance of dicliuous Exogens. They are trees, shrubs, or herbs, with opposite or alternate often stipulate leaves, and involucrate incomplete sometimes achlamydeous flowers. Perianth when present inferior, lobed, with glandular scaly or petaloid appendages; stamens definite or indefinite, separate or united in one or more bundles. Ovary one two three or more celled; ovules one or two. Fruit usually of three carpels, which separate in an elastic manner, sometimes fleshy and not opening; seeds with albumen, and often an aril; embryo with a superior radicle. The plants abound in equinoctial America; they are also found in North America, Africa, India, and Europe. They are generally acrid and poisonous, and contain much milky juice. Some yield starch, others oils and caoutchouc. *Euphorbia Lathyris*, the caper spurge, has purgative seeds, and a resinous matter having similar qualities is procured from other species, such as *E. officinarum, antiquorum* and *canariensis*. Cattimandoo, a kind of caoutchouc, is got from another species in India. Castor oil is procured from *Ricinus communis* seeds; croton oil from those of *Croton Tiglium*. The seeds of *Jatropha Curcas*, or physic-nut, are purgatives. *Stillingia sebifera* is the tallow tree of China—the fatty matter being procured from the fruit. Dyes are supplied by *Crozophora tinctoria*, the turnsole, and *Rottlera tinctoria*; the latter plant also yields kamila, a powder from its capsule, used for worms. Teak is yielded by *Oldfieldia africana*; caoutchouc by *Siphonia elastica, lutea, brevifolia, brasiliensis*, and *Spruceana*; and the poisonous manchineel by *Hippomane Mancinella*. *Janipha Manihot* or *Manihot utilissima* furnishes cassava and tapioca, which consist of the starchy matter from its root. *Aleurites triloba* yields cloe oil; *Anisophyllum laurinum* bears an acid fruit called monkey-apple in Sierra Leone. *Colliguaja odorifera* has peculiar jumping or moving seeds, owing to their becoming the habitation of the larva of an insect. Boxwood is the produce of *Buxus sempervirens*. *Hura crepitans*, the sandbox tree, has a fruit consisting of numerous carpels which, when dry, separate with a loud report. Species of *Euphorbia* abound in Africa, and some of them attain a height of thirty or forty feet, with a diameter of two feet at the base of the stem. There are 230 known genera, and about 2,500 species. Examples: *Euphorbia, Hura, Mercurialis, Acalypha, Siphonia, Jatropha, Ricinus, Andrachne, Xylophylla, Phyllanthus*. [J. H. B.]

EUPHORBIA. The Spurge genus, which gives its name to the order *Euphorbiaceæ*, comprises a very large number of species distributed throughout almost the whole world, and varying exceedingly in general or outward appearance, but corresponding closely in the structure of the flowers. All have to a greater or less extent a milky juice. In the temperate regions of the northern hemisphere the species are for the most part herbaceous; in warmer countries, especially those of the southern hemisphere, they have a shrubby or even tree-like habit. Many of the South African kinds, as well as those of other countries, possess succulent spiny leafless stems like *Cacti*. Variable as is the appearance of these plants as regards their stems and leaves, their flowers are all arranged on the same plan. The flowers are monoecious, collected into heads, surrounded by bracts constituting an involucre; these flowerheads are placed in umbels variously branched aggregated into clusters round the top of the stem. The involucre is more or less cup-shaped, four or five-toothed, the lobes or teeth alternating with a number of glands of various form. Within the involucre are a number of stamens surrounding a stalked ovary, hence giving the appearance of a single flower; but this is not really so, as each stamen represents a single male flower, because it is jointed in the middle, and has at its base a separate scale. There are really several monandrous male flowers surrounding a central stalked female, which latter consists of a three-celled ovary, with a three-cleft style. The fruit consists of three single-seeded carpels.

A comparatively small number of kinds are cultivated in this country, either for their beauty or as objects of curiosity; of the former *E. punicea, E. splendens, E. fulgens, E. prunifolia*, and *E. Bojeri* may be mentioned. These are all remarkable for the brilliant scarlet colour of the bracts

VEGETATION OF TENERIFFE WITH SUCCULENT EUPHORBIAS.

of the involucre, and as they flower in winter time and remain long in bloom, they are deservedly great favourites.

Those cultivated for their singular and grotesque appearance are such as have succulent prickly cactus-like stems, and are for the most part without leaves. Among the best known are *E. grandidens*, a tall-growing kind, sending out whorls of branches like those of a candelabrum; *E. officinarum*, *E. antiquorum*, and *E. canariensis*, all somewhat similar; *E. Hystrix*, which has long spines and lance-shaped leaves at the top part of the stem, the lower portion being destitute of them; and *E. meloformis*, a dwarf species, in shape like a melon or an *Echinocactus*, but without spines. Several of them are so like cactuses that they are frequently mistaken for them, especially as the flowers are comparatively rarely seen; a slight puncture with a pin or the point of a knife will, however, immediately decide the matter, as, should the plant be a *Euphorbia*, a milky fluid will ooze out.

In some districts the succulent Spurges are found in great abundance, as they are able to thrive where little else can grow. Thus in the Canary Islands and Teneriffe *E. canariensis* grows in great abundance in arid rocky districts. Professor Smythe speaks of this plant as attaining a height of ten or more feet, while the branches spread on all sides over twenty feet. The stems are erect, stiff, leafless, prismoidal and ill-favoured, 'the product of light and raw heat, salt land, and no shade or genial moisture.' In some parts of South Africa, too, the tall columnar species constitute the characteristic feature of the landscape — *E. grandidens* for one, being said to attain a height of thirty feet and upwards.

The milky juice, which forms one of the constant characteristics of these plants, contains active medicinal properties. Hence in all countries where they grow, some of them have been, or are, employed medicinally. The most important of them for this purpose are some of the succulent ones, which furnish the drug known as Euphorbium. The exact kind which supplies this resinous substance is not precisely known. *E. canariensis*, *E. officinarum*, *E. antiquorum*, and *E. tetragona* have all been mentioned. In commerce Euphorbium exists in the form of small irregular yellowish lumps, pierced with one or more holes, in which are often found the remains of the prickles of the stem from which the resin exuded. The drug is procured from Barbary, where the natives are said to make incisions into the branches, in consequence of which the milky juice exudes. This is so acrid, that it excoriates the hand when applied to it. The juice is allowed to dry and harden on the stem, and after a time the lumps fall off and are collected with caution, the collectors being obliged, says Mr. Jackson in his account of Morocco, to tie a cloth over their mouth and nostrils, to prevent the small dusty particles from annoying them, as they produce incessant sneezing. Euphorbium is an intensely acrid substance, which severely affects the eyes, nose, and lungs of those who come in contact with the drug in its powdered state, if the greatest precautions be not taken. It is said also to induce delirium. From its violent action, it is now rarely if ever used in medicine, but it was employed as an emetic, purgative, etc., and externally as a rubefacient.

The natives of India are said to use the juice of *E. antiquorum* as an external application in rheumatism and neuralgia, and when diluted as a purgative. *E. Nivulia* is used for similar purposes. The juice of *E. heptagona* and other African species is employed by the natives to poison their arrows, while the same purpose is effected in Brazil by the juice of *E. cotinifolia*; that of *E. ligularia* is used in India for the removal of warts; the root, moreover, of the Indian species first named is employed both internally and externally in cases of snake-bite. So also many of the leafy species in which the stem is not fleshy, are considered valuable as purgatives in many parts of the world. Others are esteemed for the cure of syphilis; while some are employed to poison fish. *E. hiberna* is said to have been used in Ireland for this purpose. The roots of some species are emetic, such as those of *E. Gerardiana*, as well as those of *E. Pithyusa* in the south of Europe, and of *E. Ipecacuanha* in America, but they are not to be relied on, as they are apt to produce dangerous purging. Nor are astringent and aromatic properties wanting, for *E. thymifolia*, an Indian plant, has these qualities, as also *E. hypericifolia*, a native of tropical America.

The poisonous principle pervading these plants is more or less dissipated by heat, and hence we hear of a few of them forming articles of diet; thus *E. edulis* is mentioned as a pot herb, so also *E. balsamifera*; the juice too of the latter is said when concentrated (by heat?) to furnish an edible jelly, which is eaten by the natives of the Canary Islands; and Berthollet even mentions the natives of Teneriffe as being in the habit of removing the bark from *E. canariensis*, and then sucking the inner portion of the stem, in order to quench their thirst. This is indeed not so improbable as at first appears, as it is the limpid watery ascending sap which is taken, while the acrid milky descending sap is removed with the rind of the tree, which it percolates. The juice of *E. Cattimandoo*, a native of the Madras presidency, furnishes caoutchouc of a quality which is such as to enable it to be put to a variety of uses; some of it was favourably reported on in the Jury Reports of the Great Exhibition of 1851. Dr. Wight, in his *Icones Plantarum Indiæ orientalis*, gives the following information on the authority of Mr. Elliot : — 'The milk is obtained by cutting off the branches, when it flows freely. It is collected and boiled on the spot, at which time it is very elastic, but after being formed into cakes or cylinders, it becomes resinous or brittle, in which state it is sold in the bazaars, and employed as a cement for fixing knives into handles, and other similar purposes, which are effected by heating it. It is also employed medicinally as an outward application in cases of rheumatism. The gum has not the valuable property, like gutta percha, of being ductile at all times. It can be moulded to any shape when first boiled, but as far as we know not afterwards, though some plan may be found for rendering it subsequently pliable.' Dr. Wight further remarks that when exposed to the heat of a fire or lamp it rapidly softens, and becomes as adhesive to the hand as shoemaker's wax, but when soaked for some time in warm water, it slowly softens, becomes pliable and plastic, and in that state takes any required form. Another of the Indian species, *E. Tirucalli*, is frequently used in Coromandel, for making hedges, as animals for the most part will not touch it, though goats will eat it in spite of the acrid juice, which latter is used medicinally by the natives. It goes by the name of Milk Hedge. *E. phosphorea* derives the name from the fact of its sap emitting a phosphorescent light on a warm night in the Brazilian forests. There are several British species, which partake more or less of the acrid properties so general in this group. [M. T. M.]

EUPHORBIALES. One of Lindley's alliances, including *Euphorbiaceæ*, and a few small related groups.

EUPHORBIUM. A gum resin obtained from certain succulent species of *Euphorbia*.

EUPHRASIA. The Eye-bright is a small annual belonging to the *Scrophularineæ*, distinguished by a tubular four-cleft calyx, a two-lipped corolla, the upper lip two-lobed, the lower three-lobed, and an oblong compressed capsule, containing numerous pendulous ribbed seeds. Eye-bright is a common plant in heaths and dry meadows, growing to the height of six to twelve inches, with small sessile leaves arranged in opposite pairs, and several flowers near the ends of the branches, white spotted with yellow and purple. From the frequent mention of Euphrasy by the poets, it would appear to have been formerly held in high repute for its medical virtues, a view which is confirmed by the statements of the old herbalists, who recommended its use both outwardly and inwardly, in powder and in decoction, for complaints of the eyes. It is still a rustic remedy as an eye-water, but is said by some to be injurious rather than beneficial. French, *Eufraise*; German, *Augentrost*. [C. A. J.]

EUPHROSYNE. A genus of *Compositæ*, composed of two annual Mexican weeds, which grow from one to two feet high, and have alternate twice or thrice pinnatisected leaves, and terminal panicles of little white flower-heads, each about the size of a pea. The flowers have much resemblance to *Ambrosia*, but differ in having male and female flowers in the same capitulum. [A. A. B.]

EUPLOCA. A North American genus of *Ehretiaceæ*, probably not distinct from *Messerschmidtia*, and consisting of herbs furnished with rough leaves, and funnel-shaped flowers. [J. T. S.]

EUPOMATIA. A genus of laurel-like shrubs, natives of the eastern extratropical parts of New Holland. It forms at present a kind of botanical puzzle, being evidently allied to *Anonaceæ*, and yet differing very materially from the other genera of that family. *E. laurifolia* was discovered by the late Robert Brown, and though it is cultivated in conservatories, it has not produced its flowers. The stamens are perigynous, and the inner ones sterile, petal-like; the tube of the calyx is coherent with the ovaries, while the limb separates by a transverse slit, like a lid, from the tube. In these flowers the access of the pollen to the stigmas appears to be completely cut off by the number and disposition of the internal barren petal-like stamens; but the communication is restored, says the learned botanist who first described the plant, by certain minute insects eating the petal-like filaments, while the antheriferous stamens, which are either expanded or reflected, and appear to be even slightly irritable, remain untouched. Recently a second species has been discovered, and named by Dr. Mueller *E. Bennettii*, in compliment to Dr. Bennett, the well-known Australian naturalist. This species has produced flowers in this country, and is figured in the *Botanical Magazine* (t. 4848), under the name of *E. laurina*, its distinctness from that species not having been at first detected. [M. T. M.]

EUPTERIS. A sectional name applied by Agardh to the normal forms of *Pteris*, as distinguished from the groups represented by *P. aquilina* and *Litobrochia vespertilionis*. It is applied by Newman as a generic name to the common Bracken, *Pteris aquilina*.

EUROTIA. A genus of *Chenopodiaceæ*, found in Russia and Siberia, and consisting of annuals with numerous branches, narrow entire leaves, and male flowers four or five together at the apices of small branches. The female flowers are solitary in the axils of the leaves, with the fruiting perianth limb growing out in the shape of two horns. [J. T. S.]

EURYA. A genus of *Ternstromiaceæ*, composed of a number of evergreen shrubs or small trees found in India, China, and the adjacent islands, one species extending to the Feejee Islands. Their very minute and unisexual flowers, together with their small berried fruits, serve to distinguish them. The leaves are not unlike those of the tea plant, and the small white flowers are arranged in bundles in their axils. [A. A. B.]

EURYALE. Before the discovery of the *Victoria regia*, the Indian aquatic herb bearing the above name took rank, perhaps, as the noblest aquatic plant in cultivation, at least as to its leaves, which are much like those of the *Victoria*, but smaller, of a circular form, with very prominent and spiny veins on the rich purple under-surface: the upper surface being covered with rounded eminences, and of a dark green colour; and the size varying from one to four feet in diameter. The flowers are stalked, and have a calyx which is adherent below to the ovary, but above it is divided into four segments; the whole outer surface of this calyx is covered with strong spines; the petals are from twenty to thirty, smaller than the calyx, and of a purple colour; the stamens numerous, detached, all fertile. Fruit a round many-seeded berry, crowned by the persistent calyx. By these characters *Euryale* is distinguished from *Victoria*: both belong to the *Nymphæaceæ*. *E. ferox* is a native of the eastern part of Bengal and other quarters of India, where also it is frequently cultivated. Its seeds are floury, and after being baked in sand are eaten by the natives. The Hindoo physicians, moreover, say that they possess medicinal virtues. The plant is also grown in China for the sake of its seeds. [M. T. M.]

EURYBIA. A genus of trees or shrubs of the composite family, confined to Australia, Tasmania, and New Zealand, and numbering upwards of sixty species. In many respects it is allied to *Aster* of the northern hemisphere, but the flower-heads do not contain nearly so many florets. The genus *Olearia* is their closest relationship, but there the pappus is double, while here it is single. They are very variable in appearance, some being large trees, others heath-like shrubs; and the daisy-like flower-heads are either solitary or numerous and panicled at the ends of the branches.

The silver-leaved Musk tree, *E. argophylla*, is a Tasmanian plant, attaining a height of twenty to twenty-five feet, with a girth of three feet. It is often seen in greenhouses as a shrub, where it is cultivated for the musky odour of its leaves. The wood of the tree is hard and takes a good polish. The larger-leaved species, which are chiefly confined to New Zealand and Tasmania, are much like this in appearance. The smaller-leaved ones are more common on the continent. Amongst these latter is *E. Dampieri*, found in great abundance by Dampier on one of the islands on the north-west coast, which now bears his name, and called by him Rosemary, from its resemblance to that plant. The name Daisy-tree is given to *E. lirata* in Tasmania. The name of the genus is that of the mother of the stars in Greek mythology. [A. A. B.]

EURYCLES. A genus of amaryllids, of the pancratiform group, found in the Eastern Archipelago and in New Holland. The bulbs are ovate; the leaves are broad and petiolated; and the scape, which scarcely precedes the leaves, supports a many-flowered umbel of flowers, of which the tube is cylindrical, the limb regular, with equal segments, and the cup frequently imperfect. *E. amboinensis* is a stove bulb. *E. australasica*, or *Cunninghami*, a smaller species from Australia, is called the Brisbane Lily. [T. M.]

EURYCOMA. A genus of shrubs from Sumatra, referred by Planchon to the *Simarubaceæ*. The leaves are compound, and clustered at the extremity of the branches; the flowers are panicled, of a purple colour, and on some plants unisexual, on others perfect. *E. longifolia* is stated to be a valuable febrifuge. [M. T. M.]

EURYLOBIUM. A genus of *Stilbaceæ*, a family of corollifloral dicotyledons, consisting of shrubs furnished with rough linear leaves; and flowers of which the calyx is five-toothed, the corolla tubular, five-cleft, its two upper lobes larger and connate, the tube hairy within, and the stamens four. *E. serrulatum*, the only species which is known, is a native of South Africa. [J. H. B.]

EURYNEMA. A small annual Arabian plant belonging to the *Zygophyllaceæ*. The flowers are on long stalks, which are bent in the middle; stamens five, their filaments dilated at the base, shorter than the anthers; ovary on a short stalk, with several ovules in each of its five compartments. Fruit capsular. [M. T. M.]

EURYOPS. A genus of *Compositæ*, nearly allied to *Senecio*, but differing in the nature of the involucre, which is composed of one series of scales having their margins more or less united, so that the involucre has the appearance of a toothed cup; while in *Senecio* the scales are free. The hairs of the short woolly pappus are curiously bent in a zig-zag manner, and their outer row is often deflexed. Of about thirty known species, one is found in Arabia, another in Abyssinia, and the remainder in S. Africa. They are handsome often resinous under-shrubs, very variable in appearance, some having the leaves smooth undivided pine-like, while in others they are wedge-shaped and toothed, and in a goodly number are pinnately-lobed and cut. The yellow-rayed flower-heads are small disposed in corymbs, or large and solitary. *E. speciosissimus* is called Resinbush by the colonists, because of a gummy exudation often seen on the stem and leaves. The handsome flower-heads are nearly four inches across. [A. A. B.]

EUSCAPHIS *staphyleoides* represents a genus of *Staphyleaceæ*, found in Japan, the Corea, and the Loo Choo Islands. It resembles the common *Staphylea pinnata*, but it is easily recognised by its fruits, which are composed of three distinct bladdery carpels; while in *Staphylea* the carpels are united so as to form one bladdery capsule. The plant grows to a bush of about twelve feet high, and is furnished with opposite pinnate smooth leaves, which are a good deal like those of the elder; and so also are the little white or yellowish flowers, which are numerous and disposed in terminal panicles. According to Siebold, the plant is a favourite in Japanese gardens from its neat habit and its pretty bladdery fruits, which are of a red colour when ripe, and remain on the bush till winter approaches. It is prized also for its medical properties. The inner bark of the root is bitter and astringent, and is given in infusion in cases of dysentery and chronic diarrhœa. The leaves are not so efficacious, and have when fresh a disagreeable fishy smell. The name of the genus has reference to the pretty fruits, which open along the inner surface into the form of a little boat. [A. A. B.]

EUSTACHYS. A genus of grasses belonging to the tribe *Chlorideæ*; now included in *Chloris*. [D. M.]

EUSTEGIA. A genus of South African *Asclepiadaceæ*, containing a few species of decumbent branching perennial herbs, with opposite hastate leaves, and sub-umbellate flowers, the calyx of which is five-parted, and the corolla rotate and five-cleft, with a triple staminal crown: the divisions of the two inner whorls of the latter alternating with the lobes of the outer whorl and with the anthers. [W. C.]

EUSTIGMA *oblongifolium*. A small tree of South China, forming a genus of *Hamamelideæ*, distinguished chiefly by its long broad flat stigmas. The flowers are in small loose heads, without any petals except five small gland-like scales, and have five stamens, with obtuse two-valved anthers, and a half-inferior ovary.

EUSTOMA. A gentianaceous annual plant, native of North America, the flowers of which are white, with a deeply five-cleft calyx, a funnel-shaped corolla, with a contracted tube into the middle of which the stamens are inserted, and a large two-lobed stigma. [M. T. M.]

EUSTREPHUS. A genus of *Liliaceæ*, consisting of twining woody-stemmed plants, from the warmer parts of Australia, with elliptical or lanceolate leaves, and aggregate, pedicellate flowers, from the axils of the leaves and the end of the stem. The flowers are purplish, with a six-parted spreading perianth, the inner divisions fringed. [J. T. S.]

EUTASSA. A genus of coniferous plants, sometimes considered as a section of *Araucaria*, and including those species which have been called needle-leaved, such as *A. excelsa*, *Cunninghami*, and *Cookii*. [See ARAUCARIA.] The species included under *Eutassa* are found in Australia, as at Norfolk Island, Moreton Bay, New Holland, and New Caledonia. As a sub-genus of *Araucaria*, sometimes called *Eutacta*, it is thus defined: scales of the cone broadly winged; a distinct basilar appendage to the seed; anthers six to ten-celled; cotyledons four. [J. H. B.]

EUTAXIA. A genus of pretty leguminous bushes found in Australia, but chiefly confined to the western portions. There are about a dozen species. They have much the appearance of *Pultenæa* or *Dillwynia*, so well known as greenhouse plants, and differ from the former in the standard being about as broad as it is long, not broader; from the latter in the wings being shorter than the keel, not equal to it in length; and from both in having opposite leaves. They are for the most part much-branched low-growing bushes, with small often heath-like leaves, and axillary golden-yellow pea-flowers, two or three together. *E. myrtifolia* is a well-known greenhouse plant, whose slender stems are often seen thickly covered in the spring and summer months with the pretty yellow blossoms. [A. A. B.]

EUTERPE. A genus of palms of extremely graceful habit, having slender almost cylindrical stems, sometimes nearly a hundred feet in height, surmounted by a tuft of pinnate leaves, the leaflets of which are narrow, very regular and close together, and generally hang downwards. The bases of the leaf-stalks are dilated, and form cylindrical sheaths round a considerable portion of the upper part of the stem, giving it a woollen appearance. Ten species are known, all natives of the forests of tropical South America, where they grow together in large masses; some inhabiting moist swampy places on the banks of rivers, and others extending a considerable height up the sides of mountains. Their flower-spikes, which grow out horizontally from the stem below the swelling of the leaf-stalks, are simply branched, and the flowers are seated in little furrows upon the branches, with bracts at their base: the males and females being in pairs on the lower parts of the branches, while the males are most numerous on the upper parts. The fruits are of a dark purple colour, with a thin fibrous fleshy rind, enclosing a single seed.

E. edulis, the Assai Palm of Pará, grows in swampy places, particularly upon the banks of rivers within the tidal limits, where it attains a height of thirty or forty feet, and has a stem about as thick as a man's arm, slightly bulged out at the base, and generally curved or leaning over. Its fruit, which resembles a sloe in size and colour, has a thin coating of clotted fibrous flesh, from which the inhabitants of Pará manufacture a beverage called Assai. This is prepared by throwing the ripe fruits into a vessel containing warm water, and allowing them to soak for about an hour, and then, the water being partly poured off, kneading them thoroughly with the hands, fresh cold water being occasionally added, until all the pulp is detached, when the liquid is separated by straining, and is then fit for use. It is of a thick creamy consistence, and of a fine plum colour; and when sweetened with sugar, and thickened with cassava farina, it is very nutritious, and forms the greater part of the daily food of a large number of the inhabitants of Pará, with whom Assaï is a great favourite.

E. montana, a West Indian species, is cultivated in the hothouses of this country. It attains a height of about twenty feet, and has the base of its stem much swollen or bulged out. The central portion of the upper part of the stem, including the leaf-bud, of this and the other species is eaten either when cooked as a vegetable or pickled; but the tree must be destroyed in order to obtain it. [A. S.]

EUTHALES. A goodeniaceous genus, native of the south-west coast of Australia. It bears a tubular unequally five-cleft inferior calyx, a corolla cleft on one side at the apex with a bilabiate limb, free anthers, an undivided style, with the indusium of the stigma bilabiate, and a four-valved capsule. They are stemless herbs with long-stalked nearly entire leaves, and yellow flowers. [R. H.]

EUTHEMIS. A genus placed by some authors in *Saurogesiaceæ*, and by others in *Ochnaceæ*; differing from any in the former in its berried fruit, and from any in the latter in the fruit being composed of five united carpels, instead of numerous *free* carpels seated on a rounded and thickened receptacle. It is composed of a few beautiful little shrubs of the Malayan Archipelago, having smooth rounded stems furnished with alternate, elliptical or lance-shaped leaves beautifully and minutely serrulate at the margins, and the glossy blades exquisitely marked with a great abundance of parallel nerves running at right angles to the midrib, the spaces between them forming a beautiful network of veinlets. The flowers are white or tinged with purple, and disposed in axillary or terminal compound racemes. [A. A. B.]

EUTHYSTACHYS. A genus of *Stilbaceæ*, entirely confined to S. Africa, and differing from its nearest ally, *Campylostachys*, in its straight, not curved, flower-spikes, whence the name of the genus, and in the little funnel-shaped corollas, which have a five-lobed instead of a four-lobed border. The only known species, *E. abbreviata*, is a smooth shrub, with heath-like four-ranked leaves thickly set on the *stems*, which terminate in a bracted spike of flowers. From the other genera this differs in having a calyx two of whose narrow segments are free, and three are united by their margins nearly to the summit. [A. A. B.]

EUTOCA. Annual herbs belonging to the *Hydrophyllaceæ*, of an erect habit, with rough leaves, and clusters of showy flowers. They are natives of North America, especially California, and are often grown in European gardens. The species mostly cultivated are *E. Menziesii* or *multiflora*, about eighteen inches high, with downy narrow leaves, either entire or lobed, and blue flowers. *E. viscida* is much branched, with heart-shaped deeply-cut toothed clammy leaves, and elongated racemes of blue flowers with a rose-coloured tube.

All the species are elegant and hardy. French, *Eutoque*. [C. A. J.]

EUTRIANA. A genus of grasses belonging to the tribe *Chlorideæ*. The inflorescence is for the most part in short racemose spikes; spikelets one-sided, alternately sessile, two to three-flowered; glumes two, keeled, the exterior larger; pales two, of thickish texture, the inferior one three-cleft, the superior two-keeled. The score of known species are nearly all natives of South America. [D. M.]

EUTROPIS. An imperfectly described genus of *Asclepiadaceæ*, containing a single species, abundant in the Punjab, and forming a low twining fleshy lance-leaved undershrub. Its position is between *Calotropis* and *Paratropis*, having the angular and saccate sinuose corolla, membrane-lipped anthers and corona of the former; and the coronal leaflets cleft, and the pollen masses oval and ventricose, as in the latter. [W. C.]

EUXENIA. A genus of opposite-leaved Chilian shrubs belonging to the composite family, and distinguished amongst its allies by each yellow flower-head being entirely composed of unisexual florets, all of which are tubular and five-toothed. The leaves, somewhat rough to the touch, are broadly oval or lance-shaped; and the globose yellow flower-heads usually grow two or three together, and are stalked at the ends of the twigs, and about half an inch across; the achenes are four-sided, without pappus. There are but two species. *E. grata*, with broadly oval leaves, is called by the Chilians Palo Negro; the other, *E. Matiqui*, with lance-shaped leaves, is called Matiqui. In both the leaves have a pleasant aromatic scent. [A. A. B.]

EVANESCENTI-VENOSE. When lateral veins disappear within the margin.

EVAX. A genus of *Compositæ*, found in the Mediterranean region and in California, and composed of a few minute tufted annual herbs, having all their parts clothed with white wool like many of the cudweeds. In some species, as in *E. eriosphæra*, the whole plant does not exceed a quarter of an inch in diameter, and looks like a little ball of wool, whence the specific name. None of them exceed four inches in height, and if branched the branches are not more than two inches long, and terminate in a sessile flower-head surrounded by a rosette of oblong woolly leaves. The genus is chiefly distinguished among its allies by the elongated cone-shaped receptacle on which the florets are seated, and by the achenes being destitute of pappus. [A. A. B.]

EVELYNA. A numerous genus of South American epiphytal orchids, found growing on stems and trunks of trees, and readily known by their habit. They have erect wiry stems, one to three feet high, furnished with lance-shaped strongly ribbed leaves, and terminating in a few-flowered spike, the flowers enveloped by long coloured bracts. The anther is two-celled, with eight pollen masses attached in fours to a very short caudicle with a triangular gland. *E. Caravata*, from the West Indies and French Guiana, is in cultivation. It has hispid stems a foot high, bearing long lance-shaped rough leaves, and bright yellow flowers with a beautifully fringed lip, almost hidden from view by long pink bracts. The genus bears the name of John Evelyn, an eminent patriot of the seventeenth century. *Evelinanthus* is a synonym. [A. A. B.]

EVENING FLOWER. *Hesperantha*.

EVENNESS. An absence of elevations or depressions of the surface of any part or organ.

EVERGREEN. Continuing to bear green leaves all the year round.

EVERNIA. A small genus of lichens belonging to the nameold tribe of *Parmeliaceæ*, differing from *Ramalina* in their having a distinct under-surface to the flat erect branched fronds. They are sometimes prettily coloured, *E. flaricoma* and *vulpina* being of a beautiful yellow. *E. prunastri* is common in almost every thicket, and was formerly ground down with starch to make hair powder. It was used, at the instigation of Lord Dundonald, as a substitute for gum in cotton-printing. The yellow species contain two distinct colouring principles, and of these *E. vulpina* is said to be poisonous to wolves. *E. flavicans* occurs in the south of England, but prefers warmer regions. [M. J. B.]

EVIA. A genus of Indian trees belonging to the *Anacardiaceæ*, and, judging from description, so closely allied to *Spondias* as hardly to be distinguished from it. In *Evia* the filaments are awl-shaped, in *Spondias* thread-like. The fruits of *Evia* are edible. [M. T. M.]

EVITTATE. Not striped; destitute of vittæ.

EVODIA. A genus of small rutaceous trees or shrubs, natives of tropical New Holland and the Indian Archipelago. The flowers are disposed in a panicled manner, and the flower-stalks are jointed in the middle. The parts of the flower are four-fold; the calyx persistent; the petals and stamens inserted at the base of a cup-shaped sinuous disc, which encircles the lower part of the four ovaries; the styles are four, becoming after a time fused into one. The fruit consists of four carpels which separate. *E. triphylla* is a stove-shrub with white flowers. [M. T. M.]

EVODIANTHUS. A genus of *Pandanaceæ*, consisting of climbing somewhat palm-like plants, with cleft leaves, and monœcious flowers arranged on stalked spadices, which are protected by three bracts. The perianth of the male flower, in which the distinguishing characters of the genus reside, is tubular, funnel-shaped and curved, the lower portion fleshy and triangular, the upper part bell-shaped, somewhat four-cornered, the limb very short and divided into several lobes, which are arranged in two rows, those of the outer row detached one from the other, those of the inner confluent, and provided with two teeth. The species are natives of Costa Rica, and greatly resemble those of *Carludovica*. [M. T. M.]

EVOLUTIO. The act of developement.

EVOLVULUS. A considerable genus of *Convolvulaceæ*, containing nearly sixty described species, natives chiefly of tropical America, but with one or two species from the warmer regions of the Old World. They are annual herbs, or have a perennial sometimes woody stock, and bear entire usually small nearly sessile leaves, and small flowers on axillary peduncles, or in terminal spikes or racemes, with the corolla campanulate or funnel-shaped, and angular or lobed. [W. C.]

EVONYMUS. *Euonymus*.

EVOSMIA. Tropical American shrubs or small trees, belonging to the *Cinchonaceæ*, and having red flowers on slender axillary stalks, the corolla wheel-shaped, the stamens short. The fruit is a four-celled berry, crowned by the limb of the calyx, and having an agreeable odour. Sir R. Schomburgk says that cases of poisoning among the Indians have arisen from their using the wood of one of these plants, *E. corymbosa*, as a spit whereon to cook meat. [M. T. M.]

EWALDIA. A genus of begoniads, consisting of villous shrubby plants found in Brazil. Their staminate flowers have four, and the pistillate five sepals; anthers oblong, with united filaments; style persistent, its branches surrounded by a continuous papillose band, which makes two spiral turns; placentas undivided, their transverse sections being ovate. There are two known species, *E. ferruginea* and *E. lobata*; both of them formerly included under *Begonia*. The genus is named in honour of Dr. Ewald, of the Berlin Academy. [J. H. B.]

EX. See E. But *exo* signifies outwards or external, as in exogens and exintine, *quasi* exointine.

EXACUM. Erect branched annual herbs, with opposite sessile leaves and showy, blue, yellow, or white flowers, belonging to the *Gentianaceæ*. The calyx is bell-shaped and four-cleft; the corolla salver-shaped, four-cleft, with an inflated tube; the capsule globose, two-celled, many-seeded, and splitting; the seeds minute. The plant described by Sir J. Smith under the name of *Exacum filiforme* is the *Gentiana filiformis* of Linnæus and the *Cicendia filiformis* of modern botanists. French, *Gentianelle*; German, *Kugelruhre*. [C. A. J.]

EXADENUS. Tropical American annuals of the gentian family, with linear leaves and four-parted flowers, the corolla wheel-shaped, four-cleft, persistent, and each of its four segments provided on the outside near the base with a sessile or stalked gland; capsule two-seeded. [M. T. M.]

EXALBUMINOSE. Having no albumen.

EXANTHEMATA. Skin diseases, blotches of leaves, &c.

EXAREOLATE. Not spaced out.

EXARISTATE. Destitute of an arista, awn, or beard.

EXASPERATE. Covered with hard short stiff points.

EXCENTRIC. Out of the centre.

EXCIPULE. That part of the thallus of a lichen which forms a rim and base to the shield. Also a similar part in certain fungals.

EXCŒCARIA. A small genus of spurgeworts consisting of about eighteen species, five or six of which belong to India, while the remainder are natives of the West Indies and Brazil. Most of them are woody shrubs, but a few form small trees. Their leaves are usually alternate, and either entire or with their margins toothed. The flowers are produced in catkins, some species having the males and females on distinct trees, and others bearing them in different parts of the same catkin. The individual flowers have neither calyx nor corolla, but their place is occupied by a variable number of little bracts. The fruit is three-celled.

E. Agallochum was at one time supposed to yield the fragrant resinous Indian wood called Agallochum, Aloes or Eagle wood, which is now, however, known to be the produce of *Aquilaria Agallochum*, a plant belonging to a totally different natural order. It is a native of India, where it is commonly found growing in salt marshes, and is sometimes employed for strengthening the banks of rivers in places within the influence of the sea water. It forms a small crooked tree or large branching shrub, with egg-shaped leaves, having round blunt teeth along their edges. The different sexes of the flowers are borne on distinct trees, the male catkins being very long, and either solitary or in pairs, while the females are much shorter, and sometimes three together. When the tree is wounded, a white milky juice flows from it, which is of a very acrid nature, producing inflammation and ulceration if allowed to come in contact with the skin. If it gets into the eyes it causes blindness. The wood is used for charcoal and firewood, but the smoke from it is said to cause intolerable pain in the eyes. [A. S.]

Gussonia of Sprengel from Brazil, and *Gymnanthes* of Swartz from the W. Indies, both monœcious, are included by modern authors in the present genus.

EXCRETION. Any superfluous matter thrown off by the living plant externally.

EXCURRENT. Running out. When a stem remains always central, all the other parts being regularly disposed round it, as in the stem of a fir tree.

EXEMBRYONATE. A name given to cryptogams in consequence of their spores not containing an embryo like the seeds of phænogams. Though, however, the spores contain no embryo in the higher cryptogams, the archegonia contain a cell which goes through the same process of cell-division as the embryonic cell in phænogams, sometimes producing a distinct plant, sometimes only fruit. [M. J. B.]

EXINDUSIATE. Not having an indusium.

EXINTINE. The middle coat of a pollen grain, or if three or four coatings are present, then that which is next the intine.

EXOCARPUS. A genus of *Thymelaceæ* or *Daphnaceæ*, though some refer it to a separate order, *Anthoboleæ*. The flowers are sometimes perfect, at other times incomplete. The perianth is four to five-parted; stamens four to five, inserted on the base of the perianth, the filaments short; ovary free, one-celled; the style very short, and the stigma capitate. Fruit, a single-seeded nut, supported on an enlarged berried peduncle. Trees and shrubs of New Holland; found also sparingly at the Moluccas. They have scattered, often minute, leaves, which have no stipules; flowers small in axillary spikes, with caducous bracts, the flower-stalk enlarging after fertilisation. There are four known species. [J. H. B.]

EXOCHORDA. A beautiful Chinese bush of the rose family, cultivated in England and quite hardy. It is remarkable for the structure of its fruits, which consist of five small compressed bony carpels adhering round a central axis in a star-like manner. From the axis or growing point stand five erect placentary cords, which enter the carpels on their inner face near the top, suspending from the apex two thin seeds. These cords remain after the carpels have fallen, and have suggested the name of the genus. The only species, *E. grandiflora*, is a smooth bush with alternate nearly lance-shaped entire leaves, the stems terminated by racemes of handsome white flowers, which appear in May, and are nearly as large as those of the mock-orange; they have a bell-shaped calyx with a five-parted border, five rounded petals, and fifteen to twenty stamens. The plant bears also the name of *Spiræa grandiflora*. [A. A. B.]

EXOGENS. A name given to one of the great classes of the vegetable kingdom, corresponding with the Dicotyledons. The name Exogen is derived from the Greek words signifying 'outwards' and 'to grow,' meaning growing outwardly, and has reference to the mode in which the woody circles are produced, viz. from the centre outwardly towards the circumference. The age of an exogenous tree, particularly in temperate climates, may be determined by counting the number of zones or circles in the woody stem, each circle marking one year's growth, and the last-formed circle being external. All the native trees of Britain are exogenous. The characters of the class are given under the head DICOTYLEDONS. [J. H. B.]

EXOGENOUS. Growing by addition to the outer parts of the stem.

EXOGONIUM. A genus of *Convolvulaceæ* very closely allied to *Convolvulus* and *Ipomœa*, from both which it is distinguished by its stamens projecting from the tube of the corolla; and from the former by its button-like stigma. *E. Purga*, a Mexican climbing plant, with salver-shaped purplish flowers, furnishes the true Jalap tubers of commerce. These are roundish, of variable size, the largest

Exogonium Purga.

being about as large as an orange, and of a dark colour. They owe their well-known purgative properties to their resinous ingredients, and hence worm-eaten tubers are more valued than sound ones, as the insects eat the farinaceous and woody portions of the tuber and leave the resin. Various species of *Ipomœa* are also said to furnish a spurious kind of jalap. [M. T. M.]

EXORHIZÆ. A name given to exogenous or dicotyledonous plants, from the mode in which the young root sprouts when the seed is placed in the ground. The term is derived from the Greek *exo* outwardly, and *rhiza* a root, meaning root pushing outwardly, in allusion to it pushing out directly in a tapering manner, and not coming out in the form of numerous rootlets through sheaths as in the *Endorhizæ*, or monocotyledons. [J. H. B.]

EXORHIZAL. That kind of germination in which the point of the radicle itself becomes the first root.

EXOSMOSE. That force which causes a viscid fluid lying on the outside of an organic membrane to attract watery fluid through it.

EXOSTEMMA. A genus of tropical trees or shrubs of the *Cinchona* family. They have whitish or pink flowers of a funnel-like form, the segments of the limb linear and rolled back; the five stamens project to a considerable distance from the corolla, hence the name of the genus. The ovary is two-celled, with a long style, and almost undivided stigma; capsule two-seeded: some of the kinds are in cultivation. The barks of the West Indian species possess febrifugal qualities, as in the closely allied *Cinchona*. [M. T. M.]

EXOSTOME. The aperture in the outer integument of an ovule.

EXOSTOSIS. A name given to a diseased condition in plants, in which hard masses of wood are produced, projecting like warts or tumours from the main stem or roots. Most cases seem to arise from tissues developed round adventitious buds which do not properly break through the bark. These are sometimes completely concealed, as the knaurs in beech, which are often quite free. Sometimes there is a continued multiplication of fresh buds, and in proportion as these are more or less developed, we have the besom-like bodies on birch, or the rough tumours on elms. Cypress knees, which sometimes grow to a great size, and when hollowed are used for beehives in the United States, grow by a similar disease on the roots of *Taxodium*. Fine specimens may be seen at Sion. The tumour at the junction of a graft with its stock seems to arise from some different cause, which is not at present ascertained. [M. J. B.]

EXSERTED. Projecting beyond the orifice of an organ.

EXSUCCOUS. Juiceless.

EXTINE. The outer coat of a pollen grain.

EXTRA. On the outside of, or beyond; as *Extra-axillaris*, beyond the axil; *Extra-foliarius*, beyond a leaf; *Extra-medianus*, beyond the middle.

EXTRORSE. Turned outwards from the axis of growth of the series of organs to which it belongs.

EYE. A term in gardening for a leaf-bud; also for the centre or the central markings of a flower.

EYEBRIGHT. *Euphrasia*.

EYSENHARDTIA. A genus of *Leguminosæ* nearly related to *Amorpha* and *Dalea*, but differing from the former—which has only one petal, and that the standard—in its corolla of five petals, and from the latter, in its little sabre-shaped pod being much longer than the calyx.

E. amorphoides is a much-branched shrub or small tree, five to twenty feet high, found in Texas and Mexico; its slender ash-coloured branches are furnished with an abundance of pinnate leaves, and the little white pea-flowers are very numerous, and disposed in dense racemes at the ends of the twigs, succeeded by thin sabre-shaped pods. The only other species, *E. spinosa*, also a Mexican bush, has the ends of its flower spikes hardened into spiny points after the flowers have fallen. The genus bears the name of C. W. Eysenhardt, once professor in the university of Konigsberg. [A. A. B.]

FAAM, or FAHAM. *Angræcum fragrans*.

FABA. The typical genus of the order *Fabaceæ* or *Leguminosæ*. It consists of annual plants rising from two to four feet high, having smooth quadrangular hollow stems; alternate pinnated leaves, formed of from two to four pairs of entire oval leaflets; and numerous large white or violet highly fragrant blossoms, marked with dark violet-coloured veins and blotches on the petals. The seeds are produced within a long green pod, or legume, and are roundish kidney-shaped, and more or less depressed or flattened.

The common Bean, *F. vulgaris*, is a hardy annual, generally believed to be a native of the shores of the Caspian Sea, as well as of Egypt and other parts of the East. It is a vegetable of very great antiquity, and is noticed in sacred history upwards of a thousand years before the Christian era (2 Samuel xvii. 28). The earlier Greeks and Athenians are stated to have cultivated beans, and offered them as a sacrifice to their gods—a practice which, according to Pliny, was in later times followed by the Romans. One of the noblest families of ancient Rome—the Fabii—derived its name from its ancestors having been celebrated for the great success which attended their culture of beans. Yet, strange to say, the most superstitious notions were entertained respecting their composition, and fitness for being used as food for man, so that some of the ancient philosophers enjoined their followers to abstain from eating them. They appear to have been known in this country from time immemorial; when, or how, they were introduced we have no information; it is, however, generally supposed to have been by the Romans.

Among the industrious classes, beans when full grown are a favourite vegetable, and considered to be very nutritious to persons with strong constitutions, but to those of a delicate habit they are not to be recommended unless in a very young state, when, if properly dressed, they form an excellent dish. There are several varieties in cultivation, which chiefly differ from one another in being tall or dwarf, early or late; or in the colour of the beans being brownish red, or green. [W. B. B.]

FABACEÆ. The bean or leguminous family, a natural order of calyciflorial dicotyledons, better known by the name *Leguminosæ*, under which head their peculiar characteristics are described. The plants are distinguished either by their papilionaceous (pea-like) flowers, or by their fruit being a legume; in other words, a pod like that of the pea or bean. [J. H. B.]

FABAGELLE. (Fr.) *Zygophyllum*.

FABIANA. A genus of South American shrubs, belonging to the *Solanaceæ*. They have alternate scattered or overlapping leaves, and extra-axillary flower-stalks, bearing a single flower, with a tubular five-cleft calyx, and funnel-like corolla, whose tube is gradually dilated upwards, and whose limb is divided into five short lobes. The five stamens are included, and of unequal length; the anthers open by slits; the capsule is two-celled, included within the persistent calyx, and divided by two valves. *F. imbricata* is a neat half-hardy shrub, of fastigiate habit, with white flowers, and has much of the general appearance of a heath. [M. T. M.]

FABRICIA. A genus of *Myrtaceæ*, consisting of New Holland shrubs, with broad oblong glaucous dotted leaves, and solitary axillary white or yellow flowers, with a bell-shaped adherent calyx-tube, and a five-cleft deciduous limb; the five petals roundish, attached to the throat of the calyx; numerous stamens, inserted with the petals, and shorter than they; and a partly-adherent many-celled ovary, each compartment containing several ovules. The fruit is a capsule opening at the top through the backs of the valves. Two or three species are in cultivation. [M. T. M.]

FABRICOUPIER, or FALABRIQUIER. (Fr.) *Celtis australis*.

FACELIS. A little annual composite plant found in Chili, and also on the opposite side of the continent. It resembles a cudweed in appearance, and differs from its allies in having the tubular ray-florets female and in many series, and those of the disk fewer in number, more slender, and perfect. The weak stems seldom exceed eight inches high, and are furnished with numerous narrow somewhat wedge-shaped leaves; and the little narrow flower-heads, containing pink-tipped florets, are clustered at the ends of the stem. The achenes are silky, and crowned with a pappus of one series of feathery hairs. [A. A. B.]

FACIES. The general appearance of a plant.

FADYENIA. A curious West Indian aspidioid fern, remarkable for having its small sterile recumbent fronds broader than the fertile, and attenuated and proliferous at the point, the fertile being erect and blunt, almost covered by the two rows of sori. The only species is *F. prolifera*, a dwarf plant but a few inches in stature, both forms of fronds being simple. The fronds have netted veins, and are remarkable for the large size of the sori, and the very much elongated sinus of the indusium, which is reniform.

The name *Fadyenia* has also been proposed by Endlicher, for the *Garrya Fady-cuii* of Hooker. [T. M.]

FAFEEIL. One of the Arabian names of *Papyrus*.

FAGARASTRUM. A genus of *Amyridaceæ*, consisting of certain shrubs, natives of the Cape of Good Hope and of tropical Africa, having hermaphrodite flowers with a short three to four-parted calyx, three to four petals, and twice as many stamens, the alternate ones shorter than the rest, the filaments thickened above the base, and the anthers large. Both petals and stamens are inserted into a kind of stalk, supporting the three to four-celled ovary, in each compartment of which are two ovules suspended from the top. The fruit is as yet unknown. [M. T. M.]

FAGE. (Fr.) *Fagus sylvatica*.

FAGELIA. A genus of *Leguminosæ*, composed of a few twining herbs found in South Africa and Abyssinia. They are more or less clothed with yellowish clammy hairs, and have ternate leaves somewhat like those of *Phaseolus multiflorus*, but smaller, with nearly triangular leaflets. Their pretty yellow pea blossoms are borne on long axillary racemes. The chief distinguishing characters of the genus are, the deeply divided calyx, the obtuse keeled petal longer than the wings, and the two-seeded turgid pods, about half an inch long. [A. A. B.]

FAGHUREH of Avicenna. *Xanthoxylon hastile*.

FAGOPYRUM. The common Buckwheat and a few other species, of Asiatic origin, are included in this genus of *Polygonaceæ*. They are herbaceous plants, with erect branching stems, and heart-shaped or halbert-shaped leaves. The perianth is cut into five equal divisions, and does not increase in size along with the fruit, like that of some allied plants; and the eight stamens alternate with eight round glands. The fruit is three-sided, and not enveloped in the perianth, like that of *Polygonum*; the seed is mealy.

The common Buckwheat, or Brank as it is sometimes called, *F. esculentum*, is an annual plant with a branched stem, growing two or three feet high. It is a native of central Asia, but has been so long extensively cultivated, that it has become naturalised in various parts of Europe. In this country it is only grown to a small extent, and principally for the purpose of affording food for pheasants. On the continent, however, and also in some parts of the United States, Buckwheat is largely employed for human food; and the thin cakes made of it are said to be very delicious. As a food, its nutritious properties are greatly inferior to wheat, but it ranks much higher than rice. In France it is called Sarrasin and Blé noir. The plant is still sometimes called *Polygonum Fagopyrum*. [A. S.]

FAGRÆA. A genus of Asiatic or Polynesian *Loganiaceæ* consisting of thick-leaved trees or shrubs, sometimes found growing on loose mould that may have gathered on the stems or forks of other trees. Their chief distinguishing features consist in the border of the tubular corolla being five, rarely six to seven-lobed, the lobes twisted in the bud; and in the fruit being a two-celled berry. From most genera in the family they are readily recognised by the remarkably thick and leathery texture of their smooth and entire, usually elliptical or lance-shaped leaves. The flowers are white or cream-coloured, and often fragrant; in some very large, and thick in texture, with a trumpet-like tube, two to five inches long (in *F. auriculata*, one of the largest-flowered species, with a border six inches across); in others, where the flowers are very numerous and disposed in terminal corymbs, the corollas are much smaller. The flowers are succeeded by berries, which in the larger-flowered species are of the size of a duck's egg, and contain numerous seeds. Altogether they have much the appearance of *Gardenia*, and are chiefly distinguished by their ovaries being superior. Upwards of thirty species are known. The name *Cyrtophyllum* is sometimes given to some of the smaller-flowered species. [A. A. B.]

FAGUS. A genus of *Corylaceæ*, distinguished by having triangular nuts enclosed within a spiny capsule or husk. The most important of the few species is *F. sylvatica*, the Common Beech, a well-known European tree, and a native also of Armenia, Palestine, and Asia Minor. It forms a large and very handsome tree, especially when growing on chalky hills; and though its timber is not of the best quality, it is found extremely useful for a variety of purposes, and is also one of the best kinds of wood for fuel. The nuts or mast are, like acorns, much sought after by swine; and in some parts, where the tree abounds, the animals are driven into the beech-woods in autumn. A useful oil is also expressed from the nuts. For a full account of the uses of the Beech, the reader is referred to Loudon's *Arboretum Britannicum*.

There are some very ornamental varieties of the common Beech to be met with in cultivation; as, for example, the Purple Beech, with purple leaves; Copper Beech, with copper-coloured leaves; and Fern-leaved Beech, with the leaves variously cut into narrow segments resembling the fronds of a fern. [T. M.]

FAIR MAID OF FRANCE. *Ranunculus aconitifolius fl. pleno*.

FAIRY RINGS. Green circles or parts of circles in pastures produced by various species of agarics and other *Fungi*. They appear to be generated in the following manner:—A patch of spawn, according to the fashion of many *Fungi*, spreads centrifugally in every direction, and produces a crop at its extreme edge. The soil in the inner part of the disc is exhausted, and the spawn there dies or becomes effete. The crop of fungi meanwhile perishes and supplies a rich manure to the grass, which is in consequence of a vivid green; the parts within the ring, in consequence of former exhaustion, looking dry and parched, and those beyond less luxuriant from comparative want of manure. Thus, year after year, the ring increases in diameter till it attains dimensions of many yards across. If any accident happens to the spawn in the first instance, a part only of the circle may be developed. Rings of fungi often occur in woods, but as they grow amongst decayed leaves, the circles are seldom observed by any except professed mycologists. *Marasmius oreades*, *Agaricus gambosus*, and *A. arvensis* are amongst the most prominent inhabitants of Fairy Rings. [M. J. B.]

FALCATE, FALCIFORM. Plane and curved in any degree, with parallel edges, like the blade of a reaper's sickle; as the pod of *Medicago falcata*.

FALCONERIA. The name of a few Indian trees of the spurgewort family, very nearly related to *Sapium*, and chiefly differing in having the male and female flowers on different instead of on the same tree. The species are trees of considerable dimensions, sometimes attaining a height of sixty feet, the stems abounding in a milky juice, the branches furnished with stalked smooth leaves, and the inconspicuous green flowers arranged in axillary tufted, erect or drooping spikes. The fruits are about the size of a pea. The genus bears the name of Dr. H. Falconer, an English botanist and zoologist, distinguished for his discoveries in fossil zoology. It is referred to the *Stilaginaceae* by Lindley, but that family is now pretty generally acknowledged to be a mere group of *Euphorbiaceae*. [A. A. B.]

FALKIA. A genus of *Convolvulaceae*, containing two species, one of which is scattered pretty generally over the world, and the other confined to Mexico. They are small creeping pubescent herbs, without milky juice, and have reniform petiolate entire leaves, and ebracteate oneflowered axillary peduncles. The calyx is five-parted, and the corolla campanulate and five-cleft. [W. C.]

FALLING STARS. The popular name in many districts of the common *Nostoc*, which often surprises by its sudden appearance on gravel walks, after a shower, where it was unnoticed just before. Dryden alludes to this substance when singing of fairies in the following lines, more fanciful than truly poetical:—

And lest our leap from the sky prove too far,
We slide on the back of a new falling star,
And drop from above
In a jelly of love. [M. J. B.]

FALL POISON. *Amianthium muscaetoxicum*.

FALSE BARK. That layer on the outside of the stem of an Endogen, which consists of cellular tissue into which fibrous tissue passes obliquely.

FALSE-NERVED. When veins have no vascular tissue, but are formed of simple elongated cellular tissue; as in mosses, seaweeds, &c.

FAN-SHAPED. Plaited like a fan; as the leaf of *Borassus flabelliformis*.

FAN-VEINED. When the veins or ribs are disposed like those of a fan.

FARIAM. In rows; thus *bifariam*, in two rows; *trifariam*, in three rows, &c.

FARINACEOUS. Having the texture of flour, as the albumen of wheat.

FARINOSE. Covered with a white mealy substance; as the leaves of *Primula farinosa*.

FAROUCHE. (Fr.) *Trifolium incarnatum*.

FARRO. Polish wheat, *Triticum polonicum*.

FARSETIA. A genus of *Cruciferae*, allied to *Alyssum*, differing by the oblong pouch containing numerous seeds which have the funicle free from the body of the seed. They are natives of the Mediterranean region and temperate Asia. *Berteroa* is scarcely different, the chief distinction being that the partition of the pouch is destitute of the nerve which occurs in *Farsetia*; and *Aubrietia* has as little claim to be separated on account of its seeds not being margined. Small plants, often shrubby at the base, with white, yellow, or purple flowers. [J. T. S.]

FASCIA (adj. FASCIATE). A cross band of colour.

FASCIATED. When a stem becomes much flattened instead of retaining its usual cylindrical figure, as in the cockscomb, &c.

FASCIARIUS. Narrow; very long, with the two opposite margins parallel, as the leaves of the seawrack.

FASCICLE, FASCICLED, FASCICULATED. When several similar things proceed from a common point, as the leaves of the larch, or the tubers of a dahlia.

FASCICULATO-RAMOSE. When branches or roots are drawn closely together so as to be almost parallel.

FASTIGIATE. Tapering to a narrow point, pyramidal; as the branches of the Lombardy poplar.

FAT PORK. *Clusia flava*.

FAU. (Fr.) *Fagus sylvatica*.

FAUREA. A genus of *Proteaceae*, containing a single species, *F. saligna*, distinguished by having a club-shaped tubular silky calyx with a four-cleft limb; four stamens, with short filaments, attached to the segments of the calyx; and an ovary covered with silky hairs, and crowned with a filiform style and oblong stigma. The fruit is a bearded nut, tipped with the permanent style. It is a small tree of South Africa, with alternate lanceolate acute subfalcate shining leaves, and bearing its flowers in solitary terminal densely crowded spikes. [B. H.]

FAUSSE-AIRELLE. (Fr.) *Gaylussacia.*—CAMPANULE. *Michauxia campanuloides.*—GESSE. *Vicia lathyroides.*—IRIS. *Morœa iridioides.*—JOUBARBE *Gregoria Vitaliana.*—LYCHNIDE. *Nyctcrinia Lychnidea.*—PAQUERETTE. *Bellium bellidioides.*—RENONCULE. *Anemone ranunculoides.*—VIPERINE. *Onosma echioides.*

FAUX. The orifice of a calyx or corolla.

FAUX-ACACIA. (Fr.) *Robinia.*—ARMERIA. *Armeria Pseudo-Armeria.*—BAGUENAUDIER. *Coronilla Emerus.*—COTONNIER. *Gomphocarpus fruticosus.*—DRAGONNIER. *Yucca Draconis.*—EBENIER. *Cytisus Laburnum.*—FRAISIER. *Potentilla Fragariastrum.*—HELIOTROPE. *Tournefortia.*—INDIGO. *Amorpha fruticosa.*—JALAP. *Mirabilis Jalapa.*—LISERON. *Polygonum Convolvulus.*—MUSCARI. *Muscari monstrosum.*—NARCISSE. *Narcissus Pseudo-Narcissus.*—NEFLIER. *Pyrus Chamæmespilus.*—NERPRUN. *Hippophae rhamnoides.*—PARTHENIUM. *Anthemis parthenioides.*—PERSIL. *Æthusa Cynapium.*—PIMENT. *Solanum Pseudocapsicum.*—PISTACHIER. *Staphylea pinnata.*—PLATANE. *Acer Pseudo-Platanus.*—SAFRAN. *Carthamus tinctorius.*—SAPIN. *Abies excelsa.*—SENE. *Colutea arborescens.*—TEUCRIUM. *Verbena teucrioides.*—THUIA. *Cupressus thyoides.*—TREMBLE. *Populus tremuloides.*—TURBITH. *Thapsia villosa.*

FAVA DE S. IGNACIO. *Anisosperma Passiflora.*

FAVEOLATE. Honeycombed. The same as Favose.

FAVILLÆ. A term applied by algologists to those capsules in *Algæ* in which the nucleus, consisting of many spores, is formed within a single mother-cell, as in *Ceramium.* When several contiguous cells are fertile, the group is called a *favillidium.* Sometimes a coccidium, when enclosing a multitude of nuclei, or favillæ, is called a favillidium. [M. J. B.]

FAVIOLLE À BOUQUETS. (Fr.) *Phaseolus multiflorus.*

FAVOSE. Excavated in the manner of a section of honeycomb, as the receptacle of many composites.

FAVOSO-AREOLATE. Divided into spaces resembling the cavities of honeycomb.

FAVOSO-DEHISCENT. Appearing honeycombed after dehiscence, as the anther of *Viscum.*

FAYARD. (Fr.) *Fagus sylvatica.*

FEA-BERRY. The Gooseberry, *Ribes Grossularia.*

FEATHERFOIL. An American name for *Hottonia.*

FEATHER-VEINED. Having veins which proceed from a midrib at an acute angle.

FEATHERY. Consisting of long hairs which are themselves hairy, as the pappus of *Leontodon Taraxacum.*

FEDIA. A genus of small succulent annuals belonging to the *Valerianaceæ*, distinguished from *Valeriana* by having the fruit crowned with unequal teeth instead of a feathery pappus. There are several British species all of similar habit, growing from six to eight inches high, with slender repeatedly-forked stems, oblong spathulate leaves, and very minute whitish flowers, some few of which are solitary in the upper forks of the stem, the rest crowded into terminal leafy heads. *F. olitoria,* Corn Salad, or Lamb's Lettuce, is the most frequent, and is a common weed in cornfields and other cultivated ground. Under the names of *Mache, Boursette, Doucette,* and *Blanchette,* this species is still commonly cultivated on the continent; as well as another species with large leaves called *Mache d'Italie* or *Régence.* The genus *Fedia* is included by some botanists under *Valerianella.* [C. A. J.]

FÉEA. A small genus of hymenophylloid ferns, separated from *Trichomanes* by their dimorphous fronds, and from *Hymenostachys,* which has dimorphous fronds, by their free veins. They are dwarf tropical subpellucid plants, with the sterile fronds pinnatifid or pinnate, and the fertile ones reduced to a mere spike with marginal cysts containing the spore-cases. [T. M.]

FEELER-WORT. *Catasetum.*

FELICIA. A genus of *Compositæ*, separated from *Aster* chiefly by its short uniserial withering pappus, the hairs of which are filiform, flexuose, and serrulate. They are herbs or suffruticose plants of the Cape of Good Hope, with branching stems, narrow alternate leaves, and flower-heads with usually white or blue rays. One of the species, *F. tenella,* is sometimes cultivated among annuals under the name of *Aster tenellus.* [T. M.]

FELLEUS. Bitter as gall.

FELWORT. *Swertia;* also an old name for *Gentiana lutea.*

FELOUGNE. (Fr.) *Chelidonium majus.*

FENDLERA. The shrub so named in honour of a well-known botanical collector, belongs to the order *Philadelpheæ.* The tube of the calyx is marked with eight ridges, the limb four-parted; petals four, deltoid, stalked, irregularly notched; stamens eight, the filament prolonged beyond each side of the anther, into a linear lobe, and the anthers provided with a small spiny point; styles more or less consolidated, permanent on the four-celled capsular fruit. [M. T. M.]

FENESTRA (adj. FENESTRATE). An opening through a membrane, like a window in a wall.

FENNEL. *Fœniculum vulgare.* —, AZOREAN. *Fœniculum dulce.* —, GIANT. *Ferula.* —, HOG'S, or SOW. *Peucedanum officinale.* —, SWEET. The Finocchio, *Fœniculum dulce.*

FENNEL-FLOWER. *Nigella.*

FENOUIL. (Fr.) *Fœniculum.* —, BÂTARD. *Anethum graveolens.* — DE MER. *Crithmum maritimum.* — DE PORC. *Peucedanum officinale.*

FENUGREEK. *Trigonella Fœnum græcum.*

FÉNU-GREC. (Fr.) *Trigonella Fœnum græcum.*

FENZLIA. A genus of tropical New Holland shrubs, belonging to the *Melastomaceæ.* They are covered by bran-like scales, and have thick entire leaves, and rose-coloured flowers on short axillary stalks. The calyx has two bracts at the base, a globose tube, and a limb of five acute, spreading segments; petals five; stamens numerous, shorter than the petals, with globular anthers, whose two cells are separated by a thickened connective, and which open by long clefts. The fruit is a berry, crowned by the calyx-limb, one-celled, one-seeded by abortion. [M. T. M.]

The name *Fenzlia dianthiflora* is applied in gardens to a beautiful dwarf Californian annual belonging to the *Polemoniaceæ,* which in cultivation forms a closely ramified spreading tuft, bearing a profusion of its delicate rosy-tinted flowers with a yellow throat, surrounded by five dark-coloured dots. These flowers have a tubulose-campanulate deeply five-cleft calyx, and a funnel-shaped corolla, with broad spreading obovate dentate limb segments. This plant is more correctly called *Gilia dianthoides.* [T. M.]

FER, FERUS. A Latin termination signifying the carrying of something, as *florifer,* the carrier of flowers.

FERRUGINOUS. Light brown, with a little mixture of red.

FERDINANDA. A genus of yellow-flowered Mexican bushes of the Composite family, nearly allied to *Heliopsis,* from which they are easily recognised by having numerous small flower-heads arranged in corymbs at the ends of the branches, instead of single and large flower-heads. The alternate or opposite leaves are rough; and the flower-heads have an involucre of one to three series of narrow scales, the outer row of florets being strap-shaped and female, the inner tubular and perfect. The achenes are four-sided, each embraced by a chaffy scale, and seated on a conical receptacle; the pappus is entirely absent, or present in the outer florets in the form of two to five scales. [A. A. B.]

FERDINANDEZIA. A genus comprising ten species of epiphytal orchids from tropical America. They differ more in habit than character from *Oncidium*; and have slender stems thickly covered with overlapping triangular leaves, the edges, instead of the flattened portion, pointing upwards; the flowers are small, yellow, and disposed in axillary racemes or panicles from the axils of the upper leaves; the two pollen-masses are pear-shaped, without a caudicle, and attached to a small ovate gland. The species have much in common. *Lockhartia* is another name for the genus. [A. A. B.]

FERDINANDUSA, or FERDINANDIA. These names both refer to the same genus of *Cinchonaceæ,* which consists of Brazilian trees, with leathery leaves; and the flowers in panicles, with a funnel-shaped corolla, whose limb is divided into four ovate notched revolute segments, and four stamens slightly protruding from the corolla. The fruit is a two-celled capsule bursting by two valves, which separate also from the calyx, which is cleft lengthwise; the seeds are winged. [M. T. M.]

FER-À-CHEVAL. (Fr.) *Hippocrepis unisiliquosa.*

FERN, BEECH. *Polypodium Phegopteris.* —, BLADDER. *Cystopteris.* —, BRISTLE. *Trichomanes.* —, BUCKLER. *Lastrea.* —, CINNAMON. *Osmunda cinnamomea.* —, CLIMBING. *Lygodium.* —, FEMALE. *Athyrium Filix-fœmina*; also *Lastrea Thelypteris,* and *Pteris aquilina.* —, FILM. *Hymenophyllum.* —, FLOWERING. *Osmunda,* and also *Anemia.* —, HARD. *Blechnum Spicant.* —, HARE'S-FOOT. *Davallia canariensis.* —, HOLLY. *Polystichum Lonchitis.* —, LADY. *Athyrium Filix-fœmina.* —, LIP. *Cheilanthes.* —, MAIDEN-HAIR. *Adiantum Capillus Veneris.* —, MALE. *Lastrea Filix-mas.* —, MARSH. *Lastrea Thelypteris.* —, MOUNTAIN. *Lastrea montana.* —, OAK. *Polypodium Dryopteris.* —, OSTRICH. *Struthiopteris.* —, PARSLEY. *Allosorus crispus*; also sometimes applied to *Athyrium Filix-fœmina crispum.* —, POD. *Ceratopteris thalictroides.* —, RATTLESNAKE. *Botrychium virginicum.* —, ROYAL. *Osmunda regalis.* —, ROYAL of Calabar. *Litobrochia Currori.* —, SCALE. *Ceterach.* —, SENSITIVE. *Onoclea sensibilis.* —, SHIELD. *Aspidium.* —, STONE. *Allosorus crispus.* —, SUN. *Polypodium Phegopteris.* —, SWEET. *Lastrea fragrans* and *montana.* —, SWORD. *Xiphopteris.* —, TARA. *Pteris esculenta.* —, WALKING. *Camptosorus rhizophyllus.* —, WALL. *Polypodium vulgare.* —, WATER. *Osmunda.* —, WOOD. A name applied to the American *Lastreas.*

FERN-BUSH, SWEET. An American name for *Comptonia asplenifolia.*

FERN-ROOT, TASMANIAN. The caudex of *Pteris esculenta.*

FERNS. The highest of the sub-groups of Acrogens, technically called FILICES: which see.

FERONIA. The Wood-apple or Elephant-apple tree of India, *F. elephantum*, is the only species belonging to this genus of *Aurantiaceæ*. It is common throughout India, Ceylon, and Burmah, and forms a large tree, yielding a hard, heavy wood, of great strength but not durable. When wounded, there flows from it a transparent gum, which is mixed with other gums and sent to this country under the name of East Indian Gum Arabic. The tree has pinnate leaves composed of shining stalkless leaflets, and the flowers are arranged in racemes, containing a mixture of male, female, and perfect blossoms; these have a flat five-toothed calyx, five (occasionally four or six) white spreading petals, ten stamens, and a five-celled ovary. The fruit, which is about the size of an apple, has a very hard, rough, woody rind, and contains a pulpy flesh with numerous seeds imbedded in it. This pulp is eatable, and, like that of the Bengal quince, which is the fruit of a closely allied tree, it exerts a beneficial action in cases of dysentery and diarrhœa; a jelly resembling black currant jelly is also prepared from it. The leaves have an odour like that of anise, and the native Indian doctors employ them as a stomachic and carminative. [A. S.]

FERRARIA. A genus of Cape *Iridaceæ*, with tuberous rhizomes, simple or paniculately branched stems, two-ranked ensiform thick nervose glaucous leaves, and very fleeting flowers, which consist of a six-parted perianth, with oblong undulated spreading or reflexed segments, the exterior ones being broader than the others; three stamens, with the filaments connate into a tube; and a three-celled many-seeded ovary, surmounted by a filiform style, and three dilated petaloid multifid stigmas. The flowers are highly curious, but dingy, and very fugacious. [T. M.]

FERTILE. Having the power of producing perfect seeds; or fertilised; or producing a large quantity of seeds.

FERULA. A genus of *Umbelliferæ* characterised by the presence of compound umbels, variable involucres, a five-toothed calyx, ovate pointed petals, and compressed fruits; each half of which is surrounded by a membranous border, and has three thread-like ridges, the two lateral ones losing themselves in the wing-like margin. There are three or more channels for oil in the furrows between the ridges, and four on the surface that touches the other half of the fruit.

The species are natives of the Mediterranean and Persian regions, with tall-growing pithy stems and deeply-divided leaves, the segments of which are frequently linear. *F. communis* attains sometimes in English gardens a height of fifteen feet, and is known under the name of Giant Fennel. It is a common plant in Sicily, where the pith in the interior of the stem is used for tinder. *F. persica*, a dwarf species, was formerly supposed to be the source of asafœtida, but the greater portion of this drug is the produce of *Narthex Asafœtida*. *F. orientalis* and *F. tingitana* are said to yield African Ammoniacum, a gum-resin like asafœtida, but less powerful. Sagapenum, a similar drug, is supposed likewise to be the produce of some species of this genus, but great uncertainty prevails on the subject. [M. T. M.]

FESTUCA. A very extensive genus of grasses, typical of the tribe *Festuceæ*. The species have either a panicled or racemed inflorescence, with flattened spikelets, which are two to many-flowered; glumes two, unequal, thinner than the pales, which latter are ribbed, rounded on the back, acute, with the setæ terminal or nearly so; stamens three, rarely one to two; styles two, short; stigmas feathery. The genus embraces about 200 species, which have a wide geographical range over nearly the whole surface of the globe, and are divided into four sections, namely, *Nardurus*, *Schlerocloa*, *Vulpia*, and true *Festuceæ*. There are nine of the species natives of Britain, and among them some of our most valuable meadow and pasture grasses. *F. pratensis*, the Meadow Fescue, and *F. duriuscula*, the Hard Fescue, are both excellent kinds, and highly prized for agricultural purposes. *F. ovina*, the Sheep's Fescue, is important for subalpine pastures, where it grows freely, and is much relished by sheep. It is also useful for forming lawns, where the grass is required to be kept short and neatly dressed. Many of the foreign species are also useful for the same purposes, especially *F. heterophylla*, *Halleri*, and *ralesiaca*. Although the Fescue grasses are rather remarkable among the family, for the large quantity of saccharine matter in their composition, one species, *F. quadridentata*, is said to be poisonous in Quito, where it is called Pigouil. See *Lindley's Vegetable Kingdom*, p. 113. [D. M.]

FETID. Having a disagreeable smell of any kind.

FÉTUQUE. (Fr.) *Festuca*. — DES BREBIS, *Festuca ovina*. — TRAÇANTE, *Festuca rubra*.

FEUILLÆA. A genus of tropical American *Cucurbitaceæ*, belonging to the small section of that order which is distinguished by the anthers not being sinuous. The species are perennial herbaceous plants, with rather woody stems, climbing up trees to a great height and supporting themselves by means of tendrils, which are said to proceed from the axils of the leaves, instead of from the sides as in the common gourd. They have large, roundish, smooth leaves, frequently lobed, and the male and female flowers are borne on distinct plants, both having a five-lobed calyx, and a wheel-shaped corolla with five divisions. The fruit is globular and has a woody shell, marked with a scar which forms a zone round it, and shows the division between the enlarged calyx and the shell of the fruit; it contains a number of large flat seeds embedded in solid flesh, and does not split open when ripe.

F. cordifolia is the Sequa or Cacoon Antidote of Jamaica, where it is a common plant in shady woods, climbing to a great height up the trunks of trees. The fruits are four or five inches in diameter, and contain from twelve to fifteen large flat seeds, which possess purgative and emetic properties and have an intensely bitter taste. In Jamaica the negroes employ them as a remedy in a variety of diseases, and consider them to be an antidote against the effects of poison; they also obtain a large quantity of semi-solid fatty oil, which is liberated by pressing and boiling them in water.

The seeds of an allied species called Abilla in Peru, contain so much oil that the Peruvians use them for making candles. These are made by cutting cubical pieces of the seed and stringing them upon a thin piece of stick, the point of which is lighted. The candles thus rudely constructed, burn well, with a tolerably clear light, and, not being readily extinguished by wind, are commonly used in the open-air processions of the Roman Catholic Church. Another curious use is made of these Abilla seeds: the shell is lined with a soft felt-like substance, which when dry forms an excellent tinder, and the Indian, by rapidly twirling a pointed stick upon it, soon obtains a light; thus the same seed furnishes him with his candle, and with tinder for lighting it. [A. S.]

FEVE, or F. DE MARAIS. (Fr.) *Faba vulgaris*.

FEVER-BUSH. An American name for *Benzoin*.

FEVERFEW. *Pyrethrum Parthenium*. —, BASTARD. *Parthenium Hysterophorus*.

FÉVEROLLE. (Fr.) *Faba vulgaris*.

FEVERWORT. *Triosteum*.

FÉVIER. (Fr.) *Gleditschia*. — D'AMÉRIQUE. *Gleditschia triacanthos*.

FIBRE, ELEMENTARY. That thread which is turned round the interior of the tubes that are called spiral vessels, or of any similar kind of tissue.

FIBRILLÆ (adj. FIBRILLOSE). The roots of lichens; any kind of small thread-shaped root; also applied occasionally among fungals to the stipe.

FIBROUS. Containing a great proportion of woody fibre; as the rind of a cocoa-nut.

FIBRO-VASCULAR. Consisting of woody tissue and spiral or other vessels.

FICAIRE. (Fr.) *Ficaria ranunculoides*.

FICARIA. This genus is distinguished from *Ranunculus* by its having three deciduous instead of five persistent sepals, and nine petals instead of five; in all other respects it is a true crowfoot; indeed, our native species, *F. ranunculoides*, is not unfrequently described under the name of *Ranunculus Ficaria*. Though called Small Celandine and Lesser Celandine, it is totally distinct from the true celandine (*Chelidonium*). Being one of the earliest of British flowering plants, and its petals being of a beautiful golden-yellow, and its leaves a glossy green, it is a general favourite. Its roots consist of a number of small fleshy tubers, which store up nourishment like bulbs during the whole of the summer and autumn. Gerarde's description of its duration is worth quoting for its accuracy and quaintness: ' It commeth forth about the calends of March, and floureth a little after; it beginneth to fade away in Aprill, it is quite gone in May, afterwards it is hard to be found, yea, scarcely the root.' This might be taken for an allegorical epitome of the life of man. The young leaves of *Ficaria*, according to Linnæus, are sometimes used as greens in Sweden. A variety with double flowers is occasionally cultivated. French, *Petite Chélidoine*; German, *Feigenranunkel*. [C. A. J.]

The trivial name of Pilewort has been bestowed upon this plant from the structure of its tubercles, which grow in bundles of small tubers, so like the shape of those excrescences which occur in the more distressing cases of piles (hæmorrhoids), that our forefathers, who chose their medicines, not from a knowledge of the properties and qualities of the plants, but from a kind of fancy as to Nature's external impress indicating innate virtues, adopted it as a remedy for this malady. Culpepper is most enthusiastic in describing its virtues: ' Here is another secret for my countrymen and women—a couple of them together. Pilewort made into an oil, ointment, or plaster, readily cures both the piles, or hæmorrhoids, and the king's-evil. The very herb borne about one's body next the skin helps in such diseases, though it never touched the place grieved. Let poor people make much for their uses. With this I cured my own daughter of the king's-evil.' Confident as are these assertions, yet the use of the plant is all but discontinued in the present day, medical practitioners very properly looking for sounder principles than those derived from the doctrine of similitudes.

In *Green's Universal Herbal* we find the following observations:—'The particular form of the roots probably recommended this plant as a cure for the piles; and this fancied quality was the origin of the English name. The roots are sometimes washed bare by the rains; and this induced the ignorant and superstitious to imagine that it rained wheat, to which the uncovered tubercles bear a little resemblance.' That this plant, from these and other reasons, was long considered as a 'herb of grace,' there can be no doubt; however, it is at present looked upon principally as a weed which can best be got rid of, when troublesome, by opening drains and thinning out trees or thickets. [J. B.]

FICINIA. A genus of cyperaceous plants belonging to the tribe *Scirpeæ*. The inflorescence is either in solitary spikes or in conglomerated heads of spikes. Scales imbricated, some of the lower empty; styles three-cleft, rarely two-cleft; ovary with a fleshy disc; achenes sharply pointed or muticous. There are upwards of forty species, nearly all of which are natives of South Africa. [D. M.]

FICOIDALES. One of Lindley's alliances of perigynal Exogens, represented by *Mesembryanthemum*.

FICOIDE. (Fr.) *Mesembryanthemum*.

FICOIDE.Æ, or Fig-Marigold family. A natural order of calycifloral dicotyledons, the type of Lindley's ficoidal alliance. The order is better known as *Mesembryaceæ* or *Mesembryanthemaceæ*. [J. H. B.]

FICUS. A genus of *Moraceæ*, including the cultivated Fig. The flowers are usually incomplete, collected on axillary receptacles, which are either stalked or sessile, pear-shaped or globular, with three bracts at the base. There is a four to six-leaved perizone; in the staminate flowers one to six stamens; and in the pistillate a one-celled ovary. The fruit consists of globose or angular achenes, with a dry thin rarely pulpy pericarp. They are erect or creeping trees or shrubs, found in Southern Europe and Africa, and in large numbers in the warm parts of India, and in the islands of the Indian Sea and of the Southern Ocean. They have alternate rarely opposite entire or lobed leaves. There are nearly 160 known species. Of the cultivated Fig there are a vast number of varieties. The part eaten is the hollow receptacle which contains the flowers. The achenes, or, as they are commonly called, seeds, are ultimately immersed in the pulpy mass of the receptacle. Turkey figs are imported from Smyrna in small boxes called drums. From the old genus *Ficus*, Miquel has separated the genera *Urostigma*, *Pharmacosycea*, *Pogonotrophe*, *Sycomorus*, *Covellia*, and *Synœcia*. See Plate 6, figs. a, f; and Plate 10, fig. b. [J. H. B.]

The Fig of our gardens is the *F. Carica* of botanists. The name *Ficus* applied to this very anciently known fruit, is most probably derived from Peg, its Hebrew name; that of Carica is from Caria in Asia Minor, where fine varieties of it have long existed. According to various authors, it is a native of Western Asia, Northern Africa, and the south of Europe, including Greece and Italy. It is certainly indigenous to Asia Minor; but it may have been thence introduced and naturalised in the islands of the Mediterranean, and the countries near its shores, both in Europe and Africa.

The Fig is a deciduous tree, fifteen to twenty or even thirty feet high in favourable climates. The alternate leaves are cordate, more or less deeply three to five lobed, and rough. The fruit is generally shortly turbinate, but some varieties are of an elongated pyriform shape; the skin soft, with shallow longitudinal furrows; the colour yellowish-white, greenish-brown, purplish-brown, violet, or dark purple. It consists of a hollow fleshy receptacle with an orifice in the top, which is surrounded and nearly closed by a number of imbricated scales—as many as 200, according to Duhamel. The flowers, unlike those of most fruit-trees, make no outward appearance, but are concealed within the fig on its internal surface: they are male and female, the former situated near the orifice, the latter in that part of the concavity next the stalk. On cutting open a fig, when it has attained little more than one-third its size, the flowers will be seen in full development, and, provided the stamens are perfect, fertilisation takes place at that stage of growth. But it often happens that the stamens are imperfect, and no seeds are formed; nevertheless the fruit swells and ripens.

Under favourable circumstances, a fruit or two is formed along the shoots at the base of almost every leaf. Of these the quantity that sometimes attains maturity is enormous; but frequently, from vicissitudes of cold in some climates and heat in others, much of the fruit drops prematurely. It may not do so at the time when dryness prevails, but at some future period when moisture is sufficiently abundant: in fact, the injury caused by drought to this fruit becomes most apparent after moisture has started the tree into vigorous growth, and hence the true but remote cause of failure in the crop is apt to be overlooked. And if this be sometimes the case now, it was much more likely to be generally so in former times, when there was amongst cultivators but little intelligence as regards tracing effects to their causes. Accordingly, to prevent the fruit of the Fig tree from dropping prematurely, and to hasten its ripening, the process of caprification was resorted to. This consisted in placing the fruit of a wild sort, called the Caprifig, amongst the cultivated ones. An insect of the gnat family infests the former, which it leaves to attack the latter, entering to the interior of the fruit by the orifice. It is a very ancient practice, for it is mentioned by the earliest Greek writers on natural history, and is even minutely described by Theophrastus. It appears to have originated in Greece. Pliny remarks that it was only used in the islands of the Archipelago; that, in his time, it was entirely unknown to the Italians; and that there was no tradition of its ever having been introduced to Syria or Palestine. Its utility was doubted by some authors, and among others by the celebrated Duhamel. He thought it questionable whether by caprification the maturity of the fruit was hastened, except in the same way as apples and pears are when attacked by the grub. Professor Gasparrini, in an essay written for the Royal Academy of Sciences of Naples, details a number of experiments which he had made, and repeated in different years. Their results led to the conclusion that caprification is useless for the setting and ripening of the fruit, and instead of making the figs remain on the tree, it either causes or facilitates their fall, especially when the insect has penetrated into the inside, and produced decay by its own death. According to Gasparrini, the practice of caprification ought to be abolished, as it entails expense, and deteriorates the flavour of the figs. The French naturalist, Olivier, says it is being abandoned in some islands of the Archipelago where it was formerly practised, but in which excellent figs are still produced. We have thought it necessary to briefly notice the operation, as so much has been written with regard to its presumed advantageous effects; but from what has been stated, it will be seen that, according to the investigations of modern science, it is proved to be not only unnecessary, but positively injurious.

Figs have been used in the east as an article of food from time immemorial. They were amongst the fruits brought back from Canaan by the Israelites sent by Moses to report on the productions of that land. We read of a present having been made to David of 200 cakes of figs. They were probably used chiefly in the dried state. The drying is easily effected in a warm climate by exposure to the sun's rays, in the same way as those grapes are dried, which are called from that circumstance raisins of the sun. Like the grape, the substance of the fig abounds in what is termed grape sugar. In drying, some of this exudes and forms that soft white powder which we see on the imported dried figs. They are thus preserved in their own sugar, and rendered fit for storing up as an article of food.

Figs were considered of such necessity by the Athenians that their exportation from Attica was prohibited. Those who informed against persons violating this law were called 'Sycophantai,' from two Greek words signifying the discoverers of figs. These informers appear to have been especially disliked, for their name gave rise to the term sycophant, used for designing liars and impostors generally, as well as flatterers.

The Figs of Athens were celebrated for their exquisite flavour; and Xerxes was induced by them to undertake the conquest of Attica. The African figs were also much admired at Rome, although Pliny says, 'it is not long since they began to grow figs in Africa.' Cato, in order to stimulate the Roman senators to declare war against Carthage, showed them a fig brought from thence. It was fresh and in good condition, and all agreed that it must have been quite recently pulled from the tree. 'Yes,' said Cato, 'it is not yet three days since this fig was gathered at Carthage; see by it how near to the walls of the city we have a mortal enemy.' This argument determined the senate to commence the third Punic war, the result of which was that Carthage, the rival of Rome, was utterly destroyed.

Only six varieties of Figs were known in Italy in the time of Cato. Others were introduced from Negropont and Scio, according to Pliny, who gives a catalogue of thirty sorts. The fig may have been introduced into Britain, along with the vine, by the Romans, or subsequently by the monks. But if it had, it seems to have disappeared till brought from Italy by Cardinal Pole, either when he returned from that country in 1525, or after his second residence abroad in 1548. In either case the identical trees which he brought, and which were planted in the garden of the Archiepiscopal Palace at Lambeth, have certainly existed for more than 300 years. This proves that the fig lives to a great age, even under less favourable circumstances than it enjoys in its native country. Another tree, brought from Aleppo by Dr. Pocock, was planted in the garden of one of the colleges at Oxford in 1648. Having been injured by fire in 1809, the old trunk decayed and was removed, but fresh shoots sprang up, some of which in 1819 were twenty-one feet high. In this country a chalk subsoil, and a climate like that near the south coast, appear to suit the fig best. There the trees grow and bear as standards. They are liable, however, to be killed to the ground in winters of excessive severity; but they spring up afresh from the roots. There was an orchard, not exceeding three-quarters of an acre, at Tarring, near Worthing, in Sussex, containing 100 standard fig-trees. About 100 dozen of ripe figs were usually gathered daily from these trees during August, September, and October. By selecting similarly favourable spots, it may be fairly concluded that this country could supply itself with abundance of fresh figs. As for dry ones, they are obtained in large quantities from Turkey, the Mediterranean, and other countries; but the supply for centuries back has chiefly been from Turkey. The import has been as much as 1,000 tons a year; and now that the duty is taken off, the quantity imported will doubtless be much greater. The wood of the Fig is soft and spongy; and as it can in consequence be easily charged with oil and emery, it is used in some countries by locksmiths and armourers for polishing. (R. T.)

FIDDLE-SHAPED. Obovate, with one or two deep recesses or indentations on each side, as the leaves of the fiddle-dock, *Rumex pulcher*.

FIDDLEWOOD. *Citharexylon*.

FIDYS, FISSUS. Divided *half-way* into two or more parts.

FIELDIA. A genus of Australian *Gesneraceæ*, having only a single species, *F. australis*. It has a five-parted calyx with blind spathaceous bracts; a tubular swollen corolla, with a five-parted slightly two-lipped limb; five stamens, four of which are fertile; and a style scarcely as long as the stamens, terminating in a bilamellar stigma. The fruit is an ovate many-seeded berry. The plant has opposite, remote, shortly-stalked broadly lanceolate leaves, and axillary, solitary pendulous flowers of a pale green colour. It is a climber, with a rooting stem, attaching itself to the trunk of tree ferns, &c. The name is sometimes applied to certain *Vandas*. [R. H.]

FIG. *Ficus*. —, ADAM'S. *Musa paradisiaca*. —, BARBARY. *Opuntia vulgaris*. —, COMMON. *Ficus Carica*. —, DEVIL'S or INFERNAL. *Argemone mexicana*. —, HOTTENTOTS. *Mesembryanthemum edule*. —, INDIAN. *Opuntia*, especially *O. vulgaris*; also a general name for the *Cactaceæ*. —, KEG, of Japan. *Diospyros Kaki*. —, PHARAOH'S. *Sycomorus antiquorum*. —, SACRED. *Ficus religiosa*.

FIG-MARIGOLD. *Mesembryanthemum*.

FIGUE BANANE. (Fr.) *Musa sapientum*. — CAQUE. *Diospyros Kaki*. — MODIQUE. *Clusia flava*.

FIGUIER. *Ficus*. — COMMUN. *Ficus Carica*. — D'ADAM. *Musa paradisiaca*. — D'INDE. *Opuntia vulgaris*.

FIG-WORT. *Scrophularia*. The term Figworts has also been applied to the scrophulariaceous order.

FILAMENT. The stalk of the anther. Any kind of thread-shaped body.

FILAGO. A genus of small herbaceous *Compositæ*, distinguished by their chaffy receptacle, the absence of a pappus, and by the female florets being mixed among the scales of the imbricated involucre. They are mostly annuals of low stature, having the stems and leaves hoary with cottony down, and inconspicuous flowers of the texture popularly known as everlasting. The commonest British species are *F. minima*, a hoary little plant three to four inches high, with erect stems, very narrow leaves, and brownish-yellow flowers; and *F. germanica*, a plant of similar habit, six to eight inches high, with an erect stem terminating in a globular head of small conical flowers, from the base of which usually spring two or three horizontal branches terminating in like manner. This curious mode of growth occasioned the term *Herba impia* to be applied by the old botanists to this plant, as if the offspring were unduly elevating themselves above the parent. None of the foreign species are worthy of especial notice. French, *Cotonnière*; German, *Filzkraut*. [C. A. J.]

FILBERT. *Corylus Avellana*. —, WEST INDIAN. *Entada scandens*.

PILPIL BURREE. An Indian name for the fruits of *Vitex trifolia*.

FILICES. One of the principal groups of cryptogams, some of the leading peculiarities of which will be found explained in the article ACROGENS. They are commonly called Ferns, and consist of arborescent or herbaceous perennial, very rarely annual plants; those of arborescent habit having a trunk varying from two or three to sixty or eighty feet in height, and formed of the consolidated bases of the fronds, surrounding a soft central mass of tissue; those of herbaceous habit either having a caudex formed on a plan similar to the arborescent kinds, but on a smaller scale, the young fronds forming the growing point, or having a more or less fleshy rhizome whose growing point is in advance of the development of the fronds, which are produced from its sides instead of its apex. Arborescent Ferns are represented in Plates 2, fig. *e*, 9, and 12; and a simple-fronded Fern in Plate 12.

All true Ferns, under which name are included nearly all the ferns that are known, may be recognised by the circinate growth of their young leaves, and by their hypophyllous fructification. The fronds are very various in regard to size and form, some being simple, others many times cut or divided; while some measure but an inch, and others many feet, in length. In the majority of instances there is no material difference of aspect between those fronds which are fertile and those which are sterile; but in others, including whole groups, the *Acrosticheæ* for example, there is a manifest contraction of the fertile fronds, which are sometimes reduced to mere ribs and spikes clustering with masses of the spore-cases.

The spore-cases, which are collected into heaps called sori, consist of little one-celled vesicles, girt either longitudinally, vertically, or obliquely by a jointed ring, which nearly, or in some cases completely, surrounds them. This ring is elastic, and by its contraction disrupts the spore-case and scatters the contained spores—mere dust-like atoms, invisible except in a mass to the naked eye. The sorus, or heap of spore-cases, is in some groups naked, but in others covered while young by a membrane called the indusium.

The spores of Ferns are produced by cell-division within the spore-cases, and are consequently unattached, and variously shaped and sculptured. They consist of two coats containing a grumous mass. On germination the outer coat bursts, and the inner is elongated and protruded, and by cell-division becomes converted into a thin marchantiform frond or prothallus. On the under-surface of the prothallus, two kinds of bodies are borne, one of which, the antherid, produces spiral ciliated spermatozoids, while the other, which forms the archegone or female cell, is sunk in the tissue. The cell at the base of the archegones, after impregnation, gives rise to a new plant, which is gradually developed, and is of different duration in different species, producing successive crops of fronds and spore-cases.

Many schemes have been proposed for the classification of Ferns, but that seems to be preferable, which is based on the modifications of the vascular system taken in conjunction with the fructification. All Ferns are referrible to one of the groups *Ophioglosseaceæ*, *Marattiaceæ*, or *Polypodia-*

VEGETATION OF JAVA: TREE FERNS IN THE FOREGROUND;
A FOREST OF AMENTACEÆ IN THE DISTANCE.
(AFTER BLUME)

ceæ, of which the two first, sometimes called pseudo-Ferns, are very limited, while the latter, containing the true Ferns, includes the greater portion of all the known species.

The three groups just named are distinguished from each other by the nature and structure of their spore-cases. The presence of the annulus or ring around the spore-case, in some form, either completely surrounding it, or in a more or less rudimentary condition, is the distinctive peculiarity of the *Polypodiaceæ*; while the *Marattiaceæ* and the *Ophioglossaceæ* are separated from it by the absence of any such ring, rudimentary or otherwise, and are distinguished from each other by the obvious characters that the *Marattiaceæ* have their sori dorsiferous, that is, on the back or under surface of their fronds, as is commonly the case among true Ferns, while the *Ophioglossaceæ* have their sori marginal, the spore-bearing or fertile fronds being contracted. The *Ophioglossaceæ* are few in number, and present little difference of structure; the *Marattiaceæ*, however, form three small tribes, of which the *Marattineæ* have their sori ranged in two lines facing each other, forming distinct oblong masses; the *Kaulfussineæ* have distinct circular sori, the spore-cases of each sorus being concrete into a single annular series, and furnished with openings towards the centre; and the *Danæineæ* have their sori connate over the whole under surface, which then shows long parallel lines of small round cavities.

The *Polypodiaceæ* offer so much variety of structure that it becomes necessary to subdivide them, and for this purpose the peculiarities in the form of the spore-cases, or in their number and position, or in the structure and development of the annulus or ring, are most relied on. This gives the following groups:— *Polypodineæ*, the most extensive of all, with spore-cases almost equally convex, having a vertical and nearly complete ring, and bursting transversely at a part on the anterior side, called the stoma, where the striæ of the ring become dilated into elongate parallel cells. *Cyatheineæ*, with spore-cases sessile or nearly so, seated on an elevated receptacle, oblique-laterally compressed, the nearly complete ring being, in consequence, more or less obliquely vertical, that is, vertical below, curving laterally towards the top, bursting transversely; they approach very near the *Polypodineæ*, through some species of *Alsophila*, in which the characteristic obliquity of the ring is little apparent. *Matonineæ*, a single species only, with spore-cases sessile, bursting horizontally, not vertically, the ring broad, sub-oblique, and nearly complete, the sori dorsal and oligocarpous, covered by umbonato-hemispherical indusia, which are peltate or affixed by a central stalk. *Gleicheineæ*, with the ring complete transverse, either truly or obliquely horizontal, the spore-cases globose-pyriform, forming oligocarpous sori, i.e. sori consisting of but few spore-cases (two or four to ten or twelve) situated at the back of the frond, sessile or nearly so, and bursting vertically; fronds rigid and opaque, and usually dichotomously-branched. *Trichomanineæ*, with the ring resembling that of the *Gleicheineæ*, but the spore-cases lenticular, clustered on an often exserted receptacle, which is a prolongation of the vein beyond the ordinary margin of the frond, so that the sori become extrorse marginal or projected outwards, as well as opening outwardly; fronds pellucid-membranaceous. *Schizæineæ*, with the ring horizontal or transverse, situated quite at the apex of the oval spore-case, which is, in consequence, said to be radiate-striate at the apex; the spore-case also sometimes resupinate, or turned upside down, so that the true apex is below; habit sometimes scandent. *Ceratopteridineæ*, one or two aquatic species, the spore-cases sometimes furnished with a very rudimentary ring, reduced, as in *Osmundineæ*, to a few parallel striæ, sometimes furnished with a very broad and more lengthened ring; spores bluntly triangular, marked with three series of concentric lines. *Osmundineæ*, with the spore-cases two-valved, bursting vertically at the apex, the ring very rudimentary, reduced to a few parallel vertical striæ on one side near the apex of the spore-case. In all but the last of these groups, the spore-cases are not valvate, and consequently, when they open for the liberation of the spores, they burst partially or irregularly, and do not split at the top in two equal divisions, as occurs in the *Osmundineæ*.

These primary and secondary groups will be more readily comparable in the following summary:—

Spore-cases ringless.

1. OPHIOGLOSSACEÆ—Fructifications marginal on rachiform fronds.
2. MARATTIACEÆ—Fructifications dorsal on flat leafy fronds.
 § *Marattineæ*—Sori oblong, distinct, longitudinally biserial.
 § *Kaulfussineæ*—Sori circular, distinct; spore-cases annularly concrete.
 § *Danæineæ*—Sori connate throughout.

Spore-cases having a jointed ring.

3. POLYPODIACEÆ—Spore-cases not valvate; rarely somewhat two-valved vertically.
 § *Polypodineæ*—Ring vertical, nearly complete; spore-cases usually stalked, gibbous; receptacles superficial or immersed.
 § *Cyatheineæ*—Ring obliquely vertical, nearly complete, narrow; spore-cases crowded, sessile or subsessile, oblique-laterally compressed; receptacles elevated.
 § *Matonineæ*—Ring sub-oblique, nearly complete, broad; spore-cases few, sessile, gibbous; sori oligocarpous.
 § *Gleicheineæ*—Ring zonal, horizontally or obliquely transverse, complete; spore-cases sessile or subsessile, ver-

tically compressed; sori dorsal; fronds rigid.

§ *Trichomaninew*—Ring and spore-cases as in *Gleicheniew*; sori extrorse-marginal; fronds pellucid.

§ *Schizæinew*—Ring apical, complete, horizontally transverse; spore-cases sessile or subsessile, oval, crowned by the convergent striæ of the ring, sometimes resupinate.

§ *Ceratopteridinew*—Ring rudimentary or more or less incomplete, very broad, flat, obliquely-vertical; spore-cases sessile, globose.

§ *Osmundinew*—Spore-cases vertically two-valved; ring rudimentary, transverse.

The *Polypodiew* are further divided into lesser groups characterised by the form, position, and vestiture of the sori. There is little difference of opinion amongst pteridologists as to the three principal divisions, but a good deal of diversity as to the value of the minor ones. [T. M.]

FILICOLOGY. That part of Botany which treats of Ferns.

FILICALES. That alliance of Acrogens to which the Ferns belong.

FILIFORM, FILIFORMIS. Slender, like a thread.

FILIPENDULE. (Fr.) *Spiræa Filipendula*.

FIMBRIA. A fringe. An elastic toothed membrane, situated beneath the operculum in urn-mosses.

FIMBRIATE. Having the margin bordered by long slender processes, forming a fringe.

FIMBRIATO-LACINIATE. Having the edge cut up into divisions which are fimbriated.

FIMBRILLIFEROUS. Bearing many little fringes, as the receptacle of some composites.

FIMBRISTYLIS. A genus of cyperaceous plants, belonging to the tribe *Scirpew*, having the inflorescence in spikes, solitary, in pairs, or in crowded heads, many-flowered; scales imbricated all round, the lower larger, one or two of them barren; style compressed and fringed, its base enlarged, adhering to the ovary. This genus embraces nearly 200 species, which have a wide geographical range, though most of them are natives of rather warm countries. [D. M.]

FIMETARIOUS. Growing on or amidst dung.

FINCKEA. A genus of the heathworts, having the corolla cylindrical and tubular, with a four-toothed border. The name was given by Klotzsch in honour of a botanist named Finck. The species are Cape shrubs, the leaves three or four together and hairy; corolla hairy and about as long as the calyx. [G. D.]

FINGERED. The same as Digitate.

FINGERHUTHIA. A genus of grasses belonging to the tribe *Phalaridew*. Glumes two, equal, with bristly points, keeled and membranaceous; lower flowers fertile, as long as the glumes; pales rigid, the lower rather the longest, keeled, with a short bristle at the point, five to seven nerved, the upper shorter and slightly compressed; stamens three, with bearded anthers; styles two; upper flowers imperfect. Only two species are described, both of which are natives of South Africa. [D. M.]

FIN HOUSSY. (Fr.) *Trifolium repens*.

FINLAYSONIA. A genus of *Asclepiadaceæ*, containing a single species, native of India. It is a twining glabrous plant, yielding a milky juice, and having opposite obovate leaves, and numerous small flowers arranged in interpetiolar corymbs. The calyx is small, five-cleft; and the corolla rotate. The staminal crown, which rises from the throat consists of five delicate white slightly converging threads, each bent back at the apex so as to form a small hook; the stamens are distinct, with short filaments, and large anthers adhering to the stigma; and the divaricate follicles contain numerous large flat obovate seeds with a few silky fibres. [W. C.]

FINOCCHIO. *Fœniculum dulce*. —, ASSES'. *Fœniculum piperitum*.

FIORIN. *Agrostis stolonifera*, and *Agrostis alba*.

FIORIN DES ANGLAIS. (Fr.) *Agrostis stolonifera*.

FIR. A general name for the trees referred to the coniferous genera *Pinus*, *Abies*, *Larix*, &c. —, BALM OF GILEAD, *Abies balsamea*. —, HEMLOCK SPRUCE, *Abies canadensis*. —, PARASOL. *Sciadopitys verticillata*. —, PLUM. *Prumnopitys elegans*. —, SCOTCH. *Pinus sylvestris*. —, SILVER. *Abies pectinata*; also a general name for the species sometimes referred to *Picea*. —, SPRUCE. *Abies excelsa*; also a general name for the species of true *Abies*.

FIREWEED. An American name for *Erechtites hieracifolia*.

FIR-RAPES. Lindley's name for the *Monotropaceæ*.

FIRS, JOINT. A name proposed by Lindley for the order *Gnetaceæ*.

FISCHERIA. A genus of *Asclepiadaceæ*, containing about ten species from the West Indies and Central America. They are twining hairy shrubs, with opposite cordate leaves, and many flowers in racemes on long interpetiolar peduncles, which thicken upwards and are scarred by the deciduous pedicels. The calyx is five-parted; the corolla is rotate and five-cleft, the divisions having a curled indentation at their apices; the staminal crown is simple or double; the pentagonal stigma covers the pollen-masses; and the follicles are ovoid and fleshy. [W. C.]

FISH-POISON. *Lepidium Piscidium.* —, JAMAICA. *Piscidia Erythrina.*

FISSENIA. A genus of *Loasaceæ*, found in Arabia and the interior of South Africa, remarkable as being the only representative of the family in the eastern hemisphere. It differs from other genera in having a three-celled fruit, with one seed in each cell. The only species, *F. spathulata*, is a branching bush with straw-

Fissenia spathulata.

coloured stems, alternate stalked lobed leaves not unlike those of the gooseberry but larger, and pale green flowers four to six together at the ends of the twigs; the flowers have ten petals, five large and rounded, and five small and narrow, very numerous stamens, and three styles. The little ten-ribbed fruits or nuts crowned with the five long narrow calyx lobes, look like miniature shuttlecocks. [A. A. B.]

FISSICALYX. A tree from Venezuela, with pinnate leaves and terminal panicles of yellow flowers, forming a genus of *Leguminosæ* of the tribe *Dalbergieæ*, distinguished from all others by the irregularly split calyx; by the anthers opening in terminal pores; and by the fruit being surrounded by a broad membranous wing proceeding from the centre instead of the edges of the valves, thus giving the fruit the appearance of that of *Guaiacum*.

FISSIDENS. A pretty genus of mosses, containing both acrocarpous and cladocarpous species, and at once distinguished by their peculiar habit arising from the flat broad-keeled two-ranked leaves with a sheathing base. The peristome is single, and the sixteen teeth of which it is composed deeply cleft. The species grow on banks, on stiff soil, or near watercourses, and vary from a line to two inches in length. They occur in both hemispheres and in various climates, the species of very distant countries being frequently identical. The shoots sometimes bear reproductive bodies at their apex, distinct from the proper fruit. These are occasionally close to the male organs. [M. J. B.]

FISSIDENTEÆ. A natural order of mosses which are remarkable for their peristome being like that of *Dicranum* or almost rudimentary, accompanied by a totally different habit due to the flat broad-keeled sheathing leaves. *Fissidens* has already been noticed. *Drepanophyllum*, from which the order is sometimes called *Drepanophylleæ*, is a magnificent moss abounding in Cayenne, with a tawny tint, the habit of a *Jungermannia*, and a nearly naked peristome. The tips of the male plants bear, in close connection with the antheridia, tufts of jointed fusiform purplish gemmæ. *Conomitrium* is an aquatic genus, and has irregular unequally split often truncate teeth without any central line, and a mitræform veil. The species grow in running water, and one only has at present been found in Europe. [M. J. B.]

FISSIPAROUS. Propagating by a subdivision of the interior of a cell into two or more other cells, by the production of a membranous partition or septum, from the lining of the mother cell.

FISSUS. Divided *halfway*; usually into a determinate number of segments. We say *bifidus*, split in two; *trifidus*, in three, and so on; or *multifidus*, when the segments are very numerous.

FISTULAR, FISTULOUS. This is said of a cylindrical or terete body which is hollow, but closed at each end, as the leaves and stems of the onion.

FISTULINA. A genus of pore-bearing *Fungi*, sometimes wrongly associated with *Hydnum*. It is distinguished from the fleshy *Polypori* by the free tubes which are at first closed and look like little stellate pink or cream-coloured pimples. *F. hepatica*, our only species, grows on the trunks of old oaks, and acquires sometimes a very large size. When divided it looks like beetroot, and drips with red juice. It is not unwholesome, but in our opinion not a pleasant article of food, however disguised with lemon juice, cayenne pepper, or other condiments. [M. J. B.]

FITCH. The Vetch or Tare.

FITCHIA. A genus of arborescent cichoraceous *Compositæ* founded on a single species, *F. nutans*, from Elizabeth Island in the South Pacific. It is a noble plant nearly related to *Rea*, and has thick woody stems, opposite broad ovate-cordate leaves, and large terminal drooping heads of flowers, hanging by longish stalks. The involucre is broadly campanulate, composed of about three series of orbicular scales, which enclose numerous ligulate male flowers; the females are unknown. The filaments and style are very much exserted. The achenes are compressed, clothed with silky hairs, and terminated by a pair of elongated hairy setæ. It is named in compliment to Mr. W. Fitch, a clever botanical artist, by whom the drawings of the plants figured in this work have been executed. [T. M.]

FITTWEED. *Eryngium fœtidum.*

FITZROYA. A genus of coniferous or cone-bearing plants belonging to the sub-order *Cupressineæ*. It was named by Dr. Hooker in honour of Captain Fitzroy, who first discovered the tree. The fruit is in small starlike cones which consist of nine scales, three in each whorl; the lower three and upper three are barren, while the intermediate three are fertile, and bear each three winged seeds. The leaves are in threes, sometimes twos or fours, ovate-oblong, flat, without stalks. There is one species, *F. patagonica*, an evergreen tree growing to the height of 100 feet, with slender spreading branches which curve at the extremities. The tree, which is found on the mountains of Patagonia, bears the ordinary winters of Britain. [J. H. B.]

FIVE-FINGERS. *Potentilla reptans* and *canadensis*.

FLABELLATUS, FLABELLIFORMIS. Fan-shaped.

FLACCID. Wilted, or relaxed in consequence of the loss of moisture.

FLACOURTIACEÆ. (*Bixaceæ, Blinds.*) A natural order of thalamifloral dicotyledons belonging to Lindley's violal alliance of hypogynous Exogens. They are shrubs or small trees with alternate leaves having no stipules, often marked with round transparent dots. Sepals and petals four to seven, the latter sometimes wanting; stamens same number as petals or a multiple thereof; ovules attached to parietal placentas. Fruit one-celled, either fleshy and indehiscent, or a four to five valved capsule containing pulp, in which numerous albuminous seeds are enveloped. The plants are natives of the East and West Indies, and of Africa. Two or three species are found at the Cape of Good Hope, and one or two in New Zealand. Some of the plants are bitter and astringent; others yield edible fruits. Arnotto is the orange-red pulp of the fruit of *Bixa Orellana*; it is used as a dye, for staining cheese, and in the manufacture of chocolate. Some *Flacourtias* yield subacid fruit. There are thirty-five known genera, including *Flacourtia, Prockia, Bixa, Azara, Erythrospermum*; and about 100 species. [J. H. B.]

FLACOURTIA. The typical genus of *Flacourtiaceæ*, characterised by having a succulent fruit and several stigmas. It has distinct male and female apetalous flowers, usually borne on different plants; the males have a great number of stamens crowded together upon the dilated receptacle, but not surrounded by glands like those of *Bosmea*; the females have an ovary crowned with from four to nine narrow radiating stigmas. The species are mostly shrubs, but some few, however, attain a height of twenty or thirty feet, and nearly all are armed with thorns. They are found in tropical Asia, Africa, and America.

The young shoots and leaves of *F. cataphracta* are used medicinally by the native Indian doctors, who prescribe them in diarrhœa, and also as an infusion to remove hoarseness; they are astringent and stomachic. *F. Ramontchi* is a small tree, native of Madagascar and India, producing a dark violet or black fruit about the size and shape of a plum, and having a sharp but sweet taste. *F. sepiaria*, a bushy shrub, is used in some parts of India for making hedges, its spiny nature rendering it peculiarly suitable for that purpose. The fruits are sold in the markets, and, like those of *F. Ramontchi* and *sepida*, have a pleasant subacid flavour when perfectly ripe, but the unripe fruit is extremely astringent. The Indian doctors use a liniment made of the bark in cases of gout, and an infusion of it as a cure for snakebites. [A. S.]

FLAG. *Iris.* —, **CAT-TAIL.** An American name for *Typha*. —, **CORN.** *Gladiolus.* —, **SWEET.** *Acorus Calamus.* —, **YELLOW.** *Iris Pseud-acorus.*

FLAGELLARIA. A genus of *Commelynaceæ*, but referred to *Juncaceæ* by some authors. Natives of India and Australia, with lanceolate leaves, sheathing the stem at the base, and terminating in a spiral tendril; flowers paniculate, bracteolate, the perianth six-cleft coloured persistent, with the three inner segments largest; stamens six, with simple glabrous filaments; fruit a pea-like drupe containing a single seed. [J. T. S.]

FLAGELLIFORM. Long, taper, and supple, like the thong of a whip, as the runners of many plants.

FLAGELLUM. A twig, or small branch; also a runner like that of the strawberry.

FLAMBE. (Fr.) *Iris germanica.* — **PETITE.** *Iris pumila.*

FLAMBOYANTE. (Fr.) *Tulipa turcica.*

FLAME-COLOURED, FLAMMEUS. Very lively scarlet; fiery red.

FLAME TREE. *Brachychiton acerifolium.*

FLAMME. (Fr.) *Iris germanica.*

FLAVEDO. Yellowness; a disease in plants in which the green parts assume that colour.

FLAVESCENS, FLAVIDUS, FLAVUS. A pure pale yellow.

FLAVO-VIRENS. Green, much stained with yellow.

FLAVERIA. An herbaceous biennial composite, distinguished by having the common involucre imbricated with unequal scales, and the partial of two to five leaves containing as many florets, a naked receptacle, and no pappus. *F. Contrayerba* is a native of Peru, and derives its name from its being used to dye yellow. It grows to the height of eighteen inches, with lanceolate serrated sharp-pointed leaves, and terminal heads of yellow flowers. [C. A. J.]

FLAX. The common name for *Linum*; also the fibre obtained from the stems of *Linum usitatissimum*. —, FALSE. An American name for Camelina. —, NEW ZEALAND. *Phormium tenax*. —, TOAD. *Linaria*.

FLAX-BUSH. *Phormium tenax*.

FLAX-SEED. *Radiola*.

FLAX-STAR. *Lysimachia Linum-stellatum*.

FLAXWORTS. A name for the order *Linaceæ*.

FLEABANE. *Conyza*; also *Pulicaria vulgaris* and *dysenterica*; also *Erigeron viscosum*, *graveolens*, and *acre*. —, AFRICAN. *Tarchonanthus*. —, MARSH. *Pluchea*.

FLEA-SEED. The seed of *Plantago Psyllium*.

FLEAU DES PRÉS. (Fr.) *Phleum pratense*.

FLEAWORT. *Pulicaria vulgaris*; also *Plantago Psyllium*.

FLÉCHIÈRE. (Fr.) *Sagittaria sagittifolia*.

FLEMINGIA. A genus of erect, prostrate, or sometimes twining plants of the pea family, and nearly allied to *Rhynchosia*, but differing in having a turgid and two-seeded, instead of a flattened and many-seeded pod. Most of the twenty known species are found in India, a few extending to the northern and eastern portions of Australia, and one, *F. betulifolia*, occurring in W. Africa. The stems are furnished with simple or trifoliate stalked leaves, often having glandular dots; in some species the stipules are large and chaffy. The small vetch-like flowers are purple, white with pink lines, or yellow, and disposed in axillary compound racemes or panicles.

One of the most elegant of them, *F. strobilifera*, is remarkable for its drooping catkin-like racemes, furnished with large, pale yellow kidney-shaped bracts, each of which encloses a fascicle of white flowers marked with pink lines. The leaves are simple, ovate and acute, and vary much in size. The plant has been in cultivation.

A beautiful purple-flowered species, *F. vestita*, is cultivated in many parts of N. W. India for the sake of its edible tuberous roots, which are nearly elliptical, and about an inch long. The plant is prostrate, with weak stems, and hairy clover-like leaves, formed of rounded hairy leaflets. The purple blossoms are larger than in any other of the genus, and are remarkable for being placed two or three together on the apex of a slender axillary flower stalk, those of the other species being disposed in racemes. The genus bears the name of Dr. J. Fleming, an Indian botanist. [A. A. B.]

FLÉOLE. (Fr.) *Phleum pratense*.

FLESH of vegetable bodies. The soft parts.

FLEUR DE COUCOU. (Fr.) *Lychnis Flos-cuculi*; also *Primula veris*, and *Narcissus Pseudo-Narcissus*. — DE CHAPAU'D. *Stapelia variegata*. — DES DAMES. *Heliotropium peruvianum*. — DE JUPITER. *Lychnis Flos Jovis*. — DE LA PASSION. *Passiflora*. — DE LA TRINITÉ. *Viola tricolor*. — DE LIS. *Phalangium Liliago*; also *Iris germanica* and other European species. The Fleur de lis representing the Iris is the emblem of France, and was called by old English authors Flower de Luce. — DE MIEL. *Melianthus major*. — D'OR ET D'ARGENT. *Lonicera confusa*. — DE PAON. *Poinciana pulcherrima*. — DE PAQUES. *Bellis perennis*. — DE PARADIS. *Poinciana pulcherrima*. — DE QUATRE HEURES. *Mirabilis dichotoma*. — DES VEUVES. *Scabiosa atropurpurea*. — DU DIABLE. *Iris susiana*. — D'UNE HEURE. *Hibiscus vesicarius* and others. — DU GRAND-SEIGNEUR. *Amberboa moschata*. — DU SOLEIL. *Helianthus annuus*. — EN CASQUE. *Aconitum Napellus*.

FLEURYA. A genus of *Urticaceæ* composed of a number of annual or perennial weeds, found in the tropics of both hemispheres. They are much like common nettles in appearance, and some of the species are furnished with stings, but they may be readily distinguished from them by their alternate (not opposite) leaves, as well as by the narrow bifid stipules which accompany them. From other allied genera they differ in their little, oblong or rounded and compressed achenes having concave depressions on both surfaces. [A. A. B.]

FLEXUOSE. Zig-zag; having a wavy direction, gently bending alternately inwards and outwards.

FLINDERSIA. A genus of *Cedrelaceæ*, having a calyx of five short teeth; five

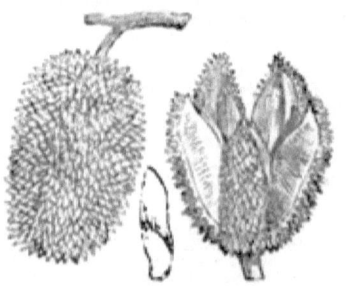

Flindersia australis.

white, ovate, plane petals, slightly hairy on the exterior; ten stamens of which only five are fertile, the alternate ones being sterile; and a simple erect obtusely five-angled style, with a peltate five-lobed stigma. The capsule is woody, oblong, obtuse, five-valved, the exterior thickly covered with sharp-pointed tubercles. They are lofty trees, having alternate pinnate leaves; found in New South Wales and the Moluccas. The natives of these islands use the rough tuberculated fruit as rasps in preparing roots &c. for food. [R. H.]

FLIX-WEED. *Sisymbrium Sophia.*

FLOCCI. Woolly threads, found mixed with sporules in fungals; also any wool-like hairs.

FLOCCOSE. Covered with close woolly hairs, which fall away in little tufts.

FLORAL. Of or belonging to the flower. — ENVELOPES. The calyx and corolla, one or both.

FLOR DE AROMA. *Acacia Farnesiana.* — DE ISABEL. *Barkeria spectabilis.* — DE JESUS. *Lelia acuminata.* — DE MAYO. *Lelia majalis.*

FLORESTINA. A genus of Mexican composite herbaceous plants, with the habit of *Stevia*, to which they were formerly referred. They are covered with small appressed white hairs, and have entire or pedately divided leaves, and flowers borne in loose corymbs surrounded by an involucre of eight short bracts. The corollas have all a short tube, and a whitish or purple limb divided into five segments. The achenes are somewhat four-cornered, surmounted by a pappus of eight or twelve membranous scales. [M. T. M.]

FLORETS. When many small flowers are collected in clusters or heads, each flower is called a floret. The *florets of the disk* are those which occupy the centre of the head of a composite; while *florets of the ray* occupy the circumference.

FLORIDEÆ. A name given to the rose-spored *Algæ*, in consequence of many of them exhibiting the rosy tints of flowers. They are now more generally known as Rhodosperms. [M. J. B.]

FLORIFER. Flower-bearing.

FLORIPAROUS. Producing flowers; also a monstrosity consisting in the production of other flowers instead of fruit.

FLORIPONDIO. *Datura sanguinea.*

FLORKEA. A North American annual herb, referred to the limnantheous division of *Tropæolaceæ*. It is a marsh plant, with slender decumbent stems, and alternate pinnately-parted leaves, bearing solitary small white flowers in their axils. It is distinguished from *Limnanthes* chiefly by its having trimerous instead of pentamerous flowers. [T. M.]

FLORUS. In composition = flowered: thus *uniflorus* is one-flowered; *biflorus*, two-flowered; *triflorus*, three-flowered; *multiflorus*, many-flowered, &c.

FLOS. A flower. — COMPOSITUS. An old name for the capitulum.

FLOS ADONIS. *Adonis autumnalis.* — AËRIS. *Aërides Arachnites.* — CARDINALIS. *Quamoclit vulgaris.* — CUCULI. *Cardamine pratensis*; also *Lychnis Flos-cuculi.* — GLOBOSUS. *Gomphrena globosa.* — JOVIS. *Lychnis Flos Jovis.* — MARTINI. *Alstromeria Flos Martini.* — PASSIONIS. *Passiflora cærulea* and others. — SOLIS. *Helianthus annuus*; also *Helianthemum vulgare.* — SUSANNÆ. *Plantanthera Susannæ.* — TRINITATIS. *Viola tricolor.*

FLOSCOPA. A genus of *Commelynaceæ* more usually known under the name of *Dithyrocarpus*, and distinguished by having nearly regular flowers, with six stamens, all fertile, with parallel anther-cells; and a two-celled ovary and capsule, with one ovule and seed in each cell. There are but very few species, natives of the tropical regions both of the new and the old world. The most common species, *F. paniculata*, is abundant in Asia, and is also found in South Africa and Brazil. It is a herb of two or three feet in height, with acuminate leaves, and a dense hirsute terminal panicle of small blue flowers.

FLOSCULI (adj. FLOSCULOSE). The same as Florets.

FLOTOVIA. A genus of spiny S. American bushes belonging to the *Compositæ*, and nearly related to *Barnadesia*, from which it differs in its regularly five-parted, not two-lipped corollas. The numerous florets of the flower-heads, and the insertion of the stamens on the middle of the corolla tube, are the characters which separate the genus from its other allies. Upwards of twenty species are known, the greater portion found in Brazil, a few in the Peruvian Andes, and one or two in Chili. For the most part they are bushes of four to six feet, with alternate leaves, accompanied by two straight slender spines. The pink flower-heads in some are solitary at the ends of the branches, but more commonly numerous and small; the silky achenes are crowned with a pappus of one series of beautiful feathery awns. *F. argentea*, a pretty Andean species, is by some referred to *Dasyphyllum*. [A. A. B.]

FLOURENSIA. A genus of resiniferous composite shrubs found in New Mexico and Chili, and nearly allied to *Helianthus*, differing chiefly in the tongue-like branches of the style, which are obtuse and not tipped with an awl-shaped appendage. The leaves are usually covered with a gummy exudation. The yellow rayed flower-heads are usually large and solitary or two to three together at the ends of the twigs. The most handsome species is *F. thurifera*, which grows to a bush of four to six feet in Chili, and has its twigs as well as its broadly lance-shaped leaves covered with a resinous substance, which is collected and burnt as incense in the churches; its fine yellow flower-heads are single at the ends of the twigs, and more than two inches across. The Chilians call the plant Maravilla, or Maravilla del Campo. Four species are known. [A. A. B.]

FLOUVE. (Fr.) *Anthoxanthum.* — DES BRESSANTS. *Anthoxanthum odoratum.*

FLOWER. That assemblage of organs in a plant, of which the stamens, or pistils, or both, form part.

FLOWER DE LUCE. An old English name for the common species of *Iris — germanica, florentina,* &c.

FLOWER OF CRETE. *Mesembryanthemum Tripolium.*

FLOWER OF JOVE. *Lychnis Flos Jovis.*

FLOWER-FENCE. *Poinciana.* —, BARBADOS. *Poinciana pulcherrima.* —, BASTARD. *Adenanthera.*

FLOWER-GENTLE. *Amaranthus.*

FLUELLIN. *Linaria Elatine* and *spuria;* also *Veronica officinalis.*

FLÜGGEA. A genus of *Euphorbiaceæ* nearly related to *Phyllanthus,* from which it may be recognised by the sterile flowers having three stamens surrounding a rudimentary ovary, there being no rudimentary ovary in those of *Phyllanthus.* The genus consists of several much branched smooth and entire-leaved bushes found in most tropical countries of the eastern hemisphere. The leaves are obovate or ovate. The flowers, sterile and fertile on different plants, are minute, green, and disposed in fascicles or cymes in the axils of the leaves. The fruits are dry capsules or berries about the size of a pea or smaller. The bark of *F. virosus,* according to Roxburgh, is strongly astringent, and possesses the property of intoxicating fish when thrown in the water, thus rendering them easily caught. The white berries of *F. leucopyrus* in India, and *F. abyssinica* in Eastern tropical Africa, are eaten by the natives. A Chinese species, *F. suffruticosa,* has long been known as *Geblera suffruticosa.* [A. A. B.]

FLUITANS. Floating upon the surface of water.

FLUVIAL, FLUVIATILE. Of or belonging to the water.

FLUTEAU. (Fr.) *Alisma Plantago.*

FLUVIALES. A natural order of monocotyledonous aquatics established by Ventinat. The plants are now included in NAIADACEÆ: which see. [J. H. B.]

FLY-AGARIC. The common name of *Agaricus muscarius,* a splendid scarlet species studded with white or yellow warts, which is common in birch woods, and is used to make a decoction for destroying flies. Its narcotic properties are so strong that in some countries it is employed, mixed with the juice of cranberries, to produce intoxication, the dry plant being more efficacious for this purpose than the fresh. The effects are somewhat different from those of other narcotics, being characterised by extreme stimulation of the muscles. The nervous system is at times excited to such a degree as to produce the most ludicrous actions. It is a curious fact that the urine of persons who have partaken of the fungus acquires the same narcotic properties; or, in other words, that the narcotic principle, like some other substances, passes through the urine without change. (See E. A. Parkes on the *Composition of Urine,* 1860.) In excess it is doubtless a dangerous poison, and we have known temporary intoxication arise from its accidental use. [M. J. B.]

FLY-POISON. *Amianthium muscætoxicum.*

FLY-TRAP. *Apocynum androsæmifolium.* —, VENUS'S. *Dionæa muscipula.*

FLY-WORT. A name applied to those species of *Catasetum* formerly called *Myanthus.*

FŒMINEUS. Female, that is, bearing pistils only.

FŒNICULUM. A genus of umbelliferous plants, with finely dissected leaves, no involucres, and yellow flowers. It is distinguished from *Anethum,* to which it is very closely allied, by the fruits being somewhat compressed from side to side, and not from back to front. *F. vulgare,* the common Fennel, is a native of temperate Europe and Western Asia; in this country it is usually found in dry chalky soil, at no great distance from the sea. The Sweet Fennel, *F. dulce,* is by some considered as only a variety of the preceding; but it differs in being a smaller plant, in the stem being compressed at the base, not round, in the smaller number of rays to the umbel, in the greater size of the fruit, in flowering earlier, &c. It is grown in this country as a potherb and for garnishing; its fruit supplies an aromatic oil, which is carminative like dill. [M. T. M.]

FŒNUM GRÆCUM. *Trigonella Fœnum græcum,* so called because formerly made into hay in Greece. It was also cultivated by the Romans, and is still occasionally employed in the agriculture of the south of Europe. The plant and seeds are strongly scented, with the new hay-like odour of coumarin.

FŒTIDIA. A genus of shrubby plants doubtfully placed in *Myrtaceæ,* the leaves being alternate and without dots, and the flowers destitute of petals. The three known species are natives of the Mauritius and Madagascar, attaining a height of thirty or forty feet, the ash-coloured branches furnished at their extremities with smooth entire, ovate or lance-shaped leaves. The flowers are axillary and solitary, and have a three or four-sided calyx tube, with a border of four triangular lobes, and very numerous stamens. The fruits are woody, four-sided, four-celled, somewhat top-shaped nuts. *F. mauritiana,* known as ' Le Bois puant' in the Mauritius, furnishes good furniture wood. According to Roxburgh, the bark is very tough, red within, bitter and astringent. It is used by woodmen for bundling up the wood, instead of cord. The common appellation of Stinkwood has no doubt suggested the name of the genus. [A. A. B.]

FOG-FRUIT. An American name for *Lippia nodiflora.*

FOLIACEOUS. Having the texture or form of a leaf, as the branches of *Xylophylla.*

FOLIA MALABATHRI. The aromatic dried leaves of *Cinnamomum nitidum.*

FOLIAR. Inserted upon, or proceeding from the leaf; thus a *cirrhus foliaris* is a tendril growing from a leaf.

FOLIATION. The act of leafing.

FOLIATE. Clothed with leaves.

FOLLICULARES. A suborder of plants belonging to the *Proteaceæ*, and referred by Decandolle to the monochlamydeous dicotyledons. They are distinguished by their coriaceous or woody follicles, which contain one or many seeds. The general characters of the order are given under PROTEACEÆ. The follicular division contains such genera as *Grevillea, Hakea, Lambertia, Rhopala, Knightia, Telopia, Lomatia, Banksia,* and *Dryandra.* [J. H. B.]

FOLIIPAROUS. Producing leaves only, as leaf-buds.

FOLIOLE (adj. FOLIOLATE). A leaflet; the secondary divisions of a compound leaf.

FOLIOSE. Covered closely with leaves.

FOLLE-AVOINE. (Fr.) *Avena fatua.*

FOLLETTE. (Fr.) *Atriplex hortensis.*

FOLLICLE. A kind of fruit, consisting of a single carpel, dehiscing by the ventral suture only, as in *Delphinium, Asclepias, Apocynum,* &c.

FONTANESIA. A Syrian shrub of the olive family, named in honour of M. Desfontaines, an eminent French botanist. It has lance-shaped leaves, ciliated at the margins; and white clustered flowers with a persistent four-cleft calyx, and four petals united into pairs at the base by the adhesion of the two stamens. The ovary is two to three-celled, with two suspended ovules, the style short, the stigma cleft into two threadlike divisions. The capsule is thin, notched, indehiscent, surmounted by the stigmas, and having a membranous margin. [M. T. M.]

FONTINALEI, FONTINALIS. A small section and genus of pleurocarpous mosses, distinguished by their aquatic habit, the nearly sessile capsule immersed in the perichætial leaves, and the cancellated inner peristome. The principal genus, *Fontinalis*, contains a few species confined to temperate regions of the northern hemisphere. The leaves are curiously compressed, so that the stems have a triquetrous outline. There is, however, no nerve as in the *Drepanophylli*, and they are frequently split down the middle, each division looking like a separate leaf. *F. antipyretica* is common about millwheels, on stones, roots, &c., in running streams; and does not fructify unless the plant is exposed or the ends of the branches come up to the surface. The dried plant is used by the Laplanders to stuff the space between the chimneys and the wall, to prevent fire, as it does not easily ignite. *Fontinalis* has a mitræform calyptra, but in the neighbouring genus *Dichelyma* the calyptra is dimidiate, while the capsules are more exserted. [M. J. B.]

FONTINALIS, FONTANUS. Growing in or near a spring of water.

FOOL'S-STONES. *Orchis mascula* and *Morio.*

FORAMEN. An aperture. The foramen of an ovule is an aperture through the integuments, allowing the passage of the pollen tubes to the nucleus.

FORAMINULE. The ostiolum of certain fungals.

FORBIDDEN FRUIT. *Citrus Paradisi,* — (of London). A variety of the shaddock, *C. decumana.* — (of Italy). The Pomme d'Adam, a variety of *C. Limetta.* — (of Paris). The sweet-skinned orange, a variety of *C. Aurantium.*

FORCIPATE. Forked, like a pair of pincers.

FORGET-ME-NOT. *Myosotis palustris.* —, ANTARCTIC. *Myosotidium nobile.*

FORFICARIA *graminifolia* is a terrestrial tuberous-rooted orchid of South Africa, with narrow, rigid, grassy leaves shorter than the stem, which is one to two feet high, and terminating in a loose raceme of eight to ten flowers, each supported by a short membranaceous bract. The upper sepal is boat-shaped, the two lateral ones larger, keeled and acute, the petals bristle-like and hairy, and the lip very short, fleshy and pubescent. There is only one species. [A. A. B.]

FORNICATE. Arched.

FORNIX. Little arched scales in the orifice of some flowers.

FORRESTIA. A genus of *Commelynaceæ* found in New Guinea. Herbs with elliptical lanceolate glabrous leaves having hairy sheaths, and red flowers in dense heads, with six stamens, the filaments of which are glabrous. [J. T. S.]

FORSKOLEA. A genus of *Urticaceæ*, found in various parts of Africa, the Mediterranean region, Arabia, and North-West India. It belongs to a small tribe characterised by the male flowers having but one stamen; and is distinguished from its near allies by its minute flowers being enclosed in a two or many-leaved involucre. The five known species are branching herbs, with tough somewhat woody stems clothed with rigid stinging hairs, furnished with lance-shaped or ovate leaves, the little flowers densely packed in their axils, males and females in the same involucre. *F. angustifolia* is said to be used in the Canary Islands to promote perspiration. The genus bears the name of M. P. Forskal, a botanist of the last century who wrote a flora of Egypt and Arabia Felix. [A. A. B.]

FORSTERA. A genus of *Stylidiaceæ* with a calyx of from three to six lobes; corolla white, campanulate, with a four to nine-lobed spreading limb; stamens and style united into a central erect column surmounting the ovary, having at its base two large erect subulate or lunate glands. The anthers, on the top of the column, are sessile, bursting transversely, the two halves hooded, the upper turned back; stigma two-lobed, feathery, and spreading in the female flowers; fruit a membranous capsule. The stems are simple or branched, two to three inches in height; the leaves more or less closely imbricated, alternate; the flowers monœcious or diœcious. Very remarkable Alpine plants, natives of Tasmania, New Zealand, and Fuegia. [R. H.]

FORSYTHIA. A genus of shrubs belonging to the *Oleaceæ*, having a four-parted calyx, a shallow bell-shaped four-cleft corolla, a two-lobed stigma, and capsular fruit. *F. viridissima* is very ornamental in March, with its numerous tufts of rather large pendulous bright greenish-yellow flowers, which grow two or three together from all parts of the rod-like branches. After these have faded, the slightly aromatic shining oblong lanceolate leaves make their appearance; they remain till late in autumn, turning yellow or purple before they fall off. The shrub then has somewhat the habit of a willow, but the stems are four-angled and studded with a number of large prominent buds. It is perfectly hardy. [C. A. J.]

FORTUNÆA. A genus of *Juglandaceæ* found in North China and Japan, and represented by a single species, *F. chinensis*, a smooth pinnate-leaved bush with the aspect of a *Sumach*, its leaves being composed of five to seven pairs of lance-shaped unequal-sided sharply serrated leaflets with an odd one. The branches are terminated by a cluster of slender drooping catkins of green male flowers somewhat like those of a willow, and a solitary cone-like and erect female catkin, made up of a number of hard-pointed bracts closely overlapping each other, and each bearing in its axil a little one-seeded, two-winged nut. According to Mr. Fortune, who first gathered the plant, and whose name it bears, the Chinese use the fruits to dye the black colour of their clothes. Its cone-like female catkins suffice to distinguish it from other genera. [A. A. B.]

FORTUYNIA. A genus of *Cruciferæ* allied to *Raphanus*, but with the two lower cells of the pod empty, the two upper one-seeded and indehiscent. The pod is flattened, broadly winged, resembling that of *Isatis*: indeed, the genus is founded on the *I. Garcini*, an eastern plant. [J. T. S.]

FOTHERGILLA. A genus of the witch-hazels, named in honour of Dr. Fothergill, a London physician and patron of Botany of the last century. The corolla is wanting; the stamens usually twenty-four in number, their filaments long and clubbed. The species are dwarf deciduous shrubs, natives of North America, having white and sweet-scented flowers. [G. D.]

FOUGÈRE COMMUNE or GRANDE. (Fr.) *Pteris aquilina*. — FEMELLE. *Athyrium Filix-fœmina*, also *Pteris aquilina*. — FLEURIE. *Osmunda regalis*. — MÂLE. *Lastrea Filix-mas*. — MUSQUÉE. *Scandix odorata*.

FOUNTAIN TREE. A popular name for *Cedrus Deodara*.

FOURCROYA. An amaryllidaceous genus closely related to *Agave*, and like it having long-lived massive stems, great fleshy leaves, and a very tall pyramidal terminal inflorescence produced after the lapse of many years. Herbert speaks of *F. longæva* as the most magnificent plant in the order, beyond all comparison: its stem forty feet high; its leaves less rigid and erect than in *Yucca*; its inflorescence thirty feet high, the lower branches of the terminal pyramid twelve to fifteen feet long; and its white flowers innumerable. These flowers consist of a six-parted perianth, with regular nearly patent segments, subulate conniving filaments with versatile anthers, a straight hollow triangular style enlarged below, and a triangular fringed stigma. The species are rather widely dispersed, occurring in South America, Mexico, West Indies, New Holland, and Madagascar. *F. gigantea* was formerly called *Agave fœtida*, and is a smaller plant than the foregoing. [T. M.]

FOUR O'CLOCK FLOWER. *Mirabilis dichotoma*.

FOUROUCHE. (Fr.) *Trifolium incarnatum*.

FOUTEAU. (Fr.) *Fagus sylvatica*.

FOVEA (adj. FOVEATE, dim. FOVEOLATE). A small excavation or pit: hence pitted.

FOVEOLE. The perithecium of certain fungals.

FOVEOLARIA. A genus of the *Styrax* family, peculiar to Peru, and represented by a single species, *F. ferruginea*, so named because of the copious rusty down which clothes the branches, flower racemes, and under surface of the leaves. It is a tall bush, with alternate elliptical entire leaves; and the little white flowers, which are somewhat like those of the *Styrax*, but smaller, are borne in axillary racemes, and have a five-toothed calyx, five oblong petals, and ten stamens adhering by their stalks into a tube so as to be monadelphous; this latter character being the chief distinguishing feature. The fruits are little ovoid berries. [A. A. B.]

FOVILLA. The imaginary fluid or emanation which it was formerly thought that the pollen discharged when performing the act of fertilisation. The fluid actually contained in the pollen-grain.

FOXBANE. *Aconitum Vulparia.*

FOX-CHOP. *Mesembryanthemum vulpinum.*

FOXGLOVE. *Digitalis.* —, DOWNY FALSE. An American name for *Gerardia flava.* —, LADIES'. *Verbascum Thapsus.*

FOX-GRAPE. *Vitis vulpina, Labrusca,* &c.

FOX-TAIL. *Lycopodium clavatum.*

FRACID. Of a pasty texture; between fleshy and pulpy.

FRAGARIA. A genus of *Rosaceæ*, distinguished by its ten-cleft calyx, its five petals, and its seeds inserted on a fleshy receptacle. This fleshy receptacle is the fruit known as the Strawberry. The name *Fragaria* is derived from *fragrans*—the fruit, as is well known, being peculiarly perfumed. The common name of Strawberry has been given, according to Sir Joseph Banks and others, on account of straw having been laid to prevent the fruit from getting soiled in wet weather.

There are several species, of which the principal are, *F. vesca*, to which belong the wood and Alpine varieties; *F. viridis*, the green; *F. elatior*, the Hautbois; *F. virginiana*, the Virginian or scarlet; *F. grandiflora*, the pine; and *F. chilensis*, the Chilian strawberry.

Previous to 1629, the date assigned to the introduction of the Scarlet Strawberry from Virginia, the Wood Strawberry must have been the sort generally gathered for sale in this country. 'Strabery ripe,' together with 'Gode Peascode,' and 'Cherrys in the ryse,' were mentioned as some of the London cries by Lidgate in a poem which he wrote, probably 400 years ago, or nearly, for he died in 1483. Peas must have been then cultivated for sale; strawberries may have been partially so, or they may have been chiefly gathered for the purpose where found growing in their wild state. But Hollinshed mentions the fact that Gloucester asked the Bishop of Ely for strawberries when contemplating the death of Hastings; and the circumstance has been dramatised by Shakespeare:—

'My lord of Ely, when I was last in Holborn,
I saw good strawberries in your garden there.'

The palace and garden of the Bishop were situated in that part of London now called Ely Place; and the grounds sloping to the then open stream or rivulet of Holborn must have been well adapted for the growth of strawberries.

The green strawberry is a European plant, but rarely met with. The fruit is small, very abundant, roundish; the flesh has a greenish tinge, solid, and juicy; and having somewhat of the pineapple flavour, something good might result from crossing it with other sorts.

The Hautbois is a native of England and the continent of Europe. It has not been found in a wild state so plentifully as the Wood or Alpine kinds. The cultivated varieties of this species sometimes bear most abundant crops; occasionally, however, the plants only produce sterile flowers, yet perhaps in another season the same plants again bear profusely, sterility being induced by circumstances which occasion a too rapid development of the parts of fructification, and their consequent imperfection. This led to the supposition that the Strawberry was a diœcious plant; but it is not so, for the rudiments of stamens and pistils, more or less perfect, can always be detected. The Hautbois have plicated, rugose leaves, and the fruit has a musky flavour, which many persons greatly prefer.

The Virginian or Scarlet Strawberry has many varieties, of various forms, round, conical, and oblong, some of them sugary and mild, but most of the scarlets have a brisk acidity. The Old Scarlet still ranks amongst the earliest ripening sorts. Although it was almost the only scarlet known for nearly 200 years after its introduction, and a shy bearer, no attempts to change it by crossing appear to have been made till within the last fifty years. By accident some good varieties of it were obtained; now, by artificial crossing, they are exceedingly numerous. They cross readily amongst themselves, and likewise with the pines.

The Pine Strawberries have generally large flowers and fruit, with foliage of a darker green and thicker substance than that of the scarlets, and the leaves are not so sharply serrated. According to some authorities, the species is a native of Surinam, but the true Old Pine was doubtless obtained from Carolina. This sort is very much superior to many formerly cultivated under the denomination of Pine Strawberries, and which, although most abundant bearers, have been abandoned for new sorts that are both prolific and good in quality. Still, as regards richness of flavour, the true Old Pine or Carolina maintains its high character. Instead of being hollow and spongy, it is solid and juicy—so much so, that a basket of it may be detected among other sorts by its greater weight.

The Chili Strawberry is, as its name implies, a native of Chili; but Prof. Decaisne, in his splendid work, the *Jardin Fruitier du Musée*, states that it is not confined to that part of South America, but has also been found on the western coast of North America, in California, and Oregon. The whole plant is covered with silky hairs, those on the scapes and peduncles spreading horizontally. It is of vigorous growth where the climate is suitable for it;

but it was found rather tender for our severe winters, and it does not succeed in the climate of Paris, but in the south-west of France it thrives admirably. It was introduced to Marseilles from Chili, in 1712, by Frazier, a French officer of marine artillery. Five plants were all that survived the voyage; but in 1857 about 450 acres were occupied with this sort alone in the neighbourhood of Brest, where the mildness of the winters and moisture of the air are favourable to its growth. It was cultivated in this country by Philip Miller in 1727; afterwards it appears to have been lost, till reintroduced by the Horticultural Society. Though not itself adapted for our climate, very beneficial results have followed its reintroduction, for, by crossing, a very large variety called Wilmot's Superb was raised; and though this had too much of the tenderness of its parent, yet, by repeated crossing, others less tender and of better quality have been obtained, and among them that so extensively grown and so well known by the name of the British Queen, one of our most valuable sorts.

Formerly strawberries were chiefly carried to the London market by women in head-load baskets. These women came mostly from Wales and Shropshire, and returned after the fruit-gathering season was over. They often made two journeys from Isleworth or Twickenham to London, thus walking between thirty and forty miles daily, with heavy loads on their heads for half that distance. Such labour is now almost entirely done away with, and spring vans are employed for the conveyance of strawberries to the markets. [R. T.]

FRAGILARIA. A genus of *Diatomaceæ* in which the frustules adhere intimately to each other, so as to form long ribbon-like threads which are narrower at one end than the other, probably from the rupture of the thread in the centre. *F. hiemalis* is very common in little pools and runlets in early spring, and is always a pleasing microscopic object. [M. J. B.]

FRAGON. (Fr.) *Ruscus.*

FRAISIER. (Fr.) *Fragaria.* — À CHÂSSIS. *Fragaria minor.* — BRESLINGE. *Fragaria collina.* — BUISSON. *Fragaria efflagellis.* — D'ANGLETERRE. *Fragaria minor.* — DE L'INDE. *Fragaria indica*, sometimes called *Duchesnea fragarioides.* — DES BOIS. *Fragaria sylvestris.* — DE TOUS LES MOIS. *Fragaria semperflorens.* — FRESSANT. *Fragaria hortensis.* — STÉRILE. *Potentilla Fragariastrum.* — DE VERSAILLES. *Fragaria monophylla.*

FRAMBOISIER. (Fr.) *Rubus Idæus.* — DU CANADA. *Rubus odoratus.*

FRANCISCEA. A genus of Brazilian shrubs, included among the *Scrophulariaceæ*, and closely allied to *Brunsfelsia*, from which it differs in the tube of the corolla being curved and dilated at its summit. There are several species in cultivation as stove plants, most of them having very showy salver-shaped purple flowers. The root of *F. uniflora*, and, to a less extent, the leaves, are used in Brazil in syphilitic complaints: hence the plant is called by the Portuguese Vegetable Mercury. It is bitter, purgative, emetic, and is poisonous in large doses. [M. T. M.]

FRANCISIA. A genus of *Chamælauciæ*, having a calyx with five short teeth, the tube of which is cylindrico-pentagonal and partially united with the ovary; a corolla of five suborbiculate converging petals inserted in the throat of the calyx; twenty stamens, those opposite the petals trifid, the middle lobes being antheriferous, and those alternate with the petals simple; and a filiform style with a hooked stigma. It is a slightly branched shrub, with crowded linear triquetrous leaves full of pellucid dots, and terminal clustered flowers. Native of New South Wales. [B. H.]

FRANCOACEÆ. (*Francoads.*) A natural order of calycifloral dicotyledons belonging to Lindley's erical alliance of hypogynous Exogens. Stemless herbs, with lobed or pinnate exstipulate leaves, and scape-like stalks bearing racemes of flowers; the calyx four-cleft; petals four, persistent for a long time; stamens about sixteen, attached to the lower part of the calyx, the alternate ones abortive. Ovary free, four-celled; ovules numerous; no style; stigma four-lobed. Fruit a four-valved capsule; seeds numerous, with a minute embryo and fleshy albumen. They are natives of Chili. Their qualities are astringent and slightly sedative. There are two genera, *Francoa* and *Tetilla*; and five species. [J. H. B.]

FRANCOA. A genus of perennial Chilian herbs, typical of the *Francoaceæ*, having lyrately pinnatifid leaves which are nearly all radical, and flowers in simple or branched racemes, the pedicels bearing single flowers, and having a persistent bract at the base. The calyx is four-parted, the petals four, the stamens eight fertile alternating with eight sterile, and the ovary free, four-celled, with numerous ovules. There are four or five species. Their juice is said to be sedative, and the roots of some are used for dying black. [J. H. B.]

FRANGIPANE. *Plumieria rubra.*

FRANGIPANIER. (Fr.) *Plumieria.*

FRANKENIACEÆ. (*Frankeniads.*) A natural order of thalamifloral dicotyledons belonging to Lindley's violal alliance of hypogynous Exogens. Herbs or undershrubs with branching stems, and opposite exstipulate leaves with a sheathing base. Flowers sessile, imbedded in the leaves; sepals four to five, united in a furrowed tube, persistent; petals alternate with sepals, often with scaly appendages; stamens four to five or twice these numbers, the anthers roundish, versatile, opening lengthwise. Ovary superior, with a slender cleft style, and numerous anatropal ovules attached to parietal placentas. Fruit a one-celled capsule, enclosed by the calyx; seeds very small. Chiefly natives of North Africa and the south of Europe; a few have been found in South Africa, South America, the temperate parts of Asia, and Australia. They have scarcely any properties of importance. The leaves of *Beatsonia portulacæfolia* are used in St. Helena as a substitute for tea. There are six genera, including *Frankenia*, *Beatsonia*, and *Hypericopsis*, and upwards of thirty species. [J. H. B.]

FRANKENIA. A genus of small heath-like herbs or sub-shrubs giving name to the order *Frankeniaceæ*, distinguished by having the petals furnished with claws which are equal in length to the tube of the calyx, six stamens, a three-cleft stigma, and a three-celled many-seeded capsule. *Frankenia* is represented in Great Britain by *F. lævis*, a procumbent plant, with numerous narrow oblong leaves which grow in tufts, and flowers rising from the forks of the stems or from the axils of the upper leaves. It grows in muddy marshes by the seaside in many parts of Europe and the Canary Islands; in England chiefly on the eastern coast. Other species come from the shores of the Mediterranean, the Cape of Good Hope, North America, and New Holland. French, *Franquenia*. [C. A. J.]

FRANKINCENSE. The odoriferous resin called Olibanum obtained from *Boswellia* —, EUROPEAN. A resinous exudation of the spruce fir. The name is also applied to *Pinus Tæda*.

FRANKLANDIA. A proteaceous genus containing only one species, *F. fucifolia*, a small upright shrub, very remarkable in its appearance, having scattered filiform dichotomous leaves, covered with orange-coloured glands and warts. The flowers have a salver-shaped calyx with a straight slender cylindrical tube and four-cleft deciduous limb; four stamens included within the tube of the calyx; and a filiform ovary, with spindle-shaped style, and inversely conical stigma. The fruit is a small nut with a single seed. It is a native of South-west Australia. [R. H.]

FRASERA. A North American genus of the gentian family, consisting of biennial herbs with axillary stalked flowers, having a wheel-shaped four-cleft corolla, whose segments have in the middle a glandular depression, protected by a fringed scale. *F. carolinensis*, or *F. Walteri* as it is also called, is a curious little plant found in the morasses of North America. It furnishes a fine gentian-like bitter, and when fresh is said to be emetic and cathartic. The roots have been imported under the name of American Calumba. [M. T. M.]

FRAXINEÆ. The ash tribe, a suborder of *Oleaceæ*. It includes those genera which have a winged fruit or samara, with one or more seeds. Among these are comprised the common ash (*Fraxinus*), and the manna ash (*Ornus*). [J. H. B.]

FRAXINELLA. *Dictamnus albus*.

FRAXINELLE. (Fr.) *Dictamnus albus*.

FRAXINUS. The Ash, a familiar tree belonging to the *Oleaceæ*, well distinguished by its fruit, which is dry and indehiscent, two-celled, two-seeded, compressed, and ending in a leaf-like expansion (samara). *F. excelsior* is indigenous throughout the greater part of Europe, the north of Africa, and some parts of Asia. Not remarkable for robustness, grandeur, or longevity, it rests its claim to distinction among European trees on qualities scarcely less striking. In height, gracefulness of form, and elegance of foliage, it has no superiors, scarcely any competitor. 'Its branches at first keep close to the trunk, and form acute angles with it; but as they begin to lengthen, they generally take an easy sweep, and the looseness of the leaves corresponding with the lightness of the spray, the whole forms an elegant depending foliage.' (*Gilpin*.) The 'sweep' described by Gilpin is especially remarkable in old trees, the lower pendent branches of which are curved upwards at the extremities in a way which quite typifies the tree. In early spring the spray assumes a characteristic appearance, occasioned by the numerous clusters of flowers which appear at the extremities of the branches, at least a month before the leaves. These flowers are minute and remarkably simple in their structure, being destitute both of calyx and corolla; but, being exceedingly numerous, and of a dark purple colour, they are very conspicuous. They grow in dense clusters on the extremities of those branches which were produced in the former year, and eventually become diffuse, and are finally succeeded by bunches of pendent seeds not inappropriately called keys. The foliage of the ash is very late in making its appearance, and it takes its departure among the first, though the precise time at which it sheds its leaves varies much in different individuals. The leaves are composed of about five pairs of acute notched leaflets with a terminal odd one, which last is occasionally abortive. A variety named *heterophylla* has most of the leaves simple. Another variety is well known as the Weeping Ash, all the existing specimens of which were originally derived from a tree discovered about the middle of the last century growing near Wimpole in Cambridgeshire.

As a timber tree the Ash is exceedingly valuable, on account of its quick growth, and the toughness and elasticity of the wood, in which latter quality it surpasses every European tree. In its younger stages (when it is called Ground-Ash), it is used for walking-sticks, hoops, and hop-poles. The matured timber is converted into ploughs, axle-trees, harrows, oars, carts, ladders, handles for tools and a variety of other implements; and as fuel it is unrivalled. Several American species of ash resemble the European ash in general appearance and qualities. The Ash is the badge of the clan Menzies. French, *Frêne*; German, *Esche*. [C. A. J.]

The Common Ash has perhaps a greater number of superstitions connected with it than almost any other tree, for it would seem that in England it unites the honours usually attributed to the rowan tree, or mountain ash of Scotland, with those peculiar to itself. Or perhaps the supposed powers of keeping witches at a respectful distance of the mountain ash—

'Rowan tree and Red Thread
Keep the witches at their speed—

have been attributed to it from the similarity of the leaves of the one to those of the other, thus giving rise to the name of ash for very dissimilar trees.

One of the most remarkable, and perhaps the most ancient, usages to which the Ash was appropriated, was that of passing children who were ruptured through a cleft in the bole of a young tree. Evelyn says: 'I have heard it affirmed with great confidence, and upon experience, that the rupture to which many children are obnoxious is healed by passing the infant through a wide cleft made in the bole or stem of a growing ash tree; it is then carried a second time round the ash, and caused to repass the same aperture as before. The rupture of the child being bound up, it is supposed to heal as the cleft of the tree closes and coalesces.' In this case, where both parents were living, the father presented the child, and the mother received it. In the Museum of Natural History in Worcester is a portion of a young ash which was probably submitted to this operation not many years since, and which did not heal as it grew, but retained an oval aperture in the stem. That this superstition lingered until very recently we know, as the Rev. T. Bree describes a case as having occurred in Warwickshire.

A superstition prevailed among the old leeches that a shrewmouse, on creeping over the limbs of man or the lower animals, was the cause of cramp and paralysis. To cure this, a hole was made with an auger in the bole of an ash tree, and a poor live shrew was fastened in with the plug of wood that had been abstracted. It is even now a not quite exploded belief that a shrewmouse running over the foot, will cause lameness, the antidote for which was the application of a twig of 'shrew ash.' Thus Gilbert White says: 'We have several persons now living in the village, who, in their childhood, were supposed to be healed by this superstitious ceremony, derived down, perhaps, from our Saxon ancestors, who practised it before their conversion to Christianity.' The same author describes the preparation of the 'shrew ash' as follows:—At the south corner of the plestor, or area, near the church, there stood, about twenty years ago, a very old, grotesque, hollow pollard-ash, which for ages had been looked upon with no small veneration as a shrew ash. Now a shrew ash is an ash whose twigs or branches, when gently applied to the limbs of cattle, will immediately relieve the pains which a beast suffers from the running of a shrewmouse over the part affected; for it is supposed that a shrewmouse is of so baneful and deleterious a nature, that whenever it creeps over a beast, be it horse, cow, or sheep, the suffering animal is afflicted with cruel anguish, and threatened with the loss of the use of the limb. Against this accident, to which they were continually liable, our provident forefathers always kept a shrew ash at hand, which, when once medicated, would maintain its virtue for ever. The manner of preparing the shrew ash was by means of the shrewmouse as already described, in which doubtless some strange invocations were used; but as we do not know them in these degenerate days, we may suppose the charm is lost. Not so, however, that attributed to the even-leaf from the Ash, that is, where the leaf terminates with two opposite pinnæ instead of the usual single terminal leaflet. In Wiltshire and Gloucestershire it is not at all uncommon for the lucky finder of the often much-coveted even-leaf to invoke it as follows:—

'Even-ash, I do thee pluck,
Hoping thus to meet good luck;
If no luck I get from thee,
Better far be on the tree.'

This simple charm keeps away witches; and we can only say that in our younger days we have travelled with an even-leaved ash on many an eerie night, and we never saw a witch.

Evelyn says that 'the chymists exceedingly recommend the seed of ash to be an admirable remedy for stone. But whether by power of magic or nature, I determine not'—doubtless from the power of its roots to rive rocks, and the facility with which this tree will grow in stony places. Be this as it may, it is, though a very old remedy, now discarded; and, indeed, of the many virtues the Ash was once supposed to possess (and we have not named them all), it now boasts none but the utilitarian one of being a most useful timber tree. However, in this relation we must not forget to mention that the root of the ash yields a most curious veined or cameloted wood in which superstition, ere now, has traced extraordinary figures. Thus Evelyn quotes one Jacobus Gaffereinus for the assertion, in his book of *Unheard-of Curiosities*, that 'of a tree found in Holland, which being cleft, had, in several slivers, the figures of a chalice, a priest's alb, his stole, and several other pontifical vestments.'   [J. B.]

FREE. Not adhering to anything else; not adnate to any other body.

FREMONTIA. A remarkable and beautiful Californian bush, belonging to the *Sterculiaceæ*. Along with the hand-plant of Mexico (*Cheirostemon*), it differs from the others in that group in the flowers having no petals; and from the latter it is readily recognised by the bell-shaped calyx, which remains attached, and does not fall away when the flower withers.

F. californica was first discovered by Col. Fremont (whose name it bears), in one of his Californian expeditions in the northern part of the Sierra Nevada. It forms a deciduous bush four to ten feet high, having much the aspect of an ordinary fig-tree. The rounded five to seven-lobed leaves, however, are smaller than those of the fig, and clothed with rusty hairs underneath.

Fremontia californica.

The handsome yellow flowers are produced singly on the ends of short spur-like branches, and consist of a broadly bell-shaped calyx of five spreading divisions, clothed sparsely with cinnamon-coloured down outside; five stamens having their stalks united below into a cup; and an ovoid ovary surrounded by the staminal cup, and terminating in a simple style. The fruits are oval capsules, which, when ripe, split into five woody portions, each of which contains a few black seeds. [A. A. B.]

FRENCH BERRIES. The fruits of *Rhamnus infectorius*, *saxatilis*, *amygdalinus*, &c.

FRENELA. A genus of *Coniferæ* of the tribe *Cupressineæ*. The flowers are incomplete, the staminate and pistillate ones on the same plant: the former in cylindrical catkins, with numerous stamens, imbricated in six rows; the latter in globular or conical cones of six scales. The seeds are numerous, winged on both sides. Resinous trees or shrubs of New Holland, with cylindrical or three-angled branches, and ternate, scale-like, persistent, and decurrent leaves. They are two years in ripening their seeds. The name was given by Mirbel after M. Frenel. There are twenty known species. [J. H. B.]

FRÊNE. (Fr.) *Fraxinus excelsior.* — À FLEURS. *Ornus europœa.* — À LA MANNE. *Fraxinus rotundifolia.*

FRESENIA. A South African genus of *Compositæ* characterised by its yellow flower-heads containing about fifteen florets, all of which are tubular and perfect; and by its achenes crowned with a double pappus, the exterior abort and chaffy, the interior of rough hairs. *F. leptophylla* and *scaposa*, the only species, are dwarf undershrubs, the former with opposite linear smooth, the latter with alternate downy leaves, and both with small terminal flower-heads. [A. A. B.]

FRESHWATER SOLDIER. *Stratiotes aloides.*

FREYCINETIA. A genus of *Pandanaceæ* consisting of climbing or scrambling trees, natives of the Indian Archipelago, Norfolk Island, New Zealand, &c. They have the habit of *Pandanus*, from which they are distinguished by having their male flowers upon an unbranched spadix, by their female flowers having abortive stamens, and by the ovaries having numerous ovules placed on three parietal placentæ. The fruit consists of numerous fleshy drupe-like carpels, completely or partially fused together, so that there are many or only one cell containing the numerous seeds. Examples of this genus are shown in Plate 10, fig. c, and Plate 14, fig. c. [M. T. M.]

FREYERA. A genus of umbellifers, having the fruit flattened laterally, each half with five sharp wing-like ridges; and comprehending a single herbaceous plant, a native of Illyria, the stem of which is slightly branched, with compound leaves twice divided, the divisions one or three-lobed; flowers white; fruit black. [G. D.]

FREZIERA. A South American genus of *Ternstrœmiaceæ*, chiefly found in the temperate regions of the Andes. They are evergreen bushes or trees of considerable magnitude, often with the aspect of laurels, but their leaves are usually covered beneath with close-pressed silky down. The little white flowers are usually two or three together, sometimes solitary in the axils of the leaves; they have a five-leaved calyx, five rounded petals, numerous stamens, and a three to five-celled ovary, which becomes a small dry berry with numerous seeds. *Cleyera*, to which this genus is most nearly allied, has few seeds in each cell. The wood of *F. chrysophylla* is preferred to many others in the Peruvian Andes for making charcoal; its lance-shaped leaves are clothed beneath with golden down. *F. theoides*, a common West Indian species, has smooth leaves like those of the tea, said to be astringent and to have a similar taste. [A. A. B.]

FRIAR'S-COWL. *Arisarum vulgare.*

FRIESIA. The name formerly given to the Tasmanian species of *Aristotelia*.

FRIJOLES. A Spanish name for various kinds of pulse.

FRINGED. The same as Fimbriate.

FRINGE-MYRTLES. A name given by Lindley to the *Chamælauciaceæ*.

FRINGE-TREE. *Chionanthus.*

FRITILLARIA. A genus of liliaceous plants of ornamental character, found in the south of Europe and in Asia. They are perennials, furnished with bulbs, and have erect annual stems with alternate or somewhat whorled often glaucous leaves, nodding bell-shaped flowers, sometimes solitary and terminating the stem, sometimes disposed in the form of a raceme in the axils of the upper leaves, or sometimes collected into a whorl beneath a terminal leafy tuft, this last being the arrangement in the Crown Imperial, *F. imperialis*, one of the most stately of the species. The perianth is six-parted, and each of its segments has a honey-pore near its base; within this are six stamens, and a three-celled ovary crowned by a three-parted style. In several of the species, especially in the native one, *F. Meleagris*, the colours of the flower are chequered, whence it is said the name is derived, from *fritillus*, assumed to mean a chess-board. [T. M.]

FRITZSCHIA. A genus of *Melastomaceæ*, composed of a few dwarf perennial Brazilian herbs, having much the aspect of the common thyme. Their minute leaves are smooth and marked with glandular dots, which is unusual in this family, and the slender twigs are terminated by solitary small purple flowers, which have a tubular calyx, four elliptical petals, eight straight stamens, with ovoid anthers united to their filaments by a short connective which has on its inner face two tubercles or short spurs. The tetramerous structure distinguishes them from some, and the nature of the stamens from others, of their allies. [A. A. B.]

FRÖLICHIA. A genus of *Amaranthaceæ* nearly allied to *Gomphrena*, from which it differs in having a tubular perianth five-cleft at the apex, and stamens with the filaments united into a long tube. They are natives of tropical America, one species reaching as far north as Illinois; and consist of hairy or woolly herbs, with opposite sessile leaves and spiked flowers, each with three scarious bracts. [J. T. S.]

FROG-BIT. *Hydrocharis morsus ranæ*. —, AMERICAN. *Limnobium*.

FROG-CHEESE. A name applied occasionally to the larger puff-balls when young. [M. J. B.]

FROLE. (Fr.) *Arbutus Unedo*.

FROMAGEON. (Fr.) *Malva rotundifolia*.

FROMAGER. (Fr.) *Bombax*. — ÉPINEUX. *Bombax Ceiba*.

FROMENTAL. (Fr.) *Avena elatior*.

FROMENT. (Fr.) *Triticum*. — CULTIVÉ. *Triticum vulgare*. — DES HAIES. *Triticum caninum*. — LOCULAR. *Triticum monococcum*.

FROND, FRONS. A combination of leaf and stem, as in many algals and liverworts; also improperly applied to a leaf which bears reproductive bodies, as that of doraiferous ferns. Linnæus applied it to palm leaves, and so destroyed its meaning.

FRONDOSE. Covered with leaves; bearing a great number of leaves.

FRONDIPAROUS. A monstrosity, consisting in the production of leaves instead of fruit.

FROPIERA. A small tree from the Mauritius with alternate evergreen entire leaves and small flowers in axillary clusters or short racemes, forming a very distinct genus, whose immediate affinities have not been ascertained. The dotted leaves and most points of structure are those of *Myrtaceæ*, but the ovary is entirely superior, and the stamens definite.

FROSTED. A term applied to surfaces in which a dewy appearance is opaque, as if the drops were congealed.

FROST-WEED. *Helianthemum canadense*.

FRUCTIFICATION. The parts of the flower; or more properly the fruit and its parts.

FRUCTIPAROUS. A monstrosity, consisting in the production of several fruits, instead of the one which is metamorphosed.

FRUIT. That part of a plant which consists of the ripened carpels, and the parts adhering to them. —, SPURIOUS. Any kind of inflorescence which grows up with the fruit, and forms one body with it, as a pine cone.

FRULLANIA. A large genus of *Jungermanniaceæ*, distinguished by its numerous archegones and complicated leaves. The species occur in all parts of the world, but are far more common in tropical or sub-tropical countries than in Europe. *F. tamarisci* is almost universally distributed, and is found abundantly in rather mountainous heathy districts, where it is conspicuous for its purple hue. The leaves in this genus are remarkable for the inflated lobes on their under side. [M. J. B.]

FRUSTULES. The joints into which the brittleworts separate.

FRUSTULOSE. Consisting of small fragments.

FRUTA DE BURRO, of Carthagena. A poisonous plant supposed to be a species of *Capparis*. —, of Humboldt. The fruit of *Xylopia grandiflora*. — DE PARAÓ. The fruit of *Schmidelia edulis*.

FRUTEX (adj. FRUTICOSE, FRUTESCENT). A shrub; a woody plant which does not form a trunk, but divides into branches nearly down to the ground.

FRUTICULUS. A small shrub.

FUCACEÆ. A natural order of dark-spored *Algæ*, consisting of olive-coloured inarticulate seaweeds whose spores are contained in spherical cavities in the frond. Most of them are large species of a tough leathery substance, and assume a dark colour when dried. Many of them have distinct leafy and even two-ranked appendages, while others are destitute of any distinction whatever between stem and receptacle. In *Himanthalia* the frond is a small cup-shaped body, the receptacle being repeatedly forked and many feet long. In many cases the receptacles form little pod-like solitary or fasciculate bodies projecting from the stem, while in others they are merely slight swellings. Another cause of variety of aspect arises from the different nature of the air-bladders by which they are sustained in the water. These are sometimes entirely wanting, sometimes simple, and sometimes compound or arranged in necklace-like rows. In all alike, whatever the habit may be, the spores are contained in cavities resembling the cells of *Dothidea*.

Fucaceæ exist in all parts of the world calculated for the growth of seaweeds, and, though much more abundant as regards species in warm than in temperate regions, have numerous representatives in the latter. All are probably occasionally attached, but they may exist for centuries as floating masses, as in the instance of the Gulf weed. *Durvillæa utilis*, which is remarkable for its having the habit of a *Laminaria*, though belonging truly to this order, and distinguished by the large cells like those of a honeycomb contained in its frond, is used in Chili and elsewhere for thickening soup. The greater part of these plants contain a great quantity of carbonate of soda, which was once procured from them in considerable quantities in the form of kelp, and they in common with some other melanosperms are a fertile source of iodine, one of the most important medicines in the Pharmacopœia. [M. J. B.]

FUCHSIA. A genus of *Onagraceæ* characterised by having a funnel-shaped coloured deciduous four-parted calyx, sometimes with a very long tube; four petals set in the mouth of the calyx-tube, and alternating with its segments; eight exserted stamens; and a long style with a capitate stigma. The flowers are succeeded by oblong bluntly four-cornered berries.

A plausible story has been often printed which attributes the introduction of the *Fuchsia* into England to a sailor, whose wife or mother was induced to sell it to Mr. Lee, a nurseryman, who in the course of the following summer made a profit of 300 guineas by the transaction. This is said to have happened about the close of the last century. It was, however, a hundred years before this time that a monk named Father Plumier discovered the first specimen of the family, which he dedicated to the memory of Melchior Adam Fuchs. This first species was named *Fuchsia triphylla flore coccineo*, and a description of it is to be found in the works of Plumier, published in 1703. With the exception of *F. excorticata* and *F. procumbens*, which are natives of New Zealand, all the species belong to the central and southern regions of America, in shady moist places, in forests, or on the lofty mountains of Mexico, Peru, and Chili. The number of distinct species at present known is more than fifty, which have been introduced from time to time since the beginning of the present century; but the varieties most prized by florists date only from the year 1837, when *F. fulgens* was introduced. The introduction of this species, and soon afterwards of *F. corymbiflora*, *cordifolia*, and *serratifolia*, gave to horticulturists the opportunity of hybridising these long-flowered species with the globose kinds, and the result has been the annual appearance of varieties which, from a garden point of view, have surpassed their predecessors, to be themselves eclipsed in their turn. [C. A. J.]

FUCHSIA, AUSTRALIAN or NATIVE. A colonial name for *Correa*.

FUCUS. A name formerly applied indiscriminately to almost all the more solid *Algæ*, though now confined to a single genus of *Fucaceæ*. Attempts have been made to subdivide even this from slight differences in the spores or in the disposition of the male organs. *Fucus*, as now generally restricted, comprises those social seaweeds which have a flat or compressed forked frond, the air-vessels when present formed by the occasional swelling of the branches, or in their substance, and also have receptacles filled with mucus, traversed by a network of jointed filaments. *Fucus*, in fact, contains such species as *serratus* and *nodosus*, which are as common on our coasts as grass in the fields. The antherids are produced either on the same or on different plants, and their spermatozoids have been proved to have active functions from their effects on the spores, which, without their access, are not capable of reproducing the species, though they commence an imperfect germination.

Many of the species are more or less exposed at low water. *F. canaliculatus*, however, which is referred by some to a distinct genus, *Pelvetia*, is remarkable for its amphibious habit, growing as it does frequently on large boulders, where it is dried up by the sun into a hard brown mass. This, however, recovers its usual appearance entirely with the first return of the tide, and is so little incommoded by the change, that it even brings fruit to perfection in such situations. As, however, there is a point beyond which endurance is impossible, it is not known on the coasts of the United States, where the hot turning sun would completely destroy vitality before the return of the tide. The peculiar leathery texture of the frond seems to enable it to bear considerable change without inconvenience.

These plants afford a considerable proportion of the seaweed thrown up upon our shores and collected for manure, as it was formerly for making kelp. Cattle also occasionally browse upon them, or they are boiled and given with coarse meal as food.

The gelatinous receptacles are sometimes used as applications to scrofulous swellings. Any benefit which results must depend on the small quantity of iodine which they contain.

Most of the species are confined to the Northern seas. *F. vesiculosus*, though so common under a variety of forms both in the North Atlantic and Pacific, does not exist in the Mediterranean except in floating masses carried in through the Straits of Gibraltar. *F. nodosus* occasionally exists in similar floating masses, and then assumes curious forms which have been registered as distinct species distinguished by their mode of branching and other characters. *F. vesiculosus* is the badge of the M'Neills. [M. J. B.]

FUGACIOUS, FUGAX. Falling off, or perishing very rapidly.

FUGOSIA. A genus of *Malvaceæ*, consisting of shrubs, natives of tropical America, Africa, and Australia. Their flowers are surrounded by an outer calyx or involucel of six or more leaves, within which is a five-cleft calyx dotted over with black spots, and five oblique petals. The capsule is three to four-celled, opening through the backs of the carpels. [M. T. M.]

FUIRENA. A genus of cyperaceous plants belonging to the tribe *Scirpeæ*, having the inflorescence in solitary spikes, in spikes of three, or in crowded heads of spikes, many-flowered; scales imbricated, the outer frequently empty; stamens three; styles three-cleft; achenes triangular, with the bases of the styles adhering. There are about forty species, mostly natives of the warmer parts of the globe, chiefly in the southern hemisphere. [D. M.]

FULCIENS. Supporting or propping up anything; said of one organ which is placed beneath another.

FULCRA (adj. **FULCRATE**). Additional organs, such as pitchers, stipules, tendrils, spines, prickles, hairs, &c.

FULCRACEOUS. Of or belonging to the fulcra.

FULIGINOUS, FULIGINOSE. Dirty brown, verging upon black.

FULVOUS. Dull yellow, with a mixture of grey and brown.

FULWA. A solid buttery oil obtained from *Bassia butyracea*.

FUMARIACEÆ. (*Fumeworts.*) A natural order of thalamifloral dicotyledons, belonging to Lindley's herberal alliance of hypogynous Exogens. Herbs with brittle stems, watery juice, alternate cut exstipulate leaves, and irregular unsymmetrical flowers. Sepals two, deciduous; petals four, cruciate, irregular, one or two of them often saccate or spurred, and the two inner frequently cohering at the apex so as to include the anthers and stigma; stamens either four and free, or six and diadelphous, each bundle being opposite the outer petals, the central anther two-celled, and the two outer one-celled. Fruit a round and indehiscent achene, or a one-celled and two-valved pod; seeds crested with a minute embryo. Natives chiefly of the temperate regions of the northern hemisphere; a few occur at the Cape of Good Hope. They possess slight bitterness and acridity. *Dielytra spectabilis* has very showy flowers. There are about 160 species, distributed in eighteen genera, of which *Hypecoum, Fumaria, Corydalis, Dielytra,* and *Platycapnos* are examples. [J. H. B.]

FUMARIA. The Fumitory, a genus of herbaceous plants giving name to the order *Fumariaceæ*, among which they are distinguished by having one of the petals swollen or spurred at the base, and a one-seeded capsule which does not open. The species vary but little in habit, being small slender herbs with weak climbing or straggling stems, decompound leaves, and clusters or spikes of small tubular irregular flowers of a pinkish hue tipped with purple, or rarely white. Several kinds of Fumitory are common weeds in cornfields and other cultivated ground, varying in luxuriance according to the richness of the soil. *F. officinalis* is said to be a common weed throughout the world, and has been long esteemed for its medicinal virtues, the juice having been recommended as a purifier of the blood, and an infusion of the leaves as a cosmetic. Though now not valued in England, it occurs in lists of French medicinal plants as a depurative. French, *Fumeterre*; German, *Erdrauch*. [C. A. J.]

The Fumitory is essentially an agrarian plant, tracking both garden and field culture over a great part of Asia as well as Europe. It is probably from this cause that the species are so variable, or perhaps we should say, that so many varieties occur; and being sown with different kinds of seeds, such as clover, flax, and other crops, which may be obtained from different parts of the world, we need not wonder if a variable mode of growth should be the consequence of the varying conditions which plants so circumstanced must encounter. The typical species is *F. officinalis*, which was formerly in repute for a variety of diseases. Its generic name, indeed, is said to be derived from the Latin *fumus*, smoke, which, Pliny tells us, was given because the juice of the plant brought on such a flow of tears that the sight became dim as in smoke, and hence its reputed use in affections of the eye. It is now no longer employed medicinally, although a volume might be written of what has been said of its virtues and the many diseases in which it was held as a remedy by a host of physicians from Dioscorides to Cullen. [J. B.]

FUMETERRE. (Fr.) *Fumaria.* — **BULBEUSE.** *Corydalis bulbosa.*

FUMOUS, FUMOSE. Grey, changing to brown; smoke-coloured.

FUMEWORTS. The plants of the order *Fumariaceæ.*

FUMITORY. *Fumaria* —, **BULBOUS** *Corydalis bulbosa*, also *Adoxa Moschatellina*.

FUNALIS. Formed of coarse fibres resembling cords.

FUNARIEI, FUNARIA. A small natural order and genus of acrocarpous mosses with a pear-shaped capsule, and the calyptra much inflated and vesicular below, and subulate above. The peristome is either double, single, or altogether wanting, the vesicular calyptra being the point of greatest importance. *Funaria hygrometrica* grows in all parts of the world, and is extremely common in this country, especially on charred or burnt soil, and is conspicuous from its large calyptra and cernuous heads. [M. J. B.]

FUNDAMENTAL. Constituting the essential part of anything, in a plant, the axis and its appendages.

FUNDI or **FUNDUNGI.** The Hungry Rice, *Paspalum exile*.

FUNDUS PLANTÆ. The collar, or place of junction of root and stem.

FUNGALES, FUNGALS. A name intended to include under one head *Fungi* and *Lichens*, the latter of which are so closely allied that it is often difficult to tell to which division some given species may belong. [M. J. B.]

FUNGI. A large class of cryptogams distinguished from *Algæ* more by habit than by any general character. They agree with them in their cellular structure, which is void of anything like vascular tissue except in a very few cases, while they differ in their scarcely ever being aquatic, in deriving nutriment from the substance on which they grow, and in the far lower degree of development of the organs of impregnation — the impregnating cells, where they really possess a sexual function, being extremely simple, void of cilia, and therefore possessed of nothing more than molecular motion, the only exception being that of *Leptomitus* and its allies, which seem to be almost intermediate. The myxogastrous *Fungi*, whose spores produce a body resembling certain Infusoria, are wholly exceptional, and the indications of animal life which they exhibit point in another direction.

Minute and abstruse as are these differences, it is almost impossible to distinguish certain *Fungi* and *Algæ* without them. Take, for instance, a *Peronospora* and a *Chroolepus*. Both exhibit erect branched threads, from the upper part of which cells are produced containing a thick grumous matter. At first sight no one would think they could belong to very different sections of the vegetable kingdom. When, however, we look more closely, we find first, that the one is a true parasite, the other growing indifferently on bark or stone, and deriving its nourishment from the surrounding air; and then when we turn from the habit to intimate structure, we find that the spores of the *Peronospora* fall off and germinate at once, while the analogous bodies in the *Chroolepus* burst and send out a multitude of minute reproductive bodies moving about for a time by means of long lash-shaped cilia. A second form of fruit which occurs in *Peronospora* shows a greater difference between the two as genera, but not as regards important sectional character. A similar parallel might be made in other cases.

Popularly speaking, *Fungi* may be recognised either as the creatures of corruption — springing, that is, from various bodies, whether animal or vegetable, in a more or less advanced stage of decomposition — or as parasites on living bodies, producing an injurious change. The ephemeral toadstools of the hotbed, the mushrooms of our rich pastures, the sap-balls on decaying trees, the moulds which infest our food and even the tissues of living animals, the mildew bunt and smut of our corn-crops, with many other more or less familiar objects, are so many *Fungi*, all agreeing in the main particulars which we have indicated, and so differing from the green scum of our brooks, and the weeds of the sea, though distinguished from each other by essential differences of structure. In some, no indications of sexual differences have been found, while in others bodies occur, which in all probability have an especial sexual function, though at present we are without actual proof of the fact.

Fungi are divided into two great sections, characterised by the mode in which the reproductive bodies are formed. In the one, they are simply the terminal joint or joints of the component threads or cells, altered in form from those which precede them, and at length falling off and reproducing the plant, in which case they are called spores. In the other they are formed from the contents of certain sacs or asci, and are usually definite in number, and multiples of four, where they are not reduced below that number; in this case they are called sporidia. Both spores and sporidia may be multicellular, and in germination give rise to as many threads of spawn as there are cells. In many species of the latter division, a second form of fruit occurs, which is naked as in the first; and in every division two or more kinds of fruit are frequently produced by the same species, a fact which takes from the mathematical precision of the two great divisions, though it does not interfere with their natural affinities.

Fungi may be divided into six principal classes, the first four of which bear naked spores, the two latter sporidia: —

1. HYMENOMYCETES, in which the fructifying surface is at length exposed, if not so in its first origin. Mushrooms and sap-balls are well-known examples.

2. GASTEROMYCETES, in which the fructifying surface is always enclosed at first, and is never completely exposed, except in old age or decay, in consequence of its sinuous intricate character, even

when the peridium bursts. Puff-balls are a familiar example.

3. CONIOMYCETES, in which the spawn or vegetative part is reduced to a minimum, and the abundant spores at length form a dusty or more rarely a gelatinous mass. The rust and bunt of corn afford ready instances.

4. HYPHOMYCETES, in which the vegetative part consists mostly of threads which are not woven into a solid mass except in a few cases which border on *Hymenomycetes*. The naked-seeded monlds belong to this division.

5. ASCOMYCETES, in which the sacs or asci which contain the sporidia are either packed into an exposed hymenium, or line the interior of the fruit-bearing cysts. Morels afford an example of the first, and the insect *Sphæria* of the second.

6. PHYSOMYCETES, in which the component threads are more or less free as in *Hyphomycetes*. The common bread mould is an excellent example.

Each of these divisions is again divided. In a few instances the bodies which at first sight seem to be the spores or ultimate fruit, are in reality a sort of prothallus. Sometimes a third evolution takes place before the ultimate spore is formed. The truly parasitic fungi of the third division give us examples.

The uses of *Fungi* are various. To enumerate them here would be merely to go over ground which must be again travelled under individual orders and species. It is sufficient to say that they afford excellent and abundant food, valuable medicines, besides less important assistance in domestic economy. Their office in the organised world is to check exuberant growth, to facilitate decomposition, to regulate the balance of the component elements of the atmosphere, to promote fertility, and to nourish myriads of the smaller members of the animal kingdom. They occur in every part of the world where the cold is not too intense to destroy their spawn, or where there is sufficient moisture, though they abound the most in moist temperate regions where the summer is warm. There are but few certain traces of them in antediluvian strata, and those only in the more recent. Most of them, however, are too soft and fugitive to make it likely that they should have been preserved. [M. J. B.]

FUNGIFORM, FUNGILLIFORM. Cylindrical, having a rounded convex overhanging extremity.

FUNGINOUS. Of or belonging to a fungus.

FUNICULUS, FUNICLE. The cord or thread which sometimes connects the ovule or seed to the placenta.

FUNILIFORM. Formed of cord-like fibres.

FUNKIA. A genus of *Liliaceæ* found in China and Japan, having fasciculate roots, the leaves usually all radical, stalked, ovate or cordate, acuminate and plaited, the caulline ones, when present, sessile. The flowers grow in racemes and are blue or white, with a tubular six-parted perianth, and the style and stamens bent down. The seeds have a black membranous coat, produced into a wing at the apex. A few species are known, and they are mostly introduced to our gardens. [J. T. S.]

FUNNEL-SHAPED. A calyx or corolla, or other organ, in which the tube is obconical, gradually enlarging upwards into the limb, so that the whole resembles a funnel, as in the *Convolvulus*.

FURBIURNE. An Arab name for *Euphorbia officinarum*.

FURCATE. Having long terminal lobes, like the prongs of a fork, as *Ophioglossum pendulum*.

FURCELLARIA. A genus of rose-spored *Algæ* belonging to the natural order of *Cryptonemiaceæ*, with a forked cylindrical fastigiate frond, having the capsules lodged in the pod-like branches. *F. fastigiata*, the only known species, which is widely distributed in the Northern Atlantic, is one of the commonest sea-weeds on our coast. It is so like *Polyides rotundus* that it is very difficult to distinguish them except when in fruit; the sponge-like masses in which the capsules of *Polyides* are immersed, afford, however, a marked distinction. [M. J. B.]

FURFURACEOUS. Scurfy; covered with soft scales, which are easily displaced.

FURROWED. Marked by longitudinal channels, as the stem of the parsnep.

FURZE. The gorse or whin, *Ulex europæus*.

FUSAIN. (Fr.) *Euonymus europæus*.

FUSANUS. A genus of sandalworts, having flowers of mixed character, some being perfect, having stamens and pistils; others with stamens or with pistils only. The border of the calyx is deeply divided into four pieces, which spread horizontally like the spokes of a wheel, but ultimately fall off; the stamens are four in number. The species are small trees or shrubs, natives of the Cape of Good Hope, and of the southern parts of New Holland. Dr. Lindley states that 'the fruit of the Quandang nut (*F. acuminatus*) is as sweet and useful to the New-Hollanders as almonds are to us.' [G. D.]

FUSARIUM. A genus of moulds closely resembling *Fusisporium*, but consisting of *Fungi* which burst forth from beneath the cuticle of the plants on which they grow, in little gelatinous spots. *F. heterosporium* has affected rye in the south of England during hot seasons, and *F. Mori*, a species first described by Léveillé, is the pest of the white mulberry crops cultivated for silkworms, forming on the leaves brown gelatinous specks which exhaust their nutritive qualities. The other species are of

[FUSE] The Treasury of Botany. 614

little importance from an economical point of view. [M. J. B.]

FUSETTE. The Spanish name for *Rhus Cotinus*.

FUSISPORIUM. A genus of moulds with septate spindle-shaped spores springing from free mucedinous threads, and at length forming a gelatinous mass. It is distinguished from *Fusarium* by its not bursting forth from beneath the cuticle as in that genus. Several of the species are destructive to vegetables, such as turnips, beet-root, gourds, &c. *F. Solani* is extremely injurious to potatos, and in company with *Peronospora infestans* hastens the decomposition which is due to that parasite, or converts the tubers into a hard dry innutritious mass. The flocci are, however, too much developed to make this a typical *Fusisporium*. [M. J. B.]

FUSCOUS. Brown, with a greyish or blackish tinge.

FUSIFORM. Thick, tapering to each end; as the root of the long radish. Sometimes conical roots are called fusiform.

FUSTET. (Fr.) *Rhus Cotinus*.

FUSTIC. A dye-stuff, consisting of the wood of *Maclura tinctoria*. —, YOUNG. The wood of *Rhus Cotinus*.

G.ERDTIA. A genus of *Begoniaceæ* having the staminate and pistillate flowers on the same plant, arranged in dichotomous cymes. The staminate flowers have a white four-leaved perianth, and twenty to thirty stamens; and the pistillate ones, which also have a white four-leaved perianth, have a three-winged ovary with three central bifid placentas. They are Brazilian shrubby plants with smooth shining jointed stems and branches, semicordate leaves, and large shining deciduous stipules. The four known species are included by most authors under *Begonia*. [J. H. B.]

G.ERTNERA. A genus of opposite-leaved bushes or small trees of the *Logania* family, differing from most of the genera, in the fruits being two-celled berries with one or rarely two instead of numerous seeds in each cell. The erect and not lateral attachment of the seeds serves to distinguish them from their nearest allies. The greater number of the thirty known species are found in Mauritius and Madagascar, the remainder in W. Africa and the Malayan peninsula and islands. The smooth entire leaves are lance-shaped, ovate or elliptical, and the flowers are white, green, or rose-coloured; in some species not unlike those of the common privet and arranged in a similar manner, in others disposed in compact terminal heads, and in a goodly number in corymbs. The calyx is usually very minute, but in *G. calycina*, a Mauritian species, it is enlarged, bell-shaped, and coloured. The corolla tube has a flat border of five narrow lobes, and bears on its inner face five stamens. The ovary becomes, when ripe, a white, black, or blue berry about the size of a pea, with two seeds. [A. A. B.]

GAGEA. An extensive genus of *Liliaceæ* formerly included in *Ornithogalum*, from which it is easily distinguished by the seeds having a yellowish (not black) seed-coat, and the stamens adhering more distinctly to the segments of the perianth. The species are natives of Europe, temperate Asia, and northern Africa, and resemble each other closely in having linear root-leaves, and a scape with a terminal bracteated umbel or corymb of greenish-yellow flowers rather large for the size of the plants. The perianth is persistent, of six patent nearly equal divisions; the stamens six; the style terminated by a three-lobed stigma; the capsule three-celled and three-valved. *G. lutea* is a British species, though rather rare; it is distinguished from allied European species by having no accessory bulb included in the common envelope. [J. T. S.]

GAGLEE. *Arum maculatum*.

GAIAC. A name applied in French Guiana to the wood of *Dipteryx odorata*.

GAILLARDIA. A genus of handsome annual or perennial North American herbs of the composite family, chiefly found in the Southern States, some extending to Oregon, and *G. aristata* reaching across the Rocky Mountains to the Winipeg Valley. The chief features of the genus are the slender bristles instead of chaffy scales of the receptacle, the long and filiform styles, the neuter ray florets, and the villous achenes crowned with a pappus of six to ten membranaceous one-nerved scales, which are prolonged into an awn. The leaves are sometimes pinnatifid, but more generally entire or obscurely toothed, lance-shaped and rough, the cauline ones sessile. The flower-heads, about two inches across, are single and supported on naked stalks, the strap-shaped ray florets three to five-toothed, sometimes brick-red or purple below, sometimes altogether yellow. The slender hairs of the stems and leaves are seen to be curiously jointed when looked at through a lens. Six species are known, all of them pretty border plants. [A. A. B.]

GAILLET. (Fr.) *Galium*.

GAIMARDIA. A genus of *Desvauxiaceæ*, differing from the rest of the order by having two instead of only one stamen. It contains a small tufted herb from the Maclovian Islands, with erect stems branched at the apex and densely leafy; the branches with scattered leaves; the leaves imbricated, bayonet-shaped, with sheathing bases; the flower-spike solitary terminal, with one-flowered spikelets. [J. T. S.]

GAINIER COMMUN. (Fr.) *Cercis Siliquastrum*.

GAIROUTTE. (Fr.) *Lathyrus Cicera*.

GAITRES BERRIES. The fruits of *Cornus sanguinea*.

GALA, GALACTO. In Greek compounds = milk or white as milk.

GALACTITES. A genus of *Compositæ* peculiar to the Mediterranean region and the Canary Islands. The three known species have much the aspect of, and are nearly allied to, *Cnicus*, differing chiefly in the outer florets of the flower-head being sterile and larger than the others, as in *Centaurea*. The stems seldom exceed two feet high; the leaves are pinnatifid with spiny-pointed segments, spotted with white above, and covered with cottony down below, the bases of the upper ones decurrent, and forming a wing to the stems. The flower-heads, which contain numerous white or pink florets, are either clustered and sessile on the ends of the branches, or grow simply on long stalks. *G. tomentosa* is remarkable among the thistles for having a milky juice like that so common in the chicory group. [A. A. B.]

GALACTODENDRON. A generic name given by some authors to the celebrated Cow-tree or 'Palo de Vaca' of South America, now more generally referred to BROSIMUM; which see. [A. S.]

GALAM BUTTER. A reddish-white solid oil obtained from *Bassia butyracea*.

GALANE. (Fr.) *Chelone*.

GALANGAL or GALANGALE. The aromatic *Alpinia Galanga*; also *A. racemosa*, *Allughas*, and *pyramidalis*; in Sweden it is called *Galgant*. Also a common name for *Kæmpferia*.

GALANT DE JOUR. (Fr.) *Cestrum diurnum*. — DE SOIR. *Cestrum vespertinum*. — DE NUIT. *Cestrum nocturnum*.

GALARDIENNE. (Fr.) *Gaillardia*.

GALANTHUS. A genus of *Amaryllidaceæ* characterised by having a six-leaved bell-shaped perianth, the exterior segments concave and spreading, the interior shorter, erect, and emarginate; six stamens inserted on an epigynous disk, with very short filaments and erect convergent anthers; a straight filiform style with simple acute stigma; and a three-celled ovary with numerous ovules. *G. nivalis* is the common Snowdrop, a dwarf bulbous plant found in some parts of England, and having a pair of narrow linear glaucous leaves, and drooping white flowers dotted with green on the inner segments, and generally solitary at the top of the short scape. *G. plicata*, the Crimean Snowdrop, is similar, but larger and handsomer, with the leaves broad linear and plicate. Our English Snowdrop is welcomed as one of the earliest floral harbingers of spring, the 'first pale blossom of the unripened year,' and a double-flowered variety is much cultivated. [T. M.]

GALATELLA. A genus of perennial herbs of the composite family, numbering about twenty species, found in the temperate parts of Asia, one species only occurring in the United States. They have much the appearance of *Aster*, and only differ in the ray florets being neuter; while from *Linosyris* they differ in the ray florets being white or purple, never yellow. The stems are simple below, branching above, and furnished with narrow entire leaves, and numerous flower-heads arranged in terminal corymbs. The ray florets are white or blue, those of the disk yellow; and the achenes are hairy or villous and crowned with a pappus consisting of numerous rigid and filiform rough bristles. [A. A. B.]

GALAX. The name of a genus of wintergreens distinguished by having the filaments united to form a tube, with ten teeth at the end, the five teeth opposite to the petals having no anthers, and the other five bearing perfect anthers. The name is derived from the Greek word signifying 'milk,' and probably refers to the colour of the numerous small flowers. The only species is *G. aphylla*, a tufted herbaceous plant, with scaly creeping root-stocks, and a native of open woods in the southern parts of the United States, extending northwards to Virginia. [G. D.]

GALAXIA. A genus of Cape *Iridaceæ*, forming dwarf plants with bulb-tuberous rhizomes, short stems bearing a terminal cluster of narrow leaves and handsome flowers, consisting of a funnel-shaped perianth, with a slender terete tube, and six-parted equal limb of oblong wedge-shaped spreading segments, the outer of which have a nectariferous cavity at the base. They have three stamens, with the filaments connate into a short tube, and the arrow-shaped anthers affixed by their base; a filiform triquetrous club-shaped style with three fringed convolute stigmas; and a three-celled ovary containing many ovules. There are some five or six species; of which *G. ovata* grows three or four inches high, and has ovate-oblong plicate ciliated leaves, and large bright yellow flowers. [T. M.]

GALBA. A durable Indian wood produced by *Calophyllum Calaba*.

GALBANUM. A Persian umbelliferous plant, the fruit only of which is known, has been described under this name, from the supposition that it was the source of the drug galbanum; a supposition, however, that is at present unsupported by evidence. The fruits of *G. officinale* are elliptical and flattened from back to front; each half-fruit has seven elevated bluntly keeled ridges; the intervening channels are broad, and have no vittæ or reservoirs for oil, but on the commissure or surface by which the two halves of the fruit are in contact, there are two vittæ. [M. T. M.]

The name Galbanum is also applied to a balsamic gum-resin, of which that obtained from Persia is ascertained to be produced by *Opoidia galbanifera*; its properties are similar but inferior to those of asafœtida. It is supposed to be also yielded by other umbellifers.

GALBULUS. A strobilus, whose scales

GALE]   **The Treasury of Botany.**   516

are fleshy, and combined into a uniform mass; as the fruit of the juniper.

GALE, SWEET. *Myrica Gale.*

GALEA. The helmet or arched part of a flower, always placed at the back, that is, next to the axis.

GALEANDRA. This was formerly recognised as a distinct genus of orchids, but is now referred to *Eulophia* by Dr. Blume. The Mexican *G. Baueri*, frequently cultivated by orchid growers, is epiphytal, with cylindrical stems bearing several lance-shaped nerved leaves, and beautiful drooping racemes of yellow flowers, the lip having parallel purple lines near the apex, which has wavy margins. *G. Devoniana* is another handsome species, with large chocolate-coloured flowers, having a funnel-shaped white lip marked with pink lines. [A. A. B.]

GALEARIA. A genus of handsome laurel-leaved bushes found in the Malay Peninsula and Archipelago, referred by some authors to the *Stilagineæ*, and by others to the *Euphorbiaceæ*; from the former of which it differs in the flowers having petals, and from any genus in the latter by its solitary and terminal, often drooping flower-spikes, which are sometimes more than a foot long. The leaves are accompanied by minute stipules; and the minute green flowers are diœcious, the males with a five-parted calyx, five concave petals, and ten free stamens; the females with similar calyx and corolla, and an ovary crowned with three or five minute stigmas. The fruits are rounded, fleshy, about the size of a pea when only one cell is perfected, larger and two or three-lobed when two or three are perfected: each cell containing one seed. The names *Bennettia* and *Cremostachys* have been given to some of these plants. [A. A. B.]

GALEGA. A genus of smooth erect perennial herbs of the leguminous family, having pinnate leaves, arrow-headed stipules, and long axillary racemes of pretty lilac or white pea-flowers. The few known species are found in the Mediterranean region, and extend eastward to Persia. They are nearly related to *Glycyrrhiza*, but the pods are narrow and smooth, and contain numerous seeds, while those of the liquorice are broad, usually rough externally, and one to four-seeded. The roots have a sweetish taste. The stems are furnished with unequally pinnate leaves made up of eight to ten pairs of ovate lance-shaped or linear leaflets. *G. officinalis*, the Goat's Rue, was at one time in repute as a cordial in fevers and convulsions, but it has long fallen into disuse. The generic name, derived from the Greek signifying milk, refers to its supposed property of increasing the milk of animals which feed upon the plants. [A. A. B.]

GALENIA. A genus of *Tetragoniaceæ* consisting of herbs or shrubs from the Cape of Good Hope, usually much branched, hairy or papillose, with alternate or opposite entire fleshy leaves and sessile flowers, generally cymose or paniculate. Calyx deeply four or five-cleft, coloured within; corolla absent; stamens eight or ten; ovary two to five-celled; capsule woody or corky, varying in shape according to the number of cells in the ovary. [J. T. S.]

GALEOBDOLON. The name of a section of *Lamium* distinguished by having the corolla tube obliquely annulate within, contracted below, and dilated and subventricose above the annulus, where it is also somewhat recurved and lengthened out; and by the helmet being elongated and narrowed at the base. The principal species, *Lamium Galeobdolon*, our native Archangel, is sometimes separated under the name of *G. luteum*. [T. M.]

GALEOGLOSSA. The name of certain Ferns, otherwise referred to *Niphobolus*.

GALEOPSIS. A genus of labiates, called Hemp-nettles, distinguished by their equally five-toothed calyx, by the two lower stamens being longer than the other pair, by the two-lipped corolla, of which the upper lip is arched, the lower three-lobed, and by the diverging anther-cells, which open longitudinally. The commonest species is *G. Tetrahit*, an annual weed frequently met with in cultivated ground. It grows to the height of a foot or more, and is well marked by its hispid stem, which is singularly swollen beneath the joints, by the very long rigid calyx teeth, and by the purple, sometimes white, flowers. *G. Ladanum* has the stems less hairy than the last, and the stem is not swollen beneath the joints; it grows principally on a limestone or chalk soil. *G. versicolor* approaches in habit to *G. Tetrahit*, from which it may be distinguished by its more showy yellow flowers having a blotch of purple on the lower lip; this is found in several parts of England, but is most abundant in Scotland, especially in cultivated fields among the Highlands. *G. ochroleuca*, with large pale yellow flowers without spots, grows in sandy cornfields, but is rare. French, *Galéope*; German, *Taube Nessel*. [C. A. J.]

GALEOTTIA. This name has been given to an obscure Mexican orchid supposed to be closely allied to *Batemannia*, but to differ in having a large ovate gland and short caudicle, *Batemannia* having no caudicle. It has besides been applied to a genus of *Acanthaceæ*, which has also been called *Glockeria*, and is related to *Stenostephanus*, from which it differs in its bilabiate corolla. This latter is a Mexican shrub, with nutant crimson flowers in terminal panicles. [T. M.]

GALEWORTS. Lindley's name for the *Myricaceæ*.

GALIACEÆ. (*Stellates, Madderworts.*) A natural order of calycifloral dicotyledons belonging to Lindley's cinchonal alliance of epigynous Exogens. The order has been sometimes called *Stellatæ* from the star-like arrangement of the leaves; and by many it is reckoned as a suborder of *Ru-*

biaceæ, which is thus made to include both *Cinchonaceæ* and *Galiaceæ*. Herbs with whorled exstipulate leaves, and angular stems. Calyx superior, the limb obsolete, four to five or six-lobed; corolla gamopetalous, rotate or tubular, regular, divided like the calyx; stamens equal in number to the corolline lobes and alternate with them. Ovary two-celled, with solitary erect ovules; styles two; stigma undivided. Fruit two-celled, with two seeds; embryo in the axis of horny albumen. Natives of the northern parts of the northern hemisphere, and of high mountains in South America and Australia. The order contains some plants used for dyeing and some having tonic qualities. The horny albumen of goose-grass or cleavers (*Galium Aparine*) has been used as a substitute for coffee. The root of madder (*Rubia tinctorum*) is employed as a dye, and supplies the Turkey-red; that of *Rubia cordifolia* furnishes the dye called munjeet in India. The leaves of woodruff (*Asperula odorata*) are fragrant when dried. There are ten known genera and about 340 species. Examples: *Galium*, *Rubia*, *Asperula*. [J. H. B.]

GALIMETA WOOD. The timber of *Bumelia salicifolia*.

GALINGALE. *Cyperus*, especially *C. longus*.

GALININGUE. (Fr.) A kind of olive.

GALINSOGA. A genus of annual South American weeds of the composite family, furnished with opposite ovate three-nerved nettle-like leaves, and small axillary or terminal stalked flower-heads having an involucre of three to five ovate scales, enclosing four or five white or purple ray florets with pistil only, and numerous yellow tubular perfect disk florets; the unsexed achenes are crowned with a pappus of lacerated chaffy scales, and seated on a conical chaffy receptacle. *G. parviflora*, a species with smooth leaves, white ray florets, and a habit like that of the annual mercury, is naturalised in many countries, and has lately become a pest in the market gardens around London. [A. A. B.]

GALIOTE. (Fr.) *Geum urbanum*.

GALIPEA. A genus of rutaceous shrubs or small trees, natives of tropical America, the flowers of which have a salver-shaped corolla with spreading acute lobes; four to seven stamens, somewhat adherent to the petals, sometimes all fertile, but usually only two of them antheriferous; a cup-shaped disk; five styles, becoming ultimately fused into one, with a four to five-grooved stigma; and five or fewer carpels. The bark of one or more of the species, such as *G. officinalis* and *G. Cusparia*, is used in medicine as an aromatic or stimulant tonic. Dr. Hancock, who had large experience of its use in tropical South America, even preferred it to cinchona in the treatment of fever. In this country it is but little used, being deemed inferior to other remedies, and possibly from the fact that a false Angostura bark was at one time, through inadvertence or cupidity, substituted for the genuine bark. This false bark occasioned several dangerous accidents, which led some of the continental governments to prohibit the use of Angostura or Cusparia bark. The spurious bark proved to have been really derived from the deadly nux-vomica tree. This nux-vomica bark, it appears, was also sold in Calcutta for the harmless bark of *Soymida febrifuga*; and a preparation of the former, to be used instead of quinine by the Indian army, was made under the impression that it was a valuable and harmless remedy. Dr. O'Shaughnessy fortunately discovered the error in time to prevent the dreadful consequences which might have ensued from the employment of this preparation. The reader is referred to *Pereira's Materia Medica* (ll. part II. p. 1915) for full details as to the means, chemical and otherwise, of distinguishing the true from the false Angostura barks, the most readily recognisable features of the true bark being, that it occurs in pieces which are not so much twisted or bent as the nux-vomica bark, that it has a disagreeable odour which is not noticed in the false bark, and from being lighter is more readily broken or cut. It is stated that the natives employ the true Angostura bark to stupefy fishes, in the same way that cinchona bark is said to be used by the Peruvians. [M. T. M.]

GALIUM. The typical genus of *Galiaceæ*, consisting of numerous herbaceous plants, distinguished by having a minute almost obsolete calyx, a four-lobed wheel-shaped almost tubeless corolla, and a fructification consisting of two seed-vessels, each containing a single dry seed. Upwards of 160 species are described, of which fourteen are British. They all agree in having square stems and whorled leaves; and the roots of most afford a purple dye. Some are perennials, others annual. The predominating colour of the flowers is white; and the number of leaves in a whorl varies from four to ten. Of the British species, *G. verum*, Bedstraw, and *G. cruciatum*, Cross-wort, are perennial, and bear yellow flowers. *G. Aparine*, Goose-grass, derives its English name from the avidity with which the young stems and leaves are eaten by geese; it is called Cleavers on account of the tenacity with which the fruit adheres to any rough and soft substance. It is a long straggling annual plant, abundant in hedges and among bushes, through which it climbs, supporting itself by the hooked prickles with which it is copiously invested. The globular seeds covered with hooked prickles, found on the dress of persons who walk through bushy places in autumn, are derived from this plant. *G. saxatile* is the pretty little species, only a few inches high, which is so frequently seen in heathy places, associated with wild thyme, bird's-foot trefoil, and tormentil; its flowers are of a brilliant white, and are succeeded by reddish fruit which is conspicuous by its abundance. French, *Gaillet*; German, *Labkraut*. [C. A. J.]

GALL OF THE EARTH. *Mulgedium floridanum*, or, according to Dr. Asa Gray, *Nabulus Fraseri.*

GALLESIA. A genus of *Phytolaccaceæ*, a large Brazilian tree, with alternate stalked ovate or oval entire pellucid-dotted, smooth leaves, small tubercular stipules, and a many-flowered terminal panicle of sessile flowers, each with three bracteoles, and having a four-parted calyx and numerous stamens in two rows. The fruit is a samaroid achene, with a large scimetar-shaped wing at the apex. [J. T. S.]

GALLINHA CHOCA. *Erythroxylon suberosum.*

GALLS. Excrescences of various kinds and forms produced in plants by the presence of the larvæ of different insects. The forms which they assume are multitudinous, and the changes produced in the tissues various. They occur on all parts of the plant, and sometimes in great quantities, but they appear in general to do little harm if they do not attack the parts of fructification. It is probable that the change of growth depends in the first place upon some acrid fluid discharged together with the egg. The process of caprification, in which figs are stimulated to generate juicy instead of dry tissues, is strictly analogous, though there is no external alteration of form. The rootlike galls in grasses are produced by larvæ between the sheath and the stem, and not penetrating the substance. [M. J. B.]

GALPHIMIA. An anagram of *Malpighia*, applied to a genus of Mexican malpighiaceous shrubs, some of which are cultivated as evergreens in our stoves. They have a five-parted calyx whose segments are mostly destitute of glands; five stalked petals generally ribbed on their outer surface; ten stamens slightly adherent at the base; and a three-lobed, three-celled ovary with a solitary pendulous ovule in each compartment. The fruit consists of three two-valved carpels. [M. T. M.]

GALUNCHA. An Indian febrifuge prepared from the stems of *Tinospora verrucosa* and *cordifolia.*

GAMASS. The Squamash or Biscuitroot, *Camassia esculenta.*

GAMBIR. A powerful astringent obtained from *Uncaria Gambir*, and employed as a substitute for catechu.

GAMBOGE, AMERICAN. The juice of *Vismia guianensis.* —, CEYLON. A gum-resin obtained from *Garcinia Cambogia*, also called *Cambogia Gutta* and *Hebradendron gambogioides.* —, MYSORE. The gum-resin of *Garcinia pictoria*, otherwise *Hebradendron pictorium.* —, SIAM. A gum-resin supposed by some to be the produce of *Garcinia cochinchinensis*, and by others that of *G. Cambogia.*

GAMO. In Greek compounds = united by the edges; thus *gamophyllus* signifies leaves united by their edges, while *gamosepalous* means monosepalous, and *gamopetalous*, monopetalous.

GAMOLEPIS. A small genus of South African *Compositæ*, having smooth entire three-lobed or pinnatifid leaves, and terminal solitary or corymbose, long-stalked flower-heads containing numerous florets. They are nearly related to *Leucanthemum*, but differ in the scales of the involucre being in one series, and more or less united by their margins so as to form a cup. The ray florets are strap-shaped, and contain only a pistil, the disk florets being tubular and perfect; while the achenes are smooth wingless and destitute of pappus. [A. A. B.]

GAMOPLEXIS *orobanchoides* is the name of a tuberous-rooted North-west Indian orchid which is destitute of leaves, and has the aspect of an *Orobanche*. It is notable for its parasitism, which is rare amongst endogenous plants. Dr. Falconer states that the tuberous rhizome emits no root-fibres by which to fix itself on other plants, but is itself matted over by their slender rootlets, giving rise to the appearance of the plant being the subject of a parasitical growth rather than a parasite itself. The stem is one to two feet high, pale straw colour, terminating in a long raceme of flowers. The lip is combined with the sepals and petals to form a tubular perianth, whence the name of the genus; and the pollen is not waxy or powdery, but granular as in *Gastrodia*, which differs in the lip being free, instead of connate with the tube of the perianth. [A. A. B.]

GANDASULI. (Fr.) *Hedychium.* — À BOUQUETS. *Hedychium coronarium.*

GANGLIA. The mycelium of certain fungals.

GANGRENE. A disease ending in putrid decay.

GANITRE. (Fr.) *Elæocarpus.*

GANNE. (Fr.) *Molinia cærulea.*

GANT DE NOTRE DAME. (Fr.) *Campanula Trachelium*; also *Aquilegia vulgaris*, and *Digitalis purpurea.*

GANTELÉE. (Fr.) *Campanula Trachelium*; also *Digitalis purpurea.*

GANTIÈRE. (Fr.) *Digitalis purpurea.*

GANYMEDES. A name proposed for a few species of *Narcissus*, e. g. *N. triandrus*, *pulchellus*, *nutans*, &c. They are called Rush Daffodils from the rush-like leaves. The perianth has a slender drooping tube and reflexed limb, the cup or coronet is equal to or shorter than the limb, the sepaline stamens are prolonged, and the style is straight and slender. [T. M.]

GARANCE. (Fr.) *Rubia tinctorum.*

GARB. *Salix babylonica.*

GARBANZOS. The Spanish name of the Gram, *Cicer arietinum.*

GARCINIA. A genus of *Clusiaceæ*, consisting of several opposite-leaved trees whose stems yield, in greater or less quantity, a yellow resinous juice which in *G. Cambogia* is known as Ceylon Gamboge. The chief features of the genus are: unisexual or rarely perfect flowers, having a calyx of four rounded leaves, and four petals of similar form; in the males numerous stamens which are free or united into one or four parcels; and in the females a few barren stamens surrounding a globose ovary which is from two to ten-celled with one ovule in each cell, and is crowned by a shield-like entire or lobed stigma.

The greater portion of the species are found in India and the Malay Archipelago. All have glossy laurel-like leathery leaves. The flowers are either white tinged with pink, or yellow, and arranged in clusters in the axils of the leaves or in panicles at the ends of the twigs. The species from which the Gamboge or Cambogo of commerce is obtained in largest quantity are commonly known under the name CAMBOGIA; which see. *G. pictoria*, which is found in the Coorg district of Malabar, yields a gamboge said by Dr. Christison and others to have properties similar to those of the Ceylon and Siam gamboge, being 'excellent as a pigment, efficient as a purgative, and equal to the gamboge in common use.' It does not, however, appear to be imported in any quantity, by far the greater portion brought to this country being sent from Siam to Singapore and shipped from that port. This plant is a tall tree with elliptical leaves, small yellow axillary solitary flowers, and berries about the size of a cherry, with four one-seeded cells.

The Mangostan or Mangosteen (*G. Mangostana*), so well known for its luscious fruit, is found in the Malay Islands, where it grows to a tree of middling stature with a conical head, the branches furnished with glossy leathery elliptical-oblong pointed leaves, and the flowers single and nearly sessile at the ends of the twigs, of a dull red colour, and as large as dog-roses. Dr. Abel, writing of the fruits of Batavia, says: 'First in beauty and flavour was the celebrated Mangostan. This, so often eulogised by travellers, certainly deserves much of the praise bestowed upon it. It is of a spherical form, of the size of a small orange, when ripe reddish-brown, and when old of a chestnut-brown colour. Its succulent rind is nearly the fourth of an inch in thickness. It contains a very powerful astringent juice, and in wet weather exudes a yellow gum which is a variety of gamboge. On removing the rind, its esculent substance appears in the form of a juicy pulp having the whiteness and solubility of snow, and of a refreshing, delicate, delicious flavour. We were all anxious to carry away with us some precise expression of its qualities; but after satisfying ourselves that it partook of the compound taste of the pine-apple and peach, we were obliged to confess it had many other equally good but utterly inexpressible qualities.' Any amount of the fruit may be eaten without injury, and it is said to be given to those afflicted with fever along with the sweet orange. The Chinese use the bark as a basis for a black dye, and it is also used in dysentery. In 1855 it first produced its blossom and fruit in this country in the gardens of the Duke of Northumberland at Syon, from whence it was figured by Sir W. J. Hooker in the *Botanical Magazine* (t. 4847). It has been cultivated in the south-

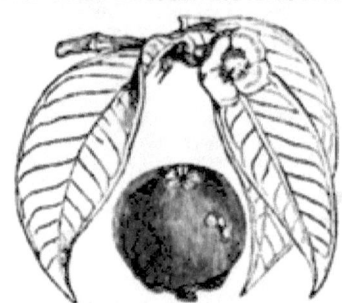

Garcinia Mangostana.

ern and eastern parts of India, but does not there attain the same perfection as it does in the Malay Archipelago. A few small-flowered species, with the stamens in four parcels, and the two-celled ovary crowned with a shield-like stigma, are by some authors kept as a separate genus under the name *Discostigma*. [A. A. B.]

GARDENER'S GARTERS. *Digraphis arundinacea variegata.*

GARDENIA. A genus of *Cinchonaceæ*, and a favourite with cultivators on account of the fragrance and beauty of its flowers. It consists of trees or shrubs, frequently spiny, and indigenous in tropical Asia and Africa, as well as at the Cape of Good Hope. The calyx tube is ovate, the limb variously divided; corolla white or yellowish, more or less funnel-shaped, with its limb divided into from five to nine somewhat twisted segments, and its tube considerably longer than the calyx; anthers five to nine, sessile on the throat of the corolla, from which they project to a short distance; ovary partially five-celled, with numerous ovules; fruit berry-like, crowned by the limb of the calyx. *G. gummifera* and *G. lucida*, East Indian species, yield a fragrant resin something like Elemi, which might be turned to some account. The fruit of *G. campanulata* is stated by Roxburgh to be used by the natives as a cathartic, and also to wash out stains in silk. Several kinds of this beautiful genus are in cultivation. The Cape Jasmine, *G. florida* and *G. radicans*, both found with double flowers of exquisite fragrance, and *G. Stanleyana*, with long trumpet-shaped blossoms, are among the most beautiful species. [M. T. M.]

GARDENIOLA. The name of a Brazilian shrub of the *Cinchona* family, having numerous flattened branches, with polygamous flowers; the males in groups of three

GARD]

or fire, sessile on the ends of the branches; and the females in similar groups, but stalked. The corolla is salver-shaped, with a short and slightly inflated tube, hairy throat, and four-lobed limb. The fruit is berry-like, black, and two-celled. [M. T. M.]

GARDE-ROBE. (Fr.) *Artemisia Abrotanum.*

GARDNERIA. A genus of *Loganiaceæ* composed of two scandent opposite-leaved bushes found in India and Japan. They differ from most in the family in having little berried two-celled fruits, with one or two instead of many seeds in each cell; and the lateral attachment of these in a shield-like manner, not erect from the base, serves to distinguish the plants from their nearest allies. The narrow or broadly lance-shaped leaves are smooth and entire, and the small yellow blossoms are disposed in loose panicles which arise from the axils of the leaves, and in size and form are not unlike those of the privet. The ripe fruits are scarlet berries, about the size of a large red currant, with two seeds. [A. A. B.]

GARDOQUIA. The name of a genus of labiates, having the teeth of the calyx short, straight, and nearly of the same size; the corolla with a long almost straight tube, its upper lip notched, the lower in three lobes, the middle one of which is broadest; style bifid at the end, the divisions small and equal in length. The name was given in honour of Gardoqui, a Spanish financier who promoted the publication of a Flora of Peru. The species are low shrubs or undershrubs chiefly natives of Peru and Chili, rare in North America, and having showy usually pink flowers. [G. D.]

GARGET. An American name for *Phytolacca decandra.*

GARIDELLA. A genus of *Ranunculaceæ* consisting of a few herbs from the Mediterranean region and temperate Asia, resembling *Nigella*, but more slender in habit, the ovary containing two or three, not five or more carpels, and the styles very short. They are erect annuals with dissected leaves, and small solitary whitish flowers at the extremity of long peduncles. *G. Nigellastrum*, which is found in the south of France, Spain, and Crete, is the most frequent. [J. T. S.]

GARLAND FLOWER. A common name for *Hedychium*; also applied to *Daphne Cneorum, Pleurandra Cneorum*, and *Erica persoluta.*

GARLIC, or GARLICK. *Allium sativum.* —, HEDGE. *Sisymbrium Alliaria.* —, HONEY. *Nectaroscordum.*

GARLIC SHRUB. *Bignonia alliacea*; also *Petiveria alliacea.*

GARLICKWORT. *Sisymbrium Alliaria.*

GARO DE MALACA. (Fr.) *Aquilaria ovata.*

GAROU. (Fr.) *Daphne Mezereum.*

The Treasury of Botany.

520

GAROUPE. (Fr.) *Cneorum tricoccum.*

GAROUSSE. (Fr.) *Lathyrus Cicera.*

GARNETBERRY. *Ribes rubrum.*

GARRYACEÆ. (*Garryads.*) A natural order of monochlamydeous dicotyledons belonging to Lindley's garryal alliance of diclinous Exogens. Shrubs with opposite exstipulate leaves and catkin-bearing imperfect flowers surrounded by united bracts. The staminate flowers have a four-leaved perianth, alternating with four stamens. The pistillate flowers have an adherent two-toothed perianth. Ovary one-celled; styles two; ovules two, pendulous with long cords. Fruit a two-seeded berry; embryo minute in the base of fleshy albumen. The wood is not arranged in circles, and there is an absence of dotted vessels. They are natives chiefly of the temperate parts of America. The few species are distributed between the genera *Garrya* and *Fadyenia*, the latter of which, however, is not generally adopted. [J. H. B.]

GARRYA. The only admitted genus of *Garryaceæ*, composed of opposite-leaved evergreen bushes, found in California, Mexico, Cuba, and Jamaica. *G. elliptica* is one of the most desirable evergreens we have in our gardens, especially as it produces its pale greenish-yellow blossoms in the spring months when little else is in bloom. It was sent from California by the lamented Douglas in 1818. This bush, with much the aspect of an evergreen oak, may be seen sometimes eight to ten feet high, its branches clad with dark green elliptical leaves. The flowers are male and female on different plants, arranged in elegant drooping necklace-like catkins which proceed from near the apex of the shoots, and are often from four to seven inches long. In the male plant (which alone is in cultivation) they are clothed with silky hairs, and a plant covered with these tassels of pale yellow flowers waving in the wind, has a singularly graceful appearance. Each link of the necklace (if we may so speak) is composed of a cup-shaped bract enclosing three flowers, each having a calyx of four divisions and four stamens. In the female the disposition of the flowers is the same. The fruit is a two-seeded berry somewhat like that of the hawthorn in size and shape. The calyx of some species is destitute of the two teeth which are seen in *G. elliptica*, and the tips of the calyx-leaves in the male flowers remain united to each other; these are separated by some authors, who give to them the name of *Fadyenia*, but such differences are not distinctive. The flowers of some species are in compound instead of simple racemes, but none can be compared to *G. elliptica* for beauty. [A. A. B.]

GARUGA. An Indian name applied to a genus of *Amyridaceæ*, consisting of trees whose flowers have a five-cleft bell-shaped calyx; five petals inserted between the notched and glandular lobes of a fleshy disk; and a pulpy fruit with five or fewer bony one-seeded stones. *G. pinnata*, an

Indian species, and *G. madagascariensis* are occasionally met with as stove shrubs, with fine pinnated foliage and panicles of yellow flowers. [M. T. M.]

GARVANCE. (Fr.) *Cicer arietinum*.

GASTERIA. The name of certain species of *Aloe*, which are regarded by some as being distinct from that genus. They are mostly dwarf stemless plants with the thick succulent spotted or warted tongue-shaped leaves often rigidly two-ranked, and the long arching spikes of green-tipped red flowers freely produced. The curvature and bellying of the flower-tubes has suggested the name; the distinguishing character of the group is indeed furnished by the curved tube of the perianth swollen at the base, by the stamens being adglutinated to the perianth at the base, and by the capsule being subcostate. They are mostly ornamental plants, and, like the other aloes, natives of the Cape of Good Hope. [T. M.]

GASTEROMYCETES. One of the six great divisions of *Fungi*, containing those genera with naked spores in which the fruit-bearing surface is either permanently concealed in a surrounding peridium, or in which, when the peridium bursts, the hymenium is complicated like the crumb of a loaf, so that a small portion only is exposed. In *Montagnites*, however, the hymenium consists of true gills. The genera are divisible into six natural groups as follows:—*Podaxinei*: mostly clavate; hymenium sinuous, enclosed at first in a volva-like peridium, and exposed partially by its rupture, withering or entirely drying up so as to form a dusty mass. *Hypogaei*: subterraneous; peridium seldom distinct. *Phalloidei*: hymenium at first enclosed in a gelatinous volva, at length diffluent. *Nidulariacei*: peridium mostly cup-shaped, enclosing several sporangia. *Trichogastres*: subglobose, not having a distinct volva; hymenium at first cellular, at length leaving a dusty mass of threads and spores. *Myxogastres*: hymenium and mycelium at first gelatinous. [M. J. B.]

GASTONIA. The name of a genus of Ivyworts, distinguished by having the corolla with five or six petals; the stamens ten to twelve, attached to the petals, and apparently in pairs opposite to them; the fruit a dry berry with eighteen cells, each of which contains one seed. The name was given by Commerson in honour of Gaston de Bourbon, son of Henry IV. The only species, *G. palmata*, is a native of Mauritius. [G. D.]

GASTRANTHUS. A genus of *Gesneraceae* containing two species from South America. They are undershrubs with opposite oblong crenate leaves, and few umbellate flowers. The divisions of the calyx are lanceolate; the corolla oblique and shortly spurred, with the limb cut into five unequal small roundish lobes; the four didynamous stamens included; the disk very small, but swelled on one side into a large gland half covering the ovary; and the apex of the style cup-shaped, and slightly bilobed. [W. C.]

GASTRIDIUM. A genus of grasses of the tribe *Agrostideae*, consisting of a single species, *G. lendigerum*, or, as it is sometimes called, *G. australe*, one of our rarer British species, and very common in the Mediterranean region. It is an elegant erect-growing annual plant, six or eight inches high, with the panicle contracted into a loose tapering spike two to three inches long, of a pale green, and shining with a satiny lustre. It has been separated from *Agrostis* on account of the polished enlarged base of the outer glumes. [T. M.]

GASTROCHILUS. A genus of *Zingiberaceae*, whose flowers have a tubular calyx, and a corolla with a long tube, the outer segments of the limb equal, the inner ones unequal, the two lateral wide, united at the base with the filament to form a kind of tube, the middle segment or lip large and distended, whence the name. *G. pulcherrima*, a native of Rangoon, and one or two other Indian species, are occasionally met with in cultivation, and are very ornamental. [M. T. M.]

GASTRODIA. This is the genus which gives its name to a small tribe (*Gastrodieae*) of the orchid family characterised by the granular instead of waxy or powdery pollen-masses. There are two known species, *G. Cunninghamii* from New Zealand, and *G. sesamoides* from Tasmania and Australia, both leafless parasites with the aspect of *Orobanche*, and like that found growing on the roots of other plants. The whole plant is of a uniform pale brown colour, the stems one to three feet high, furnished with a few obtuse bracts, and terminating in a long raceme of flowers, the sepals and petals united so as to form a tubular perianth, but the lip free and not connate with the perianth as in *Gamoplexis*. The root of the New Zealand species is eaten by the natives, who call it Peri; it is about eighteen inches long, as thick as the finger, and full of starch. [A. A. B.]

GASTROLOBIUM. An extensive genus of the pea family, peculiar to the south-western portions of Australia. It is known by the two-lipped and five-toothed calyx without bracts; the pea-flower corolla with petals nearly equal in length; and the stalked two-seeded ventricose or inflated pods, seldom larger than a pea. *Pultenaea* differs in having sessile pods, as well as heath-like foliage. Most of the Gastrolobes are bushes of two to four feet high, with twiggy stems furnished with opposite often whorled leaves varying much in form, and pretty yellow blossoms, sometimes in twos in the axils of the leaves, but more usually in short racemes arising from near the apex of the twigs. A number of the species of this and of allied genera are known in Western Australia as Poison-plants; and farmers lose annually a large number of cattle through their eating the foliage. Mr. James Drummond, in *Hooker's*

Journal of Botany (ii. p. 352), says: 'The finest and strongest animals are the first victims: a difficulty of breathing is perceptible for a few minutes, when they stagger, drop down, and it is all over with them. After the death of the animal, the stomach assumes a brown colour, and is tenderer than it ought to be; but it appears to me the poison enters into the circulation and altogether stops the action of the lungs and heart. The raw flesh poisons cats, and the blood, which is darker than usual, dogs; but the roasted or boiled flesh is eaten by the natives and some of the settlers without their appearing to suffer any inconvenience.' The poisonous effects were attributed by Mr. Drummond, at the time he wrote this, to a species of *Lobelia*, but he afterwards found out that they were due to the plants of this and allied genera. Dr. Harvey says the worst of the Poisonplants is *G. bilobum*. This plant has oblong nearly smooth slightly two-lobed leaves, placed four in a whorl round the stem, and terminal umbels of pretty yellow flowers, the keel and wing petals marked with purple. *G. spinosum* has similar properties. Altogether there are about a dozen species in cultivation in greenhouses. The generic name has reference to the bellied form of the pods. [A. A. B.]

GASTRONEMA. A small genus of South African *Amaryllidaceae*, closely allied to *Cyrtanthus*, and not unfrequently united therewith. The perianth tube is slender below, curved and widely campanulate above, the limb short and reflexed: of the six stamens, which have decurrent conniving filaments and short anthers, the three upper are longer and incurved, the petaline ones inserted at the top, and the sepaline ones near the tube; the style is declinate. *G. clavatum*, the original species, is a pretty little bulb, with slender deciduous leaves and one or two white flowers striped with red. [T. M.]

GATEN, GATTEN, GATTER, or GATTERIDGE TREE. *Cornus sanguinea*; also *Euonymus europaeus*, and *Viburnum Opulus*.

GATILIER. (Fr.) *Vitex Agnus castus*.

GATTIE. An Indian gum obtained from the Baboul, *Acacia arabica*.

GAUB. An Indian name for the astringent medicinal fruit of *Diospyros Embryopteris*.

GAUDE or VAUDE. (Fr.) *Reseda Luteola*.

GAUDICHAUDIA. A genus of Mexican climbing shrubs, belonging to the *Malpighiaceae*, and remarkable for producing constantly two kinds of flowers, the most numerous and perfect of which have a five-cleft glandular calyx; five stalked toothed petals; five stamens, two of which are usually sterile; three ovaries united at their inner edge; and a fruit winged at the sides and back. The more imperfect flowers have a calyx without glands; no petals, or only rudimentary ones; and two ovaries with imperfect styles. The flowers are yellow. [M. T. M.]

GAUDINIA. A genus of grasses of the tribe *Aveneae*, now generally regarded as forming a section of *Avena*. [D. M.]

GAULTHERIA. A large genus of stiff branching ericaceous shrubs or small trees with evergreen leaves, principally inhabiting the American continent, extending from Magellaens' Strait in the south as far north as Canada and Vancouver's Island. A few are found in Asia, principally in the Himalayas and the mountainous parts of Java; and five or six occur in Tasmania and New Zealand. The leaves are leathery, smooth and shining, and in many species the young branches are covered with bristly hairs. The flowers are small, ovate, with a contracted mouth, and enclose ten stamens: they are white, scarlet, or rose-coloured, and produced singly or in racemes at the ends or from the sides of the branches. The five-lobed calyx frequently increases in size after the flowering period, and sometimes becomes fleshy. The anthers open by pores at the top, and terminate by two bristles. The fruit is small and nearly globular, and when ripe splits open through the middle of each of the five cells.

G. procumbens, a little creeping plant, of the Northern United States and Canada, grows about five or six inches high. The erect stiff branches bear tufts of shining, evergreen oval leaves at their summits; and the drooping white flowers, produced singly from the bases of the leaves, are succeeded by fleshy bright red berries, formed by the enlargement of the calyx which encloses the true fruit. All parts of this plant, which is commonly called Wintergreen in the United States, possess a rather pleasant peculiar aromatic odour and flavour, due to the presence of a volatile oil, which, when separated by distillation, is known as Wintergreen oil. It is of a pale green colour, having the same composition as birch-bark oil, and is employed medicinally as a cordial stimulant. The leaves also possess a considerable degree of astringency, and their tincture is useful in diarrhoea. The berries are known by various names, such as Partridge-berry, Chequer-berry, Deer-berry, Tea-berry, Boxberry, &c., and afford winter food to partridges, deer, and other animals. The plant is likewise called Mountain Tea, its leaves being used as a substitute for tea or for flavouring genuine tea.

The Shallon or Salal of the north-west coast of America, *G. Shallon*, is a small shrubby plant, growing about a foot and a half high, flourishing in shady pine forests where few other plants will live. Its dark purple fleshy berries, which are produced in great abundance, have a very agreeable flavour and make excellent tarts; they are much eaten by the natives, who prepare a kind of bread by mashing them together and drying them in the sun. [A. S.]

GAULTHÉRIE DU CANADA. (Fr.) *Gaultheria procumbens.*

GAURA. A genus of onagrads, in which the tube of the calyx is long and three or four-angled below; the corolla of four, rarely three petals, turned to the upper side; the stamens eight, rarely six, those opposite the petals shortest; the fruit a hard woody nut, with three or four prominent angles, and usually four-celled. The name, from the Greek signifying superb, is not generally applicable to the species. The plants are natives of North America, and have alternate leaves varying in outline, and the flowers in spikes, white or rose-coloured, rarely yellow, turning to reddish when fading. [G. D.]

GAYA. Tropical American herbs, belonging to the mallow family, having solitary yellow flowers, whose structure is that of the closely allied *Sida*, from which, however, the present genus is distinguished by the capsule, which consists of several one-seeded carpels, opening along the back by two valves, and thus allowing of the protrusion of an inner strap-shaped valve-like appendage. [M. T. M.]

GAYAC OFFICINAL. (Fr.) *Guaiacum officinale.*

GAYAL. An Indian name for *Agave vivipara.*

GAYBINE. *Pharbitis.*

GAYLUSSACIA. A genus of tropical American shrubs, belonging to the *Vacciniaceæ*, and named in honour of the celebrated French chemist M. Gay-Lussac. The leaves are terminated by a hard spine; the corolla is tubular, distended at the base; and the stamens are inserted into the calyx, the anthers being without horns. The ovary is inferior, and the fruit succulent, crowned by the limb of the calyx, with ten one-seeded stones. *G. Pseudo-Vaccinium* is a greenhouse shrub with pretty red flowers. [M. T. M.]

GAZANIA. A genus of low-growing herbs of the composite family, peculiar to Southern Africa. The greater proportion are stemless, with a rosette of pinnatifid leaves having linear segments, generally white with close-pressed silky down beneath. In the caulescent species, the leaves are mostly narrow oblong or lance-shaped, glossy green above, white beneath. The flower-heads are large and handsome, with yellow strap-shaped ray florets, and tubular disk florets usually of a darker colour. The principal characters of the genus are: an involucre of many scales, whose margins are united nearly to the summit, so as to form a sort of cup; neuter ray florets; perfect disk florets; and wingless achenes clothed with silky hairs, which nearly hide the double pappus of thin and delicate hairs. The double pappus serves to distinguish this from *Gorteria*, a South African genus of very similar appearance. One of the most handsome and best known of the pinnatifid-leaved species is *G. Pa-*

vonia, which has long been in cultivation as a greenhouse plant, and is a beautiful object when its large dark-centred orange-coloured flower-heads, nearly three inches across, are expanded. The plant is said to be one of the greatest ornaments of the waysides in its native country, opening its blossoms only in sunshine. Upwards of forty species are enumerated. [A. A. B.]

GAZLES. *Ribes rubrum.*

GAZON D'ESPAGNE, or **D'OLYMPE.** (Fr.) *Armeria maritima.* — TURC. *Saxifraga hypnoides.*

GEAN. The wild Cherry, *Cerasus Avium.*

GEASTER. A genus of puffballs distinguished by the outer coat or peridium being perfectly distinct from the inner, which contains the spores, and splitting ultimately into several divisions, so as to have the appearance of a star, whence the name of Earth-star. Sometimes the outer peridium consists of two separable coats, of which the inner becomes at length inverted, so that it is lifted up and supported by the tips of its lobes upon those of the outer coat, which gave rise to the Man Fungus of the older herbalists. The inner peridium is either sessile or stipitate, and sometimes without any trace of an aperture for the dispersion of the spores, while in several species there is a distinct orifice which is variously fringed, folded, &c. In *G. coliformis* there are numerous orifices, and many confluent stems. In a young state the hymenium, as in *Lycoperdon*, looks like the crumb of bread, and in that condition it has the same structure as the gills of an agaric, though afterwards it dries up, leaving behind a mass of threads and spores. In general each peridium springs from its own mass of spawn, but in a fine species which occurs in Cuba, Ceylon, and Japan, there is a common expanded mycelium. Some of the species, as *G. hygrometricus*, are extremely sensitive of moisture, and are driven about by the wind as shapeless masses, till the first shower expands them like the fruit of the *Mesembryanthemum*. Others, on the contrary, expand when dry, and contract when moist.

The Earth-stars are amongst our rarer, or at least more local fungi, and are found on leaves in shady places, or on exposed banks and sands. They are more common in the south-eastern and southern parts of England than in other parts of Great Britain. Species occur in all warmer latitudes, but do not ascend very high northwards, or if they occur at all it is only in small quantities. [M. J. B.]

GEBLERA. The name given by Fischer and Meyer to a Chinese herb of the spurge-wort family, now referred to FLÜGGEA: which see. [A. A. B.]

GEIGERA. The name of a shrub of the rue family, native of tropical Australia, with five-parted flowers, having the stamens inserted beneath a fleshy five-lobed disk, in the centre of which the five ovaries are placed. The fruit consists of from one to three carpels which are adherent at their base. [M. T. M.]

GEISSOIS. A genus of *Cunoniaceæ*, native of New Caledonia, distinguished by having a calyx of four leathery ovate sepals with shaggy hairs on the inside, no corolla, eight to ten stamens with rich crimson filaments an inch long, and a style bearing two stigmas. The seed-vessel is two-celled and two-valved, containing many compressed winged seeds. It consists of a small tree, bearing closely packed flowers in long racemes on the old wood, and opposite leaves with five slightly serrated leaflets. A plant of this genus has been lately introduced which is possibly distinct from the original species, *G. racemosa*, described by Labillardiere. [R. H.]

GEISSOLOMA. The name applied to a South African shrub, referred to the *Penæaceæ*, and distinguished from *Penæa* by the imbricated arrangement of the lobes of the perianth; by the presence of eight stamens, the anthers of which have not a fleshy connective; and by the pendulous ovules: thus affording a singular illustration of the great difference existing between some plants in certain cases, where nevertheless it is not considered advisable to place them in different groups, because, in spite of their numerous points of diversity, they are yet more closely allied one to the other than to anything else. *G. marginata*, a greenhouse shrub, has red flowers surrounded by a number of scale-like bracts. [M. T. M.]

GEISSOMERIA. A genus of *Acanthaceæ*, containing nine species from Brazil. They are undershrubs, with a tetragonous stem, oval or oblong leaves, and long red, often velvety flowers, in many-flowered spikes. These have a five-parted calyx, a tubular corolla dilated upwards, four stamens inserted near the base of the corolla tube, the filaments hairy at the base, and the one-celled anthers acute at both ends. The fruit is oval, and four-seeded. [W. C.]

GEISSORHIZA. A genus of South African *Iridaceæ*, one species of which has been found in Abyssinia. The plants have bulb-tuberous rhizomes, narrow setaceous or sword-shaped leaves, and a simple or branched stem bearing the large showy flowers in one-sided spikes. The perianth is funnel-shaped, with a short tube, and an ample six-parted nearly equal limb, the segments of which bear a nectariferous pore at the base; the three stamens are incinded; the style is filiform and declinate, with three linear wedge-shaped conduplicate stigmas; and the ovary is three-celled, with numerous ovules arranged in two rows in the central angles of the cells. The rhizomes are covered by the crustaceous or scarious remains of the bases of the leaves, which lie over each other like the tiles of a roof, and hence the name of Tile-root has been given to the plants. The Ixia-like flowers are very showy, and various in colour. [T. M.]

GELA. *Entada Purætha*.

GELASINE. A genus of *Iridaceæ* closely allied to *Trichonema*, with which it is united by many botanists. *G. azurea*, a dwarf bulbous plant from the Rio Grande in South America, is the type. [T. M.]

GELIDIACEÆ. A natural order of rose-spored *Algæ* belonging to the group which bears necklaces of spores (*Desmiospermeæ*), and amongst these distinguished by the placenta being axial or suspended by filaments in the cavity of the external or half-immersed capsules. It contains many very beautiful *Algæ*, especially in warmer latitudes, amongst which the *Hypneæ* are conspicuous, on almost every tropical coast, for the hooked tips of the fronds. *Gelidium corneum*, one of our commonest and most variable seaweeds, with its rigid compressed more or less repeatedly pinnate frond, occurs almost everywhere in some form or other. [M. J. B.]

GELINEÆ. Cells in algals secreting vegetable jelly.

GELL, or GILL. *Glechoma hederacea*.

GELSEMIUM. A genus of *Loganiaceæ*, consisting of an evergreen lactescent climbing shrub, found in the vicinity of rivers in the southern states of America. It has opposite lance-shaped shining leaves with small axillary glands, and few-flowered axillary fascicles of sweet-scented yellow flowers, which have a small five-lobed calyx, and a large funnel-shaped corolla, with a five-cleft almost equal limb. The fruit is composed of two separable jointed follicles containing numerous flat seeds. *G. nitidum* is called the Carolina Jasmine. [T. M.]

GEMINATE. United or collected in pairs.

GEMINI. Two together.

GEMINIFLOROUS. When two flowers grow together.

GEMMA. A leaf-bud; leaf-buds are sometimes also called *folifera gemmæ*, and flower-buds (alabastri), *floriferæ gemmæ*. The term Gemmæ is also applied to certain small reproductive bodies found in some liverworts, which are regarded as analogous to leaf-buds.

GEMMATIO. The act of budding; the manner in which young leaves are folded up in the bud prior to its unfolding.

GEMMULE. The plumule; also the ovule.

GEN. Persian manna, an exudation caused by insects on the stems of *Tamarix*, according to some authorities; but according to others it is produced by *Alhagi Maurorum*.

GENDARUSSA. A genus of *Acanthaceæ*, containing a single species, growing everywhere in India. It is a shrub with narrow leaves, and spicate flowers on axillary pedicels furnished with small bracts and bracteoles. The calyx is regularly five-parted; the corolla tube is short, and its limb two-lipped, with the upper lip arching; there are two stamens, with two-celled anthers; and a slender rigid four-seeded capsule. The leaves and stalk of *G. vulgaris* have, when rubbed, a strong and not unpleasant smell, and are, after being roasted, prescribed in India in cases of chronic rheumatism attended with swelling of the joints. [W. C.]

GENESTROLLE. *Genista tinctoria.*

GENET. *Genista.* — BLANC. *Cytisus albus.* — D'ESPAGNE. *Genista juncea.* — ÉPINEUX. *Ulex europæus.*

GENETTE. (Fr.) *Narcissus Pseudo-Narcissus.*

GENETYLLIS. A small genus of *Chamælauciaceæ*, distinguished by having sessile flowers with two small bracts, a calyx of five short obtuse entire lobes, five ovate slightly acute petals, twenty short stamens, the alternate ones sterile, a filiform bearded style, and a one-celled seed-vessel with four or five seeds. The flower-heads in one section of the genus are enclosed in large coloured ovate or oblong involucres about an inch in length, generally of a reddish hue, or white striped with red, and on this account they are exceedingly handsome plants, particularly *G. tulipifera*, *G. macrostegia*, *G. speciosa*,

Genetyllis tulipifera.

and *G. Hookeriana*. The leaves are either lanceolate, spathulate, or linear and three-angled, full of glandular dots, the upper ones generally membranaceous. The name of *Hedaroma* was given by Dr. Lindley to certain of the involucrate species from South-West Australia. The species are remarkable for the exquisite sweetness of their foliage, which with the half-ripe fruit retain their fragrance for such a length of time that they possibly might be considered worth collecting for the perfumer. They are mostly heath-like shrubs,

and are natives of the south and south-west parts of Australia. [R. H.]

GENEVRETTE. A wine made from juniper berries.

GENÉVRIER. *Juniperus communis.* — À L'ENCENS. *Juniperus thurifera.* — CADE. *Juniperus Oxycedrus,* — FEMELLE. *Juniperus tamariscifolia.* — MÂLE. *Juniperus cupressifolia.*

GENICULATE. Bent abruptly like a knee; as the stems of many grasses.

GENICULUM. The node of a stem.

GENIPA. A genus of tropical American trees of the cinchona family, nearly allied to *Gardenia*, but differing in the tube of the corolla, which is much shorter than in that genus, so that the five to six ovate segments of the limb are longer than the tube. The fruit is succulent, with a rather thick rind, crowned by the calyx, and tapering at each end. Some of the species furnish edible fruits. Thus *G. americana* produces the Genipap fruit, as large as an orange, and with an agreeable flavour. In Surinam the same fruit is called the Marmalade Box. The fruit of *G. brasiliensis* furnishes a violet dye. A few of the species are in cultivation as evergreen stove plants. [M. T. M.]

GENIPAP. The fruit of *Genipa americana.*

GENIPI BLANC. *Artemisia Mutellina.* — NOIR. *Artemisia spicata.*

GENIP TREE. *Genipa*; also an old West Indian name for *Melicocca bijuga.*

GENISTA. An extensive genus of leguminous plants, including the *Planta Genista* or *Plante genêt* of the French, from which a celebrated race of English kings, the Plantagenets, took their name, in consequence of their wearing a sprig of the plant as a distinctive badge. The genus contains more than a hundred species, chiefly abounding in the countries bordering on the Mediterranean, in Western Asia, and in the Canary Islands, three being indigenous to Britain. They are all small branching shrubs, sometimes armed with spines, seldom growing higher than five or six feet, and often not more than a foot, bearing simple or trifoliolate leaves, and abundant yellow flowers, which are produced either singly or in racemes or clusters from the angles of the leaves or at the ends of the branches. They have a five-toothed calyx; a papilionaceous corolla, of which the keel becomes curved backwards after flowering; ten stamens united into an entire sheath, bearing long and short anthers alternately; and a flattened or sometimes roundish pod constricted between the seeds.

The Petty Whin, *G. anglica*, is a small prickly struggling English shrub, with numerous decumbent stems, bearing small lance-shaped leaves and yellow flowers, and armed with sharp spines, whence the plant is frequently called Needle Greenweed. *G. tinctoria*, a native of Central and Southern Europe, common in England, is a low bushy tufted shrub, producing numerous woody unarmed stems, which send forth stiff erect angular green branches, clothed with leaves varying from narrow lance-shaped to broadly-elliptical, and bearing short racemes of yellow flowers at the ends of the branches. This was formerly of some importance as a dye plant, but it has long been superseded by dyes of foreign origin. It is commonly known under the name of Woadwaxen or Dyer's Greenweed, but the colour derived from it was a bright yellow, and it was only by afterwards dipping the yellow yarn or cloth into a blue solution of woad (*Isatis*) that the green tint was obtained. This was the process by which was obtained the once celebrated Kendal green, so called from the town of Kendal in Westmoreland, in the vicinity of which the plant was abundant, and where also the process was first introduced by Flemish emigrants in the reign of Edward III. The plant thrives upon very poor soil, and is regarded by agriculturists of the present day as an indication of the poverty of the land where it abounds. [A. S.]

GENISTELLA. *Genista anglica*, and others; also *Aspalathus spinosa*.

GENTIAN. *Gentiana*. The gentian root of the druggists is furnished by *G. lutea*.
—, BASTARD. *Hypericum Sarothra*, or *Sarothra gentianoides*. —, HORSE. *Triosteum*. —, SPURRED. *Halenia*.

GENTIANACEÆ. (*Gentianworts*.) A natural order of corollifloral dicotyledons belonging to Lindley's gentianal alliance of perigynous Exogens. Herbs, rarely shrubs, with opposite entire exstipulate, usually ribbed leaves, and showy flowers. Calyx divided, persistent; corolla persistent, imbricate or induplicate, and often twisted in æstivation; stamens alternate with the corolline segments; ovary of two carpels, placed to the right and left of the axis, one-celled with two, parietal often introflexed placentas; style one; stigmas two. Fruit a capsule or berry; seeds numerous, with fleshy albumen and a minute embryo. They are found in almost all parts of the world, some at high elevations, and others in hot tropical plains. Bitterness prevails generally in the order. Some of the plants have emetic and narcotic qualities. The root of the yellow gentian of the Alps (*Gentiana lutea*) is used medicinally as a tonic. In the Himalaya, chirata (*Agathotes Chirayta*) is employed as a bitter tonic. *Exacum bicolor* and *Ophelia elegans* are similarly used. There are about 70 known genera, and upwards of 500 species. Examples: *Gentiana, Swertia, Chironia, Erythræa, Chlora, Lisianthus, Menyanthes, Villarsia, Limnanthemum*. [J. H. B.]

GENTIANA. A large genus of herbaceous plants, giving their name to the order *Gentianaceæ*. The Gentians are perennial plants, with opposite ribbed leaves; a calyx of four or five valvate segments;
a four to five or occasionally ten-parted corolla; four to five stamens; and a one-celled ovary with two stigmas either separate and rolled back, or contiguous and funnel-shaped. The fruit is a two-valved, one-celled, many-seeded capsule. While blue is the most frequent colour, yet white, yellow, and even red flowers are met with. The red-flowered species are almost confined to the Andes; while blue-flowered species ascend the Himalayas to the height of 16,000 feet. The great majority are found in hilly or mountainous districts, in the northern hemisphere both in the old and new world, extending also to the tropics. Most of the genera which inhabit the elevated regions of the temperate or tropical zones are likewise found in the arctic or antarctic regions in great abundance, but this does not appear to be the case with *Gentiana*.

The abundance and beauty of the Gentians on the European Alps never fail to arrest the attention and demand the admiration of the traveller, who knows not whether most to admire the noble appearance presented by some of the taller more stately-looking species, such as *G. lutea*, or the intensely brilliant blue colour of some of the more lowly-growing species, such as *G. verna* or *acaulis*, and which may be found growing in profusion on little swards environed on all sides by ice-clad rocks and mighty glaciers. A few species are native in the British islands; the most frequently met with being *G. Amarella*, an erect branching annual, sometimes attaining to the height of ten or twelve inches, the flowers panicled, of a pale purple colour, the calyx with five segments, and the corolla with a fringe of hairs at the throat. *G. campestris* much resembles this, but has the parts of the flower in fours, not in fives, and has two of the lobes of its calyx larger than the other two, which they overlap. *G. Pneumonanthe*, a rare English perennial species, has a stem nearly a foot in height, the upper leaves linear, and the corolla an inch and a half long, bright blue without hairs in its throat, but with five greenish lines on the exterior. *G. verna* is a low-growing perennial, growing in dense tufts with very short flower-stalks, terminated by a single bright blue flower. It is a rare plant in this country, but abundant in mountainous meadows in Central and Southern Europe. *G. nivalis* is only found in Britain on some of the Scotch mountains: it is a slender branching annual, each branch being terminated by a blue flower about half an inch in length. Several of the species are in cultivation, such as *G. lutea, G. purpurea, G. acaulis* the Gentianella of gardens, and *G. cruita*, a North American kind, with the four lobes of the corolla fringed at the margin.

It is not only for their beauty that these plants are prized, but for their medicinal properties. All the species are, to a greater or less extent, pervaded by a pure bitter principle, which confers valuable tonic virtues on them, not always unaccompanied by some slight degree of narcotic or acrid effect. The roots of *G. lutea* are principally used in this country; they are collected in Switzerland and the Tyrol. The roots of *G. purpurea*, *G. punctata*, and *G. pannonica* are used for like purposes; indeed, almost any species might be employed that could be obtained in sufficient quantity. *G. cruciata* has been invested with imaginary virtues, because its leaves grow in the form of a cross; it has been recommended in hydrophobia. The Swiss make a liqueur from some of the species. Some of the Himalayan and North American kinds are used, like the European ones, as tonics. [M. T. M.]

GENTIANE. (Fr.) *Gentiana lutea*.

GENTIANELLA. A common name for *Gentiana acaulis*; also *Cryphiacanthus barbadensis*.

GENTIANELLE. (Fr.) *Exacum*.

GENTIANWORTS. Lindley's name for the order *Gentianaceæ*.

GEOCARYUM. The name of a genus of *Umbelliferæ* closely allied to *Bunium*, but distinguished from it by the styles being more erect, and by the structure of the fruits, which have five ridges, with as many wide oil-channels in the interspaces. In *Bunium*, on the contrary, there are two or three such channels in each interspace. *G. capillifolium* is a native of the south of Europe, Barbary, &c., and has a bulb-like stock, whence the name of the genus, which signifies earth-nut. [M. T. M.]

GEOCOCCUS. A diminutive cruciferous annual stemless herb of Western Australia, throwing out from the neck a series of pinnatifid leaves, from whose axils emerge the minute flowers. After flowering, the peduncles become deflexed, and bury the small seed-vessels about an inch beneath the surface of the ground. [T. M.]

GEODORUM. The generic name of a few terrestrial East Indian orchids of the tribe *Vandeæ*. They have tuberous roots, radical lance-shaped or elliptical leaves six to eighteen inches long, and lateral flower-scapes terminating in a nodding spike of flowers, which in some are of a pale green colour, the lip white and veined with yellow or purple lines, and in others are blush with a yellow spot on the lip. The sepals and petals are free and connivent; the lip hooded, sessile, and not jointed with the very short column; there are two bilobed pollen-masses with a short caudicle and a transverse gland. *G. dilatatum* and *G. citrinum* are in cultivation. [A. A. B.]

GEOFFROYA. A genus of pinnate-leaved South American *Leguminosæ* of the tribe *Dalbergieæ*, and differing from most in that group in having drupaceous fruits instead of thin dry pods. From *Andira*, to which it is most nearly allied, it differs in the flowers being in simple racemes instead of panicles, and usually yellow with a fetid smell, instead of purple and smelling sweetly; the calyx, also, is distinctly or deeply instead of obscurely toothed. There are five species enumerated.

G. superba is a tree of eighteen to twenty-five feet, with the habit of a tamarind, and is found in Brazil and Venezuela. Its pinnate leaves are four to six inches long; and the yellow fetid pea-flowers are in simple racemes the length of the leaves. The fruits, about the size and form of a walnut, have a greenish-yellow downy rind, a fleshy pulp, and a hard nut or stone enclosing a single seed. Humboldt, in his *Plantes Équinoxiales*, says this is a truly magnificent tree, from the disposition of its branches clothed with beautiful green leaves, as well as from the great abundance of its yellow flowers. According to the same authority, the wood is hard, susceptible of a fine polish, and useful for building purposes, while the fruits are not agreeable, but are eaten by children and much sought after by various animals. Mr. Gardner says the fruits are called Mari in various parts of Brazil. He found them to be the principal food of the inhabitants of the Ilha de St. Pedro, who boil them, eating the fleshy portion first, and then the kernel. Amendora, or Almond, is a name given to the tree on the Amazon. [A. A. B.]

GEOGLOSSUM. A genus of ascomycetous *Fungi* which in outward aspect has the appearance of *Clavaria*, and in consequence has been wrongly associated with the clavate *Fungi*. The whole plant is club-shaped, with the hymenium covering the entire surface of the club except at the base, the distinction between head and stem being generally only slightly marked. There are two distinct groups, those which are black or brown, and those which are green, purple, &c. In the former the sporidia are septate and much elongated, in the latter minute and simple. The species occur in closely-shaven lawns, in grassy pastures, and amongst *Sphagnum* or rottenwood, &c. Occasionally the stem is either viscid and scaly, or densely velvety. No species appears to be esculent. Earth-tongues occur in most temperate parts of the world, but they are more frequent in Europe than elsewhere. [M. J. B.]

GEONOMA. A genus of palms closely resembling *Chamædorea* in general appearance, and like these confined to the tropics of the western hemisphere, where they form part of the underwood of dense forests. There are about forty known species, a few of which are stemless, but the generality have slender reed-like polished stems, marked with rings or scars of fallen leaves, and bearing at their summit a tuft of large leaves, which are usually quite entire when young, but afterwards split so as to become irregularly pinnate. The male and female flowers are borne on the same plant, but are sometimes, though not always, on distinct spikes; each spike is enclosed in a double spathe, and is either unbranched or variously branched, the small yellow or purple flowers being seated in little hollows, the males in clusters of two or three, and the females solitary. The fruits are very small, and contain a single horny seed. None of the species possess any particular features of interest; the only useful purposes to which they are applied being that of supplying leaves for thatching huts, and flexible stems for walking-sticks. [A. S.]

GEOPHILA. A genus of *Cinchonaceæ*, called 'earth-loving' from the creeping habit of the plants. The species are natives of tropical America and the East Indies, and somewhat resemble violets in their mode of growth. They are nearly allied to *Psychotria*, from which they are distinguished by their flowers being in heads surrounded by a few bracts, and by the calyx being more deeply divided into five linear spreading segments. [M. T. M.]

GEORCHIS. A small genus of orchids found growing among moss, &c., in the damp woods of India and Java. The species have all the habit of *Goodyera*, and, according to Dr. Lindley, differ from that genus in the very sharp-pointed anthers and stigma, the latter of which splits into two long bristle-like arms. The slender stems throw out roots at intervals, and bear a number of ovate or heart-shaped leaves one to three inches in length, while the small white or pink flowers are disposed in terminal spikes. [A. A. B.]

GEORGINA. A name given by Willdenow to the genus *Dahlia*, but not generally adopted.

GEOTHERMOMETER. A thermometer constructed especially for determining the temperature of the earth.

GERANIACEÆ. (*Cranesbills*.) A natural order of. thalamifloral dicotyledons, characteristic of Lindley's geranial alliance of hypogynous Exogens. Herbs or shrubs with swollen joints, and opposite or alternate leaves, which are usually palmately veined and lobed, often stipulate. Sepals five imbricate, one of them sometimes spurred; petals five, with claws, contorted in bud; stamens usually ten, united by their filaments, some occasionally sterile; ovary of five bi-ovular carpels placed round an elongated axis, to which the styles adhere. Fruit formed of five one-seeded carpels, which finally separate from the base of the central axis or beak, and curve upwards by means of the attached styles; seed exalbuminous; embryo curved and doubled up, with plaited cotyledons. The plants are distributed over various parts of the world. The species of *Pelargonium* abound at the Cape of Good Hope; those of *Geranium* and *Erodium* are chiefly natives of Europe, North America, and Northern Asia. They have astringent and aromatic qualities, many of them are fragrant, and some have a musky odour. They are sometimes tuberous, and the tubers are eaten. There are numerous hybrids among the plants of this order, and it is not easy to determine the exact number of species, but about 640 are recorded. *Erodium*, *Geranium*, *Pelargonium*, and *Monsonia*, are examples. [J. H. B.]

GERANIUM. The Cranesbills, whose name is derived from the long central beak of the fruit, form a large genus of the *Geraniaceæ*, distinguished by having regular flowers, ten stamens with the filaments united at the base, and five carpels each tipped by a long glabrous awn ('the persistent style'), which becomes recurved when it separates from the central axis, not spirally twisted as in *Erodium*. They are herbs, very rarely undershrubs, growing in all temperate climates, having stems with enlarged joints and palmately lobed cleft or divided leaves, the lower ones stalked, the upper sessile. The one or two-flowered peduncles have small bracts at the base of the pedicels, and the flowers are often large and brightly coloured. Most of the species are astringent, particularly the North American *G. maculatum* or Alumroot, the rhizome of which is used in its native country instead of kino. The Tasmanian *G. parviflorum* is there called the Native Carrot, and its tubers used as food. There are about a dozen British species. The genus *Pelargonium*, to which belong the popular Geraniums of our gardens, is distinguished from the Cranesbills, by its irregular flowers, and adherent calycine spur. [J. T. S.]

GERANIUM, INDIAN. A term used by perfumers for *Andropogon Nardus*. —,
NETTLE. A popular name for *Coleus fruticosus*.

GERANION. (Fr.) *Geranium*, including *Pelargonium* and *Erodium*.

GERARDIA. A genus of *Scrophulariaceæ*, consisting of annual or perennial herbs, rarely shrubby at the base, and most if not all the species probably more or less parasitical on the roots of other plants. The leaves are opposite, or the upper ones alternate, all entire or very rarely cut. The flowers, sessile or pedicellate in the upper axils, or forming short terminal racemes, are usually purple or pink, and downy outside. The calyx is campanulate and five-toothed; the corolla obliquely tubular or campanulate, with five broad more or less spreading lobes; the stamens four, didynamous, not longer than the corolla, with two-celled anthers. The capsule opens loculicidally in two valves, and contains numerous small seeds. There are about two dozen species known, natives of North or South America, and most of them very handsome. All attempts to cultivate them have, however, failed. The dried specimens usually turn quite black.

GERBE-D'OR. (Fr.) *Solidago canadensis*.

GERBERA. A genus of that group of the *Compositæ* called *Mutisiaceæ*, in which all the florets are two-lipped. The genus is almost entirely African, and is represented in greatest numbers at the Cape. Upwards of a dozen species are known, all stemless perennial herbs, with their leaves usually clothed beneath with white cottony down; some, as *G. asplenifolia*, have pinnatifid leaves, calling to mind the fronds of *Asplenium Trichomanes*, but the larger proportion have the leaves oval or oblong and entire. The flower-stalk which rises from the crown bears on its apex a single large flower-head one to two inches across, the ray florets in which are yellow, purple, or blood colour, disposed in a single or double row, and containing a pistil only, while the disk florets are usually yellow and perfect. The achenes are cylindrical or flattened, smooth, beaked, and crowned with a pappus of two or more series of rough hairs. The name *Lasiopus* is by some authors given to those species which have a double row of ray florets. [A. A. B.]

GERMANDER. *Teucrium Chamædrys*; also a common name for the genus *Teucrium*. —, WATER. *Teucrium Scordium*. —, WOOD. *Teucrium Scorodonia*.

GERMANDRÉE. (Fr.) *Teucrium*. — AQUATIQUE. *Teucrium Scordium*. — FEMELLE. *Teucrium Botrys*. — MARITIME. *Teucrium Marum*. — OFFICINALE. *Teucrium Chamædrys*. — SAUVAGE. *Teucrium Scorodonia*.

GERMAN TINDER. The Soft Amadou, *Polyporus fomentarius*.

GERMEN. The ovary.

GERMINATION. The first act of growth by an embryo plant, connected with the absorption of oxygen and the extrication of carbonic acid. Germination ceases when the latter begins to be decomposed.

GERNOTTE. (Fr.) *Bunium Bulbocastanum*.

GÉROFLE, or GÉROFLIER. (Fr.) *Caryophyllus aromaticus*.

GERONTOGEOUS. Of or belonging to the Old World.

GEROPOGON. A genus of the composite family nearly related to *Tragopogon*, and consisting of an annual glabrous herb of the south of Europe, having simple stems, subamplexicaul entire elongated leaves, and capitules of purplish flowers solitary at the thickened apex of the stems. It differs from *Tragopogon* in having hair-like scales on the receptacle, and in the nature of the pappus. [T. M.]

GERVAO. A Brazilian name for *Stachytarpha jamaicensis*.

GESNERACEÆ. (*Cyrtandraceæ, Didymocarpeæ, Gesnerworts*.) A natural order of corolliform dicotyledons belonging to Lindley's bignonial alliance of perigynous Exogens. Herbs or shrubs often growing from scaly tubers, with wrinkled usually opposite leaves and showy flowers; calyx half-adherent five-parted; corolla more or less irregular, five-lobed; stamens two, or four and didynamous with the rudiment of a fifth, the anthers often combined. Ovary one-celled, surrounded at the base by glands or a ring. Fruit capsular or succulent, one-celled, with parietal placentas to the right and left of the axis. Natives of various parts of the world, chiefly the warmer regions of America. The succulent fruits are occasionally edible, and some of the species yield a dye. The leaves of some of them produce buds when laid on the soil. There are upwards of 80 genera, and nearly 300 species. *Gesnera*, *Gloxinia*, *Achimenes*, *Streptocarpus*, and *Cyrtandra* furnish examples. [J. H. B.]

GESNERA. The typical genus of *Gesneraceæ*, consisting of numerous tropical South American species, some of which are amongst the most beautiful of the herbaceous plants cultivated in our stoves. It has, like some other genera of the order, been much broken up by modern botanists. Some of the species are referred to a division called *Brachylomateæ*, in which there are squamose catkin-like stolones, as in *Achimenes*; others to the *Eugesnereæ*, which have tuberous rhizomes; and others to the *Rhytidophylleæ*, which are shrubby or subshrubby in habit. In the modern restricted form, *Gesnera* itself consists of plants with depressed tubers, a racemose corymbose or panicled inflorescence, and somewhat two-lipped corollas, much longer than the calyx, and with a short galea or upper lip, differing in the latter particular from *Dircæa*, another of the *Eugesnereæ*, in which the upper lip is very much elongated and fornicate. The flowers have also from two to five conspicuous glands. Of the five groups into which the restricted genus is divided, the following species are examples: *G. discolor*, *macrostachya*, *tuberosa*, *Sceptrum*, and *punctata*. The most familiar of the separated genera in the several tribes are the following, the first three of which belong to the *Brachylomateæ*, the next two to the *Eugesnereæ*, and the remainder to the *Rhytidophylleæ*:—

Nægelia: with a campanulate-cylindrical corolla tube, inflated beneath, and short subbilabiate limb, a five-crenate perigynous glandular ring, and a stomatomorphous stigma: ex. *G. zebrina*.

Kohleria: with an oblique corolla having a cylindrical or tumid deflexed tube, and ringent limb, five subequal glands, and a bifid stigma: ex. *G. Seemanni*.

Cryptoloma: with a straight corolla tube, and very short limb, five subequal glands, and a bifid stigma: ex. *G. hondensis*.

Rechsteineria: with a subbilabiate tubular corolla, two large dorsal glands and three smaller ventral ones, and a stomatomorphous stigma: ex. *G. allagophylla*.

Dircæa: with a gaping tubular corolla, the upper lip elongate-fornicate, the lower truncate, two connate dorsal glands, and a stomatomorphous stigma: ex. *G. bulbosa*.

Houttea: with a long corolla tube, and short spreading limb, and five glands, of which the dorsal are larger and connate: ex. *G. pardina*.

Moussonia: with a shorter subinflated corolla tube, and scarcely spreading limb, and five subequal glands: ex. *G. elongata*.

Herinequia; with a very long slightly curved corolla tube, and straight limb, and a five-lobed toothed ring: ex. *G. libanensis*. [T. M.]

GESNÉRIE BIZARRE. (Fr.) *Tydæa picta*.

GESNERWORTS. A name proposed by Lindley for the *Gesneraceæ*.

GESSE. (Fr.) *Lathyrus*. — CHICHE. *Lathyrus Cicera*. — DE PRÉS. *Lathyrus pratensis*. — GRANDE. *Lathyrus latifolius*. — SAUVAGE. *Lathyrus sylvestris*. — VELUE. *Lathyrus hirsutus*.

GESSETTE. (Fr.) *Lathyrus Cicera*.

GETHYLLIS. A small genus of Cape *Amaryllidaceæ*, allied to *Sternbergia* and *Oporanthus*, and consisting of dwarf bulbous plants, with linear leaves, and short one-flowered flower-scapes. The perianth tube is long cylindrical, the limb of six segments regular and spreading; the stamens inserted in the mouth of the tube, and sometimes by superficially doubled or trebled or multiplied numerously, with erect anthers; and the style connate with the perianth tube, free and exserted at top, with a capitately trigonous stigma. The capsule is berry-like and succulent, and is said to be esculent. *G. undulata* has the leaves remarkably waved at the edge, and ciliated with strong bristles. [T. M.]

GEUM. A genus of perennial *Rosaceæ*, deriving its generic name from the Greek word *geuo*, which signifies to have an agreeable taste, on account of the slightly aromatic flavour of the roots of some of the species. The main characters of the genus reside in the calyx, whose limb is five-cleft, with five little bracts on the exterior, and in the carpels which are dry with hardened hooked styles forming collectively a kind of burr. Two species are natives of Britain, *G. urbanum* and *rivale*. The former, known as Avens or Herb Bennett, has an erect slightly branched stem; the lower leaves deeply divided in a pinnate manner, with a large terminal lobe, the side lobes in pairs, some of them much smaller than the rest; the flowers yellow, with small spreading petals. The root of this plant, called by the old herbalists Clove-root, *radix caryophyllata*, has an aromatic clove-like odour, and, as it possesses astringent properties, it has been used in diarrhœa, dysentery, intermittent fevers, &c. It was formerly put into ale to give it a clove-like flavour and prevent it turning sour, and has been recommended to be chewed when the breath is foul. The Water-avens, *G. rivale*, has the leaves more hairy, the flowers much larger, drooping, and of a dull purple colour, and the head of fruits separated from the calyx by a short stalk. This plant is frequently found in a prolified state, that is, with a branch or a second flower in the centre of the original one. Other species of this genus are widely diffused over the temperate regions of the northern hemisphere. *G. canadense* is found in Canada and the United States, where it is known by the name of Chocolate or Blood root, and is used as a mild tonic. Several species are cultivated in this country: among the handsomest is *G. coccineum*, with scarlet flowers. [M. T. M.]

GEVIN. (Fr.) *Quadria*.

GHEKOOL, or GHET-KOL. An Indian name for the acrid tubers of *Typhonium trilobatum*.

GHETCHOO. An Indian name for *Aponogeton monostachyon*, the tubers of which are used like potatos.

GIBBER. A pouch-like enlargement of the base of a calyx, corolla, &c.

GIBBEROSE, GIBBOUS, or GIBBOSE. More convex or tumid in one place than another.

GIESEKIA. A genus of *Phytolaccaceæ*, containing tropical or subtropical annual herbs from Asia and Africa, with prostrate dichotomous stems, linear-oblong or spathulate entire fleshy leaves, rough with subcutaneous glands, and small greenish flowers, often becoming purple, in small umbellate or contracted cymes opposite the leaves. The fruit consists of three to five rough utricles. The name is sometimes written *Gisekia*. [J. T. S.]

GIESLERIA. A gesneraceous plant of herbaceous habit, now included in *Tydæa*.

GIGARTINA. A genus of the large natural order of rose-spored *Algæ*, called *Cryptonemiaceæ*. The capsules, which are globose and external, contain several roundish masses of spores; the frond is flat or cylindrical and mostly branched, composed of innumerable longitudinal and horizontal threads in a firm pellucid jelly; and the tetraspores are collected in little heaps or sori. The genus is very nearly allied to *Iridæa* and *Chondrus*. Many of the species are covered with projecting tubercular or spine-shaped processes, so as to make the frond rough like a rasp. *G. mamillosa* is often found amongst caragreen. *G. speciosa*, the Jelly-plant of the Australian colonists, is now referred to *Eucheuma*. [M. J. B.]

GIGOT. (Fr.) *Iris fœtidissima*.

GILIA. A genus of pretty American *Polemoniaceæ*. The calyx is bell-shaped five-cleft; the corolla funnel-shaped or approaching to bell-shaped; the stamens five in number, inserted at the throat of the corolla; and each cell of the capsule contains several angular seeds. The species have been distributed into several subgenera, of which the most familiar are: *Ipomopsis*, represented by *G. coronopifolia*; *Leptosiphon*, by *G. androsacea*; *Dianthoides*, by *G. dianthoides*, or, as it is also called, *Fenzlia dianthiflora*; *Leptodactylon*, by *G. californica*; and *Eugilia*, of which *G. tricolor*, *achilleæfolia*, and *capitata* are well-known illustrations. Most of them are pretty garden flowers, the different groups being for the most part regarded as distinct families by cultivators. *G. coronopifolia*, *elegans*, and *aggregata*, known in gardens as species of *Ipomopsis*, are biennials; the *Leptodactylons* are perennials; and the rest are mostly annuals. [C. A. J.]

GILIBERTIA. A genus of ivyworts, characterised by having the corolla with five to ten petals; stamens five to ten, attached to the petals and alternate with them; style short, ending in five to ten stigmas, which are at first erect, and then spreading. The genus was named in honour of J. E. Gilibert, a French botanist. The only species is a small shrub of Peru, having alternate oblong acute leaves, slightly toothed; and flowers in terminal compound umbels. [G. D.]

GILL, or GELL. *Nepeta Glechoma.*

GILLENIA. A genus of perennial herbaceous plants belonging to the *Rosaceæ* and allied to *Spiræa*, from which, however, it is well distinguished by its funnel-shaped calyx, very short stamens, and five carpels combined into a five-celled capsule, with two seeds in each cell. The leaves are ternate, with stalked serrated leaflets; the flowers whitish or red, axillary and terminal, on long flower-stalks. The roots are medicinal, possessing in a mild degree the properties of ipecacuanha. Two species only are described by botanists, both natives of North America: *G. trifoliata*, distinguished by its very narrow pointed stipules; and *G. stipulacea*, the stipules of which are large, ovate and deeply cut. [C. A. J.]

GILLIESIACEÆ. (*Gilliesiads.*) A natural order of hypogynous monocotyledons belonging to Lindley's lilial alliance of Endogens. Bulbous plants, with grass-like leaves, and umbellate flowers enclosed in a spathe. Perianth of two portions, the outer petaloid and herbaceous, six-leaved, the inner minute often five-toothed; stamens six, three sometimes sterile. Capsule three-celled, three-valved, many-seeded, opening in a loculicidal manner; covering of seed black and brittle; embryo curved; albumen fleshy. Natives of Chili. The genera are *Gilliesia* and *Miersia*, comprising about half a dozen species. [J. H. B.]

GILLIESIA. A genus of Chilian bulbous herbs, belonging to *Gilliesiaceæ*. They have linear flaccid root-leaves, and subdecumbent scapes, the flowers cernuous, greenish, inconspicuous, in an umbel with two leaf-like bracts at the base. The exterior involucre is five-leaved, with the two lateral interior ones much smaller than the others; the interior involucre many-leaved, surrounding a slipper-like perianth lobe; the stamens are united into a cup, the three posterior ones sterile. [J. T. S.]

GILLIFLOWER. A name corrupted from the French *Girofleé*: also written Gilloflower and Gillyflower, and further corrupted into July-flower; that of the old writers was *Dianthus Caryophyllus*, of the moderns, *Matthiola*. —, CLOVE. *Dianthus Caryophyllus*. —, MARSH. *Lychnis Flos cuculi*. —, QUEEN'S. *Hesperis matronalis*.

—, ROGUE'S. *Hesperis matronalis*. —, SEA. *Armeria vulgaris*. —, STOCK. *Matthiola incana, annua, &c.* —, WALL. *Cheiranthus Cheiri*. —, WATER. *Hottonia palustris*. —, WINTER, *Hesperis matronalis*.

GILLS. The lamellæ or plates growing perpendicularly from the cap or pileus of an agaric.

GILVUS. Dull yellow, with a mixture of grey and red.

GINGELLY OIL. The oil of *Sesamum orientale*.

GINGEMBRE. (Fr.) *Zingiber*.

GINGER-GRASS OIL. An essential oil obtained from *Andropogon Nardus*.

GINGER. *Zingiber officinale*. The ginger of the shops is the dried rhizomes of this plant; black or East Indian ginger is the unscraped rhizome prepared by scalding; white or Jamaica is the scraped rhizome dried in the sun. —, AMADA. *Curcuma Amada*. —, EGYPTIAN, *Colocasia esculenta*. —, INDIAN. *Asarum canadense*. —, MANGO. *Curcuma Amada*. —, RED. The same as East Indian ginger. —, WILD. *Asarum canadense*. —, WOOD. An old name for *Anemone ranunculoides*.

GINGERBREAD-TREE. The Doom Palm, *Hyphæne thebaica*; also *Parinarium macrophyllum*.

GINGERWORTS. A popular name for the *Zingiberaceæ*.

GINGILIE OIL. The oil of *Sesamum orientale*.

GINGO, or GINKGO. The aboriginal Japanese name of *Salisburia adiantifolia*.

GINSEN. (Fr.) *Panax*.

GINSENG. The root of one or more species of *Panax*. It is also called Ginschen. Pereira gives *P. quinquefolium* as American Ginseng, and *P. Schinseng* as Asiatic Ginseng.

GIPSYWORT. *Lycopus europæus*.

GIRANDOLE. (Fr.) *Coburgia*.

GIRARDINIA. A genus belonging to a small group of the nettle family, characterised by its stinging properties. From *Urtica* itself it differs in having alternate instead of opposite leaves, and from other allies in the calyx of the fertile flowers being two-parted, one of the segments being much the larger and three-toothed, the other small, linear, or sometimes abortive. The species, three of which are East African and three East Indian, are tall annual or perennial herbs, having all their parts clothed with long and sharp white stinging hairs. The stalked leaves, which are accompanied by large stipules, are sometimes nearly a foot in length, variously lobed and coarsely toothed, some like those of the hemp, others like those of the maple in form. The small green flowers, like those of a nettle, are unisexual; the males in racemes and the females in compact cymes, densely clothed with stinging hairs. *G. heterophylla*, one of the commonest Himalayan species, has three to seven lobed leaves, five inches to a foot long. Dr. Hooker, in his *Himalayan Journals*, alluding to this plant, says: 'The quantity of gigantic nettles growing on the border of maize fields was quite wonderful; their long white stings look most formidable, but, though they sting powerfully, the pain only lasts half an hour or so.' According to the same authority, a sort of cloth, and also a sort of cordage, are made from fibre furnished by the stems of this plant in Sikkim. In Southern India, the stems of *G. Leschenaultiana*, which is closely allied to the preceding, yield a good silken fibre which is made into thread. The process of separation, in some places, is performed by boiling the stems; in others, by steeping them in water for twelve days or so, when the outer or fibrous portion is readily removed, and afterwards spun into a beautiful soft thread. The fibre exists in large proportions, and the tow bears great resemblance to sheep's wool. [A. A. B.]

GIRAUMONT. (Fr.) *Cucurbita Pepo*. The seeds of some cucurbitaceous plant, bearing the name of Giraumont seeds, are used to destroy tape-worm.

GIREOUDIA. A genus of *Begoniaceæ*, named after M. Gireoud, a Berlin horticulturist. Flowers monœcious: the staminate ones with two obovate petals, numerous stamens of nearly equal length, and oblong anthers opening laterally; the pistillate ones with two petals, a trigonal inferior three-celled ovary which is unequally winged, and crescentic stigmas surrounded at the margin by a papillose band. The capsule is triquetrous and top-shaped, opening by curved chinks at the origin of the wings. They are fleshy undershrubs, erect or creeping, found in Central America and in Mexico, and have usually entire lobed leaves with long petioles and large stipules. The flowers are in axillary dichotomous cymes. There are about thirty species, all of which were formerly included in *Begonia*. [J. H. B.]

GIROFLÉE. (Fr.) *Cheiranthus*. — DES JARDINS. *Matthiola incana*. — DE MAHON. *Malcolmia maritima*. — DE MURAILLE. *Cheiranthus Cheiri*. — JAUNÂTRE. *Cheiranthus ochroleucus*. — JAUNE. *Cheiranthus Cheiri*. — QUARANTAINE. *Matthiola annua*. — VIOLETTE. *Cheiranthus Cheiri*.

GIROFLIER, or G. AROMATIQUE. (Fr.) *Caryophyllus aromaticus*.

GIROLLES. (Fr.) *Sium Sisarum*.

GIROSELLE. (Fr.) *Dodecatheon*.

GITH. The Corn Cockle, *Agrostemma Githago*.

GITHAGINEUS. Greenish-red.

GITHAGO. The name of one of the groups included in *Lychnis*, and represented by the *Agrostemma Githago* of Linnæus. The lamina of the petals is entire and without appendages. [T. M.]

GITHOPSIS. A genus of *Campanulaceæ*, nearly allied to *Specularia*, but differing in the narrow-campanulate, not rotate, corolla, in the filaments without hairs, and in the capsule opening in terminal pores. It comprises two Californian annuals, with small blue flowers, scarcely showing between the long segments of the calyx.

GLABER, GLABRATE, GLABROUS. Smooth; having no hairs.

GLACIALE. (Fr.) *Mesembryanthemum crystallinum*.

GLADDON, GLADEN, or GLADER. *Iris fœtidissima*.

GLADIATE. Sword-shaped; the same as Ensiform.

GLADIOLE, WATER. *Butomus umbellatus*.

GLADIOLUS. An extensive and very beautiful genus of *Iridaceæ*, found sparingly in the warmer parts of Europe and in the Mediterranean region, and much more abundantly in South Africa. They form fleshy corms, from which grow the erect stems, terminating in a spike of flowers of greater or less length, the leaves being distichous and equitant, and either narrow and grass-like or rush-like, or broader and sword-shaped. The flowers consist of an irregular perianth, with a terete tube, and six-parted bilabiate limb; three stamens inserted in the tube; and an obtusely three-cornered three-celled ovary, containing numerous ovules in several rows in the central angle of the cells. The ovary is crowned by a filiform style, with three petaloid stigmas. There is great variety amongst the species, not only in aspect, but also in size, and in the form of the flowers. Certain of them, chiefly *G. natalensis*, *floribundus*, and *cardinalis*, have, by cross-breeding and continued seeding, yielded a race of half-hardy so-called bulbs, which rank amongst the most ornamental of our popular garden flowers, and of which new varieties are raised annually in large quantities. These are all stately plants, growing from three to six feet in height, and producing long spikes of large blossoms of the most varied and striking colours. The European species are hardy garden flowers in favourable situations. [T. M.]

GLADWYN. *Iris fœtidissima*.

GLAIVANE. (Fr.) *Xiphidium*.

GLAND DE TERRE. (Fr.) *Lathyrus tuberosus*.

GLANDACEUS. Yellowish-brown, the colour of an acorn.

GLANDS, GLANDULES. Wart-like swellings found on the surface of plants, or at one end of their hairs. They are extremely various in form.

GLANDULOSE, GLANDULIFEROUS. Bearing glands.

GLANDULOSO-SERRATE. Having serratures tipped by glands.

GLANDULAR. Covered with hairs bearing glands upon their tips; as the fruit of roses, the pods of *Adenocarpus*, &c.

GLANS. An inferior fruit, one-celled by abortion, not dehiscing, containing one or two seeds, and seated in a cupule; as in the acorn.

GLAPHYRIA. A genus of myrtaceous shrubs, natives of the Malayan Islands, &c. The limb of the calyx is five-lobed; petals five; fruit succulent, with five many-seeded compartments.' *G. nitida* is called by the Malays the Tree of Long Life, probably because it is enabled to grow at greater elevations than other forest trees. The leaves are used as a substitute for tea. [M. T. M.]

GLAREOSE. Growing in gravelly places.

GLASSWORT. *Salicornia.* —, PRICKLY. *Salsola Kali.* —, WHITE. *Suæda maritima.*

GLAUCESCENT. Dull green, passing into greyish-blue.

GLAUCIUM. A genus of herbaceous plants belonging to the *Papaveraceæ*, well marked by their very long pod-like two-valved and two-celled capsule. The Yellow Horn Poppy, *G. luteum*, is a common plant on the sandy sea-shore of Europe and some parts of North America, where it may be detected even in winter by its large, rough, deeply-cut leaves of a decided glaucous hue. In summer it attains the height of about two feet, and is made conspicuous not only by the white hue of its foliage, but by its large flowers of four delicate pale-yellow petals, which last only for a day, and are succeeded by very long curved pods, which are rough with tubercles. *G. phœniceum*, a smaller species, with scarlet flowers, and a black spot at the base of each petal, is said to have been found in England, but is not considered a native. Several other species, with yellow or scarlet flowers, are cultivated, and are considered ornamental plants; they are either annual or biennial, and abound in a copper-coloured acrid juice, which is said to be poisonous and to occasion madness. German, *Gehörnte schollkraut.* [C. A. J.]

GLAUCOUS. Covered with a fine bloom, like the plum or the cabbage-leaf.

GLAUX. A pretty little herbaceous perennial, belonging to the *Primulaceæ*. The flower is destitute of a corolla, but the bell-shaped calyx is coloured and five-lobed; the capsule is globose, five-valved, and contains about five seeds. *G. maritima*, the only species, grows abundantly on most parts of the sea-coast, just above high-water mark, and in salt marshes. The roots, which creep extensively, are composed of long zigzag fibres, and send up numerous matted stems, four to five inches high, and densely clothed with oblong fleshy smooth entire leaves, which are pale underneath and salt to the taste. The flesh-coloured flowers are solitary, nearly sessile, and axillary. The glaucous hue of the leaves sufficiently accounts for the systematic name; but whence it derived one of its English names, Sea Milkwort, is not so plain. Another name, which is appropriate enough, is Black Saltwort. French, *Glauce*; German, *Milchkraut.* [C. A. J.]

GLAYEUL. (Fr.) *Gladiolus communis.* — PUANT. *Iris fœtidissima.*

GLEBA, GLEBULA. The peridium or the fleshy part of certain fungals.

GLEBULÆ (adj. GLEBULOSE). Little roundish elevations of the thallus of lichens; also the spores of certain fungals.

GLECHOMA. The Linnæan generic name of *Nepeta Glechoma* and other allied species of *Nepeta*. *G. hederacea*, Ground Ivy, is a well-known trailing herbaceous plant, with kidney-shaped crenate leaves and violet-purple flowers; formerly much esteemed for its supposed medicinal virtues. Its leaves are slightly bitter and aromatic, on which account it was used to give a flavour to ale; hence its old names Ale-hoof and Tun-hoof. The juice was recommended to be dropped into the ears to cure singing in that organ; mixed with wine, and dropped into the eyes, it was supposed to cure inflammation; taken as snuff, it was a specific for a head-ache; and an extract or decoction, mixed with honey or sugar-candy, was a favourite remedy in complaints of the chest. Village herbalists still hold it in repute, and use it, when dried, as a substitute for tea. Gerarde enumerates among its other virtues, that, 'boiled in mutton-broth, it helpeth weake and aking backs;' a prescription which many modern physicians would no doubt endorse, if administered with the same accompaniment. French, *Terrete*; German, *Gundelreben.* [C. A. J.]

GLECHON. A genus of plants of the labiate order, distinguished by the tube of the corolla being as long as the calyx; the two lower stamens only present and fertile; and the style bifid at the apex, the upper lobe very short, the lower long and compressed. The species are Brazilian shrubs of humble growth, having the leaves usually small, the flowers in groups varying from two to six, and red, blue, or yellowish in colour. [G. D.]

GLEDITSCHIA. A small genus of thorny leguminous trees, inhabiting various parts of North America and China. They have once or twice pinnated leaves, and small dense spikes of inconspicuous greenish flowers, some of which are perfect, while others are of one sex only. The pod is flat, and contains either one or several flat seeds, surrounded by a sweet pulpy substance, and separated from each other by transverse divisions. The three-thorned Acacia, or Honey-locust tree, *G. triacanthos*, is a native of the United States, and is commonly cultivated, both there and in this country, as an ornamental tree. It grows from fifty to eighty feet high, sending forth large spreading branches, and while young both stem and branches are formidably armed with stout, usually triple, thorns, tapering to very sharp points; but as the tree increases in size these thorns are principally confined to the smaller branches. In the autumn the trees bear numerous long thin and flat pendulous pods, which are usually curved and often twisted, and have been compared to 'large apple-parings, pendent from the branches.' They are sometimes as much as a foot and a half long, and contain numerous seeds, enveloped in a sweet pulpy substance, from which a kind of sugar is said to have been extracted. The wood is coarse-grained, very hard, and splits easily, but is not much employed except for fences and similar purposes. *G. monosperma*, the Water-locust of the Southern United States, is a very large tree, closely resembling the last in general appearance, but its flat pods are small and nearly oval, and contain only one seed. [A. S.]

GLEICHENIACEÆ. A group or suborder of Ferns, remarkably distinct in aspect from other ferns. They belong to the long series which is distinguished by the spore-cases having a jointed ring, and bursting irregularly instead of being valvate; but the spore-cases are sessile, and the ring is more or less strictly horizontal, and consequently the fissure made by their bursting takes a vertical instead of the more usual horizontal direction. The additional features of rigid opaque fronds and oligocarpous dorsal sori complete the distinctive marks of the group, of which *Gleichenia* is the principal genus. [T. M.]

GLEICHENIA. A genus of polypodiaceous Ferns, typical of the tribe *Gleichenineæ*. They are furnished with creeping rhizomes, rigid usually often repeatedly dichotomously forked fronds, with the ultimate branches pinnatifid, and either bearing small rounded or ovate segments, or larger linear ones resembling the teeth of a comb. The sori are naked, sometimes placed in a hollow space, oligocarpous, that is, consisting of but few spore-cases, the number varying from two to four in one group, and from eight to twelve in another. The latter series, which agrees with that having the linear segments, forms the group *Mertensia*, which some pteridologists regard as a distinct genus. There are many species scattered widely in the tropics both of the Old and New World, and extending to Chili and the Australasian region. [T. M.]

GLI. An intoxicating liquor prepared by the Hottentots from *Lichtensteinia pyrethrifolia*.

GLIDEWORT. *Galeopsis Tetrahit*.

GLINUS. A genus usually referred to *Mesembryaceæ*, but considered by A. Richard, Endlicher, and others, as belonging to the *Caryophyllaceæ*, tribe *Mollugineæ*. They are annual prostrate branched herbs, rarely undershrubs, growing in tropical and subtropical regions, a single species occurring in the south of Europe. Leaves alternate or falsely verticillate; flowers inconspicuous, with a five-cleft calyx; the corolla absent, or with numerous very narrow strap-shaped petals, and three to twenty stamens. [J. T. S.]

GLOBBA. A genus of tropical Asiatic herbaceous plants belonging to the *Zingiberaceæ*, and having a terminal loosely-clustered inflorescence, the flowers of which have a three-cleft tubular calyx; a corolla with a slender tube and a six-parted limb, the three outer divisions equal, and two of the inner ones narrow or very small, while the remaining one, or lip, is large, undivided, and partly united with the filament in a tubular manner. The ovary is one-celled. They are handsome plants, with singular-looking yellow or pinkish flowers; some of them grown in this country as stove plants. The fruit of *G. uniformis* is said to be edible. [M. T. M.]

GLOBE-FLOWER. *Trollius*; also *Gomphrena globosa*.

GLOBOSE. Forming nearly a true sphere.

GLOBULAIRE. (Fr.) *Globularia vulgaris*.

GLOBULARIACEÆ. A natural order of corollifloral dicotyledons, belonging to Lindley's echial alliance of perigynous Exogens. Lindley unites *Selaginaceæ* with this order, but De Candolle separates them. A small group of herbaceous or shrubby plants with alternate exstipulate smooth leaves, and capitate flowers surrounded by an involucre. Calyx five-divided with quincuncial æstivation; corolla tubular, lipped, five-lobed; stamens four, inserted into the upper part of the tube of the corolla, the anther becoming one-celled, and opening by a single longitudinal slit. Ovary free, one-celled; ovule one, pendulous anatropal. Fruit an achene, enclosed by the calyx; seed with fleshy albumen; embryo with a superior radicle. Natives of Europe, and of the parts of Asia and Africa nearest Europe. There are but few species, distributed in the genera *Carradoria* and *Globularia*. [J. H. B.]

GLOBULARIA. A genus containing a few species of herbs, shrubs, or undershrubs, natives of the countries bordering on the Mediterranean Sea. They have alternate entire spathulate leaves, and flowers collected upon a common receptacle, and surrounded by a many-leaved involucre. The calyx is unequally five-cleft; the corolla tubular, with the limb two-lipped, the upper lip bipartite and smaller than the lower, which is trifid; and there are four stamens, inserted at the top of the corolla tube. The ovary is composed of a single carpel, and contains a single pendulous anatropal ovule. This genus differs from the *Selaginaceæ*, in having the ovary formed of a single carpel, as well as in its habit, in both of which respects it agrees with *Dipsaceæ*, but the plants of this order have an inferior ovary. [W. C.]

GLOBULEA. Succulent plants, natives of the Cape, with flat or sickle-shaped leaves, arranged in a rosette. The flowers are small, arranged in dense clusters, and have five petals bent inwards, each of them tipped with a little globule of waxy matter, whence the name of the genus, which differs little from *Crassula*, save in the direction of the petals. Several kinds are in cultivation. [M. T. M.]

GLOBULINE. Elementary cells; starch grains.

GLOBULUS. A kind of perithecium occurring among fungals; the antheridium of *Chara*; also, a round deciduous shield, found in such lichens as *Isidium*, formed of the thallus, and leaving a hole where it falls off.

GLOCHIDION. A genus of the spurgewort family, comprising upwards of fifty species of shrubs or small trees, for the most part found in India, a few extending eastward to Japan; others occurring in tropical Australia and the adjacent islands, and three being natives of West Africa. From *Phyllanthus*, to which they are closely allied, they differ in the flowers being destitute of a glandular disk, and generally in the more numerous cells of the ovary. Their alternate leaves are often arranged in a distichous manner; the blades of some of them have a metallic lustre, while others are clothed with soft short down. Their inconspicuous yellow or green flowers are male and female on the same plant, disposed in axillary clusters, the males usually occupying the circumference and the females the centre, both having a five or six-parted calyx. The fruits are globular or depressed capsules, sometimes covered with a thin and fleshy red coat, but more often quite dry; when ripe they split into three to ten portions. The bark of *G. nitida* is said by Roxburgh to be astringent. *Bradleia* and *Gynoon* are now referred to this genus. [A. A. B.]

GLOCHIS (adj. GLOCHIDATE). Hooked back at the point, like a fish-hook.

GLOIOCARP. The quadruple spore or tetrachocarp of some algals.

GLOMERATE. Collected into close heads or parcels.

GLOMERULI. The same as Soredia.

GLOMERULUS. A cluster of capitules enclosed in a common involucre, as in *Echinops*.

GLORIOSA. The name of a group of remarkably handsome hothouse herbaceous-stemmed climbers, more correctly called METHONICA; which see. [T. M.]

GLORY-TREE. *Clerodendron*.

GLOSSAPSIS *tentaculata* is a tuberous-rooted orchid, peculiar to the island of Hong-kong and the adjacent mainland. According to Mr. Bentham, it has the habit and characters of the small-flowered *Habenarias*, except that the terminal glands of the caudicles of the pollen-masses are received into distinct cells of the stigma. The root is an ovoid tuber; the stem, including the slender erect spike of small green flowers, eight to twelve inches high; the leaves three to four, oblong or lance-shaped, and one to three inches in length; the lip is deeply three-lobed, the lobes long and thread-like, somewhat resembling the antennæ of an insect. [A. A. B.]

GLOSSOCOMIA. A genus of bellworts distinguished by having the calyx five-lobed, reflexed; the corolla bell-shaped, five-lobed; the stigmas three, ovate; and the fruit three-celled. *G. ovata* is a hardy downy perennial, a native of Northern India, with ovate heart-shaped leaves and showy bell-like flowers. [G. D.]

GLOSSODIA. A small genus of Australian orchids, belonging to the tribe *Arethuseæ*, and most nearly related to *Caladenia*, of which, according to Dr. Hooker, they may be considered a mere section, with no glands on the disk of the lip, and a solitary bifid long appendage at the base of that organ, somewhat resembling a serpent's tongue, whence the generic name. They have tuberous roots; a solitary lance-shaped or oblong leaf, about three inches long; and a slender, erect, nearly naked stem, six inches to a foot high, bearing at its apex from one to three extremely pretty blue flowers, sometimes beautifully speckled with white, and about an inch in diameter. The flowers are nearly regular, the lip undivided, the column winged, and the anthers terminal, with four powdery compressed pollen-masses. [A. A. B.]

GLOSSOLOGY. That part of Botany which teaches the meaning of technical terms.

GLOSSONEMA. A genus of *Asclepiadaceæ*, containing three species natives of Arabia and North-Eastern Africa. They are hoary perennial branching herbs, with opposite linear leaves, and small flowers on short interpetiolar peduncles. The calyx is five-parted; the corolla campanulate and five-cleft, with a tubercle on the inner surface of each lobe towards the apex; and the staminal crown is made up of five lobes alternating with those of the corolla, and having a dilated emarginate apex, with a contorted filament rising from the indentation. The stigma is slightly two-lobed; the follicles smooth, or covered with spines; and the seeds comose. [W. C.]

GLOSSOPETALUM. A small Mexican bush referred to the *Celastraceæ*, from the other genera in which, it is easily recognised by having ten stamens instead of five, and a simple instead of a compound pistil. *G. spinescens* is much branched, two to four feet high, having stiff rounded twigs, which end in spiny points, and are furnished with little alternate entire leaves, those on the flowering twigs being reduced to scale-like processes. The small white flowers are axillary and stalked, and are succeeded by a little ovoid seed-vessel containing one seed, which is furnished with a minute arıl. (A. A. B.)

GLOSSOSTEMON. A genus of *Byttneriaceæ* nearly related to *Abroma*, and like it having pretty purple blossoms, but differing in the greater number of stamens and the form of the barren filaments. The stamens are thirty-five in number, disposed in five parcels, each parcel consisting of six anther-bearing stamens and a central barren one, which is much broader, longer, and tongue-like, whence the name of the genus. *G. Bruguieri*, the only species, is found in various parts of Persia. It is a low-growing plant, with a perennial rootstock, from which arise a few unbranched stems furnished with large soft leaves somewhat like those of a hollyhock. The stems and leaves, which are of a pale straw colour, are covered with starry hairs. Each stem terminates in a corymb of elegant dark purple flowers. (A. A. B.)

GLOSSOSTIGMA. A genus of *Scrophulariaceæ*, consisting of minute tufted mosslike creeping herbs, much resembling small specimens of our *Limosella*, but the valves of the capsule bear the dissepiments in their centre, instead of being parallel to the dissepiment. There are two species, one a native of India, the other of New Zealand and Australia. The flowers in both are very minute.

GLOUTERON. (Fr.) *Lappa communis*; also *Xanthium Strumarium*.

GLOXINIA. A genus of gesnerworts, distinguished by its corolla approaching to bell-shaped, with the border oblique, the upper lip shortest and two-lobed, the lower three-lobed with the middle lobe largest; and also by the summit of the style being rounded and hollowed. The name was given in honour of Gloxin, a botanical author of the last century. The species are natives of tropical America, and have opposite stalked leaves of rather thick texture, and axillary flowers, usually single or a few together, large, nodding, and of various colours (white violet red or greenish yellow), sometimes variegated with spots. The species are among the greatest ornaments of our hothouses, their richly-coloured leaves, and their ample, graceful, and delicately-tinted flowers, having gained for them a prominent place among introduced plants. Here, as in many other instances, the process of hybridising has been resorted to with the best results; the older kinds with drooping flowers, have of late been giving place to forms with the corolla almost regular and nearly erect —the latter peculiarity having this recommendation, that the border and throat of the corolla, to which parts much of the beauty of the flower is owing, are presented to the eye. *Gloxinias* may be propagated by their leaves. (G. D.)

GLUMALES. An alliance of Endogens,
comprising the grasses, sedges, and a few minor groups.

GLUMA, GLUME. The exterior series of the scales which constitute the flower of a grass.

GLUMELLA. That part of the flower of a grass now called the Palea or Pale. Also, in the language of Richard, one of the hypogynous scales in such a plant.

GLUMELLULA. The hypogynous scale in the flower of a grass.

GLUTA. A Javanese tree with the appearance of the mango, and flowers in panicles resembling those of *Clematis Flammula*. The calyx is tubular and deciduous; petals four five or six, spreading, longer than the calyx, attached, as also are the stamens, to a stalk supporting the ovary; style lateral; fruit succulent, one-seeded. The genus belongs to the order *Anacardiaceæ*. [M. T. M.]

GLUTINIUM. The flesh of certain fungals.

GLUTINOSE. Covered with a sticky exudation.

GLUTTIER. (Fr.) *Sapium*. — DES OISELEURS. *Sapium aucuparium*.

GLYCE. A genus of *Cruciferæ*, generally called *Königa*, and now reunited to *Alyssum* by many authors. It has the pouch nearly ovate, with flattish valves, the cells one or few-seeded, the funiculus of the seed adhering to the back of the septum, and the seeds wing-margined. The calyx is spreading, the petals entire, white or yellow, the hypogynous glands eight, and the filaments without basal appendages. *G. maritima*, the Sweet Alyssum of gardens, is found in some places in Britain, but only imperfectly naturalised where escaped from gardens. [J. T. S.]

GLYCERIA. A genus of grasses belonging to the tribe *Festuceæ*, distinguished principally from *Poa* by having the florets in more linear subcylindrical spikelets. The outer glumes and pales are membrano-herbaceous, with sharply prominent nerves, and a scarious margin. Steudel describes thirty-seven species, which are chiefly natives of the colder and more temperate parts of the world. The best known species is *G. fluitans*, or Manna-grass, which grows in most watery places in Britain. The long floating stems spread over pools of water and ditches, where cattle may frequently be seen wading to considerable depths to eat them. The seeds of some of the species are greedily fed on by ducks and other aquatic birds. (D. M.)

GLYCINE. A small genus of *Leguminosæ*, all, excepting one, being slender decumbent or twining herbs, with alternate stalked leaves made up of three to seven leaflets varying much in form, and bearing axillary racemes or fascicles of small yellow or violet pea-flowers. The genus belongs to the tribe *Phaseoleæ*, and is most nearly allied to *Teramnus*, from which it is distinguished by its pods being destitute of the hardened hooked style seen in the latter, and by the ten stamens, which are united into a tube, being all, instead of the alternate ones only, anther-bearing. The species are pretty equally distributed through tropical Asia, Africa, and Australia, where a few inhabit extratropical regions. The Sooja of the Japanese, *G. Soja*, the only erect species of the genus, a dwarf annual hairy plant, a good deal like the common dwarf kidney or French bean (*Phaseolus vulgaris*), has small violet or yellow flowers, borne in short axillary racemes, and succeeded by oblong two to five-seeded hairy pods. The seeds, like kidney beans in form but smaller, are called Miso by the Japanese, and are made into a sauce which they call Sooja or Soy. The manner of making it is said to be by boiling the beans with equal quantities of barley or wheat, and leaving it for three months to ferment, after which salt and water is added, and the liquid strained. The sauce is said to be used by them in many of their dishes, and they use the beans in soups. The Chinese cook the beans also in various ways, and the plant is cultivated for the sake of them in various parts of India and its Archipelago. Mr. Bentham groups the species in three sections, which some regard as genera: *Soja*, with flowers fascicled on the racemes, and falcate pods with depressions but not transverse lines between the seeds; *Johnia*, with flowers similarly arranged, and straight pods with transverse lines between the seeds; and *Leptocyamus*, with solitary flowers on the racemes, and straight pods. The *Glycine* or *Wistaria* of gardens is now referred to *Millettia*. [A. A. B.]

GLYCOSMIS. A name indicative of the sweetly-smelling flowers in the genus to which it refers, which consists of tropical Asiatic trees or shrubs, belonging to the *Aurantiaceae*, and closely allied to *Limonia*, but differing in the absence of spines, in the eight stamens being alternately long and short, in the short thick conical style, &c. *G. pentaphylla* is a common undershrub in the uncultivated districts of Coromandel. *G. citrifolia* is remarkable for the delicious flavour of its fruits. [M. T. M.]

GLYCYRRHIZA. The best known plant of this genus is that which reputedly furnishes Spanish Liquorice, *G. glabra*—though possibly other species may be employed for the same purpose. *G. glabra* is an herbaceous perennial, with pinnate leaves and bluish flowers, and is cultivated in this country for the sake of its root, which contains a peculiar sugar-like substance, giving to the extract its flavour and slight demulcent property. To make the extract the root is sliced and boiled in water; after a time the liquor is strained and allowed to evaporate till it becomes of a proper consistence. Large quantities of this extract are imported from Spain, whence the term Spanish Liquorice; much is also imported from Italy, where it is prepared from the root of *G. echinata*. It is imported in rolls five or six inches long, about the thickness of a man's thumb, and is packed in the leaves of the sweet bay. What is called refined liquorice is common liquorice dissolved in water, and again evaporated. It is said that both kinds are adulterated to a considerable extent, and that copper is often to be detected in them—probably from the extract having been made in an unclean copper vessel. Liquorice extract is demulcent in colds and coughs, but it is most extensively employed by the large porter brewers. The genus belongs to the *Leguminosae*, and is characterised by the presence of a tubular five-cleft two-lipped calyx; an ovate straight standard, a keel of two straight pointed petals; stamens in two parcels; style thread-like; pod ovate, compressed, one to four-seeded. [M. T. M.]

GLYPHÆA. A genus of *Tiliaceae*, of which *G. grewioides*, the only species, is a West African bush, furnished with smooth, alternate, papery, three-ribbed, toothed leaves, varying from lance-shaped to oblong, and bearing yellow flowers in axillary umbels. They have a calyx of five narrow sepals, five petals; numerous stamens; and an ovary tipped with a simple style. The fruits are many-furrowed, spindle-shaped, three to five-celled, many-seeded, the seeds one above another, and separated by a thin cellular partition. [A. A. B.]

GLYPHOSPERMUM. A name applied to a genus of *Gentianaceae*, on account of the seeds, which are pitted. They are Peruvian shrubs, with small purple polygamous flowers, having a five-cleft tubular corolla, a one-celled ovary, no style, and a button-shaped two-lobed stigma. [M. T. M.]

GLYPHOTÆNIUM. A name proposed by J. Smith for *Goniopteris crispata*.

GLYPTOSTROBUS, or Embossed Cypress, is a genus of coniferous plants, allied to *Taxodium*. The name is derived from the Greek words 'glyptos,' carved or engraved, and 'strobos,' a cone, from the embossing on the scales. The flowers are monœcious. The cones grow at the end of lateral branches, and are ovate or oblong, consisting of several unequal leathery scales, which rise from the same point at the base; each scale covers two seeds, which are erect, ovate, and compressed. They are trees or shrubs, found in China, with straight or pendulous branches, and scattered, linear awl-shaped, three-angled leaves. *G. heterophyllus*, a small tree eight to ten feet high, is the Chinese Water Pine, planted along the margins of rice-fields near Canton, and found also in other parts of China. [J. H. B.]

GMELINA. A genus of *Verbenaceae*, consisting of a number of East Indian trees or shrubs, characterised by their cup-shaped minutely four to five-toothed calyx; tubular corollas, with the tube narrow below, somewhat bell-shaped above, and spreading and two-lipped at the border; and drupe-like two to four-celled fruits with one seed in each cell. The leaves are simple, entire, and generally oval and pointed; and the handsome yellow blossoms are disposed in raceme-like panicles, the branches of which are clothed with short yellow down. *G. arborea*, a large timber tree of the mountainous parts of India, affords a good wood useful for many purposes. According to Roxburgh, that of such trees as will square into logs from eighteen to twenty-four inches, bears much resemblance to teak, with the same colour, a closer grain, as light if not lighter, and easily worked. He found the wood to resist the effects of the sun and water better than teak, and remarks that the decks of pinnaces are made of this wood at Chittagong, &c., because it resists the weather better than any other, and does not shrink or warp. Of *G. Rheedii*, a Ceylon tree, producing large and numerous tawny-yellow flowers in the summer months, the bark and roots, as well as those of *G. asiatica*, are used medicinally by the Cingalese. [A. A. B.]

GNAPHALIUM. The Everlasting: a genus of plants belonging to the *Compositæ*, distinguished from *Antennaria* by having the heads all alike and the receptacle naked, and from *Filago* by having the receptacle flat and not conical. The involucre or common calyx, in all the species, is of the peculiar character termed scarious or everlasting; hence the English name. Many of them, with white, yellow, or pink flowers, are natives of the Cape of Good Hope. The foliage is usually thickly invested with white woolly down, and the flower-heads are remarkable for the permanence of their form and colour. *G. luteo-album* is the only British species which has any pretensions to beauty; it has only been found wild in one or two places in England, but is more frequent in Jersey. *G. uliginosum*, a minute tufted plant, with narrow cottony leaves, and numerous heads of small yellowish-brown flowers, is very common on damp heaths and in places where water has stood during winter. French, *Gnaphale*; German, *Ruhrpflanze*. [C. A. J.]

GNAPHALODES. Three little Australian weeds, belonging to the composite family, and in appearance much like our own cudweeds (*Filago*), being clothed with cottony wool; they are, moreover, nearly allied to them, but differ in all the florets being perfect, instead of the outer ones being female and the inner perfect. The flower-heads have an involucre of numerous scales; and the achenes, seated on a cone-shaped naked receptacle, are smooth, and crowned with a pappus of five narrow and rigid ciliate scales. [A. A. B.]

GNAVELLE. (Fr.) *Scleranthus*.

GNAWED. The same as Erose.

GNETACEÆ. (Joint Firs.) A natural order of monochlamydeous dicotyledons, belonging to Lindley's class of Gymnogens. Small trees or creeping shrubs, not resinous, with jointed stems and branches, and opposite reticulated, sometimes scaly leaves. Flowers monœcious or diœcious, arranged in catkins or heads, surrounded by opposite scales which unite more or less completely. The staminate flowers have a one-leaved perianth, and one-celled anthers, opening by pores; the pistillate ones either have no covering, or are enclosed by two scales. Ovules usually considered naked, one of their coats being protruded through the hole so as to form a long style-like process; seed with a succulent covering; embryo with a long twisted suspensor. Natives of temperate as well as warm regions in Europe, Asia, and South America. The seeds of some of them are eaten. There are two genera, *Ephedra* and *Gnetum*, and about thirty species. [J. H. B.]

GNETUM. A genus of plants typical of the order *Gnetaceæ*. The flowers are produced in cylindrical jointed catkins, the staminate ones having a membranaceous perianth, a single stamen, and an anther opening by a pore; and the pistillate ones being without any proper covering. The ovule is solitary and orthotropal; and the seed has an outer succulent coat. Trees or creeping shrubs found in tropical Asia and Guiana. They have jointed knotty branches, opposite, exstipulate, entire, smooth leaves, and axillary or terminal stalked catkins. There are some half-dozen species. The outer covering of the seeds of *G. urens* is lined with stinging hairs. The seeds of *G. Gnemon* and other species are roasted and eaten. [J. H. B.]

GNIDIA. A genus of *Thymelaceæ*, bearing complete tetramerous flowers, whose calyx is coloured, funnel-shaped, with a regular four-divided limb; scales four to eight, inserted into the upper part of the calycine tube and projecting beyond it; anthers eight, in two rows, attached to the tube of the calyx; ovary sessile; style lateral, equalling the tube of the calyx; stigma capitate and papillose. The fruit is a nut, enclosed by the persistent calyx. Shrubs or undershrubs found in the southern and eastern tropical parts of Africa. They are heath-like plants, with slender branches, scattered or opposite leaves, terminal usually capitate flowers, which are of a white, yellow, reddish, or lilac colour, and are mostly pubescent externally. There are fifty known species. The bark of *G. daphnoides* supplies ropes in Madagascar. [J. H. B.]

GNOMONICAL. Bent at right angles.

GOATBUSH. *Castela Nicolsoni*.

GOATROOT. *Ononis Natrix*.

GOATWEED. *Capraria biflora*.

GOATS-BANE. *Aconitum Tragoctonum*.

GOATS-BEARD. *Tragopogon*; also *Spiræa Aruncus*.

GOATS-FOOT. *Oxalis caprina*.

GOATS-HORN. *Astragalus Ægiceras*.

GOATS-THORN. *Astragalus Tragacantha*, and *A. Poterium*.

GOBBO. *Abelmoschus esculentus*.

GOBE-MOUCHE. (Fr.) *Silene muscipula*; also *Dracunculus crinitus*, *Apocynum androsæmifolium*, &c.

GOBET. (Fr.) *Ceranus vulgaris*.

GOBLET-SHAPED. The same as Crateriform.

GOBO. The Japanese name of *Arctium Lappa*.

GOCKROO. *Ruellia longifolia*, an Indian drug.

GODET. (Fr.) *Narcissus Pseudo-Narcissus*.

GODETIA. A genus of ornamental annuals, belonging to the *Onagraceæ*, and closely allied to Evening Primroses (*Œnothera*), from which they may be known by bearing flowers of a purple or pink hue, never yellow. The true Evening Primroses, as their name implies, do not open their flowers in the sunshine, but the *Godetias* are subject to no such rule. The majority of the species are natives of America, and are much grown in English gardens for the sake of their showy flowers. Some of them are remarkable for the brilliant colour of their anthers, and others for the deep purple spots on their petals. They are all very similar in habit, upright more or less branching herbs, with the broad four-petaled flowers in the upper leaf-axils. [C. A. J.]

GODOYA. A genus of tropical American trees, of doubtful position, but referred by Lindley to *Ochnaceæ*. The leaves are shining, thick, marked with very numerous transverse striæ or veins. The flowers are yellow, disposed in clusters, the calyx consisting of several series of overlapping coloured leaves; the five petals convolute; and the stamens numerous, the outermost of them sterile, free or united into five distinct bundles alternating with the petals, the inner ones free and fertile. The capsule woody, three to five-celled, bursting by five valves. *G. gemmiflora* is a stove plant of elegant appearance. [M. T. M.]

GOD'S FLOWER. *Helichrysum Stœchas*.

GOD-TREE. *Eriodendron anfractuosum*.

GŒPPERTIA. A genus of Brazilian and West Indian trees, of the laurel family, having a six-parted wheel-shaped perianth, nine fertile stamens in three rows, the innermost row provided with glands, the anthers opening by two or four valves. Fruit placed within the hardened tube of the perianth. [M. T. M.]

GOGANE. (Fr.) *Fritillaria Meleagris*.

GOLDBACHIA. A genus of *Cruciferæ*, consisting of annuals found in the Levant and in the Caspian Desert. They have alternate oblong leaves, and racemes of flowers, small, white or lilac, opposite the leaves. The pod is short, breaking transversely, when mature, into two one-seeded joints. [J. T. S.]

GOGO. *Entada Purseatha*.

GOKOKF. A collective Japanese name for bread-stuffs and pulse.

GOLD-CUPS. *Ranunculus bulbosus, acris*, &c.

GOLD-DUST. A popular name for *Alyssum saxatile*.

GOLDE. *Calendula officinalis*.

GOLDEN-CHAIN. *Cytisus Laburnum*.

GOLDEN-CLUB. *Orontium*.

GOLDEN-CROWN. *Chrysostemma*.

GOLDEN-FLOWER. *Chrysanthemum*.

GOLDEN-HAIR. *Chrysocoma Coma aurea*.

GOLDEN-PERT. *Gratiola aurea*.

GOLDEN-ROD. The common name for *Solidago*; also *Leontice Chrysogonum*. —, RAYLESS. An American name for *Bigelovia*. — TREE. *Iowa Coromera*.

GOLD-FLOWER. *Helichrysum Stœchas*.

GOLDFUSSIA. A considerable genus of *Acanthaceæ*, containing twenty-four species, natives of India. They are shrubs with serrate penninerved leaves, having all the nerves directed upwards, but not reaching the apex. The flowers have two deciduous bracts, and are arranged in a head or spike, which after the fall of the bracts becomes very loose; there is an unequally five-parted calyx, a funnel-shaped corolla with an equally five-cleft limb, four didynamous included stamens with nodding anthers, and a subulate irritable stigma. [W. C.]

GOLDILOCKS, or GOLDYLOCKS. *Helichrysum Stœchas*; also *Ranunculus auricomus*, *Hymenophyllum tunbridgense*, and a common name for *Chrysocoma*.

GOLDINS. *Chrysanthemum segetum*.

GOLD-KNOBS, or GOLD-KNOPPES. *Ranunculus acris, bulbosus*, &c.

GOLD OF PLEASURE. *Camelina sativa*.

GOLD-SHRUB. *Palicourea speciosa*.

GOLD-THREAD. *Coptis trifolia*.

GOI-KAKRA. *Momordica mixta*.

GOMART. (Fr.) *Bursera*.

GOMBAUT, or GOMBO. *Abelmoschus esculentus*. Gombo is also used for the fibre-yielding *Hibiscus cannabinus*.

GOMMA DA BATATA. A purgative drug obtained from *Ipomœa operculata*.

GOMPHIA. A genus of tropical ochnaceous trees or shrubs, most abundant in Brazil. They bear panicles of handsome yellow flowers, having the following structure: sepals five, coloured, deciduous; petals five, generally stalked; stamens ten, the anthers opening by pores; ovary five to six-lobed, placed on a thickened receptacle; style very short; fruit succulent, placed on the enlarged receptacle. Some of them are grown in this country as evergreen stove shrubs; they are for the most part West Indian. [M. T. M.]

GOMPHOCARPUS. A rather considerable genus of *Asclepiadaceæ*, containing fifty species of shrubs, natives of Southern and North-Eastern Africa and Arabia, and with one species common throughout the warmer regions of the world. They have opposite rarely whorled leaves, and generally showy flowers on many-flowered interpetiolar peduncles; the calyx five-parted, the corolla rotate or reflexed and five-parted; the staminal crown inserted on the top of the gynostegium, and consisting of five conduplicate leaflets, the pollen-masses attenuated upwards; and the smooth or echinate ventricose follicles containing many comose seeds. The leaves of *G. fruticosus*, the Arghel of Syria, are employed for adulterating senna: this plant is sometimes referred to *Solenostemma*. [W. C.]

GOMPHOGYNE. The name of a Himalayan climbing plant, belonging to the *Cucurbitaceæ*. Its flowers are unisexual: the males with five sepals, five fringed petals, and five stamens which are united at the base; and the females with a one-celled ovary containing three ovules. The fruit is capsular. [M. T. M.]

GOMPHOLOBIUM. A genus of elegant leguminous undershrubs belonging to the tribe *Podalyrieæ*, in which the ten stamens are free. It comprises about thirty species, all of them found in South and West Australia. They are readily distinguished by their spherical or oblong many-seeded pods, and by their compound alternate leaves, made up of a varying number of leaflets, which are often heath-like. Many of them are cultivated in greenhouses, where they produce their blossoms in the spring and summer months. One of the most beautiful is *G. venustum*, a plant with slender flexuose branches furnished with smooth pinnate leaves of four to eight pairs of narrow linear leaflets, the stem terminating in a corymb of beautiful rose-purple pea-flowers. The largest-flowered species is *G. barbigerum*, so named because of the keel-petal being fringed; it is a smooth bush whose angular stems are furnished with trifoliolate sessile leaves, of narrow flax-like leaflets, the pale yellow pea-flowers being solitary in the axils. Amongst a goodly proportion with heath-like leaves, *G. uncinatum* is noteworthy as being in South Australia very hurtful to sheep that may eat of it; the leaves are sessile and composed of three narrow leaflets hooked at the point; the flowers yellow, axillary, and solitary. [A. A. B.]

GOMPHONEMA. A genus of *Diatomaceæ*, distinguished by its forked permanent stems and wedge-shaped frustules, which are often contracted near the apex, and sometimes also towards the base when seen laterally. *G. Berkeleri*, which is synonymous with the old *Meridion vernale*, occurs in every brook in spring, forming brown cushion-like gelatinous masses adhering to stones, leaves of aquatic plants, &c. *G. geminatum* is less generally diffused, but not uncommon, and is remarkable for its very large frustules, which form a magnificent microscopic object. The species were formerly confounded with *Vorticella*, a genus of undoubted animals, and well known to every student of freshwater *Algæ* by the curious motions of the stem. [M. J. B.]

GOMPHOSIA. A genus of cinchonaceous shrubs, natives of Peru and New Granada, having flowers whose calyx is provided with minute glands like those on the stipules. The corolla is salver-shaped, with a long tube, and a four to five-lobed limb; the stamens of unequal length, but all projecting beyond the corolla, the lobes of the anthers bent back, and connected by a very broad connective. The capsule is few-seeded, and bursts from above downwards into two valves. The seeds are winged. [M. T. M.]

GOMPHOSTEMMA. The generic name of plants belonging to the labiate order, having the corolla with its tube dilated upwards, and its border with two nearly equal lips; and the style with a bifid stigma, the two halves equal and awl-shaped. The species are herbs, natives of India, with simple, usually erect, rarely procumbent stems, the leaves large and shaggy with hairs. [G. D.]

GOMPHRENA. A genus of *Amarantaceæ*, in which the flowers are sometimes incomplete as regards stamens and pistils. There is a perianth of five leaves, very rarely five-cleft, five stamens united into a tube, the filaments dilated, with a trifid apex, the intermediate segment bearing a one-celled anther. The ovary is one-celled with a single ovule; the fruit one-seeded included within the perianth. They are undershrubs or herbs with opposite often semi-amplexicaul leaves, and flowers in lax spikes or panicles, or in globular heads. They abound in tropical America, and are rare in Asia and Australasia. There are ninety known species. [J. H. B.]

GOMUTI, or **GOMUTO.** An Eastern palm, *Saguerus saccharifer*, which yields a bristly fibre, called Gomuto or Gomuti fibre.

GONAKIE. An African name for *Acacia Adansonii*, which yields good building timber.

GONATANTHUS. The name of an Indian herbaceous plant, of the *Arum* family, with a tuberous rootstock, peltate leaves, and a very long leathery spathe, rolled round at the base and prolonged into a long point at the other extremity. The spadix is short, bearing stamens above, ovaries in the middle, and rudimentary flowers at the lower part; anthers numerous, six-celled, the cells adhering in a whorl to the peltate thick connective, and opening by pores. Ovaries numerous, detached. [M. T. M.]

GONGONHA. *Ilex Gongonha*, the leaves of which, like those of Maté, *I. paraguayensis*, are used for making tea.

GONGORA. A singular genus of orchids found growing on tree-stems in tropical America. They have oblong, grooved, two-leaved pseudobulbs, the leaves broadly lance-shaped, plaited and a foot or more in length; and, growing from the base of the pseudobulb, drooping flower-racemes which are sometimes two feet long. The lateral sepals are free and spreading, the upper one remote and connate with the back of the lengthened, arched, hammer-headed column; the petals small and adnate to the middle of the column; while the curious clawed lip is continuous with the base of the column, and contracted in the middle, the lower portion being furnished on each side with a bristle-like horn, and the terminal part vertical and pointed, with the opposite faces folded together. The anther is two-celled, with two linear pollen-masses on the end of a narrow caudicle fixed at the base to a small gland. Upwards of a dozen species are known. *G. atropurpurea*, from Trinidad, has long pendent racemes of curiously formed purple flowers, reminding one of some insect. In *G. maculata*, from Demerara, they are yellow marked with blood-red spots. The structure of the flowers of these curious plants is very singular, and well repays examination. Those of *G. galeata*, better known in gardens under the name of *Acropera Loddigesii*, and especially those of a closely related plant called *Acropera luteola*, have been subjected to a close examination by Mr. Darwin, the result of which is that he believes some orchids to be unisexual, although both male and female organs are present in each flower. See Darwin, *On Orchid Fertilisation*, p. 21. (A. A. B.)

GONGRONEMA. A genus of *Asclepiadaceæ*, nearly allied to *Gymnema*, containing a few species of twining shrubs, natives of India. They have opposite coriaceous and glabrous leaves, and small flowers in large lax compound corymbs, except in one species in which they are arranged in a simple umbel. The calyx is five-parted, and the rotate corolla five-cleft, with the throat and tube naked; there is no staminal corona, but the gynostegium has small fleshy glands at its base. The follicles are smooth. (W. C.)

GONGYLODES. Having an irregular roundish figure.

GONGYLOSPERMEÆ. A division of rose-spored *Algæ*, containing those genera in which the spores are collected without order in a mucous or membranaceous mother cell. The nucleus is sometimes compound. The filamentous *Ceramiaceæ*, and the solid *Rhodymeniaceæ* and *Cryptonemiaceæ*, belong to this division. (M. J. B.)

GONGYLUS. The spores of certain fungals. Also a round, hard, deciduous body connected with the reproduction of certain seaweeds.

GONIDIA. A name applied to the green spherical cells in the thallus of lichens which are the distinctive mark between these plants and *Fungi*. They assume different types in different divisions. In most lichens they are of a pure green, and are developed from the tips of the constituent threads singly or in tufts; in *Collema* they are less highly coloured, and form moniliform threads resembling those of *Nostoc*; in *Pauha* they are large and gelatinous, increasing by cell-division as in some *Palmelia*; while in *Ephebe* they are quadripartite, and resemble *Bamatococcus* in their developement. (M. J. B.)

GONIOMA. A genus of dogbanes, having the tube of the corolla angular at the upper part, the interior being hairy, and the border five-cleft; and two seed-vessels rough on the outside, the seeds having a long wing. *G. Kamassi* is a native of the Cape, in the form of a shrub, with branches swollen at the points where leaves arise: the latter are in pairs below, in threes above; flowers small and yellow. [G. D.]

GONIOPHLEBIUM. A genus of polypodiaceous ferns, having the naked globose sori of *Polypodium*, and forming one of the genera of the polypodineous group with netted veins. The peculiar characteristic of *Goniophlebium* amongst these, is that the veins are forked or pinnate from a central costa, the lower anterior branches being usually free and fertile at the apex, and the rest angularly or arcuately anastomosing, and producing from their angles free excurrent veinlets, which

Goniophlebium glaucophyllum.

are often fertile, the marginal veinlets being free. There are often several series of anastomosing veinlets, but sometimes only one. The free (and in mature specimens usually fertile) veinlet produced within the basal areole distinguishes this group specially from its allies. There are a considerable number of species, found abundantly in South America and the West Indies, and in India and the Eastern and Pacific Islands, more rarely in tropical Africa, the Mascarene Islands, and Madagascar. A few are simple-fronded species, with a creeping ivy-like habit, and contracted fertile fronds; but they have mostly stoutish slow-creeping rhizomes, and large pinnate or pinnatifid fronds, often of pendulous habit, and sometimes several feet in length, as in *G. subauriculatum*, a very handsome Javanese species, in which, as in a few other allied kinds, the sori are sunk in little hollows which form excrescence-like knobs on the upper surface. [T. M.]

GONIOPTERIS. A genus of polypodiaceous ferns, having round naked sori, and connivently anastomosing veins, in which latter peculiarity they differ from *Polypodium*. They have a short erect or decumbent caudex, and herbaceous or suboriaceous pinnatifid, pinnate, or pinnato-pinnatifid fronds, the latter having some resemblance in aspect to our common male fern. The species are not numerous, but widely dispersed, being found in the West Indies and South America, in tropical Africa and Madagascar, and in India, the Pacific Islands, Australia, and New Zealand. [T. M.]

GONIOSTEMMA. A genus of *Asclepiadaceæ*, containing a single species, a native of India. It is a twining shrub, with opposite elliptical-oblong and glabrous leaves, and small flowers in lax many-flowered panicles like axillary cymes; they have rotate five-cleft corollas, and the staminal crown is gamophyllous, tubular, five-angled and five-lobed, and adherent to the base of the gynostegium. This genus is separated from *Secamone* and *Toxocarpus*, its nearest allies, by the structure of the staminal crown, and by habit. [W. C.]

GONOCALYX. A very beautiful vacciniaceous plant discovered by Schlim at an elevation of 7,000 feet in New Grenada, in the provinces of Pamplona and Ocaña. It forms a shrub of erect bushy habit, thickly clothed with small nearly orbicular leaves, and bearing fine bright red tubular flowers. The young leaves and shoots are of a purplish-rose colour. The only species has been called *G. pulcher*. [T. M.]

GONOGONO. *Myristica spuria*.

GONOLOBUS. A large genus of *Asclepiadaceæ*, natives of North America, consisting of twining herbaceous or shrubby plants, with opposite heart-shaped leaves, and greenish or dingy purple flowers in racemes or corymbs on interpetiolar peduncles. With a five-parted calyx, they have a rotate or reflexed and spreading corolla, the limb of which is five-parted, the staminal crown forming a small fleshy wavy-lobed ring in the throat. The follicles, which are turgid, more or less ribbed, and armed with soft warty processes, contain many comose seeds. Upwards of sixty species have been described. [W. C.]

GONOPHORUM. A short stalk which bears the stamens and carpels in such plants as anonads, &c.

GONOSTEMON. A section of *Stapelia*, characterised by having the outer of the two whorls of the staminal crown composed of five ligulate leaflets, and the interior of as many simple hooked spines. [W. C.]

GONYANTHES. A genus of *Burmanniaceæ*, consisting of two or three species from tropical Asia, differing from *Burmannia* chiefly in the capsule, which opens by transverse fissures opposite the cells. They are all slender leafless herbs, a few inches in height, with small terminal flowers, either solitary or few together in a little cyme.

GONYSTYLUS *Miquelianus* is the name given by Miquel to the tree that produces the fragrant wood called Kaju Garu by the Malays. It is very much like eaglewood, or *Aquilaria Agallochum*.

GONZALEA. A genus of South American shrubs belonging to the *Cinchonaceæ*. The tube of the calyx is somewhat globular, its limb four-parted; corolla funnel-shaped or salver-shaped, hairy externally, stamens four, included within the corolla; stigmas four; seeds minute. [M. T. M.]

GOODENIACEÆ. (*Goodenoriæ, Scævoinceæ, Goodeniadæ*.) A natural order of calycifloral dicotyledons, belonging to Lindley's campanal alliance of epigynous Exogens. Herbs, rarely shrubs, not milky, with scattered exstipulate leaves and distinct flowers. Calyx usually superior, three to five-divided; corolla more or less superior, usually irregular, with a split tube and a five-parted lipped limb; æstivation conduplicate; stamens five separate; ovary one to two-celled; placentas free central; stigma surrounded by an indusium. Fruit capsular or drupaceous; seeds albuminous. Natives chiefly of Australia and the islands of the Southern Ocean. Some of the plants are used as esculent vegetables, and their pith is employed for economical purposes. *Scævola Tuccada* furnishes the rice-paper of the Malay Archipelago; the leaves of the plant are eaten as a pot-herb, and its fruit is succulent. There are about two dozen genera, and nearly two hundred species. Examples: *Scævola, Goodenia, Velleja, Leschenaultia*. [J. H. B.]

GOODENIA. A genus of *Goodeniaceæ*, distinguished by having a superior calyx with a five-parted limb, the corolla generally two-lipped, with the tube cleft at the back; five stamens, with distinct anthers cohering before expansion; and a simple style, the stigma with a cup-shaped indusium. The capsule is two rarely four-celled. Herbaceous plants, or a few of them small shrubs, with alternate entire or toothed leaves, sometimes covered with white silky down, and axillary or terminal flowers usually yellow, rarely blue or purplish. Natives of Australia, Tasmania, and New Zealand; one species, *G. repens*, being also found in South America. [R. H.]

GOODIA. A genus of the pea family, consisting of three species, two of which are common to Tasmania and South-Eastern Australia, the other confined to Western Australia. All are handsome erect much-branched bushes, with alternate trifoliolate leaves like those of the birdsfoot-trefoil — whence the name *lotifolia* applied to one of the species. The branches bear towards their apex racemes of golden yellow flowers, like those of a laburnum, but smaller. The chief features of the genus are the trifoliolate leaves; two-lipped calyx, the lips not deeply divided; stamens all united into a sheath; and thin and flat veined pods. Its nearest ally is *Bossiæa*, from which the compound leaves distinguish it. *G. lotifolia* and *G. pubescens* are both in cultivation in greenhouses. The genus commemorates the name of Peter Good, a collector for Kew Gardens, who died in Australia. [A. A. B.]

GOOD KING HARRY. *Chenopodium* (or *Blitum*) *Bonus Henricus*.

GOODYERA. A genus of terrestrial orchids with small flowers like those of *Spiranthes*, but the spike is not spiral, and the lip does not embrace the column, has no callosities at the base, and is contracted at the top into a recurved point. It consists of very few species, all from the northern hemisphere, and mostly from high latitudes or mountain ranges. *G. repens*, generally found in moist woods, is widely spread over Northern Europe, Asia, and America, but in Britain only occurs in the highlands of Scotland. It has a creeping rootstock and an erect flowering stem of six inches to a foot, with a few ovate leaves near the base. The flowers are of a greenish white, in a slender one-sided terminal spike.

GOOGUL. *Balsamodendron Mukul.*

GOOLS. Various Marigolds, as *Calendula officinalis*, *Caltha palustris*, and *Chrysanthemum segetum*.

GOOMALA. *Batatas edulis.*

GOOMPANY. The wood of *Odina Wodier*, used in India for railway sleepers.

GOONCH. A Hindoo name for the seeds of *Abrus precatorius*.

GOONSOORA. An Indian fibre-yielding *Hibiscus*.

GOORA NUTS. The seeds of *Cola acuminata*.

GOORGOORA. *Beptonia buxifolia.*

GOOSE and GOSLINGS. *Orchis Morio.*

GOOSEBERRY. *Ribes Uva crispa*, often called *R. Grossularia*. —, BARBADOS. *Pereskia aculeata.* —, COROMANDEL. *Averrhoa Carambola.* —, TAHITI. *Cicca disticha.*

GOOSEFOOT. A common name for *Chenopodium*; also *Aspalathus Chenopoda*.

GOOSETONGUE. *Achillea Ptarmica.*

GOOSESHARE. *Galium Aparine.*

GOOWA. The Betel nut, *Areca Catechu*.

GORDONIA. A genus of *Ternstromiaceæ*, natives of North America and of the Alps of tropical and sub-tropical Asia, consisting of shrubs with alternate coriaceous entire leaves, and solitary one-flowered peduncles. The calyx is persistent, of five nearly equal concave sepals; the corolla of five petals alternate with the sepals, imbricate in æstivation; the stamens numerous, hypogynous. The fruit is a four to five-celled capsule, with two to four pendulous seeds in each cell. There are seven known species. [J. H. B.]

GORSE. The Common Furze, *Ulex europæus*.

GORTERIA. A small genus of dwarf annual herbs of the composite family peculiar to South Africa, their stems and linear or oblong-lanceolate leaves more or less hispid, and the latter clothed underneath with a close-pressed white down; and the twigs terminated by solitary yellow flower-heads nearly an inch across, and somewhat like those of the common marigold. These have an involucre of many series of narrow scales with (eventually) hardened tips; when the flowers wither, these involucres contract at the top, so that the seeds cannot escape: the latter, therefore, when they germinate, push their stems upwards and their roots downwards through the spiny nut-like involucres, which remain attached at the collar of the root, and have the appearance of a spiny tuber of the size of a hazel-nut. The ray florets are strap-shaped neuter, those of the disk tubular and perfect; the achenes are villous at the summit only, and surmounted with a short crown-like pappus composed of a single series of scales, these characters of the fruit distinguishing the genus from *Gazania*, to which it is nearly allied. [A. A. B.]

GORY-DEW. *Palmella cruenta.*

GOSSYPIANTHUS. A genus of *Amaranthaceæ*, containing perennial North American herbs with woolly procumbent stems, elongate spathulate root-leaves, those of the stem much smaller, opposite, nearly sessile, ovate, and entire, more or less densely covered with silky wool. The flowers are axillary, densely aggregated, covered with wool, and have a five-leaved perianth, five stamens with free filaments, and one-celled anthers without intermediate teeth. [J. T. S.]

GOSSYPIUM. This small genus of *Malvaceæ* is one of the most important of the whole vegetable kingdom, for to it we are indebted for the valuable and well-known article Cotton, which occupies such a prominent place in the manufacturing industry of this and other countries, and which gives employment to so large a proportion of our mercantile marine. The number of species of *Gossypium* is extremely uncertain. Between twenty and thirty have been described and named by botanists, but the characters on which they are founded are so slight and variable, that probably they may be reduced to five or six, three of which yield the Cotton of commerce. The genus is indigenous to both the Asiatic and American continents, but it has been so extensively spread by means of cultivation that it is now found throughout all parts of the world, within the limits of 35° north and south of the equator. All the species and varieties form herbaceous or shrubby perennial plants, varying in height according to the climate and soil in which they grow, some not exceeding two or three feet, while others reach a height of fifteen or twenty feet. Annual cotton plants are frequently spoken of, but, although generally treated as such, none of them are really annuals properly so called. Their leaves grow upon stalks placed alternately upon the branches, and are generally heart-shaped, and most commonly either three or five-lobed, with the lobes sharp or rounded; they generally have one or more glands upon the under side of the principal veins near the stalk. The flowers are usually large and showy, and grow singly upon stalks in the axils of the leaves. They have a cup-shaped shortly five-toothed calyx, surrounded by a larger outer calyx or involucel of three broad deeply cut segments, joined together and heart-shaped at the base; a corolla of five petals; many stamens united into a central column; and a three or five-celled ovary. The fruit is a three or five-celled capsule, which bursts open through the middle of each cell when ripe, exposing the numerous seeds covered with the beautiful cellular filaments known under the name of cotton. The seeds themselves contain a considerable quantity of bland oil, which has been brought greatly into use during the last few years; and the cake formed by pressing the decorticated seeds has proved a valuable food for cattle.

G. barbadense is the species cultivated

Gossypium barbadense.

in the United States, where two well-marked varieties are recognised. First, the Sea Island or long-staple cotton, which was introduced from the Bahamas in 1785,
and is only grown on the low islands and sea-coast of Georgia and South Carolina; it is the most valuable kind, having a fine, soft, silky staple from an inch and a half to an inch and three-quarters long, and is easily separated from the seed. Second, Upland, Georgian, Bowed, or short-staple cotton, which forms the bulk of American cotton, and is the produce of the upland or inland districts of the Southern States; the staple is only an inch or an inch and a quarter long, and it adheres firmly to the seed, which is also covered with short down. Egyptian cotton, and the kind called Bourbon, are likewise referable to this species.

G. herbaceum is the indigenous Indian species, and yields the bulk of the cotton of that country; it is also grown in the south of Europe and other countries bordering on the Mediterranean, Persia, &c. Its seeds are woolly and yield a very short-stapled cotton. *G. peruvianum* yields the cottons imported from Pernambuco, Bahia, and other parts of Brazil, from Peru, &c. It is sometimes called kidney cotton, on account of its seeds adhering firmly together in the form of a kidney.

The use of cotton dates from prehistoric ages, both in the Old World and the New. It is frequently mentioned in the *Institutes of Manu*, a work written eight centuries before the Christian era. Upon the discovery of America it was found in common use among the inhabitants, and cotton cloth has since been found in the tombs of the Incas of Peru. From India the plant spread into Persia and Arabia. Pliny, early in the Christian era, mentions that it grew in 'Upper Egypt, on the side of Arabia,' where robes for the Egyptian priests were made of the cotton. It was brought to Spain by the Mahometan conquerors of that country, and from thence it spread through other parts of Southern Europe, but it has never formed an article of much importance in the agriculture of those countries. India supplied by far the largest part of the cotton fabrics used in Europe until the rise of the English manufacture in the latter half of last century. The introduction of this important manufacture into England took place about the close of the sixteenth century, when, in consequence of religious persecution, a number of Flemings fled to this country, and established it at Bolton and Manchester. But previously to the brilliant inventions of Hargreaves, Arkwright, Crompton, and others, it was merely a domestic manufacture, and the cotton was only used for the weft of the cloth. At first our supply of raw cotton was obtained from Southern Europe and the Levant, and later from the West Indies and South America, and in smaller quantities from India and Bourbon. Towards the end of last century, however, the great and increasing demand caused the Americans to turn their attention to its production in the Southern States; and such has been their success that, till their fratricidal war broke out, they sup-piled four-fifths of the enormous quantity annually consumed in this country. Some idea of the rapid increase of the English cotton manufactures may be gained from the fact that in the year 1751, previous to the introduction of spinning by machinery, our imports of raw cotton amounted to only 2,976,610 lbs., while in 1800 they had risen to 56,010,732 lbs.; and in 1860, the enormous quantity of 1,390,939,725 lbs. was imported, of which the United States supplied no less than 1,115,890,608 lbs.—a remarkable fact when we consider that the cotton plant is not indigenous to those States, and that its cultivation for exportation only commenced between seventy and eighty years ago.

The harvest in America commences in August and lasts till December. After being picked and dried, the cotton is separated from the seeds by means of machines called gins, and is then tightly compressed into bales averaging about 430 lbs. in weight. Two kinds of gins are used in America, the saw-gin and the roller-gin—the first, consisting of numerous circular saws revolving between iron grids, being used for the short-staple variety; and the latter, which is merely a pair of rollers, for the long-staple.

The value of English cotton manufactures in 1860 was estimated at 121,304,458*l.*, being the product of 33,000,000 spindles, giving employment, directly and indirectly, to one million men, women, and children, and requiring a capital of not less than 150,000,000*l.* sterling. [A. S.]

GO-TO-BED-AT-NOON. *Tragopogon pratense.*

GOUANIA. A genus of *Rhamnaceæ*, consisting of large rambling climbing shrubs inhabiting the forests of tropical America and Asia, but principally the former. They have alternate leaves with veins running straight from the midrib to the margin; and some of their smaller branches are generally transformed into tendrils, which serve to support them. The flowers are usually produced in clusters along leafless branches, forming long slender spikes; the lower part or tube of their calyx adhering to the ovary, while the upper part is divided into five spreading segments alternating with five petals, each of which is partly rolled round a stamen, or has a stamen lying in a hollow formed by it. The fruit usually has three wings or sharp angles, but in some species it is nearly globular and without wings.

There are upwards of twenty species of this genus, the most interesting being *G. domingensis*, a common creeper in the West Indies and Brazil. In Jamaica it is called Chaw-stick, on account of its thin flexible stems being chewed as an agreeable stomachic; tooth-brushes are also made by cutting pieces of chaw-stick to a convenient length and fraying out the ends; and a tooth powder is prepared by pulverising the dried stems. It is said to possess febrifugal properties; and on account of its pleasant bitter taste is commonly used for flavouring different cooling beverages. [A. S.]

GOUDOTIA. A genus referred to *Juncaceæ*, founded on a curious little plant from the Andes, which has stems growing in dense tufts, with short distichous closely imbricated leaves, and stalked scarious flowers, diœcious by abortion. [J. T. S.]

GOUET. (Fr.) *Arum maculatum.* — À CAPUCHON. *Arisarum vulgare.* — CHEVELU, or GOBE-MOUCHE. *Dracunculus crinitus.* — EN CAPUCHON. *Arisæma ringens.* — SERPENTAIRE. *Dracunculus vulgaris.*

GOUI. *Adansonia digitata.*

GOURD. The common name for *Cucurbita*; the varieties of the common Gourd, *C. Pepo*, and of a few other species, are very numerous. —, BITTER. *Citrullus Colocynthis.* —, BOTTLE, CLUB, or TRUMPET. Different forms of *Lagenaria vulgaris.* —, COLOCYNTH. *Citrullus Colocynthis.* —, ETHIOPIAN SOUR. *Adansonia digitata.* —, GOOSEBERRY. *Momordica echinata.* —, ORANGE. *Cucurbita aurantia.* —, SNAKE. *Trichosanthes.* —, SPANISH. *Cucurbita maxima.* —, SOUR. *Adansonia Gregorii.* —, SQUASH. *Cucurbita Melopepo.* —, WHITE, of India. *Benincasa cerifera.*

GOURDE. (Fr.) *Lagenaria vulgaris.*

GOURLIEA. A genus of *Leguminosæ* related to *Sophora*, but the pods, instead of being long and constricted between the seeds, are elliptical with a somewhat fleshy rind, of the size and form of a plum-stone when mature, and usually perfecting but one seed. There are but two species known, natives of Chili and Buenos Ayres. They are bushes or small trees, with pale smooth bark, and short spine-like lateral branches, from which arise racemes or fascicles of small yellow pea-flowers. At the time of flowering, the hoary leaves, which consist of about four pairs of oblong leaflets and an odd one, are not fully developed. Chanar or Chanal is the name given to the bushes in Chili and Buenos Ayres; and, according to Tweedie, the pulp of the fruit is used in flavouring sweet wines in the latter place, and at Entre Rios. The name of Mr. Robert Gourlie, who gathered plants at Mendoza and died there, is perpetuated in the genus. [A. A. B.]

GOUTTE DE LIN. (Fr.) *Cuscuta europæa.* — DE SANG. *Adonis autumnalis.*

GOUTWEED, or GOUTWORT. *Ægopodium Podagraria.*

GOUTY-STEMMED TREE. An Australian name for *Delabechea.*

GOUYAVIER, or GOYAVIER. (Fr.) *Psidium.*

GOVENIA. A genus of terrestrial orchids peculiar to the moist woods of tropical America. The leaves are radical, broadly lance-shaped or oblong, plaited, and from one to two feet in length. The rect flower-scape also is radical, and terminates in a spike or raceme of medium-sized flowers, each supported by a narrow bract; they are usually white or cream-coloured, but in some yellow, with or without blood-red spots. The sepals and petals are free and of nearly equal length; the lip much shorter, without spur, entire, and jointed to the base of the column; and the anther contains four solid pollen-masses fixed to a short caudicle with a small triangular gland. There are sixteen species known, seven of which have been cultivated, but none are remarkable for their beauty. The genus is named after J. R. Gowen, Esq. (A. A. B.)

GOWAN. In Scotland, the Daisy, *Bellis perennis*; but appertaining rather to *Caltha*, *Calendula*, and *Chrysanthemum*, from 'gowlan,' a corruption of 'golden;' see GOOLS, and GOLDINS.

GOZELL. The Gooseberry, *Ribes uva-crispa*.

GRABOWSKIA. This ill-sounding name is applied, in honour of a Silesian botanist, to a curious Brazilian shrub which has been referred to *Solanaceæ*, but seems more closely allied to *Ehretiaceæ*. *G. boerhaaviæ-folia* is much-branched, with axillary spines, and solitary flowers opposite the leaves or grouped in panicles at the end of the branches. Its flowers have five stamens projecting from the tubular corolla, their filaments hairy in the middle; and a four-celled ovary. The fruit is succulent, enclosed within the calyx, having two woody stones, each divided into two compartments containing a single seed. (M. T. M.)

GRACILARIA. A genus of rose-spored *Algæ* belonging to the natural order *Sphærococcoideæ*, amongst which it is distinguished by its cylindrical compressed or flat frond with oblong cruciate tetraspores dispersed among the superficial cells of the branches and branchlets. It is the same with *Plocaria*, and therefore furnishes the Corsican and Ceylon moss. (M. J. B.)

GRACILIS. Slender; applied to parts which are long and narrow.

GR.ELLSIA. A genus of *Cruciferæ* found in Persia, and represented by *G. saxifragæfolia*, a perennial herb, with a habit like that of *Saxifraga granulata*. The leaf-stalks remain attached to the short root-stock after the smooth rounded notched blades wither; the flower-stalk bears a number of white racemed flowers not unlike those of the cuckoo-flower; while the fruits are small oblong much-compressed silicles, ripening but a single seed. (A. A. B.)

GRAHAMIA. A genus of *Portulaceæ*, consisting of a small Chilian shrubby plant, with alternate fleshy oblong terete leaves, and solitary flowers at the extremity of the branches, the calyx having eight or nine imbricated bracts, the white petals five in number, and the stamens numerous, with the filaments united at the base. [J. T. S.]

GRAINS OF PARADISE. The seeds of *Amomum Grana Paradisi*; also called Guinea Grains.

GRAINES D'AMBRETTE. (Fr.) *Abelmoschus moschatus.* — D'AVIGNON. *Rhamnus infectorius*, *saxatilis*, &c. — DE CANARY. *Phalaris canariensis*. — DES MOLUQUES. *Croton Tiglium.* — MUSQUÉES. *Abelmoschus moschatus.* — D'OISEAU. *Phalaris canariensis.* — DE PERROQUET. *Carthamus tinctorius.* — DE TILLY. *Croton Tiglium.*

GRAM. The Chick Pea, *Cicer arietinum.* —, BLACK. *Phaseolus Mungo melanospermus.* —, GREEN. *Phaseolus Mungo chlorospermus*, and *P. radiatus.* —, HORSE. *Dolichos uniflorus.* —, TURKISH. *Phaseolus aconitifolius.* —, RED. *Dolichos Catjang.* —, WHITE. *Soja hispida.*

GRAMEN FLEURI. (Fr.) *Stellaria Holostea.* — TREMBLANT. *Briza media.*

GRAMIGNA. The underground stems of *Triticum repens*, used in Italy as food for horses.

GRAMINACEÆ. (*Gramineæ*, *Grasses*.) A natural order of glumiferous monocotyledons belonging to Lindley's glumal alliance of Endogens. Herbaceous plants with round usually hollow jointed stems; narrow alternate leaves, having a split sheath and often a ligule at its summit; and flowers arranged in spikes or panicles, perfect or imperfect. The flowers are composed of a series of leaves or bracts—the outer, called glumes, alternate, often unequal, usually two, sometimes one, rarely none; the rest, called pales or glumelles, usually two, alternate, the lower or outer one being simple, the upper or inner having two dorsal or lateral ribs, and supposed to be formed of two pales united; sometimes one or both are wanting. The glumes enclose one or more flowers, and among the flowers there are often abortive florets. Stamens hypogynous, one to six, usually three; anthers versatile. Ovary superior, one-celled, with two (rarely one or none) scales called lodicules; ovule one; styles two or three, rarely united; stigmas often feathery. Fruit a caryopsis; embryo lenticular, lying on one side at the base of farinaceous albumen. Grasses are widely distributed over the world, forming about one twenty-second of all known plants, according to Schouw. They are social plants, forming herbage in temperate climates, and becoming arborescent in tropical countries. The order is a very important one, as supplying food for man and animals. The various cultivated grains and the pasture grasses belong to it. It is said that darnel grass (*Lolium temulentum*) has poisonous qualities, and some think that it is the tares of Scripture. Several species of *Andropogon* yield fragrant oils, such as kum-kus, roussa oil, and citronelle. The bamboo (*Bambusa arundinacea*) is one of the most useful grasses in warm countries; the sugar-cane (*Saccharum officinarum*) is another valuable grass from a commercial point of view. Among the cereal grasses cultivated for food may be enumerated:— wheat, barley, oats, rye, rice, Indian corn, millets, Guinea corn, and swamp rice. The grains of *Coix Lachryma* are used as beads under the name of Job's tears. The tussac grass of the Falkland Islands is the *Dactylis cæspitosa*. Some grasses are useful in binding the loose sand of the sea-shore. There are about 300 genera of grasses and 4,000 species. Examples: *Oryza, Zea, Phleum, Panicum, Anthoxanthum, Poa, Dactylis, Festuca, Bromus, Bambusa, Lolium, Triticum, Hordeum, Saccharum.* [J. H. B.]

GRAMMADENIA. A small genus of *Myrsinaceæ*, found in the West Indies and the adjoining mainland, related to *Myrsine*, but having the flowers in racemes instead of fascicles; and to *Cybianthus*, but having a five to six-parted instead of four-parted calyx and corolla. Their stems are abundantly furnished with sessile, lance-shaped, entire leaves, marked with curious linear glands, thus suggesting the name of the genus. The very minute flowers are succeeded by a globose ovary, which becomes when ripe a round berry the size of a small pea, with few seeds. [A. A. B.]

GRAMMANTHES. Succulent herbaceous plants, natives of the Cape of Good Hope, forming a genus of *Crassulaceæ*, nearly allied to *Crassula*, but distinguished from it by the corolla, which is tubular, with a limb divided into five or six oval lobes, and by the absence of scales at the base of the ovary. They are pretty little plants as seen during sunshine. [M. T. M.]

GRAMMATOCARPUS. A genus of *Loasaceæ* found in Chili and Peru, and nearly allied to *Loasa*, differing chiefly in its slender twisted capsular fruits, which are one to two inches long, and not much thicker than their stalks. The Chilian species, *G. volubilis*, is a slender twining annual herb, with opposite twice pinnatifid leaves, and stalked yellow cup-shaped flowers, solitary in the axils of the leaves, and nearly an inch across; they have a calyx border of five linear segments; ten petals, five large and somewhat spurred at the base, and five smaller three-awned at the apex; and numerous stamens, the fertile ones in five bundles. [A. A. B.]

GRAMMATOPHYLLUM. The few species which make up this genus of orchids are amongst the most choice in cultivation. *G. speciosum* has been called the Queen of Orchidaceous plants. This superb species, a native of Java and the adjacent islands, has stout stems from six to ten feet long, bearing a number of strap-shaped leaves one to two feet in length, arranged in a two-ranked manner. The flower-scape arises from the base of the stem, and is sometimes six feet in length, the flowers numerous but distant on the panicle, each borne on a stalk (ovary) about six inches long, this being also the diameter of the fully expanded flowers, which are of a bright yellow colour, spotted and blotched with deep purple; the lip is trilobed and comparatively small. From Manilla we have *G. multiflorum*, a plant with pseudobulbs instead of lengthened stems, producing from its apex three or four long strap-shaped leaves, and from its base a raceme nearly two feet long of yellow flowers beautifully painted over with blood-red stains of grotesque form. *G. Ellisii*, another pseudobulbous species, was introduced from Madagascar, and has the sepals and petals yellow and beautifully barred transversely with dark lines, while the petals and lip are of a pale pink colour. This plant is considered by Reichenbach to form a distinct genus, to which he gives the name *Grammangis*. The genus is nearly related to *Cymbidium*—the principal difference being, according to Dr. Lindley, that in the latter the gland of the pollen-masses is triangular, while in this it is crescent-shaped, with one pollen-mass at each extremity of the crescent. There is also a shallow sac at the base of the column and lip, not noted in *Cymbidium*. [A. A. B.]

GRAMMATOTHECA. A genus of slender branching herbs, natives of the Cape of Good Hope, and belonging to the *Lobeliaceæ*. They are distinguished mainly by their corolla, which is tubular below, with a five-parted limb in two divisions, the lower lip consisting of three pendent segments, larger than the two constituting the upper lip; the style is concealed within the corolla, and bears a two-lobed stigma whose lobes are widely separate one from the other. The genus is closely related to *Clintonia*. [M. T. M.]

GRAMMITIS. A genus of polypodiaceous ferns, producing oblique naked oblong or elliptic sori, and having free simple or forked veins. The group is often restricted to certain small simple-fronded plants, of which *G. Billardieri* may be taken as the type; but to these are sometimes added a few larger compound-fronded species, more closely resembling *Gymnogramma* in habit, but having simple oblong instead of forked sori. [T. M.]

GRAMMICUS. When the spots upon a surface assume the form and appearance of letters.

GRAMON DE MONTAGNE. (Fr.) *Smilax aspera.*

GRANA MOLUCCANA. The seeds of *Croton Tiglium* and *Parana.* —PARADISI. The seeds of *Amomum Granum Paradisi.* —SAGU. The granulated Sago of commerce. —TETRASTICHA. The spores of certain fungals. — TIGLIA or TILLA. The seeds of *Croton Tiglium.*

GRANADILLA. *Passiflora quadrangularis, maliformis, laurifolia, incarnata, edulis,* &c., which bear edible fruits.

GRAND BAUME. (Fr.) *Pyrethrum Tanacetum.* — GENTIANE. *Gentiana lutea.* — MILLET. *Sorghum vulgare.* — MO-

NARQUE. *Narcissus concolor.* — ORCHIS MILITAIRE. *Orchis fusca.* — PIN. P*inus Pinaster.* — PLANTAIN. *Plantago major.* — RAIFORT. *Cochlearia Armoracia.* — SCEAU DE SALOMON. *Convallaria multiflora.* — TRÈFLE ROUGE. *Trifolium pratense.* — VALÉRIANE. *Valeriana Phu.*

GRANDE CAPUCINE. (Fr.) *Tropæolum majus.* — CIGÜE. *Conium maculatum.* — CONSOUDE. *Symphytum officinale.* — DOUVE. *Ranunculus Lingua.* — ÉCLAIRE. *Chelidonium majus.* — ÉPIAIRE. *Stachys sylvatica.* — ÉSULE. *Euphorbia Lathyris.* — FOUGÈRE. *Pteris aquilina.* — GESSE. *Lathyrus latifolius.* — LUNAIRE. *Lunaria biennis.* — MAUVE. *Malva sylvestris.* — ORTIE. *Urtica dioica.* — OSEILLE. *Rumex Acetosa.* — PATIENCE DES EAUX. *Rumex Hydrolapathum.* — PERVENCHE. *Vinca major.* — RENOUÉE. *Polygonum orientale.* — SAUGE. *Salvia officinalis.* — VRILLÉE BÂTARDE. *Polygonum dumetorum.*

GRANGEA. A few small prostrate or erect weeds of the chamomile group of the composite family, most nearly related to *Cotula*, differing chiefly in the broadly three-toothed ray florets, and in the presence of a small cup-shaped fringed pappus. The species are widely diffused over the tropics of both hemispheres; and have pinnatifid leaves, and solitary terminal yellow flower-heads, much like those of a chamomile divested of its white ray florets; all the florets are tubular, the outer bearing pistil only, the inner perfect. *G. maderaspatana*, a very common weed all over India, occurs in Brazil, growing abundantly in sandy plains, and is used, according to Mr. Gardner, all over the country instead of chamomile, for which it is said to be an excellent substitute. It is known by the name of Marcella. [A. A. B.]

GRANGERIA. A genus of *Chrysobalaneæ*. *G. borbonica*, the only species, is a common bush or small tree of the Mauritius, where it is known as Arbre de Buis (box tree). It has glossy green coriaceous leaves, in form like those of the common box but somewhat larger; and the small white flowers are disposed in short racemes, and have a five-parted calyx, five rounded petals, fifteen stamens, and a style arising from the base of a woolly ovary which, when ripe, becomes a three-sided pyriform drupe, with a single seed. The genus is nearly related to the American *Hirtellas*, but differs in the stamens being regularly disposed, and not all arising from one side of the flower. [A. A. B.]

GRANITICUS. Growing in granitic soil.

GRANTIA. A genus of Persian herbaceous succulent-leaved composite plants. The involucre consists of two rows of somewhat leafy bracts; the outer florets are strap-shaped and neuter, the inner ones tubular and perfect, placed upon a pitted receptacle, with membranous scales between the pits; the branches of the style are elongated and cylindrical. The fruits are somewhat cylindrical, ribbed, crowned by a pappus, of which the outer row consists of a few narrow scales, the inner of hairy bristles. [M. T. M.]

GRANULA. Large spores contained in the centre of many algals, as *Gloionema*. Among fungals it sometimes expresses a spore-case.

GRANULAR, GRANULATE. Divided into little knobs or knots, as the roots of *Saxifraga granulata*.

GRANULES. Any small particles; grains; the hollow shells which constitute pollen.

GRAPE. The well-known fruit of the vine, *Vitis vinifera.* —, BEARS. *Vaccinium Arctostaphylos* and *Arctostaphylos Uva ursi.* —, CHICKEN. *Vitis cordifolia.* —, CORINTH. The fruits of the Black Corinth variety of *Vitis vinifera*, which when dried form the currants or corinths of the shops. —, FOX. *Vitis vulpina* and *Vitis Labrusca.* —, FROST. *Vitis cordifolia.* —, SEA. *Ephedra distachya*; also *Sargassum bacciferum.* —, SEASIDE. *Coccoloba*, especially *C. uvifera.* —, WILD, of Peru. *Chondrodendrum convolvulaceum.* —, WINTER. *Vitis cordifolia.*

GRAPE FLOWER. *Muscari racemosum.*

GRAPHIDEI. A natural order of lichens, distinguished by the disk of the fruit being linear and either simple or branched. There is generally a distinct receptacle, though this is sometimes wanting. It is exactly analogous to *Hysterium* amongst *Fungi*. Many fine species occur in tropical countries, but temperate regions produce a great many, and we have many striking representatives in the genus *Opegrapha*, which adorn the trunks of trees in our forests. In *Sclerophyton* the fruit is collected in linear elevations of the crust, so that it is parallel with *Trypethelium* amongst *Verrucariei*. Though *Opegrapha* is so common in the northern hemisphere, it does not occur at all in New Zealand. In *Arthronia*, which is one of the lowest genera of lichens, the receptacle vanishes altogether. [M. J. B.]

GRAPPLE PLANT. The colonial (Cape) name of *Uncaria procumbens*.

GRAPTOPHYLLUM. A genus of *Acanthaceæ* containing a single species, a native of India, but having escaped from gardens it has been diffused over the tropical regions of both the Old and New worlds. It is a shrub with oblong or ovate variegated leaves, and flowers in terminal racemes; they have an equally five-parted calyx, a ringent corolla with the upper lip arched and the lower trifid, and two stamens with sagittate anthers. The capsule is rostrate. [W. C.]

GRASS. A general name for all graminaceous plants. —, ARROW. *Triglochin*. —, ARTIFICIAL. A name given by agriculturists to various fodder plants, as clover, lucerne, sainfoin, &c. —, AWNED HAIR. *Muhlenbergia capillaris*. —, BALLOCK. *Orchis*. —, BARLEY. *Hordeum*. —, BARNYARD. *Panicum Crus galli*. —, BASTARD KNOT. *Corrigiola littoralis*. —, BASTARD MILLET. *Paspalum*. —, BEAR. *Yucca filamentosa*. —, BEARD. *Andropogon*. —, BENT. *Agrostis*; also applied to any wiry-stemmed grass growing on a bent or common. —, BERMUDA. *Cynodon*. —, BLACK. *Alopecurus agrestis*. —, BLACK OAT. *Stipa arenacea*. —, BLUE. *Poa compressa*. —, BLUE-EYED. An American name for *Sisyrinchium*. —, BOTTLE. *Setaria glauca*. —, BOTTLE-BRUSH. An American name for *Elymus hystrix*. —, BRISTLE-TAILED. *Chæturus*. —, BRISTLY FOXTAIL. *Setaria*. —, BROME. *Bromus*. —, BURDOCK. *Lappago racemosa*. —, BURR. *Cenchrus*. —, CAPON'STAIL. *Festuca Myurus*. —, CANARY. *Phalaris canariensis*, the grain of which is the canary seed of the shops. —, CARNATION. *Carex glauca*, and others. —, CATSTAIL. *Phleum*. —, CHINA. The fibre of the Rheea, *Bœhmeria nivea*. —, CLAVER. An old name for Clover, *Trifolium pratense*. —, COCK'SCOMB. *Cynosurus echinatus*. —, COCK'SFOOT. *Dactylis glomerata*. —, COMB-FRINGE. *Dactyloctenium*. —, CORD. *Spartina stricta*. —, COTTON. *Eriophorum*. —, COUCH. *Triticum (Agropyrum) repens*. —, COW. *Trifolium pratense*; also *Polygonum aviculare*. —, CRAB. *Digitaria sanguinalis*; also an American name for *Eleusine*; also *Salicornia herbacea*. —, CRESTED HAIR. *Kœleria cristata*. —, CUCKOO. *Luzula campestris*. —, DARNEL. *Lolium*; also especially *Lolium temulentum*. —, DEER. *Rhexia*. —, DEW. *Dactylis glomerata*. —, DITCH. An American name for *Ruppia*. —, DOG. *Triticum caninum*. —, DOG'S-TAIL. *Cynosurus*. —, DOG'STOOTH. *Triticum caninum*. —, DOOB. *Cynodon Dactylon*. —, DROP-SEED. An American name for *Sporobolus* and *Muhlenbergia*. —, EEL. An American name for *Zostera* and *Vallisneria*. —, ELEPHANT'S. *Typha elephantina*. —, FALSE RED-TOP. *Poa serotina*. —, FEATHER. *Stipa pennata*. —, FESCUE. *Festuca*. —, FINGER. *Digitaria*. —, FIORIN. *Agrostis vulgaris*; now more commonly applied to *A. alba* and *stolonifera*. —, FIVE-LEAVED. *Potentilla reptans*. —, FLEA. *Carex pulicaris*. —, FLOTE, or FLOAT. *Glyceria fluitans*. —, FODDER. *Chilochloa*. —, FOUR-LEAVED. *Paris quadrifolia*. —, FOXTAIL. *Alopecurus*. —, FRENCH. *Onobrychis sativa*. —, FRENCH SPARROW. *Ornithopalum pyrenaicum*. —, FROG. *Salicornia herbacea*. —, GALLOW. *Cannabis sativa*. —, GAMA. *Tripsacum dactyloides*, an esteemed fodder grass in North America and Mexico. —, GHORONA. A reputedly poisonous Indian grass, supposed to be *Paspalum scrobiculatum*. —, GINGER. *Andropogon Nardus*. —, GOATSBEARD. *Ægopogon*. —, GOOSE. *Galium Aparine*; also *Potentilla anserina*; also an American name for *Polygonum aviculare*. —, GREAT GOOSE. *Asperugo procumbens*. —, GREEN. *Chloris*. —, GRIP. *Galium Aparine*. —, GUINEA. *Panicum jumentorum*, also known as *P. maximum*. —, HAIR. *Aira*; also *Trichochloa*; also an American name for *Agrostis scabra*. —, HARD. *Sclerochloa*; also *Ægilops*; also *Dactylis glomerata*. —, HARE'STAIL. *Lagurus*. —, HASSOCK. *Aira cæspitosa*. —, HEATH. *Triodia decumbens*. —, HEDGEHOG. *Echinochloa*; also applied in America to *Cenchrus*. —, HERD. *Agrostis dispar*. —, HERD'S, of New England. *Phleum pratense*. —, HERD'S, of Pennsylvania. *Agrostis vulgaris*. —, HOLY. *Hierochloa borealis*. —, HORN. *Cratochloa*. —, HORN OF PLENTY. *Cornucopiæ cucullatum*. —, INDIAN. An American name for *Sorghum nutans*. —, INDIAN DOOB. *Cynodon Dactylon*. —, KANGAROO. *Anthistiria australis*. —, KNOT. *Triticum repens*; also *Illecebrum* and *Polygonum aviculare*. —, KNOT, of Shakspeare. *Agrostis stolonifera*. —, LEMON. *Andropogon Schœnanthus*. —, LOB, or LOP. *Bromus mollis*. —, LONG. *Harrochloa*. —, LOVE. *Eragrostis*. —, LYME. *Elymus*. —, MAIDENHAIR. *Briza media*. —, MANNA. *Glyceria fluitans*. —, MARL. *Trifolium pratense*, or, according to some authorities, *T. medium*. —, MARRAM. *Elymus arenarius*; also *Ammophila arenaria*. —, MARSH. An American name for *Spartina*. —, MARSH HEDGEHOG. *Carex flava*. —, MAT. *Nardus stricta*; also *Ammophila arenaria*. —, MEADOW. *Poa*. —, MELIC. *Melica*. —, MILLET. *Milium*; also *Sorghum vulgare*, *Panicum miliaceum*, *Setaria italica*, &c. —, MONKEY. A commercial name for the whalebone-like fibre of *Attalea funifera*. —, MOOR. *Sesleria cœrulea*. —, MOUNTAIN, of Jamaica. *Andropogon bicornis*. —, MOUSE-EAR SCORPION. *Myosotis palustris*. —, MOUSETAIL. *Festuca Myurus*; also *Alopecurus agrestis*. —, MYRTLE. *Acorus Calamus*. —, NAKED-BEARD. *Gymnopogon*. —, NIT. *Gastridium*. —, NUT. *Cyperus Hydra*. —, OAT. *Arrhenatherum avenaceum*; also various species of *Avena*; also *Bromus mollis*. —OF PARNASSUS. *Parnassia*. —, ONE-GLUMED. *Monachne*. —, ORANGE. *Hypericum Sarothra*. —, ORCHARD. *Dactylis glomerata*. —, PAMPAS. *Gynerium argenteum*. —, PANIC. *Panicum*; also *Ehrharta panicea*. —, PARA. A commercial name of the Piassaba fibre of *Attalea funifera*. —, PENNY. *Rhinanthus Crista galli*. —, PEPPER. *Pilularia globulifera*; also an American name for *Lepidium*. —, PIGEON'S. *Verbena officinalis*. —, POVERTY. *Aristida dichotoma*. —, PRICKLY. *Echinochloa*. —, PUDDING. *Mentha Pulegium*. —, QUAKE, or QUAKING. *Briza*. —, QUICK, or QUITCH. *Triticum repens*. —, RATTLESNAKE. *Glyceria canadensis*. —, RAY. *Lolium perenne*. —, RED-TOP. *Uralepis cuprea*. —, REED. *Arundo*; also *Calamagrostis*, and *Phalaris*. —, REED BENT. *Calamagrostis*. —, RIB. *Plantago lanceolata*. —,

GRAS] **The Treasury of Botany.** 550

RIE. *Hordeum pratense*; also *Lolium perenne*. —, RIBBON. *Ingraphis arundinacea variegata*. —, RICE CUT. An American name for *Leersia oryzoides*. —, ROPE. *Restio.* —, ROT. *Pinguicula vulgaris.* —, ROUGH. *Dactylis glomerata.* —, RUSH. An American name for *Vilfa.* —, RUSH SALT. *Spartina juncea.* —, RYE. *Hordeum pratense* and *murinum*; also *Secale* and *Lolium.* —, SAND. *Craleps purpurea.* —, SCORPION. *Myosotis.* —, SCOTCH, of Jamaica. *Panicum molle.* —, SCURVY. *Cochlearia officinalis.* —, SEA. *Zuppia maritima.* —, SEA HARD. *Ophiurus.* —, SEA LYME. *Elymus.* —, SEA MAT. *Ammophila arenaria.* —, SEA SPUR. *Glyceria distans.* —, SENECA. *Hierochloa borealis.* —, SESAME. *Tripsacum.* —, SHAVE. *Equisetum hyemale.* —, SHELLY. *Triticum repens.* —, SHERE. *Carex.* —, SHORE. *Littorella lacustris.* —, SHRUBBY. *Thamnochortus.* —, SILK. *Eriocoma cuspidata.* —, SLENDER. *Leptochloa.* —, SMALL. *Hierochloa.* —, SOFT. *Holcus.* —, SOUR. *Panicum leucophaeum.* —, SPARROW. *Asparagus officinalis.* —, SPEAR. *Poa.* —, SPIKE. *Uniola.* —, SPIKED. *Triglochin.* —, SPIKED QUAKING. *Brizopyrum.* —, SPRING. *Anthoxanthum.* —, SPURT. *Scirpus maritimus.* —, SQUIRREL-TAIL. *Hordeum jubatum.* —, STANDER. *Orchis mascula.* —, STAR. *Callitriche*; also an American name for *Hypoxys* and *Aletris.* —, STRIPED. *Digraphis arundinacea variegata.* —, SWEET. *Glyceria.* —, SWINES. *Polygonum aviculare.* —, SWORD. *Gladiolus*; also *Arenaria segetalis*, and *Melilotus segetalis.* —, THIN. *Agrostis elata* and *perennans.* —, THREE-LEAVED. *Trifolium.* —, TIMOTHY. *Phleum pratense.* —, TRIPLE-AWNED. *Aristida.* —, TOAD. *Juncus bufonius.* —, TURTLE. *Zostera marina.* —, TUSSAC, or TUSSOCK. *Dactylis cespitosa.* —, TWIG. *Rhabdochloa.* —, TWO-PENNY. *Lysimachia Nummularia.* —, VANILLA. *Hierochloa borealis.* —, VELVET. *Holcus lanatus.* —, VERNAL. *Anthoxanthum odoratum.* —, VIPER'S. *Scorsonera.* —, WATER SCORPION. *Myosotis palustris.* —, WATER STAR. *Lytanthus graminens.* —, WHEAT. *Triticum.* —, WHITE. *Leersia virginica.* —, WHITLOW. *Draba*, especially *Draba verna*; also *Saxifraga tridactylites.* —, WILD OAT. *Danthonia.* —, WIND. *Apera Spica venti.* —, WIRE. *Eleusine indica*, and *Poa compressa.* —, WIRE BENT. *Nardus stricta.* —, WOOD. *Sorghum (Andropogon) nutans*; also *Luzula sylvatica.* —, WOOD REED. *Cinna.* —, WOOLLY. *Lasiagrostis.* —, WOOLLY-BEARD. *Erianthus.* —, WORM. *Spigelia*; also *Sedum album.* —, YARD. An American name for *Eleusine.* —, YELLOW-EYED. *Xyris.*

GRASS-CLOTH PLANT. *Boehmeria nivea.*

GRASSETTE. (Fr.) *Pinguicula vulgaris.*

GRASS-GREEN. Clear, lively green, without any mixture.

GRASS OIL. An oil obtained from *Andropogon Iwarancusa.*

GRASS-TREE. *Xanthorrhœa*; also *Richea dracophylla*, and *Kingia australis.*

GRASS-WRACK. *Zostera marina.*

GRATIA DEI. *Gratiola officinalis.*

GRATIOLA. A genus of Scrophulariaceæ, consisting of perennial herbaceous plants, found wild in central Europe, North America, and extra-tropical Australia. The flowers have a calyx of five equal divisions, a tubular corolla whose limb is two-lipped, the upper lip notched or cleft into two divisions, the lower three-cleft; four stamens, two of which are sterile and longer than the fertile; and a capsular fruit. *G. officinalis*, the Hedge Hyssop of the herbalists, was in former times called *Gratia Dei*, on account of its active medicinal properties. It is a bitter purgative and emetic, and is even poisonous in large doses. It is not used in medical practice in this country, but is said to have formed the basis of a famous nostrum for gout called Eau médicinale. Haller says that the abundance of this plant in some of the Swiss meadows renders it dangerous to allow cattle to feed in them. *G. peruviana* has similar properties. [M T M.]

GRATIOLE. (Fr.) *Gratiola officinalis.*

GRATTERON. (Fr.) *Galium Aparine.*

GRAVELIN. (Fr.) *Quercus pedunculata.*

GRAVEL-ROOT. *Eupatorium purpureum.*

GRAVEOLENS. Strong-scented; having a smell which is unpleasant because of its intensity.

GRAVESIA. A genus of Madagascar *Melastomaceæ*, of which *G. bertolonioides* is a nearly stemless hairy herb, with opposite ovate five to seven ribbed crenelled leaves, and flower-stalks arising from the axis bearing an umbel of flowers which have a top-shaped five-toothed hairy calyx, five ovate petals, and ten stamens of equal length, with the connective, produced below into an obtuse spur-like appendage. This latter character serves to distinguish the genus among its near allies. [A. A. B.]

GRAWATHA. Curra-tow, the fibre of *Bromelia* (or *Ananassa*) *Sagenaria*, which is twisted into ropes.

GRAYA. A genus of Chenopodiaceæ, comprising a North American erect branched spiny shrub with solitary or fascicled oblong-lanceolate entire fleshy leaves and diœcious flowers. The male flowers have a regular five-parted perigone and five stamens; the females a monosepalous perigone, compressed and winged, notched at the apex, and bulging above the middle within, and a subulate style with two filiform stigmas. [J. T. S.]

GREEDS. *Potamogeton.*

GREENBRIER. An American name for *Smilax.*

GREENGAGE. A delicious variety of plum.

GREENHEART TREE. *Nectandra Rodiæi.*

GREEN-MAN. *Acerus anthropophora.*

GREENS. The familiar domestic name for open-hearted Cabbages, Kale, and other leafy esculents; also applied to *Lemna.*

GREENWEED, or **GREENWOOD.** *Genista tinctoria* and *pilosa.*

GREENWITHE. *Vanilla claviculata.*

GREGGIA. A genus of *Cruciferæ* from New Mexico, discovered by Dr. Gregg, who died in California through over-exertion in scientific pursuits. This plant, called *G. camporum* from its growing on the Campos or plains, has the habit of a wallflower, and all its parts clothed with hoary pubescence. The stems are furnished with alternate spathulate sinuate leaves, and the pink or white flowers, somewhat like those of the Brompton stock but smaller, are disposed in loose terminal racemes. The narrowed pods (siliques) are about an inch long and flattened laterally, so that the valves are boat-shaped. The genus differs from others of the *Lepidium* group, by its long pods, and from its nearest allies in having incumbent cotyledons. [A. A. B.]

GREGRE TREE. *Erythrophleum guineense.*

GREGORIA. A genus of primworts, having a five-cleft bell-shaped calyx; a salver-shaped corolla, its tube dilated at the upper end, with a border of five spreading lobes, and five ovules, two only of which reach maturity. The only species is a small herb formerly known as *Aretia Vitaliana,* a native of the Pyrenees. [G. D.]

GRÉMIL. (Fr.) *Lithospermum officinale.*

GREMILLET. (Fr.) *Myosotis.*

GRENADIER. (Fr.) *Punica Granatum.*

GRENADILLE. (Fr.) *Passiflora.*

GRENADIN. (Fr.) *Dianthus Caryophyllus.*

GRENIERA. A genus of *Caryophyllaceæ* which it has been proposed to separate from *Alsine* on account of the seeds being much compressed, with a transparent wing round the back, and a thin layer of albumen above the peripherical embryo. *G. Douglasii* and *tenella* are slender herbs from California and Arkansas, with the habit of *Alsine verna* and *tenuifolia.* [J. T. S.]

GRENOUILLETTE. (Fr.) *Ranunculus acris,* and others.

GREVILLEA. A genus of *Proteaceæ,* distinguished by having apetalous flowers; a calyx which is either four-cleft or has four linear sepals broadish at the end; four ovate sessile anthers, one of which is attached to the concave apex of each sepal; and an elongated curved style, with the stigma either lateral or oblique, plane or concave. The seed-vessel, called a follicle, is woody or leathery, containing one or two occasionally winged oval seeds. This is the most extensive and also the handsomest genus of the order. It contains every variety of form, from lofty trees a hundred feet in height, with a girth of eight feet, as in *G. robusta,* the Silk Oak of the colonists, to humble procumbent shrubs, as in *G. lancifolia.* The foliage is equally varied: in *G. juniperina, ericifolia, &c.,* it is needle-shaped; in *G. glabella* and *juncifolia,* it is filiform; in *G. obliqua, polystachya,* and *Leucadendron,* it is linear, twelve to eighteen inches in length; in *G. asplenifolia* and *mimosoides,* it is linear and serrated; in *G. laurifolia* it is ovate and entire; in *G. angulata* and *agrifolia* it is rounded at the apex, wedge-shaped and serrated; in *G. ilicifolia, acanthifolia,* and *Cunninghamii,* it is deeply cut, with sharp prickly teeth; in *G. cinerea* and *buxifolia,*

Grevillea acanthifolia.

box-leaved; in *G. anethifolia* and *triternata,* triternate; in *G. Gaudichaudii, Aquifolium, Sturtii, &c.,* pinnatifid; in *G. Banksii, Caleyi, robusta, &c.,* pinnate or bipinnatifid. The inflorescence is in spikes generally of a deep rich red, occasionally yellow as in *G. sulphurea, Banksii,* and *Chrysodendron.* In the latter species the flower-spikes exceed one foot in length, and are extremely beautiful. In *G. Dryandri, asplenifolia, Caleyi,* and *robusta,* the flowers are also in long spikes, of a deep red colour. The seed-vessels in the following species are of a hard woody substance, nearly spherical, from an inch to two inches in diameter, viz. *G. refracta, mimosoides,* and *Leucadendron,* and especially *G. gibbosa;* these are all either tropical or subtropical plants. The genus is spread over every portion of Australia, and two species, *G. australis* and *G. Sturtii,* are found in Tasmania. [B. H.]

GREWIA. An extensive genus of *Tiliaceæ,* consisting of shrubs or small trees, with simple usually serrated leaves, natives of the tropical and subtropical regions of the Asiatic and African continents, and also of the islands of the Malayan Archipelago, the Fijis, &c., but not found on the

American continent. The flowers have five sepals, which are coloured (not green) on the inside and often hairy outside; and five petals, each with a gland or hollow at the base inside, and inserted at the bottom of the stalk-like receptacle of the three to four-celled ovary, while the numerous stamens are inserted round its summit. The fruit consists of from one to four stones, each containing one or two seeds. Upwards of eighty species of this genus are described.

G. asiatica and *sapida* have both small red fruits, which, on account of their pleasant acid taste, are commonly used in India for flavouring sherbets. The wood of the Dhamnoo, *G. elastica*, a species common in the Himalayas, is very strong and elastic, and is consequently much prized by the natives for making their bows, besides which it is used for carriage-shafts and other purposes where elasticity is requisite. At the Cape of Good Hope, the elastic wood of *G. occidentalis*, called Kruysbesje, is used for similar purposes. Most of the species have a fibrous inner bark, which is commonly employed by the natives for making fishing-nets, ropes, twine, &c. [A. S.]

GRIAS. A genus of *Barringtoniaceæ* peculiar to the West Indies and the adjoining mainland. The Anchovy Pear of Jamaica, *G. cauliflora*, has long been cultivated in plant stoves for the sake of its magnificent foliage. It is a slender tall unbranched tree, furnished at top with a large crown of drooping glossy-green alternate lance-shaped or spathulate entire leaves, which are sometimes upwards of three feet long. The flowers (not well known) are said to be large, white, arranged in clusters which arise from the old wood, and consisting of a superior four-toothed calyx, four rounded petals, numerous stamens in five rows with their stalks united at the base, and an ovary tipped with a cruciform sessile stigma. The fruits are said to be russet-brown drupes, and to be pickled and eaten like the mango, having a similar taste. *G. Fendleri*, found in Panama, with equally handsome leaves, has its flowers in short racemes arising from the trunk, yellow, and one to two inches across. [A. A. B.]

GRIFFINIA. A small genus of South American *Amaryllidaceæ*, consisting of dwarfish bulbous plants, with broad oblong petiolated nervose leaves, and a many-flowered umbel of handsome purplish flowers. The perianth has a short cylindrical declinate tube, and unequal reflexed limb of six segments, the lower of which are divaricate, and the lowest stretched forward; there are six stamens with thread-shaped filaments, one of them ascending, the rest declinate; and a three-celled ovary, containing two collateral ovules in each cell, and tipped by a three-furrowed style, and an undivided or obsoletely three-lobed stigma. *G. hyacinthina*, the best known species, grows in woods on the hills behind Rio Janeiro, and is a very ornamental species. [T. M.]

GRIFFITHIA. An Indian shrub of the *Cinchona* family, with glandular leaves and spiny stems; flowers white, in terminal clusters, with a funnel-shaped corolla, whose throat is hairy, and whose limb is divided into five oblong acute segments; ovary two-celled, surmounted by a fleshy disk; stigma undivided striated. The fruit is succulent and reddish. [M. T. M.]

GRIGG. *Calluna vulgaris.*

GRIGNON. (Fr.) The wood of *Bucida Buceras.*

GRIGRI. A name in Trinidad for the wood of *Astrocaryum aculeatum.*

GRIMMIA. A genus of acrocarpous mosses, distinguished, as now reduced, by the columella not adhering to the lid, the short even tip of the veil which is entire and not lacerated at the base, and the generally exserted capsule. The peristome, when present, consists of sixteen large lanceolate convex teeth, which are split once or twice. *G. pulvinata*, remarkable for its curved peduncle, from whence it obtained formerly the name of the Swan's-neck Bryum, forms cushion-like tufts, hoary with the long white hair-points of the leaves, and thickly studded with fruit. The other British species are either Alpine or subalpine. [M. J. B.]

GRIMMIEI. A natural order of mosses, with an equal often sessile capsule, a single peristome, a mitræform calyptra, and leaves of a dark green, always terminated by a white hair, and formed of punctiform cells. *Schistidium*, in which the columella is adnate with the lid, and the capsules are immersed; *Grimmia*, with its free lid; and *Racomitrium*, with its straggling habit, confirmed by the awl-shaped granulated beak of the veil, are the British genera. *Driptodon* differs from *Racomitrium* merely in its forked stems and fastigiate innovations, and is generally united with that genus. They are found in various climates, *Schistidium apocarpum*, which is one of our more common mosses, appearing also both in Asia and South America. [M. J. B.]

GRINDELIA. A genus of *Compositæ* numbering upwards of a dozen species. The prairies of the Saskatchawan are their northern limit, Patagonia the southern, and they are found in greatest plenty in Texas and Mexico. Their chief distinguishing feature is the pappus, which consists of from two to eight rigid narrow awns, which fall early. They are biennial or perennial suffruticose plants, with branching stems, spathulate radical leaves, and sessile or clasping cauline ones, and yellow flower-heads, solitary at the ends of the twigs, and from one to two inches across. Most of the species have all their parts more or less covered with a glutinous varnish when young. [A. A. B.]

GRIOT, GRIOTTE, or GRIOTTIER. (Fr.) Names applied to varieties of *Cerasus vulgaris.*

GRISAILLE, or GRISARD. (Fr.) *Populus canescens.*

GRISEBACHIA. A genus of heathworts, distinguished by the following marks:—calyx bell-shaped, and slightly four-angled; corolla scarcely longer than the calyx; filaments covered with stiff hairs; the style ending in a small very blunt stigma; seed-vessel compressed, two-celled, two-seeded. The genus was named in honour of Grisebach, a German botanist. The species are heathlike shrubs, natives of the Cape. [G. D.]

GRISELINIA *lucida* is an evergreen shrub forming a genus of *Cornaceæ* nearly allied to *Aucuba*, which it also resembles in habit. The leaves are of a bright shining green, alternate and quite entire; the flowers small, diœcious, in terminal panicles, the males with five stamens, the females with an inferior ovary of one or two cells, but with three stigmas. The fruit is a berry with a single pendulous seed. It is in cultivation in our botanical gardens.

GRISET (Fr.) *Hippophaë rhamnoides.*

GRISETS. Pure grey, a little verging to blue.

GRISLEA. A genus of *Lythraceæ* consisting of a few handsome opposite-leaved bushes or small trees. *G. tomentosa*, a very common East Indian species, has sessile lance-shaped entire leaves clothed with white down underneath, and pretty scarlet fuchsia-like blossoms arranged in axillary cymes, and consisting of a tubular coloured calyx with a four to six-toothed border, and a like number of green glands in the clefts, four to six small narrow petals, eight to twenty stamens protruded beyond the calyx tube, and an ovary tipped with a simple style. According to Roxburgh, the calyx tube, which closely invests the ripe capsules, does not lose its colour when withered, and thus the shrub has a gaudy appearance even when in fruit. The flowers, mixed with those of *Morinda*, are used as a dye known as Dhaeo in India. One African and one American species are known. [A. A. B.]

GRIT-BERRY. *Comarostaphylis.*

GROATS, or GRITS. The grain of the oat deprived of its husks.

GROBYA. A genus of epiphytal orchids of Brazil, having ovate pseudobulbs, with a few grassy ribbed leaves at their apex, and a drooping flower-scape proceeding from the base of the pseudobulb, and ending in a short raceme of yellow or greenish flowers tinged and spotted with purple. The lower connate crescent-shaped sepals are larger than the upper, the petals broader, forming a sort of helmet overhanging the lip, which is small and five-lobed at the apex, and the two bilobed pollen-masses have each a distinct caudicle attached to an oval gland. *G. Amherstiæ* and *G. galeata* are the two known species, both in cultivation. The genus is named in compliment to Lord Grey of Groby. [A. A. B.]

GROMWELL, or GROMELL. *Lithospermum officinale.* —, **FALSE.** *Onosmodium.*

GRONOVIA. A genus usually placed in *Loaseaceæ*, from most of the genera in which it differs in the flowers having five instead of numerous stamens, and the ovary one instead of many ovules. The only known species, *G. scandens*, found in Mexico and New Grenada, is a scandent herb very like the common bryony of our hedges. Its small yellow flowers have a funnel-shaped calyx with a five-toothed border, and near its base an accessory calyx of five small bracts, the five small petals are inserted on the calyx tube, and the fruit is a little indehiscent capsule, with one seed. [A. A. B.]

GROS BLÉ. (Fr.) *Triticum turgidum.* — **GOBET.** *Cerasus vulgaris.*

GROSEILLIER. (Fr.) *Ribes.* — À MAQUEREAUX. The cultivated varieties of *Ribes Uva crispa.* — ÉPINEUX SAUVAGE. The wild Gooseberry, *Ribes Uva crispa.*

GROSIER. The Scotch name of the Gooseberry.

GROSSAILLE. (Fr.) *Triticum.*

GROSSE GRIOTTE. (Fr.) *Cerasus vulgaris.* —JONQUILLE. *Narcissus odorus.*

GROSSIFICATION. The swelling of the ovary after fertilisation.

GROSSULARIACEÆ. (*Grossularieæ, Ribesiaceæ, Currantworts.*) A natural order of calycifloral dicotyledons characterising Lindley's grossal alliance of epigynous Exogens. Shrubs often spiny, with alternate palmately-lobed leaves, without true stipules. Calyx superior, limb four to five-lobed; petals small, five; stamens five; ovary one-celled with two parietal placentas; styles more or less united. Fruit a berry, crowned with the remains of the flower; seeds numerous, albuminous. Natives of the temperate parts of Europe, Asia, and America. Wholesome plants, often supplying edible fruits, such as the gooseberry, red currant, and black currant. Some of the plants are showy garden shrubs. There appear to be only two known genera, *Ribes* and *Polyosma*, and about a hundred species. [J. H. B.]

GROSSUS. Coarse; larger than usual; thus *grosse crenatus* = coarsely crenated; *grosse serratus* = coarsely serrated.

GROUNDHEELE. *Veronica officinalis.*

GROUNDSEL. *Senecio*, especially *S. vulgaris*; also *Hyoscyamus Senecionis.* — TREE. *Baccharis halimifolia.*

GROWING POINT. The soft centre of a bud, over which the nascent leaves are formed; and all modifications of it.

GRUBBIACEÆ. A natural order of monochlamydeous dicotyledons, containing only the genus *Grubbia*, and referred by Lindley and others to the *Bruniaceæ* in the umbellal alliance of epigynous Exogens.

Some regard it as an order which should be placed between *Santalaceæ* and *Bruniaceæ*, from the former of which it differs in habit and inflorescence, in the lobes of the stamens scarcely adhering at the base, in the form of the anthers, and in the bilocular ovary; and from the latter, in the want of lobes either to the calyx or corolla, the valvate æstivation, and the long embryo. [J. H. B.]

GRUBBIA. A genus of bruniads, distinguished by having hermaphrodite flowers in the axils of single bracts, and grouped in small heads with a two-leaved involucre; style very short, truncate at the end, which is slightly three-lobed. The species are Cape shrubs, with four-angled branches, and having the leaves in pairs, shortly stalked, narrow, acute, with their margins rolled back. [G. D.]

GRUGRU. A Trinidad name for *Astrocaryum vulgare*, and also *Acrocomia sclerocarpa*.

GRUMIXAMEIRA. One of the edible-fruited *Eugenias* of Brazil.

GRUMOUS. Divided into little clustered grains; as the fæcula in the stem of the sago palm.

GRUVELIA. A genus of *Boragineæ* from Chili, with the fruit as in *Cynoglossum*, but the corolla tubular, five-toothed at the apex, and scarcely exceeding the calyx. It has the habit of an *Arenaria*, and slender leaves, the lower and middle ones opposite, the upper alternate. [J. T. S.]

GUABINOBA. The berries of certain Brazilian species of *Psidium*.

GUACO. *Aristolochia Guaco*. Besides this, which is the true Guaco, *Mikania Guaco* and *Aristolochia anguicida* have had the reputation of yielding this South American alexipharmic. —, MEXICAN. A poison obtained from a species of *Convolvulus*.

GUAIACUM. A genus of *Zygophylleæ*, consisting of West Indian and South American trees, noted for the resin which they secrete, and the extreme hardness of their wood. They have pinnate leaves and blue flowers, which have a calyx of five unequal segments, five stalked petals, ten stamens, and a stalked five-celled five-angled capsule, sometimes by abortion two to three-celled.

G. officinale is an ornamental tree with pretty blue flowers. Its trunk yields the greenish-brown hard heavy wood, called by turners *lignum vitæ*, which is used for blocks and pulleys, rulers, skittle-balls, and other purposes where hardness is required and weight is not an objection; the logs are imported from Jamaica. As is also the case with the laburnum, there is great difference in the colour of the old or heart wood and that of the young or sap wood, which is of a light yellow colour. The fibres of this wood are cross-grained. The resin, commonly called gum guaiacum, exudes from the stem, and is also obtained by jagging or notching the stem and allowing the exuding juice to harden; or by boring holes in logs of the wood and then placing them on a fire, so that the resin is melted and runs through the hole into a calabash put to receive it; or in small quantities by boiling the chips in salt and water, when the resin floats on the top and may be removed. Guaiacum is greenish-brown, with a balsamic fragrance, and is remarkable for the changes of colour which it undergoes when brought into contact with various substances. Gluten gives it a blue tint, and hence guaiacum has been proposed as a test of the goodness of

Guaiacum officinale.

wheaten bread, which contains gluten. Gum arabic, milk, various roots, &c., as those of the carrot, potato, colchicum, and horseradish, possess a similar property. Nitric acid and chlorine change guaiacum successively to green, blue, and brown. These changes in colour are said to be due to the absorption of oxygen by guaiacic acid, the active principle of guaiacum. The resin, as well as the bark and wood, are used medicinally as stimulants in chronic rheumatism, skin diseases, and other complaints. *G. sanctum* is used for like purposes in the West Indies, where also the leaves are used as a substitute for soap, having strong detersive properties. [M. T. M.]

GUAIABARA. *Coccoloba uvifera*.

GUAIAVA. *Psidium*.

GUALLAGA. A West Indian name for *Zamia media*.

GUANDEE. *Cajanus indicus*.

GUAO. A West Indian name for *Comocladia dentata*.

GUARANA. A substance prepared in South America from the seeds of *Paullinia sorbilis*, which are pounded into a paste called guarana bread, and hardened in the sun. It is used as a remedy for various diseases, as well as to form a most refreshing beverage.

GUARANENE. A white crystalline

bitter substance, obtained from guarana, nearly identical with theine and caffeine.

GUAREA. The vernacular name, in Cuba, of a meliaceous tree, the flowers of which are in axillary clusters, with the stamens united into a cylindrical or somewhat prismatic tube, the free margin of which is entire or slightly waved, the anthers being enclosed within it. The ovary is four-celled, placed on a stalk-like disk, and the capsule is four-valved, with four or eight seeds. The trees of this genus are more or less purgative and emetic in their effects. *G. trichilioides* and other species have a musk-like perfume. Some of them present a peculiarity in the growth of their leaves which are pinnate; after a while the lower leaflets fall, and young ones grow at the end of the same leaf-stalk, which elongates, the lower older portion becoming woody, with an outer bark and a semblance of pith within—assuming in fact the characters of a branch. [M. T. M.]

GUATTERIA. A genus of *Anonaceæ*, named in honour of an Italian botanist, and consisting of trees or shrubs with lateral or terminal inflorescence. The flowers have six petals in two rows, flat oblong or linear, and all of the same form, and the carpels are distinct, each containing a single erect seed. *G. virgata* is said to yield some of the light wood used by coachbuilders under the name of Lancewood; see also DUUGETIA. *G. longifolia* is an ornamental tree, commonly planted by roadsides in Bengal. *G. suberosa*, which has cork-like bark, is a native of Ceylon and various parts of India. [M. T. M.]

GUAVA. *Psidium pyriferum, pomiferum*, &c.

GUAZA. The narcotic tops of the Indian hemp, *Cannabis sativa indica.*

GUAZUMA. A genus of shrubs or small trees of the *Byttneria* family, nearly allied to *Theobroma*, but differing in their woody tubercular fruits of the size of a hazel-nut, the entire instead of two-lobed appendage at the ends of the petals, and in their whole appearance. They are found in the East Indies and the islands of Eastern Africa, but are most frequent in tropical America. The leaves are like those of the elm, and their small white pink or yellow flowers are borne in axillary cymes. *G. tomentosa* is common in India and America. The French colonists in the West Indies call it Orme d'Amerique, from its resemblance to the elm. According to M'Fadyen, it grows in Jamaica to a height of twenty to twenty-five feet, and is allowed to grow in pasture lands, not only for the sake of its shade, but because the cattle feed and thrive on the foliage and fruit. The latter, coarsely bruised, are given to horses as a substitute for corn, their nutritive properties being attributed to the mucilage which abounds in them, and also in the inner bark. This mucilage is given out abundantly on infusion or decoction in water, and, according to the same authority, has been employed as a substitute for gelatine or albumen, in clarifying cane juice in the manufacture of sugar. A like infusion is given internally as a remedy for cutaneous diseases. The timber is light, splits readily, and is employed for the staves of sugar hogsheads. The plant is known by the name of Bastard Cedar to English colonists in Jamaica. A strong fibre is obtained from the young shoots of the same species in India. Cord made from it was found by Dr. Roxburgh to break at 100 lbs. when dry, and at 140 lbs. when wet. [A. A. B.]

GUÈDE. (Fr.) *Isatis tinctoria.*

GUÊPES VÉGÉTANTES. A name applied to a species of wasp in the West Indies, when affected by *Cordiceps sphærocephala.* The parasite has a long cylindrical curved stem with a club-shaped head, and at length weighs down and kills the wasp. The accounts of earlier observers, who affirmed that they had seen the wasps flying about with their heavy burden, were long disbelieved, but they have been confirmed by more recent authorities. The fungus does not seem to fructify till after the death of the insect. We have at least seen no perfect individuals. [M. J. B.]

GUERNÉSIENNE. (Fr.) *Nerine sarniensis.*

GUETTARDA. A genus of shrubs or small trees, natives of tropical America and Asia, and belonging to the *Cinchonaceæ*. The corolla is salver-shaped, with a cylindrical tube, and a limb divided into four to nine oblong segments; anthers four to nine, sessile, concealed within the corolla; ovary with from four to nine compartments, each containing a single erect ovule. The fruit is succulent, with a bony four to nine-celled stone. [M. T. M.]

GUEULE DE LION, or DE LOUP. (Fr.) *Antirrhinum majus.*

GUI. (Fr.) *Viscum album.*

GUIGNE ROUGE, or GUIGNIER. (Fr.) *Cerasus avium.*

GUILANDINA. A small genus of leguminous plants found in nearly every tropical country, particularly upon the seashore, its extensive distribution being caused by the transportation of its seeds (which have an exceedingly hard impervious shell) from one country to another by means of oceanic currents. There are three or four species, which form prickly trailing shrubs ten or twelve feet or more in height, having twice pinnated leaves, the stalks covered with short down and bearing recurved prickles on the under side. The flowers are of a rusty yellow colour, and are borne in racemes; they have a five-parted calyx with a short tube, and a corolla of five nearly equal-sized petals, the stamens being ten in number, distinct, and hairy at the base. The pods, which are about two or three inches long, flattened, but bulged out in the centre, and covered with prickles, contain one, two, or three large bony seeds. *G. Bonduc* has solitary prickles on the leaves, and the seeds are yellow. *G. Bonducella* differs by its prickles being in pairs, and its seeds lead-coloured. The seeds of both are very hard, and beautifully polished, and are called Nicker nuts or Bonduc nuts, the latter word being derived from the Arabic, *Bondoq*, signifying a necklace, the seeds being commonly strung into necklaces, bracelets, rosaries, &c. The kernels have a very bitter taste, and are employed by Indian doctors as a tonic and febrifuge. The roots also are said to possess similar properties: indeed, the Singhalese employ every part of these plants medicinally. The oil obtained from the seeds is supposed to be useful in convulsions and palsy.   [A. S.]

GUILDINGIA. A group of melastomads now referred to *Monriria*.

GUILIELMA. A genus of palms confined to the tropical regions of South America, and containing three species, which have tall slender trunks marked with circular scars and armed with exceedingly sharp black spines. The large pinnate leaves have spiny leaflets and footstalks. The flower spikes are simply branched, and bear male and female flowers mixed together. The fruit is large and egg-shaped, containing a single seed.

G. speciosa, the Peach Palm, a native of Venezuela and Guiana, is cultivated on the banks of the Amazon and Rio Negro. It grows sixty or eighty feet high, and has its stems armed with rings of long sharp needle-like spines. The fruits, which are borne in large drooping bunches, are about the size of apricots, and of a bright scarlet colour at the top passing into bright orange below; their fleshy outer portion (sarcocarp) contains a large quantity of starchy matter, which forms a considerable portion of the food of the natives. They are either boiled or roasted, and when eaten with salt resemble a potato in flavour; or they are sometimes eaten with molasses. A beverage is also prepared by fermenting them in water; and the meal obtained from them is made into cakes. The wood of old trees is black, and so exceedingly hard that it turns the edge of an ordinary axe. [A. S.]

GUILNO. (Fr.) *Bromus catharticus*.

GUIMAUVE. (Fr.) *Althœa officinalis*. — EN ARBRE. *Hibiscus syriacus*.

GUINCHE. (Fr.) *Molinia cœrulea*.

GUINDOLLE, or GUINDOUX. (Fr.) *Cerasus vulgaris*.

GUINEA-HEN FLOWER. *Fritillaria meleagris*.

GUINEA-HEN WEED. *Petiveria alliacea*.

GUIRILA. The Persian insect-powder, prepared from *Pyrethrum carneum*, and *P. roseum*.

GUIZOTIA. A small genus of annual opposite-leaved composite herbs found in Abyssinia and India, and nearly related to *Heliopsis*, differing chiefly in the presence of a ring of thick jointed hairs outside the corolla tubes near the base. *G. oleifera*, a plant with the habit of *Bidens cernua*, has lance-shaped stem-clasping leaves, and solitary stalked yellow-rayed flower-heads about an inch and a half across at the ends of the twigs; the ray florets female, the disk florets perfect; the achenes smooth and destitute of pappus. The plant is cultivated in Abyssinia and in India for the sake of a bland oil like that of *Sesamum*, which is expressed from the seeds, and is commonly used in India as a lamp-oil and as a condiment. The plant is sown in the Mysore districts in the autumn months, perfecting its seeds in about twelve weeks after it is sown. The yield is said to be about two bushels an acre. The oil is sweet-tasted, and is known in India as Ram-til oil.   [A. A. B.]

GUJ-PIPPUL. *Scindapsus officinalis*.

GULF WEED (called also by voyagers Sea-lentils, Sea-grasses, and Sargazo) is the celebrated *Sargassum bacciferum*, which occupies a more or less interrupted space between the 20th and 45th parallels of north latitude, extending over more than a quarter of a million of square miles. It was first discovered by Columbus, unless indeed the Phœnicians fell in with it during their early voyages, as seems possible from a passage in Aristotle. The seaweed floats on the surface, being propagated from age to age by buds, and never in that situation yielding fruit, which when produced consists of little bundles of receptacles in the axils of the leaves. The area occupied by the seaweed is determined by the course of the currents in the Atlantic, and occasionally a few stragglers are carried northward by the Gulf Stream, and are thrown even upon our own coasts. The origin of this mass of seaweed has not been determined. Its increase in deep water is, however, the less surprising if we remember that the root of seaweeds merely performs the office of a holdfast, and has not the function of a true root.   [M. J. B.]

GUM, ACAROID. A resinous product of *Xanthorrhœa hastilis* or *arborea*. — AMMONIACUM. The gum-resin of *Dorema ammoniacum*. — ANIME or ANIMI. A resinous product of *Hymenœa Courbaril*; also Indian Copal, the produce of *Valeria indica*. — ARABIC. The gummy product of various *Acacias*, as *vera*, *arabica*, *Verek*, *Seyal*, *Senegal*, *tortilis*, &c. —, ARTIFICIAL. Dextrine, obtained from potato starch. —, AUSTRALIAN. A kind of gum arabic. —, BABOOL. The gum of *Acacia arabica*. —, BARBARY. The gum of *Acacia gummifera*. —, BASSORA. A gum whose origin is unknown; it is supposed to be the produce of a *Cactus* or a *Mesembryanthemum*. —, BLACK-BOY, or BOTANY-BAY. A fragrant resinous product of *Xanthorrhœa arborea* or *hastilis*. —, BRITISH. A preparation of roasted starch. —, BUTEA. Bengal Kino, the gum-resin of *Butea frondosa* and *superba*. —, CAPE. The gum of *Acacia Karroo* or capensis. —, CARANA. The gum-resin of *Icica Carana*. —, CASHEW. The gum of *Anacardium occidentale*. —, CEDAR. A gum-resin resembling olibanum, obtained in the Cape Colony from *Widdringtonia juniperoides*. —, CHERRY-TREE. The gum produced from the stems of *Cerasus avium* and *vulgaris*, *Prunus domestica*, and other drupaceous trees. — COPAL. A gum-resinous product of *Trachylobium Martianum*, and *Gærtnerianum*. —, DOCTOR'S. The gum-resin of *Rhus Metopium*. — DRAGON. The gum-resin of *Pterocarpus Draco*; also the name sometimes given to gum tragacanth in the shops. —, EAST INDIAN. The gum of *Acacia arabica*, probably. — ELASTIC. Caoutchouc, the product of *Siphonia elastica*. — ELEMI. The gum-resin of *Amyris Plumieri* and *hexandra*. —, GRASS-TREE. The resinous product of *Xanthorrhœa australis*. — GUTTA, American. The gum-resin of *Vismea guianensis*. — GUAIACUM. The gum-resin of *Guaiacum officinale*. —, HOG. A gum-resinous juice variously ascribed to *Moronobea coccinea*, *Rhus Metopium*, *Clusia flava*, and *Hedwigia balsamifera*. — JUNIPER. The resin of *Callitris quadrivalvis*. — KINO. The gum of *Pterocarpus erinaceus*, and according to some of *Pterocarpus Marsupium*; also a similar product of *Eucalyptus resinifera*. — KUTEERA. The gum of *Cochlospermum Gossypium*, or according to others of *Sterculia urens*, or of *Acacia leucophlœa*. — LAC. The gummy product of *Erythrina monosperma*, and in Ceylon of *Aleurites laccifera*; a similar product is yielded by *Ficus indica*, *benghalensis*, &c. — LADANUM. The gum-resinous product of *Cistus creticus*; also of *C. ladaniferus* and *Ledon*. — LEDON. *Cistus Ledon*. —, MOROCCO. The gum of *Acacia gummifera*. — MYRRH. The gum-resin of *Balsamodendron Myrrha*. — OPOCALPASUM. The gum of *Acacia gummifera*. — ORENBERG. A gummy exudation of the larch, *Abies Larix*. — SANDARACH. The resin of *Callitris quadrivalvis*. — SASSA. A kind of false tragacanth obtained from *Inga Sassa*. — SENEGAL. The gum of *Acacia Senegal*, *Seyal*, *Verek*, *Adansonii*, &c. —, SOUDAN. A kind of gum arabic. — SUCCORY. The gummy juice of *Chondrilla juncea*. —, SWEET. *Liquidambar styraciflua*. — THUR. A kind of gum arabic. — TRAGACANTH. The gummy exudation of *Astragalus gummifer*, *strobiliferus* or *Dicksoni*, *verus*, *creticus*, and *aristatus*. — TRAGACANTH of Sierra Leone. The gum of *Sterculia Tragacantha*. —, TURKEY. The true white gum arabic. —, WATTLE. The gum of *Acacia mollissima*. —, YELLOW. A resinous product of *Xanthorrhœa hastilis* or *arborea*.

GUM-TREE. *Eucalyptus*; also *Xanthorrhœa*. — of Jamaica. *Hippomane biglandulosa*. —, BLACK. *Nyssa villosa*. —, BLUE. *Eucalyptus globulus*. —, RED. *Eucalyptus resinifera*. —, SOUR. *Nyssa villosa* and *biflora*. —, WHITE. *Eucalyptus resinifera*. —, YELLOW. *Nyssa villosa*.

GUM-ARABIC TREE. *Acacia Verek*. —, RED. *Acacia Adansonii*.

GUM-WOOD. The timber of *Eucalyptus*.

GUMBO MUSQUÉ. The seeds of *Abelmoschus moschatus*.

GUNDALI. An Indian name for *Pæderia fœtida*.

GUNGUN, or GUNJUN. A balsamic product of *Dipterocarpus lævis*.

GUNJA. *Abrus precatorius*.

GUNJAH. The dried Indian Hemp plant, *Cannabis sativa indica*.

GUNNALA. An Indian name for *Cassia fistula*.

GUNNERA. A genus of ivyworts, having the following characters: perigone or outer part of the flower usually adherent to the ovary, its border in four pieces, two of which are tooth-like, the other two resembling petals but sometimes wanting; stamens two, alternate with the two small lobes of the perigone; stigmas two, plumose. The name was given by Linnæus in honour of Ernest Gunner, a bishop of Norway, who published a flora of that country. The plants are herbaceous, natives of South America and the Sandwich Isles. One of them, *G. scabra*, is known in gardens by its coarse rough rhubarb-like lobed leaves, and its singular elongated conical inflorescence. [G. D.]

GUNNIA *australis* is the only epiphytal orchid of Tasmania, where it is found growing on the stems of trees and shrubs. It is a little plant, hardly a span high, with wiry roots, a few lance-shaped leaves three to four inches long, and flower-racemes of about the same length. The flowers, which smell like honeysuckle, are small and green except the clawed lip, which is marked with lilac lines on a white and yellow ground. As a genus it is hardly different from *Sarcochilus*, with which, indeed, it is united by some authors. Named in honour of Mr. R. C. Gunn of Tasmania, who is well known in connection with the botany of that island. [A. A. B.]

GUNNY. A coarse kind of cloth made from jute, the fibre of *Corchorus capsularis*, and sunn, the fibre of *Crotalaria juncea*.

GURLTIA. A genus of *Begoniaceæ*, consisting of erect branching shrubs, natives of Brazil. The flowers are cymose, the staminate and pistillate ones in the same plant; the former have four white perianth leaves, and numerous stamens; the latter five unequal perianth leaves, and a three-winged ovary, with a bifid stigma surrounded by a papillose band which is twice spirally twisted. There are four species. [J. H. B.]

GUSTAVIA. A genus of *Barringtoniaceæ* peculiar to tropical America, and consisting of trees or shrubs with large handsome alternate ovate or spathulate glossy leaves, and showy white flowers sometimes five or six inches across, tinged with pink, not unlike those of some *Magnolias*, and disposed in racemes or umbels at the ends of the twigs. They consist of a top-shaped calyx with an entire or lobed border, four to eight rounded or oval petals, very numerous stamens whose filaments are united below into a ring; and a four to six-celled ovary tipped with a short conical style and sulcate stigma. The fruits are somewhat fleshy and apple-like. The wood of *G. urceolata*, used for making hoops, is called Bois puant in Cayenne, because it becomes very fœtid after exposure to the air. The small fruits of *G. speciosa*, according to Humboldt, when eaten, have the singular property of causing the body to assume a yellow colour, which, however, leaves it in the course of a day or two without any application. The bruised leaves of *G. brasiliana* are said by Martius to have an unpleasant smell, and are used in cases of indurated liver. The roots are acrid, aromatic, and bitter; and the emetic fruit intoxicates fish. [A. A. B.]

GUTHNICKIA. The name of a few species separated from *Achimenes*. It forms one of the genera with a perignnous and nearly entire thickened ring, and a stomatomorphous stigma. Among these it is known by the long gaping corolla, the tube of which is straight and subcylindrical, and by the stamens being adnate with the lower part of the corolla tube. They are hairy leafy Mexican herbs, with solitary axillary scarlet flowers. [T. M.]

GUTTA PERCHA. The gum-resin of *Isonandra Gutta*. — TRAP. The inspissated sap of *Artocarpus*.

GUTTATUS. Spotted; that is, when colour is disposed in small spots.

GUTIERREZIA. A small genus of composite plants, of the same group as *Solidago*, and differing from its near allies in the achenes of the disk and ray florets being fertile and furnished with a pappus of several linear or oblong chaffy scales. They are peculiar to America, and extend from the prairies of the Red River to Mexico, a few occurring in Chili and the extreme south of the continent. For the most part they are branching herbs one to three feet high, with slender twiggy stems furnished with linear entire gummy leaves, and small yellow flower-heads very numerous, arranged in corymbs at the ends of the twigs. *G. gymnospermoides*, the only species with any pretensions to beauty, has flowers very much larger than the others, and not unlike those of *Pulicaria dysenterica*. [A. A. B.]

GUTTIFERÆ. (*Clusiaceæ, Guttiferæ*.) A natural order of thalamifloral dicotyledons, belonging to Lindley's guttiferal alliance of hypogynous Exogens. Trees or shrubs with a resinous juice, opposite leathery entire leaves, and often incomplete flowers; sepals and petals two four five six or eight, the former often unequal, the latter equilateral; stamens numerous, often united; disk fleshy; ovary one or many-celled; stigma usually sessile and radiate. Fruit dry or succulent, one or many-celled; seeds exalbuminous, often immersed in pulp. Natives of humid and hot places in tropical regions, chiefly in South America. Several are found in India, a few in Madagascar and the continent of Africa. The plants are generally acrid, and yield a yellow gum-resin. Gamboge is produced by *Cambogia Gutta*, *Garcinia cochinchinensis*, *G. elliptica*, and *G. tinctoria*. The famous mangosteen fruit is procured from *Garcinia Mangostana*. The American mammee apple is the produce of *Mammea americana*. Keyna oil is obtained from species of *Calophyllum*. The *Clusias* are handsome trees. *Pentadesma butyracea* is the butter and tallow tree of Sierra Leone; its fruit yields fatty matter. There are 32 known genera and upwards of 150 species. Examples: *Clusia, Garcinia, Cambogia, Calophyllum*. [J. H. B.]

GUYONIA. A genus of *Melastomaceæ*, bearing pentamerous flowers, having the teeth of the calyx acute, the petals ovate-lanceolate, the stamens ten, equal, with ovoid blunt anthers, and the ovary five-celled. They are tender smooth herbs with prostrate and ascending stems, small rhomboid-ovate leaves, and small solitary rose-coloured flowers. *G. tenella* inhabits moist ground on the banks of the Senegambia rivers. [J. H. B.]

GUZMANNIA. A genus of tropical American herbs, belonging to the *Bromeliaceæ*, and having an inferior calyx of three equal segments cohering at the base and spirally twisted, three petals rolled together into a tube, the anthers also cohering so as to form a tube. The seeds are numerous, provided with hairs, and enclosed in a three-celled three-valved capsule. *G. tricolor* is a pretty species with flowers on a spike, concealed by the bracts, the lowermost of which are green, while the upper are scarlet. [M. T. M.]

GYMNADENIA. A genus of terrestrial orchids, founded on the *Orchis conopsea*, which has the pollen-masses not enclosed in any process of the stigma. Several other species of European and North American orchids have been associated with it by some botanists, whilst others retain them in *Orchis* or in *Habenaria*.

GYMNANDRA. A genus of *Selaginaceæ*, containing six species of herbaceous plants, natives of Siberia, Arctic America, and the mountains of India. The flowers grow in long spikes at the apex of an erect scape. The calyx is spathe-like, with a fissure in front, and two or three-lobed behind; the tubular corolla is two-lipped; there are two stamens; the free bilocular ovary bears a long exserted style and a stigma with two capitate lobes; and the fruit is surrounded by the withered bracts and calyx, and consists of two achenes each containing a cylindrical pendulous seed. The structure of the fruit of this genus separates it from *Scrophulariaceæ*, to which it is otherwise nearly related. On the whole, it seems to belong to *Selaginaceæ*, though differing remarkably from the other genera of the order. [W. C.]

GYMNANTHERA. A genus of *Asclepiadaceæ*, containing a single species from New Holland, a twining glabrous shrub with milky juice, opposite leaves, and whitish-green flowers on lateral peduncles, having a salver-shaped corolla, and a staminal crown of five awned scales inserted at the summit of the tube and alternating with the corolla lobes. The capsule is cylindrical divaricate, and containing many comose seeds. [W. C.]

GYMNOBALANUS. Tropical American trees, constituting a genus of *Lauraceæ*, and known by the following characters:— The perianth in both male and female flowers is divided into six nearly equal segments, and is deciduous; the male flowers have nine stamens in three rows, the inner row bearing sessile glands, a long style, and an abortive ovary; the females nine sterile stamens, a one-celled ovary, and a short style. The fruit is succulent, on a thickened flower-stalk. [M. T. M.]

GYMNOCARPIUM. *Polypodium.*

GYMNOCIDIUM. The swelling occasionally found at the base of the sporecase of urn-mosses.

GYMNOCLADUS. A large American tree, *G. canadensis*, called the Kentucky Coffee-tree, is the sole representative of this genus of leguminous plants. It is common throughout the Northern United States, and in Canada, where it is called Chicot; and is frequently cultivated, either as an ornamental tree, or for its timber, which is strong and of a compact fine grain, and used for building purposes, common cabinet-making, &c. It attains a height of fifty or sixty feet, the trunk being frequently destitute of branches for the first thirty feet, but seldom more than twelve or fifteen inches in diameter. Its leaves are twice pinnate, and sometimes as much as three feet in length, consisting of a main stalk with several pairs of secondary stalks bearing numerous oval dull-green leaflets, except the lowest pair, which have a single leaflet. The flowers are whitish and borne in racemes from the angles of the leaves, the separate sexes being on different trees; their calyx is tubular and five-cleft, and the corolla of five equal-sized petals inserted into the top of the calyx tube, along with the ten short distinct stamens. The fruit is a hard flattened pod, from six to ten inches long, containing several flattish seeds imbedded in a mass of pulp. The common American name of this tree was given to it in consequence of the early settlers in Kentucky having made use of its seeds as a substitute for coffee, at a time when they could not procure the genuine article. The rough bark of the tree is excessively bitter, and contains *saponine*, a substance which, like soap, forms a lather in water. [A. S.]

GYMNOGENS, or GYMNOSPERMS. Naked-seeded plants, forming a division of dicotyledons or Exogens considered by Lindley as a distinct class. It includes the *Coniferæ* or pines and firs, *Taxaceæ* or yews, *Gnetaceæ* or joint-firs, and *Cycadaceæ* or cycads. These orders are usually called naked-seeded because there is no proper ovary, the seeds being fertilised by the pollen coming into direct contact with the foramen of the ovule without the intervention of a stigma. Some authors have of late doubted the correctness of this statement, and have considered the so-called seeds as a bicarpellary ovary containing one seed. Gymnospermous plants are represented largely in the fossil flora of the secondary strata. [J. H. B.]

GYMNOGRAMMA. A genus of polypodiaceous ferns, having free forked veins and linear sori which are more or less frequently forked—that is to say, the sporecases, being distributed along a portion of the veins, are continued both above and below the points where the veins fork. The sori in some species are very much elongated, and form contiguous narrow lines over a great portion of the fertile fronds. In other species the under surface, and sometimes also the upper, is farinosely ceraceous, and usually either of a white or yellow colour, these being the Gold and Silver Ferns so frequently seen in cultivation on account of the beauty of their coloured fronds. This genus contains two of the very few known annual ferns, *G. chærophylla* and *G. leptophylla*, the first a West Indian plant, the second scattered over nearly the whole of the temperate regions of the globe from Jersey to New Zealand, and found also in the Neilgherries and Cuba. The other species are widely dispersed, but chiefly found in tropical countries. [T. M.]

GYMNOGYNE *cotuloides*, which is the only known species of the genus, is a slender composite annual weed, peculiar to West Australia. Its unbranched stems, furnished with grassy leaves, are terminated by a single flower-head the size of a pea, and somewhat like those of *Cotula*. The outer florets are female and destitute of corolla (whence the name); and the inner male, with a tubular four-toothed corolla, and free anthers. The genus is related to *Euphrosyne*, differing in the many series of female florets, the four-toothed males, and the imbricated achenes. [A. A. B.]

GYMNOGYNOUS. Having a naked ovary.

GYMNOMESIUM. A genus differing but little from *Arum*, except in the presence of rudimentary flowers above the stamens, and not elsewhere. The ovaries contain several ovules. *G. pictum* is frequently met with in gardens under its former name of *Arum*. It is a native of Corsica, Sardinia, &c. [M. T. M.]

GYMNOPETALUM. The name of a genus of *Cucurbitaceæ*, allied to *Bryonia*, and having a calyx with a contracted throat.

a five-parted corolla, anthers cohering into a cone, and an ovate beaked few-seeded fruit. [M. T. M.]

GYMNOPOGON. A genus of grasses belonging to the tribe *Chlorideæ*, having the inflorescence in panicles, with simple alternate branchlets; spikelets two-flowered; glumes two, keeled, nearly equal, or the lower shortest. There are about half a dozen species, all natives of Brazil excepting *G. racemosus*, which is North American. [D. M.]

GYMNOPSIS. A genus of composite plants, comprising about a dozen species, spread over America from Texas to Brazil. They are herbaceous or somewhat shrubby, with opposite three-nerved toothed nettle-like rough leaves, and axillary or terminal stalked yellow-rayed heads of numerous florets, those of the ray neutral, of the disk perfect. The achenes are seated on a convex receptacle, each enveloped in a chaffy scale and surmounted by a minutely toothed pappus-crown. The presence and nature of the pappus, together with the opposite triple-nerved leaves, are the chief features. [A. A. B.]

GYMNOPTERIS. A genus of polypodiaceous ferns belonging to the tribe *Pleurogrammeæ*, in which the linear-oblong sori are placed close to and parallel with the costa. In this group it is distinguished by its compoundly anastomosing veins, and by the fructification occupying distinct contracted fronds. The species have sometimes been referred to the *Acrosticheæ*, from which, however, the definite linear sori, confined to the receptacular veins, seem to separate them. They are all Eastern tropical plants. [T. M.]

GYMNOS. In Greek compounds = naked, or uncovered.

GYMNOSCHŒNUS. A genus of cyperaceous plants belonging to the tribe *Rhynchosporeæ*. The inflorescence is in distichous two-flowered spikelets. Only one species is described, namely, *G. adustus*, a native of Van Diemen's Land. [D. M.]

GYMNOSIPHON. A small slender leafless herb, from the Indian Archipelago, forming a genus of *Burmanniaceæ*, scarcely differing from the tropical American genus *Dictyostegia*.

GYMNOSPERMA. A genus of *Compositæ*, receiving its name from its chief distinguishing feature, namely, the naked achenes (without pappus). The two species, found in Texas and Mexico, are smooth glutinous shrubby plants, with twiggy stems furnished with linear entire leaves, and terminating in corymbs of numerous small yellow flower-heads, each containing from eight to fourteen florets, those of the ray strap-shaped and pistillate, and those of the disk tubular and perfect. [A. A. B.]

GYMNOSTACHYS. An East Australian perennial, with a thick rootstock, and grassy leaves, from among which rises a two-edged stalk or scape, bearing towards its summit a number of clustered slender spikes or spadices, each having at its base a short keeled spathe; stamens four; ovary one-celled; fruit succulent, blue, one-seeded. *G. anceps*, the only species, is in cultivation. The genus is nearly allied to *Acorus*, and is included with it among *Orontiaceæ*. [M. T. M.]

GYMNOSTACHYUM. A genus of *Acanthaceæ* with the habit of *Eranthemum* and *Cryptophragmium*. It comprises dwarf herbs, with spreading often variegated leaves, and erect spike-like racemes of tubular flowers. The calyx is five-parted nearly equal; the corolla two-lipped, with the upper lip bidentate; the stamens two, with the anther-cells parallel; the stigmas bifid; and the capsule columnar and four-cornered. *G. ceylanicum*, a Ceylon species, is a pretty stove-herb, with the leaves variegated with white along the course of the veins; and *G. Verschaffeltii*, from Para, is a still prettier plant with the numerous reticulated veins coloured red, as occurs in *Hæmadictyon venosum*. They are chiefly interesting on account of their prettily marked foliage. [T. M.]

GYMNOSTEPHIUM. A genus of composite plants peculiar to South Africa. The three known species are herbaceous or somewhat shrubby plants, with alternate linear entire leaves, and small solitary stalked flower-heads terminating the twigs, the ray florets strap-shaped and blue, the disk tubular and yellow. The achenes of the ray being without pappus, and those of the disk being sterile, suffice to distinguish these plants from daisies and their other allies. [A. A. B.]

GYMNOSTOMUM. A genus of acrocarpous mosses, formerly containing almost every moss destitute of a peristome, but now restricted to those species which differ only in this character from *Weissia*. The species mostly, though not exclusively, inhabit temperate regions. [M. J. B.]

GYMNOSTYLIS. Under the name of *G. anthemifolia* is sometimes cultivated in botanic gardens a stemless South American herb with chamomile-like leaves, and clusters of small woolly flower-heads sitting in their midst. The plant is usually placed in *Soliva*. [A. A. B.]

GYMNOTETRASPERMOUS. Having such a four-lobed ovary as is found in labiates, which was formerly thought to consist of four naked seeds.

GYMNOTHECA. A small genus of marattiaceous ferns, separated from *Marattia* on account of the absence of an involucre beneath the sorus. They are the same in habit and in general aspect, having large globose scaly rhizomes, and ample bipinnate fronds, with articulated pinnules. The typical species is *G. cicutæfolia*. [T. M.]

The name has also been given to a Chinese herbaceous plant with the habit of *Saururus*, and referred to the *Saururaceæ*. The flowers are small, placed on spikes, in the axils of small somewhat fleshy bracts;

stamens six, inserted on the top of the ovary (?), which is inferior, one-celled, with four parietal placentæ; ovules numerous; styles four. (*Decaisne.*) [M. T. M.]

GYMNOTHRIX. A genus of grasses belonging to the tribe *Paniceæ*, joined by Steudel with *Pennisetum*. [D. M.]

GYNAION. A genus of *Cordiaceæ*, containing a single species from the Himalayas. It is a woody plant with alternate or sub-opposite elliptical entire leaves, and flowers in cymes at the ends of the branches. The calyx is unequally four to six-lobed; the corolla funnel-shaped, with the limb divided into four to six obovate oblong lobes; the five stamens have hairy filaments; and the four-celled ovary is globose and glabrous, perforated at the apex, without style or stigma. This remarkable plant has the habit of *Varronia rotundifolia*, but differs from it in the structure of the flower, which seems to be the monstrous flower of a *Cordia* wanting the style and stigma; it differs from *Cordia* in the structure of the calyx and corolla. [W. C.]

GYNANDROUS. Having the stamens and style and ovary all blended into one common body, as in orchids, *Aristolochia*, &c.

GYNERIUM. A genus of grasses belonging to the tribe *Arundineæ*, and distinguished chiefly by the species having showy flower-panicles, the spikelets of which are two-flowered, the male and female flowers on distinct plants. Steudel describes six species, which are natives of Brazil and Chili, save one, *G. zelandicum*. That which is best known and cultivated in Britain is *G. argenteum*, the Pampas grass, so called from its being a native of the vast plains of South America called Pampas. This splendid grass has proved sufficiently hardy to withstand the rigours of our winters in Britain without protection; and blossoming as it does in October, when most other flowers are past, its value is much enhanced. Few plants produce a finer effect than good tufts of this grass, either when cultivated singly on lawns, or in the front of shrubberies, where evergreen plants afford a dark background, as a contrast to its large silvery-white feather-like panicles. Under favourable circumstances, the culms rise from ten to twelve feet high or upwards, forty, fifty, or occasionally more, springing from one plant. This Pampas grass was first introduced to Europe in 1843, through seeds sent from Buenos Ayres by Mr. Tweedie to the Glasnevin Botanic Garden, and it is now cultivated in most gardens of note throughout Great Britain, as well as on the continent of Europe, and in Australia. [D. M.]

GYNIXUS, or GYNIZUS. The depressed stigmatic surface of orchids.

GYNOBASE. The growing point inserted between the base of carpels in a conical manner, so as to throw them into an oblique position.

GYNOCARDIA. A genus of *Pangiaceæ*, differing from others in the very numerous stamens (about a hundred), and in the sterile flowers, whose anthers are fixed by the base. *G. odorata*, the only known species, is a handsome East Indian tree, abundant in hot valleys in the Sikkim Himalaya and Khasya, comparable to the common sycamore for size, having glossy entire alternate leaves, and yellow sweet-scented flowers an inch and a half across, growing in clusters generally from the old wood, the different sexes on separate plants. They have four or five calyx leaves, a like number of petals and scales opposite them, numerous stamens in the sterile flowers, and a few abortive ones in the fertile, surrounding an ovary which is tipped with five short styles. The fruits are globular ash-coloured berries the size of a shaddock, and enclose numerous seeds, imbedded in pulp. According to Roxburgh, the seeds contain an oil; and are beaten up with clarified butter, and used by the natives as a remedy for cutaneous diseases. [A. A. B.]

GYNOCEPHALIUM. A Japanese climbing shrub, belonging to the *Artocarpaceæ*. The leaves are heart-shaped and undivided; the female flowers are globular, the male in panicled heads. The genus is closely allied to *Conocephalus*, but is distinguished from it by the four-parted perianth of the male flower, the cleft style, and the crumpled seed-leaves. [M. f. M.]

GYNŒCIUM. The pistil and all that belongs to it.

GYNOPHORE. The stalk of the ovary, within the origin of the calyx.

GYNOPLEURA. A genus of Chilian herbs, with entire or lobed leaves, and yellow flowers in terminal clusters or tufts. They belong to the *Malesherbiaceæ*, and are known by their bell-shaped calyx, five petals inserted outside an annular membranous slightly toothed corona, which is attached to the throat of the calyx; and five stamens, inserted at the base of a short stalk that supports the one-celled ovary. [M. T. M.]

GYNOSTEMIUM. The column of orchids; that is to say, the part formed by the union of stamens, style, and stigma.

GYNOSTEMMA. A genus of *Cucurbitaceæ*, natives of Java, having entire or lobed leaves, unisexual flowers in panicles, a two-ranked calyx, no corolla, five monadelphous stamens, and a half-inferior ovary. The fruit is succulent, with three or four one-seeded stones. [M. T. M.]

GYNOXIS. A genus of composite plants confined to America, and there found from Guatemala southward to Peru. They are nearly related to *Senecio*, and chiefly differ in the style-branches being prolonged into conical hispid points, instead of being obtuse. Some are trailing bushes with alternate lance-shaped or ovate leaves; and the yellow flower-heads few and large, or numerous small and arranged in corymbs. Another and larger group, which is restricted to the Andean regions of Equador and New Grenada, are erect oppositeleaved bushes or small trees, with white or yellow rayed or rayless flower-heads, resembling those of our own groundsels. This latter group M. Weddell keeps alone in the genus, and places the former with true groundsels. To the latter belongs *G. fragrans*, the only species known in cultivation, a scandent bush with somewhat fleshy ovate leaves, and pale yellow fragrant flowers, the heads about an inch across and disposed in loose corymbs at the ends of the twigs. About twenty species are enumerated. [A. A. B.]

GYNURA. A genus of the composite family, containing upwards of twenty species, all found in the tropics of the eastern hemisphere, occurring in greatest numbers in India and its Archipelago. From *Senecio*, to which it is closely allied, it differs in the style-branches having long protruding points. Many of the species are coarse perennial weeds, with distantly toothed or pinnatifid leaves, the angular stems terminating in corymbs of rich yellow flower-heads, whose florets are all tubular. Other species are scrambling undershrubs. The rootstocks in some are thick and fleshy, and not unfrequently the leaves are of a fine purple colour underneath, which is the case with the *G. bicolor*, a species from the Moluccas, cultivated in hothouses for the sake of its leaves, and having rich orange-coloured flower-heads disposed in loose corymbs. [A. A. B.]

GYPSOCALLIS. *Erica*.

GYPSOPHILA. A genus of annual or perennial evergreen herbaceous plants belonging to the *Caryophyllaceæ*, in the alsineous division of which they are distinguished by having the calyx campanulate, angular, somewhat five-lobed, the margins membranous; five petals without claws; ten stamens; two styles; and a one-celled capsule. The species, which are numerous, have leaves like those of the pink, and small white or pink flowers, which are usually disposed in diffuse panicles. They inhabit various parts of Europe and Asia, growing mostly in rocky or stony places, especially in a limestone soil. Some of them are occasionally cultivated as border plants, or on rockeries. [C. A. J.]

GYRANDRA. A Mexican perennial with the habit of a *Chironia*, and forming a genus of *Gentianaceæ*. The flowers have a five-parted wheel-shaped purple corolla, into the throat of which are inserted the stamens, whose showy yellow twisted anthers give a distinguishing character and name to the genus. [M. T. M.]

GYRATE. The same as Circinate; curled inwards like a crozier.

GYRINOPSIS. An aquilariaceous shrub of the Philippine Islands. The perianth is coloured and funnel-shaped, with a five-cleft limb; and there are ten short hairy scales placed in pairs opposite the lobes of the perianth. [M. T. M.]

GYROCARPUS. A genus of apetalous Exogens, consisting of trees having polygamous flowers, natives of the East Indies and tropical parts of America. The leaves are alternate, undivided or lobed, and the flowers are collected in dense panicles. The calyx in the hermaphrodite flowers is superior, and four to eight-lobed; the stamens four, with glands interposed, and the anthers remarkable for opening by valves which turn upwards. The ovary, which is completely adherent to the tube of the calyx, is one-celled with one pendulous ovule, the style slender, and the stigma obtuse. The fruit is nut-like, two-winged at the apex, from two of the lobes of the calyx enlarging while the others fall off. The male flowers have the same lobed calyx and stamens as the hermaphrodite. This genus is very near *Illigera*, from which it differs in its fruit being winged at the apex, not on the sides, and in common with it is nearly allied to *Laurineæ* and *Combretaceæ*, with the latter of which families it has been combined. It is sometimes considered as the type of a distinct family, and separated under the name of *Gyrocarpeæ*. [B. C.]

GYROMA, or GYRUS. The ring or articulated circle which surrounds the sporecases of ferns; also a button-like shield, such as is found among lichens in the genus *Gyrophora*.

GYROPHORA. A genus of lichens belonging to the order *Pyxinei*, distinguished by its curiously convoluted fruit, a number of disks being produced in a proliferous manner within the original fruit. The species grow on rocks and large boulders, and are remarkable as supplying the Tripe de Roche of the Arctic voyagers, so called from the bullate dilated frond. The bitter principle is so strong in these plants, that, though they have considerable nutritive qualities, they do not agree as an article of food with many constitutions. *Umbilicaria*, which is distinguished by the more simple disks, supplies also a part of the Tripe, which is collected without much discrimination of species. [M. J. B.]

GYROSE. Bent backwards and forwards as the anthers of cucurbits.

GYROSELLE. (Fr.) *Dodecatheon*.

GYROSTEMON. A genus of *Gyrostemoneæ*, consisting of small branched shrubs from South-western Australia with alternate linear semi-cylindrical mucronate leaves, and solitary axillary stalked diœcious flowers, with a six or seven-lobed calyx; the males have numerous stamens, in several rows, the females many carpels placed round a thick central axis. The fruit is obovate, of many membranaceous cocci, in a single row. This latter character distinguishes it from *Codonocarpus*, which has the cocci arranged in more than one row. [J. T. S.]

HAAGEA. A genus of *Begoniaceæ*, called after Haage, an Erfurt horticulturist. The flowers are rose-coloured, monœcious, with two perianth leaves: the staminate ones with numerous stamens; and the pistillate ones with an inferior three-celled equally three-winged ovary, a three-parted smooth persistent style, and broadly expanded stigmas, surrounded by a papillose band twisted once spirally. The only species, *H. dipetala*, a shrub with semicordate petiolate leaves, and pendulous floral cymes, native of the East Indies, was formerly called *Begonia dipetala*. [J. H. B.]

HABEL-ASSIS. (Fr.) *Cyperus esculentus*.

HABENARIA. A well-known extensive genus of terrestrial tuberous-rooted orchids, more or less generally distributed, though most numerous in India and Africa. It is represented in Britain by *H. bifolia* and *H. chlorantha*, known respectively as the small and large Butterfly Orchis. They are both very similar in aspect, having a stem a foot or more in length, with two oblong obtuse leaves at the base, above that a few narrow green bracts, and then an erect terminal spike of very fragrant long-spurred white flowers. The difference between the two is, that in *H. chlorantha* the flowers are larger, the throat or the nectary or spur much wider, and the two pollen-masses more distant from each other. For a most interesting account of the mode of fertilisation in these two plants, see Mr. Darwin's book, *On the Fertilisation of Orchids*. Some of the Indian species are notable for the length of spur, as in the appropriately named *H. longecalcarata*, where, with flowers an inch across, the spur is four inches in length. The habit of most of the species is similar to that of our native *Orchis*, to which they are closely related, differing chiefly in the two glands of the pollen-masses being inserted into separate pouches instead of into a common one. The flowers vary much in colour, some being green, others rose, a goodly number golden yellow, but the greater part white, and usually very fragrant. [A. A. B.]

HABINE. (Fr.) *Dolichos melanophthalmus*.

HABIT. The general appearance of a plant; its manner of growth, without reference to details of structure.

HABITAT. The situation in which a plant grows in a wild state.

HABLITZIA. A perennial Caucasian twining herb, of the order *Amaranthaceæ*, with a turnip-shaped root and furrowed stem, large alternate long-stalked glabrous cordate-acuminate entire leaves, and flowers in small cymes collected into dense axillary panicles: the perigone green and five-cleft, the stamens five. [J. T. S.]

HABRACANTHUS. A genus of *Acanthaceæ*, containing three species from Mexico, herbs or shrubs with oblong or oval leaves, and white or red flowers in terminal panicles or in few-flowered axillary cymes. The calyx is deeply five-parted, and the corolla ringent, with the upper lip falcate and entire, and the lower three-parted, there are two exserted diverging stamens, and the ovary is surrounded by a broad disk at the base, has four ovules near the middle, and is surmounted by an acute stigma. [W. C.]

HABRANTHUS. A genus of hippeastriform *Amaryllidaceæ*, distinguished in that group, which has a narrow-mouthed perianth tube, by the perianth being declinate, but not convolute into a tube-like form as in the allied *Phycella*, and by the faucial membrane being annular. They consist of handsome South American bulbs, found principally in Chili, Monte Video, and Buenos Ayres. The plants have narrow two-ranked flaccid leaves, and a precocious scape which either is single-flowered or bears an umbel of few or many flowers of a crimson, scarlet, rose, purple, whitish, yellow, or red and yellow colour. The perianth is subcampanulate with a short tube, the limb more or less spreading, the stamens unequal inserted at the mouth of the tube, the faucial membrane annular, and the stigma three-lobed. [T. M.]

HABROTHAMNUS. A genus of beautiful Mexican shrubs, belonging to the *Solanaceæ*. The flowers have a bell-shaped five-toothed calyx, a club-shaped tubular corolla, with the limb contracted and five-toothed; five stamens concealed within the corolla; and a button-shaped stigma. The fruit is succulent, surrounded by the calyx, two-celled, each cell containing a few seeds. The panicles of red or purple flowers are borne in abundance, and justify the name applied to them—which signifies graceful branch. [M. T. M.]

HABROZIA. A genus of *Scleranthaceæ*, differing from the rest of the order in having the utricle adhering to the seed, and the calyx tube not constricted at the throat. It is a small annual oriental herb, with slender stems, setaceous leaves, and many-flowered terminal cymes. [J. T. S.]

HABZELIA. A small genus of *Anonaceæ*, having a very wide geographical distribution, two species being found in Malaya, two on the western coast of tropical Africa, and the remainder in Guiana and Cuba. It belongs to the *Xylopieæ*, and is distinguished from its allies by the torus being flat, instead of drawn up into a cone or hollowed out. The flowers are three-sided, having three sepals joined together at the bottom, and six petals arranged in two series, the inner ones being rather smaller than the outer. The fruit consists of numerous long nearly cylindrical pods, separate from each other, and containing a number of oblong seeds. The plants are either tall shrubs or trees about twenty or thirty feet high, and have long simple leaves of a leathery texture, from the base of which the flowers are produced either singly or in clusters. *H. æthiopica*, a tall shrub, with pointed egg-shaped leaves, covered with whitish down underneath, but smooth and green above, is a native of Western Africa, where its fruit, which consists of a number of smooth pod-like carpels about the thickness of a quill, and two inches long, is dried and used instead of pepper, whence it is often called Negro-pepper, Guinea pepper, or Ethiopian pepper, and by old authors *Piper æthiopicum.* The fruits of *H. undulata* are used in the same way, as also are those of *H. aromatica*; indeed, it is probable that the fruits of all the species possess similar aromatic and pungent properties. [A. S.]

HACKBERRY. *Celtis crassifolia*; also *C. occidentalis.*

HACKMATACK. The American Larch, *Abies pendula.*

HACQUETIA. A genus of umbellifers, distinguished by having the limb of the calyx leaf-like and persistent, forming a crown to the fruit, which is contracted at the sides, each half having five narrow ribs. The genus was named in honour of Hacquet, who published an account of the Alpine plants of Carniola. *H. Epipactis*, the only species, is a small herbaceous perennial plant resembling an *Astrantia*, having digitate three-lobed leaves, and a single umbel of yellow flowers. [G. D.]

HADSCHY. The narcotic Hashish, *Cannabis sativa.*

HÆMADICTYON. A genus of dogbanes, distinguished by the border of the corolla having five equal broad bent lobes, its tube having on its inside five small scales alternate with the lobes; the style ending in a head-like summit; and by five small glands being situate at the base of the seed-vessel, alternate with the divisions of the calyx. The name is from the Greek, and indicates the crimson tint of the leaf-veins. The species are climbing shrubs, natives of tropical America. [G. D.]

HÆMANTHUS. A genus of Amaryllidaceæ, consisting for the most part of South African bulbs, some few species being found in tropical Africa. It belongs to the amaryllidiform group, that with a solid scape, and the stamens not connected by a cup. The special characteristics are a regular perianth with straight tube, and a valveless fruit with a pulpaceous middle coat. The species are rather numerous. They have tunicated bulbs with the scales often two-ranked, and few leaves, often only two, which are thickish and plane, erect or lying flat on the ground. The scape is short, often with a pair of bracts at the base, sometimes coloured and terminating in an umbel of many crowded flowers, usually with a many-leaved spathe, the leaflets of which are erect, often coloured, and longer than the flowers. The flowers are red or white, sometimes very showy. The perianth is six-cleft with erect or spreading segments, and a short tube; the stamens six, exserted; the style filiform, with a simple or obsoletely three-lobed stigma; and the ovary three-celled, becoming a globose or oblong berry. [T. M.]

HÆMARIA *discolor*, or, as it has been called, *Goodyera discolor*, is a small herbaceous orchid of South China, having creeping succulent stems throwing out roots at intervals, bearing towards the apex a few ovate leaves, and ending in an erect flower-spike a few inches in length furnished with a number of crimson bracts. The leaves are nearly three inches long, green above, and crimson underneath; and the flowers are white and three-quarters of an inch across. The plant has altogether the appearance of a *Goodyera*, and differs chiefly from that genus in the sepals being like the petals and not herbaceous. It is cultivated in gardens for the sake of its beautiful leaves. [A. A. B.]

HÆMATITICUS. Dull red, with a slight mixture of brown.

HÆMATOCOCCUS. A genus of chlorospermis, in which, however, some of the species have red, and some green spores, and probably for this reason the word *Glœocapsa* has been substituted for it. The plants consist of a shapeless gelatinous mass made up of vesicles containing a variable number of spores which propagate the plant by cell-division, new cells being formed from the divided spores within the mother cell. The species are numerous, and are important as illustrating under a simple form the great principle of increase by cell-division. The individual plants closely resemble the gonidia of *Ephebe*. Some of the species are, it is believed, merely an early stage of certain lichens. [M. J. B.]

HÆMATORCHIS. *Cyrtosia.*

HÆMATOSTAPHIS. A small glabrous tree from tropical Africa, with pinnate leaves and long axillary panicles of small white flowers, forming a genus of *Anacardiaceæ*, allied to *Odina, Schinus*, &c., but differing from all in its trimerous irregular flowers, and in habit. The drupes of a deep crimson, collected in bunches resembling grapes, are eatable, and although acid are not unpleasant.

HÆMATOXYLON. The tree yielding the well-known Logwood of commerce is the sole representative of this genus of *Leguminosæ Cæsalpineæ*. It is a native of the Bay of Campeachy in Yucatan—whence it is named *H. campechianum*—and also of other parts of Central America, and has been introduced into, and is now naturalised in, many of the West Indian islands. The tree is one of medium size, seldom exceeding forty feet in height, with a trunk about a foot and a half in diameter, and having its smaller branches covered with white bark, often spiny. The leaves are pinnate, consisting of from three to four pairs of small smooth obversely egg-shaped leaflets; and the yellow flowers are produced in racemes from the bases of the leaves. The pod is flat, tapered to both ends, and contains two seeds, but instead of splitting open along the edges, as many other pods do when ripe, its thin sides burst irregularly and allow the escape of the seeds.

Logwood, the produce of this tree, was one of the valuable commodities introduced into Europe by the Spaniards, during the early part of the century following the discovery of America by Columbus. Its use in England dates from the time of Queen Elizabeth, but the dyers of that period were so little acquainted with the chemical principles involved in the art of dyeing, that they failed to render its colours sufficiently permanent, and the prejudice against it consequently became so strong, that, in the twenty-third year of Elizabeth's reign, an act of parliament was

Hæmatoxylon campechianum.

passed prohibiting its use, and ordering it to be burned wherever found within the realm; and although it was subsequently surreptitiously introduced under the name of Black-wood, this law was not repealed until the time of Charles II., nearly a century afterwards. At the present day it is largely employed by calico-printers and cloth-dyers, and also by hat-makers, who use it, in combination with indigo and certain mordants, for imparting the fine black to silk hats. It likewise forms an ingredient in some of the commoner descriptions of writing-ink. Its properties depend upon the presence of a colouring principle termed *hæmatoxylin* or *hæmatin* by Chevreul.

Logwood occurs in commerce in logs about three feet long, and consists of the heart-wood of the tree, from which the sapwood, which is light-coloured and valueless, has been removed. It is of a deep dull brownish-red colour, and very hard and heavy; and, for the convenience of dyers, it is cut into chips by means of powerful machinery. Our imports in 1860 amounted to 26,938 tons, the greater and most valued portion of which was the produce of Central America; the remainder being from the West Indian Islands. [A. S.]

HÆMODORACEÆ. (*Velloziew, Bloodroots.*) A natural order of epigynous monocotyledons belonging to Lindley's narcissal alliance of Endogens. Perennial plants with fibrous roots, and sword-shaped equitant leaves, and bearing woolly hairs or scurf on their stems and flowers. Perianth tubular, six-divided; stamens three, placed opposite the segments of the perianth, or six; anthers introrse. Ovary three-celled, sometimes one-celled; style and stigma simple. Fruit usually capsular, opening by valves, covered by the withered perianth; embryo in cartilaginous albumen. Natives of America, the Cape, and New Holland. The roots of some of the plants yield a red colour—hence the name of the order. Bitterness exists in some of them. There are about a dozen genera, and fifty species. Examples: *Hæmodorum, Alstris, Vellozia, Barbacenia.* [J. H. B.]

HÆMODORUM. A genus of *Hæmodoraceæ*, consisting of perennial glabrous Australian herbs having fasciculate tubers, simple leafy stems, with half-sheathing plane or somewhat terete averse leaves, and corymbs or branched spikes of flowers, the perianth of which has a tube connate with the base of the ovary, and a limb of six narrow persistent segments, three stamens, and a three-celled ovary with filiform style and simple stigma. [T. M.]

HAGBERRY. *Cerasus Padus;* also *Celtis crassifolia.*

HAIMARADA. *Vandellia diffusa.*

HAIR, AFRICAN. The fibre of the leaves of the Palmetto, *Chamærops humilis.*

HAIRBELL. *Campanula rotundifolia.*

HAIR-BRANCH TREE. *Trichocladus crinitus.*

HAIR-POINTED. Terminating in a very fine weak point.

HAIR-SHAPED. The same as Filiform, but more slender, so as to resemble a hair; it is often applied to the fine ramifications of the inflorescence of grasses.

HAIRS. Small delicate transparent conical expansions of the epidermis, consisting of one or more cells.

HAIRY. Covered with short weak thin hairs.

HAIR-TRIGGER FLOWER. *Stylidium graminifolium.*

HAI-TSAI. A transparent gluten much used in China; the chief ingredient is supposed to be *Plocaria tenax*, a small seaweed.

HAKEA. A large proteaceous genus containing above one hundred species. It is distinguished by having a calyx irregularly four-cleft or with four linear or spathulate sepals, each lobe or sepal bearing on its concave apex an ovate sessile anther; a filiform style, and terminal or oblique stigma; and a one-celled seed-vessel (follicle), which is generally woody, ovate or oblong and swollen, rarely globose, smooth or tuberculated, and often with two spurs. The foliage is extremely variable: in *H. aricularis, propinqua, pugioniformis, longicuspis, Cunninghamii, lorea,* &c., it is simple, filiform, occasionally furrowed, and the points mostly very sharp, the leaves in the last-named species being from eighteen inches to two feet in length; in *H. lasiocarpha, trifurcata,* &c., it is very narrow and flat; in *H. linearis, florida, ilicifolia, prostrata,* &c., it is linear-lanceolate or ovate, with more or less spiny margins; in *H. cucullata, conchifolia,* and *Victoriæ,* it is broadly heart-shaped, with toothed margins; in *H. arborescens, Leucadendron, pandanicarpa, dactyloides,* &c., it is linear-spathulate and of a very leathery texture. The fruit of *H. pandanicarpa* is very large, and covered with conical tubercles. The leaves in *H. mimosoides, saligna, oleifolia,* &c., are either lanceolate or ovate. The genus consists generally of tall shrubs, or occasionally of small trees, as *H. lorea, Preissii, arborescens,* &c. Some of the species have been found in every portion of Australia and Tasmania that has yet been visited. [R. H.]

HALBERD-WEED. *Neurolæna.*

HALBERT-HEADED. Abruptly enlarged at the base into two diverging lobes, like the head of a halbert.

HALEDSCH. *Balanites ægyptiaca.*

HALENIA. A genus of Siberian herbaceous plants of the gentian family. Their flowers have a four-parted calyx; a four-cleft corolla whose segments are prolonged at the base into a spur; four stamens; and a one-celled ovary with a two-lobed stigma. *H. heterantha* is remarkable from its lower flowers having no spurs, while the upper ones are provided with them. [M. T. M.]

HALESIACE.E. One of the names of the order *Styracaceæ.*

HALESIA. A genus of *Styracaceæ*, differing from the others in its two to four-winged fruits, which are one to two inches long, with a bony one to four-celled kernel. The Snowdrop or Silver-bell trees, as the species are commonly called, are natives of the United States. They are deciduous shrubs or small trees with alternate stalked ovate-oblong toothed leaves, an inch or two long when the plant is in flower, but much larger when mature. The flowers bear much resemblance to snowdrops, and are supported on slender drooping stalks, two or three together, arising from the buds of the preceding year. A fine tree about thirty feet high of *H. tetraptera* may be seen in the Arboretum in Kew Gardens, flowering in May and June. The genus bears the name of Dr. Stephen Hales. [A. A. B.]

HALF. Sometimes used in the sense of one-sided; as *half-cordate*, which signifies cordate on one side only. — MONOPETALOUS. Having the petals united, but so slightly that they easily separate. — NETTED. When of several layers of anything, the outer one only is netted; as in the roots of *Gladiolus communis.* — STEM-CLASPING. Clasping the base in a small degree. — TERETE. A long narrow body, flat on one side, convex on the other.

HALVED. When the inequality of the two sides of an organ is so great that one half of the figure is either wholly or nearly wanting, as the leaf of many *Begonias.*

HALIANTHE. (Fr.) *Arenaria peploides.*

HALIDRYS. A generic name given to the old *Fucus siliquosus*, which is a frequent inhabitant of our coasts, and distinguished at once from all other native *Algæ* by the pod-like jointed air-bladders. The only other representative of the genus, *H. osmundacea*, is found on the northwest coast of America. [M. J. B.]

HALIMEDA. A genus of calcareous green-spored *Algæ*, with the habit of the Indian fig, belonging to the natural order *Siphoneæ.* The frond is composed, like *Caulerpa*, of a branched thread which traverses every part of the plant, but never has any articulations. The endochrome is at length resolved into minute zoospores. The species are all inhabitants of warm seas. *H. Opuntia* is widely diffused, and is found in the Atlantic and Pacific oceans, and also in the Mediterranean and Red seas. The plants grow in sand or amongst fragments of shells, being attached by a mass of fine thread-like roots, which grasp the particles of sand &c., and form a little ball. [M. J. B.]

HALIMUS. A group of *Chenopodiaceæ*, allied to *Atriplex*, but now sunk under the genus *Obione*; it is, however, retained as a section of the latter, distinguished by having the perigone surrounding the fruit closed, and joined by the whole length of their sides. *Obione pedunculata*, an annual, found, though rarely, in salt marshes in the south-east of England, belongs to this section; it has alternate obovate or oblong slightly fleshy leaves, with a mealy covering, and axillary glomerules of small flowers arranged in interrupted spikes. The fruit is remarkable from the pedunculated obcordate fruiting bracts. [J. T. S.]

HALLERIA. A genus of *Scrophulariaceæ*, consisting of erect glabrous shrubs, with opposite ovate evergreen leaves, and showy scarlet flowers, solitary or clustered in the upper axils. The shape of the corolla is nearly that of a *Pentstemon*, but the calyx is broad and cup-shaped; there is no rudimentary fifth stamen, and the fruit is a berry. There are three species known, all natives of the Cape Colony in South Africa.

HALLIA. A genus of *Leguminosæ* peculiar to South Africa, consisting of a few erect or decumbent perennial herbs, with slender angled or winged stems, simple heart-shaped or lance-shaped leaves, and in their axils solitary or twin stalked flowers, nearly the size of those of a vetch, and violet-coloured. The genus is most nearly related to *Alhagi*; but they are spiny erect bushes, with more than one seed to the pod, while here the habit is very different, and the minute compressed pods have but one seed. Linnæus named the genus after Berger Martin Hall, one of his pupils. [A. A. B.]

HALOCNEMUM. A genus of *Chenopodiaceæ*, allied to *Salicornia*, but having the perigone of three scale-like leaves, not monophyllous. They are small leafless jointed-stemmed plants, with the flowers collected into terminal spikes, much as in *Salicornia*. They occur chiefly in Southern Russia, Siberia, &c. [J. T. S.]

HALODENDRON SATINÉ. (Fr.) *Halimodendron argenteum*.

HALODULE. A genus of *Naiadaceæ* allied to *Zannichellia*, of which it has the habit, but with diœcious flowers; the leaves resemble those of *Zostera* in miniature. The plant grows in estuaries in Madagascar. [J. T. S.]

HALOGETON. A genus of *Chenopodiaceæ*, allied to *Salsola*, but having the seed vertical instead of horizontal. They are herbs or small shrubs found in Southern Russia, Siberia, Persia, &c., with alternate or opposite fleshy semi-cylindrical leaves, and axillary glomerules of flowers, of which the perigone is furnished with transverse wings when in fruit. The seeds of *H. tamariscifolium*, a Spanish species, are called Spanish Wormseed from their anthelmintic properties. [J. T. S.]

HALOPHILA. A genus of small herbaceous plants growing in salt marshes in Madagascar and elsewhere, usually referred to the *Podostemaceæ*, but excluded by Tulasne in his elaborate treatise on that order. They are plants of little general interest, having unisexual flowers, with a two-leaved perianth, and three stamens; and in the female ones, a stalked one-celled ovary. [M. T. M.]

HALORAGACEÆ. (*Halorageæ, Hippurideæ, Cercodianeæ, Hydrocaryes, Hippurids.*) A natural order of calyciflorial dicotyledons, belonging to Lindley's myrtal alliance of epigynous Exogens. Herbs or undershrubs, often aquatic, with alternate opposite or whorled leaves, and small frequently incomplete flowers. Calyx adherent, with a minute limb; petals inserted into the upper part of the calyx, or absent; stamens attached to the calyx; ovary with one or more cells; ovules pendulous anatropal. Fruit dry, not opening; seeds solitary, pendulous. The plants may be regarded as an imperfect form of *Onagraceæ*. They are found in damp places, ditches, and slow streams in all parts of the world. Some yield edible seeds. The kernels of *Trapa natans* and *T. bicornis*, called water chestnuts, and of *T. bispinosa*, singhara nuts, are used as articles of diet. The fruit of these plants has a peculiar horned aspect. *Hippuris vulgaris* is the common mare's-tail of our ponds. There are ten genera and about eighty species. Examples: *Hippuris, Myriophyllum, Haloragis, Trapa*. [J. H. B.]

HALORAGIS. A genus of *Haloragaceæ*, differing from the greater number of plants of this order, in not being aquatic. They occur in tropical Asia, and more abundantly in Australia and New Zealand. Their lower leaves are opposite, the upper often alternate; and the flowers are axillary, solitary or aggregated, combined into spikes, racemes, or even panicles. *H. citriodora*, the Piri-piri of the New Zealanders, has scented leaves. [J. T. S.]

HAMADRYAS. A genus of *Ranunculaceæ* from the Antarctic regions. More or less silky herbs, with palmately-parted or undivided leaves, and scapes with one to three flowers, the calyx and corolla externally hairy, the former with five or six sepals, the latter with ten or twelve long linear subulate petals. The flowers are diœcious by abortion, and the female ones have an ovate globose head of pistils terminated by hooked styles. [J. T. S.]

HAMAMELIDACEÆ. (*Witch Hazels.*) A natural order of calyciflorial dicotyledons, belonging to Lindley's umbellal alliance of epigynous Exogens. Trees or shrubs with alternate feather-veined leaves having deciduous stipules. Calyx four to five-divided; petals four, five, or wanting; stamens eight, the anthers introrse; ovary two-celled, inferior; ovules solitary or several; styles two. Fruit two-valved; seeds pendulous, albuminous. In some of the plants there are circular disk-like markings in the woody tubes. Natives of North America, Asia, and Africa. There are thirteen known genera, including *Hamamelis, Rhodoleia*, and *Bucklandia*. [J. H. B.]

HAMAMELIS. A genus of the witch-hazel order, distinguished by its calyx being four-parted; the corolla of four petals; the stamens four, alternate with the petals, and having four scale-like bodies—rudimentary stamens—opposite the petals. The name was adopted from a Greek term used to indicate resemblance to an apple tree, a comparison which is scarcely applicable. The species are shrubs of North America and China, with alternate leaves, and usually yellow flowers. *H. virginica* has been long known in cultivation. It has obovate toothed leaves, and is widely diffused in North America, attaining a height of ten or twelve feet, its yellow flowers appearing in the fall of the year, and its fruit ripening in spring. Its seeds contain a quantity of oil, and are edible; while the leaves and bark are astringent. It is employed as a remedy in various ways by the aborigines. [G. D.]

HAMATO-SERRATUS. When serratures have a somewhat hooked form.

HAMELIA. Tropical American shrubs, forming a genus of *Cinchonaceæ*, named in honour of M. du Hamel, a noted French vegetable physiologist. The flower orange-coloured and tubular; stamens five, concealed within the corolla; ovary five-celled, surmounted by an epigynous disk; style simple; stigma undivided; fruit succulent, five-celled, with numerous seeds in each compartment. *H. patens* and other species are in cultivation as stove plants, and have handsome flowers. [M. T. M.]

HAMELINIA. A genus founded by Richard on imperfect female specimens of *Astelia Banksii* or *Solandri*, and consequently not adopted by other botanists who have had better opportunities of examining these species. [J. T. S.]

HAMI (adj. **HAMATE, HAMOSE**). Hooks, hairs, or small spines which are hooked at the point.

HAMILTONIA. Indian shrubs with fragrant flowers, constituting a genus of *Cinchonaceæ*. The flowers have a funnel-shaped corolla with a long tube and a limb divided into five oblong lobes; stamens five, concealed within the tube of the corolla; ovary inferior, five-celled; style simple; stigma with five acute segments; capsule one-celled, with five one-seeded stones. *H. suaveolens* and *H. scabra* are cultivated in stoves, for the sake of their white fragrant flowers. [M. T. M.]

HAMMERSEDGE. *Carex hirta.*

HAMPEA. A genus of the *Bombax* family peculiar in the nature of its fruits, which are rusty-coloured capsules of the size of a cherry, bursting into two or three portions, each portion containing a single seed with a fleshy aril at its base. There are but two species, one a Mexican bush, the other a tree of New Grenada. Both have alternate long-stalked leaves, like those of the common poplar but larger; and bear on the same tree sterile and fertile white flowers, about half an inch across, solitary or two to three together in the axils of the leaves. They have a bell-shaped calyx with an entire border, five narrow petals slightly united at the base, and numerous stamens of unequal length; or, in the fertile flowers, a few barren stamens united into a ring inserted on the base of the petals, and surrounding the ovary. [A. A. B.]

HAMULOSE. Covered with little hooks.

HAMULUS. A kind of hooked bristle found in the flower of *Uncinia*. Schleiden regards it as a third glume, free from the two which form the flask.

HANBURIA *mexicana* is the sole representative of a genus of *Cucurbitaceæ* peculiar to the mountains near Cordoba, Mexico, and named after Daniel Hanbury, a distinguished London pharmacologist. It is a climber, having a pentagonal stem, furnished with simple tendrils, cordate leaves, axillary or terminal white and monœcious flowers, the males being arranged in racemes, whilst the females are solitary in the axils of the leaves. The calyx and corolla are bell-shaped; the ovary is four-celled, each cell containing one seed; and the fruit is oval, covered with long spines, and bursting open like that of *Momordica*, propelling the flat circular seeds (resembling those of *Feuillea*) to some distance. The Mexican squirrels are fond of eating the seeds, but, being unable to open a fruit so well protected by spines, they wait in the morning for the time when the first rays of the sun fall upon the ripe ones and cause them to burst. The Mexicans term the plant Chayotilla, from the close resemblance of its fruit to that of the cayoti or chayota (*Sechium edule*). [B. S.]

HANCHINOL. The Mexican name for *Heimia salicifolia.*

HANCORNIA. A small genus of *Apocynaceæ*, found in Brazil, and forming small trees or shrubs, abounding in all parts with a viscid milky juice, which is one of the sources of caoutchouc. They have entire opposite leaves, marked with pellucid veins; and sweet-smelling flowers resembling those of the jasmine. The calyx is five-parted, without glands; the corolla has a long narrow tube, hairy inside, and the five segments spread out when the flower opens, but are previously rolled round each other; the five stamens are inserted into the middle of the tube; and the ovary is divided into two cells, and has a long thread-like smooth style, and a forked stigma. The fruit is a large globular or pear-shaped fleshy berry, exuding a milky juice when wounded, and containing numerous hard seeds lying in pulp.

H. speciosa is a small tree somewhat resembling the weeping-birch in habit, with drooping branches, and small oblong leaves, sharp at the base, and rounded but with a short point at the apex. It is called Mangaiba or Mangawa, and bears a most delicious fruit, which is a great favourite with the Brazilians, but is only fit to eat when perfectly ripe, or after being kept for a short time. It is about the size of a plum, of a yellow colour marked with red spots or streaks. The milky juice of the tree, when exposed to the air, hardens into a kind of caoutchouc. [A. S.]

HAND-PLANT. *Cheirostemon platanoides.*

HANNEBANNE. (Fr.) *Hyoscyamus niger.*

HANNOA. The name of a Senegambian tree, forming a genus of *Simarubaceæ*. The flowers are unisexual, the males with the sepals combined into a somewhat two-lipped calyx, and the rudimentary ovaries concealed within a large disk. The female flowers are not known. [M. T. M.]

HAPLANTHUS. A genus of *Acanthaceæ*, containing three species, natives of India. They are erect branching herbs, with ovate petiolate leaves, and flowers in few-flowered terminal racemes furnished with small bracts. The calyx is five-parted, the corolla funnel-shaped, with an unequally five-cleft limb; the two stamens are included, and the capsule is linear and flattened, with several seeds. [W. C.]

HAPLODESMIUM. A genus of *Melastomaceæ*, consisting of a shrubby branching small-leaved plant, with elliptical leaves and tetramerous flowers. The calyx is campanulate, its teeth nearly equalling the tube; the petals oblong-ovate, blunt; the stamens eight equal; and the ovary free, four-celled. The fruit is a four-valved capsule crowned by the persistent teeth of the calyx. *H. Lindenianum*, a native of the Andes about Truxillo, grows at an elevation of from 4,000 to 12,000 feet. [J. H. B.]

HAPLOL.ENE.E. A tribe of frondose *Jungermanniaceæ*, characterised by a one-leaved involucre without any true perianth (the sheathing tube being merely the veil), a spherical capsule, and dichotomous ribbed fronds. Sometimes the rib is confluent with the margin. This tribe contains some of the finest of the frondose liverworts, vying with the smaller *Hymenophylla* in beauty and delicacy of frond (see SYMPHYOGYNA). *Pellia epiphylla* is a well-known British representative. [M. J. B.]

HAPLOPAPPUS. An American genus of *Compositæ*, distinguished from its allies in the *Solidagineæ* by the oblong or top-shaped more or less silky achenes being crowned with a pappus of rigid (not capillary) bristles. Some are North American, but the greater number are Chilian, and some of them inhabit the high Andean regions. They are mostly perennials, with alternate lance-shaped or oblong leaves, and twigs terminated by yellow-rayed flower-heads, though in some the heads are without rays. A few are nearly stemless, with leaves like those of the daisy but sharply toothed, while others have pinnatifid downy leaves. A shrubby Chilian species, *H. Baylahuen*, with glutinous stems, and spathulate unequally-toothed leaves embracing the stem by their narrowed base, is used by the Chilians, according to M. Gay, in the treatment of various diseases in their domestic animals, and is called by them Baylahuen. [A. A. B.]

HAPLOPHLEBIA. *Alsophila*.

HAPLOPETALUM. A genus of *Legnotideæ*, which tribe Mr. Bentham refers to the order *Rhizophoraceæ*. The genus is thus characterised:— Calyx four-parted; petals four entire; stamens four or five times as numerous as the petals, inserted on the margin of a very short disk; lower part of the ovary, which alone contains the ovules, adherent to the calyx, the upper part detached. The species is a Feejean plant. [M. T. M.]

HAPLOPHYLLUM. A genus of perennial plants or undershrubs, natives of Southern Europe &c., and distinguished from *Ruta* by their simple leaves, and five to six-parted flowers, the filaments hairy on their inner surface, and the style thickened towards the top. [M. T. M.]

HAPLOPTERIS. A genus of polypodiaceous ferns of the group *Pterideæ*, having simple coriaceous fasciculate fronds, on which the sori are linear continuous and marginal, with a broad firm marginal inflexed indusium opening along the inner edge. The veins are simple from a central costa. *H. scolopendrina*, the only species, a native of Bourbon, has quite the aspect of a broad-fronded species of *Vittaria* or *Tæniopsis*. [T. M.]

HAPLOSCIADIUM. An Abyssinian umbellifer with radical twice-pinnated leaves, like those of a *Meum*, and simple umbels on simple or scarcely branched scapes. It is supposed to constitute a distinct genus, but the fruit is not sufficiently known to characterise it with certainty.

HAPLOSTEMMA. A name proposed by Endlicher to receive a plant which Decaisne has referred to *Vincetoxicum*, from which it does not differ materially. [W. C.]

HAPLOSTYLUS. *Rhynchospora*.

HAPLOTAXIS. The same as *Aplotaxis*.

HARDENBERGIA. A genus of *Leguminosæ* found in Southern and Western Australia, and consisting of a few slender woody climbers, very similar in appearance, and all desirable as greenhouse plants from the profusion of their flowers. They are most nearly related to *Kennedya*, from which they are readily distinguished by their flowers being small and numerous, arranged in stalked racemes, instead of few and nearly as large as those of a pea. *H. monophylla*, a common greenhouse climber, has alternate smooth leaves, bearing a single lance-shaped or oblong leaflet two to three inches long, and prominently nerved. The racemes vary in length, but are generally longer than the leaves, and bear numerous usually blue flowers. The long carrot-shaped somewhat woody root of this plant is called by the colonists Sarsaparilla, and, according to Mr. Adamson, is used by the goldminers in infusion as a substitute for that root. Other species have three leaflets instead of one. The genus bears the name of Frances Countess Hardenberg, sister of Baron Hugel the eminent German traveller. [A. A. B.]

HARDHACK. *Spiræa tomentosa*.

HARDHAY. *Hypericum quadrangulare*.

HARDHEADS. *Centaurea nigra*.

HARDOCK, or HARLOCK. Probably the Burdock, *Arctium Lappa*.

HARDWICKIA. A small genus of East Indian trees, belonging to the *Cæsalpinia* group of *Leguminosæ*, and nearly related to the copaiva-balsam trees of South America. The abruptly pinnate leaves in *H. binata* are composed of one, and in *H. pinnata* of three pairs of opposite unequal-sided somewhat oval leaflets; and the minute dull yellow flowers are arranged in a spiked manner in axillary or terminal panicles. Each flower consists of four or five sepals; eight to ten stamens, the alternate ones shorter; and an ovary crowned with a short style, and a shield-like stigma. The lance-shaped pods are two to three inches long, compressed and one-seeded. Both species are trees of considerable size, and *H. binata* is said to yield a good timber suitable for many purposes. [A. A. B.]

HAREBELL. *Hyacinthus nonscriptus*. The name is also sometimes applied to the Hairbell, *Campanula rotundifolia*.

HAREBURR. *Arctium Lappa*.

HARE'SBANE. *Aconitum Lagoctonum*.

HARE'SBEARD. *Verbascum Thapsus*.

HARE'SEAR. *Bupleurum*; also *Erysimum austriacum* and *orientale*. —, BASTARD. *Phyllis Nobla*.

HARE'SFOOT. *Ochroma Lagopus*; also *Trifolium arvense*.

HARE'S-LETTUCE, or **HARE'S-PALACE.** *Sonchus oleraceus*.

HARE'STAIL. *Lagurus ovatus*.

HARETHISTLE. *Sonchus oleraceus*.

HARICOT. (Fr.) *Phaseolus*. The ripe seeds of *P. vulgaris* and other species of kidney-bean are cooked under the general name of Haricots. — À PIEDS. *Phaseolus nanus*. — A RAMER, BLANC, or COMMUN. *Phaseolus vulgaris*. — DE HOLLANDE. *Phaseolus compressus*. — DE LA JAMAIQUE. *Phaseolus lathyroides*. — D'ESPAGNE. *Phaseolus multiflorus*. — DE PRAGUE. *Phaseolus sphaericus*. — DE SOISSONS. *Phaseolus compressus*. — DE TONQUIN. *Phaseolus tunkinensis*. — EN ARBRE. *Bictoria frutescens*. — EN TOUFFE. *Phaseolus nanus*. — EN ZIGZAG. *Phaseolus Mungo*. — FLAGEOLET NAIN. *Phaseolus tumidus*. — LIMAÇON. *Phaseolus Caracalla*. — NAIN. *Phaseolus nanus*. — PRINCESSE. *Phaseolus tumidus*. — ROUGE D'ORLEANS. *Phaseolus vulgaris*.

HARIF, or **HEIRIFF.** *Galium Aparine*.

HARINA. A genus of East Indian palms, previously described under the name of *Wallichia*, by which they are most generally known. *Harina*, however, forms a section of the genus, characterised by having the male and female flowers upon the same plant, the males being in dense masses, and having an undivided calyx, and six stamens. [A. S.]

HARLANDIA. The glabrous climbing plant described under this name, and native at Hong Kong, is stated by Mr. Bentham to belong to the cucurbitaceous genus *Kariria*. [M. T. M.]

HARLOCK. Probably Burdock, *Arctium Lappa*.

HARPALYCE. A small genus of handsome erect pinnate-leaved bushes of Mexico and Brazil, belonging to the *Galega* group of the *Leguminosae*, and differing from its allies in the calyx being cleft nearly to the base and consisting of but two narrow and entire segments nearly as long as the corolla. *H. brasiliana*, a bush of four to eight feet high, clothed with a reddish velvety down, bears handsome scarlet pea-flowers disposed in a panicled or racemed manner towards the ends of the twigs. The Mexican species are smooth, and have purple flowers. In all, five of the ten stamens, which are united into a sheath, are shorter than the others, and have small rounded anthers. The pods are coriaceous, somewhat flattened, and many seeded, and, as in *Cassia*, the seeds are separated from each other by transverse partitions. [A. A. B.]

HARPANEMA. A genus of *Asclepiadaceae*, containing a single species, a native of Madagascar. It is a climbing shrub with opposite glabrous coriaceous leaves, and small flowers in compound axillary cymes. The calyx is five-parted; the corolla is rotate and five-cleft; the staminal corona consists of five linear blind hooked processes alternating with the lobes of the corolla; the anthers have a fleshy apex bent down upon the stigma; and the pollen-masses are attached by fours to the stigmatic corpuscles. [W. C.]

HARRISONIA. The name of a shrub with prickly branches, found in the island of Timor, and referred to the *Simarubaceae*, among which it is known by the staineous being attached to hairy two-lobed scales; by the four-lobed ovary; and by the four styles, separate at the base, but united above. The same name has been applied to an asclepiad with scarlet flowers, now included under *Bartera*. [M. T. M.]

HARSTRONG or **HORESTRONG.** *Peucedanum officinale*.

HARTIGHSEA. A small genus of *Meliaceae* confined to the islands of the Indian Archipelago, New Zealand, the east coast of New Holland, and Norfolk Island. They are trees of moderate height, with large pinnate leaves, and long panicles of smallish flowers, which have a small four or five-lobed calyx, five narrow petals joined together by their bases, the tube of the stamens cylindrical and fleshy, with eight or ten rounded notches at the apex, and the three-celled ovary included within a tubular disk occupying the centre of the flower. *H. spectabilis*, a native of New Zealand, forms a tree forty or fifty feet high. Its drooping panicles of pale-coloured flowers measure from eight to twelve inches in length, and grow from the main trunk or older branches. The New-Zealanders call the tree Kohe or Wahahe. Its leaves have a bitter taste, and are employed as a substitute for hops, and a spirituous infusion of them as a stomachic medicine. [A. S.]

HARTOGIA. A genus of *Celastraceae* peculiar to South Africa, and represented by a single species, *H. capensis*, a small much-branched tree, with opposite lance-shaped serrated leaves, and small white numerous flowers in axillary cymes or panicles. The fruits are dry elliptical two-celled two-seeded drupes, as large as a good-sized pea. The seeds being destitute of albumen, and not surrounded by an aril, are the distinguishing characters. John Hartog, whose name is commemorated in this genus, was an early Dutch traveller in South Africa and Ceylon. [A. A. B.]

HART'SBALLS. *Elaphomyces.*

HART'SHORN. *Plantago Coronopus.*

HARTSTONGUE. *Scolopendrium;* also *Olfersia cervina.*

HARTWEGIA *purpurea.* An epiphytal orchid of Mexico and Guatemala, with a short stem bearing a single lance-shaped leaf covered with brownish spots, and an erect wiry flower-scape a foot in length, with a few small bright pink flowers at the apex. It is closely related to *Epidendrum;* and is named after Mr. Theodor Hartweg, once collector in South America for the Royal Horticultural Society. [A. A. B.]

HARTWORT. *Tordylium.*

HARVEST-BELLS. *Gentiana Pneumonanthe.*

HARVEYA *capensis* is an erect simple herb, a parasite on the roots of heaths at the Cape of Good Hope; and constitutes a genus of *Scrophulariaceæ* of the tribe *Gerardieæ.* It is nearly allied to *Aulaya,* and, like that genus, has four didynamous stamens, all bearing anthers, with one fertile ovate awned cell, and the other cell long and subulate but empty; it differs chiefly in its large inflated herbaceous calyx.

HASHISH. The Arabian name of the narcotic *Cannabis sativa.*

HASKWORT. *Campanula latifolia.*

HASSAGAY TREE. *Curtisia faginea.*

HASSELQUISTIA. A genus of umbellifers distinguished by the petals of the central flowers being inversely ovate and slightly notched at the end, those in the circumference of the umbel spreading and two-cleft; by half of each fruit produced by the central flowers being abortive, the other partly folded round it; and by the fruits at the outer part of the umbel being flat with a thick winged border, slightly wrinkled. The genus was named by Linnæus in honour of Hasselquist, a well-known Eastern traveller. The species are annual herbs, natives of Syria, and have the stems hairy. [G. D.]

HASSKARLIA. The name of a genus of Indian and Javanese *Pandanaceæ.* The fruits consist of three to five or rarely more ovaries united together, each one-seeded, the seeds being like those of the allied genus *Freycinetia.* [M. T. M.]

HASTATE. Shaped like the head of a halbert.

HATHER. The common Heath or Heather.

HAUSTORIUM. A small root which attaches itself to the surface of some other plant, and lives by sucking it. A sucker, as in dodder, ivy, &c.

HAUTBOIS. A kind of Strawberry, *Fragaria elatior.*

HAUTBOIS. (Fr.) *Sambucus nigra.*

HAVER. The Wild Oat, *Avena fatua.*

HAW. The fruit of the hawthorn, *Cratægus Oxyacantha.* —, BLACK. *Viburnum prunifolium.*

HAWKBIT. *Apargia;* also *Hieracium.*

HAWKNUT. *Bunium flexuosum.*

HAWK'SBEARD. *Crepis.*

HAWKWEED. *Hieracium.*

HAWORTHIA. One of the subdivisions of the genus *Aloe,* consisting of small curious-looking and extremely interesting succulent herbs of South Africa, distinguished by having erect flowers, the perianth with a straight tube and two-lipped limb, the stamens adherent to the base of the perianth, and the capsule ribbed. Some of the species are remarkable for the translucent substance of their leaves, or for their elegant reticulated markings. [T. M.]

HAWTHORN. *Cratægus Oxyacantha.* —, INDIAN. *Rhaphiolepis.*

HAYLOCKIA. One of the hippeastriform *Amaryllidaceæ* referred to a separate genus. It is a small bulb, with biennial very narrow linear leaves, and autumnal one-flowered concealed scapes bearing a solitary white flower stained with purple. This has a cylindrical tube enlarged at the mouth, and a regular limb, funnel-shaped below and partially spreading above; the filaments of alternate lengths, conniving, the sepaline inserted at the base of the limb, the petaline higher; and the style erect, with a three-cleft stigma. It is allied to *Zephyranthus,* and is found in Uruguay. [T. M.]

HAYMAIDS. *Glechoma.*

HAZEL. *Corylus Avellana.* —, WITCH. *Hamamelis.*

HAZELWORT. *Asarum europæum.*

HEAD-ACHE TREE. *Premna integrifolia.*

HEART, FLOATING. An American name for *Limnanthemum.*

HEART'S-EASE. *Viola tricolor.*

HEART-SEED. *Cardiospermum.*

HEART-SHAPED. The same as Cordate.

HEART-WOOD. The central part of the timber of Exogens, hardened or altered by age.

HEATH. *Erica.* —, BERRIED. *Empetrum.* —, IRISH. *Menziesia* (or *Dabœcia*) *polifolia.* —, MOOR. *Gypsocallis.* —, ST. DABEOC'S. *Menziesia polifolia.* —, SEA. *Frankenia.*

HEATHER. *Calluna vulgaris.* —, HIMALAYAN. *Andromeda fastigiata.*

HEATHWORTS. Lindley's name for the *Ericaceæ*.

HEAUMIER. (Fr.) *Cerasus vulgaris*.

HEBECLADUS. A genus of *Solanaceæ*, closely allied to *Atropa*. The name is given in allusion to the downy branches of the species. The corolla is funnel-shaped, with a large tube, longer than the calyx, the limb spreading, wavy, five-cleft, frequently with small teeth intermediate between the lobes. The species are natives of tropical America, and one of them, *H. biflorus*, with yellow flowers, is cultivated in our greenhouses. [M. T. M.]

HEBECLINIUM. A South American genus of *Compositæ*, closely allied to *Eupatorium*, differing chiefly in the elevated and villous instead of flat and naked receptacles on which the florets are seated. The species are herbaceous or somewhat shrubby plants with opposite leaves, and the twigs are terminated by corymbs of numerous white, purple, or rose-coloured flower-heads containing tubular florets with protruding styles. One of the most handsome is the Mexican *H. ianthinum*, a good-sized bush, having the stems and branches clothed with rusty down, and the numerous flower-heads disposed in corymbs, and of a fine mauve colour. [A.A.B.]

HEBENSTREITIA. A genus of *Selaginaceæ*, containing sixteen species, natives of the Cape of Good Hope. They are undershrubs with alternate or scattered leaves, and membranaceous bracts surrounding the flowers. The calyx is monosepalous, and the corolla tubular at the base, with a somewhat one-lipped limb; there are four exserted stamens; and the deflexed style passes through a fissure of the corolla. [W. C.]

HECASTOPHYLLUM. The same as *Ecastaphyllum*.

HECUBÆA. A genus of *Compositæ* peculiar to Mexico, and there represented by a single species, *H. scorzonerœfolia*, a smooth unbranched herb about a foot high, furnished with a few alternate entire lance-shaped leaves, and terminal solitary long-stalked yellow-rayed flower-heads an inch or more across. The ray florets are strap-shaped and female, the strap deeply divided into three or five parts; and the disk florets are tubular, five-toothed, and perfect. The relationship of the genus is with the North American *Helenium*, from which it differs in the achenes being destitute of pappus. The analogy of the two genera is curiously expressed in the names they bear: *Hecuba* was the daughter of Dimas, king of Thrace, and *Helenus* (*Helenium*) was one of her sons. [A. A. B.]

HEDAROMA. A name sometimes given to some involucrate species of *Genetyllis*.

HEDEOMA. A genus of labiates almost confined to the American continent, but found in various countries from Brazil to Canada. They are annual or perennial herbs or dwarf shrubs, with small leaves and whorls of flowers borne towards the tops of the branches. The genus is principally distinguished from its allies by having only the two lower stamens fertile, the two upper ones being either short and sterile, or altogether wanting; and by the corolla being short and never of a scarlet colour. *H. pulegioides*, the Penny-royal of America, is an annual, with numerous branches, small opposite egg-shaped leaves, and small pale-blue flowers. It is found in the United States from Carolina to Canada, and is extensively used for medical purposes, particularly in domestic practice, large quantities of it being brought to the markets for sale. An infusion or tea of it is a popular remedy for colds and pains in the legs. The whole plant has a strong pungent but pleasant scent, and a mint-like taste. [A. S.]

HEDERACEÆ. Another name for the order *Araliaceæ*.

HEDERA. A genus of *Araliaceæ*, consisting of evergreen climbing shrubs, with simple extipulate leaves, and an umbellate inflorescence. The margin of the calyx is elevated and five-toothed, the petals five, not cohering at the apex, the stamens five, the style single with five obscure stigmas, and the berries five-celled. The common Ivy, *H. Helix*, one of our wild plants, is the hedge of the Gordons. This well-known evergreen climber, which mantles and canopies the picturesque ruin, adorns in winter the bare trunks of deciduous timber trees, clothes the hedge-row banks of our rural lanes, is admitted to various uses in the decoration of our gardens, and is made by poets the emblem of friendship. Its stems cling by means of little rootlets to the walls or tree-trunks with which they come in contact, throwing out right and left their shining five-angled leaves, but after they have reached the summit of the object to which they cling, they branch out into woody bushy heads with simple leaves, bearing at the end of every twig a little umbel of yellowish flowers succeeded by dark-coloured berries. The plant is liable to much variation, and many interesting varieties are in cultivation. Many tropical species once referred here now form the genera *Oreopanax*, *Dendropanax*, *Agalma*, *Scindpsphyllum*, &c. An ivy-clad ruin is shown in Plate 20. [T. M.]

HEDGEBELLS. *Calystegia sepium*.

HEDGEBERRY. *Cerasus avium*.

HEDGEHOG. *Medicago intertexta*.

HEDGEMAIDS. *Glechoma hederacea*.

HEDWIGIA. A West Indian tree, abounding in resin. It forms a genus of *Amyridaceæ*, among which it may be distinguished by its four-parted flowers, and by its fruit, which is fleshy externally, furrowed, with four one-seeded stones in the interior. [M. T. M.]

HEDYCHIUM. The handsome and fra-

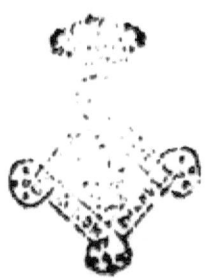

grant flowers of some of the species of this genus of *Zingiberaceæ* render them great favourites in the bothouse. They are plants with tuberous roots, herbaceous stems, clasping leaves, and a terminal spicate inflorescence. The corolla consists of six segments in two rows, five nearly equal in size, the sixth or lip large notched or more deeply divided; the filaments thread-like; and the fruit capsular. The species are natives of tropical Asia. [M. T. M.]

HEDYOSMUM. A genus of fragrant resin-bearing shrubs belonging to the *Chloranthaceæ*. They have unisexual flowers, the males in close spikes without bracts, the females solitary or in groups of four, sessile, provided with bracts; ovary triangular. The species are natives of Brazil and other districts of tropical America. Some of them are used medicinally as antispasmodics &c. [M. T. M.]

HEDYOTIS. A genus of *Cinchonaceæ*, comprising a number of herbaceous or somewhat shrubby plants, dispersed throughout the tropics. The floral whorls are arranged in fours; the corolla is funnel-shaped or wheel-shaped; the ovary has two compartments, and is surmounted by an epigynous disk and cleft stigma; the fruit is a capsule. *H. umbellata* supplies a valuable red dye in Coromandel and other parts of India where it is cultivated. The Chay-root, as it is called, is the dye employed for producing the durable red colour for which the chintzes of India are noted (*Simmonds*). Wild chay-roots are preferred to cultivated ones, and licenses to dig the former are granted in Ceylon. The colouring matter resides in the rind and outer portions of the root. The leaves of this plant are also used by the natives as expectorants.

Some of the species, especially those formerly included under the genus *Houstonia*, are cultivated in gardens, their low stature, elegant appearance, and pretty flowers rendering them desirable plants for rock-work &c. The colour of the flowers varies from white to scarlet blue and purple. [M. T. M.]

HEDYPNOIS. A name given by Pliny to a kind of wild endive, said to have medicinal virtues, being astringent and useful in dysentery. By modern botanists the name is applied to a genus of uninteresting annual herbaceous plants with diffuse stems, toothed leaves, and yellow flowers, belonging to the *Cichoraceæ*. The generic characters are:— Receptacle naked; involucre furnished with small bracts; florets of the disk furnished with a double pappus, the outer bristly, the inner chaffy; pappus of the ray a membranous finely-toothed margin. [C. A. J.]

HEDYSARUM. A family of herbaceous or somewhat shrubby leguminous plants, distinguished by the peculiar structure of the seed-pod, which is composed of numerous even one-seeded joints convex on both sides. The leaves are pinnate, with an odd leaflet; and the flowers axillary, in stalked clusters or spikes, purple, white, or cream-coloured. A large number of species are known, many of which are handsome plants, and some are valuable for their nutritive properties as fodder. *H. Alhagi*, sometimes described under the name of *Alhagi Maurorum*, is a thorny shrub, common in the East, and produces a substance called manna from its supposed resemblance to the 'manna' of the Israelites. *H. gyrans* is remarkable for the property possessed by its leaves of setting up a spontaneous motion, independent, as far as observation reaches, of all external impressions. Without being touched and without being excited by heat, light, wind, or rain, sometimes a single leaflet, sometimes a whole leaf, oscillates or gyrates, continuing to move for an indefinite time, and ceasing without known cause. *H. coronarium* is the plant commonly known in English gardens under the inappropriate name of French Honeysuckle, it being a native of Italy, and having no affinity with the honeysuckle (*Lonicera*). Its latter name it owes no doubt to its similarity to red clover, often called honeysuckle by country children from the use which they make of its sweet flower-tubes. It is a native of Spain and Italy, where it is gathered in great quantities as food for cattle. French and German, *Sulla*. [C. A. J.]

HEGBERRY. *Cerasus avium*.

HEGEMONE. A genus of *Ranunculaceæ*, allied to *Trollius*, found in the Altai near the limit of perpetual snow. The species on which the genus is founded, *H. lilacina*, has an erect stem, leafy at the base, the leaves palmately five-parted, those of the stem similar. The flower is solitary and terminal, pale lilac, with fifteen or twenty persistent petaloid sepals and about ten small irregular petals, having an oblong limb and short tubular base; carpels numerous sessile. [J. T. S.]

HEIMIA. A genus of *Lythraceæ* in which it is remarkable for its yellow flowers, blue or purple being the prevailing colour in the family. The two known species, *H. salicifolia* and *H. grandiflora*, are both smooth erect bushy shrubs, the former common to Texas, Mexico, and Buenos Ayres, the latter confined to Buenos Ayres. The willow-like leaves are opposite below and alternate above, and the yellow flowers, placed singly in the axils of the leaves, have great superficial resemblance to those of *Lysimachia vulgaris*, but in structure are widely different. According to Mr. Tweedie, both species are common in pasture lands about Buenos Ayres, and, as the cattle do not browse upon them, there is always an abundance of their gay yellow blossoms, which are called *abro sol*, 'it is open sun.' The twigs are strewed on floors to drive away fleas, of which there are abundance. The willow-leaved species is said to excite violent perspiration. The Mexicans consider it a potent remedy for venereal diseases, and call it Hanchinol (*Lindley*). The genus is named in honour of Dr. Heim, a distinguished physician of Berlin. [A. A. B.]

HEINSIA. A shrub of the *Cinchona* family, native of Sierra Leone. It has spiny branches, and white flowers in clusters of three or four at the ends of the branches. The calyx has a five-parted limb with leafy segments; the corolla is salver-shaped, its tube longer than the calyx, very hairy within; anthers sessile, concealed by the hairs of the corolla; ovary two-celled; fruit dry, hard, with two indehiscent compartments; seeds numerous. *H. jasminiflora* is an evergreen stove shrub. [M. T. M.]

HEINTZIA. A genus of *Gesneraceæ*, containing a single species from central America, an undershrub, with erect stem, opposite fleshy leaves, and axillary umbellate inflorescence. The flowers have a free five-parted calyx; a funnel-shaped corolla, tomentose on the outside, the limb cut into five roundish segments; four didynamous stamens; and a one-celled ovary surrounded by a disk, and having two bilobed parietal placentæ, with numerous anatropal ovules; the simple style has a funnel-shaped stigma. The fruit is fleshy and one-celled. [W. C.]

HEISTERIA. A genus of small trees found in some of the West Indian islands, and also in Guiana and Brazil. They belong to the *Olacaceæ*, which has very few other representatives in the western hemisphere. The flowers are produced either singly or in little clusters at the bases of the leaves. The calyx is small and five-cleft, but increases greatly in size, spreading out after flowering, and ultimately surrounding the ripe fruit; there are five petals, and ten fertile stamens. The fruit is olive-shaped, enclosed in the enlarged fleshy calyx, and contains a single seed. *H. coccinea* forms a tree fifteen or twenty feet high, with shining oblong leaves, and small white flowers borne singly on short stalks. It is a native of the West Indian islands, particularly of Martinique, where the French call it Bois perdrix, which is a corruption of Pois perdrix, signifying partridge pea, the fleshy red fruits forming a favourite food of pigeons and other birds. The corrupt French name Bois perdrix, however, has led to the supposition that the prettily marked wood called 'Partridge wood' by cabinet-makers was derived from it; but such is not the case, the source of that wood remaining unknown. [A. S.]

HELCIA *sanguinolenta* is a pretty cultivated terrestrial orchid from the Peruvian Andes, having the habit of *Trichopilia*, and differing from that genus, according to Dr. Lindley, in that the column, instead of being rolled up in the lip, stands erect and clear of it, the anther two instead of one-celled, and the anther bed with a deep fringed border instead of two lacerated processes. The plant has elongated ovate pseudobulbs, a single undulate leaf, and a one-flowered peduncle shorter than the pseudobulb. The sepals and petals are olive-coloured, marked with crimson spots, and the lip white with crimson and yellow streaks. About the middle the lip contracts, and has two fleshy lobes standing erect on each side of the column, without however touching it; the space between these lobes, forming the base of the lip, is a deep hairy pit. [A. A. B.]

HELDE. *Tanacetum vulgare*.

HELENIUM. A genus of herbaceous perennials belonging to the corymbiferous tribe of compound flowers. The characters are:—Receptacle of the disk naked, of the ray chaffy; pappus five-awned; involucre one-leaved, many-parted; florets of the ray three-cleft. The species are all natives of America, and bear yellow flowers. French, *Hélénie*. [C. A. J.]

HELIAMPHORA. A genus of plants described by Bentham, belonging to the *Sarraceniaceæ*. Perennial herbaceous plants found in muddy places in Guiana with radical leaves, the petiole of which is tubular and in the form of a pitcher with an

Heliamphora nutans.

oblique mouth; and an erect scape with nodding white or pale rose-coloured flowers. The perianth consists of four to five hypogynous imbricated parts; the stamens are indefinite and hypogynous; and the ovary is three-celled, with numerous ovules on an axile placenta. The pitchers are lined with hairs of a peculiar nature. The only species is *H. nutans*. [J. H. B.]

HELIANTHEMUM. A genus of low mostly prostrate shrubby or subshrubby plants, closely related to *Cistus*, from which they differ in having imperfectly three-celled, instead of five or ten-celled capsules. They are most plentiful in the warmer and temperate parts of Europe, and in North Africa, but occur also in Egypt, in Arabia, in the Canaries and adjacent isles, in North America, and even in Brazil. They are showy plants, with simple subevergreen leaves, and five-petaled fugacious flowers, mostly in terminal racemes, and having a calyx of from three to four sepals, a capitate stigma, a triquetrous ovary, and a three-valved capsule. Unlike their allies, the Cistuses, they do not appear to have any active properties, but it is stated of the common species, *H. vulgare*, that the stamens, if touched during sunshine, spread slowly, and lie down upon the petals. Many double-flowered varieties of the cultivated species have been originated in gardens. [T. M.]

HELIANTHUS. A genus of *Compositæ* or *Asteraceæ*, consisting of coarse tall-growing herbs, with large rough leaves and yellow flowers. The greater portion are natives of North America.

The only species grown for culinary purposes is *H. tuberosus*, the Jerusalem Artichoke, which, although stated to be a native of Brazil, is a hardy perennial attaining the height of six or eight feet, and, with its large rough alternate heart-shaped somewhat pointed leaves, has considerable resemblance in habit and appearance to the common sunflower. The name of Jerusalem Artichoke is considered to be a corruption of the Italian *Girasole Articocco*, or Sunflower Artichoke, under which name it is said to have been originally distributed from the Farnese garden at Rome soon after its introduction to Europe in 1617.

The roots are creeping, and towards the close of autumn produce, like the potato, a number of round irregular reddish or yellow tubers, clustered together and of considerable size. They are used either boiled and mashed with butter, or baked in pies, and when nicely cooked are not only well flavoured, but considered to be both wholesome and nutritious—more so even than the potato, as they may be eaten by invalids when debarred from the use of other vegetables. On the continent they are in considerable demand for soups, and before the potato became plentiful, they were a good deal in use in this country Parkinson, writing in 1629, says they were then so common in London 'that even the vulgar began to despise them: they were baked in pies with marrow, dates, ginger, sack, &c., and, being so plentiful and cheap, rather bred a loathing than a liking for them.' Hence it appears that, as the culture of the potato extended, it gradually displaced the Jerusalem Artichoke, and at the present time the latter is only grown to a very limited extent in first-class gardens. Since the failure of the potato crops, the Jerusalem Artichoke has been strongly recommended as a substitute for that vegetable; but notwithstanding all that has been said and written in its favour, it is still far from common, and by no means esteemed so much as it deserves to be. [W. B. B.]

HELICHRYSUM. A genus of herbaceous or shrubby plants belonging to the corymbiferous tribe of *Compositæ*, and of which the characters are:—'Receptacle naked; pappus hairy or feathery; involucre imbricated, radiate, scariose; ray coloured.

Most of the species are natives of the Cape of Good Hope. As the name 'gold of the sun' indicates, the flower-heads are beautifully radiated, and while some species are of a brilliant yellow, others are white, pink, or crimson. In all, the radiating involucre is very conspicuous, and retains much of its elegant form and brilliant colour when dried. *H. macranthum*, an Australian species, when first introduced, bore only white flower-heads slightly tinged with red outside, but varieties have now been raised which have exchanged the primitive white hue for numerous shades of red, orange, or rose-colour. Thus the plant, originally worthy of note for the large size of its heads, has acquired a new interest in horticulture. *H. orientale*, a native of Crete and Africa, is the Immortelle of the French. The flower-heads of this species are yellow, but are often dyed green, orange, or black, and are much employed in the making of wreaths intended to be votive offerings to the dead. In drying the flowers of these plants, they should be suspended head downwards. German, *Strohblume*. [C. A. J.]

HELICIA. A genus of *Proteaceæ* having a cylindrical club-shaped calyx with four slightly spathulate sepals, each of which bears a nearly sessile anther a little below its apex. The seed-vessel is a single-seeded follicle which does not open by valves. The leaves are ovate or oblong, five to ten inches in length, simple, scattered, sometimes opposite, herbaceous or leathery in texture, entire or toothed. The flowers grow in axillary or terminal racemes. The species form lofty trees or large shrubs of tropical Asia; one of them, *H. australasica*, has been found in Victoria. [R. H.]

HELICOGYRATE. Having a ring or gyrus carried obliquely round it; as in the spore-cases of *Trichomanes*.

HELICOID. Twisted like the shell of a snail.

HELICONIA. A fine genus of herbaceous plants, belonging to the *Musaceæ*, and inhabiting tropical America. They are distinguished from their congeners by their fruit, which is capsular, separating into three one-seeded compartments. The shoots of *H. psittacorum* are eaten in the West Indies, as also are the fruits of *H. Bihai*. [M. T. M.]

H. Maria Alexandrowna, named after the Empress of Russia, a remarkable New Grenada species, with the habit of *Musa*, produces a useful fibre. Its trunk attains twelve to fifteen feet in height, and is formed of the sheaths of the leaf-stalks. The peduncles project beyond the leaves, and curving downwards bear a narrow flattened spike two and a half feet long, the red flowers of which are almost concealed by the spathe and white bracts.

HELICTERES. A genus of *Sterculiaceæ*, containing upwards of thirty species, mostly natives of the tropics of the Western hemisphere. They are shrubs, usually covered with rusty stellate down, the leaves simple, heart-shaped with the basal lobes unequal; the flowers in little clusters in the angles of the leaves, five-petaled, with the stamens united into a long column surrounding the stalk of the ovary, but separating at the summit into from five to fifteen filaments, partly sterile. The fruit consists of five carpels, which are generally twisted together in a screw-like manner. *H. Isora* is a native of Southern India, where its singular twisted screw-like fruit, about two inches in length, is called 'twisted stick,' 'twisted horn,' or 'twisty,' and, on account of its shape and name, is supposed to be a sovereign remedy against colic or twistings of the bowels. [A. S.]

HELIOCARPUS. A genus of *Tiliaceæ*, found in Mexico, Central America, and New Grenada, readily recognised among its allies by the fruits, which are thin nearly circular bodies a quarter of an inch in diameter, beautifully ciliated round the margin with a row of radiating bristles. The resemblance of the fruits to little suns is expressed in the generic name. The species, all very similar in appearance, are shrubs, or some of them forest trees of considerable size, furnished with alternate long-stalked heart-shaped usually three-lobed leaves. The minute densely clustered yellow or green flowers are disposed in panicles or cymes terminating the branches. They consist of four sepals, four petals, twelve to twenty stamens, and a bifid style surmounting a two-celled ovary, which when ripe becomes a two-seeded fruit. [A. A. B.]

HELIOPHILA. A large genus of *Cruciferæ*, with twice-folded cotyledons. All the species are from the Cape of Good Hope, and are annual herbs or undershrubs, with branched stems, and racemes of yellow white rose-coloured or more frequently blue flowers. They have a more or less elongated pod with two flat or (in the elongated pods) slightly compressed valves. The calyx is equal at the base, which distinguishes it from the allied genus *Chamira*. [J. T. S.]

HELIOPSIS. A perennial herbaceous plant belonging to the corymbiferous tribe of *Compositæ*. The involucre is imbricated, the florets of the ray long and narrow, the receptacle chaffy, and the fruit four-cornered without a pappus. *H. lævis*, the only species, is an American plant attaining a height of five or six feet, with rather broad serrated leaves, and large yellow flowers. [C. A. J.]

HELIOSIS. A term applied to the spots produced upon leaves by the concentration of the rays of the sun through inequalities of the glass of conservatories, or through drops of water resting upon them. In the latter case the destruction is seldom so complete as in the former, and the chlorophyll is merely altered, especially in the circumference, and not destroyed. Such spots sometimes, on the contrary, arise from the congelation or low temperature of the drops. They afford a nidus for minute fungi, which are not in consequence to be considered as the cause. [M. J. B.]

HELIOSPERMA, or HELICOSPERMA. A proposed genus of *Caryophyllaceæ*, which may, however, be rather taken to represent a section of *Silene*. The flowers are solitary or cymose, long-stalked, with a clavate campanulate calyx. The capsule is one-celled, containing lenticular compressed seeds, having a series of prominent points round the back. *S. alpestris* and *quadrifida*, natives of central and southern Europe, belong to the section thus defined, which is by no means a natural one. [J. T. S.]

HELIOTROPE. *Heliotropium*, especially in a popular sense. *H. peruvianum*. —, WINTER. *Nardosmia fragrans*.

HÉLIOTROPE. (Fr.) *Heliotropium*. — D'HIVER. *Nardosmia fragrans*.

HELIOTROPIACEÆ. A group of corolliflorai dicotyledons, considered by most botanical writers as a suborder of *Ehretiaceæ*. The plants have a circinate inflorescence, regular symmetrical flowers, five stamens, and four united achenes forming the fruit. They are found in Europe and South America. See EHRETIACEÆ. [J. H. B.]

HELIOTROPIUM. The Heliotrope or Turnsole, is a large genus of *Ehretiaceæ*, differing from the greater number of genera in having exalbuminous seeds; from *Schleidenia*, by having a salver-shaped, not funnel-shaped corolla; and from *Tiaridium*, by the fruit not being two-lobed. They are herbs or undershrubs found chiefly in tropical and subtropical regions, but a few species reach Europe, and one, *H. europæum*, is distributed over the greater part of southern and central Europe. They are furnished with strigose hairs, entire oval oblong or lanceolate leaves, and terminal or lateral one-sided usually circinate racemes of small white or lilac flowers. The fruit is separable into four nuts, or drupes, having a thin fleshy covering. Some of the species are sweet-scented, as the *H. peruvianum*, which is much cultivated on that account; on account of their agreeable scent, its flowers get the popular name of Cherry-pie. [J. T. S.]

HELIPTERUM. A considerable genus of *Compositæ*, separated from *Helichrysum*, to which a large proportion of what are commonly known as everlasting flowers belong, by having the hairs of the pappus feathery (plumose) instead of rough (pilose). They are annuals or perennials found in South Africa, Australia, and Tasmania, commonly furnished with lance-shaped or linear leaves, thickly clothed with short white wool, and usually having each twig terminated by a single flower-head, though in a few species the heads are numerous and corymbose. The thin dry papery scales of the involucre, pink, yellow, or white in colour, give beauty to these flower-heads. The inner series of scales are often spread out into a flat border so as to have the appearance of ray florets; but the florets are all tubular and minute, yellow or purple, usually perfect, a few of the outer ones sometimes female.

H. humilis, well known as *Aphelexis humilis* in greenhouses, is one of the most handsome South African species. Its much-branched whip-like stems, clothed with compressed leaves, are terminated by a large handsome deep rose-coloured flower-head, expanding only in sunshine. Another remarkable African species, *H. eximium*, has sessile elliptical leaves clothed like the stems with close cottony wool, and having the consistence of the ears of some animal; and its flower-heads, disposed in corymbs at the ends of the branches, are of a vivid purple, not unlike those of the globe amaranth. *H. incanum* is a beautiful little Australian species a foot high growing in tufts, the flower-heads having the outer scales purple, and the inner ones white; this plant is known as Native Amaranth in Tasmania.

The name 'Everlasting flower' is promiscuously applied to the plants of this genus and their allies. Bouquets of them are sometimes seen, and when well selected and tastefully arranged, they look extremely beautiful, preserving their colour for a long period, especially if kept from dust by a glass shade. [A. A. B.]

HELLEBORE. *Helleborus*. —, AMERICAN WHITE. *Veratrum viride*. —, BLACK. *Helleborus niger*. —, BLACK, of the ancients. *Helleborus officinalis*. —, FALSE. An American name for *Veratrum*. —, STINKING. *Helleborus fœtidus*. —, SWAMP. *Veratrum viride*. —, WHITE. *Veratrum album*. —, WINTER. *Eranthus hyemalis*.

HELLÉBORE À FLEURS ROSE. (Fr.) *Helleborus niger*.

HELLEBORINE. *Epipactis*.

HELLEBORINE. (Fr.) *Serapias Lingua*.

HELLEBORUS. A Latinised form of an old Greek name applied to some plants of this genus, and significant of their injurious or poisonous effects when eaten. The genus is included among the *Ranunculaceæ*, and consists of perennial low-growing plants with palmate or pedate leathery leaves, five persistent sepals, eight to ten tubular petals two-lobed at the top, and several carpels each with many seeds.

The species, for the most part, are found in Southern Europe and Central Asia. Among the best known is the Christmas Rose, *H. niger*, a common plant in gardens, where it blooms in winter and early spring. Its leaves are pedate, dark, shining, and smooth, and the flower-stalk rises directly from the root, bearing one or two flowers and as many bracts; the sepals are large, white or pinkish, and petal-like, the true petals being greenish and tubular. The plant probably derives its name of Black Hellebore from its dark-coloured rootstock and the numerous fibres proceeding from it. These roots are occasionally used in medicine as a powerful cathartic, but its violent narcotic and acrid properties preclude its general use. The Black Hellebore used by the Greeks has been determined by Dr. Sibthorp to be *H. officinalis*, a handsome

Helleborus niger (flower).

plant with a branching stem, bearing numerous serrated bracts, and three to five whitish flowers. It is a native of Greece, Asia Minor, &c. According to Pliny, Black

Helleborus niger (leaf).

Hellebore was used as a purgative in mania by Melampus, a soothsayer and physician, 1,400 years before Christ; hence the name *Melampodium* has been applied to the Hellebores.

Two species are found wild in many parts of England, especially on a limestone soil, though it is a matter of doubt whether they may not have been introduced at some former time. *H. fœtidus*, the Bear'sfoot, has numerous flowers in a large loosely spreading panicle, with numerous bracts frequently exhibiting every intermediate form between the ordinary divided leaf of the plant and the ovate undivided light green bract. The flowers are globular, from the sepals converging at their extremities; their sepals are green edged with pink. It is a handsome plant, and finds a place in shrubberies from its ornamental character. *H. viridis*, the Green Hellebore, is a smaller plant with fewer flowers; the sepals are

HELL] The Treasury of Botany. 573

spreading and of a yellowish-green colour. There is reason to believe that the last-named species, and probably also the others, do not flower every year or even send up many leaves, but that in certain

Helleborus fœtidus (flowering branch).

seasons, and under favonrable circumstances, the growth of the plant is more luxuriant than in others. The writer of this notice has seen the Green Hellebore in abundance one season, and found little or none of it in the following one, though to all appearance the locality had not been disturbed. On the other hand, when the plants have been purposely uprooted, as was the case in a copse near Oxford to which cattle had access, though the extirpation seemed complete, yet in two years an abundance of the plant sprang up—a fact first made known by the illness, if not the death, of some of the cattle. [M. T. M.]

HELLWEED. Cuscuta.

HELMET. The same as Galea.

HELMET-FLOWER. Scutellaria; also Aconitum and Coryanthes.

HELMINTHIA. A common wayside composite weed of the cichoraceous group, well marked by its double involucre, the inner one of which is composed of eight to ten close scales, the outer of several large loose leafy bracts. It has hispid almost prickly stems, and leaves of the same character, the lower ones lanceolate, the upper heart-shaped embracing the stem. The flowers are in small terminal heads, of a dull yellow hue and uninteresting. The fruit, which is beaked and singularly corrugated, bears some resemblance to 'a little worm,' which is the meaning of the systematic name. The English name, Ox-tongue, has reference to the shape and roughness of the leaves. [C. A. J.]

HELMINTHOSPORIUM. A large genus of the dark-threaded moulds (Dematiei), characterised by their more or less elongated septate spores, which are dark like the mother threads. Many of the species are common on decayed wood; and it is conjectured that some are mere conditions of higher fungi. The genus is very close to *Cladosporium*, which differs principally in its threads being less carbonised and its fruit less complicated. [M. J. B.]

HELMINTHOSTACHYS. A genus of *Ophioglossaceæ* consisting of a single species, *H. zeylanica*, a pseudofern, with stout horizontal rhizomes and somewhat coriaceous fronds, which are divided into a trifoliately digitato-pedate sterile branch, and a simple spicate fertile branch, on which the glomerate verticillate pedicellate tufts of spore-cases are distichously arranged, each whorl being terminated by a crest-like appendage. The veins are forked from a central costa, with the branches free. Besides Ceylon it is found in India and the islands of the Archipelago. [T. M.]

HELONIAS. A genus of *Melanthaceæ* found in North America. They have broadly lanceolate root-leaves, from a tuberous rootstock; and a bracteated scape, bearing a dense raceme of nearly sessile flowers, which are perfect, with a perianth of six oblong persistent leaves, six long slender filaments, three revolute styles, and a three-lobed pod. *H. bullata* is found in the United States, and produces, in early spring, a short raceme of purplish flowers turning green when fading. This genus has been till lately one of the most heterogeneous; but by separating *Chamælirion*, *Schœnocaulon*, and *Aminanthium*, it has assumed a more natural aspect. [J. T. S.]

HELOSCIADIUM. A genus of low umbelliferous aquatics, inhabiting various parts of the world, and represented in Britain by two species, of which *H. nodiflorum* is the most common. This plant is frequently found growing with watercresses, for which it is sometimes gathered; it may, however, be distinguished, not only by its umbellate flowers, but by its serrated lanceolate leaves. No serious consequences need be apprehended from eating the leaves, as its properties are antiscorbutic, and by no means violent in their effects. [C. A. J.]

HELOSIS. A genus of parasitical plants inhabiting the tropical and subtropical regions of the American continent, and belonging to the *Balanophoraceæ*. They have a cylindrical branched rootstock from which proceed numerous flower-stalks, bearing ovoid or globose heads of unisexual flowers: the males with a three-parted perianth and united stamens; the females with two styles. Some of the species are used as styptics. Dr. Hooker remarks that the flowers are rarely, if ever, self-fertilised, but that this process is effected by the agency of insects. [M. T. M.]

HELOTHRIX. A small Tasmanian genus of cyperaceous plants, belonging to the tribe *Scirpeæ*, and distinguished chiefly by the inflorescence being in distichous spikelets. The two lower scales are barren, the two upper produce perfect flowers; perigones with four bristles; stamens three; styles bifid. [D. M.]

HELVELLEI, or ELVELLACEI. An

order of ascomycetous *Fungi*, distinguished by the hymenium being more or less exposed, though sometimes covered at first by a veil, or the inflexed border of the receptacle. Many of the species are large, and afford good articles of food, while, on the contrary, many are small and mere botanical curiosities. It includes the esculent *Helvella*, the morels, &c., besides a multitude of species varying greatly in colour, texture, and form. In a large portion of these the receptacle is depressed, to form a cup or disk, but in others it is so raised that it becomes pileiform; the borders are then more or less closely attached to the stem, till at last they are quite confluent with it, so as to form a club-shaped body with scarcely any distinct stem, as in *Geoglossum difforme*. [M. J. B.]

HELVELLA. A fine genus of ascomycetous *Fungi*, distinguished by the pileate receptacle, which is hollow and barren below, and whose borders hang down on the stem, to which they are either slightly attached or quite free. The fructifying surface is even and free from pits; the asci contain large elliptic sporidia with one or two nuclei. The stem is sometimes simple, but it is also at times so deeply grooved that it appears as if it were made of many confluent stems. The cinereous-black *H. lacunosa*, and the pallid *H. crispa*, are our most common species, and both of them are esculent, and when well stewed form an acceptable dish. *H. esculenta*, which has been found abundantly in pine woods at Weybridge by Mr. Currey, is now referred to *Gyromitra*, in consequence of the hymenium having many gyrose raised ribs, and is known by this character and its brown tint. It is much eaten on the continent; but in some conditions appears to be dangerous. [M. J. B.]

HELVOLUS. Greyish-yellow, with a little brown.

HELWINGIACEÆ, HELWINGIA. A natural order and a genus of monochlamydeous dicotyledons, included in Lindley's garryal alliance of diclinous Exogens. A shrub with the leaves alternate, and the flowers clustered on the midrib of the leaves. The flowers are staminate and pistillate; perianth three to four-parted, with ovate spreading segments; æstivation valvate; stamens three to four, alternate with the segments of the perianth; ovary adherent to the perianth, crowned with an epigynous disk, three to four-celled with a pendulous ovule in each cell; stigmas three to four diverging. Fruit drupaceous, crowned by the remains of the styles and disk. It comes from Japan, and has alternate petiolate acuminate stipulate leaves, and small flowers. The young leaves of *Helwingia ruscifolia* are used in Japan as an esculent vegetable. The genus is by some placed in *Araliaceæ*. [J. H. B.]

HÉMÉROCALLE BLEUE. (Fr.) *Funkia ovata*. — DU JAPON. *Funkia subcordata*.

HEMEROCALLIDEÆ. The *Hemerocallis* family, a subdivision of the natural order *Liliaceæ*, which belongs to the hypogynous monocotyledons or Endogens. They are showy plants, bearing umbellate or racemose flowers, white, yellow, red, or blue. *Phormium tenax* yields New Zealand flax. *Sansevieria cylindrica* yields fibres for cordage in Africa. Examples occur in *Hemerocallis*, *Funkia*, *Agapanthus*, and *Tritonia*: see LILIACEÆ. [J. H. B.]

HEMEROCALLIS. The Day Lily, a genus of *Liliaceæ*, differing from the other tubero-fasciculate rooted lilies, by having the segments of the perianth united into a tube, and by their larger yellow or orange flowers. The leaves are all radical, very long or broadly linear, keeled, the scape branched at the top with few flowers, and a shortly trumpet-shaped perianth. They are chiefly natives of temperate Asia and Eastern Europe, though the two commonest species, *H. flava* and *H. fulva*, occur even in France. [J. T. S.]

HEMESTHEUM. *Lastrea*.

HEMI. In Greek compounds = half, or halved.

HEMIANATROPOUS. An ovule which is anatropal, with half the raphe free.

HEMIANDRA. A genus of labiates, having the calyx bell-shaped and two-lipped, the stamens four, the filaments smooth, one half of each anther alone producing pollen. The name indicates the last character above mentioned, viz. the imperfect state of an anther. The species of this genus are erect or decumbent shrubs, natives of the south-eastern parts of Australia, with narrow stiff entire leaves, bearing in their axils the solitary flowers. [G. D.]

HEMIANTHUS *micranthemoides* is a minute North American annual, constituting a genus of *Scrophulariaceæ*, scarcely differing from *Micranthemum*, by the calyx being toothed only and not lobed, and by a more irregular corolla.

HEMICARPHA. A genus of cyperaceous plants belonging to the tribe *Hypolitreæ*, distinguished chiefly by the inflorescence being in solitary many-flowered spikes; scales imbricated, obovate-cuneate, and deciduous; stamen one; styles cleft; achenes elliptic-oblong. Steudel describes five species, which are natives of warm climates in Africa and South America. [D. M.]

HEMICHROA. A genus of *Amarantaceæ*, consisting of small undershrubs from the shores of South Australia. They have alternate semi-terete exstipulate leaves, and solitary sessile axillary bibracteated flowers, with a five-leaved calyx coloured within, and two to five stamens united at the base. [J. T. S.]

HEMICLIDIA. A South-west Australian proteaceous genus containing a single species, *H. Baxteri*, a shrub growing about five feet in height, clothed with rigid edge-shaped pinnatifid leaves having sharp-pointed lobes. The involucre is imbricated, and the flowers consist of a four-left calyx, the concave segments of which each bear an anther. The seed-vessel is airy, of a crustaceous texture, containing single wingless seed. [It. H.]

HEMICRAMBE. A genus of *Cruciferæ* from North Africa, with the habit of *rassica*, having lyrate leaves and yellow flowers. The pod has two joints, the lower one being pear-shaped, empty or with one or two seeds, the terminal one sword-shaped, three-nerved, three or four-seeded, its beak winged, without seeds. [J. T. S.]

HEMICYCLIA. A genus of *Euphorbiaceæ*, consisting of a few trees or shrubs natives of East India, Java, and North Australia. Most genera of spurgeworts have three-celled ovaries, but the ovary in these plants is one-celled with two ovules, thus showing in a measure the intimate connection of spurgeworts with antidesmads. The species are smooth trees or bushes, with alternate ovate or lance-shaped entire coriaceous leaves, and minute green or white flowers in clusters in their axils, the males and females on different plants. The fruits are oval drupes not much larger than a pea, usually ripening but one seed. [A. A. B.]

HEMIDESMUS. A genus of *Asclepiadaceæ*, containing three species of twining plants, natives of India and the Moluccas. They have opposite leaves, and small flowers on interpetiolar cymes; the calyx five-parted; the corolla rotate, with five fleshy roundish scales inserted in the throat below the sinuses and forming the staminal crown. The stamens are united at the base, free above, inserted in the tube of the corolla. The apiculate anthers four-celled; the stigma large, peltate and glabrous. The follicles are cylindrical, smooth, and very much divaricated, with ominous seeds. The roots of *H. indicus* are largely employed in India as a substitute for sarsaparilla; its diuretic effect is remarkable; it acts equally well as a diaphoretic and tonic. [W. C.]

HEMIDICTYUM. A genus of polypodiaceous ferns belonging to the *Asplenieæ*, among which it is distinguished by having the veins parallel and not joined near the costa, but reticulated near the margin, finishing off by a straight or arcuate connecting veinlet at the edge. The typical species, *H. marginatum*, in which the marginal veinlet is straight, is a large tropical fern, with pinnate fronds of a light green colour and delicate texture, widely distributed over South America and the West Indies. [T. M.]

HEMIDYSTROPHIA. A term applied to express the partial nourishment of trees from the unequal distribution of their roots or from the encroachment of other trees. Trees on a wall are necessarily in this condition. [M. J. B.]

HEMIGENIA. A genus of labiates,

having the calyx somewhat bell-shaped, deeply five-cleft, the divisions equal; stamens four, one cell of each anther bearing pollen, the other abortive, the upper anthers hairy or bearded. They are Australian shrubs of little interest. [G. D.]

HEMIGRAPHIS. A genus of *Acanthaceæ*, containing two species, natives of India, perennial branching villous herbs, with alternate oblong serrate leaves, and axillary flowers, either solitary or aggregated in terminal spicate heads. The calyx is unequally five-parted; the corolla funnel-shaped and resupinate, with an unequally five-lobed limb. The stamens four didynamous; the stigma is simple and pubescent. The capsule is seedless above, but contains below from six to eight echinate seeds. [W. C.]

HEMIGYRUS. The same as Follicle.

HEMIMERIS. A genus of *Scrophulariaceæ*, consisting of small much-branched spreading annuals, with opposite leaves, and small yellow flowers in the upper axils, or clustered at the ends of the branches. The calyx is five-cleft, the corolla spreading, four-lobed, and slightly two-lipped, with two deeply-coloured depressions at the base of the lower lip. There are only two stamens, with one-celled anthers. The capsule is globular, more or less opening septicidally in two valves. There are three species known, all natives of the Cape Colony in South Africa.

HEMIONITIS. A genus of polypodiaceous ferns containing a few simple-fronded species found in the tropics of both the old and new worlds. The fronds are cordate sagittate or palmate, often proliferous, and the fertile ones generally taller. These latter are clothed with a network of closely reticulated lines of naked sporecases, which is the characteristic of the genus. The veins are reticulated just like the sort. [T. M.]

HEMIPHRAGMA *heterophyllum* is a prostrate herb, often spreading to a great extent, a native of the Himalayas, forming a genus of *Scrophulariaceæ*. The principal leaves along the wiry branches are small rounded and cordate, with dense clusters of short subulate secondary leaves in their axils. The flowers are small and pink, usually sessile and solitary, with a campanulate five-lobed corolla and four stamens. The fruit is a succulent capsule, almost a berry, but opening in two bifid valves.

HEMIPHUES. A small densely-tufted Alpine plant from Tasmania, constituting a genus of *Umbelliferæ*, remarkable for the fruit, which contains only a single cell and seed. The leaves are radical, spathulate, on short pedicels, the flowers in simple umbels on short simple scapes.

HEMIPOGON. A genus of *Asclepiadaceæ*, containing two species from Brazil. They are cæspitose undershrubs with rigid subulate glabrous sessile leaves in opposite pairs or in whorls, and solitary or ternate subsessile extra-axillary flowers. The calyx consists of five acute rigid sepals. The corolla is campanulate, the limb cut into five acute erect lobes; and there is no staminal crown. [W. C.]

HEMISTEGIA. *Hemitelia.*

HEMISTEMMA. A small genus of *Dilleniaceæ*, in which the stamens are situated upon only one side of the flower. The species are natives of Madagascar and the northern part of Australia; they are all small twiggy plants with yellow flowers, and resemble the rock-roses of Europe, their leaves being small, entire, and of a leathery texture, smooth above, but covered with white woolly hairs underneath. The calyx consists of five permanent sepals, the corolla of five petals; the stamens indefinite, a portion of them being sterile and resembling scales; and the two distinct ovaries are terminated by thin thread-like styles. [A. S.]

HEMITELIA. A genus of tree-ferns of the polypodiaceous order and the tribe *Cyatheineæ.* The fronds are large herbaceo-coriaceous, pinnate, bipinnate, or sometimes decompound, the veins parallel-forked or pinnate from a central costa, the basal ones arcuately anastomosing, forming elongated costal areoles from the outer side of which free veinlets are given off. This venation, taken together with the presence of a half cup-shaped involucre investing the sorus, characterizes the group, except in the case of *H. speciosa*, in which the costal arc is only here and there developed. They are South American or West Indian plants. [T. M.]

HEMITERIA. A monstrosity of elementary organs, or of appendages of the axis.

HEMITRICHOUS. Half covered with hairs.

HEMITROPAL. A slight modification of the anatropal ovule, in which the axis of the nucleus is more curved.

HEMLOCK. *Conium maculatum.* —, GROUND. *Taxus canadensis.* —, WATER. *Phellandrium aquaticum;* also *Cicuta virosa* and *maculata.*

HEMLOCK SPRUCE. *Abies canadensis.*

HEMP. The name of various valuable fibres employed for manufacturing purposes; and also of the plants which produce them. Common Hemp is *Cannabis sativa.* —, AFRICAN. *Sanseviera zeylanica* and others. —, BASTARD. *Datisca cannabina.* —, BENGAL or BOMBAY. *Crotalaria juncea.* —, BOWSTRING. *Sanseviera zeylanica* and others. —, BOWSTRING, of India. *Calotropis gigantea.* —, BROWN. *Crotalaria juncea.* —, BROWN INDIAN. *Hibiscus cannabinus.* —, INDIAN. *Apocynum cannabinum.* —, JUBBALPORE. *Crotalaria tenuifolia.* —, MADRAS. *Crotalaria juncea.* —, MANILLA. *Musa textilis.* —, SISAL. *Agave Sisalana.* —, SUNN. *Crotalaria juncea.* —, VIRGINIAN or WATER. *Acnida cannabina.*

HEMP-WEED, CLIMBING. An American name for *Mikania.*

HEMPWORTS. Lindley's name for the *Cannabinaceæ.*

HEN AND CHICKEN. The name given to a proliferous variety of the Daisy, *Bellis perennis;* also *Sempervivum soboliferum.*

HENBANE. *Hyoscyamus niger.*

HENBIT. *Veronica hederifolia;* also *Lamium amplexicaule.*

HENDERSONIA. One of the most striking genera of those *Coniomycetes* whose spores spring from the walls of a perithecium. The spores are always more or less articulated, and afford many exquisite objects for the microscope. Most of the species are, however, in all probability, mere states of different *Sphæriacei.* The most striking perhaps is one which occurs on dead seeds, the elongated spores of which have many transverse divisions, each articulation containing a large nucleus. *H. polycystis,* however, carries the division of the spores still further, having many vertical as well as transverse septa, and being moreover elegantly coated with a thick gelatinous stratum. [M. J. B.]

HENFREYA. A genus of *Acanthaceæ,* named in honour of the late Professor Henfrey. It is a climber, differing in this respect from most plants of the order; and is also distinguished by its anthers, which have awn-like processes at the base, and by the small two-lobed stigma. There seems, however, little to distinguish the genus from *Asystasia. H. scandens,* a native of Sierra Leone, is an elegant stove climber. [M. T. M.]

HENNÉ. *Lawsonia inermis.*

HENRIQUEZIA. A genus of handsome bignoniaceous trees of Brazil and Venezuela, exceptional in having a calyx whose tube is adherent to instead of free from the ovary, its border four instead of five-toothed; in having five perfect stamens instead of four; and in the presence of stipules to the leaves. They have oblong or obovate entire leaves placed in whorls of three to five round the stem. The handsome tubular five-lobed pink or white flowers, like those of some *Bignonia,* are disposed in dense panicles at the ends of the branches. The fruits, not the least curious part of the plant, are flat hard-shelled bodies of the shape of a bean, two-celled, opening transversely by two valves, each cell containing four seeds. The latter germinate while still in the fruit. [A. A. B.]

HENRYA. A genus of *Acanthaceæ,* containing two species, natives of Central America. They are shrubs, with hairy glandulose petiolate and ovate leaves, and spicate flowers in an involucre composed of two bracts, but apparently monophyllous from the two neighbouring margins of the bracts being united on the one side while they are free on the other. The calyx is small and five-parted; the corolla two-lipped, the upper lip deeply bifid, and the lower cut into two spathulate lobes. [W. C.]

HENSCHELIA. The name applied to a shrub, native of the Philippine Islands, and of uncertain position. It is of climbing habit with trifoliolate leaves, greenish flowers in axillary panicles; calyx of ten sepals in two rows; petals ten; stamens five, placed in front of the five outer sepals; ovary one-celled, with two ovules; stigmas five radiating. By Miers it is placed in the order *Phytocrenaceæ*. [M. T. M.]

HEN'SFOOT. *Caucalis daucoides*.

HENSLOVIACEÆ, HENSLOVIA. A natural order and a genus of calycifloral dicotyledons, belonging to Lindley's saxifragal alliance of perigynous Exogens. Trees with opposite entire leathery exstipulate leaves, and minute diœcious racemose flowers. Perianth five-parted, lined with a woolly disk, the æstivation valvate; stamens five, alternate with the segments of the perianth, inserted on a glandular perigynous disk; ovary superior, two-celled; ovules numerous, anatropal. Fruit a capsule opening by two valves; seeds numerous, minute, exalbuminous. They are natives of the tropical parts of India. There are three or four species of *Henslovia*, the only known genus, which was named after the late Professor Henslow of Cambridge. [J. H. B.]

HENSLOVIAN MEMBRANE. The cuticle; so called because Professor Henslow was one of its discoverers.

HENSLOWIA. A genus of *Santalaceæ*, having monœcious flowers, the perianth adherent to the ovary, with a five-cleft limb; and the stamens inserted at the base of the segments of the perianth, and shorter than them, with awl-shaped filaments, and introrse two-celled anthers. The ovary is inferior, unilocular, covered by a disk, and containing two pendulous ovules. Fruit drupaceous, one-seeded. Shrubby plants of the Indian Archipelago, with alternate nearly sessile leaves, and small greenish flowers. There are eight known species. [J. H. B.]

HENWARE. *Alaria esculenta*.

HEP, or HIP. The fruit of the Dog Rose, *Rosa canina*.

HEPATICA. A subgenus or section of *Anemone*, marked by having the carpels without tails, and the involucre of three simple leaves close to the flower so as to resemble a calyx. The common *H. triloba* of gardens is a native of continental Europe. In a wild state the flowers are generally blue, more rarely rose-colour or white, but in cultivation many other tints are to be found. The three-lobed leaves were fancied to resemble the liver—whence the name. [J. T. S.]

HEPATICÆ. The cryptogams belonging to this curious section, known popularly under the name of Liverworts, though confounded with lichens, differ from the mosses, to which they are closely allied, in their capsule, whether opening definitely or indefinitely, never having a distinct lid, and consequently in the total absence of a peristome. In many genera there is no stem, but the leafy shoots are replaced by an expanded membranous frond which may be quite simple or repeatedly forked, while it is sometimes irregularly lobed or laciniate. Sometimes it is crisped and plicate, and sometimes furnished with gill-like plates above. Below it is generally attached to a substance on which it grows by slender delicate rootlets. In the leafy species, the leaves have rarely the same lanceolate outline so common in mosses, and they are often accompanied by stipules or lobes which give them a habit which is very distinct from that of most mosses, though the *Hypopterygii* amongst them show something of the same structure. The section comprises three distinct natural orders as follows:—

1. RICCIACEI, in which the capsules are valveless, and either sunk in the frond or seated on its surface. The spores are not mixed with the spiral threads called elaters.
2. MARCHANTIACEI, with valvate capsules seated on the under side of a stalked target-shaped disk. Spores mixed with elaters.
3. JUNGERMANNIACEI, with solitary fruit splitting into four equal valves. Spores mixed with elaters.

The development of the fruit and the manner of impregnation are the same in these as in mosses. They are also extensively propagated by gemmæ. [M. J. B.]

HEPATICUS. Dull brown with a little yellow.

HÉPATIQUE. (Fr.) *Marchantia*. — BLANCHE. *Parnassia palustris*. — DES JARDINS. *Hepatica triloba*. — DORÉE. A common name applied to several species of *Saxifraga*. — ÉTOILÉE. *Marchantia polymorpha*; also *Asperula odorata*. — PRINTANIÈRE. *Hepatica triloba*.

HEPTA. In Greek composition = seven.

HERACLEUM. A genus of umbellifers, distinguished by having the fruit compressed from the back, each half of it with three dorsal slender ribs, and one at each marginal line, one oil-vessel in each furrow, and generally two in the commissure.

The generic name is derived from Hercules, probably in reference to the properties of some, or the size of others. The number of described species is considerable, and they are somewhat difficult to distinguish. They are widely diffused, occurring in different parts of India, in Europe and America. Several have been long known in cultivation, but are not possessed of any very special recommendations. One species has of late years been a very general object of culture on account of its large size and commanding appearance, viz. *H. giganteum*, a native of Siberia. This species is easily raised, and flowers the year after being sown, or sometimes a year later still, the latter being usually the more vigorous and attaining larger size. Individuals ten to twelve feet high are common, with a circumference of stem equal to about as many inches.

Some of the species are turned to various useful purposes. Our native *H. Sphondylium* is used for feeding pigs, and in Scania, according to Linnæus, is employed as a domestic remedy. A Kamtschatkan species is had recourse to by the natives; the footstalks of the lower leaves, when properly treated, yield a sweet exudation which is employed in the preparation of a distilled spirit. The roots and stems of *H. lanatum* are eaten by some of the native tribes of North America. The young shoots of *H. pubescens* contain a sweet and aromatic juice, and are used as food in some parts of the Caucasus. [G. D.]

HERB BENNET. *Geum urbanum*; also *Conium maculatum*, and *Valeriana officinalis*. — CHRISTOPHER. *Actæa spicata*; also *Osmunda regalis*, and *Pulicaria dysenterica*. — GERARD. *Ægopodium Podagraria*. — IMPIOUS. *Filago germanica*. — IVE, or IVY. *Ajuga Iva*; also *Coronopus Ruellii*, and *Plantago Coronopus*. — MARGARET. *Bellis perennis*. — OF GRACE. *Ruta graveolens*. — PARIS. *Paris quadrifolia*. — PETER. *Primula veris*. —, POOR-MAN'S. *Gratiola officinalis*. — ROBERT. *Geranium Robertianum*. — TRUELOVE. *Paris quadrifolia*. — TRINITY. *Viola tricolor*; also *Hepatica triloba*. — TWOPENCE. *Lysimachia Nummularia*.

HERBACEOUS. Merely green, or thin green and cellular, as the tissue of membranous leaves. Also producing an annual stem from a perennial root.

HERBA ADMIRATIONIS. *Leucas zeylanica*. — ARTICULARIS. *Silene inflata*. — BARONIS. *Thymus Herba barona*. — IMPIA. *Filago germanica*. — INDICA. *Ionidium enneaspermum*. — MERORIS. *Phyllanthus Urinaria*. — PARIS. *Paris quadrifolia*. — ROTA. *Ptarmica Herba rota*. — SANCTI JACOBI. *Senecio Jacobæa*. — SANCTI STEPHANI. *Circæa*. — SENTIENS. *Oxalis sensitiva*. — STELLA. *Plantago Coronopus*. — SUPPLEX. *Cymbidium ovatum*. — VIVA. *Oxalis sensitiva*. — VULNERATA. *Bupleurum falcatum*.

HERBARIUM. A collection of dried plants systematically arranged.

HERBE λ CENT GOÛTS. (Fr.) *Artemisia vulgaris*. — λ CLOQUES. *Physalis Alkekengi*. — λ COTON. *Asclepias Cornuti*. — À ÉCURER. *Chara*. — À ÉTERNUER. *Ptarmica vulgaris*. — À GÉRARD. *Ægopodium Podagraria*. — À JAUNIR. *Genista tinctoria*, and *Reseda luteola*. — À L'ARAIGNÉE. *Anthericum ramosum*, and *Nigella damascena*. — À LA COU-
PÉE. *Sedum Telephium*. — À LA MANNE. *Glyceria fluitans*. — À LA RATE. *Scolopendrium vulgare*. — À LA REINE. *Nicotiana Tabacum*. — À LA TAUPE. *Datura Tatula*. — À LA VIERGE. *Narcissus poeticus*. — À L'ÉPERVIER. *Hypochæris radicata*. — À L'ESQUINANCIE. *Asperula cynanchica*, and *Geranium Robertianum*. — À L'HIRONDELLE. *Passerina Stellera*. — A PRINTEMPS. *Chenopodium Botrys*. — À MILLE FLORINS. *Erythræa Centaurium*. — À OUATE. *Asclepias Cornuti* — λ PARIS. *Paris quadrifolia*. — À PAUVRE HOMME. *Gratiola officinalis*. — À ROBERT. *Geranium Robertianum*. — AU CANCER. *Herniaria glabra*. — AU CHANTRE. *Sisymbrium officinale*. — AU CHARPENTIER. *Achillea Ageratum*. — AU LAIT DE NOTRE-DAME. *Pulmonaria officinalis*. — AU NOMBRIL. *Cynoglossum linifolium*. — AU VENT. *Anemone Pulsatilla*. — AUX ÂNES. *Onothera biennis*. — AUX BŒUFS. *Chelidonium majus*. — AUX CENT MIRACLES. *Ophioglossum vulgatum*. — AUX CHARPENTIERS. *Achillea Millefolium*, and *Sedum Telephium*. — AUX CHATS. *Nepeta Cataria*, and *Teucrium Marum*. — AUX CINQ COUTURES. *Plantago lanceolata*. — AUX CUILLERS. *Cochlearia officinalis*. — AUX CUREDENTS. *Ammi Visnaga*. — AUX ÉCUS. *Lysimachia Nummularia*, and *Lunaria biennis*. — AUX FEMMES BATTUES. *Tamus communis*. — AUX GOUTTEUX. *Ægopodium Podagraria*. — AUX GUEUX. *Clematis Vitalba*. — AUX HÉMORRHOÏDES. *Ficaria ranunculoides*. — AUX MAGICIENNES. *Circæa Lutetiana*. — AUX MAMELLES. *Lapsana communis*. — AUX MASSUES. *Lycopodium clavatum*. — AUX MITES. *Verbascum Blattaria*. — AUX PANTHÈRES. *Doronicum Pardalianches*. — AUX PERLES. *Lithospermum officinale*. — AUX POUMONS. *Pulmonaria officinalis*. — AUX POUX. *Delphinium Staphisagria*, and *Pedicularis palustris*. — AUX PUCES. *Plantago Psyllium*. — AUX SERPENTS. *Trichosanthes anguina*. — AUX SONNETTES. *Fritillaria imperialis*. — AUX SORCIÈRES. *Circæa Lutetiana*. — AUX TEIGNEUX. *Tussilago Petasites*. — AUX VERRUES. *Chelidonium majus*, and *Heliotropium europæum*. — AUX VIPÈRES. *Echium vulgare*. — BLANCHE. *Didis candidissima*. — CACHÉE. *Lathræa clandestina*. — CANICULAIRE. *Hyoscyamus niger*. — CHASTE. *Vitex Agnus castus*. — CŒUR. *Pulmonaria officinalis*. — D'AMOUR. *Reseda odorata*. — DE GUINÉE. *Panicum altissimum*. — DE L'HIRONDELLE. *Chelidonium majus*. — DE LA BAIE D'HUDSON. A kind of *Poa*. — DE LA SAINT JEAN. *Hypericum perforatum*. — DE LA TRINITÉ. *Hepatica triloba*. — DE SAINTE APOL-

LINE. *Hyoscyamus niger*. — DE SAINTE BARBE. *Barbarea vulgaris*. — DE SAINT ÉTIENNE. *Circæa Lutetiana*. — DE SAINT FIACRE. *Heliotropium europæum*. — DE SAINT INNOCENT. *Polygonum Hydropiper*. — DE SAINT JOSEPH. *Scabiosa succisa*. — DE SAINT ROCHE. *Inula dysenterica*. — DES FEMMES BATTUES. *Bryonia dioica*. — DES MAGICIENS. *Datura Stramonium*. — DU BON HENRI. *Blitum Bonus Henricus*. — DU CARDINAL. *Symphytum officinale*. — DU DIABLE. *Datura Stramonium*, and *Plumbago scandens*. — DU GRAND-PRIEUR. *Nicotiana Tabacum*. — DU SIÈGE. *Scrophularia aquatica*. — DU VENT. *Anemone Pulsatilla*. — EMPOISONNÉE. *Atropa Belladonna*. — MAUCHE. *Reseda odorata*. — MORE. *Solanum nigrum*. — MUSQUÉE. *Adoxa Moschatellina*. — SACRÉE. *Melittis Melissophyllum, Nicotiana Tabacum,* and *Verbena officinalis*. — ST. CHRISTOPHE. *Actæa spicata*. — ST. PIERRE. *Crithmum maritimum*. — SANS COUTURE. *Ophioglossum vulgatum*.

HERBERTIA. A genus of dwarf bulbous Iridaceous perennials from Texas and Chili, one species found in Brazil. They have narrow acute radical leaves, and a short scape bearing at top several pretty blue or yellow flowers, which have a short-tubed six-parted perianth, with the outer segments triangular, acute, and reflexed, and the shorter inner ones rounded and erect, three monadelphous stamens inserted at the base of the exterior segments, and a three-celled ovary, crowned with three trifid stigmas having recurved petaloid branches. The genus, which is allied to *Cypella* and *Iris*, is named in honour of the late Dean of Manchester, who was a high authority on all matters relating to bulbous plants. [T. M.]

HERCULES' CLUB. *Xanthoxylon Clava Herculis*.

HÉRISSONNE. (Fr.) *Erinacea pungens*.

HERITIERA. A genus of *Sterculiaceæ*, containing two trees of considerable magnitude, found on the coasts of India, Africa, and many islands of the eastern hemisphere; in a cultivated state only in the West Indies. They are pyramidal trees with large handsome stalked entire alternate leaves of a silvery white underneath, this silvery appearance giving rise to the name of 'Looking-glass tree,' sometimes applied to them. The blades in *H. macrophylla* are eight to fourteen inches long by four to six broad. The fine foliage and symmetrical habit of this species render it a beautiful object in a plant stove where it has space to grow. The minute reddish-coloured unisexual flowers are disposed in terminal panicles; they have a five-lobed or toothed calyx; the sterile with five sessile anthers united into a tube, and the fertile with five sessile ovaries which become, when ripe, hard nearly boat-shaped carpels. They usually ripen but one seed, and do not open when ripe, in this respect differing from *Sterculia*, as well as in their less numerous stamens. C. L. L'Héritier, whose name is here perpetuated, was a distinguished French botanist. [A. A. B.]

HERMANNIEÆ. A section of the order *Byttneriaceæ*, distinguished by the following characters: — Petals flat; stamens monadelphous at the base, equal to the petals in number and opposite to them, all fertile; ovary one or many-celled, with two or many ovules in each cell. They are herbs or shrubs found in intertropical regions, but most abundant at the Cape of Good Hope. The group includes the genera *Waltheria, Melochia, Riedlea, Physodium, Hermannia,* and *Mahernia*; see BYTTNERIACEÆ. [J. B. B.]

HERMANNIA. An extensive genus of *Byttneriaceæ*, including about eighty species. The chief features of the genus are: — A bell-shaped five-cleft calyx; five clawed petals, the claws hollowed; five stamens, with their filaments flattened, but not dilated above the middle in the form of a + as in *Mahernia*; and a five-celled ovary, which, when ripe, is a five-angled capsule with many seeds. The species are twiggy undershrubs, having the stems and leaves, especially the latter, which are often accompanied with leaf-like stipules, more or less clothed with starry hairs. The pretty nodding sometimes sweet-scented flowers are pale yellow, orange, or reddish-coloured, disposed in dense clusters or loose racemes or panicles at the ends of the twigs. The genus bears the name of Paul Hermann, once professor of botany at Leyden. [A. A. B.]

HERMAPHRODITE. Containing both stamens and pistil.

HERMAS. A genus of umbellifers, characterised by the calyx having a five-parted persistent border; and the fruit ovate, each half with five ribs, the middle one prominent, those on each side of it larger, the other two smaller. The species are small Cape herbs, with soft downy undivided leaves. The outer flowers of the umbels have stamens only, the others have both stamens and pistil. [G. D.]

HERMINIERA. A genus of tropical African trees, of the leguminous family, having thorny branches, abruptly pinnate leaves, and large orange-coloured flowers, succeeded by linear oblong compressed legumes, which become at length spirally twisted. The wood of *H. elaphroxylon*, the only species, is very white, remarkably soft, having the appearance of a mass of pith, with the medullary rays and annual rings almost imperceptible. The natives apply it to various uses. [T. M.]

HERMINIUM. A genus of terrestrial orchids, with small flowers very nearly allied to those of *Orchis*, but the perianth has no spur, and the anther-cells are distant at the base, the glands of the stalks of the pollen-masses protruding below the cells. There are but very few species, all natives of the northern or Alpine regions of Europe and Asia. *H. Monorchis*, the Musk Orchis, the most common and widely-spread species, is occasionally found in southern and eastern England. It has globular tubers like those of an *Orchis*, but the new one is always produced at some distance from the stem at the end of a thickish fibre, so that the plant moves each year to a distance of one or more inches from the spot it previously occupied. The stem is slender, three to six inches high, with two or three narrow leaves near its base. The flowers, in a terminal spike, are small, of a yellowish green, with narrow sepals and petals.

HERMIONE. One of the divisions of the genus *Narcissus*, kept separate by some botanists, and consisting mainly of the plants which in gardens bear the name of *Polyanthus Narcissus*. According to Herbert, the distinctions are: that the cup is shorter than the slender cylindrical tube of the flower; the stamens with conniving filaments, situate unequally near the mouth of the tube, and free only at the curved point; and incumbent acute-oval anthers attached by the middle; and the straight slender style. Most of the *Narcissi* imported along with hyacinths from Holland, for spring flowering in gardens, are of this group. [T. M.]

HERMODACTE. (Fr.) *Iris tuberosa.*

HERMODACTYLUS. The name of a few Eastern plants often included in *Iris*, but sometimes regarded as distinct. They have fleshy tubers, glaucescent quadrangular leaves much longer than the stem which supports the curious black and green velvety flower, very small inner perianth-segments, and an oblong ovary narrowed to each end. *Iris tuberosa*, the typical species, is often called the Snake's-head Iris. [T. M.]

HERNANDIA. A genus of apetalous Exogens, the station of which in the natural system is regarded as doubtful: by some it has been separated as the type of a distinct family, the *Hernandiaceæ*. It consists of three or four or perhaps more species, tropical trees inhabiting both the East and West Indies and Guiana. The leaves are cordate, peltate, and smooth; and the flowers, which are monœcious, are in panicled masses, having a yellowish appearance from the sepals being petaloid. The male flower has six sepals, and three stamens opposite the three outer sepals; between the bases of the stamens are three pairs of glands. The anthers open by two valves, reflected laterally. The female flower, the structure of which has hitherto been imperfectly understood, proves on examination to have the ovary inferior, and at its base external to the calyx it is enclosed by a cup-like involucre, which in the male flower is wanting. The sepals are eight, or sometimes in imperfectly hermaphrodite flowers nine, and it has four barren stamens which are like the glands of the male flower. The ovary is one-celled, containing one pendulous ovule; the style is short, furrowed on one side, and the stigma is broad and lobulated. The seed, in which the radicle is superior, contains no albumen, and the embryo has a crumpled appearance, in addition to which each cotyledon is three-lobed at its base. By its valvular anthers it is nearly related to *Lauraceæ*, but in its inferior ovary it is nearer *Combretaceæ*, and its station consequently is near *Gyrocarpus* and *Illigera* in the latter family, the flowers of these genera having no petals, and their anthers opening by valves. The bark, seed, and young leaves of *H. sonora* are slightly purgative. It is said that the fibrous roots chewed and applied to wounds caused by the Macassar poison form an effectual cure. The juice of the leaves is a powerful depilatory, destroying hair wherever applied without pain. The wood is light; that of *H. guianensis* takes fire so readily from a flint and steel, that it is used as amadou. [B. C.]

HERNANT SEEDS. The commercial name for the seeds of *Hernandia ovigera*, used for dyeing.

HERNIARIA. A genus of *Illecebraceæ*, found in barren places in the temperate regions of Europe, Asia, and Africa. They are small annuals or undershrubs with oval, oblong, or linear leaves, and small scarious stipules. The minute flowers, in lateral clusters generally arranged in an interrupted leafy spike, have a five-parted calyx, five petals reduced to mere threads, five stamens, two stigmas, and a membranous utricular fruit. *H. glabra* is a native of Britain, and not unfrequent in the south-western counties. [J. T. S.]

HERNIOLE. (Fr.) *Herniaria glabra.*

HERON'S BILL. *Erodium.*

HERPESTIS. A genus of *Scrophulariaceæ*, allied to *Gratiola*, and having, like that genus, didynamous stamens, with two-celled anthers, and a capsule opening septicidally in two entire or bifid valves. It is, however, readily known by the calyx consisting of five distinct very unequal sepals, the lowest outer one always much larger than the others, and the two innermost often very narrow. There are above forty species known, natives of various parts of America, Africa, Australia, or southern Asia. They are all herbs, mostly procumbent or prostrate, more rarely erect, with rather small flowers usually yellow or pale blue. The most common are, *H. Monnieria*, a small creeping glabrous plant, with rather thick entire leaves, and a pale blue or nearly white flower, very abundant in almost all hot countries in moist situations; and *H. chamædryoides*, a much-branched spreading species with ovate toothed leaves and yellow flowers, common in the mountainous districts of America from South Brazil to Mexico.

HERRERIA. A genus of *Liliaceæ* of doubtful affinity, having the habit of the

Asparageæ, but in structure resembling the *Anthericeæ*. They are undershrubs found in Brazil and Chili, with tuberous rootstock, climbing stems, whorled-fascicled lanceolate or linear leaves, and small scented flowers in many-flowered axillary racemes. The perianth is herbaceous, six-parted, persistent; the stamens six; the capsule membranaceous, three-winged, and three-celled. [J. T. S.]

HERSCHELIA *cœlestis* is the name of a terrestrial orchid of South Africa, with a stem a foot high, bearing at the base a number of narrow grassy leaves, and ending in a raceme of pretty flowers an inch across and of an intense sky-blue colour — therefore most appropriately named by Dr. Lindley in honour of Sir John Herschel, the celebrated astronomer. The upper sepal is helmet-shaped, spurred near the base, larger than the lower ones, and hiding the petals. The beak is trilobed, and between it and the anthers is a curious forked linear appendage. [A. A. B.]

HESPERANTHA. A genus of Cape *Iridaceæ* closely allied to *Ixia*, the species remarkable for expanding their sweet-scented flowers in the evening — whence the name. They are bulb-tuberous plants with sword-shaped leaves; and the flowers, which grow in loose spikes, have a long-tubed hypocrateriform perianth with six equal spreading limb-segments, three stamens inserted in the perianth tube, and three stigmas, which are elongate narrow-linear and conduplicate. The flowers are mostly white, sometimes stained outside with some dark colour. [T. M.]

HESPERIDEÆ. A name given by Linnæus to a natural order comprising the genera *Citrus*, *Styrax*, and *Garcinia*. It has sometimes been applied to the orange family. Endlicher gives the name *Hesperides* to one of his classes embracing the orders *Humiriaceæ*, *Olacineæ*, *Aurantiaceæ*, *Meliaceæ*, and *Cedrelaceæ*. It is thus defined :— Trees or shrubs with alternate exstipulate usually compound leaves. Calyx free, imbricate in æstivation; corolla with petals equal in number to the segments of the calyx, valvate or convolute in æstivation; stamens twice or four times the number of the petals, free, monadelphous or polyadelphous; carpels numerous, united into a one or many-celled ovary; ovules solitary or many, usually anatropal; embryo very often exalbuminous; cotyledons mostly fleshy. [J. H. B.]

HESPERIDIUM. A many-celled superior indehiscent fruit, pulpy within, and covered by a separable rind; as the orange.

HESPERIS. The Rocket, a genus of *Cruciferæ*, belonging to the section having the radicle of the seed bent over the back of one of the flat cotyledons. It is distinguished from *Malcolmia* by the blunt not sharp-pointed lobes of the stigma at the end of the long cylindrical pod. They are biennial or annual (rarely perennial) herbs with somewhat the habit of the stock, but usually with less stellate pubescence. The flowers are large, purple, lilac, white, or dirty yellow; in some of the species sweet-scented in the evening, whence the generic name. The common garden Rocket, or Dame's Violet, is *H. matronalis*, a native of Europe, but probably not indigenous to Britain; many varieties exist in cultivation, with white, purple, variegated, or double flowers. [J. T. S.]

HESPEROMELES. The name of a few shrubs or trees of considerable size belonging to the *Pomaceæ*, and found at elevations of eight to thirteen thousand feet on the Andes of Peru and New Grenada. They have alternate stalked coriaceous ovate or oblong leaves, and white or pink flowers much like those of the hawthorn in size and disposition. From this genus they chiefly differ in the ovaries, five in number, having each but one instead of two ovules. The fruits are also like those of the hawthorn. *H. lanuginosa* grows to a large tree in New Grenada; Mr. Purdie remarks that it forms the entire forest, beginning at ten thousand and reaching to fourteen thousand feet, or near the perpetual snow limit. *Hesperomeles* signifies Western Apple. [A. A. B.]

HESPEROSCORDON. A genus of *Liliaceæ*, differing from *Brodiæa* by having all the six stamens anther-bearing, and the ovary sessile. They are herbs found in western North America, having much the habit of some species of *Allium*, and with large white or bluish flowers. [J. T. S.]

HESSEA. A small genus of *Amaryllidaceæ*, characterised by having a bifid spathe, a short-tubed regular-limbed perianth, equal subulate filaments becoming reflexed and bearing short anthers, a filiform style, and a trifid fimbriated stigma. It is represented by the *Amaryllis stellaris* of Jacquin. The name *Hessea* has also been given to the genus *Carpolyza*. [T. M.]

HETÆRIA. A small Australian marsh plant, belonging to the *Philydraceæ*, and differing from *Philydrum* by its kidney-shaped anther lobes, its central placenta ultimately detached from the three valves of the capsule, and by its smooth seeds. *H. pygmæa* is a small rush-like plant with a spike of flowers of a yellow colour and invested by bracts. [M. T. M.]

HETERANTHERA. A genus of *Pontederaceæ*, consisting of small aquatic herbs with roundish long-stalked or linear leaves, and one or two small white or blue flowers produced from a spathe in the axil of a sheathing leaf-stalk. The perianth is salver-shaped, with a long slender tube and a spreading six-lobed limb. *H. reniformis*, the Mud Plantain, with roundish kidney-shaped leaves and white flowers, is not unfrequent by the muddy banks of streams in the Southern United States. [J. T. S.]

HETEROCARYUM. A genus of *Boragineæ*, natives of temperate Asia, resembling *Omphalodes*, but having the calyx segments caducous, the column of styles adhering to the nuts as far as their middle, and the peduncles thickened. [J. T. S.]

HETEROCEPHALOUS. Bearing in the same individual, heads of entirely male flowers, and others entirely female.

HETEROCHÆNIA. A genus of hellworts, having the tube of the calyx obconical, deeply five-cleft, with the lobes ciliated; seed-vessel three-celled, opening first by three valves at the summit, subsequently by rupture of other parts. The genus was founded by De Candolle, to include the Mascaren plant formerly called *Wahlenbergia ensifolia*. [G. D.]

HETEROCODON. An annual from the Oregon territory in North America, distinguished as a genus of *Campanulaceæ* by Nuttall, on account of the lower flowers having no corolla; but it is probably only a form or variety of *Specularia perfoliata*.

HETERODON. A genus of brunlads, distinguished by the calyx having ten teeth, five of which are short and blunt, and five elongated. The only species is a shrub, a native of the Cape, having semicylindrical leaves, which are hairy, ending in awn-like points. [G. D.]

HETEROGAMOUS. When in a capitulum the florets of the ray are either neuter or female, and those of the disk male.

HETEROIDEOUS. Diversified in form.

HETEROLÆNA. A subdivision of *Pimelea* in which the capitula are terminal, and the involucre formed of four rarely five to eight leaves, and these leaves are unlike the foliage of the branches, differing in magnitude or in form and texture, often coloured. They are shrubby plants of New Holland and Tasmania, with opposite leaves. There are thirty-eight species of *Pimelea* in this subdivision. [J. H. B.]

HETEROLEPIS. A small genus of *Compositæ*, nearly related to *Gazania*, and found in South Africa. The species differ from this and their other allies, in having the hairs of the pappus (which are of unequal length and ciliated) in two or three series. All are branching bushes, with rosemary-like leaves, and handsome flower-heads with the florets all yellow. [A. A. B.]

HETEROMORPHA. A genus of umbellifers, distinguished from its congeners by its peculiar fruit, which is apparently five-winged owing to the different aspect of its two halves, the outer being provided with two wing-like ridges, the inner with three. The species are natives of the Cape of Good Hope. [G. D.]

HETERONEMEÆ. A name applied to the higher cryptogams by Fries to express the fact of the more complicated germination than in the lower cryptogams. The production of the pseudocotyledons in ferns appears to be what he had more especially in view. It may, however, be objected that in *Puccinia* and some other fungi there is a decided prothallus preceding the formation of true fruit. [M. J. B.]

HETERONEURON. *Pœcilopteris*.

HETEROPAPPUS. The name formerly given to a few *Compositæ* of North China and Japan, with flower-heads like *Aster*. They are now known to belong to the genus *CALIMERIS*; which see. [A. A. B.]

HETEROPHRAGMA. A genus of *Bignoniaceæ*, containing a single species from India. It is a large tree with opposite or ternate impari-pinnate leaves, and whitish flowers in dense terminal downy panicles. The calyx is campanulate and three-lobed; the corolla equally five-parted, with the margins of the divisions waved; there are four fertile stamens; the ovary is surrounded by a purple disk, and surmounted by a simple style and a two-cleft stigma; the capsule is long and pointed; and the seeds have a broad wing. [W. C.]

HETEROPOGON. A genus of grasses belonging to the tribe *Andropogoneæ*, now included in *Andropogon*. They are mostly natives of Mexico. [D. M.]

HETEROPSIS. A genus of Brazilian plants, of the family *Araceæ*, deriving its name from the fact that the appearance of the plant is different from that of most of its congeners. The stem is woody and branching, with lance-shaped leaves; the spathe hooded, deciduous; spadix blunt, covered with male and female flowers, intermixed; the anthers are two-celled and gaping; ovaries two-celled, with two ovules in each cell. [M. T. M.]

HETEROPTERIS. A genus of American climbing shrubs, with yellow or bluish flowers, belonging to the *Malpighiaceæ*. Several are cultivated as evergreen stove climbing plants; their flowers have a calyx with eight glands; stamens all fertile; styles three; fruit with a wing thickened on the lower margin. [M. T. M.]

HETEROS. In Greek compounds = variable, or various.

HETEROSTEMMA. A small genus of *Asclepiadaceæ*, natives of India and the Moluccas. They are glabrous twining shrubs, with opposite membranaceous leaves, and flowers in few-flowered interpetiolar umbels. The calyx consists of five ovate sepals; the corolla is rotate and five-parted, with spreading lobes. The five-leaved staminal crown is very variable, differing in each species. The follicles are smooth and divaricate, and contain about twenty comose seeds. [W. C.]

HETEROTOMA. The name of a Mexican herbaceous plant, constituting a genus of *Lobeliaceæ*. It has a two-lipped calyx; a tubular corolla, the tube of which is irregularly dilated at the base into a spur-like form; anthers cohering, the two lower ones hairy; ovary with two compartments; stigma two-lobed. The flowers are large, purple, arranged in racemes. *H. lobelioides* is the Bird-plant of Mexico. [M. T. M.]

HETEROTROPA. The name applied to a genus of *Aristolochiaceæ*, represented by a Japanese herb, with a coloured pitcher-

-shaped perianth contracted at the throat, where it is provided with a plicated ring or 'corona.' The anthers are twelve in number, arranged in two rows; the outer, on triangular filaments, open inwardly, and are partially united together; the inner ones are sessile, open outwardly, and have their connective prolonged into a lance-shaped point. The plant has the appearance of *Asarum*, from which genus the above characters amply distinguish it. Its leaves are heart-shaped, marked with white spots. [M. T. M.]

HETEROTROPAL. Lying parallel with the hilum. A term applied only to the embryo.

HÊTRE. (Fr.) *Fagus sylvatica*.

HEUCHERA. A genus of perennial herbaceous plants of elegant appearance, natives of North America and Siberia, and included in the *Saxifragaceæ*. The petals are five, inserted into the upper part of the tube of the calyx, of a linear form and slightly unequal; stamens five, inserted with the petals; ovary one-celled, with two parietal placentæ; styles elongated, divergent; fruit bursting between the styles. The flowers are borne in clusters which rise from a number of lobed toothed leaves. Several of the species are grown in English gardens. The root of *H. americana* is so astringent that it is called Alum root. [M. T. M.]

HEWARDIA. A genus of polypodiaceous ferns, agreeing with *Adiantum* in all the essential points of fructification, but distinguished from it by having the veins reticulated. They have linear continuous sori, as in *Adiantum Wilsoni* and its allies, and are pinnate, bipinnate, or pedately tripinnate plants of South America. It is named after Mr. R. Heward, an amateur pteridologist, and one of the contributors to this work. The name has also been given to a melanthaceous stemless herb from Tasmania, having ensiform distichous leaves, and star-shaped purple flowers, and the habit of an *Iris* or *Sisyrinchium*; but for this the name of *Isophysis* has been suggested. [T. M.]

HEXA. In Greek compounds = six. Thus: *Hexalepidous*, consisting of six scales: *Hexapterous*, having six wings or membranous expansions; *Hexapyrenous*, having six stones; *Hexapetaloid*, consisting of six coloured parts, like petals; *Hexarinous*, having six stamens.

HEXACENTRIS. A small genus of *Acanthaceæ*, containing three species from India. They are climbing shrubs with dentate leaves, and purple or yellow flowers in axillary and terminal many-flowered racemes. The small calyx is unequally toothed, and is surrounded by two small bracts. The corolla has a short tube and an oblique five-cleft limb. The four didynamous stamens have erect two-celled anthers, which, in the shorter pair, have both cells spurred, and the longer pair have a spur on one only. A short subulate sterile fifth stamen is present. The stigma is bifurcate. *H. mysorensis* is very ornamental. [W. C.]

HEXADESMIA. A few epiphytal orchids of Central America, differing from *Epidendrum* in having six instead of four pollen-masses; whence the generic name. They are tufted plants a few inches high, with narrow oblong pseudobulbs, a few short grassy leaves, and a number of inconspicuous green or white flowers in a terminal raceme. [A. A. B.]

HEXAGONIA. A fine genus of pore-bearing *Fungi*, distinguished by its large angular pores, which resemble the cells of a honeycomb. Most of the species are hard and woody, but one or two are thin and flexible as paper. They are, with but one or two exceptions, inhabitants of tropical countries. We have no species in Great Britain, but *H. sericea* is found in the forests above Canada. One or two species are found on gum trees in Australia. In some Indian species the pores are one-sixth of an inch across. [M. J. B.]

HEXALOBUS. A genus of anonaceous shrubs, inhabiting Senegal and Madagascar. They have a six-cleft corolla, with the spreading segments in two rows; numerous club-shaped stamens, attached to the sides of a convex receptacle; and numerous ovaries with sessile stigmas; fruit of several few-seeded berries. [M. T. M.]

HEYNEA. A genus of Indian trees belonging to the *Meliaceæ*, among which they are distinguished by the tube formed by the union of the stamens, which is deeply five-cleft, the segments being also cleft; the anthers are ten, sharply pointed; ovary two-celled, imbedded in a fleshy disk, and ripening into a somewhat fleshy capsular fruit, which is one-celled by abortion and single-seeded. [M. T. M.]

HIANS. Gaping; opening by a long narrow fissure cut across the shorter axis.

HIBBERTIA. A genus of *Dilleniaceæ* confined to Australia and Tasmania, comprising about fifty species. They usually form little heath-like tufted shrubs, or their slender stems trail along the ground, but occasionally they grow several feet in length and climb upon other shrubs. Their flowers are yellow, borne at the ends of the branches, and generally give out a very unpleasant odour; they have five thick leathery permanent sepals, and five thin fugaceous petals; the stamens are very numerous, entirely free or united at their bases into several bundles; and the one-celled ovaries, two to five in number, are terminated by a diverging style. The fruit consists of two or more carpels splitting open down the inner edge, and containing one or several roundish shining seeds, each partly surrounded by an aril.

H. dentata, a climbing species, is one of the most showy, and grows six or eight feet high. *H. grossulariæfolia* is another of the climbing kinds, having leaves somewhat resembling those of the common gooseberry bush, its trailing stems tinged with red, and its flowers produced in great abundance at the ends of little side branches. *H. volubilis*, the largest species of the genus, has a stiff climbing stem, and pale yellow flowers two inches across, but most disagreeably scented. [A. S.]

HIBERNACULUM. The poetical name of a bud or bulb.

HIBERNAL. Of or belonging to winter.

HIBISCUS. The Rose-mallow family, a very large genus of *Malvaceæ*, characterized by their large showy flowers being borne singly upon stalks towards the ends of the branches; by having an outer calyx or involucel composed of numerous leaves, and an inner or true calyx cut into five divisions at the top, which does not fall away after flowering; by having five petals broad at top and narrow towards the base, where they unite with the tube of the stamens; and by the latter forming a sheath round the five-branched style, and emitting filaments bearing kidney-shaped anthers throughout the greater part of its length. The fruit is five-celled, with numerous seeds. The majority of the species are tropical, but a few are found in temperate regions, and one, *H. Trionum*, occurs in the South of Europe and also in New Zealand. Most of them are shrubs, but a few form moderately high trees. All possess the mucilaginous properties common to the order, and several are eaten as potherbs, while their inner bark yields more or less fibre.

H. cannabinus has a prickly stem, six or eight feet high, and deeply-parted leaves somewhat resembling those of hemp. The flowers are pale yellow with a dark purple blotch at the bottom of each petal. This is a native of the East Indies, where it is cultivated on account of the fibre contained in its stems, the seeds being sown thickly so as to induce the plants to grow up tall, straight, and unbranched. The fibre, like that of other malvaceous plants, bears more resemblance to jute than to hemp, though it is sometimes called Indian Hemp. It comes to this country in small quantities, and is sometimes called Bastard Jute. In Western India the plant is called Ambarec, and its leaves are eaten as a pot-herb, and an oil is extracted from its seeds.

H. Rosa sinensis, a well-known ornament of our hothouses, is a native of India, China, and other parts of Asia. It is a tree of twenty or thirty feet high; and has very variable flowers—double, single, red, dark purple, yellow, white, or variegated, according to the particular variety. These flowers contain a quantity of astringent juice, and when bruised rapidly turn black or deep purple; they are used by the Chinese ladies for dyeing their hair and eyebrows, and in Java for blacking shoes, whence the plant is frequently called the Shoe-black Plant.

H. syriacus, commonly called Althæa frutex, is a hardy deciduous shrub, with large showy flowers, produced in great profusion in the autumn months. [A. S.]

HIBISCUS, BASTARD. *Achania Malvaviscus*.

HICKORY. *Carya.*

HIDDEN-VEINED. Having the veins so buried in the parenchyma, that they are not visible upon external inspection.

HIERACIUM. A large and exceedingly difficult genus of cichoraceous plants, mostly with yellow flowers, inhabiting the temperate countries of the eastern hemisphere, and distinguished among allied genera by having a brown brittle pappus and no beak to the fruit. From twenty to thirty species are indigenous to Britain, growing in hedges, woods, and mountains. One of the best known and most attractive of these is *H. Pilosella*, common on heaths and in dry pastures, a dwarf plant with creeping leafy scions, elliptical leaves clothed above with scattered long hairs, and bearing on leafless stalks a single brilliant light yellow flower. Other common species are *H. sylvaticum* and *H. umbellatum*, tall weeds with leafy stems and uninteresting yellow flowers. Several others are more or less frequent, but can only be discriminated by the application of much patient care. *H. aurantiacum*, called Grim-the-collier from the black hairs which clothe the flower-stalk and involucre, is an ornamental plant with orange-coloured flowers, often cultivated in flower gardens. The systematic name *Hieracium*, the English Hawkweed, the French *Éperviére*, and the German *Habichtskraut*, all have reference to an ancient belief that birds of prey made use of the juice of these plants to strengthen their vision. [C. A. J.]

HIEROCHLOA. A genus of grasses belonging to the *Phalarideæ*, and consisting of several species spread over the colder parts of both hemispheres. They have loose spreading or narrow crowded panicles; three-flowered spikelets, the two lower flowers being males with three stamens, and the upper one smaller with two stamens and hermaphrodite; the glumes are scarious, boat-shaped, and pointed. One native species, *H. borealis*, found near Thurso, occurs in mountain pastures in Northern Europe, Asia, and America, and also in New Zealand. The name *Hierochloa*, sometimes written *Hierochloë*—whence Holy-grass—refers to the practice, adopted in some parts of Germany, of strewing it before the doors of churches on festival days. [T. M.]

HIGGINSIA. A genus of small Peruvian shrubs, belonging to the *Cinchonaceæ*. The parts of the flower are arranged in fours; the corolla is somewhat bell-shaped, with a short tube, concealing the stamens within it; the ovary has two compartments; the ovules are numerous, the style short, and the stigma cleft and projecting. The fruit is berry-like, and two-celled. See CAMPYLOBOTRYS. [M. T. M.]

HIG-TAPER. *Verbascum Thapsus.* The name is, according to Dr. Prior, often incorrectly spelt High-taper.

HIGHWATER SHRUB. An American name for *Iva*.

HILIFER. Bearing a hilum upon its surface.

HILLIA. The memory of Sir John Hill, a writer on various branches of botany, is held in little respect in this country, owing to some unseemly disputes with some of his contemporaries and with the Royal Society; nevertheless a genus of plants has been named in his honour, consisting of small tropical American shrubs reported to grow upon the trunks of trees, and belonging to the *Cinchonaceæ*. They have somewhat fleshy leaves; an involucre of three or four bracts outside the calyx, the limb of which is divided into two to four narrow segments; a salver-shaped corolla, with a long tube, distended at the throat, and concealing four to six sessile anthers; a thread-like style, and thick stigma. The fruit is a long pod-like two-celled two-valved capsule, with numerous seeds which are provided with a loose integument prolonged at one end into a long brush-like appendage. [M. T. M.]

HILUM. The scar produced by the separation of a seed from its placenta. Also used to indicate any point of attachment; and the apertures in the extine of pollen grains.

HIMANTHALIA. A genus of olive-spored *Algæ*, remarkable for the large immensely elongated forked receptacles, and the little cup-shaped frond which scarcely exceeds an inch in diameter. The plant is common on some parts of our coast, though rather local. The fronds when young sometimes become detached and form little bladders which make a loud report when trod upon. The only species, *H. lorea*, is known by the name of Sea-thongs from the strap-like appearance of the receptacles. The plant is biennial, the receptacle not being produced till the second year. It extends southward as far as Spain, but prefers rather cold waters. It is very rare, if found at all, on the coast of America. [M. J. B.]

HIMATANTHUS. A Brazilian tree, constituting a genus of *Cinchonaceæ*. The flowers are arranged in spikes which are covered by a large spathe-like bract, falling off before the flowers expand. The parts of the flower are arranged in fives; the corolla is very long and funnel-shaped, concealing within it the stamens; the style is somewhat club-shaped; and the ovary has two compartments. The fruit is unknown. [M. T. M.]

HIMERANTHUS. A genus of *Solanaceæ*. The flowers are placed singly on long stalks, and have a bell-shaped corolla, to the base of which, internally, the stamens with strap-shaped filaments are attached. The ovary is two-celled, the fruit fleshy, many-seeded, supported by the persistent calyx. It is found in Uruguay; and it is singular that a plant so nearly allied to the true mandrake should be supposed by the natives of that country to possess the same power of exciting the passions as was attributed to the mandrake, in Greece &c., by the ancients, and even by mediæval writers. [M. T. M.]

HINA. The Pacific Island name for a Gourd.

HINAU, or HINO. *Elæocarpus Hinau*, the bark of which is used for dyeing in New Zealand.

HINDA. An Indian name for the Wild Date, *Phœnix sylvestris*.

HINDBERRY. *Rubus Idæus*.

HINDHEAL. *Chenopodium Botrys*.

HINDSIA. A genus of cinchonaceous shrubs, natives of Brazil. The flowers have a calyx with unequal linear segments, sometimes dilated in a leaf-like manner; a funnel-shaped corolla with a long tube, somewhat dilated at the upper part; anthers on very short stalks at the top of the tube of the corolla; and a style divided at its upper part into two long linear compressed hairy branches. The capsule bursts by two valves, and contains numerous seeds. *H. violacea* is a stove plant of great beauty, with large deep blue flowers. [M. T. M.]

HING. The Indian name for Asafœtida.

HINOID. When veins proceed entirely from the midrib of a leaf, and are parallel and undivided; as in ginger-worts.

HIP-TREE. *Rosa canina*, the fruits of which are called Hips.

HIPWORT. *Cotyledon Umbilicus*.

HIPPEASTRUM. The Knight's Star Lily, a genus of *Amaryllidaceæ*, consisting of South American and West Indian bulbs, remarkable for their showy flowers, and comprising most of the plants cultivated in hothouses under the name of *Amaryllis*, these being for the most part hybrids, which are very freely produced in the genus. The leaves, which are vernal, are bifarious, and precede or accompany the flowers; the latter usually grow several together at the top of a hollow scape, and are large and in most cases very handsome, the somewhat funnel-shaped declinate perianth having an abbreviated and narrow-mouthed tube with the faucial membrane deficient on the lower side; and a very irregular limb, the upper sepaline being wider, and the lower petaline narrower than the other segments. The filaments are declinate, curved, unequal, and unequally inserted into the throat; and the style is three-lobed or three-cleft. The flowers of some of the species, as *aulicum*, *equestre*, and *reginum*, are crimson, scarlet, or orange-red, with a green star; of *vittatum* white striped with red; and of *reticulatum* purplish-red, beautifully veined with deeper red, and with a white central star. [T. M.]

HIPPIA. A genus of South African

Compositæ, consisting of slender herbs or small branching shrubs, with leaves and flower-heads something like chamomile. The leaves are pinnatifid; the flowers are minute yellow rayless, disposed in corymbs at the ends of the twigs, and not unlike those of *Artemisia*, to which the genus is allied. The outer florets have pistils only, the inner stamens; and the orbicular compressed achenes have slightly winged margins, and no pappus. (A. A. B.)

HIPPOBROMUS *alatus*, the only representative of a genus of *Sapindaceæ*, is a South African tree of considerable size, with alternate unequally-pinnate leaves, bearing in their axils short velvety clusters of small reddish flowers. The leaves are made up of four to six pairs of unequalsided serrate leaflets; and the flowers are unisexual, the sterile with five sepals, five petals, and eight stamens, the fertile with a like calyx and corolla, and a few barren stamens surrounding a three-celled ovary tipped with a short style. The genus differs from *Sapindus*, in the petals being destitute of a scale or tuft of hairs on their inner surface, as well as in the round berried fruits the size of a pea accompanied by the remaining calyx. The colonial name of the tree is Paardepis. (A. A. B.)

HIPPOCASTANEÆ. A group of hypogynous Exogens, forming a subdivision of the order SAPINDACEÆ; which see.

HIPPOCRATEACEÆ. A natural order of thalamifloral dicotyledons, included in Lindley's rhamnal alliance of perigynous Exogens. Shrubby plants with opposite simple leaves having deciduous stipules; sepals and petals five imbricate; stamens three monadelphous. Fruit either consisting of three-winged carpels, or laccate. The prominent character of the order is the ternary stamens, and pentamerous sepals and petals. The plants are chiefly natives of South America, but some are found in Africa and Asia. The nuts of *Hippocratea comosa* are oily and sweet. The fruit of *Tontelea pyriformis* is eaten in Sierra Leone. There are seven genera and about ninety species. Examples: *Hippocratea*, *Tontelea*, and *Salacia*. (J. H. B.)

HIPPOCRATEA. A genus of the small order *Hippocrateaceæ*, consisting of upwards of thirty species, the greater part natives of the tropics of the western hemisphere, the remainder found principally in Western Africa, India, and the Island of Timor. They are climbing shrubs, with opposite entire or toothed usually smooth leaves, and panicles of small inconspicuous flowers, produced from the axils of the leaves, and characterized by the anthers of their stamens consisting of single cells, which burst open transversely. Their fruit also differs considerably from those of the allied genera, being composed of three (occasionally only one or two) separate flattened leathery carpels, which split down the middle into two halves when ripe, each half resembling in shape a little boat. (A. S.)

HIPPOCREPIFORM. Horseshoe-shaped.

HIPPOCREPIS. The Horseshoe Vetch, a genus of herbaceous or somewhat shrubby leguminous plants, so called from the peculiar form of their seed-vessels, which are long and jointed, each joint being oneseeded and curved into a shape somewhat resembling that of a horseshoe. In all the species the leaves are pinnate, with a terminal leaflet. The flowers are yellow, in some species solitary in the axils of the leaves, but more frequently collected into simple umbels on slender axillary stalks. The only British species, *H. comosa*, is a low trailing plant with much of the habit of the common bird's-foot trefoil, but differs both in the shape of its leaves and pods. It is not uncommon on sunny banks of chalk or limestone. Several other species, some of which are annuals, inhabit the south of Europe. French, *Hippocrèpe*; German, *Hufeisenpflanze*. [C. A. J.]

HIPPOMANE. The celebrated poisonous Manchineel or Manzanillo tree of tropical South America (*Hippomane Mancinella*) is the only species of this genus of spurgeworts. It is a tree forty or fifty feet high, common in many of the West Indian Islands and in Venezuela and Panama, usually growing on sandy sea-shores. Its leaves are stalked, shining green, egg-shaped or elliptical, with the edges cut into saw-like teeth, having a single gland on the upper side at the junction of the stalk and leaf. Its flowers are very small and inconspicuous, and of separate sexes, borne on long slender spikes, the females few placed singly at the base of the spike, the males in little clusters occupying the upper part. The calyx of the males is two-parted, and that of the females three-parted, the male containing two or four stamens joined together by their filaments, and the females a many-celled ovary, crowned with from four to eight styles and reflexed stigmas. Its fruit is a roundish fleshy yellowishgreen berry.

The virulent nature of the juice of the Manchineel tree has given rise, in the western hemisphere, to nearly as wonderful stories as those associated with the upas tree in the eastern; but although there can be no doubt that it possesses extremely poisonous properties, its powers have been greatly exaggerated, and many of the tales must be regarded as fabulous. Among the statements referable to the latter class may be included the assertions that grass will not grow under it, that mere sleeping in its shade causes death, that its juice raises blisters difficult to heal when applied to the skin, and others of a like nature. It is certain, however, that the juice, which resembles pure white milk, does possess a considerable amount of acridity, and that some persons suffer great pain from incautiously handling it, while others again do not experience the slightest inconvenience from it, its effects, as in the case of the poisonous *Rhus* of North America, depending upon peculiarities in the constitutions of different individuals. Perhaps its most dangerous property is that of causing blindness, if by chance the least drop of the milk, or the smoke of the burning wood, comes in contact with the eyes. Dr. Seemann, in his *Narrative of the Voyage of H.M.S. Herald,* states that at Veraguas some of the ship's carpenters were blinded for several days from the juice getting into their eyes whilst cutting down Manchineel trees; while he himself suffered temporary loss of sight from merely gathering specimens; and that the same accident happened to a boat's crew from using the wood for making a fire. Salt water is said to be an efficacious remedy. The fruit also abounds in a similar acrid milky juice, and, from its tempting appearance, is sometimes bitten by those who are unaware of its deleterious properties, but its burning effect upon the lips soon causes them to desist. It is commonly asserted that the Indians use the juice for poisoning the barbs of their arrows, but, from its excessively volatile nature, this is improbable. [A. S.]

Hippomane Manclnella.

HIPPOPHAË. A shrub or low tree of the order *Elaeagnaceæ*, distinguished by bearing the male flowers (with four stamens) in catkins, and the female in the axils of the leaves, on separate plants; the calyx tubular, finally assuming the character of a berry containing a single seed. *H. rhamnoides,* Sea Buckthorn or Sallow Thorn, is a native of many parts of the coast of Europe, including England, preferring a sandy soil, but sometimes found on the cliffs. In its native haunts it is usually a thick bush with numerous branches terminating each in a thorn. The leaves are narrow, of a peculiar leaden green above, silvery and scaly below. The berries, which are produced in great abundance, are yellow and of an acid flavour. The Tartars, it is said, make a jelly of them, and the fishermen of the Gulf of Bothnia prepare from them a fish-sauce, but in England they appear to be neglected. French, *Argousier*; German, *Haftdorn.* [C. A. J.]

HIPPURIDEÆ. A natural group described by Link, now included under HALORAGACEÆ: which see.

HIPPURIS. Aquatic herbaceous plants with whorls of narrow leaves, and inconspicuous flowers (also in whorls) of very simple structure. There are only two or three species, all much alike. They grow either wholly or partially submersed in ditches and canals, sending up from their creeping roots numerous unbranched erect stems, having at short intervals whorls of linear leaves, in the axils of which are the small inconspicuous flowers, each of which contains a single stamen, but no petals, and an ovary with a single seed. The most abundant species is *H. vulgaris,* the common Mare's-tail, plentiful not only in Great Britain, but throughout Europe and North America. There is some resemblance in habit between these plants and *Equisetum,* but in all essential characters they are perfectly distinct. French, *Pesse d'eau*; German, *Schafthalm.* [C. A. J.]

HIPTAGE. A genus of climbing shrubs belonging to the *Malpighiaceæ.* The flowers have a calyx provided with one large gland; unequal fringed petals; ten stamens, all fertile, and one larger than the rest; one style; and a fruit of three or fewer carpels, each provided with three wings. They are of a white or yellowish colour, are fragrant. [M. T. M.]

HIREA. A genus of *Malpighiaceæ,* distinguished chiefly by its ten stamens, all of which are fertile and slightly united at the base; by its three styles compressed at their summits, with truncate two-coloured stigmas; and by its fruits with lateral wings. [M. T. M.]

HIRCINOUS. Smelling like a goat.

HIRONDINAIRE. (Fr.) *Vincetoxicum officinale.*

HIRSE. A kind of *Panicum* or Millet.

HIRSUTE, HIRTUS. Hairy; covered by long tolerably distinct hairs.

HIRTELLA. A genus of tropical American shrubs or small trees, of the order *Chrysobalanaceæ,* differing from its allies in its five-petaled flowers with from three to fifteen long protruding stamens arising from one side of the flower. Upwards of thirty species are known, all of them with alternate shortly-stalked leaves accompanied by stipules; the flowers small, white or purplish, disposed in axillary or terminal racemes, and remarkable for their protruding stamens, which are usually much longer than the corollas. The fruits are pear-shaped furrowed drupes nearly an inch long, with one seed. *H. silicea* is a tree of Trinidad, where its bark, which is rich in siliceous matter, is said by M. Cruger to be used by the Indians in making pottery. *H. physophora,* a Brazilian species, is exceptional, in having on each side of the short leafstalks a leafy bladder-like process as large as a good-sized pea, with an opening at top. [A. A. B.]

www.ingramcontent.com/pod-product-compliance
Lightning Source LLC
Chambersburg PA
CBHW021225300426
44111CB00007B/429